T0192865

Habitats and Biota of the Gulf of Mexico: Before the Deepwater Horizon Oil Spill

Volume 2: Fish Resources, Fisheries, Sea Turtles, Avian Resources, Marine Mammals, Diseases and Mortalities

Habitats and Biota of the Gulf of Mexico: Before the Deepwater Horizon Oil Spill

Volume 2: Fish Resources, Fisheries, Sea Turtles, Avian Resources, Marine Mammals, Diseases and Mortalities

Edited by

C. Herb Ward

Rice University, Houston, TX, USA

Authors

Joanna Burger

Yong Chen

William E. Hawkins

Kym Rouse Holzwart

Walter R. Keithly, Jr.

Robin M. Overstreet

Kenneth J. Roberts

Roldán A. Valverde

Bernd Würsig

Editor
C. Herb Ward
Department of Civil
   and Environmental Engineering
Rice University
Houston, TX, USA
wardch@rice.edu

ISBN 978-1-4939-8055-0   ISBN 978-1-4939-3456-0   (eBook)
DOI 10.1007/978-1-4939-3456-0

© The Editor(s) (if applicable) and The Author(s) 2017. This book is an open access publication.
Softcover reprint of the hardcover 1st edition 2017
Open Access. This book is licensed under the terms of the Creative Commons Attribution-NonCommercial 2.5 International License (http://creativecommons.org/licenses/by-nc/2.5/), which permits any noncommercial use, sharing, adaptation, distribution and reproduction in any medium or format, as long as you give appropriate credit to the original author(s) and the source, provide a link to the Creative Commons license and indicate if changes were made.
The images or other third party material in this book are included in the book's Creative Commons license, unless indicated otherwise in a credit line to the material. If material is not included in the book's Creative Commons license and your intended use is not permitted by statutory regulation or exceeds the permitted use, you will need to obtain permission directly from the copyright holder.
This work is subject to copyright. All commercial rights are reserved by the Publisher, whether the whole or part of the material is concerned, specifically the rights of translation, reprinting, reuse of illustrations, recitation, broadcasting, reproduction on microfilms or in any other physical way, and transmission or information storage and retrieval, electronic adaptation, computer software, or by similar or dissimilar methodology now known or hereafter developed.
The use of general descriptive names, registered names, trademarks, service marks, etc. in this publication does not imply, even in the absence of a specific statement, that such names are exempt from the relevant protective laws and regulations and therefore free for general use.
The publisher, the authors and the editors are safe to assume that the advice and information in this book are believed to be true and accurate at the date of publication. Neither the publisher nor the authors or the editors give a warranty, express or implied, with respect to the material contained herein or for any errors or omissions that may have been made. The publisher remains neutral with regard to jurisdictional claims in published maps and institutional affiliations.

Cover Design: Map data from National Oceanic and Atmospheric Administration National Centers for Environmental Information ETOPO1 Global Relief Model and ESRI USA Base Layer. Map Design by Christopher Dunn, Ramboll Environ, Portland, ME.

Printed on acid-free paper

This Springer imprint is published by Springer Nature
The registered company is Springer Science+Business Media LLC
The registered company address is: 233 Spring Street, New York, NY 10013, U.S.A.

Preface

The Deepwater Horizon accident and oil spill in the Gulf of Mexico from the Macondo well began on April 20, 2010. Oil flowed into the Gulf for 87 days until the well was capped on July 15, 2010, and declared sealed on September 19, 2010. The United States (USA) Government initially estimated that a total oil discharge into the Gulf of 4.9 million barrels (210 million U.S. gallons) resulted from the spill; however, the estimate was challenged in litigation, reduced to 3.19 million barrels by a trial court, and remains in dispute. A massive cleanup, restoration, and research program followed and continues to the present, mostly funded by BP Exploration & Production Inc. (BP).

The Deepwater Horizon accident and oil spill quickly polarized factions of both the government regulatory and scientific communities, which resulted in a continuing barrage of conflicting opinions and reports in the media and at scientific meetings. In the aftermath of the oil spill, it quickly became apparent that much of the differences in opinion being expressed about biological and ecological effects was based on individual perceptions of the status and health of the Gulf of Mexico before the spill. Because of the very large differences between the Deepwater Horizon oil spill and the next largest oil spill in the Gulf (Ixtoc 1), few comparisons of pre-spill conditions and post-spill effects could be made.

BP funded cooperative research with government agencies on the effects of the Gulf oil spill and external competitively awarded independent research through their $500 million Gulf of Mexico Research Initiative (GoMRI) program. However, little of the research addressed the status and ecological health of the Gulf of Mexico *before* the Deepwater Horizon accident to serve as baseline to help assess post-spill effects.

Perhaps because of my 30-year background as the founding Editor in Chief of *Environmental Toxicology and Chemistry*, in teaching oil spill cleanup courses in the 1980s, in editing the *The Offshore Ecology Investigation* volume, and my work on tar ball formation from oil spilled in the Gulf, BP asked me to identify potential authors with appropriate expertise to research and write baseline white papers on the status and ecosystem health of the Gulf of Mexico before the Deepwater Horizon accident. Dozens of potential authors were identified and vetted for conflicts. Those selected as authors of white papers were given complete freedom to research and write their papers. I worked with the authors much in the mode of a journal editor to help them develop advanced drafts of their papers suitable for external peer review. As editor I researched and selected the peer reviewers for each paper and worked with the authors to address peer reviewer comments, which at times required preparation of additional text, figures, and tables. Author coordination meetings were held at the James A. Baker III Institute for Public Policy at Rice University.

After most of the white papers had been written, edited, and vetted by peers, BP proposed to publish them as a *SpringerOpen* two-volume series under the Creative Commons License for noncommercial use to promote wide distribution and free access.

In organizing and editing this two-volume series on baseline conditions in the Gulf of Mexico before the Deepwater Horizon oil spill, I have been assisted by Diana Freeman and Mary Cormier at Rice University; Alexa Wenning, Michael Bock, Laura Leighton, Jonathan Ipock[a], and Richard Wenning at Ramboll Environ; and Catherine Vogel who prepared the text and figures for preparation of page proofs by Springer. All involved in writing and editing this book series have been compensated for their time and efforts.

C. Herb Ward, Series Editor

A.J. Foyt Family Chair of Engineering, Professor of Civil and Environmental Engineering, and Professor of Ecology and Evolutionary Biology Emeritus and Scholar in Environmental Science and Technology Policy, Baker Institute for Public Policy, Rice University, Houston, TX.

---

[a]The late Jonathan "Jon" Ipock (1986-2015) tragically died too young. While working with Ramboll Environ, Inc., he tirelessly obtained documents, compiled data and references, and prepared maps and graphs for Chapter 7 (Offshore Plankton and Benthos of the Gulf of Mexico), Chapter 9 (Fish Resources of the Gulf of Mexico), and Chapter 11 (Sea Turtles of the Gulf of Mexico). During his short career Jon worked at two environmental consulting firms for more than eight years, first as a volunteer student intern, then as an associate ecologist. Jon's thirst for ecology was endless; he eagerly learned all he could and was one of ecology's rising stars.

About the Editor

C. Herb Ward

C. Herb Ward is Professor Emeritus at Rice University. He held the A. J. Foyt Family Chair of Engineering and was Professor of Civil and Environmental Engineering in the George R. Brown School of Engineering and Professor of Ecology and Evolutionary Biology in the Weiss School of Natural Sciences. He is now a Scholar in Environmental Science and Technology Policy in the James A. Baker III Institute for Public Policy at Rice University. He received his B.S. (1955) in Biology and Agricultural Science from New Mexico College of Agriculture and Mechanical Arts; his M.S. (1958) and Ph.D. (1960) in Microbial Diseases, Physiology, and Genetics of Plants from Cornell University; and the M.P.H. (1978) in Environmental Health from the University of Texas School of Public Health. He is a Registered Professional Engineer in Texas and a Board-Certified Environmental Engineer by the American Academy of Environmental Engineers. He was the founding Chair of the Department of Environmental Science and Engineering, Chair of the Department of Civil and Environmental Engineering, and the inaugural Director of the Energy and Environmental Systems Institute at Rice University. He also served as Director of the U.S. Environmental Protection Agency (USEPA)-sponsored National Center for Ground Water Research and the U.S. Department of Defense (DoD)-sponsored Advanced Applied (Environmental) Technology Development Facility. Dr. Ward was a member of the USEPA Science Advisory Board and served as Chair of the Scientific Advisory Board of the DoD Strategic Environmental Research and Development Program (SERDP). He is the founding Editor in Chief of the scientific journal *Environmental Toxicology and Chemistry* and led the development of the journal of *Industrial Microbiology and Biotechnology*. He is a Fellow in the American Academy of Microbiology (AAM), Society of Industrial Microbiology and Biotechnology (SIMB), and the Society of Environmental Toxicology and Chemistry (SETAC). Dr. Ward received the Pohland Medal for Outstanding Contributions to Bridging Environmental Research, Education, and Practice and the Brown and Caldwell Lifetime Achievement Award for Environmental Remediation in 2006, the Water Environment Federation McKee Medal for Achievement in Groundwater Restoration in 2007, and the SIMB Charles Thom Award for Bioremediation Research in 2011 and was recognized as a Distinguished Alumnus by New Mexico State University in 2013. He was a coauthor of the 2011 AAM report, *Microbes and Oil Spills*.

About the Authors

Joanna Burger

Joanna Burger is Distinguished Professor of biology at Rutgers University. She received her B.S. from the State University of New York at Albany, M.S. from Cornell University, a Ph.D. from the University of Minnesota, and an honorary Ph.D. from the University of Alaska at Fairbanks. Her main interests are behavioral ecology of vertebrates, environmental monitoring, ecotoxicology, and ecological and human health risk assessment. She has conducted behavioral ecology studies on all continents and has a 40-year study of behavior, population dynamics, and contaminant levels of colonial-nesting seabirds in New Jersey. She has examined the effects of heavy metals, oil, and other environmental factors on behavioral development in the field and laboratory. For 20 years she has worked with the Consortium for Risk Evaluation with Stakeholder Participation to examine ecological and human risk on Department of Energy lands, including the value of ecological resources. She led a biological expedition to the Aleutians to examine seafood safety because of possible radionuclide exposure from the underground nuclear tests performed in the 1970s. Dr. Burger has published over 600 peer-reviewed papers and more than 20 books, including *Behavior of Marine Animals, Seabirds and Other Marine Vertebrates, The Common Tern, The Black Skimmer: Social Dynamics of a Colonial Species, Before and After an Oil Spill: The Arthur Kill, Oil Spills, Protecting The Commons: A Framework for Resource Management in the Americas, The Biology of Seabirds*, and *A Visual Guide to Birds*. She served on several editorial boards, including *Environmental Research, Environmental Monitoring and Assessment, Journal of Toxicology and Environmental Health, Science and the Total Environment*, and *Environmental Bioindicators*. She served on the National Academy of Sciences' Commission of Life Sciences, Board of Biology, Board of Environmental Science and Toxicology, and several National Research Council committees. She served on several other national and international committees for the U.S. Environmental Protection Agency, U.S. Fish & Wildlife Service, and the Scientific Committee on Problems of the Environment (SCOPE). She is a Fellow of the American Ornithologists' Union, the International Union of Pure and Applied Chemistry, and the American Association for the Advancement of Science. She received the Brewster Medal in ornithology, and the Lifetime Achievement Award of the Society of Risk Analysis.

Yong Chen

Yong Chen is Professor of fisheries science in the School of Marine Science at the University of Maine in Orono, Maine. He received his bachelor of agriculture degree (1983) in fisheries science from the Ocean University of China and M.S. (1991) and Ph.D. (1995) degrees in zoology (fisheries ecology) from the University of Toronto in Canada. He started his career as a fisheries stock assessment scientist in the New South Wales Fisheries Department in Sydney, Australia, in 1995. In 1997, he was appointed Assistant Professor and Canadian Natural Science and Engineering Research Council Associate Chair in fisheries conservation at the Memorial University of Newfoundland. He moved to the University of Maine in 2000 and was promoted to the rank of Professor in 2007.

Dr. Chen's research is focused primarily on fisheries ecology, fish population dynamics, and stock assessment. He has extensive expertise working on the spatiotemporal dynamics of fish populations and communities of commercial and recreational importance in the Gulf of Maine (e.g., American lobster, sea urchin, demersal and pelagic finfish), Gulf of Mexico (tilefish habitat), Newfoundland (groundfish fisheries), Oman (finfish surveys), Kuwait (impacts of the Gulf War on shrimp fisheries), New Zealand (abalone fisheries), Australia (spiny lobster, abalone, and finfish), and China (oceanic squid and tuna fisheries). He has published 135 peer-reviewed papers in scientific journals and many technical and fish stock assessment reports and has received over $4.5 million in competitive research support. He is a Center for Independent Expert reviewer who is regularly invited to review National Oceanic and Atmospheric Administration stock assessment work. He frequently serves as a scientific consultant and reviewer on fisheries-related programs in the United States, Canada, China, New Zealand, Oman, and Kuwait. He has advised or co-advised 25 M.S. and 11 Ph.D. students, 9 postdoctoral research associates, and 8 short-term visiting Ph.D. students. Dr. Chen is actively involved in community and professional service, including peer reviewer for scientific journals and funding agencies. He is currently a member of the New England Fisheries Management Council's Scientific and Statistical Committee. He is Co-Editor of the *Canadian Journal of Fisheries and Aquatic Sciences*.

William E. Hawkins

William E. Hawkins is Professor Emeritus in the Department of Coastal Sciences at the Gulf Coast Research Laboratory (GCRL) of the University of Southern Mississippi. He retired from GCRL in 2011 as Director of the institution, which he served for more than 30 years. His undergraduate degree was from Mississippi State University and graduate degrees from the University of Mississippi Medical Center. He conducted his dissertation research at GCRL, and after receiving his doctorate he went to teach in the Department of Medical Anatomy and Embryology at the State University of Utrecht in the Netherlands and conduct research in the Centre for Electron Microscopy. On returning to the United States, he taught in the College of Medicine of the University of South Alabama then rejoined the Gulf Coast Research Laboratory in 1979 where he remained until he retired. At GCRL his research focused on the pathobiology of marine organisms, particularly the use of fish in studying disease processes. Technologies using the Japanese medaka for cancer research developed by him and colleagues at GCRL received millions of dollars in competitive grant support. He was instrumental in the development of the Thad Cochran Marine Aquaculture Center at GCRL, a more than $40M research facility built on the GCRL Cedar Point Campus. He led the recovery of GCRL from the effects of Hurricane Katrina and taught a graduate level course on the effects of that storm on the Mississippi Gulf Coast and the responses of individuals and institutions such as GCRL to the disaster.

Kym Rouse Holzwart

Kym Rouse Holzwart was a Senior Science Advisor and Certified Senior Ecologist at ENVIRON International Corporation in Tampa, Florida, when the chapter on sea turtles was written. She is now a senior environmental scientist with the Southwest Florida Water Management District. She has more than 30 years of experience as an ecologist and limnologist designing, managing, and participating in applied research projects. She has B.S. degrees in biology (1985) and limnology (1986) and an M.S. in biological sciences (1993) from the University of Central Florida. She worked for the St. Johns River Water Management District (SJRWMD), in Orlando, Florida (1987–1993), on a wide variety of tasks from permitting to enforcement and compliance, water quality sampling, and evaluating wetlands. Her M.S. thesis research, funded by the Water Management District, addressed the bioaccumulation of heavy metals by fish inhabiting storm water treatment ponds. She was a research ecologist at Oak Ridge National Laboratory (1993–1996) and participated in large environmental and ecological risk assessment projects.

Ms. Rouse Holzwart has experience in aquatic ecosystem assessment; environmental toxicology; water quality, sediment, and biological sampling; environmental permitting; and ecosystem services valuation. She has been involved in numerous ecological risk assessments evaluating the effects of contaminants, introduced species, and other disturbances on mammals, birds, fish, reptiles, amphibians, invertebrates, and plants. In addition, she has designed, managed, and conducted numerous environmental toxicology studies in aquatic and terrestrial ecosystems, including studies on reptiles and fish, with particular emphasis on the accumulation and effects of heavy metals. Her research has focused on aquatic ecosystems in the southeast United States and Florida. She has published more than 50 papers in the scientific literature and has coauthored hundreds of technical reports. She currently serves on the Ecological Society of America Board of Professional Certification, the Society of Environmental Toxicology and Chemistry Ecotoxicology of Amphibians and Reptiles Advisory Group, the Council of the Florida Academy of Sciences as Biological Sciences Section chair and councilor at large, and the Florida Lake Management Society Board of Directors.

Walter R. Keithly Jr.

Walter R. Keithly earned a doctoral degree in food and resource economics at the University of Florida in 1985 specializing in environmental and natural resource economics. He then accepted and currently holds a faculty appointment in the Department of Agricultural Economics and Agribusiness at Louisiana State University. Since joining Louisiana State University, he has authored or coauthored more than 100 papers covering a wide range of topics relevant to fisheries and coastal management. Recent fisheries-related research interests have focused on:

1. Effects of increasing shrimp imports on prices received by Gulf of Mexico shrimp fishermen
2. Factors determining location choice by Gulf of Mexico shrimp fishermen
3. The influence of *Vibrio vulnificus* on dockside oyster prices in the Gulf of Mexico and other producing regions of the United States
4. Analysis of quality and its influence on seafood demand
5. Industry perceptions related to the Gulf of Mexico red snapper catch share program

Other research interests have included (1) analysis of factors determining wetland values in Louisiana, (2) analysis of incentives that can be provided to wetland owners to encourage protection/rehabilitation of coastal wetland properties, and (3) the influence of bounties on the harvest of nutria. He frequently gives presentations at academic and industry-oriented meetings and serves on numerous federal committees, including as chair of the Standing Scientific and Statistical Committee and the Socioeconomic Committee of the Gulf of Mexico Fishery Management Council. He has also served on the Caribbean Standing and Scientific Statistical Committee of the Caribbean Fishery Management Council.

Dr. Keithly often serves as consultant to state, federal, and international organizations, including the Food and Agricultural Organization of the United Nations, the National Oceanic and Atmospheric Administration, and, most recently, the state of Maryland where he provided an economic review of the introduction of nonindigenous oysters to the Chesapeake Region. He has been an expert witness on numerous occasions where he has estimated financial losses due to natural resources damage.

Robin M. Overstreet

Robin M. Overstreet is Professor Emeritus at the University of Southern Mississippi (USM) in the Department of Coastal Sciences on the Gulf Coast Research Laboratory (GCRL) campus. He graduated with a B.A. in biology from the University of Oregon in 1963 and an

M.S. and Ph.D. in marine biology from the Institute of Marine Sciences, University of Miami (now the Rosenstiel School of Marine and Atmospheric Science), in 1966 and 1968, respectively, followed by a National Institutes of Health postdoctoral fellowship in parasitology at Tulane Medical School. He spent 45 years of his professional career at the GCRL, which became part of USM in 1998 where he was a Full Professor. He garnered almost $20M in extramural support for his own research, collaborated on a total of about $50M in research funding for GCRL and USM, and was advisor to numerous graduate students from USM and several other universities. He published over 300 peer-reviewed research papers in his primary fields of research (1) parasitology and diseases, (2) aquaculture and fisheries science, and (3) environmental biology and neoplasms. He has three *Digenea* genera named after him and 27 species of patronymics. Since 1969, he has studied parasites, including viral agents, in penaeid shrimps and other crustaceans as well as other invertebrates, fishes, other lower vertebrates, and warm-blooded animals. He still has students, technicians, space, and an active research program including work on crustacean viral infections, but focusing on parasites. Robin has received several awards including Fellow of the American Academy of Microbiology, academic and applied research awards from USM, and leadership positions and awards from different parasitological and fisheries societies. He also was a Visiting Professor at the University of Queensland, Curtin University, and University of Rome. He enjoys tennis, old Fords, and photography.

Kenneth J. Roberts

Kenneth J. Roberts is Associate Vice-Chancellor Emeritus of the Louisiana State University AgCenter. He received B.S. (1966) and M.S. (1968) degrees in agricultural economics from Louisiana State University. He earned a Ph.D. (1973) in natural resource economics from Oregon State University where he also taught as an Instructor. His academic and professional career continued on the faculty of Clemson University where he was Associate Professor with tenure in the Department of Agricultural Economics. He was an Assistant Program Manager for the National Sea Grant Program in the National Oceanic and Atmospheric Administration while on sabbatical leave from Clemson University. In 1978 Dr. Roberts accepted a joint marine economics research and extension service position at Louisiana State University with the Louisiana State University Sea Grant College Program. With an interest in leadership development among Louisiana's agricultural and fisheries food producers, processors, and service industries, he became Director of the endowed Agricultural Leadership Development Program at the university. He was appointed Associate Vice-Chancellor of the Louisiana State University AgCenter, which he held until retirement in 2008.

Dr. Roberts' professional career has included both extension service and research in the fields of fisheries economics, seafood processing and marketing, and the economics of pond and recirculating aquaculture. He published frequently in these fields, but his most cherished accomplishments were those related to service on public agency and policy-shaping groups such as the South Atlantic Fishery Management Council and the Gulf of Mexico Fishery Management Council (GMFMC). Dr. Roberts became Vice-Chairman and then Chairman of the GMFMC. Collectively, he has been a member of the Councils' Scientific and Statistical Committees for over three decades. He has also provided expert advice via the GMFMC's Socioeconomics Scientific and Statistical Advisory Committee and controlled access committees regarding red snapper and grouper fishery management plan amendments. The U.S. Secretary of Commerce appointed Dr. Roberts to the Marine Fisheries Advisory Committee (MAFAC) to participate in the shaping of recommendations on marine aquaculture, controlled access to fisheries, and other emerging issues. A year 2020 plan for the National Marine Fisheries Service was developed.

Roldán A. Valverde

Roldán A. Valverde is Professor in the Department of Biology at Southeastern Louisiana University. He received his B.S. degree (1985) from the Universidad Nacional of Costa Rica and his Ph.D. in zoology (1996) from Texas A&M University. During his doctoral studies, he studied endocrine stress responses of arribada olive ridleys at Nancite Beach. Following his doctoral degree, he worked as leader of the Green Sea Turtle Tagging Program at Tortuguero, Costa Rica. He then pursued postdoctoral training in the Department of Biology, University of Michigan, studying the neuroendocrine mechanisms of amphibian metamorphosis. He accepted an academic position at Xavier University of New Orleans in 2001 and moved to his current position at Southeastern Louisiana University in 2004. Dr. Valverde is primarily a sea turtle biologist with training in comparative and integrative endocrinology. His main academic interests focus on the endocrine stress response of turtles and on the nesting ecology of sea turtles. His research combines laboratory experimental work with captive-raised and wild red-ear slider (*Trachemys scripta*) turtles examining the effects of organic pollutants on turtle physiology. His sea turtle work includes nesting ecology and reproductive and stress physiology, mainly of olive ridley sea turtles. Much of his work on olive ridley nesting ecology is conducted with collaborators in Costa Rica.

Dr. Valverde has 21 peer-reviewed publications, including one book chapter, over 40 coauthored presentations at professional meetings, 15 invited presentations, and several sea turtle workshops and training courses geared to Latin Americans. He has directed 21 graduate students and 12 undergraduates at Southeastern Louisiana University or Latin American institutions. Over the years, he has received 13 extramural research grants and 10 intramural grants to support his research and graduate training program. He has reviewed manuscripts for several scientific journals, served as a reviewer for National Science Foundation grant programs (UMEB, GK-12, and REU), and served on the editorial board of the *Marine Turtle Newsletter*. Dr. Valverde is a member of the International Sea Turtle Society, for which he served as a board member since 2010 and as President in 2014.

Bernd Würsig

Bernd Würsig holds the George P. Mitchell '40 Chair of Sustainable Fisheries and is Regents' Professor at Texas A&M University (TAMU). He was recently honored by appointment as University Distinguished Professor at TAMU. He received his B.S. (1971) in zoology from Ohio State University and Ph.D. (1978) in an interdisciplinary doctoral program at Stony Brook University. After going through the professor ranks at Moss Landing Marine Laboratories in California (1981–1989), Dr. Würsig became the inaugural Chair of the Texas A&M System's Marine Biology Graduate Program, an interdisciplinary entity among three campuses. Würsig teaches undergraduate and graduate courses on marine mammalogy, specializing in behavior and behavioral ecology. He has published widely in the popular literature as part of teaching endeavors, such as for the *Journal of Natural History* and *Scientific American*, and he has been advisor to numerous movies made for television on nature interpretation, as well as the IMAX movie *Dolphins* in 2000 that was nominated for an Academy Award for Best Documentary Short Subject. He also leads field courses on marine birds and mammals in New Zealand.

Professor Würsig has published more than 150 peer-reviewed papers in scientific journals and 6 books and has been senior advisor to 70 graduate students. He and his students and postdoctoral fellows have studied cetaceans, pinnipeds, and sea birds on all continents, with present focus primarily on social strategies of dusky dolphins in New Zealand, western gray whales in far east Russia, Indo-Pacific humpback dolphins in Hong Kong, humpback whales in Puerto Rico, small cetaceans of the Mediterranean Sea, diversity of cetaceans of southern

California, and bottlenose dolphins in the Gulf of Mexico. Recent books are *The Encyclopedia of Marine Mammals*, 2nd Edition (with Bill Perrin and Hans Thewissen, 2009), and *The Dusky Dolphin*: *Master Acrobat Off Different Shores* (with Melany Würsig, 2010). Bernd and Melany enjoy their gardens in New Zealand, the Arizona desert, and coastal South Texas, three marvelously different biomes.

External Peer Reviewers

Charles M. Adams
Food and Resource Economics
University of Florida
Gainesville, FL, USA

John B. Anderson
Sedimentology and Earth Science
Rice University
Houston, TX, USA

Susan S. Bell
Marine and Restoration Ecology
University of South Florida
Tampa, FL, USA

William F. Font
Fish Ecological Parasitology
Southeastern Louisiana University
Hammond, LA, USA

Mark A. Fraker
Marine Mammal Ecology
TerraMar Environmental Research LLC
Ashland, OR, USA

Jonathon H. Grabowski
Ecology and Fisheries Biology
Northeastern University
Boston, MA, USA

Frank R. Moore
Bird Migration and Ecology
University of Southern Mississippi
Hattiesburg, MS, USA

Pamela T. Plotkin
Sea Turtle Behavioral Ecology
Texas A&M University
College Station, TX, USA

Steve W. Ross
Marine Fish Ecology
University of North Carolina
Wilmington, NC, USA

Roger Sassen
Marine Geochemistry
Texas A&M University
College Station, TX, USA

Greg W. Stunz
Marine Biology and Fisheries
Texas A&M University
Corpus Christi, TX, USA

John H. Trefry
Chemical Oceanography
Florida Institute of Technology
Melbourne, FL, USA

Edward S. Van Vleet
Chemical Oceanography
University of South Florida
St. Petersburg, FL, USA

Contents

VOLUME 2

VOLUME 1

List of Figures

List of Tables

CHAPTER 9

FISH RESOURCES OF THE GULF OF MEXICO

Yong Chen[1]

[1]School of Marine Sciences, University of Maine, Orono, ME 04469, USA
ychen@maine.edu

9.1 INTRODUCTION

The Gulf of Mexico, surrounded on three sides by continental landmass, is the nineth largest waterbody in the world; it is semi-enclosed with its east connecting to the Atlantic Ocean through the Straits of Florida and its south to the Caribbean Sea through the Yucatán Channel. The Gulf of Mexico basin resembles a large crater with a wide shallow rim. Approximately 38 % of Gulf waters are shallow, intertidal areas. The continental shelf (<200 meters (m) or <656 feet [ft]) and continental slope (200–3,000 m or 656–9,843 ft) represent 22 and 20 % of the Gulf of Mexico basin, respectively, and abyssal regions deeper than 3,000 m (9,843 ft) comprise the remaining 58 % (USEPA 1994). The Sigsbee Deep in the southwestern Gulf of Mexico is the deepest region at 4,384 m (14,383 ft). The average water depth of the Gulf is about 1,615 m (5,299 ft). The boundary of the Gulf of Mexico used in this evaluation follows that defined in McEachran and Fechhelm (2005), which does not exclude the Florida Keys and the northeastern coast of Cuba.

The ichthyofaunal community in the Gulf of Mexico is dynamic and varies greatly, both spatially and temporally, because of fish movement/migration, diversified life-history strategies, fishing pressure, and varying hydrographic, oceanographic, and geographic conditions. It consists of a large number of reef-dependent, demersal species (e.g., snappers and groupers); coastal demersal species (e.g., drums and mullets); demersal species (e.g., tilefishes and porgies); coastal pelagic species (e.g., herrings and jacks); highly migratory, pelagic species (e.g., tunas and billfishes); small and large coastal sharks; and pelagic sharks (McEachran and Fechhelm 2005; Parsons 2006). Because of its unique oceanographic and hydrographic conditions, geological location, and availability of a great diversity of habitats, the Gulf of Mexico ecosystem has a relatively high biodiversity, with a large number of fish and shark species compared to other areas in the United States (Chesney et al. 2000).

Finfish and sharks, both as prey and predators, play significant roles in regulating the dynamics of the Gulf of Mexico ecosystem and the energy flows between organisms of different trophic levels (Hoese and Moore 1998; McEachran and Fechhelm 2005; Parsons 2006). Small coastal pelagic forage fishes, such as herrings and anchovies, filter feed on plankton and play a critical role in transferring primary productivity into fish biomass that is useable for other fish species of higher trophic levels or that directly supports commercial and recreational fisheries (McEachran and Fechhelm 2005; Anderson and McDonald 2007). These fish species form the forage base in the Gulf of Mexico ecosystem. Fish species of higher trophic levels usually prey on forage species, juvenile fish, and other organisms, such as squids, crabs, and shrimps. Many finfish and sharks are apex predators in the Gulf of Mexico ecosystem and are important in regulating the dynamics of their prey species (Hoese and Moore 1998; McEachran and Fechhelm 2005; Parsons 2006). Large oceanic pelagic species, such as tunas, billfishes, and sharks, tend to have few predators and prey on many finfish and

© The Author(s) 2017
C.H. Ward (ed.), *Habitats and Biota of the Gulf of Mexico: Before the Deepwater Horizon Oil Spill*,
DOI 10.1007/978-1-4939-3456-0_1

invertebrate species. Substantial changes in the dynamics of populations and communities of key forage and apex predator species can have significant cascading effects on Gulf of Mexico ecological processes (Anderson and McDonald 2007).

In the Gulf of Mexico, finfish and sharks support important commercial and recreational fisheries, two of the most important industries in the region, as well as one of the most productive fisheries in the world (Chesney et al. 2000). Overall, approximately 25 % of the U.S. commercial fish landings and 40 % of the recreational harvest occur in the Gulf of Mexico. Commercial landings of finfish and shellfish in the Gulf of Mexico totaled over 590 million kilograms (kg) [1.3 billion pounds (lb)], valued at $661 million in 2008, and 8 of the top 20 fishing ports by value and 4 of the top 7 fishing ports by weight in the United States are located in the Gulf of Mexico (NMFS 2009a). More than 24.1 million recreational fishery trips were made in 2008 in the Gulf of Mexico, resulting in a catch of 190 million fish (NMFS 2009a). Therefore, the economic and social values of fisheries in the Gulf of Mexico are huge and should not be underestimated.

The Gulf of Mexico provides a wide range of habitats for its ichthyofaunal community, but long-term anthropogenic and natural stressors and perturbations, such as rapid coastal development, pollution, overfishing, and natural disasters, have altered the Gulf of Mexico ecosystem and the dynamics of its fish community and populations (O'Connell et al. 2004). However, it is difficult to quantitatively assess and separate the impacts of human and natural perturbations on the resilience of the Gulf of Mexico ecosystem because of the limitations of available data.

The Gulf of Mexico receives about 50 % of all watershed discharge in the United States, and more than 3,100 point-source outfalls discharge into the Gulf of Mexico. Pesticides and nutrients used in the watersheds of the U.S. states bordering the Gulf exceed those used in any of the other coastal zones in the United States. The entire U.S. Gulf of Mexico coastline has been under fish consumption advisory for mercury since 1994 (USEPA 1994). Fifty-nine percent of the estuarine areas of the U.S. Gulf of Mexico, which are essential nursery and spawning grounds for many finfish and sharks, assessed from 1997 through 2000, were considered impaired or threatened (USEPA 2004). Coastal wetlands and nearshore seagrass beds are critical nursery and spawning grounds for many finfish and sharks; however, Lewis et al. (2007) estimated that 78 square kilometers (km^2) (30 miles^2) of wetlands were being lost annually and that 20–100 % of the seagrasses have been destroyed in some areas of the Gulf of Mexico. The deterioration and even total loss of these critical habitats may greatly reduce the carrying capacity of the Gulf of Mexico for many fish and shark species that depend on these areas as their critical habitat. Overfishing and shrimp fishery bycatch have substantially reduced the population abundance of many fish and shark species of commercial and recreational importance, resulting in some important species being classified as in the status of overfishing and/or being overfished (SEDAR 31 2009; NMFS 2012a).

The objective of this chapter is to provide an overview, synthesis, and evaluation of the life histories, population and community structures, and population dynamics, distribution, and abundance of fish representative of the species and habitat diversity in the Gulf prior to the Deepwater Horizon event. The primary focus is on information believed critical to the overall understanding of the spatiotemporal dynamics and habitat needs of key finfish, shark, and ray species and the major anthropogenic and environmental drivers that influence their conditions in the Gulf of Mexico.

Hoese and Moore (1998) and McEachran and Fechhelm (2005) documented 1,443 finfish species in 223 families in the Gulf of Mexico. A representative subset of 100 key families of finfish were evaluated for their distribution and habitat needs in the Gulf of Mexico (Table 9.1). Finfish families with high to medium importance to commercial and recreational fisheries in the Gulf of Mexico were identified (Table 9.2). Ten finfish families were selected for evaluation

Table 9.1. Summary of the Key Finfish Families, Their Distributions, and Preferred Habitats in the Gulf of Mexico

Family	Common Name	Number of Species	Distribution and Preferred Habitat
Dactylopteridae	Flying gurnards	1 species	Benthic in shallow to moderate depths, sandy bottom
Scorpaenidae	Scorpionfishes	20–21 species in 9 genera	Sedentary benthic from the intertidal zone to 2,200 m
Triglidae	Searobins	14–15 species in 2 genera	Benthic on sandy to muddy bottom on continental and slopes
Peristediidae	Armored searobins	8–12 species in 1 genus	Benthic on continental and insular slopes
Centropomidae	Snooks	6 species in 1 genus	Catadromous
Moronidae	Temperate basses	3 species in 1 genus	Stenohaline or euryhaline and anadromous
Acropomatidae	Temperate ocean-basses	4 species in 2 genera	In water column between 87 and 910 m
Howellidae	Not Available	1 species in 1 genus	Pelagic or benthopelagic over outer continental shelves and slopes
Serranidae	**Seabasses**	61–62 species in 20 genera	Benthic up to depths of 500 m, some hard bottom, and some soft bottom and sea-grass beds
Grammatidae	Basslets	2–3 species in 1–2 genera	Near ledges and drop-offs on deep reefs
Opistognathidae	Jawfishes	7–9 species in 2 genera	Sandy to muddy bottom near reefs from nearshore to 375 m deep
Priacanthidae	Bigeyes	3 species in 3 genera	Associated with reefs on continental shelves
Apongonidae	Cardinalfishes	15 species in 3 genera	Associated with rocky and coral reefs and sandy/weedy areas
Epigonidae	Deepwater cardinalfishes	6 species in 1 genus	Benthic pelagic in depths from 75 to 3,700 m over continental and insular slopes
Malacanthidae	**Tilefishes**	6 species in 3 genera	Burrow in bottom from shoreline to 500 m
Pomatomidae	Bluefishes	Monotypic	Continental shelves
Echeneidae	Remoras	8 species in 4 genera	Attachment to sharks, billfishes, rays, whales, dolphins, seabasses, jacks, and cobia
Rachycentridae	Cobia	Monotypic	Pelagic, but also associated with coral reefs and man-made surface structures
Coryphaenidae	**Dolphinfishes**	2 species in 1 genus	Epipelagic in oceanic waters and over continental shelves, associated with surface structures

(continued)

Table 9.1. (continued)

Family	Common Name	Number of Species	Distribution and Preferred Habitat
Carangidae	**Jacks**	28–29 species in 14–15 genera	Pelagic over continental and insular shelves
Bramidae	Pomfrets	5 species in 4 genera	Most pelagic between the surface and 600 m
Caristidae	Manefishes	1 species	Epipelagic to bathypelagic from 100 to 2,000 m
Emmelichthyidae	Rovers	2 species in 2 genera	Benthopelagic often found over drop-offs nearby islands and deep reefs
Lutjanidae	**Snappers**	16–17 species in 6 genera	Most associated reefs on continental and insular shelves and slopes
Symphysanodontidae	Not available	1–2 species in 1 genus	Between depths of 50–500 m over continental and insular shelves and slopes
Lobotidae	Tripletails	1 species	Benthic and associated with coastal waters and estuaries
Gerreidae	Mojarras	12 species in 4 genera	Sandy to muddy bottoms in coastal waters and estuaries
Haemulidae	Grunts	18 species in 5 genera	Associated with coral reefs in coastal waters
Inermiidae	Bonnetmouths	1–2 species in 2 genera	Pelagic over continental and insular shelves
Sparidae	Porgies	16 species in 6 genera	Benthic on continental and insular shelves, coral reefs
Polynemidae	Threadfins	2–3 species in 1 genus	Shallow sandy to muddy bottom
Sciaenidae	**Drums**	25–29 species in 14–15 genera	Sandy to muddy bottom in the coastal waters
Mullidae	Goatfishes	4 species in 4 genera	Benthic on continental and insular shelves
Pempheridae	Sweepers	1 species	On coral reefs and in caves or other cavities
Bathyclupeidae	Not available	2 species in 1 genus	mesopelagic and bathypelagic
Chaetodontidae	Butterflyfishes	6 species in 2 genera	Associated with coral reefs in shallow waters
Pomacanthidae	Angelfishes	6 species in 3 genera	Associated with coral reefs in shallow waters
Kyphosidae	Sea chubs	2 species in 1 genus	Associated with coral reefs and rocky areas in shallow waters
Cirrhitidae	Hawkfishes	1 species	Benthic in shallow waters
Pomacentridae	Damselfishes	14 species in 4 genera	Associated with coral reefs

(continued)

Table 9.1. (continued)

Family	Common Name	Number of Species	Distribution and Preferred Habitat
Labridae	Wrasses	17–19 species in 8 genera	Most associated with coral reefs
Scaridae	Parrotfishes	14 species in 4 genera	Around coral reefs in shallow waters
Zoarcidae	Eelpouts	3 species in 3 genera	Benthic or benthopelagic or mesopelagic
Chiamodontidae	Not available	Possibly 9–10 species in 4 genera	Mesopelagic and bathypelagic
Percophidae	Flatheads	2–3 species in 1 genus	Benthic from the outer continental shelf to the upper slope
Uranoscopidae	Stargazers	3 species in 3 genera	Benthic on the continental and insular shelves
Tripterygiidae	Not Available	4 species in 1 genus	Benthic and cryptic in shallow water
Dactyloscopidae	Sand stargazers	4–6 species in 3 genera	Benthic in sandy and reef habitats on continental and insular shelves
Labrisomidae	Scaly blennies	19–20 species in 5 genera	Benthic in coral and rocky reefs in shallow water
Chaenopsidae	Tube blennies	11–12 species in 6 genera	Benthic in rocky and coral reefs
Blenniidae	Combtooth blennies	14 species in 8 genera	Benthic in shallow marine water
Gobiesocidae	Clingfishes	2–3 species in 1–2 genera	From nearshore to 200 m, attaching to hard substrates and plants
Callionymidae	Dragonets	4 species in 3 genera	Benthic associated with sandy to muddy bottom and seagrass beds, some with coral reefs
Draconettidae	Draconetts	1 species	Outer continental and insular shelves and upper slopes
Eleotridae	Sleepers	5 species in 5 genera	Most benthic in fresh and brackish waters, some on coral reefs
Gobiidae	Gobies	58–62 species in 26 genera	Most benthic, some free-swimming, from shore to depths of 500 m, coral reefs
Microdesmidae	Wormfishes	4 species in 2 genera	Burrow into soft muddy and sandy bottom
Ephippidae	Spadefishes	1 species	Associated with coral reefs, artificial reefs, and rocky area
Luvaridae	Louvars	1 species	In oceanic waters between 200 and 600 m
Acanthuridae	Surgeonfishes	3 species in 1 genus	In coral and rocky reefs to a depth of about 100 m

(continued)

Table 9.1. (continued)

Family	Common Name	Number of Species	Distribution and Preferred Habitat
Scombrolabracidae	Not Available	Single species	Found in depths from 560 to 1,340 m in the northern and southern Gulf of Mexico
Sphyraenidae	Barracudas	3–4 species in 1 genus	Pelagic in neritic waters and associated with reefs and sea-grass beds
Gempylidae	Snake mackerels	9 species in 9 genera	Pelagic or benthopelagic in oceanic waters from 200 to 1,000 m
Thichiuridae	Cutlassfishes	5 species in 5 genera	Pelagic or benthopelagic in oceanic waters from the surface to 1,000 m
Scombridae	**Mackerels and tunas**	14–15 species in 8 genera	Epipelagic in marine ecosystem
Xiphiidae	**Billfishes**	5 species in 4 genera	Epipelagic and mesopelagic, highly migratory
Centrolophidae	Medusafishes	2–3 in 1–2 genera	Epipelagic to demersal over continental shelves, some are pelagic on the high seas
Nomeidae	Diftfishes	7 species in 3 genera	In mid-water or demersal over continental shelves and oceanic waters
Ariommatidae	Not Available	3 species in 1 genus	In deepwater near continental and insular shelves
Tetragonuridae	Not Available	1 species	In oceanic waters at epipelagic and mesopelagic depths
Stromateidae	Butterfishes	2 species in 1 genus	Along continental margins
Bothidae	Lefteye founders	7 species in 5 genera	Benthic and associated with soft bottoms on continental shelves
Paralichthyidae	Not available	22–23 species in 8 genera	Benthic and associated with soft bottoms on continental shelves
Achiridae	Not available	5–6 species in 3 genera	Benthic on inner continental and insular shelves
Cynoglossidae	Tonguefishes	11–12 species in 1 genus	Benthic on continental shelves and the upper slopes
Triacanthodidae	Spikefishes	3 species in 2 genera	Near the bottom between 46 and 900 m
Balistidae	Triggerfishes	6 species in 4 genera	Benthic
Monacanthidae	Filefishes	10 species in 4 genera	Coral or rocky reefs and sea-grass beds
Ostraciidae	Cowfishes	5 species in 3 genera	Benthic and associated with coral and rocky reefs, and sea-grass beds
Tetraodontidae	Puffers	9 species 3 genera	In shallow depths

(continued)

Table 9.1. (continued)

Family	Common Name	Number of Species	Distribution and Preferred Habitat
Diodontidae	Porcupinefishes	6–7 species in 2 genera	Benthic and associated with floating seaweed
Molidae	Ocean sunfishes	3 species in 3 genera	Pelagic
Clupeidae	**Herrings**	At least 12 species in 8 genera	Pelagic and schooling
Engraulidae	Anchovies	At least 5 species in 2 genera	Pelagic
Synodontidae	Lizardfishes	At least 7 species 3 genera	Benthic
Mugilidae	**Mullets**	At least 4 species in 2 genera	Estuaries and freshwater
Batrachoididae	Toadfishes	At least 3 species in 2 genera	Benthic
Aentennariidae	Frogfishes	At least 4 species in 2 genera	In coral or sponge-encrusted substrates, middle shelf
Ogcocephalidae	Batfishes	At least 4 species in 3 genera	In the shelf and deeper areas, near bottom
Bregmacerotidae	Codlets	1 species	On the middle shelf
Steindachneridae	Not available	1 species	In deeper waters, muddy bottom
Phycidae	Hakes	At least 3 species in 1 genus	Offshore in deep and cold waters
Ophidiidae	Cusk-eels	At least 7 species in 4 genera	Deep-water fishing crevices or in burrows in the mud
Exocoetidae	Flyingfishes	12 species in 9 genera	Frequently jump from the water and skip over the surface
Belonidae	Needlefishes	4 species in 4 genera	In inshore Gulf and bays
Fundulidae	Killifishes	6 species in 3 genera	In inshore, coastal, estuaries, and bays
Altherinidae	Silversides	4 species in 2 genera	In estuarine and coastal areas
Holocentridae	Squirrelfishes	7 species in 3 genera	On offshore reefs
Syngnathidae	Pipefishes	10 species in 4 genera	In vegetated areas
Muraenidae	Moray eels	5 species in 1 genus	In continental shelves, associated with reefs

Families in *bold* were selected for evaluation. Data compiled based on McEachran and Fechhelm (2005) and Hoese and Moore (1998)

Table 9.2. Key Finfish Families with High to Medium Value to Recreational and Commercial Fisheries in the Gulf of Mexico

Fisheries Values	Family	Species Selected
Finfish families with high to medium commercial values	**Serranidae (Seabasses)**	Red grouper (*Epinephelus morio*)
	Carangidae (Jacks)	Greater amberjack (*Seriola dumerili*)
	Lutjanidae (Snappers)	Red snapper (*Lutjanus campechanus*)
	Scombridae (Mackerels and tunas)	Bluefin tuna (*Thunnus thynnus*), king mackerel (*Scomberomorus cavalla*)
	Clupeidae (Herrings)	Gulf menhaden (*Brevoortia patronus*)
	Mugilidae (Mullets)	Striped mullet (*Mugil cephalus*)
	Moronidae (Temperate basses)	No species selected
	Malacanthidae (Tilefishes)	Tilefish (*Lopholatilus chamaeleonticeps*)
	Coryphaenidae (Dolphinfishes)	Dolphinfish (*Coryphaena hippurus*)
	Sparidae (Porgies)	No species selected
	Stromateidae (Butterfishes)	No species selected
	Balistidae (Triggerfishes)	No species selected
Finfish families with high to medium recreational values	**Serranidae (Seabasses)**	Red grouper (*Epinephelus morio*)
	Priacanthidae (Bigeyes)	No species selected
	Pomatomidae (Bluefishes)	No species selected
	Coryphaenidae (Dolphinfishes)	Dolphinfish (*Coryphaena hippurus*)
	Carangidae (Jacks)	Greater amberjack (*Seriola dumerili*)
	Lutjanidae (Snappers)	Red snapper (*Lutjanus campechanus*)
	Sciaenidae (Drums)	Red drum (*Sciaenops ocellatus*)
	Scombridae (Mackerels and tunas)	Bluefin tuna (*Thunnus thynnus*), king mackerel (*Scomberomorus cavalla*)
	Xiphiidae (Billfishes)	Atlantic sailfish (*Istiophorus albicans*), Atlantic blue marlin (*Makaira nigricans*), Atlantic swordfish (*Xiphias gladius*)
	Sphyraenidae (Barracudas)	No species selected
	Paralichthyidae	No species selected
	Balistidae (Triggerfishes)	No species selected
	Mugilidae (Mullets)	Striped mullet (*Mugil cephalus*)

Families in *bold* were selected for evaluation

based on information in Tables 9.1 and 9.2 and the following criteria: (1) relative importance to the ecosystem of the Gulf of Mexico; (2) importance to commercial and/or recreational fisheries; (3) abundance (high and low population sizes) and range of fish distributions (e.g., coastal waters and estuaries versus open ocean) in the Gulf of Mexico; (4) diversity of life histories (e.g., long-lived versus short-lived, slow growing versus fast growing, and early mature versus late mature); (5) movements (e.g., sedentary/inactive versus highly migratory); and (6) habitat needs (e.g., low salinity versus high salinity, low temperature versus high temperature, habitat generalist versus habitat specialist). The ten finfish families selected included Lutjanidae (snappers), Clupeidae (herrings), Serranidae (seabasses), Scombridae (mackerels and tunas), Xiphiidae (billfishes), Sciaenidae (drums), Malacanthidae (tilefishes), Coryphaenidae

Table 9.3. Key Finfish Species of High Commercial and/or Recreational Importance in the Gulf of Mexico Listed by Habitat

Habitat	Finfish Species
Benthic	Rock hind grouper (*Epinephelus adscensionis*), Yellowfin grouper (*Mycteroperca venenosa*), Scamp grouper (*Mycteroperca phenax*), Red hind (*Epinephelus guttatus*), Atlantic goliath grouper (*Epinephelus itajara*), Nassau grouper (*Epinephelus striatus*), **Red grouper (*Epinephelus morio*)**, Gag grouper (*Mycteroperca microlepis*), Yellowedge grouper (*Hyporthodus flavolimbatus*), Mutton snapper (*Lutjanus analis*), Blackfin snapper (*Lutjanus buccanella*), **Red snapper (*Lutjanus campechanus*)**, Lane snapper (*Lutjanus synagris*), Silk snapper (*Lutjanus vivanus*), Yellowtail snapper (*Ocyurus chrysurus*), Vermillion snapper (*Rhomboplites aurorubens*), **Tilefish (*Lopholatilus chamaeleonticeps*)**, Blueline snapper (*Lutjanus kasmira*), Golden snapper (*Lutjanus inermis*), **Red drum (*Sciaenops ocellatus*)**, Black drum (*Pogonias cromis*), Bluefish (*Pomatomus saltatrix*), Common snook (*Centropomus undecimalis*), Crevalle jack (*Caranx hippos*), Spotted seatrout (*Cynoscion nebulosus*), and **Striped mullet (*Mugil cephalus*)**
Pelagic and highly migratory	Skipjack (*Katsuwonus pelamis*), Albacore (*Thunnus alalunga*), Bigeye (*Thunnus obesus*), **Atlantic bluefin tuna (*Thunnus thynnus*)**, Yellowfin tuna (*Thunnus albacores*), Small tunas, **Atlantic blue marlin (*Makaira nigricans*)**, White marlin (*Tetrapturus albidus*), **Atlantic sailfish (*Istiophorus albicans*)**, and **Atlantic swordfish (*Xiphias gladius*)**
Pelagic	**Dolphinfish (*Coryphaena hippurus*)**, Spanish mackerel (*Scomberomorus maculatus*), Cobia (*Rachycentron canadum*), Atlantic thread herring (*Opisthonema oglinum*), **King mackerel (*Scomberomorus cavalla*)**, Spanish sardine (*Sardinella aurita*), **Menhaden (*Brevoortia* spp.)**, and **Greater amberjack (*Seriola dumerili*)**

Species *highlighted* were selected for evaluation

(dolphinfishes), Mugilidae (mullets), and Carangidae (jacks) (Table 9.2). Based on their distribution, habitat needs, and commercial and recreational importance, 13 representative species of finfish were selected from the ten families for detailed evaluation in this chapter (Table 9.3). Species selected include red snapper (*Lutjanus campechanus*); menhaden, including Gulf menhaden (*Brevoortia patronus*), finescale menhaden (*Brevoortia gunteri*), and yellowfin menhaden (*Brevoortia smithi*); red grouper (*Epinephelus morio*); Atlantic bluefin tuna (*Thunnus thynnus*); Atlantic blue marlin (*Makaira nigricans*); Atlantic swordfish (*Xiphias gladius*); Atlantic sailfish (*Istiophorus albicans*); red drum (*Sciaenops ocellatus*); tilefish (*Lopholatilus chamaeleonticeps*); king mackerel (*Scomberomorus cavalla*); dolphinfish (*Coryphaena hippurus*); striped mullet (*Mugil cephalus*); and greater amberjack (*Seriola dumerili*). These are representative species that are demersal and reef-dependent (red snapper and red grouper); offshore demersal (tilefish); coastal demersal (red drum and striped mullet); highly migratory and pelagic (Atlantic bluefin tuna, Atlantic blue marlin, Atlantic swordfish, and Atlantic sailfish); offshore pelagic (dolphinfish); and coastal pelagic (menhaden, king mackerel, and greater amberjack). Although many finfish species of great ecological, commercial, and recreational importance, such as many species in the families of snappers, seabasses, tunas, and jacks, were not selected (Table 9.3), they are well represented by the above 13 species with respect to spatiotemporal distributions, life histories, fisheries, and habitat needs.

The status and management of the four groups of shark species in the Gulf of Mexico, Small Coastal Sharks, Large Coastal Sharks, Pelagic Sharks, and Sharks Prohibited from Fisheries, were also evaluated. All of the four species in the Small Coastal Sharks group (Atlantic sharpnose shark, blacknose shark, bonnethead shark, and finetooth shark) were evaluated. Two of the 11 species in the Large Coastal Sharks group (sandbar shark and blacktip shark) were selected for evaluation because they are two of the most abundant and most commercially and recreationally important shark species, and they are widely distributed in the Gulf of Mexico. Rays and skates were also evaluated with three species (giant manta ray, cownose ray, and smalltooth sawfish) being selected because of their abundance and distribution.

Stock assessments to estimate stock abundance and determine stock status are only conducted for a very small number of marine organisms in the Gulf of Mexico (e.g., overfished and/or overfishing). A recent study indicates that of about 60 fish stocks managed in the Gulf of Mexico, information to determine their status is only available for fewer than half (Karnauskas et al. 2013). No formal stock assessments had been done for the vast majority of fish species in the Gulf of Mexico prior to the Deepwater Horizon event, and currently there is limited knowledge about the status of most fish species that live in and/or use the Gulf in part of their lifecycle.

9.2 OVERVIEW OF THE GULF OF MEXICO ECOSYSTEM FOR FINFISH

The Gulf of Mexico provides a wide variety of habitats for finfish and sharks (McEachran and Fechhelm 2005), ranging from coastal marsh, seagrasses, mangroves, river mouths, and reefs to man-made structures such as oil and gas platforms, continental shelf, slope, and deepwaters (Figure 9.1). There is large spatiotemporal variability in oceanographic conditions, with the Gulf of Mexico influenced greatly by inflows and discharges from rivers and other land-based sources, including the Mississippi River, and by large-scale oceanographic features, such as the Loop Current and associated core eddies of different thermal conditions (Govoni

Figure 9.1. Coral reefs, such as this one on the Flower Garden Banks, are one of a wide variety of habitats available to finfish and sharks in the Gulf of Mexico (photograph by Emma Hickerson, Flower Garden Banks National Marine Sanctuary) (from NMS 2013).

and Grimes 1992; Sturges and Leben 2000). Combined, these factors result in large spatiotemporal variability in physicochemical conditions, causing primary production to vary markedly within and across areas of different oceanographic conditions in the Gulf of Mexico (Grimes and Finucane 1991; Biggs 1992). Physical–chemical variability affects the distribution, growth, and mortality of pelagic larvae of many fish species in the Gulf of Mexico (Govoni et al. 1989; DeVries et al. 1990; Lang et al. 1994). Higher abundance, increased growth, and reduced mortality have been observed for larvae within frontal features created by riverine discharge and hydrodynamic convergence (Lang et al. 1994; Hoffmayer et al. 2007). The great spatiotemporal variability in oceanographic and physicochemical conditions provides a large diversity of habitat for fish species that often require different habitats in their different stages of life history.

The general movement patterns and key habitat requirements for fish of different life-history stages are responsible for the formulation of fish community structure (O'Connell et al. 2004). Environmental variables, such as temperature, primary production, current, salinity, depth, dissolved oxygen, water clarity, substrate, and geographic area, have been found important in regulating the spatiotemporal dynamics of fish communities (McEachran and Fechhelm 2005). Because of the large spatiotemporal variability in these environmental variables, fish community structure varies temporally among seasons and years and spatially over estuarine categories and geographic areas, such as in areas east and west of the Mississippi River (Hoese and Moore 1998; McEachran and Fechhelm 2005).

For a given fish population, the dynamics of distribution, abundance, and life-history processes are greatly influenced by abiotic factors, such as water temperature, salinity, dissolved oxygen, and substrate, as well as a variety of biotic factors, such as food availability, intra- and interspecific competition, and predator abundance (Briggs 1974; Richards et al. 1989; Ahrenholz 1991). The most profound impacts of these factors on the dynamics of a fish population usually occur during their early life-history stages, when their survival rates are most sensitive to the change in biotic and abiotic environments (Gallaway et al. 2009). For marine fish species that tend to have planktonic early life-history stages (eggs and larvae), their survival rates during the planktonic life stage are usually a function of parental abundance and fecundity and their complex interactions with predation, oceanographic processes, and prey abundance (Richards and Lindeman 1987). Parental abundance can be greatly affected by the level of fishing mortality. The process of fish growing from early life-history stages to catchable sizes, or becoming catchable in commercial and recreational fisheries, is often referred to as recruitment, which consists of largely distinct ecological processes including survival of a cohort of planktonic eggs and larvae, spatiotemporal patterns of demersal settlement of free-swimming juveniles, and natural and fishing mortality of adults and juveniles before they reach the catchable sizes defined by fishing gear selectivity or minimum legal size requirements (Gallaway et al. 2009). Spatiotemporal variations in recruitment, which can be affected by environmental variables and commercial and recreational fisheries, contribute to variability in fish populations and community structure in the Gulf of Mexico ecosystem (O'Connell et al. 2004; McEachran and Fechhelm 2005).

9.2.1 Key Environmental Variables Influencing Spatiotemporal Dynamics of Fish Populations

Several natural environmental gradients result in the diversity of habitats, which contributes to the relatively high species richness in the Gulf of Mexico and spatiotemporal distribution of finfish species (McEachran and Fechhelm 2005). The first gradient is salinity, which tends to increase from west to east along the coastline as a result of spatial variability in rainfall, river output, and temperature. Bottom composition is the second gradient (McEachran and Fechhelm 2005). Large amounts of fine-grained sediments exist in the Gulf of Mexico along the East Texas and Louisiana coasts as a result of large riverine inputs. Bottom sediments become coarse-grained and sandy off the arid South Texas coast and less sandy and muddier away from the barrier islands. Rocky reefs appear on the 40-fathom contour off Texas and on the continental shelf off Louisiana, providing a hard bottom substrate habitat suitable for species of tropical reef fish not typically found in the inshore shallow waters. From east of the Mississippi Delta, the shelf tends to have coarse-grained sandy sediment with large areas of hard bottom and accumulations of shells, which differs greatly from that of most of the western Gulf of Mexico (McEachran and Fechhelm 2005). The Florida West coast mainly has limestone and detrital-derived sediments, which provide suitable habitat for the spread of many coral reef fishes northward. The spatial variation in sediments contributes greatly to the diversity of habitat for fish requiring specific bottom substrates in different stages of their life history in the Gulf of Mexico (McEachran and Fechhelm 2005).

A third gradient is the spatial variability in depth from the shore to the edge of the continental shelf, resulting in large spatiotemporal variability in the temperature regime, which provides habitat diversity for different fishes (McEachran and Fechhelm 2005). The Gulf of Mexico has greater seasonal changes in thermal habitat than regions to the south or east (Backus et al. 1997); these changes provide a diversity of habitat niches for fish species with differing thermal habitat requirements.

Currents play a central role in regulating the sources of fish recruitment, as well as the transportation and distribution of fish larvae, which can have great impacts on the dynamics of fish populations (Richards et al. 1989; DeVries et al. 2006). Many fish species spawning in the Gulf of Mexico depend on seasonal and often wind-driven currents to transport their larvae into estuarine nursery areas. The Loop Current, which enters the Gulf of Mexico from the Yucatán Channel and begins the Gulf Stream, dominates the Gulf of Mexico oceanographic features. It contains a rich variety of larval tropical fishes that grow and settle on the reefs of the eastern Gulf of Mexico, while eddies transport additional species into the western Gulf of Mexico (McEachran and Fechhelm 2005). The Loop Current may further act as an important geographic isolating mechanism that separates inshore fish populations of the eastern and western Gulf of Mexico (Govoni and Grimes 1992; Sturges and Leben 2000). This may result in the degree of endemism found in the western Gulf of Mexico (Shipp 1992). Thus, the spatial structure of Gulf of Mexico fishes is influenced greatly by the Loop Current and its associated anticyclonic rings (Kleisner et al. 2010).

Approximately 4,000 oil and gas platforms exist in the northern Gulf of Mexico, acting as one of the most extensive man-made reef structures in the world. Many of these petroleum platforms have existed for more than 40 years and have greatly affected spatiotemporal distributions of pelagic fish species (Franks 2000). These platforms vary greatly in size and structural complexity from small, single-well platforms to large, multi-well platforms with complex structures that are installed in both inshore shallow waters and in waters more than 250 km (155 miles) offshore and deeper than 2,000 m (6,562 ft) (Cranswick and Regg 1997; Franks 2000). These platforms form additional new habitat in the northern Gulf of Mexico that attracts pelagic and mid-water fish species to form a distinctive ichthyofaunal community different from the faunal assemblage in the surrounding natural habitat (Gallaway and Lewbel 1982; Franks 2000).

Extreme conditions of environmental variables, such as temperature, salinity, and dissolved oxygen, as well as the existence of natural and human-induced toxic substances, can result in significant temporary or even permanent loss of habitats that can lead to die-offs of fishes in the affected areas (McEachran and Fechhelm 2005). Drastic events, such as red tides (Riley et al. 1989), brown tides (Buskey and Hyatt 1995), and extreme freezes (McEachron et al. 1994), in the Gulf of Mexico can significantly increase fish mortality and cause large-scale die-offs. Subtle and long-term changes can cause a gradual shift of the fish community from more temperate species to more tropical species or vice versa for a given region in the Gulf of Mexico (O'Connell et al. 2004). For example, tropical fish tend to be rare inshore, but are commonly found along the South Texas coast (Hoese and Moore 1998). Climatic events, such as hurricanes and floods, can also affect fish community and population dynamics in the Gulf of Mexico; for example, storms are believed to enhance red drum recruitment (Matlock 1987).

9.2.2 The Fish Community in the Gulf of Mexico

The ichthyofaunal community of the Gulf of Mexico has features similar to those of both warm temperate and tropical waters. The Gulf of Mexico has a relatively rich fish fauna for its size and has nearly 10 % of the world's known marine fish species (Nelson 2006). McEachran and Fechhelm (2005) suggest that the species richness and composition in the Gulf of Mexico is largely similar to that in the West-Central Atlantic region (Cape Hatteras, North Carolina to the equator). Previous studies have documented 1,443 finfish species in 700 genera, 223 families, and 45 orders in the Gulf (Hoese and Moore 1998; McEachran and Fechhelm 2005), which is 200 more species and 54 more genera than what occurs in the eastern Atlantic Ocean between the Arctic and the southern coast of Morocco, including the Mediterranean. This is equal to 64.2 % of the species, 81.6 % of the genera, 92 % of the families, and all of the orders of fish in the West-Central Atlantic Ocean (McEachran and Fechhelm 2005). Species of fish in the West-Central Atlantic Ocean that are not found in the Gulf of Mexico are mostly deep-sea and oceanic fish; temperate fishes rarely occur to the south of Cape Hatteras. Tropical fishes are rare north of Central America or west of the Bahamas or Great Antilles. Relatively large seasonal temperature changes and the lack of extensive reef habitat may exclude species that are not adapted to seasonal changes in thermal habitat and are reef-dependent (McEachran and Fechhelm 2005).

According to McEachran and Fechhelm (2005), only 4.6 % of the 1,443 species (66 species) can be defined as endemic to the Gulf of Mexico. Of these, only nine species are omnipresent and distributed throughout the Gulf of Mexico. The majority of the endemic species are distributed in one or two of the three subregions in the Gulf of Mexico (eastern, northwestern, and southern). Five species of fish that are widely distributed along the U.S. east coast, including Atlantic sturgeon (*Acipenser oxyrhynchus*), striped bass (*Morone saxatilis*), black sea bass (*Centropristis striata*), banded drum (*Larimus fasciatus*), and shelf flounder (*Etropus cyclosquamus*), have or have had isolated populations in the northern Gulf of Mexico, suggesting that the Gulf populations are or were remnants of western extremes of once continuous populations (Smith et al. 2002; McEachran and Fechhelm 2005). However, species that limit their distribution in the southern subregion of the Gulf of Mexico tend not to be endemic to the Gulf of Mexico. This indicates that the Yucatán Peninsula, unlike the Florida Peninsula, is not a biogeographic barrier (Smith et al. 2002). The Gulf of Mexico has deep sills in the Straits of Yucatán and in the Straits of Florida, which may allow for easy movement of fishes between the Gulf of Mexico and other areas.

The Gulf of Mexico cannot be defined as a biogeographic province, which requires that more than 10 % of all the species be endemic (Briggs 1974). However, it can be considered a unique biogeographic region because of its high fish species richness and unique community of warm temperate and tropical fish species (McEachran and Fechhelm 2005). This may be due to a combination of diversity of habitats, geological and oceanographic conditions, and geographic location, which makes it accessible to warm temperate and tropical shore fishes and most deep-sea pelagic and benthic fish species (McEachran and Fechhelm 2005).

Table 9.4. Summary of Finfish Spatial Distributions within the Gulf of Mexico (data from McEachran and Fechhelm 2005)

Distribution Region	Percent of Total Species (%)	Species
Entire Gulf of Mexico	48.8	Wide-ranging epipelagic, mesopelagic, and benthic fish species
Eastern subregion of Gulf of Mexico	14.6	Mainly fish species in the families: Ophichthidae, Alepocephalodae, Melanostomiidae, Notosudidae, Paralepididae, Syngnathidae, Opistognathidae, Apogonidae, Chaenopsidae, and Gobiidae
Northwestern subregion of Gulf of Mexico	3.6	Mainly fish species endemic to the Gulf of Mexico or found in the southeastern United States and some deep-sea fish species, such as narrownose chimaera, bigeye sand tiger, fangtooth snake-eel, snipe-eel, and blue slickhead
Southern subregion of Gulf of Mexico	6.4	Fish species that also occur in the Caribbean Sea
Northern subregion of Gulf of Mexico (including both eastern and northwestern subregions)	17.5	Most fish species also occur along the eastern seaboard of the United States
Western subregion of Gulf of Mexico (including both northwestern and southern subregions)	3	Some fish species have disjunct populations along the eastern seaboard of the United States
Both eastern and southern subregions of Gulf of Mexico	5.3	Fish species tend to be associated with reefs and can often be found in the Florida Keys, the Bahamas, and the Greater and Lesser Antilles

Almost half of the 1,443 species occurring throughout the Gulf of Mexico can be considered ubiquitous within their respective depth (Table 9.4). These species include wide-ranging epipelagic fishes, e.g., blacktip shark (Carcharhinidae), Gulf menhaden (Clupeidae), Atlantic needlefish (Belonidae), Atlantic flyingfish (Exocoetidae), and common halfbeak (Hemiramphidae); mesopelagic fishes, e.g., Garrick (Gonostomatidae), hatchetfish (Sternoptychidae), lightfish (Phosichthyidae), and smallfin lanternfish (Myctophidae); benthic fishes of the continental shelf, e.g., squirrelfish (Holocentridae), red grouper (Serranidae), red snapper (Lutjanidae), and red drum (Sciaenidae); and benthic fishes of the slope, e.g., blackfin spiderfish (Ipnopidae), Western Atlantic grenadier (Macrouridae), and beardless codling (Moridae) (McEachran and Fechhelm 2005). These families of species also tend to be distributed in other regions of the Atlantic, Pacific, and Indian Oceans. The remaining 51.2 % of the 1,443 fish species mainly limit their spatial distributions within a subregion of the Gulf of Mexico (Table 9.4). For example, a total of 211 species (14.6 %) can be found only in the eastern subregion of the Gulf of Mexico (Table 9.4), and most of these species are mesopelagic fishes that may reflect intrusion of the Loop Current into the eastern Gulf of Mexico. The distribution patterns reflect spatial variability in geological and oceanographic conditions and other habitat variables (Hoese and Moore 1998; McEachran and Fechhelm 2005). For example, benthic species that prefer terrigenous substrates are mainly found in the northern and western Gulf of Mexico, and benthic fishes associated with calcareous substrates tend to be found in the calcareous shelves of Florida and the Yucatán; species preferring warm temperate habitats are usually found in the northern Gulf of Mexico, while those preferring tropical habitats tend to occur in the southern Gulf of Mexico (McEachran and Fechhelm 2005). Although the results may be biased by the difference in sampling efforts, the eastern Gulf of Mexico (Florida Bay to Pensacola, Florida or Mobile Bay, Alabama) appears to have the highest number of species (1,259), followed by the western Gulf of Mexico (Pensacola or Mobile Bays to Cape Rojo, Mexico, 1,056 species), with the southern Gulf of Mexico (Cape Rojo to Cape Catoche, Mexico) having the lowest species diversity (916) (Table 9.4).

More than 1,112 species of finfish, sharks, and rays in the Gulf of Mexico were included in the FishBase database developed by the World Fisheries Center (Froese and Pauly 2009). Although this is not a complete list [the number of finfish species alone is 1,443 as suggested by McEachran and Fechhelm (2005)], the species included in FishBase represent a majority of the finfish, sharks, and rays in the Gulf of Mexico. Based on habitat needs and distribution in the water column, these 1,112 species are divided into seven groups in FishBase: reef-associated, bathydemersal, bathypelagic, benthopelagic, demersal, pelagic-neritic, and pelagic-oceanic. Of the 1,112 fish species in FishBase, more than one-third are reef-associated species, and the benthopelagic and pelagic-oceanic fish species have the lowest species diversity (Table 9.5). The trophic level of fishes associated with each habitat tends to vary greatly (Figure 9.2). The pelagic-oceanic species tend to have the highest average trophic level, while the reef-associated fish species tend to have the widest distribution of trophic levels (Table 9.5; Figure 9.2). The maximum size also varies greatly within each habitat group, with the pelagic-oceanic group having the largest average maximum size and the bathypelagic group having the smallest average size (Table 9.5).

Although various fishery-dependent and fishery-independent monitoring programs have been developed and implemented (McEachran and Fechhelm 2005) and some species, such as red snapper and Gulf menhaden, are well researched, many fish populations in the Gulf of Mexico are still not well understood compared to those of other marine ecosystems in the United States (Rowe and Kennicutt 2009). Therefore, large uncertainty still remains on the dynamics and conditions of many Gulf of Mexico fish populations of commercial and recreational importance (NMFS 2012a).

Table 9.5. Summary of Average Trophic Level, Number of Species, and Average Maximum Size Calculated for Each Habitat Group for the 1,112 Finfish, Shark, and Ray Species in the Gulf of Mexico (data from FishBase 2013)

Habitat	Trophic Level[a]				Size (cm)	
	Mean	Coefficient of Variation	# of Species	% of Species	Mean	Coefficient of Variation
Reef-associated	3.470	0.181	384	34.5	53.73	1.27
Bathydemersal	3.671	0.097	131	11.8	50.91	1.16
Bathypelagic	3.695	0.123	158	14.2	32.44	1.25
Benthopelagic	3.745	0.142	66	5.9	90.92	1.45
Demersal	3.568	0.126	260	23.4	56.26	1.62
Pelagic-neritic	3.389	0.167	46	4.1	46.10	0.97
Pelagic-oceanic	3.970	0.121	67	6.0	216.21	1.53

[a]Trophic level measures the number of steps the fish, shark, or ray is from the start of the food chain: 1 = primary producers that make their own food, such as plants and algae; 2 = primary consumers, such as herbivores consuming primary producers; 3 = secondary consumers, such as carnivores eating herbivores; 4 = tertiary consumers, such as carnivores eating other carnivores; and 5 = apex predators that are at the top of the food chain with no predators (FishBase 2013)

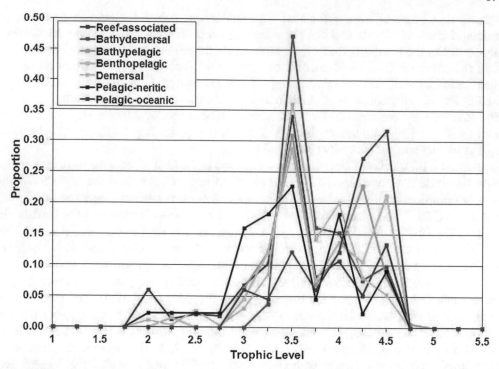

Figure 9.2. The distribution of trophic levels for fish, shark, and ray species of different habitats in the Gulf of Mexico. Trophic level measures the number of steps the fish, shark, or ray is from the start of the food chain: 1 = primary producers that make their own food, such as plants and algae; 2 = primary consumers, such as herbivores consuming primary producers; 3 = secondary consumers, such as carnivores eating herbivores; 4 = tertiary consumers, such as carnivores eating other carnivores; and 5 = apex predators that are at the top of the food chain with no predators (data from FishBase 2013).

9.3   POPULATION DYNAMICS OF KEY FINFISH SPECIES OF ECOLOGICAL, COMMERCIAL, AND RECREATIONAL IMPORTANCE

Many fish species support highly valued commercial and recreational fisheries. These species range from reef-dependent snappers and groupers, to highly migratory tuna and billfish, to coastal pelagic menhaden and mackerel, and coastal demersal drums and jacks (Hoese and Moore 1998; McEachran and Fechhelm 2005). They differ greatly in their ecological roles, life histories, habitat needs, and contributions to commercial and recreational fisheries. As described earlier, 13 representative species have been selected from ten families for evaluation in this chapter: red snapper; menhaden, including Gulf menhaden, finescale menhaden, and yellowfin menhaden; red grouper; Atlantic bluefin tuna; Atlantic blue marlin; Atlantic swordfish; Atlantic sailfish; red drum; tilefish; king mackerel; dolphinfish; striped mullet, and greater amberjack (Table 9.3). Because of their ecological and fisheries significance in the Gulf of Mexico, snapper and grouper species were also evaluated as families for their distribution, life history, fisheries, and habitat needs.

9.3.1 Snappers (Family Lutjanidae)

The family Lutjanidae, or snappers, is composed of 17 genera and about 100 species of mostly reef-dwelling marine fishes that are divided into four subfamilies (Allen 1985). Snappers are confined mostly to tropical and subtropical regions of all oceans, while three species occur in freshwater; juveniles of many snapper species inhabit brackish mangrove estuaries and the lower reaches of freshwater streams. Snappers occur in four discrete geographic faunas, and snappers that occur in the western Atlantic Ocean are not found in any other region (Allen 1985). Snappers that occur in the Gulf of Mexico region include 16–17 species in six genera from the family Lutjaninae (Table 9.1) (McEachran and Fechhelm 2005).

Snappers have separate sexes, sexual differentiation remains constant throughout their life span, and sexual dimorphism is rare (Martinez-Andrade 2003). A key reproductive strategy utilized by many species of inshore-dwelling snappers is an extensive migration to selective offshore areas along outer reefs to form seasonal spawning aggregations in the week or so prior to the full moon (Martinez-Andrade 2003). Snapper larvae are most common relatively close to shore, in waters over the continental shelf, or in large coral reef lagoons; they are relatively rare in the more offshore areas at the edge of the shelf and in oceanic waters (Allen 1985). Snappers can grow to about 1 m (3.3 ft) in length, and the typical maximum life span of snappers has been estimated between 4 and 21 years. Most snappers occur in shallow to intermediate depths to 100 m (328 ft), although some are largely confined to deepwater (100–500 m or 328–1,640 ft) (Allen 1985).

Snappers are active predators feeding mostly at night on a variety of prey (Allen 1985). Fishes dominate the diet of most species, and other common prey include crabs, shrimps, other crustaceans, gastropods, cephalopods, and planktons. Generally, the larger, deep-bodied snappers feed on other fishes and large invertebrates on or near the surface of the reef and are usually equipped with large canine teeth adapted for seizing and holding their prey.

Landings of snappers are of significant volume and economic value because of the excellent quality of their meat and high demand, making them some of the most desirable species in the market (Martinez-Andrade 2003). The Gulf of Mexico Fishery Management Council (GMFMC) manages snappers under the Reef Fish Fishery (GMFMC 2004a). The Reef Fish Management Plan (FMP) currently includes 42 species, and snapper species managed under this FMP include red snapper, queen snapper (*Etelis oculatus*), mutton snapper (*Lutjanus analis*), schoolmaster (*Lutjanus apodus*), blackfin snapper (*Lutjanus buccanella*), cubera snapper (*Lutjanus cyanopterus*), gray or mangrove snapper (*Lutjanus griseus*), dog snapper (*Lutjanus jocu*), mahogany snapper (*Lutjanus mahogoni*), lane snapper (*Lutjanus synagris*), silk snapper (*Lutjanus vivanus*), yellowtail snapper (*Ocyurus chrysurus*), wenchman (*Pristipomoides aquilonaris*), and vermilion snapper (*Rhomboplites aurorubens*).

Because of its recreational and commercial importance as a prized food fish in the Gulf of Mexico, red snapper was selected as the representative snapper species for evaluation (Figure 9.3). Key life-history parameters for red snapper are summarized in Table 9.6 and discussed in the sections below. A summary of red snapper habitat information is presented in Table 9.7, while Table 9.8 includes stock and fisheries information for the red snapper; this information is also discussed in more detail in the following sections.

Figure 9.3. Red snapper (*Lutjanus campechanus*) on a coral reef in the Gulf of Mexico (from von Brandis 2013).

Table 9.6. Summary of Life-History Information for Red Snapper (*Lutjanus campechanus*)

Parameter	Value	Reference
von Bertalanffy growth model parameters[a]	$L_\infty = 876.9$ mm (34.5 inches [in]) fork length (FL)[2]	SEDAR 7 (2005)
	$L_\infty = 876.9$ mm (34.5 in.) FL	Nelson and Manooch (1982)
	$K = 0.22$ per year	SEDAR 7 (2005)
	$K = 0.17$ per year	Nelson and Manooch (1982)
	$t_0 = 0.37$ years	SEDAR 7 (2005)
	$t_0 = -0.1$ years	Nelson and Manooch (1982)
Age at first maturity	1 year	Cook et al. (2009)
	2 years	Fitzhugh et al. (2004), Woods et al. (2007)
Length at first maturity	Smallest females showing evidence of recent spawning: 196 mm (7.7 in.) and 216 mm (8.5 in.) FL	Cook et al. (2009)
	296 mm (11.6 in.) FL	Fitzhugh et al. (2004)
	285 mm (11.2 in.) FL	Woods et al. (2007)
Spawning season	April through September, peaks June through August	Bradley and Bryan (1975), Futch and Burger (1976), Render (1995), Collins et al. (1996)
Spawning location	Spawn offshore on the shelf and upper continental slope over sand and mud bottom areas away from reefs, highest abundances occur in the Northern Gulf of Mexico off central and western Louisiana	Szedlmayer and Furman (2000), Collins et al. (2001), Woods (2003), Fitzhugh et al. (2004), GMFMC (2004a), Lyczkowski-Shultz and Hanisko (2007)
Common prey of juveniles	Diet comprised primarily of fish and invertebrates from reef and soft bottom habitat. Fishes include blennies, *Halichoeres* sp., Serranidae (*Serranus* sp., *Centropristes* sp.). Invertebrates include shrimps (mantis shrimp, rock shrimp, Alpheidae, Hippolytidae), squid, octopuses, and crabs	Bradley and Bryan (1975), Beaumariage and Bullock (1976), Futch and Burger (1976), Szedlmayer and Lee (2004)

(continued)

Table 9.6. (continued)

Parameter	Value	Reference
Common prey of adults	Soft bottom prey are a major component of the diet, but reef associated fishes are taken when abundant. Fishes include gulf pipefish, shoal flounder, puffer family, striped mullet, sea robin family, rough scad, butterfish family, sand perch, and clupeids. Invertebrates include mantis shrimp, crabs, gastropods, and zooplankton	McCawley and Cowan (2007), Addis et al. (2011)
Common prey of large adults	For large adults, feeding is independent of reef habitat and includes a wide variety of prey from reef, soft bottom, pelagic, and *Sargassum* habitats	Gallaway (1981)
Common predators	Data not available	

Note: *mm* millimeter(s), *in.* inch(es)

[a] The von Bertalanffy growth model describes how fish length changes with age and can be written as $L_t = L_\infty(1 - e^{-K(t - t0)})$, where L_t is fish length at age t, L_∞ is the maximum attainable length, K is the growth coefficient describing how fast fishes approach L_∞, and t_0 is a theoretical age at which fish size is 0 (Ricker 1975)

[b] Fork length (FL) is the length from the tip of the snout to the end of the middle caudal fin rays (fork of the tail fin) (FishBase 2013)

Table 9.7. Summary of Habitat Information for Red Snapper (*Lutjanus campechanus*)

Parameter	Value	Reference
Habitat preferences and temporal/spatial distribution of juveniles	Around age 6 months, juveniles recruit to structured habitat and reefs with medium relief, structures about 1 m³ (35.3 ft³) in size; salinity approximately 35 ppt; temperature from 24 to 26 °C; dissolved oxygen at least 5 mg/L; depth from 18 to 64 m (59–210 ft); highest distribution of juveniles are found from Alabama to southern Texas	Gallaway et al. (1999), Szedlmayer and Lee (2004)
Habitat preferences and temporal/spatial distribution of adults	Around age of 1.5 years, adults start recruiting to large reefs, natural rock outcroppings, offshore petroleum platforms, wrecks, and large artificial reefs across the continental shelf to the shelf edge	Stanley (1994), Gallaway et al. (1999), Patterson et al. (2001), Nieland and Wilson (2003)
Habitat preferences and temporal/spatial distribution of spawning adults	Older fish, age 8+ years, reach sizes that render them largely invulnerable to predation, and spend a larger portion of their time over soft bottoms; highest abundances occur in the northern and western Gulf over mud bottoms with depressions or lumps; depth from 55 to 92 m (180–302 ft)	Boland et al. (1983), Render (1995), Nieland and Wilson (2003), Mitchell et al. (2004)
Designated Essential Fish Habitat for juveniles and adults	All estuaries in the U.S. Gulf of Mexico; the U.S./Mexico border to the boundary between the areas covered by the GMFMC and the SAFMC from estuarine waters out to depths of 100 fathoms	GMFMC (2005)

Note: °C degrees Celsius, *GMFMC* Gulf of Mexico Fishery Management Council, *mg/L* milligram(s) per liter, *ppt* part(s) per thousand, *SAFMC* South Atlantic Fishery Management Council

Table 9.8. Summary of Stock and Fisheries Information for Red Snapper (*Lutjanus campechanus*)

Parameter	Value	References
General geographic distribution	Gulf of Mexico, Caribbean Sea, and U.S. Atlantic coast to northern South America. Greatest abundance occurs in the northern Gulf off southwestern Louisiana and Alabama, as well as on the Campeche Banks off of Mexico	Patterson et al. (2001), GMFMC (2004a), SEDAR 31 (2009), Walter and Ingram (2009)
Commercial importance	High, commercial landings highest in the western Gulf of Mexico	SEDAR 7 (2005)
Recreational importance	High, recreational landings highest in the western Gulf of Mexico	SEDAR 7 (2005)
Management agency	NMFS, GMFMC	SEDAR 31 (2009)
Management boundary	Mexico-Texas boarder to west of the Florida Keys (GMFMC boundaries)	SEDAR 31 (2009)
Stock structure within the Gulf of Mexico	Managed as one stock, but assessed as two subunits (east/west of the Mississippi River)	SEDAR 31 (2009)
Status (overfished/ overfishing)	Overfished 2001–2012; overfishing 2001–2011	NMFS (2012a)

Note: NMFS National Marine Fisheries Service

9.3.1.1   Key Life-History Processes and Ecology

Red snapper are distributed throughout the Gulf of Mexico and the U.S. Atlantic coast (Figure 9.4). Genetic studies support the hypothesis of a single red snapper stock in the northern Gulf of Mexico (Gold et al. 1997; Heist and Gold 2000).

Larval abundance is directly related to adult abundance (Lyczkowski-Shultz et al. 2005). During the peak spawning months, the highest larval density is found off the Louisiana coast at depths of 50–100 m (164–328 ft) (Table 9.6), and abundance tends to be lower east of the Mississippi River compared to west of the Mississippi River. According to fall plankton surveys, red snapper larvae can be found less frequently and in lower abundance in the eastern Gulf of Mexico than in the western Gulf. Larvae were also found between the 100 and 200 m (328 and 656 ft) depth contours throughout the Gulf of Mexico, indicating that red snapper spawn from the mid-shelf to the continental slope.

After reaching 50 mm (1.9 in.) total length [TL refers to the length from the tip of the snout to the tip of the longer lobe of the caudal fin or tail, usually measured with the lobes compressed along the midline (FishBase 2013)], these age-0 red snapper are taken as bycatch in the Gulf of Mexico penaeid shrimp fishery and continue to be taken as bycatch as age-1 red snapper. The highest density of age-0 to -1 red snapper is found in the northern Gulf of Mexico at depths between 18 and 55 m (59 and 180 ft) from the Alabama–Florida border to the Texas–Mexico border (Gallaway et al. 1999). They tend to prefer shell and sand substrates (Szedlmayer and Howe 1997). Studies suggest an ontogenetic shift from low-relief to higher-relief habitat with size and age (Szedlmayer and Lee 2004; Wells 2007). The newly settled fish smaller than 40 mm (1.6 in.) TL mostly occur in open habitat, but begin moving onto the reefs as their sizes approach 100 mm (3.9 in.) TL (Table 9.7). They tend to have a high degree of fidelity to these habitats (Workman et al. 2002).

Red snapper enter the targeted commercial and recreational fisheries at age 2 for the rest of their life span (Wilson and Nieland 2001). They can be found across the shelf to the shelf

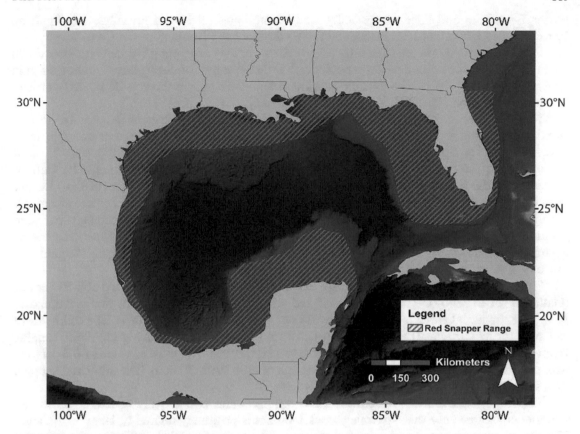

Figure 9.4. Range of the red snapper (*Lutjanus campechanus*) in the Gulf of Mexico and Florida east coast (from USGS 2010a).

edge and show an affinity for vertical structure (Table 9.7) (Patterson et al. 2001), especially from 2 to 10 years of age. Red snapper older than 8–10 years are no longer totally dependent on structured habitats and are capable of foraging over open habitat (Table 9.7). A National Marine Fisheries Service (NMFS) bottom-longline survey suggests that red snapper tend to be most abundant at depths from 55 to 92 m (180–302 ft) and that older and larger red snapper are found more frequently in the western Gulf of Mexico, while younger and smaller fish are found in the eastern Gulf (Mitchell et al. 2004). Adult red snapper tend to experience a seasonal depth-related movement toward shallower water (inner-mid shelf) in the spring/summer months and offshore (mid-outer shelf) in the winter months (Bradley and Bryan 1975). This movement may be related to spawning-related activity (SEDAR 7 2005).

Red snapper have some rather unique life-history traits (Table 9.6). In the Gulf of Mexico, they can reach maturity at young ages but have a long life span of more than 50 years (Szedlmayer and Shipp 1994; Wilson and Nieland 2001). Red snapper are batch spawners, with an estimated spawning duration of 180 days and a mean spawning frequency of 3.0 in the eastern Gulf of Mexico and 2.9 in the western Gulf of Mexico (SEDAR 7 2005). Lyczkowski-Shultz and Hanisko (2007) suggest that red snapper tend to spawn over a wide depth range from the mid-shelf to the continental slope. The eggs are pelagic and hatch in about 20–27 h after fertilization (Minton et al. 1983). The larvae remain pelagic for about 26–30 days until metamorphosis and settlement (Rooker et al. 2004). After the completion of the pelagic larval stage, red snapper settle and move to structured habitat, such as low-relief, relic-shell habitat (Workman and

Foster 1994; Piko and Szedlmayer 2007), and become post-settlement juveniles, ranging from 19 to 50 mm (0.75–1.9 in.) TL in size and 29–66 days in age (Szedlmayer and Conti 1999).

Red snapper experience high rates of growth when they are young but begin to slow down when they reach the age of 8–10 years. There is little evidence for strong sexual dimorphism in growth (Goodyear 1995). The average maximum attainable size in the von Bertalanffy growth equation is less than 900 mm (35.4 in.) TL (Table 9.6).

Females tend to mature at relatively smaller sizes and earlier ages in the eastern Gulf of Mexico compared to those in the western Gulf of Mexico (SEDAR 7 2005). For example, in an analysis done for Southeast Data, Assessment and Review (SEDAR) 7 (2005), over 75 % of females were mature by 300 mm (11.8 in.) FL for samples taken from the eastern Gulf of Mexico, but the proportion in the west was still below 75 % even at 350 mm (13.8 in.) FL. For both regions, all females were mature after reaching 650 mm (25.6 in.) FL. The red snapper is highly fecund and, on average, a female of age 10 can produce over 60 million eggs per year. Fecundity-at-length data can be best quantified with power or exponential functions, but an asymptotic function provides a better fit for fecundity-at-age data, suggesting that fecundity is more dependent upon length, rather than age (SEDAR 7 2005).

Natural mortality[1] (M) during the egg stage of the red snapper is estimated at 0.50 per day (Gallaway et al. 2007). The mortality of red snapper larvae is high, and the accumulative M during the larval stage is estimated at 6.76 per year (Gallaway et al. 2007). The estimates of red snapper M for ages 0 and 1 varied greatly among studies (Gallaway et al. 2009), ranging from 0.98 to 3.7 and 0.6 to 1.4 for age-0 and age-1 fish, respectively. An M value of 0.6 per year was used for age-1 fish in recent stock assessments (SEDAR 7 2005). The adult red snapper M was assumed to be 0.1 per year in the assessment.

The newly hatched larval density in the water column is positively related to adult fish abundance, suggesting that spawning stock biomass is positively related to larval abundance. The abundances of age-0 and age-1 red snapper are poorly correlated, indicating the existence of density-dependent mortality in early life history. The availability of low relief, natural habitat for the post-settlement of red snapper (ages 0 and 1) is suggested as a major limiting factor in the observed level of recruitment (Gallaway et al. 2009). However, Cowan et al. (2011) suggest that age-1 red snapper are more vulnerable to shrimp trawl bycatch as compared to age-0 fish, weakening the above argument about the role of low-relief habitats in the shallow Gulf of Mexico. They further state that habitat limitation is not a strong factor in regulating recruitment dynamics of red snapper (Cowan et al. 2011).

The number of recruits, measured as the number of red snapper at age 1, estimated for the eastern U.S. Gulf of Mexico in the stock assessment (SEDAR 7 2009) is much higher than that for the western U.S. Gulf of Mexico (Figure 9.5). The recruitment of red snapper in both the western and eastern Gulf has fluctuated over time. In the eastern Gulf of Mexico, recruitment reached one of the highest values in 2003, but continued to decline from 2003 through 2008, with the recruits in 2008 being less than half of the recruits in 2003.

Red snapper from Alabama tend to mature at smaller sizes and younger ages than those from Louisiana (Woods et al. 2007). Differences in maturation are also found between the eastern Gulf of Mexico (Mississippi, Alabama, and Florida west coast) and western Gulf of

---

[1] Mortality is usually measured as an instantaneous rate. Total mortality (Z) is the sum of fishing mortality (F) and natural mortality (M). The proportion of fish dead as a result of Z can be calculated as $1 - \exp(-Z)$. Thus, in the absence of fishing mortality ($F = 0$), an M of 0.50 per day for red snapper eggs is equivalent to 39.3 % of red snapper eggs dying per day and an accumulative M of 6.76 per year during the larval stage is equivalent to an annual mortality rate of 99.9 %.

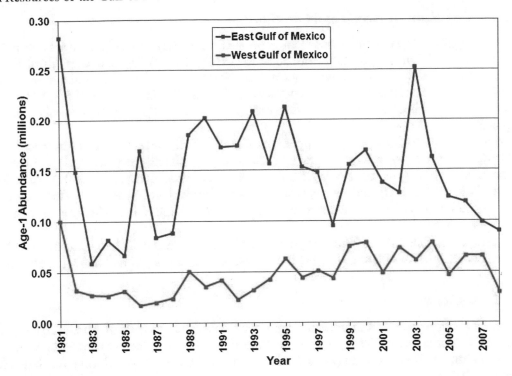

Figure 9.5. Abundance of red snapper (*Lutjanus campechanus*) recruits measured as age-1 fish for the eastern and western U.S. Gulf of Mexico from 1981 to 2008 (data from SEDAR 7 Update 2009).

Mexico (Louisiana and Texas) (SEDAR 7 2005). Young (to age 8) red snapper in the eastern Gulf of Mexico tend to have a higher reproductive output at age compared to those in the western Gulf. A single stock of red snapper in the Gulf of Mexico has been suggested by genetic studies (Camper et al. 1993; Gold et al. 1997; Heist and Gold 2000), which may result from the lack of sufficient time since the Pleistocene epoch for spatially separated substocks of red snapper in the Gulf of Mexico to have become genetically distinct, or from enough mixing to maintain homogeneity in the population. However, phenotypic differences have been identified in growth, maturation, abundance, age/size compositions, prey compositions, and fishery dynamics between the eastern and western Gulf of Mexico. To account for such differences between the two areas in the stock assessment, the Gulf of Mexico red snapper stock is considered to consist of the two substocks. Although there is evidence of large differences in life history and population dynamics at fine spatial scales, such as among different reefs (Gallaway et al. 2009), more studies are needed to evaluate the potential existence of metapopulation structure.

9.3.1.2 Predators and Prey

Juvenile red snapper prey mainly on fishes and invertebrates from reefs and soft bottom habitats (Table 9.6). A diet shift from open-water prey to reef prey was observed by Szedlmayer and Lee (2004) as fish moved from open to reef habitat, suggesting that reef habitat provides not only protection from predation but also additional food sources. The diet of adult red snapper also includes many species of fishes and invertebrates (Table 9.6).

9.3.1.3   Key Habitat Needs and Distribution

Red snapper eggs are pelagic and float to the surface. Newly hatched larvae are also pelagic and are found to be most abundant from 50 to 100 m (164–328 ft) depths in the Gulf of Mexico west of the Mississippi River. After they reach 16–19 mm (0.6–0.7 in.) TL in about 26–30 days of age, they settle to the bottom. The newly settled fish smaller than 40 mm (1.6 in.) TL mostly occur in open habitat, but begin moving onto the reefs as their sizes approach 100 mm (3.9 in.) TL.

Prior to 8–10 years of age, red snapper tend to prefer shell and sand substrates (Szedlmayer and Howe 1997), are attracted to natural and artificial (e.g., oil and gas platforms) reef habitats, and have a high degree of fidelity to these habitats (Workman et al. 2002). Additional characteristics of juvenile red snapper habitat are described in Table 9.6. After they reach age 8–10, they tend to be less attached to reef habitats and spend most of their time in open waters (Table 9.6).

Essential fish habitat has been designated for Reef Fish, which includes juvenile and adult red snapper. Reef fish essential fish habitat is described in Table 9.7 and shown in Figure 9.6.

9.3.1.4   Fisheries

Red snapper support an important commercial fishery in the Gulf of Mexico. The fishery began in Pensacola about 150 years ago (Bortone et al. 1977) and then expanded to the waters off Galveston, Texas, the Campeche Banks, and the Dry Tortugas during the late 1800s (Goodyear 1995).

Commercial landings in the United States are divided into four separate fisheries based on fishing gear (headline and longline) and fishing location (eastern and western Gulf of Mexico): (1) handline east, (2) handline west, (3) longline east, and (4) longline west (Figure 9.7). Most of the catch was landed with handline in the western Gulf of Mexico (Figure 9.7). The total landings tend to have a decreasing trend and reached the lowest value around 1992. The catch doubled for the next 10–12 years, but decreased drastically after 2006 as a result of a large decrease in the western Gulf of Mexico (Figure 9.7).

Red snapper bycatch in the shrimp fishery, mainly consisting of fishes of ages 0 and 1, dominate the catch in numbers of fish (SEDAR 7 2009). The number of red snapper discarded as bycatch has fluctuated between 10 and 60 million fish in most years since the 1970s, and is the lowest in recent years (Figure 9.8). The recreational and commercial fisheries combined take roughly 3–4 million red snapper annually. Targeted commercial and recreational red snapper fisheries dominate removals in weight, accounting for about 4 million kg (9 million lb) in recent years. The annual weight of the shrimp bycatch discarded was estimated to be roughly 1–3 million kg (2–3 million lb) of red snapper (SEDAR 7 2005).

Estimates of the recreational catch for red snapper in the Gulf of Mexico since 1981 are obtained from three surveys: (1) the Marine Recreational Fishery Statistics Survey conducted by the NMFS, (2) the Texas Marine Sport-Harvest Monitoring Program by the Texas Parks and Wildlife Department, and (3) the Headboat Survey conducted by the NMFS, Southeast Fisheries Science Center. The estimated recreational landings of red snapper show a decreasing trend over time since 1981 in the U.S. Gulf of Mexico (Figure 9.9). However, it appears to be relatively stable around one half million kg (1.1 million lb) since 2000.

The recreational fishery of Gulf of Mexico red snapper is managed with a size limit, daily bag limit, seasonal length, and allocation quota. For the 2009 recreational fishing season, the size limit was 40.6 cm (16 in.) TL, the daily bag limit was two fish, the fishing season was from June 1 to August 15 (75 days), and the annual quota allocation was 1.11 million kg (2.45 million lb) (SEDAR 7 2009).

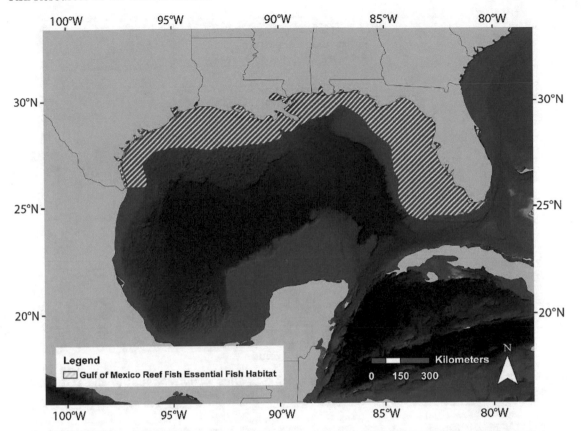

Figure 9.6. The Gulf of Mexico Fishery Management Council's Gulf of Mexico Reef Fish Essential Fish Habitat for queen snapper (*Etelis oculatus*), mutton snapper (*Lutjanus analis*), blackfin snapper (*Lutjanus buccanella*), red snapper (*Lutjanus campechanus*), cubera snapper (*Lutjanus cyanopterus*), gray or mangrove snapper (*Lutjanus griseus*), lane snapper (*Lutjanus synagris*), silk snapper (*Lutjanus vivanus*), yellowtail snapper (*Ocyurus chrysurus*), wenchman (*Pristipomoides aquilonaris*), vermilion snapper (*Rhomboplites aurorubens*), speckled hind (*Epinephelus drummondhayi*), yellowedge grouper (*Epinephelus flavolimbatus*), goliath grouper (*Epinephelus itajara*), red grouper (*Epinephelus morio*), warsaw grouper (*Epinephelus nigritus*), snowy grouper (*Epinephelus niveatus*), black grouper (*Mycteroperca bonaci*), yellowmouth grouper (*Mycteroperca interstitialis*), gag (*Mycteroperca microlepis*), scamp (*Mycteroperca phenax*), yellowfin grouper (*Mycteroperca venenosa*), goldface tilefish (*Caulolatilus chrysops*), blueline tilefish (*Caulolatilus microps*), tilefish (*Lopholatilus chamaeleonticeps*), greater amberjack (*Seriola dumerili*), lesser amberjack (*Seriola fasciata*), almaco amberjack (*Seriola rivoliana*), banded rudderfish (*Seriola zonata*), gray triggerfish (*Balistes capriscus*), and hogfish (*Lachnolaimus maximus*) (from GMFMC 2004b).

The recreational fishery of Gulf of Mexico red snapper is managed with a size limit, daily bag limit, seasonal length, and allocation quota. For the 2009 recreational fishing season, the size limit was 40.6 cm (16 in.) TL, the daily bag limit was two fish, the fishing season was from June 1 to August 15 (75 days), and the annual quota allocation was 1.11 million kg (2.45 million lb) (SEDAR 7 2009).

The red snapper stock is a single management unit in the Gulf of Mexico extending from the U.S.–Mexico border in the west through the northern Gulf waters and west of the Dry Tortugas and the Florida Keys. The assessment assumes there are two sub-units of the red snapper stock within this region, separated roughly by the Mississippi River (SEDAR 7 2009). The GMFMC is responsible for assessing the red snapper stock status in the Gulf of Mexico under Section 303 of the Magnuson-Stevens Act.

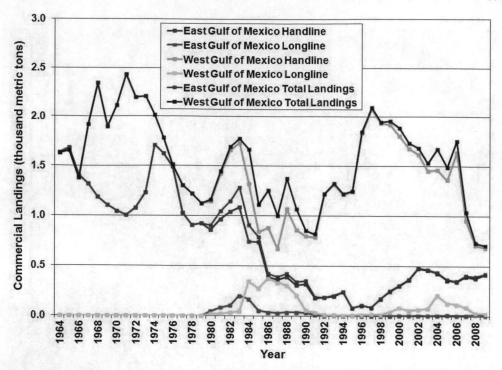

Figure 9.7. Commercial landings of red snapper (*Lutjanus campechanus*) in the U.S. Gulf of Mexico from 1964 through 2009 (data from SEDAR 7 Update 2009).

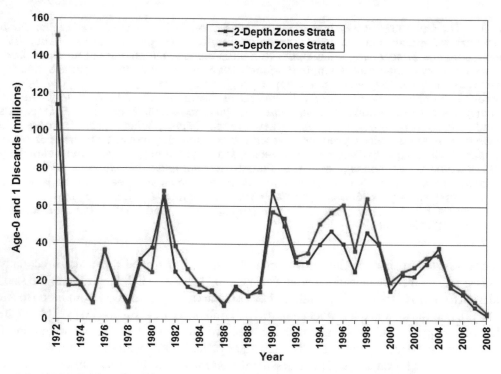

Figure 9.8. Estimated median number of young red snapper (*Lutjanus campechanus*) (ages 0–1) discarded in the shrimp fishery in the U.S. Gulf of Mexico using the 2-depth (0–10, 10+ fathoms) and 3-depth (0–10, 10+, 30+ fathoms) zones strata models from 1972 through 2008 (data from SEDAR 7 Update 2009).

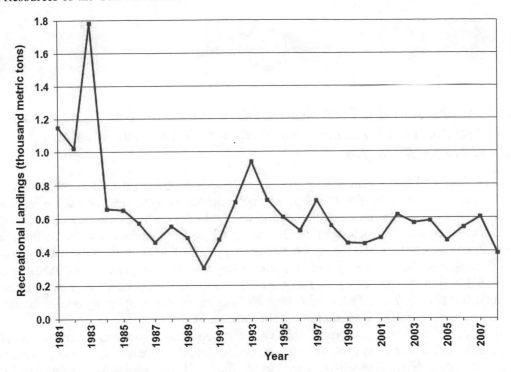

Figure 9.9. Recreational landings of red snapper (*Lutjanus campechanus*) in the U.S. Gulf of Mexico from 1981 through 2008 (data from SEDAR 7 Update 2009).

In the assessment of the Gulf of Mexico fish stocks, a fishery with fishing mortality higher than the maximum fishing mortality threshold (MFMT) is defined as in the status of "overfishing," meaning that fish stocks cannot be sustained under such a level of fishing mortality. A fish population with its biomass lower than the minimum spawning stock threshold (MSST) is defined "overfished," meaning that the stock biomass is too low and reproductive potentials are severely depleted. For the Gulf of Mexico red snapper stock, the MFMT is defined as $F_{SPR26\%}$ (SEDAR 7 2009), a fishing mortality (F) at which the reproductive potential is only 26 % of the maximum reproductive potential in the absence of fishing mortality and was estimated at 0.53 per year. The average F from 2006 through 2008 was 1.00, almost two times as high as the MFMT. The fishing mortality of optimal yield (F_{OY}) was determined at 75 % of the MFMT (e.g., 0.39). Thus, recent fishing mortality was much higher than the F_{OY} and the MFMT, suggesting that overfishing occurred in the Gulf of Mexico red snapper fishery prior to 2010 (SEDAR 7 2009). The stock biomass at which maximum sustainable yield (MSY) is achieved (S_{MSY}) was estimated at 4.6 million kg (10.16 million lb), and for the Gulf of Mexico red snapper, the MSST was calculated as $(1 - M)S_{MSY}$, where M has a value of 0.1 per year, which yields a MSST of 4.1 million kg (9.14 million lb). The biomass as of 2008 was only 1.78, much lower than the MSST (<20 % of the MSST), suggesting that the red snapper stock biomass was severely overfished prior to 2010 (SEDAR 7 2009). Thus, the Gulf of Mexico red snapper stock was overfished, and overfishing occurred in the Gulf of Mexico prior to 2010 (Table 9.8). An early stock assessment (SEDAR 7 2005) also suggests that Gulf of Mexico red snapper were grossly overfished through 2003, and the estimated spawning potential ratio (SPR) was less than 5 %.

9.3.2   Menhaden: Gulf Menhaden (*Brevoortia patronus*), Finescale Menhaden (*Brevoortia gunteri*), and Yellowfin Menhaden (*Brevoortia smithi*)

In the Gulf of Mexico, menhaden play a critical role in linking plankton with upper level predators. Because of their filter feeding abilities, menhaden can consume and redistribute a significant amount of primary production and energy in the Gulf of Mexico. They are small, marine, filter feeding fish belonging to the family Clupeidae (herrings, shads, sardines, hilsa and menhadens). Gulf menhaden are considered the Gulf of Mexico complement to Atlantic menhaden (*Brevoortia tyrannus*) based on morphological and genetic analyses (Dahlberg 1970; Anderson 2007). Both species support large-scale, commercial *reduction* fisheries (not directly consumed but used to make fish products), with Gulf menhaden supporting one of the largest fisheries, by weight, in the United States (Pritchard 2005).

Menhaden abundance can greatly influence the population dynamics of many predatory fish species, such as tunas, drums, and sharks; in addition, they are also a very important food source for many birds (Overstreet and Heard 1982). Three species of menhadens, Gulf menhaden, finescale menhaden, and yellowfin menhaden, are distributed in the Gulf of Mexico. Key life-history parameters, habitat preferences and distribution, and general information on the menhaden stock and fishery in the Gulf of Mexico are presented in the tables and paragraphs that follow (Tables 9.9, 9.10, and 9.11).

9.3.2.1   Key Life-History Processes and Ecology

Menhaden are flat and dull silver with a greenish back, have soft flesh, and a deeply forked tail. A prominent black spot is found behind the gill cover, followed by a row of smaller spots. The three species of menhaden, Gulf menhaden, finescale menhaden, and yellowfin menhaden, are distributed in the Gulf of Mexico from estuarine waters outwards to the continental shelf, although they are most likely distributed in less saline waters of estuaries and can be found in bays, lagoons, and river mouths (Table 9.10 and Figure 9.10). Gulf menhaden tend to have larger scales than yellowfin menhaden, and finescale menhaden lack the row of smaller spots that occur on Gulf menhaden. All three species have yellowish fins (McEachran and Fechhelm 2005).

The Gulf menhaden occurs throughout the Gulf of Mexico but is mainly distributed in nearshore waters (Table 9.10 and Figure 9.10). Yellowfin menhaden mainly inhabit estuarine or nearshore areas and do not seem to have seasonal migratory behavior. Finescale menhaden also occur in estuarine or nearshore areas. No evidence suggests that finescale menhaden are subject to any systematic seasonal migration, but there appears to be a seasonal shift of larger finescale menhaden between Texas bays (Gunter 1945). In the southern Gulf of Mexico, the range of Gulf menhaden overlaps that of the finescale menhaden (Anderson and McDonald 2007), and it appears that these two species may engage in resource partitioning, a process whereby closely related or trophic-overlapped species occurring in close proximity results in subtle differences in ecological niches (Castillo-Rivera et al. 1996). In the eastern Gulf, the range of Gulf menhaden overlaps that of the yellowfin menhaden, and there is evidence of hybridization between the two species (Anderson and Karel 2007).

Table 9.9. Summary of Life-History Information for Gulf Menhaden (*Brevoortia patronus*), Finescale Menhaden (*Brevoortia gunteri*), and Yellowfin Menhaden (*Brevoortia smithi*)

Parameter	Value	Reference
von Bertalanffy growth model parameters (see Table 9.6 for explanation)	$L_\infty = 225.9$ mm (8.9 in.) FL	Vaughan et al. (2000)
	$L_\infty = 212.2$ mm (8.3 in.) FL	SEDAR 27 (2011)
	$K = 0.56$ per year	Vaughan et al. (2000)
	$K = 0.69$ per year	SEDAR 27 (2011)
	$t_0 = -0.43$ years	Vaughan et al. (2000)
	$t_0 = -0.31$ years	SEDAR 27 (2011)
Age at first maturity	2 years	Lewis and Roithmayer (1981), Nelson and Ahrenholz (1986), Vaughan et al. (2000, 2007)
Length at first maturity	183.1 mm (7.2 in.) FL	Lewis and Roithmayer (1981), SEDAR 27 (2011)
Spawning season	October through March, peaks December through February	Christmas and Waller (1975), Lewis and Roithmayer (1981)
Spawning location	High salinity, offshore, open Gulf waters; highest abundances occur from Texas to Alabama, concentrated near the Mississippi Delta	Lewis and Roithmayer (1981)
Common prey of juveniles and adults	Zooplankton, phytoplankton, and detritus	Reintjes and Pacheco (1966), Deegan (1985), Ahrenholz (1991)
Common predators	Brown pelicans, osprey, common loons, mackerel, bluefish, blue runner, ladyfish, sharks, white and spotted seatrout, longnose and alligator gars, and red drum	Reintjes (1970), Etzold and Christmas (1979), Overstreet and Heard (1982), Spitzer (1989)

All menhaden species are estuarine-dependent and marine migratory species (Anderson and McDonald 2007). In general, spawning usually takes places in the offshore marine environment during winter (Table 9.9) (Gunter 1945; Simmons 1957; Dahlberg 1970; Houde and Swanson 1975). Egg hatch and early growth of larvae usually occur when currents from offshore spawning grounds transport them to low-salinity estuary nursery grounds (Minello and Webb 1997). This process usually takes 1–2 months. The transported larvae enter estuarine bays, sounds, and streams and metamorphose into juveniles. Menhaden juveniles inhabit estuarine areas until the following fall or early winter, when they migrate offshore (Table 9.10). Adults are usually distributed in large schools in nearshore oceanic waters and large estuarine systems. Because the Gulf menhaden has a similar life-history process and is much more abundant and widely distributed than the other two menhaden species, the following discussion is focused on Gulf menhaden.

The spawning season estimated for the Gulf menhaden differs among studies, varying from December through February and October through March (Table 9.9) (Suttkus 1956; Combs 1969; Christmas and Waller 1975; Shaw et al. 1985a, b). This might reflect impacts of environmental conditions, which vary from year to year (SEDAR 27 2011). Gulf menhaden are multiple and intermittent spawners with ova being released in batches over a protracted spawning season (Combs 1969; Lewis and Roithmayer 1981). Spawning can occur from nearshore to 60 miles offshore along the entire U.S. Gulf coast (Table 9.9) (Christmas and Waller 1975). However, Fore (1970) analyzed the distributions of eggs and concluded that

Table 9.10. Summary of Habitat Information for Gulf Menhaden (*Brevoortia patronus*), Finescale Menhaden (*Brevoortia gunteri*), and Yellowfin Menhaden (*Brevoortia smithi*)

Parameter	Value	Reference
Habitat preferences and temporal/spatial distribution of juveniles	Early juveniles settle in shallow (0–2 m or 0–6.6 ft deep), quiet, low salinity areas nearshore during late winter to spring; estuarine marsh edge habitat (also coastal rivers, streams, bays, bayous, and other quiet, low salinity, nearshore habitat) provides adequate forage and protection from predators; salinity from 0 to 15 ppt; temperature from 5 to 35 °C; bottom depth $\leq$2 m ($\leq$6.6 ft); juveniles migrate offshore during winter and move back to coastal waters the following spring as age-1 adults	Christmas and Gunter (1960), Reintjes (1970), Perret et al. (1971), Fore and Baxter (1972), Christmas and Waller (1973), Copeland and Bechtel (1974), Etzold and Christmas (1979), Christmas et al. (1982), Addis et al. (2011), SEDAR 27 (2011)
Habitat preferences and temporal/spatial distribution of adults	Non-gravid, developing adults associated with mid-range salinities of estuary; salinity from 5 to 25 ppt; temperature from 5 to 35 °C; maturing juveniles and adults are typically found in open bay and Gulf waters with non-vegetated bottoms and emigrate from estuarine to open Gulf waters from mid-summer through winter; following overwintering or spawning in offshore waters, all surviving age classes migrate back to estuaries in March and April	Reintjes (1970), Christmas and Waller (1973), Etzold and Christmas (1979), Lassuy (1983), Addis et al. (2011)
Habitat preferences and temporal/spatial distribution of spawning adults	Gravid adults generally associated with higher-salinity, open bay and open Gulf waters; spawning typically takes place over the continental shelf during winter; salinity from 15 to 36 ppt; temperature from 14 to 25 °C; depth from 8 to 70 m (26–230 ft)	Turner (1969), Fore (1970), Christmas and Waller (1975), Christmas et al. (1982), Lassuy (1983), Shaw et al. (1985a, b), SEDAR 27 (2011), Addis et al. (2011)
Designated essential fish habitat for juveniles and adults	None designated because not federally managed	

spawning of Gulf menhaden occurred mainly over the continental shelf between Sabine Pass, Texas and Alabama, with the greatest concentrations being found in waters between the 8 and 70 m (26.2 and 230 ft) contours off Texas and Louisiana and near the Mississippi Delta.

The eggs of Gulf menhaden are planktonic and drift with prevailing currents for almost 48 h before hatching. Early larvae also drift with the current and feed on phytoplankton. Currents transport Gulf menhaden larvae into low-salinity estuaries for early growth (Minello and Webb 1997). This transportation from spawning grounds to estuarine nursery grounds is critical for the survival of Gulf menhaden larvae. As they grow larger and become able to swim, they shift their diet to zooplankton. After developing gill rakers, they filter-feed on plankton, typically near the surface. In fresh and brackish estuaries and rivers, they grow rapidly in spring and summer, and by fall, they migrate to high-salinity offshore waters no deeper than 100 m (328 ft). No east–west component of annual migration was found for Gulf menhaden in tagging studies (Kroger and Pristas 1975; Pristas et al. 1976); however, Gulf menhaden from the eastern and western extremes of their ranges tend to move toward the center of their range with age (Ahrenholz 1991).

Table 9.11. Summary of Stock and Fisheries Information for Gulf Menhaden (*Brevoortia patronus*), Finescale Menhaden (*Brevoortia gunteri*), and Yellowfin Menhaden (*Brevoortia smithi*)

Parameter	Value	Reference
General geographic distribution	Coastal Gulf of Mexico with highest abundances occurring from Texas to Alabama, concentrated near the Mississippi Delta; Gulf menhaden: from Yucatán Peninsula in Mexico, across the western and northern Gulf to Tampa Bay, Florida; nearshore marine and estuarine waters from Cape Sable, Florida to Veracruz, Mexico, with centers of abundance off Louisiana and Mississippi; finescale menhaden: from Mississippi Sound southwestward to the Gulf of Campeche in Mexico; yellowfin menhaden: from Chandeleur Sound, Louisiana southward to the Caloosahatchee River, Florida (presumably around the Florida peninsula) to Cape Lookout, North Carolina	Hildebrand (1948), Christmas and Gunter (1960), Lassuy (1983)
Commercial importance	High, Gulf menhaden support second largest single fishery in the United States by weight	Lassuy (1983)
Recreational importance	Low, important as a bait fish	Addis et al. (2011)
Management agency	GSMFC, respective Gulf state marine agencies	SEDAR 27 (2011)
Management boundary	The menhaden fishery generally operates in state waters; the respective state marine agencies are responsible for regulating and monitoring the gulf menhaden fishing in their waters. The Gulf states cooperate with each other through the GSMFC to enact multi-state cooperative management of gulf menhaden, without relinquishing their individual state authorities.	VanderKooy and Smith (2002), SEDAR 27 (2011)
Stock structure within the Gulf of Mexico	Gulf menhaden comprise >99 % of the annual catch from the menhaden fishery; the management unit is defined as the total population of Gulf menhaden in the U.S. Gulf of Mexico	Ahrenholz (1981), SEDAR 27 (2011)
Status (overfished/ overfishing)	Through 2004: not overfished, no overfishing occurring; successfully managed under a regional Fisheries Management Plan since 1978; as of 2011, the stock is not overfished, and no overfishing is occurring	Vaughan et al. (2007), SEDAR 27 (2011)

Note: GSMFC Gulf States Marine Fisheries Commission

Few Gulf menhaden spawn in their first winter, but almost all fish are mature by their second winter when they reach age 1+. Female Gulf menhaden are generally mature after they reach about 150 mm (5.9 in.) FL and larger (Table 9.9) (Lewis and Roithmayer 1981). The life span of Gulf menhaden is about 5–6 years. The maximum size observed for Gulf menhaden is 223 mm (8.8 in.) FL (Ahrenholz 1991).

Limited information on age and size at maturity is available for finescale and yellowfin menhaden. Female finescale menhaden were found to be mature at the size of 150 mm (5.9 in.) TL (Gunter 1945), and female yellowfin menhaden were found to be mature at 186 mm (7.3 in.) FL (Hellier 1968). The maximum size reported is 281 mm (11 in.) FL for yellowfin menhaden and 289 mm (11.4 in.) FL for finescale menhaden (Ahrenholz 1991).

Younger fish are thought to be more vulnerable to predation, and thus M may decline with size or age (SEDAR 27 2011). In addition to varying with size or age, M also tends to vary from year to year, reflecting annual variability of habitat variables (Figure 9.11).

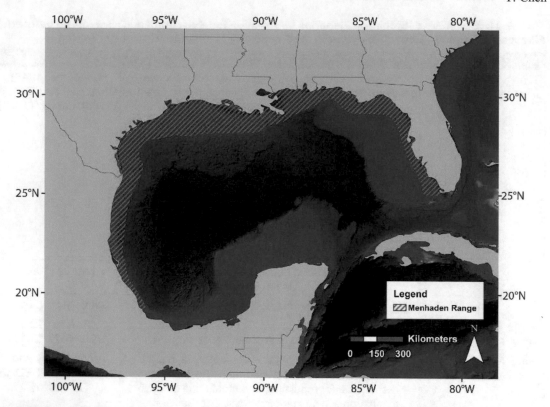

Figure 9.10. Range of menhaden, including Gulf menhaden (*Brevoortia patronus*), finescale menhaden (*Brevoortia gunteri*), and yellowfin menhaden (*Brevoortia smithi*), in the Gulf of Mexico (USGS 2010b).

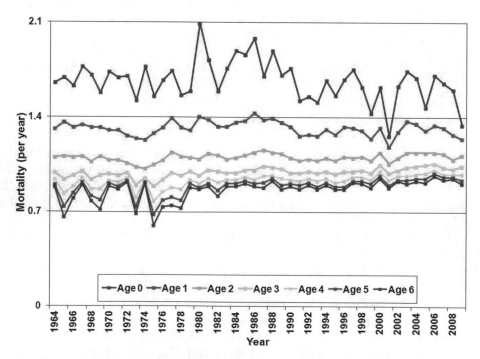

Figure 9.11. Annual natural mortality (*M*) for different age groups of Gulf menhaden (*Brevoortia patronus*) from 1964 through 2009 (data from SEDAR 27 2011).

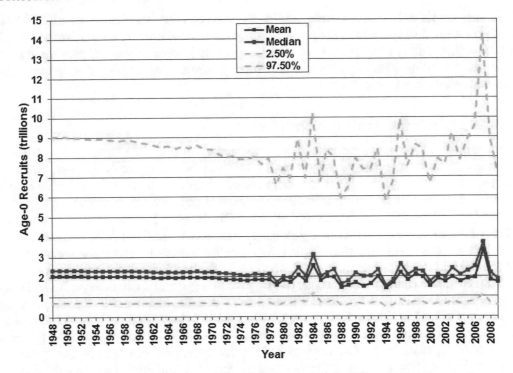

Figure 9.12. Estimates of annual recruitment of Gulf menhaden (*Brevoortia patronus*), measured as the number of age-0 fish, from 1948 through 2009 (data from SEDAR 27 2011).

Recruitment tends to fluctuate over time without a clear temporal trend. However, large uncertainty appears to be associated with the Gulf menhaden recruitment estimates (Figure 9.12).

Populations of Gulf menhaden throughout the Gulf of Mexico are generally thought to comprise a single genetic stock (SEDAR 27 2011). No evidence supports the existence of multiple stocks for finescale menhaden and yellowfin menhaden within the Gulf of Mexico. There is no strong evidence supporting the existence of metapopulations (groups of spatially separated populations of the same species that interact at some level). However, there appears to be large spatial variability in key life-history parameters, such as M, growth, and maturation. Stock structure also varies over time and space (SEDAR 27 2011).

9.3.2.2 Predators and Prey

Menhaden are omnivorous filter feeders that remove food resources from the water column via their gill rakers while swimming (Table 9.9). Their filtration efficiency is largely a function of branchio-spicule spacing of the gill rakers changing allometrically as menhaden grow (Friedland et al. 2006). Small Gulf menhaden larvae primarily feed on large phytoplankton (e.g., dinoflagellates) and some zooplankton (Govoni et al. 1983). As the larvae grow, phytoplankton become less important in the diet, and large zooplankton, especially copepods, become more important. After metamorphosis into juveniles, Gulf menhaden become filter-feeding omnivores (Table 9.10). However, some of the phytoplankton that the juvenile Gulf menhaden consume is an order of magnitude smaller than the smallest phytoplankton consumed at larval stages (Chipman 1959; June and Carlson 1971). Menhaden may also feed on their own eggs (Nelson et al. 1977), as well as eggs and larvae of other fishes and invertebrates (Peck 1893; McHugh 1967).

Because of their high abundance and schooling behavior, menhaden of all life-history stages from eggs through adults are potential prey for a large number of piscivorous fish and birds (Table 9.9). Many invertebrate predators, especially in oceanic waters, prey upon menhaden larvae, including chaetognaths (arrow worms), squids (mollusks), ctenophores (comb jellies), and jellyfishes (coelenterates).

9.3.2.3   Key Habitat Needs and Distribution

Larvae and early juveniles are often found associated with estuarine marsh edges for forage and protection from predators, and juveniles and adults are typically in open water with non-vegetated bottoms (Table 9.10). Offshore spawning ensures that Gulf menhaden eggs and larvae are euryhaline. Most Gulf menhaden eggs occur in waters with salinities over 25 parts per thousand (ppt) (Fore 1970; Christmas and Waller 1973). Eggs and larvae are found throughout the Gulf of Mexico waters with salinity ranging from 20.7 to 36.6 ppt (Christmas et al. 1982). As the larvae move inshore, they require low salinity waters to complete metamorphosis. The entrance of larvae into estuaries with abundant food and lower salinities may be essential to their survival and to their metamorphosis into juveniles (June and Chamberlin 1959). Temperature may be more critical to egg development than to juveniles and to adults that are distributed widely in the Gulf of Mexico with large spatial variability in temperature. Eggs and larvae have been observed in waters with temperatures ranging from 11 °C (February) to 18 °C (March) in northern Florida, from 16 °C (January) to 23 °C (March) in southern Florida, and from 10 °C (January) to 15 °C (December) in the Mississippi Sound. Menhaden may be subject to cold mortality under freezing winter conditions, especially in narrow or shallow tidal areas. Large fish kills may also occur during the summer, as a result of plankton blooms and low dissolved oxygen or hypoxic conditions (Christmas and Waller 1973; Etzold and Christmas 1979).

Menhaden tend to have high habitat elasticity to adapt to changes in their habitats. In a study examining fish assemblages in an estuary from 1950 to 2000 (O'Connell et al. 2004), Gulf menhaden were found to change little in their frequency or position within the estuarine ecosystem even though the estuary had deteriorated substantially in environmental quality and the fish assemblage shifted from a croaker-dominated complex to an anchovy-dominated complex. This indicates that Gulf menhaden are elastic in their ability to adapt to short- or long-term changes in environmental conditions (O'Connell et al. 2004). Because menhaden are not federally managed, no essential fish habitat has been designated (Table 9.10).

9.3.2.4   Fisheries

The Gulf menhaden fishery has great ecological, economic, and social importance. Although menhaden are bony, oily, and usually not directly consumed by humans, they are an important source of fishmeal and fish oil. Both of these reduction products are used as feed for livestock and aquaculture, such as for salmon, shrimp, tilapia, and catfish. Fish oil made from menhaden is also used as a dietary supplement and as a raw material for products, such as lipstick. Menhaden is one of the best baitfish available. Fresh or frozen menhaden are commonly used as whole or cut bait for snapper and king mackerel fishing (SEDAR 27 2011).

Gulf menhaden supports one of the largest fisheries in the United States (Table 9.11), which dates back to the 1800s. On average, 400–600 kilotons of Gulf menhaden are extracted and used for reduction annually (Figure 9.13), with a much smaller amount being captured for use as bait. Landings have had a decreasing trend since the 1980s, when they were the highest (Figure 9.13). Most of the Gulf menhaden landed in the reduction fishery was ages 1 and 2, representing 57 and 38 % of the annual catch on average, respectively. Commercial reduction

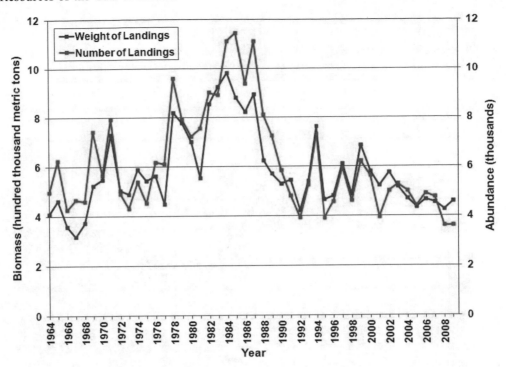

Figure 9.13. Landings of Gulf menhaden (*Brevoortia patronus*) in the reduction fishery from 1964 through 2009 (data from SEDAR 27 2011).

fishery catches are landed from areas ranging from Florida to Texas, with the majority of recent catches coming from Louisiana waters (SEDAR 27 2011).

The Gulf menhaden reduction fishery has been managed under a regional FMP since 1978. Management of the Gulf menhaden fishery is through partnerships among the NMFS Beaufort Laboratory, the state marine agencies, the menhaden industry, and the Gulf States Marine Fisheries Commission (GSMFC) (Table 9.11). It is one of the most detailed and data-rich fisheries currently operated in the Gulf of Mexico. A statistical catch-at-age model, the Beaufort Assessment Model (BAM), was used as the base model for the most recent stock assessment (SEDAR 27 2011). The BAM model assumes one coast-wide population of Gulf menhaden in the Gulf of Mexico. The BAM model for Gulf menhaden uses annual time steps, including landings data from 1948 to 2010. The 1948 data are from close to the beginning of the fishery and, thus, tend to represent unfished conditions for Gulf menhaden. The BAM model incorporates various fishery-dependent and fishery-independent data, including abundance indices and age compositions derived from various survey programs, commercial catch-at-age data, and biological information on growth, maturation, and M.

Total egg production, a more accurate quantification of population reproductive potential than spawning stock biomass, was estimated in the most recent stock assessment (Figure 9.14) (SEDAR 27 2011). It appears to have a decreasing trend prior to the mid-1980s, but shows an increasing trend since the late 1980s. The total egg production estimates for recent years tend to be higher than those for most years since the mid-1970s, suggesting that the stock is in good condition (Figure 9.14). Based on the most recent stock assessment (SEDAR 27 2011), the Gulf of Mexico Gulf menhaden population was not overfished and overfishing did not occur in 2010 (Table 9.11).

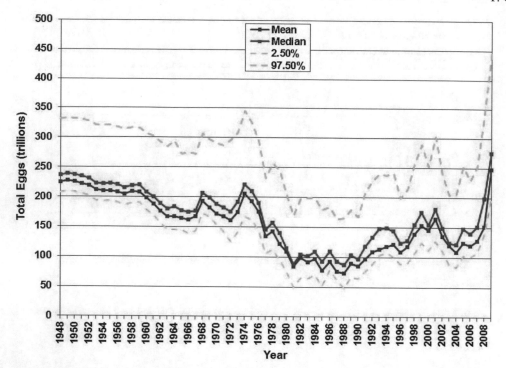

Figure 9.14. Estimates of Gulf menhaden (*Brevoortia patronus*) total egg production from 1948 through 2009 (data from SEDAR 27 2011).

9.3.3 Groupers (Family Serranidae, Subfamily Epinephelinae)

The subfamily Epinephelinae of the family Serranidae consists of about 160 species of marine fishes in 15 genera that are commonly known as the groupers, rockcods, hinds, and seabasses (Heemstra and Randall 1993). Groupers are bottom-associated fishes found in the tropical and subtropical waters of oceans and are of considerable ecological and economic value. Groupers are generally associated with hard or rocky bottoms, and most species occur on coral reefs, occupying caves, ledges, and crevices (Figure 9.15) (Jory and Iverson 1989; Heemstra and Randall 1993). Some species occur in depths of 100–200 m (328–656 ft), with the majority inhabiting depths less than 100 m (328 ft). Most grouper species apparently migrate vertically as they grow, with larger fish living at progressively deeper depths (Jory and Iverson 1989).

As the major predators of coral reef ecosystems, most groupers feed on a variety of fishes, large crustaceans, and cephalopods (Heemstra and Randall 1993). Most groupers are ambush predators, hiding among the coral and rocks until a fish or crustacean goes by, and catch their prey with a quick rush and snap of their powerful jaws. The large head and mouth of the typical grouper enables it to suck in a large volume of water and its prey in less than one second (Heemstra and Randall 1993).

Groupers are typically solitary fishes except for occasional spawning aggregations and are generally resident on a particular reef for many years; this site specificity and their relatively slow growth rate makes them vulnerable to overfishing (Heemstra and Randall 1993). Most groupers are protogynous hermaphrodites, meaning that all fish are first females and then change into males at a certain age/size (Jory and Iverson 1989).

Figure 9.15. Two yellowmouth groupers (*Mycteroperca interstitialis*) eye one another near a large brain coral in Flower Garden Banks National Marine Sanctuary (from NMS 2013).

Fifteen species of groupers are managed under the Reef Fish Fishery by the GMFMC (GMFMC 2004b). The fishery is divided into shallow-water and deep-water grouper complexes (SEDAR 12 Update 2009). Species in the shallow-water complex include red grouper, gag grouper (*Mycteroperca microlepis*), black grouper (*Mycteroperca bonaci*), scamp (*Mycteroperca phenax*), yellowfin grouper (*Mycteroperca venenosa*), yellowmouth grouper (*Mycteroperca interstitialis*) (Figure 9.15), rock hind (*Epinephelus adsensionis*), and red hind (*Epinephelus guttatus*). The deep-water grouper complex includes snowy grouper (*Epinephelus niveatus*), yellowedge grouper (*Epinephelus flavolimbatus*), speckled hind (*Epinephelus drummondhayi*), warsaw grouper (*Epinephelus nigritus*), and misty grouper (*Hyporthodus mystacinus*). Nassau grouper (*Epinephelus striatus*) and goliath grouper (*Epinephelus itajara*) (Figure 9.16) are managed as individual species and are prohibited from being harvested.

Red grouper are among the most abundant, popular, and important commercial fish in the Gulf of Mexico; therefore, this species was selected as the representative species of grouper for evaluation (Figure 9.17). Key life-history parameters for red grouper are summarized in Table 9.12 and discussed in detail in the following paragraphs. In addition, information on habitat preferences and distribution of the red grouper stock and fishery is presented in Tables 9.13 and 9.14 and discussed below.

9.3.3.1   Key Life-History Processes and Ecology

In the Gulf of Mexico, red grouper are distributed along the continental shelf, and the center of distribution along the U.S. coast is in the eastern Gulf of Mexico (Table 9.13; Figure 9.18). Genetic differences within the Gulf of Mexico tend to be small, suggesting a single population within the Gulf. This may have resulted from historic bottlenecks in population abundance that helped maintain the most common genotypes (Richardson and Gold 1997).

Red grouper spend their larval phase in the plankton. Juveniles occupy nearshore reefs and seagrass beds; adult red grouper leave nearshore reefs and move offshore to rocky bottom habitat (Tables 9.13 and 9.14). Red grouper are usually solitary until spawning time.

Figure 9.16. Goliath grouper (*Epinephelus itajara*) is one of the species of grouper prohibited from being harvested in the Gulf of Mexico (from Puntel 2016).

Figure 9.17. Red grouper (*Epinephelus morio*) on a coral reef in the Gulf of Mexico (from Dombrowski 2012).

Most red grouper exhibit limited movement throughout their life span and can exhibit high site fidelity at older ages upon reaching mid- to outer-shelf depths, which may result from the species habitat-structuring and haremic (territorial) mating behavior (Coleman and Koenig 2010). Limited movement shown by most red grouper throughout their lives could give rise

Table 9.12. Summary of Life-History Information for Red Grouper (*Epinephelus morio*)

Parameter	Value	Reference
von Bertalanffy growth model parameters—Gulf of Mexico (see Table 9.6 for explanation)	$L_\infty = 808$ mm (31.8 in.) TL	Goodyear (1995)
	$L_\infty = 854$ mm (33.6 in.) TL	SEDAR 12 (2006), SEDAR 12 Update (2009)
	L_∞ (West Florida) = 792 mm (31.2 in.) TL	Data from Moe (1969) converted by Lombardi-Carlson et al. (2002)
	L_∞ (Northern West Florida) = 800.1 mm (31.5 in.) TL	Lombardi-Carlson et al. (2008)
	L_∞ (Southern West Florida) = 863.1 mm (33.9 in.) TL	Lombardi-Carlson et al. (2008)
	$K = 0.21$ per year	Goodyear (1995)
	$K = 0.16$ per year	SEDAR 12 (2006), SEDAR 12 Update (2009)
	K (West Florida) = 0.18 per year	Data from Moe (1969) converted by Lombardi-Carlson et al. (2002)
	K (Northern West Florida) = 0.23 per year	Lombardi-Carlson et al. (2008)
	K (Southern West Florida) = 0.15 per year	Lombardi-Carlson et al. (2008)
	$t_0 = -0.3$ years	Goodyear (1995)
	$t_0 = 0.19$ years	SEDAR 12 (2006), SEDAR 12 Update (2009)
	t_0 (West Florida) = -0.45 years	Data from Moe (1969) converted by Lombardi-Carlson et al. (2002)
	t_0 (Northern West Florida) = 1.12 years	Lombardi-Carlson et al. (2008)
	t_0 (Southern West Florida) = 0.05 years	Lombardi-Carlson et al. (2008)
Age at female maturity (50 %)	4–6 years	Moe (1969), Beaumariage and Bullock (1976), Brule et al. (1999)
	2.4 years	Burgos (2001)
	≥2 years	NMFS (2002a)
	2 years (definitely mature model)	Fitzhugh et al. (2006)
	3.5 years (effectively mature model)	Fitzhugh et al. (2006)
	3 years	SEDAR 12 (2006), SEDAR 12 Update (2009)
Length at female maturity (50 %)	450 mm (17.7 in.) SL	Moe (1969)
	485 mm (19.1 in.) FL	Moe (1969)
	509 mm (20 in.) FL	Brule et al. (1999)
	487 mm (19.2 in.) TL	Burgos (2001)
	280 mm (11 in.) TL (definitely mature model)	Fitzhugh et al. (2006)
	380 mm (14.9 in.) TL (effectively mature model)	Fitzhugh et al. (2006)

(continued)

Table 9.12. (continued)

Parameter	Value	Reference
Age at transition from female to mature male (50 %)	5–10 years	Moe (1969), Beaumariage and Bullock (1976)
	7–14 years	Brule et al. (1999)
	7.2 years	Burgos (2001)
	16 years	NMFS (2002a)
	13 years	Collins et al. (2002)
	10.5 years	Fitzhugh et al. (2006)
	11 years	SEDAR 12 (2006), SEDAR 12 Update (2009)
Length at transition from female to mature male (50 %)	275–500 mm (10.8–19.7 in.) SL	Moe (1969) Beaumariage and Bullock (1976)
	597 mm (23.5 in.) FL	Brule et al. (1999)
	≥584 mm (≥22.9 in.)	Brule et al. (1999)
	690 mm (27.2 in.) TL	Burgos (2001)
	800–900 mm (31.5–35.4 in.) TL	Collins et al. (2002)
	765 mm (30.1 in.) TL	Fitzhugh et al. (2006)
Spawning season	January through June, peaks March through May	Moe (1969), Johnson et al. (1998), Collins et al. (2002), Fitzhugh et al. (2006)
Spawning location	Offshore waters, do not aggregate to spawn	Coleman et al. (1996), Brule et al. (1999)
Common prey of juveniles and adults	Snappers, sea breams, porgies, and many small fish species; portunid and calappid crabs, octopuses, squids, stomatopods and other shrimps, panulirid and scyllarid lobsters, and amphipods	Gudger (1929), Longley and Hildebrand (1941), Moe (1969), Jory and Iverson (1989), Bullock and Smith (1991)
Common predators	Larger groupers and piscivorous fishes, sandbar shark, and great hammerhead shark	Smith (1961), Moe (1969), Compagno (1984)

Table 9.13. Summary of Habitat Information for Red Grouper (*Epinephelus morio*)

Parameter	Value	Reference
Habitat preferences and temporal/spatial distribution of juveniles	Juveniles <5 years of age inhabit shallow, nearshore reefs and seagrass beds; depths of 3–18 m (9.8–59 ft)	Moe (1966), Beaumariage and Bullock (1976), Bullock and Smith (1991)
Habitat preferences and temporal/spatial distribution of adults	Age 4–6 years, coinciding with sexual maturity, adults leave nearshore reefs and move offshore; mainly inhabit rocky bottoms at depths of 36–122 m (118–400 ft); frequently occupy crevices, ledges, and caverns in limestone reefs; depths of 36–189 m (118–620 ft); temperatures from 15 to 30 °C	Cervigon (1966), Moe (1969), Roe (1976), Beaumariage and Bullock (1976), Fischer et al. (1978), Bullock and Smith (1991)
Habitat preferences and temporal/spatial distribution of spawning adults	Offshore waters; do not aggregate to spawn	Coleman et al. (1996), Brule et al. (1999)
Designated Essential Fish Habitat for juveniles and adults	All Gulf of Mexico estuaries; the U.S./Mexico border to the boundary between the areas covered by the GMFMC and the SAFMC from estuarine waters out to depths of 100 fathoms	GMFMC (2005)

Table 9.14. Summary of Stock and Fisheries Information for Red Grouper (*Epinephelus morio*)

Parameter	Value	Reference
General geographic distribution	Massachusetts to Brazil; especially abundant in the Gulf of Mexico and on the Yucatán Peninsula shelf; center of abundance is in the Florida shelf and the eastern Gulf of Mexico	Roe (1976), Bullock and Smith (1991)
Commercial importance	High	
Recreational importance	High	
Management agency	GMFMC	SEDAR 12 (2006)
Management boundary	All U.S. federal waters in the Gulf of Mexico within the GMFMC boundaries; U.S./Mexico border through the northern Gulf of Mexico waters to the Florida Keys; the Gulf of Mexico and South Atlantic Stocks are divided along U.S. Highway 1 in the Florida Keys	SEDAR 12 (2006), SEDAR 12 Update (2009)
Stock structure within the Gulf of Mexico	Managed as a single stock in Gulf of Mexico	SEDAR 12 (2006c)
Status (overfished/ overfishing)	Overfishing 2000–2004, overfishing not occurring 2005–2008, not overfished 2005–2008; but overfishing might occur and local populations might be overfished in some areas in northeastern and southern Gulf of Mexico	NMFS (2004, 2007, 2012a), SEDAR 12 (2006)

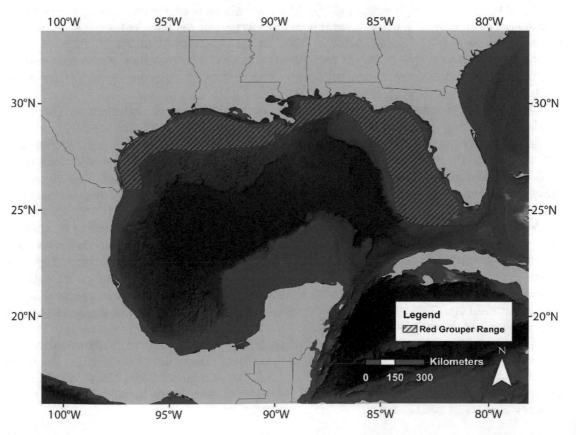

Figure 9.18. Range of the red grouper (*Epinephelus morio*) in the Gulf of Mexico (from NOAA 2013a).

to complex sub-stock structure. However, tagging studies have shown that some red grouper taken inshore during summer feeding and cohort migrations moved in large numbers in response to unusual events, such as hurricanes. For example, following Hurricane Lily in 2002, both juvenile and adult red grouper were abundant on artificial reefs and petroleum platforms off Mississippi where they previously had been absent (Franks 2003). However, since 2002, red grouper off Mississippi have become scarce.

The red grouper is a protogynous hermaphrodite, with all fish beginning life as females. Most of the females transform to males between 7 and 14 years of age after reaching at least 58.4 cm (23 in.) in length (Table 9.12) (Moe 1969; Brule et al. 1999). Females become sexually mature at 4–6 years of age and at a size of 39.9 cm (15.7 in.) standard length [SL, the length of a fish measured from the tip of the snout to the posterior end of the last vertebra, which excludes the tail] (Table 9.12) (FishBase 2013; Bullock and Smith 1991), while males become reproductively significant at age 10 and older. However, in a recent stock assessment (SEDAR 12 Update 2009), 50 % of females were found to be mature at 3 years of age and becoming males, and 50 % of males became mature at 11 years old. Red grouper have a life span of approximately 25–30 years (SEDAR 12 2006) and can grow up to 125 cm (49.2 in.) in length (McGovern et al. 2002). Although abundance of red grouper has changed substantially over time, the sex ratio of the population has not changed greatly since 1975 (Coleman et al. 1996). Peak spawning occurs in late spring, during March and May in the eastern Gulf of Mexico, but spawning may occur from January through June in the Gulf of Mexico and Caribbean Sea (Tables 9.12 and 9.13) (Johnson et al. 1998). Red grouper are indeterminate batch spawners (Johnson et al. 1998; Collins et al. 2002). Fecundity is related to size and ranges from 312,000 to 5,735,700 eggs.

Spawning red grouper release their sperm and eggs in offshore waters (Table 9.13). The fertilized eggs require high salinity (32 ppt) to maintain their buoyancy. The eggs hatch into larvae approximately 30 h after spawning and live as part of the zooplankton that drifts with the ocean currents. The larvae settle to the bottom substrate at about 35–50 days after hatching and reach a size of 20–25 mm (0.8–0.9 in.) SL. The duration of the red grouper larval stage is within the range of 31–66 days for other grouper species (Lindeman et al. 2000).

A constant M rate of 0.2 per year was used in early stock assessments (Schirripa et al. 1999). Using different models (e.g., Jensen 1996; Quinn and Deriso 1999), M was estimated to range from 0.14 to 0.24. Based on a recent estimate of maximum age (29 years) for the Gulf of Mexico red grouper (SEDAR 12 2006), M was estimated to be 0.14 for all age classes using the regression model developed by Hoenig (1983). However, the assumption of having the same M across all the age groups may not be realistic. An age-varying M approach was, thus, developed (Lorenzen 1996), which inversely relates the M-at-age to the mean weight-at-age by a power function incorporating a scaling parameter. Lorenzen (1996) provided point estimates and 90 % confidence intervals of the power and scaling parameters for oceanic fishes, which are used for initial parameterization. The estimated M using the Lorenzen method varies with age and is considered more biologically plausible than a fixed M for all ages (SEDAR 12 2006). The estimate was then re-scaled to the oldest observed age (29 years) so that the cumulative M through this age was equivalent to that of a constant M ($M = 0.14$) for all ages.

The distribution of major red grouper fishing grounds and the limited movement shown in tagging studies indicate that the spatial distribution of recruitment varies greatly. The Big Bend region of Florida (DeVries et al. 2006; SEDAR 12 2006) and the shallow (<20 m or <65.6 ft) areas off Southwest Florida (Pinellas and Charlotte Counties) were hypothesized to be two primary sources of recruitment.

Significant differences in size and age structure and in growth rates of red grouper were found north and south of 28°N latitude (Lombardi-Carlson et al. 2006). A tagging study conducted by Mote Marine Laboratory strongly suggested that red grouper (age 2–4 years) had limited range. This tendency could contribute to future stock separation given enough time. The large spatial variability in growth and age structure of red grouper also supports the existence of a more complex subpopulation structure that is not genetically distinctive but functionally independent (Fischer et al. 2004).

9.3.3.2 Predators and Prey

In their early juvenile stages, red grouper feed primarily on demersal crustaceans in seagrass beds. As the juveniles become sexually mature, they move out to deeper rocky bottoms and feed on small fishes, such as snappers and porgies, and invertebrates, such as shrimps and crabs (Table 9.12). The proportion of the diet consisting of fish increases with red grouper size. Top predators, such as sharks, prey on juvenile and adult red grouper (Table 9.12). Red grouper are known to be susceptible to red tide poisoning (SEDAR 12 Update 2009).

9.3.3.3 Key Habitat Needs and Distribution

The fertilized eggs of red grouper require high salinity (32 ppt) to maintain their buoyancy. Red grouper larvae are pelagic and are transported by ocean currents from spawning grounds to settlement grounds. Juveniles occupy nearshore reefs and move offshore when they become adults (Table 9.13). Adult red grouper are non-migratory and are often seen resting on the bottom substrate. The designated essential fish habitat for juvenile and adult red grouper, as well as many other managed grouper species, is included in the Reef Fish FMP (Table 9.13 and Figure 9.6).

9.3.3.4 Fisheries

Red grouper is the most abundant grouper species in the Gulf of Mexico, which helps explain its status as the primary commercial grouper species by weight and second most recreationally caught grouper species (GMFMC 2011). Red grouper are managed as a single management unit in the Gulf of Mexico extending from the U.S.–Mexico border in the west through the northern Gulf waters and west of the Dry Tortugas and the Florida Keys (Table 9.14) (SEDAR 12 Update 2009). Landings are regulated through the implementation of allowable biological catch (ABC), size limits, trip limits, quotas, seasonal closures, area closures, and gear restrictions. These regulations have been constantly adjusted over time based on improved understanding of the population dynamics of red grouper and stock status.

Red grouper total landings in the United States are taken from four fleets: longline, commercial handline, commercial trap, and recreational. These combined fleets fluctuated with an overall declining trend, falling from almost 3.9 million gutted kg (8.7 million gutted lb) in 1986 to about 2.1 million gutted kg (4.6 million gutted lb) in 1998 (SEDAR 12 Update 2009). Total landings then increased sharply, reaching almost 3.2 million gutted kg (7.1 million gutted lb) in 1999, while stabilizing at an average of 3.4 million gutted kg (7.5 million gutted lb) until 2005 and nearing the estimated optimal yield (OY) of 3.4 million gutted kg (7.6 million gutted lb) (SEDAR 12 2006; SEDAR 12 Update 2009). Total landings began a decreasing trend in 2006, and reached 2.5 million gutted kg (5.6 million gutted lb) in 2008 (SEDAR 12 Update 2009).

Commercial longline landings in the United States from 1986 to 2005 showed a gradual increase with a range of 0.9–2.0 million gutted kg (2.0–4.3 million gutted lb), while commercial handline landings declined considerably from 1.7 to 0.5 million gutted kg (3.7–0.9 million gutted lb) before stabilizing in 2000 at 0.8 million gutted kg (1.8 million gutted lb) (SEDAR 12 2006). The commercial trap fishery contributed less than either the commercial handline or longline, while only landing about 0.5 million gutted kg (1.1 million gutted lb) annually in 1995 and 2000. Recreational landings, including all components, were equal to a third of all commercial landings from 1986 to 2008.

The annual estimated rate of total fishing mortality (landings and discards combined) for directed fleets increased steadily from 0.25 in 1986 to a peak of 0.29 in 1993, before falling steadily to 0.16 in 1998. The rate of fishing mortality increased slightly in 1999 to around 0.2, followed by a decreasing trend to 0.18 for 2005 (SEDAR 12 Update 2009). Discard mortality is typically 10 % of the landings attributed to directed fleets (SEDAR 12 2006).

The recreational fishery of the Gulf of Mexico red grouper is managed with size limits, daily bag limits, seasonal length, and allocation quotas. For the 2009 recreational fishing season, the size limit was 40.6 cm (16 in.) TL, the daily bag limit was two fish, seasonal length was from June 1 to August 15 (75 days), and the annual quota allocation was 1.1 million gutted kg (2.43 million lb) (SEDAR 12 2006). Some of the regulations implemented have been questioned for their unintended biological implications. For example, Goodyear (1995) raised concerns about the use of a high minimum size limit (50.8 cm or 20 in TL) on red grouper that show great variation in growth, suggesting that the disproportionally high harvest rate of faster growing red grouper may select for the heritable trait for slow growth.

Total stock abundance averaged 27.4 million fish and varied with little trend between 1986 and 1999. However, abundance jumped sharply in 2000 to 39.5 million fish as the strong 1999-year class entered the estimated population at age 1 (SEDAR 12 Update 2009). Total abundance tapered off gradually thereafter to the terminal estimate of 31.2 million fish in 2008 (SEDAR 12 Update 2009). An analysis of stock recruitment and abundance-at-age data from 1986 to 2005 indicated a maturing stock primarily consisting of individuals approximately 10 years old, while older individuals declined in abundance from 1986 to the mid-1990s (SEDAR 12 Addendum 1 2007). Spawning stock is measured as total female gonad weight. Estimated spawning stock gradually improved over the assessment period, from an average of 460 metric tons (1 metric ton = 1.102 U.S. short ton) of eggs in the late 1980s to an average of almost 680 metric tons in the last few years, which included the observed high of 713 metric tons of eggs in 2008 (SEDAR 12 Update 2009). Estimated recruitment at age 1 indicated two notably strong year classes (1996 and 1999), while exhibiting a slightly increasing trend from 1986 to 2005. Recruitment over those years averaged 9.7 million fish, with peak values of 13.2 million in 1997 and 21.1 million in 2000 (SEDAR 12 Update 2009).

Both the 2006 and 2009 updated stock assessments concluded that Gulf of Mexico red grouper stocks were neither overfished nor experiencing overfishing and almost approached OY based on data through 2005 and 2008 (Table 9.14) (SEDAR 12 2006; SEDAR 12 Update 2009). However, the 2009 red grouper stock assessment did indicate a stock decline since 2005, but an episodic 20 % stock mortality event was attributed as the primary source for the decline in concurrence with typical fishing and natural mortality (GMFMC 2011). Successful management and the 50 % U.S. harvest reduction in the last 55 years have encouraged rebounding stocks and allowed the GMFMC to set the 2011 Total Allowable Catch (TAC) at 5.68 million lb gutted weight based on March 2010 projections (GMFMC 2011).

The large variability in the spatial distribution of the red grouper stock within the Gulf of Mexico due to the distribution of suitable habitats, larval transportation patterns, and lack of movement must be considered for these results. Furthermore, both fishery-dependent and fishery-independent monitoring programs clearly have shown that red grouper in the Gulf are characterized by periodic strong year classes, the latest being 1996, 1999, and possibly 2002

Figure 9.19. Atlantic bluefin tuna (*Thunnus thynnus*) in a net (from DeepAqua 2010).

(DeVries et al. 2006; SEDAR 12 Update 2009). Understanding the red grouper's unique life history and continued landings monitoring are critical to management towards OY of this ecologically, socially, and economically important Gulf of Mexico stock.

9.3.4 Atlantic Bluefin Tuna (*Thunnus thynnus*)

The Atlantic bluefin tuna is the largest member of the family Scombridae (mackerels and tunas); fishes in this family are generally predators in pelagic ecosystems, are fast swimming, and are some of the most important and familiar food and sport fish species (Figure 9.19). Atlantic bluefin tuna are highly migratory and experience large-scale, transoceanic movements between foraging and spawning grounds over a wide range of pelagic environments from warm tropical to subpolar waters of the North Atlantic Ocean (Figure 9.20) (Mather et al. 1995; Collette et al. 2001; Fromentin and Powers 2005), and the northern Gulf of Mexico is one of the spawning locations of Atlantic bluefin tuna (Table 9.15). Based on genetic and tagging studies, two separate stocks are defined with their separate spawning grounds in the Gulf of Mexico (western stock) and Mediterranean Sea (eastern stock), respectively (Block et al. 2005; Boustany et al. 2007; Carlsson et al. 2007). Information for the western stock or western Atlantic population of bluefin tuna is summarized in the tables and text that follow (Tables 9.15, 9.16, and 9.17).

9.3.4.1 Key Life-History Processes and Ecology

The western Atlantic bluefin tuna stock, with the Gulf of Mexico as its main spawning grounds, is much smaller than the eastern Atlantic bluefin tuna stock; in addition, its spawning stock biomass has declined by over 90 % in the last 30 years (ICCAT 2012a). The International Commission for the Conservation of Atlantic Tunas (ICCAT) is responsible for the assessment and management of the two Atlantic bluefin tuna stocks. Because the focus of this chapter is the Gulf of Mexico, the western Atlantic bluefin tuna stock has been selected for evaluation (Figure 9.20 and Table 9.17).

The timing and distance traveled to spawning grounds varies among spawning adults of different origins in the eastern and western Atlantic. Individuals of western stock origin move

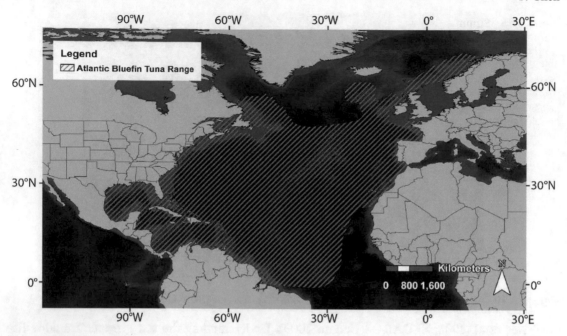

Figure 9.20. Range of the Atlantic bluefin tuna (*Thunnus thynnus*) (modified from Maguire et al. 2006).

directly from foraging grounds in the western and central North Atlantic to the Gulf of Mexico in the late winter and early spring. Ovaries of western Atlantic bluefin tuna are well developed in April and May (Baglin 1982), and spawning occurs from mid- to late May (Brothers et al. 1983). Most individuals are present on the spawning grounds from March to early July, but the spawning period varies (Table 9.15). A fraction of the stock moves into the highly productive waters of the Gulf of Maine, Scotian Shelf, central North Atlantic, and east of the Flemish Cap before returning to the Gulf of Mexico to spawn. Evidence supports site fidelity to natal areas for fish after reaching reproductive maturity.

Spawning in the Gulf of Mexico occurs in the northern areas, primarily in waters west of the Loop Current in the northern slope waters from 85 and 95°W (Table 9.15) (Block et al. 2005; Teo et al. 2007b). The location and intensity of spawning is influenced by the spatial and temporal variability in the location of major oceanographic features (fronts) and environmental conditions (e.g., waters with sea-surface temperatures above the 24 °C threshold). Thus, changes in the location of the Loop Current from year to year lead to changes in the distribution of bluefin tuna eggs and larvae in the Gulf of Mexico.

Western Atlantic bluefin tuna exhibit distinct behaviors during the three phases (entry, breeding, and exit phases) on their spawning grounds with changes in diving time, depths, and thermal biology (Block et al. 2001a; Teo et al. 2007b). As the bluefin tuna enter and exit the Gulf of Mexico, they tend to dive to significantly deeper daily maximum depths (>500 m or >1,640 ft) and exhibit directed movement paths going to and leaving spawning areas. In the breeding phase, which lasts for about 20 days (Block et al. 2001a; Teo et al. 2007b), the fish exhibit significantly shallower daily maximum depths, perform shallow oscillatory dives, and have movement paths that are significantly more residential and sinuous (Teo et al. 2007b).

High concentrations of Atlantic bluefin tuna larvae have been found in a broad region of the northern Gulf of Mexico, with peaks near the continental shelf break (e.g., 26–28°N latitude, 85–94°W longitude) (Richards 1976, 1997; Turner et al. 1996; Nishida et al. 1998).

Table 9.15. Summary of Life-History Information for Western Atlantic Bluefin Tuna (*Thunnus thynnus*)

Parameter	Value	Reference
von Bertalanffy growth model parameters (see Table 9.6 for explanation)	$L_\infty = 382$ cm (150.4 in.)	Turner and Restrepo (1994), ICCAT (2010)
	$L_\infty = 315$ cm (124 in.)	ICCAT (2012a)
	$K = 0.08$ per year	Turner and Restrepo (1994), ICCAT (2010)
	$K = 0.09$ per year	ICCAT (2012a)
	$t_0 = -0.71$ years	Turner and Restrepo (1994), ICCAT (2010)
	$t_0 = -1.13$ years	ICCAT (2012a)
Age at maturity	9 years	Baglin (1982), ICCAT (2010, 2012a)
	10 years	Magnuson et al. (1994)
	11 years	Block et al. (2005)
Length at maturity	200 cm (78.7 in.) curved fork length (CFL)[a]	Magnuson et al. (1994)
	241 ± 28 cm (94.9 ± 11 in.) CFL	Block et al. (2005)
Spawning season	April to mid-June	Mather et al. (1995), Fromentin and Powers (2005), Rooker et al. (2007), ICCAT (2010)
Spawning location	Northern Gulf of Mexico in waters along the continental shelf break and slope	Richards (1976), Richards and Potthoff (1980), Mather et al. (1995), Turner et al. (1996), Richards (1997), Nishida et al. (1998), Block et al. (2001a, b, 2005), Fromentin and Powers (2005), Rooker et al. (2007), Teo et al. (2007a), ICCAT (2010)
Common prey of juveniles	Small fishes, fish larvae, and zooplankton	Uotani et al. (1981, 1990), Miyashita et al. (2001), Rooker et al. (2007)
Common prey of adults	Atlantic herring, Atlantic mackerel, bluefish, sand lances, silver hake, spiny dogfish, demersal fishes, krill, squids, and crustaceans	Nichols (1922), Crane (1936), Bigelow and Schroeder (1953), Dragovich (1970), Mason (1976), Matthews et al. (1977), Holliday (1978), Eggleston and Bochenek (1990), Chase (2002), Sarà and Sarà (2007)
Common predators of juveniles	Larger fishes and gelatinous zooplankton	McGowan and Richards (1989)
Common predators of adults	Toothed whales, swordfish, and sharks	Tiews (1963), Chase (2002)

[a]Curved fork length (CFL) is the measurement of the length of a tuna taken in a line tracing the contour of the body from the tip of the upper jaw to the fork of the tail, which abuts the upper side of the pectoral fin and the upper side of the caudal keel (FishBase 2013)

Atlantic bluefin tuna larvae also occur from the southern Gulf of Mexico to the Yucatán Channel (Richards and Potthoff 1980; McGowan and Richards 1986) and from the Straits of Florida to the Bahamas (Brothers et al. 1983; McGowan and Richards 1989).

Juveniles leave spawning grounds in the Gulf of Mexico in June to begin migration to nursery areas located off the North Carolina and Massachusetts coasts from Cape Hatteras to Cape Cod in waters over the continental shelf (Table 9.16). From June to March, adults inhabit feeding grounds in the central and northern Atlantic (Table 9.16).

Table 9.16. Summary of Habitat Information for Western Atlantic Bluefin Tuna (*Thunnus thynnus*)

Parameter	Value	Reference
Habitat preferences and spatial/ temporal distribution of juveniles	In June, juveniles leaving spawning grounds in the Gulf of Mexico begin migration to nursery areas located between Cape Hatteras, North Carolina and Cape Cod, Massachusetts in waters over the continental shelf for the summer and farther offshore in the winter	McGowan and Richards (1989), Mather et al. (1995)
Habitat preferences and spatial/ temporal distribution of adults	Epipelagic and oceanic, coming inshore seasonally to feed; feeding typically at depths <200 m (<656 ft) and >12 °C in waters above the thermocline; June through March: adults inhabit foraging grounds along the east coast of North America in waters over the continental shelf and in the central North Atlantic; April through June: non-mature adults inhabit waters over the continental shelf along the southeastern U.S. coast	Tiews (1963), Collette and Nauen (1983), Block et al. (2001a), Stokesbury et al. (2004), De Metrio et al. (2005)
Habitat preferences and spatial/ temporal distribution of spawning adults	April–June: migrate to spawning grounds in the northern Gulf of Mexico where spawning occurs along the continental slope in waters between the 200- and 3,000-m (656 and 9,843-ft) contours; prefer waters with moderate eddy kinetic energy, low surface chlorophyll concentrations, moderate wind speeds, and temperatures from 22.6 to 27.5°C; June–March: migrate through the Straits of Florida to foraging grounds off the Northeast U.S. and Canadian coasts; foraging grounds include waters overlying North American continental shelf, slope, Gulf Stream waters, the South and Mid-Atlantic Bight, the Gulf of Maine, and the Nova Scotia Shelf; larger individuals move into higher latitudes than smaller fish; occasionally forage in the central North Atlantic crossing the 45°W meridian, moving into the Eastern Atlantic and back before returning to spawning areas in the Gulf of Mexico	Mather et al. (1995), Block et al. (2001a, 2005), Karakulak et al. (2004a, b), Garcia et al. (2005), Teo et al. (2007a, b), Rooker et al. (2007)
Designated essential fish habitat for juveniles	Waters off North Carolina, south of Cape Hatteras, to Cape Cod	NMFS (2009b)
Designated essential fish habitat for adults	Pelagic waters of the central Gulf of Mexico and the mid-east coast of Florida; North Carolina from Cape Lookout to Cape Hatteras; New England from Connecticut to the mid-coast of Maine	NMFS (2009b)
Designated essential fish habitat for spawning adults	In the Gulf of Mexico, from the 100 m (328 ft) depth contour to the Exclusive Economic Zone (EEZ), continuing to the mid-east coast of Florida	NMFS (2009b)

The vertical distribution of western Atlantic bluefin tuna is often influenced by their feeding behavior and thermal biology. Atlantic bluefin tuna spend a considerable amount of time in the upper mixed layer, particularly on the inner continental shelf, where diving depths are limited by the bathymetry (Block et al. 2001b). Feeding in the mixed layer above the thermocline is common for both tropical and temperate tunas, and vertical use patterns may vary temporally as a function of shifts in prey distribution (Musyl et al. 2003; Kitagawa et al. 2006). Although Atlantic bluefin tuna spend most of their time in waters shallower than 200 m (656 ft), they are capable of diving to 1,000 m (3,281 ft) when in offshore waters

Table 9.17. Summary of Stock and Fisheries Information for Western Atlantic Bluefin Tuna (*Thunnus thynnus*)

Parameter	Value	Reference
General geographic distribution	From warm tropical waters in the Gulf of Mexico and the Caribbean to subpolar waters of the North Atlantic Ocean; Atlantic waters west of the 45°W meridian, from 55°N to 0° latitude	Collette and Nauen (1983), Mather et al. (1995), Vinnichenko (1996), Collette et al. (2001), Fromentin and Powers (2005), Rooker et al. (2007)
Commercial importance	High	
Recreational importance	High	
Management agency	NMFS, Highly Migratory Species Management Division (HMSMD); ICCAT	NMFS (2009b)
Management boundary	North Atlantic Ocean west of the 45°W meridian, including the Gulf of Mexico and Caribbean	Collette and Nauen (1983), NMFS (2009b)
Stock structure	Managed as East and West Atlantic Stocks; separated by the 45°W meridian	Rooker et al. (2007), NMFS (2009b)
Status (overfished/ overfishing)	Overfished from at least 2000–2012; overfishing from at least 2000–2012 (the conclusion could differ if a different productivity regime was assumed)	NMFS (2001, 2002b, 2003, 2004, 2005, 2006a, 2007, 2008, 2009a, 2010, 2011, 2012a)

(Block et al. 2001b; Stokesbury et al. 2004; De Metrio et al. 2005). The frequency of deep dives tends to be greatest for Atlantic bluefin tuna when they occur in the warm Gulf of Mexico waters (Block et al. 2001b; Teo et al. 2007b). Because Atlantic bluefin tuna are endothermic and can be thermally stressed in the warm Gulf of Mexico waters, the frequency of deep dives beneath the thermocline in the Gulf of Mexico may result from their efforts to avoid overheating (Block et al. 2005).

Bluefin tuna are oviparous (producing eggs that develop and hatch outside the maternal body), iteroparous (producing offspring several times over many seasons), and are multiple batch spawners (Schaefer 2001). The number of eggs produced is dependent on the size of the fish. Fertilization takes place directly in the water column (Fromentin 2009), and hatching occurs after 2 days. Atlantic bluefin tuna larvae are pelagic and reabsorb the yolk sac within a few days (Fromentin and Powers 2005).

Juvenile bluefin tuna grow rapidly. Growth tends to be linear during the larval phase (2–10 days) at a rate of 0.3–0.4 mm (0.012–0.016 in.)/day (Scott et al. 1993), similar to those reported for other tuna species from temperate and tropical regions, e.g., Pacific bluefin tuna (*Thunnus orientalis*), 0.33 mm (0.013 in.)/day (Miyashita et al. 2001); yellowfin tuna (*Thunnus albacores*), 0.47 mm (0.018 in.)/day (Lang et al. 1994); and southern bluefin tuna (*Thunnus maccoyii*), 0.28–0.36 mm (0.11–0.14 in.)/day (Jenkins and Davis 1990; Jenkins et al. 1991). A growth rate of 1.4 mm (0.06 in.)/day was reported for juveniles in the western Atlantic (267–413 mm or 10.5–16.3 in FL; from 70 to 200 days) (Brothers et al. 1983).

The mean observed length of Atlantic bluefin tuna at ages 1 and 2 in the western Atlantic was 53 and 75 cm (20.9 and 29.5 in.) FL, respectively (Atlantic Bluefin Tuna Status Review 2011). Estimated lengths of Atlantic bluefin tuna at ages 4 and 5 were 118 cm (46.5 in.) and 139 cm (54.7 in.) FL, respectively. Growth trajectories of Atlantic bluefin tuna are similar for young fish (ages 1–5) between eastern and western Atlantic stocks but diverge for older individuals, with size at age being greater for the western Atlantic stock. After age 5, growth trajectories of Atlantic bluefin tuna show marked differences between the eastern and western Atlantic, with the length at age being greater in the western Atlantic than in the eastern Atlantic (Atlantic Bluefin Tuna Status Review 2011). For example, at age 10, mean size in the western Atlantic was 212 cm (83.5 in.) FL, compared to 200 cm (78.7 in.) FL for the eastern Atlantic bluefin tuna. The general trend of greater length at age in the western Atlantic is exhibited in the growth models used for ICCAT assessments in the east (Cort 1991) and west (Turner and Restrepo 1994).

The western spawning stock in the Gulf of Mexico is comprised of large, late-maturing individuals. The estimated age at maturity ranges from 7 to 12 years, with the most commonly used age and size at maturity for the Gulf of Mexico Atlantic bluefin tuna being age 10 and 200 cm (78.7 in.) curved fork length (CFL), the measurement of the length of a tuna taken in a line tracing the contour of the body from the tip of the upper jaw to the fork of the tail, which abuts the upper side of the pectoral fin and the upper side of the caudal keel (FishBase 2013) (Table 9.15). However, Atlantic bluefin tuna reach sexual maturity as early as age 3 or 4 in the eastern Atlantic. Sex-specific differences in growth occur, with males growing slightly faster than females and reaching slightly larger sizes by age 10. Bluefin tuna are a long-lived species, with a life span of about 40 years.

The M of Atlantic bluefin tuna during early life-history stages mainly results from starvation and predation. Daily mortality during the larval stage has been estimated at 0.20 per day for the western stock. This estimate is lower than values reported for more tropical tunas during comparable periods: yellowfin tuna ($M = 0.33$ per day; Lang et al. 1994) and southern bluefin tuna ($M = 0.66$ per day; Davis 1991). The mortality of tunas during the juvenile phase is largely a function of size or age rather than species or habitat (Hampton 2000). In the most recent stock assessment, the M rate has been set at 0.14 per year and assumed to be age-independent (NMFS 2012b).

Large uncertainty is associated with the recruitment dynamics estimated in the most recent stock assessment (NMFS 2012a). Two levels of recruitment dynamics were considered in the stock assessment, low and high productivity. These levels could yield very different conclusions about the status of the Atlantic bluefin tuna stock and fishery.

Seasonal differences in growth occur for Atlantic bluefin tuna. The existence of a slowdown in growth during the winter has been confirmed for both juveniles (Mather and Schuck 1960; Furnestin and Dardignac 1962; Cort 1991) and adults (Tiews 1963; Butler et al. 1977). Large differences in growth, maturation, stock structure, and movement have been identified between the eastern and western Atlantic bluefin tuna. Genetic differentiation and natal homing behavior, observed in genetic and archival tagging studies, provide strong evidence for independence of the Gulf of Mexico and Mediterranean Sea Atlantic bluefin tuna stocks (Block et al. 2005; Carlsson et al. 2007; Boustany et al. 2008).

The stock structure is complicated because some fraction of the stock undertakes trans-Atlantic migration annually and/or ontogenetically, but migrants return to their natal sites to spawn. Although resident subpopulations exist in the eastern Atlantic bluefin tuna stock (De Metrio et al. 2005), there is no strong evidence for subpopulations in the western Atlantic bluefin tuna stock.

9.3.4.2 Predators and Prey

Atlantic bluefin tuna are opportunistic feeders and consume a wide variety of prey. As larvae and small juveniles, their diet is comprised primarily of zooplankton, with copepods as the main stomach item (Table 9.15) (Uotani et al. 1981, 1990). Their larvae are capable of feeding on other fish larvae (Miyashita et al. 2001). The diet of older juveniles and adults consists mainly of fishes, cephalopods (mostly squid), and crustaceans (Table 9.15).

Demersal fishes and invertebrates are also found in the stomachs of Atlantic bluefin tuna, especially for those that inhabit nearshore environments. Although no single taxon dominates, as a group, demersal organisms may comprise as much as 20 % of the stomach contents by number (Chase 2002). Large Atlantic bluefin tuna (e.g., >230 cm or >90.5 in CFL) may consume large individual prey items, such as bluefish (*Pomatomus saltatrix*) and spiny dogfish (*Squalus acanthias*) (Table 9.15) (Matthews et al. 1977; Chase 2002). The trophic level of adult Atlantic bluefin tuna is one level higher than those of other congeners. Predators of Atlantic bluefin tuna include swordfish, sharks, and whales (Table 9.15).

9.3.4.3 Key Habitat Needs and Distribution

Oceanographic conditions appear important for bluefin tuna spawning, and the actual location of spawning within each basin likely represents a balance between habitat requirements of larvae and the physiological limitations of adults. Key variables include bathymetry, sea surface temperature, eddy kinetic energy, surface chlorophyll concentration, and surface wind speed; sea surface temperature is the most important parameter. The sea surface temperatures reported for Atlantic bluefin tuna on putative spawning grounds in the Gulf of Mexico ranged from approximately 22.6–27.5 °C (Teo et al. 2007b). Because the northern slope waters of the Gulf of Mexico are above the purported 24 °C spawning threshold in early spring (Block et al. 2001a, b, 2005; Teo et al. 2007b), it is not surprising that Atlantic bluefin tuna begin spawning earlier in the Gulf of Mexico. In a study by Teo et al. (2007b), Atlantic bluefin tuna exhibited significant preference for areas with continental slope waters (2,800–3,400 m or 9,186–11,155 ft), moderate sea surface temperatures (24–25 and 26–27 °C), moderate eddy kinetic energy (251–355 cm^2/s^2), low surface chlorophyll concentrations (0.10–0.16 mg/m^3), and moderate wind speeds (6–7 and 9–9.5 m/s or 19.7–22.9 and 29.5–31.2 ft/s).

Temperature and depth are important factors influencing the distribution of Atlantic bluefin tuna in different life-history stages (Table 9.16). Essential fish habitat has been designated for different life-stages of Atlantic bluefin tuna, including eggs, larvae, juveniles, adults, and spawning adults (Table 9.16; Figures 9.21, 9.22, 9.23, and 9.24). In addition, a Habitat Area of Particular Concern has been designated for bluefin tuna (Figure 9.25).

9.3.4.4 Fisheries

Atlantic bluefin tuna are very valuable and highly prized; they support an important commercial and recreational fishery in the United States. The total catch for western Atlantic bluefin tuna peaked at 18,671 metric tons in 1964 as a result of the Japanese longline fishery for large fish off Brazil and the U.S. purse seine fishery for juvenile fish (NMFS 2012b). Landings dropped sharply thereafter with the collapse of these two fisheries, but increased again to average over 5,000 metric tons (11 million lb) in the 1970s due to the expansion of the Japanese longline fleet into the Northwest Atlantic and Gulf of Mexico and increased efforts in the purse seine fishery targeting larger fish for the sashimi market. The total catch for western Atlantic bluefin tuna, including discards, has generally been relatively stable since 1982 due to the imposition of quotas (Figure 9.26) (NMFS 2012b). Recent changes in landings mainly result from annual changes in the catch quota. The decline through 2007 was primarily due to considerable reductions in catch levels for U.S. fisheries. The majority of the western Atlantic bluefin tuna catch in recent years is from the commercial longline and sport fisheries (Figure 9.26).

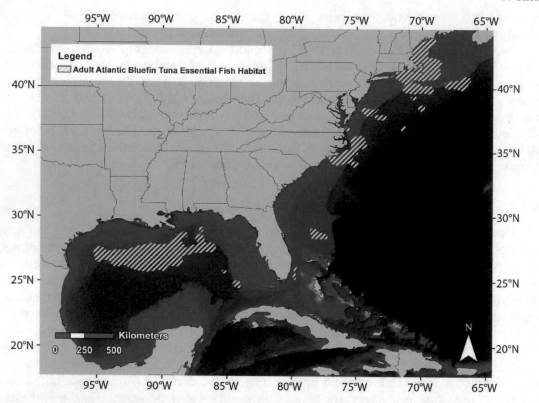

Figure 9.21. Essential fish habitat for adult Atlantic bluefin tuna (*Thunnus thynnus*) (from NOAA Fisheries Office of Sustainable Fisheries 2009a).

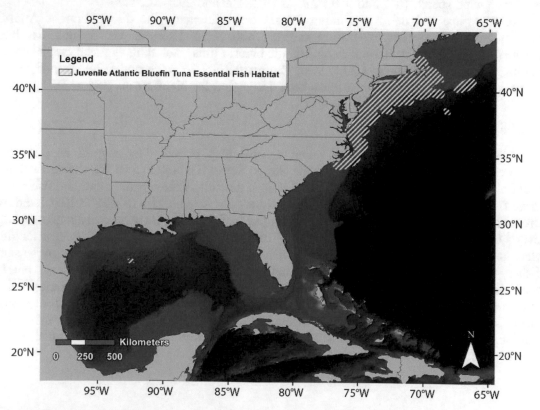

Figure 9.22. Essential fish habitat for juvenile Atlantic bluefin tuna (*Thunnus thynnus*) (from NOAA Fisheries Office of Sustainable Fisheries 2009a).

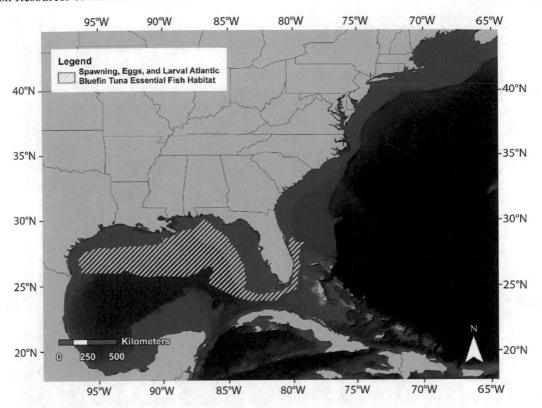

Figure 9.23. Essential fish habitat for spawning, eggs, and larval Atlantic bluefin tuna (*Thunnus thynnus*) (from NOAA Fisheries Office of Sustainable Fisheries 2009a).

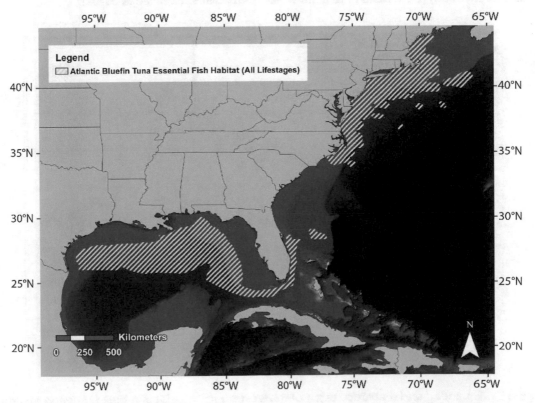

Figure 9.24. Essential fish habitat for all lifestages of Atlantic bluefin tuna (*Thunnus thynnus*) (from NOAA Fisheries Office of Sustainable Fisheries 2009a).

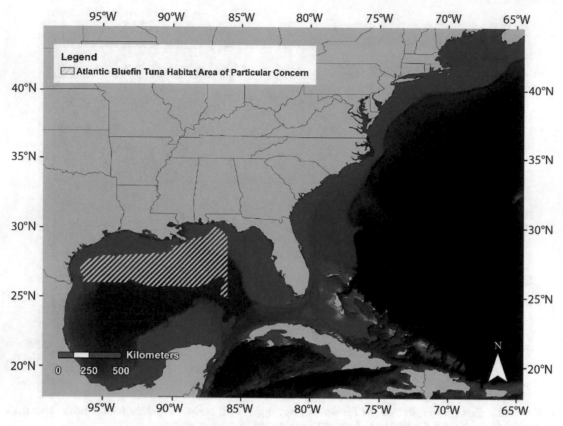

Figure 9.25. Highly migratory species habitat area of particular concern for Atlantic bluefin tuna (*Thunnus thynnus*) (from NOAA Fisheries Office of Sustainable Fisheries 2009a).

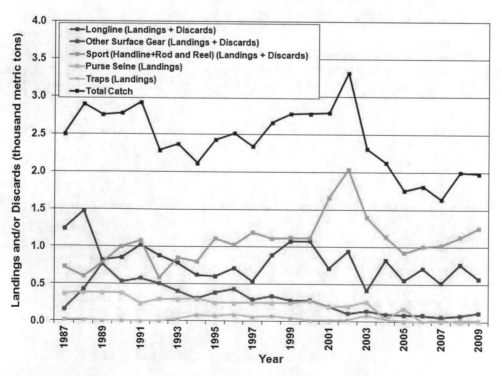

Figure 9.26. Landed and discarded catch of the western Atlantic bluefin tuna (*Thunnus thynnus*) stock for different gears from 1987 through 2009 (data from ICCAT 2012a).

Atlantic bluefin tuna are managed domestically by the NMFS's Highly Migratory Species Management Division (HMSMD) and internationally by the ICCAT (Table 9.17). The spawning stock biomass of the western Atlantic bluefin tuna has declined substantially over the past few decades and is at a very low level despite more than 20 years of strict regulations on the western Atlantic bluefin tuna fishery (NMFS 2012b).

Large uncertainty is associated with the most recent Atlantic bluefin tuna stock assessment, in particular with the estimated recruitment. The status of the population and fishery are dependent on the assumptions made on recruitment dynamics. For the high productivity scenario, the western Atlantic bluefin tuna stock is considered overfished (e.g., population level is too low) and the fishery is in the status of overfishing (e.g., fishing mortality is too high) (Table 9.17) (NMFS 2012a). However, for the low productivity scenario, the western Atlantic bluefin tuna stock is not overfished and the fishery is not in the status of overfishing (NMFS 2012a). Because of the limited information available, it is not clear which scenario more realistically describes the dynamics of Atlantic bluefin tuna recruitment.

Despite the uncertainty in the stock assessment, the stock biomass of the western Atlantic bluefin tuna has decreased greatly since the 1970s, mainly as a result of overfishing (NMFS 2012a). Overfishing over the last several decades has greatly reduced the spawning stock biomass and stock reproductive potential, likely resulting in poor recruitment and current low stock biomass of the Atlantic bluefin tuna. However, the western Atlantic bluefin tuna stock appears to be stable or even slightly increasing over the last 10 years, perhaps resulting from conservation measures and regulations (NMFS 2012a).

9.3.5 Atlantic Blue Marlin (*Makaira nigricans*)

The Atlantic blue marlin, a species of marlin endemic to the Atlantic Ocean, is widely distributed throughout the tropical and temperate waters of the Atlantic Ocean and Gulf of Mexico and is considered to be a single stock in the Atlantic Ocean (Figure 9.27). The Atlantic blue marlin is an apex predator and is considered a highly prized species in sport fisheries in the Gulf of Mexico. Recent stock assessments of Atlantic blue marlin by the ICCAT suggest that stocks are well below the level to support the MSY. Because of its economic and ecological importance, the Atlantic blue marlin was selected as a representative species to be evaluated in this chapter.

9.3.5.1 Key Life-History Processes and Ecology

As an apex predator, the Atlantic blue marlin plays a critical role in the ocean ecosystem (ICCAT 2012b). The Atlantic blue marlin is the most tropical of the billfishes and is a blue water fish that spends most of its life in the open sea (Tables 9.18 and 9.19). They rarely aggregate in schools and are usually found as scattered single individuals. Their distributional areas range from about latitude 45°N to about latitude 35°S (Table 9.20). Blue marlin are less abundant in the eastern Atlantic, where they mostly occur off Africa between the latitudes of 25°N and 25°S (NMFS 2009b).

The distribution and movement patterns of Atlantic blue marlin within the Gulf of Mexico tend to vary among individuals. Some may spend considerable time within the Gulf of Mexico for feeding and spawning, while others move seasonally between the Gulf of Mexico and tropical areas, such as the Bahamas. A tagging study with pop-up archival transmitting tags in the Gulf of Mexico suggested that most tagged fish remained in the Gulf of Mexico, with some fish exhibiting egress into Belize (Caribbean Sea) and the U.S. Virgin Islands (Kraus et al. 2011).

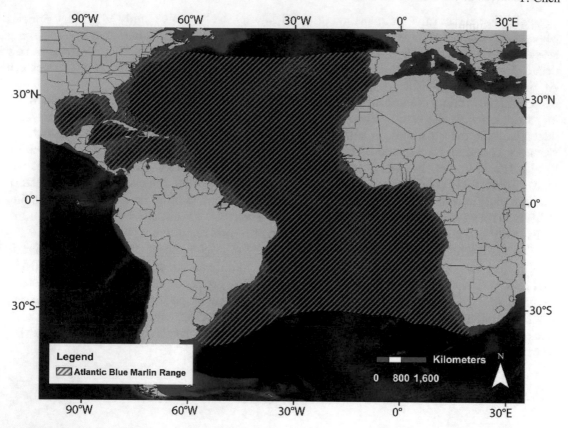

Figure 9.27. Range of the Atlantic blue marlin (*Makaira nigricans*) (modified from Maguire et al. 2006).

Table 9.18. Summary of Life-History Information for Atlantic Blue Marlin (*Makaira nigricans*)

Parameter	Value	Reference
von Bertalanffy growth model parameters	Not available	
Age at maturity	2–4 years	ICCAT (1997)
Length at maturity	Not available	
Weight at maturity	Males: 35–44 kg (77.2–97 lb) Females: 47–60 kg (103.6–132.3 lb)	NMFS (2009b)
Spawning season	July through October in the North Atlantic; February and March in the South Atlantic; May and June are peak spawning months for fish off Florida and the Bahamas	de Sylva and Breder (1997), NMFS (2009b)
Spawning location	Pelagic waters in the North and South Atlantic from a northern extreme of 32°N off of Bermuda to a southern extreme of 25°S off the Brazilian coast; corresponds to sea surface temperatures around 28 °C	Bartlett and Haedrich (1968), Serafy et al. (2003), Luckhurst et al. (2006), NMFS (2009b)
Common prey of juveniles and adults	Feed primarily on tuna-like fishes, squid, and on a wide size range of other organisms; dolphinfish, mackerels, tunas, and bonitos are important prey in the Gulf of Mexico	Rivas (1975), Davies and Bortone (1976), Nakamura (1985)
Common predators of juvenile and adults	Very little is known	ICCAT (2012b)

Fish Resources of the Gulf of Mexico

Table 9.19. Summary of Habitat Information for Atlantic Blue Marlin (*Makaira nigricans*)

Parameter	Value	Reference
Habitat preferences and temporal/spatial distribution of juveniles and adults	Epipelagic and oceanic, generally found in blue waters with a temperature range of 22–31 °C; January to April in the Southwest Atlantic from 5°S to 30°S, and from June to October in the Northwest Atlantic between 10°N and 35°N; May, November, and December are transitional months; seasonal movements related to changes in sea surface temperatures; in the northern Gulf of Mexico they are associated with low productivity blue waters and the Loop Current	Rivas (1975), NMFS (2009b)
Habitat preferences and temporal/spatial distribution of spawning adults	Pelagic waters in the North and South Atlantic from a northern extreme of 32°N off Bermuda to a southern extreme of 25°S off the Brazilian coast; sea surface temperatures around 28 °C; May and June are peak spawning months for fish off Florida and the Bahamas	Bartlett and Haedrich (1968), de Sylva and Breder (1997), Serafy et al. (2003), Luckhurst et al. (2006), NMFS (2009b)
Designated essential fish habitat for juveniles	In the central Gulf of Mexico, from southern Texas to the Florida Panhandle; through the Florida Keys to southern Cape Cod; Puerto Rico and the Virgin Islands	NMFS (2009b)
Designated essential fish habitat for adults	In the central Gulf of Mexico, from southern Texas to the Florida Panhandle; through the Florida Keys to southern Cape Cod; Puerto Rico and the Virgin Islands	NMFS (2009b)
Designated essential fish habitat for spawning adults	Mid-east coast of Florida through the Florida Keys; waters off the northwest coast of Puerto Rico.	NMFS (2009b)

Table 9.20. Summary of Stock and Fisheries Information for Atlantic Blue Marlin (*Makaira nigricans*)

Parameter	Value	Reference
General geographic distribution	Tropical and subtropical waters of the Atlantic Ocean; ranging from 45°N to 35°S	NMFS (2009b)
Commercial importance	No commercial U.S. fishery	NMFS (2009b)
Recreational importance	High	NMFS (2009b)
Management agency	NMFS, HMSMD; ICCAT	NMFS (2009b)
Management boundary	Atlantic Ocean	
Stock structure	Single Atlantic-wide stock	ICCAT (2001)
Status (overfished/ overfishing)	Overfished from at least 2000–2011; overfishing from at least 2000–2011	NMFS (2001, 2002b, 2003, 2004, 2005, 2006a, 2007, 2008, 2009a, 2010, 2011)

However, tagged fish showed highly variable movement patterns, regardless of tagging location, season, or egress status. Seasonal changes in distribution suggested a north–south cyclical movement pattern within the Gulf of Mexico that supported a new perspective on Atlantic blue marlin, in which the Gulf of Mexico provides suitable year-round habitat that is utilized by a subset of the Atlantic population. An analysis of otolith chemistry of Atlantic blue marlin also suggested that movement out of the Gulf of Mexico for Atlantic blue marlin may be more limited, as compared to other regions (Wells et al. 2010).

Atlantic blue marlin in the Gulf of Mexico tend to remain in offshore waters (Table 9.19). However, they may move close to the coast from July to September. They spawn in the Gulf of Mexico as early as May and continue to spawn throughout the summer (Table 9.18). Atlantic blue marlin that spawn on spawning grounds off Texas and Louisiana during the summer remain in the Gulf through the fall and winter. Blue marlin tag/recapture data from the Gulf of Mexico indicate that seasonal movements may occur between the Gulf of Mexico (summer) and the Bahamas (winter). Several data sources indicate that the Gulf of Mexico may serve as important spawning and/or nursery habitat for blue marlin (Brown-Peterson et al. 2008; Rooker et al. 2012). Blue marlin larvae were found in a 2005 fishery-independent survey in the areas from 27 to 28°N to 90 to 94°W in July (Rooker et al. 2012). This seems to suggest that blue marlin can spawn in the northern Gulf of Mexico (Brown-Peterson et al. 2008; Kraus et al. 2011). However, larvae mainly are present near the western margin of Loop Current on the continental shelf in relatively warm waters (>27 °C). The presence of young blue marlin larvae along the boundary of the Loop Current may be a result of transport from Caribbean/Straits of Florida spawning events (Kraus et al. 2011; Rooker et al. 2012). Because no spawning-capable adults have been captured in this region, it is unlikely that blue marlin spawn in the Loop Current in the northeastern Gulf of Mexico (Brown-Peterson et al. 2008). Strong histological evidence supports the lack of spawning in the northern Gulf of Mexico east of the Mississippi River, which is augmented by the failure to capture blue marlin larvae in areas not associated with the Loop Current (Kraus and Rooker 2007; Kraus et al. 2011; Rooker et al. 2012). Thus, the likelihood of blue marlin spawning in the northern Gulf of Mexico is slim, although the northern Gulf of Mexico supports an active recreational fishery for blue marlin from May through September (Brown-Peterson et al. 2008).

Limited published information is available on blue marlin biology and life history from the Gulf of Mexico (Table 9.18) (De Sylva et al. 2000). Females are batch spawners and can spawn as many as four times in a spawning season (Brown-Peterson et al. 2004). They often release more than seven million eggs at once, each approximately 1 mm (0.04 in.) in diameter (Brown-Peterson et al. 2008). The larvae may grow as much as 16 mm (0.63 in.) in a day (Brown-Peterson et al. 2008). Males may live for 18 years and females up to 27 years. Females can grow up to four times the weight of males (Wilson et al. 1991; ICCAT 1997).

The M estimated using the Hoenig method (Hoenig 1983) at a maximum age of 30 years is 0.14 (Hill et al. 1989), which was used in the most recent stock assessment. The estimated blue marlin recruitment fluctuates over time and has been low in recent years (Figure 9.28). However, there is great uncertainty associated with the estimated recruitment, which results mainly from uncertainty in the quality of fishery and biological data, as well as the assumed stock structure and population dynamics.

Given the large distributional area that the Atlantic blue marlin occupies and the existence of multiple spawning grounds, blue marlin in different areas may be subject to different environmental stressors and prey availability. This may result in large spatial variability in key life-history parameters, such as growth and maturation.

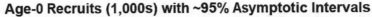

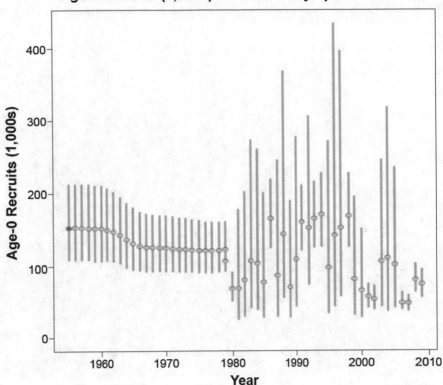

Figure 9.28. Temporal variability in Atlantic blue marlin (*Makaira nigricans*) recruitment estimated with the fully integrated stock assessment model (redrawn from Figure 30, ICCAT 2012b).

Although there is evidence indicating that some Atlantic blue marlin may be able to complete most of their life cycle from spawning to feeding within the Gulf of Mexico, many studies suggest that Atlantic blue marlin larvae are not produced within the Gulf of Mexico; rather, they are transported via the Loop Current from tropical areas. The evidence for the existence of multiple spawning grounds suggests that the stock structure may be more complicated than a one-unit stock assumed in the stock assessment. More evidence is needed to test the hypothesis that the Gulf of Mexico provides suitable year-round habitat that is utilized by a subset of the Atlantic blue marlin population (e.g., existence of a substock of Atlantic blue marlin in the Gulf of Mexico), given the uncertainty regarding whether the Atlantic blue marlin larvae come from within the Gulf of Mexico or originate in Caribbean waters.

9.3.5.2 Predators and Prey

Atlantic blue marlin larvae feed on a variety of zooplankton, along with drifting fish eggs and other larvae. Juvenile and adult Atlantic blue marlin typically feed near the surface but sometimes travel to great depths in search of prey, and feed opportunistically on a wide variety of fish and invertebrates (Table 9.18). Blue marlin have been documented to take prey as large as white marlin, yellowfin, and bigeye tuna in the 45 kg (100 lb) range and are also capable of feeding on small but numerous prey, such as filefish and snipefish. The Atlantic blue marlin has few predators apart from humans (McEachran and Fechhelm 2005).

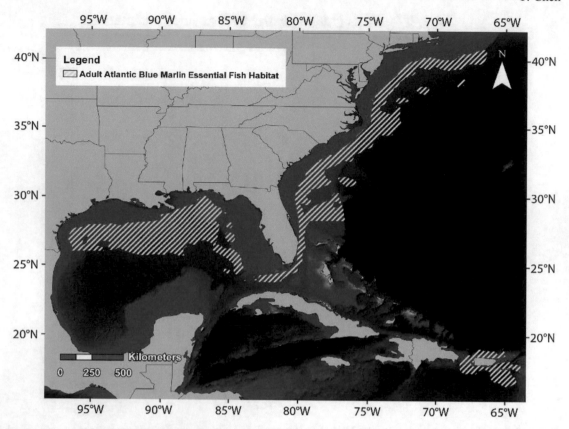

Figure 9.29. Essential fish habitat for adult Atlantic blue marlin (*Makaira nigricans*) (from NOAA Fisheries Office of Sustainable Fisheries 2009b).

9.3.5.3 Key Habitat Needs and Distribution

Atlantic blue marlin usually inhabit waters warmer than 24 °C, but have been found at surface water temperatures as high as 30.5 °C and as low as 21.7 °C (Table 9.19). Because Atlantic blue marlin prefers blue water, the clarity of water is also an important factor influencing its distribution (NMFS 2009b). Essential fish habitat has been designated for eggs, larvae, juveniles, adults, and spawning adults of Atlantic blue marlin (Table 9.19; Figures 9.29, 9.30, 9.31, 9.32).

9.3.5.4 Fisheries

Because of their relative rarity, beauty, and sporting qualities, Atlantic blue marlin are considered one of the most prestigious catches in recreational fisheries, and they support a multi-million dollar industry that includes hundreds of companies and thousands of jobs for boat operators, boat builders, marinas, dealerships, and fishing tackle manufacturers and dealers in the Gulf of Mexico region. The Atlantic blue marlin catch increased abruptly in the early 1960s nearing 9,000 metric tons, but dropped quickly. The catch has been quite stable since the late 1960s, varying between 2,000 and 5,000 metric tons for most years during this time period (Figure 9.33).

The management of Atlantic blue marlin is subject to domestic and international regulations (Table 9.20). The current Atlantic blue marlin stock assessment indicates that the stock level was low in 2009, fishing mortality was high, and the catch level of 3,431 metric tons in 2010

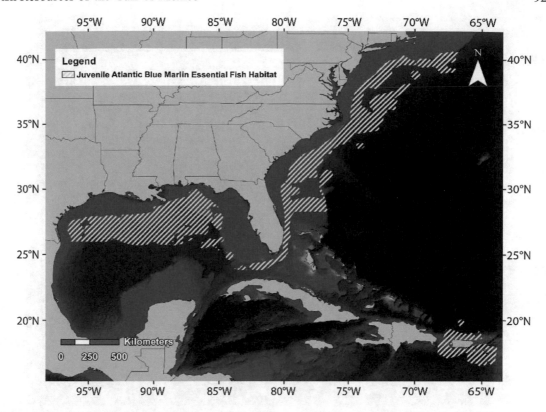

Figure 9.30. Essential fish habitat for juvenile Atlantic blue marlin (*Makaira nigricans*) (from NOAA Fisheries Office of Sustainable Fisheries 2009b).

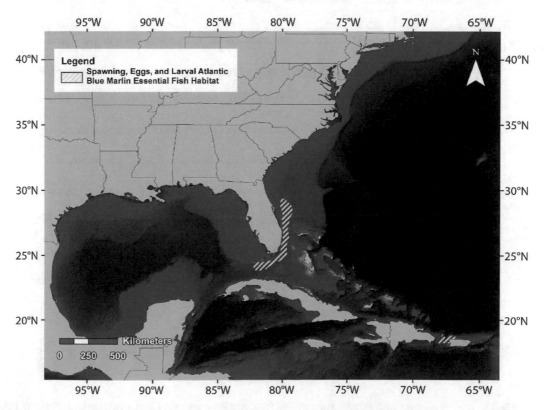

Figure 9.31. Essential fish habitat for spawning, eggs, and larval Atlantic blue marlin (*Makaira nigricans*) (from NOAA Fisheries Office of Sustainable Fisheries 2009b).

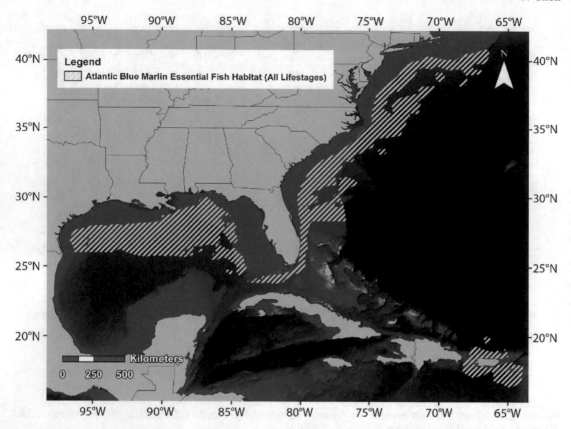

Figure 9.32. Essential fish habitat for all lifestages of Atlantic blue marlin (*Makaira nigricans*) (from NOAA Fisheries Office of Sustainable Fisheries 2009b).

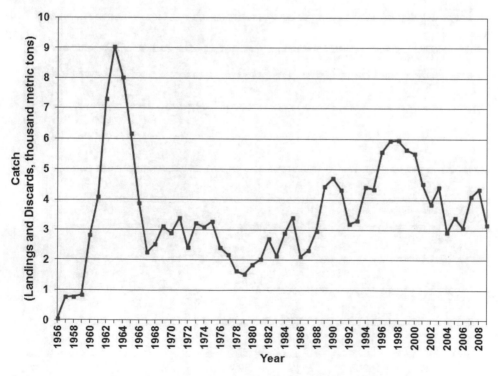

Figure 9.33. Landed and discarded catch of the Atlantic blue marlin (*Makaira nigricans*) from 1956 through 2009 (data from ICCAT 2012b).

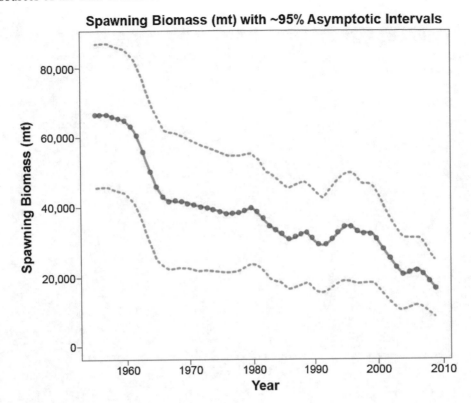

Figure 9.34. Temporal variability in Atlantic blue marlin (*Makaira nigricans*) spawning stock biomass estimated with the fully integrated stock assessment model (redrawn from Figure 29, ICCAT 2012b).

would likely result in continuing stock decline. A rebuilding plan needs to be developed for the stock of Atlantic blue marlin to reduce the annual total catch below 2,000 metric tons to allow the stock to increase. The Atlantic blue marlin population declined greatly during the last century (Figure 9.34); overfishing is currently occurring, and the stock is overfished (Table 9.20). The International Union for the Conservation of Nature currently considers it a threatened species due to overfishing and the substantially reduced stock abundance.

The spawning stock biomass of Atlantic blue marlin has decreased greatly since the 1960s (Figure 9.34). The recent stock biomass is approximately 25 % of the biomass that existed in the 1950s. However, there is large uncertainty associated with the estimates. This uncertainty mainly results from uncertainty in the quality of fishery and biological data, as well as the assumed stock structure and population dynamics.

9.3.6 Atlantic Swordfish (*Xiphias gladius*)

The Atlantic swordfish is a highly migratory and circumglobal species; it is widely distributed in the Atlantic Ocean, including tropical, temperate, and some cold water regions from 50°N to 45°S in the western Atlantic and 60°N to 50°S in the eastern Atlantic (Figure 9.35 and Table 9.21) (Palko et al. 1981; Nakamura 1985; NMFS 2009b). Currently, the ICCAT considers the existence of three distinct management units: North Atlantic, South Atlantic, and Mediterranean Sea. The North Atlantic stock is separated from the South Atlantic stock

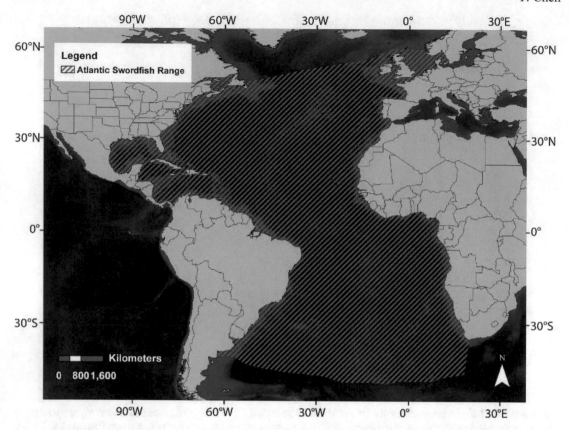

Figure 9.35. Range of the Atlantic swordfish (*Xiphias gladius*) (modified from Maguire et al. 2006).

Table 9.21. Summary of Stock and Fisheries Information for Atlantic Swordfish (*Xiphias gladius*)

Parameter	Value	Reference
General geographic distribution	Circumglobal species; range includes tropical, temperate, and some cold water regions from 50°N to 45°S in the western Atlantic, and 60°N to 50°S in the eastern Atlantic	Nakamura (1985), NMFS (2009b)
Commercial importance	High	
Recreational importance	Medium	
Management agency	NMFS, Highly Migratory Species Management Division; International Commission for the Conservation of Atlantic Tunas	NMFS (2009b)
Management boundary	North and South Atlantic stocks are separated at 5°N	NMFS (2009b)
Stock structure within the Gulf of Mexico	Part of the North Atlantic stock	NMFS (2009b)
Status (overfished/ overfishing)	Overfished prior to 2000–2002; overfishing prior to 2000–2002	NMFS (2001, 2002b, 2003, 2004, 2005, 2006a, 2007, 2008, 2009a, 2010, 2011)

at 5°N. The results of biological (Tserpes and Tsimenides 1995), genetic (Chow and Takeyama 2000; Kasapidis et al. 2007), and tagging (García-Cortés et al. 2003; Neilson et al. 2007) studies clearly supported this delineation of population structure, although intermixing among the three stocks was found in some studies (Alvarado-Bremer et al. 2007). The North Atlantic stock of the Atlantic swordfish was evaluated because it is an apex predator that plays an important role in its marine ecosystems, it supports an important fishery in the United States, and the Gulf of Mexico is an important Atlantic swordfish nursery, feeding, and spawning ground.

9.3.6.1 Key Life-History Processes and Ecology

Because swordfish are difficult to age, there is a lot of uncertainty about some of their basic life-history processes, such as growth and maturation (Table 9.22). In general, juvenile swordfish grow rapidly, reaching about 140 cm (55.1 in.) lower-jaw fork length [LJFL, which is

Table 9.22. Summary of Life-History Information for Atlantic Swordfish (*Xiphias gladius*)

Parameter	Value	Reference
von Bertalanffy growth model parameters	Not available	
Age at first maturity	Females: 4–5 years Males: 1.4 years	Palko et al. (1981), Nakamura (1985), Arocha (1997), NMFS (2009b)
Weight at first maturity	Females: 74 kg (163.1 lb) Males: 21 kg (46.3 lb)	Palko et al. (1981), Nakamura (1985), Arocha (1997), NMFS (2009b)
Length at first maturity (50 %)	Females: 179–182 cm (70.5–71.7 in.) lower jaw fork length (LJFL)[a] Males: 112–129 cm (44.1–50.8 in.) LJFL	Palko et al. (1981), Nakamura (1985), Arocha (1997), NMFS (2009b)
Spawning season	December through June in the western North Atlantic and northern Caribbean; April through August off of the southeast coast of the United States	Arocha (1997)
Spawning location	Between 15°N and 35°N, west of 40°W meridian; major spawning grounds in the Straits of Yucatán, the Straits of Florida, and in the vicinity of the northernmost arc of the Gulf Loop Current	Grall et al. (1983), Arocha (1997), Govoni et al. (2003)
Common prey of juveniles	Squids, fishes, and pelagic crustaceans	Palko et al. (1981)
Common prey of adults	Small tunas, dolphinfishes, lancetfish, snake mackerels, flyingfishes, barracudas, mackerels, herrings, anchovies, sardines, sauries, needlefishes, hakes, pomfrets, cutlass fish, lightfishes, hatchet fishes, redfish, lanternfishes, and cuttlefishes, octopus, and squids, such as *Ommastrephes*, *Loligo*, and *Illex*	Toll and Hess (1981), Nakamura (1985)
Common predators of juveniles	Sharks, tunas, billfishes, and adult swordfish	Palko et al. (1981)
Common predators of adults	Sperm whales, killer whales, and large sharks, such as mako sharks	NMFS (2009b)

[a]Lower jaw fork length is from the tip of the lower jaw to the fork in the tail (FishBase 2013)

from the tip of the lower jaw to the fork in the tail (FishBase 2013)] by age 3. The growth rate decreases after age 3, perhaps as a result of maturation. There is sexual dimorphism, with females growing faster and reaching larger maximum sizes than males (Table 9.22). Tagging studies have shown that some swordfish can live up to 15 years.

Juvenile Atlantic swordfish of the North Atlantic stock occur year-round in the Gulf of Mexico, the Florida Atlantic coast, and waters near the Charleston Bump (Table 9.23) (Palko

Table 9.23. Summary of Habitat Information for Atlantic Swordfish (*Xiphias gladius*)

Parameter	Value	Reference
Habitat preferences and temporal/spatial distribution of juveniles	The Gulf of Mexico, the Atlantic coast of Florida, and waters near the Charleston Bump	Palko et al. (1981), Cramer and Scott (1998)
Habitat preferences and temporal/spatial distribution of adults	Epipelagic to meso-pelagic; temperature range from 18 to 22 °C; concentrate along boundary currents of the Gulf Stream and the Gulf of Mexico Loop Current; some move northeastward along U.S. continental shelf in summer and return southwestward in autumn; another group moves from deepwater westward toward the continental shelf in summer and back into deepwater in autumn	Palko et al. (1981), Nakamura (1985), Arocha (1997), Govoni et al. (2003), NMFS (2009b)
Habitat preferences and temporal/spatial distribution of spawning adults	Between 15°N and 35°N, west of 40°W meridian; most spawning takes place in waters with surface temperatures above 20–22 °C; major spawning grounds thought to occur in the Straits of Yucatán, the Straits of Florida, and in the vicinity of the northernmost arc of the Gulf Loop Current; move to warmer waters for spawning and cooler waters for feeding; south of the Sargasso Sea and in the upper Caribbean, spawning occurs from December through March; off the U.S. southeast coast, spawning occurs from April through August	Palko et al. (1981), Grall et al. (1983), Nakamura (1985), Arocha (1997), Govoni et al. (2003), NMFS (2009b)
Designated essential fish habitat for juveniles	In the central Gulf of Mexico, from southern Texas through the Florida Keys; Atlantic east coast from South Florida to Cape Cod; Puerto Rico and the Virgin Islands	NMFS (2009b)
Designated essential fish habitat for adults	In the central Gulf of Mexico, from southern Texas to the Florida Panhandle and western Florida Keys; Atlantic east coast from southern Florida to the mid-east coast of Florida and Georgia to Cape Cod; Puerto Rico and the Virgin Islands	NMFS (2009b)
Designated essential fish habitat for spawning adults	From off Cape Hatteras, North Carolina extending south around Peninsular Florida through the Gulf of Mexico to the U.S./Mexico border from the 200 m (656 ft) isobath to the EEZ boundary; associated with the Loop Current boundaries in the Gulf and the western edge of the Gulf Stream in the Atlantic; also, all U.S. waters of the Caribbean from the 200 m (656 ft) isobath to the EEZ boundary	NMFS (2009b)

et al. 1981; Cramer and Scott 1998). Adult Atlantic swordfish tend to concentrate along boundary currents of the Gulf Stream and the Gulf of Mexico Loop Current (Table 9.23). They are subject to seasonal movement: one group moves northeastward along the U.S. continental shelf in summer and returns southwestward in autumn, and another group moves from deepwater westward toward the continental shelf in summer and back into deepwater in autumn (Palko et al. 1981; Arocha 1997; NMFS 2009b).

Atlantic swordfish tend to move to warmer waters for spawning and cooler waters for feeding. They tend to migrate to the preferred temperatures or areas for spawning during the peak of a spawning season (Palko et al. 1981; Tserpes et al. 2008). Atlantic swordfish appear to spawn throughout the year, and spawning timing tends to vary among different spawning areas (Tables 9.21 and 9.22). Seasonal latitudinal migrations of swordfish, which may result from seasonal changes in sea surface temperature, are well documented (Nakamura 1985; Seki et al. 2002; Takahashi et al. 2003; Neilson et al. 2009).

Although Atlantic swordfish have evolved a specialized muscle that functions like a brain heater and enables them to tolerate a wide range of temperatures and move rapidly between warm surface waters and cold waters at great depths (Carey 1990), their vertical distribution is generally limited by the depth of the thermocline (Block et al. 1992). Takahashi et al. (2003) also indicated that the vertical swimming behavior of swordfish changes in response to near-surface water temperatures.

Limited information is available on the M of the Atlantic swordfish. In the assessment based on the results of the virtual population analysis (VPA) model, M was assumed to be 0.2 per year (Scott and Porch 2007). However, no information or evidence is presented to justify the choice of this value.

9.3.6.2 Predators and Prey

Atlantic swordfish are diurnal feeders rising close to the mixed surface layer at night and descending to deeper waters during the day to feed on pelagic fishes and squids (Carey 1990). Swordfish mainly feed on prey concentrations associated with vertical density discontinuities (Carey and Robison 1981), such as the thermocline (Draganik and Cholyst 1988). Juvenile and adult Atlantic swordfish predate on squids, tunas, dolphinfishes, mackerels, and pelagic crustaceans (Table 9.21). Sperm whales, killer whales, and large sharks prey on swordfish (Table 9.21).

9.3.6.3 Key Habitat Needs and Distribution

Oceanographic variables that may influence the distribution and abundance of Atlantic swordfish include sea surface temperature; depth of the thermocline (Carey 1990); sea surface height anomaly, which is a good indicator of possible oceanographic activities, such as gyres and eddies (Seki et al. 2002; Tserpes et al. 2008); existence of thermal fronts, frontal zones, and eddy fields that can produce locally elevated chlorophyll concentrations and zooplankton abundance that stimulate feeding conditions (Podestá et al. 1993; Logerwell and Smith 2001); and chlorophyll concentrations, which regulate the distribution and abundance of the prey of swordfish (Tserpes et al. 2008; Yáñez et al. 2009). The spatial distribution and abundance of swordfish also may be determined by other factors, such as distinct bathymetric features. Many studies have indicated that the distribution of swordfish is also associated with bottom topographic structures and thermal fronts, such as submarine canyons or hummocky bumps (Carey and Robison 1981; Carey 1990; Podestá et al. 1993; Sedberry and Loefer 2001; Damalas et al. 2007). The average temperature preferred by swordfish during the day can be as low as

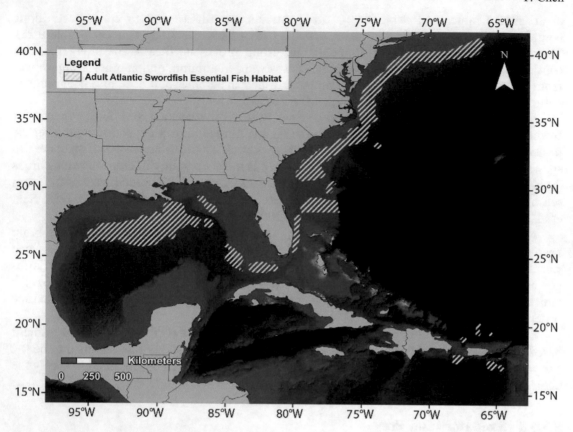

Figure 9.36. Essential fish habitat for adult Atlantic swordfish (*Xiphias gladius*) (from NOAA Fisheries Office of Sustainable Fisheries 2009c).

10 °C, while it is 28 °C at night when they move up to the near-surface waters (Sedberry and Loefer 2001).

The spatial distributions of essential fish habitat that have been designated for various lifestages of Atlantic swordfish in the Gulf of Mexico, along the U.S. east coast, and around Puerto Rico are shown in Figures 9.36, 9.37, 9.38, and 9.39. Table 9.22 includes the definitions of essential fish habitat that have been established for juvenile, adult, and spawning adult Atlantic swordfish.

9.3.6.4 Fisheries

Atlantic swordfish support an important commercial and recreational fishery in the United States, including the Gulf of Mexico (NMFS 2012a). Canada, Spain, and the United States have operated a targeted pelagic longline Atlantic swordfish fishery since the late 1950s or early 1960s in the North Atlantic (NMFS 2009a, 2012a). The harpoon fisheries have existed at least since the late 1800s in the Northwest Atlantic Ocean. In addition, some driftnet activities for swordfish occur around the Straits of Gibraltar area and in other Atlantic areas (e.g., off the coast of West Africa). The primary fisheries that take swordfish as bycatch are tuna fishing fleets from Taiwan, Japan, Korea, and France (Collette et al. 2012). The tuna longline fishery has operated throughout the Atlantic since 1956, with substantial catches of swordfish as bycatch in some years.

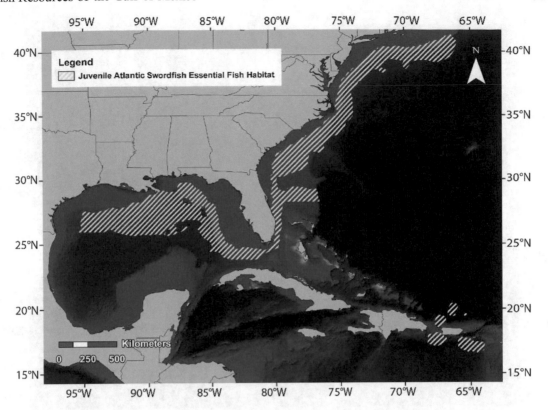

Figure 9.37. Essential fish habitat for juvenile Atlantic swordfish (*Xiphias gladius*) (from NOAA Fisheries Office of Sustainable Fisheries 2009c).

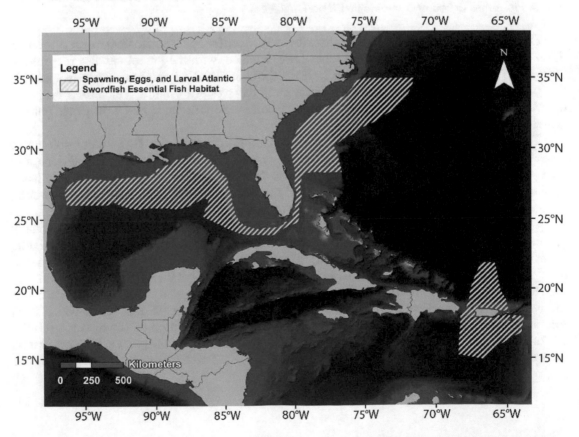

Figure 9.38. Essential fish habitat for spawning, eggs, and larval Atlantic swordfish (*Xiphias gladius*) (from NOAA Fisheries Office of Sustainable Fisheries 2009c).

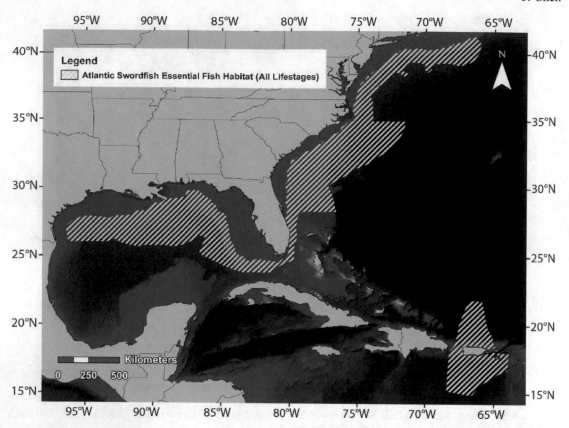

Figure 9.39. Essential fish habitat for all lifestages of Atlantic swordfish (*Xiphias gladius*) (from NOAA Fisheries Office of Sustainable Fisheries 2009c).

U.S. catches (landings plus dead discards) of swordfish peaked in 1990, with a total of 5,519 metric tons. Since then, U.S. catches have declined, with the lowest catches reported in 2006 (2,057 metric tons). Most (93 % in 2008) of the U.S. swordfish catches have been landed in the pelagic longline fishery operated throughout the western Atlantic, including the Gulf of Mexico and Caribbean Sea, that targeted both yellowfin tuna and swordfish (NMFS 2009a, 2012a).

The U.S. pelagic longline fleet has decreased substantially in size from about 400 active vessels in the 1990s to about 120 vessels in 2008 as a result of regulations, market conditions, and fuel prices. Atlantic swordfish also support a small recreational fishery, which currently lands only a small proportion of total U.S. landings (75 metric tons in 2008). This fishery has, however, expanded in the last few years and is projected to continue to grow (NMFS 2009a, 2012a).

The Atlantic swordfish fishery is managed both domestically and internationally (Table 9.23), and catch limits are one of the regulations used in managing the North Atlantic swordfish stock (NMFS 2010, 2011, 2012a). The total allowable catch (TAC) of 14,000 metric tons per year during the 2007–2009 period was reduced to 13,700 metric tons in 2010 and 12,836 metric tons in 2011. Minimum size limits are also used to manage the fishery. There are two minimum size options that are applied to the entire Atlantic: 125 cm (49.2 in.) LJFL with a 15 % tolerance or 119 cm (46.8 in.) LJFL with zero tolerance and evaluation of the discards (NMFS 2012a). A number of time/area closures went into effect in 2001 for the pelagic longline vessels operating within the U.S. Exclusive Economic Zone (EEZ). Two permanent closures, one in the

Gulf of Mexico and the other along the Florida east coast, were established to reduce the bycatch of undersized swordfish. Circle hooks became mandatory for the U.S. pelagic longline fleet in 2004 to reduce mortality of discarded bycatch species, including swordfish (NMFS 2009a, 2012a).

The North Atlantic swordfish fishery is considered to be fully rebuilt, and overfishing is not occurring (Table 9.21) (NMFS 2012a). The latest Standing Committee on Research and Statistics stock assessment indicates that the North Atlantic swordfish stock has greater than 50 % probability to be at or above biomass at MSY (NMFS 2009a). The estimated relative biomass trend shows a consistent increase since 2000. Fishing mortality has been below fishing mortality at MSY since 2005; therefore, the rebuilding objective has been achieved. However, great uncertainty is associated with the stock assessment resulting from the quality of fisheries data and biological parameters (e.g., M and growth), suitability of stock assessment models, lack of understanding of some key life-history process, and assumed stock structure. More data are needed for improved understanding of key biotic and abiotic factors influencing the recruitment dynamics of the North Atlantic swordfish.

9.3.7 Atlantic Sailfish (*Istiophorus albicans*)

The Atlantic sailfish is a pelagic-oceanic and highly migratory species (Figure 9.40). It is distributed in tropical and temperate waters about 40°N in the Northwest Atlantic, 50°N in the Northeast Atlantic, 40°S in the Southwest Atlantic, and 32°S in the Southeast Atlantic (Figure 9.41 and Table 9.24). Although the importance of sailfish in commercial fisheries is limited, this species plays an important role in recreational fisheries (Table 9.24). In addition, because it is so important in recreational fisheries, the Atlantic sailfish is the official saltwater fish of Florida. As one of the top predator species that are highly migratory and distributed widely, Atlantic sailfish play an important ecological role in Gulf of Mexico ecosystems.

Figure 9.40. Atlantic sailfish (*Istiophorus albicans*) feeding (from NaluPhoto 2012).

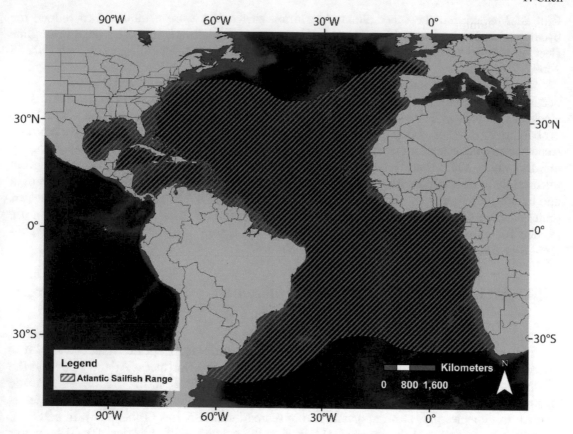

Figure 9.41. Range of the Atlantic sailfish (*Istiophorus albicans*) (modified from Maguire et al. 2006).

Table 9.24. Summary of Stock and Fisheries Information for Atlantic Sailfish (*Istiophorus albicans*)

Parameter	Value	Reference
General geographic distribution	Circumtropical distribution; range from 40°N to 40°S in the western Atlantic and 50°N to 32°S in the eastern Atlantic	NMFS (2009b)
Commercial importance	Low, commercial landings prohibited in the United States	ICCAT (2010)
Recreational importance	High	
Management agency	NMFS, Highly Migratory Species Management Division; ICCAT	NMFS (2009b)
Management boundary	40°W north of 5°N; 30°W from 5°N to the equator; 20°W south of the equator	ICCAT (2011a)
Stock structure within the Gulf of Mexico	Part of the Western Atlantic stock	ICCAT (2010)
Status (overfished/ overfishing)	Overfished prior to 2000–2008; overfishing prior to 2000–2011	NMFS (2001, 2002b, 2003, 2004, 2005, 2006a, 2007, 2008, 2009a, 2010, 2011)

Table 9.25. Summary of Life-History Information for Atlantic Sailfish (*Istiophorus albicans*)

Parameter	Value	Reference
von Bertalanffy growth model parameters	Not available	
Age at first maturity	3 years	de Sylva and Breder (1997)
Weight at first maturity	Males: 10 kg (22 lb); females: 13–18 kg (28.7–39.7 lb)	de Sylva and Breder (1997)
Length at first maturity	Not available	
Spawning season	Multiple spawners; in the western Atlantic, spawning activity moves northward as summer progresses; in the northern Gulf of Mexico, spawning occurs from May to September; from Cuba to the Carolinas, spawning occurs from April to September	Bumguardner and Anderson (2008), NMFS (2009b)
Spawning location	Shallow waters around Florida from the Keys to Palm Beach on the east coast; in the northern Gulf of Mexico, including off Texas; offshore from Cuba to the Carolinas	Bumguardner and Anderson (2008), NMFS (2009b)
Common prey of juveniles and adults	Little tunny, halfbeaks, cutlassfish, rudderfish, jacks, pinfish, bullet tuna, sea robin, Atlantic moonfish, squids, shrimps, and gastropods	Beardsley et al. (1975), Davies and Bortone (1976), Nakamura (1985)
Common predators of adults	Killer whales, bottlenose dolphin, and sharks	Beardsley et al. 1975

Key information for the Atlantic sailfish is summarized in Tables 9.24, 9.25, and 9.26 and discussed in detail in the following paragraphs.

9.3.7.1 Key Life-History Processes and Ecology

Atlantic sailfish usually occur in the upper layers of warm water above the thermocline offshore (NMFS 2009b), but can also descend to deepwater and often migrate into nearshore shallow waters. They occasionally form schools or smaller groups of 3–30 individuals but more frequently appear in loose aggregations over a wide area. Atlantic sailfish are distributed throughout the Gulf of Mexico (Figure 9.41); they can be found year-round in the southern Gulf of Mexico but move into the northern Gulf only during the summer season (Table 9.26). No transatlantic or transequatorial movements have been documented using tag-recapture methods (Orbesen et al. 2010).

Juvenile and adult Atlantic sailfish spend winters in warm waters, often occurring in small schools, and spread out during the summer (Table 9.26). However, there appears to be a year-round Florida east coast population (Beardsley et al. 1975; Nakamura 1985; Bayley and Prince 1993; NMFS 2009b; Orbesen et al. 2010). Atlantic sailfish often move northward in early summer in the western Atlantic to engage in spawning activity (NMFS 2009b). Spawning can begin as early as April, but occurs mainly in summer (Table 9.25). Atlantic sailfish can spawn in various oceanographic conditions from offshore in deepwater to inshore shallow waters near the surface in the warm season (Tables 9.25 and 9.26). The Gulf of Mexico has been identified as an important and critical spawning ground for this species, and large concentrations of sailfish larvae have been found in the Gulf of Mexico during the summer, which suggests that July is the peak of the spawning season for Atlantic sailfish in the Gulf of Mexico (Simms 2009).

Table 9.26. Summary of Habitat Information for Atlantic Sailfish (*Istiophorus albicans*)

Parameter	Value	Reference
Habitat preferences and temporal/spatial distribution of juveniles and adults	Mainly oceanic, but migrate into shallow coastal waters; in the southern Gulf of Mexico, usually found above the thermocline at depths of <20 m (<65.6 ft), repeatedly making short duration dives below the thermocline to depths of 50–150 m (164–492 ft); in some areas of their range, the thermocline occurs at depth of 200–250 m (656–820 ft); preferred temperature range of 21–29 °C; Winter: small schools around the Florida Keys and off eastern Florida, in the Caribbean, and in offshore waters throughout the Gulf of Mexico; summer: spread out along the U.S. east coast as far north as Maine, although there is a year-round Florida east coast population; no transatlantic or transequatorial movements have been documented using tag-recapture methods	Beardsley et al. 1975, Nakamura (1985), Bayley and Prince (1993), NMFS (2009b), Orbesen et al. (2010), Kerstetter et al. (2010)
Habitat preferences for spawning adults	Shallow waters, 9–12 m (29.5–39.4 ft) deep; around Florida from the Keys to Palm Beach on the east coast; in the northern Gulf of Mexico, including off Texas; offshore beyond the 100 m (328 ft) isobath from Cuba to the Carolinas; spawning activity moves northward in the western Atlantic as summer progresses	Bumguardner and Anderson (2008), NMFS (2009b)
Designated essential fish habitat for juveniles	In the central Gulf of Mexico, off southern Texas, Louisiana, and the Florida Panhandle; Atlantic east coast from the Florida Keys to mid-coast of South Carolina; the Outer Banks of North Carolina and Maryland; eastern Puerto Rico and the Virgin Islands	NMFS (2009b)
Designated essential fish habitat for adults	In the central Gulf of Mexico, off southern Texas, Louisiana, and the Florida Panhandle; Atlantic east coast from the Florida Keys to northern Florida, off of Georgia, and Cape Hatteras; also around the Virgin Islands	NMFS (2009b)
Designated essential fish habitat for spawning adults	Off the Southeast Florida coast to Key West; associated with waters of the Gulf Stream and Florida Straits from 5 miles offshore out to the EEZ boundary	NMFS (2009b)

Atlantic sailfish grow fast, reaching 137 cm (53.9 in.) in length and 3 kg (6.6 lb) in weight in 6 months and 183 cm (72 in.) and 9 kg (19.8 lb), respectively, in just 1 year. Growth then slows down, and like other billfish, female sailfish grow to be larger than males (Table 9.25). A large female sailfish may release as many as 4.5 million eggs. The M tends to be high during early life-history stages but becomes relatively stable for juvenile sailfish (Luthy et al. 2005; Richardson et al. 2009).

Large variability exists in life-history parameters over the distributional areas of Atlantic sailfish (e.g., East Atlantic versus West Atlantic). However, the growth of juveniles was found to be almost uniform within the Gulf of Mexico (Simms 2009). More studies are needed to evaluate the possible spatial and temporal variability in key life-history parameters. The Atlantic sailfish within the Gulf of Mexico is considered part of the Western Atlantic stock (ICCAT 2010). While studies have indicated the presence of a year-round Florida east coast stock, it is not clear if the Gulf of Mexico has a year-round population.

9.3.7.2 Predators and Prey

Juvenile and adult Atlantic sailfish feed primarily on small pelagic fishes, such as tunas and jacks; they also feed on shrimps, cephalopods, and gastropods (Table 9.25). Feeding can occur at the surface and in mid-water, along reef edges, or along the sea floor. Atlantic sailfish predators include killer whales, bottlenose dolphin, and sharks (Table 9.25).

9.3.7.3 Key Habitat Needs and Distribution

Water temperature and, in some cases, wind conditions are important habitat variables influencing the distribution of Atlantic sailfish (Table 9.26). At the northern and southern extremes of their distribution, Atlantic sailfish occur only in the warmer months. The seasonal changes in distribution may be linked to prey migrations.

The rates of Atlantic sailfish bycatch in the pelagic longline fisheries are two times higher in the Gulf of Mexico than in other areas of the North Atlantic during the spawning season, from May through September (De Sylva and Breder 1997), suggesting that spawning biomass in the Gulf of Mexico tends to be higher than spawning biomass in other areas of the Atlantic. This suggests that the Gulf of Mexico provides an important spawning and larval habitat for Atlantic sailfish.

Essential fish habitat has been designated for juvenile, adult, and spawning Atlantic sailfish. These habitats are described in Table 9.26 and shown in Figures 9.42, 9.43, 9.44, and 9.45.

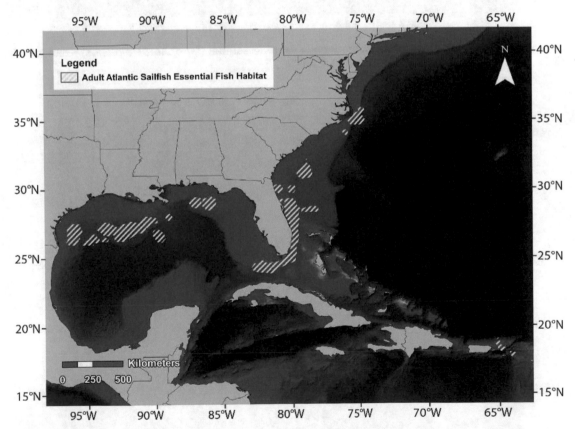

Figure 9.42. Essential fish habitat for adult Atlantic sailfish (*Istiophorus albicans*) (from NOAA Fisheries Office of Sustainable Fisheries 2009d).

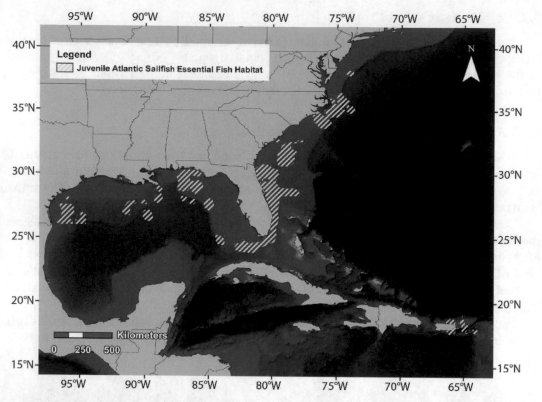

Figure 9.43. Essential fish habitat for juvenile Atlantic sailfish (*Istiophorus albicans*) (from NOAA Fisheries Office of Sustainable Fisheries 2009d).

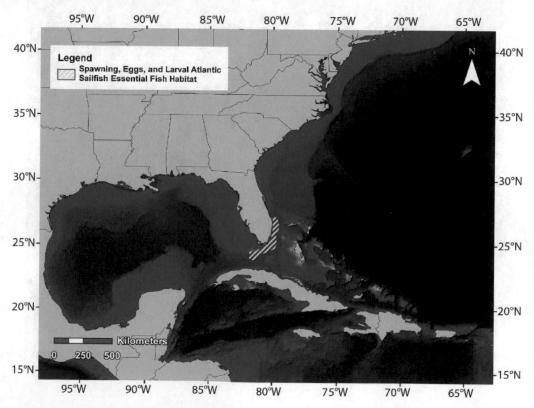

Figure 9.44. Essential fish habitat for spawning, eggs, and larval Atlantic sailfish (*Istiophorus albicans*) (from NOAA Fisheries Office of Sustainable Fisheries 2009d).

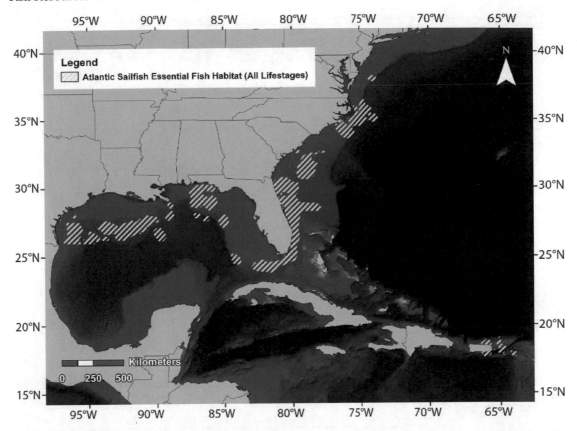

Figure 9.45. Essential fish habitat for all lifestages of Atlantic sailfish (*Istiophorus albicans*) (from NOAA Fisheries Office of Sustainable Fisheries 2009d).

9.3.7.4   Fisheries

The United States has historically landed a large quantity of Atlantic sailfish (Figure 9.46). Prior to the mid-1990s, the U.S. share of landings from the West Atlantic sailfish stock varied between 20 and 60 %, with an average of approximately 40 % (Figure 9.47). Beginning around 2000, landings in the United States and the U.S. share of landings dropped dramatically (Figures 9.46 and 9.47). This may reflect the fact that a targeted commercial fishery for Atlantic sailfish is prohibited in the Gulf of Mexico and other U.S. waters. The current Atlantic sailfish commercial catch is bycatch in pelagic longlines, which are commonly used in the Gulf of Mexico to target swordfish and yellowfin tuna.

Atlantic sailfish are subject to domestic and international management regulations (Table 9.24). For the West Atlantic sailfish stock, the most recent stock assessment suggests that overfishing is probably occurring, and the stock may be overfished (Table 9.24) (ICCAT 2011b). However, because of the large uncertainty associated with the data and stock structure, the results are not conclusive. The recent stock assessment (ICCAT 2011b) suggests that the West Atlantic stock suffered great declines in abundance prior to 1990. However, since 1990, different abundance indices tend to suggest conflicting trends, with some indicating declines and others indicating increases or no trends (ICCAT 2011b).

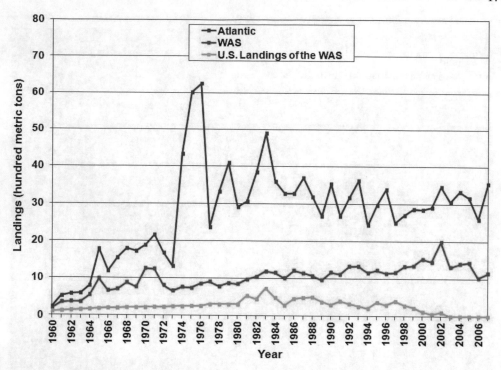

Figure 9.46. Landings of the Atlantic sailfish (*Istiophorus albicans*) stock, West Atlantic sailfish (*Istiophorus albicans*) stock (WAS), and U.S. landings of the WAS from 1960 through 2007 (data from NMFS 2012b).

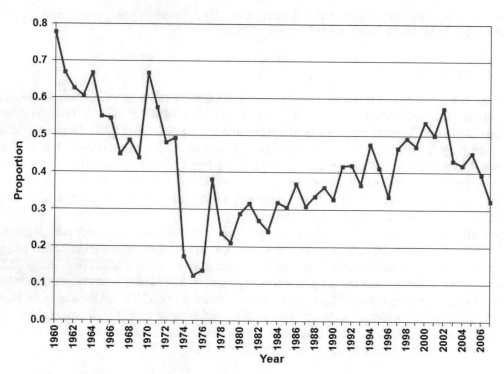

Figure 9.47. Proportion of the U.S. share of the total landings of the West Atlantic sailfish (*Istiophorus albicans*) stock from 1960 through 2007 (data from ICCAT 2011b).

Figure 9.48. Red drum (*Sciaenops ocellatus*) in grass flats (from Ftlaudgirl 2016a).

9.3.8 Red Drum (*Sciaenops ocellatus*)

Red drum, an estuarine-dependent species, is widely distributed in various habitats throughout the Gulf of Mexico and plays an important ecological role in Gulf of Mexico coastal ecosystems (Figure 9.48) (Powers et al. 2012). In the Gulf of Mexico, red drum occur from northern Mexico into extreme Southwest Florida (Figure 9.49). The overall Gulf of Mexico stock was considered overfished, and overfishing was occurring in the early 2000s (Table 9.29). A harvest moratorium has been implemented on red drum in federal waters since 2007. Thus, there is currently no U.S. commercial fishery targeting this species. This fishing moratorium in federal waters is considered to be one of the main reasons for the recent recovery of red drum abundance. Because the red drum is a highly prized sportfish and supports an important recreational fishery in the Gulf of Mexico and because the fishery is a good example to demonstrate potential impacts of management regulations in federal waters on fish population dynamics, it was selected for evaluation in this chapter. Information on red drum, such as life-history parameters, habitat information, and stock and fisheries information, is summarized in Tables 9.27, 9.28, and 9.29 and discussed in detail in the following paragraphs.

9.3.8.1 Key Life-History Processes and Ecology

Life-history parameters of red drum, such as growth and maturation, vary greatly with location and time; growth rates are likely higher in more southerly estuaries (Table 9.27) (Powers et al. 2012). Depending on their habitat, red drum can be from 271 to 383 mm (10.7–15.1 in.) in size at the end of the first year, and red drum growth is rapid through the ages of 4–5 years. Males tend to mature at younger ages than females (Table 9.27). The maximum age of red drum is around 40 years in Florida (Murphy and Taylor 1990); however, red drum as old as 60 years have been reported in North Carolina waters (Ross et al. 1995).

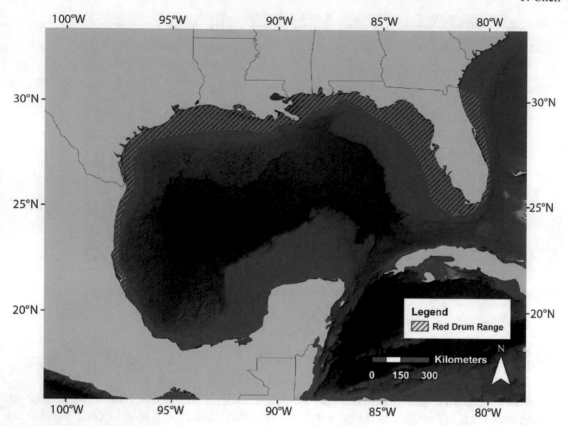

Figure 9.49. Range of the red drum (*Sciaenops ocellatus*) in the Gulf of Mexico and along the Florida east coast (from USGS 2010c).

The fecundity of red drum depends on fish size, and a female red drum can lay eggs ranging from 200,000 to more than three million per batch. Red drum eggs tend to be subject to high mortality (Peters and McMichael 1987; Goodyear 1989). Larval red drum use vertical migrations to ride high salinity tidal currents into tidal creeks and shallow salt marsh nursery habitats (Wenner 1999). They are transported or move to quiet, shallow water with grassy or muddy bottoms to feed on detritus or dead and decomposing organisms (Buckley 1984; Pattillo et al. 1997). A rapid decline in water temperature can cause large mortalities of early juvenile red drum. Tagging studies suggest that they remain in the same area and generally move less than three miles from where they were tagged.

There is large uncertainty associated with the estimation of M of red drum. The M estimated using the observed maximum age ranges from 0.10 to 0.33 per year, and the estimated M based on growth parameters tends to be higher, from 0.42 to 0.92 per year. The estimated M based on age composition data was 0.20 per year, which supports the M estimated from the observed maximum ages (Goodyear 1989).

The distribution of juvenile red drum is typically limited to inshore waters in the Gulf of Mexico, except during fall and winter (Table 9.28). Adult red drum spend less time in bays and estuaries and more time in open Gulf of Mexico waters (Table 9.28). Spawning red drum can be found in both open and nearshore waters in the Gulf of Mexico and tend to spawn near shorelines during late summer and fall (Tables 9.27 and 9.28). There is little evidence of seasonal migration of red drum, and they have been found in rivers and tidal creeks during

Table 9.27. Summary of Life-History Information for Red Drum (*Sciaenops ocellatus*)

Parameter	Value	Reference
von Bertalanffy growth model parameters (see Table 9.6 for explanation)	L_∞ (Texas, age <3.8 years) = 982 mm (38.7 in.)	Porch (2000)
	L_∞ (Texas, age ≥3.8 years) = 982 mm (38.7 in.)	
	L_∞ (Louisiana/Mississippi/Alabama, age <3.3 years) = 1,017 mm (40 in.)	
	L_∞ (Louisiana/Mississippi/Alabama, age ≥3.3 years) = 1,017 mm (40 in.)	
	L_∞ (Florida Gulf coast) = 935 mm (36.8 in.)	Murphy and Taylor (1990)
	L_∞ (Florida, age <2.8 years) = 1,019 mm (40.1 in.)	Porch (2000)
	L_∞ (Florida, age ≥2.8 years) = 1,019 mm (40.1 in.)	
	K (Texas, age <3.8 years) = 0.31 per year	
	K (Texas, age ≥3.8 years) = 0.15 per year	
	K (Louisiana/Mississippi/Alabama, age <3.3 years) = 0.41 per year	
	K (Louisiana/Mississippi/Alabama, age ≥3.3 years) = 0.11 per year	
	K (Florida Gulf coast) = 0.46 per year	Murphy and Taylor (1990)
	K (Florida, age <2.8 years) = 0.40 per year	Porch (2000)
	K (Florida, age ≥2.8 years) = 0.19 per year	
	t_0 (Texas, age <3.8 years) = −0.18 years	
	t_0 (Texas, age ≥3.8 years) = −4.78 years	
	t_0 (Louisiana/Mississippi/Alabama, age <3.3 years) = 0.06 years	
	t_0 (Louisiana/Mississippi/Alabama, age ≥3.3 years) = −8.39 years	
	t_0 (Florida Gulf coast) = 0.029 years	Murphy and Taylor (1990)
	t_0 (Florida, age <2.8 years) = −0.04 years	Porch (2000)
	t_0 (Florida, age ≥2.8 years) = −3.06 years	
Age at first maturity	Male: 1–3 years Female: 3–6 years	Murphy and Taylor (1990), Addis et al. (2011)
Length at first maturity	Gulf of Mexico (Sexes combined) = 740–750 mm (29.1–29.5 in.)	NMFS, SERO (1986)
	Male: 411–791 mm (16.2–31.1 in.) TL; 50 % at 552 mm (21.7 in.)	Murphy and Taylor (1990)
	Female: 629 to 900 mm (24.8 to 35.4 in.) TL; 50 % at 874 mm (34.4 in.)	
Spawning season	Late summer and early fall; peak September through October	Wilson and Nieland (1994), Addis et al. (2011)
Spawning location	Open Gulf of Mexico waters, inlets, within estuaries, or in nearshore shelf waters	Pearson (1929), Yokel (1966), Jannke (1971), Loman (1978), NMFS, SERO (1986)
Common prey of juveniles	Copepods, mysid shrimp, and amphipods	Peters and McMichael (1987)
Common prey of adults	Menhadens, anchovies, lizard fish, mullets, pinfish, sea catfish, spot, Atlantic croaker, mollusks, crabs, and shrimps	Boothby and Avault (1971), Bass and Avault (1975)
Common predators of juveniles	Amberjack, large piscivorous fishes, sharks, and birds; typically not normal part of diet of any common estuarine predator	Overstreet (1983), Porch (2000)
Common predators of adults	Sharks; not a normal part of the diet of any common estuarine predator	

Table 9.28. Summary of Habitat Information for Red Drum (*Sciaenops ocellatus*)

Parameter	Value	Reference
Habitat preferences and spatial/temporal distribution for juveniles	Typically limited to rivers, bays, bayous, canals, tidal creeks, boat basins, and passes within estuaries; also within seagrass beds and over oyster bars, mud flats, and sand bottoms; salinity of 5–35 ppt; temperature from 5 to 35 °C; older juveniles may move into open Gulf of Mexico waters during fall and winter	Pearson (1929), Kilby (1955), Perret et al. (1971), Matlock and Weaver (1979), Peters and McMichael (1987), Osburn et al. (1982), NMFS, SERO (1986)
Habitat preferences and spatial/temporal distribution for adults	Along coastal beaches and in nearshore shelf waters; move farther into open Gulf of Mexico waters and spend less time in bays and estuaries as they mature; optimum salinity range of 30–35 ppt; temperatures from 3 to 35 °C; depths from 40 to 70 m (131.2–229.7 ft); in eastern and western Gulf of Mexico, including South Florida and South Texas, typically inhabit bays and near Gulf waters; in northern Gulf of Mexico, from the Florida Panhandle to North Texas, may move farther offshore, especially in the area between Mobile Bay, Alabama and the area east of the Mississippi Delta	Springer (1960), Simmons and Breuer (1962), Beaumariage and Wittich (1966), Beaumariage (1969a), Moe (1972), Heath et al. (1979), Overstreet (1983), NMFS, SERO (1986), Peters and McMichael (1987), Addis et al. (2011)
Habitat preferences and spatial/temporal distribution for spawning adults	Open Gulf of Mexico waters, near passes and inlets, within estuaries, or in nearshore shelf waters; temperatures from 22 to 26 °C; salinity around 30 ppt	Pearson (1929), Yokel (1966), Jannke (1971), Christmas and Waller (1973), Johnson (1978), Loman (1978), Roberts et al. (1978), Holt et al. (1981), NMFS, SERO (1986), Murphy and Munyandorero (2009)
Designated essential fish habitat	All estuaries in the Gulf of Mexico; Vermilion Bay, Louisiana to the eastern edge of Mobile Bay, Alabama, out to depths of 25 fathoms; Crystal River, Florida to Naples, Florida, between depths of 5 and 10 fathoms; and Cape Sable, Florida, to the boundary between the areas covered by the GMFMC and the SAFMC, between depths of 5 and 10 fathoms (1 ftm = 1.8 m = 6 ft)	GMFMC (2005)

the winter. Tides and water temperatures influence daily movement from shallow to deepwaters. During the fall, especially during stormy weather, adult red drum can move to the beaches in the Gulf of Mexico.

Genetic studies have concluded that Atlantic and Gulf of Mexico red drum are two distinct subpopulations, likely resulting from oceanographic and geographic conditions in South Florida, which limits genetic exchange between the two coastal groups (Gold and Richardson 1991; Gold et al. 1993; Seyoum et al. 2000). Population structure within the Gulf of Mexico is complicated because red drum have limited coastal movement and migrate back to a natal estuary (Gold et al. 1999; Gold and Turner 2002). Genetic studies indicate significant patterns of heterogeneity in Gulf of Mexico red drum, suggesting that the genetic difference increases

Table 9.29. Summary of Stock and Fisheries Information for Red Drum (*Sciaenops ocellatus*)

Parameter	Value	Reference
General geographic distribution	In the Gulf of Mexico from northern Mexico along the Gulf coast into extreme Southwest Florida; along the Atlantic coast from Key West, Florida to New Jersey; occasionally as far north as the Gulf of Maine	Yokel (1966), Lux and Mahoney (1969), Castro Aguirre (1978), NMFS, SERO (1986), Porch (2000)
Commercial importance	Low	
Recreational importance	High	
Management agency	GMFMC; respective Gulf state marine agencies	NMFS, SERO (1986)
Management boundary	GMFMC boundaries; respective state marine agencies are responsible for regulating and monitoring the red drum fishing in their waters	
Stock structure within the Gulf of Mexico	Single Gulf of Mexico stock	
Status (overfished/overfishing)	Overfished from prior to 2000–2005, overfished condition undefined 2006–2011; overfishing occurring prior to 2000 and from 2001 to 2003	NMFS (2001, 2002b, 2003, 2004, 2005, 2006a, 2007, 2008, 2009a, 2010, 2011)

with the distances between the estuaries (Gold et al. 1993, 1999). Tagging studies suggest that juvenile red drum have limited dispersal but that adults can travel considerable distances in the Gulf of Mexico (Osburn et al. 1982; Overstreet 1983). Metapopulation structure may exist for the red drum in the Gulf of Mexico, and despite the likely complex spatial structure of the stock, red drum in the Gulf of Mexico is considered as a single stock, which implicitly assumes no spatial heterogeneity in the Gulf of Mexico red drum population. The impacts of this unrealistic assumption on the stock assessment and management of red drum are unknown.

9.3.8.2 Predators and Prey

Red drum generally are bottom feeders, but can feed in the water column when the opportunity arises. Juveniles feed on invertebrates, while adults feed on many species of fish, including menhadens and mullets, as well as invertebrates, including crabs and shrimps (Table 9.27). Red drum predators include piscivorous fishes, sharks, and birds (Table 9.27).

9.3.8.3 Key Habitat Needs and Distribution

The larvae of red drum prefer vegetated muddy bottom. Juvenile red drum prefer rivers, bays, canals, tidal creeks, passes in estuaries, seagrass beds, oyster bars, mud flats, and sand bottom (Table 9.28). As they mature, red drum move farther into the open Gulf of Mexico, and adults can be found along coastal beaches and nearshore shelf waters (Table 9.28). Essential fish habitat, which is shown in Figure 9.50 and described in Table 9.28, has been designated for the red drum.

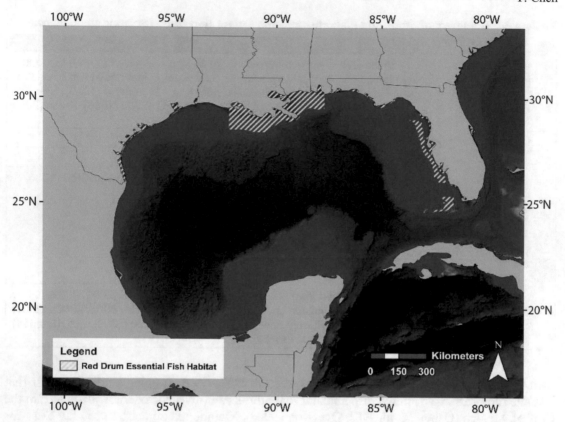

Figure 9.50. The Gulf of Mexico Fishery Management Council's essential fish habitat for red drum (*Sciaenops ocellatus*) (from GMFMC 2004c).

9.3.8.4 Fisheries

The red drum was designated as a protected game fish in 2007 under Executive Order 13449. The order prohibits the sale of red drum caught in U.S. waters, resulting in the elimination of the commercial fishery targeting red drum in federal waters and in most state waters. In Florida, the recreational hook-and-line fishery has been the sole source of red drum landings since 1988. The Florida landings were about 230,000 kg (0.5 million lb) in 1988, but quickly increased to an average of about 771,000 kg (1.7 million lb) during the 1990s and stabilized in the 2000s at close to 900,000 kg (2 million lb) on average (Figure 9.51).

Red drum in the Gulf of Mexico are managed by the GMFMC and relevant Gulf state marine resource management agencies (Table 9.29). No Gulf-wide formal stock assessment is available for red drum. However, a stock assessment was conducted in 2009 for red drum in Florida waters. The assessment was done separately for the Florida Gulf and Atlantic coasts. For red drum along the Florida Gulf coast, stock abundance increased substantially over the time after elimination of the commercial fishery. However, recruitment is relatively stable (Figure 9.52).

The overall Gulf of Mexico red drum stock was considered overfished, and overfishing was occurring in the early 2000s (Table 9.29). However, because of the harvest moratorium on red drum in federal waters, fishery-dependent data are not available in federal waters and fishery-independent data are limited, which makes it difficult to conduct a comprehensive stock

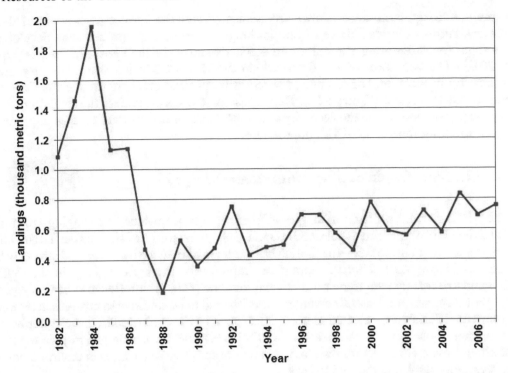

Figure 9.51. Total red drum (*Sciaenops ocellatus*) landings along the Florida Gulf coast from 1982 through 2007. There were no commercial landings after 1988 (data from Murphy and Munyandorero 2009).

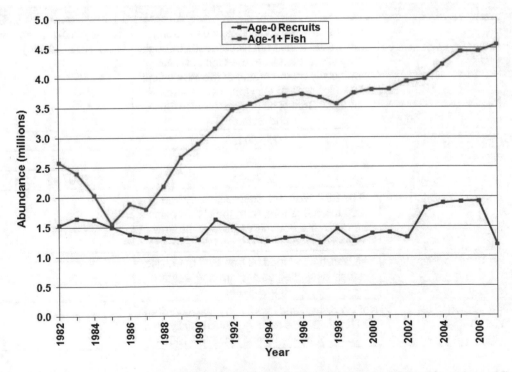

Figure 9.52. Estimated red drum (*Sciaenops ocellatus*) recruitment abundance (number of fish at age 0) and stock abundance (total number of fish age 1 or older) along the Gulf coast of Florida from 1982 through 2007 (data from Murphy and Munyandorero 2009).

assessment. Clearly, large uncertainties are associated with the current assessment of the red drum stock status described above. A bottom longline survey program has been developed to collect data for monitoring the red drum stock dynamics in the Gulf of Mexico (Powers et al. 2012). The condition of red drum within the Gulf of Mexico may also vary greatly from location to location. For example, the red drum population along the Florida Gulf coast appears to be recovered (Figure 9.52). Regardless of the uncertainties, the evolution of this fishery clearly demonstrates the necessity and importance of appropriate management regulations in improving the status of fish populations.

9.3.9 Tilefish (*Lopholatilus chamaeleonticeps*)

The tilefish, often referred to as golden tilefish, is a deepwater fish ranging from Nova Scotia to the Gulf of Mexico (Table 9.30) (Dooley 1978); the range of the tilefish in the Gulf of Mexico is shown in Figure 9.53. The tilefish has a unique burrowing behavior and strong habitat preferences (Table 9.31). Tilefish support an important commercial fishery in the Gulf of Mexico and are mainly caught by handline and longline (Table 9.30). Because of their specific habitat requirements and lack of movement, tilefish tend to be sensitive to changes in their local environment. Tilefish was selected as one of the species to be evaluated in this chapter because they represent those benthic demersal species (Table 9.1) that have wide geographic separation and limited movements, require distinct habitats, are sensitive to changes in environment, and support an important commercial fishery.

Table 9.30. Summary of Stock and Fisheries Information for Tilefish (*Lopholatilus chamaeleonticeps*)

Parameter	Value	Reference
General geographic distribution	In the western Atlantic, along the outer continental shelf from Nova Scotia through Key West, Florida; in the Gulf of Mexico, particularly off the mouth of the Mississippi River in De Soto Canyon, Texas, and the Campeche Banks; off of Venezuela to Guyana and Surinam	Dooley (1978), Lombardi et al. (2010)
Commercial importance	Medium	
Recreational importance	Low	
Management agency	GMFMC	SEDAR 22 (2011)
Management boundary	The EEZ, from the state boundary line to 200 miles offshore, from the U.S./Mexico border to the boundary between the areas covered by the GMFMC and the SAFMC	
Stock structure within the Gulf of Mexico	All tilefish combined as one Gulf of Mexico stock; assessed as eastern and western populations	
Status (overfished/ overfishing)	Not overfished and overfishing not occurring in 2010; stock size reduced substantially as a result of heavy fishing since the 1960s	

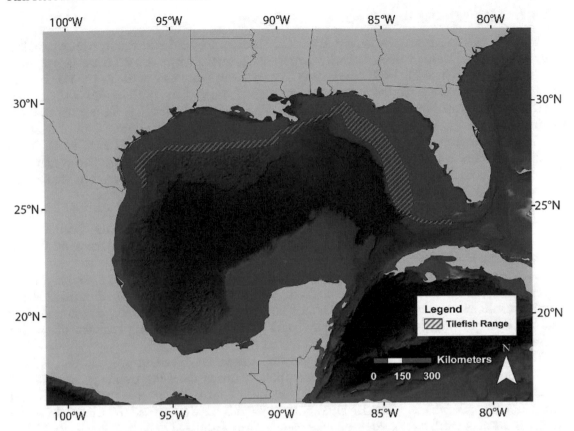

Figure 9.53. Range of the tilefish (*Lopholatilus chamaeleonticeps*) in the Gulf of Mexico (from NOAA 2003).

Table 9.31. Summary of Habitat Information for Tilefish (*Lopholatilus chamaeleonticeps*)

Parameter	Value	Reference
Habitat preferences and spatial/temporal distribution	Inhabit burrows along the continental slope and shelf with distinct sediment, depth, and temperature preferences; burrows excavated from firm mud, silt, sand, and clay sediment along the continental slope; prefer depths from about 120–360 m (393.7–1,181 ft); bottom temperatures from about 9 to 18 °C; tagging results suggest adult movements are minimal; Gulf of Mexico populations are most dense off the mouth of the Mississippi River in Desoto Canyon, Texas, and the Campeche Banks	Nelson and Carpenter (1968), Dooley (1978), Able et al. (1982), Grimes et al. (1983), Katz et al. (1983), Lombardi et al. (2010), SEDAR 22 (2011), Walter et al. (2011)
Designated essential fish habitat	All estuaries in the U.S. Gulf of Mexico; the U.S./Mexico border to the boundary between the areas covered by the GMFMC and the SAFMC from estuarine waters out to depths of 100 fathoms	GMFMC (2005)

9.3.9.1 Key Life-History Processes and Ecology

Tilefish are the largest and longest lived of the tilefish species in the family Malacanthidae. They grow slowly and exhibit sexually dimorphic growth, with males having larger sizes (Table 9.32) (Turner et al. 1983; Grimes and Turner 1999; Lombardi et al. 2010). Tilefish can live for more than 40 years, and maximum sizes range from 96.5 to 111.9 cm (37.9–44 in.). Their age at maturity varies greatly over their distributional areas, with tilefish in the northern waters maturing late and at a large size compared to tilefish in the South Atlantic and Gulf of Mexico. Tilefish mature at age 5 in the North-Mid-Atlantic region (Grimes et al. 1988), age 3 in the South Atlantic, and age 2 in the Gulf of Mexico (Table 9.32). It appears that the age/size at maturity for tilefish has declined over time (Palmer et al. 1998). Compared to other species of similar life history and life span, tilefish tend to mature at younger ages and smaller sizes.

The spawning season for tilefish varies greatly among regions and is typically from January to June (Table 9.32). Tilefish are batch spawners and spawn multiple times throughout a spawning season (Palmer et al. 1998). Annual fecundity increases with size from 195,000 to 8 million eggs per female (Grimes et al. 1988; Palmer et al. 1998).

Tilefish in the Gulf of Mexico have demonstrated some evidence of sequential hermaphrodism, suggesting that tilefish tend to be protogynous (Lombardi et al. 2010; SEDAR 22 2011), but the results are not conclusive. Males may be more vulnerable to fishing pressure as they tend to be larger than females, which may result in disrupting spawning behavior of tilefish (Grimes and Turner 1999).

Because of the long life span, slow growth, a complex breeding process, and habitat specificity and limitations, tilefish are susceptible to mass mortality events as a result of sudden changes in their local environment, such as the intrusion of cold water (Harris and Grossman 1985; Barans and Stender 1993). Many methods have been used to estimate M in the stock assessment (SEDAR 22 2011), and the mean M was estimated to be 0.11 per year, which is comparable to other fish species of similar life history.

The number of recruits estimated for tilefish tended to increase gradually over time prior to the mid-1990s for both the East and West U.S. Gulf of Mexico (Figure 9.54). A more than threefold increase in recruitment was believed to occur in 1997, followed by a large decline back to the levels of the 1980s and 1990s. Recruitment continued to decline after 2000, but has recovered slightly since 2005; this temporal pattern is the same for both the East and West Gulf of Mexico (Figure 9.54).

Wide geographic separation and restricted movements limit possible adult exchanges between the Gulf of Mexico and other regions, which may require the Gulf of Mexico tilefish to be a separate stock in assessment and management. Even within the Gulf of Mexico, because of patchy distribution and the likely lack of movement, tilefish may have much more complex spatial structure, which has not been considered in the current stock assessment and management.

9.3.9.2 Predators and Prey

Tilefish prey on a wide variety of invertebrates and fish, including decapod crustaceans, squids, bivalve mollusks, sea cucumbers, spiny dogfish, and eels (Table 9.32). Sharks, large tilefish, and other predatory fish are the main predators of tilefish.

9.3.9.3 Key Habitat Needs and Distribution

The restrictions of habitat in sediment type, depth, and temperature by adult tilefish may prevent them from moving long distances (Table 9.31). This was shown in tagging studies,

Table 9.32. Summary of Life-History Information for Tilefish (*Lopholatilus chamaeleonticeps*)

Parameter	Value	Reference
von Bertalanffy growth model parameters (see Table 9.6 for explanation)	L_∞ (All data combined) = 830 mm (32.7 in.)	Palmer et al. (1998), Lombardi et al. (2010), SEDAR 22 (2011)
	L_∞ (Males) = 767 mm (30.2 in.)	Lombardi et al. (2010), SEDAR 22 (2011)
	L_∞ (Females) = 613 mm (24.1 in.)	
	L_∞ (East Gulf of Mexico Population) = 878 mm (34.6 in.)	SEDAR 22 (2011), Walter et al. (2011)
	L_∞ (West Gulf of Mexico Population) = 773 mm (30.4 in.)	
	K (All data combined) = 0.13 per year	Palmer et al. (1998), Lombardi et al. (2010), SEDAR 22 (2011)
	K (Males) = 0.15 per year	Lombardi et al. (2010), SEDAR 22 (2011)
	K (Females) = 0.13 per year	
	K (East Gulf of Mexico Population) = 0.11 per year	SEDAR 22 (2011), Walter et al. (2011)
	K (West Gulf of Mexico Population) = 0.17 per year	
	t_0 (All data combined) = −2.14 years	Palmer et al. (1998), Lombardi et al. (2010), SEDAR 22 (2011)
	t_0 (Males) = −1.46 years	Lombardi et al. (2010), SEDAR 22 (2011)
	t_0 (Females) = −4.56 years	
	t_0 (East Gulf of Mexico Population) = −2.86 years	SEDAR 22 (2011), Walter et al. (2011)
	t_0 (West Gulf of Mexico Population) = −2.36 years	
Age at first maturity (50 %)	Females: 2 years	Lombardi et al. (2010), SEDAR 22 (2011)
Length at first maturity (50 %)	Females: 344 mm (13.5 in.) TL Transition to male: 564 mm (22.2 in.) TL (assuming protogyny occurs)	
Spawning season	January to June, peak in April; extended season of 9 months or longer may be possible	
Spawning location	Not available	
Common prey of juveniles and adults	Decapod crustaceans, squids, salps, bivalve mollusks, annelids, sea cucumbers, actinians, eels, spiny dogfish, and other fish species	Linton (1901), Dooley (1978)
Common predators of juveniles	Large tilefish and other fish species	Freeman and Turner (1977)
Common predators of adults	Sharks	Able et al. (1982)

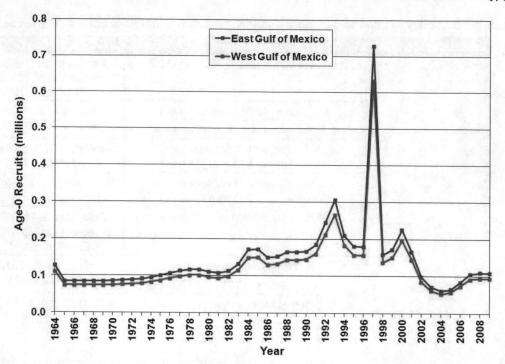

Figure 9.54. Estimated recruitment abundance, measured as age-0 fish, for tilefish (*Lopholatilus chamaeleonticeps*) in the East and West U.S. Gulf of Mexico from 1964 through 2009 (data from SEDAR 22 2011).

which suggested that the movement of tilefish was minimal (Grimes et al. 1983; Katz et al. 1983). This implies that the suitability of local habitat is critical for tilefish. Tilefish are included as one of the Reef Fish species for which essential fish habitat has been designated; this designated habitat is shown in Figure 9.6 and described in Table 9.31.

Tilefish eggs often occur in late spring and summer in the upper water column near the edge of the continental shelf in the Gulf of Mexico (Fahay and Berrien 1981; Fahay 1983; Erickson et al. 1985; Grimes et al. 1988). They hatch in about 40 h at temperatures from 22 to 24.6 °C.

Larval tilefish are pelagic and can be found during the summer in Gulf of Mexico offshore waters (Fahay and Berrien 1981; Fahay 1983; Turner et al. 1983). Early juveniles are still pelagic but start to settle to the bottom at a size of 9–15.5 mm (0.35–0.61 in.) SL (Fahay 1983). The benthic juveniles burrow and occupy simple vertical shafts in the substrate (Able et al. 1982). In the Gulf of Mexico, adults inhabit burrows excavated from firm mud, silt, sand, and clay along the continental slope and shelf, with distinct depth and temperature preferences (Table 9.31).

9.3.9.4 Fisheries

Prior to 1980, tilefish landings were low, but the commercial fishery took off in 1980, reaching the highest level at around 430 metric tons in 1988, which was immediately followed by a large decline (Figure 9.55). Since 1990, tilefish landings have fluctuated between 100 and 250 metric tons.

Tilefish in the Gulf of Mexico are managed under the FMP for the Reef Fish Fishery, which was implemented in 1984. The FMP was developed to: (1) rebuild declining reef fish stocks

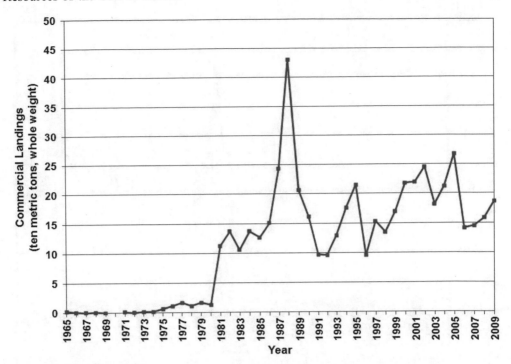

Figure 9.55. Vertical and longline commercial landings of tilefish (*Lopholatilus chamaeleonticeps*) in the U.S. Gulf of Mexico from 1965 through 2009. No data are available for 1970 (data from SEDAR 22 2011).

wherever they occur within the fishery; (2) establish a fishery reporting system for monitoring the Reef Fish Fishery; (3) conserve reef fish habitats, increase reef fish habitats in appropriate areas, and provide protection for juveniles while protecting existing new habitats; and (4) minimize conflicts between user groups of the resource and conflicts for space (SEDAR 22 2011).

Tilefish fishing mortality rates were low prior to 1980, increased quickly after that and reached the highest level in 1988 (Figure 9.56); 1988 was also the year of the highest tilefish landings to date (Figure 9.55). Fishing mortality has decreased since 1988 to around 0.10 during the 1990s and 0.15 during the 2000s (Figure 9.56).

The stock biomass of tilefish in both the eastern and western U.S. Gulf of Mexico has declined substantially since the 1960s (Figure 9.57). The rate of decline in stock biomass for the western Gulf of Mexico is higher than that for the eastern Gulf. The tilefish stock biomass for the eastern Gulf of Mexico has been higher than that for the western Gulf over most of the years included in the stock assessment. However, in the last 2 years in the assessment, the western Gulf of Mexico tilefish stock biomass was higher than that in the East, which might have resulted from higher landings in the eastern Gulf of Mexico (SEDAR 22 2011). The stock assessment results, however, need to be interpreted cautiously.

Most of the tilefish samples were taken from relatively shallow waters, while the stock assessment also covered deep offshore waters from which few samples were taken (SEDAR 22 2011). This inconsistency may result in large uncertainty in the estimation of key life-history parameters, including growth and M, and subsequently the stock dynamics.

Most of the tilefish samples were taken from relatively shallow waters, while the stock assessment also covered deep offshore waters from which few samples were taken (SEDAR

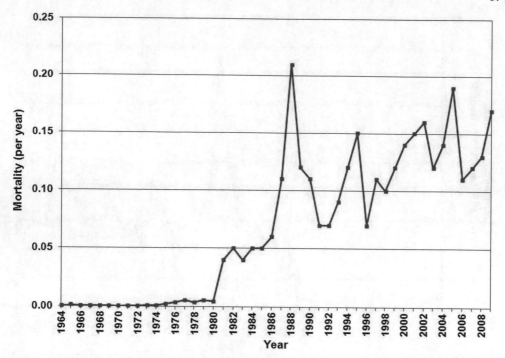

Figure 9.56. Estimated fishing mortality for Gulf of Mexico tilefish (*Lopholatilus chamaeleonticeps*) from 1964 through 2009 (data from SEDAR 22 2011).

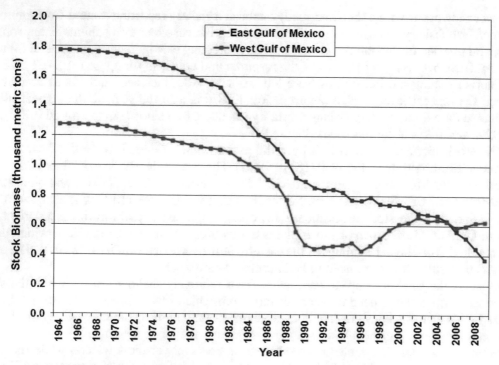

Figure 9.57. Estimated stock biomass for tilefish (*Lopholatilus chamaeleonticeps*) in East and West U.S. Gulf of Mexico from 1964 through 2009 (data from SEDAR 22 2011).

22 2011). This inconsistency may result in large uncertainty in the estimation of key life-history parameters, including growth and M, and subsequently the stock dynamics.

The recent stock assessment suggests that tilefish in the Gulf of Mexico are not overfished (Table 9.30). Most scenarios evaluated in the assessment also suggest that overfishing is not occurring for the Gulf of Mexico tilefish stock. However, at least one scenario considered in the assessment suggests that the Gulf of Mexico tilefish stock is subject to overfishing.

9.3.10 King Mackerel (*Scomberomorus cavalla*)

The king mackerel, a subtropical species of mackerel in the family Scombridae, is mainly distributed in tropical and subtropical waters (Table 9.33) (Beaumariage 1973). It is a migratory species of mackerel that occurs in the open waters of the western Atlantic Ocean and Gulf of

Table 9.33. Summary of Stock and Fisheries Information for King Mackerel (*Scomberomous cavalla*)

Parameter	Value	Reference
General geographic distribution	Western Atlantic from Massachusetts to Rio de Janeiro, Brazil, including waters of the Gulf of Mexico and Caribbean Sea. The coastal area from Florida to Massachusetts is inhabited only during the warmer months of the year	Collette and Nauen (1983), Collette and Russo (1984)
Commercial importance	Medium	
Recreational importance	High	
Management agency	GMFMC and SAFMC	SEDAR 16 (2009), Addis et al. (2011)
Management boundary	Managed as a Gulf of Mexico population in U.S. waters from Texas to Florida and an Atlantic population from the Florida east coast to the Carolinas. During the winter (November 1– March 31), the Flagler-Volusia County line in Florida separates the Gulf of Mexico and Atlantic groups; in the summer (April 1–October 31), the Monroe-Collier County line in Florida separates the two groups.	SEDAR 16 (2009)
Stock structure within the Gulf of Mexico	Current management defines two migratory units, Gulf of Mexico and South Atlantic. Mixing of the two stocks occurs in the region delimited by the Flagler-Volusia and Monroe-Collier County lines on the Florida coast during the winter months. A third group may be found in the western Gulf of Mexico in Mexico, Texas, and seasonally, in Louisiana. There may also be a well-defined group on the Campeche Banks in the southern Gulf of Mexico that mixes to a low degree with other western and northern Gulf of Mexico stocks.	Grimes et al. (1987), Johnson et al. (1994), Arrenguín-Sánchez et al. (1995), DeVries and Grimes (1997), Roelke and Cifuentes (1997), SEDAR 16 (2009), Addis et al. (2011)
Status (overfished/ overfishing)	Overfished from prior to 2000–2003; declared rebuilt in 2008; overfishing occurred prior to 2000	NMFS (2001, 2002b, 2003, 2004, 2005, 2006a, 2007, 2008, 2009a, 2010, 2011)

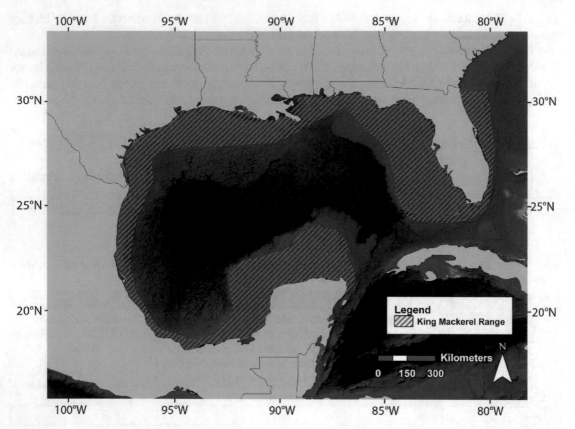

Figure 9.58. Range of the king mackerel (*Scomberomorus cavalla*) in the Gulf of Mexico and along the Florida east coast (from USGS 2010d).

Mexico (Figure 9.58). Because king mackerel are opportunistic and voracious carnivores and are among the most sought-after gamefish throughout their distributional range (Table 9.33) (Beaumariage 1973), they are the representative mackerel species selected for evaluation. In addition, king mackerel support both commercial and recreational fishing industries in the Gulf of Mexico (SEDAR 16 2009). General stock and fishery information, habitat preferences and life-history parameters for king mackerel are summarized in Tables 9.33, 9.34, and 9.35 and discussed in detail in the paragraphs below.

9.3.10.1 Key Life-History Processes and Ecology

The king mackerel inhabits coastal areas, usually in waters less than 73 m (239 ft) deep, and coral reefs, offshore currents, tide rips, and large bays. Two migratory groups of king mackerel have been identified to exist in U.S. waters: the Gulf of Mexico group, ranging from the Texas coast in summer to the middle-east coast of Florida from November through March; and the Atlantic group off North Carolina to southeast Florida that migrates in spring and fall (Figure 9.59). King mackerel spawn from May through October in the coastal waters of the northern and southern Gulf of Mexico in depths ranging from 35 to 183 m (115–600 ft) (Tables 9.34 and 9.35).

Depending on its size, a female may lay from 50,000 to several million eggs over a spawning season (Addis et al. 2011). Eggs of king mackerel are fertilized in the water column

Table 9.34. Summary of Habitat Information for King Mackerel (*Scomberomorus cavalla*)

Parameter	Value	Reference
Habitat preferences and temporal/spatial distribution for juveniles	Epipelagic, neritic tropical, subtropical, and temperate waters; depths from 6 to 46 m (19.7–151 ft); limited by a minimum water temperature of 20 °C; mostly small, young fish <6 years old, migrate south along the Florida Peninsula in late fall and overwinter off South Florida; in spring, as water temperatures warm, fish migrate northward and return to summer spawning grounds. Summer and fall months, inhabit the northern Gulf of Mexico and off the Carolinas	Beaumariage (1969b), Powers and Eldridge (1983), Collette and Russo (1984), Finucane et al. (1986), Fable et al. (1987), Sutter et al. (1991), Schaefer and Fable (1994), SEDAR 16 (2009)
Habitat preferences and temporal/spatial distribution for adults	Epipelagic, neritic tropical, subtropical, and temperate waters; depths from 6 to 46 m (19.7–151 ft); north–south migrations tend to follow the 20 °C isotherm; adults follow the same migration patterns as juveniles; however, older, larger fish may inhabit the northern Gulf and waters off the Carolinas year-round	Beaumariage (1969b, 1973), Manooch and Laws (1979), Powers and Eldridge (1983), Collette and Russo (1984), Finucane et al. (1986), Fable et al. (1987), Sutter et al. (1991), Schaefer and Fable (1994), SEDAR 16 (2009), Addis et al. (2011)
Habitat preferences and temporal/spatial distribution for spawning adults	Waters 35–183 m (115–600 ft); over the middle and outer continental shelf in the northeastern and northwestern Gulf of Mexico in spring and summer	Wollam (1970), McEachran et al. (1980), Finucane et al. (1986)
Designated Essential Fish Habitat	All estuaries in the U.S. Gulf of Mexico; the U.S./Mexico border to the boundary between the areas covered by the GMFMC and the SAFMC from estuarine waters out to depths of 100 fathoms	GMFMC (2005)

and hatch in about 24 h. Little is known about young-of-the-year (YOY) king mackerel (SEDAR 16 2009).

A typical age-1 fish can reach an average weight of 1.4–1.8 kg (3.1–3.9 lb) and a FL of 60 cm (23.6 in.) (SEDAR 16 2009). Female king mackerel can grow much larger than males, and few male king mackerel weigh more than 7 kg (15.4 lb) (Johnson et al. 1983; Finucane et al. 1986). For example, at age 7, females reach an average weight of 9.5 kg (20.9 lb), while males typically weigh 5 kg (11 lb). There is temporal and spatial variation, as well as differences between males and females, in the growth and maturation of king mackerel (Table 9.35).

In the recent stock assessment (SEDAR 16 2009), the M was set at 0.16 and 0.17 for South Atlantic and Gulf of Mexico king mackerel, respectively. Age-specific M was based on a scaled Lorenzen curve. Two migratory units, Gulf of Mexico and South Atlantic, are currently managed (Table 9.33).

Table 9.35. Summary of Life-History Information for King Mackerel (*Scomberomorus cavalla*)

Parameter	Value	Reference
von Bertalanffy growth model parameters (see Table 9.6 for explanation)	L_∞ (Males, western Gulf of Mexico) = 102.9 cm (40.5 in.) FL	DeVries and Grimes (1997)
	L_∞ (Males, eastern Gulf of Mexico) = 102.6 cm (40.4 in.) FL	
	L_∞ (Males, eastern Gulf of Mexico) = 93 cm (36.6 in.) FL	Shepard et al. (2010)
	L_∞ (Females, western Gulf of Mexico) = 134.1 cm (52.8 in.) FL	DeVries and Grimes (1997)
	L_∞ (Females, eastern Gulf of Mexico) = 137.9 cm (54.3 in.) FL	
	L_∞ (Females, eastern Gulf of Mexico) = 124.5 cm (49 in.) FL	Shepard et al. (2010)
	L_∞ (Combined sexes, Gulf stock) = 122.4 cm (48.2 in.) FL	Ortiz and Palmer (2008), Ortiz et al. (2008), SEDAR 16 (2009)
	L_∞ (Females, Gulf Stock) = 132.8 cm (52.3 in.) FL	
	L_∞ (Males, Gulf Stock) = 100 cm (39.4 in.) FL	
	K (Males, western Gulf of Mexico) = 0.20 per year	DeVries and Grimes (1997)
	K (Males, eastern Gulf of Mexico) = 0.25 per year	
	K (Males, eastern Gulf of Mexico) = 0.35 per year	Shepard et al. (2010)
	K (Females, western Gulf of Mexico) = 0.15 per year	DeVries and Grimes (1997)
	K (Females, eastern Gulf of Mexico) = 0.17 per year	
	K (Females, eastern Gulf of Mexico) = 0.26 per year	Shepard et al. (2010)
	K (Combined sexes, Gulf stock) = 0.18 per year	Ortiz and Palmer (2008), Ortiz et al. (2008), SEDAR 16 (2009)
	K (Females, Gulf Stock) = 0.17 per year	
	K (Males, Gulf Stock) = 0.23 per year	
	t_0 (Males, western Gulf of Mexico) = −2.7 years	DeVries and Grimes (1997)
	t_0 (Males, eastern Gulf of Mexico) = −1.8 years	
	t_0 (Males, eastern Gulf of Mexico) = −0.17 years	Shepard et al. (2010)
	t_0 (Females, western Gulf of Mexico) = −2.7 years	DeVries and Grimes (1997)
	t_0 (Females, eastern Gulf of Mexico) = −1.8 years	
	t_0 (Females, eastern Gulf of Mexico) = −0.17 years	Shepard et al. (2010)
	t_0 (Combined sexes, Gulf stock) = −2.6 years	Ortiz and Palmer (2008), Ortiz et al. (2008), SEDAR 16 (2009)
	t_0 (Females, Gulf Stock) = −2.5 years	
	t_0 (Males, Gulf Stock) = −2.6 years	
Age at first maturity	Females: 5–6 years	Johnson et al. (1983)
	Females: 4 years	Beaumariage (1973), Gesteira and Mesquita (1976), Finucane et al. (1986)
	Males: 3 years	Beaumariage (1973)

(continued)

Table 9.35. (continued)

Parameter	Value	Reference
Length at first maturity	Females: Before reaching 86.1 cm (33.9 in.) FL	Johnson et al. (1983)
	Females: 60.2 cm (23.7 in.) FL; most >70.0 cm (>27.6 in.) FL	Fitzhugh et al. (2008)
	Females, first occurrence: 45.0–49.9 cm (17.7–19.6 in.) FL	Finucane et al. (1986)
	Females, 50 %: 55.0–59.9 cm (21.6–23.6 in.) FL	
Spawning season	May through October; peak May through July	Fitzhugh et al. (2008), Addis et al. (2011)
Spawning location	Coastal waters of the southern and northern Gulf of Mexico and off the South Atlantic coast	Burns (1981), Grimes et al. (1990)
Common prey of juveniles and adults	Schooling fishes including: Spanish sardine, scaled sardine, Atlantic thread herring, round scad, blue runner, Atlantic bumper, weakfish, cutlassfish, flying fish, striped anchovy, and scombrids; shrimps and squids	Beaumariage (1973), Saloman and Naughton (1983)
Common predators	Pelagic sharks, little tunny, and dolphins	GMFMC and SAFMC (2011)

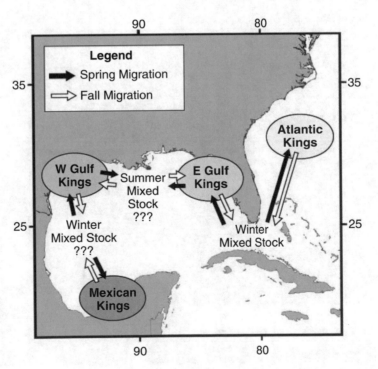

Figure 9.59. Seasonal migratory pattern of king mackerel (*Scomberomorus cavalla*) in the Gulf of Mexico hypothesized based on tagging data (Figure 4.2 redrawn from SEDAR 16 2009).

9.3.10.2 Predators and Prey

King mackerel are opportunistic carnivores (Table 9.35). They eat a wide variety of schooling pelagic fishes, including sardines, herrings, and anchovies, as well and shrimp and squid (Table 9.35) (Beaumariage 1973; Saloman and Naughton 1983). Predators include sharks and dolphins (Table 9.35).

9.3.10.3 Key Habitat Needs and Distribution

King mackerel commonly occur in depths of 12–45 m (39.4–147.6 ft) (Table 9.34), where the principal fisheries occur. Both juvenile and adult king mackerel prefer epipelagic, neritic tropical, subtropical, and temperate waters (Table 9.34). Larger fish (heavier than 9 kg or 19.8 lb) often occur inshore in the mouths of inlets and harbors; occasionally, they are found at depths of 180 m (590 ft) at the edge of the Gulf Stream. King mackerel prefer water temperatures in the range of 20–29 °C; their distribution may be limited by a minimum water temperature tolerance of 20 °C (Table 9.34). The king mackerel is included as one of the Coastal Migratory Pelagics for which essential fish habitat has been designated (Figure 9.60 and Table 9.34).

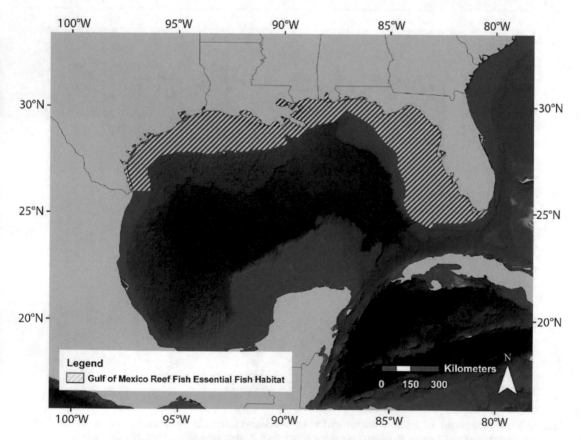

Figure 9.60. Essential fish habitat for king mackerel (*Scomberomorus cavalla*), Spanish mackerel (*Scomberomorus maculatus*), cobia (*Rachycentron canadum*), cero (*Scomberomorus regalis*), little tunny (*Euthynnus alletteratus*), dolphinfish (*Coryphaena hippurus*), and bluefish (*Pomatomus saltatrix*) (from GMFMC 2004d).

9.3.10.4 Fisheries

In the king mackerel recreational fishery, gear used includes trolling with various live and dead baitfish, spoons, jigs, and other artificial lures (SEDAR 16 2009). Gear used in the king mackerel commercial fishery includes run-around gill nets, trolling with large planers, and heavy tackle and lures similar to those used by sport fishers. The recreational fishery lands more king mackerel than the commercial fishery (Figure 9.61). Fishing mortality for Gulf of Mexico king mackerel also includes discarded bycatch of king mackerel in the Gulf of Mexico shrimp fishery, with most discards being YOY fish, and discarded dead fish in the recreational fishery (Figure 9.62). The number of dead king mackerel discarded in the recreational fishery is much smaller than the bycatch in the shrimp fishery (Figure 9.62).

King mackerel in the Gulf of Mexico are managed by the GMFMC and SAFMC under the FMP for Coastal Migratory Pelagic Resources of the Gulf of Mexico and South Atlantic, which was approved in 1982 and implemented in 1983 (Table 9.33). The limit reference points are $0.80 * B_{MSY}$ (stock biomass that can produce MSY) for biomass, which is baseline to determine if fish stock is overfished, and F_{MSY} for fishing mortality, which is the baseline to determine if overfishing occurs (SEDAR 16 2009). Overfishing has occurred in the past (Table 9.33), but the Gulf of Mexico migratory group of king mackerel was not overfished and was not experiencing overfishing in fishing year 2006/07 (Figure 9.63).

The recruitment of Gulf of Mexico king mackerel has fluctuated but has had an increasing trend since the early 1980s (Figure 9.64). The stock abundance of all fish age 1 or older in the Gulf of Mexico group also increased over time. A large increase in stock abundance appeared to occur in fishing year 2003/04 (Figure 9.64). However, the estimation of the recent stock abundance and recruitment tends to be subject to large uncertainty and even biases (most often

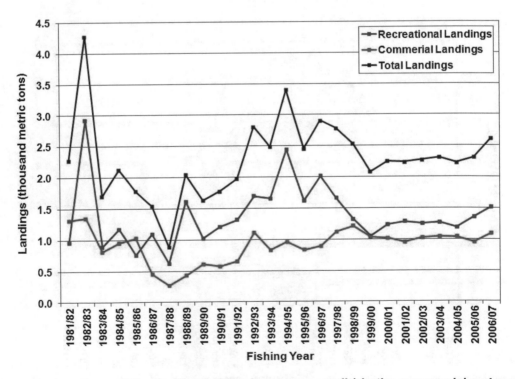

Figure 9.61. Landings of king mackerel (*Scomberomorus cavalla*) in the commercial and recreational fisheries in the U.S. Gulf of Mexico from 1981 through 2007 (data from SEDAR 16 2009).

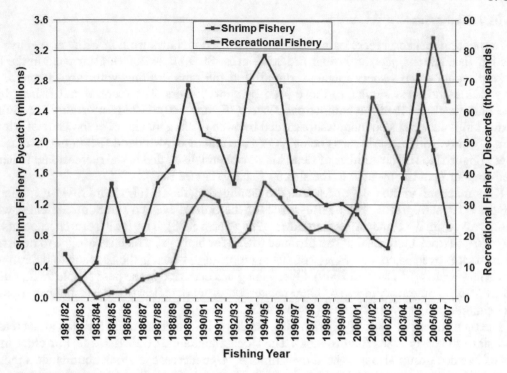

Figure 9.62. King mackerel (*Scomberomorus cavalla*) bycatch in the shrimp fishery and discarded (dead) in the recreational fishery for the migratory group in the Gulf of Mexico from 1981 through 2007 (data from SEDAR 16 2009).

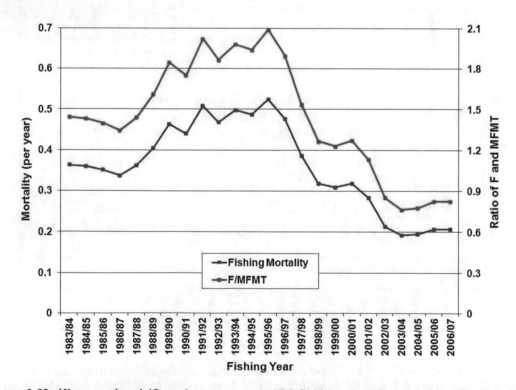

Figure 9.63. King mackerel (*Scomberomorus cavalla*) fishing mortality and the ratio of current fishing mortality (*F*) versus the maximum fishing mortality threshold (MFMT), which is used to determine if the fishery is subject to overfishing of the migratory group in the Gulf of Mexico, from 1983 through 2007 (data from SEDAR 16 2009).

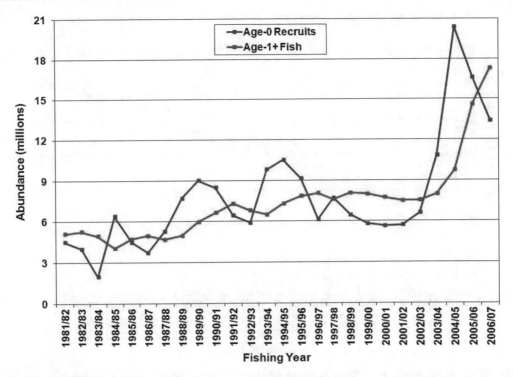

Figure 9.64. Recruitment measured as the abundance of king mackerel (*Scomberomorus cavalla*) at age 0 and stock abundance measured as number of fish age 1 or older for the migratory group in the Gulf of Mexico from 1981 through 2007 (data from SEDAR 16 2009).

overestimation) because of retrospective errors in stock assessment (Mohn 1999). Therefore, the large increase in stock abundance and recruitment in recent years should be interpreted cautiously.

The king mackerel spawning stock has also increased over time. The ratio of spawning stock to the minimum spawning stock threshold has become larger than one in recent years (Figure 9.65), suggesting that the Gulf of Mexico king mackerel is not overfished.

9.3.11 Dolphinfish (*Coryphaena hippurus*)

Dolphinfish, often referred to as mahi-mahi or dorado, are large, fast-swimming, and surface-dwelling fishes found in off-shore temperate, tropical, and ocean waters worldwide (Figure 9.66) (NMFS 2009c). The range of dolphinfish in the Gulf of Mexico is shown in Figure 9.67. As one of the top coastal pelagic predators, dolphinfish play an important role in the Gulf of Mexico ecosystem; in addition, they are a significant species in both commercial and recreational fisheries in the Gulf of Mexico (NMFS 2009c). While details are included in the paragraphs that follow, life history, habitat, and stock and fisheries information for dolphinfish are summarized in Tables 9.36, 9.37, and 9.38.

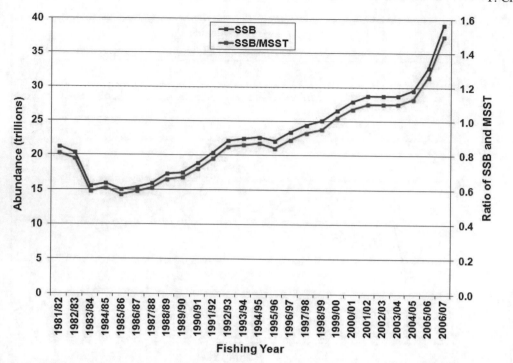

Figure 9.65. King mackerel (*Scomberomorus cavalla*) spawning stock measured as the abundance of hydrated eggs (SSB) and the ratio of spawning stock versus minimum spawning stock threshold (MSST), which is a pre-set spawning stock biomass used to determine if the stock is overfished for the migratory group in the Gulf of Mexico from 1981 through 2007 (data from SEDAR 16 2009).

Figure 9.66. Dolphinfish (*Coryphaena hippurus*) underwater off the Florida coast (from Ftlaudgirl 2016b).

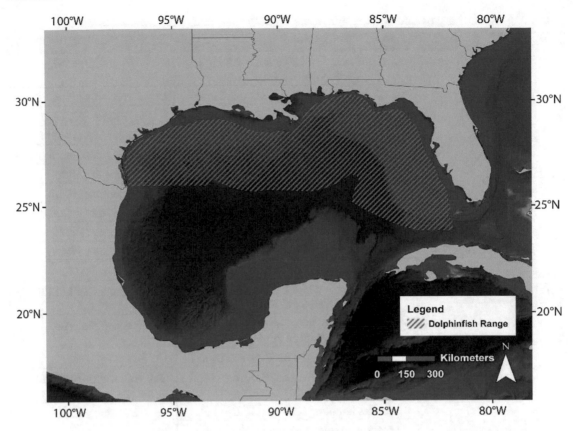

Figure 9.67. Range of the dolphinfish (*Coryphaena hippurus*) in the Gulf of Mexico (from NOAA 2013b).

9.3.11.1 Key Life-History Processes and Ecology

Females may spawn 2–3 times per year, and produce between 80,000 and 1,000,000 eggs per spawning event, depending on their sizes (Beardsley 1967). Dolphinfish spawn in warm ocean currents (water temperatures greater than 24 °C) throughout much of the year (Tables 9.36, 9.37, and 9.38), and spawning occurs in the open water when the water temperature rises. The spawning season varies greatly with latitude. In the northern Gulf of Mexico, spawning occurs at least from April through December, while spawning takes place throughout the year in the southern Gulf of Mexico, and oceanic waters are preferred (Tables 9.36 and 9.37).

Larval and juvenile dolphinfish are commonly found in floating seaweed (Gibbs and Collette 1959; Beardsley 1967; Rose and Hassler 1974). In waters above 34 °C, dolphinfish larvae are found year-round, with greater numbers detected in spring and fall. In a study conducted in the northern Gulf of Mexico, 70 % of the youngest larvae collected were found at a depth greater than 180 m (590 ft).

Dolphinfish are among the fastest-growing fish, and they become sexually mature within a few months (Table 9.36). The length at maturity varies spatially, ranging from 35 to 53 cm (13.8–20.9 in.) FL (Table 9.36).

Table 9.36. Summary of Life-History Information for Dolphinfish (*Coryphaena hippurus*)

Parameter	Value	Reference
von Bertalanffy growth model parameters (see Table 9.6 for explanation)	L_∞ (Caribbean) = 120.8 cm (47.6 in.) SL; 155.9 cm (61.4 in.) TL	Oxenford (1985)
	L_∞ (Puerto Rico) = 142.7 cm (56.2 in.) FL	Rivera and Appeldoorn (2000)
	L_∞ (South Florida) = 171.0 cm (67.3 in.) FL	Prager (2000)
	K (Caribbean) = 3.5 per year	Oxenford (1985)
	K (Puerto Rico) = 2.2 per year	Rivera and Appeldoorn (2000)
	K (South Florida) = 0.58 per year	Prager (2000)
	t_0 (Caribbean) = 0.05 years	Oxenford (1985)
	t_0 (Puerto Rico) = −0.05 years	Rivera and Appeldoorn (2000)
	t_0 (South Florida) = 0.7 years	Prager (2000)
Age at first maturity	Sexes combined: 6–7 months	Beardsley (1967)
	Females: 3–4 months/Males: 4 months	Bentivoglio (1988)
Length at first maturity	Gulf of Mexico, sexes combined: 53.0 cm (20.9 in.) FL	GMFMC and SAFMC (2011)
	Gulf of Mexico, females: 49.0–52.0 cm (19.3–20.5 in.) FL	Bentivoglio (1988)
	Gulf of Mexico, males: 52.8 cm (20.8 in.) FL	
	Florida, sexes combined: 35.0 cm (13.8 in.) FL	GMFMC and SAFMC (2011)
	Florida, females: 35.0 cm (13.8 in.) FL	Beardsley (1967)
	Florida, males: 42.7 cm (16.8 in.) FL	
Spawning season	Spawning season varies with latitude: year-round in the Florida Current, peak from November through July; year-round in southern Gulf of Mexico; at least April through December in northern Gulf of Mexico, peaks in spring and early fall; from June through July in the Gulf Stream near North Carolina	Gibbs and Collette (1959), Beardsley (1967), Powels and Stender (1976), Ditty et al. (1994), GMFMC and SAFMC (2011)
Spawning location	Waters warmer than 24 °C in the Atlantic along the Southeast United States, the Gulf of Mexico, Puerto Rico, and Barbados; prefers oceanic waters rather than shelf waters in the Gulf of Mexico	Beardsley (1967), Ditty et al. (1994)
Common prey of juveniles and adults	Small fishes, crabs, and shrimps associated with *Sargassum*; small oceanic pelagic species, such as flying fishes, halfbeaks, mackerels, man-o-war fish, Sargassum fish, and rough triggerfish; juveniles of large oceanic pelagic species, including tunas, billfishes, jacks, and dolphinfish; pelagic larvae of neritic, benthic species, including flying gurnards, triggerfishes, pufferfishes, and grunts; invertebrates such as cephalopods, mysids, and scyphozoans	Manooch et al. (1984); SEFSC (1998), Oxenford (1999), GMFMC and SAFMC (2011)
Common predators of juveniles and adults	Large tunas, sharks, marlins, sailfishes, and swordfishes	Oxenford (1999), GMFMC and SAFMC (2011)

Table 9.37. Summary of Habitat Information for Dolphinfish (*Coryphaena hippurus*)

Parameter	Value	Reference
Habitat preferences and spatial/temporal distribution for juveniles and adults	Tropical and subtropical waters; closely associated with floating objects and *Sargassum*; able to tolerate salinities from 16 to 26 ppt; typically restricted to waters warmer than 20 °C, but can tolerate temperatures from 15 to 29.3 °C; December through February off Puerto Rico; April through May in the Bahamas; May through June off the Florida east coast and Georgia; June through July off the Carolinas coast; July through August off Bermuda; April through August in the Gulf of Mexico	Gibbs and Collette (1959), Beardsley (1967), Rose and Hassler (1974), Hassler and Hogarth (1977), Oxenford and Hunt (1986), SEFSC (1998), Oxenford (1999), GMFMC and SAFMC (2011)
Habitat preferences and spatial/temporal distribution for spawning adults	Waters warmer than 24 °C in the Atlantic along the Southeast United States, Gulf of Mexico, Puerto Rico, and Barbados; prefers oceanic waters rather than shelf waters in the Gulf of Mexico; spawning season varies with latitude; year-round in the Florida Current, peak from November through July; year-round in southern Gulf of Mexico; at least April through December in northern Gulf of Mexico, peaks in spring and early fall; from June through July in the Gulf Stream near North Carolina	Gibbs and Collette (1959), Beardsley (1967), Powels and Stender (1976), Ditty et al. (1994), GMFMC and SAFMC (2011)
Designated essential fish habitat	All estuaries in the U.S. Gulf of Mexico; the U.S./Mexico border to the boundary between the areas covered by the GMFMC and the SAFMC from estuarine waters out to depths of 100 fathoms (180 m; 600 ft)	GMFMC (2005)

Table 9.38. Summary of Stock and Fisheries Information for Dolphinfish (*Coryphaena hippurus*)

Parameter	Value	Reference
General geographic distribution	Broadly distributed in tropical to warm-temperate waters of the Atlantic, Pacific, and Indian Oceans; in the North Atlantic, from New England to Brazil, including the Gulf of Mexico and Caribbean	NMFS (2009c)
Commercial importance	High	
Recreational importance	High	
Management agency	South Atlantic, Mid-Atlantic, and New England Fishery Management Councils; included in the GMFMC Coastal Pelagics Fishery, but not the management unit	NMFS (2009c)
Management boundary	U.S. Atlantic waters; southern boundary at the border between the GMFMC and SAFMC	GMFMC and SAFMC (2011)
Stock structure within the Gulf of Mexico	Single stock in the Atlantic, U.S. Caribbean, and Gulf of Mexico	
Status (overfished/ overfishing)	No overfishing occurring 2000–2011; not overfished 2000–2011	NMFS (2001, 2002b, 2003, 2004, 2005, 2006a, 2007, 2008, 2009a, 2009c, 2010, 2011)

The estimated M ranged from 0.68 to 0.80 per year in a previous stock assessment (Prager 2000). This range is consistent with the values used for yellowfin tuna that are also a wide-ranging, fast-growing, and predatory species found in similar warm ocean waters.

Some studies suggest that there might be multiple dolphinfish stocks based on the analysis of biological and morphological variables (Oxenford and Hunt 1986; Duarte-Neto et al. 2008; Lessa et al. 2009). However, genetic connectivity was found between migratory groups in the Atlantic, Caribbean, and Gulf of Mexico, which leads to the definition of a single stock in the Atlantic, U.S. Caribbean, and Gulf of Mexico for dolphinfish (GMFMC and SAFMC 2011). Impacts of uncertainty regarding stock structure were evaluated in an assessment by the Caribbean Regional Fisheries Mechanism (CRFM 2006).

9.3.11.2 Predators and Prey

Dolphinfish are carnivorous and feed on a variety of fish and invertebrates; examples include crabs and shrimps associated with *Sargassum* and juvenile tunas, billfishes, jacks, and dolphinfish (Table 9.36). Predators of dolphinfish include large tunas, marlins, sailfishes, and swordfishes, as well as sharks (Table 9.36).

9.3.11.3 Key Habitat Needs and Distribution

Both juvenile and adult dolphinfish prefer tropical and subtropical oceanic waters and are closely associated with floating objects and *Sargassum* (Table 9.37). While juveniles are restricted to waters that are warmer than 20 °C, adults can tolerate temperatures ranging from 15 to 29.3 °C.

Essential fish habitat has been designated for seven species managed as Coastal Migratory Pelagics, and dolphinfish is included as one of the species. Table 9.37 contains a description of the designated habitat for dolphinfish; it is shown in Figure 9.60, with that of several other Gulf of Mexico fish species.

9.3.11.4 Fisheries

The dolphinfish supports an important recreational and commercial fishery in the Gulf of Mexico (Table 9.38). From 1998 to 2006, on average, 6,240 metric tons (13.8 million lb) of dolphinfish were landed in the recreational fishery, which consisted of 94 % of the total dolphinfish landings; commercial fishermen landed 415 metric tons (914,909 lb) (NMFS 2009c). The total landings of dolphinfish increased from 2,100 metric tons (4.6 million lb) in 1981 to a peak of 11,300 metric tons (24.9 million lb) in 1997. Dolphinfish landings decreased to 5,800 metric tons (12.8 million lb) in 2006 (NMFS 2009c). Multiple councils manage dolphinfish (Table 9.38).

A time series of relative abundance index data was developed based on U.S. longline fishery data, which was then used for the assessment of dolphinfish using a surplus production model (Prager 2000). The estimated MSY was about 12,000 metric tons (26.5 million lb)/year, and the estimated fishing mortality that yielded MSY (F_{MSY}) was about 0.5 per year. The estimated stock biomass in 1998 was above B_{MSY}, suggesting that the stock was not overfished in 1998. This assessment suggested some increase in stock size relative to previous estimates and that the fishery was sustainable (Prager 2000). Although a large uncertainty may exist in the assessment as a result of the quality and quantity of data and uncertainty about the stock structure, a recent assessment suggested that there was no decline in catch per unit effort (CPUE) indices and that the current fishing mortality level appears to be sustainable (Collette et al. 2012). The life history of fast growth, early maturation, and high M suggests that dolphinfish may be able to withstand a relatively high exploitation rate. Overfishing appears not to be occurring, and the dolphinfish stock was not overfished from 2000 through 2011 (Table 9.38).

9.3.12 Striped Mullet (*Mugil cephalus*)

The striped mullet is a cosmopolitan species distributed worldwide throughout estuarine, coastal tropical, and warm temperate waters (Figure 9.68) (Addis et al. 2011); its distribution throughout the Gulf of Mexico is shown in Figure 9.69. Striped mullet are catadromous, which means they spawn in saltwater, but return to freshwater to feed and grow (De Silva 1980).

As a widely distributed, abundant, and low trophic level fish, striped mullet play an important role in Gulf of Mexico coastal ecosystems (McEachran and Fechhelm 2005). It captures and transfers food and energy that cannot be utilized by other finfish species of higher trophic levels and is an important prey species for many finfish and sharks. In addition, striped mullet support one of the most important inshore commercial finfish fisheries in Florida (Mahmoudi 2000, 2005, 2008). In the tables (Tables 9.39, 9.40, and 9.41) and text that follow, life history, habitat, and stock and fisheries information for striped mullet are summarized.

9.3.12.1 Key Life-History Processes and Ecology

The movement of spawning adult to offshore spawning areas may be linked to lunar or tidal cycles (Rivas 1980). In the northern Gulf of Mexico, peak spawning occurs in November and December; spawning occurs slightly later in the more southern areas of the eastern and western Gulf of Mexico (Table 9.39).

The fecundity of a female depends on its size and ranges from 250,000 to 2.2 million eggs. Striped mullet appear to spawn only once each year, and eggs are small, non-adhesive, and pelagic (Collins 1985; Greeley et al. 1987). Fertilization is external in the water column (Ross 2001), and fertilized eggs hatch in about 48 h. Nocturnal spawning, followed by the rapid

Figure 9.68. A school of striped mullet (*Mugil cephalus*) swim along the bottom of Fanning Springs, Florida (from Wood 2016).

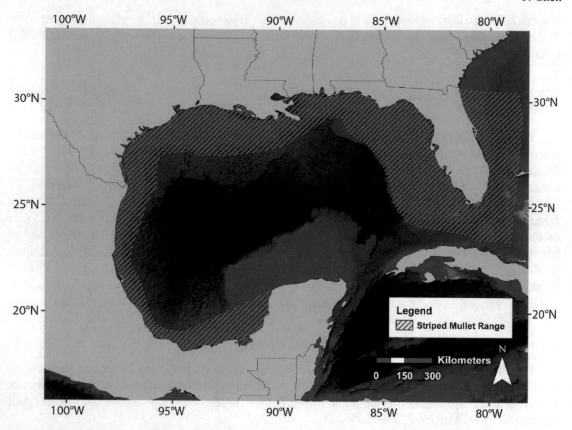

Figure 9.69. Range of the striped mullet (*Mugil cephalus*) in the Gulf of Mexico and along the Florida east coast (from USGS 2010b).

development of fertilized eggs, may reflect possible adaptations minimizing the probability of eggs being exposed to heavy waves (Martin and Drewry 1978).

Juvenile and adult striped mullet can be found in freshwater, as well as shallow marine and estuarine waters (Table 9.40). Both males and females approach sexual maturity in freshwater (Ross 2001), which occurs after 2–3 years of age (Table 9.39). The life span of striped mullet is about 5 or 6 years, but few striped mullet live past 4 years (Rivas 1980).

The M of striped mullet was assumed to be 0.3 per year, constant for all age groups and for all years, in the Florida stock assessment (Mahmoudi 2005). However, the reliability of this assumed M remains unknown.

Striped mullet distributed in the U.S. Gulf of Mexico are considered a unit stock. However, because of limited movements and wide distribution in various habitats of the Gulf of Mexico, their key life-history parameters, such as growth and maturation, may have large spatial variability, and they tend to be assessed and managed under regional or state-specific management programs (Leard et al. 1995).

9.3.12.2 Predators and Prey

The striped mullet is a detritivore/invertivore and a filter feeder (Goldstein and Simon 1999), and common food sources for juveniles and adults include microalgae, diatoms, dinoflagellates, plant detritus, and organic sediments (Table 9.39). They usually feed at surface

Table 9.39. Summary of Life-History Information for Striped Mullet (*Mugil cephalus*)

Parameter	Value	Reference
von Bertalanffy growth model parameters (see Table 9.6 for explanation)	L_∞ (Florida Central West coast, females) = 45.2 cm (17.8 in.) TL	Mahmoudi (1991)
	L_∞ (Florida Central West coast, males) = 36.3 cm (14.3 in.) TL	
	L_∞ (Apalachicola Bay, Florida, females) = 36.1 cm (14.2 in.) TL	
	L_∞ (Apalachicola Bay, Florida, males) = 32.8 cm (12.9 in.) TL	
	L_∞ (Pensacola Bay, Florida, females) = 42.2 cm (16.6 in.) TL	
	L_∞ (Pensacola Bay, Florida, males) = 36.1 cm (14.2 in.) TL	
	L_∞ (Veracruz, Mexico, females) = 62.2 cm (24.5 in.) TL	Ibañez Aguirre et al. (1999)
	L_∞ (Veracruz, Mexico, males) = 60.2 cm (23.7 in.) TL	
	K (Florida Central West coast, females) = 0.385 per year	Mahmoudi (1991)
	K (Florida Central West coast, males) = 0.66 per year	
	K (Apalachicola Bay, Florida, females) = 0.85 per year	
	K (Apalachicola Bay, Florida, males) = 1.07 per year	
	K (Pensacola Bay, Florida, females) = 0.42 per year	
	K (Pensacola Bay, Florida, males) = 0.65 per year	
	K (Veracruz, Mexico, females) = 0.11 per year	Ibañez Aguirre et al. (1999)
	K (Veracruz, Mexico, males) = 0.11 per year	
	t_0 (Florida Central West coast, females) = -0.13 years	Mahmoudi (1991)
	t_0 (Florida Central West coast, males) = 0.003 years	
	t_0 (Florida Central West coast, males) = −0.11 years	
	t_0 (Apalachicola Bay, Florida, males) = −0.17 years	
	t_0 (Pensacola Bay, Florida, females) = −0.13 years	
	t_0 (Pensacola Bay, Florida, males) = −0.26 years	
	t_0 (Veracruz, Mexico, females) = −2.67 years	Ibañez Aguirre et al. (1999)
	t_0 (Veracruz, Mexico, males) = −2.98 years	
Age at first maturity	2–3 years	Broadhead (1953, 1958), Rivas (1980), Thompson et al. (1989), Mahmoudi (2000)
Length at first maturity	29–38 cm (11.4–14.9 in.) FL	Mahmoudi (2000)
Spawning season	Mid-October through late January; peak spawning occurs in November and December for the northern Gulf of Mexico and slightly later in the more southern areas in the eastern and western Gulf of Mexico	Thompson et al. (1989), Mahmoudi (1991), Ditty and Shaw (1996)

(continued)

Table 9.39. (continued)

Parameter	Value	Reference
Spawning location	Typically occurs near the surface in offshore marine waters	Ditty and Shaw (1996)
Common food sources for juveniles and adults	Epiphytic and benthic microalgae, benthic diatoms and dinoflagellates, plant detritus, and organic sediments	Odum (1970), Collins (1981), Addis et al. (2011)
Common predators	Common snook, spotted seatrout, red drum, hardhead catfish, southern flounder, bull shark, alligator gar, sea birds, and marine mammals	Gunter (1945), Breuer (1957), Simmons (1957), Darnell (1958), Thomson (1963)

Table 9.40. **Summary of Habitat Information for Striped Mullet (*Mugil cephalus*)**

Parameter	Value	Reference
Habitat preferences and spatial/temporal distribution of juveniles	Nursery areas are thought to be secondary and tertiary bays; salinities ranging from 0 to 35 ppt; temperature from 5 to 34.9 °C; juveniles spend the rest of their first year of life in coastal waters, salt marshes, and estuaries; often move to deeper water in the fall when the adults migrate offshore to spawn; large numbers of immature mullet overwinter in estuaries	Perret et al. (1971), Nordlie et al. (1982), Collins (1985), Mahmoudi (2000)
Habitat preferences and spatial/temporal distribution of adults	Reside in fresh waters and shallow marine and estuarine waters nearshore, including open beaches, flats, lagoons, bays, rivers, salt marshes, and grass beds; prefer soft sediments, such as mud and sand, containing decaying organic detritus, but also occur over fine silt, ground shell, and oyster bars; salinities ranging from 0 to 35 ppt; temperatures from 5 to 34.9 °C; do not move or migrate extensively outside of estuaries, except to spawn	Gunter (1945), Broadhead and Mefford (1956), Simmons (1957), Arnold and Thompson (1958), Perret et al. (1971), Moore (1974), Nordlie et al. (1982), Collins (1981, 1985), Mahmoudi (2000)
Habitat preferences and spatial/temporal distribution of spawning adults	In the fall, large schools of adult mullet gather near the lower parts of rivers and the mouths of bays in preparation for traveling to the open sea; fall and winter: migrate out of bays and estuaries to spawn in deep open water; may also spawn inshore, near passes along outside beaches, and in the ocean near inlets; distances of 8–32 km (4.9–19.9 miles) offshore and in water deeper than 40 m (131.2 ft); spawning has been observed 65–80 km (40.4–49.7 miles) offshore over water 1,000–1,800 m (3,281–5,905 ft) deep in the Gulf of Mexico; in Florida, spawning migrations are typically southward along the east coast and the west coast from Cedar Key to Homosassa; migrations from Tampa Bay are usually northward; return to the estuaries and ascend toward freshwater after the spawning season	Breder (1940), Gunter (1945), Taylor (1951), Broadhead (1953), Anderson (1958), Arnold and Thompson (1958), Futch (1966, 1976), Finucane et al. (1978), Collins (1985), Mahmoudi (1993, 2000), Leard et al. (1995), Ditty and Shaw (1996)
Designated essential fish habitat	None designated because not federally managed	

Table 9.41. Summary of Stock and Fisheries Information for Striped Mullet (*Mugil cephalus*)

Parameter	Value	Reference
General geographic distribution	Distributed worldwide inhabiting estuaries and coastal waters in all oceans between latitudes of 42°N and 42°S; in the western Atlantic from Brazil to Nova Scotia; most abundant at sub-tropical latitudes	Thomson (1963), Hoese and Moore (1998), Addis et al. (2011)
Commercial importance	Medium	
Recreational importance	High	
Management agency	The GSMFC; individual Gulf States are directly responsible for management	Leard et al. (1995)
Management boundary	State jurisdictional waters	
Stock structure within the Gulf of Mexico	The total population of striped mullet occurring in the U.S. Gulf of Mexico is considered a unit stock. However, due to limited movements, populations may be managed under regional or state-specific management programs.	
Status (overfished/ overfishing)	Florida stocks not subject to overfishing from 1995 to 2007; not overfished from 1995 to 2007	Mahmoudi (2000, 2005, 2008)

boundaries by sucking up mud surfaces or grazing on diatoms or algae attached to rock or plant surfaces (Odum 1970; Ross 2001). Larvae tend to feed on microcrustaceans, such as copepods and insect larvae (Etnier and Starnes 1993). When striped mullet reach 40 mm (1.6 in.) SL, their feeding shifts from grazing on surface/subsurface materials to digging into bottom sediments. Fish reaching 110 mm (4.3 in.) SL can dig 5–7 mm (0.2–0.3 in.) into the sediment, and a striped mullet of 200 mm (7.9 in.) SL may filter over 450 kg (992 lb) of bottom sediment in a year. Sand grains can consist of 50–60 % of the diet of fish larger than 40 mm (1.6 in.) SL (Odum 1970; Eggold and Motta 1992). Adult striped mullet may feed opportunistically on animal prey when highly abundant (e.g., spawning aggregations of marine bristleworms) (Bishop and Miglarese 1978). Bacteria may be important in the diet of striped mullet in muddy areas (Moriarity 1976). Feeding becomes active during the daytime, peaking near midday, and starts to decline in the afternoon. Digestion rates were found to be lower for fish inhabiting freshwater, compared to those in saltwater (Perera and De Silva 1978).

Striped mullet have many predators (Table 9.39). They include snooks, seatrouts, drums, catfishes, flounders, sharks, and gars, as well as sea birds and marine mammals.

9.3.12.3 Key Habitat Needs and Distribution

Striped mullet often enter and inhabit estuaries and freshwater environments (McEachran and Fechhelm 2005). Adult mullet can be found in waters ranging from 0 to 75 ppt salinity, but juveniles cannot tolerate such wide salinity ranges. The habitat preferences of juvenile and adult striped mullet include shallow estuarine and marine waters, as well as contiguous freshwaters (Table 9.40). Because it is not federally managed, essential fish habitat has not been designated for striped mullet in the Gulf of Mexico.

9.3.12.4 Fisheries

The commercial fishery for ripe striped mullet increased significantly in the early 1980s as a result of development of the roe export market (Mahmoudi 2000, 2005, 2008). Various regulations have been implemented in the management of the striped mullet fishery since 1989, and commercial fishing has been strictly restricted since 1995, when Florida prohibited the use of gill and other entangling nets in state waters. This caused a rapid decline in landings and fishing effort since 1995, especially on the Florida Gulf coast. Important regulations developed for managing the mullet fishery included seasonal closures in the early 1950s, a minimum size in 1989, gear restrictions and temporal closures in the early 1990s, and the elimination of the use of gill nets in 1995 (Mahmoudi 2000, 2005, 2008).

The striped mullet is a very important species targeted by the recreational fishery for food and bait in the Gulf of Mexico (Mahmoudi 2000, 2005, 2008). Cast nets are used almost exclusively in the striped mullet recreational fishery. In Florida, landings in the recreational fishery were less than 14 % of the total statewide striped mullet landings from 1998 through 2001 and fluctuated widely from year to year. Since 1995, annual recreational harvests have averaged 356,909 fish (169,250 kg or 373,132 lb) and 425,055 fish (352,713 kg or 777,600 lb) on the Northwest and Southwest Florida Gulf coasts, respectively.

The striped mullet fishery is managed by multiple entities, and the Florida striped mullet stocks have not been overfished and were not subject to overfishing in recent years (Table 9.41). No formal stock assessment has been conducted for striped mullet in other parts of the Gulf of Mexico.

The fishing mortality rate of striped mullet in Florida waters has declined significantly since the net ban was implemented on both coasts of Florida in 1995. The recent fishing mortality rates were below the management target levels (Mahmoudi 2008). This has resulted in a gradual increase of the spawning stock biomass especially along the Florida Gulf coast, where over 85 % of striped mullet are landed. The current striped mullet stocks appear to be healthy, and current levels of fishing effort appear to be sustainable (Mahmoudi 2008).

9.3.13 Greater Amberjack (*Seriola dumerili*)

The greater amberjack is the largest genus in the family Carangidae, with a maximum length of 200 cm (78.7 in.) (Figure 9.70) (Murie and Parkyn 2010). It is a popular fish targeted in recreational fisheries, as well as in commercial fisheries and was selected as the representative jack species for evaluation in this chapter.

Greater amberjack are widely distributed in the Gulf of Mexico (Figure 9.71). In the southern Gulf of Mexico, they sometimes move to nearshore waters (Harris et al. 2007). Greater amberjack are often found near reefs, including artificial reefs, floating wrecks, and offshore oil and gas platforms in the northern Gulf of Mexico. Information regarding the life history, habitat preferences, and stock and fisheries for the greater amberjack is summarized in the tables (Tables 9.42, 9.43, and 9.44) and paragraphs below

9.3.13.1 Key Life-History Processes and Ecology

Greater amberjack spawn from March through June, and little is known regarding the spawning aggregations of the Gulf of Mexico population. The age and length of maturity of female greater amberjack is variable and ranges from 1 to 6 years (Table 9.42).

The daily instantaneous M was estimated at 0.005 for YOY greater amberjack from 40 to 130 days old, resulting in a cumulative M of 36 % for a 100-day period (Wells and Rooker 2004b). Greater amberjack in the Gulf of Mexico tend to have a life span of at least 15 years,

Figure 9.70. School of greater amberjack (*Seriola dumerili*) around a shipwreck (from semet 2013).

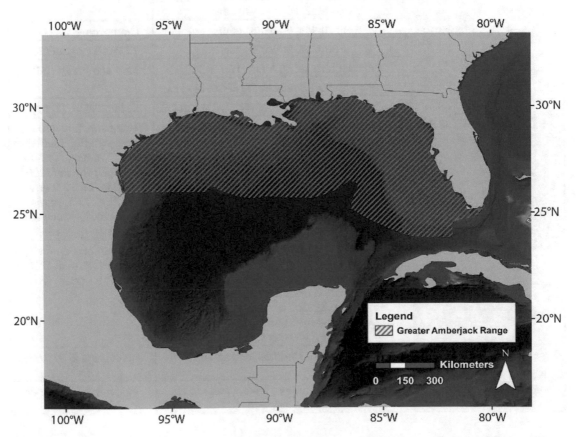

Figure 9.71. Range of the greater amberjack (*Seriola dumerili*) in the Gulf of Mexico (from NOAA 2013c).

Table 9.42. Summary of Life-History Information for Greater Amberjack (*Seriola dumerili*)

Parameter	Value	Reference
von Bertalanffy growth model parameters (see Table 9.6 for explanation)	L_∞ (Combined sexes, Gulf of Mexico) = 111 cm (43.7 in.) FL	Manooch and Potts (1997a)
	L_∞ (Combined sexes, Gulf of Mexico) = 171.2 cm (67.4 in.) FL	Schirripa and Burns (1997)
	L_∞ (Combined sexes, Gulf of Mexico) = 138.9 cm (54.7 in.) FL	Beasley (1993), Thompson et al. (1999)
	K (Combined sexes, Gulf of Mexico) = 0.23 per year	Manooch and Potts (1997a)
	K (Combined sexes, Gulf of Mexico) = 0.26 per year	Schirripa and Burns (1997)
	K (Combined sexes, Gulf of Mexico) = 0.25 per year	Beasley (1993), Thompson et al. (1999)
	t_0 (Combined sexes, Gulf of Mexico) = −0.72 years	Manooch and Potts (1997a)
	t_0 (Combined sexes, Gulf of Mexico) = −0.04 years	Schirripa and Burns (1997)
	t_0 (Combined sexes, Gulf of Mexico) = −0.79 years	Beasley (1993), Thompson et al. (1999)
Age at first maturity	Females: 1–6 years	Harris et al. (2007)
	Females, 50 %: 3–4 years	
	Females, 50 %: 3 years	Thompson et al. (1991)
	Females, 50 %: 4 years	SEDAR 9 Update (2011)
Length at first maturity	Smallest female: 50.1 cm (19.7 in.) FL	Murie and Parkyn (2010)
	Females, 50 %: 85–90 cm (33.5–35.4 in.) FL	
	Females, 50 %: 71.9–74.5 cm (28.3–29.3 in.) FL	Harris et al. (2007)
Spawning season	March through June, peak around April through May	Thompson et al. (1991), Beasley (1993), McClellan and Cummings (1997), Wells and Rooker (2003, 2004a, b), Sedberry et al. (2006), Harris et al. (2007), Murie and Parkyn (2010)
Spawning location	In the Atlantic, from North Carolina to the Florida Keys, concentrated in areas off South Florida and the Florida Keys; it is not known if the Gulf of Mexico population utilizes the spawning area off South Florida; Gulf of Mexico spawning aggregations have not been discussed in the literature	McClellan and Cummings (1997), Harris et al. (2007), SEDAR 9 Update (2011)
Common prey of adults	Bigeye scad, sardines, and squids	Andovora and Pipitone (1997)
Common predators	Yellowfin tuna, European hake, brown noddy, and sooty tern	Andovora and Pipitone (1997)

Table 9.43. Summary of Habitat Information for Greater Amberjack (*Seriola dumerili*)

Parameter	Value	Reference
Habitat preferences and temporal/spatial distribution of juveniles	Associated with pelagic *Sargassum* mats until 5–6 months of age, after which juveniles transition to adult habitat, including reefs, rock outcrops, and wrecks; YOY are most common during May and June in offshore waters of the Gulf of Mexico	Bortone et al. (1977), Manooch and Potts (1997b), Wells and Rooker (2004a, b), Ingram (2006)
Habitat preferences and temporal/spatial distribution of adults	Pelagic and epibenthic; congregate around reefs, rock outcrops, and wrecks in depths ranging from 18 to 72 m (59–236.2 ft); tagging studies of the Gulf of Mexico population demonstrated no trends in movement; in the northern Gulf of Mexico, movements appear random; some fish from West-Central Florida move to South Florida, where some evidence of stock mixing occurs	McClellan and Cummings (1997), Manooch and Potts (1997a), Carpenter (2002), SEDAR 9 (2006a), Harris et al. (2007)
Habitat preferences and temporal/spatial distribution of spawning adults	In the South Atlantic, known to spawn over both the middle and outer shelf, as well as on upper-slope reefs from 45 to 122 m (147.6–400.3 ft), with bottom temperatures around 24 °C; during the winter, individuals from the Atlantic population move into Florida's Atlantic waters for spring spawning, which primarily occurs off South Florida and the Florida Keys during April and May; some fish from West-Central Florida showed movement to South Florida, where some evidence of stock mixing occurs	McClellan and Cummings (1997), Lee and Williams (1999), Sedberry et al. (2006), Harris et al. (2007)
Designated essential fish habitat	All Gulf of Mexico estuaries; the U.S./Mexico border to the boundary between the areas covered by the GMFMC and the SAFMC from estuarine waters out to depths of 100 fathoms	GMFMC (2005)

Table 9.44. Summary of Stock and Fisheries Information for Greater Amberjack (*Seriola dumerili*)

Parameter	Value	Reference
General geographic distribution	A pelagic and epibenthic, reef-associated species with circumglobal distribution in warm-temperate waters; in the western Atlantic, ranges from Nova Scotia to Brazil, including Bermuda, the Gulf of Mexico, and the Caribbean	Manooch (1984), Harris et al. (2007)
Commercial importance	Medium	
Recreational importance	Medium–High	
Management agency	GMFMC (Gulf stock) and SAFMC (Atlantic stock)	SEDAR 9 (2006)
Management boundary	The geographic boundary of the Gulf and Atlantic management units occurs from approximately the Dry Tortugas through the Florida Keys and to the mainland of Florida	
Stock structure within the Gulf of Mexico	Gulf stock inhabits the northern Gulf of Mexico and along Southwest Florida; Atlantic stock inhabits South Florida, the Florida Keys, and the U.S. South Atlantic region	Gold and Richardson (1998), SEDAR 9 (2006)
Status (overfished/ overfishing)	Overfishing occurring from 2004 to 2011; overfished from 2001 to 2011	NMFS (2001, 2002b, 2003, 2004, 2005, 2006a, 2007, 2008, 2009a, 2010, 2011)

based on age samples available (Manooch and Potts 1997b; Thompson et al. 1999). Using the method of Hoenig (1983), this yields a value for M of 0.28. The M used in the stock assessments is 0.25 (Turner et al. 2000; SEDAR 9 2006).

9.3.13.2 Predators and Prey

Juvenile greater amberjack feed mainly on planktonic decapods and other small invertebrates (Andovora and Pipitone 1997). Adult greater amberjack are opportunistic predators, feeding on benthic and pelagic fishes and invertebrates, such as scads, sardines, and squids (Table 9.42). Predators of the greater amberjack in the Gulf of Mexico often include larger fishes, such as tunas, and seabirds, such as brown noddys and sooty terns (Table 9.42).

9.3.13.3 Key Habitat Needs and Distribution

Juvenile greater amberjack are associated with pelagic *Sargassum* mats until about 6 months of age (Table 9.43). The habitat of adults includes pelagic and epibenthic waters, and greater amberjack congregate around reefs, rock outcrops, and wrecks (Ingram, 2006). Some greater amberjack are full-time residents along the Florida Gulf and Atlantic coasts, while others may migrate from the South Atlantic Bight into inshore waters during certain times of the year. Greater amberjack tend to congregate in schools when they are young; however, the schooling behavior changes as the fish grow older, and old fish are primarily solitary.

The greater amberjack is managed under the Reef Fish FMP by the GMFMC (GMFMC 2004a). Essential fish habitat that has been designated for Reef Fish is shown in Figure 9.6 and described in Table 9.43.

9.3.13.4 Fisheries

Greater amberjack are caught primarily with hydraulic reels, handlines, rods-and-reels, longlines, and traps. For stock assessment and management, greater amberjack are considered as two stocks in the United States (Table 9.44). The Gulf of Mexico stock inhabits the northern Gulf of Mexico and along the Southwest Florida coast, while the South Atlantic stock inhabits South Florida, the Florida Keys, and the U.S. South Atlantic region (Gold and Richardson 1998; SEDAR 9 2006).

Various management regulations have been in place since 1998 within the Gulf of Mexico fishery (SEDAR 9 2006). The bag limit is one fish per day in the recreational fishery, with a 71.1 cm (28 in.) minimum legal length. In the commercial fishery, the minimum legal size is 91.4 cm (36 in.), and there is a seasonal closure from March through May when greater amberjack spawn. The majority of greater amberjack commercial landings are from handline (Figure 9.72). The landings increased dramatically during the 1980s, but have exhibited a decreasing trend since the late 1980s. Since 1998, the landings have fluctuated from year to year, but are relatively stable compared to the variability in landings observed in the 1980s and 1990s (Figure 9.72).

The abundance of greater amberjack in the Gulf of Mexico decreased in the late 1980s through the mid-1990s (Figure 9.73). However, the stock assessment conducted in 2006 (SEDAR 9 2006) indicated that the stock abundance increased after 1998, reaching a high level in 2000, but then decreased again (Figure 9.73). Spawning stock fecundity had similar temporal trends as recruitment prior to 1998 (Figure 9.74) (SEDAR 9 2006). However, after 1998, the stock fecundity had an opposite temporal trend compared to the recruitment (e.g., an

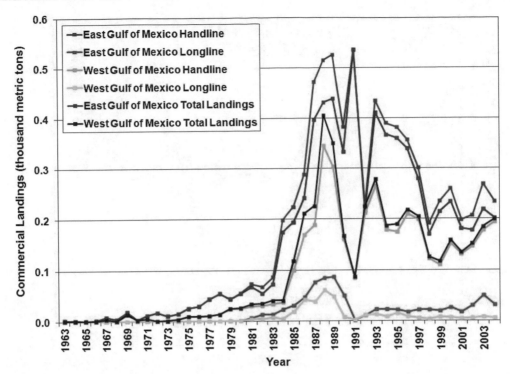

Figure 9.72. Landings of greater amberjack (*Seriola dumerili*) by gear type (longline and handline) and area (western Gulf of Mexico and eastern Gulf of Mexico) in the Gulf of Mexico from 1963 through 2004 (data from SEDAR 9 2006).

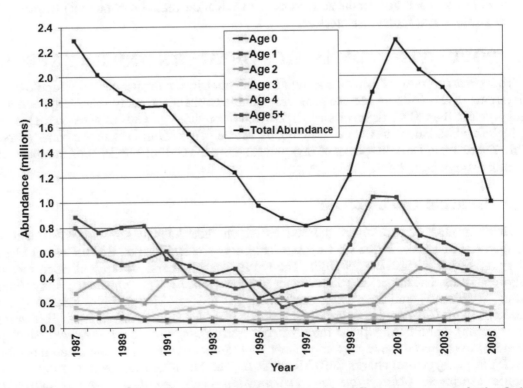

Figure 9.73. Greater amberjack (*Seriola dumerili*) abundance for different age groups in the Gulf of Mexico from 1987 through 2005 (data from SEDAR 9 2006).

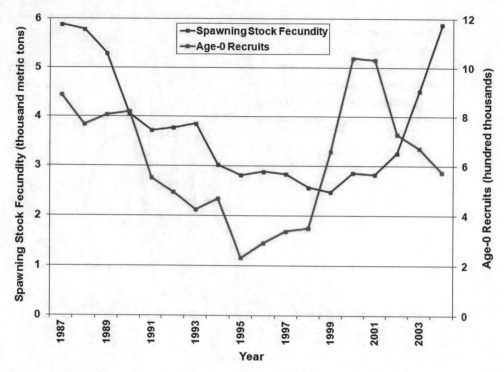

Figure 9.74. Greater amberjack (*Seriola dumerili*) recruitment (age-0 fish) and spawning stock fecundity in the Gulf of Mexico from 1987 through 2004 (data from SEDAR 9 2006a).

increased stock fecundity versus an increased and then decreased recruitment) (Figure 9.74). The cause for this difference is unclear.

9.4   POPULATION DYNAMICS OF SHARK AND RAY SPECIES

The great diversity of oceanographic and bathymetric conditions, geology, topology, and ecosystems in the Gulf of Mexico provides suitable habitats for many shark, ray, and skate species. More than 51 shark species are distributed throughout the Gulf of Mexico region, with the highest abundances in the central Gulf of Mexico from Louisiana to Alabama (Parsons 2006). There are at least 49 species of rays and skates in the Gulf of Mexico, and six species are endemic (McEachran 2009).

9.4.1   General Introduction

Three groups of sharks are defined based on their habitats for the assessment and management of sharks in the Gulf of Mexico: Large Coastal Sharks, Small Coastal Sharks, and Pelagic Sharks (SEDAR 29 2012). The Large Coastal Sharks group is divided into two subgroups: Ridgeback Species, which include sandbar (*Carcharhinus plumbeus*), silky (*Carcharhinus falciformis*), and tiger (*Galeocerdo cuvier*) sharks; and Non-Ridgeback Species, which include blacktip (*Carcharhinus limbatus*), spinner (*Carcharhinus brevipinna*), bull (*Carcharhinus leucas*), lemon (*Negaprion brevirostris*), nurse (*Ginglymostoma cirratum*), scalloped hammerhead (*Sphyrna lewini)*, great hammerhead (*Sphyrna mokarran*), and smooth hammerhead (*Sphyrna zygaena*) sharks (SEDAR 29 2012. The Small Coastal Sharks group includes Atlantic sharpnose (*Rhizoprionodon terraenovae*), blacknose (*Carcharhinus acronotus*), bonnethead (*Sphyrna tiburo*), and finetooth (*Carcharhinus isodon*) sharks (SEDAR 29 2012). The Pelagic Sharks group includes blue (*Prionace glauca*), oceanic whitetip (*Carcharhinus longimanus*), porbeagle (*Lamna nasus*), shortfin mako (*Isurus oxyrinchus*), and common thresher (*Alopias vulpinus*) sharks (SEDAR 29 2012).

In addition to these three groups of sharks, because of very low population biomass and poor stock conditions (e.g., overfished), the following 19 shark species are listed as commercially and recreationally prohibited species: sand tiger (*Odontaspis taurus*), bigeye sand tiger (*Odontaspis noronhai*), whale (*Rhincodon typus*), basking (*Cetorhinus maximus*), great white (*Carcharodon carcharias*), dusky (*Carcharhinus obscurus*), bignose (*Carcharhinus altimus*), Galapagos (*Carcharhinus galapagensis*), night (*Carcharhinus signatus*), Caribbean reef (*Carcharhinus perezi*), narrowtooth (*Carcharhinus brachyurus*), Atlantic angel (*Squatina dumerili*), Caribbean sharpnose (*Rhizoprionodon porosus*), smalltail (*Carcharhinus porosus*), bigeye sixgill (*Hexanchus nakamurai*), bigeye thresher (*Alopias superciliosus*), longfin mako (*Isurus paucus*), sevengill (*Heptranchias perlo*), and sixgill (*Hexanchus griseus*) sharks. The first five species were part of the Large Coastal Sharks group until 1997, the second nine species were part of Large Coastal Sharks group until 1999, and the last five species were part of the Pelagic Sharks group until 1999. All of these species are prohibited in commercial and recreational fisheries (NMFS 2006b, 2009b).

The abundance of a given shark species and species composition tends to vary spatially and temporally, corresponding to spatial and temporal variability in their habitats (Parsons 2006). For example, blacktip shark, spinner shark, Atlantic sharpnose shark, and bull shark are abundant in coastal waters. Adult blacktip shark is more abundant in the central Gulf of Mexico than any other region. Tiger shark is not reported to utilize coastal nursery areas; however, their young are distributed offshore. Whale shark is distributed along much of the Gulf of Mexico, with highest concentrations off the Louisiana Delta; their distribution can be both near coastal areas and well offshore (Parsons 2006). In the southern Gulf of Mexico and along the Florida Gulf coast, the most common coastal shark is the bonnethead shark; blacktip, blacknose, and lemon sharks are also abundant. In the northern Gulf of Mexico, the Atlantic sharpnose is the dominant species, and blacktip and finetooth sharks are also common. The bull shark is most common in and around the marshes of Louisiana (Parsons 2006). Many coastal sharks experience seasonal inshore–offshore movements to avoid unfavorable thermal habitats. For example, in the northern Gulf of Mexico, coastal shark species tend to move offshore in the fall and winter into warmer and deeper offshore waters, and then return to inshore nursery areas for the spring and summer months. Such seasonal inshore–offshore movements are less clear in the southern Gulf of Mexico, where water temperatures are less variable seasonally (Parsons 2006).

In general, sharks tend to have a life history with slow growth, late maturity, and low reproductive rates, making them vulnerable to exploitation. Heavy exploitation can greatly reduce the biomass of shark stocks, resulting in overfished populations. Baum and Myers (2004) suggested that fishing might drive some shark populations in the Gulf of Mexico to extremely low abundances. However, their analyses and conclusions were criticized as flawed. Burgess et al. (2005) suggested that Baum and Myers (2004) overstated the severity of low shark population levels in the Gulf of Mexico; however, even though they questioned the severity of overfishing, they agreed that many shark stocks in the Gulf of Mexico had been overfished. Of 39 species included in the shark FMP in the Gulf of Mexico, 19 species are listed as commercially and recreationally prohibited species. In addition, the Large Coastal Sharks, Small Coastal Sharks, and Pelagic Sharks groups are now subject to strict management regulations for both commercial and recreational fisheries, including limitations on the type of fishing gear that can be used, size limits, temporally and/or spatially allocated catch quota, requirements for landing conditions to prevent shark finning and species identification, requirements for reporting to improve catch data quality, license requirements, and restrictions of catch and fishing times/locations for research.

9.4.2 Stock Assessment and Management History

The FMP was developed in 1993 for sharks of the Atlantic Ocean (NMFS 1993). It includes the following major management measures: (1) establishing a fishery management unit (FMU) consisting of 39 frequently caught species of Atlantic sharks, separated into three groups for assessment and regulatory purposes (Large Coastal Sharks, Small Coastal Sharks, and Pelagic Sharks); (2) developing assessment protocols for determining annual quotas and other management regulations for commercial fisheries for the Large Coastal Sharks and Pelagic Sharks groups; and (3) defining management regulations for the recreational shark fisheries. The 1993 plan also identified 34 additional species of sharks that were not included in the FMU but were included in the fishery for data reporting purposes (NMFS 1993).

The Large Coastal Sharks group was determined to be overfished based on a 1992 stock assessment and a rebuilding plan was developed, which forms the basis for determining subsequent annual catch quotas for the Large Coastal Sharks stocks. The 1996 stock assessment suggested that Large Coastal Sharks stocks were not on the path for rebuilding (SEFSC 1996). In 1996, the NMFS developed a new rebuilding plan for the Large Coastal Sharks and Small Coastal Sharks stocks to be consistent with the revised definition of overfishing and establishment of new provisions for rebuilding overfished stocks, minimizing bycatch mortality, and protecting essential fish habitat in the amendments to the Magnuson-Stevens Act.

In 62 FR 16648, April 7, 1997, the NMFS issued the final rule prohibiting the directed commercial fishing for, landing of, or sale of five species of sharks from the Atlantic Large Coastal Sharks group (NMFS 1999). These five species were placed in a new Prohibited Species group that included sand tiger, bigeye sand tiger, whale, basking, and great white sharks. These shark species were excluded from directed fishing as a precautionary measure to prevent directed fisheries and markets from developing (50 CFR Part 678, proposed rule). Of the five prohibited species, only sand tiger and bigeye sand tiger sharks were exploited commercially, accounting for less than 1 % of the landings in the directed Large Coastal Sharks fishery (50 CFR Part 678, proposed rule). Sand tiger and bigeye sand tiger sharks were determined to be highly vulnerable to overfishing due to a maximum litter size of only two pups (SEFSC 1998; NMFS 1999). Whale and basking sharks were particularly vulnerable to indiscriminate mortality due to their habit of swimming near the surface (50 CFR Part 678, proposed rule). Great white shark was determined to be susceptible to overfishing due to low reproductive potential, although limited information was available at the time. Because a recreational fishery already existed for the great white shark in parts of its range, the fishery was restricted to catch and release only (50 CFR Part 678, final rule) (NMFS 2003).

In April 1999, the NMFS published the Final FMP for Atlantic Tuna, Swordfish, and Sharks (NMFS 1999). A court order prevented implementation of shark specific rules until a settlement agreement was reached resolving several 1997 and 1999 lawsuits in 2000. The settlement agreement did not address any regulations affecting prohibited shark species (NMFS 2006b). Differing from the previous legislation that prohibited the possession of species known to be vulnerable to fishing pressures, this legislation allowed possession of only those species with stock sizes known to be able to withstand fishing mortality (NMFS 1999). This 1999 FMP increased the total number of prohibited shark species to 19, which included whale, basking, sand tiger, bigeye sand tiger, great white, dusky, night, bignose, Galapagos, Caribbean reef, and narrowtooth sharks from the Large Coastal Sharks group; Caribbean sharpnose, smalltail, and

Atlantic angel sharks from the Small Coastal Sharks group; and longfin mako, bigeye thresher, sevengill, sixgill, and bigeye sixgill sharks from the Pelagic Sharks group (NMFS 1999). The goal of this action was to prevent the development of directed fisheries or markets for uncommon or seriously depleted species, as well as those thought to be highly susceptible to exploitation (NMFS 2006b). This FMP defined a new Deepwater and Other Sharks management group to extend the protection of the finning prohibition to all species of sharks, including the 34 species previously included in the fishery in 1993 only for data collection purposes. The 1999 FMP also included life-history information and designated essential fish habitat for highly migratory species, including many shark species within the FMU; however, limited life-history information for some shark species prevented the definition of essential fish habitat at that time (NMFS 1999). Based on a stock assessment conducted for the Large Coastal Sharks and Small Coastal Sharks stocks in 2002, Amendment 1 to the 1999 FMP (for tunas, billfish, and sharks) was added in 2003, which included: (1) aggregating the Large Coastal Sharks group; (2) using MSY as a basis for setting commercial quotas; (3) eliminating the commercial minimum size; and (4) developing various management regulations, including area-specific catch quotas and temporal and spatial closures to reduce fishing and bycatch mortality (NMFS 2003).

The 2006 Consolidated FMP required that the owners and operators using pelagic and demersal longline gear take mandatory workshops and certifications to reduce bycatch mortality and that all federally permitted shark dealers be trained in the identification of shark carcasses (NMFS 2006b). This FMP also included a plan for preventing the overfishing of finetooth sharks by expanding observer coverage and collecting more information on finetooth shark catch. A stock assessment was conducted in 2006 on the Large Coastal Sharks group, which included sandbar, blacktip, porbeagle, and dusky sharks (SEDAR 11 2006). The assessment suggested that dusky and sandbar sharks were overfished, with overfishing still occurring, and that porbeagle sharks were overfished. Amendment 1 to the Consolidated FMP of 2006 updated and expanded upon the life-history information and essential fish habitat for sharks within the FMU (NMFS 2009b). Amendment 2 was added to the 2006 Consolidated FMP for developing rebuilding plans for overfished shark species. Amendment 3 was added to the 2006 Consolidated FMP to address issues raised in the Small Coastal Sharks stock assessment in 2007, which assessed finetooth, Atlantic sharpnose, blacknose, and bonnethead sharks separately. Blacknose sharks were considered overfished with overfishing occurring; however, Atlantic sharpnose, bonnethead, and finetooth sharks were not overfished and overfishing was not occurring (NMFS 2009b).

9.4.3 Small Coastal Sharks Group

The Small Coastal Sharks group currently includes four species: Atlantic sharpnose, blacknose, bonnethead, and finetooth sharks (SEDAR 29 2012). They are widely distributed in coastal waters of the Gulf of Mexico and experience seasonal inshore–offshore movements, usually leaving inshore waters in October and November for warm offshore waters and moving back into inshore waters in spring. These species tend to be smaller than 150 cm (59 in.) TL and have a maximum life span of less than 12 years. They become sexually mature at relatively young ages (2–3 years old), with males often maturing sooner than females. The reproductive cycle is usually annual within the Gulf of Mexico, with an average litter size ranging from 3 to 10 pups. One single stock is assumed in the assessment. Tagging studies provide little evidence to support mixing between the sharks of this group in the South Atlantic and Gulf of Mexico, which suggests that most small coastal sharks complete their life cycles within the Gulf of Mexico. Species in the Small Coastal Sharks group support important commercial and recreational fisheries and are often taken as bycatch in finfish fisheries (SEDAR 13 2007). In the most recent stock assessment (SEDAR 13 2007), sharpnose, finetooth, and bonnethead sharks were determined healthy, with no overfishing occurring and stocks were not overfished. However, the blacknose shark was considered to be overfished, with overfishing still occurring. Detailed descriptions of the distributions and life histories of these species are provided below.

9.4.3.1   Atlantic Sharpnose Shark

The Atlantic sharpnose shark is mainly distributed in waters from the Bay of Fundy to the Straits of Florida and the Gulf of Mexico (SEDAR 13 2007). Even though no genetic differences were found (Heist et al. 1996), based on tagging and life-history studies, a two-stock hypothesis has been proposed: an Atlantic stock distributed from North Carolina to the Straits of Florida and a Gulf of Mexico stock from the Florida Keys throughout the Gulf of Mexico. Little mixing was found in tagging studies between these two stocks (SEDAR 13 2007), suggesting that the two stocks are rather independent. However, large differences have been observed in the life histories between samples collected from the two areas, which might have resulted from differences in the sampling times, locations, and habitats. Most catch of this species is from the Florida east coast, and a single working stock was assumed for the assessment (SEDAR 13 2007).

The Atlantic sharpnose shark is the most common shark in the northern Gulf of Mexico (SEDAR 13 2007). They tend to engage in seasonal inshore–offshore movement, leaving the coast in October and November for warmer offshore waters and returning in March and April. Most adult females tend to be found just offshore in deepwaters. It appears that there is sex and size segregation in the distribution of the population in the Gulf of Mexico (Parsons 2006). Young-of-the-year, juvenile, and adult Atlantic sharpnose sharks prefer sandy and seagrass bottoms, but also can be found on muddy grounds (Bethea et al. 2009). The shallow waters in the extensive barrier islands of the northern Gulf of Mexico are important Atlantic sharpnose shark pupping and nursery grounds.

Age and growth of the Atlantic sharpnose shark in the Atlantic Ocean and Gulf of Mexico have been based on vertebral age analysis (Parsons 1983; Branstetter 1987; Carlson and Baremore 2003). The Atlantic sharpnose shark has a maximum length of about 107 cm (42.1 in.) and a maximum age of 11 years (Loefer and Sedberry 2003). Tagging studies, however, suggest that the maximum age should be 12 years. Nearly all females and males become sexually mature at the age of 2.5 years in the Gulf of Mexico and at age 3.5 in the South Atlantic. Peak mating activity occurs in June and July, and the gestation period is 10–12 months. Reproductive periodicity is annual for both the Gulf of Mexico and the Atlantic Ocean. Fecundity is 4.1 pups per year, with pupping occurring in June. The annual survival rate is about 0.7 for age-1 sharks and slightly higher (around 0.75) for adults (SEDAR 13 2007).

9.4.3.2   Blacknose Shark

Blacknose sharks occur from North Carolina to Brazil, including the Gulf of Mexico (SEDAR 13 2007). They can usually be found in inshore shallow waters in the northern Gulf of Mexico, although not very abundant. Genetic studies suggest that the reproductive cycles differ by basin, but tagging data show no mixing, so they are considered as one unit stock in the assessment (SEDAR 13 2007).

The blacknose shark is small, with a maximum size around 150 cm (59 in.) TL (Parsons 2006). Age and growth was studied for the blacknose shark in the Atlantic Ocean and Gulf of Mexico (Carlson et al. 1999; Driggers et al. 2004). Males mature at about 100–110 cm (39.4–43.3 in.) TL and females at 110–115 cm (43.3–45.3 in.) TL. Mating occurs in late summer or early fall. The gestation period is about 9–10 months. The fecundity is about 3–6 pups, with pupping months in May and June. The reproductive periodicity in the Gulf of Mexico is annual, while the periodicity is considered biennial in the South Atlantic. The annual survival rates are 0.72 for age-1 sharks and 0.76–0.83 for adults. Their main prey includes fish, squid, shrimp, and other invertebrates.

9.4.3.3 Bonnethead Shark

The bonnethead shark is distributed from New England to south of Brazil and commonly occurs in the Gulf of Mexico (SEDAR 13 2007). Bonnethead sharks are considered in the most recent stock assessment a single stock from North Carolina through the Straits of Florida and the Gulf of Mexico (SEDAR 13, 2007). However, there are no data supporting this single-stock hypothesis. In the Gulf of Mexico, it is especially abundant east of Mobile Bay and is the dominant shark species in the shallow coastal waters of the Florida Gulf coast. Like most coastal sharks in the Gulf of Mexico, bonnethead sharks are also subject to seasonal inshore–offshore movements, leaving the coastal waters in October and November for warm offshore waters and moving back to inshore areas in the spring.

Bonnethead sharks are small, with a maximum size of about 109 cm (42.9 in.) TL for males and 124 cm (48.8 in.) TL for females. Age and growth have only been studied for bonnethead sharks in the eastern Gulf of Mexico (Parsons 1993; Carlson and Parsons 1997; Lombardi-Carlson et al. 2003). The maximum age is estimated at 7.5 years based on vertebral age analysis. However, tagging studies suggest that a maximum age of 12 years is a more reasonable estimate. Males become sexually mature at around 2 years of age and females at 2.5 years. They may mate in the fall, with the mated females storing sperm until the following spring when their eggs ovulate for fertilization. Gestation is 4–5 months, the shortest of any placental viviparous (give birth to young) shark species. Their reproductive cycle is annual, with pupping time in August (Parsons 1993; Carlson and Parsons 1997; Lombardi-Carlson et al. 2003). The average size of a litter is about 10 pups.

Juvenile bonnethead sharks tend to be associated with sandy and seagrass bottoms, and adults can be found on muddy, sandy, and seagrass bottoms. Although they feed mainly on blue crabs, shrimp, and squid, occasionally, fish can be found in their stomachs. The first-year survival rate is about 0.66 per year, and survival rates of adults range from 0.66 to 0.81 per year.

9.4.3.4 Finetooth Shark

Finetooth sharks are distributed in the western Atlantic from New York to southern Brazil (SEDAR 13 2007). They are abundant in the northern Gulf of Mexico. Finetooth sharks from North Carolina through the Straits of Florida and into the Gulf of Mexico are considered a single stock because of the lack of genetic differences. However, there is a low exchange of individuals between the Gulf of Mexico and South Atlantic (SEDAR 13 2007). They are one of the most abundant species in inshore waters of the northern Gulf of Mexico.

The finetooth shark is a medium-sized shark, reaching a maximum size of 180 cm (70.9 in.) TL and a maximum age of 12 years (Parsons 2006; SEDAR 13 2007). Males and females become sexually mature at 120 and 137 cm (47.2 and 53.9 in.) TL, respectively, and at ages of 6–7 years old (Parsons 2006; SEDAR 13 2007). Mating occurs in late spring and early summer. The reproductive cycle is biennial, with pupping in June and an average litter size of 3–4 pups (Neer and Thompson 2004). The gestation period likely lasts 11–12 months. They mainly feed on finfish, including mackerel, whiting, and sea trout. Young-of-the-year finetooth sharks prefer muddy bottoms; juveniles also mainly exist on the muddy bottom but can also be found on sandy and seagrass bottoms, while adults usually are associated with seagrass and sandy bottoms (Bethea et al. 2009).

9.4.4 Large Coastal Sharks Group

Currently, the Large Coastal Sharks group consists of 11 shark species that are widely distributed throughout the world (SEDAR 29 2012). In the western Atlantic Ocean, they can be found from along the U.S. Atlantic coast all the way to the south of Brazil. All of the Large Coastal Sharks can be found in the Gulf of Mexico (SEDAR 29 2012). They are considered either as part of the South Atlantic stock or as an independent stock in the assessment and management. Most species of Large Coastal Sharks use the inshore shallow waters of the northern Gulf of Mexico as their spawning and nursery grounds. They tend to move into the inshore shallow waters in the Gulf of Mexico during the spring to give birth to their offspring. The inshore shallow waters provide refuges for their newborn offspring from potential predators (usually large sharks). The young sharks spend summers in the inshore waters for feeding. The preferred bottoms range from sand to mud to seagrass. Young-of-the-year sharks tend to occur more frequently in shallower water with higher temperatures, lower salinities, and more turbid conditions compared to the habitats for juveniles and adults. Small and young sharks may select these habitats as a refuge from larger and more active predators (Bethea et al. 2009). Most sharks move into warmer offshore waters in the fall. Compared to species of Small Coastal Sharks, most of the Large Coastal Sharks tend to become sexually mature at a later age.

The Large Coastal Sharks group supports several important commercial and recreational fisheries in the Gulf of Mexico (SEDAR 29 2012). Although most species are not overfished and overfishing is not occurring, most stock abundances have been reduced over time. The catch quota for the Large Coastal Sharks stocks has been reduced continuously over time since the 1990s but has become relatively stable since the mid-2000s (Figure 9.75), perhaps reflecting a stabilized Large Coastal Sharks group.

Sandbar and blacktip sharks are two of the most abundant and most commercially and recreationally important shark species in the Gulf of Mexico (SEDAR 29 2012). They both belong to the Large Coastal Sharks group and are widely distributed in the Gulf of Mexico. As top predators that are abundant in Gulf of Mexico coastal ecosystems, sandbar and blacktip sharks play an important role in regulating the ecosystem dynamics of the Gulf of Mexico; therefore, they were selected as representative species for evaluation.

9.4.4.1 Sandbar Shark

Sandbar sharks, one of the largest coastal sharks in the world, can be found in the subtropical waters of the western Atlantic from southern Massachusetts in the United States to southern Brazil, including the Gulf of Mexico between 44°N and 36°S (SEDAR 21 2011). They usually prefer waters ranging from 23 to 27 °C in temperature. Sandbar sharks occur over muddy or sandy bottoms in shallow coastal waters, such as estuaries, bays, river mouths, and harbors, and on continental and insular shelves. They spend most of the time in waters 20–65 m (65.6–213 ft) deep; they also occur in deeper waters (200 m or 656 ft or more), as well as intertidal zones. However, they tend to avoid the surf zone and beach areas. Sandbar sharks usually swim alone or aggregate in sex-segregated schools varying in size.

Sandbar sharks are viviparous (SEDAR 21 2011). Males reach maturity between 1.3 and 1.8 m (4.3–5.9 ft) in size, while females mature at 1.45–1.8 m (4.8–5.9 ft). Birth sizes of pups range from 55 to 70 cm (21.6–27.6 in.) long. Mating occurs in the spring or early summer (May through

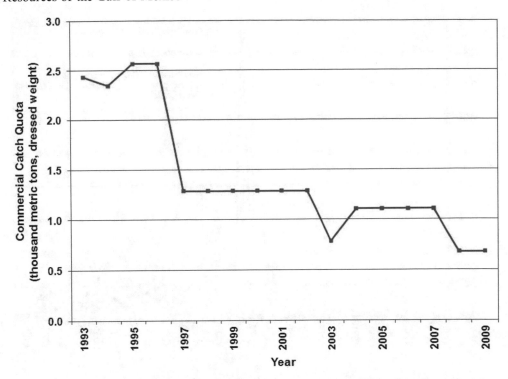

Figure 9.75. Annual commercial catch quota for the Large Coastal Sharks group in the Atlantic Ocean and Gulf of Mexico from 1993 through 2009. The catch quota data do not include bycatch and discards (data from Table 8 in SEDAR 29 2012).

June). Once fertilization occurs, the gestation period is 8–9 months in the western Atlantic population, where pups are born between June and August. The female has a triennial reproductive cycle. The litter size is typically between 6 and 13 pups, depending on the size of the mother. In the northern Gulf of Mexico, an important nursery area exists around Cape San Blas, Florida. Juvenile sandbar sharks are also captured off Mississippi and Alabama, suggesting the existence of other nursery grounds. Females give birth in shallow water nursing grounds so that YOY and juveniles sharks can be protected from predation by larger sharks, such as bull sharks. Juveniles remain in or near the nursery grounds until late fall after which they form schools and migrate to deeper waters. They return to the nursery grounds during warmer months. After reaching the age of 5 years, they begin to follow the wider migrations of adults. Sandbar sharks are opportunistic bottom feeders preying on bony fishes, smaller sharks, rays, cephalopods, gastropods, crabs, and shrimps. Sandbar sharks feed throughout the day but become more active at night. Predators of sandbar sharks include tiger sharks and great white sharks, on occasion. Sandbar shark M was assumed to be 0.14 in the most recent stock assessment (SEDAR 21 2011).

Sandbar sharks in the South Atlantic and Gulf of Mexico are assessed and managed as a single stock (SEDAR 21 2011). Mexican fisheries and U.S. recreational fishing dominated the catch prior to the mid-1980s (Figure 9.76). After 1985, the commercial catch in the Gulf of Mexico increased quickly and comprised almost half of the total catch between 1985 and 1995. After the mid-1990s, catch in the Gulf of Mexico decreased rapidly (Figure 9.76). Sandbar shark stock abundance has decreased substantially since 1960, and stock abundance in 2009, which is the most recent year covered in the most recent stock assessment (SEDAR 21 2011), is only about 25 % of the stock biomass in the 1960s. Spawning stock fecundity (calculated as numbers × proportion mature × fecundity in numbers) describing the stock reproductive potential also has the same trend as stock abundance (Figure 9.77).

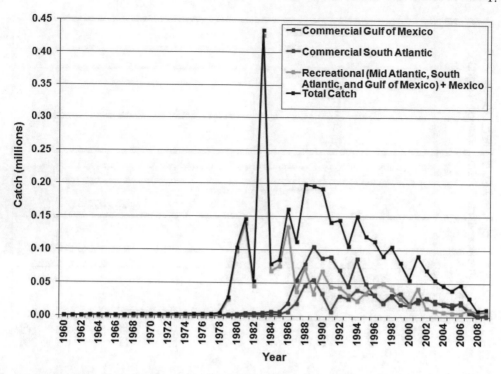

Figure 9.76. Sandbar shark (*Carcharhinus plumbeus*) catch by recreational and commercial fisheries in the Mid-Atlantic, South Atlantic, and Gulf of Mexico from 1960 through 2009 (data from SEDAR 21 2011b).

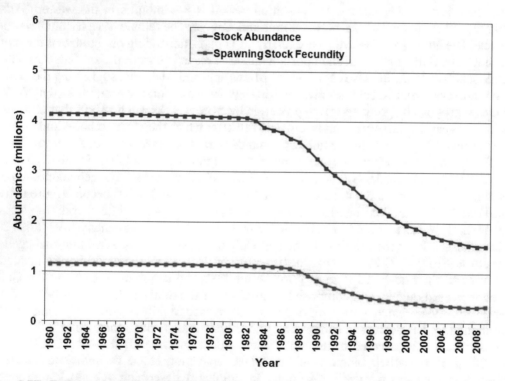

Figure 9.77. Estimated stock abundance and spawning stock fecundity for the sandbar shark (*Carcharhinus plumbeus*) in the Gulf of Mexico from 1960 through 2009 (data from SEDAR 21 2011b).

Figure 9.78. Blacktip shark (*Carcharhinus limbatus*) (from Block 2011).

The most recent stock assessment suggests that the sandbar shark stock was overfished and, therefore, subject to rebuilding. However, in the base run and in most sensitivity runs, the stock was found not to be currently subject to overfishing (F2009/FMSY ranges from 0.29 to 0.93) (SEDAR 21 2011). Overfishing was found to be occurring (F2009/F_{MSY} of 2.62) only for the low productivity scenario (SEDAR 21 2011).

9.4.4.2 Blacktip Shark

The blacktip shark (Figure 9.78), a fast-swimming and highly active shark species, is widely distributed in coastal tropical and subtropical waters around the world, including brackish habitats (SEDAR 29 2012). In the western Atlantic, their distribution ranges from southern New England to southern Brazil, including the Gulf of Mexico and the Caribbean. The blacktip shark is one of the most abundant shark species in the Gulf of Mexico. Based on tagging and genetic studies, two stocks are defined in the stock assessment: the Atlantic stock distributed from Delaware to the Straits of Florida and the Gulf of Mexico stock. Although adult blacktip sharks are highly mobile and often disperse over long distances, tagging studies provide little evidence to support mixing between the two stocks (Keeney et al. 2005; SEDAR 11 2006). They are philopatric (behavior of remaining in, or returning to, their birthplace) and return to their original nursery areas to give birth, which can result in subgroups of genetically distinct breeding stocks that overlap in geographic distributional ranges (Keeney et al. 2003, 2005).

Blacktip sharks are targeted as a prized and high quality food fish, and are captured in targeted commercial and recreational fisheries (SEDAR 29 2012). The majority of landings are from the demersal longline fishery. Another major source of mortality in the Gulf of Mexico is discards in the Gulf of Mexico menhaden fishery. The landings of blacktip shark in the Gulf of Mexico increased rapidly in the late 1980s but have decreased substantially since 1990 (Figure 9.79). The lowest catch level occurred in 2008, and landings have increased slightly since then (Figure 9.79).

Female blacktip sharks can reach up to 200 cm (78.7 in.) TL, while males can reach up to about 180 cm (70.9 in.) TL (Parsons 2006). Maximum ages found in the most recent stock

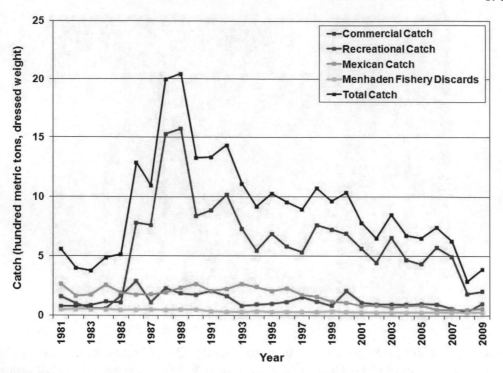

Figure 9.79. Blacktip shark (*Carcharhinus limbatus*) catch in the Gulf of Mexico in recreational and commercial fisheries from 1981 through 2009 (data from SEDAR 29 2012).

assessment (SEDAR 29 2012) were 18.5 years for females and 23.5 years for males, a significant increase of 6 years and 12 years for females and males, respectively, compared to ages observed by Carlson et al. (2006). The M is assumed to be between 0.1 and 0.2, decreasing with age (Figure 9.80).

The blacktip shark is a synchronous, seasonally reproducing species with reproductive activity (e.g., mating and parturition) mainly occurring in March through May. Length and age at 50 % maturity are 105.8 cm (41.6 in.) FL and 4.8 years for males and 119.2 cm (46.9 in.) FL and 6.3 years for females, respectively. Female blacktip sharks have a biennial ovarian cycle. The gestation period ranges from 10 months (Parsons 2006) to approximately 12 months (SEDAR 29 2012); the average fecundity is 4.5 pups (ranging from 1 to 10 pups), with the average size at birth at 38 cm (14.9 in.) FL (or about 60 cm or 23.6 in TL). Fecundity was found to increase with both maternal size and age. Females are also capable of asexual reproduction in the absence of males.

Blacktip sharks mainly feed on fishes, squids, and sometimes crustaceans. In the Gulf of Mexico, the most important prey of the blacktip shark is the Gulf menhaden, followed by the Atlantic croaker (*Micropogonias undulatus*) (Barry 2002). Juveniles may be prey of other large sharks, but adults have no known predators.

Blacktip sharks do not inhabit oceanic waters, although some individuals may be found some distance offshore (Compagno 1984). Most blacktip sharks are found in water less than 30 m (98.4 ft) deep over continental and insular shelves; though, they may dive to 64 m (210 ft) (Froese and Pauly 2009). Their favored habitats include muddy bays, island lagoons, and the

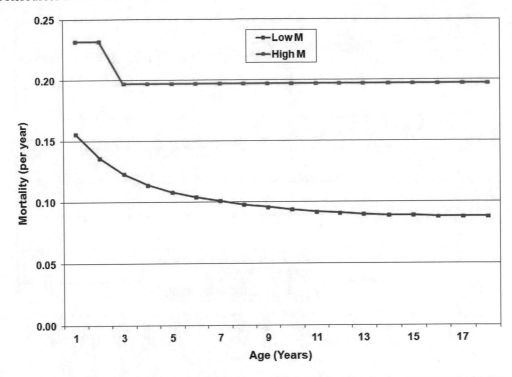

Figure 9.80. Age-specific natural mortality (*M*) assumed in the assessment of the Gulf of Mexico blacktip shark (*Carcharhinus limbatus*) stock (data from SEDAR 29 2012).

drop-offs near coral reefs. Juvenile blacktip shark abundance showed significant correlation with turbidity/water clarity (Bethea et al. 2009). They can also be tolerant of low salinity, moving into estuaries and mangrove swamps. Seasonal migration has been documented to avoid unfavorable thermal habitats, usually moving into warm waters during the fall and returning to inshore feeding/nursery grounds in the spring. Newborn and juvenile blacktip sharks can be found on muddy/sandy/seagrass grounds in inshore shallow nurseries in late spring and early summer, and grown females tend to return to the nurseries where they were born to give birth. Young blacktip sharks are most likely to form aggregations in early summer to avoid predators (Heupel and Simpfendorfer 2005). There tends to be segregation by sex and age, with adult males and nonpregnant females being found apart from pregnant females; juveniles are separated from both groups in the winter (Castro 1996).

The abundance of blacktip sharks in the Gulf of Mexico has had a relatively small decrease since the 1980s and seems to have stabilized or slightly increased since 2000 (Figure 9.81). The spawning stock fecundity that describes stock reproductive potential has the same trend as stock abundance. The most recent stock assessment concluded that the Gulf of Mexico blacktip shark stock was not overfished, and overfishing was not occurring (SEDAR 29 2012). This conclusion is robust with respect to all of the uncertainty in data quality and quantity and assumptions considered in the assessment (SEDAR 29 2012).

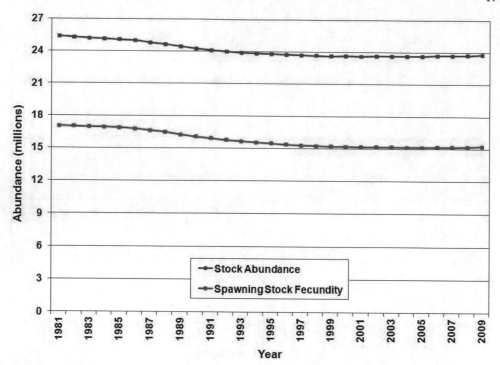

Figure 9.81. Estimated blacktip shark (*Carcharhinus limbatus*) abundance and spawning stock fecundity for the Gulf of Mexico from 1981 through 2009 (data from SEDAR 29 2012).

9.4.5 Pelagic Sharks Group

The Pelagic Sharks group was initially identified in the 1993 FMP and included the following ten species: shortfin mako, longfin mako, porbeagle, common thresher, bigeye thresher, blue, oceanic whitetip, sevengill, sixgill, and bigeye sixgill (NMFS 1993). Since 1993, five species have been moved to the group of sharks that are prohibited from fishing because of low population levels. Therefore, the Pelagic Sharks group currently includes blue, oceanic whitetip, porbeagle, shortfin mako, and common thresher sharks.

Sharks included in the Pelagic Sharks group are transoceanic, cosmopolitan species that, in general, are highly migratory. In the Western Atlantic, most of these species can be found from Maine to Argentina, including the Gulf of Mexico and the Caribbean. They tend to stay in oceanic deepwater areas but sometimes come close to shore. They can be found most frequently from the surface to depths of at least 200 m (656 ft), but also appear at depths over 1,000 m (3,281 ft). This group of sharks tends to have the largest body sizes. For example, the common thresher can be as large as over 700 cm (275.6 in.) TL (FishBase 2013).

Pelagic sharks are the top predators in the marine ecosystem, feeding mostly on oceanic bony fishes, but also on threadfins, stingrays, sea turtles, sea birds, gastropods, squids, crustaceans, mammalian carrion, tunas, and dolphinfish. Like other shark species, they are viviparous and may be subject to partial segregation by size and sex in some areas.

Pelagic sharks are often caught as bycatch in the North Atlantic Ocean by fishing fleets from several nations. In the U.S. Atlantic, Gulf of Mexico, and the Caribbean, the combined commercial and recreational catch and discards tend to fluctuate greatly over time (Figure 9.82). Since the mid-1990s, the catch has been fairly stable, remaining around 20,000 sharks per year. For the most recent year included in the time series (e.g., 2006), the majority of the catch was from the recreational fishery.

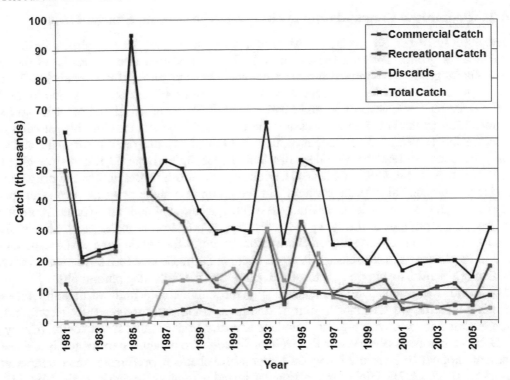

Figure 9.82. Catch of the Pelagic Shark group in the U.S. Atlantic, Gulf of Mexico, and Caribbean from 1981 through 2006 (data from Cortés 2008).

9.4.6 Prohibited Sharks

Many species of sharks are prohibited from being commercially and recreationally fished because of low populations and poor stock conditions. These sharks, which are now in the Prohibited Species group, were formerly in other managed groups and are described below.

9.4.6.1 Prohibited Sharks Formerly in the Small Coastal Sharks Group

The 1993 FMP for Sharks of the Atlantic Ocean listed seven species of sharks in the Small Coastal Sharks group. Of these seven species, three were moved into the newly created Prohibited Species group in the 1999 FMP for Atlantic Tuna, Swordfish, and Sharks. Since 1999, commercial and recreational fishermen have been prohibited from possessing these three species, which include the Atlantic angel, Caribbean sharpnose, and smalltail sharks. Atlantic angel, Caribbean sharpnose, and smalltail sharks occupy shallow coastal waters and estuaries from the Gulf of Mexico south throughout the Caribbean. The Atlantic angel shark can be found in waters as far north as New England in the western North Atlantic, and the Caribbean sharpnose inhabits waters between 24°N and 35°S. The Atlantic angel shark is dorsoventrally flattened, resembling a ray. Angel sharks reproduce biennially, bearing as many as 16 pups per litter (Castro 1983). The Caribbean sharpnose shark is closely related to the Atlantic sharpnose, with similar biology and life history, differing only in the number of precaudal vertebrae and geographic range (Springer 1964). All of these shark species have long gestation periods of about 10 months (Carlson et al. 2004).

9.4.6.2 Prohibited Sharks Formerly in the Large Coastal Sharks Group

The 1993 FMP for Sharks of the Atlantic Ocean listed 22 species of sharks in the Large Coastal Sharks group. Through legislation in 1997 by the NMFS, five species from the Large Coastal Sharks group were moved into the Prohibited Species group. Of the remaining 17 species in the Large Coastal Sharks group, six were added to the Prohibited Species group in the 1999 FMP for Atlantic Tuna, Swordfish, and Sharks. Since 1999, commercial and recreational fishermen have been prohibited from possessing these 11 shark species: basking, bigeye sand tiger, bignose, Caribbean reef, dusky, Galapagos, narrowtooth, night, sand tiger, great white, and whale sharks. These sharks tend to be widely distributed globally and can be found throughout the western North Atlantic, Gulf of Mexico, Caribbean, and south to Brazil. The majority of these 11 species inhabit coastal to pelagic waters, but the group also includes a few deepwater species. These sharks tend to mature late, and many have long gestation periods and biennial reproductive cycles. Feeding strategies in this group range from apex predator, as in the great white shark, to generalist feeders and scavengers, and filter feeders, such as the basking and whale sharks. Great white and whale sharks were selected as representative Large Coastal Sharks that were moved to the Prohibited Sharks group and are briefly discussed in the following paragraphs.

The great white shark occurs sporadically throughout cold and warm temperate seas (Figure 9.83). In the western North Atlantic, the great white shark ranges from Newfoundland to the Gulf of Mexico, with highest abundances in the Mid-Atlantic Bight region (Casey and Pratt 1985). It has been observed in the Gulf of Mexico from January to September. Seasonal movements appear to be related to water temperature changes, preferring water temperatures of 12–25 °C (Miles 1971). Higher proportions of juvenile great white sharks in the Mid-Atlantic Bight region suggest that this area may serve as a nursery area (Casey and Pratt 1985). Great white sharks are an apex predator feeding primarily on fish as juveniles and switching to primarily marine mammals after reaching a length of over 300 cm (118 in.) (Klimley 1985; McCosker 1985). Little is known about great white shark reproduction, as few gravid females have been examined. Great white sharks carry 7–10 embryos and are thought to reach maturity at 9–10 years (Cailliet et al. 1985; Francis 1996; Uchida et al. 1996). Small localized populations,

Figure 9.83. Great white shark (*Carcharodon carcharias*) (from Dascher 2013).

Figure 9.84. Whale shark (*Rhincodon typus*) (from crisod 2013).

susceptibility to longlines, and limited reproductive potential contribute to making the great white shark vulnerable to overfishing (Strong et al. 1992).

The whale shark is the largest fish in the ocean, reaching lengths of over 12 m (39.4 ft) (Figure 9.84). It is a slow-moving filter feeder distributed throughout the world in tropical seas (Castro 1983). The range of the whale shark includes the northern Gulf of Mexico, and they appear to be more abundant in the western Gulf of Mexico than the eastern Gulf (Burks et al. 2006). Whale sharks sometimes form large feeding aggregations near the surface, and as many as 100 individuals or more join these aggregations. Very little is known about whale shark reproduction. One gravid female has been described, carrying 300 young in various stages of development. Due to its wide range, whale shark populations may have to be managed as an ocean-wide population. The whale shark has been demonstrated to be susceptible to overfishing based on records of the Taiwanese fishery.

9.4.6.3   Prohibited Sharks Formerly in the Pelagic Sharks Group

Ten species of sharks were included in the Pelagic Sharks group in the 1993 FMP for Sharks of the Atlantic Ocean (NMFS 1993). Of these 10 species, five were moved into the Prohibited Species group in the 1999 FMP for Atlantic Tuna, Swordfish, and Sharks (NMFS 1999). Since 1999, commercial and recreational fishermen have been prohibited from possessing the following five shark species: longfin mako, bigeye thresher, sevengill, sixgill, and bigeye sixgill sharks. These sharks tend to occur in waters 100s to 1,000s of meters deep and have very wide global distributions, including the Gulf of Mexico, Caribbean, and western North Atlantic. They tend to have a generalist feeding strategy, preying on various bony and cartilaginous fishes, squids, and crustaceans, as well as scavenging carrion. This group tends to have slow growth rates, a late age at maturity, and small litter sizes, with the exception of the sixgill shark, which can have as many as 20–100 pups in a single litter. The longfin mako and bigeye sixgill sharks were not described by science until the 1960s, and very little is known about them. The pelagic sharks of the Prohibited Species group are susceptible to bycatch in fisheries utilizing bottom trawls and longlines, such as those used in the tuna, swordfish, and tilefish fisheries.

Figure 9.85. Southern sting rays (*Dasyatis americana*) often rest in the valleys between pinnacles and Stetson Bank, which is located within the Flower Garden Banks National Marine Sanctuary (photograph by Emma Hickerson) (from NMS 2013).

Due to low fecundity, slow maturation rates, and likelihood of bycatch mortality, these sharks are highly susceptible to overfishing (NMFS 2006b, 2009b).

9.4.7 Rays and Skates

Rays and skates are a diverse group of cartilaginous fishes and inhabit marine ecosystems in the Gulf of Mexico, primarily along the bottom, ranging from shoreline depths to 2000 m (6,562 ft) deep (Figure 9.85) (McEachran 2009). Only three species (Family Mobulidae) inhabit the oceanic surface and epipelagic zone. They are primarily specialized for a bottom-dwelling lifestyle, feeding primarily on benthic invertebrates. This lends to their unique physiology, which includes teeth resembling flat-crowned plates for crushing shells and exoskeletons, a highly protrusible mouth advantageous for the suction of benthic invertebrates from the substrate, and varying degrees of dorsoventral flattening conducive to camouflage and ambushing prey (Pough et al. 2008). Similar to what has been suggested for the skull structure of a hammerhead shark, this flattening increases the distribution of ampullae of Lorenzini across a larger surface area, which may be more conducive to seeking out prey along the bottom (Pough et al. 2008). However, some rays are highly specialized zooplankton strainers; one species even uses an elongated snout with tooth-like structures for slashing at schools of fish (McEachran and de Carvalho 2002). Rays and skates can vary greatly in size from 13 cm (5.1 in.) to 7 m (23 ft), and can weigh from 10 g (35 oz) to more than 2,700 kg (6,000 lb). Rays

Figure 9.86. Manta ray (*Manta birostris*) (from Scubaguys 2016).

and skates are separated for both their reproductive strategy and tail differences. Rays are viviparous, while skates are oviparous and lay collagenous egg cases that are commonly called "mermaid's purses." A skate tail is typically long, thick, and finned, while a typical ray tail is more whip-like and often contains a specialized venomous or serrated dorsal barb. No direct fishery exists for rays and skates in the Gulf of Mexico; however, intensive bycatch during shrimp trawling and bottom longlining can negatively impact rays and skates (Sheperd and Myers 2005). Three representative species of rays and skates were selected for further discussion in the paragraphs that follow.

The giant manta ray (*Manta birostris*) is the world's largest ray reaching 1,814 kg (4,000 lb), with an average wingspan of 6.7 m (22 ft) (Figure 9.86). It primarily inhabits pelagic waters from 0 to 100 m (328 ft) deep, but can also be found near reefs, over deeper waters, as well as in muddy, intertidal habitats (McEachran 2009; FFWCC 2013a). Giant manta rays are considered highly transient, migratory, and circumglobal; however, some debate exists on whether records from other oceans may be for different species (McEachran and de Carvalho 2002). The giant manta ray belongs to Family Mobulidae, which is unique from other ray families because they feed almost exclusively on zooplankton while slowly swimming in the epipelagic and oceanic surface zones using funneling fins near the mouth called rostra and specialized gill rakers (FFWCC 2013a). This family has a low reproductive potential; giant manta rays can viviparously produce up to two pups per litter, although one is considered more typical, with a gestation period of 10–14 months, and a possible life span of approximately 20 years (Bigelow and Schroeder 1953; Homma et al. 1999; Ebert 2003).

The cownose ray (*Rhinoptera bonasus*) is a semi-pelagic species that can form large schools anywhere from 100s to 1,000s of individuals (Neer 2005; FFWCC 2013b). Cownose rays have an average wingspan of 0.9 m (3 ft) and are often seen actively swimming or leaping out of the water. They primarily inhabit pelagic waters from 0 to 25 m (82 ft) deep in bays, estuaries, river mouths, and even in the open ocean (McEachran 2009; FFWCC 2013b). In the Western Atlantic, cownose rays can be found from southern New England to Argentina (McEachran and de Carvalho 2002; McEachran 2009). Cownose rays are considered migratory but can be found in some estuaries throughout the year (McEachran and de Carvalho 2002). Their diet consists of bivalve mollusks, crustaceans, and polychaetes (FFWCC 2013b). Like other rays and skates, cownose rays have a low reproductive potential and can viviparously produce up to 2–6 pups per litter, although one may be considered more typical (McEachran and de Carvalho 2002; FFWCC 2013b).

The smalltooth sawfish (*Pristis pectinata*) is a recognizable species given its elongated snout that can measure up to one-quarter of its total length and be lined with 24–28 unpaired teeth (FFWCC 2013c). It can grow up to 5.5 m (18 ft) and primarily inhabits shallow coastal waters near river mouths, estuaries, bays, or depths up to 125 m (410 ft) (Bigelow and Schroeder 1953; Simpfendorfer and Wiley 2005; McEachran 2009). Historically, in the Western Atlantic, small-tooth sawfish ranged from New York to Brazil, but there has been significant population reduction and range contraction. Today, it is found for the most part only between the Caloosahatchee River in Florida and the Florida Keys (NMFS 2009d). The diet of the smalltooth sawfish consists mostly of small schooling fishes, such as anchovies or mullets, which are injured or killed as the snout is slashed through the school (Bigelow and Schroeder 1953). Smalltooth sawfish have a low reproductive potential, but can potentially produce up to 20 pups per litter, with an approximate gestation period of 5 months (Bigelow and Schroeder 1953; NMFS 2009d).

9.5 SUMMARY

The Gulf of Mexico is one of the most productive marine ecosystems in the world, with high fish species richness and high fishery productivity supported by a great diversity of habitat types. Finfish and shark species play critical roles in the Gulf of Mexico ecosystem and in the spatiotemporal dynamics of their populations. They are strongly influenced by habitat quality and biotic and abiotic factors, including hydrographic and geographic conditions, predation, food supply, fishing, natural weather, geochemical cycles, and the impacts of human activity and coastal development. However, heavy fishing over the last several decades and the long-term effects of anthropogenic and natural stressors on the finfish and shark species and their habitats in the Gulf of Mexico have resulted in populations of great commercial and recreational importance being defined as being overfished and/or undergoing overfishing (Table 9.45).

This chapter has evaluated the distribution, life history, habitat needs, fisheries, and population status of some of the most important finfish and shark species in the Gulf of Mexico prior to the Deepwater Horizon event and has attempted to analyze factors that have most influenced their health and productivity. The 13 finfish species selected for detailed evaluation in this chapter are representative of the Gulf of Mexico. They vary greatly in life history and distribution, are important to commercial and recreational fisheries, and consist of fish species of almost all habitat types. A summary of the status of the 13 fish species selected for evaluation is presented in Table 9.45. Some important general conclusions for finfish and shark in the Gulf of Mexico include the following:

- Inshore estuaries in the Gulf of Mexico are critical habitats because they are ideal nursery and feeding grounds for most finfish and shark species, providing food and refuge from juvenile predation.

- The Gulf of Mexico provides critical spawning grounds for many highly migratory fish and shark species of great ecological, commercial, and recreational importance.

- Only a small fraction (4.6 %) of finfish is endemic to the Gulf of Mexico, although some fish species of commercial and recreational importance can complete their entire life cycle within the Gulf of Mexico ecosystem.

- Almost half of finfish species in the Gulf of Mexico are omnipresent and can be found throughout the Gulf; the other half is limited to parts of the Gulf of Mexico.

Table 9.45. Summary of the Status of 13 Finfish Species Selected for Evaluation

Species	Fishery status	Key habitat	Other important information
Red snapper (*Lutjanus campechanus*)	Overfishing occurring 2001–2012; stock overfished 2001–2012	Structured habitat and reefs across shelf to shelf edge	Found throughout Gulf of Mexico, long life span, mature at young ages
Menhaden, including Gulf menhaden (*Brevoortia patronus*), finescale menhaden (*Brevoortia gunteri*), and yellowfin menhaden (*Brevoortia smithi*)	Overfishing not occurring; stock not overfished in the 2000s	Estuaries and other quiet low salinity nearshore habitat for juveniles, open bay and Gulf waters with vegetable bottom for adults	Found throughout coastal Gulf of Mexico, high habitat elasticity to adapt to changes in habitat
Red grouper (*Epinephelus morio*)	Overfished 2000–2002; overfishing 2002; not overfished and no overfishing 2005–2008, but some local subpopulations in northeastern and southern portion of the Gulf of Mexico overfished and overfishing occurring	Shallow nearshore reefs and seagrass beds for juveniles, shore rocky bottom ledges and caverns in limestone reefs for mature adults	Limited movement, large spatial variability in life history, complex stock structure and possible existence of local subpopulations
Atlantic bluefin tuna (*Thunnus thynnus*)	Large uncertainty in status depending on the assumption on stock productivity; most likely overfishing and overfished prior to the 2000s	Epipelagic and oceanic, feeding inshore, spawn in northern Gulf of Mexico	Highly migratory, strictly regulated, highly priced
Atlantic blue marlin (*Makaira nigricans*)	Overfishing and overfished 2000–2011	Epipelagic and oceanic, blue waters associated with the Loop Current	No commercial fishery in the United States, highly migratory
Atlantic swordfish (*Xiphias gladius*)	Overfished and overfishing 2000–2002; considered rebuilt and not overfished; no overfishing since 2002	Pelagic-oceanic	Highly migratory, valuable sport fishery
Atlantic sailfish (*Istiophorus albicans*)	Overfished 2000–2005; overfishing occurring from 2001 to 2003; not overfished from 2006 to 2011	Oceanic, but migrate into shallow waters	Highly migratory, tropic and temperate waters
Red drum (*Sciaenops ocellatus*)	Overfished 2000–2005; overfishing 2000–2003; not overfished 2006–2011	Found in coastal beaches and nearshore shelf waters	Complex stock structure, commercial fishery prohibited in federal waters, valuable sport fishery
Striped mullet (*Mugil cephalus*)	No Gulf-wide assessment; Florida stock not overfished 1995–2007; overfishing not occurring 1995–2007	Nursery grounds in secondary and tertiary bays, shallow inshore marine and estuarine waters, soft sediment	Inshore fishery, important recreational fishery

(continued)

Table 9.45. (continued)

Species	Fishery status	Key habitat	Other important information
Greater amberjack (*Seriola dumerili*)	Overfished 2001–2011; overfishing occurring 2004–2011	Pelagic *Sargassum* mats until 5–6 months old, reefs, rock outcrops and wrecks, pelagic and epibenthic	Like congregation
Tilefish (*Lopholatilus chamaeleonticeps*)	Not overfished and no overfishing; large uncertainty in the status	Demersal living in burrows in mud, silt and clay sediments along the continental slopes	Long lived, slow growth, complex breeding process, habitat specificity, sensitivity to changes in habitat
King mackerel (*Scomberomorus cavalla*)	Overfishing occurring prior to 2000; overfished 2000–2003; declared rebuilt 2008	Epipelagic tropic, subtropic and temperate waters	Opportunistic carnivores
Dolphinfish (*Coryphaena hippurus*)	Overfishing not occurring 2000–2011; not overfished 2000–2011	Tropical and warm-temperate waters	One of the fastest growing species, high commercial and recreational values

- Many fish species in the Gulf of Mexico experience inshore–offshore movement in response to changes in environmental conditions (e.g., thermal habitat) and/or needs of life-history processes (e.g., spawning).

- Fish species differ widely in their ability to adapt to changes in the biotic and abiotic environment in the Gulf of Mexico; omnipresent species are the most robust in adapting to changes in habitat.

- There is large spatiotemporal variability in key life-history parameters (e.g., growth and maturation) for many Gulf of Mexico fish species.

- Many fish species in the Gulf are often considered a unit fish stock, which implicitly assumes adequate Gulf-wide mixing for fisheries stock assessment and management, even though evidence suggests complex stock structure within the Gulf of Mexico (e.g., existence of meta-population or local populations as a result of spatial isolations).

- A wide variety of short-term and long-term anthropogenic and natural stressors, such as rapid coastal development, pollution, heavy fishing, climate change, and natural disasters, have reduced the resilience and robustness of the Gulf of Mexico ecosystem with respect to human and natural perturbations.

- Different fish species tend to respond to fishing and changes in habitat in different ways. Some species are more sensitive, and others are more robust with respect to changes in fishing mortality and habitat.

- High fishing pressure and degraded environment have changed key life-history parameters of important fish species in the Gulf of Mexico, e.g., reduced size at age and earlier maturation, reduced stock reproductive potential, increased temporal fluctuation of recruitment, and impaired ability of fish stocks to recover from low stock abundance.

- Many finfish and shark species of great ecological, commercial, and recreational importance have been subject to overfishing and being overfished. Their present population abundances tend to be much lower compared to historic levels, which may be the result of a combination of factors, including high fishing pressure, large bycatch mortality, and recruitment failure likely due to low spawning stock biomass and/or unfavorable environmental conditions.

- Recent management regulations appear to have been effective in improving the population levels of some important fish species, which have recovered or are moving towards recovery from overfishing and/or overfished status.

- No formal stock assessment had been done for the vast majority of fish species in the Gulf of Mexico immediately prior to the Deepwater Horizon event, and subsequently, there is limited knowledge about the status of these fish species (e.g., if they are overfished and/or if overfishing is occurring).

As discussed above, numerous long-term anthropogenic stressors, including fishing pressure, as well as a variety of natural stressors, affect finfish in the Gulf of Mexico. Of the 13 finfish species evaluated in this chapter, five species were being overfished and/or were in the status of overfishing immediately prior to the Deepwater Horizon event. The five species included red snapper, red grouper (some local subpopulations), Atlantic bluefin tuna (most likely but the uncertainty is high), Atlantic blue marlin, and greater amberjack. In addition, many shark species were overfished or were in the status of overfishing immediately before or around April 2010. Of 39 shark species included in the shark Fisheries Management Plan in the Gulf of Mexico, 19 species are listed as commercially and recreationally prohibited species because of very low population biomass and poor stock conditions. Finfish species evaluated in this chapter that were determined to have healthy stocks in the Gulf of Mexico immediately before the Deepwater Horizon event included menhaden, Atlantic swordfish, Atlantic sailfish, red drum, striped mullet, tilefish, king mackerel, and dolphinfish.

ACKNOWLEDGMENTS

BP sponsored the preparation of this chapter. This chapter has been peer reviewed by anonymous and independent reviewers with substantial experience in the subject matter. I thank the peer reviewers, as well as others, who provided assistance with research and the compilation of information. Completing this chapter would not have been possible without the tireless work of Kym Rouse Holzwart, Jonathan Ipock, and Paul Irvin, ENVIRON International Corporation, in obtaining documents, compiling data and information, and preparing maps, graphs, and lists of references.

Small fish images used throughout Chapter 9 are from GulfFINFO (http://gulffishinfo.org/) with the exception of the following: (1) Yellowfin Menhaden (from DM Nelson and ME Pattillo (1992) Distribution and abundance of fishes and invertebrates in Gulf of Mexico estuaries. NOAA National Ocean Service, Rockville, MD, USA. Image available at https://www.flickr. com/photos/internetarchivebookimages/20786380240/in/photolist-xXtgv2-xV8FtW-xEPCEd, accessed 14 December 2016); (2) Finescale Menhaden (from http://txmarspecies.tamug.edu/ fishdetails.cfm?scinameID=Brevoortia%20gunteri, accessed 13 December 2016); (3) Atlantic Bluefin Tuna (NOAA FishWatch, http://www.fishwatch.gov/profiles/western-atlantic-bluefin-tuna, accessed 14 December 2016); (4) Blue Marlin (Oceloti, 2014, iStock image at http://www. istockphoto.com/vector/blue-marlin-fish-gm505255597-44750310?clarity=false, accessed 5

December 2016); and (5) Atlantic Sailfish (Szabo D, 2012, iStock image at http://www.istock-photo.com/vector/atlantic-sailfish-gm156019592-13474335?clarity=false; accessed 13 December 2016).

REFERENCES

Able KW, Grimes CB, Cooper RA, Uzmann JR (1982) Burrow construction and behavior of tilefish, *Lopholatilus chamaeleonticeps*, in the Hudson Submarine Canyon. Environ Biol Fish 7:199–205

Addis D, Chagaris D, Cooper W, Mahmoudi B, Muller RG, Munyandorero J, Murphy MD, O'Hop J (2011) Florida's inshore and nearshore species: 2010 status and trends report. Florida Fish and Wildlife Conservation Commission, Fish Wildlife Research Institute, St. Petersburg, FL, USA. March. 334 p

Ahrenholz DW (1981) Recruitment and exploitation of Gulf menhaden, *Brevoortia patronus*. Fish Bull 79:325–335

Ahrenholz DW (1991) Population biology and life history of the North American menhadens, *Brevoortia* spp. Mar Fish Rev 53:3–19

Allen GR (1985) FAO species catalogue: An annotated and illustrated catalogue of Lutjanid species known to date. FAO fisheries synopsis no. 125, vol 6. Food and Agriculture Organization of the United Nations, Rome, LZ, Italy, 208 p

Alvarado-Bremer JR, Mejuto J, Gomez-Marquez J, Pla-Znuy C, Vinas J, Marques C, Grieg TW (2007) Genetic population structure of the Atlantic swordfish: Current status and future directions. Collect Vol Sci Pap ICCAT 6:107–118

Anderson WW (1958) Larval development, growth, and spawning of striped mullet (*Mugil cephalus*) along the South Atlantic Coast of the United States. Fish Bull 58:501–519

Anderson JD (2007) Systematics of the North American menhadens: Molecular evolutionary reconstructions in the genus *Brevoortia* (Clupeiformes: Clupeidae). Fish Bull 105:368–378

Anderson JD, Karel WJ (2007) Genetic evidence for asymmetric hybridization between men-hadens (*Brevoortia* spp.) from Peninsular Florida. J Fish Biol 71:235–249

Anderson JD, McDonald DL (2007) Morphological and genetic investigations of two western Gulf of Mexico menhadens (*Brevoortia* spp.). J Fish Biol 70:139–147

Andovora F, Pipitone C (1997) Food and feeding habits of the amberjack, *Seriola dumerili*, in the central Mediterranean Sea during the spawning season. Cah Biol Mar 38:9196

Arnold EL Jr, Thompson JR (1958) Offshore spawning of the striped mullet (*Mugil cephalus*) in the Gulf of Mexico. Copeia 1958:130–132

Arocha F (1997) The reproductive dynamics of swordfish *Xiphias gladius* L. and management implications in the Northwestern Atlantic. PhD Thesis, University of Miami, Coral Gables, FL, USA, 383 p

Arrenguín-Sánchez F, Cabrera MA, Aguilar FA (1995) Population dynamics of the king mackerel (*Scomberomorus cavalla*) of the Campeche Bank, Mexico. Sci Mar 59:637–645

Atlantic Bluefin Tuna Status Review Team (2011) Status review report of Atlantic bluefin tuna (*Thunnus thynnus*). Report to National Marine Fisheries Service, Northeast Regional Office, March 22, 104 p

Backus S, Durfee CG III, Mourou G, Kapteyn HC, Murnane MM (1997) 0.2-TW laser system at 1kHz. Opt Lett 22:1256–1258

Baglin RE Jr (1982) Reproductive biology of western Atlantic bluefin tuna (*Thunnus thynnus*). Fish Bull 80:121–134

Barans CA, Stender BW (1993) Trends in tilefish distribution and relative abundance off South Carolina and Georgia. Trans Am Fish Soc 122:165–178

Barry KP (2002) Feeding habits of Blacktip Sharks, *Brevoortia gunteri*, and Atlantic Sharpnose Sharks, *Rhizoprionodon terraenovae*, in Louisiana Coastal Waters. MS Thesis, Louisiana State University, Baton Rouge, LA, USA, 72 p

Bartlett MR, Haedrich RL (1968) Neuston nets and South Atlantic blue marlin *Makaira nigricans*. Copeia 1968:469–474

Bass RJ, Avault JW Jr (1975) Food habit, length-weight relationship, condition factor, and growth of juvenile red drum, *Sciaenops ocellatus*, in Louisiana. Trans Am Fish Soc 104:35–45

Baum JK, Myers RA (2004) Shifting baselines and the decline of pelagic sharks in the Gulf of Mexico. Ecol Lett 7:135–145

Bayley RE, Prince ED (1993) A review of tag release and recapture files for Istiophoridae from the Southeast Fisheries Center's cooperatives game fish tagging program, 1954 to present. Collect Vol Sci Pap ICCAT 41:527–548

Beardsley GL Jr (1967) Age, growth, and reproduction of the dolphin, *Coryphaena hippurus*, in the Straits of Florida. Copeia 1967:441–451

Beardsley GL Jr, Merrett NR, Richards WJ (1975) Synopsis of the biology of the Sailfish *Istiophorus platypterus* (Shaw and Nodder, 1791). In: Proceedings, International Billfish Symposium, Kailua-Kona, HI, USA, August, Part 3, pp 95–120

Beasley M (1993) Age and growth of greater amberjack, *Seriola dumerili*, from the northern Gulf of Mexico. MS Thesis, Louisiana State University, Baton Rouge, LA, USA, 85 p

Beaumariage DS (1969a) Returns from the 1965 Schlitz Tagging Program including a cumulative analysis of previous results. Florida Board of Conservation, Marine Research Lab, Technical Series No. 59. Florida Department of Natural Resources, Division of Marine Resources, St. Petersburg, FL, USA, 38 p

Beaumariage DS (1969b) Current status of biological investigations of Florida's Mackerel Fisheries. In: Proceedings, 22nd Annual Meeting of the Gulf and Caribbean Fisheries Institute, Coral Gables, FL, USA, May, pp 79–86

Beaumariage DS (1973) Age, growth, and reproduction of King Mackerel, *Scomberomorus cavallas*, in Florida. Florida Marine Research Lab, Publication No. I. Florida Department of Natural Resources, Division of Marine Resources, St. Petersburg, FL, USA, 45 p

Beaumariage DS, Bullock LH (1976) Biological research on snappers and groupers as related to fishery management requirements. In: Bullis HR Jr, Jones AC (eds) Colloquium on snapper-grouper fishery resources of the western Central Atlantic Ocean. Florida Sea Grant College Report No. 17. St. Petersburg, FL, USA, pp 86–94

Beaumariage DS, Wittich AC (1966) Returns from the 1964 Schlitz tagging program. Florida Board of Conservation, Marine Research Lab, Technical series no. 47. Florida Department of Natural Resources, Division of Marine Resources, St. Petersburg, FL, USA, 50 p

Bentivoglio AA (1988) Investigations into the growth, maturity, mortality rates and occurrence of the Dolphin (*Coryphaena hippurus*) in the Gulf of Mexico. MS Thesis, University College of North Wales, Bangor, Wales, UK, 37 p

Bethea DM, Hollensead L, Carlson JK, Ajemian MJ, Grubbs RD, Hoffmayer ER, Del Rio R, Peterson GW, Baltz DM, Romine J (2009) Shark nursery grounds and essential fish habitat studies: Gulfspan Gulf of Mexico FY'08 Cooperative Gulf of Mexico States shark pupping and nursery survey. Contribution report PCB-08/02 to the National Oceanic and Atmospheric Administration, Highly Migratory Species Division, Narragansett, RI, USA, 64 p

Bigelow HB, Schroeder WC (1953) Fishes of the Gulf of Maine. Fish Bull 54:1–577

Biggs DC (1992) Nutrients, plankton, and productivity in a warm-core ring in the western Gulf of Mexico. J Geophys Res 97:2143–2154

Bishop JM, Miglarese JV (1978) Carnivorous feeding in adult striped mullet. Copeia 1978:705–707

Block B (2011) Blacktip reef shark. iStockphoto, Calgary, Alberta, Canada. http://www.istockphoto. com/stock-photo-7471636-blacktip-reef-shark.php?st=808bdc3. Accessed 5 December 2016

Block BA, Booth DT, Carey FG (1992) Depth and temperature of the blue marlin, *Makaira nigricans*, observed by acoustic telemetry. Mar Biol 114:175–183

Block BA, Dewar H, Blackwell SB, Williams TD, Prince ED, Boustany A, Farwell CJ, Dau DJ, Seitz A (2001a) Archival and pop-up satellite tagging of Atlantic bluefin tuna. In: Sibert JR, Nielsen JL (eds) Electronic tagging and tracking in marine fisheries. Kluwer Academic, Dordrecht, The Netherlands, pp 65–88

Block BA, Dewar H, Blackwell SB, Williams TD, Prince ED, Farwell CJ, Boustany A, Teo SLH, Seitz A, Walli A, Fudge D (2001b) Migratory movements, depth preferences, and thermal biology of Atlantic bluefin tuna. Science 293:1310–1314

Block BA, Teo SLH, Walli A, Boustany A, Stokesbury MJW, Farwell CJ, Weng KC, Dewar H, Williams TD (2005) Electronic tagging and population structure of Atlantic Bluefin Tuna. Nature 34:1121–1127

Boland GS, Gallaway BJ, Baker JS, Lewbel GS (1983) Ecological effects of energy development on Reef Fish of the Flower Garden Banks. Report No. NOAA-83120106. NTIS PB84-128289, Contract No. NA80-GA-C-00057. National Marine Fisheries Service, Galveston, TX, USA, 466 p

Boothby RN, Avault WJ Jr (1971) Food habits, length-weight relationship, and condition factor of the red drum (*Sciaenops ocellata*) in southeastern Louisiana. Trans Am Fish Soc 100:290–295

Bortone SA, Hastings PA, Collard SB (1977) The pelagic-*Sargassum* ichthyofauna of the eastern Gulf of Mexico. Northeast Gulf Sci 1:60–67

Boustany AM, Reeb CA, Teo SL, De Metrio G, Block BA (2007) Genetic data and electronic tagging indicate that the Gulf of Mexico and Mediterranean Sea are reproductively isolated stocks of bluefin tuna (*Thunnus thynnus*). Collect Vol Sci Pap ICCAT 60:1154–1159

Boustany AM, Reeb CA, Block BA (2008) Mitochondrial DNA and electronic tracking reveal population structure of Atlantic bluefin tuna (*Thunnus thynnus*). Mar Biol 156:13–24

Bradley E, Bryan CE (1975) Life history and fishery of the red snapper (*Lutjanus campechanus*) in the Northwestern Gulf of Mexico. In: Proceedings, 27th annual Gulf and Caribbean Fisheries Institute and the 17th annual international game fish research conference, Miami, FL, USA, November, pp 77–106

Branstetter S (1987) Age and growth estimates for blacktip, *Brevoortia gunteri*, and spinner, *C. brevipinna*, sharks from the northwestern Gulf of Mexico. Copeia 1987:964–974

Breder CM Jr (1940) The spawning of *Mugil cephalus* on the Florida West Coast. Copeia 1940:138–139

Breuer JP (1957) An ecological survey of Baffin and Alazan Bays, Texas. Publ Inst Mar Sci Univ Texas 4:134–155

Briggs JC (1974) Marine zoogeography. McGraw-Hill Companies, New York, NY, USA, 475 p

Broadhead GC (1953) Investigations of the Black Mullet, *Mugil cephalus* L., in Northwest Florida. Florida Board of Conservation, Marine Research Lab, Technical Series No. 7. Florida Department of Natural Resources, Division of Marine Resources, St. Petersburg, FL, USA, 33 p

Broadhead GC (1958) Growth of the Black Mullet, (*Mugil cephalus* L.) in West and Northwest Florida. Florida Board of Conservation, Marine Research Lab, Technical Series No. 25.

Florida Department of Natural Resources, Division of Marine Resources, St. Petersburg, FL, USA, 29 p

Broadhead GC, Mefford HP (1956) The migration and exploitation of the Black Mullet, *Mugil cephalus*, Linnaeus, in Florida as determined from tagging during 1949–1953. Florida Board of Conservation, Marine Research Lab, Technical Series No. 18. Florida Department of Natural Resources, Division of Marine Resources, St. Petersburg, FL, USA, 31 p

Brothers EB, Prince ED, Lee DW (1983) Age and growth of young-of-the-year Bluefin Tuna, *Thunnus thynnus*, from Otolith Microstructure. In: Proceedings, International Workshop on Age Determination of Oceanic Pelagic Fishes: Tunas, Billfishes and Sharks, Miami, FL, USA, February, pp 49–59

Brown-Peterson NJ, Franks JS, Comyns BH, Hendon LA, Hoffmayer ER, Hendon JR, Waller RS (2004) Aspects of the reproduction of large pelagic fishes in the Northern Gulf of Mexico. In: Proceedings, 55th Annual Meeting of the Gulf and Caribbean Fisheries Institute, Xel Ha, Quintana Roo, Mexico, November, pp 1016–1017

Brown-Peterson NJ, Franks JS, Comyns BH, McDowell JR (2008) Do Blue Marlin spawn in the Northern Gulf of Mexico. In: Proceedings, 60th Annual Meeting of the Gulf and Caribbean Fisheries Institute, Punta Cana, Dominican Republic, November, pp 372–378

Brule T, Deniel C, Colas-Marrufo T, Sanchez-Crespo M (1999) Red grouper reproduction in the southern Gulf of Mexico. Trans Am Fish Soc 128:385–402

Buckley J (1984) Habitat suitability index models: Larval and juvenile red drum. FWS/OBS 82/10.74, U.S. Department of the Interior, Fish and Wildlife Service, Washington, DC, USA, 15 p

Bullock LH, Smith GB (1991) Seabasses (Pisces: Serranidae), Memoirs of the Hourglass Cruises, vol VIII, Part II. Florida Marine Research Institute, Department of Natural Resources, St. Petersburg, FL, USA, 243 p

Bumguardner BW, Anderson JD (2008) Age and growth, reproduction and genetics of Billfish in Gulf of Mexico Waters off Texas. In: Proceedings, Atlantic Billfish Research Program Symposium-Gulf States Marine Fisheries Commission Spring Meeting, Galveston, TX, USA, March, pp 52–66

Burgess GH, Beerkircher LR, Cailliet GM, Carlson JK, Cortés E, Goldman KJ, Simpfendorfer CA (2005) Is the collapse of shark populations in the Northwest Atlantic Ocean and Gulf of Mexico real? Fish 30:19–26

Burgos JM (2001) Life history of the Red Grouper (*Epinephelus morio*) off the North Carolina and South Carolina Coast. University of Charleston, Charleston, SC, USA, 90 p

Burks CM, Driggers WB III, Mullin KD (2006) Abundance and distribution of whale sharks (*Rhincodon typus*) in the northern Gulf of Mexico. Fish Bull 104:570–584

Burns KM (1981) Seasonal and areal distribution of scombrid larvae in the vicinity of Palm Beach, Florida. MA Thesis, University of South Florida, Tampa, FL, USA, 66 p

Buskey EJ, Hyatt CJ (1995) Effects of The Texas (USA) "brown tide" alga on planktonic grazers. Mar Ecol Prog Ser 126:285–292

Butler MJA, Caddy JF, Dickson CA, Hunt JJ, Burnett CD (1977) Apparent age and growth, based on otolith analysis of giant bluefin tuna (*Thunnus thynnus*) in the 1975–1976 Canadian catch. Collect Vol Sci Pap ICCAT 5:318–330

Cailliet GM, Natanson LJ, Welden BA, Ebert DA (1985) Preliminary studies on the age and growth of the white shark, *Carcharodon carcharias*, using vertebral bands. Mem South Calif Acad Sci 9:49–60

Camper JD, Barber RC, Richardson LR, Gold JR (1993) Mitochondrial DNA variation among red snapper (*Lutjanus campechanus*) from the Gulf of Mexico. Mol Mar Biol Biotechnol 2:154–161

Carey FG (1990) Further acoustic telemetry observations of swordfish. Mar Recreat Fish 13:103–122

Carey FG, Robison BH (1981) Daily patterns in the activities of swordfish, *Xiphias gladius*, observed by acoustic telemetry. Fish Bull 79:277–292

Carlson JK, Baremore IE (2003) Changes in biological parameters of Atlantic sharpnose shark *Rhizoprionodon terraenovae* in the Gulf of Mexico: Evidence for density-dependent growth and maturity? Mar Freshw Res 54:227–234

Carlson JK, Parsons GR (1997) Age and growth of the bonnethead shark, *Sphyrna tiburo*, from Northwest Florida, with comments on clinal variation. Environ Biol Fish 50:331–341

Carlson JK, Cort E, Johnson AG (1999) Age and growth of the blacknose shark, *Carcharhinus acronotus*, in the eastern Gulf of Mexico. Copeia 1999:684–691

Carlson JK, Bethea DM, Middlemiss A, Baremore IE (2004) Shark nursery grounds and essential fish habitat studies, Gulfspan Gulf of Mexico-FY04. An internal report to NOAA's Highly Migratory Species Office, Sustainable Fisheries Division Contribution No. PCB-04/06. Panama City Beach, FL, USA, 21 p

Carlson JK, Sulikowski JR, Baremore IE (2006) Do differences in life history exist for blacktip sharks, *Carcharhinus limbatus*, from the United States South Atlantic Bight and eastern Gulf of Mexico? Environ Biol Fish 77:279–292

Carlsson J, McDowell JR, Carlsson JE, Graves JE (2007) Genetic identity of YOY bluefin tuna from the eastern and western Atlantic spawning areas. J Hered 98:23–28

Carpenter KE (2002) FAO species identification guide for fishery purposes: The living marine resources of the Western Central Atlantic: Bony fishes, Part 2 (Opistognathidae to Molidae), sea turtles and marine mammals. American Society of Ichthyologists and Herpetologists Special Publication No. 5, vol 3. Food and Agriculture Organization of the United Nations, Rome, LZ, Italy, pp 1375–2127

Casey JG, Pratt HL Jr (1985) Distribution of the white shark, *Carcharodon carcharias*, in the western North Atlantic. Mem South Calif Acad Sci 9:2–14

Castillo-Rivera M, Kobelkowsky A, Zamayoa V (1996) Food resource partitioning and trophic morphology of *Brevoortia gunteri* and *B. patronus*. J Fish Biol 49:1102–1111

Castro JI (1983) The sharks of North American Waters. Texas A&M University Press, College Station, TX, USA, 180 p

Castro JI (1996) Biology of the blacktip shark, *Brevoortia gunteri*, off the southeastern United States. Bull Mar Sci 59:508–522

Castro Aguirre JL (1978) Catalogo Sistematico Peces Marinos Pentran a los Advas Contenentals de Mexico con Aspectos Geograficos Etologicos. Cientifica No. 19, Instituto Nacional de Pesca, Benito Juárez, Mexico, 298 p

Cervigon F (1966) Los Peces Marinos de Venezuela. Fundacion La Salle de Ciencias Naturales. Caracas, Venezuela, 951 p

Chase BC (2002) Differences in the diet of Atlantic bluefin tuna (*Thunnus thynnus*) at five seasonal feeding grounds on the New England continental shelf. Fish Bull 100:168–180

Chesney EJ, Baltz DM, Thomas RG (2000) Louisiana estuarine and coastal fisheries and habitats: Perspectives from a fish's eye view. Ecol Appl 10:350–366

Chipman WA (1959) Use of radioisotopes in studies of the foods and feeding activities of marine animals. Pubbl Stn Zool Napoli II 31(Suppl):154–175

Chow S, Takeyama H (2000) Nuclear and mitochondrial DNA analyses reveal four genetically separated breeding units of the swordfish. J Fish Biol 56:1087–1098

Christmas JY Jr, Gunter G (1960) Distribution of menhaden, genus *Brevoortia*, in the Gulf of Mexico. Trans Am Fish Soc 89:338–343

Christmas JY, Waller RS (1973) Mississippi: Estuarine vertebrates. In: Christmas JY (ed) Cooperative Gulf of Mexico Estuarine Inventory and Study, Mississippi. Gulf Coast Research Laboratory, Ocean Springs, MS, USA, pp 320–434

Christmas JY, Waller RS (1975) Location and time of menhaden spawning in the Gulf of Mexico. Gulf Coast Research Laboratory, Ocean Springs, MS, USA, 20 p

Christmas JY, McBee JT, Waller RS, Sutter FC, III (1982) Habitat suitability index models: Gulf menhaden. FWS/OBS 82/10.23. U.S. Department of Interior, Fish and Wildlife Service, Washington, DC, USA, 23 p

Coleman FC, Koenig CC (2010) The effects of fishing, climate change, and other anthropogenic disturbances on red grouper and other reef fishes in the Gulf of Mexico. Integr Comp Biol 50:201–212

Coleman FC, Koenig CC, Collins LA (1996) Reproductive styles of shallow-water groupers (Pisces: Serranidae) in the eastern Gulf of Mexico and the consequences of fishing spawning aggregations. Environ Biol Fish 47:129–141

Collette BB, Nauen CE (1983) FAO species catalogue: An annotated and illustrated catalogue of tunas, mackerels, bonitos and related species known to date. FAO Fisheries Synopsis No. 125, vol 2. Food and Agriculture Organization of the United Nations, Rome, LZ, Italy, 137 p

Collette BB, Russo JL (1984) Morphology, systematics, and biology of the Spanish mackerels (*Scomberomorus*, Scombridae). Fish Bull 82:545–689

Collette BB, Reeb C, Block B (2001) Systematics of the tuna and mackerels (Scombridae). In: Block BA, Stevens ED (eds) Tuna: Physiology, ecology, and evolution. Academic Press, San Diego, CA, USA, pp 5–35

Collette B, Acero A, Amorim AF, Boustany A, Canales RC, Cardenas G, Carpenter N, Di Natale A, Fox W, Fredou FL, Graves J, Viera Hazin FH, Juan Jorda M, Minte Vera C, Miyabe N, Cruz R, Nelson R, Oxenford H, Schaefer K, Serra R, Sun C, Teixeira RP, Pires F, Travassos PE, Uozumi Y, Yanez E (2012) *Coryphaena hippurus*. In: IUCN (International Union for the Conservation of Nature). 2012 IUCN red list of threatened species, Version 2012.2. http://www.iucnredlist.org/. Accessed 31 May 2013

Collins MR (1981) The feeding periodicity of striped mullet *Mugil cephalus*, Linnaeus, in two Florida habitats. J Fish Biol 19:307–315

Collins MR (1985) Species profiles: Life histories and environmental requirements of coastal fishes and invertebrates (South Florida), striped mullet. U.S. Fish and Wildlife Service Biology Report 82(11.34), TR EL-82-4. U.S. Army Corps of Engineers, Coastal Ecology Group, Waterways Experiment Station, Vicksburg, MS, USA. 11 p

Collins LA, Johnson AG, Keim CP (1996) Spawning and annual fecundity of the red snapper (*Lutjanus campechanus*) from the northeastern Gulf of Mexico. In: Arreguín-Sánchez F, Munro JL, Balgos MC, Pauly D (eds) Biology, fisheries and culture of tropical groupers and snappers. ICLARM Conference Proceedings 48, Campeche, Mexico, pp 174–188

Collins LA, Fitzhugh GR, Mourand L, Lombardi LA, Walling WT Jr, Fable WA, Burnett MR, Allman RJ (2001) Preliminary results from a continuing study of spawning and fecundity in the red snapper (Lutjanidae: *Lutjanus campechanus*) from the Gulf of Mexico, 1998–1999. In Proceedings, 52nd Annual Meeting of the Gulf and Caribbean Fisheries Institute, Key West, FL, USA, November, pp 34–47

Collins LA, Fitzhugh GR, Lombardi-Carlson LA, Lyon HM, Walling WT, Oliver DW (2002) Characterization of Red Grouper (Serranidae: *Epinephelus morio*) reproduction from the Eastern Gulf of Mexico. Panama City Laboratory Contribution Series 2002-07, National Marine Fisheries Service, Southeastern Fisheries Science Center, Panama City Beach, FL, USA, 20 p

Combs RM (1969) Embryogenesis, histology and organology of the ovary of *Brevoortia patronus*. Gulf Res Rep 2:333–434

Compagno LJV (1984) FAO species catalogue: An annotated and illustrated catalogue of shark species known to date, Part 2: Carcharhiniformes. FAO Fisheries Synopsis No. 12, vol 4. Food and Agriculture Organization of the United Nations, Rome, LZ, Italy, pp 251–655

Cook M, Barnett BK, Duncan MS, Allman RJ, Porch CE, Fitzhugh GR (2009) Characterization of red snapper (*Lutjanus campechanus*) size and age at sexual maturity for the 2009 Gulf of Mexico SEDAR Update, Draft working document. Panama City Laboratory Contribution Series 2009–16, National Marine Fisheries Service, Southeastern Fisheries Science Center, Panama City Beach, FL, USA, August, 16 p

Copeland BJ, Bechtel TJ (1974) Some environmental limits of six Gulf Coast estuarine organisms. Contrib Mar Sci 18:169–203

Cort JL (1991) Age and growth of the bluefin tuna, *Thunnus thynnus* (L.) of the Northeast Atlantic. Collect Vol Sci Pap ICCAT 35:213–230

Cortés E (2008) Catches of pelagic sharks from the western North Atlantic Ocean, including the Gulf of Mexico and Caribbean Sea. Collect Vol Sci Pap ICCAT 62:1434–1446

Cowan JH, Grimes CB, Patterson WF, Walters CJ, Jones AC, Lindberg WJ, Rose KA (2011) Red snapper management in the Gulf of Mexico: Science-or faith-based? Rev Fish Biol Fish 21:87–204

Cramer J, Scott G (1998) Summarization of catch and effort in the pelagic longline fishery and analysis of the effect of two degree square closures on swordfish and discards landings. Sustainable Fisheries Division Contribution MIA-97/98-17, Southeast Fisheries Science Center, Miami, FL, USA, 22 p

Crane J (1936) Notes on the biology and ecology of giant tuna *Thunnus thynnus*, L., observed at Portland, Maine. Zoologica 21:207–212

Cranswick D, Regg J (1997) Deep-water in the Gulf of Mexico: America's new frontier. OCS Rep. 97-0004. U.S. Department of the Interior, Minerals Management Service, New Orleans, LA, USA, 41 p

CRFM (Caribbean Regional Fisheries Mechanism) (2006) Report of the Second Annual Scientific Meeting—Port of Spain, Trinidad and Tobago, 13–22 March 2006, vol 1, Suppl 1. CRFM Secretariat, Belize and St. Vincent and the Grenadines, 48 p

crisod (2013) White shark. iStockphoto, Calgary, Alberta, Canada. http://www.istockphoto.com/stock-photo-17993543-white-shark.php?st=33b18. Accessed 5 December 2016

Dahlberg MD (1970) Atlantic and Gulf of Mexico menhadens, genus *Brevoortia* (Pisces: Clupeidae). Bull Fla State Mus Biol Sci 15:91–162

Damalas D, Megalofonou P, Apostolopoulou M (2007) Environmental, spatial, temporal and operational effects on swordfish (*Xiphias gladius*) catch rates of eastern Mediterranean Sea longline fisheries. Fish Res 84:233–246

Darnell RM (1958) Food habits of fishes and larger vertebrates of Lake Pontchartrain, Louisiana, an estuarine community. Publ Inst Mar Sci Univ Texas 5:354–416

Dascher C (2013) Great white shark [Internet]. iStockphoto, Calgary, Alberta, Canada. http://www.istockphoto.com/stock-photo-7436967-great-white-shark.php?st=290490c. Accessed 5 December 2016

Davies JH, Bortone SA (1976) Partial food list of 3 species of Istiophoridae (Pisces) from the northeastern Gulf of Mexico. Fla Sci 39:249–253

Davis TL (1991) Advection, dispersion and mortality of a patch of southern bluefin tuna larvae *Thunnus maccoyii* in the East Indian Ocean. Mar Ecol Prog Ser 73:33–45

De Metrio G, Arnold GP, de la Serna JM, Block BA, Megalofonou P, Lutcavage M, Oray I, Deflorio M (2005) Movements of bluefin tuna (*Thunnus thynnus* L.) tagged in the Mediterranean Sea with pop-up satellite tags. Collect Vol Sci Pap ICCAT 58:1337–1340

De Silva SS (1980) Biology of juvenile grey mullet: A short review. Aquaculture 19:21–36

De Sylva DP, Breder PR (1997) Reproduction, gonad histology, and spawning cycles of North Atlantic billfishes (Istiophoridae). Bull Mar Sci 60:668–697

De Sylva DP, Richards WJ, Capo TR, Serafy JE (2000) Potential effects of human activities on billfishes (Istiophoridae and Xiphiidae) in the western Atlantic Ocean. Bull Mar Sci 66:187–198

Deegan LA (1985) The population ecology and nutrient transport in Gulf Menhaden in Fourleague Bay, Louisiana. PhD Thesis, Louisiana State University, Baton Rouge, LA, USA, 134 p

DeepAqua (2010) Bluefin tuna in net. iStockphoto, Calgary, Alberta, Canada. http://www.istockphoto.com/photo/bluefin-tuna-in-net-gm95662678-11554516?clarity=false. Accessed December 4, 2016

DeVries DA, Grimes CB (1997) Spatial and temporal variation in age and growth of king mackerel, *Scomberomorus cavallas*, 1977–1992. Fish Bull 95:694–708

DeVries DA, Grimes CB, Lang KL, White BDW (1990) Age and growth of king and Spanish mackerel larvae and juveniles from the Gulf of Mexico and U.S. South Atlantic Bight. Environ Biol Fish 29:135–143

DeVries DA, Brusher JH, Fitzhugh GR (2006) Spatial and temporal patterns in demographics and catch rates of red grouper from a fishery-independent trap survey in the Northeast Gulf of Mexico, 2004–2005. SEDAR 12-DW-08, Panama City Laboratory Contribution Series 2006-13, National Marine Fisheries Service, Southeastern Fisheries Science Center, Panama City Beach, FL, USA, 7 p

Ditty JG, Shaw RF (1996) Spatial and temporal distribution of larval striped mullet (*Mugil cephalus*) and white mullet (*Mugil curema*, Family: Mugilidae) in the northern Gulf of Mexico, with notes on mountain mullet, *Gonostomus monticola*. Bull Mar Sci 59:271–288

Ditty JG, Shaw RF, Grimes CB, Cope JS (1994) Larval development, distribution, and abundance of common dolphin, *Coryphaena hippurus*, and pompano dolphin, *C. equiselis* (Family: Coryphaenidae), in the northern Gulf of Mexico. Fish Bull 92:275–291

Dombrowski B (2012) Red grouper swimming in coral reef underwater. iStockphoto, Calgary, Alberta, Canada. http://www.istockphoto.com/photo/red-grouper-swimming-in-coral-reef-underwater-gm139547530-377657. Accessed December 4, 2016

Dooley JK (1978) Systematics and biology of the tilefishes (Perciformes: Branchiostegidae and Malacanthidae), with description of two new species. NOAA Technical Report Circular No. 411. National Oceanic and Atmospheric Administration, National Marine Fisheries Service, Washington, DC, USA, 78 p

Draganik B, Cholyst J (1988) Temperature and moonlight as stimulators for feeding activity by swordfish. Collect Vol Sci Pap ICCAT 27:305–314

Dragovich A (1970) The food of bluefin tuna (*Thunnus thynnus*) in the western North Atlantic Ocean. Trans Am Fish Soc 99:723–731

Driggers W, Carlson J, Cullum B, Dean J, Oakley D (2004) Age and growth of the blacknose shark, *Carcharhinus acronotus*, in the western North Atlantic Ocean with comments on regional variation in growth rates. Environ Biol Fish 71:171–178

Duarte-Neto P, Lessa R, Stosic B, Morize E (2008) The use of sagittal otoliths in discriminating stocks of common dolphinfish (*Coryphaena hippurus*) off northeastern Brazil using multi-shape descriptors. ICES J Mar Sci 65:1144–1152

Ebert DA (2003) Sharks, rays and chimaeras of California. University of California Press, Berkeley, CA, USA, 287 p

Eggleston DB, Bochenek EA (1990) Stomach contents and parasite infestation of school bluefin tuna, *Thunnus thynnus*, collected from the Middle Atlantic Bight, Virginia. Fish Bull 88:389–395

Eggold BT, Motta PJ (1992) Ontogenetic dietary shifts and morphological correlates in striped mullet, *Mugil cephalus*. Environ Biol Fish 34:139–158

Erickson DL, Harris MJ, Grossman GD (1985) Ovarian cycling of tilefish, *Lopholatilus chamaeleonticeps* Goode and Bean, from the South Atlantic Bight, USA. J Fish Biol 27:131–146

Etnier DA, Starnes WC (1993) The fishes of Tennessee. University of Tennessee Press, Knoxville, TN, USA, 681 p

Etzold DJ, Christmas JY (1979) A Mississippi marine finfish management plan. MASGP-78-046, Mississippi-Alabama Sea Grant Consortium, Ocean Springs, MS, USA, 36 p

Fable WA Jr, Trent L, Bane GW, Ellsworth SW (1987) Movements of king mackerel, *Scomberomorus cavallas*, tagged in Southeast Louisiana, 1983–1985. Mar Fish Rev 49:98–101

Fahay MP (1983) Guide to the early stages of marine fishes occurring in the western North Atlantic Ocean, Cape Hatteras to the southern Scotian Shelf. J Northw Atl Fish Sci 4:1–423

Fahay MP, Berrien P (1981) Preliminary description of larval tilefish (*Lopholatilus chamaeleonticeps*). Rapp PV Cons Int Explor Mer 178:600–602

FFWCC (Florida Fish and Wildlife Conservation Commission) (2013a) Eagle rays: Cownose rays. http://myfwc.com/research/saltwater/sharks-rays/ray-species/cownose-ray/. Accessed 31 May 2013

FFWCC (2013b) Manta rays: Giant Manta. http://myfwc.com/research/saltwater/sharks-rays/ray-species/giant-manta/. Accessed 31 May 2013

FFWCC (2013c) Sawfishes: Smalltooth Sawfish. http://myfwc.com/research/saltwater/sharks-rays/ray-species/smalltooth-sawfish/. Accessed 31 May 2013

Finucane JH, Collins LA, Barger LE (1978) Spawning of the striped mullet, *Mugil cephalus*, in the northwestern Gulf of Mexico. Northeast Gulf Sci 2:148–151

Finucane JH, Collins LA, Brusher HA, Saloman CH (1986) Reproductive biology of king mackerel, *Scomberomorus cavallas*, from the southeastern United States. Fish Bull 84:841–850

Fischer W, Bianchi G, Scott WB (eds) (1978) FAO species identification sheets for fishery purposes: Western Central Atlantic, Fishing Area 31, vol IV. Food and Agriculture Organization of the United Nations, Rome, LZ, Italy

Fischer AJ, Baker MS Jr, Wilson CA (2004) Red snapper (*Lutjanus campechanus*) demographic structure in the northern Gulf of Mexico based on spatial patterns in growth rates and morphometrics. Fish Bull 102:593–603

FishBase (2013) http://fishbase.org/. Accessed 31 May 2013

Fitzhugh GR, Duncan MS, Collins LA, Walling WT, Oliver DW (2004) Characterization of red snapper (*Lutjanus campechanus*) reproduction for the 2004 Gulf of Mexico SEDAR. SEDAR 7-DW-35, Panama City Laboratory Contribution Series 2004-01. National Marine Fisheries Service, Southeastern Fisheries Science Center, Panama City Beach, FL, USA, 29 p

Fitzhugh GR, Lyon HM, Walling WT, Levins CF, Lombardi-Carlson LA (2006) An update of Gulf of Mexico red grouper reproductive data and parameters for SEDAR 12. SEDAR 12-DW-04, Panama City Laboratory Contribution Series 2006-14, National Marine Fisheries Service, Southeastern Fisheries Science Center, Panama City Beach, FL, USA, 18 p

Fitzhugh GR, Levins CF, Walling WT, Gamby M, Lyon H, DeVries DA (2008) Batch fecundity and an attempt to estimate spawning frequency of king mackerel (*Scomberomorus cavallas*) in U.S. Waters. SEDAR 16-DW-06, Southeast Data, Assessment and Review, North Charleston, SC, USA, 17 p

Fore PL (1970) Oceanic distribution of eggs and larvae of the Gulf menhaden. Report of the Bureau of Commercial Fisheries Biology Laboratory, Beaufort, North Carolina, for the Fiscal Year Ending June 30, 1968, U.S. Fish and Wildlife Service Circular 341, pp 11–13

Fore PL, Baxter KN (1972) Diel fluctuations in the catch of larval Gulf menhaden, *Brevoortia patronus*, at Galveston Entrance, Texas. Trans Am Fish Soc 101:729–732

Francis M (1996) Observations of a pregnant white shark with a review of reproductive biology. In: Klimley AP, Ainley DG (eds) Great White Sharks: The biology of *Carcharodon carcharias*. Academic Press, San Diego, CA, USA, pp 157–172

Franks JS (2000) A review: Pelagic fishes at petroleum platforms in the Northern Gulf of Mexico; diversity, interrelationships, and perspective. Pêche thonière et dispositifs de concentration de poissons, Caribbean-Martinique, 15–19 Oct 1999

Franks JS (2003) First record of goliath grouper, *Epinephelus itajara*, in Mississippi coastal waters with comments on the first documented occurrence of red grouper, *Epinephelus morio*, off Mississippi. In: Proceedings, 56th Annual Meeting of the Gulf Caribbean Fisheries Institute, Tortola, British Virgin Islands, November, pp 295–306

Freeman BL, Turner SC (1977) Biological and fisheries data on tilefish, *Lopholatilus chamaeleonticeps* Goode and Bean. Sandy Hook Laboratory Technical Series Report No. 5. Northeast Fisheries Science Center, National Marine Fisheries Service, Sandy Hook, NJ, USA, 41 p

Friedland KD, Ahrenholz DW, Smith JW, Manning M, Ryan J (2006) Sieving functional morphology of the gill raker feeding apparatus of Atlantic menhaden. J Exp Zool A Comp Exp Biol 305:974–985

Froese R, Pauly D (eds) (2009) FishBase 2009: Concepts, design and data sources. www.fishbase.org. Accessed 31 May 2013

Fromentin JM (2009) Lessons from the past: Investigating historical data from bluefin tuna fisheries. Fish 10:197–216

Fromentin JM, Powers JE (2005) Atlantic bluefin tuna: Population dynamics, ecology, fisheries and management. Fish 6:281–306

Ftlaudgirl (2016a) Redfish is swimming in the grass flats ocean. Bigstockphoto, New York, NY, USA. http://www.bigstockphoto.com/image-41736652/stock-photo-redfish-is-swimming-in-the-grass-flats-ocean. Accessed 5 December 2016

Ftlaudgirl (2016b) Mahi mahi swimming underwater in blue ocean. Bigstockphoto, New York, NY, USA. http://www.bigstockphoto.com/image-33299180/stock-photo-mahi-mahi-swimming-underwater-in-blue-ocean. Accessed 5 December 2016

Furnestin J, Dardignac J (1962) Le thon rouge du maroc atlantique (*Thunnus thynnus* Linn,). Rev Trav Inst Marit 26:382–398

Futch C (1966) Lisa—The Florida black mullet. Florida Board of Conservation, Marine Research Laboratory, Leaflet Series 6, St. Petersburg, FL, USA, 6 p

Futch C (1976) Biology of striped mullet. In: Cato JC, McCullough WE (eds) Economics, biology, and food technology of mullet. Florida Sea Grant Program Report 15, Gainesville, FL, USA, pp 63–69

Futch RB, Burger GE (1976) Age, growth, and production of red snapper in Florida waters. In: Bullis HR Jr, Jones AC (eds) Colloquium on Snapper-Grouper fishery resources of the Western Central Atlantic Ocean. Florida Sea Grant Program Report 17, Gainesville, FL, USA, pp 165–184

Gallaway BJ (1981) An ecosystem analysis of oil and gas development on the Texas-Louisiana Continental Shelf. FWS/OBS-81/27, U.S. Fish and Wildlife Service, Office of Biological Services, Washington, DC, USA, 89 p

Gallaway BJ, Lewbel GS (1982) The ecology of petroleum platforms in the northwestern Gulf of Mexico: A community profile. FWS/OBS 82/27, U.S. Fish and Wildlife Service, 92 p

Gallaway BJ, Cole JG, Meyer R, Roscigno P (1999) Delineation of essential habitat for juvenile red snapper in the northwestern Gulf of Mexico. Trans Am Fish Soc 128:713–726

Gallaway BJ, Gazey WJ, Cole JG, Fechhelm RG (2007) Estimation of potential impacts from offshore liquefied natural gas terminals on red snapper and red drum fisheries in the Gulf of Mexico: an alternative approach. Trans Am Fish Soc 136:655–677

Gallaway BJ, Szedlmayer ST, Gazey WJ (2009) A life history review for red snapper in the Gulf of Mexico with an evaluation of the importance of offshore petroleum platforms and other artificial reefs. Rev Fish Sci 17:48–67

Garcia A, Alemany F, de la Serna JM, Oray I, Karakulak S, Rollandi L, Arigo A, Mazzola S (2005) Preliminary results of the 2004 bluefin tuna larval surveys off different Mediterranean sites (Balearic Archipelago, Levantine Sea, and the Sicilian Channel). Collect Vol Sci Pap ICCAT 58:1420–1428

García-Cortés B, Mejuto J, Quintans M (2003) Summary of swordfish (*Xiphias gladius*) recaptures carried out by the Spanish surface longline fleet in the Atlantic Ocean: 1984–2002. Collect Vol Sci Pap ICCAT 55:1476–1484

Gesteira TCV, Mesquita ALL (1976) Epoca de reproducao tamanho e idade na primeira desova da cavala e da serra, na costa do Estado do Ceara (Brasil). Arq Cienc Mar 16:83–86

Gibbs RH, Collette BB (1959) On the identification, distribution, and biology of the dolphins, *Coryphaena hippurus* and *C. equiselis*. Bull Mar Sci Gulf Caribb 9:117–152

GMFMC (Gulf of Mexico Fishery Management Council) (2004a) Amendment 22 of the Reef Fish Fishery Management Plan to Set Red Snapper Sustainable Fisheries Act Targets and Thresholds, Set a Rebuilding Plan, and Establish Bycatch Reporting Methodologies for the Reef Fish Fishery, Tampa, FL, USA, 76 p

GMFMC (2004b) EFH map for reef fish in the final environmental impact statement for the generic amendment to the fishery management plans of the Gulf of Mexico. http://www.habitat.noaa.gov/protection/efh/newInv/index.html/. Accessed 31 May 2013

GMFMC (2004c) EFH Map for red drum in the final environmental impact statement for the generic amendment to the fishery management plans of the Gulf of Mexico. http://www.habitat.noaa.gov/protection/efh/newInv/index.html/. Accessed 31 May 2013

GMFMC (2004d) EFH map for coastal migratory pelagics in the final environmental impact statement for the generic amendment to the fishery management plans of the Gulf of Mexico. http://www.habitat.noaa.gov/protection/efh/newInv/index.html/. Accessed 31 May 2013

GMFMC (2005) Generic Amendment Number 3 for Addressing Essential Fish Habitat Requirements, Habitat Areas of Particular Concern, and Adverse Effects of Fishing in the Following Fishery Management Plans of the Gulf of Mexico: Shrimp Fishery of the Gulf of Mexico, United States Waters; Red Drum Fishery of the Gulf of Mexico; Reef Fish Fishery of the Gulf of Mexico; Coastal Migratory Pelagic Resources (Mackerels) in the Gulf of Mexico and South Atlantic; Stone Crab Fishery of the Gulf of Mexico; Spiny Lobster in the Gulf of Mexico and South Atlantic; Coral and Coral Reefs of the Gulf of Mexico. Tampa, FL, USA, 104 p

GMFMC (2011) Final regulatory amendment to set 2011–2015 total allowable catch and adjust bag limit for red grouper. Tampa, FL, USA, 46 p

GMFMC and SAFMC (South Atlantic Fishery Management Council) (2011) Final Amendment 18 to the Fishery Management Plan for Coastal Migratory Pelagic Resources in the Gulf of Mexico and Atlantic Region Including Environmental Assessment, Regulatory Impact Review, and Regulatory Flexibility Act Analysis. GMFMC, Tampa, FL, USA and South Atlantic Fishery Management Council, North Charleston, SC, USA, 373 p

Gold JR, Richardson LR (1991) Genetic studies in marine fishes. IV. An analysis of population structure in the red drum (*Sciaenops ocellatus*) using mitochondrial DNA. Fish Res 12:213–241

Gold JR, Richardson LR (1998) Population structure in greater amberjack, *Seriola dumerili*, from the Gulf of Mexico and the western Atlantic Ocean. Fish Bull 96:767–778

Gold JR, Turner T (2002) Population structure of red drum (*Sciaenops ocellatus*) in the northern Gulf of Mexico, as inferred from variation in nuclear-encoded microsatellites. Mar Biol 140:249–265

Gold JR, Richardson LR, Furman C, King TL (1993) Mitochondrial DNA differentiation and population structure in red drum (*Sciaenops ocellatus*) from the Gulf of Mexico and Atlantic Ocean. Mar Biol 116:175–185

Gold JR, Sun F, Richardson LR (1997) Population structure of red snapper from the Gulf of Mexico as inferred from analysis of mitochondrial DNA. Trans Am Fish Soc 126:386–396

Gold JR, Richardson LR, Turner TF (1999) Temporal stability and spatial divergence of mitochondrial DNA haplotype frequencies in red drum (*Sciaenops ocellatus*) from coastal regions of the western Atlantic Ocean and Gulf of Mexico. Mar Biol 133:593–602

Goldstein RM, Simon PT (1999) Toward a united definition of guild structure for feeding ecology of North American freshwater fishes. In: Simon TP (ed) Assessing the sustainability and biological integrity of water resources using fish communities. CRC Press, Boca Raton, FL, USA, pp 123–202

Goodyear CP (1989) Status of the red drum stocks of the Gulf of Mexico report for 1989. CRD 88/89-14, Southeast Fisheries Science Center, Miami Laboratory, Coastal Resources Division, Miami, FL, USA, 64 p

Goodyear CP (1995) Mean size at age: An evaluation of sampling strategies using simulated red grouper data. Trans Am Fish Soc 124:746–755

Govoni JJ, Grimes CB (1992) The surface accumulation of larval fishes by hydrodynamic convergence within the Mississippi River plume front. Cont Shelf Res 12:1265–1276

Govoni JJ, Hoss DE, Chester AJ (1983) Comparative feeding of three species of larval fishes in the northern Gulf of Mexico: *Brevoortia patronus*, *Leiostomus xamhurus*, and *Micropogonius undulates*. Mar Ecol Prog Ser 13:189–199

Govoni JJ, Hoss DE, Colby DR (1989) The spatial distribution of larval fishes about the Mississippi River plume. Limnol Oceanogr 34:178–187

Govoni JJ, Laban EH, Hare JA (2003) The early life history of swordfish (*Xiphias gladius*) in the western North Atlantic. Fish Bull 101:778–789

Grall C, de Sylva DP, Houde ED (1983) Distribution, relative abundance, and seasonality of swordfish larvae. Trans Am Fish Soc 112:235–246

Greeley MS Jr, Calder DR, Wallace RA (1987) Oocyte growth and development in the striped mullet, *Mugil cephalus*, during seasonal ovarian recrudescence: Relationship to fecundity and size at maturity. Fish Bull 85:187–200

Grimes CB, Finucane JH (1991) Spatial distribution and abundance of larval and juvenile fish, chlorophyll and macrozooplankton around the Mississippi River discharge plume, and the role of the plume in fish recruitment. Mar Ecol Prog Ser 75:109–119

Grimes CB, Turner SC (1999) The complex life history of tilefish, *Lopholatilus chamaeleonticeps*. Am Film 23:17–26

Grimes CB, Turner SC, Able KW (1983) A technique for tagging deepwater fish. Fish Bull 81:663–666

Grimes CB, Johnson AG, Fable WA Jr (1987) Delineation of King Mackerel (*Scomberomorus cavalla*) stocks along the U.S. East Coast and in the Gulf of Mexico. In: Proceedings, Stock Identification Workshop, Panama City Beach, FL, USA, November, pp 186–187

Grimes CB, Idelberger CF, Able KW, Turner SC (1988) The reproductive biology of tilefish, *Lopholatilus chamaeleonticeps* Goode and Bean, from the United States Mid-Atlantic Bight, and the effects of fishing on the breeding system. Fish Bull 86:745–762

Grimes CB, Finucane JH, Collins LA, DeVries DA (1990) Young king mackerel, *Scombero-morus cavallas*, in the Gulf of Mexico, a summary of the distribution and occurrence of larvae and juveniles, and spawning dates for Mexican juveniles. Bull Mar Sci 46:640–654

Gudger EW (1929) On the morphology, coloration and behavior of seventy teleostean fishes of Tortugas, Florida. Carnegie Inst Wash Pap Tortugas Lab 26:149–204

Gunter G (1945) Studies on marine fishes of Texas. Publ Inst Mar Sci Univ Texas 1:1–190

Hampton J (2000) Natural mortality rates in tropical tunas: Size really does matter. Can J Fish Aquat Sci 57:1002–1010

Harris MJ, Grossman GD (1985) Growth, mortality, and age composition of a lightly exploited tilefish stock off Georgia. Trans Am Fish Soc 114:837–846

Harris PJ, Wyanski DM, White DB, Mikell PP, Eyo PB (2007) Age, growth, and reproduction of greater amberjack off the southeastern U.S. Atlantic coast. Trans Am Fish Soc 136:1534–1545

Hassler WW, Hogarth WT (1977) The growth and culture of dolphin (*Coryphaena hippurus*) in North Carolina. Aquaculture 12:115–122

Heath SR, Eckmayer WJ, Wade CW, Trimble WC, Tatum WM (1979) Research and management of Alabama coastal fisheries. Annual Progress Report PL 88-309, Project 2-330-R-1. Alabama Marine Resources Division, Gulf Shores, AL, USA, 70 p

Heemstra PC, Randall JE (1993) FAO Species catalogue: An annotated and illustrated catalogue of the grouper, Rockcod, Hind, Coral Grouper, and Lyretail Species known to date. FAO Fisheries Synopsis No. 125, vol 16. Food and Agriculture Organization of the United Nations, Rome, LZ, Italy, 382 p

Heist EJ, Gold JR (2000) DNA microsatellite loci and genetic structure of red snapper in the Gulf of Mexico. Trans Am Fish Soc 129:469–475

Heist EJ, Musick JA, Graves JE (1996) Genetic population structure of the shortfin mako (*Isurus oxyrinchus*) inferred from restriction fragment length polymorphism analysis of mitochondrial DNA. Can J Fish Aquat Sci 53:583–588

Hellier WF Jr (1968) Artificial fertilization among yellowfin and Gulf menhaden (*Brevoortia*) and their hybrid. Trans Am Fish Soc 97:119–123

Heupel MR, Simpfendorfer CA (2005) Quantitative analysis of aggregation behavior in juvenile blacktip sharks. Mar Biol 147:1239–1249

Hildebrand SF (1948) A review of the American menhaden, genus *Brevoortia*, with a description of a new species. Smithson Misc Coll 107:1–39

Hill KT, Cailliet GM, Radtke RL (1989) A comparative analysis of growth zones in four calcified structures of Pacific blue marlin, *Makaira nigricans*. Fish Bull 87:829–843

Hoenig JM (1983) Empirical use of longevity data to estimate mortality rates. Fish Bull 8:898–903

Hoese HD, Moore RH (1998) Fishes of the Gulf of Mexico-Texas, Louisiana, and adjacent waters, 2nd edn. Texas A&M University Press, College Station, TX, USA, 422 p

Hoffmayer ER, Franks JS, Driggers WB, Oswald KJ, Quattro JM (2007) Observations of a feeding aggregation of whale sharks, *Rhincodon typus*, in the North Central Gulf of Mexico. Gulf Caribb Res 19:69–74

Holliday M (1978) Food of Atlantic Bluefin Tuna, *Thunnus thynnus* (L.), from the coastal waters of North Carolina to Massachusetts. Long Island University, Long Island, NY, USA, 31 p

Holt J, Godbout R, Arnold CR (1981) Effects of temperature and salinity on egg hatching and larval survival of red drum, *Sciaenops ocellatus*. Fish Bull 79:569–573

Homma K, Maruyama T, Itoh T, Ishihara H, Uchida S (1999) Biology of the manta ray, *Manta birostris* Walbaum, in the Indo-Pacific. In: Proceedings, 5th Indo-Pacific Fisheries Conference, Noumea, New Caledonia, November, pp 209–216

Houde E, Swanson LJ Jr (1975) Description of eggs and larvae of yellowfin menhaden, *Brevoortia smithi*. Fish Bull 73:660–673

Ibañez Aguirre AL, Gallardo-Cabello M, Chiappa CX (1999) Growth analysis of striped mullet, *Mugil cephalus*, and white mullet, *M. curema* (Pisces: Mugilidae), in the Gulf of Mexico. Fish Bull 97:861–872

ICCAT (International Commission for the Conservation of Atlantic Tunas) (1997) Report of the ICCAT SCRS bluefin tuna stock assessment session. Collect Vol Sci Pap ICCAT 46:1–186

ICCAT (2001) Report of the fourth ICCAT billfish workshop. Collect Vol Sci Pap ICCAT 53:1–130

ICCAT (2010) Report of the 2009 sailfish stock assessment. Collect Vol Sci Pap ICCAT 65:1507–1632

ICCAT (2011a) ICCAT geographical delimitations (2011 version). http://www.iccat.int/Data/ICCATMaps2011.pdf. Accessed 31 May 2013

ICCAT (2011b) Report of the 2010 Atlantic bluefin tuna stock assessment session. Collect Vol Sci Pap ICCAT 66:505–714

ICCAT (2012a) Report of the 2012 Atlantic bluefin tuna stock assessment session. Document No. SCI-033/2012. International Commission for the Conservation of Atlantic Tunas, Madrid, Spain, 124 p

ICCAT (2012b) Report of the 2011 blue marlin stock assessment and white marlin data preparatory meeting. Collect Vol Sci Pap ICCAT 68:1273–1386

Ingram GW Jr (2006) Data summary of gray triggerfish (*Balistes capriscus*), vermilion snapper (*Rhomboplites aurorubens*), and greater amberjack (*Seriola dumerili*) collected during small pelagic trawl surveys, 1988–1996. SEDAR 9-DW-22, Southeast Data, Assessment and Review, North Charleston, SC, USA, 13 p

Jannke TE (1971) Abundance of young sciaenid fishes in Everglades National Park, Florida, in relation to season and other variables. University of Miami, Coral Gables, FL, USA, 128 p

Jenkins GP, Davis TLO (1990) Age, growth rate, and growth trajectory determined from otolith microstructure of southern bluefin tuna *Thunnus maccoyii* larvae. Mar Ecol Prog Ser 63:93–104

Jenkins GP, Young JW, Davis TL (1991) Density dependence of larval growth of a marine fish, the southern bluefin tuna, *Thunnus maccoyii*. Can J Fish Aquat Sci 48:1358–1363

Jensen AL (1996) Beverton and Holt Life history invariants result from optimal trade-off of reproduction and survival. Can J Fish Aquat Sci 53:820–822

Johnson GD (1978) Development of fishes of the Mid-Atlantic Bight: An atlas of egg, larval and juvenile stages, Part IV: Carangidae through Ephippidae. FWS/OB5-78/12, U.S. Fish and Wildlife Service, Office of Biological Sciences, Washington, DC, USA, 314 p

Johnson AG, Fable WA Jr, Williams ML, Barger LE (1983) Age, growth, and mortality of king mackerel, *Scomberomorus cavalla*, from the southeastern United States. Fish Bull 81:97–106

Johnson AG, Fable WA Jr, Grimes CB, Trent L, Perez JV (1994) Evidence for distinct stocks of king mackerel, *Scomberomorus cavallas*, in the Gulf of Mexico. Fish Bull 92:91–101

Johnson AK, Thomas P, Wilson RR Jr (1998) Seasonal cycles of gonadal development and plasma sex steroid levels in *Epinephelus morio*, a protogynous grouper in the eastern Gulf of Mexico. J Fish Biol 52:502–518

Jory DE, Iverson ES (1989) Species profiles: Life histories and environmental requirements of coastal fishes and invertebrates (South Florida): Black, red, and Nassau Groupers. U.S. Fish and Wildlife Service Biology Report 82(11.110), TR EL-82-4. U.S. Army Corps of Engineers, Coastal Ecology Group, Waterways Experiment Station, Vicksburg, MS, USA, 21 p

June FC, Carlson FT (1971) Food of young Atlantic menhaden, *Brevoortia tyrannus*, in relation to metamorphosis. Fish Bull 68:493–512

June FC, Chamberlin JL (1959) The role of the estuary in the life history and biology of Atlantic menhaden. In: Proceedings, 11th Annual Meeting of the Gulf and Caribbean Fisheries Institute, Coral Gables, FL, USA, November, pp 41–45

Karakulak S, Oray I, Corriero A, Deflorio M, Santamaria N, Desantis S, De Metrio G (2004a) Evidence of a spawning area for the bluefin tuna (*Thunnus thynnus* L.) in the eastern Mediterranean. J Appl Ichthyol 20:318–320

Karakulak S, Oray I, Corriero A, Aprea A, Spedicato D, Zubani D, Santamaria N, De Metrio G (2004b) First information on the reproductive biology of the bluefin tuna (*Thunnus thynnus*) in the eastern Mediterranean. Collect Vol Sci Pap ICCAT 56:1158–1162

Karnauskas M, Schirripa MJ, Kelble CR, Cook GS, Craig JK (2013) Ecosystem status report for the Gulf of Mexico. NOAA Technical Memorandum NMFS-SEFSC-653. NOAA, Washington, DC, USA, 52 p

Kasapidis P, Mejuto J, Tserpes G, Antoniou A, Garcia-Cortes B, Peristeraki P, Magoulas A (2007) Genetic structure of the swordfish (*Xiphias gladius*) stocks in the Atlantic using microsatellite DNA analysis. Collect Vol Sci Pap ICCAT 61:89–98

Katz SJ, Grimes CB, Able KW (1983) Delineation of tilefish, *Lopholatilus chamaeleonticeps*, stocks along the United States East Coast and in the Gulf of Mexico. Fish Bull 81:41–50

Keeney DB, Heupel M, Hueter RE, Heist EJ (2003) Genetic heterogeneity among blacktip shark, *Brevoortia gunteri*, continental nurseries along the U.S. Atlantic and Gulf of Mexico. Mar Biol 143:1039–1046

Keeney DB, Heupel MR, Hueter RE, Heist EJ (2005) Microsatellite and mitochondrial DNA analyses of the genetic structure of blacktip shark (*Brevoortia gunteri*) nurseries in the northwestern Atlantic, Gulf of Mexico, and Caribbean Sea. Mol Ecol 14:1911–1923

Kerstetter DW, Bayse SM, Graves JE (2010) Sailfish (*Istiophorus platypterus*) habitat utilization in the southern Gulf of Mexico and Florida Straits with Implications on vulnerability to shallow-set pelagic longline gear. Collect Vol Sci Pap ICCAT 65:1701–1712

Kilby JD (1955) The fishes of two Gulf coastal marsh areas of Florida. Tulane Stud Zool 2:175–247

Kitagawa T, Kimura S, Nakata H, Yamada H (2006) Thermal adaptation of Pacific bluefin tuna *Thunnus orientalis* to temperate waters. Fish Sci 72:149–156

Kleisner KM, Walter JF, Diamond SL, Die DJ (2010) Modeling the spatial autocorrelation of pelagic fish abundance. Mar Ecol Prog Ser 411:203–213

Klimley AP (1985) The areal distribution and autoecology of the white shark, *Carcharodon carcharias*, off the west coast of North America. Mem South Cali Acad Sci 9:15–40

Kraus RT, Rooker JR (2007) Patterns of vertical habitat use by Atlantic blue marlin (*Makaira nigricans*) in the Gulf of Mexico. Gulf Caribb Res 19:89–97

Kraus RT, Wells RJD, Rooker JR (2011) Horizontal movements of Atlantic blue marlin (*Makaira nigricans*) in the Gulf of Mexico. Mar Biol 158:699–713

Kroger RL, Pristas PJ (1975) Movements of tagged juvenile menhaden (*Brevoortia patronus*) in the Gulf of Mexico. Tex J Sci 26:473–477

Lang KL, Grimes CB, Shaw RF (1994) Variations in the age and growth of yellowfin tuna larvae, *Thunnus albacares*, collected about the Mississippi River plume. Environ Biol Fish 39:259–270

Lassuy DR (1983) Species profiles: Life histories and environmental requirements (Gulf of Mexico) Gulf Menhaden. U.S. Fish and Wildlife Service Biology Report FWS/OBS-82(11.2), TR EL-82-4, U.S. Army Corps of Engineers, Coastal Ecology Group, Waterways Experiment Station, Vicksburg, MS, USA, 13 p

Leard R, Mahmoudi B, Blanchet H, Lazauski H, Spiller K, Buchanan M, Dyer C, Keithly W (1995) The striped mullet fishery of the Gulf of Mexico, United States: A regional management plan. Report number 33, Gulf States Marine Fisheries Commission, Ocean Springs, MS, USA, 196 p

Lee TN, Williams E (1999) Mean distribution and seasonal variability of coastal currents and temperature in the Florida Keys with implications for larval recruitment. Bull Mar Sci 64:35–56

Lessa RP, Monteiro A, Duarte-Neto PJ, Vieira AC (2009) Multidimensional analysis of fishery production systems in the state of Pernambuco, Brazil. J Appl Ichthyol 25:256–268

Lewis RM, Roithmayer CM (1981) Spawning and sexual maturity of Gulf menhaden *Brevoortia patronus*. Fish Bull 78:947–951

Lewis MA, Dantin DD, Chancy CA, Abel KC, Lewis KG (2007) Florida seagrass habitat evaluation: A comparative survey for chemical quality. Environ Pollut 146:206–218

Lindeman KC, Pugliese R, Waugh GT, Ault JS (2000) Developmental patterns within a multispecies reef fishery: Management applications for essential fish habitats and protected areas. Bull Mar Sci 66:929–956

Linton E (1901) Fish parasites collected at Woods Hole in 1898. Bull US Fish Comm 19:267–304

Loefer JK, Sedberry GR (2003) Life history of the Atlantic sharpnose shark (*Rhizoprionodon terraenovae*) (Richardson, 1836) off the southeastern United States. Fish Bull 101:57–88

Logerwell EA, Smith PE (2001) Mesoscale eddies and survival of late stage Pacific sardine (*Sardinops sagax*) larvae. Fish Oceanogr 10:13–25

Loman M (1978) Other finfish. In: Christmas JY (ed) Fisheries assessment and monitoring—Mississippi. Completion report, PL 88-309, 2-215-R. Gulf Coast Research Laboratory, Ocean Springs, MD, USA, pp 143–147

Lombardi LA, Fitzhugh G, Lyon H (2010) Golden tilefish (*Lopholatilus chamaeleonticeps*) age, growth, and reproduction from the Northeastern Gulf of Mexico: 1985, 1997–2009. SEDAR 22-DW-01, Southeast Data, Assessment and Review, North Charleston, SC, USA, 35 p

Lombardi-Carlson LA, Fitzhugh GR, Mikulas JJ (2002) Red grouper (*Epinephelus morio*) age-length structure and description of growth from the Eastern Gulf of Mexico: 1992–2001. Panama City Laboratory Contribution Series 2002-06. National Marine Fisheries Service, Southeastern Fisheries Science Center, Panama City Beach, FL, USA, 42 p

Lombardi-Carlson LA, Cort E, Parsons GR, Manire CA (2003) Latitudinal variation in life-history traits of bonnethead sharks, *Sphyrna tiburo* (Carcharhiniformes: Sphyrnidae), from the eastern Gulf of Mexico. Mar Freshw Res 54:875–883

Lombardi-Carlson LA, Palmer C, Gardner C, Farsky B (2006) Temporal and spatial trends in red grouper (*Epinephelus morio*) age and growth from the northeastern Gulf of Mexico: 1979–2005. SEDAR 12-DW-03, Panama City Laboratory Contribution Series 2006-09. National Marine Fisheries Service, Southeastern Fisheries Science Center, Panama City Beach, FL, USA, 43 p

Lombardi-Carlson LA, Fitzhugh GR, Palmer C, Gardner C, Farsky R, Ortiz M (2008) Regional size, age and growth differences of red grouper (*Epinephelus morio*) along the West Coast of Florida. Fish Res 91:239–251

Longley WH, Hildebrand SF (1941) Systematic catalogue of the fishes of Tortugas, Florida, with observations on colour, habits and local distributions. Carnegie Inst Wash Pap Tortugas Lab 34:1–331

Lorenzen K (1996) A simple von Bertalanffy model for density-dependent growth in extensive aquaculture, with an application to common carp (*Cyprinus carpio*). Aquaculture 142:191–205

Luckhurst BE, Prince ED, Llopiz JK, Snodgrass D, Brothers EB (2006) Evidence of blue marlin (*Makaira nigricans*) spawning in Bermuda waters and elevated mercury levels in large specimens. Bull Mar Sci 79:691–704

Luthy SA, Cowen RK, Serafy JE, McDowell JR (2005) Toward identification of larval sailfish (*Istiophorus platypterus*), white marlin (*Tetrapturus albidus*), and blue marlin (*Makaira nigricans*) in the western North Atlantic Ocean. Fish Bull 103:588–600

Lux FE, Mahoney JV (1969) First record of the channel bass, *Sciaenops ocellata* (Linnaeus), in the Gulf of Maine. Copeia 1969:632–633

Lyczkowski-Shultz J, Hanisko DS (2007) A time series of observations on red snapper larvae from SEAMAP surveys, 1982–2003: Seasonal occurrence, distribution, abundance, and size. In: Patterson WF, Gowan JH Jr, Fitzhugh GR, Nieland DL (eds) Red Snapper ecology and fisheries in the U.S. Gulf of Mexico, vol 60, American Fisheries Society Symposium Series. American Fisheries Society, Bethesda, MD, USA, pp 3–24

Lyczkowski-Shultz J, Hanisko DS, Ingram GW (2005) The potential for incorporating a larval index of abundance for stock assessment of red snapper, *Lutjanus campechanus*. SEDAR 7-DW-14. National Marine Fisheries Service, Miami, FL, USA, 10 p

Magnuson JJ, Block BA, Deriso RB, Gold JR, Grant WS, Quinn TJ, Saila SB, Shapiro L, Stevens ED (eds) (1994) An assessment of Atlantic Bluefin Tuna. National Academy Press, Washington, DC, USA, 148 p

Maguire JJ, Sissenwine M, Csirke J, Grainger R, Garcia S (2006) The state of world highly migratory, straddling and other high seas fishery resources and associated species. FAO fisheries technical paper, no. 495, Rome, LZ, Italy, 84 p

Mahmoudi B (1991) Population assessment of black mullet (*Mugil cephalus*) in the Eastern Gulf of Mexico. Final report of cooperative agreement (MARFIN) NA86-WC-H-06138. Florida Department of Environmental Protection, St. Petersburg, FL, USA, 78 p

Mahmoudi B (1993) Update on black mullet stock assessment. Final report submitted to the Florida Marine Fisheries Commission, Tallahassee, FL, USA, 38 p

Mahmoudi B (2000) Status and trends in the Florida Mullet Fishery and an updated stock assessment. Florida Fish and Wildlife Commission. Florida Marine Research Institute, St. Petersburg, FL, USA, 48 p

Mahmoudi B (2005) The 2005 update of the stock assessment for striped mullet, *Mugil cephalus*, in Florida. Florida Fish and Wildlife Conservation Commission. Fish and Wildlife Research Institute, St. Petersburg, FL, USA, 43 p

Mahmoudi B (2008) The 2008 update of the stock assessment for striped mullet, *Mugil cephalus*, in Florida. In-House Report IHR2008. Florida Fish and Wildlife Conservation Commission, Fish and Wildlife Research Institute, St. Petersburg, FL, USA, 114 p

Manooch CS III (1984) Fisherman's guide to the fishes of the Southeastern United States. North Carolina Museum of Natural History, Raleigh, NC, USA, 362 p

Manooch CS III, Laws ST (1979) Survey of the charter boat troll fishery in North Carolina, 1977. Mar Fish Rev 41:1–11

Manooch CS III, Potts JC (1997a) Age, growth, and mortality estimates of greater amberjack, *Seriola dumerili*, from the U.S. Gulf of Mexico headboat fishery. Bull Mar Sci 61:671–683

Manooch CS III, Potts JC (1997b) Age, growth, and mortality estimates of greater amberjack from the southeastern United States. Fish Res 30:229–240

Manooch CS III, Mason DL, Nelson RS (1984) Food and gastrointestinal parasites of dolphin *Coryphaena hippurus* collected along the southeastern and Gulf Coasts of the United States. Bull Jap Soc Sci Fish 50:1511–1525

NMFS (National Marine Fisheries Service) (1993) Fishery management plan for sharks of the Atlantic Ocean. NOAA, Office of Sustainable Fisheries, Highly Migratory Species Management Division, Silver Spring, MD, USA, 287 p

NMS (National Marine Sanctuaries) (2013) Fish photos. National Oceanic and Atmospheric Administration, Silver Spring, MD, USA. https://marinelife.noaa.gov. Accessed 31 May 2013

Martin FD, Drewry GE (1978) Development of fishes of the Mid-Atlantic Bight, an Atlas of egg, larval, and juvenile stages, vol 6. U.S. Fish and Wildlife Service, Washington, DC, USA, 416 p

Martinez-Andrade F (2003) A comparison of life histories and ecological aspects among snappers (Pisces: Lutjanidae). PhD Thesis, Louisiana State University, Baton Rouge, LA, USA, 201 p

Mason JM (1976) Food of Small, Northwestern Atlantic Bluefin Tuna, *Thunnus thynnus* (L.). MS Thesis, University of Rhode Island, Kingston, RI, USA, 31 p

Mather FJ III, Schuck HA (1960) Growth of bluefin tuna of the western North Atlantic. Fish Bull 179:39–52

Mather FJ, Mason JM Jr, Jones A (1995) Historical document: Life history and fisheries of Atlantic bluefin tuna. NMFS-SEFSC-370, NOAA technical memorandum. National Marine Fisheries Service, Southeast Fisheries Science Center, Miami, FL, USA, 165 p

Matlock GC (1987) The role of hurricanes in determining year class strength. Contrib Mar Sci 30:39–47

Matlock GC, Weaver JE (1979) Fish tagging in Texas Bays during November 1975–September 1976. Management data series no. 1. Texas Parks and Wildlife Department, Coastal Fisheries Branch, Austin, TX, USA, 136 p

Matthews FD, Damaker DM, Knapp LW, Collette BB (1977) Food of Western North Atlantic tunas (*Thunnus*) and lancetfish (*Alepisaurus*). NOAA technical report 706. National Oceanic and Atmospheric Administration, National Marine Fisheries Service, Special Scientific Report on Fisheries, Washington, DC, USA, 26 p

McCawley JR, Cowan JH (2007) Seasonal and size specific diet and prey demand of red snapper on artificial reefs. In: Patterson WF, Cowan JH Jr, Fitzhugh GR, Nieland DL (eds) Red snapper ecology and fisheries in the U.S. Gulf of Mexico. American Fisheries Society symposium series 60. American Fisheries Society, Bethesda, MD, USA, pp 77–104

McClellan DB, Cummings NJ (1997) Preliminary analysis of tag and recapture data of the greater amberjack, *Seriola dumerili*, in the Southeastern United States. In: Proceedings, 49th Annual Meeting of the Gulf and Caribbean Fisheries Institute, Dover, Christ Church, New Zealand, November, pp 25–45

McCosker JE (1985) White shark attack behavior: Observations of and speculations about predator and prey strategies. Mem South Calif Acad Sci 9:123–135

McEachran JD (2009) Fishes (Vertebrata: Pisces) of the Gulf of Mexico. In: Felder DL, Camp DK (eds) Gulf of Mexico: Origin, waters, and biota, vol 1, Biodiversity. Texas A&M University Press, College Station, TX, USA, pp 1223–1317

McEachran JD, de Carvalho MR (2002) Batoid fishes. In: Carpenter KE (ed) FAO species identification guide for fishery purposes: The living marine resources of the Western Central Atlantic, introduction, molluscs, crustaceans, hagfishes, sharks, batoid fishes, and chimaeras, vol 1. American Society of Ichthyologists and Herpetologists Special Publication No. 5. Food and Agriculture Organization of the United Nations, Rome, LZ, Italy, pp 507–586

McEachran JD, Fechhelm JD (2005) Fishes of the Gulf of Mexico, vol 2, Scorpaeniformes to Tetraodontiformes. University of Texas Press, Austin, TX, USA, 1004 p

McEachran JD, Finucane JH, Hall LS (1980) Distribution, seasonality and abundance of king and Spanish mackerel larvae in the northwestern Gulf of Mexico (Pisces: Scombridae). Northeast Gulf Sci 4:1–16

McEachron LW, Matlock GC, Bryan CE, Unger P, Cody TJ, Martin JH (1994) Winter mass mortality of animals in Texas bays. Northeast Gulf Sci 13:121–138

McGovern JC, Burgos JM, Harris PJ, Sedberry GR, Loefer JK, Pashuk O, Russ D (2002) Aspects of the life history of Red Grouper, *Epinephelus morio*, along the Southeastern United States. MARFIN final report NA97FF0347, South Carolina Department of Natural Resources, Charleston, SC, USA, 59 p

McGowan MF, Richards WJ (1986) Distribution and abundance of bluefin tuna (*Thunnus thynnus*) larvae in the Gulf of Mexico in 1982 and 1983 with estimates of the biomass and population size of the spawning stock for 1977, 1978, and 1981–1983. Collect Vol Sci Pap ICCAT 24:182–195

McGowan MF, Richards WJ (1989) Bluefin tuna, *Thunnus thynnus*, larvae in the Gulf Stream off the southeastern United States: Satellite and shipboard observations of their environment. Fish Bull 87:615–631

McHugh JL (1967) Estuarine nekton. In: Lauff GH (ed) Estuaries. American Association for the Advancement of Science Publication No. 83. Washington, DC, USA, pp 581–620

Miles P (1971) The mystery of the great white shark. Oceans 4:51–59

Minello TJ, Webb JW Jr (1997) Use of natural and created *Spartina alterniflora* salt marshes by fishery species and other aquatic fauna in Galveston Bay, Texas, USA. Mar Ecol Prog Ser 151:165–179

Minton RV, Hawke JP, Tatum WM (1983) Hormone induced spawning of red snapper, *Lutjanus campechanus*. Aquaculture 30:363–368

Mitchell KM, Henwood T, Fitzhugh GR, Allman RJ (2004) Distribution, abundance, and age structure of red snapper (*Lutjanus campechanus*) caught on research longlines in U.S. Gulf of Mexico. Gulf Mex Sci 22:164–172

Miyashita S, Sawada Y, Okada T, Murata O, Kumai H (2001) Morphological development and growth of laboratory-reared larval and juvenile *Thunnus thynnus* (Pisces: Scombridae). Fish Bull 99:601–616

Moe MA Jr (1966) Tagging fishes in Florida offshore waters. Florida Board of Conservation, Marine Research Lab, Technical Series No. 49. Florida Department of Natural Resources, Division of Marine Resources, St. Petersburg, FL, USA, 40 p

Moe MA Jr (1969) Biology of the red grouper *Epinephelus morio* (Valenciennes) from the Eastern Gulf of Mexico. Florida Board of Conservation, Marine Research Lab, Technical Series No. 10. Florida Department of Natural Resources, Division of Marine Resources, St. Petersburg, FL, USA, 95 p

Moe MA Jr (1972) Movement and migration of South Florida fishes. Florida Board of Conservation, Marine Research Lab, Technical Series No. 69. Florida Department of Natural Resources, Division of Marine Resources, St. Petersburg, FL, USA, 25 p

Mohn R (1999) The retrospective problem in sequential population analysis: An investigation using cod fishery and simulated data. ICES J Mar Sci 56:473–488

Moore RH (1974) General ecology, distribution, and relative abundance of *Mugil cephalus* and *Mugil curema* on the South Texas Coast. Contrib Mar Sci 18:241–255

Moriarity DJW (1976) Quantitative studies on bacteria and algae in the food of the mullet *Mugil cephalus* L. and the prawn *Metapenaeus bennettae* (Racek and Dall). J Exp Mar Biol Ecol 22:131–143

Murie DJ, Parkyn DC (2010) Age, growth and sex maturity of greater amberjack (*Seriola dumerili*) in the Gulf of Mexico, MARFIN Grant No. NA05NMF4331071. In: Proceedings, 18th Annual Marine Fisheries Research Initiative Program (MARFIN) Conference, St. Petersburg, FL, USA, April, pp 28–29

Murphy MD, Munyandorero J (2009) An assessment of the status of red drum in Florida waters through 2007. IHR 2008-008, Florida Fish and Wildlife Conservation Commission, Fish and Wildlife Research Institute, St. Petersburg, FL, USA, April, 106 p

Murphy MD, Taylor RG (1990) Reproduction, growth, and mortality of red drum *Sciaenops ocellatus*, in Florida waters. Fish Bull 88:531–542

Musyl MK, Brill RW, Boggs CH, Curran DS, Kazama TK, Seki MP (2003) Vertical movements of bigeye tuna (*Thunnus obesus*) associated with islands, buoys, and seamounts near the main Hawaiian Islands from archival tagging data. Fish Oceanogr 12:152–169

Nakamura I (1985) FAO species catalogue: An annotated and illustrated catalogue of marlins, sailfishes, spearfishes and swordfishes known to date. FAO Fisheries Synopsis No. 125, vol 5. Food and Agriculture Organization of the United Nations, Rome, LZ, Italy, 65 p

NaluPhoto (2012) Sailfish with bait ball. iStockphoto, Calgary, Alberta, Canada. http://www.istockphoto.com/photo/sailfish-with-bait-ball-gm138186926-16170177. Accessed December 5, 2016

Neer JA (2005) Aspects of the life history, ecophysiology, bioenergetics, and population dynamics of the cownose ray, *Rhinoptera bonasus*, in the Northern Gulf of Mexico. PhD Thesis, Louisiana State University, Baton Rouge, LA, USA, 124 p

Neer JA, Thompson BA (2004) Aspects of the biology of the finetooth shark, *Carcharhinus isodon*, in Louisiana waters. Gulf Mex Sci 22:108–113

Neilson JD, Paul SD, Smith SC (2007) Stock structure of swordfish (*Xiphias gladius*) in the Atlantic: A review of the non-genetic evidence. Collect Vol Sci Pap ICCAT 61:25–60

Neilson JD, Smith S, Royer F, Paul SD, Porter JM, Lutcavage M (2009) Investigations of horizontal movements of Atlantic swordfish using pop-up satellite archival tags. Rev Methods Technol Fish 9 Biol Fish 9 (Part 1, Part 3):145–159

Nelson JS (2006) Fishes of the world, 4th edn. John Wiley and Sons, New York, NY, USA, 601 p

Nelson WR, Ahrenholz DW (1986) Population and fishery characteristics of Gulf menhaden, *Brevoortia patronus*. Fish Bull 84:311–325

Nelson WR, Carpenter JS (1968) Bottom longline explorations in the Gulf of Mexico. Comp Fish Rev 30:57–62

Nelson RS, Manooch CS III (1982) Growth and mortality of red snappers in the West-Central Atlantic Ocean and northern Gulf of Mexico. Trans Am Fish Soc 111:465–475

Nelson WR, Ingham MC, Schaaf WE (1977) Larval transport and year-class strength of Atlantic menhaden, *Brevoortia tyrannus*. Fish Bull 75:23–41

Nichols JT (1922) Color of the tuna. Copeia 1922:74–75

Nieland DL, Wilson CA (2003) Red snapper recruitment to and disappearance from oil and gas platforms in the northern Gulf of Mexico. In: Stanberg DR, Scarborough-Bull A (eds) Fisheries, reefs, and offshore development, vol 36, American Fisheries Society Symposium Series. American Fisheries Society, Bethesda, MD, USA, pp 73–81

Nishida T, Tsuji S, Segawa K (1998) Spatial data analyses of Atlantic bluefin tuna larval surveys in the 1994 ICCAT BYP. Collect Vol Sci Pap ICCAT 48:107–110

NMFS (National Marine Fisheries Service) (1993) Fishery management plan for sharks of the Atlantic Ocean. NOAA, Office of Sustainable Fisheries, Highly Migratory Species Management Division, Silver Spring, MD, USA, 287 p

NMFS (1999) Amendment 1 to the Atlantic Billfish fishery management plan including: Revised final supplemental environmental impact statement, regulatory impact review final regulatory flexibility analysis, and social impact assessment/fishery impact statement. NOAA, Office of Sustainable Fisheries, Highly Migratory Species Management Division, Silver Spring, MD, USA, 387 p

NMFS (2001) Annual report to congress on the status of U.S. Fisheries-2000. U.S. Department of Commerce, NOAA, Silver Spring, MD, USA, 122 p

NMFS (2002a) Status of red grouper in United States Waters of the Gulf of Mexico During 1986–2001. Sustainable Fisheries Division Contribution No. SFD-01/02-175. NOAA, Southeast Fisheries Science Center, Miami, FL, USA

NMFS (2002b) Annual report to congress on the status of U.S. Fisheries-2001. U.S. Department of Commerce, NOAA, NMFS, Silver Spring, MD, USA, 142 p

NMFS (2003) Annual report to congress on the status of U.S. Fisheries-2002. U.S. Department of Commerce, NOAA, Silver Spring, MD, USA, 156 p

NMFS (2004) Annual report to congress on the status of U.S. Fisheries-2003. U.S. Department of Commerce, NOAA, Silver Spring, MD, USA, 24 p

NMFS (2005) Annual report to congress on the status of U.S. Fisheries-2004. U.S. Department of Commerce, NOAA, Silver Spring, MD, USA, 20 p

NMFS (2006a) Annual report to congress on the status of U.S. Fisheries-2005. U.S. Department of Commerce, NOAA, Silver Spring, MD, USA, 20 p

NMFS (2006b) Final consolidated atlantic highly migratory species fishery management plan including: Final environmental impact statement, final regulatory impact review, final regulatory flexibility analysis, final social impact assessment, framework actions, and the 2006 stock assessment and fishery evaluation report. NOAA, Office of Sustainable Fisheries, Highly Migratory Species Management Division, Silver Spring, MD, USA, 629 p

NMFS (2007) Annual report to congress on the status of U.S. Fisheries-2006. U.S. Department of Commerce, NOAA, Silver Spring, MD, USA, 28 p

NMFS (2008) Annual report to congress on the status of U.S. Fisheries-2007. U.S. Department of Commerce, NOAA, Silver Spring, MD, USA, 23 p

NMFS (2009a) Annual report to congress on the status of U.S. Fisheries-2008. U.S. Department of Commerce, NOAA, Silver Spring, MD, USA, 23 p

NMFS (2009b). Final amendment 1 to the 2006 consolidated atlantic highly migratory species fishery management plan, essential fish habitat. U.S. Department of Commerce, NOAA, Office of Sustainable Fisheries, Highly Migratory Species Management Division, Silver Spring, MD, USA, 395 p

NMFS (2009c) Our living oceans, report on the status of U.S. living marine resources. 6th ed. NOAA Technical Memorandum NMFS-F/SPO-80. U.S. Department of Commerce, NOAA, Washington, DC, USA, 369 p

NMFS (2009d) Recovery plan for smalltooth sawfish (*Pristis pectinata*). Prepared by the Smalltooth Sawfish Recovery Team for the National Marine Fisheries Service, Silver Spring, MD, USA, 102 p

NMFS (2010) Annual report to congress on the status of U.S. Fisheries-2009. U.S. Department of Commerce, NOAA, NMFS, Silver Spring, MD, USA, 20 p

NMFS (2011) Annual report to congress on the status of U.S. Fisheries-2010. U.S. Department of Commerce, NOAA, Silver Spring, MD, USA, 21 p

NMFS (2012a) Annual report to congress on the status of U.S. Fisheries-2011. U.S. Department of Commerce, NOAA, Silver Spring, MD, USA, 20 p

NMFS (2012b) Species information system public portal, NOAA Fisheries Service. https://www.st.nmfs.noaa.gov/sisPortal/sisPortalMain.jsp. Accessed 31 May 2013

NMFS, SERO (Southeast Regional Office) (1986) Final secretarial fishery management plan regulatory impact review regulatory flexibility analysis for the red drum fishery of the Gulf of Mexico. NMFS Miami, FL, USA. December

NOAA (National Oceanic and Atmospheric Administration) (2003) Tilefish distributions in the Gulf of Mexico. http://www.ncddc.noaa.gov/website/CHP. Accessed 31 May 2013

NOAA Fisheries Office of Sustainable Fisheries (2009a) HMS EFH 2009 Bluefin Tuna. http://www.nmfs.noaa.gov/sfa/hms/EFH/shapefiles.htm/. Accessed 31 May 2013

NOAA Fisheries Office of Sustainable Fisheries (2009b) HMS EFH 2009 Blue Marlin. http://www.nmfs.noaa.gov/sfa/hms/EFH/shapefiles.htm/. Accessed 31 May 2013

NOAA Fisheries Office of Sustainable Fisheries (2009c) HMS EFH 2009 Swordfish. http://www.nmfs.noaa.gov/sfa/hms/EFH/shapefiles.htm/. Accessed 31 May 2013

NOAA Fisheries Office of Sustainable Fisheries (2009d) HMS EFH 2009 Sailfish. http://www.nmfs.noaa.gov/sfa/hms/EFH/shapefiles.htm/. Accessed 31 May 2013

NOAA (2013a) Red grouper distributions in the Gulf of Mexico. http://www.ncddc.noaa.gov /website/CHP. Accessed 31 May 2013

NOAA (2013b) Dolphin distributions in the Gulf of Mexico. http://www.ncddc.noaa.gov /website/CHP. Accessed 31 May 2013

NOAA (2013c) Greater amberjack distributions in the Gulf of Mexico. http://www.ncddc.noaa.gov/website/CHP. Accessed 31 May 2013

Nordlie FG, Szelistowshi WA, Nordlie WC (1982) Ontogenesis of osmoregulation in the striped mullet *Mugil cephalus* Linnaeus. J Fish Biol 20:79–86

O'Connell MT, Cashner RC, Schieble CS (2004) Fish assemblage stability over fifty years in the Lake Pontchartrain estuary: Comparisons among habitats using canonical correspondence analysis. Estuaries 27:807–817

Odum WE (1970) Utilization of the direct grazing and plant detritus food chains by the striped mullet, *Mugil cephalus*. In: Steele JJ (ed) Marine food chains. Oliver and Boyd, Edinburgh, Scotland, UK, pp 222–240

Orbesen ES, Snodgrass D, Hoolihan JP, Prince ED (2010) Updated U.S. conventional tagging data base for Atlantic sailfish (1956–2009), with comments on potential stock structure. Collect Vol Sci Pap ICCAT 65:1692–1700

Ortiz M, Palmer C (2008) Review and estimates of von Bertalanffy growth curves for the King Mackerel Atlantic and Gulf of Mexico stock units. SEDAR 16-DW-12. Southeast Data, Assessment and Review, North Charleston, SC, USA, 20 p

Ortiz M, Methot R, Cass-Calay SL, Linton B (2008) Preliminary report King Mackerel stock assessment results 2008. SEDAR 16-AW-08. Southeast Data, Assessment and Review, North Charleston, SC, USA, 75 p

Osburn HR, Matlock GC, Green AW (1982) Red drum (*Sciaenops ocellatus*) movement in Texas bays. Contrib Mar Sci 25:85–97

Overstreet RM (1983) Aspects of the biology of the red drum, *Sciaenops ocellatus*, in Mississippi. Gulf Res Rep Suppl 1:1–43

Overstreet RM, Heard RW (1982) Food contents of six commercial fishes from Mississippi Sound. Gulf Res Rep 7:137–149

Oxenford HA (1985) Biology of the Dolphin, *Coryphaena hippurus*, and its implications for the Barbadian Fishery. University of the West Indies, Cave Hill, St. Michael, 366 p

Oxenford HA (1999) Biology of the dolphinfish (*Coryphaena hippurus*) in the western Central Atlantic: A review. Sci Mar 63:277–301

Oxenford HA, Hunt W (1986) A preliminary investigation of the stock structure of the dolphin, *Coryphaena hippurus*, in the western Central Atlantic. Fish Bull 84:451–460

Palko BJ, Beardsley GL, Richards WJ (1981) Synopsis of the biology of the Swordfish, *Xiphias gladius* Linnaeus. FAO Fisheries Synopsis No. 127. NOAA Technical Report, NMFS Circular 441. U.S. Department of Commerce, National Oceanic and Atmospheric Administration, National Marine Fisheries Service, Washington, DC, USA, 21 p

Palmer SM, Harris PJ, Powers PT (1998) Age, growth, and reproduction of tilefish, *Lopholatilus chamaeleonticeps*, along the southeast coast of the United States 1980–87 and 1996–98. SEDAR 4-DW-18, Southeast Data, Assessment and Review, North Charleston, SC, USA, 34 p

Parsons GR (1983) The reproductive biology of the Atlantic sharpnose shark, *Rhizoprionodon terraenovae* (Richardson). Fish Bull 81:61–74

Parsons GR (1993) Geographic variation in reproduction between two populations of the bonnethead shark, *Sphyrna tiburo*. Environ Biol Fish 38:25–35

Parsons GR (2006) Sharks, skates, and rays of the Gulf of Mexico, A field guide. University Press of Mississippi, Jackson, MS, USA, 165 p

Patterson WF, Watterson JC, Shipp RL, Cowan JH Jr (2001) Movement of tagged red snapper in the northern Gulf of Mexico. Trans Am Fish Soc 130:533–545

Pattillo ME, Czapla TE, Nelson DM, Monaco ME (1997) Distribution and abundance of fishes and invertebrates in Gulf of Mexico Estuaries, vol II: Species life history summaries. ELMR Report No. 11. NOAA, National Ocean Service, Strategic Environmental Assessments Division, Estuarine Living Marine Resources Program, Silver Spring, MD, USA, 377 p

Pearson JC (1929) Natural history and conservation of the redfish and other commercial sciaenids on the Texas Coast. Bull US Bur Fish 44:129–214

Peck J (1893) On the food of the menhaden. Bull US Fish Comm 13:113–126

Perera PAB, De Silva SS (1978) Studies on the biology of the young grey mullet (*Mugil cephalus*) digestion. Mar Biol 44:383–387

Perret WS, Latapie WR, Pollard IF, Mock WR, Adkins BG, Gaidry W, White CJ (1971) Fishes and invertebrates collected in trawl and seine samples in Louisiana estuaries. In: Section I, Cooperative Gulf of Mexico Estuarine Inventory and Study, Louisiana, Phase I, Area Description and Phase IV, Biology, Louisiana Wildlife and Fisheries Commission, Baton Rouge, LA, USA, pp 39–105

Peters KM, McMichael RG Jr (1987) Early life history of *Sciaenops ocellatus* (Pisces: Sciaenidae) in Tampa Bay, Florida. Estuaries 10:92–107

Piko AA, Szedlmayer ST (2007) Effects of habitat complexity and predator exclusion on the abundance of juvenile red snapper. J Fish Biol 70:758–769

Podestá GP, Browder JA, Hoey JJ (1993) Exploring the association between swordfish catch rates and thermal fronts on U.S. longline grounds in the western North Atlantic. Cont Shelf Res 13:253–277

Porch CE (2000) Status of the red drum stocks of the Gulf of Mexico, Version 2.1. SFD-99/00-85, Southeast Fisheries Science Center, Miami Laboratory, Miami, FL, USA, 43 p

Pough HF, Janis CM, Heiser JB (2008) Vertebrate life, 8th edn. Benjamin Cummings Publishing Company, San Francisco, CA, USA, 752 p

Powels H, Stender BW (1976) Observations on composition, seasonality and distribution of ichthyoplankton from MARMAP cruises in the South Atlantic Bight in 1973. South Carolina Marine Research Center Technical Report Series No. 11, Charleston, SC, USA, May

Powers JE, Eldridge P (1983) A preliminary Assessment of king mackerel resources of the Southeast United States. Unpublished report, NOAA, NMFS, Southeast Fisheries Science Center, Miami, FL, USA, 38 p

Powers SP, Hightower CL, Drymon JM, Johnson MW (2012) Age composition and distribution of red drum (*Sciaenops ocellatus*) in offshore waters of the North Central Gulf of Mexico: An evaluation of a stock under a federal harvest moratorium. Fish Bull 110:283–292

Prager MH (2000) Exploratory assessment of dolphinfish, *Coryphaena hippurus*, based on U.S. landings from the Atlantic Ocean and the Gulf of Mexico. NOAA, NMFS, Southeast Fisheries Science Center, Beaufort, NC, USA, 18 p

Pristas PJ, Levi EJ, Dryfoos RL (1976) Analysis of returns of tagged Gulf menhaden. Fish Bull 74:112–117

Pritchard ES (ed) (2005) Fisheries of the United States 2004. National Marine Fisheries Service, Office of Science and Technology, Silver Spring, MD, USA, 19 p

Puntel LF (2016) Goliath Grouper and shipwreck. iStockphoto, Calgary, Alberta, Canada. http://www.istockphoto.com/photo/goliath-grouper-and-shipwreck-gm518386616-90009669?st=_p_goliath%20grouper. Accessed 13 December 2016

Quinn TJ II, Deriso RB (1999) Quantitative fish dynamics. Oxford University Press, New York, NY, USA, 546 p

Reintjes JW (1970) The Gulf menhaden and our changing estuaries. In: Proceedings, 22nd Annual Meeting of the Gulf and Caribbean Fisheries Institute, Coral Gables, FL, USA, May, pp 87–90

Reintjes JW, Pacheco AL (1966) The relation of menhaden to estuaries. In: Smith RF, Swartz AH, Massmann WH (eds) A Symposium on Estuarine Fisheries. American Fisheries Society Special Publication No. 3. American Fisheries Society, Bethesda, MD, USA, pp 50–58

Render JH (1995) The life history (age, growth, and reproduction) of red snapper (*Lutjanus campechanus*) and its affinity for oil and gas platforms. PhD Thesis, Louisiana State University, Baton Rouge, LA, USA, 76 p

Richards WJ (1976) Spawning of bluefin tuna (*Thunnus thynnus*) in the Atlantic Ocean and adjacent seas. Collect Vol Sci Pap ICCAT 5:267–278

Richards WJ (1997) Report on U.S. collections from the Gulf of Mexico, 1994. Collect Vol Sci Pap ICCAT 46:186–188

Richards WJ, Lindeman KC (1987) Recruitment dynamics of reef fishes: Planktonic processes, settlement and demersal ecologies, and fishery analysis. Bull Mar Sci 41:392–410

Richards WJ, Potthoff T (1980) Distribution and abundance of bluefin tuna larvae in the Gulf of Mexico in 1977 and 1978. Collect Vol Sci Pap ICCAT 9:433–441

Richards WJ, Leming T, McGowan MF, Lamkin JT, Kelley-Fraga S (1989) Distribution of fish larvae in relation to hydrographic features of the Loop Current boundary in the Gulf of Mexico. ICES Mar Sci Symp 191:169–176

Richardson LR, Gold JR (1997) Mitochondrial DNA diversity in and population structure of red grouper, *Epinephelus morio*, from the Gulf of Mexico. Fish Bull 95:174–179

Richardson DE, Cowen RK, Prince ED, Sponaugle S (2009) Importance of the Straits of Florida spawning ground to Atlantic sailfish (*Istiophorus platypterus*) and blue marlin (*Makaira nigricans*). Fish Oceanogr 18:402–418

Ricker WE (1975) Computation and interpretation of biological statistics of fish populations. Fish Res Board Can Bull 191:382

Riley CM, Holt SA, Holt GJ, Buskey EJ, Arnold CR (1989) Mortality of larval red drum (*Sciaenops ocellatus*) associated with a *Ptychodiscus brevis* red tide. Publ Inst Mar Sci Univ Texas 31:137–146

Rivas LR (1975) Synopsis of biological data on blue marlin, *Makaira nigricans* Lacepede, 1802. In: Shomura RS, Williams F (eds) Proceedings of the International Billfish Symposium, Kailua-Kona, HI, USA, August, Part 3, pp 1–16

Rivas LR (1980) Synopsis of knowledge on the taxonomy, biology, distribution and fishery of the Gulf of Mexico mullets (Pisces: Mugilidae). In: Proceedings, Workshop for Potential Fishery Resources of the Northern Gulf of Mexico, New Orleans, LA, USA, pp 34–53

Rivera GA, Appeldoorn RS (2000) Age and growth of dolphin, *Coryphaena hippurus*, off Puerto Rico. Fish Bull 98:345–352

Roberts DE Jr, Harpster BV, Henderson GE (1978) Conditioning and spawning of the red drum (*Sciaenops ocellatus*) under varied conditions of photoperiod and temperature. In: Proceedings, 9th Annual Meeting of the World Mariculture Society, Atlanta, GA, USA, January, pp 311–332

Roe RB (1976) Distributions of snappers and groupers in the Gulf of Mexico as determined from exploratory fishing data. In: Proceedings, Colloquium on Snapper-Grouper Fishery Resources of the Western Central Atlantic Ocean, Pensacola Beach, FL, USA, October, pp 129–164

Roelke LA, Cifuentes LA (1997) Use of stable isotopes to assess groups of king mackerel, *Scomberomorus cavallas*, in the Gulf of Mexico and southeastern Florida. Fish Bull 95:540–551

Rooker JR, Landry AM, Geary BW, Harper JA (2004) Assessment of a shell bank and associated substrates as nursery habitat of post settlement red snapper. Estuar Coast Shelf Sci 59:653–661

Rooker JR, Alvarado Bremer JR, Block BA, De Metrio G, Corriero A, Kraus RT, Prince ED, Rodriguez-Marin E, Secor DH (2007) Life history and stock structure of Atlantic bluefin tuna (*Thunnus thynnus*). Rev Fish Sci 15:265–310

Rooker JR, Simms JR, Wells RJD, Holt SA, Holt GJ, Graves JE, Furey NB (2012) Distribution and habitat associations of billfish and swordfish larvae across mesoscale features in the Gulf of Mexico. PLoS One 7(4):e34180. doi:10.1371/journal.pone.0034180

Rose CD, Hassler WW (1974) Food habits and sex ratios of dolphin captured in the western Atlantic Ocean off Hatteras, North Carolina. Trans Am Fish Soc 103:94–100

Ross ST (2001) The inland fishes of Mississippi. University Press of Mississippi, Jackson, MS, USA, 624 p

Ross JL, Stevens TM, Vaughan DS (1995) Age, growth, mortality, and reproductive biology of red drums in North Carolina waters. Trans Am Fish Soc 124:37–54

Rowe GT, Kennicutt MC II (2009) Northern Gulf of Mexico Continental Slope habitats and benthic ecology study: Final report. OCS Study MMS 2009-039. U.S. Department of the Interior, Minerals Management Service, Gulf of Mexico OCS Region, New Orleans, LA, USA, 456 p

SAFMC (South Atlantic Fishery Management Council) (1998) Dolphin/Wahoo workshop report. Charleston, SC, USA, May

Saloman CH, Naughton SP (1983) Food of king mackerel, *Scomberomorus cavallas*, from the Southeastern United States including the Gulf of Mexico. NOAA Technical Memorandum

NMFS-SEFC-126. U.S. Department of Commerce, NOAA, NMFS, Washington, DC, USA, 25 p

Sarà G, Sarà R (2007) Feeding habits and trophic levels of bluefin tuna (*Thunnus thynnus*) of different size classes in the Mediterranean Sea. J Appl Ichthyol 23:122–127

Schaefer KM (2001) Reproductive biology of tunas. Fish Physiol 19:225–270

Schaefer HC, Fable WA Jr (1994) King mackerel, *Scomberomorus cavallas*, mark-recapture studies off Florida's East Coast. Mar Fish Rev 56:13–23

Schirripa MJ, Burns KM (1997) Growth estimates for three species of reef fish in the eastern Gulf of Mexico. Bull Mar Sci 61:581–591

Schirripa MJ, Legault CM, Ortiz M (1999) The red grouper fishery of the Gulf of Mexico: Assessment 3.0. SEDAR 12-RD05, 98/99-56. Sustainable Fisheries Division, Southeast Fisheries Science Center, Miami, FL, USA, 58 p

Scott G, Porch C (2007) Additional VPA analysis of northern Atlantic swordfish. Collect Vol Sci Pap ICCAT 60:2069–2076

Scott GP, Turner SC, Churchill GB, Richards WJ, Brothers EB (1993) Indices of larval bluefin tuna, *Thunnus thynnus*, abundance in the Gulf of Mexico: Modelling variability in growth, mortality, and gear selectivity. Bull Mar Sci 53:912–929

Scubaguys (2016) Manta flying thru water. Bigstockphoto, New York, New York, USA. http://www.bigstockphoto.com/image-5985799/stock-photo-manta-flying-thru-water. Accessed 5 December 2016

semet (2013) School of amberjacks with a shipwreck behind. iStockphoto, Calgary, Alberta, Canada. http://www.istockphoto.com/photo/school-of-amberjacks-with-a-shipwreck-behind-gm182746082-12690060. Accessed December 5, 2016

SEDAR 7 (Southeast Data, Assessment and Review) (2005) SEDAR 7, stock assessment report, Gulf of Mexico Red Snapper. North Charleston, SC, USA, 480 p

SEDAR 9 (2006) SEDAR 9, stock assessment report, Gulf of Mexico Greater Amberjack. North Charleston, SC, USA, 178 p

SEDAR 11 (2006) SEDAR 11, stock assessment report, large coastal shark complex, blacktip and sandbar shark. Panama City, FL, USA, 387 p

SEDAR 12 (2006) SEDAR 12, Stock assessment report, Gulf of Mexico red grouper. North Charleston, SC, USA, 358 p

SEDAR 13 (2007) SEDAR 13, stock assessment report, small coastal complex, Atlantic sharpnose, blacknose, bonnethead, and finetooth shark. Panama City, FL, USA, 395 p

SEDAR 7 (2009) SEDAR 7 update, stock assessment report, Gulf of Mexico Red Snapper. Miami, FL, USA, 224 p

SEDAR 12 (2009) SEDAR 12 update, stock assessment report, Gulf of Mexico Red Grouper. North Charleston, SC, USA, 143 p

SEDAR 16 (2009) SEDAR 16, stock assessment report, South Atlantic and Gulf of Mexico King Mackerel. North Charleston, SC, USA, 484 p

SEDAR 31 (2009) SEDAR 31, stock assessment report, Gulf of Mexico red snapper. Pensacola, FL, USA, 224 p

SEDAR 9 Update (2011) SEDAR 9 update, stock assessment report, Gulf of Mexico greater amberjack. North Charleston, SC, USA, 190 p

SEDAR 21 (2011) SEDAR 21, stock assessment report, HMS sandbar shark stock assessment report. Annapolis, MD, USA, 459 p

SEDAR 22 (2011) SEDAR 22, stock assessment report, Gulf of Mexico tilefish. North Charleston, SC, USA, 467 p

SEDAR 27 (2011) SEDAR 27, stock assessment report, Gulf of Mexico Menhaden. North Charleston, SC, USA, 492 p

SEDAR 29 (2012) SEDAR 29, stock assessment report, HMS Gulf of Mexico blacktip shark. North Charleston, SC, USA, 197 p

SEDAR 12 Addendum 1 (2007) SEDAR 12, stock assessment report, Gulf of Mexico Red Grouper. Miami, FL, USA, 64 p

Sedberry GR, Loefer J (2001) Satellite telemetry tracking of swordfish, *Xiphias gladius*, off the eastern United States. Mar Biol 139:355–360

Sedberry GR, Pashuk O, Wyanski DM, Stephen JA, Weinbach P (2006) Spawning locations for Atlantic reef fishes off the Southeastern U.S. In: Proceedings, 57th Annual Meeting of the Gulf and Caribbean Fisheries Institute, St. Petersburg, FL, USA, November, pp 463–514

SEFSC (Southeast Fisheries Science Center) (1996) Report on the shark evaluation workshop. NMFSC, Sustainable Fisheries Division, Miami, FL, USA, 80 p

SEFSC (1998) Report on the shark evaluation workshop. NMFSC, SEFSC, Southeast Fisheries Science Center, Sustainable Fisheries Division, Miami, FL, USA, 109 p

Seki MP, Polovina JJ, Kobayashi DR, Bidigare RR, Mitchum GT (2002) An oceanographic characterization of swordfish (*Xiphias gladius*) longline fishing grounds in the springtime subtropical North Pacific. Fish Oceanogr 11:251–266

Serafy JE, Cowen RK, Paris CB, Capo TR, Luthy SA (2003) Evidence of blue marlin, *Makaira nigricans*, spawning in the vicinity of Exuma Sound, Bahamas. Mar Freshw Res 54:299–306

Seyoum S, Tringali MD, Bert TM, McElroy D, Stokes R (2000) An analysis of genetic population structure in red drum, *Sciaenops ocellatus*, based on mtDNA control region sequences. Fish Bull 98:127–138

Shaw RF, Cowan JH Jr, Tillman TL (1985a) Distribution and density of *Brevoortia patronus* (Gulf menhaden) eggs and larvae in the continental shelf waters of Western Louisiana. Bull Mar Sci 36:96–103

Shaw RF, Wiseman WJ Jr, Turner RE, Rouse LJ Jr, Condrey RE, Kelly FJ Jr (1985b) Transport of larval Gulf menhaden *Brevoortia patronus* in continental shelf waters of western Louisiana: a hypothesis. Trans Am Fish Soc 114:452–460

Shepard KE, Patterson WF, DeVries DA, Ortiz M (2010) Contemporary versus historical estimates of king mackerel (*Scomberomorus cavallas*) age and growth in the U.S. Atlantic Ocean and Gulf of Mexico. Bull Mar Sci 86:515–532

Sheperd TD, Myers RA (2005) Direct and indirect fishery effects on small coastal elasmobranchs in the northern Gulf of Mexico. Ecol Lett 8:1095–1104

Shipp RL (1992) Introduction: biogeography of Alabama's marine fishes. In: Boschung HT Jr (ed) Catalog of freshwater and marine fishes of Alabama. Alabama Museum of Natural History Bulletin No. 14. University of Alabama, Tuscaloosa, AL, USA, pp 7–9

Simmons EG (1957) Ecological survey of the Upper Laguna Madre of Texas. Publ Inst Mar Sci Univ Texas 4:156–200

Simmons EG, Breuer JP (1962) A study of redfish (*Sciaenops ocellatus* Linnaeus) and black drum (*Pogonias cromis* Linnaeus). Publ Inst Mar Sci Univ Texas 8:184–211

Simms J (2009). Early life ecology of sailfish, *Istiophorus platypterus*, in the Northern Gulf of Mexico. MS thesis, Texas A&M University, College Station, TX, USA, 49 p

Simpfendorfer CA, Wiley TR (2005) Determination of the distribution of Florida's remnant sawfish population and identification of areas critical to their conservation, final report. Florida Fish and Wildlife Conservation Commission, Tallahassee, FL, USA, 40 p

Smith CL (1961) Synopsis of biological data on groupers (*Epinephelus* and Allied Genera) of the Western North Atlantic. FAO Fisheries Biology Synopsis No. 23. Food and Agriculture Organization of the United Nations, Rome, LZ, Italy, 61 p

Smith ML, Carpenter KE, Waller RW (2002) An introduction to the oceanography, geology, biogeography, and fisheries of the tropical and subtropical Western Central Atlantic. In:

Carpenter KE (ed) FAO species identification guide for fishery purposes: The living marine resources of the Western Central Atlantic, introduction, molluscs, crustaceans, hagfishes, sharks, batoid fishes, and chimaeras, vol 1. American Society of Ichthyologists and Herpetologists Special Publication No. 5. Food and Agriculture Organization of the United Nations, Rome, LZ, Italy, pp 1375–1398

Spitzer PR (1989) Osprey. In: Proceedings, Northeast Raptor Management Symposium Workshop, National Wildlife Federation, Washington, DC, USA, March, pp 22–29

Springer VG (1960) Icthyological surveys of the lower St. Lucie and Indian Rivers, Florida East Coast. Mimeo file report no. 60-19. Florida State Board of Conservation Marine Laboratory, St. Petersburg, FL, USA, 22 p

Springer VG (1964) A revision of the carcharhinid shark genera *Scoliodon*, *Loxodon*, and *Rhizoprionodon*. Proc US Nat Mus 115:559–632

Stanley DR (1994) Seasonal and spatial abundance and size distribution of fishes associated with a petroleum platform in the Northern Gulf of Mexico. PhD Thesis, Louisiana State University, Baton Rouge, LA, USA, 123 p

Stokesbury MJW, Teo SLH, Seitz A, O'Dor RK, Block BA (2004) Movement of Atlantic bluefin tuna (*Thunnus thynnus*) as determined by satellite tagging experiments initiated off New England. Can J Fish Aquat Sci 61:1976–1987

Strong WR Jr, Murphy RC, Bruce BD, Nelson DR (1992) Movements and associated observations of bait-attracted white sharks, *Carcharodon carcharias*: a preliminary report. Aust J Mar Freshw Res 43:13–20

Sturges W, Leben R (2000) Frequency of ring separations from the Loop Current in the Gulf of Mexico: A revised estimate. J Phys Oceanogr 30:1814–1819

Sutter FC, Williams RO, Godcharles MF (1991) Movement patterns and stock affinities of king mackerel in the southeastern United States. Fish Bull 89:315–324

Suttkus RD (1956) Early life history of the Gulf menhaden, *Brevoortia patronus*, in Louisiana. Trans N Amer Wildl Conf 21:290–307

Szedlmayer ST, Conti J (1999) Nursery habitats, growth rates, and seasonality of age-0 red snapper, *Lutjanus campechanus*, in the Northeast Gulf of Mexico. Fish Bull 97:626–635

Szedlmayer ST, Furman C (2000) Estimation of abundance, mortality, fecundity, age frequency, and growth rates of red snapper, *Lutjanus campechanus*, from a fishery independent, stratified random survey, final report for 1999. Gulf and South Atlantic Fisheries Development Foundation, Tampa, FL, USA, 43 p

Szedlmayer ST, Howe JC (1997) Substrate preference in age-0 red snapper, *Lutjanus campechanus*. Environ Biol Fish 50:203–207

Szedlmayer ST, Lee JD (2004) Diet shifts of juvenile red snapper (*Lutjanus campechanus*) with changes in habitat and fish size. Fish Bull 102:366–375

Szedlmayer ST, Shipp RL (1994) Movement and growth of red snapper, *Lutjanus campechanus*, from an artificial reef area in the northeastern Gulf of Mexico. Bull Mar Sci 55:2–3

Takahashi M, Okamura H, Yokawa K, Okazaki M (2003) Swimming behaviour and migration of a swordfish recorded by an archival tag. Mar Freshw Res 54:527–534

Taylor HF (1951) Survey of marine fisheries of North Carolina. University of North Carolina Press, Chapel Hill, NC, USA, 555 p

Teo SLH, Boustany A, Block BA (2007a) Oceanographic preferences of Atlantic bluefin tuna, *Thunnus thynnus*, on their Gulf of Mexico breeding grounds. Mar Biol 152:1105–1119

Teo SLH, Boustany A, Dewar H, Stokesbury M, Weng K, Beemer S, Seitz A, Farwell C, Prince ED, Block BA (2007b) Annual migrations, diving behavior and thermal biology of Atlantic bluefin tuna, *Thunnus thynnus*, on their Gulf of Mexico breeding grounds. Mar Biol 151:1–18

Thompson BA, Render JH, Allen KL, Nieland DL (1989) Life history and population dynamics of commercially harvested striped mullet *Mugil cephalus* in Louisiana. Final report to the Louisiana Board of Regents. LSU-CFI-89-01. Coastal Fisheries Institute, Louisiana State University, Baton Rouge, LA, USA, 80 p

Thompson BA, Wilson CA, Render JH, Beasley M, Cauthron C (1991) Age, growth, and reproductive biology of greater amberjack and cobia from Louisiana waters. MARFIN final report NA90AAH-MF722. Marine Fisheries Research Initiative Program, National Marine Fisheries Service, St. Petersburg, FL, USA, 77 p

Thompson BA, Beasley M, Wilson CA (1999) Age distribution and growth of greater amberjack, *Seriola dumerili*, from the North-Central Gulf of Mexico. Fish Bull 97:362–371

Thomson JM (1963) Synopsis of biological data on the grey mullet *Mugil cephalus*, Linnaeus 1758. Australia CSIRO Division of Fisheries and Oceanography, Fishery Synopsis No. 1. Hobart, Tasmania, Australia, 68 p

Tiews K (1963). Synopsis of biological data on the bluefin tuna *Thunnus thynnus* (Linnaeus) 1758 (Atlantic and Mediterranean). In: Rosa H Jr (ed) Proceedings of the World Scientific Meeting on the Biology of Tunas and Related Species. FAO Fisheries Synopsis No. 6. Food and Agriculture Organization of the United Nations, Rome, LZ, Italy, pp 422–481

Toll RB, Hess SC (1981) Cephalopods in the diet of the swordfish, *Xiphias gladius*, from the Florida Straits. Fish Bull 79:765–774

Tserpes G, Tsimenides N (1995) Determination of age and growth of swordfish, *Xiphias gladius* L., 1758, in the eastern Mediterranean using anal-fin spines. Fish Bull 93:594–602

Tserpes G, Peristeraki P, Valavanis VD (2008) Distribution of swordfish in the eastern Mediterranean, in relation to environmental factors and the species biology. Hydrobiologia 612:241–250

Turner WR (1969) Life history of menhadens in the Eastern Gulf of Mexico. Trans Am Fish Soc 98:216–224

Turner SC, Restrepo VR (1994) A review of the growth rate of West Atlantic bluefin tuna, *Thunnus thynnus*, estimated from marked and recaptured fish. Collect Vol Sci Pap ICCAT 42:170–172

Turner SC, Grimes CB, Able KW (1983) Growth, mortality, and age/size structure of the fisheries for tilefish, *Lopholatilus chamaeleonticeps*, in the middle Atlantic-Southern New England region. Fish Bull 81:751–763

Turner SC, Arocha F, Scott GP (1996) U.S. swordfish catch at age by sex. Collect Vol Sci Pap ICCAT 45:373–378

Turner SC, Cummings NJ, Porch CE (2000) Stock assessments of Gulf of Mexico greater amberjack using data through 1998. SFD 99/00-100. Southeast Fisheries Science Center, National Marine Fisheries Service, Miami, FL, USA, 27 p

Uchida S, Toda M, Teshima K, Yano K (1996) Pregnant white sharks and full-term embryos from Japan. In: Klimley AP, Ainley DG (eds) Great White Sharks: The biology of *Carcharodon carcharias*. Academic, Academic Press, San Diego, CA, USA, pp 139–155

Uotani I, Matsuzaki K, Makino Y, Noda K, Inamura O, Horikawa M (1981) Food habits of larvae of tunas and their related species in the area northwest of Australia. Bull Jap Soc Sci Fish 47:1165–1172

Uotani I, Saito T, Hiranuma K, Nishikawa Y (1990) Feeding habit of bluefin tuna *Thunnus thynnus* larvae in the western North Pacific Ocean. Nippon Suisan Gakkaishi 56:713–717

USEPA (U.S. Environmental Protection Agency) (1994) Living aquatic resources action agenda for the Gulf of Mexico, first generation—management committee report. EPA 800-B-94-007. U.S. Environmental Protection Agency, Office of Water, Gulf of Mexico Program, Stennis, MS, USA, 168 p

USEPA (2004) National coastal condition report II. EPA 620/R-03/002, U.S. Environmental Protection Agency, Office of Research and Development, Washington, DC, USA, 286 p

USGS (U.S. Geological Survey) (2010a). Gulf Coast wildlife habitat ranges: Scamp grouper, mangrove snapper and red snapper. http://www.usgs.gov/oilspill/sesc/maps.asp#ranges/. Accessed 31 May 2013

USGS (2010b) Gulf Coast wildlife habitat ranges: Bay anchovy, menhaden and striped mullet. http://www.usgs.gov/oilspill/sesc/maps.asp#ranges/. Accessed 31 May 2013

USGS (2010c) Gulf Coast Wildlife habitat ranges: Red drum, black drum, sand seatrout, spotted seatrout, Atlantic croaker and southern kingfish. http://www.usgs.gov/oilspill/sesc/maps.asp#ranges/. Accessed 31 May 2013

USGS (2010d) Gulf Coast Wildlife habitat ranges: King mackerel, Spanish mackerel and sailfish. http://www.usgs.gov/oilspill/sesc/maps.asp#ranges/. Accessed 31 May 2013

VanderKooy SJ, Smith JW (eds) (2002) The Menhaden Fishery of the Gulf of Mexico, United States: A regional management plan, 2002 revision. Report No. 99. Gulf States Marine Fisheries Commission, Ocean Springs, MS, USA, 143 p

Vaughan DS, Smith JW, Prager MH (2000) Population characteristics of gulf menhaden, *Brevoortia patronus*. NOAA technical report NMFS 149. National Marine Fisheries Service, Seattle, WA, USA, 19 p

Vaughan DS, Shertzer KW, Smith JW (2007) Gulf menhaden (*Brevoortia patronus*) in the U.S. Gulf of Mexico: Fishery characteristics and biological reference points for management. Fish Res 83:263–275

Vinnichenko VI (1996) New data on the distribution of some species of tuna (Scombridae) in the North Atlantic. J Ichthyol 36:679–681

von Brandis R (2013) Red snapper profile. iStockphoto, Calgary, Alberta, Canada. http://www.istockphoto.com/stock-photo-20240311-red-snapper-profile.php. Accessed 4 December 2016

Walter J, Ingram W (2009) Exploration of the NMFS bottom longline survey for potential deepwater cryptic biomass. National Marine Fisheries Service, Southeast Fisheries Science Center, 18 p. http://ftp.gulfcouncil.org/. Accessed 31 May 2013

Walter JF, Cook M, Linton B, Lombardi L, Quinlan JA (2011) Explorations of habitat associations of yellowedge grouper and golden tilefish. SEDAR 22-DW-05. Southeast Data, Assessment and Review, North Charleston, SC, USA, 26 p

Wells RJD (2007) The effects of trawling and habitat use on red snapper and the associated community. PhD Thesis, Louisiana State University, Baton Rouge, LA, USA, 179 p

Wells RJD, Rooker JR (2003) Distribution and abundance of fishes associated with *Sargassum* mats in the NW Gulf of Mexico. In: Proceedings, 54th Annual Meeting of the Gulf and Caribbean Fisheries Institute, Providenciales, Turks & Caicos Islands, November, pp 609–621

Wells RJD, Rooker JR (2004a) Distribution, age, and growth of young-of-the-year greater amberjack (*Seriola dumerili*) associated with pelagic *Sargassum*. Fish Bull 102:545–554

Wells RJD, Rooker JR (2004b) Spatial and temporal patterns of habitat use by fishes associated with *Sargassum* mats in the northwestern Gulf of Mexico. Bull Mar Sci 74:81–99

Wells RJD, Rooker JR, Prince ED (2010) Regional variation in the otolith chemistry of blue marlin (*Makaira nigricans*) and white marlin (*Tetrapturus albidus*) from the western North Atlantic Ocean. Fish Res 106:430–435

Wenner C (1999) Red drum: Natural history and fishing techniques in South Carolina. Educational report no. 17. Marine Resources Research Institute, Marine Resources Division, South Carolina Department of Natural Resources, Charleston, SC, USA, 40 p

Wilson CA, Nieland DL (1994) Reproductive biology of red drum, *Sciaenops ocellatus*, from the neritic waters of the northern Gulf of Mexico. Fish Bull 92:841–850

Wilson CA, Nieland DL (2001) Age and growth of red snapper, *Lutjanus campechanus*, from the northern Gulf of Mexico off Louisiana. Fish Bull 99:653–664

Wilson CA, Dean JM, Prince ED, Lee DW (1991) An examination of sexual dimorphism in Atlantic and Pacific blue marlin using body weight, sagittae weight, and age estimates. J Exp Mar Biol Ecol 151:209–225

Wollam MB (1970) Description and distribution of larvae and early juveniles of king mackerel, *Scomberomorus cavallas* (Cuvier), and Spanish mackerel, *Scomberomorus maculatus* (Mitchill); (Pisces: Scombridae); in the Western North Atlantic. Florida Board of Conservation, Marine Research Lab, Technical Series No. 61. Florida Department of Natural Resources, Division of Marine Resources, St. Petersburg, FL, USA, 35 p

Wood M (2016) A school of striped mullet swim along the bottom of Fanning Springs, Florida. http://www.gettyimages.com/photos/a-school-of-striped-mullet-swim?excludenudity=true &family=creative&page=1&phrase=a%20school%20of%20striped%20mullet%20swim& sort=best#license. Accessed 13 December 2016.

Woods MK (2003) Demographic differences in reproductive biology of female red snapper (*Lutjanus campechanus*) in the Northern Gulf of Mexico. MS Thesis, University of South Alabama, Mobile, AL, USA, 129 p

Woods MK, Cowan JH Jr, Nieland DL (2007) Demographic difference in northern Gulf of Mexico red snapper reproduction maturation: Implication for the unit stock hypothesis. In: Patterson WF, III, Cowan JH Jr, Fitzhugh GR, Nieland DL (eds) Red snapper ecology and fisheries in the U.S. Gulf of Mexico. American Fisheries Society symposium series 60. Bethesda, MD, USA, pp 217–277

Workman IK, Foster DG (1994) Occurrence and behavior of juvenile red snapper, *Lutjanus campechanus*, on commercial shrimp fishing grounds in the northeastern Gulf of Mexico. Mar Fish Rev 56:9–11

Workman IK, Shah A, Foster D, Hataway B (2002) Habitat preferences and site fidelity of juvenile red snapper (*Lutjanus campechanus*). ICES J Mar Sci 59(Suppl):S43–S50

Yáñez E, Silva C, Barbieri MA, Órdenes A, Vega R (2009) Environmental conditions associated with swordfish size compositions and catches off the Chilean coast. Latin Am J Aquat Res 37:71–81

Yokel BJ (1966) A contribution to the biology and distribution of the red drum, *Sciaenops ocellatus*. University of Miami, Coral Gables, FL, USA, 160 p

Open Access This chapter is licensed under the terms of the Creative Commons Attribution-NonCommercial 2.5 International License (http://creativecommons.org/licenses/by-nc/2.5/), which permits any noncommercial use, sharing, adaptation, distribution and reproduction in any medium or format, as long as you give appropriate credit to the original author(s) and the source, provide a link to the Creative Commons license and indicate if changes were made.

The images or other third party material in this chapter are included in the chapter's Creative Commons license, unless indicated otherwise in a credit line to the material. If material is not included in the chapter's Creative Commons license and your intended use is not permitted by statutory regulation or exceeds the permitted use, you will need to obtain permission directly from the copyright holder.

CHAPTER 10

COMMERCIAL AND RECREATIONAL FISHERIES OF THE GULF OF MEXICO

Walter R. Keithly Jr.[1] and Kenneth J. Roberts[1]

[1]Louisiana State University, Baton Rouge, LA, USA
walterk@lsu.edu

10.1 INTRODUCTION

The users of the Gulf of Mexico living marine resources are as diverse as the species and habitats. Depicting the economic components both annually and over time generally is based on agency-collected data primarily focused on landings. The revenue element of use being well documented serves commercial industry analyses partially and leaves a void that confronts recreational industry researchers. Missing critical elements for depicting economic conditions include, but are not limited to, production costs, expenditures by anglers, site-specific data, marketing and processing prices, and margins. Research at universities, by consultants, and within agencies on various economic issues occurs on a project basis. Project studies do not occur consistently enough over time on any species, much less a large enough component of Gulf of Mexico species, to be relied upon for the increasingly complex mix of decisions faced by agencies. Agencies in turn must be responsive to harvesters and increasingly strong regional and national nongovernmental organizations (NGOs). Agencies, users groups, and NGOs face decisions that include habitat protection, avoidance of indirect impacts of harvest gear, access, determining initial catch shares, allocations, law enforcement, and juxtaposition with other agency regulations. The existing data reporting system relied on for this chapter cannot be expected to adequately serve economic researchers addressing the range of inquiries associated with commerce in fisheries. Special projects of short duration from various funding sources most likely will be necessary to meet the needs of participants in the decision-making process. This chapter makes use of the data reporting systems maintained by agencies. State agencies in the Gulf of Mexico are generally unified in their reporting via agreements founded by the Gulf States Marine Fisheries Commission (GSMFC). This congressionally authorized commission has an increasing presence in organizing fisheries data and providing Internet access in a timely manner. Of particular interest is the GSMFC's role in specific analyses focused to fill special needs. The most recent example is commitment to a multiyear economic study of the inshore commercial shrimping sector. This economic analysis fills a void and has added value as it can be coupled with findings of National Oceanic and Atmospheric Administration (NOAA) Fisheries' research. Beginning in 2006, NOAA Fisheries began an annual economic survey of federal Gulf of Mexico shrimp permit holders that provides valuable insight over time of the region's largest commercial fishery. Essentially all other economic perspective of Gulf of Mexico commercial fisheries must be ascertained from annual NOAA and GSMFC reports interspersed with irregularly funded special projects.

When addressing the complexities of the angler harvest of Gulf of Mexico species, economists are no richer in terms of data sources. The core source of most reports is the Marine Recreational Fisheries Statistics Survey (MRFSS), which was later renamed the

© The Author(s) 2017
C.H. Ward (ed.), *Habitats and Biota of the Gulf of Mexico: Before the Deepwater Horizon Oil Spill*,
DOI 10.1007/978-1-4939-3456-0_2

Marine Recreational Information Program (MRIP). Established in 1979, the MRFSS evolved over the years into a system reflective of the difficulties associated with estimating (1) catch by species, (2) participants, (3) fishing by location, (4) target species, (5) fishing mode, and (6) expenditures by anglers. The use of the database was undertaken with knowledge of changes made over time to improve not only the representativeness of the data but also access. It is noteworthy that the state of Texas does not participate in the annual MRFSS/MRIP survey. Consequently, all discussion of catch by species, participation, and trips made by anglers are exclusive of Texas. However, there is Texas data on angler expenditures and related multipliers included from other sources to make that section as complete as possible. The recreational fisheries are addressed on the basis of economic activities associated with the pursuit of fish. Expenditures and associated indirect impacts springing from multiplier effects must serve as both the cost of angling and the base from which gross benefits can be estimated.

This chapter deals with the complexity of angling with attention to the Gulf of Mexico and state levels inclusive of species-specific findings to give the best possible descriptive background of the marine recreational fisheries. With the understanding that the commercial harvest of Gulf of Mexico fish species is a capture and sale process, there can be minimal comparability with the pursuit of recreational fisheries in terms of economics. Decisions on the use of Gulf of Mexico marine fish species will remain an interesting public process as data improves and economic analyses become more numerous with attention to both descriptive and analytical needs. Beginning with a review of federal, regional, and state management, a review of the commercial and recreational fishing industry in the Gulf of Mexico will be presented in general and for specific, commercially and recreationally important marine species. With respect to the commercial sector, emphasis is given to analysis of the shrimp, crab, menhaden (*Brevoortia patronus*), oyster, and reef fish industries. Recreationally important marine species for which special emphasis is given include spotted seatrout (*Cynoscion nebulosus*), red drum (*Sciaenops ocellatus*), groupers, snappers, and coastal pelagics. This review also includes estimates of expenditure and cost multipliers associated with input–output analyses. This assessment will focus first on the commercial fishing industry followed by the recreational angler-based industry. The chapter ends with a review of the Florida, Mississippi, Alabama, and Louisiana harvests since they represent major recreational fishing foci. A summary of the results of this review is presented in the final section of this chapter.

10.2 THE MANAGEMENT PROCESS AT ITS BASE

The mobility of most living marine resources pursued for harvest results in three levels of public entities—federal, state, and regional—being involved in management for the sustainable flow of benefits. Federal, state, and regional responsibilities established by law are approached by entities with similar but not uniform authorizations. Often, agencies charged with the management of fisheries resources in the Gulf of Mexico evolve with expanded abilities to influence the use of marine species. Criteria for guiding the public use of fishery resources can be found in legislation but more frequently in regulations promulgated by agencies. It is beyond the needs of this document to detail the regulations and authority by which agencies act to move resources toward sustainability. Agency websites can be searched for insight to the origin of authorizing legislation and current status of species-specific management activities.

10.2.1 Federal Oversight: National Oceanic and Atmospheric Administration

Of the agencies, the federal level is the most subject to change. Passage of the Fishery Conservation and Management Act (FCMA) in 1976 began an increased level of oversight at the federal level. The passage was associated with many prior years of numerous nations extending fisheries oversight to 200 miles (mi) (322 kilometers [km]). Fishery management councils were authorized around the nation. The membership of the Gulf of Mexico Fishery Management Council (GMFMC) included (1) state fishery agency representatives from Florida, Alabama, Mississippi, Louisiana, and Texas, (2) citizens appointed by the U.S. Secretary of Commerce from nominations by the region's governors, and (3) NOAA Fisheries' regional director. GMFMC develops fishery management plans for species common to the federal Exclusive Economic Zone (EEZ). Plan development evolves from guidelines established by federal legislation with frequent amendments necessary due to changing (1) use patterns, (2) technologies of fish harvesting, (3) legislation, (4) data, and (5) analysis methodologies. NOAA has final authority to approve, modify, or deny any amendment to a fishery management plan emanating from the GMFMC.

10.2.2 State Agency Management

The five states with Gulf coastal borders have authority to manage fishery resources on the basis of their preferred regulatory approaches to achieving goals. All have similar goals regarding conserving living resources for sustainable use over time. Though the focus is on state waters, there is the need for substantial interaction and cooperation with other states and the GMFMC. The movement of many species at critical life phases to waters of other Gulf States and waters seaward of state coastal boundaries necessitates formal working relationships to assure oversight throughout the various habitats. Seaward coastal boundaries vary from 9 mi (14.5 km) in Texas and the west coast of Florida to the traditional 3 mi (4.8 km) for the other three states on the Gulf. State agencies have designees on the GMFMC to convey local regulatory perspectives in the federal fishery plan development process. When species are totally within state waters or move laterally along the coast, coastal state regional coordination is authorized through the GSMFC.

The shrimp fisheries exemplify complexity for the management structure in the Gulf. The shrimp industry in Louisiana waters produces the Gulf's largest landings in pounds. Agency management approaches involve a large inshore fishery and harvest of smaller shrimp sizes (i.e., a larger number of shrimp to the pound at harvest). The management from Texas' state agency, Texas Parks and Wildlife, is for a lessened inshore catch and cooperative management with the GMFMC for larger-sized shrimp (i.e., fewer shrimp to the harvested pound).

Texas is unique among the states in that it has a voluntary commercial fishing license buyback program. The license buyback programs for bay shrimp, blue crab (*Callinectes sapidus*), and finfish seek to stabilize fishing efforts through time in order to promote healthy fisheries stocks. Funds for the buyback come from a surcharge on related commercial fishing licenses and a saltwater fishing stamp endorsement to recreational licenses.

10.2.3 Gulf States Marine Fisheries Commission

Acknowledging the joint interest of coastal states to achieve multiple goals for management of mobile fishery resources, Congress authorized the formation of multistate commissions in 1949. Utilization of fishery resources to meet food, employment, economic, and recreation needs of citizens was reasoned to be facilitated by use based on conservation and a multistate oversight. The GSMFC includes 15 commissioners to oversee the implementation and evaluation of efforts to coordinate management among Gulf States. Each governor appoints a commissioner and each state legislature appoints one as well. The other five commissioners are the state fishery agency directors.

Though the GSMFC does not have direct regulatory authority, it clearly has been successful in stimulating deliberations leading to cooperative planning, data programs, and research. An understanding of the key role that fishery data improvement plays in goal achievement for the Gulf has been a visible part of GSMFC actions. While there are many GSMFC programs, the creation of fishery-independent and fishery-dependent data collection programs serve to prove the value of regional cooperation. The former is termed the Southeast Area Monitoring and Assessment Program (SEAMAP). The latter comprises two elements: (1) Commercial Fisheries Information Network (ComFIN) and (2) Recreational Fisheries Information Network (RecFIN). A common element of both the ComFIN and RecFIN programs is an emerging program to administer collection of economic information on Gulf fisheries.

Following the active hurricane year of 2005, Congress assigned the GSMFC a leadership role in recovery programming. A 5-year program began in 2006 to oversee rehabilitation and recovery efforts. This emergency assistance to Gulf States established a format for action that resulted in valuable experience on enabling fisheries agencies to respond with coordinated programs.

10.3 GULF OF MEXICO COMMERCIAL FISHERIES IN AGGREGATE

10.3.1 Gulf of Mexico Landings

The capture of free ranging marine species for commercial use occurs from a large area subject to both within-year and between-year variability in environmental and economic conditions. Environmental conditions including water temperature, salinity, and turbidity—in conjunction with the life cycles of many of the species that inhabit the Gulf—all contribute to availability. Species availability, in conjunction with those economic conditions that determine whether a trip will be profitable, including the price received for the harvested product and the cost of inputs used in the harvesting process, provide signals to the harvesting units as to whether a trip will be financially viable. This viability along with the multitude of regulations that can also govern fishing patterns influences fishing effort, and ultimately the catch. Considering this, landings of a specific year cannot be descriptive of Gulf fisheries from either a biological or economic viewpoint. For this chapter, the 20-year period from 1990 to 2009 was chosen as inclusive of (1) pre- and post-management agency changes, (2) active tropical storm periods, (3) challenging production cost situations, and (4) high and low points in the national economy. This approach acknowledges that a species' stock level and economic conditions of inputs and demand play roles in landings levels. This perspective conveys a need to avoid reference to beginning-year and end-year comparisons. Rather, a 3-year average was used to depict landings and associated value as the beginning and end focus of comments. There is a distinction between location of landings and location of catch. This is particularly the case for the shrimp fisheries and most finfish. Location of catch is best documented for Gulf shrimp fisheries by offshore zones east to west across the Gulf and by inshore versus offshore. When data are available to differentiate landings from catch, that data is reported in the sections dealing with key species. Data by state are also reported in the key species sections.

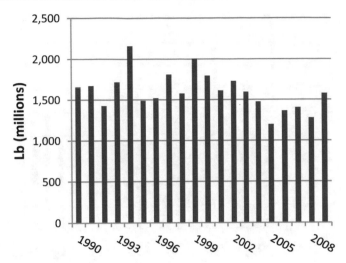

Figure 10.1. Gulf of Mexico commercial fishery landings, 1990–2009 (NMFS FSD; data accessed 2012—see Appendix A) (*Note*: 1 lb = 0.454 kg).

In the latest 3-year period, 2007–2009, landings of all species combined were 1.4 billion pounds (Figure 10.1). This was 10 % lower than the initial 3-year period (1990–1992). With respect to the nation's total fisheries, Gulf landings were near a 16 % share at the start and end of the 20-year period. Both U.S. and Gulf landings fell over the 20-year period to leave the Gulf shares essentially unchanged.

The Gulf landings share for the key species—menhaden, brown shrimp, white shrimp, blue crab, and oysters—demonstrate the national significance of the region's fisheries. These key species accounted for 94 % of Gulf landings in the latest 3-year period. While other species, primarily finfish, are harvested, their trends do not convey overall change in Gulf landings.

10.3.1.1 Menhaden Landings

Gulf Menhaden

The menhaden fishery landings for the 20-year period ranged from a high of 1.7 billion pounds in 1994 to a low of 0.8 billion pounds in 2005 (Figure 10.2). The average was a 21 % decrease for the nation. The resulting Gulf share of national landings was 69 %. Essentially all menhaden landings occur in Louisiana (80 %) and Mississippi (19 %) for the industrial production of fish meal and oils. However, this is a case where there is some divergence due to catch location. Some Louisiana landings occasionally are caught off Texas. Mississippi landings can originate from Louisiana and vice versa.

10.3.1.2 Brown Shrimp Landings

Brown Shrimp

Landings of brown shrimp (whole weight) ranged from a high of 168 million pounds in 1990 to the period low of 79 million pounds in 2008 (Figure 10.3, left panel). The average landings on

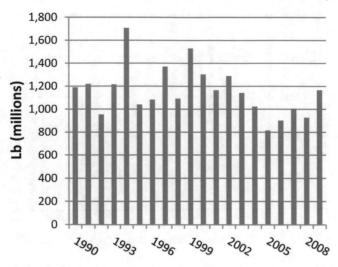

Figure 10.2. Gulf of Mexico commercial menhaden landings, 1990–2009 (NMFS FSD; data accessed 2012—see Appendix A) (*Note*: 1 lb = 0.454 kg).

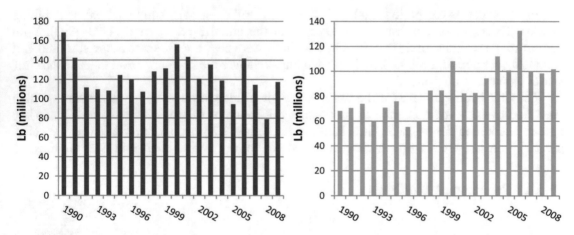

Figure 10.3. Gulf of Mexico brown shrimp (*left panel*) and white shrimp (*right panel*) landings, 1990–2009 (NMFS FSD; data accessed 2012—see Appendix A) (*Note*: 1 lb = 0.454 kg).

the basis of the 3-year groupings decreased 27 %. With Gulf landings accounting for 95 % of U.S. production, national landings were then down 27 %.

10.3.1.3   White Shrimp Landings

White Shrimp

Annual white shrimp landings ranged from the period high of 132 million pounds in 2006 to the period low of 55 million pounds in 1996 (Figure 10.3, right panel). The average annual landings, on the basis of the 3-year groupings, increased 41 %. U.S. landings showed a smaller increase (30 %) when the non-Gulf landings decrease (20 %) was included. The Gulf's increased white shrimp production for the period almost negated the lower production from the brown shrimp fishery, which left total shrimp landings essentially unchanged.

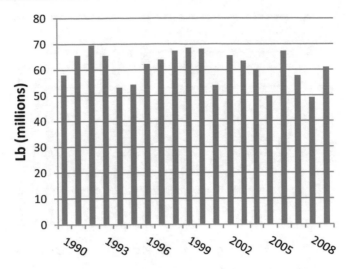

Figure 10.4. Gulf of Mexico commercial blue crab landings, 1990–2009 (NMFS FSD; data accessed 2012—see Appendix A) (*Note*: 1 lb = 0.454 kg).

10.3.1.4   Blue Crab Landings

Blue Crab

Statistics are reported for three blue crab products: (1) hard blue crab, (2) peeler crab, and (3) soft crab. Hard blue crab is, by far, the target of harvesters. Peeler is a designation for a crab in molt stage that results in a soft crab that can be marketed. Only hard blue crab landings are addressed herein, because it is the largest commodity form and also would reflect changes in levels of the other forms (Figure 10.4). The Gulf crab fishery accounts for 35 % of domestic landings with the remaining landings from Chesapeake and South Atlantic areas. Gulf landings, examined in 3-year intervals, began the period of analysis at almost 65 million pounds (i.e., 1990–1992 average) and ended the period at 56 million pounds (i.e., 2007–2009 average) for a 14 % decrease (Figure 10.4). National landings fared worse with a 26 % decrease.

10.3.1.5   Oyster Landings

Eastern Oyster

U.S. landings of eastern oysters (*Crassostrea virginica*) were essentially unchanged for the period at 24 million pounds of meat. The initial 3-year average was 23.9 million pounds of meat, and the final 3-year period average was 24.4 million pounds of meat. Gulf oyster harvesters produced 13.7 million pounds in the initial period but the average for the final 3-year period rose to 22 million pounds (61 % increase) (Figure 10.5). The 22-million-pound level for the Gulf represents 90 % of the country's eastern oyster landings.

10.3.1.6   Landings of All Other Species

Dozens of species have not been covered in the aggregate discussion of the Gulf. Although comprising approximately 6 % of total landings, many of the species are the focus of GMFMC

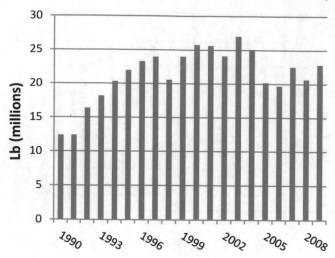

Figure 10.5. Gulf of Mexico commercial oyster landings, 1990–2009 (NMFS FSD; data accessed 2012—see Appendix A) (*Note*: 1 lb = 0.454 kg).

regulations. The reef fish complex of species includes many that are subject to technically defined designations of "subject to overfishing" and/or "overfished." As of 2009, gag grouper (*Mycteroperca microlepis*), gray triggerfish (*Balistes capriscus*), greater amberjack (*Seriola dumerili*), and red snapper (*Lutjanus campechanus*) were being managed so designated. Given the overfishing or overfished designation associated with these species, landings are constrained by regulation and significant changes in landings of these species are unlikely in the absence of a change in regulation. Changes in regulations generally reflect updated stock assessments indicating improvements/deteriorations in the health of the stock. Reef fish complex species generally entail involvement of commercial and recreational harvesters. This adds a complexity to the understanding of Gulf fisheries not present in the previously presented key species. There are small recreational harvests of oysters, blue crab, and shrimp in relation to commercial landings that are not problematic. Anglers for Gulf reef fish species are major participants in quota sharing and likely have a wider distribution throughout the Gulf landing sites than the far smaller number of commercial harvesters. More detailed discussions of the commercial harvest of reef fish species and the recreational harvest of reef fish and other species are given in subsequent sections of this chapter.

10.3.2 Aggregate Landings by State

The finfish and shellfish landings attributed to the states fluctuate as expected, yet the ranking of the states within the Gulf does not change much (Figure 10.6). Louisiana ranks first due to landings in five major species: (1) menhaden, (2) brown shrimp, (3) white shrimp, (4) blue crab, and (5) oysters. Landings are commonly above a billion pounds with menhaden accounting for 80 %. Mississippi attains the second highest landings also fueled by the menhaden fishery with a 94 % component. Most recently the west coast of Florida ranks fourth after historically holding the third spot. Landings in Texas placed third at the end of the 1990–2009 period. Alabama began and ended the period in fifth place. Differences by species among the states are presented in the sections dealing with individual key species.

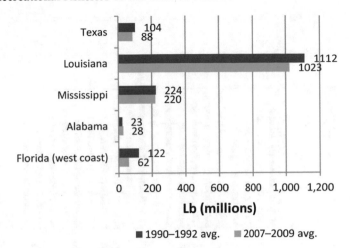

Figure 10.6. Average annual landings by state, 1990–1992 and 2007–2009 (NMFS FSD; data accessed 2012—see Appendix A) (*Note*: 1 lb = 0.454 kg).

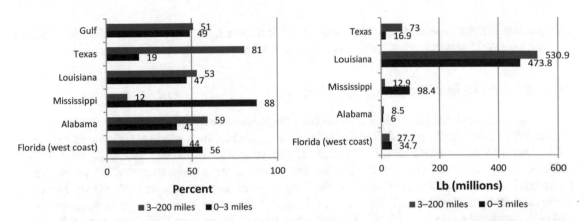

Figure 10.7. Gulf of Mexico commercial catch by distance from shore, by state, in percentage (*left panel*) and pounds (*right panel*), 1990–2009 (NMFS, Fisheries Statistics Division with percentage calculations by authors; data accessed 2012—see Appendix A) (*Note*: 1 lb = 0.454 kg).

10.3.3 Catch by Distance from Shore

The diversity of species in the Gulf subject to commercial harvest results in many being caught either totally or partially in state waters. State waters is reported in the NOAA Fisheries as 0–3 mi (0–4.8 km) offshore even though Florida has a 9 mi (14.5 km) state limit on its west coast as does Texas throughout its Gulf border. Total catch for the Gulf can be portrayed as near a 50 %–50 % split between state and federal waters (Figure 10.7). Mississippi receives the highest level of state water catch at 88 %. At the other extreme, Texas receives 81 % from the 3–200-mi (4.8–322-km) zone, largely because of a large offshore shrimp component. Louisiana, Alabama, and Florida (west coast) were nearer to receiving equal shares from state waters and offshore zones. The Gulf's large menhaden fishery generally conveys a shallow water image consistent with state waters. This accurately fits for Mississippi with 88 % of the state's catch coming from state waters. The situation is not so described in neighboring Louisiana even

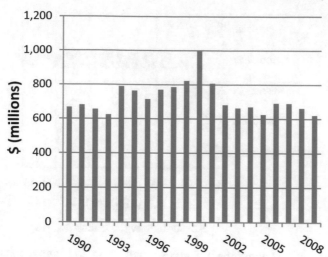

Figure 10.8. Gulf of Mexico dockside value of commercial landings, 1990–2009 (NMFS FSD; data accessed 2012—see Appendix A).

though the harvest methods are the same. Louisiana, with 53 % of total catch from federal waters, can only attain such a level if menhaden comprises a large part of the catch.

10.3.4 Dockside Value of Landings

Gulf fisheries brought in $658 million (i.e., dockside value), on average, for the last 3 years of the 1990–2009 study period. The first 3-year average for the period was $568 million for an 18 % increase in nominal terms. The high single year was 2000 with value at $997 million (Figure 10.8). The last year of the period had value at its lowest over the 20 years. Value increased while landings decreased 10 %. Key species values were mixed: (1) oyster value increased 89 % under increased supplies of 61 %, (2) blue crab landings were 14 % lower with a value increase response of 59 %, (3) menhaden landed value was 15 % higher on 9 % lower landings, (4) white shrimp value was up 24 % on much higher landings of 41 %, and (5) brown shrimp was 37 % lower on a drop of 27 % in landings. Recall that these are for 3-year averages at the start and end of the 1990–2009 period.

NOAA Fisheries maintains an ex-vessel price series with 1982 as the base year (i.e., 1982 = 100). The ex-vessel price indexes for blue crab, oysters, menhaden, and Gulf and South Atlantic shrimp are good descriptors for the Gulf. However, none of the edible finfish from the Gulf have price indexes. The substitute index used herein is that of total edible finfish in the country. Edible finfish ex-vessel prices in 1990 had an index of 130 but ended at 117 in 2009. The interpretation is that overall finfish ex-vessel prices were 30 % higher in 1990 compared to 1982 but only 17 % higher by 2009. The index for blue crab was at 152 in 1990 with a large increase to 383 by 2009. Oyster harvesters were successful marketing in 1990 at prices that put the index at 228, the highest index for the key species. By 2009, the oyster index reflected more favorable conditions with an index of 273. Ex-vessel prices in the vertically integrated menhaden industry are estimated from a small number of firms. The index levels in 1990 and 2009 were 128 and 154, respectively. The situation for shrimp necessitated that all warm water shrimp be used in the calculation, not just the brown and white shrimp noted previously in this chapter. Brown and white shrimp commonly comprise over 95 % of landings. For 1990, the index was 79 signaling a 21 % decrease from the 1982 base. Although there were

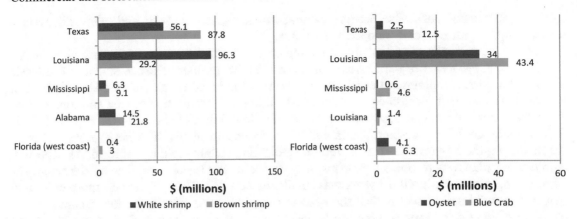

Figure 10.9. Value of commercial landings by state and species (shrimp, *left panel*; oysters and blue crab, *right panel*), (2007–2009 average) (NMFS FSD; data accessed 2012—see Appendix A).

occasional exceptions for the 20-year period, the index reflected poor ex-vessel shrimp prices. Economic conditions by 2009 were not favorable, resulting in an index of 65 (i.e., price was 35 % below 1982).

When examined at the state level, the dockside value of all landings is highly concentrated in Louisiana and Texas with shares of 43 % and 26 %, respectively. The other state achieving a double-digit contribution is Florida (west coast) at 19 %. Alabama and Mississippi range from 6 to 7 % of Gulf value. Species components of the state values are widely different. Louisiana value of individual fisheries for white shrimp, blue crab, oysters, and menhaden leads among the states. For example, the commercial dockside value of Louisiana's white shrimp landings averaged $96.3 million annually during 2007–2009, which exceeded the combined values for all other states (Figure 10.9 left panel). Similarly, the 2007–2009 annual average commercial value of Louisiana's blue crab landings ($34 million) and oyster landings ($43.4 million) exceeded the combined landings from all other Gulf states (Figure 10.9 right panel). The remaining key species, brown shrimp, is dominated by Texas landings (with an average dockside value of $87.8 million during 2007–2009), followed by Louisiana ($29.2 million) and Alabama ($21.8 million) (Figure 10.9 left panel). Key species designation of the five species fits well for all but Florida (west coast). At 19 % of total Gulf value, the area only receives 11 % of its landed value from key species. Edible finfish such as groupers and snapper bring high finfish dockside prices. These species and highly valued spiny lobster (*Panulirus argus*) and stone crab (*Menippe mercenaria*) claws push the west coast's share in the Gulf (19 %) past that depicted by key species alone (3 %).

10.3.5 Processing Plants and Related Employment

The after-landings activities necessary to convert marine shellfish and finfish into marketable consumer products in varied locations around the country are substantial. A consumer product can be as basic as one in whole form that has been washed, graded, and temperature safe to labeled frozen product at retail. With the majority of seafood consumption occurring away from home, the product processing can result in an intermediate form that allows chefs final value-added opportunities in restaurants. Estimation of total employment in such a marketing chain when imported products as well as fresh seafood imports account for large shares of supply is not attempted on a times series basis. A substitute is the use of an input/output model that accounts for activity created throughout the economy as a result of an initial sale. The next section describes economic impacts of sales, income, jobs, and value added based on an input/output model developed for NOAA.

There are minimal data available annually on the domestic processing industry. NOAA's annual report *Fisheries of the United States* includes the number of processing and wholesale plants with direct employment estimates. Indirect and induced employment estimates are not included. The state of Florida data are reported without differentiation of east and west coasts. Therefore, data to be discussed are for the non-Florida Gulf. For the 2007–2009 period, Gulf States averaged 163 processing plants and 231 wholesaling plants. The range for processing plants during the 3 years was small at 160–165 indicating stability in the near term. As expected, wholesaling plants were more numerous, in part due to the lower capital cost. The range for wholesaling plants during the 3 years was smaller at 229–232. There likely was more entry and exit in the wholesaling sector than the narrow range suggests due to the lower capital entry costs. Louisiana was home to both the largest number of processors and wholesalers (72 and 176, respectively). Mississippi had the lowest number of plants. However, in terms of employment, Mississippi led the Gulf States. Approximately one-third of the region's employment can be identified as Mississippi based. Average plant employment in Mississippi amounted to three times the level of the next highest Gulf state, Texas.

10.3.6 Economic Impact of Gulf of Mexico Commercial Fishing

Economic impacts to be portrayed include those of sales, income, and value added originating from landings and imports. The initial use of the National Marine Fisheries Service (NMFS) Fishing Industry Input/Output Model was applied to 2006. Annual analyses followed with a value-added calculation made in 2009. Thus, there are findings for the 2007–2009 period previously used to depict near term conditions with respect to landings. Separate information for the Florida west coast versus Florida east coast was not available. The Gulf economic impacts of landings had to be reported for Alabama, Mississippi, Louisiana, and Texas to avoid inconsistencies with the prior sections dealing with landings and this impact, with and without the inclusion of imports, is presented in Figure 10.10 for 2009. This represents the first year in which NMFS segmented imports from domestic product in the calculation of economic impacts.

10.3.6.1 Sales Impacts

An input/output model measures the impacts of an economic impetus, in this case the value of landings, on other sectors in a defined economy or region. Impacts estimated include the effects of domestic landings, imported seafood, wholesaling, processing, and retail on an economy. In this case the impact generated $4.6 billion from Alabama, Mississippi, Louisiana, and Texas. For the 2007–2009 period, average annual sales impacts by state, including both domestic and imported product, were (1) Alabama, $441 million; (2) Mississippi, $348 million; (3) Louisiana, $1.9 billion; and (4) Texas, $1.9 billion.[1] All four states experienced a sales impact decrease from 2007 to 2009. Using the 3-year period, landed value average results in a higher impact estimate for sales than if 2009 alone was calculated. Importers accounted for 41 % of the seafood industry's Texas sales impact. Louisiana importers had 21 % of the impact. Mississippi and Alabama seafood economy had minimal importer roles.

---

[1] The 2009 sales impact for Florida, including the east coast, equaled $13 billion. Of this total, $9.5 billion was generated by importers.

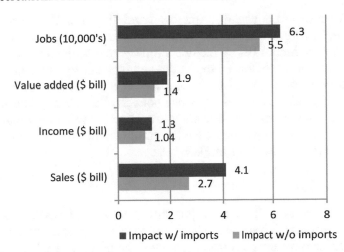

Figure 10.10. Gulf of Mexico commercial seafood industry economic impact, 2009 (U.S. Department of Commerce 2011).

10.3.6.2 Income Impacts

Income impacts are a component of sales in an economy. The income impacts for the four states associated with the use of both domestic and imported product amounted to $1.3 billion. Louisiana ($574 million) leads the Gulf and Mississippi was lowest in income generated at $113 million. Texas at $474 million was near the top and Alabama ranked third at $148 million. Each of the states experienced a reduction of income impacts from 2007 to 2009 with no change in rankings. Specifically, income impacts for the four states in 2007 were as follows: Louisiana, $1.1 billion; Mississippi, $184 million; Texas, $959 million; and Alabama, $268 million.[2]

10.3.6.3 Employment Impacts

Direct jobs in the commercial harvesting sector spur actions among companies supplying inputs and for those adding value to landings and imported product ultimately used by consumers. The four states Gulf economy averaged 92,000 seafood industry jobs during 2007–2009. Employment decreased each year from 109,000 in 2007 to the period low of 63,000 in 2009. Texas job contraction was largest at −56 % followed by −38 % in Louisiana. Alabama and Mississippi had the lower decreases with each approximately −20 %. Seafood industry jobs in 2009 were (1) Louisiana, 29,200; (2) Texas, 18,900; (3) Alabama, 8,800; and (4) Mississippi, 6,400.[3] Jobs in the retail sector comprised approximately half of the jobs over the period. As to be expected, when employment decreased the retail sector experienced the largest problems. The nation's economy began a period of slowdown that could have led to the result. However, the input/output model result of a Gulf retail sector experiencing a 60 % reduction between 2007 and 2009 is problematic in spite of Gulf landings falling 10 %.

---

[2] The income impact for Florida, including its east coast, equaled $2.4 billion in 2009 compared to $2.8 billion in 2007.

[3] The 2009 number of Florida jobs, including the east coast, equaled 64,700. The import sector accounted for more than one-half of this total.

In addition to these employment estimates from the U.S. Department of Commerce, NOAA's annual publication *Fisheries Economics of the United States* and NOAA's report *Fisheries of the United States* (FUS) include employment estimates. The later report lists employment from seafood wholesale and processing plants by state and region. With the exception of the input/output model indicating lower employment for Mississippi than the FUS report, the employment estimates are close between the reports. This closeness warrants caution because an input/output model accounts for direct employment and jobs arising from the induced effects of direct employment. So the employment estimate from the model should be higher than the direct employment in FUS.

10.3.6.4 Value Added

The value-added measure from an input/output model addresses a net concept to an industry's economic impact. Gross sales reflect that costs are associated to produce the product sold. When the transfer payments of costs for goods and services used to produce the product sold are subtracted from gross sales, a net value image emerges. Referred to as value added, the estimate yields a descriptor useful for measuring a firm's or sector's net contribution to an economy. This section continues with the Alabama, Mississippi, Louisiana, and Texas designation for the Gulf because the input/output model does not report for the Florida west coast separately. The landed value and import value of the four-state Gulf in 2009 resulted in a value added of $1.4 billion. Louisiana's post dockside firms accounted for 47 % of the total. Texas was second at a 31 % contribution to the total. Alabama at 13 % and Mississippi at 9 % had the smaller roles. There was no means by which to measure change between 2007 and 2009 because 2009 marked the first year of estimation.

10.3.6.5 Imports and Sales, Income, Employment, and Value Added Impacts

Use of imported seafood in Gulf post dockside economic endeavors can be significant to a firm's success. The *Fisheries Economics of the United States* report for 2009 includes a treatment of imports as supply that leads to economic impacts. The four economic impact measures indicate double-digit contributions by imported product: (1) 33 % of sales, (2) 21 % of income, (3) 13 % of jobs, and (4) 25 % of value added (Figure 10.10). Among states Texas' sales were 98 % higher than would have been experienced with state landings alone. Mississippi incorporated imports the least at 6 % of sales. Louisiana and Alabama used imports to gain 34 % and 16 % higher seafood industry sales, respectively.

10.3.7 Commercial Fisheries of State Managed Species

10.3.7.1 The Blue Crab Fishery

Essentially all of the nation's catch of blue crab occur in state waters. Harvesting units are small and make daily trips. These characteristics apply throughout the Chesapeake Bay, South Atlantic, and Gulf assuring that landings by state mimic catch by state. Management of the elements contributory to population levels and harvests consequently fall to state agencies. Regional cooperation via GSMFC adds another level of contribution to states achieving their goals. Gulf landings fell 14 % from the 1990–1992 base period to the end period of 2007–2009. However, the region's share of national landings increased in the comparison periods because national landings with Gulf removed fell by 32 %. Nationally, the increasing ex-vessel price for blue crab pushed dockside value up 90 %. The non-Gulf component increased over 100 %, while the Gulf increase neared only 59 %.

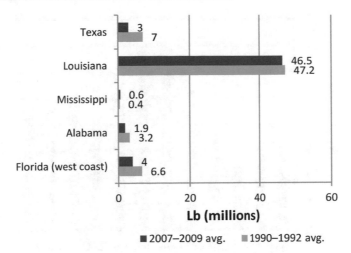

Figure 10.11. Gulf of Mexico commercial blue crab landings by state for selected periods (NMFS FSD; data accessed 2012—see Appendix A) (*Note*: 1 lb = 0.454 kg).

Within the downward landings results for the 1990–1992 period versus the 2007–2009 period, there was divergence among states. Texas (−57 %), Florida west coast (−40 %), and Alabama (−38 %) all experienced significantly lower landings (Figure 10.11). The Gulf's largest producer, Louisiana, by comparison experienced only a 2 % decline in production while the region's smallest producer, Mississippi, experienced an increase in production (Figure 10.11). Given that the Gulf blue crab production is dominated by Louisiana, the reduction in Gulf blue crab landings between 1990–1992 and 2007–2009 was minimal and largely mimicked that observed for Louisiana. These were among the lower producing states in the Gulf, but the impact with the largest producer, Louisiana, up only 2 % resulted in a decrease for the Gulf in total. Lowest producer, Mississippi, had a large percentage increase, but production approached only 500,000 lb. An important aspect of the Gulf blue crab fishery relates to the value of landings. Previously cited was the ex-vessel price performance being the best of species comprising Gulf landings. With 1982 serving as the base year for NOAA Fisheries' ex-vessel price index, the blue crab index reached 383 in 2009. In 1990, the index stood at 152 suggesting that most of the large price increase occurred from 1990 to 2009. The end period had U.S. average ex-vessel price in a small range of $0.75–$0.81 per pound with the low occurring in 2009. Gulf end-period average prices were similar at $0.73–$0.80 per pound. The national recession in 2009 must have played a role as most Gulf species attained period low levels. Exceptions were oysters and stone crab claws.

Seasonality was less of an issue with blue crab production than other species. Closed seasons were not a management approach in major producing areas. Louisiana's fishery accounts for 83 % of Gulf landings. Therefore, the occasional crab trap free periods based in avoiding gear conflicts or the facilitation of abandoned trap removal do not result in production shifts. May–September landings account for 53 % of annual landings (Figure 10.12). Winter months are lowest. Crabbers still put 4.3 million pounds on docks in the lowest month, March.

Blue crab can be graded by size with larger crabs going to live resale. Those not reaching the live resale size limit, the majority, are processed to remove the meat. However, the meat is not a uniform product; processed product is differentiated for sale as crab fingers, claw meat, white, backfin, lump, and jumbo lump. Multiple products of varied value for the human market represent perhaps the most complex of the Gulf's processing industries. Blue crab processing occurred in all five Gulf states until 2005 which marked the stoppage in Mississippi from 2006

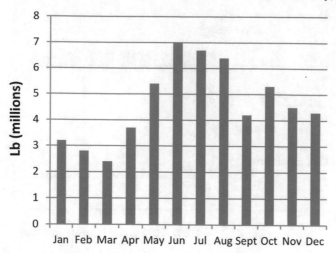

Figure 10.12. Gulf of Mexico commercial blue crab landings by month, 2007–2009 average (NMFS FSD; data accessed 2012—see Appendix A) (*Note*: 1 lb = 0.454 kg).

to the end of period used herein, 2009 (personal communication, Melissa Yencho-NOAA Fisheries). Texas processing was at the level with such a small number of firms that reporting it separately would divulge confidential data. From 2006 forward, the Texas data had to be combined with the Louisiana data to maintain confidentiality. Gulf blue crab processing data exists for Alabama, Louisiana/Texas, and Florida west coast. The 2007–2009 average Alabama processed production was six times larger than the next largest, Louisiana/Texas. Recalling the level of landings in Alabama being a 3-year average of approximately ten million pounds, points to significant cross state movement of live blue crab. The 4 % of average Gulf landings clearly would not support the Alabama processing industry's 4.2 million pounds of blue crab meat. It is an inescapable conclusion that Louisiana was the only state that could have supplied sufficient live crabs for Alabama to attain such a high processed volume.

10.3.7.2   The Menhaden Fishery

Menhaden are a small oily finfish caught in nearshore fisheries from Chesapeake Bay to the Gulf. The vast majority of landings come from catch in the 0–3 mi (0–4.8 km) coastal area. Occasionally substantial catch is from the 3–200 mi (4.8–322 km) offshore area. The prospect of offshore harvest necessitates a closer tie between state agencies and NOAA Fisheries than would be thought for a clearly nearshore focused species. The decreasing number of firms in what is a large fishery for a species used in domestic and international markets encourages close cooperation among agencies and firms. Menhaden processing results in three products: (1) fish meal for use in animal feeds, primarily poultry; (2) fish oil for mostly export markets inclusive of human food uses; and (3) soluble, which often can be an additive to the meal.

The menhaden industry is noted as vertically integrated. Processors own vessels that fish under corporate direction. Crews are compensated on the basis of shares. Reported ex-vessel price under a vertically integrated structure with a small number of firms can be expected to differ from other Gulf fisheries. The other fisheries are characterized by large numbers of harvesters operating as owner operators throughout the Gulf at all times of a year. The companies and NOAA Fisheries do generate a price so that dockside value can be reported. The index of ex-vessel price for menhaden in 1990 and 2009 was at 128 and 154, respectively. At the end of the analysis period menhaden prices were $0.06–$0.07 per pound.

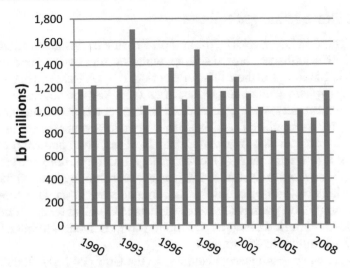

Figure 10.13. Gulf of Mexico annual menhaden landings, 1990–2009 (NMFS FSD; data accessed 2012—see Appendix A) (*Note*: 1 lb = 0.454 kg).

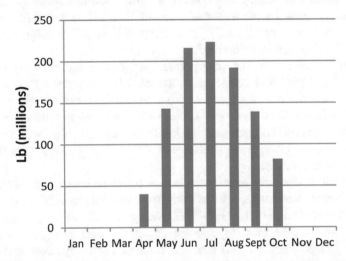

Figure 10.14. Gulf of Mexico menhaden landings by month, 2007–2009 average (NMFS FSD; data accessed 2012—see Appendix A) (*Note*: 1 lb = 0.454 kg).

Gulf landings were two-thirds of the U.S. total. On only four occasions from 1990 to 2009 did Gulf landings not reach at least one billion pounds and on only two occasions did landings exceed 1.5 billion pounds (Figure 10.13). Based on the first and last 3-year averages for the period, landings exhibited stability. Landings over the period fell only slightly with value increasing by 15 %. The number of firms over time decreased; evidently making for an increase in average landings per firm. The industrial firms are located in Mississippi and Louisiana. Much smaller firms in Florida (west coast) and Alabama focus on menhaden as bait for other fisheries such as blue crab and some recreational uses. These states land less than 1 % of the Gulf production. Landings for the industrial fishery start minimally in April, steadily increase to a peak in July, and end by October (Figure 10.14). Firms in Mississippi and Louisiana essentially fish the same times of the year.

10.3.7.3 Other State-Managed Species

The species selected as key species by the authors provide the insight needed regarding general conditions in the Gulf. Menhaden, brown shrimp, white shrimp, oysters, and blue crab combined accounted for 94 % of landings in the 2007–2009 period. NOAA Fisheries in its annual publication *Fisheries Economics of the United States* identifies Gulf key species additionally as crawfish, groupers, red snapper, mullets, stone crab, and tunas. The focus of this chapter being the northern Gulf (i.e., Alabama, Mississippi, and Louisiana) means there was no need to include crawfish and stone crab claws. The former is a freshwater species of wild and aquaculture origins found in Louisiana. The latter is overwhelmingly a Florida fishery. Like stone crab, the vast majority of Gulf striped mullet (*Mugil cephalus*) catch is Florida based; representing over 70 % of the Gulf total. Alabama and Louisiana basically account for the remainder with Alabama the larger. Total Gulf landings averaged ten million pounds of striped mullets in the most recent period. This was down from the initial 1990–1992 period average of 26 million pounds. Dockside value fell from the initial period's level of $26.4 million to $5.7 million.

Yellowfin tuna (*Thunnus albacares*) landings in the Gulf for 2007–2009 increased to 35 % of the United States. The increase did not result from increased landings compared to elsewhere in the country. Rather, Gulf landings decreased (69 %) but landings other than the Gulf fell 78 %. Prices were favorable during 1990–2009 by almost doubling nationally. Gulf yellowfin prices followed the increase by the lesser amount of 50 %. The distribution of Gulf landings was very narrow. Louisiana received 77 % of the catch in 2007–2009, which represents an increase from the 46 % share in 1990–1992.

The harvest of red snapper and grouper are subject to increasingly constraining catch regulations of the GMFMC and cooperating states. Management of commercial effort by seasonal, gear, area protections and quotas with share assignment has the near-term effect of constraining catch. Additionally, these key species have been highly prized by anglers throughout the Gulf. Commercial red snapper average landings were essentially unchanged on the basis of an initial-period versus end-period measure at 2.6 million pounds. The 1995–2006 period average was 4.5 million pounds.

Location of landings changed among the states between initial and end periods. Northern Gulf states of Alabama, Louisiana, and Mississippi experienced a 50 % decrease. Texas and Florida west coast benefitted with the 1990–1992 average of 1.2 million pounds, increasing to 1.9 million pounds by 2007–2009. Gag (*Mycteroperca microlepis*), red (*Epinephelus morio*), and warsaw grouper (*Epinephelus nigritus*) landings have consistently been attributable to Florida west coast ports. Thus, there are landings of some groupers in the northern Gulf, but these cannot be considered important compared to previously reviewed species.

10.3.8 Additional Detail on Key Commercial Species

An overview of the Gulf of Mexico commercial seafood industry, including a brief discussion of some of the key species, was provided in the previous section of this chapter. This section provides additional detail on some of these key species including shrimp, oysters, and reef fish. Shrimp is given more discussion because it is by far the largest contributor, by value, to the Gulf of Mexico seafood industry. Oysters are given additional treatment because the nature of the industry involves leasing activities, with emphasis being given to Louisiana. Reef fish species comprise a sizeable portion of commercial finfish landings and are the subject of considerable management, including recently enacted catch share programs, and are given additional consideration on this basis.

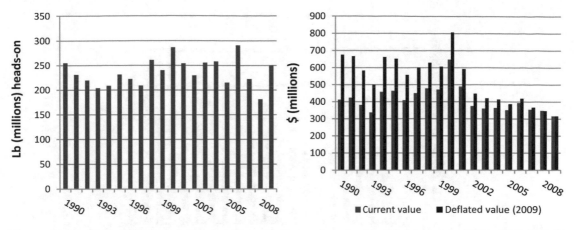

Figure 10.15. Gulf of Mexico shrimp landings (*left panel*) and value (*right panel*), 1990–2009 (NMFS FSD, data accessed 2012, with deflated values calculated by authors–see Appendix A) (*Note*: 1 lb = 0.454 kg).

10.3.8.1   The Shrimp Fishery

10.3.8.1.1   Gulf Shrimp Landings and the Relation to Imports

With a 2009 dockside value of $314 million, the shrimp fishery is the largest contributor to the $615 million (2009) Gulf of Mexico commercial fishing sector. Since it is by far the largest component of the Gulf of Mexico commercial seafood industry, it is covered in additional detail in this section.

Annual Gulf shrimp production (heads-on weight) during 1990–2009 is provided in Figure 10.15 (left panel). While exhibiting a significant amount of annual variation, the yearly changes tend to follow a random-walk process and, over time, production returns to its long-run average (while not shown in the graph, long-run production of gulf shrimp has been stable since at least the 1970s). These observed random walks are primarily the result of changes in environmental conditions that influence recruitment and growth. Since the primary species of shrimp landed in the Gulf—brown and white—are short-lived animals, with maximum age of about 1 year, any short-run deviations from the long-term average will be temporary in nature assuming environmental conditions return to normal and there is a sufficient amount of effort to harvest the available crop. Overall, annual harvest of Gulf shrimp during 1990–2009 averaged 236 million pounds with a range from 181 million pounds in 2008 to 290 million pounds in 2006. While the effort needed to harvest the aggregate shrimp crop has historically been sufficient, as addressed in subsequent sections of this chapter, changes in profitability have led to a significant decline in industry effort in recent years and an increasing concern that with further declines in effort, a portion of the annual shrimp crop may not be harvested.

While the long-run production of Gulf shrimp, in pounds, has remained stable over time, the same cannot be said about the value of landed product; especially when the influence of inflation is removed. As indicated in Figure 10.15 (right panel), the long-run dockside value of the Gulf shrimp harvest has, overall, been declining, whether considered on a current or deflated basis. This decline has been particularly pronounced since 2001. On a current dollar basis, the value of Gulf production fell from an average of just over $400 million annually during 1990–1994 to about $350 million annually during 2005–2009. After adjusting for inflation, the decline was approximately 40 %, from $617 million to $367 million (expressed in 2009 dollars).

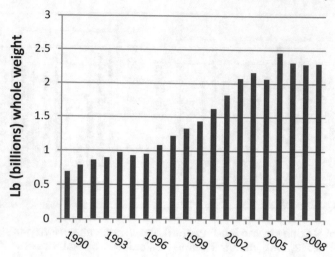

Figure 10.16. U.S. shrimp imports (whole weight), 1990–2009 (NMFS FSD, data accessed 2012, with weight conversions calculated by authors–see Appendix A) (*Note*: 1 lb = 0.454 kg).

While there are several reasons for the sharp decline in the Gulf dockside shrimp price beginning in 2001, the overriding one is that of increasing imports. The source of these imports is from more than 40 countries throughout the world with Asian countries dominating the field. As indicated by the information in Figure 10.16, import growth has been large during the considered timeframe with total imports (heads-on equivalent weight[4]) advancing from an average of 850 million pounds annually during 1990–1994 to 2.3 billion pounds annually during the 2005–2009 period. Furthermore, as indicated, much of this increase has occurred post 2000. Given the strong U.S. economy throughout the later portion of the 1990s and the concomitant increase in demand for shrimp, the increase in imports during the 1990s did not lead to any sharp decline in the Gulf of Mexico dockside value (or price). However, the large increase in imports post 2000 combined with a number of other factors, including a recession that officially began in the third quarter of 2001, resulted in a sharp and prolonged decrease in the Gulf of Mexico dockside value (via a change in price). A detailed examination of possible factors influencing this price decline can be found in Keithly and Poudel (2008).

Comparison of the information in Figure 10.15 (left panel) and Figure 10.16 clearly highlights how small Gulf landings are relative to imports. Given this and the fact that differentiation of Gulf shrimp from the imported product is minimal, one would expect changes in the Gulf and import prices to follow a similar pattern. This relationship is evident in the information in Figure 10.17. While the import price, expressed on a whole weight equivalent basis, generally exceeded the Gulf dockside price by a considerable margin during the early 1990s, this margin gradually lessened over time and had largely disappeared by the mid-2000s.[5] Furthermore, given the large share of total U.S. supply (i.e., domestic and imported product) provided by imports, along with their apparent close substitutability, one would expect that changes in

---

[4] The terms "live-weight" and "whole weight" are used interchangeably in this section.

[5] The import price, while converted to a whole weight equivalent basis, consists of different product forms and different shrimp sizes. Both of these factors will, to some extent, likely explain a portion of the price differential between import and domestic product prices. Overall, the correlation between these two price series was 0.94 during the study period.

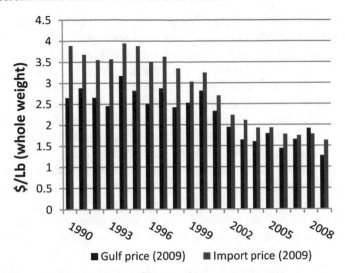

Figure 10.17. **Deflated Gulf dockside shrimp price and import price (whole weight), 1990–2009 (NMFS FSD, data accessed 2012, with weight conversions and deflated prices calculated by authors—see Appendix A) (*Note*: 1 lb = 0.454 kg).**

Gulf landings would have little or no influence on its own price. This is examined in greater detail in a subsequent section.

10.3.8.1.2   A Closer Look at Imports

As noted, a large number of countries export shrimp to the United States. Asian countries have accounted for the majority of U.S. shrimp imports since at least the early 1990s and in 2009 accounted for more than 70 % of the total (based on product weight). Thailand dominated exports to the United States in 2009 accounting for almost one-half of the Asian exports and more than one-third of total exports. Other countries of significance include Indonesia (17 % of Asian exports and 13 % of total exports to the United States), Ecuador (70 % of South American exports and 11 % of total exports to the United States), China and Vietnam (each accounting for approximately 10 % of Asian exports to the United States and 8 % of total exports to the United States), and Mexico (accounting for about 67 % of Central American exports to the United States and 7.5 % of total exports to the United States).

Employing monthly data covering the 1995–2005 period, Jones et al. (2008) examines the U.S. demand for shrimp by source in relation to prices from the sources. The analysis includes seven import sources—Mexico, Ecuador, India, Thailand, Vietnam, China, and Rest of World—and domestic (i.e., U.S.) source. Own-price elasticities for all sources were negative, as suggested by theory, and statistically significant.[6] Furthermore, the own-price elasticities were inelastic (less than − 1) for all sources implying that a 1 % increase (decrease) in price from any given source would result in a less than proportionate decrease (increase) in quantity demanded for shrimp from that source in the U.S. market. The scale elasticities, which measure the influence of a change in overall U.S. shrimp demand on the demand from the individual sources, were positive and statistically significant for all sources and ranged from a low of 0.30

---

[6] An own-price elasticity, with respect to demand, measures the change in quantity demanded of a good that will be forthcoming with respect to a 1 % change in its own price. Similarly, a cross-price elasticity measures the change in demand for a given good associated with a 1 % change in the price of a substitute (or complement) good.

(Ecuador) to a high of 1.74 (India). The scale elasticity for the U.S. production with an estimate of 0.90 indicates that the demand for U.S. produced shrimp increases by 9 % for each 10 % increase in total U.S. shrimp demand. Finally, the researchers note that "[f]or the most part, cross elasticities were negative, implying that shrimp demand exhibited a complementary relationship between countries." This finding is not easily explainable.

The large increase in U.S. shrimp imports and the resultant decline in Gulf dockside price resulted in a coalition of Southeast U.S. shrimp harvesters and processors (Gulf and South Atlantic) petitioning the U.S. International Trade Administration and the U.S. International Trade Commission for relief in the form of antidumping duties. These petitions, filed on December 31, 2003, charged six countries—China, Thailand, Vietnam, India, Ecuador, and Brazil—with unfair trade practices. These six countries exported 822 million pounds of shrimp (product weight) to the United States in 2003, which represented almost three-quarters of the total U.S. shrimp imports for that year. After an exhaustive investigation, the finding of dumping and injury was found, and duties were imposed on subject merchandise from these six countries. Details on the investigation and factors leading to the investigation are provided by Keithly and Poudel (2008) who, after analysis of the situation, conclude that these duties had only a marginal impact on limiting shrimp exports to the United States because of trade diversion effects (essentially increased shrimp imports from countries not named or merchandise not named that offset any reduction in imports from countries and merchandise named). Thus, the duties likely had only a marginal, if any, effect on increasing the price received by the domestic shrimpers for their harvested product. Furthermore, the conclusion by Keithly and Poudel (2008) would suggest that the recent stability in imports was not the result of the duties imposed on named countries and merchandise. Instead, the stability likely reflects a decline in demand in 2008 as the United States entered a deep and protracted recession. While the antidumping duties imposed on the six named countries may have had little influence on increasing the U.S. Gulf shrimp dockside price, the domestic industry did benefit significantly via funds collected from the duties and negotiated settlements to rescind reviews. Specifically, the Continued Dumping and Subsidy Offset Act of 2000 (i.e., the Byrd Amendment) provided for the annual disbursement of funds collected under the Act to the injured party (i.e., the petitioners). This disbursement totaled hundreds of millions of dollars before the Act was repealed.

As noted, Southeast U.S. processors also petitioned for relief from the growing import base. This reflected the fact that not only was the total import base increasing but the composition of the import base was also changing with value-added products comprising an increasing share of the total (Figure 10.18). Imports of peeled raw product, for example, increased from about 300 million pounds (whole weight basis) in 1990 to more than 800 million pounds in the late 2000s. Peeled cooked imports increased from about 60 million pounds (whole weight equivalent) to more than 800 million pounds. Imports of headless shell-on shrimp, by comparison, exhibited a much more modest increase—from about 325 million pounds (whole weight basis) in 1990 to 500–550 million pounds by the late 2000s.

10.3.8.1.3 A Closer Look at the Gulf Shrimp Fishery

Royal Red Shrimp

Pink Shrimp

Gulf shrimp fishermen target four species of shrimp, including brown (*Farfantepenaeus aztecus*), white (*Litopenaeus setiferus)*, pink (*Farfantepenaeus duorarum*), and royal red (*Pleoticus robustus* or *Hymenopenaeus robustus*). Other species of related organisms, such as seabobs (*Xiphopenaeus kroyeri*) and rock shrimp (*Sicyonia brevirostris*), are incidentally harvested. Of the main shrimp species, brown shrimp is the most important to offshore

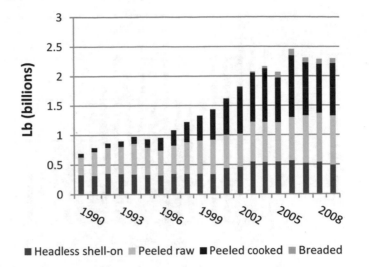

Figure 10.18. U.S. shrimp imports by product form (whole weight equivalent basis), 1990–2009 (NMFS FSD, data accessed 2012, with weight conversions calculated by authors—see Appendix A) (*Note*: 1 lb = 0.454 kg).

harvesters and is primarily caught in waters up to 40 fathoms (73.2 m) from June through October of each year. The white shrimp fishery, which approaches the importance of brown shrimp in terms of catch, typically peaks in the months of August through December. Geographically, however, white shrimp are primarily harvested from nearshore, state waters up to 20 fathoms (36.6 m), thus generally making them the target of smaller vessels. Of the remaining shrimp species, pink shrimp are primarily harvested as a distinct species off of Florida's west coast and in the Florida Keys in waters up to 30 fathoms (54.9 m). Outside Florida waters, pink shrimp are less abundant; if harvested, they tend to be caught while harvesting brown shrimp and are typically included as part of the brown shrimp harvest. Royal red shrimp, a species harvested in waters 140–275 fathoms (256–503 m) deep, are a minor component of the Gulf shrimp fishery. Unlike other shrimp species, which are relatively short-lived and thus considered to be an annual crop, royal reds have a multiple-year life span. While brown, white, and pink shrimp are all subject to capture in state and EEZ waters (depending on the time of year), royal reds are harvested exclusively in the EEZ.

Technologically, the Gulf shrimp fleet employs a wide range of both gear and vessels depending on the species and fishing area being exploited. In terms of gear, harvesters have been known to use cast nets, haul seines, stationary butterfly nets, wing nets, skimmer nets, traps, beam trawls, and otter trawls, with the otter trawl being the primary gear used in offshore and EEZ waters.

Shrimp Effort

Given the large decline in the Gulf of Mexico dockside shrimp price in conjunction with rising fuel prices, shrimp fishermen have been experiencing a cost-price squeeze for some time now. This squeeze was exacerbated in late 2001 when the dockside price fell sharply and this decline lasted for a protracted period of time (see Keithly and Poudel 2008 for additional details). Given this cost-price squeeze, it should come as no surprise that effort in the fishery has fallen. The decline in offshore effort (defined as outside the Collision Regulation [COLREG] lines), measured in terms of 24-h days fished, is given in Figure 10.19 for the 1990–2009

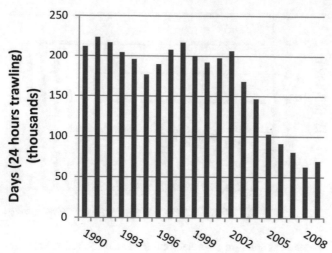

Figure 10.19. Estimated offshore effort (24-h days fished) by the Gulf of Mexico offshore shrimp fleet, 1990–2009 (NMFS Galveston Laboratory, personal communication, 2012–see Appendix A).

period. As indicated, total offshore effort approached or exceeded 200,000 days fished per year throughout the 1990s. Since 2003, however, effort has fallen sharply and in recent years, has been less than 70,000 days per year. Overall, effort in recent years has only been about one-third to one-half of the observed effort throughout the 1990s. Analysis by Nance et al. (2006) examines the relationship between catch and effort in the offshore component of the Gulf of Mexico shrimp fishery and if their analysis is valid, one can conclude that the current level of effort associated with the offshore component of the fishery is significantly less than what is required to harvest maximum yield. This conclusion, however, needs to be tempered because the treated relationship between offshore yield and effort in their analysis was considered independently of inshore shrimping activities. As the case with respect to offshore effort, inshore effort has also fallen sharply in recent years. Reduction in effort in the inshore component of the fishery would, one might hypothesize, result in increased escapement of the small shrimp to offshore waters and, hence, an increasing abundance of shrimp in the offshore waters. This increased abundance translates into a higher catch per unit of effort in the offshore waters.

A more detailed examination of effort in the two main northern Gulf of Mexico shrimp fisheries—the brown shrimp fishery and the white shrimp fishery—can be made with the aid of Figure 10.20. As indicated, total estimated effort (i.e., inshore and offshore) in the brown shrimp fishery (Grids 7–21)[7] fell from almost 200,000 days annually in the early 1990s to about 160,000 days by the late 1990s/early 2000s (effort is assumed to be directed at a particular species if at least 90 % of that trip's catch comprises that particular species). Thereafter, in association with the sharp decline in shrimp price and increasing fuel costs, effort fell precipitously to less than 50,000 days in recent years.

A somewhat different picture emerges when one examines total effort (i.e., inshore and offshore) white shrimp effort (Figure 10.20, right panel). As indicated, effort associated with this fishery showed a large increase in the mid-1990s to early 2000s with an abnormally high number of days fished being reported in 2002 (169,000 days). Thereafter, however, effort fell sharply to about 60,000 days in recent years. This decline in effort coincided with a period of increasing white shrimp harvest indicating a significant increase in the catch per unit effort.

---

[7] See Figure 10.29 for a listing of grids.

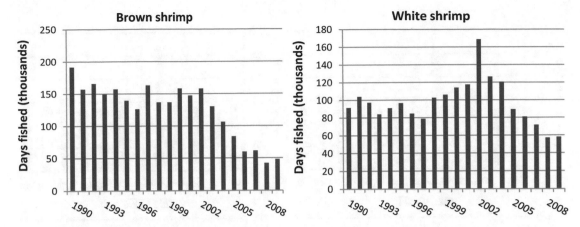

Figure 10.20. Directed shrimping effort on brown (*left panel*) and white shrimp (*right panel*) fisheries (grids 7–21), 1990–2009 (NMFS Galveston Laboratory, personal communication, 2012–see Appendix A).

As noted by Liese and Travis (2010), vessels fishing for *Penaeid* shrimp in the federal waters of the Gulf of Mexico were required to have a permit as of December 5, 2002. Subsequently, a moratorium was placed on the issuance of new permits and, according to unpublished NMFS records, a total of 1,907 vessels were permitted under the Gulf shrimp moratorium permit in 2009 (i.e., the upper-bound estimate of the number of vessels that would be legally allowed to shrimp in the federal waters of the Gulf of Mexico). Of this total, 693 of the vessels, or more than one-third of the total, were home-ported in Texas. Louisiana ranked second (545 permits; 29 %), followed by Florida (278 permits; 15 %), Mississippi (164 permits), and Alabama (149 permits). While the number of permits equaled about 1,900 in 2009, Liese and Travis (2010) report that only about 1,215 of these actively harvested shrimp in 2009.

In addition to those vessels holding a Gulf shrimp moratorium permit, which is required for shrimping in federal waters, a large number of boats shrimp only in the state waters. Based on state license sales, Miller and Isaacs (2011) estimate that the population of inshore shrimpers, excluding those that had a Gulf shrimp moratorium permit, approximated 3,765 in 2009. About 60 % of the licenses were issued in Louisiana while another 14 % and 12 % were issued in the states of Texas and Alabama, respectively.

Shrimp Size at Harvest

The size of shrimp at harvest varies significantly throughout the year and can vary over time as a result of environmental factors, dates associated with opening inshore waters, the amount of fishing pressure, where the fishing pressure is centered, or some amalgam. Cold weather, for example, can retard the growth of brown shrimp, which may yield a smaller size at harvest, all other factors being equal. Similarly, declining fishing pressure may provide the shrimp additional time to grow which would yield a larger average size at harvest (assuming all other factors are the same). The estimated average size of shrimp for four time periods—1990–1994, 1995–1999, 2000–2004, and 2005–2009—by month is given in Figure 10.21. As indicated, shrimp size is consistently smallest in May (i.e., a larger number of shrimp to the pound), associated with movement of brown shrimp from the estuaries and the opening of the inshore fishery in the northern Gulf States. The average size then increases (as the brown shrimp grows and moves offshore) until September/October when white shrimp show up in significant quantities.

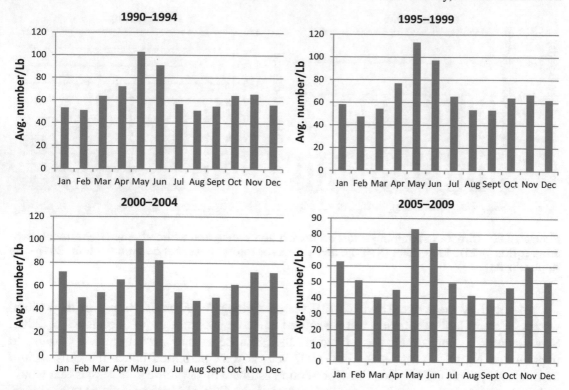

Figure 10.21. Estimated average size of shrimp at harvest (headless), by month, selected 5-year periods (NMFS Galveston Laboratory, personal communication, 2012, with calculations by authors–see Appendix A) (*Note*: 1 lb = 0.454 kg).

The apparent increase in average shrimp size (i.e., fewer shrimp to the pound), particularly after the 1995–1999 period is of interest as well. For example, the estimated average number of shrimp to the pound (headless) in May during 1990–1994 was estimated to equal 102 and increased to 113 during 1995–1999. During the May 2000–2004 period, the average declined to 98 and declined again to 85 during the 2005–2009 period. For September (roughly when white shrimp begin to move), the averages for the four 5-year periods are 55, 53, 50, and 40, respectively. The increasing shrimp size (i.e., fewer shrimp to the pound) has been particularly pronounced during the most recent 5-year period when the monthly trend held for all months but February. While not formally tested, one plausible explanation for the changing shrimp size over the period of analysis is the large reduction in effort during recent years (Figs. 10.19 and 10.20).

Size of shrimp at harvest is an important consideration for at least two reasons. First, the price the shrimper receives for his harvested product is directly related to the harvested size with smaller shrimp commanding a lower price. Second, an increase in the average shrimp size at harvest (i.e., fewer shrimp to the pound) can translate into increased harvest in the aggregate assuming natural mortality is low relative to the gains in weight that could be achieved by allowing the shrimp to grow to a larger size prior to harvest. The relationship between size of shrimp and price received on an annual basis for the 2000–2009 period is given in Figure 10.22. As is illustrated by the information in the figure, prices (undeflated) of all shrimp sizes fell during the 2000–2009 period. Furthermore, the price declines are particularly pronounced (in terms of the absolute dollar decline) for the larger-sized shrimp (i.e., smaller count to the pound). With respect to the under 15 count (i.e., less than 15 shrimp to the pound),

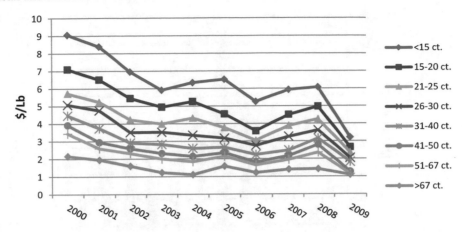

Figure 10.22. Average annual shrimp prices per pound (current) by size category (NMFS Galveston Laboratory, personal communication, 2012–see Appendix A) (*Note*: 1 lb = 0.454 kg).

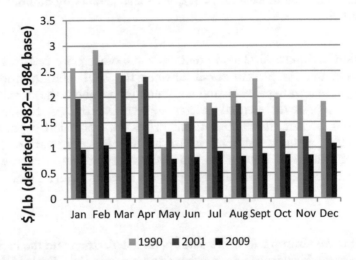

Figure 10.23. Gulf dockside price by month for selected years (prices deflated based on 1982–1984 Consumer Price Index [CPI]) (NMFS Galveston Laboratory, personal communication, 2012, with price calculations by authors–see Appendix A) (*Note*: 1 lb = 0.454 kg).

the unadjusted price fell from about $9.00 per pound to $4.00 per pound, or by about $5.00 per pound. The 51–67 count size price, by comparison, fell by only $1.75 per pound from $3.44 in 2000 to $1.69 in 2009. In all size categories, overall, the price decline between 2000 and 2009 ranged from about 45 to 55 %.

With a change in average size of shrimp harvested throughout the year comes a change in price. This is illustrated in Figure 10.23 for selected years. As indicated, price is consistently lowest in May when the average size of shrimp is smallest (see Figure 10.21) and inland waters are opened. As the brown shrimp grow and move offshore, the average price tends to increase through August. Associated with the opening of the inshore waters to white shrimp in late August, the price of shrimp begins to decline. The relatively high prices in months prior to the opening of the inshore waters to brown shrimp fishing in May (i.e., January through April) to a large extent represent the harvest of large, overwintering white shrimp.

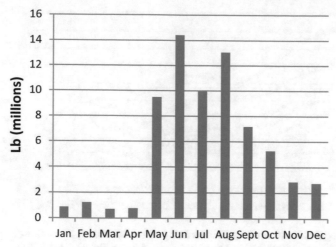

Figure 10.24. Gulf of Mexico brown shrimp harvest by month, 2005–2009 average (NMFS Galveston Laboratory, personal communication, 2012, with calculations by authors–see Appendix A) (*Note*: 1 lb = 0.454 kg).

The information in Figure 10.23 also points to some other price features meriting discussion. First, note that the sharp differential between the 2001 monthly prices and the 1990 monthly prices beginning in September and continuing throughout the remainder of the year. This sudden and sharp price differential reflects the terrorist attack of September 11, 2001 and subsequent recession. Second, the 2009 monthly prices are well below either the 1990 or 2001 deflated prices. Finally, as indicated, there is considerably less price variation by month in the 2009 prices than in either the 1990 or 2001 prices; consistent with a narrowing of the price differential between the large and small shrimp as observed in Figure 10.22.

Harvested Species

Two species, brown shrimp and white shrimp, as noted, dominate the commercial harvest of shrimp. This is particularly true in the northern and western Gulf. Both of these species tend to be seasonal in nature, and harvest is directly related to their growth and migration patterns. The seasonal nature of harvest of brown shrimp, based on the 2005–2009 period, is illustrated in Figure 10.24. Harvest tends to be small until May, which coincides with emigration of the brown shrimp from the estuaries to deeper waters and the opening of the inshore waters in the northern Gulf States. On average, 9.3 million pounds of brown shrimp were harvested in the month of May during 2005–2009, and this increased to 14 million pounds in June. Coinciding with the opening of Texas waters to shrimping, brown shrimp catch, in pounds, increased once again in August and then fell through the remainder of the year.[8]

Production (pounds) of brown shrimp by month for selected time periods during 1990–2009, expressed on a percentage basis, is given in Figure 10.25 (left panel). In general, the monthly production pattern is relatively consistent across the four 5-year time periods considered. One significant difference, however, is observed in the most recent 5-year period (2005–2009).

---

[8] In an effort to protect juvenile brown shrimp and thereby increasing shrimp yield, waters off the Texas coast and seaward to 200 mi (322 km) are closed each year from approximately May 15 to July 15.

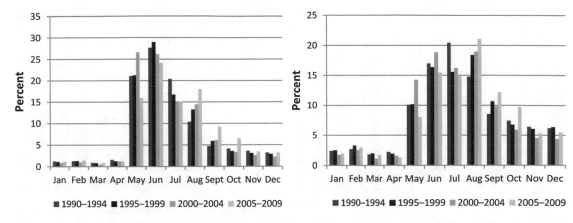

Figure 10.25. Gulf of Mexico brown shrimp harvest by month (pounds, *left panel*; value, *right panel*) expressed on a percentage basis for selected time periods, 1990–2009 (NMFS, Galveston Laboratory with calculations by authors; data 2012—see Appendix A).

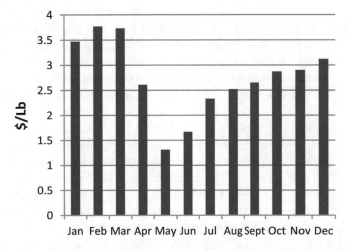

Figure 10.26. Gulf of Mexico brown shrimp price by month, 2005–2009 average (NMFS Galveston Laboratory, personal communication, 2012, with calculations by authors–see Appendix A) (*Note*: 1 lb = 0.454 kg).

In this period, the harvest of brown shrimp appears, to some extent, to be delayed. For example, May harvest, expressed on a percentage basis, was significantly lower than in other 5-year periods while harvests in the later months (August through October) were higher than in other periods. This delayed harvest may reflect the declining effort on the stock (Figure 10.20, left panel), which provides the brown shrimp stock additional time to grow.

While May and June tend to be the peak months in terms of poundage of brown shrimp harvest, peak value from the harvest tends to be in July and August (Figure 10.25, right panel). The observed difference in monthly poundage and value patterns is the result of larger brown shrimp being harvested in the later months and the increased price per pound for the harvested product. This price pattern is presented in Figure 10.26 for the 2005–2009 period. As indicated, the May brown shrimp price during 2005–2009 averaged less than $1.50 per pound. Coinciding with an increased size at harvest, the brown shrimp price increased rapidly reaching $3.00 per pound by the end of the year.

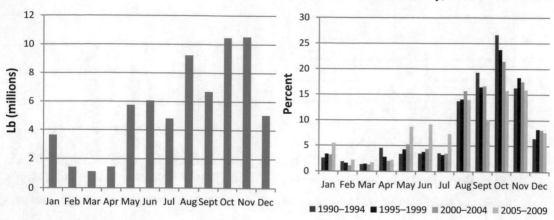

Figure 10.27. Gulf of Mexico average monthly white shrimp production for 2005–2009 (*left panel*) and production by month, expressed on a percentage basis, selected periods (*right panel*) (NMFS Galveston Laboratory, personal communication, 2012, with calculations by authors–see Appendix A) (*Note*: 1 lb = 0.454 kg).

The U.S. Gulf of Mexico white shrimp harvest has, overall, been increasing during the 1990–2009 period. For example, in 1990–1994, annual production of white shrimp averaged about 40 million pounds (heads off). By 1995–1999, the average had increased again to 42 million pounds and increased sharply to 58 million pounds in 2000–2004. Annual production of white shrimp in 2005–2009, averaging 66 million pounds, exceeded that of 1990–1994 by about 65 %.

While not as distinct as for brown shrimp, there is also a seasonal pattern to the Gulf white shrimp harvest. During 2005–2009, for example, Gulf landings of white shrimp averaged 66 million pounds (heads off). While brown shrimp catch is predominant in the 3-month period ending in August, the Gulf white shrimp catch tends to be highest in the months of August through November (Figure 10.27, left panel). This pattern is relatively consistent back to the 1990–1994 period, although the most recent 5-year period indicates a higher proportion being harvested in the May–July period at the expense of later months.

The higher proportion of white shrimp catch in the earlier months (May through July) may well reflect the increased catch of overwintering white shrimp. Specifically, with significantly less white shrimp fishing effort in recent years, an increasing proportion of the shrimp stock produced in a given year escapes catch in that year and is available for harvest in the subsequent year. This hypothesis is, to some extent, supported by examination of monthly white shrimp dockside prices (Figure 10.28). Specifically, the monthly white shrimp dockside prices tend to be relatively high in the earlier months suggesting larger shrimp that escaped harvest in the previous year. While one might argue that this price effect may be the result of low quantities being harvested in these earlier months, this argument is likely fallacious for two reasons. First, there are large quantities of brown shrimp landed in the May–July period that represent a close substitute for the white shrimp product. Second, as discussed later in this chapter, large changes in Gulf landings appear to have little influence on the Gulf dockside price due, largely, to the large import base.

Harvest by Depth and Movement of the Fleet

There are two general classes of shrimp vessels in the Gulf of Mexico—those that fish primarily in the inshore waters and those that fish primarily in the offshore waters. Smaller

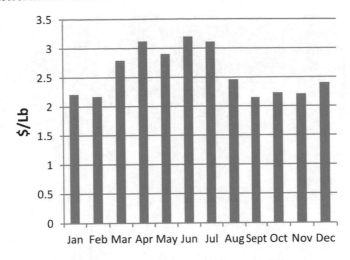

Figure 10.28. Gulf of Mexico average monthly dockside white shrimp price, 2005–2009 average (NMFS Galveston Laboratory, personal communication, 2012, with price calculations by authors– see Appendix A) (*Note*: 1 lb = 0.454 kg).

vessels, as one would expect, tend to shrimp primarily inshore while large vessels shrimp primarily offshore. The larger, offshore vessels are required to have a moratorium permit to shrimp in federal waters. According to Liese and Travis (2010), this segment of the harvesting sector accounted for two-thirds of the poundage harvested, and with the larger-sized shrimp harvested offshore, over three-quarters of the dockside revenue was generated in the fishery.

Geographical information covering the spatial distribution of catch, effort, and other critical variables for the management of the shrimp fishery are collected by the NMFS. This geographical information has three major components: a harvesting location defined on a statistical grid of longitude and latitude, a harvesting depth based on the fathom zone where harvesting was reported, and a record that identifies the port where the harvest was landed. The statistical grids are roughly defined as 1° longitudinal or latitudinal areas that project from shore out to 50 fathoms (91.4 m). Twenty-one of these grids occur in the U.S. Gulf of Mexico territorial waters. The fathom zones are defined as intervals of water depth in 5-fathom (9.1-m) increments from the U.S. shoreline out to 50 fathoms (91.4 m). Given the bathometry of the continental shelf in the northern Gulf of Mexico, the overlap of these two measures generates a maximum of 210 statistical subareas to which harvesting activity, and thus landings, are assigned during data collection (Figure 10.29).

Because the larger Gulf shrimp vessels can traverse a large geographic area in the harvesting of shrimp, the area where shrimp is caught does not necessarily reflect where it is eventually landed and thus, while landings by state were considered earlier in this chapter, it is also useful to consider catch by area. The estimated 2005–2009 annual catch by grid, expressed on a percentage basis, is given in Figure 10.30. Relatively little catch occurs along the Florida coast with the exception of grid 2 which represents the primary fishing grounds for pink shrimp. More than one-third of the total Gulf shrimp catch, by comparison, is estimated to be derived from two grids off of the coast of Louisiana (13 and 14) where both brown and white shrimp dominate the catch. All of the grids off the coast of Louisiana (13–17) account for about 60 % of the total shrimp catch, in pounds, during the 2005–2009 period. This catch, in percentage terms of the Gulf total, is about twice as high as the percentage of 2009 active shrimp moratorium permits registered to Louisiana home-ported vessels. While more than one-third

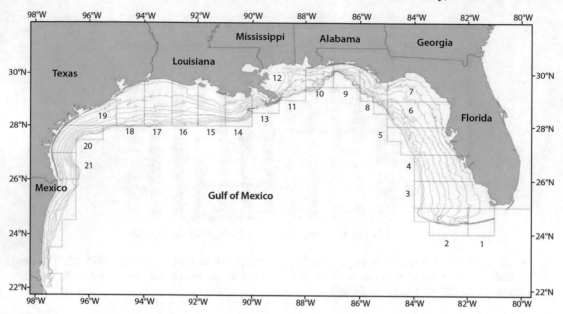

Figure 10.29. Relationship of 1° longitude/latitude statistical grids with fathom zones in the northern U.S. Gulf of Mexico (Nance et al. 2006).

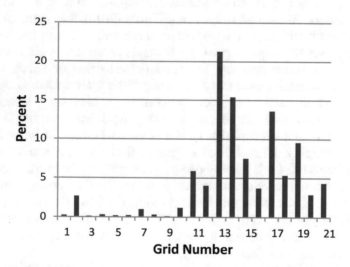

Figure 10.30. Estimated shrimp catch by grid, 2005–2009 average (NMFS Galveston Laboratory, personal communication, 2012, with calculations by authors–see Appendix A).

of the 2009 active shrimp moratorium permits were associated with vessels home-ported in Texas, catch in the grids associated with waters off the coast of Texas (grids 18–21) represented just 22 % of the total shrimp catch, in pounds, during 2005–2009.

There are several explanations as to why there is relatively high shrimp catch off the Louisiana coast, in pounds, relative to active shrimp moratorium permits issued to vessels home-ported in the state and conversely, why there is a relatively high number of active shrimp moratorium permits issued to Texas-based vessels relative to catch off of the Texas coast. The first explanation is that Texas vessels tend to be larger than Louisiana vessels and travel greater

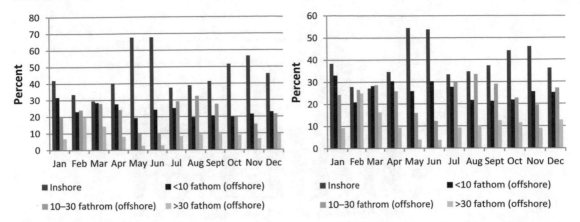

Figure 10.31. Average monthly percentage of Gulf shrimp catch from inshore and offshore waters on the basis of pounds (*left panel*) and value (*right panel*), 2005–2009 (NMFS Galveston Laboratory, personal communication, 2012, with percentage calculations by authors–see Appendix A).

distances. Ran et al. (2008) report that most of the vessels home-ported in Texas had statistical grids 14–21 as their harvesting destination, which of course includes Louisiana waters. However, according to Ran et al. (2008), Louisiana vessels only infrequently fish in the waters off the Texas coast. A second explanation is that while the catch from waters off the Texas coast, expressed in pounds, equaled only 22 % of the Gulf total catch during 2005–2009, Texas manages for a larger-sized shrimp than Louisiana which, as such, commands a price premium. Estimated catch off the Texas coast during 2005–2009, expressed on a value basis, was in excess of one-quarter of the Gulf total. Finally, while Louisiana's offshore fleet is smaller than that of Texas, in terms of the number of permitted vessels, Louisiana has a much larger inshore fleet that harvests a large amount of shrimp from its inshore waters.

Average monthly catch by depth, expressed on a poundage basis, is given for the 2005–2009 period and provided in Figure 10.31 (left panel). Similar information, expressed on a value of catch basis, is given in Figure 10.31 (right panel). As indicated, catch from inshore waters consistently represents the largest proportion of catch during each month and, in general, catch decreases with depth. Furthermore, the proportion of catch from inshore waters is directly related to the opening of the bays in association with the growth and movement of brown and white shrimp. For example, catch from inshore waters, in pounds, tends to be 40 % or less and then increases to almost 70 % in May associated with the opening of the spring season (i.e., brown shrimp season) in the northern Gulf States. As the brown shrimp grow and migrate to deeper waters, catch from inshore waters declines until the inshore waters open again in the fall for white shrimp season. By comparison, monthly shrimp catches in the less-than-10-fathom (18.3-m) offshore zone consistently fell in the narrow range of 20–30 %. Finally, with the exception of February and March, catch outside the 30-fathom (54.9-m) zone generally equaled 10 % or less.

A similar pattern to that observed for pounds emerges when one considers monthly values of catch by depth (Figure 10.31, right panel). However, the dominance of the inshore catch is lessened because the average size of shrimp caught from inshore waters tends to be smaller than that caught from offshore and, as such, commands a lower price. A comparison of inshore and offshore average monthly prices, based on 2005–2009 catches, is presented in Figure 10.32. As indicated, the May price differential ($1.28 per pound) is the largest, reflecting the opening of the inshore waters in Louisiana (which accounts for the largest proportion of the inshore catch) and the targeting of very small brown shrimp in the local bays and estuaries. The price

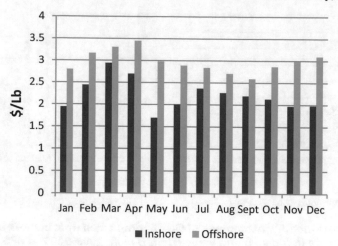

Figure 10.32. Average monthly shrimp prices for catch from inshore and offshore waters, 2005–2009 average (NMFS Galveston Laboratory, personal communication, 2012, with price calculations by authors–see Appendix A) (*Note*: 1 lb = 0.454 kg).

differential then decreases again through September after which it increases to more than $1.00 per pound by November.

Because of the diversified product (size category) mixes that bring very different market prices, shrimp harvesters can be considered as multiproduct firms. As such, one might expect the harvesters to be responsive to market signals subject to their technological and resource abundance constraints. Ran et al. (2008) examine this issue by determining to what extent harvesting effort and shrimp harvests by size category change in response to changing relative prices, while at the same time controlling for various seasonal influences that might affect the size distribution of shrimp stocks. Their analysis indicates that shrimp harvesters apparently have some ability to allocate effort across shrimp size categories in response to relative market prices, and harvesting effort is statistically targeted at low-count (large) shrimp size categories both due to their own-price and because of changing relationships with the price of other size categories. In addition, the majority of middle-sized shrimp appear to be harvested as a residual in the overall pursuit of large and small shrimp. While the harvest of these shrimp is dependent on the effort expended, the supply-response to effort changes tends to be lower than that observed for large- and small-sized shrimp. Ran et al. (2008) also found there to be some discernible differences between the supply elasticities of the nearshore waters and the deeper, offshore water fishery, with the supply generated by the deeper, offshore water fleet being more responsive to changes in effort, particularly with respect to the largest and smallest size categories.

Ran et al. (2011) examine those factors that influence location choice by vessels in the offshore fleet during two 5-year periods: 1995–1999 and 2000–2004. Factors found to influence location choice include expected revenues, attitudes towards risk, and fuel costs. The most important factor, however, is past experiences among the shrimpers at specific harvesting locations. Specifically, as noted by Ran et al. (2011) "...the behavioral inertia associated with changing fishing sites, perhaps due to lack of information or habit persistence, made harvester reluctant to change fishing location from one trip to the next (p. 41)." The authors conclude that because of changing economic conditions in the fishery (i.e., the deterioration of profits), behavior of the fleet has changed with some of the shrimpers becoming more risk averse.

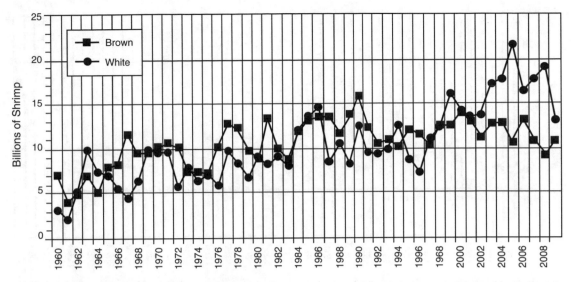

Figure 10.33. Estimated annual recruitment of brown and white shrimp into the Gulf of Mexico shrimp fishery, 1960–2009 (Nance 2011).

Variability in Shrimp Populations and Harvests

It is well known that fish populations and, to a lesser extent, subsequent harvests can vary significantly from one year to the next and that much of this variation is the result of environmental factors. Year-to-year variations can be particularly pronounced for species supported by a relatively few year classes since variations in recruitment are not *smoothed out* by older year classes. Given the short life span of brown and white shrimp, it should come as no surprise that annual populations and harvests vary substantially from year to year.

Fish populations are not directly observed which makes determination of populations difficult. One method is to estimate populations based on fishery-independent sampling of the population. Another method, which has historically been employed by the NMFS to estimate recruitment and adult shrimp populations, is based on virtual population analysis (VPA), details of which are provided by Nichols (1986). Based on this analysis, estimated Gulf of Mexico brown and white shrimp recruitment for the 1960–2009 period is given in Figure 10.33. After generally increasing from 1960 to 1990, brown shrimp recruitment fell during the next several years with the 2000 value of 14 billion recruits approximating those numbers estimated for the late 1980s. The 2009 estimated recruitment of 10.7 billion brown shrimp was approximately 15 % above the 2008 estimated recruitment of 9.25 billion shrimp. While no discernible long-term trend in estimated brown shrimp recruitment has been observed since the late 1980s, annual variation is shown to be large with year-over-year changes of 15–20 % not being uncommon.

Like brown shrimp, estimated recruitment of white shrimp showed significant variation over time with a range from 7.3 billion shrimp in 1996 to 21.5 billion shrimp in 2005. Estimated recruitment in 2008 of 19 billion shrimp exceeded the 2009 estimated recruitment of 13.1 billion shrimp by 45 %. Overall, estimated annual recruitment since the mid-1990s appears to be significantly higher than the long-term average though, as indicated, annual variation is also large.

Given the large annual variation in estimated recruitment of brown and white shrimp, it is no surprise that large variations are also observed in estimated (via VPA) parent populations

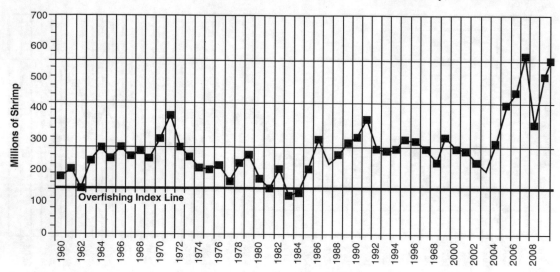

Figure 10.34. Estimated Gulf of Mexico brown shrimp parent population, 1960–2009 (Nance 2011).

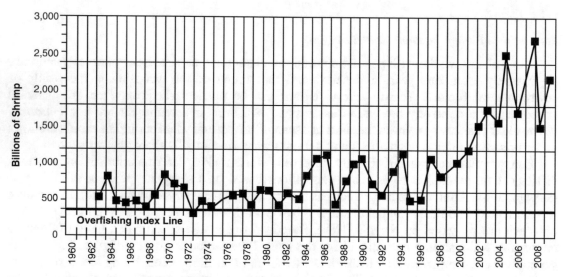

Figure 10.35. Estimated Gulf of Mexico white shrimp parent population, 1960–2009 (Nance 2011).

(Figs. 10.34 [brown shrimp] and 10.35 [white shrimp]). As indicated, the estimated brown shrimp parent population (defined as over 7 months of age during November–February) increased significantly in recent years but with large annual variations. For example, the estimated parent population in 2005 (approximately 400 million shrimp) exceeded the 2004 estimate (approximately 300 million shrimp) by 33 %. Similarly, the 2007 estimate (approximately 500 million) exceeded the 2008 estimate (approximately 350 million) by about 50 % with the 2009 estimate (approximately 500 million) exceeding the 2008 estimate by about 150 million shrimp.

The estimated white shrimp parent population clearly showed an increasing trend since the late 1990s. Nance (2011) hypothesizes that this is related to an increase in the number of overwintering white shrimp (while not stated by Nance, this is likely the result of a decline in white shrimp effort; see Figure 10.20). Like brown shrimp, the estimated population of white shrimp parents can vary substantially from one year to the next with percentage changes of more than 20 % not being uncommon, particularly in later years. For example, the estimated white shrimp parent population increased from about 1.6 billion in 2005 to more than 2.5 billion in 2006, or by more than 50 %, before falling to about 1.7 billion the following year. Similarly the change between 2008 and 2009 (from approximately 2.1–2.7 billion) represents a nearly 30 % increase.

A comparison of the information in Figure 10.33 with that in Figs. 10.34 and 10.35 gives an indication of the high shrimp natural (and/or harvest) mortality from time of recruitment until parent stage. For example, while the estimated recruitment of brown shrimp generally exceeded 10 billion shrimp in recent years (Figure 10.33), the parent population has generally fallen in the 200–500 million range (Figure 10.34). While a portion of the decline can be explained by the harvest of juvenile shrimp, the majority is undoubtedly the result of high natural mortality.

While there is no routine sampling of harvestable shrimp to determine population, the NMFS (Galveston Laboratory) forecasts western Gulf of Mexico brown shrimp production for the upcoming year (July–June) via two methods. The first method, referred to as the Baxter Bait Index, is based on the monitoring of the Galveston Bay bait shrimp fishery from late April to mid-June. The second method, referred to as The Environmental Model, uses a suite of variables (Galveston air temperature during mid-April, rainfall during early March, and bay water height during late April/early May) to predict brown shrimp production from Texas waters. The Baxter Bait Index is considered to be the more reliable of the two forecasts. Figure 10.36 shows a comparison of predicted annual harvests (July–June) based on the Baxter Bait Index and actual annual harvests for the 1980–2009 period. The correlation between the predicted and actual values (excluding 1990 for which no prediction was made) is a relatively low, 0.37. As indicated, the predicted harvest ranged from below 20 million pounds (1983) to 30 million pounds (2000) while the actual harvest ranged from less than 20 million pounds (2007) to more than 40 million pounds (1982). The average predicted harvest averaged 25 million pounds, which was also the average annual harvest during the considered period. The relatively low correlation between predicted harvest and actual harvest is likely largely driven by unpredictable changes in natural mortality and growth of the shrimp between the time of

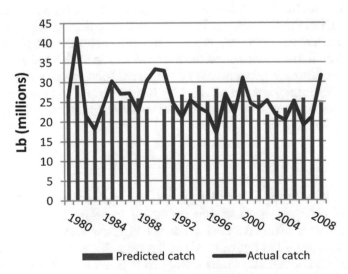

Figure 10.36. Texas offshore brown shrimp catch predictions (July–June) based on Galveston Bay bait index values in relation to actual catch, 1980–2009 [NMFS (Galveston Laboratory) 2012] (*Note*: 1 lb = 0.454 kg).

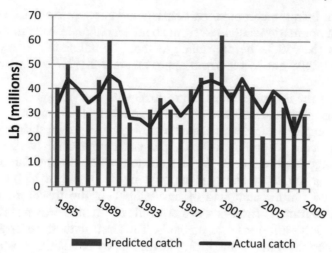

Figure 10.37. Louisiana inshore and offshore brown shrimp catch predictions (May–April) based on May catch index value and actual catch, 1985–2009 [NMFS (Galveston Laboratory) 2012] (*Note*: 1 lb = 0.454 kg).

monitoring in the bays (late April to mid-June) and the time that the shrimp move offshore and become susceptible to harvest by the Texas fleet.

Louisiana's brown shrimp catch (inshore and offshore) is also predicted for the biological year (May–April) based on catch information from Louisiana's inshore and offshore fisheries in May. These predictions for 1985–2009 along with subsequent harvests are presented in Figure 10.37. As indicated, the use of May's catch to forecast the biological year's catch is, with some notable exceptions, relatively accurate (correlation 0.72) with most turning points being correctly predicted. Thus, one can conclude that a single month's catch early in the season can provide meaningful information that can be used in predicting catch for the biological year (May–April). Finally, as indicated, there is considerable year-to-year variation in both the predicted and actual Louisiana brown shrimp catch.

The large annual fluctuations in brown and white juvenile shrimp can, of course, translate into large annual variations in harvests. However, the mapping of shrimp from the juvenile stage to either the adult stage or harvest is less than monotonic because of the large number of environmental factors that can influence the survival and growth of shrimp throughout their successive life stages. These environmental factors have been examined by a large number of researchers (see, for example, Haas et al. 2001 for brown shrimp and Diop et al. 2007 for white shrimp and references contained therein). Annual variations in harvest for the two species are clearly identified in Figure 10.38.

Catch per Unit Effort

Large variations in year-to-year and long-term shrimp abundance (Figures 10.33, 10.34, and 10.35) and long-run changes in effort (Figure 10.20) translate into short-run and long-run changes in catch per unit effort (CPUE). Annual CPUE estimates for the brown and white shrimp fisheries are presented in Figure 10.38 for the 1960–2009 period. As indicated, following abundance patterns, CPUE can vary considerably from one year to the next and has increased significantly since the early 2000s. The increased CPUE in recent years reflects, at least in part, the sharp reduction in effort (days fished) that then translates into increased shrimp availability (given the fixed short-run stock) for those trips being made.

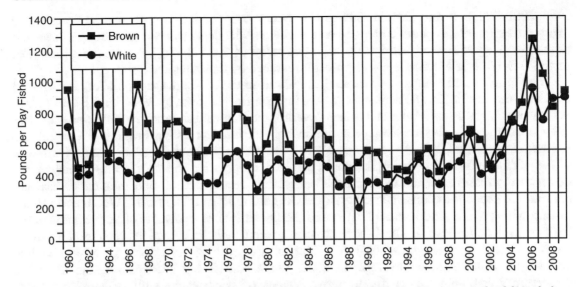

Figure 10.38. Catch per unit effort (day fished) in the Gulf of Mexico brown and white shrimp fisheries, 1960–2009 (Nance 2011) (*Note*: 1 lb = 0.454 kg).

With respect to white shrimp, CPUE was a record 931 lb per day fished in 2006 before falling to less than 750 lb per day fished in 2007. In 2008, CPUE increased again to 875 lb and equaled 882 lb in 2009. Since 2004, the CPUE associated with the white shrimp fishery has consistently been higher than in any year dating back to the 1960s. This, of course, reflects both the relatively high abundance in recent years (Figs. 10.33 and 10.35) and the relatively low level of effort targeting the species (Figure 10.20). Before the early 2000s, CPUE of less than 400 lb per day fished was not an uncommon occurrence in the white shrimp fishery.

The average CPUE for brown shrimp since 1960 has approximated 640 lb per day fished with the 2006 estimate of 1,244 lb per day fished being approximately twice the long-run average. The CPUE declined in the successive 2 years to 1,027 lb per day fished in 2007 and 821 lb per day fished in 2008 before increasing to 932 lb per day fished in 2009. Unlike the white shrimp fishery, however, CPUE associated with the brown shrimp fishery rarely (if ever) fell below 400 lb per day fished during the considered period of analysis.

Financial Condition of the Fleet

Prior to the year 2001, the U.S. shrimp industry was relatively healthy from an economic perspective. The average annual rate of return for the harvesting fleet was 12.5 % during the 1965–1995 period, even though fluctuating stocks (due to year-to-year changes in environmental conditions) led to substantial inter-year variability, including some years in which profitability was near zero or negative (Funk et al. 1998). While it is not surprising to find that profitability varies by vessel size, small vessels on average had higher rates of return, suggesting that there are decreasing returns to scale in the harvesting industry. This may be a function of ownership patterns in the industry, where smaller vessels tend to be operated by their owners and only participate in the shrimp fishery on a part-time basis when revenue and/or profit per unit of effort are high (Funk et al. 1998). For larger vessels, relatively high fixed costs and vertical integration with processors often force owners to continue harvesting regardless of the economic conditions. Over time, this leads to lower than average rates of return even though large vessels can be highly profitable when nominal dockside prices are stable and real input cost are low (as they were from

1998–2000)[9] (Travis and Griffin 2004). In the years since 2000, however, market forces have exerted tremendous economic pressure on individuals who depend on the harvesting of seafood as their primary source of income. As the largest sector of that industry by value, the shrimp fleet of the U.S. Gulf of Mexico is also the most threatened by those market forces.

The most recent analysis of financial conditions in the offshore shrimp fishery is provided by Liese and Travis (2010), while that for the inshore fleet is given by Miller and Isaacs (2011). As succinctly stated by Miller and Isaacs "overall, the financial situation in 2008 was economically unsustainable for the average active inshore shrimp harvesting business." The authors further indicate that "[t]hese results parallel similar research about the economic performance of the offshore fleet. Increasing fuel costs, increases in imported shrimp volume—which places downward pressure on domestic prices—as well as recent natural and manmade disasters continue to erode the economic vitality of the Gulf shrimp harvesting fleet." With respect to the inshore fleet, Miller and Isaacs (2011) found the net cash flow to owners of active boats in the Gulf of Mexico inshore shrimp fishery, which represents the difference between total revenues from all sources (average $45,684) and financial outlays ($39,850), to equal approximately $6,000 per fisherman, on average, in 2008. When considering all expenses, including the opportunity cost of time, profits to active owners of boats in the Gulf of Mexico inshore shrimp fishery were, on average, slightly negative in 2008. Almost 50 % of the active boat owners were found to have a negative net cash flow in 2008, and less than 10 % reported net cash flow in excess of $33,000.

With respect to the federally permitted vessels (i.e., those vessels legally allowed to shrimp in the Gulf of Mexico EEZ), Liese and Travis (2011) report that net cash flow among this segment of the shrimp harvesting sector averaged about $8,300 per active owner in 2009 based on revenues from all sources (from the sale of shrimp, disaster payments, etc.) averaging $212,000 (revenues from landed shrimp accounted for 89 % of this total) and costs averaging about $208,000. The average economic return, calculated by dividing operating revenue by the value of the vessel assets, equaled 0.3 %. This estimate stands in stark contrast to the 12.5 % return on investment reported by Funk et al. (1998) for the 1965–1995 period.

10.3.8.1.4   The Gulf of Mexico Shrimp Processing Sector

Analyses by Keithly and Roberts (1994) and Keithly et al. (2006) indicate that virtually all shrimp landed in the U.S. Gulf is processed in that region. The two primary products produced from the domestic landings, as noted by Keithly and Roberts (1994) and Keithly et al. (2006), are a headless shell-on product and a peeled-raw product. The production of these two product forms (converted to a whole weight basis) and deflated value (2009 Consumer Price Index [CPI] used as the base) are presented in Figure 10.39. Mirroring the dockside price, the price of the processed product has fallen sharply, particularly after 2000. This decline in deflated value has transpired despite long-run stability in processed poundage (the result of virtually all harvest being used in the processing sector and long-run stability in harvest; see Figure 10.15).

Comparison of the processed shrimp price (headless shell-on and peeled-raw) with the Gulf dockside shrimp price indicates that the marketing margin has significantly fallen over time

---

[9] Historically, many of the larger processors maintained their own fleets. This vertical integration was employed as a means of ensuring adequate supply of raw material for use in processing activities. These vertical integrated facilities could (and often would) absorb losses in the harvesting component of their operations in the profits generated in processing. As profitability in the processing sector eroded over time, due to competition with imported product, the ability to absorb losses in the harvesting component of the business declined. As such, vertical integration is probably not as prevalent today as in the 1980s. Unfortunately, there is little data that could be used to examine changes in vertical integration.

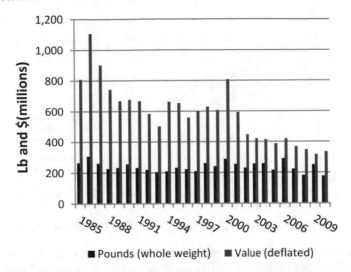

Figure 10.39. Gulf processed pounds (headless shell-on and peeled-raw products) and deflated value of processed product, 1985–2009 (NMFS Southeast Regional Office, personal communication, 2011, with deflated values calculated by authors—see Appendix A) (*Note*: 1 lb = 0.454 kg).

(Figure 10.40), particularly since 2001 and the associated rapid rise in imports of peeled-raw product (Figure 10.18). Given the reduction in margin and, one would hypothesize, associated profit per unit of output, a large proportion of the processing establishments have exited the industry while others have coped with the declining per unit profitability by increasing output.[10] The decline in number of firms in association with the declining marketing margin is given in Figure 10.41 while the increase output per firm is considered in Figure 10.42. As indicated, in association with the declining marketing margin the number of firms fell from almost 100 in the early 1990s to the mid-40s by the late 2000s. Production per firm, however, has increased, thereby mitigating, at least to some extent, the declining profitability per unit of output. Given that the long-run domestic shrimp harvest has been stable, along with the fact that existing processors use virtually all of the landings, it is apparent that the increased output per firm is the result of a reduction in the number of firms.

In general, the price received by Gulf processors for the two primary products, headless shell-on and peeled-raw, closely mirrors the import prices associated with these two products. With respect to the headless shell-on product, there are generally only small deviations between the Gulf price and import price (Figure 10.43, left panel). With respect to the peeled-raw product, the Gulf price generally exceeded the import price during the mid-1980s to early 1990s but since then the import price has consistently exceeded the domestic price (Figure 10.43, right panel). While the reason for this change is not known with certainty, it coincides with that period during which U.S. imports of farm-raised shrimp from Asian countries expanded rapidly. As such, one might hypothesize that beginning in the early 1990s, there was an increased use of this farm-raised shrimp (which is desired because of its uniform size and

---

[10] The marketing margin, by definition, reflects the difference between the processed price and the dockside price or, stated somewhat differently, the cost of inputs (including normal returns to capital and labor) to transform the product. If costs of these inputs did not significantly decline, one could state with certainty that the profit per unit output has also fallen.

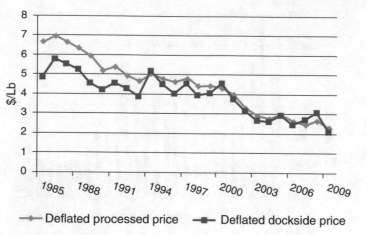

Figure 10.40. Relationship between deflated processed and dockside prices (2009 base), 1985–2009 [NMFS Southeast Regional Office (processing data), personal communication, 2011; NMFS FSD (dockside data), data accessed 2011, with deflated prices calculated by authors—see Appendix A] (*Note*: 1 lb = 0.454 kg).

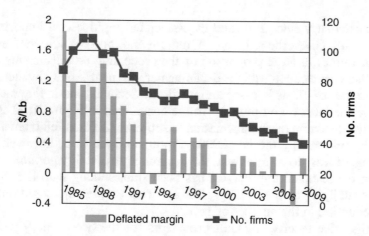

Figure 10.41. Change in number of Gulf shrimp processors in relation to change in marketing margin, 1985–2009 (NMFS Southeast Regional Office, personal communication, 2011, with deflated margins calculated by authors–see Appendix A). (*Note*: 1 lb = 0.454 kg).

year-round availability) in the raw-peeled product exported to the U.S. market. Given its desirability, a premium was likely attached to the product.

Finally, a comparison of the domestic headless shell-on price (Figure 10.43, left panel) and the domestic peeled-raw price (Figure 10.43, right panel) shows that the price received for the headless shell-on product consistently exceeds the price received for the peeled-raw product but that the price differential has been narrowing in recent years. The higher price associated with the headless shell-on product is the result of a larger-sized shrimp generally being used in the production of the headless shell-on product *vis-à-vis* the peeled-raw product. Roberts and Keithly (1991), however, document the significantly greater overall economic contribution associated with the peeled-raw product resulting from additional value-added activities.

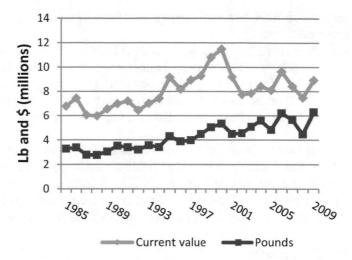

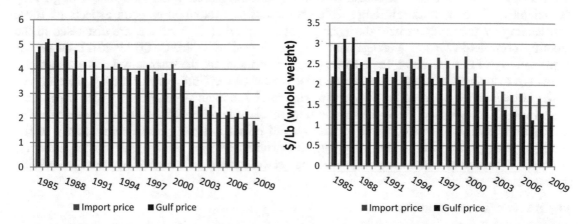

Figure 10.42. Average output per firm (headless shell-on and peeled-raw products) and current value of output per firm, 1985–2009 (NMFS Southeast Regional Office, personal communication, 2011, with calculations by authors—see Appendix A). (*Note*: 1 lb = 0.454 kg).

Figure 10.43. Relationship between the domestic processed price and import price for headless shell-on product (*left panel*) and peeled-raw product (*right panel*), 1985–2009 [NMFS Southeast Regional Office (processing data), personal communication, 2011; NMFS FSD (dockside price data), data accessed 2011, with calculations by authors—see Appendix A] (*Note*: 1 lb = 0.454 kg).

10.3.8.1.5   Impact of Gulf Shrimp Landings on Dockside Price

Arguably, the most comprehensive analysis of the impact of Gulf shrimp landings on the Gulf shrimp dockside price is that of Poudel (2008). Based on a large-scale econometric model of the world shrimp market U.S. market, the European Union [EU] market, and the Japanese market), Poudel (2008) analyzes the impacts of increased shrimp production in different regions of the world (Asia, Central America, and South America) on the Gulf of Mexico dockside price as well as the influence of changes in own landings (i.e., Gulf landings) on the dockside price. The analysis was based on quarterly data from 1990 to 2004. Overall, Poudel (2008) found the dockside price to be relatively invariant to large changes in landings with a 10 % increase (decrease) in Gulf landings resulting in a 1.7 % decline (increase) in dockside price, holding all other factors constant. The small response in price to a change in landings is not unexpected given that the U.S. shrimp supply is dominated by imports. Furthermore, given the increase in imports since 2004, one might expect that the influence of own landings on price has lessened in more recent years.

Based on monthly data from 1990 to 2008, Asche et al. (2012) use a co-integration approach to examine the relationship between the shrimp import price and the Gulf dockside price. The authors found a high degree of market integration between the imported product and domestic product and, based on this finding, conclude that large changes in Gulf of Mexico landings will lead to little change in the Gulf dockside price. Rather, imports will increase to meet domestic demand.

Together, these two studies indicate that large changes in Gulf shrimp production will result in little change in the dockside price. This finding should come as little surprise given (1) imports represent the vast majority of U.S. supply (i.e., domestic landings plus imports) and (2) there is little to differentiate the domestic product from the imported product, particularly after it enters the restaurant trade where a high percentage of the shrimp product is consumed.

10.3.8.2 The Oyster Industry

Unlike most species harvested in the Gulf of Mexico, the oyster is a sessile creature. As such, the harvesting sector can be developed around leasing operations. All Gulf States, with the exception of Alabama, maintain leases on state-regulated water bottoms, though only Louisiana and Texas maintain large-scale active leasing systems. Long-run aspects of these two leasing systems, along with the leasing systems in other Gulf States, are discussed in the recently completed Oyster Management Plan developed by GSMFC (OTTF 2012). Given the importance of Louisiana and Texas to Gulf oyster production, the leasing systems in these two states are examined in some detail after a brief review of the Gulf oyster industry. Detail given to the Louisiana segment of the industry is warranted due to its large size relative to other Gulf States and its complexity. While the lease system in Texas is somewhat less complex than that of Louisiana's, attention is also given to this system because it contributes significantly to the state's oyster production. After reviewing the production side of the Gulf oyster industry, attention is turned to examining the processing sector and the influence of harvest on dockside prices.

10.3.8.2.1 The Production Side

Gulf Production in Relation to U.S. Total

On average, the production of oysters from the Gulf of Mexico (the United States) averaged 21.3 million pounds (meat weight) annually during 1990–2009, which represented almost 60 % of the nation's 36.6 million-pound annual average production over the period. Given the large share of U.S. oyster production attributable to the Gulf region, any large changes in annual Gulf production also significantly influence U.S. production (Figure 10.44). As indicated, Gulf production, which averaged 19.3 million pounds annually during the 1990s, generally increased during much of the period with production from the region approximately doubling from the early 1990s to the late 1990s. By 2000, Gulf production reached the 25 million-pound mark, and during the decade beginning in 2000, annual production from the region averaged 23.3 million pounds. The increased Gulf production in the most recent decade has translated into an increased share of U.S. production attributable to the Gulf. Specifically, during the 1990s, the Gulf share of U.S. production equaled 54 %, and since 2000, the Gulf share of the nation's production has equaled 62 %. The Gulf has approached or exceeded the 65 % mark in most years since 2000 with the exception of the 3-year period ending in 2008 when the Gulf share fell below 60 % in each of the 3 years. Overall, U.S. production among states outside the Gulf region has averaged about 14.2 million since 2000.

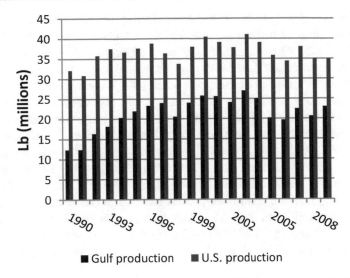

Figure 10.44. U.S. and Gulf of Mexico annual oyster production, 1990–2009 (NMFS FSD, data accessed 2012—see Appendix A) (*Note*: 1 lb = 0.454 kg).

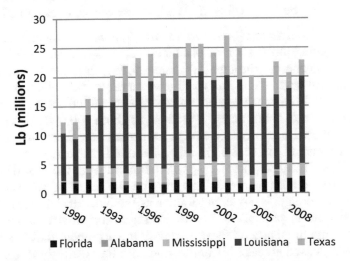

Figure 10.45. Gulf oyster production by state, 1990–2009 (NMFS FSD, data accessed 2012—see Appendix A) (*Note*: 1 lb = 0.454 kg).

Gulf Production by State

Annual oyster production for each of the five Gulf States from 1990 to 2009 is presented in Figure 10.45. As indicated, the region's production is dominated by Louisiana, which accounted for 57 % of the total during the period of analysis based on annual average production of 12.2 million pounds. Texas, with average annual production approaching 4.5 million pounds accounted for an additional 20 % of the region's total output during 1990–2009. Florida and Mississippi each contributed about 10 % to the region's total while Alabama's contribution was negligible.

Annual oyster harvest in any given state, or throughout the region, can vary significantly from one year to the next due largely to environmental perturbations. As indicated by the information in Figure 10.45, Gulf production during the 20-year period ending in 2009 fluctuated from less than 13 million pounds (1990 and 1991) to more than 25 million pounds (2000, 2001, 2003, and 2004). Abnormally low production in 1990 was most likely the result of drought conditions throughout Louisiana, which lasted for several years beginning in the mid-1980s. This drought came to an abrupt end in 1991 with a record rainfall. While initially resulting in high oyster mortality as a result of low salinity conditions throughout the state's estuary systems, the pulse of fresh water was, in the long run, beneficial to the oyster population. Gulf production was also relatively low in 2005 and 2006 due, primarily, to a reduction in harvests in Mississippi and Louisiana. This decline can be directly related to Hurricane Katrina (and to a lesser extent Hurricane Rita), which made landfall around the Mississippi/Louisiana border in August of 2005 (Mississippi was forced to close its state waters to all oyster harvesting in 2006). Similarly, when Hurricane Ike entered around Galveston Bay in 2008, there was a significant loss of infrastructure that resulted in a reduction in Texas production in that year and in 2009. This had a significant impact on production from Texas given the fact that about 80 % of the Texas production is generally taken from this one water body.

Despite some significant year-to-year variations in state annual production, each state's relative share of the region's overall total has remained extremely stable when considered in 10-year increments (Table 10.1). For example, Louisiana's share of the region's production remained at 57 % during both 10-year periods while Texas's share remained at about 21 %.

A Closer Look at the Louisiana Oyster Harvesting System: Louisiana's large annual oyster harvest is derived from a combination of production from leases and public seed grounds. By providing a stable environment through its leasing policy, the state has encouraged industry investment and has provided an impetus for the preservation, rehabilitation, and expansion of existing leases. Overall, Louisiana's leased acreage has expanded approximately fivefold since the early 1960s, from about 75,000 acres (30,350 ha) to about 400,000 acres (161,875 ha) (OTTF 2012). Despite this increase in acreage, long-run production from this leased acreage has remained relatively constant at about eight million pounds per year. Increasing leased acreage in conjunction with relatively constant long-run production from the leased acreage implies, of course, declining productivity per acre. This may be the result of several factors including (1) the recently added acreage is not as productive as the older acreage, (2) older leases are no longer as productive as in past years, (3) the average productivity of all leased acreage is declining, and (4) some amalgam of these factors. One argument that has been advanced to explain the increased leased acreage in conjunction with the relatively stable long-run production is that the increased acreage being leased is in response to wetland degradation and increasing rapid fluctuations in salinity regimes. Specifically, with the increasing exposure of

Table 10.1. 10-Year Average Annual Oyster Production in Pounds for Each Gulf State and Its Share (%) of Gulf of Mexico Production

	Florida	Alabama	Mississippi	Louisiana	Texas
1990–1999 avg.	1,964,000 (10.2 %)	594,137 (3.1 %)	1,652,250 (8.5 %)	11,118,537 (57.5 %)	4,002,534 (20.7 %)
2000–2009 avg.	2,256,747 (9.7 %)	669,528 (2.9 %)	2,172,242 (9.3 %)	13,349,786 (57.2 %)	4,895,472 (21.0 %)
1990–2009 avg.	2,110,374 (9.9 %)	631,833 (3.0 %)	1,912,246 (9.0 %)	12,234,162 (57.3 %)	4,449,003 (20.9 %)

Source: NMFS FSD, data accessed 2012—see Appendix A. *Note*: 1 lb=0.454 kg.

oyster leases to open water (due to marsh deterioration), short-term changes in the proximate reef area have become more common and with a higher magnitude of change. Hence, acreage that is productive one year may not be productive the next. As such, leaseholders may be increasingly diversifying their individual lease portfolios as a means of protecting themselves against the vagaries associated with any single lease or group of leases subject to environmental perturbations. Keithly and Kazmierczak (2006) suggest that speculation may have also contributed to the observed increase in leased acreage since the 1960s. Specifically, oil and gas activities are common in coastal Louisiana and often overlap oyster leases on a geographical basis. The researchers found that compensation for oil and gas activities is negotiated with affected lessees and may or may not be based on lease productivity. Hence, the researchers argue that considerable acreage of water-bottom is leased for the main purpose of receiving compensation rather than for the production of oysters.

Leasing activities do not operate in isolation but, instead, are intricately tied to the public grounds. Specifically, these public grounds serve as a source of seed oyster that can be transplanted to the private leases which is particularly important in those areas where natural oyster production (i.e., spat set) is limited and, as such, production from leases in these areas would be very limited in the absence of transplanting activities. An examination of Louisiana's oyster leasing activities, in recent years, and the relation between these activities and the public grounds is presented in this section.

Private Leases: Since 1999, production from private grounds has averaged eight million pounds (meats) annually (Figure 10.46, left panel). Highest observed production from private leases during the 11-year period of analysis ending in 2009 occurred in that year and equaled 11.5 million pounds. The relatively high production in the latest year may reflect, in part, the influence of the Private Oyster Lease Rehabilitation Program (POLR), which was initiated to assist leaseholders in recovery efforts after Hurricanes Katrina and Rita. Specifically, the program partially reimbursed leaseholders for (1) movement of seed from public grounds to individual leases, (2) sediment/debris removal, (3) cultch deposition, and (4) other activities. During the life of the program, which expired at the end of 2009, leaseholders were partially reimbursed for the bedding of more than 800,000 barrels of seed oysters (one barrel is equivalent to two sacks where a sack, according to the Oyster Technical Task Force (OTTF 2012), has a dimension of 1.87 cubic feet (ft^3) and supports approximately 100 lb of shell and

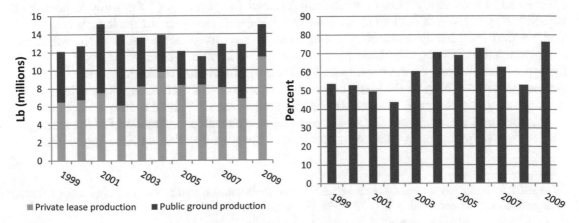

Figure 10.46. Annual oyster production from private leases and public grounds (*left panel*) and annual private lease production as a percentage of the total (*right panel*), 1999–2009 (unpublished data provided to Walter Keithly by the LDWF for years covering 1999–2008); 2009 data derived from LDWF (2010) with percentage calculations by authors; (*Note*: 1 lb = 0.454 kg).

meats on approximately 60,000 acres (24,281 ha). The bedding of these seed oysters represented approximately 40 % of the total POLR expenditures with another 40 % being paid for sediment/debris removal from private grounds. The lowest observed annual production during the 11-year period of analysis, equal to 6.2 million pounds, occurred in 2002.

Expressed on a percentage basis, production from leases as a percent of total production (i.e., leases and public grounds) equaled 60 % during the period of analysis. As indicated in Figure 10.46 (right panel), the range has been from just over 40 % (2002) to approaching 80 % (2009). As of September 21, 2010, there was a total of 384,951 acres (155,784 ha) of water bottoms being leased (personal communication with Patrick Banks, LDWF). This represents about a 5 % decline from the January 1999 leased acreage totaling 403,141 acres (163,145 ha) and about an 8 % decline from the 419,900 acres (169,928 ha) being leased in February 2001.[11] As noted by Keithly and Kazmierczak (2006), the declining acreage likely reflects a combination of the moratorium on the leasing of new acreage (this moratorium was established March 7, 2002, but excluded pending applications as of that date) and the purchase of leases by the state in furtherance of its restoration activities. Of the 392,000 acres (158,636 ha) being leased as of February 2006, more than one-third of the total (140,485 acres [56,852 ha]) was in Plaquemines Parish while an additional one-quarter of the total (91,890 acres [37,187 ha]) was Terrebonne Parish based. Other parishes contributing to the total include St. Bernard (88,139 acres [35,669 ha]), Lafourche (23,448 acres [9,489 ha]), Iberia (18,312 acres [7,411 ha]), Jefferson (18,093 acres [7,322 ha]), Vermillion (5,404 acres [2,187 ha]), and St. Mary (14 acres [5.7 ha]). In addition, some leases transverse parish borders. These include Jefferson/Lafourche (1,088 acres [440 ha]), Jefferson/Plaquemines (1,804 acres [730 ha]), Lafourche/Terrebonne (381 acres [154 ha]), Plaquemine/St. Bernard (327 acres [132 ha]), Terrebonne/St. Mary (177 acres [72 ha]), and Iberia/Vermillion (2,432 acres [984 ha]). Annual leased acreage of approximately 400,000 acres [161,874 ha] for the 11-year period ending in 2009 in conjunction with production from leased grounds during that period (averaging eight million pounds per year) yields an average annual production per acre of 20 lb. This equates to three sacks per acre based on the conversion factor of 6.47 lb of meats per sack.

While oyster yield per acre from private leases has averaged about 20 lb (meats) per year in recent years, one should recognize that all acreage is not as equally productive and some acreage is not capable of supporting oysters. With respect to the ability to support oysters, Keithly and Kazmierczak (2006), in an analysis of leasing activities, reported that 56 % of the leases considered in their study were unproductive (defined for purposes of the study as having no standing crop capable of harvest at the time the pre-impact assessment of the lease was made[12]). While some leases may have no standing crop in any given year, under more conducive environmental conditions the lease may be productive in other years. While not provided in the report, an analysis of the pre-impact assessment information collected by the researchers indicate that about 20 % of the leases had no hard bottom or shell (indicating that the lease could not support an oyster crop) while another 38 % had less than 5 % hard bottom and/or shell.

Keithly and Kazmierczak (2006) suggest that one plausible explanation for the leasing of nonproductive grounds is that of speculation. Specifically, the authors argue that the

---

[11] Detailed information on leased acreage, number of leaseholders, and number of leases for selected time periods can be obtained at http://204.196.151.247/oyster/.

[12] Much of the proposed work in the coastal region requires a Coastal Use Permit. As a part of the process in obtaining this permit, a pre-impact assessment in that area potentially impacted by the work must be conducted. This includes an assessment of oyster leases and reefs if they are located in the area of the proposed work.

juxtaposition of water-bottom leasing and oil and gas activities along the coast has likely encouraged leasing of nonproductive grounds in the expectation that compensation will be received for oil and gas activities in proximity to the lease. The authors estimated that the 2004–2005 harvesting cost per sack equaled $7.84 while the dockside price per sack equaled $16.37 yielding a profit margin of $8.53 per sack. In conjunction with average productivity per acre and number of acres, net income associated with harvesting from private leases was estimated to equal approximately $12 million in total or roughly $31 on a per acre basis. Payments to lease holders from oil and gas activities during 2004–2005, by comparison, were estimated by the authors to equal $26 to $36 per acre. Hence, the authors conclude that compensation to leaseholders from oil and gas activities equals or exceeds income derived from harvesting activities.

Other acreage, while capable of supporting a standing oyster crop, may be closed to the harvest for direct marketing on a seasonal or permanent basis as a result of health concerns. As suggested by Diagne et al. (2004), relaying of oysters from leases in harvest-limited waters to leases in approved waters, while permitted, is practiced infrequently in Louisiana with the level of activity being a function of the dockside price, availability of oysters on the public seed grounds, and the availability of oysters on private leases.

Average monthly oyster landings from private leases for the 10-year period ending in 2008 (i.e., 1999–2008) are presented in Figure 10.47 (left panel) and the same information presented on a percentage basis is given in the right panel. As indicated, production from private grounds is highest in the summer months with the 4-month period ending in August accounting for about 60 % of the total. By comparison, production during the 4-month period ending in February accounted for only about 11 % of the total production from private grounds during the 1999–2008 period.

One explanation for higher production during summer months is that the public grounds are closed throughout the summer months. The relationship between production from private leases and public grounds is illustrated in Figure 10.48. As indicated, decreases (increases) in production from the private leases can generally be associated with increases (decreases) in production from the public ground with the correlation between the two being equal to −0.795 (based on the monthly data from 1999 to 2008).

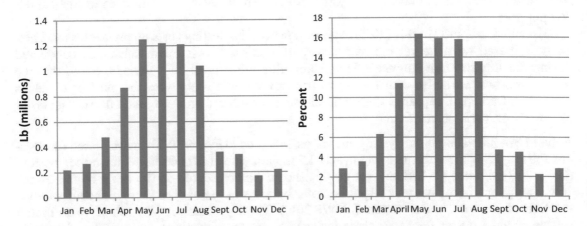

Figure 10.47. Average monthly oyster production from private leases on a poundage basis (*left panel*) and on a percentage basis (*right panel*), 1999–2008 (calculated from unpublished data provided to Walter Keithly by LDWF with percentage calculations by authors) (*Note*: 1 lb = 0.454 kg).

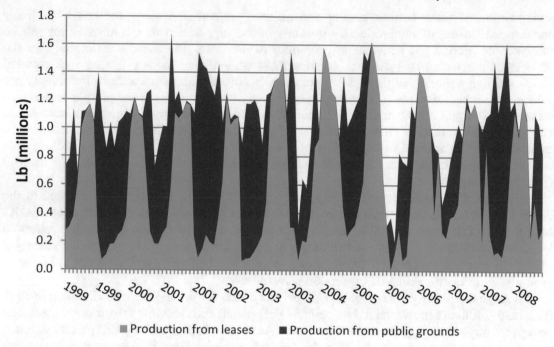

■ Production from leases   ■ Production from public grounds

Figure 10.48. Monthly oyster production from private leases and public grounds, 1999–2008 (calculated from unpublished data provided to Walter Keithly by the LDWF) (*Note*: 1 lb = 0.454 kg).

The annual value of production derived from private leases increased from about $13 million in 1999 to almost $40 million in 2009 (Figure 10.49, left panel). The relatively high dockside value from private leases in 2009 represents a combination of two factors. First, as noted, production from leases was uncharacteristically high in 2009. Second, the 2009 dockside price for oysters (meat weight) taken from private leases equaled $3.38 per pound (Figure 10.49, right panel). This price exceeded the reported dockside price in most other years by a significant margin with the 2006 and 2007 reported prices being about 8 % below the 2009 price. Using the 2006–2009 average annual dockside price (unweighted) in conjunction with the recent productivity of 20 lb per acre, annual gross oyster revenues from leasing are estimated to equal about $65 per acre.

Kazmierczak and Keithly (2005) examined per trip harvesting costs on private leases. Their analysis, based on a harvesting cost survey of Louisiana oystermen which was conducted during the July through August 2003 and June through August 2004 periods, found that the most important variables contributing to per trip variable costs were the number of sacks harvested, fuel price, captain's wage, miscellaneous costs, and crew wages (in decreasing order of impact on variable costs).

Public Grounds: The public oyster grounds, as indicated in Figure 10.50, are scattered throughout the coast. While encompassing nearly 1.7 million acres (687,966 ha), known reef bottom (about 38,000 acres [15,378 ha] though this should be considered as the lower-bound of the actual amount of reef because all public water bottoms have not been surveyed) equals only a fraction of the total water bottom (LDWF 2010). These grounds, in general, serve as both a source of seed oyster (less than three inches) and oysters for direct market (three inches or greater). They are generally open for harvest in September/October and close the following March/April.

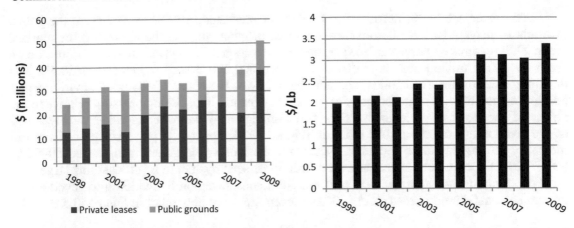

Figure 10.49. Current value of oyster production from private leases and public grounds (*left panel*) and annual dockside price for oysters taken from private leases (*right panel*), 1999–2009 (unpublished data provided to Walter Keithly by the LDWF for years covering 1999–2008 with 2009 data derived from LDWF (2010) with price calculations by authors).

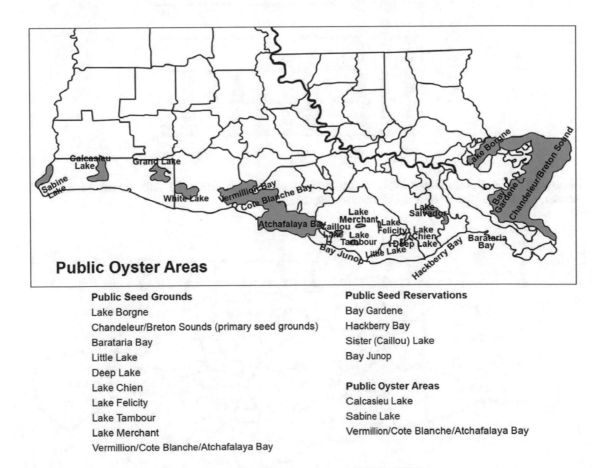

Public Seed Grounds
Lake Borgne
Chandeleur/Breton Sounds (primary seed grounds)
Barataria Bay
Little Lake
Deep Lake
Lake Chien
Lake Felicity
Lake Tambour
Lake Merchant
Vermillion/Cote Blanche/Atchafalaya Bay

Public Seed Reservations
Bay Gardene
Hackberry Bay
Sister (Caillou) Lake
Bay Junop

Public Oyster Areas
Calcasieu Lake
Sabine Lake
Vermillion/Cote Blanche/Atchafalaya Bay

Figure 10.50. Map of public oyster grounds in Louisiana (LDWF 2009).

Seed oyster, while not permitted to be directly marketed, may be moved from the public grounds to private leases where the transplanted product can later be harvested for market. Since 1999, estimated barrels of seed oyster have averaged 2.1 million annually, with a range from over five million barrels (2000) to less than 600,000 barrels (2009; Figure 10.51). The declining seed oyster availability in recent years may be the result of changing environmental conditions, the effects of numerous storms and hurricanes, high fishing pressure relative to the ability of the stock to replenish itself, or some amalgam. While the change in estimated seed oyster availability on the public grounds (Figure 10.51) appears large, estimates of natural mortality among subadult and adult oysters populations are large and can exceed 50–95 % (OTTF 2012). One would expect natural mortality of seed oysters to be at least this large.

For sampling and management purposes, the coastal region is divided into seven areas, known as coastal study areas (CSAs). These seven areas are illustrated in Figure 10.52.

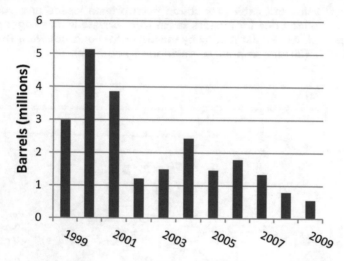

Figure 10.51. Estimated seed oyster availability on public grounds, 1999–2009 (email from Patrick Banks, LDWF, to Walter Keithly, December 29, 2009).

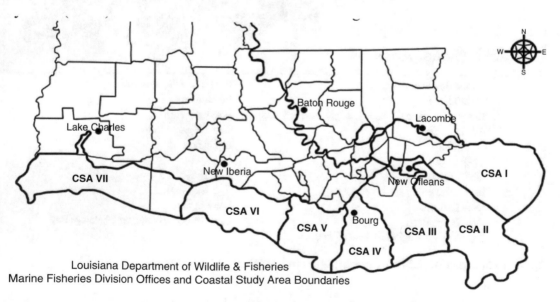

Louisiana Department of Wildlife & Fisheries
Marine Fisheries Division Offices and Coastal Study Area Boundaries

Figure 10.52. Map illustrating coastal study areas in the Louisiana (LDWF 2010).

- CSA I consists of approximately 690,000 acres (279,233 ha) of water bottoms, all east of the Mississippi River and north of the Mississippi River Gulf Outlet, of which approximately 21,000 acres (8,498 ha) represent reef.

- CSA II, which consists of approximately 17,000 acres (6,880 ha) of reefs, is character-ized by 300,000 acres (121,406 ha) of water bottoms east of the Mississippi River (and south of the Mississippi River Gulf Outlet).

- CSA III represents the Barataria Bay system and consists of approximately 140 acres (57 ha) of reef.

- CSA IV includes the Terrebonne/Timbalier Basin.

- CSA V, which includes three water bodies in Terrebonne Parish (Sister Lake, Bay Junop, and Lake Merchant), consists of approximately 13,000 acres (5,261 ha) of water bottoms of which approximately 2,500 acres (1,012 ha) is reef.

- CSA VI, found in the Vermilion/Cote Blanche/Atchafalaya Bay System, consists of about 542,000 acres (219,340 ha) of water bottoms, and the reef is an unknown portion of this total.

- CSA VII includes Calcasieu Lake and Sabine Lake; Calcasieu Lake consists of 58,290 acres (23,590 ha) of water bottoms of which 1,691 acres (684 ha) is reef.[13]

Estimated seed and sack availability from these CSAs associated with the four most recent stock assessments are presented in Table 10.2. As indicated, estimated seed variability from one year to the next in any CSA can be large. As just one example, estimated seed oyster availability in CSA I fell by more than two-thirds, from 305,000 barrels in 2008 to 83,000 barrels in 2009. While the variability from one year to the next in any CSA is important to recognize, it is also important to recognize that the direction of change across CSAs is not always consistent, even among contiguous CSAs. For example, while there was a large decline in estimated seed availability in CSA I between 2008 and 2009, estimated availability in CSA II more than doubled. The fact that the direction of change across contiguous CSAs is not consistent is not unexpected; different CSAs represent different bay systems and all are subject to their own environmental perturbations. Given the high annual natural mortality rate asso-ciated with subadult and adult oysters (i.e., 50–95 %), the high year-to-year seed oyster variability within a given CSA should come as no surprise.

The estimated harvest of seed oysters, by CSA, for the most recent 4 years is provided in Table 10.3. It is useful to consider the information in Table 10.2 in conjunction with that in Table 10.3. Estimates of seed oyster availability are generally made in July of each year and can be used to help establish what might be available for harvest in that season. Thus, the 2008 seed availability can be used to establish the seed harvest potential for the 2008–2009 public ground season. For example, estimated seed availability in CSA I for 2008 was 305,000 barrels. About 87,000 barrels of seed were subsequently transplanted to private leases during the 2008–2009 public ground harvesting season. As indicated, no transplanting from CSA VII occurs even though there is generally a significant amount of seed in that region. This is because of the long distance to leases that are primarily located in the eastern portion of the state. In examining these numbers, one should also keep in mind the incentives to transplant seed offered by the POLR program. These incentives likely increased transplanting activities during the period considered in these tables.

---

[13] Summation of total reef area by CSA exceeds the 38,000 acres (15,378 ha) previously cited. The reason(s) for this discrepancy is unknown.

Table 10.2. Estimated Seed and Market (Sack) Oyster Availability in Barrels by Coastal Study Area, 2007–2010[a]

CSA	Seed	Sack	Total	CSA	Seed	Sack	Total
	2007				2008		
I	293,219	139,136	432,355	I	305,256	750,526	1,055,782
II	451,034	309,562	760,596	II	110,751	124,393	235,144
III	10,584	2,424	13,008	III	2,036	2,949	4,985
IV	2,131	847	2,978	IV	2,277	2,267	4,544
V	96,891	127,127	224,018	V	46,863	52,237	98,100
VI	N/A[b]	N/A	N/A	VI	N/A	N/A	N/A
VII				VII	331,102	447,131	778,233
	2009				2010		
I	82,867	178,097	265,964	I	120,188	94,833	215,021
II	241,762	78,450	320,212	II	105,836	39,739	145,575
III	11,402	141	11,543	III	5,020	1,207	6,227
IV	2,236	270	2,506	IV	2,021	499	2,520
V	89,602	43,387	132,989	V	154,340	36,971	191,311
VI	N/A	N/A	N/A	VI	N/A	N/A	N/A
VII	126,047	310,503	436,550	VII	307,265	356,458	663,723

[a]There are two sacks to one barrel. Convert barrels to pounds of meat by multiplying sacks (barrels times two) by 6.47. Seed oyster availability cannot be converted to meat weight.
[b]No estimates (N/A) are given for seed and sack in CSA VI because the amount of reef area has not been determined
Source: Derived from LDWF (2007), LDWF (2008), LDWF (2009), and LDWF (2010).

Estimated statewide market oyster availability on the public grounds for the 1999–2009 period is provided in Figure 10.53. As with seed oyster availability (Figure 10.51), the estimated market oyster availability has, in general, been declining since the early 2000s. During the 11-year period ending in 2009, the average estimated market oyster availability on public grounds equaled 1.7 million barrels (3.4 million sacks) and ranged from 4.3 million barrels in 2001 to 375,000 barrels in 2006 (likely influenced in part by Hurricanes Katrina and Rita). The correlation between seed and market oyster harvests during the period of analysis equaled 0.69.

The estimated market oyster availability on public grounds by CSA is given in Table 10.2. As with seed availability, market oyster availability (i.e., sack) in any CSA can vary significantly from one year to the next. For example, estimated market availability in CSA I fell from approximately 750,000 barrels in 2008 to less than 200,000 barrels in 2009 and fell again to 95,000 barrels in 2010. Inter-year variation likely reflects a combination of changing environmental conditions that lead to changes in natural mortality and harvesting pressure.

Annual statewide harvest of market oysters from the public grounds is given in Figure 10.46 and as a percent of total production in Figure 10.54.[14] Overall, during 1999–2009, market oyster production from the public grounds averaged 5.3 million pounds, with a low of 3.1 million pounds being harvested in 2006 and a high of 7.8 million pounds being harvested in 2002.

---

[14] Landings data associated with production from the public grounds are based on trip ticket data provided by the fishermen/dealers. Information provided includes area fished. Specified areas associated with the public grounds are provided in the two figures at the end of this section.

Table 10.3. Estimated Harvests of Seed Oysters and Market Oysters in Barrels by Coastal Study Area, 2007–2010[a]

CSA	Seed	Market	Total	CSA	Seed	Market	Total
	2006–2007				2007–2008		
I	61,635	25,536	87,171	I	157,085	136,568	293,653
II	110,567	91,678	202,245	II	173,285	139,290	312,575
III	12,190	3,046	15,236	III	13,345	167	13,512
IV	1,940	0	1,940	IV	2,627	3,635	6,262
V	10	4,956	4,966	V	39,115	47,562	86,677
VI	60,390	8,884	69,274	VI	45,121	2,197	47,318
VII	0	14,171	14,171	VII	0	39,823	39,823
Total	246,732[b]	148,271	395,003	Total	430,578	369,241	799,819
	2008–2009				2009–2010		
I	87,180	85,094	172,274	I	57,055	79,014	136,069
II	77,003	132,791	209,794	II	82,688	83,807	166,495
III	1,985	1,860	3,845	III	7,885	252	8,137
IV	205	9	214	IV	0	0	0
V	600	3,502	4,102	V	4,610	6,838	11,448
VI	0	0	0	VI	0	0	0
VII	0	34,742	34,742	VII	0	68,537	68,537
Total	166,973	257,998	424,746	Total	152,238	238,448	390,686

[a]There are two sacks to one barrel. Convert market oysters in barrels to pounds of meat by multiplying sacks (barrels times two) by 6.47. Seed oyster availability cannot be converted to meat weight.
[b]Does not include relocation project.
Source: Derived from LDWF (2007), LDWF (2008), LDWF (2009), and LDWF (2010).

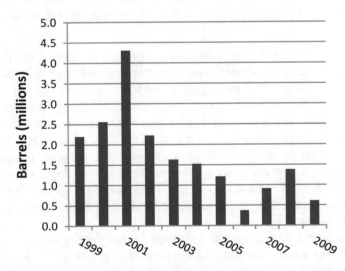

Figure 10.53. Estimated state wide market oyster availability in barrels on the state's public seed grounds, 1999–2009 (email from Patrick Banks, LDWF, to Walter Keithly, December 29, 2009).

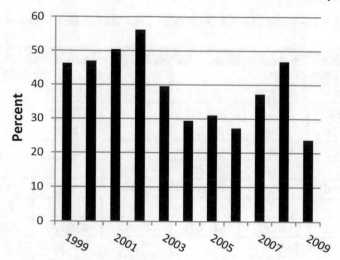

Figure 10.54. Percentage of annual harvest derived from public grounds, 1999–2009 (unpublished data provided to Walter Keithly by the LDWF for years covering 1999–2008 with 2009 data derived from LDWF (2010) with percentage calculations by authors).

On a percentage basis of the total harvest (i.e., harvest from leases and harvest of market oysters from public grounds), production from public grounds ranged from less than 30 % to more than 50 %. In general, years of high public ground production tends to correlate with low private lease production.

Estimated harvest of market oysters from the public grounds by CSA for the four most recent years is given in Table 10.3. For CSA I, although estimated 2008 market oyster availability equaled 750,000 barrels, the estimated 2008–2009 harvests equaled only 85,000 barrels. Similarly, though the 2009 market oyster availability for the region equaled 178,000 barrels, estimated 2009–2010 harvests equaled 79,000 barrels. By comparison, the 2008 estimated market oyster availability for CSA II in 2008 equaled 124,000 barrels and the 2008–2009 estimated harvest from the region equaled 133,000 barrels. Similarly, the 2009 estimated market oyster availability equaled 788,000 barrels and subsequent harvest equaled 84,000 barrels.

A Closer Look at the Texas Oyster Harvesting System: As noted by the information in Table 10.1, Texas represents the second largest oyster producing state in the Gulf of Mexico with annual production since 1990 averaging close to 4.5 million pounds. Like Louisiana, a sizeable share of the Texas oyster production is derived from leasing activities. However, leases are much more limited in Texas totaling 43 and comprising 2,321 acres (939 ha) (OTTF 2012). Furthermore, all are in Galveston Bay. As stated in OTTF (2012), "[t]he original goal of the Texas oyster lease program was to create new self-sustaining oyster producing areas under private ownership but is currently being used exclusively as depuration sites for oysters transplanted from restricted waters (pp. 8–37)." Furthermore, given that the management goals associated with the current program are currently being met, there is a moratorium on the issuance of new leases.

Given that the current leases are used exclusively for depuration sites for oysters transplanted from restricted waters, relaying of oysters from public grounds restricted waters to private beds represents the primary source of oysters that are subsequently harvested from leases. There is currently a spring (May) transplanting season and a fall (September) transplanting season with each lasting, on average, 9 days in recent years.

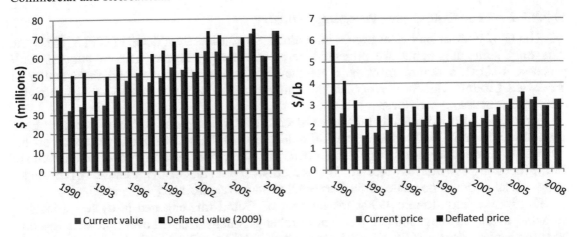

Figure 10.55. Current and deflated value of Gulf oyster production (*left panel*) and current and deflated price (*right panel*), 1990–2009 (NMFS FSD, data accessed 2012, with deflated prices calculated by authors—see Appendix A) (*Note*: 1 lb = 0.454 kg).

Since the early 1990s, production from the private leases as a percentage of total state production has fallen in the 15–30 % range annually with absolute production from the leases ranging from less than 600,000 lb in many years to close to two million pounds. The 2008 lease-based harvest equaled 535,000 lb which represented the lowest take from leases since 1993 (OTTF 2012).

Dockside Value and Price

In general, the current dockside value of Gulf oyster production, as indicated by the information in Figure 10.55 (left panel), trended upward during the 1990–2009 period. During 1990–1994, for instance, total annual Gulf value averaged about $35 million. By 2005–2009, the current value had increased to $66 million annually (or by about 90 %). Much of this increase is, of course, the result of inflation. After adjusting for inflation (expressed in 2009 dollars), the increase was much more moderate; from $54 million annually during 1990–1994 to $69 million annually during 2005–2009 (Figure 10.55, left panel). This 28 % increase in deflated value matches well with the 33 % increase in production between these two periods—15.9 million pounds to 21.2 million pounds—suggesting no long-run increase in the deflated dockside price.

The long-run constancy in deflated dockside price (meat weight) can be examined with the aid of the information in Figure 10.55 (right panel). While the 1990 deflated price was, as indicated, significantly higher than any other yearly price during the 20-year period of analysis, the deflated price fell sharply in the following three succeeding years even though the Gulf production in pounds increased after 1991 (Keithly and Diop 2001). Subsequently, Dedah et al. (2011) attribute the significant decline in price to (1) media which drew attention to the health risks associated with the consumption of raw oysters and (2) mandated labeling requirements for establishments selling raw oysters. These mandated labeling requirements were initiated in an attempt to better inform the public of the health risks associated with the consumption of raw oysters and other shellfish.

10.3.8.2.2   The Gulf Oyster Processing Industry

The Gulf of Mexico oyster processing industry is considered in detail in OTTF (2012) and thus only some highlights are presented here. The study suggests that, in general, Gulf processors shuck a minimum of 60 % of the Gulf harvested product, and the quantity of processed product closely mirrors landings in the region with increased landings implying increased processing activities.

The report also indicates that the value of Gulf oyster processing activities increased from about $30 million in 1980 to more than $60 million during 2000–2008. However, much of that increase is inflationary based and after removing inflationary effects, no growth in the value of Gulf processing activities is observed. This is consistent with the relatively long-term constancy in the deflated dockside value of the harvested product (Figure 10.55, left panel).

During the early-to-mid 1990s, the number of Gulf firms engaged in oyster processing activities averaged about 100 per year. The number gradually declined over time with less than 70 being reported since 2006. Given the long-term stability in Gulf oyster landings in conjunction with the quantity of processed product closely mirroring the Gulf landings, a declining number of firms suggests increased output per processing establishment. Overall, production per firm since 1994 has consistently exceeded 100,000 lb of oyster meats and since 2004 has exceeded 175,000 lb of meats. On average, revenues from oyster processing activities exceeded $1 million per firm for the first time in 2006, with 2007 and 2008 figures also around the $1 million figure.

10.3.8.2.3   Impact of Oyster Landings on Dockside Price

Dedah et al. (2011) is the most current and detailed analysis examining the influence of production in different regions of the United States and imports on the Gulf of Mexico dockside. The study employed a complete demand system using quarterly data covering the first quarter 1985 through the fourth quarter of 2008. Included in the analysis was Gulf oyster production. The authors found that a 10 % increase (decrease) in Gulf harvest (at its mean value) resulted in a 6.4 % decrease (increase) in the Gulf dockside oyster price. Dedah et al. (2011) also indicate that Gulf dockside price is significantly influenced by production in other regions and imports. Specifically, a 10 % increase (decrease) in Pacific production was found to result in an inverse reduction or increase in the Gulf price by 1.6 %. Finally, the authors found that a 10 % increase in all supply sources (Gulf, Chesapeake, Pacific, and imports), evaluated at the 1985–2008 mean values for all variables, results in a 9.8 % decrease in the Gulf dockside price.

10.3.8.3   The Commercial Reef Fish Sector

The Gulf of Mexico is host to a large number of reef fish species, many of which represent income generators to the commercial fishing sector. Given the susceptibility of many of the species to overfishing in conjunction with their popularity by the commercial and recreational sectors, the GMFMC is considerably involved in the management of reef fish species. Management of reef fish species with respect to the commercial sector has historically included sector quotas, size and trip limits, closed seasons, limited entry, and, more recently, the introduction of catch shares (previously called individual fishing quotas). The introduction of catch shares, as discussed in more detail below, constitutes a major shift in management regime and one that is likely to become more prevalent over time.

Following the *Gulf of Mexico Reef Fish Fishery Management Plan* (GMFMC 1981) and related amendments, Gulf of Mexico reef fish can be broadly classified into six groups: snappers, groupers, tilefish, jacks, amberjacks, and triggerfish. Annual commercial landings of these species

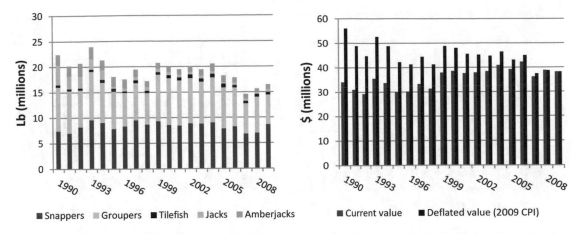

Figure 10.56. Gulf of Mexico commercial reef fish landings in pounds (*left panel*) and value (*right panel*), 1990–2009 (NMFS FSD, data accessed 2012, with deflated values calculated by authors—see Appendix A) (*Note*: 1 lb = 0.454 kg).

groups, excluding triggerfish, are presented in Figure 10.41 (left panel) for the 1990–2009 period (*Note*: commercial landings of triggerfish are minor, totaling less than 1 million pounds since 1990). Two of these species groups—snappers and groupers—dominate commercial reef fish landings with respective shares both in excess of 40 %. Overall, the total commercial landings of reef fish species declined from about 20 million pounds annually in the early to mid-1990s to about 16 million pounds annually in the later years of analysis (Figure 10.56, left panel).

The current dockside value of the commercial reef fish landings, as indicated by the information in Figure 10.56 (right panel), showed little growth during the period of analysis, and on a deflated basis (with 2009 being designated as the base year for the CPI), the value has clearly been declining. There are a number of potential reasons for this decline. First, reef fish landings have fallen marginally during the period of analysis. Second, management measures, particularly with respect to the red snapper fishery, may have contributed to lower prices than would otherwise be the case (in short, derby fishing conditions that led to market gluts). Finally, and of significant importance, imports of snappers and groupers are large; generally nearing or exceeding domestic production of these species. During 2005–2009, for example, imports of snapper averaged eight million pounds (product weight) annually, while imports of grouper averaged ten million pounds (product weight). These imports have been increasing over time and, being close substitutes for the domestic product, likely exerted downward pressure on the Gulf snapper and grouper dockside prices.

Examining snapper separately, landings have averaged about 8.4 million pounds annually since 1990 with no apparent long-run trend (Figure 10.57, left panel). While the current value of snapper landings increased from approximately $15 million per year in the early 1990s to about $20 million per year, on average, in the later years, the deflated value of Gulf of Mexico snapper landings illustrated no increase (Figure 10.57, right panel). The long-run stability in commercial snapper landings and the concomitant stability in deflated value of the landings imply, of course, long-run stability in the deflated per pound price. While the deflated dockside price exceeded $3.00 per pound in 1990 and 1991, the deflated dockside price since 1992 has fallen in the relatively narrow range of $2.40–$2.80 per pound (based on the 2009 CPI). Waters (2001) ascribes much of the decline in snapper price beginning in 1992 to management measures imposed to protect and rebuild the red snapper stock. Also, as noted, imports of snappers are large and have been increasing over time.

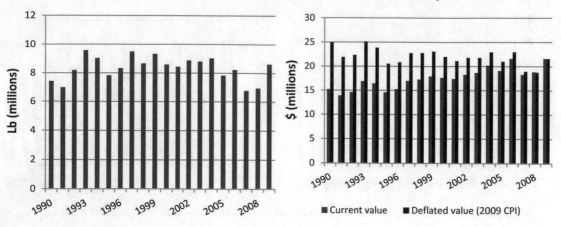

Figure 10.57. Gulf of Mexico commercial snapper landings in pounds (*left panel*) and value (*right panel*), 1990–2009 (NMFS FSD, data accessed 2012, with deflated values calculated by authors—see Appendix A) (*Note*: 1 lb = 0.454 kg).

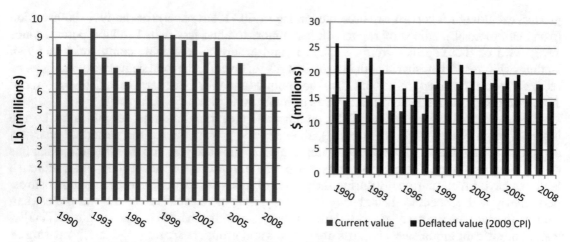

Figure 10.58. Gulf of Mexico commercial grouper landings in pounds (*left panel*) and value (*right panel*), 1990–2009 (NMFS FSD, data accessed 2012, with deflated values calculated by authors—see Appendix A) (*Note*: 1 lb = 0.454 kg).

Gulf of Mexico annual commercial grouper landings (Figure 10.58, left panel) indicate production from a low of less than six million pounds to a high of more than nine million pounds. Since 2005, landings have averaged about seven million pounds annually or about one million pounds below the eight million pound long-run average. This decline is, at least in part, the result of more stringent quotas being placed on the commercial sector in response to recent stock assessments suggesting that some species in the grouper complex are experiencing overfishing conditions. Furthermore, as was the situation with red snapper, the value of grouper landings has trended downwards (Figure 10.58, right panel) when inflationary effects are removed (based on the 2009 CPI). This is largely the result of a decline in landings in recent years given that the deflated per pound price has historically fallen in the relatively narrow range of about $2.40–$2.70 per pound with few exceptions and no apparent trend.

Two reef fish species—red snapper and vermilion snapper (*Rhomboplites auroruben*)— represent the primary species targeted by commercial fishermen in the northern Gulf of

Mexico. In the eastern Gulf, grouper dominates commercial harvest. Management of all of these species is under the purview of the GMFMC. Management measures implemented over the years to protect these stocks are numerous including minimum size restrictions, vessel quotas, closed seasons, and most recently, catch share programs for the red snapper fishery (implemented in 2007) and the grouper and tilefish fisheries (implemented in 2010). These catch share programs give harvesting rights to individuals with each individual's harvesting rights based on the total quota for the fishery and each individual's share of the total (shares will sum to 100 % of the quota). Given overall industry quotas for all of these species, landings are constrained and will expand only as the respective stocks expand.

As mentioned, the red snapper fishery is one of the two most important reef fish species targeted by the commercial sector in the northern Gulf of Mexico, and it was the first of the Gulf reef fish species to be managed under a catch share program. While originally scheduled for implementation in the mid-1990s, congressional actions delayed implementation until 2007 (see Keithly (2001) for information on the original program and congressional actions that delayed implementation). When implemented on January 1, 2007, the commercial quota was set at 2.55 million pounds (whole weight) with a quota increase of 765,000 lb later in the year (NMFS 2010). The 2007 ending quota of 3.315 million pounds was reduced to 2.550 million pounds for both 2008 and 2009. The commercial quota is of particular importance because it provides an upper boundary for commercial harvest assuming no illegal catch. In fact, the reported commercial harvest since 2007 and through 2009 has been from 96 to 97 % of the quota allocated to the sector (NMFS 2011).

Shares in the red snapper fishery can be either sold or leased.[15] As noted by Gauvin et al. (1994), if fishermen embrace the future of the catch share program and expect the program to last in perpetuity, then shares can be considered an asset worth the equivalent of the discounted stream on net income derived from that asset. According to the NMFS (2010), the mean transfer price per one-pound equivalent of shares approximated $14 in 2009, while the median price was approximately $18. The allocation price (lease price to harvest one pound in 2009) was $3.02. However, as stated in the report, the large number of transactions without reliable price information suggests that these figures should be viewed with some caution.

10.3.8.4 Menhaden

The Gulf of Mexico menhaden fishery was briefly considered in Section 10.3.1.1. The species is relatively short lived and in 2009 age-2 fish comprised an estimated 73 % of the fleet harvest and age-1 fish comprised 13 % of the harvest (NMFS, Sustainable Fisheries Branch, Beaufort, NC, 2010). Overall, the percentage of age-2 fish comprising harvest has increased over time with reasons for this increase not clearly identified. Hypotheses include (1) contraction of the fishery over time from the extremes of the species range, Florida through Texas, where smaller fish were more abundant in Mississippi and Louisiana waters and (2) a redistribution of age-1 fish to more inside waters due to deterioration of wetlands (GSMFC meeting, Orange Beach, Alabama, 2010). Given that the majority of harvest comprises only a couple of year classes, environmental factors that influence recruitment significantly influence subsequent harvest. Citing Christmas et al. (1982) and Guillory et al. (1983), Deegan (1990) suggests that low winter temperatures, high salinities, and low turbidity during the period when the menhaden are in the estuaries are correlated with poor year-classes because of their

---

[15] Details of the red snapper catch share program, including issues of transferability and leasing, can be found in Amendment 26 to the Gulf of Mexico Reef Fish Fishery Management Plan (GMFMC 2006) which is available at: http://sero.nmfs.noaa.gov/sf/pdfs/Amend_26_031606_FINAL.pdf.

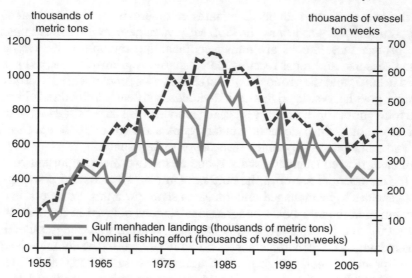

Figure 10.59. Gulf of Mexico menhaden landings and nominal effort, 1955–2009 (NMFS (Sustainable Fisheries Branch) 2010) (*Note*: 1 metric ton is equal to 2,204.6 pounds).

influence on growth and mortality rates of young-of-the-year fish. Louisiana, in fact, forecasts menhaden catch each year using a suite of factors including water temperature (off Grand Isle, Louisiana), salinity, Mississippi River discharge, juvenile menhaden catch in fishery-independent trawl samples, and expected effort.[16] Given that the fishery is largely dependent upon only 2 year classes, changes in environmental factors that influence recruitment of juveniles can significantly influence menhaden availability from one year to the next.

While environmental factors may largely drive menhaden availability, catch is determined by both environmental factors and effort employed to harvest available menhaden. The NMFS (Beaufort Laboratory) has been forecasting annual Gulf menhaden harvests based on estimates of expected fishing effort for the upcoming year (via discussion with industry). Gulf forecasts over the 1973–2009 period have differed from actual catch by an average of 15 % per year (NMFS, Beaufort Laboratory, 2010). The relationship between effort and harvest is clearly illustrated for the 1955–2009 period in Figure 10.59. While the relationship is clear, it is also apparent that annual variations in harvest are large relative to annual variations in effort. This fact is likely explained by the influence of environmental factors on the populations of the few year classes dominating harvest.

10.3.9 Additional Detail on Processing and Wholesaling

In general, there are two sources of data by which one can examine the seafood processing industry. The first source is generally referred to as *the voluntary end-of-the year processor survey*, and the data used in this survey is collected and maintained by NMFS. Data collected include detailed information (by plant) on species processed and output by product form, the value of the output, and employment. The database is very rich, and it was this database that was employed in the analysis of Gulf shrimp and oyster processing activities in this chapter. The other data source represents information collected by the U.S. Bureau of Census, and this

---

[16] http://menhaden.gsmfc.org/pdf/March%202008%20MAC%20minutes.pdf.

data source was the basis for discussion in Section 10.3.5. Both data sources have their advantages. The primary advantage of the NMFS data source is that it provides detailed information on processing activities by species and associated output value. The primary disadvantage of this source is that although detailed information by species is collected, it is not routinely published, and for reasons of confidentiality, the information is not easily accessible by the general public. The primary advantage of the data collected by the U.S. Bureau of Census is that differentiation is made between nonemployer processing firms (defined as firms that have no paid employees and are subject to federal income taxes), which tend to be small in scale, and employer establishments (which have paid employees and may, in some instances, consist of more than one firm in one or more states). In addition, payrolls are given for employer establishments. One primary limitation to this data source is the absence of detailed information on processed species and value of output (though the value of output is apparently provided for nonemployer establishments). A second limitation is that data for Florida are not differentiated between the Gulf and South Atlantic. The NMFS uses this database in describing processing activities by state in its annual *Fisheries Economics of the United States* reports, and it is this data source used in this section. All seafood wholesaling information also comes from the U.S. Bureau of Census.

Annual seafood sales and processing information for each of the five Gulf of Mexico states (which includes the east coast of Florida) for the 2000–2008 period is presented in Tables 10.4, 10.5, 10.6, 10.7, and 10.8, and 2008 information for the Gulf, in aggregate, is provided in Table 10.9. While a detailed discussion of the information in each table is beyond the scope of this chapter, a discussion of some of the primary findings is informative.

Examination of the individual state tables leads to the conclusion that there is little or no growth in the number of employer-based seafood processing establishments, and some states (particularly Florida and Mississippi) are experiencing a significant contraction in the number of establishments. Those states experiencing the largest contraction in number of establishments (on a percentage basis) also experienced a large decline in number of employees suggesting that contraction goes beyond simple consolidation, with the caveat that payroll does not appear to have fallen as much as employment.

A second noteworthy finding is that Mississippi consistently led Gulf States in terms of the number of employees among employer-based processing establishments during the period of analysis with a generally higher commensurate payroll. Given the limited commercial landings in Mississippi, it is clear that other sources of raw product, including product from other states and imports, are being used by Mississippi processors.

As shown in the tables, employer-based establishments tend to be substantially larger than the nonemployer-based firms. Specifically, payrolls among employer-based establishments in each of the Gulf States tended to exceed receipts by the nonemployer firms even though the number of employer-based establishments by state, with the exception of Mississippi, tended to be significantly less than the number of nonemployer based firms and payroll comprises only a fraction of receipts (assuming profitability).

The information in the respective tables also indicates that Florida (including east coast) and Louisiana experienced sizable increases in the number of nonemployer processing firms during the period of study with the number in Florida approximately doubling. Furthermore, receipts from the sale of prepared and packaged seafood among nonemployer firms can be highly variable from one year to the next and does not appear to strongly track changes in the number of firms.

Finally, comparing payroll from employer-based seafood wholesaling by state to seafood processing indicates that seafood wholesaling represents a major activity with payroll often exceeding that of processing (i.e., Florida and Texas). The primary exception to this finding is

Table 10.4. Florida (Including East Coast) Annual Seafood Processing and Wholesaling Activities, 2000–2008

Year		2000	2001	2002	2003	2004	2005	2006	2007	2008
Seafood sales and processing—nonemployer firms										
Seafood product prep. and packaging	Firms	102	104	116	142	177	164	174	173	202
	Receipts ($1000's)	8,330	6,350	5,064	8,047	8,652	8,756	10,184	10,497	11,065
Seafood sales and processing—employer establishments										
Seafood product prep. and packaging	Firms	41	43	33	27	24	25	22	20	23
	Employees	2,188	2,033	2,359	2,084	2,193	1,616	1,704	1,748	1,637
	Payroll ($1000's)	58,821	58,977	65,914	61,452	65,881	47,529	62,801	58,233	53,455
Seafood sales, wholesale	Firms	329	323	314	293	261	258	259	267	229
	Employees	2,915	2,670	2,395	1,835	1,948	1,883	2,091	2,308	1,913
	Payroll ($1000's)	76,363	76,717	78,160	55,874	63,276	65,339	73,897	85,019	75,203

Source: U.S. Department of Commerce (2011)

Table 10.5. Alabama Annual Seafood Processing and Wholesaling Activities, 2000–2008

Year	2000	2001	2002	2003	2004	2005	2006	2007	2008
Seafood sales and processing—nonemployer firms									
Seafood product prep. and packaging — Firms	46	39	44	36	43	40	34	47	33
Seafood product prep. and packaging — Receipts ($1000's)	3,677	2,711	3,603	1,168	3,413	3,414	1,558	1,547	1,894
Seafood sales and processing—employer establishments									
Seafood product prep. and packaging — Firms	17	21	22	24	23	26	24	23	23
Seafood product prep. and packaging — Employees	1,725	1,880	1,951	2,057	2,037	1,925	1,629	1,510	1,450
Seafood product prep. and packaging — Payroll ($1000's)	33,811	32,692	36,198	36,766	36,130	38,229	34,703	32,774	29,277
Seafood sale; wholesale — Firms	47	45	36	33	31	26	26	31	29
Seafood sale; wholesale — Employees	887	692	547	611	588	607	395	395	494
Seafood sale; wholesale — Payroll ($1000's)	10,252	9,597	7,062	6,148	6,752	6,345	6,195	6,202	8,751

Source: U.S. Department of Commerce (2011)

W.R. Keithly, Jr. and K.J. Roberts

Table 10.6. Mississippi Annual Seafood Processing and Wholesaling Activities, 2000–2008

Year		2000	2001	2002	2003	2004	2005	2006	2007	2008
Seafood sales and processing—nonemployer firms										
Seafood product prep. and packaging	Firms	10	13	15	23	18	12	22	0	17
	Receipts ($1000's)	1,300	1,186	915	1,561	1,056	1,045	1,537	ND	1,055
Seafood sales and processing—employer establishments										
Seafood product prep. and packaging	Firms	37	33	34	37	33	28	24	22	20
	Employees	4,339	4,053	3,675	4,438	3,728	3,637	3,353	3,022	3,062
	Payroll ($1000's)	73,350	65,237	70,792	80,229	66,047	63,957	60,510	60,633	61,723
Seafood sales, wholesale	Firms	30	28	29	26	29	30	23	25	18
	Employees	232	2,226	226	176	166	145	58	106	61
	Payroll ($1000's)	3,716	4,056	3,791	3,067	3,631	1,822	2,063	3,285	3,088

Source: U.S. Department of Commerce (2011)

Table 10.7. Louisiana Annual Seafood Processing and Wholesaling Activities, 2000–2008

Year		2000	2001	2002	2003	2004	2005	2006	2007	2008
Seafood sales and processing—nonemployer firms										
Seafood product prep. and packaging	Firms	39	58	66	73	75	76	99	85	77
	Receipts ($1000's)	3,466	2,918	3,006	4,678	10,097	8,513	8,179	6,253	7,365
Seafood sales, retail	Firms	172	170	185	208	204	156	181	196	182
	Receipts ($1000's)	11,806	12,586	15,201	22,637	18,148	14,585	20,046	20,932	25,900
Seafood sales and processing—employer establishments										
Seafood product prep. and packaging	Firms	56	50	50	54	54	50	40	41	36
	Employees	1,282	1,141	1,185	1,693	1,519	1,556	1,506	1,253	991
	Payroll ($1000's)	45,285	48,331	52,861	56,562	47,016	43,801	45,439	41,391	32,382
Seafood sales, wholesale	Firms	162	164	152	134	133	128	112	119	98
	Employees	1,187	1,245	1,270	1,001	975	1,037	807	954	739
	Payroll ($1000's)	21,717	23,053	22,363	19,539	19,639	17,649	21,243	21,604	18,858

Source: U.S. Department of Commerce (2011)

Table 10.8. Texas Annual Seafood Processing and Wholesaling Activities, 2000–2008

Year		2000	2001	2002	2003	2004	2005	2006	2007	2008
Seafood sales and processing—nonemployer firms										
Seafood product prep. and packaging	Firms	85	108	104	99	100	108	109	94	85
	Receipts ($1000's)	5,596	5,575	3,901	5,234	1,981	2,228	2,974	5,386	3,466
Seafood sales and processing—employer establishments										
Seafood product prep. and packaging	Firms	31	29	27	23	24	23	21	26	27
	Employees	1,305	1,506	1,453	1,274	1,177	1,288	1,155	1,207	1,169
	Payroll ($1000's)	24,374	24,507	25,772	25,426	24,394	23,842	24,302	27,813	27,045
Seafood sales, wholesale	Firms	113	129	115	99	103	97	92	104	69
	Employees	1,187	1,102	999	1,057	1,009	1,001	897	970	734
	Payroll ($1000's)	32,857	33,552	29,430	27,016	27,730	26,408	28,586	51,597	24,498

Source: U.S. Department of Commerce (2011)

Table 10.9. Gulf of Mexico (Including Florida East Coast) Seafood Processing and Wholesaling Activities, 2008

State		Florida	Alabama	Mississippi	Louisiana	Texas
Seafood sales and processing—nonemployer firms						
Seafood product prep. and packaging	Firms	202	33	17	77	85
	Receipts ($1000's)	11,065	1,894	1,055	7,365	3,466
Seafood sales and processing—employer establishments						
Seafood product prep. and packaging	Firms	23	23	20	36	27
	Employees	1,637	1,450	3,062	991	1,169
	Payroll ($1000's)	53,455	29,277	61,723	32,382	27,045
Seafood sales, wholesale	Firms	229	29	18	98	69
	Employees	1,913	494	61	739	734
	Payroll ($1000's)	75,203	8,751	3,088	15,858	24,498

Source: U.S. Department of Commerce (2011)

Mississippi (and to a lesser extent Alabama) where payroll from wholesaling is a small fraction of that associated with processing.

10.4　THE GULF OF MEXICO RECREATIONAL SECTOR

A review of the Gulf of Mexico recreational fishing sector is provided in this section. Issues to be considered include participation, number of trips, catches of various species, and expenditures and multiplier effects. Much of the information reported here is derived from MRFSS/MRIP implemented by the NMFS in 1979.[17] Collection of reliable statistics on recreational activities is notoriously elusive and, as stated by a panel convened to review the validity of the MRFSS/MRIP protocol, "[r]ecreational angling provides formidable challenges in estimating catch, effort, and economic expenditures by anglers, either regionally or nationally, due to the diversity of sites and modes of fishing available to the anglers" (National Research Council 2006).

With this in mind, it is instructive to first examine the methodology employed to collect data that are used in MRFSS/MRIP estimates of effort, participation, and catch rates. There are two independent but complementary surveys used to collect the raw data—a telephone survey and a dockside intercept survey. The telephone survey is used to determine the number of participants and trips, and the dockside intercept is used primarily for determining species caught and associated quantities.

Determining species caught and quantities are problematic for at least three reasons. First, and foremost, for reasons discussed later in this section, much of the catch is released and, as such, is not observed by the port sampler. Second, if catch by an individual is large, the port sampler may not have time to measure all fish or the angler may refuse to show all fish to the

---

[17] In early 2012, changes were made in the MRFSS estimation procedure regarding the extrapolation from the sample to the population with respect to catch. Participation estimates were not changed. Changes were made from 2004 forward. The name of the program was also changed from the Marine Recreational Fisheries Statistics Survey to the Marine Recreational Information Program. In general, changes appear in most cases to be minor. Details regarding changes, including a comparison of the MRFSS to MRIP data, is available at: http://www.countmyfish.noaa.gov/index.html.

port sampler. Finally, given limited budgets and time, port samplers are only able to interview a small proportion of anglers. As such, estimates of total catch and weight are made based on a relatively small sample and there will be imprecision in these estimates. Furthermore, the magnitude of the impression, as a percentage of the estimate, is likely to be compounded when considering infrequently caught species or when estimates are generated for a geographic region more narrowly defined than the Gulf of Mexico (e.g., state or county/parish). This later observation reflects the fact that the MRFSS/MRIP was originally designed in a manner that would yield reliable estimates of catch and effort for the Gulf of Mexico in total, but estimates at the state level would be less reliable. However, a number of states now contribute to the MRFSS/MRIP federal budget to ensure greater reliability of estimates at the state level. Other limitations associated with the MRFSS/MRIP are discussed in the National Research Council report (2006).

Given that much of the ensuing discussion regarding recreational catch (in numbers of fish) and harvest (pounds landed or released dead) is based on the MRFSS/MRIP, the reader should be cognizant that the figures given are merely estimates, and there are likely to be some errors associated with these estimates. Also, Texas is not included in the MRFSS/MRIP, and as such, all discussion of catch, harvest, trips, and participation is exclusive of Texas (though Texas is included in the Expenditures and Multipliers section of the chapter).[18] Finally more emphasis tends to be given to Louisiana and, to a lesser extent, Florida. This reflects the fact that Florida has, by far, the largest recreational fishery in the Gulf (of the four states considered) and there is currently a special interest in the Louisiana recreational fishery. In some instances, comparisons are made between the MRFSS/MRIP data and other available information, particularly regarding Louisiana.

For purposes of notation, the MRFSS/MRIP system designates fish brought into the dock and observed by the port sampler (trained interviewers) as *A*. Fish that are used for bait, released dead, or filleted (i.e., they are killed but identification is by individual anglers) are designated as *B1* (i.e., fish that are considered harvested but not seen or identified by interviewer). Finally fish claimed to be released alive by the angler (identified by individual anglers) are designated as *B2*. Given these designations, total catch is defined as A + B1 + B2. Total harvest, or removals from the stock, is defined as A + B1. Total catch (i.e., A + B1 + B2) is given only in numbers of fish because B2 is unobserved by the trained port samplers. The harvest (A + B1) is given in terms of both numbers of fish and weight.[19] These designations are used throughout the report. For purposes of this chapter, with few exceptions, analysis of catch includes fish released alive (i.e., B2). Thus, unless otherwise noted, catch will refer to the total number of fish caught, whether released or kept. Harvest (A + B1), however, is only examined in terms of pounds of fish. In general, we have attempted to provide detail on catch in a manner that the reader can also ascertain harvest (A + B1) in terms of numbers of fish.

---

[18] In addition, catch by headboats are not included in the MRFSS/MRIP data. Catch by this sector tends to be relatively limited for most species and would rarely exceed 10 % of the total estimated recreational catch of any species. Furthermore, most of the catch by this sector would be that associated with offshore species (primarily snappers and groupers).

[19] Additional detail on the sampling process can be found at http://www.st.nmfs.noaa.gov/st1/recreational/queries/glossary.html.

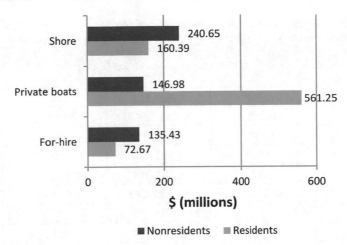

Figure 10.60. Estimated marine recreational angling trip expenditures, 2009 (NMFS 2010).

10.4.1 Expenditures and Multipliers

Each year, millions of individuals, both Gulf and non-Gulf residents, enjoy marine recreational fishing in the Gulf of Mexico. In pursuing this activity, expenditures are incurred. These expenditures can be broadly classified as either trip expenditures or durable expenditures. Furthermore, these expenditures can be incurred in association of shore-based fishing activities, private boat fishing activities, or for-hire fishing activities. The estimated 2009 trip expenditures associated with each of these activities are presented in Figure 10.60. As indicated, greatest trip expenditures were incurred in private/rental boat fishing activities, which totaled $708 million. The vast majority of this total (approximately 80 %), as might be expected, was incurred by Gulf residents. With respect to for-hire (charter) activities, by comparison, the majority of total estimated expenditures ($208 million) were incurred by nonresidents who accounted for 65 % of the total. Similarly, about 60 % of the $401 million spent on shore-based fishing activities in 2009 was incurred by nonresidents.

Expenditures incurred in the pursuit of recreational fishing activities generate jobs, sales, value added, and income in the state where these initial expenditures were incurred. NMFS has estimated economic multipliers associated with these expenditures; details of these multipliers and assumptions are given by Gentner and Steinback (2008). Focusing initially on the 2009 impacts, recreational fishing activities were estimated to generate, at a minimum, 92,000 jobs throughout the Gulf States when including indirect and induced effects associated with the initial expenditures (Table 10.10). Value added, which represents the contribution of recreational fishing to the gross domestic product of the state (region), was estimated to equal $3.3 billion. Sales, which represent the total dollar sales resulting from the initial expenditures, totaled an estimated $9.9 billion. Finally, income, which represents wages, salaries, benefits, and proprietary income resulting from the initial angler expenditures, totaled more than $5 billion.

Florida (west coast) accounted for about 50 % of the total number of generated jobs and value-added activities in 2009. Texas accounted for about 25 % of the generated jobs and value-added activities. Louisiana, though having a small population relative to Texas, accounted for about 20 % of the generated jobs and more than 15 % of the value-added activities.

Table 10.10 Economic Impacts Associated with Gulf of Mexico Angling Activities, 2006–2009

	Jobs	Sales ($1000 s)	Value Added ($1000 s)	Income ($1000 s)
2006				
Florida (West Coast)	75,257	7,823,752	4,235,087	NA
Alabama	6,572	630,181	325,523	NA
Mississippi	3,731	490,501	189,450	NA
Louisiana	26,612	2,382,034	1,199,333	NA
Texas	34,175	4,197,011	2,154,891	NA
Total	**146,347**	**15,523,479**	**8,104,284**	**NA**
2007				
Florida (West Coast)	65,799	6,829,434	3,704,818	NA
Alabama	6,759	654,353	337,493	NA
Mississippi	4,707	616,930	239,021	NA
Louisiana	27,446	2,453,392	1,234,449	NA
Texas	23,382	3,004,862	1,514,791	NA
Total[a]	**128,093**	**13,558,971**	**7,030,572**	**NA**
2008				
Florida (West Coast)	54,589	5,650,068	3,075,710	NA
Alabama	4,719	455,093	235,481	NA
Mississippi	2,930	382,778	148,837	NA
Louisiana	25,590	2,297,078	1,156,796	NA
Texas	25,544	3,288,135	1,656,545	NA
Total[a]	**113,372**	**12,073,152**	**6,273,369**	**NA**
2009				
Florida (West Coast)	42,314	4,369,022	1,532,821	2,385,738
Alabama	4,924	474,746	155,663	245,437
Mississippi	3,188	417,080	105,472	162,099
Louisiana	19,688	1,774,692	578,767	894,123
Texas	22,127	2,846,858	910,011	1,434,733
Total[a]	**92,241**	**9,900,398**	**3,282,734**	**5,122,130**

[a]*Note*: The TOTAL figures should be considered a minimum since they do not account for any trade among individual Gulf States (estimated by authors). *NA* not available. *Source*: U.S. Department of Commerce (various issues): (http://www.st.nmfs.noaa.gov/st5/publication/fisheries_economics_2009.html#).

Comparison of economic impacts across years indicates a distinct reduction in jobs, sales, and value-added activities from 2006 to 2009. For instance, while the total number of estimated jobs generated from Gulf recreational fishing activities equaled 146,000 in 2006, by 2009 the total number of estimated jobs had fallen to 92,000. Similarly, value-added activities fell from an estimated $8.1 billion in 2006 to just $3.3 billion in 2009. Much if not most of this decline can be tied to the downturn in the U.S. economy that began in 2007 and continued into 2009. Furthermore, as indicated, a large proportion of the decline can be tied to Florida (west coast), which, among the Gulf States, was particularly impacted by the most recent recession.

10.4.2 Gulf of Mexico Fishing Activities

Various issues related to the Gulf of Mexico recreational fishing sector are considered in this section of the report including an analysis of trips taken and catch (number of fish) and harvest (pounds kept). Catch and harvest are evaluated at the aggregate level and for some of the primary species groups and individual species. Missing from this section is a discussion of the number of recreational participants. This is because the MRFSS/MRIP does not provide this information at the aggregate Gulf level (in particular, estimates of nonresidents are not given because of potential *double counting* of Gulf residents fishing in more than one state). However, participation in the individual Gulf States is considered when states are evaluated on an individual basis.

10.4.2.1 Number of Angler Trips

Before considering angler trips, a definition is in order. The MRFSS/MRIP defines a single angler on a fishing trip as an angler trip. Hence, if a party of four goes out on a private/rental boat, this is considered to be four angler trips. Based on this definition, the number of saltwater angler trips, according to MRFSS/MRIP estimates, averaged just over 20 million annually during the 1990–2009 period (Figure 10.61). Throughout the 1990s, the estimated number of annual angler trips never exceeded 20 million, and the 10-year average was 17 million. Beginning in 2000, the estimated annual number of trips increased sharply with the average in the most recent decade equaling 23.1 million. During this most recent 10-year period the number of annual trips fell below 20 million only in 2002 when the total equaled 19.7 million. The maximum number of trips during the 20-year period of analysis occurred in 2004 with the estimated total exceeding 26 million.

The reason for the large increase in number of angler trips beginning in 2000 is difficult to identify. While real per capita income did rise during the late 1990s and through much of the 2000s, the increase is likely not sufficient to explain the sharp increase in estimated trips. Similarly, while population throughout the Gulf region was gradually increasing during the period of analysis, which might explain the gradual overall increase in estimated trips, there was no abrupt change in the late 1990s/early 2000s that would correlate with the large increase in trips. Similarly, after being at near-record lows (adjusted for inflation) during the 1990s, gasoline prices rose rapidly beginning around 2000 and, hence, a decline in the cost of a

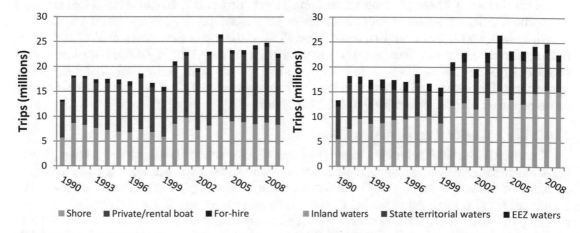

Figure 10.61. Gulf of Mexico angler trips by mode (*left panel*) and by area fished (*right panel*), 1990–2009 (NMFS FSD, data accessed 2012—see Appendix A).

recreational fishing trip does not appear to explain the observed increase in Gulf-wide recreational fishing trips. As discussed in some detail in a subsequent section, a similar trend is apparent in the number of Florida visitors (i.e., a sudden and significant increase in 2000). While it is tempting to suggest that this increase contributed to the large increase in the 2000 estimated fishing trips, one would think that increased visitation would manifest itself via an increase in for-hire trips. However, this mode did not show an appreciable increase in trips.

10.4.2.1.1   Angler Trips by Mode

Trips in the MRIP database are reported on a mode basis. Three types of modes are considered: shore-based marine fishing trips, private/rental boat-based fishing trips, and charter-based trips. The overwhelming majority of the trips, as indicated by the information in Figure 10.61 (left panel), represent shore-based fishing or fishing from private/rental boats with the former representing about 35–40 % in recent years and the later representing 55–60 %. Furthermore, virtually all growth in the total estimated number of marine angler trips during 1990–2009 was the result of increasing private/rental boat-based trips. During 1990–1999, the number of private/rental boat-based angler trips averaged 9.1 million annually, and this mode of fishing constituted 54 % of the estimated total number of trips during the period. During the most recent 10-year period of analysis, the estimated number of angler trips associated with the private/rental boat mode increased to 13.6 million annually, and its share of the total advanced to almost 60 %.

Shore-based angler trips averaged 7.9 million during 1990–2009, with a maximum estimate of 10 million being reported in 2004. During the 1990s, shore-based angler trips constituted a 42 % share of the total number based on an annual average of 7.1 million. With the large increase in private/rental boat angler trips since 2000, the share of total trips represented by the shore mode fell to 38 % based on an average of 8.7 million shore-based trips.

For-hire-based marine fishing trips in the Gulf (excluding Texas) represent a very small share of the total number of angler trips. Specifically, during the 20-year period ending in 2009, less than 4 % of the total angler trips were represented by the for-hire sector. In absolute numbers, for-hire trips peaked at just below one million in 1997, and since 2000, have averaged 786,000 annually. In a recently completed analysis, Savolainen et al. (2012) estimated that the population of for-hire operators was 3,315 in 2009. Of this total, 189 were classified as head boats operations (defined as a firm whose primary vessel carries more than six passengers on the average trip), 789 were classified as charter operations (defined as a firm whose primary vessel carries six or fewer passengers, on average, per trip), and 2,337 guide boats (defined as a firm whose primary vessel carries six or fewer passengers per trip, is approximately 28 ft in length or less, and fishes inshore on more than 75 % of the trips). Savolainen et al. (2012) report that revenues (primarily fees and tips) associated with the Gulf of Mexico for-hire sector equaled $215 million in 2009, with about one-half of the total being derived by the guide operations. The estimated revenues given by Savolainen et al. (2012) compare favorably with those reported by NMFS for 2009 (i.e., $208 million).

10.4.2.1.2   Angler Trips by Area Fished

The MRFSS/MRIP segments trips into three fishing areas: (1) inland waters (e.g., bays), (2) state territorial waters, and (3) EEZ waters, where the EEZ for Florida (west coast) is seaward of 9 nautical mi (16.7 km) and for Alabama, Mississippi, and Louisiana is seaward of 3 nautical mi (5.6 km). As indicated by the information in Figure 10.61 (right panel), the vast majority of angler trips occur in state waters (either inland or state territorial). Since 1990, 56 % of the annual total number of trips (20.1 million annual average) occurred in inland waters while an additional 35 % occurred in the state territorial waters. Angler trips in the EEZ, averaging 1.8

million annually during 1990–2009, accounted for less than 10 % of the total annual number of angler trips.

In general, a significant increase in the number of inland angler trips was apparent in the estimates while little increase was observed in either the number of territorial water trips or trips in the EEZ. For example, during 1990–1995, the number of angler trips in inland waters averaged 8.0 million annually. By 2005–2009, this number had increased to 14.3 million (about 75 %). During the same time frames, by comparison, angler trips in state territorial waters increased by less than 5 % (from an average of 7.4 million to an average of 7.6 million) while angler trips in federal waters increased by less than 15 % (from an annual average of 1.5 million to an annual average of 1.7 million). Furthermore, virtually all of the long-run increase in total angler trips, beginning in 2000, as discussed in the previous section of this chapter, appears to be attributable to an increase in number of trips being reported in inland waters.

In 2009, total Gulf of Mexico angler trips in EEZ waters (excluding Texas) equaled 1.33 million. This total represented the lowest reported figure since the early 1990s and about a 30 % decrease from the 1.87 million trips reported as recently as 2006. Similarly, angler trips in state territorial waters have fallen sharply since 2006 with the 2009 number (6.1 million) representing more than a 40 % decline when compared to the 2006 figure of 8.8 million trips. These reductions likely reflect, in part, recessionary conditions throughout the country and among Gulf States (particularly Florida, subsequently discussed, which represents the majority of recreational Gulf fishing activities, excluding Texas). Throughout the United States, per capita personal income fell from $40,900 in 2008 to $38,800 in 2009. In Florida, the decline was from $40,000 to $37,300. While angler trips in Gulf state territorial waters and federal waters have fallen in recent years, angler trips in inland waters have increased—12.6 million in 2006 and 15.2 million in 2009. It is likely that given the reduction in income and high fuel prices, some anglers substituted the less expensive fishing in inland and state territorial waters for the more expensive fishing in federal waters.

10.4.2.1.3 Angler Trips by Wave

The year-round temperate climate in the Gulf region is conducive to year-round fishing. The MRFSS/MRIP provides information on fishing activities (and catch) in 2-month increments: January/February, March/April, May/June, July/August, September/October, and November/December. While some seasonal fishing patterns in the Gulf are apparent, these patterns, as indicated by the information in Figure 10.46, are moderate in nature. Slightly more than 40 % of the reported trips during the 1990–2009 period occurred during the 4-month summer period ending in August while about 25 % of the angler trips occurred during the 4-month winter period ending in February (Figure 10.62, left panel). Furthermore, little change in seasonality pattern was evident when considering shorter time periods, such as the 2005–2009 period (Figure 10.62, right panel).

10.4.2.1.4 Angler Targeting Behavior

There are a large number of species harvested by recreational anglers in the Gulf of Mexico. When anglers are intercepted at the conclusion of a fishing trip and asked to respond to a series of questions as part of MRFSS/MRIP, one question asked as part of the survey is species targeted on that trip. The angler is allowed to list up to two individual species. Three caveats are in order when evaluating targeting activities. First, many trips contain multiple participants and only the *leader* may be interviewed. It is assumed that the *followers* are targeting the same species. Second, the interviewees may not specify (or may not have) targeted species. Hence, targeting behavior associated with any individual species represents a minimum value of all targeting behavior. Finally, interviews occur at the completion of a trip. As such,

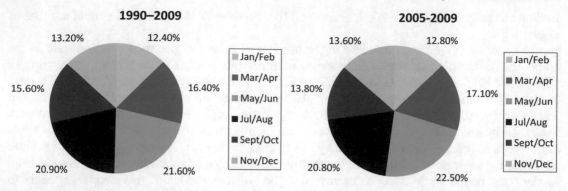

Figure 10.62. Gulf of Mexico angler trips by wave for 1990–2009 (*left panel*) and 2005–2009 (*right panel*) (NMFS FSD, data accessed 2012, with percentage calculations by authors—see Appendix A).

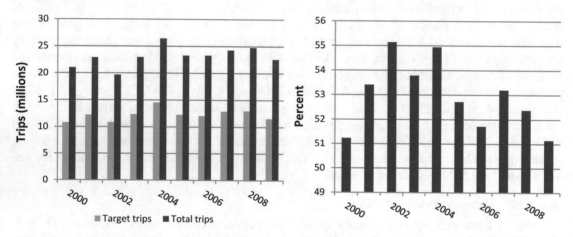

Figure 10.63. Gulf of Mexico recreational targeted trips in relation to total trips (*left panel*) and as a percentage of total trips (*right panel*) (NMFS FSD, data accessed 2012, with targeting estimates and percentages calculated by authors—see Appendix A).

there is some debate as to whether the actual catch influences how one responds to questions regarding targeting behavior.

With these caveats in mind, reported targeted angler trips in relation to total angler trips and the targeted angler trips as a percentage of total trips are illustrated in Figure 10.63. As indicated, the proportion of interviewees who list targeted species tends to be limited; ranging from about 51 to 55 % during the 10-year time period being considered (Figure 10.63, right panel). The reason(s) behind the large percentage of, and stability in, unspecified target trips is unknown but to some extent may reflect the large number of species available to recreational anglers with many of the species susceptible to harvest using the same gear. Thus, the primary purpose of a trip becomes one more of catching fish rather than catching specific species. Furthermore, one might speculate that the inability to specify targeted species may be heightened in the offshore fisheries that are largely managed by GMFMC. As discussed in a subsequent section, many of the federally managed species are subject to seasonal closures, and there may be a reluctance to indicate one is targeting a species that would need to be released during a closure of that fishery.

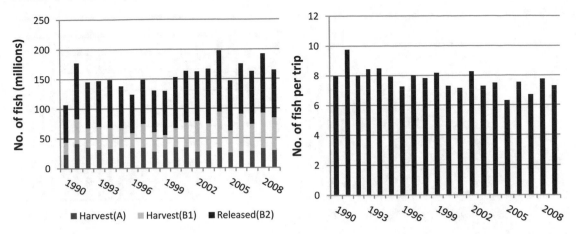

Figure 10.64. Catch in numbers of fish by recreational anglers (*left panel*) and number of fish caught per trip (*right panel*), 1990–2009 (NMFS FSD, data accessed 2012, with calculations by authors—see Appendix A).

10.4.2.1.5   Aggregate Gulf of Mexico Recreational Catch/Harvest

The estimated annual number of fish caught (A + B1 + B2) throughout the Gulf (excluding Texas) for the 1990–2009 period is given in Figure 10.64 (left panel). The total catch (by all anglers and modes) generally ranged from about 130 million fish per year to 190 million fish per year and averaged about 155 million fish per year during the 20-year period of analysis. While exhibiting considerable year-to-year variation, there is little or no long-run discernible trend in annual catch estimates in number of fish, with a possible exception of an upward and permanent shift beginning in the late 1990s. This upward shift largely corresponds with the upward movement in number of trips (see Figure 10.61).

The information in Figure 10.64 (left panel) also indicates that a large proportion of the number of fish caught is released either alive or dead. Since 1990, more than one-half of the catch in numbers of fish has been released alive (B2) with no apparent long-run trend. The high release rate is the result of a number of factors. Factors include, but are not limited to, (1) many species caught are generally considered to be undesirable (e.g., saltwater catfish), (2) most desirable species now have minimum (and sometimes maximum) size limits with catch at size below (above) that limit required to be returned to the water, (3) there are seasons for many of the desirable species and catch of that species outside the designated season must be released, and (4) retention of some species considered to be severely overfished (e.g., goliath grouper) is prohibited. While there appears to have been an upward and permanent shift in the number of fish caught beginning in the late 1990s (Figure 10.64, left panel), there appears to be no corresponding increase in the number of fish caught per trip (Figure 10.64, right panel). This is because the increased catch beginning in the late 1990s coincided with an increase in the number of angler trips. Hence, the estimated catch per angler trip remained virtually constant.

Focusing only on harvest (A + B1), the Gulf of Mexico recreational harvest (excluding Texas) averaged 73 million fish per year during the 20-year period ending in 2009 with a range from less than 50 million in 1990 to 100 million in 2006 (Figure 10.65, left panel). During the decade of the 1990s, the estimated annual harvest equaled 65 million fish per year with the figure increasing to 82 million fish per year during the most recent decade (2000–2009). With few exceptions, the number of fish harvested per trip consistently fell in the relatively narrow range of three fish to four fish with no observable trend (Figure 10.65, right panel). This would

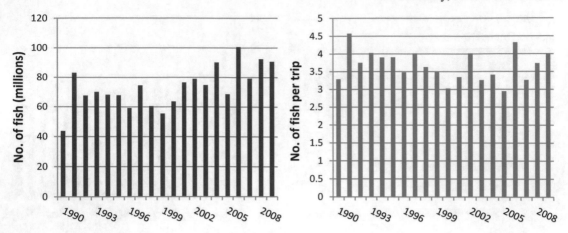

Figure 10.65. Harvest (in numbers of fish) by recreational anglers (*left panel*) and number of fish harvested per trip (*right panel*), 1990–2009 (NMFS FSD, data accessed 2012, with calculations by authors—see Appendix A).

suggest that the increasing harvest, expressed in terms of number of fish, during the most recent decade is the result of an increase in number of trips which, as noted, averaged 23 million annually during the most recent decade compared to 17 million annually during the 1990s.

Considering either catch (A + B1 + B2) or harvest (A + B1) in the aggregate has only limited value because many of the species most frequently caught are done so with the express purpose of using the catch for bait in catching more desirable species. This is clearly illustrated with the help of the information in Tables 10.11, 10.12, 10.13, 10.14, and 10.15 where each table provides the 25 most commonly caught (A + B1 + B2) and harvested (A + B1) species for a given year (1990, 1995, 2000, 2005, and 2009). As indicated, many of the species either caught (A + B1 + B2) or harvested (A + B1) tend to be used primarily as bait or released because they are not considered edible. In 1990, for example, three of the four most commonly recreationally caught species in the Gulf of Mexico (scaled sardine [*Harengula jaguana*], gizzard shad [*Dorosoma cepedianum*], and pinfish [*Lagodon rhomboids*]) represent species primarily used for bait as too are Spanish sardine (*Sardinella aurita*) and Gulf menhaden. Many other species—including hardhead catfish (*Arius felis*), gafftopsail catfish (*Bagre marinus*), and ladyfish (*Elops saurus*)—are rarely kept because they are considered *trash fish* by a large proportion of the recreational fishing population. Similarly, many of the more commonly harvested species (A + B1) in 1990 are used for bait including the three listed under the most commonly caught species. The average weight of the harvested baitfish species, as indicated, tends to be low relative to non-bait species (e.g., the average weight of scaled sardines equaled 0.0266 lb). The total estimated catch by Gulf of Mexico anglers in 1990 equaled 106 million fish (Figure 10.64, left panel). Of this total, 22.5 million, or about 20 %, represented species that would normally be used for bait. Similarly, the number of fish harvested (A + B1) in 1990 was estimated to equal 44 million (Figure 10.65, left panel). Of this total, 13.3 million, or almost 30 %, represented species primarily used for bait.

As also indicated by the information in Table 10.11, some of the species caught (A + B1 + B2) in 1990 were generally harvested (i.e., either retained or released dead) while for other species the catch and harvest figures can vary substantially. For example, an estimated 7.9 million scaled sardines were caught by recreational anglers in 1990, of which 7.4 million were harvested. By comparison, while an estimated 11.8 million spotted seatrout were caught by recreational anglers in 1990, about two-thirds of this total were released alive based on an

Table 10.11 25 Most Frequently Caught (Left Panel) and Harvested (Right Panel) Species by Gulf of Mexico Recreational Anglers, 1990

Species Name	Number of Fish Caught (A + B1 + B2)	Species Name	Number of Fish Harvested (A + B1)	Mean Weight (lbs)
Spotted Seatrout	11,782,266	Scaled Sardine	7,406,277	0.027
Scaled Sardine	7,865,548	Spotted Seatrout	3,791,206	1.297
Gizzard Shad	6,781,324	Sand Seatrout	2,785,120	0.652
Pinfish	5,758,476	Pinfish	2,352,044	0.229
Hardhead Catfish	5,335,865	White Grunt	1,961,626	0.709
Sand Seatrout	4,386,361	Atlantic Croaker	1,615,793	0.500
Atlantic Croaker	3,792,456	Spanish Mackerel	1,524,575	1.657
Spanish Mackerel	3,641,476	Sheepshead	1,363,611	2.066
Black Sea Bass	3,242,468	Striped Mullet	1,116,400	1.272
White Grunt	3,192,857	Spanish Sardine	1,046,086	0.044
Red Drum	2,267,628	Gray Triggerfish	945,723	2.372
Striped Mullet	2,253,036	Gizzard Shad	937,038	0.220
Sheepshead	2,043,894	Gulf Menhaden	890,584	0.588
Gray Snapper	1,838,071	Black Sea Bass	879,359	0.669
Gafftopsail Catfish	1,678,791	Gray Snapper	869,753	1.570
Ladyfish	1,559,712	Red Drum	813,517	6.403
Red Grouper	1,501,984	Hardhead Catfish	748,224	1.066
Sand Perch	1,270,975	Yellowtail Snapper	714,397	1.424
Spanish Sardine	1,217,217	Silver Seatrout	696,586	0.584
Southern Flounder	1,162,179	Atlantic Thread Herring	682,254	0.122
Gulf Menhaden	1,135,902	Southern Flounder	614,575	1.382
Yellowtail Snapper	1,098,766	Vermilion Snapper	549,003	0.984
Silver Seatrout	1,087,375	Dolphin	512,500	8.857
Gray Triggerfish	1,082,116	Pigfish	436,867	0.472
Blue Runner	994,028	Sand Perch	435,769	0.436

Source: (NMFS FSD, data accessed 2012, with calculations by authors—see Appendix A. Note: 1 lb = 0.454 kg)

estimated harvest figure of 3.8 million fish. Since scaled sardines are used primarily for bait, those not used on the trip are likely discarded dead. The large percentage of live-release of spotted seatrout, as discussed in greater detail later in this chapter, reflects the considerable management measures, including size limits and bag limits, implemented to protect the populations of this species throughout the Gulf of Mexico. Finally, as indicated by the information in Table 10.11, some species commonly caught by recreational anglers in the Gulf of Mexico, such as hardhead catfish, are rarely retained or released dead (A + B1). While the estimated catch of hardhead catfish in 1990 totaled 5.3 million fish which led to its 5th place ranking, hardhead catfish ranks only 17th among most commonly harvested species indicating that the vast majority of this species is reported to be released alive. Because there are virtually

Table 10.12 25 Most Frequently Caught (Left Panel) and Harvested (Right Panel) Species by Gulf of Mexico Recreational Anglers, 1995

Species Name	Number of Fish Caught (A + B1 + B2)	Species Name	Number of Fish Harvested (A + B1)	Mean Weight (lbs)
Spotted Seatrout	20,803,580	Spanish Sardine	890,074	NA
Scaled Sardine	19,570,826	Pinfish	855,943	0.220
Pinfish	12,297,105	Sand Seatrout	817,751	0.530
Red Drum	6,987,029	White Grunt	752,203	0.699
Hardhead Catfish	6,466,956	Vermilion Snapper	699,398	1.041
White Grunt	5,409,872	Southern Flounder	662,142	1.144
Sand Seatrout	4,449,631	Striped Anchovy	660,456	0.020
Sheepshead	3,573,270	Atlantic Thread Herring	637,242	0.060
Atlantic Thread Herring	3,287,447	Sand Perch	616,149	0.326
Gray Snapper	3,264,432	Southern Kingfish	612,129	0.605
Atlantic Croaker	2,409,813	Scaled Sardine	607,327	0.043
Gag	2,264,393	Pigfish	583,308	0.397
Black Sea Bass	2,168,769	Gray Triggerfish	541,630	1.893
Red Grouper	1,951,612	King Mackerel	484,248	9.199
Crevalle Jack	1,865,996	Black Sea Bass	479,187	0.591
Pigfish	1,632,945	Gulf Menhaden	456,781	0.220
Spanish Mackerel	1,572,507	Seatrout Genus	421,376	NA
Gafftopsail Catfish	1,561,431	Silver Perch	406,875	0.177
Red Snapper	1,491,284	Sheepshead	373,012	2.078
Striped Mullet	1,268,225	Yellowtail Snapper	351,082	1.212
Silver Perch	1,259,430	Hardhead Catfish	343,923	0.976
Sand Perch	1,176,826	Round Scad	327,235	NA
Blue Runner	1,108,328	Mullet Genus	294,853	NA
Black Drum	1,030,209	Bluefish	243,192	2.072
Yellowtail Snapper	966,760	Lane Snapper	228,787	1.476

Source: (NMFS FSD, data accessed 2012, with calculations by authors—see Appendix A. Note: 1 lb=0.454 kg.)

no regulations requiring the return of hardhead catfish (e.g., size limits or bag limits), the large difference between catch (A + B1 + B2) of hardhead catfish and harvest of the species (A + B1) reflects the fact that much of the recreation public considers this species to be a trash fish.

In 2009, three of the five most commonly caught species (scaled sardine, pinfish, and Atlantic thread herring [*Opisthonema oglinum*]) represent species primarily used for bait (Table 10.15). Similar findings apply if one considers harvested species. Comparison of the information in Tables 10.11 and 10.15 (as well as other selected years provided in Tables 10.12, 10.13, and 10.14) suggests that, in general, the catch and harvest of many species—such as

Table 10.13 25 Most Frequently Caught (Left Panel) and Harvested (Right Panel) Species by Gulf of Mexico Recreational Anglers, 2000

Species Name	Number of Fish Caught (A + B1 + B2)	Species Name	Number of Fish Harvested (A + B1)	Mean Weight (lbs)
Spotted Seatrout	26,859,790	Scaled Sardine	14,714,424	0.022
Scaled Sardine	17,584,629	Pinfish	5,180,685	0.401
Pinfish	9,534,737	Sand Seatrout	4,376,247	0.631
Red Drum	8,201,553	Red Drum	2,981,011	5.651
Atlantic Croaker	5,911,239	Atlantic Thread Herring	2,316,911	0.082
Sand Seatrout	5,902,934	Atlantic Croaker	1,772,120	0.456
Hardhead Catfish	4,523,357	White Grunt	1,739,314	0.875
White Grunt	4,179,199	Spanish Mackerel	1,501,056	1.986
Gray Snapper	3,907,599	Striped Mullet	1,478,051	1.145
Black Sea Bass	3,382,680	Blue Runner	1,323,223	1.097
Ladyfish	3,038,159	Sheepshead	1,258,531	2.669
Spanish Mackerel	2,897,142	Southern Kingfish	1,253,472	0.603
Sheepshead	2,837,075	Black Drum	821,396	3.342
Blue Runner	2,796,861	Spanish Sardine	816,465	0.019
Atlantic Thread Herring	2,727,189	Round Scad	775,122	NA
Gafftopsail Catfish	2,124,884	Gulf Menhaden	758,747	0.215
Crevalle Jack	2,063,277	Gray Snapper	694,916	1.612
Black Drum	1,994,186	Spotted Seatrout	660,962	1.641
Red Grouper	1,862,001	Black Sea Bass	548,872	0.815
Striped Mullet	1,841,982	Southern Flounder	546,769	1.452
Gag	1,798,006	Gag	540,122	6.820
Southern Kingfish	1,551,570	Pigfish	435,936	0.319
Pigfish	1,060,531	White Mullet	373,823	0.696
Red Snapper	998,453	Gulf Kingfish	370,069	0.663
Gulf Menhaden	950,030	Red Snapper	340,077	4.284

Source: (NMFS FSD, data accessed 2012, with calculations by authors—see Appendix A. *Note*: 1 lb = 0.454 kg.)

spotted seatrout, red drum, and scaled sardines—increased significantly between 1990 and 2009. Changes in catches/harvests of the more desired species over time are examined in greater detail later in this chapter.

The information in Tables 10.11, 10.12, 10.13, 10.14, and 10.15 leads to the conclusion that aggregate catch (A + B1 + B2) and harvest (A + B1) estimates, when analyzed in terms of numbers of fish, are significantly skewed by the inclusion of baitfish species. Given the relatively low average weight associated with the baitfish species, inclusion of these species will have little influence on aggregate poundage estimates. The large amount of catch that is released dead or otherwise not seen or identified by the interviewer (B1) is illustrated in

Table 10.14 25 Most Frequently Caught (Left Panel) and Harvested (Right Panel) Species by Gulf of Mexico Recreational Anglers, 2005

Species Name	Number of Fish Caught (A + B1 + B2)	Species Name	Number of Fish Harvested (A + B1)	Mean Weight (lbs)
Spotted Seatrout	30,194,309	Scaled Sardine	17,338,752	0.052
Scaled Sardine	18,058,990	Spotted Seatrout	10,027,333	1.470
Pinfish	9,468,582	Pinfish	5,954,236	0.281
Red Drum	8,313,858	Atlantic Thread Herring	3,915,545	0.016
Hardhead Catfish	6,103,615	Red Drum	2,316,967	6.957
Gray Snapper	5,557,647	Sheepshead	2,002,107	2.850
Sheepshead	4,341,937	Sand Seatrout	1,916,453	0.573
Atlantic Thread Herring	4,070,662	White Grunt	1,687,555	0.860
Ladyfish	3,894,394	Spanish Mackerel	1,191,652	1.729
White Grunt	3,372,101	Striped Mullet	1,080,239	1.176
Atlantic Croaker	3,344,904	Southern Kingfish	1,060,265	0.592
Gag	2,789,268	Gray Snapper	1,054,134	2.396
Red Snapper	2,738,566	Red Snapper	835,166	4.027
Sand Seatrout	2,588,201	Atlantic Croaker	770,890	0.411
Spanish Mackerel	2,497,044	White Mullet	743,687	0.665
Gafftopsail Catfish	2,163,933	Gulf Menhaden	577,043	NA
Southern Kingfish	1,643,147	Southern Flounder	541,916	1.219
Black Sea Bass	1,612,855	Gag	517,374	6.666
Red Grouper	1,460,939	Sand Perch	460,961	0.431
Common Snook	1,362,106	Black Drum	449,895	5.269
Crevalle Jack	1,346,097	Blue Runner	449,314	0.924
Black Drum	1,285,579	Menhaden Genus	390,512	NA
Striped Mullet	1,228,363	Gulf Kingfish	366,168	0.731
Blue Runner	1,047,689	Lane Snapper	349,043	1.141
Sand Perch	836,435	Round Scad	336,453	NA

Source: (NMFS FSD, data accessed 2012, with calculations by authors–see Appendix A. *Note*: 1 lb = 0.454 kg.)

Figure 10.64 (left panel). A large proportion of this total reflects the harvest of many small fish subsequently used for bait (e.g., scaled sardines and pinfish). For example, the total estimated catch in 2009 equaled 172 million fish. More than 25 % of this total number was represented by species generally used as baitfish. Similarly, about 20 % of the estimated 106 million fish caught in 1990 represent baitfish. With respect to harvested fish (A + B1), the 1990 estimate equaled 44 million (Figure 10.65, left panel). Of this total, 13.3 million, or about 30 %, represented species primarily used for bait. Likewise, about 50 % of the estimated 90 million fish harvested in 2009 represent species primarily used as baitfish.

Table 10.15 25 Most Frequently Caught (Left Panel) and Harvested (Right Panel) Species by Gulf of Mexico Recreational Anglers, 2009

Species Name	Number of Fish Caught (A + B1 + B2)	Species Name	Number of Fish Harvested (A + B1)	Mean Weight (lbs)
Scaled Sardine	31,431,676	Scaled Sardine	29,939,131	0.029
Spotted Seatrout	30,689,392	Spotted Seatrout	13,336,326	1.510
Pinfish	9,792,148	Atlantic Thread Herring	6,804,819	0.108
Red Drum	8,009,540	Pinfish	5,448,281	0.132
Atlantic Thread Herring	7,472,772	Sand Seatrout	4,200,054	0.561
Sand Seatrout	6,617,915	Red Drum	2,608,080	6.146
Hardhead Catfish	5,279,557	Sheepshead	1,573,049	2.658
Atlantic Croaker	4,897,441	Spanish Mackerel	1,503,195	1.697
Gray Snapper	4,172,791	Gray Snapper	1,300,627	1.711
Ladyfish	3,387,942	White Grunt	1,206,086	0.909
Spanish Mackerel	3,132,709	Atlantic Croaker	1,173,610	0.388
Sheepshead	2,871,863	Round Scad	1,096,334	NA
Gag	2,750,328	Ballyhoo	1,087,375	NA
Red Snapper	2,568,716	Southern Kingfish	979,390	0.575
Red Grouper	2,472,120	Red Snapper	795,585	4.885
White Grunt	2,241,227	Striped Mullet	741,904	1.172
Black Drum	1,747,954	Blue Runner	696,892	0.951
Blue Runner	1,490,693	Black Drum	664,917	5.248
Gafftopsail Catfish	1,426,345	Herring Family	647,389	NA
Southern Kingfish	1,388,023	Southern Flounder	643,630	1.358
Crevalle Jack	1,309,758	King Mackerel	509,489	8.039
Round Scad	1,128,681	Vermilion Snapper	407,787	1.030
Ballyhoo	1,088,172	Gulf Menhaden	391,449	NA
Black Sea Bass	977,919	Dolphin	341,574	08.162
Striped Mullet	957,806	Unidentified Fish	271,498	NA

Source: (NMFS FSD, data accessed 2012, with calculations by authors—see Appendix A. Note: 1 lb = 0.454 kg.)

The estimated annual harvest of fish, given on a weight basis (A + B1), for the Gulf (excluding Texas) from 1990 to 2009 is given in Figure 10.66 (left panel). As indicated, pounds harvested have historically fluctuated from about 60 million to 80 million on an annual basis with no long-run trend.

While the total harvest during the 20-year period ending in 2009 appears to be stable, there does appear to be a decline in pounds harvested (A + B1) per angler trip (Figure 10.66, right panel). During the 1990s, pounds harvested (A + B1) averaged four pounds per angler trip. During the most recent decade (2000–2009), pounds harvested had fallen by about

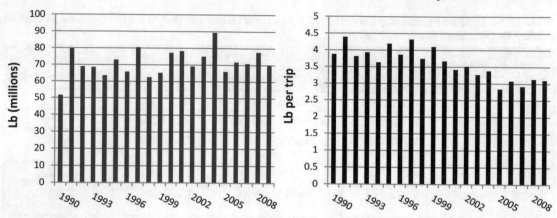

Figure 10.66. Gulf of Mexico recreational harvest in pounds (*left panel*) and harvest per trip (*right panel*), 1990–2009 (NMFS FSD, data accessed 2012, with calculations by authors—see Appendix A) (*Note*: 1 lb = 0.454 kg).

0.25–3.25 lb per angler trip, on average. This decline largely parallels the increase in angler trips in the early 2000s.

When examined by mode, fishing from private/rental boats accounted for an average of 65 % of Gulf (excluding Texas) harvest (A + B1), in pounds, during the 1990–2009 period with an annual range generally fluctuating from 60 to 70 %. The for-hire mode accounted for about 20 % of the total pounds while shore-based angling accounted for the remaining 15 %. No long-term trends were apparent in any of the three modes.

10.4.2.2 Gulf of Mexico Recreational Catch/Harvest by Species/Group

A large and diverse group of species are targeted and harvested by Gulf of Mexico recreational anglers. The species (groups) targeted and caught in the inshore waters tend to differ from those targeted and caught offshore. Analyses of primary species associated with inshore and offshore fishing activities at the Gulf level, angler trips taken in the pursuit of the catch/harvest of these species (groups), and targeting behavior are examined below.

10.4.2.2.1 Inshore Species

Two species, spotted (speckled) seatrout and red drum, dominate marine recreational angling activities in the inland and nearshore waters of the Gulf of Mexico. These two species are the preferred target of anglers throughout the Gulf and are given gamefish status in many of the Gulf States.

Spotted seatrout is managed by the individual Gulf States, and size restrictions and bag/possession limits are also determined by the individual states. These can vary significantly among the states. For example, Florida limits spotted seatrout harvest to four per harvester per day in the South region and five per harvester per day in the Northwest region. For Louisiana, the limit on spotted seatrout is 25 per person per day. VanderKooy (2010) provides a detailed listing of all laws and regulations pertaining to the recreational sector by state, and information on spotted seatrout regulations can be found in the report. While the individual Gulf States also manage red drum, management of the fishery in federal waters is under the purview of GMFMC. Since 1988, the harvest and possession of red drum from federal waters has been prohibited. By state, Florida has the most restrictive red drum bag limit (one per harvester per day) while Louisiana has the most liberal (five daily per person).

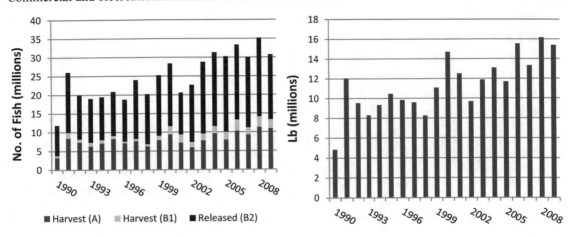

Figure 10.67. Gulf of Mexico recreational spotted sea trout catch (*left panel*) and pounds harvested (*right panel*), 1990–2009 (NMFS FSD, data accessed 2012—see Appendix A) (*Note*: 1 lb = 0.454 kg).

Spotted Seatrout

Spotted Seatrout

Estimated Gulf of Mexico catch of spotted seatrout (excluding Texas) in numbers of fish is given in Figure 10.67 (left panel) for the 1990–2009 period. In general, after falling from more than 28 million fish in 1999 to about 20 million fish in 2000, the estimated catch of spotted seatrout gradually increased during the remainder of the decade peaking at 35 million fish in 2008 and equaled 31 million in 2009. On average, approximately 60 % of the total catch is reported to be released alive (B2), and there has been no apparent trend in this average since 1990. The information in Figure 10.67 (left panel) also highlights that harvest (dead) not seen or identified by interviewer (B1) is relatively limited when considering a species not used as bait.

The estimated harvest of spotted seatrout, in pounds (A + B1), also has been gradually increasing since the 1990s with the average during the most recent 10-year period at 13.4 million pounds annually, exceeding the 1990–1999 average of 9.3 million pounds annually by about 45 % (Figure 10.67, right panel). In general, the increased spotted seatrout catch (in either number of fish or pounds landed) correlates well with the increasing number of inshore trips (see Figure 10.61, right panel). In both instances, the correlation approached or exceeded 0.80.

The role of the inshore waters to spotted seatrout catch in numbers of fish (A + B1 + B2) is clearly illustrated in Figure 10.68 (left panel). During the 1990s, about 65 % of the spotted seatrout catch (excluding Texas) was in inland waters, with the proportion increasing to more than 80 % since 2000 (the years 1990 and 1991 appear to be unexplained anomalies with respect to the percentage of spotted seatrout derived from the inshore waters). The 2009 proportion of 87 % was the highest on record. With some notable exceptions, particularly in the earlier years, less than 5 % of the catch in numbers is taken from federal waters.

Being primarily an inshore fishery, it should come as no surprise that the vast majority of recreational spotted seatrout catch comes from private/rental boats. Since 1990, the percentage of catch in numbers of fish coming from this mode (excluding Texas) has consistently equaled about 80–90 % of the total with no apparent trend (Figure 10.68, right panel). The share of

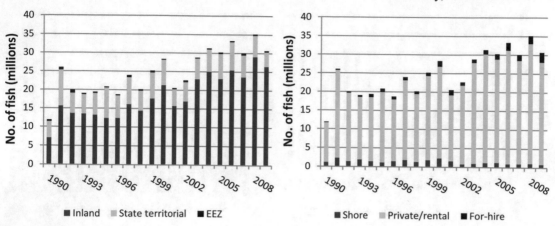

Figure 10.68. Gulf of Mexico recreational spotted seatrout catch by area fished (*left panel*) and mode (*right panel*), 1990–2009 (NMFS FSD, data accessed 2012—see Appendix A).

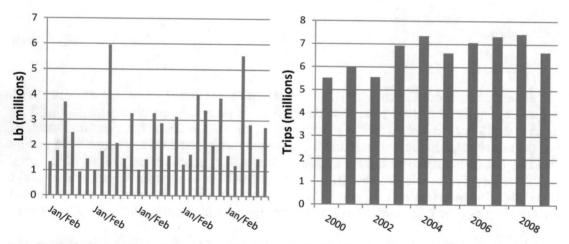

Figure 10.69. Gulf of Mexico recreational spotted seatrout harvest by wave 2005–2009 (*left panel*) and trips where spotted seatrout was caught, 2000–2009 (NMFS FSD, data accessed 2012— see Appendix A) (*Note*: 1 lb = 0.454 kg).

harvest represented by the for-hire sector has, on the other hand, been increasing particularly in recent years with the 2009 share (15 %) representing the highest on record. The recent increase in the for-hire share has come largely at the expense of the shore mode with its 2009 share (1.7 %) being the lowest on record (the shore mode represented an unweighted average of 6.5 % during the period of study).

The Gulf of Mexico recreational spotted seatrout fishery is seasonal in nature with a couple of distinct periods when examining harvest (A + B1 in pounds) in 2 month waves. Beginning in January/February, harvest tends to increase, reaching a peak in May/June of each year. It then falls but generally increases sharply again in November/December. Since 2005, nearly 33 % of the spotted seatrout harvest (A + B1), in pounds, has occurred in the May/June period with an additional 20 % being reported in both the July/August and November/December periods (Figure 10.69, left panel).

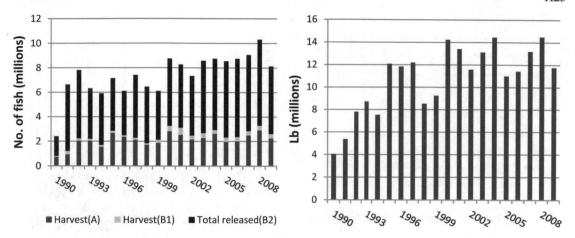

Figure 10.70. Gulf of Mexico recreational red drum catch (*left panel*) and harvested pounds (*right panel*), 1990–2009 (NMFS FSD, data accessed 2012—see Appendix A) (*Note*: 1 lb = 0.454 kg).

Gulf of Mexico (excluding Texas) angling trips resulting in the catch of spotted seatrout (A + B1 + B2) is given in Figure 10.69 (right panel) for the 2000–2009 period. As indicated, the annual number of trips reporting the catch of spotted seatrout has, in recent years, fluctuated in the 5.5–7.5 million range.

Red Drum

Red Drum

Red drum is the other popular species among inshore fishermen. Gulf of Mexico recreational anglers have caught, on average, an estimated 7.4 million red drum per year (A1 + B1 + B2) since 1990 with annual catches trending up in recent years (Figure 10.70, left panel). For example, during 1990–1994, estimated red drum catch per year averaged 5.8 million fish, while during the 1990s the average estimated catch equaled 6.2 million fish per year. Since 2000, average catch, in numbers, has equaled an estimated 8.7 million fish per year. In no year prior to 2000 did the annual estimated catch exceed eight million fish, but since 2000, annual estimated catch has not fallen below eight million fish. The impact of severe weather conditions on red drum stocks is evident in the abnormally low 1990 red drum catch (2.4 million fish) which was likely the direct result of a very hard freeze in 1989 that resulted in a high mortality in the Louisiana red drum stock. Historically, about two-thirds of the red drum catch, in numbers, have been released alive (B2) and there is no long-run apparent change to this figure. Release of this high percentage of red drum reflects, at least in part, the management measures established to protect the species from overfished conditions, particularly size limits and daily bag limits.

Similar to catch, the red drum harvest (A + B1) expressed on a weight basis increased during the mid-to-late 1990s, but since that time there has been no apparent increasing long-run trend (Figure 10.70, right panel). Since 2000, an estimated 12.9 million pounds of red drum have been harvested annually by recreational anglers throughout the Gulf of Mexico (excluding Texas). Similar to spotted seatrout, the correlation between total number of inshore trips (Figure 10.61, right panel) and total annual estimated red drum catch in numbers during the 1990–2009 period was high (0.88), as was the correlation between total number of inshore trips and the estimated harvest (A + B1) in pounds (0.83). Since 2000, no less than 80 % of the red

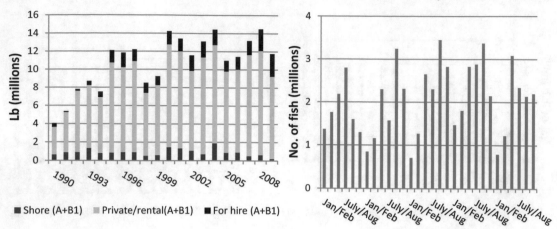

Figure 10.71. Gulf of Mexico recreational red drum harvest by mode 1990–2009 (*left panel*) and wave 2005–2009 (*right panel*) 1990–2009 (NMFS FSD, data accessed 2012—see Appendix A).

drum harvest (A + B1) in pounds has been taken from inshore waters with the total exceeding 90 % in both 2008 and 2009. Furthermore, in no year prior to 1999 did the share of red drum harvest taken from inland waters exceed 80 %, and in no year after 1999 did the share fall under 80 %.

Similar to spotted seatrout, the overwhelming majority of red drum harvest in pounds is taken from the private/rental mode with yearly estimates generally ranging from about 75 % to slightly more than 80 % (Figure 10.71, left panel) with no apparent trend. The share taken by the for-hire mode, on the other hand, has increased, especially after the mid-1990s, with the share in 2008 (16.6 %) and 2009 (21.3 %) being the largest on record. The share attributable to shore-based fishing has ranged from 16 % to less than 2 % and has averaged about 8 % during the period of analysis.

Like spotted seatrout, there is a seasonal pattern to red drum harvest with yield being lowest in January/February and gradually increasing to May/June or July/August (Figure 10.71, right panel). As with spotted seatrout, significant harvests occur in the November/December period. Spotted seatrout and red drum are often targeted on the same trip. Hence, targeting behavior for these two species combined is considered herein. During the 2000–2009 period, the number of angler trips wherein the angler reported targeting behavior for either red drum or spotted seatrout averaged 6.6 million annually with the annual estimates ranging from 5.5 million in 2000 to 8.1 million in 2004 (Figure 10.72, left panel). As a proportion of total Gulf trips, which averaged 23.1 million during 2000–2009, trips targeting red drum or spotted seatrout averaged about 30 % (Figure 10.72, right panel). This figure becomes more relevant when one considers that only about 50 % of the MRFSS/MRIP interviewees report any targeting behavior (Figure 10.63, right panel).

As indicated, the vast majority of targeting behavior for either spotted seatrout or red drum was in relation to the private/rental boat mode (Figure 10.72, left panel). In general, more than 40 % of the private/rental boat mode angler trips reported targeting either spotted seatrout or red drum with the figure consistently approximating 43 % since 2005. By comparison, only about 10 % of the anglers fishing from shore reported targeting spotted seatrout or red drum with little interyear variation. With respect to the for-hire mode, the percentage of angler trips reporting red drum/spotted trout targeting behavior ranged from about 10 % (2000) to 25 % (2009) and averaged about 15 % during the 10-year period of analysis (2000–2009).

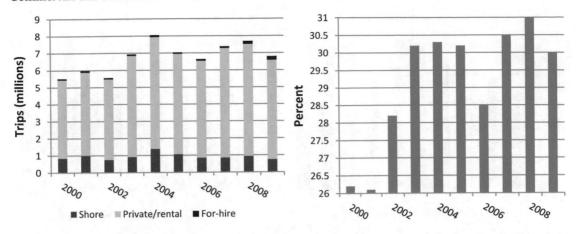

Figure 10.72. Angler trips targeting either spotted seatrout or red drum (*left panel*) and targeting percentage (*right panel*), 2000–2009 (NMFS FSD, data accessed 2012, with targeting estimates and percentages calculated by authors—see Appendix A).

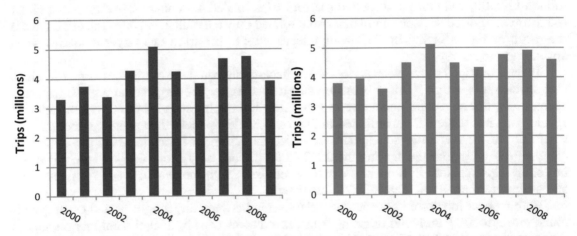

Figure 10.73. Gulf of Mexico angler trips targeting red drum (*left panel*) and Gulf of Mexico angler trips targeting spotted seatrout, 2000–2009 (NMFS FSD, data accessed 2012, with targeting estimates calculated by authors—see Appendix A).

Considering the two species separately, the number of Gulf angler trips reporting the targeting of red drum on an annual basis ranged from about 3.3 million (2000) to 5.1 million (2004) and averaged 4.1 million annually during the 10-year period of analysis (Figure 10.73, left panel). This represents about 18 % of the total 23.2 million angler trips conducted yearly, on average, throughout the Gulf (excluding Texas) during the 10-year period and almost 20 % of the 21.3 million annual trips taken in state (inland and territorial) waters.

The annual number of spotted seatrout targeting trips generally ranged from about 4 million to 5 million and averaged 4.4 million during the 10-year period ending in 2009 (Figure 10.73, right panel). This 10-year average constituted 19 % of the annual average of 23.2 million trips taken in the Gulf of Mexico (excluding Texas).

One might notice that the total targeting trips for red drum and spotted seatrout, evaluated separately (Figure 10.73), exceeds the total when combined (i.e., targeting either red drum or spotted seatrout) by a significant margin. For example, targeted red drum trips totaled 3.9

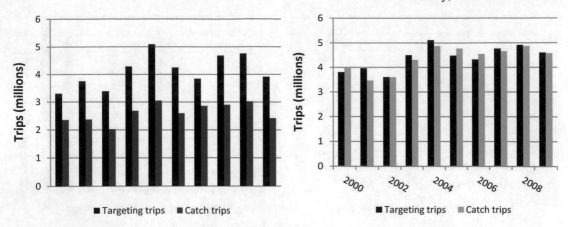

Figure 10.74. Relationship between red drum catch trips and targeting trips (*left panel*) and spotted seatrout catch trips and targeting trips (*right panel*), 2000–2009 (NMFS FSD, data accessed 2012, with targeting estimates calculated by authors—see Appendix A).

million in 2009 while targeted spotted seatrout trips totaled 4.6 million. Trips targeting either red drum or spotted seatrout, by comparison, totaled only 6.8 million in 2009. The difference is the result of the MRFSS/MRIP allowing interviewees to list up to two targeted species when answering the survey.

Targeting trips for a given species do not necessarily equal the number of trips for which that species was caught. This is because while a species may be targeted on a given trip, that species may not be caught and a given species may be caught on a given trip even though that species was not targeted. As indicated in Figure 10.74 (left panel), the annual number of trips where red drum was a targeted species exceeded the number of trips in which red drum was caught by a wide margin. During the 2000–2009 period, only about 65 % of the red drum targeting trips resulted in a positive catch of red drum with the annual range from less than 50 % to more than 70 % (Figure 10.74, left panel).

With respect to spotted seatrout, the relationship between targeting trips and catch trips is much more direct. Rarely did targeting trips exceed catch trips by a significant margin, and in many years, the number of catch trips equaled or slightly exceeded the number of targeting trips (Figure 10.74, right panel).

10.4.2.2.2   Offshore Species

Aggregate Reef Fish

Greater Amberjack

Gray Triggerfish

The Gulf of Mexico Reef Fish Fishery Management Plan includes six species groups in its management unit: triggerfishes, Jacks, wrasses, snappers, tilefish, and groupers. Based on this designation, estimated total catch of reef fish species for 1990–2009 in numbers of fish is presented in Figure 10.75 (left panel). As indicated, the total number of fish caught (A + B1 + B2) ranged from less than ten million in 1990 to more than 20 million in 1991, 2004, and 2008, with an average catch during the period equaling 15 million. While catch in numbers can exceed 20 million, the majority of this catch is released alive (B2); though much of it is subject to subsequent mortality (from hook and handling trauma or predation before the fish can recover). Since 1990, on average, almost 70 % of the reef fish catch is reportedly released alive with the figure exceeding 75 % since 2005. There are a number of reasons for the

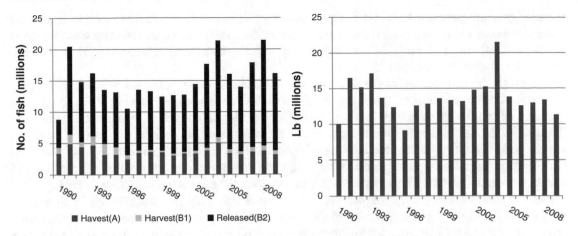

Figure 10.75. Gulf of Mexico recreational aggregate reef fish catch (*left panel*) and harvest (*right panel*), 1990–2009 (NMFS FSD, data accessed 2012—see Appendix A) (*Note*: 1 lb = 0.454 kg).

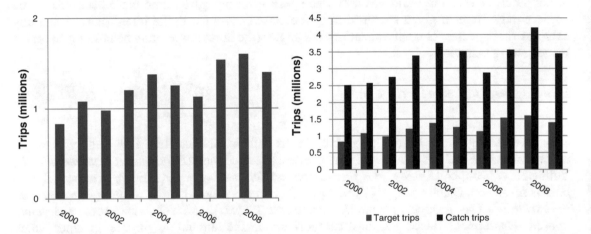

Figure 10.76. Gulf of Mexico recreational reef fish targeting trips (*left panel*) and reef fish targeting trips in relation to reef fish catch trips, 2000–2009 (NMFS FSD, data accessed 2012, with targeting estimates calculated by authors—see Appendix A).

high reef fish discard rates including regulations pertaining to minimum sizes and bag limits. Some of these regulations will be considered in more detail when considering snappers and groupers—the two primarily targeted recreational species groups within the reef fish complex.

Estimated total recreational reef fish harvest (A + B1) expressed on a weight basis is given in Figure 10.75 (right panel) for the 1990–2009 period. The aggregate harvest of reef fish averaged 13.4 million pounds during the period of analysis and ranged from about nine million pounds (1996) to more than 20 million pounds (2004). Since 2004, the annual aggregate reef fish harvest has fallen and, as discussed in the next section, much of this decline is the result of a decline in grouper harvest.

Though a considerable amount of effort is expended on the management of reef fish species by GMFMC and the regulations of managing recreational effort and take are numerous (bag limits, size limits, seasonal closures, mandatory use of circle hooks, etc.), only a small percentage of trips taken by marine anglers are in pursuit of any specific reef fish species. Since 2005, about 1.4 million trips annually (or about 6 % of the total estimated Gulf of Mexico trips) indicated a given reef fish species to be a targeted species (Figure 10.76, left panel). While a seemingly small percentage of Gulf trips target reef fish, most reef fish are caught in federal waters or state territorial waters (particularly in Florida and Texas where the state waters extend out 9 nautical mi [16.7 km]). When considering just the state territorial and EEZ waters, the proportion increases to about 15 %.

Though the number of angler trips in which a specific reef fish species is listed as being targeted is relatively limited (an average of 1.4 million trips annually during 2000–2009), the number of angler trips where a given reef fish species is caught is much more prevalent (Figure 10.76, right panel). On average, reef fish were reported caught on about 3.3 million angler trips annually during the 2000–2009 period. This represents almost 15 % of the total Gulf angler trips during the period. This of course raises the question, why is the estimated number of reef fish catch trips significantly higher than the number of reef fish targeting trips? There are at least three plausible answers to this question. First, no specific reef fish species may be targeted on a given trip even if the intent of the trip is to catch reef fish (specifically, since the MRFSS/MRIP asks for targeting behavior on specific species, one could still target reef fish without any specific species in mind). Second, many of the reef fish species, particularly red snapper, are subject to long seasonal closures, and these species may be caught in conjunction with targeting non-reef fish species. Finally, given long closed seasons, some anglers may target a given species during the closed season with the intent of releasing any catch of that species. In such a situation, the angler (the interviewee) may be hesitant to report his targeting behavior.

Individual Reef Fish Species (Groups)

Gag Grouper

Black Grouper

Groupers: The grouper family, as defined in the Gulf of Mexico Reef Fish Fishery Management Plan (GMFMC 1981), includes nine species: speckled hind (*Epinephelus drummondhayi*), Yellowedge grouper (*Epinephelus flavolimbatus*), Warsaw grouper, Snowy grouper (*Epinephelus niveatus*), Black grouper (*Mycteroperca bonaci*), Yellowmouth grouper (*Mycteroperca interstitialis*), Gag grouper, Scamp (*Mycteroperca phenax*), and Yellowfin grouper (*Mycteroperca venenosa*). Based on this categorization, the annual aggregate grouper catch (A + B1 + B2) by Gulf of Mexico recreational anglers in numbers of fish during the 1990–2009 period is presented in Figure 10.77 (left panel). As shown, aggregate catch

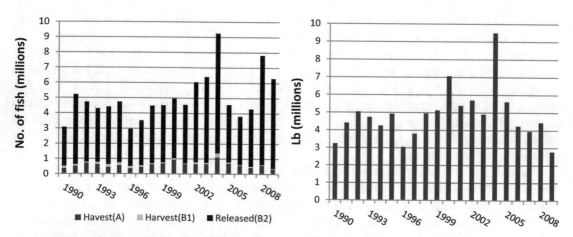

Figure 10.77. Gulf of Mexico recreational aggregate grouper catch (*left panel*) and harvest (*right panel*), 1990–2009 (NMFS FSD, data accessed 2012—see Appendix A) (*Note*: 1 lb = 0.454 kg).

(A + B1 + B2) increased rapidly after the mid-1990s peaking at about nine million fish in 2004 after which catch declined sharply. This decline, however, was transitory in nature and by 2008 the total catch in numbers of fish had again approached the eight million mark. The reasons for the precipitous decline in landings after 2004 are explored in more detail when catch of individual grouper species is considered.

On average, less than 15 % of the aggregate grouper catch by recreational anglers in the Gulf in number of fish was harvested (A + B1); the remaining (approximately) 85 % was released alive (Figure 10.77, left panel). This percentage remained extremely consistent during the period of analysis (generally 83–89 % range) with the exception of the last 2 years when the percentage exceeded 90 %. The high release rate during the period reflects aggregate bag limits, minimum size restrictions, and closed seasons (Carter et al. 2008).

Estimated aggregate harvest (A + B1) of grouper, expressed on a weight basis, is provided in Figure 10.77 (right panel) for the 1990–2009 period. As indicated, the annual harvest has fluctuated widely, ranging from less than three million pounds to more than eight million pounds. While some of the fluctuation can be explained by management actions (see individual species discussion), much of the variation likely simply reflects large annual recruitment variation. As a result of numerous environmental factors (most of which remain unknown), year-class size can vary by an order of magnitude and, as such, recruitment into the legal fishery (i.e., minimum legal size at harvest) can also vary substantially. Because of a high amount of pressure on the grouper stocks, a high percentage of the recreational grouper catch occurs shortly after the minimum legal size is reached.

Overall, about 33 % of the aggregate reef fish harvest (A + B1), expressed on a weight basis, was represented by grouper during the period of analysis. The share of aggregate reef fish harvest in pounds represented by grouper reached a maximum of 53 % in 2000 but was only about 25 % in 2009.

Grouper catch by recreational anglers consists primarily of two species—gag grouper and red grouper—which combined, account for about 95 % of the total recreational grouper harvest (A + B1) during 1990–2009. Annual harvests (A + B1) of these two species expressed on a weight basis are presented in Figure 10.78 (gag grouper, left panel; red grouper, right panel) for the 1990–2009 period. Gag grouper harvest, which accounted for about two-thirds of the total harvest of these two species, gradually increased through 2004 and declined sharply thereafter to a low of 1.5 million pounds in 2009. By comparison, recreational landings of red

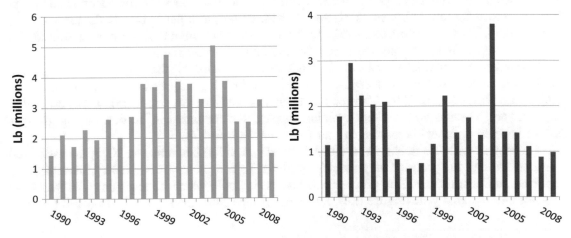

Figure 10.78. Gulf of Mexico recreational harvest of gag grouper (*left panel*) and red grouper (*right panel*), 1990–2009 (NMFS FSD, data accessed 2012—see Appendix A) (*Note*: 1 lb = 0.454 kg).

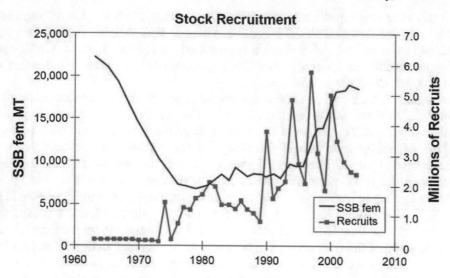

Figure 10.79. Estimated recruitment in the Gulf of Mexico gag grouper fishery (Ortiz 2006) (*Note*: SSB fem MT refers to the estimated spawning stock biomass of females).

grouper showed several cyclical trends. For example, landings during 1996–2000 were less than one-half of those during 1990–1995. After 2000, landings rose sharply, peaked in 2004, and then fell sharply. There are a couple of potential reasons why landings of both red and gag grouper fell sharply beginning in 2005 and have remained relatively low since that year. First, a large red tide event off the west coast of Florida occurred in 2005, which is believed to have led to large fish kills.[20] Furthermore, in June of 2005, an interim rule was established by GMFMC that established closed seasons for recreational grouper fishing and reduced the bag limit of red grouper to only one fish. Further restrictions were imposed in 2009 (GMFMC 2011).

Changes in recreational harvest of gag and red grouper from one year to the next or in the long run can be the result of several factors including annual variation in recruitment, long-term changes in biomass, and regulations. Changes in recruitment on subsequent harvest can be particularly pronounced in those fisheries that are heavily fished, since a large proportion of the fish that are kept are at the minimum size limit. Thus, annual variations in recruitment (year-0 fish) can have large impacts on harvest when that cohort reaches the minimum legal size at harvest. Estimated annual variation in recruitment of gag grouper is clearly demonstrated in Figure 10.79. This variation can easily translate into large annual variations in subsequent harvest at that time when that cohort reaches minimum harvest size.

The influence of a large-scale environmental perturbation on the gag grouper population can be examined with the aid of Figure 10.80. As mentioned, a large red tide event occurred off the Florida west coast in 2005. This red tide event resulted in a large reduction in the estimated gag grouper biomass which likely explains, in part, the reduction in harvest of gag grouper (as well as red grouper) beginning in that year (see Figure 10.78).

Consistent with the decline in grouper harvests (A + B1) beginning in 2005, the estimated number of targeted grouper (any species) angler trips fell by about one-half. The number of reported angler trips targeting grouper was highest in 2000, which is consistent with the above average catch in that year (Figure 10.81, left panel). Since 2005, the reported number of targeted

[20] See   http://www.sefsc.noaa.gov/sedar/download/S12RD12%20FLred%20tide%20Dec2006.pdf?id= DOCUMENT for information of the impact of this event.

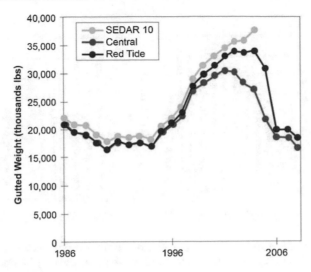

Figure 10.80. Estimated Gulf of Mexico gag grouper biomass (Anonymous 2009) (*Note*: 1 lb = 0.454 kg).

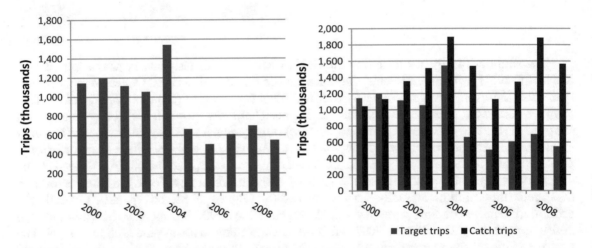

Figure 10.81. Recreational grouper targeting trips (*left panel*) and relationship between grouper catch trips and grouper targeting trips (*right panel*), 2000–2009 (NMFS FSD, data accessed 2012, with targeting estimates calculated by authors—see Appendix A).

grouper trips has averaged about 600,000 annually compared to an annual average of 1.2 million during 2000–2004. The decline in number of reported grouper targeting trips coincides with the red tide event that reportedly caused high grouper mortality (Anonymous 2009). While four named hurricanes hit Florida in 2004 (Charley, Frances, Ivan, and Jeanne), these hurricanes do not appear to have materially impacted Florida fishing trips or targeted grouper trips. In fact, some have hypothesized that the increased hurricane activity was the cause for the increased grouper catch in that year.

While the number of reported grouper angler targeting trips fell sharply beginning in 2005, the number of angler trips wherein grouper was caught fell for only a couple of years but increased sharply again in 2008 and 2009 (Figure 10.81, right panel). This is consistent with the increased grouper catches (A + B1 + B2) in number of fish as illustrated in Figure 10.77 (left panel). As these fish reach minimum harvest size, one might see a commensurate increase in harvest in pounds.

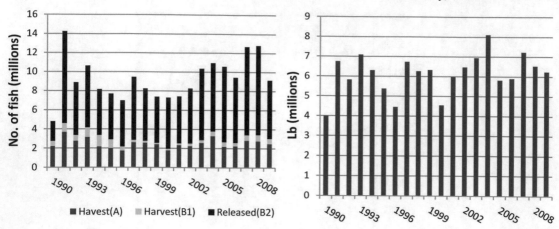

Figure 10.82. Gulf of Mexico recreational aggregate snapper catch (*left panel*) and harvest (*right panel*), 1990–2009 (NMFS FSD, data accessed 2012—see Appendix A) (*Note*: 1 lb = 0.454 kg).

Snappers

Red Snapper

Mangrove Snapper

Vermilion Snapper

The snapper family, as defined in the Gulf of Mexico Reef Fish Fishery Management Plan (GMFMC 1981) (with amendments), includes a large number of species: Queen snapper (*Etelis oculatus*), Mutton snapper (*Lutjanus analis*), Blackfin snapper (*Lutjanus buccanella*), Red snapper (*Lutjanus campechanus*), Cubera snapper (*Lutjanus cyanopterus*), Gray (mangrove) snapper (*Lutjanus griseus*), Lane snapper (*Lutjanus synagris*), Yellowtail snapper (*Ocyurus chrysurus*), Wenchman snapper (*Pristipomoides aquilonaris*), and Vermilion snapper (*Rhomboplites aurorubens*). Based on this classification, the estimated catch (A + B1 + B2) of snappers by recreational anglers in the Gulf of Mexico (excluding Texas) in number of fish averaged nine million annually during 1990–2009 (Figure 10.82, left panel). Overall, the estimated number of fish generally fell throughout the 1990s but increased throughout the 2000s approaching 13 million in 2007 and 2008 before falling to about nine million in 2009. The vast majority of snapper are released with the average approaching 70 % (with no apparent trend) during the 20-year period of analysis. The high snapper release percentage reflects the substantial regulations imposed on the primary recreational snapper species in the northern Gulf—the red snapper. These regulations include size limits, bag limits, and extended closed seasons.[21] Overall, annual snapper catch in number of fish as a proportion of aggregate reef fish catch in number of fish generally ranged from about 55 to 70 % during the 20-year period of analysis. Combined snapper and grouper catches (A + B1 + B2) consistently accounted for at least 90 % of the aggregate reef fish catch in numbers and in selected years the combined total approached or exceeded 99 % of the aggregate reef fish catch.

---

[21] A complete list of recreational red snapper size limits, bag limits, season lengths, and recreational allocation/quotas through 2005 is given by Hood et al. 2007. More recent information on recreational red snapper regulations (and other reef fish species) can be found on the Gulf of Mexico Fishery Management Council website at: http://www.gulfcouncil.org/Beta/GMFMCWeb/downloads/ Summaries_of_the_Provisions_of_FMPs.pdf

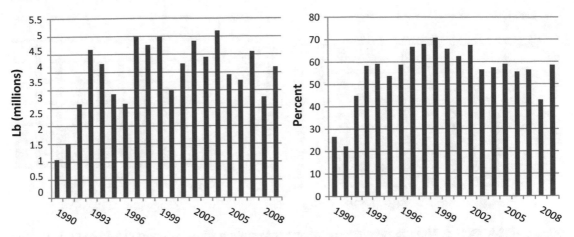

Figure 10.83. Gulf of Mexico recreational red snapper harvest (*left panel*) and as a percent of total recreational snapper harvest (right panel), 1990–2009 (NMFS FSD, data accessed 2012, with percentage calculations by authors—see Appendix A) (*Note*: 1 pound is equal to 0. 454 kilograms).

Aggregate snapper harvest by Gulf of Mexico recreational anglers (excluding Texas) in pounds (A + B1) is provided in Figure 10.82 (right panel). During the 20-year period ending in 2009, the estimated weight of aggregate snapper harvest averaged 6.1 million pounds annually with peak landings of over eight million pounds in 2004. As indicated, estimated harvest in pounds has been relatively stable since 2005.

Red snapper is by far the most popular recreational snapper species, particularly in the northern and western Gulf. Annual harvest of red snapper during 1990–2009, expressed on a weight basis, is given in Figure 10.83 (left panel), while red snapper landings in pounds as a percentage of total snapper landings are given in Figure 10.83 (right panel). As indicated, harvest of red snapper in pounds (A + B1) was very low in the early 1990s but increased from 1990 to 1993 at which point it equaled more than 4.5 million pounds. It declined again through 1996 but increased sharply in 1997. Much of the change in harvest during this period can be tied to changing regulations that are often tied to a changing red snapper population (as determined by stock assessments). Since the early 1990s, the red snapper fishery has been managed under a quota system with 51 % of the total quota given to the commercial sector and 49 % to the recreational sector. From 1996 to 2006, the recreational share of the quota equaled 4.5 million pounds a year. In 2007, the recreational quota was reduced to 3.2 million pounds and was reduced once again to 2.4 million pounds in 2008 where it remained in 2009. In an attempt to maintain the recreational harvest within its quota, GMFMC uses a combination of bag limits and fishing seasons though more often than not the final recreational harvest in a given year exceeds the quota. In 2009, for example, the recreational catch exceeded the quota by about 2.2 million pounds or almost 90 %.

Snapper angler targeting trips and snapper angler catch trips in relation to targeting trips are presented in Figure 10.84. Trips reporting the targeting of specific snapper species averaged 770,000 during the 10-year period of analysis (Figure 10.84, left panel). As was the case with grouper, snapper catch trips exceeded targeting trips by a wide margin during each of the 10 years considered (Figure 10.84, right panel). Overall, catch trips were generally about three times as large as target trips. Potential explanations for the large deviation between targeted snapper trips and catch trips were identified in the analysis of aggregate reef fish species.

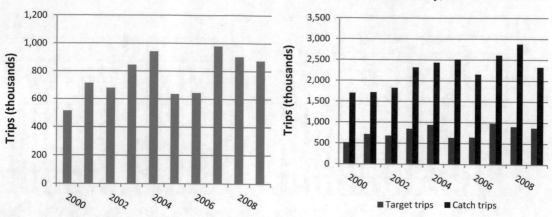

Figure 10.84. Gulf of Mexico recreational snapper targeting trips (*left panel*) and the relationship between catch trips and targeting trips (*right panel*), 2000–2009 (NMFS FSD, data accessed 2012, with targeting estimates calculated by authors—see Appendix A).

Coastal Pelagics

King Mackerel   Dolphinfish

Another group of species of high recreational interest in the Gulf of Mexico is that of coastal pelagic. There are five primary coastal pelagic species: king mackerel (*Scomberomorus cavalla*), Spanish mackerel (*Scomberomorus maculatus*), cobia (*Rachycentron canadum*), dolphinfish (*Coryphaena hippurus*), and wahoo (*Acanthocybium solandri*). Three of these species—king mackerel, Spanish mackerel, and cobia—are managed under the auspices of the GMFMC (cooperatively with the South Atlantic Management Council due to the migratory nature of the species). Recreational harvests of the other two species, while not subject to federal regulation in the Gulf, are subject to various state regulations.

Coastal Pelagics Managed by Gulf Council: Of the three coastal pelagic species managed by the GMFMC, king mackerel has historically received the most attention because it was considered to be heavily overfished. As such, the GMFMC established a total allowable catch (TAC) for the species for the 1986/1987 season (July 1 to June 30) equal to 2.9 million pounds of which 1.97 million pounds was allocated to the recreational sector.[22] With additional information and updated stock assessments, the TAC was subsequently increased to 4.25 million pounds for the 1990/1991 season (recreational allocation equal to 3.91 million pounds) and increased again to 7.8 million pounds for the 1992/1993 season (recreational allocation equal to 5.3 million pounds). In association with the recovery of the king mackerel stock, the TAC was increased again to 10.6 million pounds for the 1997/1998 season (7.2 million pound recreational quota) before being decreased marginally to 10.2 million pounds for the 2000/2001 season where it is currently maintained.

Estimated year-class strengths for the Gulf of Mexico king mackerel stock for the 1980–2000 period is presented in Figure 10.85 while the estimated biomass for the age 3+ proportion of the population (i.e., harvestable population) is given in Figure 10.86. While

---

[22] For a detailed description of historical rules and regulations, see: http://www.gulfcouncil.org/fishery_management_plans/migratory_pelagics_management.php.

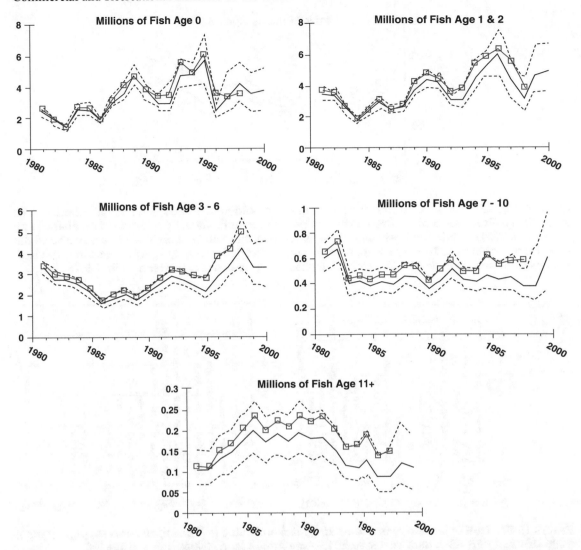

Figure 10.85. Estimated Gulf of Mexico king mackerel biomass trends, by cohort (Ortiz et al. 2002). *Note: Solid black line* represents population estimates, by age, based on analysis by Ortiz et al. (2002) while the *hashed black lines* represent the 80 % confidence interval around the population estimates. The red line represents population estimates, by age, provided in a previous assessment and provided by Ortiz et al. (2002) for purposes of comparison.

somewhat dated, the information presented in these figures clearly indicates a high amount of annual variability in recruitment, subsequent year-class strength, and harvestable biomass. As illustrated, years of strong recruitment (age 0 fish) map into larger year classes in subsequent years.

Large variations in year-class strengths and biomass, in conjunction with changing management measures, can result in large annual variations in recreational catches and harvests. This variation is evident in Figure 10.87, left panel (1990–2009 Gulf recreational catches in terms of numbers of fish) and Figure 10.87, right panel (annual recreational harvest in pounds). As indicated, the estimated Gulf recreational catch of king mackerel in numbers of fish has ranged from less than 400,000 in some years (2003, 2005, 2008) to more than 700,000 in other years (1991, 1996, 2006). Similarly, annual harvests (A + B1) have ranged from in excess of five

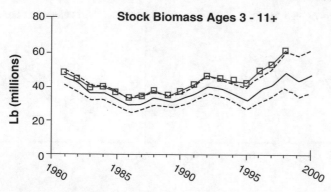

Figure 10.86. Estimated Gulf of Mexico king mackerel biomass trends, 3–11+ year cohorts (Ortiz et al. 2002). *Note*: Solid black line represents population estimates, by age, based on analysis by Ortiz et al. (2002) while the hashed black lines represent the 80 % confidence interval around the population estimates. The *red line* represents population estimates, by age, provided in a previous assessment and provided by Ortiz et al. (2002) for purposes of comparison. (1 lb = 0.454 kg).

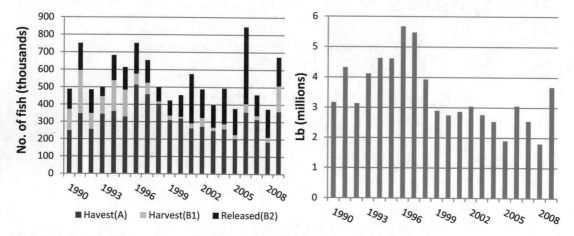

Figure 10.87. Gulf of Mexico recreational king mackerel catch (*left panel*) and harvest (*right panel*), 1990–2009 (NMFS FSD, data accessed 2012—see Appendix A) (*Note*: 1 lb = 0.454 kg).

million pounds in some years (1996, 1997) to less than two million pounds in other years (2005, 2008). In recent years, the recreational harvest has fallen far short of the approximately seven million pound recreational allocation. Almost 30 % of the king mackerel catch (A + B1 + B2) over the 1990–2009 period was reportedly released alive (B2) with the proportion approaching or exceeding 50 % in some years (e.g., 2001 and 2006).

Like king mackerel, the GMFMC management of Spanish mackerel has changed over time. In the early 1990s (1991/1992 season), the TAC was set at 8.6 million pounds with 43 % of this total allocated to the recreational sector. Bag limits varied by state with a bag limit of three fish per person per day in Texas, a bag limit of five fish per person per day in Florida, and a bag limit of ten fish per person per day in the remaining Gulf States (i.e., Alabama, Mississippi, and Louisiana). The Gulf Spanish mackerel TAC was subsequently reduced to seven million pounds for the 1996/1997 fishing year with the percentage allocation to the recreational sector remaining constant. While the recreational share remained constant, the TAC for the 1999/2000 season was increased to 9.1 million pounds and the recreational bag limit was increased to 15 fish per person per day across all Gulf States.

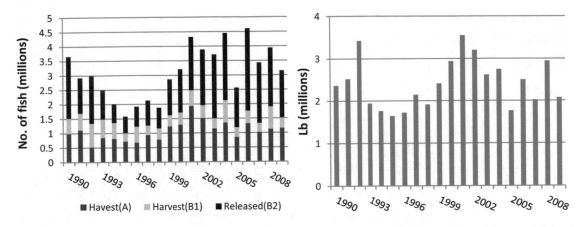

Figure 10.88. Gulf of Mexico recreational Spanish mackerel catch (*left panel*) and harvest (*right panel*), 1990–2009 (NMFS FSD, data accessed 2012—see Appendix A) (*Note*: 1 lb = 0.454 kg).

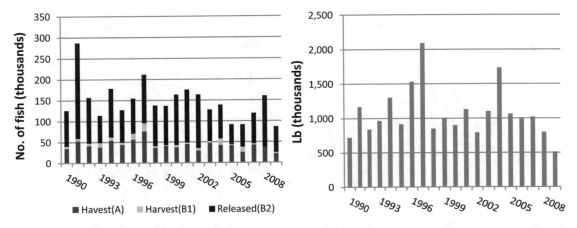

Figure 10.89. Gulf of Mexico recreational cobia catch (*left panel*) and harvest (*right panel*), 1990–2009 (NMFS FSD, data accessed 2012—see Appendix A) (*Note*: 1 lb = 0.454 kg).

Annual Gulf of Mexico recreational catches (A + B1 + B2) and harvests (A + B1) of Spanish mackerel for the 20-year period ending in 2009 are presented in Figure 10.88. Annual catch in numbers of fish is highly variable with a range from less than two million fish to more than four million fish. Similarly, the annual Spanish mackerel harvest during the 20-year period ranged from less than two million pounds in many years to 3.5 million pounds in 2001.

Cobia, as noted, is the third coastal pelagic species under the purview of the GMFMC. Annual catches and harvests of this fish are relatively limited (Figure 10.89) and are recreationally managed under a bag limit of two fish per person.

Coastal Pelagics Not Managed by Gulf Council: As noted, there are two coastal pelagic species that are harvested by recreational fishermen that are not managed by GMFMC. The first of these species, dolphinfish, are high spawners and the growth rate of the fish is very high. As such, GMFMC sees no need to manage this species (though there are some state regulations, including a Florida regulation of ten fish per person per day bag limit, not to exceed 80 per vessel per day). As indicated by the information in Figure 10.90 (left panel), annual dolphinfish catches expressed in numbers of fish can vary widely. During the period of analysis, annual catch of dolphinfish ranged from less than 400,000 fish (1992, 2002, and 2006) to more than

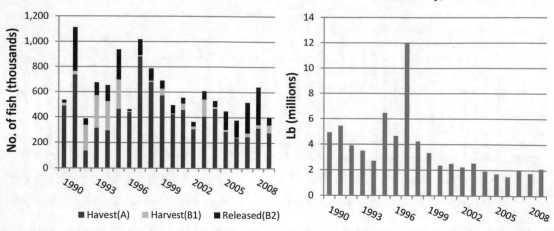

Figure 10.90. Gulf of Mexico recreational dolphinfish catch (*left panel*) and harvest (*right panel*), 1990–2009 (NMFS FSD, data accessed 2012—see Appendix A) (*Note*: 1 lb = 0.454 kg).

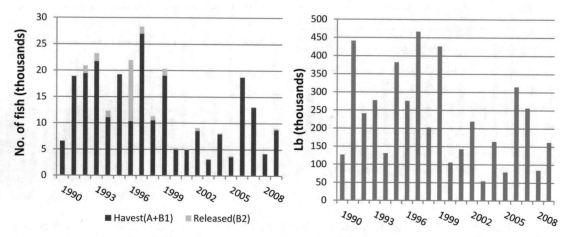

Figure 10.91. Gulf of Mexico recreational wahoo catch (*left panel*) and harvest (*right panel*), 1990–2009 (NMFS FSD, data accessed 2012—see Appendix A).

one million fish (1991 and 1997). Recreational harvest varied from a high of 12 million pounds in 1997 to less than two million pounds in a number of years (Figure 10.90, right panel). Since 2004, however, annual harvest has fallen in the relatively narrow 1.5 million pound to two million pound range.

The second coastal pelagic species not managed by the GMFMC is wahoo. Annual catches and harvests of this species are quite limited as suggested by the information in Figure 10.91.

Highly Migratory Pelagics

Swordfish

Yellowfin Tuna

Blue Marlin

A number of highly migratory species (HMS), including billfish, swordfish, tunas, and sharks, spend a portion of their respective life cycles in the Gulf of Mexico, and a high proportion of the trips targeting these species are in Florida. Estimating the recreational catch of these species is problematic for a number of reasons, and as stated in a recently completed report by the MRIP, *Highly Migratory Species Work Group & Florida Fish & Wildlife Commission* (Florida Highly Migratory Species Private Angler Survey Final Report) (FFWCC 2010):

> *Conducted by the state's Fish & Wildlife Research Institute for the past decade, the MRFSS has averaged 40,000 field intercepts annually. HMS-targeted trips comprise a small proportion of all recreational fishing trips combined, though, which makes them a `rare event' in any survey that is not directly targeting this specific segment of the recreational fishery. As a result, catch estimates for nearly all HMS species are highly imprecise due to typically low MRFSS intercept sample size.*

It is not just the low sampling rate that yields imprecise (and likely biased) estimates of catch and effort associated with HMS species, including design limitations (e.g., when sampling takes place) and coverage biases, in the MRFSS. With respect to design limitations, MRFSS intercept surveys occur only during the daytime. Completed HMS trips, given the type of fishing and the larger boats used in the activity, often arrive home at night and thus would not be sampled. With respect to coverage biases, MRFSS intercept sampling occurs only at accessible docks. A sizeable proportion of the larger recreational vessels that will, on occasion, target HMS species do not dock at public access sites. Finally, tournament caught fish are not included in MRFSS estimates because MRFSS does conduct intercept surveys at tournament sites. With these caveats in mind, a few MRFSS catch statistics are provided in this section. However, one should recognize the uncertainty with these estimates and make use of them accordingly. In an attempt to reduce potential biases and level of uncertainty, only 10-year averages are presented along with high and low catches during that interval. Estimated Gulf (excluding Texas) catch of tunas during the 2000–2009 period averaged approximately 100,000 fish per year with a low of 27,000 fish and a high of 223,000 fish. About 70 % of the total catch (A + B1 + B2) during the period was harvested (A + B1) which equaled about 1.3 million pounds per year.

Just over one million sharks per year were estimated to be caught in the Gulf (excluding Texas) during 2000–2009 with a range from about 800,000 to 1.3 million. Among HMS species, sharks have the least annual variation in catch. Of the estimated annual catch of 1.3 million sharks, less than 10 % were harvested (A + B1). The estimated number of marlin (blue [*Makaira nigricans*] and white [*Tetrapturus albidus*]) caught each year, on average, during 2000–2009 was less than 10,000; virtually none of these were kept. Finally, estimated annual swordfish catches were negligible and equal to zero in many years (likely because no intercepted anglers had caught a swordfish).

10.4.3 Recreational Activities at the State Level

10.4.3.1 Participation by State

10.4.3.1.1 Florida

The estimated annual number of marine anglers in Florida is, to a large extent, nonresident based (Figure 10.92). As discussed throughout the analysis by state, this factor alone tends to differentiate Florida from the other Gulf States. Overall, more than 50 % of the participants have historically been nonresidents with the proportion approaching 60 % during the late 1990s/early 2000s.

Furthermore, as indicated, west coast Florida participation, as measured by the number of marine anglers, increased significantly during the 1995–2009 period. Much of the growth occurred in 2000–2001 when the annual participants estimate increased to an average of about 4.2 million as compared to an average of about three million annually during

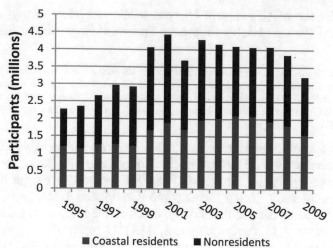

■ Coastal residents   ■ Nonresidents

Figure 10.92. Number of Florida (west coast) angler participants based on MRFSS, 1995–2009 (NMFS FSD, data accessed 2012—see Appendix A).

1998–1999. Most of the increase was the result of an increase in the number of nonresident participants.

The sharp increase in nonresident participation in the early 2000s corresponds with a sharp increase in estimated Florida visitors during this period. Specifically, from 1996 to 1998, the annual number of Florida visitors (estimated) fell in the relatively narrow range of 45 million to 49 million. In 1999, the estimate increased to 59 million and increased again to 73 million in 2000. Thereafter, the number gradually grew to almost 85 million in 2007 and 2008. In 2009, the number fell to 81 million. Corresponding to the decline in estimated visitors in 2009, the number of nonresident anglers fishing Florida waters fell from about two million in 2008 to 1.7 million in 2009. The estimated number of resident participants increased from about 1.2 million annually during the mid-1990s to more than two million by 2004 (Figure 10.92). The estimated number fell after 2006, and by 2009 it equaled only 1.6 million.

10.4.3.1.2 Alabama

The estimated number of marine angler participants in Alabama's waters is provided in Figure 10.93 for the 15-year period ending in 2009. As indicated, the total increased from less than 300,000 annually during the mid-1990s to more than 700,000 by the mid-2000s before falling to about 550,000 annually in 2008 and 2009 (the low participation rate in 2005 is undoubtedly the result of Hurricane Katrina and its impact on the marine-related infrastructure along the Alabama coast). With respect to residents (coastal and noncoastal), the estimated number of participants increased from about 180,000 annually during the mid-1990s to more than 400,000 during much of the 2000s before falling to 308,000 in 2008. The number increased to 357,000 in 2009. Overall, the average number of resident participants during 2005–2009, averaging 365,000 annually, was almost double the average number of resident participants in 1995–1999 (184,000). Coastal participants have represented about 60 % of total resident participants during the period of analysis.

The number of nonresident participants, who comprised roughly 40 % of the Alabama total during 1995–2009, grew from about 100,000 annually during the mid-1990s to 345,000 in 2004 but fell to only 160,000 in 2005, undoubtedly the influence of Hurricane Katrina. After recovering again to 320,000 in 2006, the number fell in each of the successive 3 years and

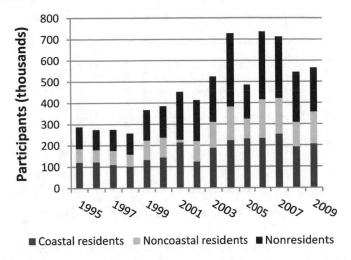

■ Coastal residents ■ Noncoastal residents ■ Nonresidents

Figure 10.93. Number of Alabama angler participants based on MRFSS, 1995–2009 (NMFS FSD, data accessed 2012—see Appendix A).

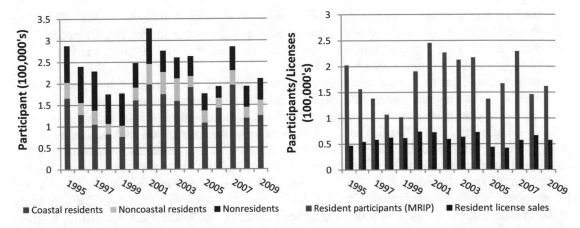

■ Coastal residents ■ Noncoastal residents ■ Nonresidents   ■ Resident participants (MRIP) ■ Resident license sales

Figure 10.94. Number of Mississippi angler participants based on MRFSS (*left panel*) and comparison of MRFSS resident participants and resident license sales (*right panel*), 1995–2009 (NMFS FSD, data accessed 2012—see Appendix A; Mississippi Department of Marine Resources, data provided by Buck Buchanan).

equaled only 210,000 in 2009. This figure represents the fewest nonresident participants since 2002 (excluding the 2005 hurricane year).

10.4.3.1.3 Mississippi

Among the Gulf States (excluding Texas), Mississippi had the fewest marine recreational participants with the estimated number over 1995–2009 averaging 235,000 annually. Unlike other Gulf States, Mississippi had no apparent growth in participation rate during the 15-year period being considered (Figure 10.94, left panel). Overall, the number of participants during 1995–1999, according the MRFSS estimates, averaged about 220,000 compared to about 210,000 during 2005–2009. While some of the decline in recent years reflects the impact on infrastructure associated with Hurricane Katrina, the downward trend in participation rate appears to have been in motion prior to 2005, though the large annual variation in number of participants makes this statement tenuous.

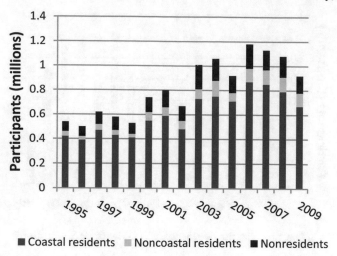

Figure 10.95. Number of Louisiana angler participants based on MRFSS, 1995–2009 (NMFS FSD, data accessed 2012—see Appendix A).

The number of resident participants, which ranged from 102,000 (1999) to 245,000 (2001), represented three-quarters of the total number of participants during the 15-year period. The estimated number of nonresident participants never exceeded the 100,000 mark and has hovered around 50,000 annually since 2006.

Information illustrating the relationship between the MRFSS annual resident participation estimates and the resident license sales is given in Figure 10.94 (right panel).[23] As indicated, the MRFSS estimates of resident participants exceed resident license sales by a wide margin with little or no apparent trend to this margin. There are a number of reasons that explain at least a portion of the differential. First, some residents are exempt from a license requirement (e.g., individuals under the age of 16; blind, paraplegic, or multiple amputee residents; member of the armed forces on active duty out of state on leave). Second, some residents required to have a license may risk fishing without it. Third, MRFSS estimates are given on a calendar year basis while license sales are given on a fiscal year basis. Finally, the MRFSS estimates are just that, estimates. As such, there is some amount of error associated with these estimates. While not illustrated, there is also a large differential between MRFSS estimates of nonresident participants and nonresident license sales. In 2009, for example, the MRFSS nonresident participation estimate was approximately 50,000 individuals compared to about 11,000 nonresident license sales (2009–2010 fiscal year).

10.4.3.1.4 Louisiana

Based on MRFSS estimates, the reported number of recreational participants fishing in Louisiana's waters increased from an average of 553,000 annually during 1995–1999 to more than one million during 2005–2009 (Figure 10.95). Coastal residents consistently represented 70–80 % of the total participants while noncoastal Louisiana residents represented 7–12 % of the total. Out-of-state participants represented 14–20 % of the total. Despite two 2005 hurricanes (Katrina and Rita) that damaged or destroyed a sizeable amount of the coastal fishing infrastructure and resulted in the dislocation of a sizeable portion of Louisiana's coastal

---

[23] While MRFSS estimates are on a calendar basis, license sales are on a June/May basis. For purposes of analysis, license sales for 1995–1996 were treated as 1995.

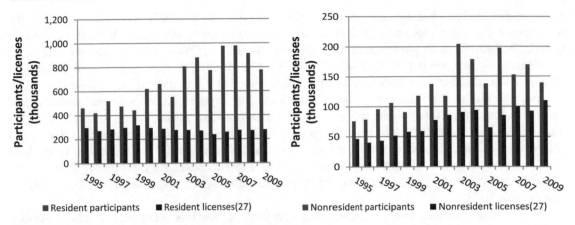

Figure 10.96. Comparison of MRFSS estimates of Louisiana participants and license sales to residents (*left panel*) and nonresidents (right panel), 1995–2009 (NMFS FSD, data accessed 2012 —see Appendix A; LDWF, http://www.wlf.louisiana.gov/licenses/statistics).

population, the 2006 MRFSS reported coastal participants (868,000) was the highest estimate during the 15-year period of analysis (ending in 2009) with the number subsequently falling in each of the 3 following years to about 670,000 in 2009. Similarly, the reported number of out-of-state participants totaled 198,000 in 2006 but then fell to 139,000 in 2009. In absolute terms, the 198,000 in 2006 ranked second only to the 204,000 (2003) reported out-of-state participants during the 15-year period of analysis ending in 2009. Combined, Louisiana coastal and noncoastal estimated number of resident participants peaked in 2006 and 2007 at about 977,000 and fell to 777,000 by 2009.

With a number of exceptions, a saltwater fishing license is required for fishing in Louisiana's waters. This requirement permits a comparison of MRFSS participation estimates with license sales. Some primary caveats are in order, however, before such a comparison is given. First, a saltwater fishing license is not required for individuals under the age of 16 or for senior citizens (currently, residents born prior to June 1, 1940, are exempt). Second, licenses expire on June 30 each year but can be purchased as early as June 1 of the previous year. Third, the MRFSS numbers are estimates extrapolated from a sample and there is likely to be some error associated with this extrapolation. Finally, the purchase of a license does not imply its use. With these caveats in mind, a comparison of annual resident saltwater licenses (privilege type 27) and MRFSS resident participation estimates is presented in Figure 10.96 (left panel). A similar comparison of nonresident saltwater license sales and MRFSS out-of-state participant estimates is given in Figure 10.96 (right panel). As indicated, without exception, the MRFSS estimates of resident participants exceed resident saltwater license sales by a wide margin, and this margin has tended to increase over time. Consistent with the significant differences in MRFSS resident (i.e., Louisiana) participant estimates and resident license sales, the correlation between the two time-series was negative and significant (−0.58).

While license gear 27 (Resident Saltwater Fishing) is the most common license required for participating in saltwater fishing activities in Louisiana, there are, in addition to those previously noted (e.g., individuals under the age of 16), several exceptions. These include residents with disabilities, active military residents, and the resident 3-day charter to name just some of the more common exemptions. While these residents are exempt from purchasing the Resident Saltwater Fishing License (27), other respective licenses are required (in many instances, these licenses are hunting and fishing combination licenses). In addition to these licenses, there is also a Resident Lifetime License and Senior Hunting/Fishing License. Inclusion of these licenses complicates analysis because some are issued only as hunting and fishing combinations and the number of Lifetime Licenses, by definition, will only increase over time. With these caveats noted, the Louisiana Department of Wildlife and Fisheries (LDWF) provides an estimate of all resident saltwater privileges. In 2009, this total was 395,000 (357,000 in 2007 and 383,000 in 2008). These numbers are still significantly below the MRFSS participation estimates.

As was the case in the comparison of resident participants, MRFSS annual estimates of out-of-state participants consistently exceeded the nonresident saltwater license sales, often by a substantial amount (Figure 10.96, right panel). The difference between the two time series as a proportion of the MRFSS estimates ranged from a low of about 20 % in 2009 to more than 50 % in several years. Despite the observed annual differences, the correlation between the MRFSS estimates of out-of-state participants and nonresident saltwater license sales was a respectable 0.78.

Yet a third source for examining participation is the U.S. Fish and Wildlife Service (USFWS) *National Survey of Fishing, Hunting, and Wildlife-Associated Recreation* (U.S. Department of Interior, Fish and Wildlife Service, and U.S. Department of Commerce, U.S. Census Bureau (2008)). This survey is conducted every 5 years with the 2006 survey being the most recently available one. For that year, the estimated number of saltwater participants (both fishing and spearing activities) in Louisiana of age 16 or greater totaled 289,000. Of this total, 248,000 were Louisiana residents and 42,000 were nonresidents. The estimated number of 2006 nonresident saltwater fishermen, however, is based on a small sample size and is thus subject to the standard caveats.[24] Estimates of number of participants from the 2001 survey yield a total of 504,000 of which 386,000 were residents and 118,000 were nonresidents. The 1996 survey yielded estimates of 255,000 resident participants (16 years of age or greater) and 90,000 nonresidents (small sample size) for a total participation estimate of 346,000. As indicated by the three USFWS surveys covering an 11-year period spanning from 1996 to 2006, the total estimated number of saltwater participants (16 years of age or more) increased from 346,000 in 1996 to 504,000 in 2001 but then fell by about 40–289,000 in 2006. Resident participants as a percentage of the total ranged from 74 % in 1996 to 86 % in 2006.

The MRFSS/MRIP survey captures only recreational finfish fishing activities and, as such, does not capture all recreational activities of potential relevance. Recreational shellfish/molluscan fishing activities also occur in Louisiana's waters. State license sales can assist in portraying these activities. One of the more common recreational shellfish fishing activities is that of crabbing. To participate in this activity, residents and nonresidents must purchase a crab trap license that allows them to employ up to 10 traps (license numbers 70/83 and 71/84). Annual sales of crab trap licenses for the period 1995–2009 are presented in Figure 10.97 (left panel). As indicated, total sales advanced from about 2800 in 1995 to more than 4,000 during each year of the 1990s and have exceeded 5,000 in 2001, 2008, and 2009. More recently, a lifetime crab trap license has been instituted by LDWF. Issuance of these licenses has increased from 1 in 2004 to 39 in 2009. Less than 1 % of the recreational crab trap licenses are issued to nonresidents.

---

[24] While the survey used to conduct the 2006 *U.S. Fish and Wildlife Service Fishing, Hunting, and Wildlife-Associated Activities* did not elicit information on individuals less than 16 years of age, the screening phase of the survey, which was initiated in April 2006, did collect information on persons 6–15 years of age and their activities in 2005. While a number of caveats associated with these estimates are given—not the least of which being potential long-term recall bias and the household participant not being the 6–15-year-old participants—an estimate of 133,000 6–15-year-old resident fishing participants in 2006 was made. Note that this estimate includes both fresh-water and saltwater participants.

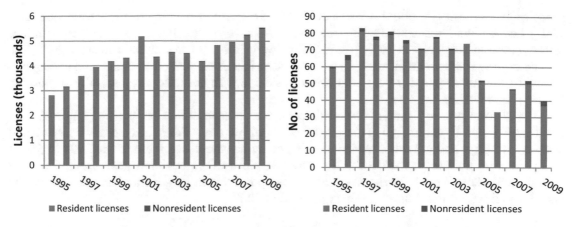

Figure 10.97. Louisiana recreational crab licenses (*left panel*) and recreational oyster licenses (*right panel*), 1995–2009 (Louisiana Department of Wildlife and Fisheries: http://www.wlf.louisiana. gov/licenses/statistics).

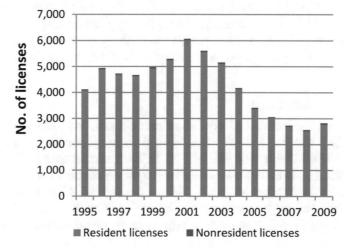

Figure 10.98. Louisiana recreational shrimp licenses (less than 16-foot trawl), 1995–2009 (LDWF: http://www.wlf.louisiana.gov/licenses/statistics).

Recreational oystering also requires a specific gear license (in addition, a basic fishing license and a saltwater fishing license is required for the recreational tonging of oysters). This license (74/89 for residents and 75/90 for nonresidents) allows one to use tongs for the recreational take of oysters. A separate tonging license is required for each tong being used, and harvesters are limited to two sacks per person per day. As indicated (Figure 10.97, right panel), the number of recreational gear licenses associated with the recreational harvest of oysters is relatively limited (never exceeding the low 80s).

Another Louisiana recreational shellfish fishing activity is that of shrimping. There are two primary gear licenses associated with this activity differing, primarily, on the size of allowable trawl and take/possession limits (in addition, the basic fishing license and saltwater fishing license are required for recreational shrimping). The first license (72/87 for residents and 73/88 for nonresidents) allows for a single trawl (up to 16 ft in length) and boat limit (not to exceed 100 lb per day heads on). As indicated by the information in Figure 10.98, sales of these licenses peaked at just over 6,000 in 2001 and thereafter declined to a low of about 2,600 in 2008 before marginally increasing to about 2,800 in 2009. More recently, a lifetime resident trawl (less than 16 ft) license has been instituted by the state. Issuance of this license to residents is minor totaling zero in 2004, one in 2005, five in 2006, five in 2007, seven in 2008, and six in 2009. The overwhelming majority of licenses giving one the privilege of using a 16-foot trawl in conjunction with the 100 lb per day boat limit are to residents with nonresidents generally accounting for less than 2 % of the total.

The second primary recreational shrimping gear license allows for a trawl up to 25 ft in length and catch limits of 250 lb (heads on) per boat per day. This is a relatively new initiated license category, and annual issuance of this license has increased from 157 in 2004 to 520 in 2009. As was the case with the smaller trawl issuance of nonresident gear licenses for the recreational use of trawls in excess of 16 ft, the up to 25 ft in length trawl licenses tend to be very limited with a maximum of 8 being issued in 2008.

10.4.3.2   Angler Trips by State

Florida, as previously discussed, dominated Gulf of Mexico marine angler participation (excluding Texas). As such, one would expect a very high proportion of the angler trips to be Florida based. In fact, Florida accounted for more than 70 % of the total estimated angler trips throughout the Gulf of Mexico (excluding Texas) during 1995–2009 (Figure 10.99, left panel). Louisiana ranked second, similar to its participation ranking, accounting for 17 % of the angler trips. Alabama and Mississippi combined contributed 12 % of the total number of trips during 1995–2009.

10.4.3.2.1   Florida (West Coast)

In association with the increasing number of participants, the estimated number of marine trips taken in Florida's waters (west coast) increased from an average of 12.3 million annually during 1995–1999 to 17 million during 2005–2009. In 2009, an estimated 15.2 million angler trips were made in Florida (west coast) waters, which represented about a 12 % decline from the 2004 peak of 17.8 million trips. Overall, the proportion of Gulf trips (excluding Texas) represented by Florida fell in the narrow range of 69–73 % with no apparent long-term trend. On average, 50 % of the Florida-based angler trips were in inland waters during 1995–2009 with

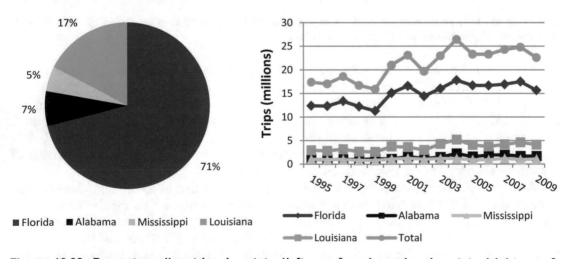

Figure 10.99. Percent angling trips by state (*left panel*) and number by state (*right panel*), 1995–2009 (NMFS FSD, data accessed 2012, with percentage calculations by authors—see Appendix A).

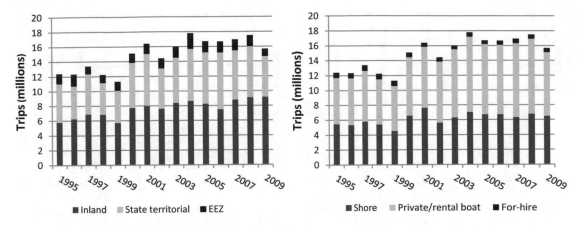

Figure 10.100. Recreational fishing trips in Florida waters by area fished (*left panel*) and by mode (*right panel*), 1995–2009 (NMFS FSD, data accessed 2012—see Appendix A).

no discernible long-run change in this proportion (Figure 10.100, left panel). Another 40 % of the trips were in state territorial waters (out to 9 nautical mi [16.7 km] in the case of Florida). About 10 % of the trips occurred in the federal waters (outside 9 nautical mi [16.7 km]).

Overall, the private/rental boat mode represented 50–60 % of the total Florida-based angler trips during 1995–2009 while the shore-based mode generally represented 45–50 % of the total angler trips (Figure 10.100, right panel). The share of total trips attributable to the for-hire sector never exceeded 6 % and in some years was as low as 3 %. Savolainen et al. (2012) estimate that the 2009 number of for-hire operations in Florida (west coast) totaled 1372 and comprised 118 head boat operations, 473 charter operations, and 781 guide boat operations. The head boat operations made, on average, 115 trips in 2009 while the charter operations and guide operations made close to 100 each. Savolainen et al. (2012) further report that the average net income to the owner of head boat operations in Florida equaled $65,000 with the owners of charter operations and guide operations netting $21,000 and $28,000, respectively.

10.4.3.2.2   Alabama

The number of marine angler trips in Alabama waters averaged 1.45 million annually during 1995–2009 and ranged from less than one million in several years (1995, 1996, 1998) to 2.3 million trips in 2004 (Figure 10.99, right panel). Since 2004, however, the estimated number of trips has fallen with the 2009 estimate equaling 1.7 million. Overall, the Gulf proportion of trips taken in Alabama waters increased during the period of analysis from 5 to 6 % during the mid-1990s to 7–8 % in more recent years.

During the mid-1990s, less than 30 % of the Alabama angler trips were in the inshore waters but by the late 2000s, this percentage had increased to more than 60 %. Conversely, whereas more than 30 % of the angler trips were in the EEZ waters in some years during the mid-1990s, this percentage has fallen sharply in later years and has averaged less than 10 % since 2006.

10.4.3.2.3   Mississippi

Angler trips in Mississippi's waters, according to MRFSS data, averaged just over one million per year with no discernible long-term trend. In 1995–1999, for example, the number of trips averaged about 925,000 annually. This figure increased only marginally to about one million annually during 2005–2009. Among the four Gulf States considered in detail in this section, Mississippi is the only one where no change is evident.

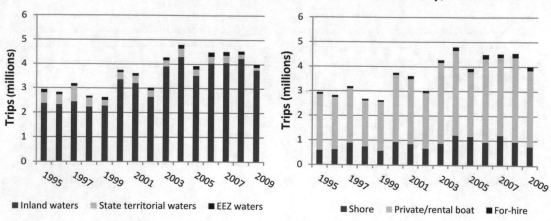

Figure 10.101. Recreational fishing trips in Louisiana waters by area fished (*left panel*) and by mode (*right panel*), 1995–2009 (NMFS FSD, data accessed 2012—see Appendix A).

10.4.3.2.4   Louisiana

During the 15-year period ending in 2009, the reported annual number of trips in Louisiana (salt) waters has ranged from 2.6 million (1999) to 5.2 million (2004) and has averaged 3.7 million (Figure 10.99, right panel). While exhibiting significant interyear variation, the number of trips, in general, exhibited an upward trend during the period of analysis. Despite two hurricanes in 2005 (Katrina and Rita) that impacted a significant portion of Louisiana's coastal infrastructure, the reported number of trips in that year fell by only about 20 % when compared to 2004 (which was the record year), and by 2006, the number of trips approached that observed pre-hurricanes. The observed number of trips in 2009 equaled 4.1 million, which represented about a 10 % decline in relation to the previous year.

Louisiana's recreational fishery is overwhelmingly inland in nature. Since 1995, recreational trips in inland waters averaged 3.2 million annually—more than 85 % of the total number of recreational trips (Figure 10.101, left panel). The percentage of trips in state territorial waters averaged 312,000 annually during 1995–2009 and ranged from a high of 20 % in 1997 to less than 4 % (2008 and 2009). In general, the proportion of total trips occurring in territorial waters has fallen during the period of analysis with a concomitant increase in the percentage of trips occurring in inland waters. During the period of analyses, the reported number of recreational trips in federal waters averaged 120,000 annually representing approximately 3 % of the total number of trips.

As indicated by the information in Figure 10.101 (right panel), 20–30 % of the total annual trips are shore based. Another 2–4 % of the trips used for-hire services. The majority of the trips, almost 75 % of the total, represent use of private/rental boats.

With respect to the Louisiana for-hire sector, Savolainen et al. (2012) estimate that the population of for-hire boats (more specifically, captains) equaled 681 in 2009 and that the number has increased substantially since the 1990s. Of this total, 100 were charter boats, 575 were guide boats, and the remaining 6 were head boats. The average number of trips made by charter vessels equaled 75, and net income to owners from charter boat operations averaged $40,000. Guide boat operations, by comparison, averaged 71 trips with net income accruing to the owner estimated at $28,000 on average.

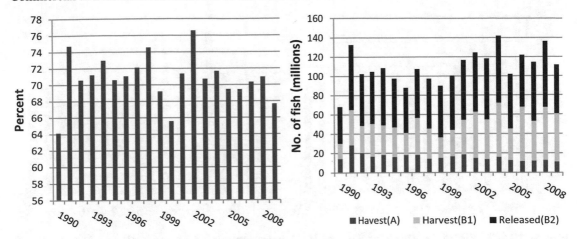

Figure 10.102. Florida's share of the Gulf aggregate recreational catch (*left panel*) and Florida aggregate recreational catch (*right panel*), 1990–2009 (NMFS FSD, data accessed 2012, with percentage calculations by authors—see Appendix A).

10.4.3.3 Catch/Harvest by State

10.4.3.3.1 Florida

Aggregate Catch/Harvest

Florida, as previously discussed, dominates the Gulf (excluding Texas) in terms of number of participants and trips. As such, it should come as no surprise that Florida also accounts for the majority of catch. Florida's share of Gulf catch in numbers of fish and harvest in pounds are given in Figure 10.102 (left panel) for the 1995–2009 period. Florida has continually maintained a 65–75 % share of Gulf catch in numbers of fish during the 15 years of analysis. This percentage of Gulf catch corresponds with the 70 % of Gulf trips (excluding Texas) taken in Florida waters. In absolute numbers, the estimated number of fish caught increased from an average of about 100 million annually during the mid-1990s to an average of about 120 million during the late 2000s, yielding a 15-year period average of about 110 million fish (Figure 10.102, right panel).

While Florida consistently accounted for about 70 % of the Gulf recreational catch in number of fish (Figure 10.102, left panel), its share, in terms of pounds harvested (A + B1), averaged only about 50 % of the Gulf total (1995–2009) during the 15-year analysis period, and its share appears to have fallen marginally since the late 1990s (Figure 10.103, left panel). In absolute numbers, Florida's annual landings in pounds of fish (A + B1), with few exceptions, have fallen in the 30–40 million pound range (Figure 10.103, right panel) with no apparent upward trend even though the number of fish caught does appear to have increased (Figure 10.102, right panel). This would suggest an increasing rate of releases or smaller fish being harvested and kept.

The reason that Florida's share of the total Gulf catch in terms of numbers exceeds its share of total Gulf harvest by weight is the inclusion of bait fish in the catch (A + B1 + B2) and harvest estimates (A + B1). Florida's catch (A + B1 + B2) and harvest (A + B1) of the top 25 species for selected years (1995, 2000, 2005, and 2009) in numbers of fish are given in Tables 10.16, 10.17, 10.18, and 10.19. As indicated, the overwhelming proportion of the state's catch/harvest in numbers of fish is represented by species generally used as baitfish. Comparison of the 2009 Florida catch/harvest estimates with those for the Gulf for the same year

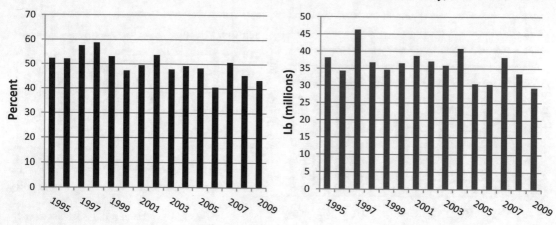

Figure 10.103. Florida's share of Gulf of Mexico recreational harvest (*left panel*) and pounds harvested (*right panel*), 1995–2009 (NMFS FSD, data accessed 2012, with percentage calculations by authors—see Appendix A) (*Note*: 1 lb = 0.454 kg).

(Tables 10.12, 10.13, 10.14, and 10.15) clearly points to the fact that virtually all of the reported Gulf of Mexico recreational baitfish catch/harvest is Florida based.

In general, there was a large amount of consistency between most commonly caught (A + B1 + B2) recreational species in Florida when comparing 1995 and 2009 (Tables 10.16 and 10.19), with no observed changes in ranking among the top five species. Furthermore, while large increases in the catches of scaled sardines and Atlantic thread herring were observed between the 2005 and 2009 periods, catches of the other three top five species (pinfish, spotted seatrout, and gray snapper) remained relatively constant. With respect to harvest (A + B1), four of the five most commonly harvested species in 1995 remained among the five most commonly harvested species in 2009 with only white grunt (*Haemulon plumierii*) falling out of the ranking and Spanish mackerel replacing it. This change in positioning reflects both a sharp decline in the estimated harvest of white grunt (from 2.8 million fish in 1995 to 1.2 million fish in 2009) and a doubling in the harvest of Spanish mackerel (from 658,000 fish to 1.4 million fish).

Analysis by Inshore and Offshore Species

Inshore Species: As in other Gulf States, the two most desirable species targeted by Florida participants in the inshore waters are spotted seatrout and red drum. Currently in Florida, the bag limit on spotted seatrout is five fish and the bag limit on red drum is one fish.

The estimated catch of spotted seatrout in Florida waters for the 1995–2009 period expressed in number of fish (A + B1 + B2) is given in Figure 10.104 (left panel), and harvest in pounds (A + B1) is given in Figure 10.104 (right panel). In terms of number of fish, catch often exceeds the ten million mark and in some years exceeds 12 million. While the number of spotted seatrout caught in Florida's waters is large, the vast majority of these fish are released alive (Figure 10.104, left panel). This finding is also confirmed by the information presented in Tables 10.16, 10.17, 10.18, and 10.19. Since 1995, approximately 85 % of the spotted seatrout caught in Florida waters has been released alive (i.e., B2) with no apparent long-run trend. Pounds harvested generally falls between two and three million pounds. The estimated weight associated with the Florida recreational spotted seatrout harvest remained virtually constant between 1995 and 2009 (with the exception of an increase in 2000) at an average of 1.45 lb per fish (Tables 10.16, 10.17, 10.18, and 10.19).

Table 10.16 25 Most Frequently Caught (Left Panel) and Harvested (Right Panel) Species by Florida Recreational Anglers, 1995

Species Name	Number of Fish Caught (A + B1 + B2)	Species Name	Number of Fish Harvested (A + B1)	Mean Weight (lbs)
Scaled Sardine	19,218,450	Scaled Sardine	15,254,950	0.042
Pinfish	11,568,160	Pinfish	15,254,950	0.185
Spotted Seatrout	8,341,695	White Grunt	15,254,950	0.696
White Grunt	5,387,420	Atlantic Thread Herring	2,637,242	0.060
Atlantic Thread Herring	3,287,447	Spotted Seatrout	1,831,312	1.452
Gray Snapper	3,161,517	Sand Seatrout	1,240,362	0.535
Hardhead Catfish	2,745,700	Sheepshead	1,237,207	1.835
Gag	2,244,954	Spanish Sardine	890,074	NA
Black Sea Bass	2,168,769	Gray Snapper	757,900	1.264
Sheepshead	2,095,996	Striped Anchovy	660,456	0.020
Red Grouper	1,951,298	Spanish Mackerel	657,562	1.690
Crevalle Jack	1,848,913	Dolphin	650,211	8.522
Pigfish	1,612,672	Sand Perch	614,892	0.326
Red Drum	1,453,207	Pigfish	574,835	0.413
Gafftopsail Catfish	1,415,476	Blue Runner	524,440	0.948
Sand Seatrout	1,410,338	Black Sea Bass	479,187	0.591
Silver Perch	1,257,550	Striped Mullet	468,509	1.056
Sand Perch	1,175,568	Vermilion Snapper	452,154	0.762
Blue Runner	1,065,211	Silver Perch	404,995	0.178
Spanish Mackerel	1,010,330	King Mackerel	403,774	9.164
Yellowtail Snapper	966,760	Seatrout Genus	395,104	NA
Spanish Sardine	890,074	Gag	390,383	6.125
Dolphin	888,518	Southern Kingfish	382,604	0.601
Seatrout Genus	837,843	Yellowtail Snapper	351,082	1.212
Grunt Genus	785,373	Round Scad	320,876	0.436

Source: NMFS FSD, data accessed 2012, with calculations by authors–see Appendix A. *Note*: 1 lb = 0.454 kg.

The catch of red drum expressed in numbers of fish (A + B1 + B2) is given in Figure 10.105 (left panel) while pounds harvested is given in Figure 10.105 (right panel). As indicated, catch of red drum in Florida waters generally fluctuated between 1.5 and 2 million fish annually until 2003 at which point catch increased significantly. Release rate equaled about 1.9 million fish per year based on total catch of 2.2 million. This indicates a release proportion of 85 % that has not changed appreciably during the period of consideration.

Offshore Species: Florida's recreational reef fish catch is large, averaging an estimated 13 million fish per year from 1995 to 2009 (Figure 10.106, left panel). This represented about 85 % of the total Gulf recreational reef fish catch (excluding Texas) in numbers of fish with the annual

Table 10.17 25 Most Frequently Caught (Left Panel) and Harvested (Right Panel) Species by Florida Recreational Anglers, 2000

Species Name	Number of Fish Caught (A + B1 + B2)	Species Name	Number of Fish Harvested (A + B1)	Mean Weight (lbs)
Scaled Sardine	17,583,932	Scaled Sardine	14,713,726	0.022
Spotted Seatrout	10,484,841	Pinfish	5,015,732	0.385
Pinfish	8,545,587	Atlantic Thread Herring	2,316,911	0.082
White Grunt	4,179,199	White Grunt	1,739,314	0.875
Gray Snapper	3,727,119	Sand Seatrout	1,620,390	0.546
Black Sea Bass	3,382,680	Spotted Seatrout	1,469,697	1.839
Ladyfish	2,758,987	Blue Runner	1,272,321	1.208
Atlantic Thread Herring	2,727,189	Spanish Mackerel	1,180,062	1.860
Spanish Mackerel	2,303,801	Striped Mullet	966,378	1.436
Blue Runner	2,262,193	Spanish Sardine	816,465	0.019
Sand Seatrout	2,189,286	Round Scad	775,122	NA
Hardhead Catfish	2,165,642	Sheepshead	697,513	2.090
Crevalle Jack	2,041,748	Gray Snapper	630,192	1.395
Sheepshead	1,938,982	Gulf Menhaden	579,657	0.223
Red Grouper	1,862,001	Black Sea Bass	548,872	0.815
Gag	1,768,555	Gag	527,939	6.732
Gafftopsail Catfish	1,741,746	Pigfish	432,774	0.314
Red Drum	1,633,350	Atlantic Croaker	404,930	0.484
Striped Mullet	1,064,127	Southern Kingfish	321,841	0.639
Pigfish	1,056,668	Silver Perch	315,476	0.199
Round Scad	915,464	Red Drum	310,044	4.614
Spanish Sardine	895,075	Striped Killifish	296,217	NA
Gulf Menhaden	702,632	White Mullet	250,926	0.729
Silver Perch	695,093	Menhaden Genus	219,017	NA
Common Snook	667,738	Red Grouper	217,853	7.179
Grunt Genus	785,373	Round Scad	320,876	0.436

Source: NMFS FSD, data accessed 2012, with calculations by authors–see Appendix A. *Note*: 1 lb = 0.454 kg.

proportion generally ranging from about 80 to 90 % (Figure 10.106, right panel). As indicated, reef fish catch in numbers of fish along Florida's west coast appears to have increased since the early 2000s. This increase closely mimics the increase in Gulf grouper catch (Figure 10.77), which is primarily a Florida fishery. Overall, the catch of reef fish species accounted for slightly more than 10 % of the total estimated angler catch (numbers of fish) of fish in Florida's waters during 1995–2009.

Estimated pounds of reef fish harvested from Florida's waters (A + B1) by recreational anglers generally ranged from about 8 to 11 million pounds, with exceptions, and averaged ten million pounds annually during the 15-year period of analysis (Figure 10.107, left panel). While variable on a year-to-year basis, no discernible trend in harvested pounds is evident. Overall,

Table 10.18 25 Most Frequently Caught (Left Panel) and Harvested (Right Panel) Species by Florida Recreational Anglers, 2005

Species Name	Number of Fish Caught (A + B1 + B2)	Species Name	Number of Fish Harvested (A + B1)	Mean Weight (lbs)
Scaled Sardine	18,055,543	Scaled Sardine	17,335,305	0.052
Spotted Seatrout	13,694,784	Pinfish	5,645,022	0.295
Pinfish	8,764,639	Atlantic Thread Herring	3,915,545	0.016
Gray Snapper	5,360,548	Spotted Seatrout	1,980,357	1.549
Atlantic Thread Herring	4,070,662	White Grunt	1,687,555	0.860
Red Drum	3,590,782	Spanish Mackerel	1,100,222	1.857
White Grunt	3,372,101	Sheepshead	1,050,108	2.121
Ladyfish	3,285,880	Gray Snapper	931,242	1.451
Sheepshead	2,869,202	Striped Mullet	806,221	1.479
Gag	2,716,307	White Mullet	722,388	0.685
Spanish Mackerel	2,314,955	Gulf Menhaden	560,549	0.000
Hardhead Catfish	2,134,717	Red Drum	501,367	3.813
Red Snapper	1,665,642	Red Snapper	491,229	3.755
Black Sea Bass	1,612,855	Gag	490,192	6.818
Red Grouper	1,453,218	Sand Perch	460,951	0.431
Gafftopsail Catfish	1,388,348	Southern Kingfish	413,214	0.674
Common Snook	1,362,106	Menhaden Genus	390,512	0.000
Crevalle Jack	1,320,171	Sand Seatrout	370,992	0.634
Striped Mullet	919,147	Round Scad	336,453	0.000
Blue Runner	871,509	Blue Runner	333,292	1.055
Sand Perch	836,425	Lane Snapper	332,042	0.753
White Mullet	732,364	Dolphin	285,999	5.951
Gulf Menhaden	598,936	Black Sea Bass	285,543	0.804
Lane Snapper	582,930	Gulf Kingfish	273,497	0.764
Southern Kingfish	553,397	Spanish Sardine	228,028	0.220

Source: NMFS FSD, data accessed 2012, with calculations by authors–see Appendix A. Note: 1 lb = 0.454 kg.

Florida's share of the annual Gulf recreational reef fish harvest (excluding Texas) generally ranged from about 60 to 80 % when evaluated on a poundage basis, though during the latest 2 years of analysis, its share fell to about 50 % (Figure 10.107, right panel).

Groupers: Groupers are strongly identified with Florida. Overall, more than 95 % of the grouper catch in terms of either number of fish (A + B1 + B2) or pounds harvested (A + B1) occurs in Florida waters and, hence, the various figures provided for the Gulf are, for all intents and purposes, equivalent to what one would see for Florida. Overall, since 1995 groupers have represented about 40 % of the reef fish caught by recreational anglers in Florida's waters in terms of number of fish (A + B1 + B2) and almost 50 % in terms of pounds landed (A + B1).

Table 10.19 25 Most Frequently Caught (Left Panel) and Harvested (Right Panel) Species by Florida Recreational Anglers, 2009

Species Name	Number of Fish Caught (A + B1 + B2)	Species Name	Number of Fish Harvested (A + B1)	Mean Weight (lbs)
Scaled Sardine	31,430,678	Scaled Sardine	29,938,133	0.028
Pinfish	9,413,639	Atlantic Thread Herring	6,804,819	0.108
Spotted Seatrout	9,032,362	Pinfish	5,384,932	0.106
Atlantic Thread Herring	7,472,772	Spanish Mackerel	1,392,399	1.620
Gray Snapper	3,998,388	Spotted Seatrout	1,370,634	1.488
Ladyfish	3,043,921	White Grunt	1,206,086	0.909
Spanish Mackerel	2,938,091	Gray Snapper	1,176,301	1.233
Gag	2,728,998	Round Scad	1,096,334	NA
Red Grouper	2,472,120	Ballyhoo	1,087,375	NA
White Grunt	2,241,227	Sand Seatrout	889,866	0.661
Hardhead Catfish	1,957,552	Blue Runner	687,199	0.948
Red Snapper	1,868,467	Sheepshead	681,263	2.193
Red Drum	1,566,251	Herring Family	647,389	NA
Sheepshead	1,466,501	Red Snapper	545,333	3.637
Blue Runner	1,446,946	Striped Mullet	490,298	1.492
Sand Seatrout	1,333,096	King Mackerel	452,892	7.101
Crevalle Jack	1,301,344	Vermilion Snapper	345,683	0.976
Round Scad	1,128,681	Gulf Menhaden	334,964	NA
Ballyhoo	1,088,172	Dolphin	334,374	7.304
Black Sea Bass	977,919	Unidentified Fish	271,498	NA
Gafftopsail Catfish	792,219	Pigfish	232,901	0.355
Pigfish	751,446	Red Drum	225,380	4.623
Common Snook	711,391	Gag	202,659	6.835
Bluefish	674,052	Ladyfish	200,634	1.208
Striped Mullet	674,022	Lane Snapper	192,094	0.882

Source: NMFS FSD, data accessed 2012, with calculations by authors—see Appendix A. *Note*: 1 lb = 0.454 kg.

Snappers

Yellowtail Snapper

The catch of snappers in Florida waters by marine recreational anglers expressed in number of fish and as a percentage of Gulf total snapper catch (excluding Texas) is given in Figure 10.108. When examined on a yearly basis, catch consistently ranged from about 6 million to 7 million fish per year during the mid-1990s until 2003 when catch increased sharply. The average estimated catch since 2003 is nine million fish annually (left panel). On a share basis,

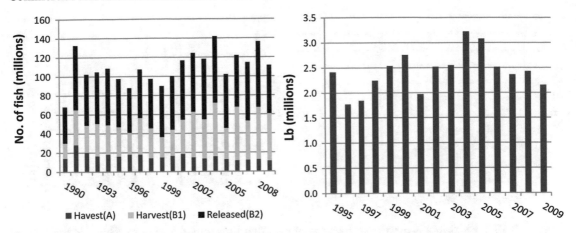

Figure 10.104. Florida recreational catch of spotted seatrout (*left panel*) and pounds harvested (*right panel*), 1995–2009 (NMFS FSD, data accessed 2012—see Appendix A) (*Note*: 1 lb = 0.454 kg).

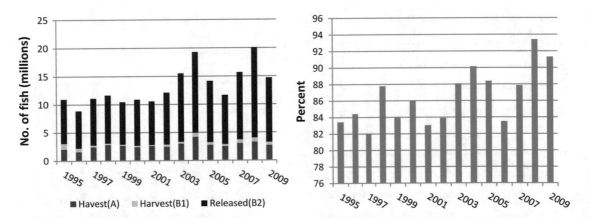

Figure 10.105. Florida recreational catch of red drum (*left panel*) and pounds harvested (*right panel*), 1995–2009 (NMFS FSD, data accessed 2012—see Appendix A) (*Note*: 1 lb = 0.454 kg).

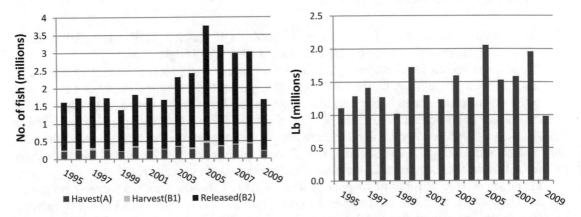

Figure 10.106. Florida's recreational catch of reef fish (*left panel*) and as a percentage of the Gulf of Mexico recreational reef fish catch, 1995–2009 (NMFS FSD, data accessed 2012, with percentage calculations by authors—see Appendix A).

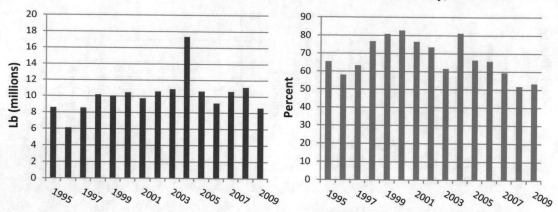

Figure 10.107. Florida's recreational harvest of reef fish (*left panel*) and percentage of the Gulf of Mexico recreational reef fish harvest (*right panel*), 1995–2009 (NMFS FSD, data accessed 2012, with percentage calculations by authors—see Appendix A) (*Note*: 1 lb = 0.454 kg).

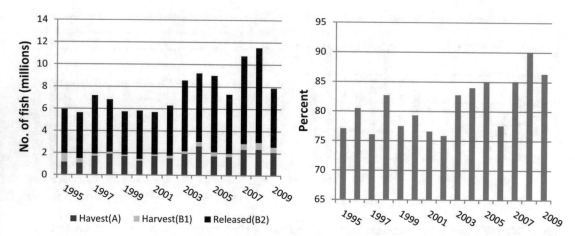

Figure 10.108. Florida's recreational catch of snappers (*left panel*) and as a percentage of the Gulf of Mexico recreational snapper catch, 1995–2009 (NMFS FSD, data accessed 2012, with percentage calculations by authors—see Appendix A).

Florida generally accounted for between 75 and 85 % of the Gulf recreational catch (A + B1 + B2) expressed in numbers of fish during 1995–2009 (right panel).

The annual recreational harvest (A + B1) of snappers from Florida waters expressed in pounds and as a percentage of the Gulf harvest is presented in Figure 10.109. As indicated, harvested pounds have been highly variable on an annual basis with a range from less than 2.5 million pounds to more than 5.5 million pounds. Since 1995, pounds of snappers harvested from Florida waters have averaged 3.8 million annually which represented almost 60 % of the total Gulf recreational snapper harvest (excluding Texas). Furthermore, as indicated, Florida's share of the Gulf's total closely mimics the absolute Florida harvest in pounds.

There are a large number of snapper species caught in Florida, with red snapper contributing the largest share of the total harvested poundage (A + B1). There has been a significant increase in the harvest of this species since 1995 (Figure 10.110, left panel) and this increase is generally attributed to growth in the stock and expansion of the stock from the northern Gulf to the Florida panhandle and down through the eastern Gulf.

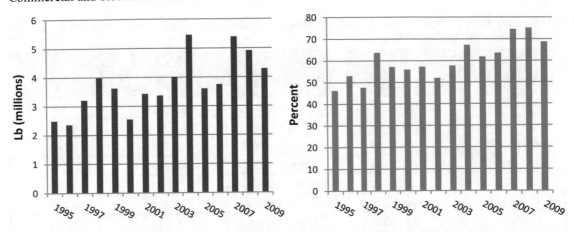

Figure 10.109. Florida's recreational harvest of snappers (*left panel*) and as a percentage of the Gulf of Mexico recreational snapper harvest, 1995–2009 (NMFS FSD, data accessed 2012, with percentage calculations by authors—see Appendix A) (*Note*: 1 lb = 0.454 kg).

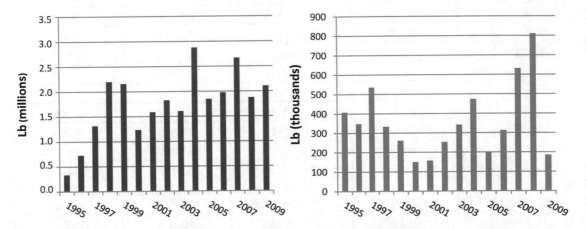

Figure 10.110. Florida's recreational harvest of red snapper (*left panel*) and yellowtail snapper (*right panel*), 1995–2009 (NMFS FSD, data accessed 2012—see Appendix A) (*Note*: 1 lb = 0.454 kg).

Yellowtail snapper is harvested largely in the Florida Keys. Recreational harvest of this species, as shown in Figure 10.110 (right panel), tends to be highly variable on a year-to-year basis with average annual landings during 1995–2009 equaling close to 400,000 lb.

10.4.3.3.2 Louisiana

Aggregate Catch

Whereas Florida's most common catches (A + B1 + B2) and harvests (A + B1) in terms of numbers of fish were baitfish, a decidedly different picture emerges when one evaluates Louisiana's catches and harvests by species for selected years ranging from 1995 to 2009 (Tables 10.20, 10.21, 10.22, and 10.23). In terms of catch, spotted seatrout dominates all other species with an estimated 11.8 million (A + B1 + B2) and 18.5 million being caught in 1995 and 2009, respectively. Red drum ranks a distant second with the estimated 1995 catch equaling 5.2

Table 10.20 Most Frequently Caught (Left Panel) and Harvested (Right Panel) Species by Louisiana Recreational Anglers, 1995

Species Name	Number of Fish Caught (A + B1 + B2)	Species Name	Number of Fish Harvested (A + B1)	Mean Weight (lbs)
Spotted Seatrout	11,772,695	Spotted Seatrout	6,884,427	1.177
Red Drum	5,217,781	Red Drum	2,449,022	5.920
Hardhead Catfish	3,457,519	Sand Seatrout	856,107	0.569
Sand Seatrout	1,116,463	Sheepshead	646,983	2.124
Atlantic Croaker	1,053,423	Largemouth Bass	427,546	1.178
Sheepshead	944,262	Atlantic Croaker	409,294	0.499
Black Drum	803,938	Red Snapper	288,484	4.245
Red Snapper	672,140	Southern Flounder	260,073	1.242
Largemouth Bass	651,826	Black Drum	230,479	4.3784
Southern Flounder	301,697	Gulf Menhaden	159,683	0.220
Pinfish	259,226	Striped Mullet	122,547	0.869
Gulf Menhaden	188,932	Hardhead Catfish	115,584	1.164
Striped Mullet	145,357	Threadfin Shad	108,903	NA
Gulf Kingfish	128,201	Silver Seatrout	99,118	1.285
Gafftopsail Catfish	114,887	Bluegill	83,914	0.205
Spot	114,332	Pinfish	78,540	0.296
Bluegill	112,658	Blue Catfish	72,358	1.332
Threadfin Shad	108,903	Gray Triggerfish	66,995	2.200
Gulf Killifish	105,952	Spot	65,346	0.364
Silver Seatrout	100,805	Gulf Killifish	64,840	NA
Blue Catfish	99,257	Sheepshead Minnow	59,909	NA
Gray Triggerfish	78,097	Gulf Kingfish	55,728	0.453
Sea Catfish Family	64,052	Dolphin	45,171	11.747
Sheepshead Minnow	59,909	Gafftopsail Catfish	42,971	2.638
Gray Snapper	52,700	Skipjack Herring	34,126	0.414

Source: NMFS FSD, data accessed 2012, with calculations by authors–see Appendix A. *Note*: 1 lb = 0.454 kg.

million fish and the 2009 catch equaling almost six million fish. Combined, these two species accounted for more than one-half of the total Louisiana recreational catch (A + B1 + B2) in 1995 and two-thirds of the state's estimated recreational catch in 2009. Overall, the ranking of the five most commonly caught species remained unchanged when comparing the information for 1995 and 2009.

Similar to the Louisiana catch statistics, two species—spotted seatrout and red drum—dominate the Louisiana recreational harvest (A + B1) statistics. These two species, combined, represented 66 % of the state's total estimated recreational harvest in 1995 and almost 80 % of the state's total 2009 harvest. Overall, the average weight of the harvested spotted seatrout

Table 10.21 25 Most Frequently Caught (Left Panel) and Harvested (Right Panel) Species by Louisiana Recreational Anglers, 2000

Species Name	Number of Fish Caught (A + B1 + B2)	Species Name	Number of Fish Harvested (A + B1)	Mean Weight (1bs)
Spotted Seatrout	15,370,235	Spotted Seatrout	8,834,473	1.310
Red Drum	6,322,698	Sand Seatrout	1,230,837	0.928
Atlantic Croaker	3,906,464	Atlantic Croaker	957,736	0.522
Hardhead Catfish	2,201,793	Black Drum	665,273	4.008
Sand Seatrout	1,836,504	Red Drum	568,100	5.311
Black Drum	1,732,633	Sheepshead	386,000	2.696
Sheepshead	671,964	Southern Flounder	373,833	1.636
Southern Flounder	444,544	Spanish Mackerel	151,080	2.039
Striped Mullet	363,547	Southern Kingfish	142,531	0.812
Spanish Mackerel	360,313	Hardhead Catfish	133,085	0.934
Pinfish	348,617	Striped Mullet	109,685	0.890
Gafftopsail Catfish	316,989	Gafftopsail Catfish	99,703	2.315
Ladyfish	2,043,894	Gulf Menhaden	95,578	NA
Southern Kingfish	1,838,071	Red Snapper	81,065	5.543
Gulf Menhaden	1,678,791	Largemouth Bass	79,170	1.390
Red Snapper	1,559,712	Pinfish	62,861	0.619
Mullet Genus	1,501,984	Gray Triggerfish	51,058	2.619
Largemouth Bass	1,270,975	Seatrout Genus	48,325	NA
Blue Runner	1,217,217	Blue Catfish	41,551	1.557
Blue Catfish	1,162,179	Gulf Kingfish	33,994	0.655
Seatrout Genus	1,135,902	Gray Snapper	33,353	2.598
Atlantic Stingray	1,098,766	Atlantic Stingray	30,581	NA
Gray Triggerfish	1,087,375	Atlantic Spadefish	23,275	3.187
Gray Snapper	1,082,116	Blue Runner	14,299	1.301
Stingray Genus	994,028	Blacktip Shark	13,574	14.364

Source: NMFS FSD, data accessed 2012, with calculations by authors—see Appendix A. *Note*: 1 lb = 0.454 kg.

remained unchanged between the 2 years (1.17 lb per fish) while the average weight of the harvested red drum fell from almost 6 lb per fish in 1995 to about 5.6 lb in 2009.

While Louisiana's recreational catch of baitfish is but a fraction of that reported for Florida (259,000 pinfish and 189,000 Gulf menhaden in 1995), some undesirable species are often caught. For example, an estimated 3.5 million and 2.7 million hardhead catfish were caught in 1995 and 2009, respectively. Harvest (A + B1) of this species in 1995, however, equaled only 116,000 and in 2009 equaled 95,000, which for each year represents less than 5 % of the catch of this species. While there are no regulations governing the harvest of this species, it is usually returned to the water alive because it is considered inedible by the majority of the recreational fishing population.

The aggregate anglers catch in Louisiana's waters for the years 1995 to 2009 expressed as a percentage of the Gulf catch and in numbers of fish (A + B1 + B2) is presented in Figure 10.111

Table 10.22 25 Most Frequently Caught (Left Panel) and Harvested (Right Panel) Species by Louisiana Recreational Anglers, 2005

Species Name	Number of Fish Caught (A + B1 + B2)	Species Name	Number of Fish Harvested (A + B1)	Mean Weight (lbs)
Spotted Seatrout	14,727,580	Spotted Seatrout	7,435,705	1.106
Red Drum	4,263,779	Red Drum	1,626,356	8.209
Hardhead Catfish	3,388,334	Sand Seatrout	973,661	0.593
Atlantic Croaker	1,333,069	Sheepshead	644,499	2.768
Sand Seatrout	1,226,313	Atlantic Croaker	442,583	0.412
Sheepshead	1,073,416	Black Drum	308,777	6.024
Black Drum	930,537	Southern Flounder	280,050	1.123
Gafftopsail Catfish	672,912	Southern Kingfish	239,777	0.538
Ladyfish	532,341	Pinfish	147,908	0.314
Southern Kingfish	410,882	Hardhead Catfish	125,832	1.11
Red Snapper	396,531	Red Snapper	110,503	4.966
Southern Flounder	355,791	Gray Snapper	107,688	4.486
Largemouth Bass	272,581	Largemouth Bass	102,857	0.986
Pinfish	232,477	Gafftopsail Catfish	86,621	3.307
Blue Catfish	189,002	Blue Catfish	78,580	1.293
Gray Snapper	155,251	Spanish Mackerel	38,785	2.133
Spanish Mackerel	75,771	Cobia	21,172	24.054
Atlantic Stingray	54,144	Channel Catfish	19,996	0.569
Bluefish	50,440	Bluegill	19,031	0.249
Blue Runner	45,997	Striped Mullet	18,046	0.595
Channel Catfish	45,894	Blackfin Tuna	15,582	22.818
Stingray Genus	35,501	Atlantic Spadefish	13,820	1.221
Atlantic Spadefish	28,914	Dolphin	13,246	3.771
Freshwater Drum	23,539	Seatrout Genus	12,446	NA
Cobia	22,362	Gulf Kingfish	11,837	0.551

Source: NMFS FSD, data accessed 2012, with calculations by authors–see Appendix A. *Note*: 1 lb = 0.454 kg.

(left panel) while the number of fish caught by anglers is given in Figure 10.111 (right panel). As indicated, the estimated annual number of fish being caught by recreational anglers, though highly variable on a year-to-year basis, increased during the period of analysis. From 1995 to 1999, the average catch expressed in number of fish equaled 27 million annually. By 2005–2009, this figure had increased to 36 million. Overall, Louisiana's share of the total Gulf catch (excluding Texas) expressed in numbers of fish averaged 21 % during the 15-year period of analysis and ranged from a low of 15 % in 2002 to a high of 26 % in 2000. As was found to be the case throughout the Gulf, a large proportion of the catch by anglers in Louisiana's waters is released with the annual estimate of 50–60 %. This high release rate reflects a combination of the catch of undesirable species (e.g., hardhead catfish) and regulations (particularly size limits and bag limits).

Table 10.23 25 Most Frequently Caught (Left Panel) and Harvested (Right Panel) Species by Louisiana Recreational Anglers, 2009

Species Name	Number of Fish Caught (A + B1 + B2)	Species Name	Number of Fish Harvested (A + B1)	Mean Weight (lbs)
Spotted Seatrout	18,532,549	Spotted Seatrout	10,557,489	1.179
Red Drum	5,959,448	Red Drum	2,236,916	5.581
Hardhead Catfish	2,693,243	Sand Seatrout	879,031	0.554
Sand Seatrout	1,726,535	Sheepshead	703,498	3.153
Atlantic Croaker	1,563,809	Black Drum	518,989	5.370
Black Drum	1,482,978	Atlantic Croaker	470,537	0.373
Sheepshead	1,174,727	Southern Flounder	285,605	1.378
Gafftopsail Catfish	536,631	Southern Kingfish	103,044	0.548
Southern Flounder	336,019	Gray Snapper	98,829	2.481
Ladyfish	329,567	Striped Mullet	97,984	0.394
Red Snapper	214,713	Red Snapper	97,250	7.075
Southern Kingfish	152,493	Hardhead Catfish	95,201	1.414
Striped Mullet	110,011	Largemouth Bass	59,344	1.343
Gray Snapper	106,433	Gafftopsail Catfish	59,194	2.056
Atlantic Bumper	87,237	Blue Catfish	51,043	2.843
Largemouth Bass	77,727	Gulf Menhaden	50,650	NA
Blue Catfish	68,697	Blackfin Tuna	47,558	18.746
Blackfin Tuna	57,048	Seatrout Genus	44,144	NA
Gulf Menhaden	50,650	Atlantic Bumper	37,076	0.110
Seatrout Genus	44,144	Channel Catfish	19,709	1.740
Pinfish	41,213	Greater Amberjack	17,277	26.761
Atlantic Stingray	33,755	Tripletail	15,580	8.799
Spanish Mackerel	29,906	Striped Bass	14,353	0.701
Channel Catfish	25,374	Spanish Mackerel	12,511	1.9870
Greater Amberjack	23,140	Pinfish	11,273	0.107

Source: NMFS FSD, data accessed 2012, with calculations by authors–see Appendix A. *Note*: 1 lb = 0.454 kg.

As was the case with the number of fish landed, the aggregate pounds of fish harvested (A + B1) from Louisiana waters have been increasing (Figure 10.112). During 1995–1999, the estimated harvest averaged about 22 million pounds annually. By 2005–2009, this annual average increased to almost 30 million pounds and would likely have been higher if not for Hurricanes Katrina and Rita in 2005 that limited fishing activities and catch in that year. Interestingly, the recreational harvest (in pounds) in 2006 was the highest observed figure during the 15-year period of analysis and may reflect an increase in species populations in that year as a result of a reduction in 2005 effort. Overall, Louisiana's share of the Gulf recreational harvest in pounds averaged 37 % during the 1995–2009 period with an annual range from about 30 % to more than 45 %.

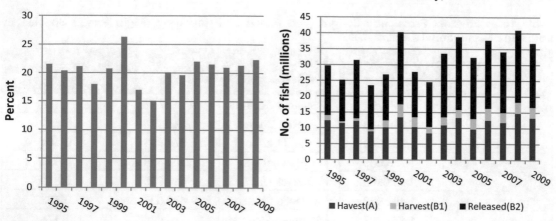

Figure 10.111. Louisiana's share of the Gulf aggregate recreational catch (*left panel*) and Louisiana recreational catch (*right panel*), 1990–2009 (NMFS FSD, data accessed 2012, with percentage calculations by authors—see Appendix A).

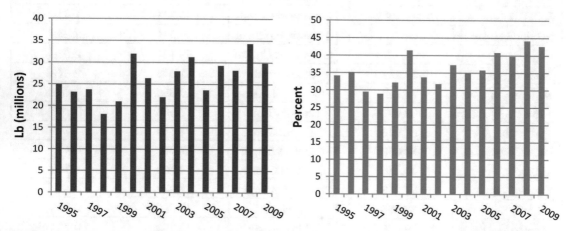

Figure 10.112. Louisiana's recreational aggregate harvest (*left panel*) and percentage of Gulf harvest (*right panel*), 1995–2009 (NMFS FSD, data accessed 2012, with percentage calculations by authors—see Appendix A) (*Note*: 1 lb = 0.454 kg).

As was the case with the number of fish landed, the aggregate pounds of fish harvested (A + B1) from Louisiana waters have been increasing (Figure 10.112). During 1995–1999, the estimated harvest averaged about 22 million pounds annually. By 2005–2009, this annual average increased to almost 30 million pounds and would likely have been higher if not for Hurricanes Katrina and Rita in 2005 that limited fishing activities and catch in that year. Interestingly, the recreational harvest (in pounds) in 2006 was the highest observed figure during the 15-year period of analysis and may reflect an increase in species populations in that year as a result of a reduction in 2005 effort. Overall, Louisiana's share of the Gulf recreational harvest in pounds averaged 37 % during the 1995–2009 period with an annual range from about 30 % to more than 45 %.

Inshore Species

Louisiana is known for its inshore fishing and is often referred to as the *Redfish Capital*. This is not surprising since the five fish bag limit is the most liberal among Gulf States, and the Louisiana red drum catch rates tend to be high. Though possibly not as well known, the same claim can be made for spotted seatrout.

Red Drum

The popularity of red drum in Louisiana becomes clear when one looks at the percentage of fishermen targeting the species on any given trip. According to the MRIP survey data, since 2000, approximately 43 % of the Louisiana interviewed respondents, on average, reported that red drum was one of the two primary species targeted. In 2001, for example, there was an estimated 3.6 million angler trips. Of this total, an estimated 1.7 million (48 %) of the total, reported red drum as one of the two primary species being targeted (Figure 10.113, left panel). This year represents the highest red drum targeting behavior during the 10-year period of analysis. Conversely, the lowest reported red drum targeting behavior (on a percentage basis) was reported in 2005 and 2006 when 1.5 million of the approximately four million angler trips (38 %) indicated red drum as one of the two primary targeted species.

While an angler may specify that he or she is targeting red drum on any given trip, it does not necessarily imply that red drum will be caught. A comparison between red drum targeted trips and trips where the catch of red drum is reported is presented in Figure 10.113 (right panel). As indicated, targeted trips consistently exceeded catch trips though the correlation between the two was a respectable 0.88.

Since 1995, the estimated number of red drum caught expressed in numbers of fish (A + B1 + B2) has averaged 5.4 million annually with an associated range of 4.1 million in 1996 to 6.6 million in 2000 (Figure 10.114). Louisiana's share of the Gulf red drum catch generally ranges from 60 % to almost 80 %. These fish can be either kept or released. As indicated, about 60 % of the red drum catch has historically been released alive with very little variation in the percentage when examined on a year-to-year basis. During the period of analysis, the correlation between the (estimated) annual number of red drum harvested (A + B1) and the number of red drum released alive (B2) was positive, equaling 0.54.

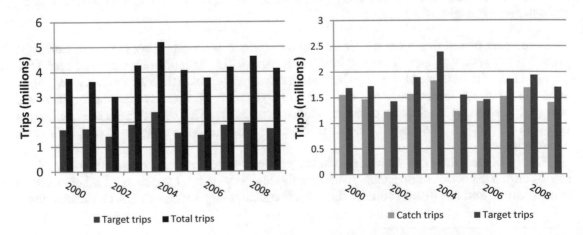

Figure 10.113. Louisiana red drum targeting trips in relation to total trips (*left panel*) and Louisiana red drum targeting trips in relation to catch trips (*right panel*), 2000–2009 (NMFS FSD, data accessed 2012, with targeting estimates calculated by authors—see Appendix A).

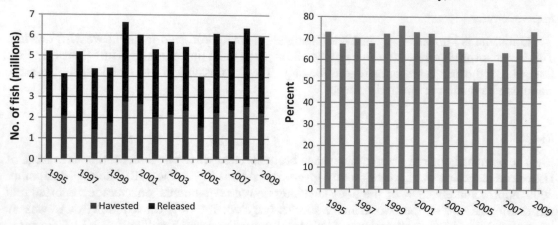

Figure 10.114. Louisiana recreational red drum catch (*left panel*) and Louisiana catch in relation to Gulf total (*right panel*), 1995–2009 (NMFS FSD, data accessed 2012, with percentage calculations by authors—see Appendix A).

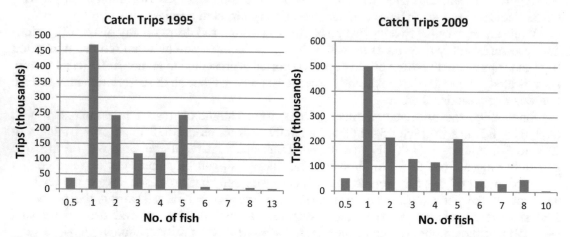

Figure 10.115. Red drum catch (number) per Louisiana angler among those trips where catch of red drum was positive, 1995 and 2009 (NMFS FSD, data accessed 2012, with estimations by authors—see Appendix A).

The catch of red drum per angler among those angler trips reporting red drum catches is provided for selected years (1995 and 2009) in Figure 10.115. As indicated, there has been little change in the distribution of the catch between the two periods with the average catch per angler exhibiting a bimodal distribution at one fish and five fish.[25] Furthermore, as shown, there are many trips where the red drum catch per angler exceeds five fish even though the bag limit is five fish per angler. The reason for this is that the catch can exceed bag limit with the excess being released.

Catch of red drum in Louisiana waters occurs overwhelmingly in inland waters (Figure 10.116). This is not unexpected given that the majority of total angler trips taken occur in inland waters and red drum is one of the most frequently targeted species. As indicated, there

---

[25] The 0.5 catch per angler is the result of parties with more than one angler and the division of the catch among the anglers.

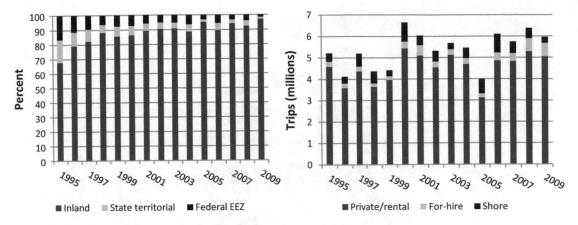

Figure 10.116. Louisiana recreational red drum catch by area (*left panel*) and mode (*right panel*), 1995–2009 (Sour NMFS FSD, data accessed 2012—see Appendix A).

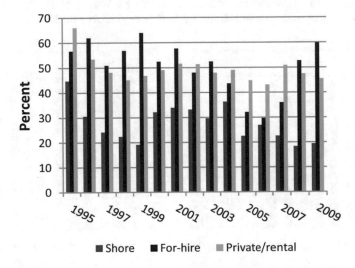

Figure 10.117. Louisiana recreational red drum targeting behavior by mode, 1995–2009 (NMFS FSD, data accessed 2012, with targeting estimates calculated by authors—see Appendix A).

has been a gradual increasing trend in the percentage of red drum catch in inland waters over the 15-year period of analysis with the inland catch representing 97 % of the total in 2009.

The overwhelming proportion of red drum catch occurs from private/rental boats with approximately 85 % of this species catch being taken by this mode since 1995 (Figure 10.116, right panel). Another 10 % of the catch has been taken from shore. In general, there was little observed change in the catch-by-mode trend during the 15-year period of analysis with the exception of the last several years when the proportion of red drum catch emanating from the for-hire mode increased at the expense of the shore mode.

In general, there is a considerable amount of red drum targeting behavior among all fishing modes (Figure 10.117). With respect to the shore-based mode, slightly less than 30 % of the angler trips reported the targeting of red drum during the 15-year period of analysis. This proportion increased to 50 % for the for-hire mode and the private/rental boat mode.

As noted, MRIP data are collected and analyzed in terms of *waves* wherein each wave represents a 2-month period. Estimated red drum catch by wave for selected periods during 1995–2009 is presented in Figure 10.118. As indicated, there is some seasonality to red drum

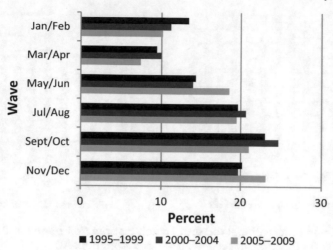

Figure 10.118. Louisiana recreational red drum catch by wave, 1995–2009 (NMFS FSD, data accessed 2012, with percentage calculations by authors—see Appendix A).

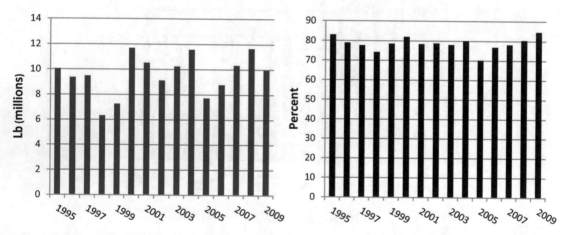

Figure 10.119. Louisiana recreational red drum harvest and Louisiana's harvest in relation to Gulf total, 1995–2009 (NMFS FSD, data accessed 2012, with percentage calculations by authors—see Appendix A) (*Note*: 1 lb = 0.454 kg).

catch, and this seasonality mimics what was observed for the Gulf. This is expected given the large proportion of the Gulf catch in number of fish is represented by Louisiana.

As previously discussed, the catch of red drum by Louisiana's anglers in numbers of fish (A + B1 + B2) accounts for the majority of the Gulf catch (excluding Texas). As such, it should come as no surprise that Louisiana's recreational harvest of red drum in pounds (A + B1) dominates the Gulf (Figure 10.119). Overall, Louisiana's annual recreational red drum harvest in pounds generally ranges from about 8 million pounds to 11 million pounds, which represents about 80 % of the Gulf total (excluding Texas). The state's share of the Gulf total in pounds harvested is marginally higher than its share in number of fish reflecting primarily larger bag limits in Louisiana and a larger sized fish being harvested.[26]

---

[26] As noted, the Gulf red drum harvest (excluding Texas) is dominated by Florida and Louisiana. In 1995, the average weight of red drum harvested in Florida equaled 4.3 lb per fish compared to 5.9 lb per fish in Louisiana. In 2009, the Florida recreationally harvested red drum averaged 1.49 lb per fish compared to 5.6 lb for fish harvested in Louisiana.

Spotted Seatrout

Along with red drum, spotted seatrout is the other species most often targeted in Louisiana. Since 1995, as indicated by the information in Figure 10.120, from about 40 % to more than 50 % of the annual trips report spotted seatrout as one of the two targeted species (average of 48 % over the 15-year period ending in 2009). The only year in which trips targeting spotted seatrout fell below 40 % was 1998 when it was marginally lower (39 %). As previously noted, the respondents to the MRIP dockside interview are allowed to state two targeting species. While not discussed here, about two-thirds of intercepted anglers in Louisiana consistently indicted that they targeted either red drum or spotted seatrout.

A close relationship exists between recreational spotted seatrout targeting trips and catch trips in Louisiana's waters (Figure 10.121). While this might suggest that the probability of

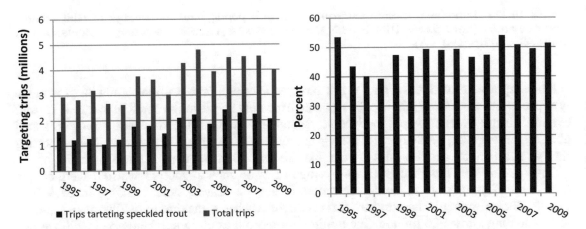

Figure 10.120. Louisiana spotted seatrout targeting trips in relation to total number of trips (*left panel*) and as a percent of total trips (*right panel*): 1995–2009 (NMFS FSD, data accessed 2012, with targeting estimates and percentages calculated by authors—see Appendix A).

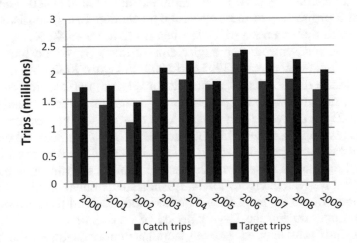

Figure 10.121. Relationship between Louisiana spotted seatrout catch trips and spotted seatrout targeting trips, 2000–2009 (NMFS FSD, data accessed 2012, with targeting estimates calculated by authors—see Appendix A).

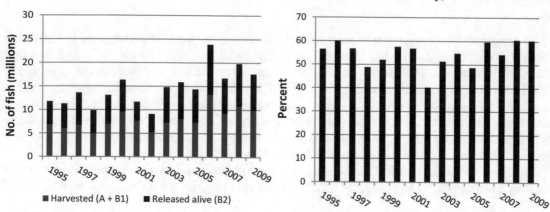

Figure 10.122. Louisiana spotted seatrout catch (*left panel*), and percentage of Gulf spotted seatrout catch (*right panel*) (NMFS FSD, data accessed 2012, with percentage calculations by authors—see Appendix A).

catching spotted seatrout is high if it is a targeted species, some caution should be used in the interpretation of this relationship. First, some trips may result in the catch of spotted seatrout even if it is not a targeted species. Second, respondents to the MRIP dockside survey are asked about their targeting behavior after the trip is concluded. There is a body of evidence suggesting that what an angler catches on a trip can bias his post-trip responses to targeting behavior.

According to MRIP statistics, an average of 14.66 million spotted seatrout were caught (A + B1 + B2) per year in Louisiana waters during the 15-year period ending in 2009 (Figure 10.122, left panel). The number of fish caught reached a maximum in 2006 with reported total equaling almost 24 million. The minimum reported catch in numbers was in 2002 at just over nine million. As with all species, spotted seatrout can be harvested (A + B1) or released alive (B2). For the 15-year period ending in 2009, an average of 55 % of the spotted seatrout catch in numbers were harvested annually with a range from about 50 % in many years to about 65 % in 2002. During the period of analysis, the correlation between the (estimated) annual number of spotted seatrout harvested (A + B1) and the number of spotted seatrout released alive (B2) was positive, equaling 0.84. When examined on an annual basis, Louisiana's catch of spotted seatrout as a proportion of the Gulf's total (excluding Texas) consistently ranged from about 50 to 60 % with the exception of 2002 when it fell to about 40 %.

The catch of spotted seatrout per angler among those trips reporting spotted seatrout catches is provided for selected years (1995 and 2009) in Figure 10.123. As indicated, there has been little change in the distribution of the catch between the two considered years.

Like red drum, the overwhelming proportion of Louisiana's recreational spotted seatrout catch (A + B1 + B2) is derived from inland waters with the percentage in recent years approaching 95 % (average for the 1995–2009 equals 90 %). Also, like red drum, about 90 % of the recreational catch of spotted seatrout in Louisiana's waters is derived from the private/rental boat mode. Louisiana's recreational harvest of spotted seatrout in pounds (A + B1) for the 1995–2009 period is given in Figure 10.124. As indicated, annual landings have ranged from less than six million pounds to more than 12 million pounds and have averaged almost nine million pounds annually during the 15-year period of consideration. This average represents about 70 % of the Gulf total spotted seatrout landings in pounds during the period.

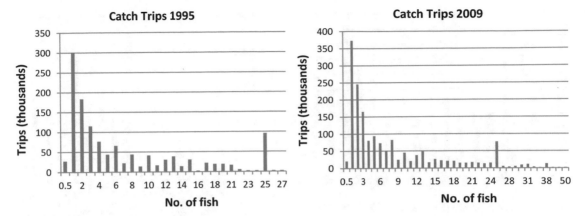

Figure 10.123. Catch of spotted seatrout per angler (in number of fish) among those trips where catch of spotted seatrout was positive, 1995 and 2009 (NMFS FSD, data accessed 2012, with estimations by authors—see Appendix A).

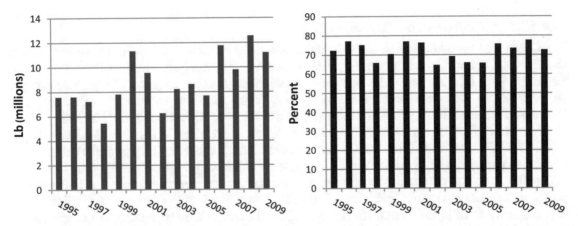

Figure 10.124. Louisiana recreational harvest of spotted seatrout (*left panel*) and percentage of Gulf total harvest (*right panel*), 1995–2009 (NMFS FSD, data accessed 2012, with percentage calculations by authors—see Appendix A) (*Note*: 1 lb = 0.454 kg).

Offshore Species

Aggregate Reef Fish: As noted when examining trips by area, Louisiana is primarily an inshore fishery. As such, it is not surprising that catch of offshore species is limited. The aggregate catch of reef fish in numbers of fish (A + B1 + B2) generally tends to be less than one million fish per year and harvest in pounds (A + B1) is generally less than two million pounds per year (Figure 10.125). Among the primary species harvested are red snapper (average annual landings of 558,000 since 1995) and greater amberjack (average landings of 231,000 lb annually since 1995).

Other Offshore Species: Other than reef fish species, two offshore species highly desired by Louisiana anglers are yellowfin tuna and blackfin tuna (*Thunnus atlanticus*). These two species are highly migratory in nature and, as such, yearly landings can fluctuate widely. Reported harvest of yellowfin tuna averaged 365,000 lb annually during 1995–2009, and blackfin tuna landings averaged 300,000 lb. Large expenditures are incurred in the harvest of these species due to the far offshore distance one must travel to catch either yellowfin or blackfin tuna and, as such, the number of trips is limited. The limited number and nature of these trips suggests caution should be exercised when assessing the reliability of these figures.

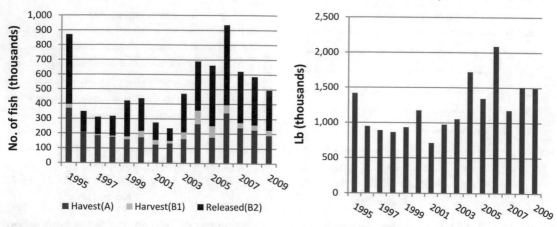

■ Havest(A)   ■ Harvest(B1)   ■ Released(B2)

Figure 10.125. Louisiana recreational aggregate reef fish catch (*left panel*) and harvested pounds, 1995–2009 (NMFS FSD, data accessed 2012—see Appendix A) (*Note*: 1 lb = 0.454 kg).

10.4.3.3.3   Alabama and Mississippi

Given that Florida and Louisiana dominate Gulf recreational catch in both numbers of fish caught (A + B1 + B2) and pounds of fish kept (A + B1), analysis given to the recreational fisheries in Alabama and Mississippi is more limited. The 25 most frequently caught (A + B1 + B2) and harvested (A + B1) species from Alabama waters for selected years between 1995 and 2009 (1995, 2000, 2005, and 2009) are identified in Tables 10.24, 10.25, 10.26, and 10.27 while comparable figures for Mississippi are given in Tables 10.28, 10.29, 10.30, and 10.31. Without going into detail, a comparison of Alabama's and Mississippi's catches and harvests in numbers of fish with that of Louisiana's would suggest that Mississippi is closer to Louisiana in terms of species caught than is Alabama. For example, while spotted seatrout represents the most frequently harvested species in numbers in both Louisiana and Mississippi, it ranks only third in Alabama. Similarly, while red drum is ranked second in Louisiana and seventh in Mississippi, it is ranked tenth in Alabama. Such a finding is not unexpected given that the coastal wetlands in Alabama are considerably more limited than in Mississippi. By comparison, many of the species most frequently harvested in Alabama represent those most often associated with offshore fishing activities (e.g., king mackerel, vermilion snapper, and gray triggerfish). This finding is consistent with fishing practices across the states. Specifically, whereas approximately 95 % of the 2009 fishing trips in both Louisiana and Mississippi were conducted in inland waters, less than 65 % of the fishing trips in Alabama were conducted in inland waters.

As the information in Figure 10.126 (left panel) indicates, recreational anglers in Alabama have, in recent years, caught about 9–12 million fish per year. Since increasing in the late 1990s, little trend is evident in recreational catch from Alabama's waters. As a proportion of the Gulf catch, in numbers of fish, Alabama has contributed as little as about 3.5 % and never more than 7 % with the 1995–2009 average equaling 5.4 % (Figure 10.126, right panel). Overall, more than 50 % of the catch during 1995–2009 was released alive (B2) with the proportion exceeding 60 % in some years.

In terms of pounds landed (A + B1), Alabama's share of the Gulf total has fallen in the relatively narrow range of 8–11 % in recent years (Figure 10.127, right panel) based on an absolute harvest that has remained stable during the 2005–2009 period ranging from about 5.7 million to 7 million pounds (Figure 10.127, left panel). As was the situation in terms of number

Table 10.24 25 Most Frequently Caught (Left Panel) and Harvested (Right Panel) Species By Alabama Recreational Anglers, 1995

Species Name	Number of Fish Caught (A + B1 + B2)	Species Name	Number of Fish Harvested (A + B1)	Mean Weight (lbs)
Sand Seatrout	1,234,756	Sand Seatrout	1,078,925	0.559
Red Snapper	567,495	Scaled Sardine	352,376	0.055
Spanish Mackerel	427,074	Red Snapper	324,633	4.010
Atlantic Croaker	406,574	Sheepshead	273,670	2.409
Scaled Sardine	352,376	Spanish Mackerel	250,118	1.691
Pinfish	344,211	Vermilion Snapper	242,816	1.170
Sheepshead	295,479	Striped Mullet	215,248	0.685
Vermilion Snapper	287,047	Gray Triggerfish	188,386	1.977
Gray Triggerfish	222,571	Atlantic Croaker	166,017	0.405
Striped Mullet	216,156	Southern Kingfish	150,153	0.671
Atlantic Spadefish	199,938	Southern Flounder	112,973	1.183
Southern Kingfish	171,281	Atlantic Spadefish	105,743	1.677
Spotted Seatrout	153,573	Bluefish	105,533	2.076
Hardhead Catfish	141,301	Spotted Seatrout	93,232	1.133
Bluefish	140,848	Red Drum	74,409	6.102
Red Drum	126,209	King Mackerel	65,071	6.6031
Southern Flounder	120,208	Pinfish	48,191	0.264
King Mackerel	84,632	Lefteye Flounder Family	28,774	NA
Tomtate	40,717	White Mullet	28,749	0.555
Greater Amberjack	40,260	Seatrout Genus	26,271	NA
Lefteye Flounder Family	35,822	Gulf Kingfish	23,503	0.551
White Mullet	28,749	Black Drum	21,412	3.084
Gulf Kingfish	27,978	Hardhead Catfish	19,223	0.907
Seatrout Genus	26,271	Greater Amberjack	16,564	20.242
Black Drum	24,039	Gray Snapper	15,478	1.081

Source: NMFS FSD, data accessed 2012, with calculations by authors–see Appendix A. Note: 1 lb = 0.454 kg.

of fish caught (A + B1 + B2), there was an apparent increase in pounds landed in the late 1990s though the reason for this increase is not obvious.

As with the number of participants and trips, Mississippi's estimated recreational catch is the lowest among the four Gulf States considered in this analysis (Figure 10.128). The observed maximum catch expressed in numbers of fish occurred in 2001 when an estimated eight million fish were caught. Anywhere from one-third to one-half of the total catch is generally released alive (B2). Overall, the recreational catch from Mississippi's waters did not exceed 5 % of the Gulf total in any of the 15 years of analysis and in some years fell as low as 2 %.

Table 10.25 25 Most Frequently Caught (Left Panel) and Harvested (Right Panel) Species By Alabama Recreational Anglers, 2000

Species Name	Number of Fish Caught (A + B1 + B2)	Species Name	Number of Fish Harvested (A + B1)	Mean Weight (lbs)
Sand Seatrout	738,646	Sand Seatrout	554,172	0.507
Atlantic Croaker	738,218	Southern Kingfish	302,402	0.534
Red Snapper	461,098	Atlantic Croaker	225,056	0.399
Southern Kingfish	458,689	Striped Mullet	170,156	0.937
Pinfish	432,737	Spanish Mackerel	162,281	2.349
Blue Runner	430,801	Spotted Seatrout	140,197	1.674
Spotted Seatrout	382,089	Sheepshead	133,462	2.977
Spanish Mackerel	218,697	Red Snapper	127,346	4.010
Sheepshead	179,962	Gulf Kingfish	125,542	0.586
Striped Mullet	173,894	White Mullet	122,897	0.623
Atlantic Spadefish	152,835	Menhaden Genus	99,330	NA
White Mullet	151,838	King Mackerel	91,576	10.916
Gulf Kingfish	133,762	Mullet Genus	89,280	NA
Red Drum	124,407	Pinfish	85,125	0.303
Menhaden Genus	124,320	Southern Flounder	63,443	1.481
King Mackerel	123,636	Bluefish	58,056	2.408
Bluefish	103,625	Red Drum	53,734	6.045
Hardhead Catfish	90,637	Gulf Menhaden	52,745	0.203
Mullet Genus	89,280	Atlantic Spadefish	35,565	2.264
Gray Snapper	86,529	Blue Runner	34,701	0.717
Southern Flounder	74,359	Black Drum	26,846	2.896
Gulf Menhaden	53,744	Gray Snapper	22,622	1.400
Little Tunny	44,632	Gray Triggerfish	15,314	2.576
Requiem Shark Family	34,849	Florida Pompano	12,757	1.561
Black Drum	29,827	Searobin Genus	12,185	NA

Source: NMFS FSD, data accessed 2012, with calculations by authors–see Appendix A. *Note*: 1 lb = 0.454 kg.

In terms of pounds harvested (A + B1), Mississippi's share fell from about 6 % in the mid-1990s to less than 3 % from 2005 to 2008 before increasing to 5 % in 2009 (Figure 10.129, right panel). This is based on harvested poundage ranging from about 1.5 million to 4.5 million (Figure 10.129, left panel). Much of the observed decline in both catch (Figure 10.128) and harvest (Figure 10.129) during the mid-2000s was undoubtedly related to the destruction in infrastructure associated with Hurricane Katrina.

Table 10.26 25 Most Frequently Caught (Left Panel) and Harvested (Right Panel) Species By Alabama Recreational Anglers, 2005

Species Name	Number of Fish Caught (A + B1 + B2)	Species Name	Number of Fish Harvested (A + B1)	Mean Weight (lbs)
Atlantic Croaker	1,683,014	Sand Seatrout	349,559	0.510
Red Snapper	650,305	Spotted Seatrout	294,437	1.860
Spotted Seatrout	617,079	Sheepshead	279,854	3.420
Sand Seatrout	612,421	Atlantic Croaker	233,043	0.428
Pinfish	467,484	Red Snapper	232,430	4.106
Southern Kingfish	409,075	Striped Mullet	221,943	0.919
Sheepshead	365,273	Southern Kingfish	191,183	0.560
Hardhead Catfish	349,698	Pinfish	158,298	0.233
Red Drum	327,984	Red Drum	153,822	7.861
Striped Mullet	254,510	Southern Flounder	150,458	1.258
Southern Flounder	230,554	Blue Runner	104,515	0.362
Blue Runner	129,795	Gray Triggerfish	82,494	2.249
Gulf Kingfish	108,247	Vermilion Snapper	74,899	1.105
Spanish Mackerel	96,234	Gulf Kingfish	71,938	0.565
Gray Triggerfish	89,455	Black Drum	68,699	8.199
Vermilion Snapper	82,812	Spanish Mackerel	45,032	1.500
Ladyfish	76,172	King Mackerel	41,509	8.108
Black Drum	75,331	Bluegill	37,084	0.388
Bluefish	72,364	Hardhead Catfish	33,459	0.750
Gafftopsail Catfish	69,927	Mullet Genus	25,055	NA
Gag	64,974	Gag	21,381	6.056
Bluegill	64,896	White Mullet	21,298	0.539
King Mackerel	54,814	Atlantic Spadefish	20,761	1.904
Atlantic Spadefish	53,477	Red Porgy	19,127	1.127
Gray Snapper	41,847	Ladyfish	16,195	1.262

Source: NMFS FSD, data accessed 2012, with calculations by authors–see Appendix A. *Note*: 1 lb = 0.454 kg.

10.5 SUMMARY

Given its diversity of species, the Gulf of Mexico offers ample opportunities to both commercial and recreational fishermen. The objective of this chapter is to provide a systematic examination of the Gulf of Mexico commercial and recreational fishing sectors focusing on a variety of topics. With respect to the commercial sector, some of the topics considered include trends in production of various species, the value of production associated with these various species, the impact of imports on dockside prices, and processing. Overall, long-term landings of most key commercial species appear to be stable and changes, where noted, appear to be tied to regulations to manage fish stocks. This is particularly true with respect to finfish stocks. Of the commercial fisheries examined, the shrimp fishery faces the greatest obstacles in terms

Table 10.27 25 Most Frequently Caught (Left Panel) and Harvested (Right Panel) Species By Alabama Recreational Anglers, 2009

Species Name	Number of Fish Caught (A + B1 + B2)	Species Name	Number of Fish Harvested (A + B1)	Mean Weight (lbs)
Sand Seatrout	2,176,890	Sand Seatrout	1,428,030	0.580
Atlantic Croaker	2,035,394	Southern Kingfish	591,217	0.595
Spotted Seatrout	1,075,150	Spotted Seatrout	318,109	2.100
Southern Kingfish	837,218	Atlantic Croaker	249,833	0.367
Red Snapper	453,175	Sheepshead	165,809	2.735
Hardhead Catfish	439,071	Southern Flounder	138,841	1.445
Pinfish	298,775	Red Snapper	138,062	5.083
Sheepshead	202,989	Spanish Mackerel	75,605	1.854
Red Drum	163,178	Vermillion Snapper	61,969	0.893
Southern Flounder	160,787	Red Drum	61,808	6.771
Spanish Mackerel	135,188	King Mackerel	52,661	9.475
King Mackerel	76,575	Pinfish	8270	0.200
Vermillion Snapper	67,768	White Mullet	42,196	0.357
White Mullet	61,976	Striped Mullet	34,979	0.854
Gray Snapper	59,930	Gray Triggerfish	34,555	2.550
Gray Triggerfish	52,989	Black Drum	28,670	6.284
Striped Mullet	52,122	Hardhead Catfish	19,564	0.825
Bluefish	43,031	Gray Snapper	18,536	2.152
Pigfish	41,901	Atlantic Spadefish	17,386	1.105
Blue Runner	38,897	Bluefish	13,985	1.804
Black Drum	38,841	Silver Perch	12,069	0.203
Gulf Flounder	34,850	Gulf Flounder	11,120	1.718
Gafftopsail Catfish	34,422	Lane Snapper	10,138	1.285
Atlantic Spadefish	26,424	Red Pongy	8,616	0.834
Ladyfish	14,414	Blue Runner	7,158	1.256

Source: NMFS FSD, data accessed 2012, with calculations by authors—see Appendix A. Note: 1 lb = 0.454 kg.

of long-run viability. Increasing imports have led to a significant decline in the price shrimpers receive for the harvested product and, in turn, a reduction in profitability. This reduction has led to a substantial downsizing of the industry with current effort in the fishery (measured in days fished) being a fraction of what it was in the 1990s. This statement applies for both the brown and white shrimp, the two species of relevance in the northern Gulf of Mexico.

Like the harvesting sector, the Gulf shrimp-processing sector has not been immune to the increasing import base. A steadily eroding marketing margin and, presumably, profit has culminated in consolidation of this sector, and remaining firms are increasing output in an attempt to counterbalance the declining marketing margin per unit of output.

Table 10.28 25 Most Frequently Caught (Left Panel) and Harvested (Right Panel) Species By Mississippi Recreational Anglers, 1995

Species Name	Number of Fish Caught (A + B1 + B2)	Species Name	Number of Fish Harvested (A + B1)	Mean Weight (lbs)
Sand Seatrout	688,074.36	Sand Seatrout	642,357.28	0.441
Atlantic Croaker	669,942.72	Striped Mullet	13,309.73	0.593
Spotted Seatrout	535,617.59	Atlantic Croaker	388,380.15	0.283
Striped Mullet	428,727.73	Southern Flounder	69,565.72	0.845
Southern Flounder	281,587.77	Spotted Seatrout	266,054.24	1.374
Sheepshead	237,534.23	Sheepshead	215,151.83	2.061
Red Drum	189,832.36	Gulf Menhaden	93,910.29	NA
Pigfish	125,507.84	Red Drum	81,965.38	7.853
Hardhead Catfish	122,435.30	Spanish Mackerel	79,882.90	1.792
Gulf Menhaden	93,910.29	Southern Kingfish	63,416.67	0.523
Spanish Mackerel	93,115.38	Red Snapper	37,535.93	3.573
Southern Kingfish	77,432.72	Atlantic Spadefish	35,548.97	1.480
Red Snapper	48,894.06	Atlantic Sharpnose Shark	32,997.40	6.070
Atlantic Sharpnose Shark	38,055.40	Pinfish	32,618.79	0.216
Atlantic Spadefish	35,548.97	Gray Snapper	22,444.24	0.809
Gafftopsail Catfish	28,706.12	Black Drum	21,236.04	3.208
Gray Snapper	27,776. 59	Gafftopsail Catfish	20,001.63	3.535
Black Drum	23,725. 02	Tripletail	19,618.58	9.467
Cobia	21,625. 78	Hardhead Catfish	16,086.65	0.734
Tripletail	19,618. 58	Blacktip Shark	13,504.86	10.149
Blacktip Shark	18,624.38	Gray Triggerfish	9,116.03	2.348
Bluefish	10,817.39	King Mackerel	7,689.57	9.574
Gray Triggerfish	10,227.84	Spot	5,040.37	0.296
King Mackerel	7,689.57	Lefteye Flounder Genus	5,039.00	NA
Requiem Shark Family	5,559.03	Bluefish	4,996.11	2.284

Source: NMFS FSD, data accessed 2012, with calculations by authors–see Appendix A. Note: 1 lb = 0.454 kg.

Direct jobs in the harvesting sector generate jobs elsewhere in the economy via companies supplying inputs and those adding value to the harvest product that is ultimately used by the consumer. For the four Gulf States considered in the analysis (Florida was excluded because the west coast could not be differentiated from the east coast), seafood industry jobs averaged 92,000 annually during 2007–2009. However, the four-state employment fell from 109,000 in 2007 to 63,000 in 2009. Income impacts for the four states equaled $1.3 billion in 2009 compared to $2.5 billion in 2007, a decline approaching 50 %.

Table 10.29 25 Most Frequently Caught (Left Panel) and Harvested (Right Panel) Species By Mississippi Recreational Anglers, 2000

Species Name	Number of Fish Caught (A + B1 + B2)	Species Name	Number of Fish Harvested (A + B1)	Mean Weight (lbs)
Sand Seatrout	1,138,498	Sand Seatrout	970,848	0.439
Atlantic Croaker	659,068	Southern Kingfish	486,699	0.512
Spotted Seatrout	622,625	Striped Mullet	231,832	0.868
Southern Kingfish	514,030	Spotted Seatrout	216,596	1.762
Striped Mullet	240,413	Atlantic Croaker	184,398	0.421
Pinfish	207,796	Southern Flounder	93,031	1.251
Red Drum	121,097	Red Drum	49,133	7.505
Southern Flounder	113,023	Sheepshead	41,556	3.422
Hardhead Catfish	65,285	Gulf Menhenden	30,768	0.203
Sheepshead	46,167	Black Drum	27,479	3.263
Gafftopsail Catfish	36,392	Gafftopsail Catfish	22,347	2.973
Gulf Menhaden	30,768	Pinfish	16,967	0.400
Black Drum	28,862	Atlantic Sharpnose Shark	11,171	5.934
Gray Snapper	23,384	Gray Snapper	8,750	0.707
Spanish Mackerel	14,331	Spanish Mackerel	7,634	1.551
Atlantic Sharpnose Shark	11,171	Gulf Kingfish	7,429	0.683
Red Snapper	9,231	Hardhead Catfish	6,494	0.999
Cobia	7,464	Red Snapper	6,379	4.750
Gulf Kingfish	7,429	Cobia	3,096	32.356
Gag	3,694	Blacktip Shark	2,797	18.257
Blacktip Shark	2,797	Tripletail	2,768	5.123
Tripletail	2,768	King Mackerel	2,305	8.043
King Mackerel	2,305	Gag	2,238	6.225
Unidentified Eel	2,273	Blue Runner	1,901	0.728
Blue Runner	1,901	Atlantic Spadefish	1,901	0.998

Source: NMFS FSD, data accessed 2012, with calculations by authors–see Appendix A. *Note*: 1 lb = 0.454 kg.

With respect to the recreational sector, topics considered include expenditures and impact, angler participation, trips, and catch and harvest. The analysis was based almost exclusively on MRFSS/MRIP statistics, the most continual and long-term monitoring program on recreational fishing patterns. Texas opted out of the program and, hence, is largely excluded from this chapter with the exception of expenditures and impacts. At the top end in terms of economic impacts, about 42,000 jobs were generated in Florida in response to recreational fishing activities with an associated $2.4 billion in income. At the bottom end, about 3,200 jobs were generated in Mississippi with additional income of $162 million. Louisiana was in the middle with the generation of almost 20,000 jobs and almost $1.0 billion in additional income.

Table 10.30 25 Most Frequently Caught (Left Panel) and Harvested (Right Panel) Species By Mississippi Recreational Anglers, 2005

Species Name	Number of Fish caught (A + B1 + B2)	Species Name	Number of Fish Harvested (A + B1)	Mean Weight (lbs)
Spotted Seatrout	1,154,866	Spotted Seatrout	316,834	1.297
Sand Seatrout	329,859	Sand Seatrout	222,240	0.492
Southern Kingfish	269,793	Southern Kingfish	216,090	0.518
Atlantic Croaker	241,377	Southern Flounder	72,485	1.231
Hardhead Catfish	230,865	Atlantic Croaker	40,813	0.256
Red Drum	131,312	Red Drum	35,422	11.521
Southern Flounder	101,119	Striped Mullet	34,028	0.886
Sheepshead	34,045	Sheepshead	27,646	4.352
Striped Mullet	34,028	Hardhead Catfish	12,174	1.335
Gafftopsail Catfish	32,746	Gulf Kingfish	8,895	0.869
Red Snapper	26,087	Spanish Mackerel	7,612	1.041
Blacktip Shark	11,162	Black Drum	6,850	1.452
Black Drum	10,136	Atlantic Sharpnose Shark	4,960	7.47
Spanish Mackerel	10,085	King Mackerel	4,940	10.431
Gulf Kingfish	8,895	Blacktip Shark	4,047	29.43
Atlantic Sharpnose Shark	4,960	Gafftopsail Catfish	4,012	3.688
King Mackerel	4,940	Pinfish	3,008	0.165
Pinfish	3,983	Tripletail	2,254	4.123
Tripletail	2,254	Cobia	1,196	32.915
Crevalle Jack	2,102	Red Snapper	1,003	2.249
Cobia	1,196	Lane Snapper	1,003	2.822
Lane Snapper	1,003	Florida Pompano	993	1.268
Florida Pompano	993	Finetooth Shark	878	10.307
Finetooth Shark	878	Blue Runner	388	0.841
Blue Runner	388	Crevalle Jack	271	1.102

Source: NMFS FSD, data accessed 2012, with calculations by authors–see Appendix A. Note: 1 lb = 0.454 kg.

Overall, marine recreational participation in three of the four states increased significantly since the mid-1990s with Mississippi being the sole exception. While participation has increased substantially, much of the growth occurred prior to the mid-2000s. It is likely that the combination of high fuel prices in recent times combined with the downturn in the U.S. economy, including Florida, negatively influenced participation and the number of trips.

While MRFSS/MRIP represents the primary data source for tracking participation over time, state-issued marine fishing license sales can also be used to track changes, subject to a number of caveats. A comparison between MRFSS/MRIP participation estimates and license sales for both Louisiana and Mississippi was made to determine whether license sales track

Table 10.31 25 Most Frequently Caught (Left Panel) and Harvested (Right Panel) Species By Mississippi Recreational Anglers, 2009

Species Name	Number of Fish Caught (A + B1 + B2)	Species Name	Number of Fish Harvested (A + B1)	Mean Weight (lbs)
Spotted Seatrout	2,049,332	Spotted Seatrout	1,090,094	1.431
Sand Seatrout	1,381,393	Sand Seatrout	1,003,126	0.441
Atlantic Croaker	1,038,030	Atlantic Croaker	339,728	0.310
Southern Flounder	328,421	Southern Flounder	209,197	1.161
Red Drum	320,663	Southern Kingfish	125,724	0.487
Hardhead Catfish	189,692	Striped Mullet	118,642	0.846
Southern Kingfish	184,865	Red Drum	83,976	8.662
Striped Mullet	121,651	Black Drum	77,811	3.685
Black Drum	112,968	Spanish Mackerel	22,680	1.458
Gafftopsail Catfish	63,073	Sheepshead	22,479	2.833
Requiem Shark Family	40,093	Atlantic Spadefish	19,978	1.153
Pinfish	38,521	Sunfish Genus	19,750	NA
Red Snapper	32,360	Requiem Shark Family	18,527	NA
Spanish Mackerel	29,523	Red Snapper	14,939	4.184
Sheepshead	27,645	Gafftopsail Catfish	7,181	2.258
Atlantic Spadefish	20,353	Gray Snapper	6,960	4.515
Sunfish Genus	19,750	Gulf Menhaden	5,763	NA
Bluegill	14,350	Hardhead Catfish	5,274	0.827
Bluefish	14,134	Bluefish	4,885	2.168
Gray Snapper	8,039	Gag	4,464	5.313
Gag	5,903	Pinfish	3,805	0.157
Gulf Menhaden	5,763	Blue Catfish	3,363	0.320
Blue Runner	4,850	Tripletail	2,963	4.668
Blue Catfish	3,363	Bluegill	2,870	0.364
King Mackerel	3,128	King Mackerel	2,850	9.668

Source: NMFS FSD, data accessed 2012, with calculations by authors–see Appendix A. *Note*: 1 lb = 0.454 kg.

MRFSS/MRIP estimates in a reasonable manner. Disturbingly, some significant differences were noted with MRFSS/MRIP estimates exceeding license sales by a large margin. While there are explanations for these observed differences (e.g., a license is not required for saltwater fishing in Louisiana if one is under the age of 16), the differences are large enough to justify further examination of MRFSS/MRIP participation data. The number of Gulf angler trips (excluding Texas) increased from about 17 million annually during the decade of the 1990s to 23 million annually during the most recent decade with a sharp increase in number of angler trips beginning in 2000. The explanation for this sharp increase in the number of angler trips is open to speculation but it coincides with a sharp increase in the number of nonresident

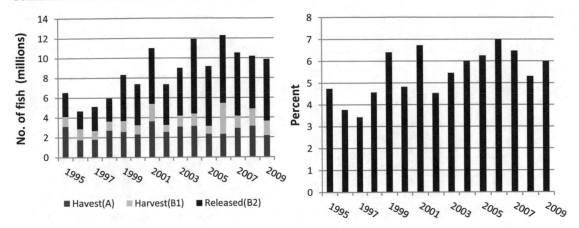

Figure 10.126. Alabama recreational catch and proportion of Gulf catch, 1995–2009 (NMFS FSD, data accessed 2012, with percentage calculations by authors—see Appendix A).

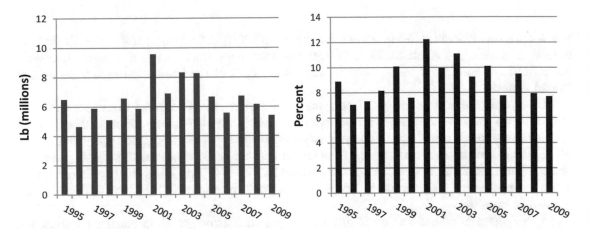

Figure 10.127. Alabama recreational harvest (*left panel*) and harvest in relation to the Gulf harvest (NMFS FSD, data accessed 2012, with percentage calculations by authors–see Appendix A) (*Note*: 1 lb = 0.454 kg).

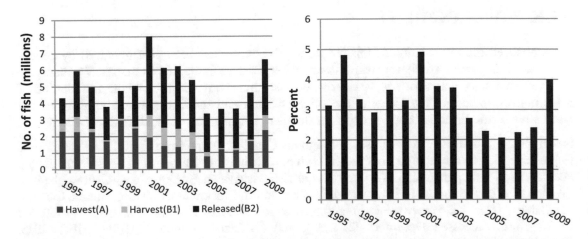

Figure 10.128. Mississippi recreational catch (*left panel*) and catch as a percent of Gulf total (*right panel*), 1995–2009 (NMFS FSD, data accessed 2012, with percentage calculations by authors–see Appendix A).

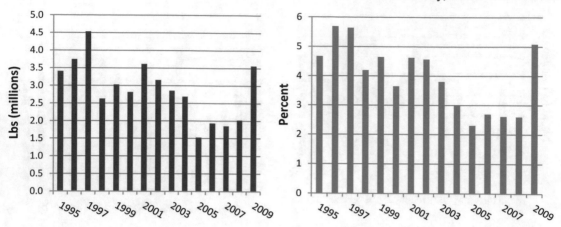

Figure 10.129. Mississippi recreational harvest (*left panel*) and Mississippi recreational harvest in relation to the Gulf harvest. (NMFS FSD, data accessed 2012, with percentage calculations by authors–see Appendix A) (*Note*: 1 lb = 0.454 kg).

participants in Florida. Florida accounted for about 70 % of the total Gulf trips during the period of analysis and about 50 % of Florida-based trips were in inland waters. Louisiana accounted for another 17 % of the total, and about 85 % of the Louisiana-based trips were in inland waters.

Given that the vast majority of Louisiana's fishing activities take place in inshore waters, it comes as no surprise that targeting behavior and catch are also largely associated with those species using inshore habitat; the two primary species are red drum and spotted seatrout. A full 50 % of all Louisiana-based angling trips target spotted seatrout and with a catch averaging about 20 million fish per year, the state accounts for 60 % of the Gulf's total spotted seatrout catch in numbers of fish. Similarly, Louisiana accounts for about 80 % of the Gulf's red drum harvest in pounds.

While there is considerable red drum and spotted seatrout catch in Florida's waters, the state can also make claim to a large offshore fishery component and reef fish is generally the target of offshore activities.

ACKNOWLEDGMENTS

BP sponsored the preparation of this chapter. This chapter has been peer reviewed by anonymous and independent reviewers with substantial experience in the subject matter. We thank the peer reviewers, as well as others, who provided assistance with research and the compilation of information, including the Fisheries Statistics Division and the Galveston Laboratory of the National Marine Fisheries Service and the Louisiana Department of Wildlife and Fisheries.

Small fish and shellfish images used throughout Chapter 10 are from GulfFINFO (http://gulf fishinfo.org/) with the exception of the following: (1) Mangrove Snapper (© 1992, Diane Rome Peebles, used with permission); (2) Grey Triggerfish (FishWatch.gov; http://www.fishwatch.gov); (3) Blue Marlin (Oceloti. 2014. Blue marlin fish illustration. iStockphoto, Calgary, Alberta, Canada. Available from http://www.istockphoto.com/vector/blue-marlin-fish-gm505255597-44750310? clarity=false. Accessed December 12, 2016); and (4) Black Grouper (THEPALMER. 2015. Black grouper engraving illustration. iStockphoto, Calgary, Alberta, Canada. Available from

http://www.istockphoto.com/vector/black-grouper-engraving-gm475800654-66202355?clarity=false.
Accessed December 12, 2016).

REFERENCES

Anonymous (2009) Stock Assessment of Gag Grouper in the Gulf of Mexico—SEDAR Update Assessment. Report of the Assessment Workshop, Miami, FL, U.S.A, March 30-April 2. NOAA (National Oceanic and Atmospheric Administration), Miami, FL, U.S.A. http://www.sefsc.noaa.gov/sedar/download/Red_Grouper_2009_Assessment_Update_Report.pdf?id=DOCUMENT. Accessed 16 Apr 2013

Asche F, Bennear LS, Oglend A, Smith MD (2012) U.S. shrimp market integration. Mar Res Econ 27:181–192

Carter DW, Agar J, Waters J (2008) Economic framework for fishery allocation decisions with an application to Gulf of Mexico Red Grouper. NOAA Technical Memorandum NMFS-SEFSC-576. NOAA Fisheries, Miami, FL, USA. 95 p

Christmas JY, McBee JT, Waller RS, Sutter FC (1982) Habitat suitability index models: Gulf Menhaden. U.S. Department of the Interior Fisheries and Wildlife Service FWS/OBS-82/10.23. U.S. Fish and Wildlife Service, National Wetlands Research Center, Lafayette, LA, USA. http://www.nwrc.usgs.gov/wdb/pub/hsi/hsi-023.pdf. Accessed 16 Apr 2013

Dedah C, Keithly WR, Kazmierczak RF (2011) An analysis of U.S. oyster demand and the influence of labeling requirements. Mar Res Econ 26:17–33

Deegan LA (1990) Effects of estuarine environmental conditions on population dynamics of young-of-the-year Gulf Menhaden. Mar Ecol Prog Ser 68:195–205

Diagne A, Keithly WR, Kazmierczak RF (2004) The effect of environmental conditions and regulatory costs on oyster relaying in Louisiana. Mar Resour Econ 19:211–224

Diop H, Keithly WR, Kazmierczak RF, Shaw RF (2007) Predicting the abundance of white shrimp (*Litopenaeus setiferus*) from environmental parameters and previous life stages. Fish Res 861:31–41

FFWCC (Florida Fish & Wildlife Conservation Commission, Fish & Wildlife Research Institute) (2010) Florida highly migratory species private angler telephone survey, Tallahassee, FL, USA. https://www.st.nmfs.noaa.gov/mdms/doc/11FL%20HMS%20PATS%20Final%20Report_Workgroup.pdf. Accessed 12 Feb 2013

Funk R, Griffin W, Mjelde J, Ozuna T, Ward J (1998) A method of imputing and simulating costs and returns in fisheries. Mar Resour Econ 13:171–183

Gauvin JR, Ward JM, Burgess EE (1994) Description and evaluation of the wreckfish (polypolon Americanus) fishery under an individual transferable quotas. Mar Resour Econ 9:99–118

Gentner B, Steinback S (2008) The economic contribution of marine angler expenditures in the United States, 2006. NOAA Technical Memorandum. NMFS-F/SPO-94. U.S. Department of Commerce, Washington, DC, USA. 301 p

GMFMC (Gulf of Mexico Fishery Management Council) (1981) Environmental impact statement and fishery management plan for the reef fish resources of the Gulf of Mexico. Gulf of Mexico Fishery Management Council, Tampa, FL, USA. http://www.gulfcouncil.org/Beta/GMFMCWeb/downloads/RF%20FMP%20and%20EIS%201981-08.pdf

GMFMC (2011) Final Reef Fish Amendment 32 (including Final Environmental Impact Statement, Regulatory Impact Review, and Regulatory, FL, USA Flexibility Analysis, Fishery

Impact Statement). NOAA, St. Petersburg, FL, USA. http://www.gulfcouncil.org/docs/amendments/Final%20RF32_EIS_October_21_2011[2].pdf Accessed 16 Apr 2013

Guillory V, Geaghan J, Roussell J (1983) Influence of environmental factors on Gulf Menhaden Recruitment. Technical Bulletin 37. Louisiana Department of Wildlife and Fisheries, Baton Rouge, LA, USA. 32 p

Haas H, Lamon EC, Rose K, Shaw R (2001) Environmental and biological factors associated with stage-specific brown shrimp abundances in Louisiana, applying a new combination of statistical techniques to recruitment data. Can J Aquat Sci 58:2258–2270

Hood PB, Stelcheck AJ, Steele P (2007) A history of red snapper management in the Gulf of Mexico. In: Patterson WF III, Cowan JH, Fitzugh GR, Nieland DL (eds) Red snapper ecology and fisheries of the U.S. Gulf of Mexico. American Fisheries Society, Bethesda, MD, USA, pp 267–284

Jones K, Harvey DJ, Hahn W, Muhammad A (2008) U.S. demand for source-differentiated shrimp: A differential approach. J Agr Appl Econ 40:609–621

Kazmierczak RF, Keithly WR (2005) The costs of harvesting oysters from private leases in Louisiana. Final Report. Louisiana Department of Natural Resources, Baton Rouge, LA, USA. 29 p

Keithly WR (2001) Initial allocation of ITQs in the Gulf of Mexico Red Snapper Fishery. In Shotten R (ed) Case studies on the allocation of transferable quota rights in fisheries. FAO Fisheries Technical Paper 411. FAO (Food and Agriculture Organization of the United Nations) Rome, Italy, pp 99–118

Keithly WR, Diop H (2001) The demand for eastern oysters, crassostrea virginica, from the Gulf of Mexico Oysters in the presence of *Vibrio vulnificus*. Mar Fish Rev 63:47–53

Keithly WR, Kazmierczak RF (2006) Economic analysis of oyster lease dynamics in Louisiana. Final Report. Louisiana Department of Natural Resources, Baton Rouge, LA, USA. 34 p

Keithly WR, Poudel P (2008) The Southeast U.S.A. shrimp industry: Issues related to trade and antidumping duties. Mar Resour Econ 23:459–483

Keithly WR, Roberts KJ (1994) Shrimp closures and their impact on the Gulf Region processing and wholesaling sector (Expanded to Include South Atlantic). Final Report. Contract NA17FF0376-01. National Marine Fisheries Service, Washington, DC, USA. 106p + Appendix

Keithly WR, Diop H, Kazmierczak RF, Travis M (2006) The impacts of imports, particularly farm-raised product, on the Southeast U.S. Shrimp processing sector. Final Report. Gulf and South Atlantic Fisheries Foundation, Tampa, FL, USA. 50 p

LDWF (Louisiana Department of Wildlife and Fisheries) (2007) Oyster stock assessment report of the public oyster areas in Louisiana seed grounds and seed reservations (Oyster Data Report Series). http://www.wlf.louisiana.gov/sites/default/files/pdf/page_fishing/32695-Oyster%20Program/2007-oyster-stock-assessment-report.pdf. Accessed 18 Apr 2013

LDWF (2008) Oyster stock assessment report of the public oyster areas in Louisiana seed grounds and seed reservations (Oyster Data Report Series). http://www.wlf.louisiana.gov/sites/default/files/pdf/page_fishing/32695-Oyster%20Program/2008-oyster-stock-assessment.pdf. Accessed 18 Apr 2013

LDWF (2009) Oyster stock assessment report of the public oyster areas in Louisiana seed grounds and seed reservations (Oyster Data Report Series). http://www.wlf.louisiana.gov/sites/default/files/pdf/page_fishing/32695-Oyster%20Program/2009_oyster_stock_assessment.pdf. Accessed 18 Apr 2013

LDWF (2010) Oyster stock assessment report of the public oyster areas in Louisiana seed grounds and seed reservations (Oyster Data Report Series). http://www.wlf.louisiana.gov/

sites/default/files/pdf/page_fishing/32695-Oyster%20Program/2010-oyster-stock-assess ment-report.pdf. Accessed 18 Apr 2013

Liese C, Travis MD (2010) The annual economic survey of Federal Gulf Shrimp permit holders: Implementation and descriptive results for 2008. NOAA Technical Memorandum NMFS-SEFSC-601. U.S. Department of Commerce, NOAA, Washington, DC, USA. http://docs.lib.noaa.gov/noaa_documents/NMFS/SEFSC/TM_NMFS_SEFSC/NMFS_SEFSC_TM_601.pdf. Accessed 16 Apr 2013

Liese C, Travis MD (2011) 2009 Economics of the federal Gulf shrimp fishery annual report. National Marine Fisheries Service, Southeast Fisheries Science Center, Miami, FL, USA. http://www.sefsc.noaa.gov/docs/2009%20Gulf%20shrimp%20econ%20report.pdf. Accessed 17 Apr 2013

Miller AL, Isaacs JC (2011) Amended February 2012. An economic survey of the Gulf of Mexico inshore shrimp fishery: Implementation and descriptive results for 2008. Gulf States Marine Fisheries Commission, Ocean Springs, MS, USA. (GSMFC publication 195). http://www.gsmfc.org/publications/GSMFC%20Number%20195.pdf Accessed 16 Apr 2013

Nance JM (2011) Stock assessment report, 2010: Gulf of Mexico shrimp fishery. Report to the Gulf of Mexico Fishery Management Council. NMFS (National Marine Fisheries Service), Southeast Fisheries Science Center, Galveston Lab, Galveston, TX, USA. 16 p

Nance JM, Keithly WR, Caillouet C, Cole J, Gaidry W, Gallaway B, Griffin W, Hart R, Travis M (2006) Estimation of effort, maximum sustainable yield, and maximum economic yield in the shrimp fishery of the Gulf of Mexico. Report to the Gulf of Mexico Fishery Management Council, Tampa, FL, USA. http://www.gulfcouncil.org/Beta/GMFMCWeb/downloads/FINAL_AdHocEffortReport_1.pdf. Accessed 16 Apr 2013

National Research Council (2006) Review of recreational fisheries survey methods. National Academies Press, Washington, DC, USA. 202 p

Nichols S (1986) Stock assessment of brown, white, and pink shrimp in the U.S. Gulf of Mexico, 1960–85. NOAA Technical Memorandum NMFS-SEFC-179. NOAA, Washington, DC, USA. https://grunt.sefsc.noaa.gov/P_QryLDS/download/TM201_TM-203.pdf?id=LDS. Accessed 16 Apr 2013

NMFS (Sustainable Fisheries Branch) (2010) Forecast for the 2010 Gulf and Atlantic Menhaden Purse-Seine fisheries and review of the 2009 fishing season. Sustainable Fisheries Branch, Beaufort, NC, USA. http://www.st.nmfs.noaa.gov/st1/market_news/menhaden%20forecast%202010.pdf. Accessed 12 Feb 2013.

NMFS (2011) Gulf of Mexico 2010 red snapper individual fishing quota annual report. SERO-LAPP-2011-09. National Marine Fisheries Service, Southeast Regional Office, Miami, FL, USA. http://sero.nmfs.noaa.gov/sf/pdfs/2010_RS_AnnualReport_Final%2010-28-11.pdf. Accessed 16 Apr 2013

NMFS (Galveston Laboratory) (2012) Forecast for the 2012 brown shrimp season in the Western Gulf of Mexico, from the Mississippi River to the U.S.—Mexico border (memo dated July 2, 2012). http://www.galvestonlab.sefsc.noaa.gov/stories/2012/Brown%20Shrimp%20Forecast/index.html

NMFS Fisheries Statistics Division (FSD) (data accessed 2012) Personal communication from the NMFS FSD, Silver Spring, MD, USA. http://www.st.nmfs.noaa.gov/st1/index.html. Accessed 12 Feb 2013

Ortiz M (2006) Status review of the Gag Grouper in the U.S. Gulf of Mexico (SEDAR 10 RW-01; also list as SEDAR 10 RW-02 on website). Presented at the SEDAR 10 Review Workshop, Atlanta, GA, USA, June 26–30. 65 p

Ortiz M, Scott GP, Cummings N, Phares PL (2002) Stock assessment analyses of the Gulf of Mexico king mackerel. MSAP/02/01. NMFS SEFSC Miami Sustainable Fisheries Division Contribution 01/02-161. National Marine Fisheries Service, Miami, FL, USA. 56 p

OTTF (Oyster Technical Task Force) (2012) The oyster fishery of the Gulf of Mexico, United States: A regional management plan—2012 revision. Publication No. 202, Gulf States Marine Fisheries Commission, Ocean Springs, MS, USA. http://www.gsmfc.org/publica tions/GSMFC%20Number%20202.pdf. Accessed 17 Apr 2013

Poudel P (2008) An analysis of the world shrimp market and the impact of an increasing import base on the Gulf of Mexico dockside price. MS Thesis, Louisiana State University, Baton Rouge, LA, USA. http://etd.lsu.edu/docs/available/etd-01242008-125555/unrestricted/pou del_thesis.pdf. Accessed 17 Apr 2013

Ran T, Keithly WR, Kazmierczak RF (2008) Modeling the intertemporal and spatial supply dynamics of the Gulf of Mexico shrimp fleet. Final Report. National Marine Fisheries Service, National Oceanic and Atmospheric Administration, Washington, DC, USA. 160 p

Ran T, Keithly WR, Kazmierczak RF (2011) Location choice behavior of Gulf of Mexico shrimpers under dynamic economic conditions. J Appl Econ 43:29–42

Roberts KJ, Keithly WR (1991) The role of small shrimp in determining economic returns. Final Report. National Marine Fisheries Service Contract NA89WC-H-MF010. NMFS, Miami, FL, USA. 51 p

Savolainen M, Caffey RH, Kazmierczak RF (2012) The recreational for-hire sector in the U.S. Gulf of Mexico: Structural and economic observations from the third decadal survey. In: Creswell RL (ed) Proceedings of the 64th Gulf and Caribbean Fisheries Institute Conference, Gulf and Caribbean Fisheries Institute, Puerto Morelos, Mexico, pp 102–113. http://procs.gcfi.org/pdf/GCFI_64-22.pdf

Travis MD, Griffin WL (2004) Update on the economic status of the Gulf of Mexico commercial shrimp fishery. SERO-ECON-04-01. http://sero.nmfs.noaa.gov/sf/socialsci/ pdfs/EconUpdateGulfShrFinal.pdf. Accessed 12 Feb 2013

U.S. Department of Commerce (2011) Fisheries economics of the United States, 2009: Economics and sociocultural status and trends series. U.S. Department of Commerce, NOAA Technical Memorandum NMFS-F/SPO-118. https://www.st.nmfs.noaa.gov/st5/publication/ index.html. Accessed 12 Feb 2013

U.S. Department of Interior, Fish and Wildlife Service, and U.S. Department of Commerce, U.S. Census Bureau (2008) 2006 National Survey of Fishing, Hunting, and Wildlife-Associated Recreation: Louisiana. (FHW/06-LA) Washington, DC, USA. 47 p + Appendixes

Van der Kooy SJ (2010) Law summary 2009. A summary of marine fishing laws and regulations for the Gulf States. Gulf States Marine Fisheries Commission, Ocean Springs, MS, USA. http://www.gsmfc.org/publications/GSMFC%20Number%20184.pdf

APPENDIX A

The National Marine Fisheries Service (NMFS) maintains a large number of databases related to the catch of commercial and recreational marine species. Many of the more frequently used databases are available to the public on line and other databases are made available upon request to the appropriate unit within NMFS. With respect to commercial fishery statistics, landings data can be accessed by logging onto the website http://www.st.nmfs.noaa. gov/commercial-fisheries/index and following the link to "Commercial Landings." Here, annual commercial landings can be downloaded by species (pounds and value) by state or region on either an annual or monthly basis. The annual databases for commercial landings extend back to 1950 while the monthly databases extend back to 1990. These databases served as the primary source for much of the commercial landings information and figures presented in this chapter. For example, annual commercial data from the website was used to generate Figures 10.1, 10.2, and 10.13 while monthly commercial data from the website was used to generate Figures. 10.12 and 10.14. In addition, this link also provides relevant information on landings by gear.

While this source provides considerable information on commercial landings by species, it is presented only at an aggregate species level and more detailed information can often be obtained via a request to the appropriate regional NMFS laboratory. For example, while shrimp landings by species (brown, white, etc.) can be downloaded from the http://www.st.nmfs.noaa.gov/commercial-fisheries/index website, many different shrimp sizes are landed and the price per pound can vary significantly depending upon size. Detailed information of this nature requires a request being made to the appropriate NMFS Laboratory, with the Galveston Laboratory maintaining the more detailed shrimp records. These records include landings by size count (e.g., Figure 10.21), harvest by area (e.g., Figure 10.30) and effort expressed in 24-h fishing days in total and by species (e.g., Figure 10.20).

When considering the U.S. commercial seafood industry, the role of imports (or exports) should be considered. Imports add to the total U.S. supply and U.S. consumption is a function of domestic landings and imports less any exports. The National Marine Fisheries Service maintains extensive databases on fishery product imports and exports differentiated by country of origin (for imports) and product forms. Data from these databases is provided on both an annual and monthly basis and can be downloaded by logging onto the website http://www.st.nmfs.noaa.gov/commercial-fisheries/index and following the link to "Foreign Trade."

A final component that should be considered when examining the commercial fishing industry is the processing sector. Processing activities, by transforming the harvested product into product forms desired by consumers, adds value to the landed product via the marketing services it provides. As discussed in Section 10.3.9 of this chapter, there are two primary data sources related to processing activities. One is referred to as *the voluntary end-of-the year processor survey*; data used in this survey is collected and maintained by NMFS. Data collected and maintained under the auspices of this survey is detailed and includes for each processing establishment: (a) processed pounds, by product form, and value associated with each species being processed by that establishment, (b) the location of the processing establishment, and (c) monthly employment. This database, which includes the use of both domestic and imported raw product, was used to generate the figures associated with shrimp processing activities (i.e., the figures in Section 10.3.8.1.4). While detailed information associated with this annual survey (e.g., processing activities for individual species or by region) is not routinely published by NMFS, specific requests can be made by contacting the NMFS Office of Science and Technology Headquarters located in Silver Spring, Maryland, USA. The second data source of relevance to processing activities is contained in annual *Fisheries Economics of the United States* reports. Information given in these reports was discussed in Section 10.3.9 of this chapter and is not repeated here.

Given the increasing economic importance of recreational marine fishing activities and the relevancy of these fishing activities in the management process, NMFS also collects and analyzes these activities. Detailed information on recreational activities, such as that included in this chapter can be viewed by logging onto http://www.st.nmfs.noaa.gov/recreational-fisheries/index and then following the link to the "Access Data" site and then to the "Run a Query" site. From there, several "pull down menus" are presented including "Select a Catch Query," "Select an Effort Query," and "Select a Participation Query." The "Select a Catch Query" menu provides the data to analyze recreational catch and harvest in aggregate, such as that presented in Figure 10.64 (left panel) and Figure 10.65 (left panel), as well as by individual species (such as,

Figure 10.67). Information can be generated in terms of either number of fish or pounds, given certain limitations, as well as by state or region. The "Select Effort Query" permits analysis of recreational activities in terms of number of trips such as that presented in Figure 10.61. Using this "pull down menu," trips by mode and area fished, such as presented in Figure 10.61, can be examined.[27] Combining information generated from the "Select a Catch Query" menu and the "Select an Effort Query" allows examination of the catch (harvest) per trip such as that presented in Figure 10.64 (right panel). The data associated with both catch and effort are collected and maintained in 2-month waves (January/February, ..., November/December) which also allows for seasonal analysis of both catch (e.g., Figure 10.69; left panel) and effort (e.g., Figure 10.62). Finally, the "Select a Participation" query gives the number of fishermen by state such as that presented in Section 10.4.3.1 of this chapter.

In addition to these "readymade" queries, more detailed data sets pertaining to MRFSS and MRIP can be downloaded from the "Access Data" site by selecting the "Download Data" option. This allows development of customized programming options and the examination of data in greater detail (e.g., county level). The data used to generate the tables reporting the 25 most commonly caught and harvested presented in this chapter, as well as targeted trips information, were derived from these databases. In addition, this site presents details regarding available information.

It is important to recognize, however, that when programs are customized for analysis, assumptions must be made at several steps of the analysis that can influence final results. One specific example related to the current analysis is that associated with targeted trip estimates given in this document. Specifically, when the document was being prepared, the website had no "readymade" query for targeted trips and the authors utilized a program originally developed by the NMFS, Southeast Regional Office (provided by Stephen Holiman) to generate the targeted trip estimates. An assumption was made in the development of this program that if, for example, only one person in a fishing party of four was interviewed and that person indicated targeting a given species than the other three members of that party would also be targeting the same species. This assumption is probably realistic in most cases but if this assumption is not made, targeted trip estimates will generally differ by a relatively small amount. Since completion of this chapter, the NMFS has added a "readymade" query for targeted trips and estimates from this query do differ (generally by a small amount) from those given.

Open Access   This chapter is licensed under the terms of the Creative Commons Attribution-NonCommercial 2.5 International License (http://creativecommons.org/licenses/by-nc/2.5/), which permits any noncommercial use, sharing, adaptation, distribution and reproduction in any medium or format, as long as you give appropriate credit to the original author(s) and the source, provide a link to the Creative Commons license and indicate if changes were made.

The images or other third party material in this chapter are included in the chapter's Creative Commons license, unless indicated otherwise in a credit line to the material. If material is not included in the chapter's Creative Commons license and your intended use is not permitted by statutory regulation or exceeds the permitted use, you will need to obtain permission directly from the copyright holder.

---

[27] Since the completion of this chapter an additional query has been added to the "Select an Effort Query" menu, that is the "Directed Trip" query. This query was not available at the time of the analysis but results presented in this paper should closely match the information given by using this query.

CHAPTER 11

SEA TURTLES OF THE GULF OF MEXICO

Roldán A. Valverde[1] and Kym Rouse Holzwart[2]

[1]Southeastern Louisiana University, Hammond, LA 70402, USA; [2]Ramboll Environ, Inc., Tampa, FL 33610, USA
roldan.valverde@selu.edu

11.1 INTRODUCTION

Five species of sea turtles are found in the Gulf of Mexico: the Kemp's ridley (*Lepidochelys kempii*), loggerhead (*Caretta caretta*), green (*Chelonia mydas*), leatherback (*Dermochelys coriacea*), and hawksbill (*Eretmochelys imbricata*). While individuals of some species of sea turtles may nest on beaches and spend nearly their entire lives in the Gulf of Mexico, such as the Kemp's ridley, others may only use the Gulf to nest, as a foraging area, or as part of their migration routes. The Gulf of Mexico provides important sea turtle nesting habitat, and many Gulf of Mexico beaches where sea turtles nest have been protected as refuges and parks. For example, protected beaches in Rancho Nuevo, Mexico and at Padre Island National Seashore (PAIS), Texas, are the major nesting beaches for the Kemp's ridley, and loggerheads and green sea turtles nest on beaches at Dry Tortugas and Everglades National Parks in Florida. Sea turtles often spend their post-hatchling and early juvenile years in the pelagic Gulf of Mexico (Witherington et al. 2012), and the nearshore waters of the Gulf provide critical foraging habitat for juvenile and adult sea turtles, as well as important mating and internesting habitat (Musick and Limpus 1997; Bolten et al. 1998; Hopkins-Murphy et al. 2003). The northern Gulf of Mexico can be divided oceanographically into the eastern half and the western half. The eastern half is influenced strongly by Caribbean inflow and has relatively clear water, while the western half is influenced by the turbid Mississippi River, incurs significant shrimp trawling pressure, and has thousands of oil and gas production structures (Hopkins-Murphy et al. 2003).

This chapter describes Gulf of Mexico sea turtle populations prior to the Deepwater Horizon incident that occurred on April 20, 2010, especially with regard to abundance and distribution. The types of information summarized for sea turtles that use the Gulf of Mexico for at least some portion of their life cycle includes life history, distribution, and abundance; location of nesting beaches and nesting numbers; and habitat use and foraging area locations for the various life stages. The natural and anthropogenic threats that affect sea turtles in the Gulf of Mexico are also discussed, including sea turtle stranding, fisheries bycatch, and other types of less common but important impacts, such as boat strikes.

Sea turtles that occur in the Gulf of Mexico are not randomly or evenly distributed spatially or temporally. They are difficult to study and monitor since they are broadly distributed, with nesting aggregations in various locations of the Gulf, have wide-ranging migrations, have long generation times and long life spans, and spend the majority of their lives at sea (Holder and Holder 2007; Witherington et al. 2009; NRC 2010). In addition, the oceanic habitat of juveniles is a major obstacle to studying immature stages of sea turtles (NRC 2010). For these reasons, the most common sea turtle population assessments have been made at nesting beaches (Schroeder and Murphy 1999). Counts of sea turtle nests provide an index of annual population productivity and an approximate index of abundance for adult females (Karnauskas

© The Author(s) 2017
C.H. Ward (ed.), *Habitats and Biota of the Gulf of Mexico: Before the Deepwater Horizon Oil Spill*,
DOI 10.1007/978-1-4939-3456-0_3

et al. 2013). However, this segment of the population represents a very small portion of the sea turtle population, and determining population sizes of sea turtles, as well as quantifying impacts on juvenile and adult males and females, is extremely challenging (NRC 2010). This type of population assessment is similar to estimating human population trends by counting women in maternity wards: while useful information is obtained, if the children were decimated from an impact, the mortality would not be detected in the adult population for decades (Bjorndal et al. 2011). The need for assessing populations of both juvenile and adult sea turtles in the water to complement assessments of nesting beaches has been widely recognized (Magnuson et al. 1990; TEWG 1998, 2000). Due to advances in genetic analyses, satellite telemetry technology and the development of new methodologies and technologies over the past 20 years, as well as the growing long-term monitoring and tagging sea turtle datasets, much has been learned regarding where sea turtles go and what they do when they are not on the nesting beach. Nevertheless, significant data gaps remain.

11.1.1   Generalized Life History of Gulf of Mexico Sea Turtles

Three basic ecosystem zones characterize the life history patterns of the five species of sea turtles that occur in the Gulf of Mexico (Figure 11.1) (NMFS et al. 2011).

1. Terrestrial Zone: the nesting beach where females lay eggs and where embryos develop.
2. Neritic Zone: the inshore marine environment from the surface to the seafloor, including bays, sounds, and estuaries, as well as the continental shelf, where water depths do not exceed 200 meters (m) (656.2 feet [ft]).
3. Oceanic Zone: the open ocean environment from the surface to the seafloor where water depths are greater than 200 m (656.2 ft).

On the nesting beach sea turtle eggs require a high-humidity environment, an incubation temperature between 25 and 35 degrees Celsius (°C), and adequate conditions for gas exchange for proper development (Ackerman 1997). The length of the incubation period varies and is inversely related to nest temperature; the warmer the sand surrounding the egg chamber, the

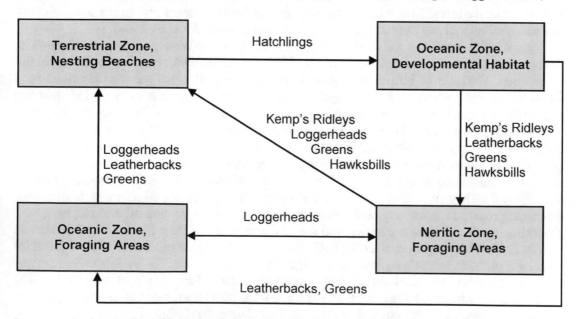

Figure 11.1. Generalized life cycle of sea turtle species that occur in the Gulf of Mexico.

faster the embryos develop (Mrosovsky and Yntema 1980; Ackerman 1997). Sex is determined by incubation temperatures prevailing during the middle third of the incubation period for all sea turtle species (Mrosovsky and Yntema 1980; Wibbels 2007). Each species has a pivotal temperature, which is the temperature at which the sex ratio is one to one. Nest temperatures higher than the pivotal temperature produce mostly females, and lower temperatures produce mostly males (Witherington et al. 2006a).

Immediately after emerging from the nest, sea turtle hatchlings begin a period of frenzied activity. During this period, they move from their nest to the surf, swim, and are swept through the surf zone (Witherington 1995; Conant et al. 2009). The hatchlings use a progression of orientation cues as they crawl to the water, swim through the surf, and migrate offshore (Lohmann and Lohmann 2003; Lohmann et al. 2012). Once they reach the oceanic zone, hatchlings spend a number of years growing and developing.

Some sea turtle species return to the neritic zone as juveniles (e.g., Kemp's ridleys, loggerheads, greens, hawksbills), while others stay in the oceanic zone (e.g., leatherbacks) (Figure 11.1). Adults of some species remain in the neritic zone their entire lives (e.g., Kemp's ridleys), and some move back and forth between the neritic and oceanic zones (e.g., logger-heads). Some sea turtles spend their entire lives in the oceanic zone (e.g., leatherbacks), with the exception of females nesting in the terrestrial zone.

While differences exist between and within species, adult females typically return to nest in the general vicinity of the beach where they hatched from eggs many years earlier and often nest at the same beach throughout their reproductive years. For example, while most logger-heads return to nest at the same beach from which they were hatched, individual loggerhead sea turtles have been known to nest on both the Atlantic and Gulf coasts of Florida (LeBuff 1974). Green turtles typically nest on the same beach where they hatched (Bowen et al. 1989; Allard et al. 1994).

11.1.2 Historical Abundance of Gulf of Mexico Sea Turtles

Sea turtles were once highly abundant throughout the Gulf of Mexico and the Caribbean. By some estimates, they may have numbered in the millions (Jackson 1997). However, since the discovery of the New World, their rookeries in the region have decreased significantly, mainly due to overexploitation of these reptiles for their meat, shell, and eggs. Because of the high quality of its meat, the impact of the direct take of juvenile and adult green turtles has historically been more pronounced than for any other sea turtle species. It has been suggested that Caribbean green turtle populations have declined as much as 99 percent (%) since the arrival of Christopher Columbus (Bowen and Avise 1995; Jackson 1997). There are numerous accounts of consumption of this species throughout the region. In fact, the name of the turtle does not derive from the outer coloration of this animal but from the color of the green fat found under the shell, for which this turtle was considered a delicacy by British royalty in the eighteenth and nineteenth centuries (Witzell 1994). Consumption of this species fueled early exploitation by local artisanal and industrial fisheries, leading to the exploitation of other sea turtle species. To appreciate the magnitude of the impacts, it is important to examine the history of green turtle exploitation in the greater Caribbean.

In his book *The Green Turtle and Man*, James Parsons (1962) documents the use of green turtles and their eggs in the region by Europeans and New World settlers. Green turtles were sought after for their calipee and calipash, the cartilage associated with the plastron and the carapace, respectively. These products were used by the English aristocracy to make turtle soup, which became a staple after the discovery of the New World. Besides supplying the English kitchens with gourmet soup, green turtle oil was also used as a substitute for butter, lamp fuel, and as a lubricant. The green turtle trade between London and the West Indies began in the mid-eighteenth century. The calipee was more abundant, with a large turtle producing between 1.1 and 1.6 kilograms (kg) (2.5 and 3.5 pounds [lb]). The English used green turtle soup as a cure for scurvy in the long transatlantic voyages and as a substitute for, or in addition to, salted beef, since turtles could be kept alive below deck on their backs for weeks. Green sea turtles were abundant and supported the exploration and settlement of the greater Caribbean, providing sailors and pioneer settlers with fresh meat. By 1878, it was estimated that some 15,000 Cayman Islands green turtles had been landed in London, ranging from 11 to 136 kg (25 to 300 lb). In 1880, imports of preserved turtles (sun-dried meat and calipee) amounted to 4,899 kg (10,800 lb). That year, a Key West factory had an estimated production of 200,000 cases of calipee. Although London was the main market, New York also constituted an important market for these products. In 1883, the largest factory of green turtle soup was Moore & Company Soups, Inc. of Newark, New Jersey. By that year, imports of live green turtles into the United States amounted to 468,646 kg (1,033,187 lb), mainly from Mexican and Nicaraguan waters. At an estimated 73 kg (160 lb) per turtle, this would be equivalent to 6,457 turtles, which did not include turtles caught in Florida and those sacrificed for calipee.

The high demand for green turtle meat and soup contributed significantly not only to the demise of many Caribbean populations, but also to those in the Gulf of Mexico. In the western Gulf, green turtles were once abundant enough to support meat and soup canneries in Texas (Groombridge and Luxmoore 1989). At least five green turtle canneries existed in the late 1800s along the coast of the northwestern Gulf of Mexico. Green and loggerhead turtle fisheries supplied these canneries (Figure 11.2). To give an idea of the magnitude of this fishery, it was reported that in 1890, a total of 265,000 kg (584,225 lb) of green turtles were caught in Texas (Hildebrand 1982). The canneries were located in Fulton, Rockport, Indianola, Point Isabel, and

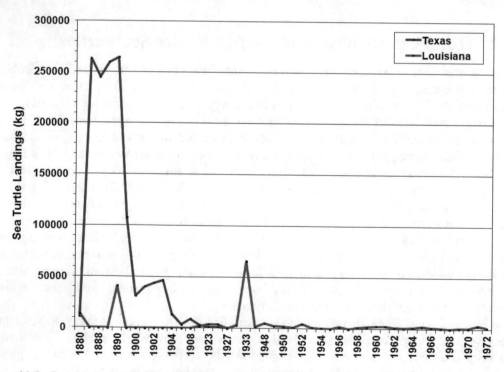

Figure 11.2. Sea turtle landings in Texas and Louisiana for available years from 1880 through 1972. The records include mainly green and loggerhead sea turtles (Rebel 1974; Cato et al. 1978; Doughty 1984).

Corpus Christi, Texas (True 1887; Doughty 1984). The Fulton cannery was the largest and operated between 1881 and 1896; in 1890, this cannery alone processed 900 green turtles (41.7 % of the turtle catch in the state) totaling 110,223 kg (243,000 lb) and produced about 40,000 0.9-kg (2-lb) cans (Doughty 1984). The green turtle cannery located in Rockport was founded around 1886. During its first 6 months of operation, the cannery processed about 3,856 kg (8,500 lb) of green turtle meat (True 1887). Green turtles caught and processed in Texas were presumed to feed in the seagrass beds located between Matagorda Bay and Laguna Madre (Hildebrand 1982). Rapid depletion of green turtle populations in Texas and elsewhere occurred because the fishery targeted juvenile turtles, a highly vulnerable stage in the life cycle of this slow-growing, late-maturing species (Crouse et al. 1987; Witzell 1994). While the green turtle fishery in Texas developed very quickly, it declined abruptly after 1892 and ended shortly thereafter, presumably due to the scarcity of turtles and to a deep freeze along the Texas coast (Hildebrand 1982). A small turtle fishery in Louisiana remained open through the early 1970s (Figure 11.2).

Prior to 1860, green turtles caught on the Florida east coast, particularly in the Indian River Lagoon, were exchanged for goods with various merchant vessels; in later years, agents purchased the catch and then shipped it mainly to New York (True 1887). By 1887, green, loggerhead, and hawksbill turtles were hunted as far north as Beaufort and Morehead City, North Carolina, where green turtles in particular were a delicacy and consumed locally (True 1887). There was no mention of Kemp's ridley sea turtles being harvested in the early records, even though this species surely was present in Florida at the time (Carr 1957). The omission appears to have occurred because Kemp's ridleys were sold as loggerheads for years (Rebel 1974; Cato et al. 1978).

Two locations on Florida's west coast—Key West and Homosassa—appear to have supported the most abundant in-water populations of green turtles in the entire Gulf of Mexico in the late 1800s. Captured turtles weighing between 18 and 45 kg (40 and 100 lb) were kept alive in small Kraals (Dutch for corral) or seawater-filled holding pens until ready to be shipped to New York (True 1887). Estimates indicate that about 50 18-kg (40-pound) turtles per week were brought to Key West throughout the year (True 1887). The green turtle fishery in the Cedar Keys area presumably arose around 1878. Fishing was concentrated in an area 32–48 kilometers (km) (20–30 miles [mi]) north and south from the main Cedar Keys port, with the shallow foraging grounds being the most productive (True 1887). Large boats brought between 1,361 and 2,268 kg (3,000 and 5,000 lb) of green turtles to port; whereas, small boats brought in only 23–363 kg (50–800 lb). Interestingly, the largest green turtle recorded at the time weighed an impressive 544 kg (1,200 lb) (True 1887). The reported weight of this turtle was questioned by Carr and Caldwell (1956), who indicated that green turtles landed in the Cedar Keys fisheries in the 1950s were no larger than 52 kg (115 lb). Alternatively, it is possible that the discrepancy reflects the impact of the decades-long turtle fishery in the region, leaving no large adult turtles in the population.

Overall, statistics show that the Florida west coast produced about 81,647 kg (180,000 lb), Louisiana about 13,608 kg (30,000 lb), and Texas approximately 24,494 kg (54,000 lb) of green turtle meat in 1880, though apparently an unspecified amount of freshwater turtle meat was also included in these records (True 1887; Rebel 1974). No mention is made of significant sea turtle fisheries for any other Gulf coast state prior to 1880 besides Florida, Louisiana, and Texas. By 1887, the most important sea turtle fisheries in the Gulf of Mexico were those of the Cedar Keys area and around Key West (True 1887; Townsend 1899). Figure 11.3 shows statistics of turtle landings on the Florida Gulf coast between 1880 and 1897 for the years for which data are available (Townsend 1899), and provides a perspective of turtle demand in the northeastern Gulf of Mexico.

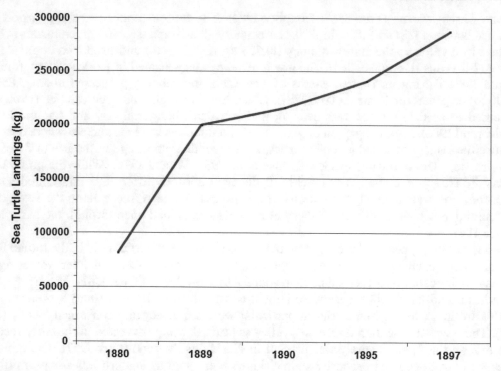

Figure 11.3. Turtle landings on the Florida Gulf coast for available years from 1880 through 1897. The data include mostly green sea turtles, along with loggerheads, hawksbills, and most likely Kemp's ridleys (from Townsend 1899).

Until 1890, the sea turtle fisheries in Florida occurred in eight counties, but by 1897 they were concentrated in four: Monroe, Levy, Franklin, and Escambia counties on the Gulf coast. The total production in 1897 alone was 287,857 kg (634,616 lb), with 86 % coming from Monroe County at the southern tip of Florida (Townsend 1899). However, an unspecified portion of the total production came from the Yucatán coast because turtles were already becoming scarce in Florida; by 1897, most of the turtles came from the Yucatán coast. Apparently, the decrease in the sea turtle populations along the Florida Gulf coast was due not only to the harvest of juveniles but also of eggs, as these were sought after eagerly by local people (Townsend 1899). Indeed, although few records of egg exploitation exist, it is believed that a large number of eggs were collected throughout the entire rim of the Gulf of Mexico (Hildebrand 1963; Witzell 1994). Turtles were so scarce by the late 1800s that Townsend (1899) called for the protection of the turtles and their eggs during their breeding season. The sea turtle fisheries continued through the early and mid-1900s on the Florida Gulf coast, though at a much lower rate (Figure 11.4), presumably due to decreases in the turtle populations (Rebel 1974). The landings records for Florida indicate a preference for green turtle meat (Figure 11.4).

In the 1900s, imports of live turtles into the United States were significant prior to the 1978 listing of sea turtles under the 1973 U.S. Endangered Species Act (ESA). The total live sea turtle imports into the United States from 1948 through 1976 amounted to 8,099,950 kg (17,857,334 lb), with a peak in 1951 and a decreasing trend to a minimum of 1,814 kg (4,000 lb) in 1975 (Figure 11.5) (Cato et al. 1978). The imported species of sea turtles included the green, olive ridley (*Lepidochelys olivacea*), loggerhead, and hawksbill, with the former two species being the most imported (Cato et al. 1978). These imports came from over 40 countries and demonstrate the demand for live sea turtles that existed in the United States until the

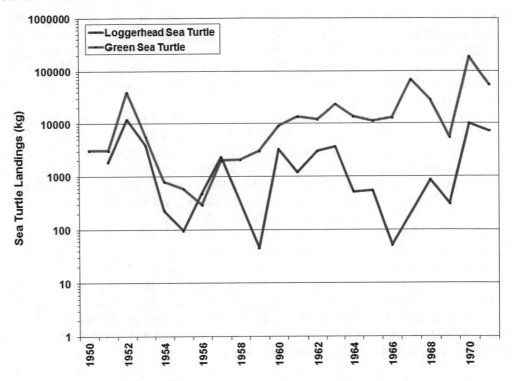

Figure 11.4. Turtle landings on the Florida Gulf coast for available years from 1950 through 1971. The data include green and loggerhead sea turtles, as well as Kemp's ridleys sold as loggerheads, and are plotted on a logarithmic scale to enhance contrast (from Rebel 1974).

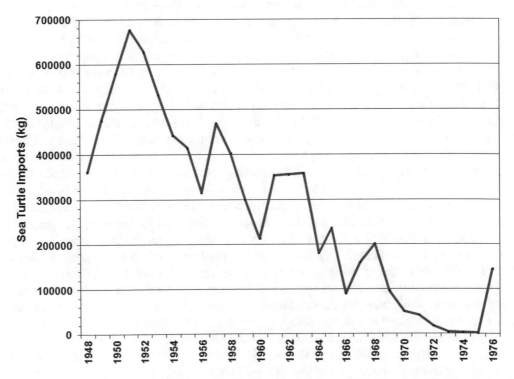

Figure 11.5. Live sea turtle imports into the United States from 1948 through 1976 (redrawn from Cato et al. 1978).

middle of the last century. The data show that demand was significant in the 1950s, but decreased steadily in the 1960s and 1970s, reaching its lowest volume from 1973 to 1975 (Figure 11.5). The spike observed in 1976 is believed to be flawed, since it does not match the overall trend up to that year (Figure 11.5) (Cato et al. 1978). Alternatively, this spike may reflect a last effort by the industry to import live sea turtles before the ban on these imports was fully implemented (Cato et al. 1978).

The green sea turtle fishery in Mexican waters of the Gulf of Mexico continued until 1990, when a total ban was imposed (DOF 1990). Some statistics are available quantifying the magnitude of this catch, which was mainly off the coasts of Quintana Roo and Campeche; for example, from 1964 through 1981, green turtles captured ranged from 14 to 74.7 % of the annual total of sea turtles captured (Márquez-M 2004). Although the commercial fishery also captured loggerheads and hawksbills, most of the take was green turtles, with an estimated average of 67.9 % from 1964 through 1981 (Márquez-M 2004).

In essence, the data provided above indicate that the sea turtle fishery that developed from the discovery of the New World through the mid-1900s was largely responsible for the decline of sea turtle populations in the Gulf of Mexico and the Caribbean. It is important to mention that the data do not include turtles and eggs consumed locally; therefore, the actual anthropogenic impact was likely of much greater magnitude on the populations of sea turtles in the Gulf. Sea turtle populations in the Gulf of Mexico have remained low to the present, mostly as a result of the impact of bycatch in various fisheries, mainly that of shrimp (McDaniel et al. 2000; Crowder and Heppell 2011; Finkbeiner et al. 2011).

11.1.3   General Nesting Abundance of Gulf of Mexico Sea Turtles

The beaches of east Texas, Louisiana, Mississippi, and Alabama in the north-central Gulf of Mexico are essentially devoid of any significant sea turtle nesting. Sea turtle nesting, in general, increases east, west, and southwest from this north-central location and reaches its zenith around the Florida and Yucatán peninsulas (Renaud 2001). Data available on nesting females show that Gulf of Mexico sea turtle populations generally exhibit a very low abundance relative to Atlantic regions outside the Gulf. This is particularly true for loggerhead sea turtles, whose nesting on Florida Gulf of Mexico beaches amounted to only 8.6 % of statewide nesting from 2001 through 2006 (Witherington et al. 2009). The only obvious exception to this general rule is the Kemp's ridley, whose main rookery is located along the beaches of Tamaulipas, Mexico, on the Mexican Gulf coast (NMFS et al. 2011). The very low nesting numbers indicate that all sea turtle populations in the Gulf of Mexico are particularly vulnerable to environmental and anthropogenic impacts, perhaps more so than populations outside the Gulf of Mexico.

An early assessment of the status of sea turtle populations in the western Gulf of Mexico was based largely on the presence/absence of turtles from Louisiana throughout the western rim of the Gulf and south to the state of Yucatán, Mexico (Hildebrand 1982). Unfortunately, no data on abundance were provided for the five species that nest in this large geographic area, which precludes establishing some sort of baseline to which current numbers can be compared. However, the review indicated that all sea turtle populations had undergone a significant decline by 1979 due to the exploitation of eggs, juveniles, and adult turtles (Hildebrand 1982).

A few years later, another attempt to assess the status of Gulf of Mexico sea turtle populations was published, which combined various sources of data, including nesting, in-water captures, aerial surveys, stranding, mortality, and bycatch data, among other datasets, providing hard numbers for the various U.S. stocks (Thompson 1988). Although the publication elicited some controversy (Dodd and Byles 1991; Thompson 1991), its message, that U.S. sea turtle stock assessments must be conducted regularly and frequently, was well taken.

In the last 30 years, many substantial efforts have been made to generate assessments of sea turtle populations on a regular basis (e.g., NMFS and USFWS 2007a, b, c, d, e, 2008; Conant et al. 2009; NMFS et al. 2011). However, most of these assessments focused on beach counts of nests and did not include in-water population assessments. This is not surprising, given the costs associated with studying highly migratory species with complex life cycles. To fully understand the health of sea turtle populations, it is imperative to generate reliable datasets of in-water turtle populations that include various demographic parameters suitable for analysis, such as age of hatchlings and juveniles and survival rates (Heppell et al. 2005), among many other parameters.

With regard to the southwestern Gulf of Mexico, all five species are present and nest on many beaches in the region (Hildebrand 1963; Sánchez-Pérez et al. 1989). On Mexican beaches of the Gulf, the Kemp's ridley and the green sea turtle are estimated to exhibit similar nesting abundances; whereas, the hawksbill and leatherback are less abundant (Márquez-M 2004). Most loggerheads nest on the Caribbean side of the Yucatán Peninsula, with low numbers of nesting occurring along the Gulf coast (Márquez-M et al. 2004). Only about ten leatherback nests were recorded on Mexican beaches of the Gulf in 2000 (Márquez-M 2004). At the level of the Gulf basin, these numbers, along with Florida numbers, confirm that the largest numbers of nesting sea turtles are located on the southwest and northeast rims of the Gulf of Mexico.

11.1.4 General In-Water Abundance of Gulf of Mexico Sea Turtles

The in-water abundance of the five species of sea turtles that inhabit the waters of the Gulf of Mexico is difficult to ascertain given the lack of long-term, systematic studies. Indeed, the Gulf may arguably be the most data-deficient basin in terms of its sea turtle populations. Efforts to determine the presence and abundance of all species in U.S. waters seem to have concentrated in Texas and Florida, likely due to the presence of nesting beaches in these states–Florida boasting by far the largest numbers (e.g., Meylan et al. 1995). Aerial surveys over Gulf of Mexico waters have been used frequently to address this deficiency of data. However, no reports exist regarding the southern Gulf of Mexico, and most of the reports available for the northern Gulf are point in time studies, over a season or a year, and lack the benefit of long-term, systematic records that could be used to establish population trends. This has led researchers to state that currently it is virtually impossible to assess and restore the sea turtle populations of the Gulf of Mexico in relation to their historical abundance (Bjorndal et al. 2011). While there are differences in the methodologies of aerial surveys that have been conducted in the Gulf and variables such as speed, altitude, visibility, and lack of consistency in the areas surveyed limit the accuracy of the observations, available aerial survey information is presented in the following paragraphs.

During aerial surveys conducted in the Gulf of Mexico from June 1980 to April 1981, loggerheads were observed nearly 50 times as often in waters off the Florida Gulf coast compared to those observed in the western Gulf (Fritts et al. 1983a). They were present throughout the year, mostly in waters less than 50 m (164 ft) deep, but the frequency of sightings was lowest during the winter. Green turtles were infrequently observed in the Gulf of Mexico. Kemp's ridleys were most frequently sighted off southwest Florida and rarely observed in the western Gulf. Leatherbacks were observed more often on the continental shelf than in deeper waters (Fritts et al. 1983a).

Differences in sea turtle distribution in the eastern and western U.S. Gulf of Mexico were found in an analysis of Sea Turtle Stranding and Salvage Network (STSSN) data (see Section 11.7.4 for an explanation of these data) collected from 1985 through 1991 (Teas 1993). Large numbers of juvenile and adult loggerheads occurred in the eastern Gulf of Mexico during the spring and summer, especially along the south Florida Gulf coast. During all seasons, juvenile and adult green and Kemp's ridley sea turtles used eastern Gulf waters extensively. Low numbers of hawksbills also used the eastern Gulf of Mexico, while leatherbacks were found in the eastern Gulf during the spring and fall as they migrated through to preferred feeding and nesting grounds (Teas 1993). The western Gulf of Mexico provided year-round habitat for juvenile loggerheads and for hatchlings to adult Kemp's ridleys. Throughout the year, juvenile green turtles used western Gulf of Mexico waters (Teas 1993). During the summer and fall when prevailing currents carried them into the western Gulf, hawksbills ranging from hatchling to juvenile were common. Leatherback sea turtles migrated through the western Gulf during the spring and fall (Teas 1993).

Surveys were conducted in 1991 and 1992 to establish relative sea turtle abundance and seasonality in the U.S. Gulf of Mexico (Braun-McNeill and Epperly 2002). The study was based on surveying fishermen along the Gulf coast from Louisiana to southern Florida year-round regarding the sighting of sea turtles. The surveys indicated that sea turtle abundance along the Gulf coast was seasonal, with turtles migrating northward in the warmer months and then migrating south in the colder months. This seasonality was in agreement with historical information obtained from turtle fisheries in the Gulf States (Stevenson 1893; Carr and Caldwell 1956). The study also demonstrated that the number of turtle sightings was significantly higher in the Florida Keys than in any other location and that, throughout the study area, most turtles tended to be located within 506 km (314 mi) from the coast (Braun-McNeill and Epperly 2002).

An analysis of National Marine Fisheries Service (NMFS) aerial survey data for September, October, and November of 1992, 1993, and 1994 was conducted to determine sea turtle spatial dynamics for the U.S. Gulf of Mexico (McDaniel 1998). The results of the study indicated that sea turtles were observed at much higher rates along the Florida Gulf coast than in the western Gulf, and the highest density of observed sea turtles occurred in the Florida Keys region (0.525 turtles per square kilometer [km^2] or 0.203 turtles per square mile [mi^2]) (McDaniel et al. 2000). These results are similar to those obtained by Fritts et al. (1983a) discussed above. Various hypotheses were proposed to explain the higher numbers of sea turtles observed in the eastern Gulf as compared to the western Gulf of Mexico; these included the following: more suitable sea turtle habitat in the eastern Gulf of Mexico, the reduction of turtles by the intense shrimp fishery in the western Gulf of Mexico, low oxygen levels off the Louisiana coast, sea turtles being attracted to shrimp vessel bycatch, and more turtles inhabiting nearshore areas compared to areas offshore (McDaniel 1998).

In the same study conducted by McDaniel et al. (2000), sea turtle abundance decreased significantly west from Florida toward the north-central Gulf of Mexico and increased 20-fold in south Texas as compared to other areas surveyed in the western Gulf, ranging from no turtles up to 0.10 turtles/km^2 (0.04 turtles/mi^2) (McDaniel et al. 2000). This is consistent with an earlier week-long aerial survey conducted in the fall of 1979 in the same south Texas area that reported a mean sea turtle density of 0.0196 turtle/km^2 (0.0076 turtles/mi^2) (Reeves and Leatherwood 1983). Interestingly, a small peak around the Chandeleur Islands in Louisiana was observed (McDaniel et al. 2000); this peak around the Chandeleur Islands is important because Louisiana waters are known foraging grounds for post-nesting Kemp's ridleys (Chávez 1968; Pritchard and Márquez-M. 1973; Ogren 1989).

In aerial surveys conducted along the U.S. Gulf Coast from September through November during 1992 through 1996, Kemp's ridleys were sighted primarily in inshore waters and most commonly occurred in the eastern Gulf of Mexico (Epperly et al. 2002). During the same surveys, loggerhead sea turtles were sighted throughout the Gulf but had a very low occurrence in offshore waters of the western Gulf. Green turtles occurred offshore and primarily were sighted in the southern portion of the Florida Gulf coast. Hawksbills occurred mainly in southwest Florida, and leatherback sea turtles were more broadly distributed and were observed predominantly in offshore waters (Epperly et al. 2002).

Loggerhead sea turtles are the most abundant sea turtle species in the Gulf of Mexico (Henwood 1987). The nearshore waters of the northwestern Gulf provide important foraging areas for loggerhead sea turtles (Plotkin et al. 1993; Plotkin 1996). Loggerhead densities of 0.04 turtles/km^2 (0.015 turtles/mi^2) were reported for the northeastern Gulf of Mexico during aerial and ship surveys conducted from 1996 through 1998 (Mullin and Hoggard 2000). In a survey of the eastern Gulf of Mexico continental shelf along a series of transects between Tampa Bay and Charlotte Harbor, Florida, conducted from November 1998 through November 2000 between the coast and the 180 m (591 ft) isobaths, the overall density of loggerhead sea turtles was estimated to be 0.013 turtles/km^2 (0.005 turtles/mi^2) (Griffin and Griffin 2003); since unidentified turtles were not included in the analyses, the abundance of loggerheads for the eastern continental shelf of the Gulf of Mexico was most likely underestimated.

11.1.5   Regulation and Protection of Gulf of Mexico Sea Turtles

Two pieces of legislation were crucial to the protection of sea turtle species around the world and in the United States: the U.S. ESA and the Convention on International Trade in Endangered Species of Wild Fauna and Flora (CITES), the latter being an international agreement regulating the international trade of endangered species. Both pieces were enacted or signed by the United States in 1973, a time when sea turtle populations, particularly those in the Gulf of Mexico and greater Caribbean, exhibited evident signs of overexploitation.

All sea turtles occurring in the Gulf of Mexico are listed under the U.S. ESA and are under the joint jurisdiction of the National Oceanic and Atmospheric Administration (NOAA), the NMFS, and the U.S. Fish and Wildlife Service (USFWS) (NOAA 2013). The USFWS has lead responsibility on the nesting beaches, while the NMFS is the lead agency in the marine environment (NMFS et al. 2011). Kemp's ridley, leatherback, and hawksbill sea turtles are listed as endangered under the ESA (NOAA 2013). The overall listing status for the loggerhead sea turtle is threatened; however, each of the nine distinct population segments (DPSs) of logger-heads has a separate listing (NOAA 2013). The Northwest Atlantic Ocean DPS of loggerheads, whose range includes the Gulf of Mexico, is listed as threatened (USFWS and NMFS 2011). Green sea turtles have two listed populations: the Florida and Mexican Pacific coast green turtle breeding colonies are listed as endangered, and green turtles in all other areas are listed as threatened (NOAA 2013).

The NMFS and USFWS established the DPS policy in 1996 (USFWS and NMFS 1996). A population is considered to be a DPS if it is both discrete and significant relative to its taxon (taxonomic group). A population may be considered discrete if it satisfies either of the following conditions: (1) it is markedly separated from other populations of the same taxon as a consequence of physical, physiological, ecological, and behavioral factors (often based on genetic evidence); or (2) it is delimited by international government boundaries within which significant differences in control of exploitation, management of habitat, conservation status, or regulatory mechanisms exist. If a population segment is considered to be discrete, the NMFS and/or the USFWS must then determine whether the DPS is significant relative to its taxon using established criteria (USFWS and NMFS 1996).

Sea turtles that occur in the Gulf of Mexico are also listed by many U.S. Gulf Coast states as threatened and endangered species. State agencies that protect, regulate, and study sea turtles along the U.S. Gulf Coast include the Florida Fish and Wildlife Conservation Commission; Alabama Department of Conservation and Natural Resources; Mississippi Department of

Wildlife, Fisheries, and Parks; Louisiana Department of Wildlife and Fisheries; and Texas Parks and Wildlife Department (ADCNR 2012; FFWCC 2012; LDWF 2012; MDWFP 2012; TPWD 2012). In Mexico, sea turtles are regulated by the Comisión Nacional de Areas Naturales Protegidas, Secretaría de Medio Ambiente y Recursos Naturales (SEMARNAT) (SEMARNAT 2012). The taking of all sea turtles in Mexico was prohibited by presidential decree in 1990 (DOF 1990), and the National Program for Protection, Conservation, Research, and Management of Marine Turtles was implemented in Mexico in 2000 (NMFS et al. 2011).

Kemp's ridley, leatherback, and hawksbill sea turtles are listed as critically endangered by the International Union for the Conservation of Nature, and loggerhead and green sea turtles are listed as endangered (IUCN 2012). All sea turtles that occur in the Gulf are listed in CITES's Appendix 1, which includes species identified as endangered and prohibits all commercial international trade (NMFS et al. 2011).

11.1.5.1   History of Kemp's Ridley Sea Turtle Protection in the United States and Mexico

Under the ESA, the Kemp's ridley sea turtle was listed as endangered throughout its range on December 2, 1970 (NMFS and USFWS 2007a). Five-year status reviews of the Kemp's ridley were conducted by the NMFS in 1985, by the USFWS in 1985 and 1991, and by the NMFS and USFWS in 1995 and 2007; no change in the Kemp's ridley endangered listing status was recommended as a result of these reviews (Mager 1985; Plotkin 1995; NMFS and USFWS 2007a). The initial recovery plan for the Kemp's ridley sea turtle, which was for all six sea turtle species occurring in the United States, was approved by the NMFS on September 19, 1984, and the first revision and separate recovery plan for the Kemp's ridley was approved by the USFWS and NMFS on August 21, 1992 (NMFS and USFWS 2007a). In 2002, the USFWS, NMFS, and Mexico's SEMARNAT initiated the process to revise the recovery plan for a second time, but this time as a binational recovery plan; the second revision of this plan was approved on September 22, 2011 (NMFS et al. 2011).

In Mexico, efforts to protect Kemp's ridleys and their nesting beaches have been ongoing since the 1960s (Márquez-M. 1994). The harvest of Kemp's ridleys in the Gulf of Mexico has been prohibited since the 1970s (Márquez-M. et al. 1989). In 1977, Rancho Nuevo, the only mass-nesting site for the Kemp's ridley, was declared a natural reserve and further protective measures were added in 1986 (DOF 1977, 1986; Márquez-M. et al. 1989). Rancho Nuevo was declared a sanctuary in 2002 and was included in the listing of Wetlands of International Importance under the Convention on Wetlands in 2004 (DOF 2002; NMFS et al. 2011).

11.1.5.2   History of Loggerhead Sea Turtle Protection in the United States

The loggerhead sea turtle was listed as threatened throughout its range under the ESA on July 28, 1978 (Conant et al. 2009). The initial recovery plan for the loggerhead sea turtle was approved by the NMFS on September 19, 1984 (NMFS et al. 2011), while the first revision of the loggerhead recovery plan, which focused on the U.S. population in the Atlantic Ocean, was approved by the USFWS and NMFS on December 26, 1991 (NMFS and USFWS 2008). No change in the loggerhead threatened listing status was recommended as a result of 5-year status reviews conducted by the NMFS in 1985 and the USFWS in 1991 (Mager 1985; USFWS 1991). While no change to the loggerhead's listing status as threatened was recommended as a result of the joint 5-year review conducted by the NMFS and USFWS in 1995, the review identified the need to conduct additional research regarding the existence of two separate nesting populations along the southeast U.S. coast: the Florida subpopulation and the subpopulation nesting from Georgia through southern Virginia (Plotkin 1995). The results of research conducted between the 1995 5-year status review and the 2007 joint review completed by the NMFS and USFWS indicated that loggerhead populations might be separated by ocean basins, and while no change to the threatened listing of the loggerhead was recommended as a result of the 2007 status review, a commitment was made to determine the applicability of the DPS policy (NMFS and USFWS 2007b).

Five recovery units/subpopulations of loggerheads within the northwest Atlantic Ocean were recognized in the second revision and most recent version of the *Recovery Plan for the Northwest Atlantic Population of the Loggerhead Sea Turtle* (NMFS and USFWS 2008). The recovery units/subpopulations include the northern subpopulation (southern Virginia through the Florida/Georgia border), Peninsular Florida subpopulation (Florida/Georgia border through Pinellas County, Florida), Dry Tortugas subpopulation (islands located west of Key West, Florida), northern Gulf of Mexico subpopulation (Franklin County, Florida through Texas), and Greater Caribbean subpopulation (Mexico through French Guiana, The Bahamas, Lesser Antilles, and Greater Antilles). A threats analysis was conducted for all loggerhead life stages in support of the recovery plan to prioritize conservation actions relative to their impact on population growth rate and to support development of management priorities (NMFS and USFWS 2008; Bolten et al. 2011). During the most recent loggerhead status review, nine loggerhead DPSs were identified (Conant et al. 2009). They include the DPSs for the North Pacific Ocean, South Pacific Ocean, North Indian Ocean, Southeast Indo-Pacific Ocean, Southwest Indian Ocean, Northwest Atlantic Ocean, Northeast Atlantic Ocean, Mediterranean Sea, and South Atlantic Ocean (Figure 11.6).

The NMFS and USFWS published a proposed rule in March 2010, in which a Northwest Atlantic Ocean DPS would be established and listed as endangered under the ESA (USFWS and NMFS 2011). However, prior to making a final determination, nesting data available after the proposed rule was published and information provided by reviewers was evaluated; it was ultimately determined that listing the Northwest Atlantic Ocean DPS as threatened was more appropriate because the nesting population was large, the overall nesting population remained

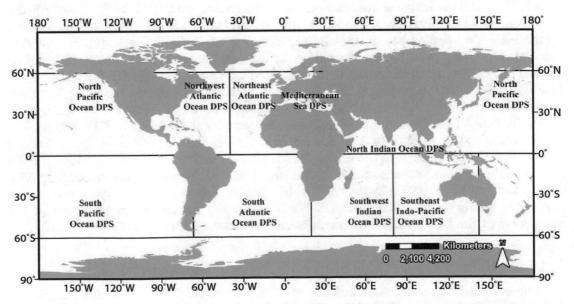

Figure 11.6. Location of the loggerhead sea turtle distinct population segment (DPS) boundaries (redrawn from USFWS and NMFS 2011).

widespread, the trend for the nesting population appeared to be stabilizing, and substantial conservation efforts were underway to address threats (USFWS and NMFS 2011).

In September 2011, the NMFS and USFWS listed nine DPSs of loggerhead sea turtles under the ESA (USFWS and NMFS 2011). Four DPSs (Southeast Indo-Pacific Ocean, Southwest Indian Ocean, Northwest Atlantic Ocean, and South Atlantic Ocean) were listed as threatened, and five DPSs (North Pacific Ocean, South Pacific Ocean, North Indian Ocean, Northeast Atlantic Ocean, and Mediterranean Sea) were listed as endangered. In addition, the rule stated that critical habitat for the two loggerhead DPSs occurring within the United States (Northwest Atlantic Ocean and Northeast Atlantic Ocean) will be proposed in future rulemaking, and information related to this effort was requested.

11.1.5.3   History of Green Sea Turtle Protection in the United States

Under the ESA on July 28, 1978, the green turtle breeding colony populations in Florida and on the Pacific coast of Mexico were listed as endangered, and green turtles in all other areas were listed as threatened (NMFS and USFWS 2007c). Five-year status reviews of the green sea turtle were conducted by the USFWS in 1983 and 1991, by the NMFS in 1985, and by the NMFS and USFWS in 1995 and 2007. In these reviews, no changes in the green turtle's listing status were recommended (Mager 1985; Plotkin 1995; NMFS and USFWS 2007c). The initial recovery plan for the green sea turtle was approved by the NMFS on September 19, 1984 (NMFS and USFWS 2007c). A recovery plan for the U.S. population of the Atlantic green turtle was approved on October 29, 1991, and recovery plans for the U.S. Pacific populations of the green turtle and the East Pacific green turtle were finalized in January 1998 (NMFS and USFWS 1991, 2007c). While the green turtle has no designated critical habitat in the Gulf of Mexico, marine critical habitat was designated in Puerto Rico on September 2, 1998 (NMFS and USFWS 2007c).

11.1.5.4   History of Leatherback Sea Turtle Protection in the United States

The leatherback sea turtle was listed as endangered throughout its range under the ESA on June 2, 1970, and the listing status has since remained unchanged (NMFS 2011b). Leatherbacks were included in the initial recovery plan for all sea turtle species in the United States approved by the NMFS on September 19, 1984 (NMFS and USFWS 2007d). The NMFS and USFWS completed recovery plans for leatherbacks in the U.S. Caribbean, Atlantic Ocean, and Gulf of Mexico in 1992 and for leatherbacks in the U.S. Pacific Ocean in 1998 (NMFS 1992; NMFS and USFWS 2007d). The leatherback has no designated critical habitat in the Gulf; however, terrestrial and marine critical habitat for the leatherback was designated on and around St. Croix, U.S. Virgin Islands in 1978 and 1979, respectively (NMFS and USFWS 2007d). The NMFS added waters adjacent to the U.S. west coast to the designated critical habitat in 2010. In 2011, waters surrounding a major nesting beach location in Puerto Rico were added (NMFS 2011b). Status reviews were conducted by the NMFS in 1985, by the USFWS in 1995 and 1991, and by the NMFS in 1995 (Plotkin 1995; NMFS 2011b). The most recent 5-year review was completed jointly by the NMFS and USFWS in 2007, and further review to determine the application of the DPS policy to leatherbacks was suggested (NMFS and USFWS 2007d).

11.1.5.5   History of Hawksbill Sea Turtle Protection in the United States

The hawksbill sea turtle was listed as endangered throughout its range under the ESA on June 2, 1970 (NMFS and USFWS 2007e). While the hawksbill has no designated critical habitat in the Gulf of Mexico, terrestrial and marine critical habitat was designated for the hawksbill in

Puerto Rico on June 24, 1982 and on September 2, 1998, respectively (NMFS and USFWS 2007e). Five-year status reviews of the hawksbill sea turtle were conducted by the NMFS in 1985, by the USFWS in 1985 and 1991, and by the NMFS and USFWS in 1995 and 2007. While no changes in the hawksbill's listing classification were recommended as a result of these reviews, a future analysis of the hawksbill was recommended in the most recent review to determine the application of the DPS policy to this species (Mager 1985; Plotkin 1995; NMFS and USFWS 2007e). The initial recovery plan for the hawksbill sea turtle was approved by the NMFS on September 19, 1984 (NMFS and USFWS 2007e). The first revision and separate recovery plan for the hawksbill in the U.S. Caribbean, Atlantic Ocean, and Gulf of Mexico was approved by the NMFS and USFWS on December 15, 1993, and a recovery plan for the U.S. Pacific populations of the hawksbill sea turtle was issued on January 12, 1998 (NMFS and USFWS 2007e).

11.2 KEMP'S RIDLEY SEA TURTLE (*LEPIDOCHELYS KEMPII*)

Along with the flatback sea turtle (*Natator depressus*), the Kemp's ridley sea turtle has the most geographically restricted distribution of all sea turtle species (Morreale et al. 2007). The smallest of the sea turtle species, the Kemp's ridley was first described in the late 1800s and named for Richard M. Kemp, a fisherman and naturalist from Key West, Florida, who submitted the type specimen (Figures 11.7 and 11.8) (NMFS et al. 2011). Unlike other species of sea turtles, which emerge individually on beaches to lay their eggs in the sand, the Kemp's ridley, as well as the closely related olive ridley, typically comes ashore in large, synchronous aggregations to lay their eggs; these events, or *arribadas*, occur at only a few beaches around the world (Plotkin 2007a). While olive ridleys typically nest at night like most sea turtles, Kemp's ridleys regularly nest during daylight hours (Safina and Wallace 2010). Another significant difference between the Kemp's ridley and the olive ridley is that the latter is the most abundant of all the sea turtle species (Valverde et al. 2012), while Kemp's ridleys are the least abundant species of sea turtles. In addition to the arribadas, the two species share the

Figure 11.7. Nesting Kemp's ridley sea turtle (from NOAA 2011).

Figure 11.8. Kemp's ridley sea turtle in the water (photograph by Kim Bassos-Hull, Mote Marine Laboratory) (NOAA 2011).

trademark *ridley dance*, in which the nesting females rock from side to side using their bodies to tamp sand on top of their nests (Safina and Wallace 2010).

The life history, particularly the nesting beach locations, of the Kemp's ridley sea turtle remained a mystery through the 1950s; some thought the Kemp's ridley was a hybrid of the loggerhead and green sea turtles (Carr 1979). However, the western Gulf of Mexico was determined to be important for this species when two Kemp's ridleys were found nesting during the day on Padre Island, Texas, in 1948 and 1950 (Werler 1951). The only mass-nesting site for the Kemp's ridley—the Rancho Nuevo area located in Tamaulipas, Mexico, already impacted from years of egg overexploitation—was not discovered by the scientific community until the early 1960s (Carr 1963; Hildebrand 1963).

Due to overexploitation and accidental mortality in fishing gear, the Kemp's ridley sea turtle came perilously close to extinction in the 1980s (Crowder and Heppell 2011). Due to the intensive, cooperative efforts by researchers and volunteers in Mexico and the United States (see Section 11.2.2), the Kemp's ridley rebounded from the brink of extinction (Heppell et al. 2007). The Kemp's ridley has recovered remarkably because conservation efforts have focused on stressors affecting all life stages, from eggs to juveniles and adults at sea. While the story of Kemp's ridley recovery is not finished, the trajectory is promising (Crowder and Heppell 2011). The combination of turtle excluder device (TED) use (see Sections 11.2.2 and 11.7.1.2), reductions in the shrimping effort, and nest protection on Mexican beaches has resulted in an unusually rapid recovery for a long-lived, slow-growing vertebrate. However, in spite of the recent gains over the lowest abundance of the 1980s and increased protection measures, Kemp's ridley populations remain significantly below historical levels.

11.2.1 Kemp's Ridley Life History, Distribution, and Abundance

The Kemp's ridley sea turtle occurs in the Gulf of Mexico and along the U.S. Atlantic Coast (Figure 11.9). The vast majority of Kemp's ridley nesting occurs on beaches in the western Gulf (Figure 11.10), and most juveniles spend time in the Gulf of Mexico oceanic zone after they

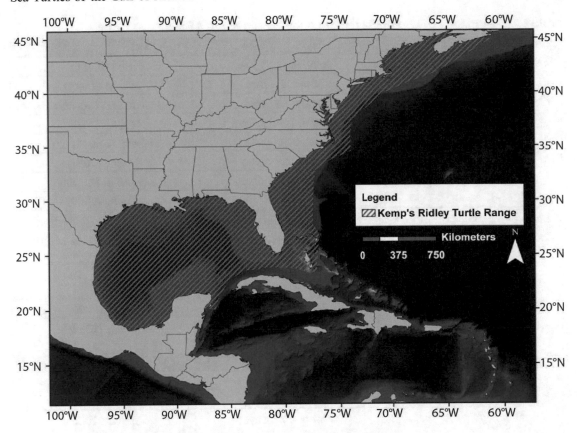

Figure 11.9. Range of the Kemp's ridley sea turtle (from NOAA 2007).

leave the beach as hatchlings (Collard and Ogren 1990; TEWG 2000; Putman et al. 2010). After the oceanic juvenile stage, juveniles recruit into the neritic zone (inshore marine environment), mostly along the Gulf continental shelf but also along the U.S. Atlantic Coast (Pritchard 1969; Ogren 1989; Schmid 1998; Witzell and Schmid 2004; Seney and Landry 2011), where they continue to feed and grow for a number of years until reaching sexual maturity. Oceanic juveniles that end up in the currents of the Atlantic Ocean move into coastal habitats along the east coast of the United States from Florida to New England, and as far north as the Grand Banks and Nova Scotia (Pritchard 1969; Ogren 1989; Morreale and Standora 1999; Watson et al. 2004; Morreale et al. 2007; Frazier et al. 2007; Landry and Seney 2008). Some Kemp's ridleys have been found in European Atlantic waters, the Mediterranean, and the Azores (Brongersma 1972; Brongersma and Carr 1983; Fontaine et al. 1989; Fretey 2001). Many of these juveniles return to the Gulf of Mexico to reproduce; for example, neritic juvenile Kemp's ridleys that were tagged along the U.S. Atlantic Coast have nested at Rancho Nuevo, Mexico (Witzell 1998; Schmid and Woodhead 2000). Adult Kemp's ridley sea turtles occur primarily in the Gulf of Mexico, typically in nearshore waters (Hildebrand 1982; Ogren 1989; USFWS and NMFS 1992; Pritchard 2007a). A summary of life history information for the Kemp's ridley specific to the Gulf of Mexico is included in Table 11.1; information available for specific Gulf beaches or regions is also included in the table.

Table 11.1. Summary of Life History Information for the Kemp's Ridley Sea Turtle

Parameter	Values	References
Nesting season: Gulf of Mexico	April through July	Hirth (1980)
Remigration interval: Rancho Nuevo, Mexico	Mean: 2 years	Márquez-M et al. (1982)
	Mean: 1.5 years	van Buskirk and Crowder (1994)
Nesting (arribada) interval: Rancho Nuevo, Mexico	Range: 20–28 days	Chávez (1969)
	Mean: 25 days	Rostal et al. (1997)
Number of nests/season: Rancho Nuevo, Mexico	Mean: 3.1 nests	Rostal (1991)
	Mean: 2.5 nests	Heppell et al. (2005)
	Mean: 3.1 nests	Rostal (2005)
Number of eggs/nest		
Rancho Nuevo, Mexico	Mean: 116 eggs, Range: 93–135 eggs	Pritchard and Márquez-M. (1973)
	Mean: 104 eggs, Range: 17–192 eggs	Márquez-M. (1994)
	Mean: 95 eggs	Coyne (2000)
Upper Texas Region	Mean: 99 eggs, Range: 71–119 eggs	Seney (2008)
Egg incubation time		
Rancho Nuevo, Mexico	Range: 50–70 days	Chávez et al. (1967)
	Range: 45–58 days	Márquez-M. (1990)
Padre Island National Seashore, Texas	Mean: 49.7 days	Shaver (2005)
Nest pivotal temperature	30.2 °C	Shaver et al. (1988)
Sex ratio of hatchlings from in situ nests (proportional female)		
Rancho Nuevo, Mexico	Mean: 0.80	Wibbels and Geis (2003)
	Mean: 0.64	T. Wibbels, UAB, unpublished data, cited in NMFS and USFWS (2007a)
Padre Island National Seashore, Texas	Mean: 0.60	Shaver (2005)
Emergence success of hatchlings from in situ nests		
Rancho Nuevo, Mexico	Mean: 0.66	USFWS (2006)
	Mean: 0.80	J. Pena, GPZ, personal communication, cited in NMFS et al. (2011)
Padre Island National Seashore, Texas	Mean: 0.62	Shaver (2005, 2006a, b, 2007, 2008), D. Shaver, PAIS, unpublished data, cited in NMFS et al. (2011)
Size of hatchlings	Mean: 4.4 cm SCL[a]	Márquez-M. (1972)
	Mean: 3.8 cm SCL	NOAA Fisheries OPR (2013)
Size of oceanic juveniles: Cedar Keys, Florida	Range: 5–19 cm SCL	Gregory and Schmid (2001)
Duration of oceanic juvenile stage: Cedar Keys and Cape Canaveral, Florida	Mean: 2 years	Schmid and Witzell (1997)
	Estimated maximum: 4 years	Putman et al. (2010)

(continued)

Table 11.1. (continued)

Parameter	Values	References
Diet of oceanic juveniles		
Lower Texas Region	Marine mollusks associated with the pelagic *Sargassum* community, including brown janthinas, *Cavolina longirostris*, *Sargassum* snails, and unidentifiable crabs, and *Sargassum*	Shaver (1991)
Texas and western Louisiana	Hardhead catfish, blue crabs, stone crabs, mottled purse crabs, and *Sargassum*	Zimmerman (1998)
Gulf Stream off Florida's Gulf coast	Marine animals associated with the pelagic *Sargassum* community, including hydroids, *Membranipora* sp., *Sargassum* anemones, serpulid polychaetes, gastropods, *Sargassum* snails, and *Sargassum* swimming crabs; *Sargassum*; and cladophora algae	Witherington et al. (2012)
Size of neritic juveniles: Sea Rim State Park, Texas to Cedar Keys, Florida	Range: 20–60 cm SCL	Ogren (1989)
Duration of neritic juvenile stage: Mississippi Sound, Mississippi to Ten Thousand Islands, Florida	Range: 8–9 years	Schmid and Barichivich (2005)
	Range: 7–8 years	Schmid and Woodhead (2000)
Diet of neritic juveniles		
Southern Texas	Speckled swimming crabs, blue crabs, mottled purse crabs, *Libinia* sp., calico crabs, surf hermits, Gulf stone crabs, bruised nassas, sharp nassas, moon snails, concentric nut clams, oysters, American stardrums, spot croakers, *Sargassum*, shoalgrass, *Gracilaria* sp., turtle grass, brown shrimp, and white shrimp	Shaver (1991)
Matagorda and Galveston Bays, Texas	Blue crabs, calico crabs, longnose spider crabs, *Ovalipes* sp., flat-clawed hermit crabs, mottled purse crabs, blood ark clams, transverse ark clams, *Anadara* sp., *Bittium* sp., angel wing clams, *Epitonium* sp., dwarf surf clams, bruised nassas, moon snails, *Terebra* sp., annelids, common sand dollars, mullet, and *Sargassum*	Seney (2008)

(continued)

Table 11.1. (continued)

Parameter	Values	References
Sabine Pass, Texas and Louisiana	Blue crabs, stone crabs, *Persephona aquilonaris*, thinstripe hermit crabs, dwarf surf clams, sharp nassas, oysters, catfish, *Sargassum*, shoalgrass, and bryozoans, including *Corallina cubensis*, common sheep's wool, and *Amathia distans*	Werner (1994)
Terrebonne Parish, Louisiana	Blue crabs, ornate blue crabs, *Nassarius* sp., and clams, including *Nuculana* sp., *Corbula* sp., and *Mulinia* sp.	Dobie et al. (1961)
Deadman Bay, Florida	Spider crabs, blue crabs, stone crabs, and mottled purse crabs	Barichivich et al. (1998)
Waccasassa Bay, Florida	Stone crabs, blue crabs, *Paguridae* sp., moon snails, bruised nassas, *Cantharus cancellarius*, eastern oysters, hooked mussels, shoalgrass, and star grass	Schmid (1998)
Charlotte Harbor Estuary, Florida	Spider crabs, mottled purse crabs, calico crabs, and blue crabs	Schmid (2011)
Gullivan Bay, Florida	Sea squirts, worm tubes, *Amathia* sp., hydroids, *Libinia* sp., mottled purse crabs, calico crabs, Atlantic horseshoe crabs, *Pitho* sp., *Hexapanopeus* sp., Florida stone crabs, giant marine hermit crabs, estuarine mud crabs, squatter pea crabs, *Marginella* sp., *Anadara* sp., *Lucina* sp., *Vermicularia* sp., turtle grass, shoalgrass, and manatee grass	Witzell and Schmid (2005)
Age at sexual maturity		
Rancho Nuevo, Mexico	Range: 5–7 years	Márquez-M (1972)
	Mean: 10 years	Coyne (2000)
Texas coast	Mean: 10 years	Caillouet et al. (1995)
	Range: 10–20 years	Shaver and Wibbels (2007)
Texas coast to southwest Florida	Range: 10–11 years	Schmid and Barichivich (2005)
Eastern Louisiana to southwest Florida	Range: 7–11 years	Schmid and Woodhead (2000)
Size of sexually mature adult females		
Rancho Nuevo, Mexico	Mean: 64 cm SCL, Range: 56–72.5 cm SCL	Burchfield et al. (1988)
	Minimum: 52.4 cm SCL	Márquez-M (1990)
Upper Texas Region to Louisiana coast	Mean: 60 cm SCL	Coyne and Landry (2000)

(continued)

Table 11.1. (continued)

Parameter	Values	References
Eastern Louisiana to southwest Florida	Mean: 60 cm SCL	Schmid and Barichivich (2005)
Diet of adults		
Lower Texas Region	Speckled swimming crabs, blue crabs, mottled purse crabs, *Libinia* sp., calico crabs, surf hermits, Gulf stone crabs, bruised nassas, sharp nassas, moon snails, concentric nut clams, oysters, star drums, spot croakers, *Sargassum*, shoalgrass, *Gracilaria* sp., turtle grass, brown shrimp, and white shrimp	Shaver (1991)
Gullivan Bay, Florida	Sea squirts, worm tubes, *Amathia* sp., hydroids, *Leptogoria* sp., *Libinia* sp., mottled purse crabs, calico crabs, Atlantic horseshoe crabs, *Pitho* sp., *Hexapanopeus* sp., Florida stone crabs, giant marine hermit crabs, estuarine mud crabs, blue crabs, squatter pea crabs, flatback mud crabs, *Nassarius* sp., *Marginella* sp., *Anadara* sp., eastern oysters, *Lucina* sp., *Vermicularia* sp., horse conches, turtle grass, shoalgrass, manatee grass, star grass, leafy caulerpa, and tonguefishes	Witzell and Schmid (2005)

[a]*SCL* straight carapace length, *cm* centimeters

11.2.1.1 Nesting Life History, Distribution, and Abundance for Gulf of Mexico Kemp's Ridleys

The single known aggregated nesting site and primary Kemp's ridley nesting beaches are located in Tamaulipas, Mexico, and include Rancho Nuevo, Tepehuajes, and Playa Dos (Figure 11.10) (Pritchard 2007b; NMFS et al. 2011). Nesting in Tamaulipas often occurs in arribadas, which may be triggered by strong onshore winds, especially north winds, as well as changes in barometric pressure (Jimenez et al. 2005). Individual nesting of Kemp's ridleys occurs from Texas to Veracruz, Mexico, and as far east as Campeche, Mexico (Figure 11.10) (Ross et al. 1989; Shaver 2005; Pritchard 2007a, b; Guzmán-Hernández et al. 2007). The majority of Kemp's ridley nesting in Texas occurs at PAIS, but low levels of nesting now regularly occur along the upper Texas coast, including in Matagorda, Brazoria, and Galveston Counties (Figure 11.10) (Shaver and Caillouet 1998; Shaver 2005; Seney 2008). The increase in nesting along the upper Texas coast represents either a northern expansion of Kemp's ridley nesting in the Gulf of Mexico or a reestablishment of its nesting range (Seney 2008).

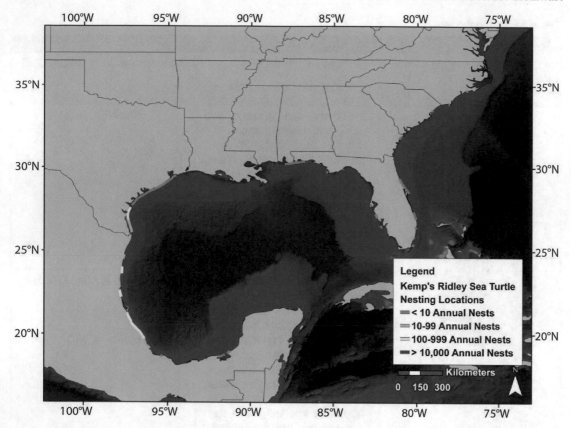

Figure 11.10. Generalized nesting beach locations of the Kemp's ridley sea turtle in the Gulf of Mexico and southeast U.S. Atlantic Coast (interpreted from Dow et al. 2007; SWOT 2010a; NMFS et al. 2011).

In addition, since the late 1980s, Kemp's ridley sea turtles have nested occasionally in Alabama, on Florida's Gulf and Atlantic coasts, and in Georgia, South Carolina, and North Carolina (Figure 11.10) (Meylan et al. 1991; Anonymous 1992; Márquez-M. et al. 1996; Johnson et al. 1999a; Williams et al. 2006).

Because of the limited nesting distribution, as well as the collaborative United States–Mexican recovery program (see Section 11.2.2), an entire time series of nesting information, beginning in 1966, is available for the Kemp's ridley for Rancho Nuevo and the adjacent beaches. This time series of information has little uncertainty after 1978, when nest protection methods became standardized and almost all nests were moved to a hatchery and recorded; however, since the mid-1990s, the level of uncertainty in estimating the population size has increased because of spatial expansion of the population and increased protection efforts (Márquez-M. et al. 1999; Heppell et al. 2007).

While more than 40,000 Kemp's ridleys were estimated to nest at Rancho Nuevo during an arribada in one day in 1947 (Carr 1963; Hildebrand 1963), only 924 nests were documented in 1978, and a low of 702 nests was recorded in 1985, representing about 300 and 228 nesting females, respectively (Figure 11.11). The number of nests observed at Rancho Nuevo and nearby beaches began to increase during the late 1980s and continued to increase at a rate of about 15 % per year (Figure 11.11) (Heppell et al. 2005; Crowder and Heppell 2011). In addition, the geographic range of nesting has expanded to the north and south of Rancho Nuevo (Heppell et al. 2007). Since 2005, the number of nests recorded in the Rancho Nuevo area each year

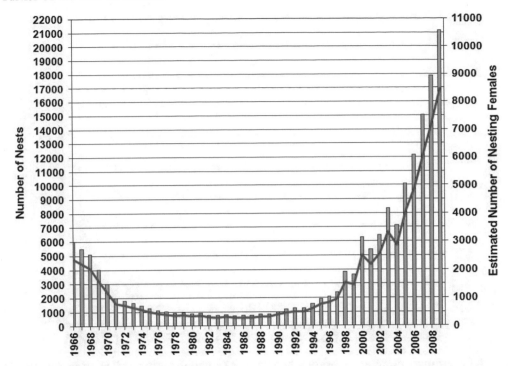

Figure 11.11. Annual number of nests (*bars*) and estimated number of nesting females (*line*), assuming 2.5 nests per female (Heppell et al. 2005), for Kemp's ridley sea turtles at Rancho Nuevo and adjacent beaches, Tamaulipas and Veracruz, Mexico from 1966 through 2009 (TEWG 2000; USFWS 2006; Alonso 2009).

consistently exceeded 10,000, indicating that at least 4,000 Kemp's ridleys were nesting each year (Figure 11.11). A record number of 21,144 nests (representing approximately 8,458 nesting Kemp's ridleys) were recorded at Rancho Nuevo and the adjacent beaches during 2009 (Figure 11.11). Approximately 13,000 nests were recorded from the Rancho Nuevo area during 2010, between 18,000 and 20,000 nests were recorded during 2011 (FuelFix 2011), and more than 21,000 nests were recorded in 2012 (NPS 2013a).

Besides its main nesting site at Tamaulipas, the beaches of Campeche, Mexico, are considered an important historic nesting site for the Kemp's ridley because regular nesting occurs, albeit at low levels (Guzmán-Hernández et al. 2007). The fact that nesting activity has been ongoing in this region, more than 1,200 km (746 mi) from Rancho Nuevo, corroborates the resilience of this highly vulnerable species and helps support the idea that Campeche was an important nesting region decades ago before the spread of overfishing and other human impacts (Márquez-M. 2004). From 1984 through 2003, 15 Kemp's ridley nests were recorded during the standardized patrolling and surveillance of sea turtle nesting beaches in Campeche. Nests were found on three beaches: ten nests on Isla Aguada, three on Isla del Carmen, and two in Sabancuy. While few in number compared to the hatchlings released at Rancho Nuevo, these nests contributed 1,109 hatchlings to the Kemp's ridley population in the Gulf of Mexico.

The number of Kemp's ridley nests along the Texas coast has increased dramatically since the late 1940s (Figure 11.12). The PAIS in Corpus Christi, Texas, is now considered a secondary nesting colony; more Kemp's ridley nests have been confirmed at PAIS than any other location in the United States during the last 50 years (Shaver 1999, 2005, 2006a). Kemp's ridleys that nest in Texas today are a mixture of head-started turtles—raised in captivity for a period of time and later released—and wild-stock turtles (Shaver 2005; Seney 2008). The larger size of the

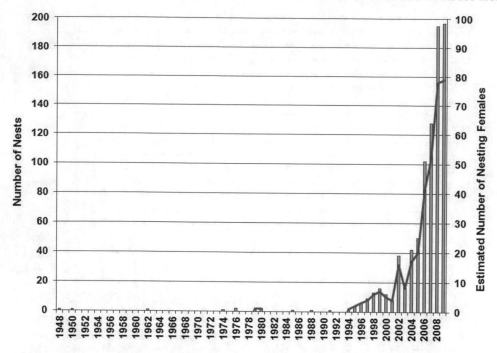

Figure 11.12. Annual number of nests (*bars*) and estimated number of nesting females (*line*), assuming 2.5 nests per female (Heppell et al. 2005), for Kemp's ridley sea turtles recorded on Texas beaches from 1948 through 2009 (Shaver and Caillouet 1998; NPS 2013b). Data were not available for 1949, 1951 through 1961, 1963 through 1973, 1975, 1977, 1978, 1981 through 1984, 1986, 1987, 1989, 1990, 1992, or 1993.

head-started turtles when released is assumed to decrease mortality (Section 11.2.2). Nesting surveys along the Texas coast, which have increased in recent years, have been extremely challenging due to the hundreds of kilometers of beaches that must be searched, limited resources, and logistical difficulties, and because of the nesting characteristics of Kemp's ridleys (e.g., nesting during the day on windy days) (Shaver and Rubio 2008). Eggs from nests found on many of the Texas beaches are moved to incubation facilities or protective corrals. While fewer than ten nests were recorded each year (for years for which data are available) prior to 1997, the number of nests on Texas beaches began to increase in 1998. Since 2006, more than 100 Kemp's ridley nests or at least 40 nesting Kemp's ridleys have been recorded for Texas beaches each year (Figure 11.12). In 2010, 141 nests were recorded along the Texas coast, and 199 nests were recorded in 2011 (NPS 2013a). A record number of 209 Kemp's ridley nests, representing about 84 nesting females, was recorded along the Texas coast in 2012 (NPS 2013b), and 153 nests were recorded during the 2013 nesting season (NPS 2013a). With increased productivity on the nesting beaches and improved survival at sea because of reductions in fishing mortality, conservation efforts boosted the current population of Kemp's ridleys from 7,000 to 8,000 breeding females nesting at multiple sites from Padre Island, Texas, to Veracruz in the southwestern Gulf of Mexico (Crowder and Heppell 2011).

11.2.1.2   Hatchling, Post-Hatchling, and Oceanic Juvenile Life History and Distribution for Gulf of Mexico Kemp's Ridleys

After the embryos have developed, the time depending on temperature and other incubation conditions (Table 11.1), Kemp's ridley hatchlings emerge from the nest *en masse* at night or

Figure 11.13. Kemp's ridley sea turtle hatchlings entering the sea after emerging from the nest (from NPS 2013c).

during the early morning and swim offshore into the oceanic zone to feed and grow (Figure 11.13) (NMFS et al. 2011). The estimated pivotal temperature for the Kemp's ridley is relatively high (30.2 °C) compared to those for other sea turtle species (29–29.6 °C) (Yntema and Mrosovsky 1982; Godfrey et al. 1999; Hulin et al. 2009); the high temperatures at the nesting beach at Rancho Nuevo appear to naturally produce a hatchling sex ratio that is female biased (Table 11.1) (Wibbels 2007).

Not much is known about the period after a Kemp's ridley hatchling leaves the beach, swims offshore, associates with boundary currents, and is transported by the open ocean currents, often known as the *lost years*. However, the Kemp's ridley lost years may be similar to what occurs for the loggerhead sea turtle (Collard and Ogren 1990; Bolten 2003; Witherington et al. 2012). The oceanic currents in the western Gulf of Mexico control Kemp's ridley hatchling transport; coastal, shelf, and offshore currents vary during the hatchling emergence period (Collard 1987). The migratory success of young turtles that quickly reach pelagic waters is highly variable and influenced by oceanic conditions across the Kemp's ridley nesting range (Putman et al. 2010).

A recent analysis of seven Kemp's ridley nesting regions indicated that Rancho Nuevo ranked highest for migratory success of hatchlings to pelagic habitat (Putman et al. 2010). The narrow continental shelf off Tamaulipas, as well as oceanic conditions offshore Tamaulipas and Veracruz, may facilitate hatchling transport to the pelagic environment within 1–4 days (Collard and Ogren 1990; Putman et al. 2010). Depending on the type, location, strength, and paths of surface currents, pelagic Kemp's ridleys may either complete the developmental phase of their life cycle in the western Gulf of Mexico or be transported to the east, entrained in the Loop Current, exit the Gulf through the Straits of Florida, and drift to the north on the western edge of the Florida Current/Gulf Stream (Collard 1987; Collard and Ogren 1990). Similar to loggerheads, post-hatchlings likely become passive migrants in oceanic currents and use the *Sargassum* community as developmental habitat (Shaver 1991; NMFS et al. 2011).

The oceanic juvenile stage can be divided into two groups: the majority that remain in the currents of the Gulf of Mexico and a smaller group that is entrained in the Florida Current and transported up the Atlantic coast by the Gulf Stream (Putman et al. 2010). Because of the variability in growth rates, there is a range in the time estimated for a hatchling to grow to a size of about 20 centimeters (cm) (7.9 in) straight carapace length (SCL), the size at which Kemp's ridleys typically transition to the next stage—the neritic juvenile stage (Table 11.1). Juvenile Kemp's ridleys in the oceanic zone feed mostly on pelagic invertebrate prey associated with the *Sargassum* community (Table 11.1).

No post-hatchling Kemp's ridleys were collected from the Gulf of Mexico off the Florida coast from 2005 through 2011 as part of a study to determine the importance of the pelagic *Sargassum*-dominated drift community to young sea turtles (Witherington et al. 2012). Hatchlings are typically around 4 cm (1.6 in) SCL in size (Table 11.1); the smallest Kemp's ridley collected by Witherington et al. (2012) was 17 cm (6.7 in) SCL. Witherington et al. (2012) sampled only in the eastern Gulf of Mexico from May through September, and post-hatchling Kemp's ridley would most likely occur in the western Gulf of Mexico off the principal nesting beaches.

Thirty-eight juvenile Kemp's ridleys, ranging in size from 17 to 28 cm (6.7 to 11 in) SCL, were captured from the *Sargassum*-dominated surface-pelagic drift community in the eastern Gulf of Mexico from 2005 through 2011 (Witherington et al. 2012). These turtles were estimated to be between 1 and 2 years old. Because they were similar in size to the lower size range observed in nearby neritic habitats (Schmid 1998; Witzell and Schmid 2004) and because most of these turtles were not found within currents that would transport them out of the Gulf, they were hypothesized to be on the cusp of recruiting into coastal habitats of the northern and eastern Gulf of Mexico. Similar to what was proposed by Collard and Ogren (1990), these data suggest that an important recruitment pulse occurs primarily in the northern and eastern Gulf of Mexico, marking the end of the oceanic juvenile stage and the beginning of the neritic juvenile stage and suggests that the open waters of the northern and eastern Gulf of Mexico are of unique importance to Kemp's ridley sea turtles (Witherington et al. 2012).

11.2.1.3 Neritic Juvenile Life History and Distribution for Gulf of Mexico Kemp's Ridleys

Juvenile Kemp's ridleys that remain in the Gulf of Mexico during the oceanic stage move into coastal waters and are known to concentrate in shallow coastal waters, bays, estuaries, and sounds of the Gulf from south Texas to southwest Florida (Ogren 1989; Rudloe et al. 1991; Schmid 1998; Witzell and Schmid 2004; Schmid and Barichivich 2005; Frazier et al. 2007; Seney 2008). Coastal developmental and foraging areas frequently used by neritic juvenile Kemp's ridleys include Bolivar Roads Channel, Sabine Pass, and Lavaca and Matagorda bays in Texas; Caillou Bay and Calcasieu Pass in Louisiana; Big Gulley in Alabama; and Charlotte Harbor, Apalachicola Bay, Apalachee Bay, Deadman Bay, Waccasassa Bay/Cedar Keys, and Gullivan Bay/Ten Thousand Islands in Florida (Landry et al. 1995, 2005; Schmid and Barichivich 2005, 2006; Renaud and Williams 1997, 2005; Eaton et al. 2008; Schmid 2011). Details regarding studies that have been conducted in these areas are presented in the following paragraphs. Juvenile neritic Kemp's ridley sea turtles have not been reported from the southern Gulf of Mexico (Carr 1984).

The initial transition, as well as subsequent movements, of juvenile Kemp's ridleys to and from these coastal habitats appears to be seasonal (NMFS et al. 2011). Data from capture-mark-recapture (CMR) and satellite telemetry studies in the Gulf of Mexico have documented that juvenile turtles leave the coastal foraging areas in the fall and move to more suitable over-wintering habitat in deeper or more southern waters and return to the same coastal feeding areas the following spring (Ogren 1989; Schmid 1998; Witzell and Schmid 2004; Landry et al. 2005; Schmid and Barichivich 2005, 2006; Renaud and Williams 2005; Schmid and Witzell

2006; Seney and Landry 2011). For example, Renaud and Williams (2005) documented the fall and winter movements of Kemp's ridleys in the Gulf of Mexico in response to changing seawater temperature using satellite telemetry. Kemp's ridleys on the Florida Gulf coast moved in a southerly direction during the months of October through January as far south as the Florida Keys; once waters began to warm, they reversed their direction of movement. Southerly and southwesterly fall and winter migrations also were observed for turtles on the central and upper Texas coast (Renaud and Williams 2005).

Satellite telemetry was also used to monitor the winter migration of six neritic juveniles on the Florida Gulf coast (Schmid and Witzell 2006). All Kemp's ridleys departed from the Cedar Keys area in late November, when the average sea surface temperature dropped from 23.6 to 17.1 °C, migrated south in December, and overwintered in offshore waters from the Anclote Keys to Captiva Island during January. In February, when water temperatures increased to an average of 16.6 °C, the turtles started moving north and began returning to the Cedar Keys area in March.

Studies conducted in the early 1990s for the U.S. Army Corps of Engineers (USACE) to collect quantitative data to assess the risks of maintenance dredging to sea turtles demonstrated that the inshore and nearshore habitats of the upper Texas and Louisiana coasts were used by Kemp's ridleys on a seasonal basis and verified that jetties and channel entrances along the Texas and Louisiana coasts served as summer developmental habitat (Landry et al. 1992, 1993, 1994, 1995; Renaud et al. 1993a, 1995a). Small turtles arrived at Sabine Pass and Calcasieu Pass in April and May, and in June, July, and August, the mean size and overall abundance of turtles increased. Turtle abundance began to decrease in September, and by November, most, if not all, of the Kemp's ridleys had left the region. Kemp's ridley abundance was highest at Sabine Pass in Texas, followed by Calcasieu Pass and Bolivar Roads Channel. Small turtles (less than 18 kg or 39.7 lb) remained nearshore from May to October and moved less than large Kemp's ridleys (greater than 24 kg or 52.9 lb). Migration patterns of Kemp's ridleys varied by season and depended on turtle size; however, the majority of tracked turtles released near Sabine Pass and Calcasieu Pass remained within a few kilometers of shore and in relatively shallow waters. In addition, most maintained strong site fidelity to the westward side of both passes, most likely because current eddies and quiet water appeared to result in more favorable habitat and accumulation of prey (e.g., blue crabs), until cold fronts forced them to migrate south along the coast. Kemp's ridleys were found in dredged channels and moved back and forth across the passes and into inshore waters through shipping channels.

The nearshore Kemp's ridley assemblages in the western Gulf of Mexico were character-ized from 1992 through 1998 by netting turtles at nine study areas from Grand Isle, Louisiana, to South Padre Island, Texas (Landry et al. 2005). The occurrence of Kemp's ridleys at Sabine and Calcasieu passes was typically limited to April through September, and no turtles were captured from December through February. The 429 Kemp's ridleys captured during the study ranged in size from 19.5 to 65.8 cm (7.7 to 25.9 in) SCL; 77 % of the turtles had an SCL of less than 40 cm (15.7 in), and about 2 % of the turtles captured were adults, with none being mature males. The results of the study indicated that nearshore Gulf of Mexico waters along the upper Texas and Louisiana coasts provide developmental habitat to neritic juveniles during late spring through summer, when blue crab abundance and discarded shrimp fishery bycatch were highest (Landry et al. 2005).

The long-term abundance and distribution of neritic juvenile Kemp's ridleys (20–40 cm [7.9–15.7 in] SCL) in the nearshore waters of the northwestern Gulf of Mexico was character-ized using 10 years of entanglement netting data (Metz 2004). The nearshore waters included beachfront sites ranging in depth from 0.6 to 2 m (1.9 to 6.6 ft), while jetty sites ranged in depth from 1.5 to 3 m (4.9 to 9.8 ft). This 10-year survey, which was conducted by the Sea Turtle and

Fisheries Ecology Research Laboratory at Texas A&M University-Galveston, is the longest of its kind in the northwestern Gulf of Mexico and was conducted at locations (index habitats) that have a consistent occurrence of juvenile through adult Kemp's ridleys. The netting surveys were conducted primarily at Sabine Pass, Texas, and at Calcasieu Pass, Louisiana, as well as secondarily near the Mermentau River, Louisiana, from April through October 1993 through 2002. During the 10-year study, 600 Kemp's ridleys were captured, ranging in size from 19.5 to 66.3 cm (7.7 to 26.1 in) SCL; all annual mean size values were between 30 and 40 cm (11.8 and 15.7 in) SCL. Of all Kemp's ridleys captured during the study, 77 % were between 20 and 40 cm (7.9 and 15.7 in) SCL, about 20 % were between 40 and 60 cm (15.7 and 23.6 in) SCL, and 2 % were larger than 60 cm (23.6 in) SCL (Metz 2004). The size of Kemp's ridleys at Sabine Pass ranged from 19.5 to 64 cm (7.7 to 25.2 in) SCL, and no turtles larger than 55 cm (21.7 in) SCL were captured after 1998. Turtles captured at Calcasieu Pass were significantly larger and ranged in size between 22.4 and 66.3 cm (8.8 and 26.1 in) SCL.

Most likely in response to rising water temperatures and seasonal occurrence of blue crab prey, the overall monthly Kemp's ridley catch per unit effort (CPUE) peaked from April through June during the 10-year study in northwestern Gulf of Mexico nearshore waters (Metz 2004). The annual mean ridley CPUE across all study areas peaked in 1994, 1997, 1999, and 2002, which suggested a 2- to 3-year cycle in abundance that could be related to temporal patterns in clutch size or hatch success at the Rancho Nuevo nesting beach resulting from variability in nesting female fecundity and the remigration interval (Metz 2004). However, there was no significant relationship between Kemp's ridley CPUE in nearshore Texas and Louisiana waters and the number of hatchlings leaving the nesting beaches at Rancho Nuevo. In fact, juvenile ridley CPUE remained relatively constant or decreased slightly, even as the number of hatchlings released from Rancho Nuevo increased exponentially. Assuming that post-hatchling mortality rates did not increase during the study period, juvenile Kemp's ridleys may have been recruiting to coastal locations outside of the northwestern Gulf of Mexico study areas (Metz 2004). The annual declines in strandings in Texas since 1994, along with the subsequent increases in Florida strandings since 1995, suggested that a shift in Kemp's ridley distribution from the western to the eastern Gulf of Mexico may have occurred in the mid-1990s, which could have been related to fluctuations in circulation patterns. Significant declines in turtle CPUE at Sabine Pass since 1997 coincided with a concurrent reduction in blue crab size; however, a similar trend was not seen at Calcasieu Pass. When evaluating various biological and abiotic factors, nesting dynamics and prey availability appeared to have had the most influence on the nearshore occurrence of Kemp's ridleys (Metz 2004).

Hook-and-line captures data, stranding and nesting records, satellite telemetry, and diet analyses were used to characterize Kemp's ridley population dynamics and movements along the Texas coast from 2003 through 2007 (Seney 2008). The results of the analyses confirmed that Kemp's ridleys use the upper Texas coast and northwestern Gulf of Mexico throughout their life and that the region was used seasonally as developmental and nesting habitat, as well as a migration and foraging corridor. Recreational hook-and-line captures, which did not include oceanic juveniles or adults, made up about one-third of non-nesting encounters along the coasts of Galveston and Jefferson counties in Texas. Juveniles demonstrated a preference for habitat type or benthic prey concentrations, rather than specific locations, in the northwestern Gulf and were found in nearshore waters along the upper Texas coast primarily during the warmer months (March through October). They also entered inshore areas, such as bays and coastal lakes, along the Texas and Louisiana coasts (Seney 2008). Adult females that nested along the upper Texas coast occupied the region during the nesting season (April through July). Juvenile and internesting adults occurred in relatively shallow Texas state waters, and post-nesting females subsequently migrated through deeper, federal waters (Seney 2008).

In a related satellite telemetry study, the inshore and continental shelf waters of the northwestern Gulf of Mexico were shown to serve as developmental and migratory habitat for the Kemp's ridley sea turtle (Seney and Landry 2011). Fifteen juveniles were fitted with transmitters and released off the upper Texas coast from 2004 through 2007. Their movements were restricted to the continental shelf from Matagorda Bay, Texas, east to waters offshore Timbalier Bay, Louisiana, and during most or all of the tracking period, the juveniles remained primarily in waters less than 5 m (16.4 ft) deep (Seney and Landry 2011). While movement patterns varied among years, the juvenile Kemp's ridleys were tracked primarily during the warmer months and preferred tidal passes, bays, coastal lakes, and nearshore waters. In addition, this investigation suggested that the preferred habitat of juvenile Kemp's ridleys may differ among years and could be related to the locations and abundances of specific prey items (Seney and Landry 2011).

The movements of juvenile Kemp's ridleys in an understudied region of the Kemp's ridley range—the north-central Gulf of Mexico—were studied by satellite tracking 12 turtles that were captured incidentally by recreational fishermen on piers or stranded live in Mississippi and Alabama, and rehabilitated and released (Lyn et al. 2012). Six turtles were released in Mississippi waters, 3.2 km (2 mi) south of East Ship Island, in November 2010, and six were released near documented feeding grounds off the Cedar Keys in Florida in April 2011. The turtles released in Mississippi migrated to warmer waters offshore (when the water temperatures decreased) and stayed in the general area of Mississippi Sound and adjacent Louisiana waters. However, within days of being released, most of the turtles released in Florida quickly began swimming up the coastline toward Alabama and Mississippi. One of the turtles released in Florida, a newly mature male, was tracked all the way to Rancho Nuevo, Mexico; it remained in this area for 2 weeks in March before returning north to waters along the Texas/Louisiana border. The results of this study indicated that releasing turtles near their hooking/stranding location is preferred over releasing them in known feeding grounds (Lyn et al. 2012).

A number of in-water tagging studies have characterized sea turtle distribution, abundance, use, and ecology in nearshore waters along the Florida Gulf coast. Nearshore waters that have been studied include Apalachee Bay (Rudloe et al. 1991; Campbell 1996), Deadman Bay (Barichivich 2006), Cedar Keys/Waccasassa Bay (Schmid and Ogren 1990; Schmid 1998; Schmid et al. 2002, 2003), Tampa Bay (Nelson 2000), Charlotte Harbor Estuary (Schmid 2011), and Ten Thousand Islands/Gullivan Bay (Witzell and Schmid 2004, 2005). Details of these investigations are presented in the following paragraphs. However, information gaps still exist since the Florida Gulf coast is extensive, and long-term, in-water studies are needed to monitor the status of juvenile sea turtles at key foraging areas (Eaton et al. 2008).

Rudloe et al. (1991) conducted a tagging study of post-oceanic juvenile Kemp's ridleys that were incidentally captured during shrimp trawling, gill netting, or fish seining in the coastal waters of the northeastern Gulf of Mexico, including Apalachee, Levy, and Dickerson bays, from 1984 through 1988. A total of 106 turtles, ranging in size from 20.3 to 57.9 cm (8 to 22.8 in) SCL (mean 36.7 cm [14.4 in] SCL), were collected over a 97 km (60 mi) stretch of the Florida Gulf coast from Shell Point, Wakulla County to St. George Island, Franklin County. While turtles were collected every month, the highest numbers of turtles were collected during May and December. Turtles obtained during December, January, and February were significantly larger than those collected in June, July, August, and September (means of 40.4 cm [15.9 in] SCL and 30 cm [11.8 in] SCL, respectively). Kemp's ridleys were collected from seagrass, sand, and mud bottom substrates at depths ranging from 0.3 m (1 ft) inshore to 32 m (105 ft) 9.7 km (6 mi) offshore, with the smallest turtles collected from depths of less than 2 m (6.5 ft). Since turtles were only recaptured within a season, the results of the study indicated a transitory Kemp's ridley population along the northwest Florida Panhandle coast. In a netting study conducted from August 1995 through July 1997, Apalachee Bay was shown to be an important developmental habitat for juvenile Kemp's ridleys; the average size of captured turtles was 34 cm (13.4 in) SCL (Campbell 1996).

The importance of Waccasassa Bay, located along the Florida Gulf coast near the Cedar Keys, as developmental habitat for neritic juvenile Kemp's ridleys was demonstrated in an investigation conducted from June 1986 through October 1995 (Schmid 1998). Turtles were captured with large-mesh tangle nets from April to November when water temperatures were greater than 20 °C and near the oyster bars of Corrigan Reef. CMR data indicated that some Kemp's ridleys remained in the vicinity of Corrigan Reef during their seasonal occurrence and returned each year, and a mean annual population size of 159 turtles, with high rates of immigration and emigration, was estimated for this area. Captured Kemp's ridleys at Corrigan Reef (253 turtles) ranged in size from 26.8 to 58.6 cm (10.6 to 23.1 in) SCL and averaged 44.5 cm (17.5 in) SCL. Turtles captured during the summer were significantly larger than those captured during the fall (means of 45.5 cm [17.9 in] SCL and 43.1 cm [17 in] SCL, respectively). With the exception of 1991 when most turtles were in the 30–40 cm (11.8–15.7 in) SCL size class, the 40–50 cm (15.7–19.7 in) SCL size class dominated the catch. This investigation, as well as earlier work (Carr and Caldwell 1956; Schmid and Ogren 1990), confirmed the occurrence of a seasonal, resident population of juvenile Kemp's ridleys in Waccasassa Bay.

In a related investigation, juvenile Kemp's ridleys were studied in Waccasassa Bay from May through August 1994 and May through November 1995 (Schmid et al. 2002, 2003). Turtles occupied foraging areas ranging from 5 to 30 km^2 (1.9 to 11.6 mi^2) in size, and they used rock outcroppings more than expected (Schmid et al. 2003). In addition, live bottom and green macroalgae habitats were used more than seagrass habitats. Turtles increased their rate of movement with the increasing velocity of the tide (Schmid et al. 2002). The rates of Kemp's ridley movement were higher and surface and submergence durations were shorter during the day; the turtles' daily activities were attributed to food acquisition and bioenergetics.

Juvenile Kemp's ridleys inhabiting the nearshore waters of the northeastern Gulf of Mexico, specifically from Apalachee Bay to Suwannee Sound, including Deadman Bay, Florida, were studied from 1995 through 1999 (Barichivich 2006). The majority of the turtles captured were from 20 to 40 cm (7.9 to 15.7 in) SCL, and most were captured in Deadman Bay (121 of 126 turtles). While fewer captures were made during the cooler months, Kemp's ridleys were captured from March through December, and turtles were captured when water temperatures were between 19.7 and 34 °C. Annual growth rates ranged from 1.25 to 8.92 cm (0.49 to 3.51 in). The large number of short-term recaptures, as well as recaptures between seasons, indicated that Deadman Bay is an important developmental habitat for Kemp's ridleys (Barichivich 2006).

To assess sea turtle occurrence relative to the channel bottom, trawl surveys were conducted within the Tampa Bay Entrance Channel during the spring, summer, and fall of 1997, as well as in the spring of 1998, by the USACE (Nelson 2000). During the surveys, two juvenile Kemp's ridleys were captured and tracked. These turtles remained in the study area for days or months at a time and eventually moved in response to changing water temperature. They either moved offshore or southward as water temperatures decreased, and when water temperatures warmed, they returned to their original location. The results of the surveys indicated that dredging activities in the Tampa Bay Entrance Channel should be conducted during extremes in water temperature during either the winter or summer (Nelson 2000).

In a study conducted from 1997 through 2004, the nearshore waters of Gullivan Bay, in the Ten Thousand Islands area off the southwest Florida Gulf coast, were determined to be important developmental habitat for juvenile Kemp's ridleys (Witzell and Schmid 2004, 2005; Eaton et al. 2008). More than 190 Kemp's ridleys, ranging in size from 21.4 to 65.2 cm (8.43 to

25.7 in) SCL and averaging 40.4 cm (15.9 in) SCL, were captured. Kemp's ridley recaptures were documented within and between sampling seasons, indicating foraging-site fidelity. Some turtles set up home ranges in this area for as long as 3 years. Kemp's ridleys preferred areas of sand substrate with plumed worm tubes and live-bottom organisms (Witzell and Schmid 2004, 2005).

From August 2009 through April 2011, monthly in-water surveys of the southeastern portion of Pine Island Sound, part of the Charlotte Harbor estuary complex on the Gulf coast in Lee County, Florida, containing live bottom habitat were conducted. These in-water surveys were a continuation of earlier tagging studies conducted in the area (Schmid 2011). Almost 70 % of the sea turtles observed during the study were Kemp's ridleys. There were 50 sightings of Kemp's ridleys, and 45 Kemp's ridleys were captured and tagged. Similar to other nearshore areas of the Florida Gulf coast, most of the Kemp's ridleys were juveniles; captured turtles ranged in size from 24.2 to 62.7 cm (9.5 to 24.7 in) minimum SCL and averaged 40.9 cm (16.1 in) minimum SCL in size (Schmid 2011). CMR data indicated both within- and between-seasons fidelity to the study area. A satellite-tracked juvenile turtle demonstrated seasonal fidelity to the area by leaving Charlotte Harbor in late fall, heading south and wintering off the Florida and Marquesas keys, and returning to within a few kilometers of its capture site in early spring (Schmid 2011). An adult-sized Kemp's ridley was also tracked by satellite; it appeared to be a transient inhabitant in the area, immediately leaving Pine Island Sound after release and moving northward to a feeding area offshore from Homosassa Bay. The results of the study reinforced the importance of Charlotte Harbor Estuary, particularly Pine Island Sound, as Kemp's ridley developmental habitat (Schmid 2011).

Most of the diet of neritic juvenile Kemp's ridley sea turtles consists of crabs (Table 11.1). However, as indicated by the lists of food items for neritic juveniles from different locations included in Table 11.1, Kemp's ridleys appear to be opportunistic foragers and readily utilize prey in a particular area. Since neritic juvenile Kemp's ridley sea turtles continue the pattern of seasonal migrations and fidelity to foraging sites for many years until maturing and moving to adult foraging areas, both the nearshore foraging grounds discussed in the paragraphs above and the offshore overwintering areas in the Gulf of Mexico are important for this life stage of Kemp's ridleys (NMFS et al. 2011).

11.2.1.4 Adult Life History and Distribution for Gulf of Mexico Kemp's Ridleys

Adult Kemp's ridley sea turtles occur primarily in nearshore waters of the Gulf of Mexico. They are occasionally found in the coastal regions of the southeast U.S. Atlantic Coast and are rare off the northeast U.S. Atlantic Coast (Hildebrand 1982; Ogren 1989; USFWS and NMFS 1992; TEWG 2000; Pritchard 2007a). Important foraging areas where adult females reside seasonally consist of biologically productive locations in the waters off the western and northern Yucatán Peninsula, including the Laguna del Carmen area off Campeche, Mexico, and the northern Gulf of Mexico from southern Texas to western Florida, such as along the Louisiana coast near the mouth of the Mississippi River (Chávez 1968; Pritchard and Márquez-M. 1973; Guzmán-Hernández et al. 2007; Shaver and Rubio 2008). While there is some geographic variation, crabs (especially portunid crabs) are the primary prey of adult Kemp's ridleys (Table 11.1).

Early on Kemp's ridley adult migration was thought to occur within the continental shelf waters of the Gulf of Mexico, mainly between nesting sites in Tamaulipas, Mexico, and coastal feeding grounds (Morreale et al. 2007). In fact, adult female Kemp's ridleys have been tracked from foraging grounds in Louisiana and Texas to the Rancho Nuevo nesting beach (Renaud et al. 1996). This pattern of moderate-distance migration was consistent with the observed high frequency of annual return of nesting females to their main nesting beach in Rancho Nuevo,

Mexico, as well as with their strategy of feeding primarily on benthic crustaceans in coastal waters (Morreale et al. 2007). Satellite tracking studies have indicated that post-nesting female Kemp's ridleys travel along coastal corridors (typically shallower than 50 m [164 ft]) along the rim of the Gulf of Mexico basin, extending from the Yucatán Peninsula to southern Florida (Byles 1989; Byles and Plotkin 1994; Renaud 1995; Renaud et al. 1996; Shaver 1999, 2001a, b; Morreale et al. 2007). Many of these turtles settled in resident feeding areas for up to several months after migrating, which demonstrated that Kemp's ridley post-nesting migrations could also be considered foraging migrations to fixed destinations (Byles and Plotkin 1994).

Between 1997 and 2006, the movements of 28 turtles (17 wild-stock and 11 head-started) that nested on North Padre Island or Mustang Island, Texas, were monitored using satellite telemetry to obtain habitat use and movement information for Kemp's ridleys that nested in south Texas (Shaver and Rubio 2008). Internesting residency was documented off south Texas, and post-nesting residency occurred in Gulf of Mexico waters from south Texas to the southern tip of Florida. After nesting for the season was complete, most of the Kemp's ridleys left south Texas and traveled, parallel to the coastline, to the northern or eastern Gulf of Mexico. This study demonstrated the importance of nearshore Gulf waters to post-nesting Kemp's ridleys, and the results were used to develop a regulation to close nearshore south Texas waters seasonally to shrimp trawling (Shaver and Rubio 2008).

The inshore and continental shelf waters of the northwestern Gulf of Mexico along the upper Texas coast were recently demonstrated to serve as migratory, internesting, and post-nesting habitat for adult Kemp's ridleys (Seney and Landry 2008, 2011). Six female Kemp's ridleys were fitted with satellite transmitters after nesting and tracked during 2005 and 2006 (Seney and Landry 2008). In a second investigation, seven adult females (six nesting and one trawl-caught) were fitted with transmitters and released off the upper Texas coast during 2004 through 2007 to characterize their movements, migration patterns, and foraging grounds in the northwestern Gulf of Mexico (Seney and Landry 2011). During both tracking studies, the females remained in the Galveston region and in the vicinity of the upper Texas coast during their internesting intervals and moved eastward along the continental shelf (20 m [66 ft] isobath) to offshore foraging areas of central Louisiana upon entering the post-nesting stage (Seney and Landry 2008, 2011).

The importance of nearshore Gulf of Mexico waters, specifically off the Louisiana coast, as critical foraging habitat for post-nesting Kemp's ridleys was recently demonstrated by Shaver et al. (2013). Satellite telemetry and switching state-space modeling was used to track 31 turtles after nesting at PAIS and Rancho Nuevo for 13 years from 1998 through 2011. Multiple turtles foraged along their migratory route before arriving at their final foraging sites. Nearshore Gulf of Mexico waters served as foraging habitat for all turtles tracked in the study, and final foraging sites were located in water less than 68 m (223 ft) deep and a mean distance of 33.2 km (20.6 mi) from the nearest mainland coast. The wide distribution of foraging sites indicates that a foraging corridor exists for Kemp's ridleys in the Gulf of Mexico (Shaver et al. 2013).

In the spring, mature female Kemp's ridleys undertake annual migrations to the western Gulf of Mexico and gather along the coast of Tamaulipas near the village of Rancho Nuevo to nest on the many kilometers of almost continuous sand beach (Pritchard 2007a, b). Reproductive females begin to arrive offshore Rancho Nuevo in March and April, with most arriving during May and June, and remain in the vicinity through the nesting season (Table 11.1) (Rostal 1991; Seney 2008). About a month before the nesting season begins, females and males aggregate to mate in nearshore waters near the beach at Rancho Nuevo (Pritchard 1969; Mendonca and Pritchard 1986).

In contrast to the pattern of female post-nesting migration, many adult male Kemp's ridleys remain in the vicinity of the nesting beach throughout the year (Shaver et al. 2005; Shaver 2006a). Shaver et al. (2005) monitored the movements of 11 adult male Kemp's ridleys captured near Rancho Nuevo, Mexico, using satellite telemetry, and while one traveled north and was last located offshore Galveston, Texas, ten remained in the vicinity of the nesting beach. As indicated by mating activities for the Kemp's ridley, which are more widespread than the nesting areas and occur in coastal and inshore waters from south Texas to Veracruz, Mexico, males that do not reside near the nesting beaches throughout the year mate with females in foraging areas or migration pathways (Shaver 1992; Morreale et al. 2007).

11.2.2 Kemp's Ridley Recovery Program

As mentioned previously, through a variety of intervention methods by Mexico and the United States that began almost 50 years ago, the Kemp's ridley has made a remarkable comeback from the brink of extinction (Márquez-M. et al., 1998, 1999; TEWG 1998; Heppell et al. 2007; Crowder and Heppell 2011). Egg protection efforts began at Rancho Nuevo in 1966 and were expanded in 1976, and a binational recovery plan was developed in 1977. Community involvement and education, which have changed attitudes regarding Kemp's ridley conservation, have been important elements of the recovery program since its implementation (Heppell et al. 2007). While no method exists to quantitatively assess the relative impact of all of the methods implemented, which is a critical issue, it is clear that conservation efforts have resulted in increased survival rates for all life stages of the Kemp's ridley (Heppell et al. 2007; Plotkin 2007b).

The Kemp's Ridley Recovery Program began in 1978, and 100 % of the nests in the Rancho Nuevo area began to be relocated to fenced corrals; these hatcheries eliminated land-based predation and human egg collection from the Kemp's ridley life cycle to ensure high egg survival rates (Cornelius et al. 2007). Over the years, nest protection efforts have expanded north and south of Rancho Nuevo, and additional corrals have been constructed. After the hatchlings emerge from the nests in these hatcheries, they are counted and released in large groups directly into the water at different locations along the beach (NMFS et al. 2011).

Currently, nests continue to be relocated to fenced corrals; however, a large number of nests are now left in situ. Many of these nests are covered to protect against predation, as was done in a 2007 arribada (NMFS et al. 2011). The slow and steady increase in the nesting population at Rancho Nuevo and the increase in numbers of females nesting in Texas have been aided by protective egg hatcheries in both Mexico and Texas (Márquez-M. et al. 1996; Heppell 1997; Shaver and Caillouet 1998).

At the same time, measures to reduce the at-sea, incidental mortality of juvenile and adult Kemp's ridleys resulting from shrimp trawling and other fishing operations began to be implemented (Heppell et al. 2007). In 1980, U.S. shrimp trawlers were excluded from Mexican waters; sailing and fishing within 6.44 km (4 mi) of the Rancho Nuevo beach was prohibited starting in the late 1980s; and by 1990, the sea turtle product trade was banned in Mexico, and TEDs were required in U.S. waters (NMFS et al. 2011). In the mid- and late-1990s, TEDs began to be required in Mexico, the fishing effort off the main nesting beaches was reduced, and a closure of the Mexican shrimping season during the primary ridley nesting period began. In 2000, an annual closure of shrimp trawling in Gulf of Mexico waters off North Padre Island, South Padre Island, and Boca Chica Beach began. Kemp's ridley feeding habitat off South Padre Island, Texas, was protected in 2002 (Márquez-M. et al., 1998, 1999; TEWG 1998; Shaver 2005, 2006b; Heppell et al. 2007).

In addition, the Kemp's Ridley Sea Turtle Head Start Experiment began in 1978, with a goal of establishing a second nesting population at PAIS in Texas where sporadic Kemp's ridley nesting had been documented in the past (Fontaine and Shaver 2005). To establish a nesting beach at PAIS, a small fraction (average of 2.8 %) of the total eggs from Rancho Nuevo was translocated for incubation, hatching, and experimental imprinting from 1978 through 1988 (Caillouet 1995). During the 23 years of the Head Start Experiment—1978 to 2000—more than 23,000 Kemp's ridley hatchlings were raised in captivity for approximately 1 year at the NMFS Laboratory in Galveston, Texas (see below; Fontaine and Shaver 2005; Shaver and Wibbels 2007). This prevented the high level of predation associated with the post-hatchling life stage in the wild, as these larger turtles would be less susceptible to predators upon their release (Shaver and Wibbels 2007). In case of extinction in the wild, a few hundred of the head-started ridleys were retained in captivity in various locations.

The majority of hatchlings from 1978 through 1988 were obtained from eggs that had been transferred from Rancho Nuevo to PAIS for experimental imprinting. Several hundred hatchlings from the 1978, 1979, 1980, and 1983 year classes were obtained directly from Rancho Nuevo (Fontaine and Shaver 2005; Shaver and Wibbels 2007). The eggs obtained from Rancho Nuevo from 1978 through 1988 (22,507 total eggs) were incubated in PAIS sand, and the hatchlings were released on the beach and allowed to swim briefly in the Gulf of Mexico (Shaver and Wibbels 2007). After the hatchlings swam approximately 5–10 m (16–33 ft), they were captured by dip net and shipped to the NMFS Laboratory in Galveston for head-starting. Approximately 77 % of the eggs incubated at PAIS produced hatchlings (15,875 total hatchlings) that were transferred to the Galveston Laboratory from 1978 through 1988 (Shaver 2005). All hatchlings reared in the Head Start Experiment were obtained directly from Rancho Nuevo from 1989 through 2000. Because of lower than pivotal incubation temperatures, males likely predominated in the 1978, 1979, 1981, and 1983 year classes. Conversely, about 78 % of the 1985 through 1988 year classes were estimated to be females given the above pivotal temperature used to incubate the eggs (overall, approximately 60 % of the 1978 through 1988 year classes were females), and over 90 % of the head-started turtles from the 1989 through 2000 year classes were likely females (Fontaine and Shaver 2005; Shaver and Wibbels 2007).

All released turtles were tagged using a variety of methods: external metal tags, *living tags* (a disk of light living tissue from the plastron transplanted to the dark carapace), internal magnetized wire tags, and passive integrated transponder tags (Fontaine and Shaver 2005; Shaver and Wibbels 2007). Except for 1983, and from 1993 through 2000, approximately 1,000–2,000 turtles were raised in captivity each year (Shaver and Wibbels 2007). Over 23 years, 27,137 Kemp's ridley hatchlings were transported to the NMFS Galveston Laboratory from either PAIS or directly from Rancho Nuevo; of those, about 88 % (23,473 hatchlings, which included 13,275 imprinted at PAIS and 10,198 imprinted at Rancho Nuevo) were successfully reared, tagged, and released into the Gulf of Mexico at sizes comparable to the juvenile oceanic or neritic stage of wild-stock Kemp's ridleys (Fontaine and Shaver 2005; Shaver and Wibbels 2007).

Head-started Kemp's ridleys, which typically weighed approximately 1 kg (2.2 lb), were released throughout the Gulf of Mexico in a variety of locations that represented habitats appropriate for late oceanic or neritic juvenile ridleys (Shaver and Wibbels 2007). In 1978 and 1979, most turtles were released off the Florida Gulf coast. However, there was concern that turtles were leaving the Gulf of Mexico and might not return to breed, since many of the released turtles were later recaptured along the Atlantic coast (Shaver and Wibbels 2007). Therefore, from 1980 through 2000, ridleys were released in the western Gulf of Mexico, primarily off the Texas coast in the waters off Padre Island.

While the site of release affected the movements and distributions of head-started turtles, overall, head-started Kemp's ridleys dispersed widely from release areas and were reported throughout the natural range and in habitats of wild-stock Kemp's ridleys (Shaver and Wibbels 2007). Growth and diet information obtained from recaptured and stranded turtles indicated that head-started ridleys adapted to feeding in the wild. Mortality rates were likely high for both wild-stock and head-started ridleys before the mandatory use of TEDs began in U.S. offshore waters in 1990, as indicated by the continued decline of Kemp's ridleys throughout the 1980s and early 1990s. In addition, because Kemp's ridley mortality was so high during the 1980s, survival to adulthood was unlikely for most turtles (Shaver and Wibbels 2007).

Although factors, such as insufficient monitoring of the nesting beaches and turtle tag loss, have affected the collection of head-started Kemp's ridley nesting data, some head-started Kemp's ridleys have nested on beaches of PAIS, North Padre Island, Mustang Island, Galveston Island, and Bolivar Peninsula in Texas, as well as in Tamaulipas, Mexico (Shaver and Wibbels 2007). The first head-started Kemp's ridley nests, laid by turtles from the 1983 and 1986 year classes, were documented at PAIS and Boca Chica Beach, Texas in 1996 (Shaver 1996; Fontaine and Shaver 2005; Shaver and Wibbels 2007). The Kemp's ridleys that have nested in Texas since 1996 have been a mixture of head-started and wild-stock turtles. However, since 2002, the majority of nesting in Texas has been by wild-stock Kemp's ridleys (NMFS et al. 2011). For example, of the 42 Kemp's ridley nests documented in Texas in 2004, 13 were from head-started turtles (Safina and Wallace 2010). Tracking studies have demonstrated that nesting head-started Kemp's ridleys have similar post-nesting movement behaviors as wild-stock ridleys (Landry and Seney 2006; Shaver and Wibbels 2007).

Because the Head Start Experiment was a large-scale experiment on the most endangered sea turtle in the world and was the subject of intense debate and controversy throughout the 1980s and early 1990s (Shaver and Wibbels 2007), it was officially terminated in 1993 but continued on a limited basis through 2000. While it is encouraging that some head-started Kemp's ridleys have survived to maturity and are nesting, it is not possible to accurately determine the effectiveness or ineffectiveness of the experiment (Shaver and Wibbels 2007). However, head-started females have nested and viable second-generation hatchlings have been produced both at PAIS and at Rancho Nuevo. In addition, a large amount of biological information was obtained as a result of the Head Start Experiment that might otherwise have never been discovered (Fontaine and Shaver 2005).

11.3 LOGGERHEAD SEA TURTLE (*CARETTA CARETTA*)

Loggerhead sea turtles are named for their large heads (Figures 11.14 and 11.15). The adults are slightly larger than hawksbills but slightly smaller than green sea turtles (Witherington et al. 2006a). Of all species of sea turtles, the life history of the loggerhead is probably the best understood (Bolten and Witherington 2003; Witherington et al. 2012). This is due to the concern regarding their decline because of the incidental capture in commercial fisheries (e.g., trawling, driftnet, longline) as well as the loss of nesting habitat caused by coastal development. In addition, loggerheads nest in areas with major conservation programs; for example, loggerheads nest on the beaches of the Archie Carr National Wildlife Refuge (ACNWR) in Florida (Figure 11.14), the most important nesting beach of loggerheads in the Western Hemisphere, and the location of a long-term research program. Compared to the other sea turtle species, the loggerhead sea turtle has the largest geographic nesting range, which includes both temperate and tropical latitudes. It is globally distributed in all temperate and tropical ocean basins, and its diet is the least specialized (Bolten and Witherington 2003).

Figure 11.14. Nesting loggerhead sea turtle on the beach at Archie Carr National Wildlife Refuge, Indian River and Brevard Counties, Florida, June 2012 (photograph by Steve A. Johnson, University of Florida, with permission).

Figure 11.15. Juvenile loggerhead sea turtle in the water (photograph by Marco Giuliano, Fondazione Cetecea) (from NOAA 2011).

11.3.1 Loggerhead Sea Turtle Life History, Distribution, and Abundance

The loggerhead sea turtle occurs throughout the temperate, subtropical, and tropical regions of the Atlantic, Pacific, and Indian oceans, and its range includes foraging areas, migration corridors, and nesting beaches (Figure 11.16) (Dodd 1988). Unlike most other sea turtle species, the loggerhead is less abundant in the tropics than it is in temperate waters, and most of its nesting beaches are located outside of the tropics (Witherington et al. 2006a). The majority of loggerhead nesting is located at the western rims of the Atlantic and Indian oceans, and only two loggerhead nesting aggregations have greater than 10,000 females nesting per year: Peninsular Florida in the United States and Masirah Island, Oman (Conant et al. 2009). Loggerhead nesting aggregations with 1,000–9,999 females nesting annually occur in Georgia through North Carolina in the United States, Quintana Roo and Yucatán in Mexico, Brazil, Cape Verde Islands, western Australia, and Japan (Márquez-M. 1990; Ehrhart et al. 2003; Conant et al. 2009). Aggregations with 100–999 females nesting annually occur in the northern Gulf of Mexico (USA), Dry Tortugas (USA), Cay Sal Bank (Bahamas), Tongaland (South Africa), Mozambique, Arabian Sea coast (Oman), Halaniyat Islands (Oman), Cyprus, Peloponnesus (Greece), Zakynthos (Greece), Crete (Greece), Turkey, and Queensland (Australia) (Conant et al. 2009).

Following are the five currently recognized life stages for the loggerhead sea turtle (TEWG 2009):

1. Year One: terrestrial zone to oceanic zone, size less than or equal to 15 cm (5.9 in) SCL.

2. Juvenile, Stage I: exclusively oceanic zone, size range of 15–63 cm (5.9–24.8 in) SCL.

3. Juvenile, Stage II: oceanic or neritic zones, size range of 41–82 cm (16.1–32.3 in) SCL.

4. Juvenile, Stage III: oceanic or neritic zones, size range of 63–100 cm (24.8–39.4 in) SCL.

5. Adult: neritic or oceanic zones, size greater than or equal to 82 cm (32.3 in) SCL.

Loggerhead sea turtles are represented by many distinct populations (Figure 11.6) (USFWS and NMFS 2011). Because the focus of this chapter is the Gulf of Mexico sea turtles, the

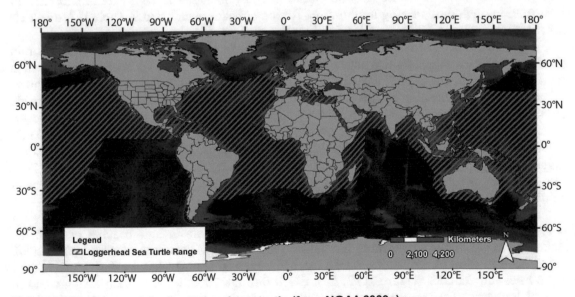

Figure 11.16. Range of the loggerhead sea turtle (from NOAA 2009a).

sections that follow focus on loggerheads that occur in the Gulf of Mexico during some portion of their life cycle. The Northwest Atlantic Ocean DPS of loggerhead sea turtles uses the Gulf of Mexico's beaches for nesting, oceanic currents for developmental habitat, and/or neritic and oceanic areas for foraging, resting, or migrating.

The Northwest Atlantic Ocean DPS of loggerhead sea turtles includes the following subpopulations (NMFS and USFWS 2008; USFWS and NMFS 2011):

- Northern Subpopulation: southern Virginia to Florida/Georgia border (rarely utilizes the Gulf of Mexico (Heppell et al. 2003)).

- Peninsular Florida Subpopulation: Florida/Georgia border south through Pinellas County, excluding the islands of Key West, Florida.

- Northern Gulf of Mexico Subpopulation: Franklin County, Florida, west through Texas.

- Dry Tortugas Subpopulation: islands west of Key West, Florida.

- Greater Caribbean Subpopulation: Mexico through French Guiana, Bahamas, Lesser and Greater Antilles.

Using nesting data from 2001 through 2010, Richards et al. (2011) estimated the Northwest Atlantic Ocean DPS adult female loggerhead population to range from 30,096 to 51,211 turtles. Individual subpopulations were estimated as follows: 258–496 adult females for Dry Tortugas, 323–634 adult females for northern Gulf of Mexico, 1,975–4,232 adult females for Greater Caribbean, and 23,655–45,058 adult females for Peninsular Florida. Richards et al. (2011) remarked that improved estimates of clutch frequency and breeding intervals, as well as better measures of temporal and spatial variation, are needed to improve population estimates based on nest counts.

11.3.1.1   Nesting Life History, Distribution, and Abundance for Gulf of Mexico Loggerheads

The generalized locations of loggerhead nesting beaches for all of the loggerhead subpopulations of the Northwest Atlantic Ocean DPS are included in Figure 11.17. Life history information, including nesting information, for loggerheads is included in Table 11.2; available information for specific Gulf of Mexico beaches or regions is also included.

The Peninsular Florida subpopulation of loggerheads is the largest nesting aggregation in the Atlantic Ocean, representing about 80 % of all nesting and about 90 % of all hatchlings in this DPS (Ehrhart et al. 2003; TEWG 2009; Witherington et al. 2009). The greatest proportion of nesting for the Peninsular Florida subpopulation occurs on the Atlantic coast in six Florida counties (Brevard, Indian River, St. Lucie, Martin, Palm Beach, and Broward); however, thousands of nests are laid each year on southwest Florida Gulf coast beaches (Figure 11.17) (TEWG 2000, 2009; Ehrhart et al. 2003; Witherington et al. 2009). In the Gulf of Mexico, loggerheads typically nest on barrier island beaches with moderate to high wave energy. They also nest on low-relief mangrove islands, such as those located in the Ten Thousand Islands in southwest Florida (Foley et al. 2000).

In 1979, approximately 10,000 loggerhead nests, or about 1,900 nesting females, were counted on beaches surveyed in Peninsular Florida (these data include beaches on both the Atlantic and Gulf coasts; separating out data for the Gulf of Mexico was not possible). In 1998, loggerhead nesting in Peninsular Florida reached a high of almost 84,600, which equals about 15,700 nesting females (Figure 11.18). From 1979 through 2000, a general increasing trend in annual loggerhead nesting occurred in Peninsular Florida; but, in 2001, annual nest counts began a decreasing trend through 2009, with a low of 44,512 nests (or about 8,243 nesting

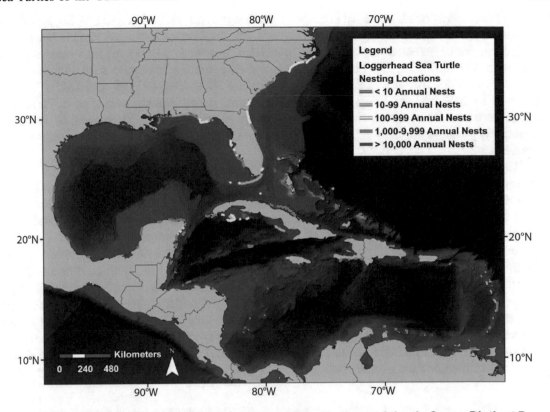

Figure 11.17. Generalized nesting beach locations of the Northwest Atlantic Ocean Distinct Population Segment of loggerhead sea turtles (interpreted from Dow et al. 2007; NMFS and USFWS 2008; SWOT 2007a).

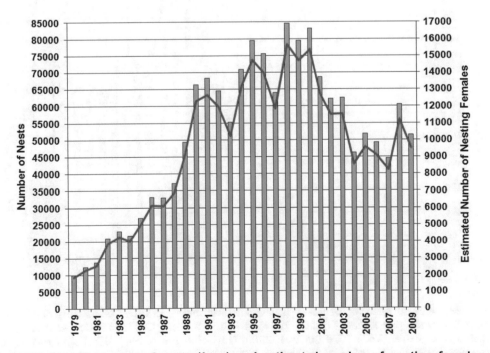

Figure 11.18. Annual number of nests (*bars*) and estimated number of nesting females (*line*), assuming 5.4 nests per female (Tucker 2010), for the Peninsular Florida subpopulation of loggerhead sea turtles from 1979 through 2009. Note that the survey effort was not consistent among years (from Meylan et al. 1995; NMFS and USFWS 2008; FFWCC FWRI 2012).

females) recorded for all surveyed Peninsular Florida beaches in 2007 (Figure 11.18). Results of a recent analysis of Florida Index Beach survey data, a subset of surveyed Florida beach data suitable for trend assessments because of consistent spatial and temporal nest counts (Witherington et al. 2009), for the Peninsular Florida subpopulation indicated a 26 % decrease in nesting from 1989 through 2008 and a 41 % decline since 1998 (NMFS and USFWS 2008). However, in 2010 and 2011, loggerhead nesting counts on all beaches surveyed in Peninsular Florida were back to numbers similar to those recorded in 2000 (73,066 nests and 67,701 nests, respectively), and a high of 97,000 nests, representing about 18,000 nesting females, was recorded in 2012 (FFWCC FWRI 2012).

The northern Gulf of Mexico subpopulation of loggerheads is one of the smallest nesting aggregations in the Atlantic and the second smallest in the western North Atlantic (TEWG 2009). The nesting beaches of this subpopulation are concentrated in the Florida Panhandle, with a consistent but small amount of nesting in other Gulf States, mostly Alabama and Texas (Figure 11.19). As part of this subpopulation, loggerhead sea turtles nest along Eglin Air Force Base on Cape San Blas and Santa Rosa Island. The number of nests laid at each location from 1994 through 1997 included the following: 53, 60, 25, and 54 nests on Cape San Blas and 32, 18, 28, and 22 nests on Santa Rosa Island (Lamont et al. 1998). Texas has almost 600 km (373 mi) of beach available to nesting sea turtles but, for unknown reasons, loggerheads do not nest regularly or in large numbers on Texas beaches (Plotkin 1989). Since 1994, annual nest counts for northern Gulf of Mexico loggerheads have consistently exceeded 600, or about 110 nesting females, with a high of 1,285 nests in 1999 and a low of 611 nests in 2007 (Figure 11.19). While it is difficult to evaluate long-term nesting trends for the northern Gulf of Mexico subpopulation

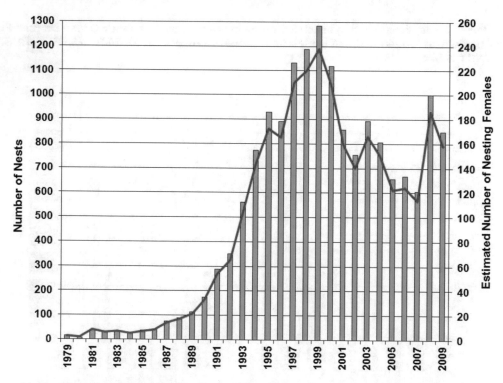

Figure 11.19. Annual number of nests (*bars*) and estimated number of nesting females (*line*), assuming 5.4 nests per female (Tucker 2010), for the northern Gulf of Mexico subpopulation of loggerhead sea turtles from 1979 through 2009. Note that the survey effort was not consistent among years (from Meylan et al. 1995; TEWG 2000; NMFS and USFWS 2008; Richards et al. 2011).

because of changed and expanded beach survey coverage, an analysis of 12 years (1995–2007) of Florida Index Beach survey data for this subpopulation indicated a significant declining trend of about 5 % per year (NMFS and USFWS 2008). In 2010 and 2011, 683 and 970 nests, respectively, were recorded for northern Gulf of Mexico loggerheads, and a high of 1,750 nests was recorded in 2012 (FFWCC FWRI 2012; Share the Beach 2013).

The Dry Tortugas subpopulation is the smallest loggerhead subpopulation of the Northwest Atlantic Ocean DPS (Figure 11.17) (TEWG 2009). This subpopulation is important not only for its genetic distinctiveness but also because it is isolated from many of the threats facing most sea turtle nesting areas (van Houtan and Pimm 2007). Loggerhead nesting activity is unevenly distributed among the seven islands within Dry Tortugas National Park (DTNP). About 90 % of loggerhead nests are laid on East Key and Loggerhead Key each year; the remaining 10 % of nesting activity occurs on Bush Key, Hospital Key, and Garden Key, and virtually no nesting occurs on Long Key and Middle Key (van Houtan and Pimm 2007). Annual nest counts for the Dry Tortugas subpopulation of loggerheads from 1995 through 2001 consistently exceeded 200 (about 40 nesting females) with a high of 340 nests recorded in 1995 (Figure 11.20). About 200 nests were also recorded for the Dry Tortugas in 2003 and 2009. A longer time series of nesting data for the Dry Tortugas subpopulation is necessary in order to detect a trend (NMFS and USFWS 2008).

The majority of nesting for the Greater Caribbean subpopulation of loggerhead sea turtles occurs in Quintana Roo, Mexico. The loggerhead nesting aggregation in Quintana Roo is the third largest in the western north Atlantic (Figure 11.17) (Ehrhart et al. 2003; TEWG 2009). Less frequent and scattered loggerhead nesting occurs along the Mexican Gulf coast from the

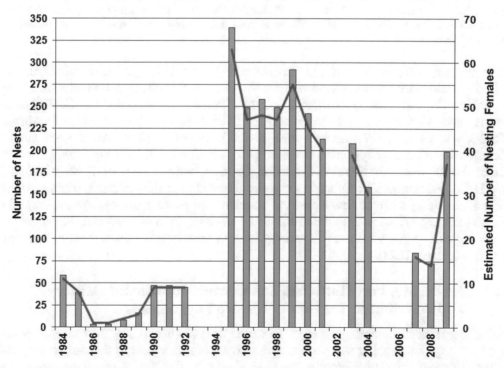

Figure 11.20. Annual number of nests (*bars*) and estimated number of nesting females (*line*), assuming 5.4 nests per female (Tucker 2010), for the Dry Tortugas subpopulation of loggerhead sea turtles from 1984 through 2009. Note that the survey effort was not consistent among years, and data were not available for 1993, 1994, 2002, 2005, or 2006 (from Meylan et al. 1995; NMFS and USFWS 2008; FFWCC FWRI 2012).

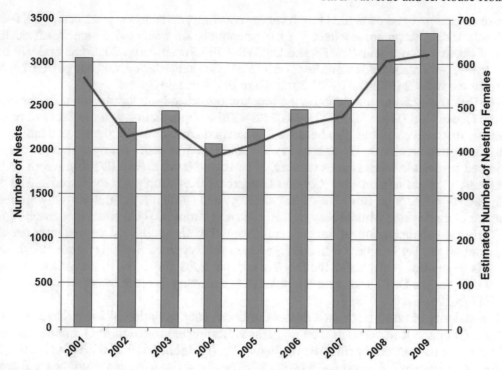

Figure 11.21. Annual number of nests (*bars*) and estimated number of nesting females (*line*), assuming 5.4 nests per female (Tucker 2010), for the Greater Caribbean subpopulation of logger-head sea turtles from 2001 through 2009. Note that the survey effort was not consistent among years (from Richards et al. 2011).

Yucatán Peninsula to as far north as Tamaulipas (Figure 11.17) (Márquez-M. 1990; Ehrhart et al. 2003). Additional nesting locations for the Greater Caribbean subpopulation of logger-heads include Cay Sal Bank in the eastern Bahamas; along the coasts of Cuba, Central America, Colombia, and Venezuela; and the eastern Caribbean Islands (Figure 11.17) (Addison and Morford 1996; Addison 1997; Ehrhart et al. 2003; Conant et al. 2009). Annual nest counts for the Greater Caribbean subpopulation of loggerheads from 2001 through 2009 consistently exceeded 2,000 nests, or about 370 nesting females, with a high of 3,349 nests in 2009 (Figure 11.21). An analysis of trends in loggerhead nesting for the Greater Caribbean subpopu-lation is challenging because few long-term standardized nesting surveys are available for the region, the survey effort at monitored beaches has changed, and scattered, low-level logger-head nesting can be found at many locations; however, nesting for this subpopulation appears to be stable (Figure 11.21).

11.3.1.2   Hatchling, Post-Hatchling, and Oceanic Juvenile Life History and Distribution for Gulf of Mexico Loggerheads

After loggerhead hatchlings emerge from the nest (Figure 11.22), they begin a period of frenzied activity, and during this period, they move from their nest to the surf, swim, and are swept through the surf zone (Witherington 1995; Conant et al. 2009). A magnetic compass and a progression of orientation cues guide the hatchlings as they swim offshore from the nesting beach (Lohmann et al. 2012). Once the swim frenzy stage ends, neonate loggerheads that have migrated offshore are mostly inactive and swim only occasionally and slowly. They begin to feed and are no longer relying on retained yolk (Witherington 2002). Pelagic post-hatchling

Figure 11.22. Loggerhead sea turtle hatchling in its frenzy stage as it approaches the sea (photograph by Burcin Tuncer) (Tuncer 2009).

northwest Atlantic Ocean loggerheads may inhabit the neritic waters just off the nesting beach for weeks to months, which may be a transition to the oceanic stage that loggerheads enter as they grow, or they may be transported by ocean currents into the Gulf of Mexico and the North Atlantic Ocean (Witherington 2002; Bolten 2003; Conant et al. 2009).

Since loggerhead hatchlings that emerge from nests on the Florida Gulf and Atlantic coasts must swim in opposite directions and search for different surface currents to migrate away from continental shelf waters, Wyneken et al. (2008) compared the pattern of swimming activity shown by the hatchlings from each coast over the first 6 days of migration. Hatchlings from both coasts were equally active during the first 24 h of swimming (the frenzy period), as well as during the daylight hours that followed (the post-frenzy period). However, the Gulf coast hatchlings were significantly more active than the Atlantic turtles during the nocturnal portion of the post-frenzy period (Wyneken et al. 2008). This difference could be related to the greater distance Gulf coast loggerhead hatchlings must negotiate to locate surface currents for transport out of the Gulf of Mexico and into the Atlantic Ocean. These behavioral differences could be determined genetically or may be due to phenotypic plasticity that occurs as the hatchlings respond to unique environmental cues on each coast (Wyneken et al. 2008).

As post-hatchlings, northwest Atlantic Ocean loggerheads inhabit areas where surface waters converge to form local downwellings, which are characterized by linear accumulations of floating material, especially *Sargassum*; these areas are common between the Gulf Stream and the southeast U.S. coast and between the Loop Current and the Florida coast in the Gulf of Mexico (Carr 1986; Witherington 2002; Witherington et al. 2012). During this time, the post-hatchlings feed on a wide variety of floating material, including organisms associated with the *Sargassum* community (Table 11.2), and are low-energy, float-and-wait foragers (Witherington 2002; Witherington et al. 2012).

In a study conducted from 2005 through 2011 to determine the importance of the pelagic *Sargassum*-dominated drift community to young sea turtles, 1,688 of 1,704 post-hatchlings that were observed were loggerheads (Witherington et al. 2012). While 30 of the post-hatchlings

Table 11.2. Summary of Life History Information for the Loggerhead Sea Turtle

Parameter	Values	References
Nesting season: Northwest Atlantic Ocean DPS	April through September	NMFS and USFWS (2008)
Remigration interval		
Quintana Roo, Mexico	Mean: 2.6 years	J. Zurita, ECOSUR, personal communication, cited in NMFS SEFSC (2009)
Casey Key, Florida	Mean: 3.7 years, Range: 1–8 years	Tucker (2010)
Cape Sable, Florida	Mean: 2 years	Davis and Whiting (1977)
Nesting interval		
Casey Key, Florida	Mean: 12 days, Range: 6–21 days	Tucker (2010)
Sanibel Island, Florida	Mean: 11 days	LeBuff (1990)
Key Island, Florida	Mean: 11 days	Addison (1996)
Cape Sable, Florida	Mean: 12 days, Range: 1–24 days	Davis and Whiting (1977)
Number of nests/season		
Casey Key, Florida	Mean: 5.4 nests, Range: 2–8 nests	Tucker (2010)
Sanibel Island, Florida	Mean: 3 nests	LeBuff (1990)
Key Island, Florida	Mean: 3.9 nests	Addison (1996)
Number of eggs/nest		
Isla Contoy, Quintana Roo, Mexico	Mean: 110 eggs, Range 71–177 eggs	Najera (1990)
Santa Rosa Island, Florida	Mean: 117 eggs, Range: 53–170 eggs	Atencio (1994)
	Mean: 116 eggs	Lamont et al. (1998)
Cape San Blas, Florida	Mean: 100 eggs	Lamont et al. (1998)
Casey Key, Florida	Mean: 102 eggs	Llew Ehrhart and Bill Redfoot, UCF, personal communication, cited in NMFS SEFSC (2009)
Cape Sable, Florida	Mean: 100 eggs, Range: 48–159 eggs	Davis and Whiting (1977)
Dry Tortugas, Florida	Mean: 102 eggs	van Houtan and Pimm (2007)
Egg incubation time		
Santa Rosa Island, Florida	Mean: 66.5 days, Range: 50–81 days	Atencio (1994)
	Mean: 54 days	Lamont et al. (1998)
Cape San Blas, Florida	Mean: 62 days	Lamont et al. (1998)
Cape Sable, Florida	Mean: 55 days	Davis and Whiting (1977)
Nest pivotal temperature	29 °C	Yntema and Mrosovsky (1982)
Sex ratio of hatchlings (proportional female)		
Sarasota, Florida	Mean: 0.71	Blair (2005)
Sanibel Island, Florida	Mean: 0.65	Blair (2005)

(continued)

Table 11.2. (continued)

Parameter	Values	References
Emergence success of hatchlings from nests		
Santa Rosa Island, Florida	Mean: 0.21	Lamont et al. (1998)
Cape San Blas, Florida	Mean: 0.27	Lamont et al. (1998)
Size of hatchlings		
Azores, Portugal	Estimated value: 15 cm SCL[a]	Bjorndal et al. (2000)
Southeastern Gulf Stream, Florida	Mean: 5.4 cm SCL, Range: 4.6–6.3 cm SCL	Eaton et al. (2008)
Size of post-hatchlings: East and west coast of Florida	Range: 3.9–7.8 cm SCL	Witherington et al. (2012)
Duration of hatchling stage: The Azores, Portugal	Estimated value: less than 1 year	Bjorndal et al. (2000)
Size of oceanic juveniles: East and west coast of Florida	Estimated range: 15–63 cm SCL[b]	Bjorndal et al. (2000), TEWG (2009)
Duration of oceanic juvenile stage: Cape Canaveral, Florida, and Madeira and the Azores, Portugal	Estimated range: 7–11.5 years	Bjorndal et al. (2003)
Diet of oceanic juveniles		
Lower Texas Region	*Sargassum*, pelagic crustaceans, and mollusks	Plotkin (1996)
East and west coast of Florida	Marine animals associated with the *Sargassum* community, including anemones, hydroids, *Aurelia* sp., and *Sargassum*	Witherington et al. (2012)
Size of oceanic juveniles at recruitment to neritic juvenile stage		
U.S. Gulf of Mexico	Range: 41.6–79.7 cm SCL[b]	Bjorndal et al. (2001)
East and west coast of Florida	Range: 31.7–98.7 cm SCL	Witherington et al. (2012)
Duration of neritic juvenile stage: U.S. Gulf of Mexico	Estimated value: 20 years	Bjorndal et al. (2001)
Diet of neritic juveniles: Lower Texas Region	Pipe cleaner sea pens, calico crabs, *Libinia* sp., blue crabs, *Persephona* sp., bivalves, gastropods, and carrion from fisheries bycatch	Plotkin et al. (1993), Plotkin (1996)
Age at sexual maturity: U.S. Gulf of Mexico	Estimated value: 27 years	Bjorndal et al. (2000, 2003)
Size of sexually mature adult females		
Quintana Roo, Mexico	Mean: 90.6 cm SCL[b], Range: 73.7–105.7 cm SCL	J. Zurita, ECOSUR, personal communication, cited in TEWG (2009)
Casey and Manasota Key, Florida	Mean: 89 cm SCL[b], Range: 74.1–105.7 cm SCL	T. Tucker, Mote Marine Laboratory, personal communication, cited in TEWG (2009)

(continued)

Table 11.2. (continued)

Parameter	Values	References
Cape Sable, Florida	Mean: 92.4 cm SCL, Range: 76.2–108 cm SCL	Davis and Whiting (1977)
U.S. Gulf of Mexico	Estimated value: 79.7 cm SCL[b]	Bjorndal et al. (2001)
Diet of adults: Lower Texas Region	Pipe cleaner sea pens, calico crabs, *Libinia* sp., blue crabs, *Persephona* sp., bivalves, gastropods, and carrion from fisheries bycatch	Plotkin et al. (1993)

[a]*SCL* straight carapace length, *cm* centimeters
[b]To convert from curved carapace length (*CCL*), the following equation was used: SCL = (0.948 × CCL) − 1.442 (Bjorndal et al. 2001)

were observed in the Gulf of Mexico, the majority of post-hatchling loggerheads were observed in the Atlantic Ocean, and only during the hatching season of adjacent Florida nesting beaches (July to October). All the loggerhead post-hatchlings were observed in both deep neritic and oceanic waters and were slightly larger than hatchlings measured on nearby nesting beaches, indicating that they had begun to feed and grow following their offshore recruitment (Witherington et al. 2012). It was not surprising that most of the observed post-hatchlings were loggerheads, given the large numbers of loggerheads that nest on Florida beaches (Figures 11.17, 11.18, and 11.19).

The oceanic juvenile life stage is better understood for loggerheads than for any of the other sea turtle species (Table 11.2) (Bolten 2003; Witherington et al. 2006a, 2012). Loggerhead sea turtle hatchlings that originate from nesting beaches in the northwest Atlantic Ocean appear to use oceanic developmental habitats and move with the North Atlantic gyre for several years before returning to their neritic foraging and nesting habitats (Bolten 2003). Using the North Atlantic gyre, these oceanic juveniles can be transported to the northeast Atlantic and Mediterranean Sea (Carr 1987; Carreras et al. 2006; Eckert et al. 2008).

Since the surface currents used by Florida Gulf coast hatchlings are unknown, hatchlings from Casey Key, Sarasota County, were evaluated to determine their likely migratory routes (Merrill and Salmon 2011). The Gulf of Mexico hatchlings were shown to possess a guidance system for responding to surface currents. A hypothesized migratory route for Casey Key hatchlings included turtles being initially carried northward by an along-shore countercurrent, then south by the eastern portion of the Loop Current. Some turtles could then exit the Gulf via the Florida Straits and become entrained within the Gulf Stream (Merrill and Salmon 2011).

The oceanic juvenile stage in the North Atlantic primarily has been studied in the waters around the Azores and Madeira. Juvenile loggerheads undergo a long period of residency around the Azores, but turtles in Madeiran waters appear to be passing through (Dellinger and Freitas 2000; Bolten 2003). While 10 % of oceanic juveniles in the Azores were determined to be from Mexico, approximately 70 % of juveniles were from the Peninsular Florida subpopulation. It could not be determined whether any of these turtles came from Gulf of Mexico nesting beaches (Bolten et al. 1998). However, none of the oceanic juveniles in the Azores were from the northern Gulf of Mexico subpopulation (Bolten et al. 1998). After many years as oceanic juveniles (Table 11.2), which could include time in the eastern Atlantic and Mediterranean Sea, northwest Atlantic Ocean loggerheads return to settle in coastal habitats as neritic juveniles (Heppell et al. 2003).

11.3.1.3 Neritic Juvenile Life History and Distribution for Gulf of Mexico Loggerheads

The neritic juvenile stage begins when loggerhead sea turtles leave the oceanic zone, and juvenile loggerheads continue to mature in the neritic zone until they reach adulthood (Bolten 2003). Juvenile loggerheads recruiting to neritic habitats in the Gulf are typically not seen until they are larger than about 30–40 cm (11.8–15.7 in) SCL (Table 11.2).

The coasts of the Yucatán Peninsula are regarded as major foraging areas for juvenile loggerheads (Ehrhart et al. 2003). Loggerheads are the most abundant sea turtle in the western Gulf of Mexico; the majority of loggerheads that occur there are neritic juveniles (Rabalais and Rabalais 1980; Plotkin 1989; Plotkin et al. 1993). In addition, large juveniles have been associated with hard substrates, such as reefs and oil production areas (Figure 11.23), and appear to use these areas for resting (Rosman et al. 1987). Renaud and Carpenter (1994) characterized the long-term movement and submergence patterns of loggerheads using satellite telemetry. Four loggerheads were captured under oil and gas platforms in the Gulf of Mexico and tracked for periods of 5–10.5 months from June 1989 through January 1991. Loggerheads spent an average of more than 90 % of their time underwater in any given season, and average submergence times ranged from 4.2 min in June to 171.7 min in January. The home ranges determined for the turtles extended from 954 to 28,833 km^2 (368 to 11,132 mi^2), while core areas ranged from 89.6 to 4,279 km^2 (35 to 1652 mi^2). The core areas included several oil and gas platforms that may have been visited on a daily, weekly, or monthly basis (Renaud and Carpenter 1994).

The Flower Garden Banks National Marine Sanctuary (FGBNMS), three offshore banks located in the northwestern Gulf of Mexico, has been determined to be the residence of a population of large juvenile loggerheads (Hickerson 2000). Underwater and above water surveys conducted by recreational scuba divers from August 1994 through April 2000 resulted in 152 sightings of sea turtles. Most of the sightings were loggerheads (87 %), but hawksbills, leatherbacks, and unidentified turtles were also observed. Six large juveniles (five females and

Figure 11.23. Loggerhead sea turtle swimming under an oil and gas platform (photograph courtesy of Ed Elfert, Chevron Corporation, photographer unknown).

one male) were captured and satellite tracked from June 1995 through September 1998. The male was recaptured three times over a 20-month period. More than 40 % of the satellite locations were within the boundaries of the FGBNMS (Hickerson 2000). An analysis using geographic information systems (GIS) indicated an average core range of the satellite-tracked loggerheads of 133.6 km^2 (52 mi^2), and an average home range of 1,074 km^2 (415 mi^2), which are similar to ranges determined for satellite-tracked loggerheads captured under oil and gas platforms in the Gulf of Mexico (Renaud and Carpenter 1994). The average core ranges of the juvenile loggerheads were within 1 km (0.6 mi) of FGBNMS boundaries, while the home range was within 30 km (18.6 mi) of the boundaries (Hickerson 2000).

During a Kemp's ridley investigation along the Texas coast from 2003 through 2007 (Seney 2008), four juvenile loggerheads, averaging 68.6 cm (23.6 in) SCL, were captured by recreational hook and line from Galveston County piers in April, August, and September. All four turtles were successfully rehabilitated and released, and none had been recaptured or stranded as of July 2008.

As part of an in-water study of juvenile Kemp's ridleys inhabiting Apalachee and Deadman bays in Florida that was conducted from 1995 through 1999, 11 loggerhead sea turtles were captured (Barichivich 2006). Loggerheads were captured in Deadman Bay from March through August; they ranged in size from 23.7 cm (9.3 in) SCL to more than 1 m (3.3 ft) SCL, with four being of adult size. Turtles were captured when water temperatures were between 20.7 and 32.7 °C. The large proportion of post-oceanic (less than 50 cm [19.7 in] SCL) loggerheads supports the hypothesis that this area may be an ejection point for turtles recruiting from the oceanic zone to the neritic zone. In addition, the observation of small loggerhead turtles suggests that they remained within the Gulf of Mexico during the oceanic juvenile stage (Barichivich 2006).

In a Kemp's ridley investigation conducted in Waccasassa Bay from June 1986 through October 1995, loggerhead sea turtles were captured from April through November (Schmid 1998). One loggerhead, measuring 86.4 cm (34 in) SCL, was netted on the seagrass shoals of Waccasassa Reefs, and 19 loggerheads, averaging 65 cm (25.6 in) SCL and ranging from 50 to 77.4 cm (19.7 to 30.5 in) SCL, were collected near the oyster bars of Corrigan Reef. Loggerhead turtles greater than 80 cm (31.5 in) SCL were caught at Corrigan Reef, but they could not be landed for data collection. Five loggerheads were recaptured, and recapture times ranged from 142 to 189 days. In an earlier survey, two loggerheads, ranging in size from 57 to 88 cm (22.4 to 34.6 in) SCL, were captured (Schmid and Ogren 1990). The results of the studies indicated the importance of oyster reefs and, to a lesser extent, seagrass beds in Waccasassa Bay as foraging habitat for juvenile and adult loggerheads (Schmid and Ogren 1990; Schmid 1998).

As a result of surveys conducted from July 1990 through December 1996, Florida Bay was shown to be an important developmental habitat for loggerhead sea turtles (Schroeder et al. 1998). The loggerheads captured averaged 80.1 cm (31.5 in) SCL and ranged in size from 48.9 to 98.7 cm (19.3 to 38.9 in) SCL; this size class represents juveniles that are just nearing maturation as well as adults.

Juvenile loggerhead sea turtles may periodically move between the neritic and oceanic zones, particularly during the winter (Bolten 2003; Morreale and Standora 2005; McClellan and Read 2007). Juvenile loggerhead sea turtles foraging in the Gulf of Mexico are primarily carnivorous but do consume some plant material in both the oceanic and neritic zones (Table 11.2). Loggerhead prey varies seasonally and geographically. For example, in the northwestern Gulf of Mexico, loggerheads feed primarily on sea pens during the spring, then primarily on crabs during the summer and fall, paralleling the annual increase in the abundance of crabs in the Gulf (Plotkin et al. 1993).

11.3.1.4   Adult Life History and Distribution for Gulf of Mexico Loggerheads

The age at which loggerhead sea turtles reach sexual maturity is variable (Bjorndal et al. 2000, 2001); however, the estimated age in the U.S. Gulf of Mexico is 27 years (Table 11.2). The duration of the adult female reproductive life stage is at least 25 years for the northwest Atlantic Ocean loggerhead nesting assemblages (Dahlen et al. 2000). While female loggerheads typically do not reproduce every year (Table 11.2), male loggerheads may breed every year (Wibbels et al. 1990). Limited studies of adult loggerheads indicate that their diet is similar to that of neritic juveniles (Table 11.2).

Essentially all shelf waters along the Gulf of Mexico shoreline are inhabited by loggerheads (Conant et al. 2009). Adult loggerheads have been associated with hard substrates, such as reefs and oil production areas, and appear to use these areas for resting (Rosman et al. 1987). Stranded adults have been observed in the Chandeleur Islands/Sound area in eastern Louisiana (Fuller 1988). During aerial surveys, loggerheads were associated with platforms off the Chandeleur Islands in the Gulf of Mexico (Lohoefener et al. 1989). Additionally, the FGBNMS coral reefs 150 km (93.2 mi) off the Louisiana/Texas coast typically have resting loggerheads present (Hickerson and Peccini 2000).

Continental shelf waters along the Florida west coast and the Yucatán Peninsula have been identified as important resident areas for adult female loggerheads from both the Peninsular Florida and northern Gulf of Mexico subpopulations (Meylan et al. 1983; Schroeder et al. 2003; Ehrhart et al. 2003; Foley et al. 2008; Conant et al. 2009; TEWG 2009). For example, approximately 98 % of the loggerheads tagged on Gulf of Mexico beaches as part of the Cooperative Marine Turtle Tagging Program, administered by the Archie Carr Center for Sea Turtle Research, from 1980 through 2007 were later recaptured in the Gulf of Mexico (TEWG 2009). Satellite telemetry and aerial/ship survey data consistently have shown that Gulf of Mexico adult female loggerheads likely remain within the Gulf or more southern geographic regions, such as Mexico and the Caribbean (TEWG 2009).

Post-nesting female loggerhead sea turtles leave the nesting beach area immediately (typically within 24 h) after the last clutch of eggs is deposited and often make directed migrations (Schroeder et al. 2003). The migratory route may be neritic or may involve crossing oceanic waters, and even if the foraging destinations are similar, turtles do not necessarily follow the same migratory routes. Ocean currents may affect migration routes; temporary course adjustments occur, and post-nesting females occasionally swim against the prevailing current. Post-nesting female loggerhead sea turtles have strong foraging area site fidelity, take up residence in discrete foraging areas on continental shelves, and may move among a number of preferred foraging sites within the larger foraging area (Schroeder et al. 2003).

During USACE trawl surveys conducted within the Tampa Bay Entrance Channel in 1997, five loggerheads that were captured—two adult males, one adult female, and two large juveniles; average size: 86.6 cm (34.1 in) SCL—were later tracked (Nelson 2000). The turtles, which were tracked from 13 to 376 days, remained in the study area for days or months at a time and eventually moved in response to changing water temperature. As water temperatures decreased, the turtles moved offshore or to the south and returned to their original location when water temperatures warmed. The results of the surveys indicated that dredging activities in the Tampa Bay Entrance Channel should be conducted during either the winter or summer when temperatures were at their extremes (Nelson 2000).

During 1998, 1999, and 2000, 38 nesting females were outfitted with transmitters and tracked by satellite after they had deposited their last clutch for the nesting season (Foley et al. 2008). Twenty-eight of the turtles were from the Peninsular Florida subpopulation (15 from the Atlantic coast and 13 from the Gulf of Mexico coast), and ten were from the northern Gulf of Mexico subpopulation. The females typically left the vicinity of the nesting beach within 24 h of nesting, and movements were usually highly directed. The post-nesting females took up residence in well-defined, relatively small (median size of 2,000 km^2 [772 mi^2]) areas on the continental shelf adjacent to Florida, Texas, Mexico, the Bahamas, and Cuba within a few weeks of departing from the nesting beaches (Foley et al. 2008). Sixty percent of the turtles (22 of 38) from both nesting assemblages took up residence off the Florida Gulf coast between the Dry Tortugas and Cape San Blas. The distribution of resident areas of female loggerheads from both nesting assemblages overlapped off the western coast of Florida, the western and northern coast of the Yucatán Peninsula, and the northern coast of Cuba (Foley et al. 2008).

Migrations of 28 post-nesting loggerheads from nesting beaches in Sarasota County—their most important Gulf of Mexico rookery—were tracked between May 2005 and December 2007 (Girard et al. 2009). Post-nesting migrations were completed in 3–68 days. Five different migration patterns were observed and included the following: six turtles remained in the vicinity of their nesting site; nine migrated to the southwestern part of the Florida Shelf; two migrated to the northeast Gulf of Mexico; five turtles migrated to the Yucatán Shelf in the southern Gulf of Mexico, Campeche Bay, or Cuba; and six loggerheads migrated to the Bahamas. Loggerheads moved along rather straight routes over the continental shelf but showed more indirect paths in oceanic waters. Smaller turtles remained on the Florida Shelf, and larger individuals showed various migration strategies, staying on the Florida Shelf or moving to long-distance foraging grounds (Girard et al. 2009).

Hart et al. (2012a) recently completed a study that identified shared regional at-sea foraging areas for female loggerheads, which was the first study to consolidate tracking data from three different nesting subpopulations in the Gulf. Ten females from nesting beaches of three subpopulations (St. Joseph Peninsula for northern Gulf of Mexico subpopulation, Casey Key for Peninsular Florida subpopulation, and DTNP for Dry Tortugas subpopulation) in the Gulf of Mexico were satellite tracked in 2008, 2009, and 2010. All turtles migrated to discrete foraging sites in two common areas located off southwest Florida and the northern Yucatán Peninsula, located 102–904 km (63.4–561.7 mi) from the nesting beaches. Within 3–35 days, turtles migrated to the foraging sites, where they all displayed high site fidelity over time. The results of the study indicated that different nesting aggregations of loggerheads in the Gulf of Mexico use common at-sea foraging areas, and these important areas should be protected (Hart et al. 2012a).

In a study of nearshore waters of Gullivan Bay, Ten Thousand Islands, that was conducted from 1997 through 2004, few loggerheads (15 turtles averaging 65.5 cm [25.8 in] SCL) were captured, possibly because they prefer deeper waters (Witzell and Schmid 2004, 2005). During surveys conducted at Key West National Wildlife Refuge (KWNWR) from 2002 through 2006, more than 470 loggerhead sea turtles were sighted, and 182 neritic juveniles and adults, ranging in size from 36.4 to 98.1 cm (14.3 to 38.6 in) SCL and averaging 75.5 cm (29.7 in) SCL, were captured, demonstrating the importance of this area as a foraging site for these size classes (Eaton et al. 2008).

A long-term study of Florida Bay demonstrated that this area provided year-round resident foraging areas for large numbers of juvenile and male and female adult loggerhead sea turtles (Schroeder et al. 2003; Eaton et al. 2008; Conant et al. 2009). From 1990 through 2006, 902 loggerheads were captured; they ranged in size from 33 to 98.7 cm (13 to 38.9 in) SCL, averaging 77.7 cm (30.6 in) SCL. Multiple recaptures of juvenile and adult loggerheads over periods of up to 10 years indicated some strong site fidelity within Florida Bay. Genetic studies have demonstrated that more than 80 % of the adult loggerheads that forage in Florida Bay are from the Peninsular Florida subpopulation, while about 10 % are from the Quintana Roo, Mexico nesting population (Heppell et al. 2003).

While the neritic zone provides important foraging, internesting, and migratory habitats for adult loggerhead sea turtles, some adults may also periodically move between the neritic and oceanic zones (Schroeder et al. 2003; Harrison and Bjorndal 2006). In addition, on a seasonal basis, loggerheads typically move from offshore to inshore and/or from south to north in the spring, reversing their direction in the fall. It is clear from many studies that water temperature is a critical environmental cue that loggerheads use to guide their movements in and out of shallow coastal waters (Hopkins-Murphy et al. 2003). Some loggerheads nesting in the Gulf may inhabit oceanic habitats and are significantly smaller than those in neritic habitats (Reich et al. 2007a).

11.4   GREEN SEA TURTLE (*CHELONIA MYDAS*)

The green sea turtle is the largest of the hard-shelled turtles (Figures 11.24, 11.25, and 11.26), weighing up to 395 kg (870 lb) (Pritchard 2010). The only other turtle that is larger than the green turtle is the leatherback sea turtle (Witherington et al. 2006b). The green turtle was named for the green cartilage found under its shell (Witzell 1994); other names given to this turtle include *edible turtle* and *soup turtle* (Hirth 1997).

The green sea turtle has been studied for centuries (Pritchard 2010). They were the first species for which a conservation and research program was established; this occurred in Tortuguero, Costa Rica, in 1955 and was the first of its kind in the world (Carr et al. 1978; Carr 1979). The Tortuguero program, which was started by the Caribbean Conservation Corporation (now known as the Sea Turtle Conservancy), continues today. Much of what is known about the biology of green sea turtles, as well as other sea turtle species, has been learned on the beaches of Tortuguero.

11.4.1   Green Sea Turtle Life History, Distribution, and Abundance

The distribution range of the green turtle includes nesting beaches, foraging areas, and migration corridors throughout the tropical and subtropical oceans of the world (Figure 11.27) (Hirth 1997). Although most green turtle populations are greatly depleted and many rookeries have been extirpated (Witherington et al. 2006b), green turtles nest in more than 80 countries and are thought to inhabit coastal areas of more than 140 countries (NMFS 2011a).

Figure 11.24. Green sea turtle nesting on the beach of Archie Carr National Wildlife Refuge, Brevard and Indian River counties, Florida, June 2012 (photograph by Steve A. Johnson, University of Florida, with permission).

Figure 11.25. Green sea turtle returning to the water after nesting at Archie Carr National Wildlife Refuge, Brevard and Indian River Counties, Florida, June 2012 (photograph by Steve A. Johnson, University of Florida, with permission).

Figure 11.26. Green sea turtle foraging on a seagrass bed (photograph by RP van Dam) (NOAA 2011).

11.4.1.1   Nesting Life History, Distribution, and Abundance for Gulf of Mexico Green Sea Turtles

Green sea turtles typically nest at night. In the tropics, nesting may span all seasons of the year, with a peak during the rainy season (Witherington et al. 2006b). The green turtle nesting life history varies throughout the Gulf of Mexico, Caribbean, and northwest Atlantic Ocean region (Table 11.3). Available green turtle life history information for specific Gulf of Mexico beaches or regions is included in Table 11.3.

Green turtles that nest on Gulf of Mexico beaches migrate to locations outside the Gulf. For example, turtles from Gulf of Mexico nesting beaches can be found foraging in the Bahamas,

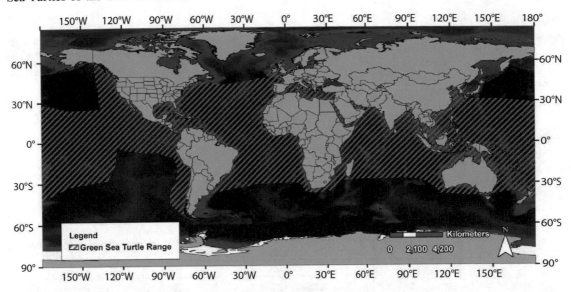

Figure 11.27. Range of the green sea turtle (from NOAA 2009b).

Table 11.3. Summary of Life History Information for the Green Sea Turtle

Parameter	Values	References
Nesting season		
Tortuguero, Costa Rica	May through September	Witherington et al. (2006b)
Santa Rosa Island, Florida	May through August	Atencio (1994)
Remigration interval		
Tortuguero, Costa Rica	Mean: 3 years	Carr et al. (1978)
Melbourne Beach, Florida	Mean: 2 years	Bjorndal et al. (1983)
Indian River Lagoon, Florida	Mean: 2 years	Witherington and Ehrhart (1989a)
Nesting interval		
Tortuguero, Costa Rica	Mean: 13 days, Range: 9–16 days	Carr and Hirth (1962)
El Cuyo, Yucatán, Yucatán Peninsula, Mexico	Mean: 12 days, Range: 10–20 days	Xavier et al. (2006)
Isla Aguada, Campeche, Yucatán Peninsula, Mexico	Mean: 11 days, Range: 8–13 days	Guzmán-Hernández et al. (2006)
Atlantic coast, Florida	Range: 9–15 days	Hirth (1997)
Number of nests/season		
Tortuguero, Costa Rica	Mean: 3 nests	Bjorndal (1982)
	Mean: 2.6 nests, Range: 2–7 nests	Bjorndal and Bolten (1992)
El Cuyo, Yucatán, Yucatán Peninsula, Mexico	Mean: 2.9 nests	Xavier et al. (2006)
Archie Carr National Wildlife Refuge, Brevard County, Florida	Mean: 3.6 nests	Johnson and Ehrhart (1994)

(continued)

Table 11.3. (continued)

Parameter	Values	References
Number of eggs/nest		
Tortuguero, Costa Rica	Mean: 104 eggs, Range: 7–178 eggs	Fowler (1979)
	Mean: 112 eggs, Range: 3–219 eggs	Bjorndal and Carr (1989)
Isla Contoy, Quintana Roo, Yucatán Peninsula, Mexico	Mean: 106 eggs, Range: 69–163 eggs	Najera (1990)
El Cuyo, Yucatán, Yucatán Peninsula, Mexico	Mean: 131 eggs	Xavier et al. (2006)
Rio Lagartos, Yucatán, Yucatán Peninsula, Mexico	Mean: 128 eggs, Range: 96–147 eggs	Najera (1990)
Santa Rosa Island, Florida	Mean: 131 eggs, Range: 76–172 eggs	Atencio (1994)
Dry Tortugas, Florida	Mean: 123 eggs	van Houtan and Pimm (2007)
Egg incubation time		
Tortuguero, Costa Rica	Mean: 56 days, Range: 48–70 days	Carr and Hirth (1962)
	Mean: 62 days, Range: 53–81 days	Fowler (1979)
Isla Aguada, Campeche, Yucatán Peninsula, Mexico	Mean: 52 days, Range: 41–66 days	Guzmán-Hernández et al. (2006)
Santa Rosa Island, Florida	Mean: 63 days, Range: 51–83 days	Atencio (1994)
Nest pivotal temperature: Tortuguero, Costa Rica	Range: 28.5–30.3 °C	Spotila et al. (1987)
Sex ratio of hatchlings from nests (proportional female): Tortuguero, Costa Rica	Range: 0.08–0.74	Spotila et al. (1987)
Emergence success of hatchlings from nests		
Tortuguero, Costa Rica	Mean: 0.51	Carr and Hirth (1962)
	Mean: 0.83	Fowler (1979)
El Cuyo, Yucatán, Yucatán Peninsula, Mexico	Mean: 0.86	Xavier et al. (2006)
Santa Rosa Island, Florida	Range: 0.13–0.48	Atencio (1994)
Size of hatchlings		
Tortuguero, Costa Rica	Mean: 5 cm SCL[a], Range: 4.6–5.6 cm SCL	Carr and Hirth (1962)
Merritt Island, Florida	Range: 4.4–5.8 cm SCL	Ehrhart (1980)
East and west coast of Florida	Range: 5.3–5.6 cm SCL	Witherington et al. (2012)
Size of oceanic juveniles		
St. Joseph Bay, Florida	Estimated mean: 20 cm SCL	Avens et al. (2012)
East and west coast of Florida	Range: 15–26.3 cm SCL	Witherington et al. (2012)

(continued)

Table 11.3. (continued)

Parameter	Values	References
Duration of oceanic juvenile stage: St. Joseph Bay, Florida	Estimated mean: 2 years	Avens et al. (2012)
Diet of oceanic juveniles: Gulf Stream off east and west coast of Florida	Marine animals related to pelagic *Sargassum*, including hydroids, *Membranipora* sp., portunid crabs, gastropods, serpulid polychaetes, *Porpita* sp., *Sargassum* nudibranchs, *Sargassum* snails, *Pyrosoma* sp.; planehead filefish; *Sargassum*; and coralline and cladophora algae	Witherington et al. (2012)
	Sargassum, *Sargassum*-affiliated invertebrates, including hydroids, bryozoans, *Porpita* sp., and *Vellela* sp.	Witherington, unpublished data, cited in Witherington et al. (2006b)
Size of oceanic juveniles at recruitment to neritic juvenile stage		
Mansfield Channel, Texas	Mean: 34.2 cm SCL, Range: 26.6–52 cm SCL	Shaver (1994)
St. Joseph Bay, Florida	Mean: 36.6 cm SCL, Range: 25–75.3 cm SCL	Foley et al. (2007)
	Mean: 36.3 cm SCL, Range: 18.1–78.5 cm SCL	Avens et al. (2012)
Cedar Key, Florida	Mean: 59.8 cm SCL	Eaton et al. (2008)
Corrigan Reef, Florida	Mean: 56.8 cm SCL, Range: 42.9–70.9 cm SCL	Schmid (1998)
Waccasassa Reef, Florida	Mean: 68 cm SCL, Range: 63–73.9 cm SCL	Schmid (1998)
Cape Sable, Florida	Mean: 40.1 cm SCL, Range: 32.8–51.9 cm SCL	Eaton et al. (2008)
Duration of neritic juvenile stage: St. Joseph Bay, Florida	Estimated range: 17–19 years	Avens et al. (2012)
Diet of neritic juveniles		
Great Inagua, Bahamas	Turtle grass, manatee grass, algae, jellyfish, sponges, and sea pens	Bjorndal (1980)
Caribbean coast of Nicaragua	Turtle grass, star grass, jellyfish, sponges, and sea pens	Mortimer (1981)
Caribbean	Turtle grass, manatee grass, and algae	Bjorndal (1985)
	Turtle grass, manatee grass, shoalgrass, star grass, eelgrass, and chicken liver sponge	Bjorndal (1997)

(continued)

Table 11.3. (continued)

Parameter	Values	References
St. Joseph Bay, Florida	Turtle grass, shoal grass, manatee grass, *Laurencia* sp., and *Entermorpha* sp.	Foley et al. (2007)
Mosquito Lagoon, Florida	Manatee grass and turtle grass	Mendonca (1981)
	Manatee grass, shoalgrass, star grass, and green and red algae	Mendonca (1983)
Age at sexual maturity		
St. Joseph Bay, Florida	Estimated range: 19–21 years	Avens et al. (2012)
Mosquito Lagoon, Florida	Estimated range: 18–27 years	Frazer and Ehrhart (1985)
Size of sexually mature adult females		
Tortuguero, Costa Rica	Mean: 100.3 cm SCL, Minimum: 69.2 cm SCL	Carr and Hirth (1962)
El Cuyo, Yucatán, Yucatán Peninsula, Mexico	Mean: 101.1 cm SCL[b]	Xavier et al. (2006)
Isla Aguada, Campeche, Yucatán Peninsula, Mexico	Mean: 101.8 cm SCL, Range: 92.3–114 cm SCL	Guzmán-Hernández et al. (2006)
Melbourne Beach, Florida	Range: 83–114 cm SCL	Witherington (1986)
Diet of adults		
Great Inagua, Bahamas	Turtle grass, manatee grass, algae, jellyfish, sponges, and sea pens	Bjorndal (1980)
Caribbean coast of Nicaragua	Turtle grass, star grass, jellyfish, sponges, and sea pens	Mortimer (1981)
Caribbean	Turtle grass, manatee grass, and algae	Bjorndal (1985)
	Turtle grass, manatee grass, shoalgrass, star grass, eelgrass, and chicken liver sponge	Bjorndal (1997)

[a]*SCL* straight carapace length, *cm* centimeters
[b]To convert from curved carapace length (*CCL*), the following equation was used: SCL = (0.9426 × CCL) − 0.0515 (Goshe 2009)

Barbados, Cuba, Puerto Rico, Venezuela, and the southeast United States (Bass et al. 2006; NMFS and USFWS 2007c). In addition, juveniles that forage in the Gulf of Mexico originate from Barbados, Costa Rica, Florida, Mexico, Venezuela, and Suriname (Bass and Witzell 2000). Tagging studies have also demonstrated that post-nesting Tortuguero green turtles migrate into the Gulf (Carr et al. 1978). Therefore, green turtle rookeries outside the Gulf of Mexico must be considered when assessing green turtles that occur in the Gulf of Mexico.

The largest green turtle rookery in the Gulf of Mexico, Caribbean, and northwest Atlantic Ocean, as well as in the entire Atlantic Ocean, is located in Tortuguero, Costa Rica (Figure 11.27) (Bjorndal et al. 1999; Witherington et al. 2006b; NMFS and USFWS 2007c). Additional major green turtle rookeries in the area include the Florida east coast, the Yucatán Peninsula, and areas along the Mexican Gulf coast (Figure 11.28). In the U.S. Gulf of Mexico, limited green turtle nesting has been documented along the Florida Gulf coast, as well as on the south Texas

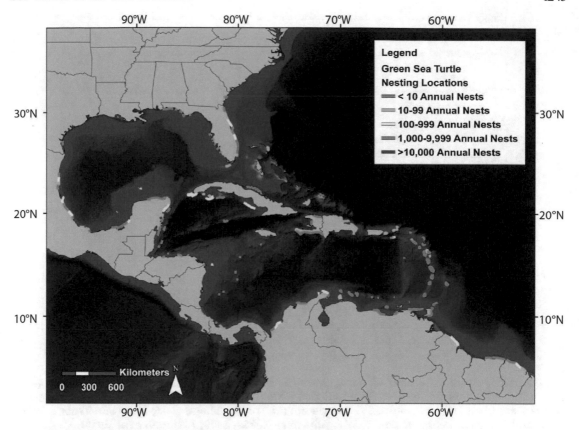

Figure 11.28. Generalized nesting locations of the green sea turtle in the Gulf of Mexico, Caribbean, and northwest Atlantic Ocean (from Dow et al. 2007; SWOT 2010b).

coast (Figure 11.28). Before 1956, no records documented green turtle nesting on Gulf of Mexico beaches or elsewhere in the continental United States However, green turtle populations are known to have been abundant historically as described in Section 11.1.2 (Meylan et al. 1995). Extensive seagrass beds in south Texas bays were once an important feeding ground for the green sea turtle (Owens et al. 1983).

A review of the green turtle nesting data through 2001 for the Tortuguero, Yucatán Peninsula, and Florida nesting beaches indicated that all three western Atlantic Ocean subpopulations were increasing (IUCN 2004). The population of nesting females for the western Atlantic Ocean and Caribbean Sea was estimated to range from 30,981 to 31,981 in 2001, a 13 to 66 % increase over past published estimates (IUCN 2004).

Green turtle nesting in Tortuguero has increased significantly since the 1970s (NMFS and USFWS 2007c). Evaluation of the trend in nesting activity on the Tortuguero beach indicated a relatively steady increase from 1971 to the mid-1980s, constant or possibly decreasing nesting during the late 1980s, and then resumption of an upward trend in the 1990s (Bjorndal et al. 1999). About 41,250 adult female green turtles emerged on beaches each year from 1971 through 1975, and from 1992 through 1996, approximately 72,200 females emerged per year (Bjorndal et al. 1999). Approximately 104,411 nests per year were laid on the beach at Tortuguero from 1999 through 2003, which corresponds to about 17,402–37,290 nesting female green turtles each year (Troëng and Rankin 2005). More than 80,000 green turtle nests were estimated for Tortuguero each year from 2003 through 2009, with a high of almost 178,000 estimated nests or about 68,000 estimated nesting females in 2007 (Figure 11.29).

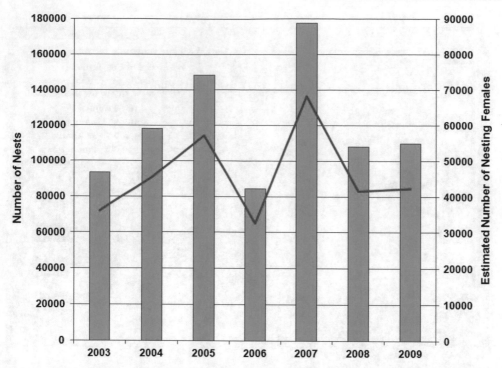

Figure 11.29. Annual number of nests (*bars*) and estimated number of nesting females (*line*), assuming 2.6 nests per female (Bjorndal and Bolten 1992), for green sea turtles in Tortuguero, Costa Rica from 2003 through 2009 (Harrison and Troëng 2004a, 2005; Haro and Troëng 2006a; Haro and Harrison 2007a; Nolasco del Aguila et al. 2008a, 2009; Atkinson et al. 2010).

In Florida, approximately 99 % of green turtle nesting occurs on the Atlantic coast, with most of the activity occurring from Brevard through Broward counties (Figure 11.28) (Witherington et al. 2006b). While the Dry Tortugas historically supported an important green turtle nesting colony of approximately 2,800 nesting females each year that was thought to be extirpated (Thompson 1988), dozens of nests have been laid on Dry Tortugas beaches in recent years (Witherington et al. 2006b). Green turtle nesting was not recorded on Florida's Gulf coast before 1987; however, green turtle nesting now occurs regularly along most of the Gulf coast, with the exception of the Big Bend area, which is the area around Apalachee Bay from Franklin County on the west end through Jefferson, Taylor, and Dixie counties on the southeast end (Figure 11.28) (Witherington et al. 2006b). Very little green turtle nesting occurs on the Texas coast; for example, one nest was documented in 1987, five nests were documented during 1998, and 15 nests were recorded in 2013 (Shaver 2000; NPS 2013a).

From 1979 through 2009, green turtle nesting in Florida has increased significantly, with a high of 12,751 nests or 3,542 nesting females in 2007 (Figure 11.30). The data for all surveyed beaches in Florida include beaches on both the Atlantic and Gulf coasts; separating out data for the Gulf of Mexico was not possible. In 2010, 13,225 green turtle nests were recorded on all surveyed beaches in Florida (FFWCC FWRI 2011a). The increasing trend in green turtle nesting in Florida is in agreement with increases observed at Tortuguero, Costa Rica and Ascension Island since the mid-1970s (Bjorndal et al. 1999; Godley et al. 2001; Troëng and Rankin 2005). Interestingly, there are significant interannual fluctuations in green turtle nesting activity in Florida, which have been noticeable only since 1990 when nesting numbers began to increase significantly (Figure 11.30). These annual nesting fluctuations are characteristic of green turtle

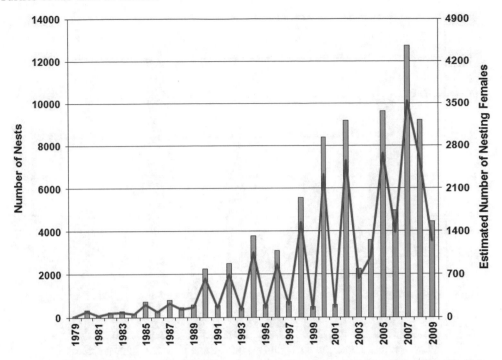

Figure 11.30. Annual number of nests (*bars*) and estimated number of nesting females (*line*), assuming 3.6 nests per female (Johnson and Ehrhart 1994), for green sea turtles for all surveyed beaches in Florida from 1979 through 2009. Note that the survey effort was not consistent among years (from FFWCC FWRI 2011a).

rookeries (Limpus and Nicholls 1988; Bjorndal et al. 1999). It has been proposed that such wide fluctuations are due to environmentally constrained primary productivity, which regulates energy budgets of this herbivorous turtle (Broderick et al. 2001).

In the Florida Panhandle, green turtles nest along Eglin Air Force Base on Santa Rosa Island; 16 and 14 nests were laid in 1994 and 1996, respectively, while no nests were recorded in 1995 or 1997 (Lamont et al. 1998). This small nesting population also displays the annual nesting fluctuations characteristic of green turtle rookeries mentioned above. Green turtles also nest on the low-relief mangrove islands located in the Ten Thousand Islands in southwest Florida (Foley et al. 2000).

From 1993 through 2002, the number of green turtle nests laid on Mexican beaches ranged from about 1,000 to over 7,000, with a high recorded in 2,000 of more than 7,200 nests, representing approximately 2,570 nesting females (Figure 11.31). Interannual fluctuations in green turtle nesting are also apparent for the Mexican Gulf coast (Figure 11.31). A summary of nesting data on Mexican beaches from 1993 through 2002 indicated that most green turtles nested in the state of Quintana Roo, followed by Veracruz, Yucatán, Campeche, and Tamaulipas, with an estimated mean annual total of 1,430, 730, 633, 535, and 141 nests, respectively (Márquez-M 2004). Daily nesting beach reconnaissance efforts along the Yucatán Peninsula indicates that green turtle nesting has increased. In the early 1980s, approximately 875 nests per year were laid; by 2000, nesting had increased to more than 1,500 nests per year (Instituto Nacional de Pesca unpublished data cited in NMFS and USFWS 2007c). By 2004, about 1,547 females were estimated to nest on Yucatán Peninsula beaches (IUCN 2004).

From 2002 through 2004, green turtle nesting activity at El Cuyo Beach, located within the Río Lagartos Biosphere Reserve on the Yucatán Peninsula in Mexico, was evaluated (Xavier

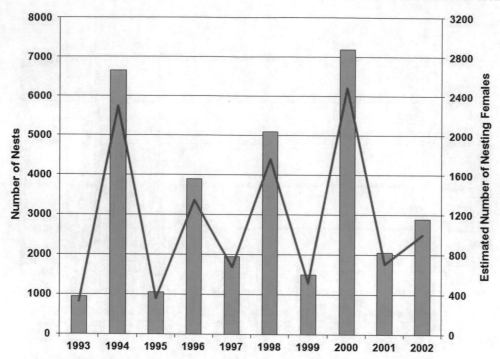

**Figure 11.31. Annual number of nests (*bars*) and estimated number of nesting females (*line*),
assuming 2.9 nests per female (Xavier et al. 2006), for green sea turtles for Mexican Gulf of Mexico
beaches from 1993 through 2002 (Márquez-M 2004).**

et al. 2006). Beach surveys were conducted from mid-April through September each year.
Green turtle nesting activity fluctuated during the study period, from a high of 390 nests in
2002, followed by 157 nests in 2003 and 172 nests in 2004. There were no differences in the size
of nesting females between nesting seasons. Green turtles exhibited high site fidelity, with an
average distance of 1.8 km (1.1 mi) between nests. Hurricane Ivan destroyed the majority of
green turtle nests in 2004 (Xavier et al. 2006). El Cuyo Beach is also one of the most important
hawksbill nesting beaches on the Yucatán Peninsula. Compared to hawksbills, green turtles had
a narrow range of preferences for beach morphological features and selected beaches with
slightly steeper slopes, mainly in the berm zone, and were clearly restricted to nesting in the
western portion of the beach (Cuevas et al. 2010).

11.4.1.2 Hatchling, Post-Hatchling, and Oceanic Juvenile Life History and Distribution for Gulf of Mexico Green Sea Turtles

Green sea turtle hatchlings emerge from the nest about four days after hatching, when sand
surface temperatures are appropriate (Figure 11.32) (Godfrey and Mrosovsky 1997), and enter
the sea, dispersing away from land and into the open ocean (Witherington et al. 2006b). While
little is known about green turtle post-hatchling ecology, some hatchlings have been observed in
convergence zones, drift lines, and *Sargassum* (Carr 1987; Witherington et al. 2006b, 2012).
Post-hatchling and young juvenile green turtles are thought to be carnivorous or omnivorous
(Table 11.3) (Hirth 1971).

The oceanic phase of juvenile green sea turtles remains one of the most poorly understood
phases of green turtle life history (NMFS and USFWS 2007c). After leaving the nesting beach as
5 cm (2 in) SCL hatchlings, green turtles disappear until they recruit to neritic habitats as

Figure 11.32. Green sea turtle hatchling moving across the beach toward the sea (photograph by Kjersti Joergensen) (Joergensen 2012).

juveniles and feed primarily on seagrasses and algae (Table 11.3) (Reich et al. 2007b). While extensive surveys in the northwest Atlantic Ocean have resulted in the sightings of thousands of loggerheads, green turtles are rarely observed (Witherington 2002; Bolten 2003; Witherington et al. 2012). A recent analysis of stable isotopes in green turtle scute tissue suggested that, before recruiting to neritic habitats, green turtles occupy similar habitats and feed at the same trophic level as oceanic-stage loggerheads for 3–5 years (Reich et al. 2007b).

Fifteen post-hatchling green turtles, with an average size of 5.4 cm (2.1 in) SCL and 44 juveniles, with an average size of 20.6 cm (8.1 in) SCL, were observed and/or captured as part of a study to determine the importance of the pelagic *Sargassum*-dominated drift community to young sea turtles in the Atlantic and Gulf of Mexico from 2005 through 2011 (Witherington et al. 2012). In contrast to loggerheads, the post-hatchling green turtles were active within the *Sargassum* or swimming. There was a size gap of 9 cm (3.5 in) between the juvenile and post-hatchling green turtles. Juveniles were estimated to be 1–2 years old, indicating that two discrete life stages of green turtles were observed in the *Sargassum* community (Witherington et al. 2012).

11.4.1.3   Neritic Juvenile Life History and Distribution for Gulf of Mexico Green Sea Turtles

Juvenile green turtles typically shift from the oceanic phase and move into neritic waters, such as protected lagoons and open coastal areas. They may remain in these areas for many years and then shift to other sites as larger juveniles (Zug and Glor 1998; Witherington et al. 2006b). Some green turtles remain in the oceanic zone for extended periods (Pelletier et al. 2003).

When they move into neritic foraging grounds, they adopt an herbivorous diet, which is unique among sea turtles (Table 11.3) (Bjorndal 1985). Green turtles utilize seagrasses as a food source by frequently grazing in the same areas, thus promoting an abundance of young grass blades with high nutritional value (Bjorndal 1980, 1985). However, juvenile, as well as adult, green turtles do consume invertebrates as part of their diet (Table 11.3).

Tracking and CMR studies on green sea turtles in south Texas, as well as sightings at jetties and channel entrances along the central and south Texas coast during the summer, suggest that these areas serve as important developmental habitats for juvenile green turtles (Shaver 1990a, 1994, 2000; Manzella et al. 1990; Renaud et al. 1992, 1993b, 1995b; Renaud and Williams 1997). For example, juvenile green turtles were studied from 1989 through 1992 in the Laguna Madre and the Mansfield Channel in Texas. Turtles, ranging initially in size from 26.6 to 52 cm (10.5 to 20.5 in) SCL, were caught during all months except January, and 42 % of the turtles were recaptured at least once; the CPUE was positively correlated with water temperature, air temperature, and water salinity (Shaver 1994).

Ten juvenile green turtles, ranging in size from 26.6 to 47.9 cm (10.5 to 18.9 in) SCL, that were tracked from July through September 1992 and nine juveniles, ranging from 29.1 to 49.7 cm (11.5 to 19.6 in) SCL, that were tracked in August and September 1992 demonstrated a preference for the jetty habitat in Brazos Santiago Pass in south Texas (Renaud et al. 1993b, 1995b). The tracking data suggested that movement behaviors of juvenile green sea turtles in the Brazos Santiago Pass area did not threaten their lives with respect to the biannual hopper dredging of the Brownsville Ship Channel because they mainly stayed in the jetty habitat and rarely entered the channel (Renaud et al. 1993b).

In addition, neritic juvenile green turtles, ranging in size from 27.5 to 29.9 cm (10.8 to 11.8 in) SCL, were tracked in 1996 and 1997 to determine the spatial and temporal distribution of green turtles in Lavaca and Matagorda bays in Texas, as well as to determine the exposure risk associated with a point-source discharge (Renaud and Williams 1997). Coincident with the distribution of seagrasses, green turtles used the southwestern portion of Lavaca Bay and the western shores of Matagorda Bay. Their home range was greater than 19.5 km^2 (7.5 mi^2), and they moved into the Gulf of Mexico during the winter months seeking warmer temperatures.

From April 1991 through March 1993, the feeding ecology of juvenile green turtles at South Padre Island was characterized by capturing turtles from jetty habitat at Brazos Santiago Pass and seagrass beds at South Bay/Mexiquita Flats (Coyne 1994). There were differences in the sizes of the turtles and feeding selectivity between the sites. Green turtles using the jetty habitat averaged 31.3 cm (12.3 in) SCL in size (range of 22.2–47.9 cm [8.74–18.9 in] SCL) and fed strictly on algae. The turtles captured from the seagrass beds ranged in size from 29.6 to 81.5 cm (11.7 to 32.1 in) SCL (mean = 44.6 cm [17.6 in] SCL); they fed primarily on seagrasses and exhibited a preference for the least abundant taxon, shoalgrass (*Halodule wrightii*). The highest growth rates were observed in spring and summer (0.62 and 0.64 cm/month [0.24 and 0.25 in/month], respectively), while turtles grew the slowest during the winter (0.14 cm/month [0.06 in/month]) (Coyne 1994). There were also seasonal differences in activity patterns, with increased movement and strong site fidelity during the warmer months.

Using stable isotope analysis of scute tissues (Gorga 2010), an intermediate stage between the shift of green turtles from the oceanic juvenile stage (when they are omnivores) to the neritic juvenile stage (when they switch to foraging on seagrass and algae) was found for juvenile green turtles inhabiting south Texas bays, such as the Lower Laguna Madre and Aransas Bay. This intermediate stage consists of an initial recruitment of neritic juveniles to jetty habitat located on the channel passes Gulf-ward of adjacent bays to forage on algae before subsequently recruiting to seagrass beds in these bays. These results and those found earlier by Coyne (1994) indicated the use of a characteristic sequence of distinct habitats by multiple life-history stages of green turtles in Texas bays (Gorga 2010).

As a result of a large hypothermic-stunning event in St. Joseph Bay along the Florida Gulf coast in Gulf County in December 2000/January 2001 that stranded 388 green turtles, information on the assemblage of green turtles along the northeastern Gulf of Mexico, which had been observed in the past (Carr and Caldwell 1956), was obtained (Foley et al. 2007). All of the green turtles were neritic juveniles, with a mean size of 36.6 cm (14.4 in) SCL (range = 25–75.3 cm [9.8–29.6 in] SCL). Genetic analyses indicated that about 81 % of the turtles were from nesting populations in Florida and the Yucatán. This assemblage is interesting because it does not have substantial representation from the nesting population in Tortuguero, Costa Rica, the Atlantic's largest green turtle nesting population (Foley et al. 2007).

Green turtle CMR data from surveys conducted in St. Joseph Bay during 2002 and 39 green turtle strandings from a small hypothermic-stunning event in 2003 indicated site fidelity to St. Joseph Bay since more than 70 % of the recaptures were originally tagged in 2001 (McMichael et al. 2003, 2008). The recapture intervals ranged from 311 to 1,193 days (mean = 636 days). Turtles ranged in size from 27.4 to 56.9 cm (10.8 to 22.4 in) SCL (mean = 37.4 cm [14.7 in] SCL). Annual growth increments ranged from 1.2 to 8.4 cm/year (0.47 to 3.3 in/year) (McMichael et al. 2008). Size-specific growth rates were as follows:

- For 30–39.9 cm (11.8–15.7 in) SCL turtles, growth rates averaged 4.7 cm/year (1.9 in/year).
- For 40–49.9 cm (15.7–19.6 in) SCL turtles, growth rates averaged 4.3 cm/year (1.7 in/year).
- For 50–59.9 cm (19.7–23.2 in) SCL turtles, growth rates averaged 4.8 cm/year (1.9 in/year).
- For 60–69.9 cm (23.6–27.5 in) SCL turtles, growth rates averaged 1.2 cm/year (0.47 in/year).

In addition, as a result of comparing green turtle mean size data throughout the Gulf of Mexico, it appears that developmental migration occurs throughout the region, which was first discussed by Carr and Caldwell (1956), with the size of turtles increasing as they move from west to east (McMichael et al. 2003). Green turtles enter St. Joseph Bay at just under 30 cm (11.8 in) SCL, and the majority of turtles remain in this habitat until they reach a size of just over 60 cm (23.6 in) SCL. The estimated mean time of residency within the bay is 7 years (±1.5 years) (Eaton et al. 2008).

As a result of a massive hypothermic-stunning event in January 2010, the population of neritic juvenile green turtles inhabiting St. Joseph Bay was characterized using necropsy and skeletochronology by evaluating more than 400 dead turtles of the more than 4,600 turtles that stranded (Avens et al. 2012). The size range of the dead green turtles was not significantly different from those that survived the hypothermic-stunning event, indicating that the sample was representative. The age of the turtles ranged from 2 to 22 years, and SCLs ranged from 18.1 to 78.5 cm (7.1 to 30.9 in). The female age distribution was significantly greater than that of males, and the mean stage duration ranged from 17 to 20 years. Growth rates of the green turtles were significantly influenced by size, age, and calendar year; however, no effect of sex, fibropapilloma status, or body condition on growth rates was found (Avens et al. 2012).

Twenty-eight neritic juvenile green turtles were also captured during an in-water study conducted from 1995 through 1999 on juvenile Kemp's ridleys inhabiting Apalachee and Deadman bays, Florida (Barichivich 2006). One green turtle, which measured 37.3 cm (14.7 in) SCL, was captured in Apalachee Bay, and 27 green turtles, ranging in size from 27.9 to 70.7 cm (11 to 27.8 in) SCL (mean of 42.2 cm [16.6 in] SCL), were captured in Deadman Bay; one green turtle was recaptured in Deadman Bay during the study. The green turtles were captured when water temperatures were between 22.2 and 32.7 °C. The large proportion of post-oceanic (less than 40 cm [15.7 in] SCL) green turtles supported the hypothesis that the Big Bend region may be an ejection point for turtles recruiting from the oceanic zone to the neritic zone. In addition, Deadman Bay, located in the largest remaining seagrass bed in North America, is an important developmental habitat for green turtles (Barichivich 2006).

Green turtles were captured from April through November in a study conducted in Waccasassa Bay from June 1986 through October 1995 (Schmid 1998). Six green turtles were netted on the seagrass shoals of Waccasassa Reefs. They ranged in size from 63 to 73.9 cm (24.8 to 29 in) SCL and averaged 68 cm (26.8 in) SCL. The four green turtles collected near the oyster bars of Corrigan Reef averaged 56.8 cm (22.4 in) SCL, with a size range of 42.9–70.9 cm (16.9–27.9 in) SCL. The results of this study, as well as an earlier survey in which nine juvenile green turtles were captured (mean SCL: 66 cm [26 in], range: 49.5–74 cm [19.5–29.1 in] SCL) (Schmid and Ogren 1990), indicated the importance of the seagrass beds in Waccasassa Bay as foraging habitat for late-stage juvenile green turtles (Schmid 1998).

As a result of surveys conducted in Florida Bay from July 1990 through December 1996 and from 2000 through 2006, the bay was determined to be an important developmental habitat for juvenile green sea turtles (Schroeder et al. 1998; Eaton et al. 2008). The green turtles captured from 1990 through 1996 ranged in size from 25.5 to 52.9 cm (10 to 20.8 in) SCL and averaged 46.2 cm (18.2 in) SCL, while the 73 juveniles captured from 2000 through 2006 averaged 45.8 cm (18 in) SCL (range of 25.5–66.1 cm [10–26 in] SCL). This size class distribution was similar to those for other nearshore developmental habitats in the Gulf of Mexico. No adult green turtles were captured or sighted (Schroeder et al. 1998). Juvenile green turtles, averaging 51.6 cm (20.3 in) SCL, also have been captured in surveys of the Ten Thousand Islands area (Witzell and Schmid 2004).

The movements of six juvenile green sea turtles, ranging in size from 33.4 to 67.5 cm (13.1 to 26.6 in) SCL captured in southwest Florida within Everglades National Park were tracked using satellite telemetry during the spring for 27 days in 2007 and 62 days in 2008 (Hart and Fujisaki 2010). These turtles were observed to be resident for several months in coastal waters ranging up to 10 m (38.2 ft) in depth near their capture and release sites. The results of this study documented habitat use by juvenile green turtles in the mangroves of southwest Florida and highlighted the need to consider the impacts of Everglades restoration activities on juvenile green turtles and their habitat (Hart and Fujisaki 2010).

11.4.1.4 Adult Life History and Distribution for Gulf of Mexico Green Sea Turtles

While growth rates vary among populations, most green turtles grow very slowly because of their low energy, mostly herbivorous diet (Bjorndal 1982). The age to maturity for green turtles ranges from less than 20 years to more than 40 years (Table 11.3) (Hirth 1997; Zug et al. 2002). Reproductive longevity for green turtles ranges from 17 to 23 years (Carr et al. 1978). After leaving the nesting beach as hatchlings and living in a variety of marine habitats for up to 40 or more years, adult female green turtles return to the same beach from where they were hatched (Bowen et al. 1989, 1992).

A recent satellite tracking study provided the first available information on green turtle migratory corridors and post-nesting foraging locations in the Yucatán (Cuevas et al. 2012). In 2011, nine post-nesting females were tracked from eight different nesting beaches. Green turtles appeared to prefer a region known at Petenes-Celestun off the northwest corner of the Yucatán Peninsula for foraging (42 % of tracked turtles), while 22 % of the tracked turtles migrated to the Florida Keys. A well-known green turtle feeding and mating area for the region, the Catoche-Contoy area off the northeast corner of the Yucatán Peninsula, was confirmed. Yucatán waters within about 15 km (9.3 mi) from shore were shown to be important migratory corridors (Cuevas et al. 2012).

Green turtle juveniles and adults are found in inshore and nearshore waters of the U.S. Gulf of Mexico from Texas to Florida (NMFS 2011a). They are known to forage in Florida's coastal waters where there is sufficient seagrass or algae. Important green turtle foraging areas along the Florida Gulf coast include the Florida Keys, Marquesas, Florida Bay, Homosassa, Crystal River, the Cedar Keys, and St. Joseph Bay (Witherington et al. 2006b; NMFS 2011a). When not nesting, adult female green sea turtles reside in Gulf of Mexico foraging areas from throughout the Florida Keys to the Dry Tortugas and waters southwest of Cape Sable (NMFS and USFWS 2007c). From 2002 through 2006, more than 900 green turtles were sighted during surveys conducted at KWNWR. In addition, almost 90 juvenile and adult turtles were captured, ranging in size from 27 to 108.5 cm (10.6 to 42.7 in) SCL and averaging 61.3 cm (24.1 in) SCL (Eaton et al. 2008). Green turtles often return to the same foraging locations after subsequent nesting migrations, and after their arrival, they typically visit specific areas for foraging and resting (Broderick et al. 2006; Taquet et al. 2006).

The coastal foraging grounds, where green turtles spend the majority of their lives, are often highly dynamic, with annual fluctuations in seawater and air temperatures, which cause the distribution and abundance of green turtle food items to vary significantly between seasons and years (Carballo et al. 2002). This variability in food item abundance may explain, in part, the significant interannual fluctuations in green turtle nesting.

11.5　LEATHERBACK SEA TURTLE (*DERMOCHELYS CORIACEA*)

First described by Vandelli in 1761 (Fretey and Bour 1980), the highly specialized leatherback sea turtle is the only living member of the family Dermochelyidae (Eckert et al. 2012). Leatherbacks are more widely distributed than any other reptile species and are the largest species of sea turtles; these gigantic turtles can measure 2 m (6.5 ft) in length and weigh up to about 900 kg (2,000 lb) (Stewart and Johnson 2006; NMFS 2011b). The largest recorded leatherback is a male that weighed 916 kg (2,019 lb), found off the shores of Wales in 1988 (Morgan 1989). Leatherbacks are one of the deepest diving vertebrates, diving to depths greater than 1,000 m (3,280 ft), surpassed only by sperm whales and elephant seals (Eckert et al. 1989a).

The leatherback sea turtle is easily identified by its unique morphology (Figure 11.33). Leatherbacks do not have hard scutes made of keratin on their carapace or plastron like the other species of sea turtles; instead, they are completely covered by a thin layer of smooth, rubbery, oily skin, which is black with white mottling dorsally and lighter colored ventrally (NMFS 2011b). Beneath the skin of the carapace lies a nearly continuous layer of small dermal bones, and the carapace is made up of seven longitudinal bony ridges, which taper to a blunt point posteriorly (NMFS 2011b). There is no sharp angle between the carapace and plastron; therefore, the body of the leatherback is barrel shaped (NMFS 2011b). Leatherbacks have a distinctive upper jaw, with two tooth-like projections that are each flanked by deep cusps, rather than a hard beak-like structure like all other species of sea turtles (NMFS 2011b). Adult leatherbacks have no scales on their head or flippers and lack claws. Their proportionately long fore flippers and streamlined body shape make the leatherback highly adapted for long migrations and deep dives in a primarily pelagic habitat (NMFS 2011b).

Figure 11.33. Leatherback sea turtle covering her eggs after nesting (photograph by Paul Mannix) (Mannix 2012).

11.5.1 Leatherback Sea Turtle Life History, Distribution, and Abundance

Leatherback sea turtles are physiologically unique among all other species of sea turtles and are capable of maintaining a core body temperature several degrees higher than the surrounding water temperature (Frair et al. 1972; Standora et al. 1984). This trait is likely due to countercurrent mechanisms in the circulatory system, peripheral insulation, regional endothermy, and large body size; this suite of adaptations is sometimes referred to as *gigantothermy*, distinct from strict ectothermy and endothermy (Greer et al. 1973; Penick et al. 1998; Eckert et al. 2012). Leatherbacks are able to forage in cold water environments without becoming hypothermic stunned and, due to their ability to tolerate cold temperatures, have the most extensive global range of any reptile (Eckert et al. 2012).

Leatherback sea turtles occur in the Pacific, Atlantic, and Indian oceans (Figure 11.34). While they nest on tropical and subtropical beaches, wide-ranging foraging areas include temperate and subarctic waters (Stewart and Johnson 2006). Adult leatherbacks have the longest migration of any reptile, greater than 5,000 km (3,107 mi), traveling between high-latitude foraging grounds and low-latitude mating and nesting areas (Pritchard 1976). Satellite-tagged leatherbacks have frequently traversed entire ocean basins (Luschi et al. 2003; Hays et al. 2004, 2006; Fossette et al. 2010). In the western Atlantic Ocean, leatherbacks have been sighted as far north as Greenland and as far south as Argentina (Carriol and Vader 2002; González-C. et al. 2011).

Leatherback sea turtles nest on every continent except Europe and Antarctica (Eckert et al. 2012), and an estimated 652 nesting sites have been documented worldwide (Wallace et al. 2010). The largest nesting assemblages in the world are along the coasts of French Guiana and Suriname (4,500–7,500 females per year) and Gabon, West Africa (1,300–2,553 females per year) (NMFS 2011b). There is no evidence of substantial declines recently at the main western

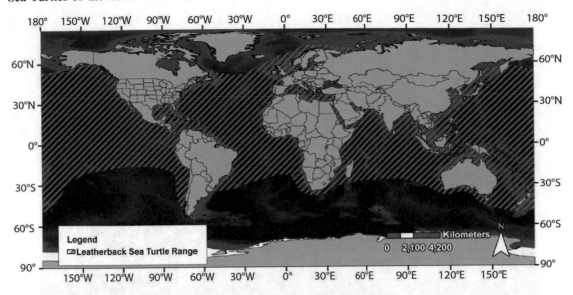

Figure 11.34. Range of the leatherback sea turtle (from NOAA 2009c).

Atlantic Ocean nesting grounds; however, significant threats are affecting populations in the eastern Atlantic, and populations in the Pacific Ocean have been decimated (Spotila et al. 2000; Eckert et al. 2012).

In 1982, 115,000 adult female leatherbacks were estimated to occur worldwide, of which 60 % nested along the Mexican Pacific coast (Pritchard 1982). Spotila et al. (1996) later estimated that only 34,500 nesting females remained worldwide in 1995. However, the most recent population estimate for the entire north Atlantic Ocean ranged from 34,000 to 94,000 adult leatherback sea turtles, and the global population was determined to be slightly female biased, ranging from 51.9 to 66.7 % female (TEWG 2007).

11.5.1.1 Nesting Life History, Distribution, and Abundance for Gulf of Mexico Leatherbacks

Compared to other species of sea turtles, genetic studies to date have revealed reduced global divergence and differentiation between leatherback nesting populations (Eckert et al. 2012). Some of the possible reasons for this include the extensive home ranges, large migrations between foraging and nesting areas, and weaker nesting beach fidelity than other sea turtle species.

The major leatherback nesting beaches in the northwest Atlantic Ocean are located along the coasts of French Guiana and Suriname, as well as along the coasts of Costa Rica and Panama (Figure 11.35). The large colony in French Guiana and Suriname appears to be stable or increasing; from 5,029 to 63,294 nests were laid each year from 1967 to 2002 (Girondot et al. 2007). The population of leatherbacks that nest on the beaches of Trinidad also appears to be stable or slightly increasing; an estimated 52,797 and 48,240 nests were laid in 2007 and 2008, respectively (Eckert et al. 2012).

In the USA, leatherback nesting occurs primarily in St. Croix in the U.S. Virgin Islands, Puerto Rico, and Florida (Figure 11.35). The first record of a leatherback nesting in Florida is from 1947 (Carr 1952), with additional reports coming after 1955 (Caldwell et al. 1955; Caldwell 1959). The highest density of leatherback nesting in Florida occurs along the Atlantic coast from Jensen Beach south to Palm Beach in Martin and Palm Beach counties (Stewart and Johnson

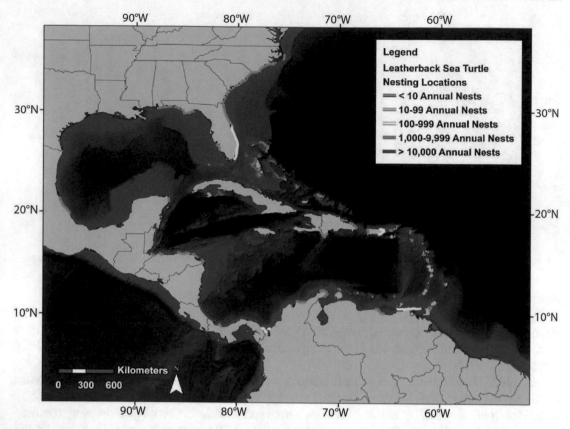

Figure 11.35. Generalized nesting beach locations of the leatherback sea turtle in the Gulf of Mexico, Caribbean, and northwest Atlantic Ocean (interpreted from Dow et al. 2007; SWOT 2007b).

2006). Low numbers of leatherback nesting also occur along the southeast Atlantic coast of the United States, from central Florida through North Carolina (Figure 11.35).

In the Gulf of Mexico, leatherback sea turtles nest at low levels along the Florida, Alabama, and Mexican coasts (Figure 11.35). Leatherback nesting was reported as early as 1962 on a beach near Destin along the Florida Gulf coast (Yerger 1965). As indicated by reports from local residents, leatherbacks may have nested on Texas beaches in the 1920s and 1930s (Hildebrand 1963). One leatherback nest was recorded at PAIS in 2008 (NPS 2013d).

Leatherback sea turtles that forage in and travel through the Gulf of Mexico originate from many different nesting beaches. For example, leatherbacks that were tagged while nesting on the Florida east coast and St. Croix were later recaptured along the Mexican Gulf coast (Eckert et al. 2012). In addition, leatherbacks that nest on Gulf of Mexico beaches migrate to and forage in waters outside the Gulf, and many satellite tracking studies have shown that leatherback sea turtles that nest in the western Atlantic migrate to Western Europe and West Africa (Eckert et al. 2012). Therefore, using data from nesting beaches of leatherbacks known to occur in the Gulf of Mexico to evaluate the population of leatherback sea turtles that use the Gulf is extremely challenging. However, to determine trends in leatherback nesting in the general western Atlantic Ocean area, data from two important nesting beaches in the area with long-term monitoring were evaluated: the Florida east coast and Tortuguero, Costa Rica. In addition, recent investigations have suggested that the Gulf of Mexico may be a significant year-round foraging ground for leatherbacks that nest along the Caribbean coast of Costa Rica, as well as Panama (Evans et al. 2007; Fossette et al. 2010).

The leatherback is a nocturnal nester and, on average, takes just over 100 min to complete the entire nesting procedure (Eckert 1987). While females generally return to their natal beaches to nest, some are known to also nest at beaches greater than 100 km (62.1 mi) apart (Eckert et al. 1989b). In comparison to the other sea turtle species, nesting leatherbacks have a warmer body temperature; therefore, they lay eggs that are at a higher temperature than eggs of other sea turtle species (Mrosovsky and Pritchard 1971). In addition, their eggs are the largest of any sea turtle species (Eckert et al. 2012). Leatherbacks are the only sea turtle species that normally lay small, irregular yolkless eggs along with viable eggs (Bell et al. 2003). The irregular eggs typically appear in a clutch during the latter half of egg laying (Eckert et al. 2012). Table 11.4 includes nesting life history information for leatherback sea turtles; since no life history

Table 11.4. Summary of Life History Information for the Leatherback Sea Turtle

Parameter	Values	References
Nesting season		
French Guiana	April through August	Hilterman and Goverse (2007)
Panama	February through August	Boulon et al. (1996)
Southeast Florida coast	March through June	Stewart and Johnson (2006)
Remigration interval		
St. Croix, U.S. Virgin Islands	Mode: 2 years, Mean: 2.2 years	Dutton et al. (2005)
Babunsanti, Samsambo, Kolukumbo, and Matapica, Suriname	Mode: 2 years	Hilterman and Goverse (2007)
Juno Beach, Florida	Mean: 2.9 years, Range: 1–6 years	Stewart (2007)
Nesting interval		
St. Croix, U.S. Virgin Islands	Mean: 9.6 days	Boulon et al. (1996)
Babunsanti, Samsambo, Kolukumbo, and Matapica, Suriname	Mean: 9.6 days	Hilterman and Goverse (2007)
Juno Beach, Florida	Mean: 10 days	Stewart and Johnson (2006)
Number of nests/season		
St. Croix, U.S. Virgin Islands	Mean: 5.3 nests	Boulon et al. (1996)
Babunsanti, Samsambo, Kolukumbo, and Matapica, Suriname	Mean: 4.6 nests	Hilterman and Goverse (2007)
Juno Beach, Florida	Estimated mean: 4.1 nests	Stewart (2007)
Number of eggs/nest		
St. Croix, U.S. Virgin Islands	Mean: 116.1 eggs	Boulon et al. (1996)
Babunsanti and Matapica, Suriname	Mean: 115.8 eggs	Hilterman and Goverse (2007)
Juno Beach, Florida	Mean: 98 eggs	Stewart and Johnson (2006)
Egg incubation time		
St. Croix, U.S. Virgin Islands	Mean: 63.2 days, Range: 57–76 days	Boulon et al. (1996)
Babunsanti, Samsambo, Kolukumbo, and Matapica, Suriname	Mean: 64 days	Hilterman and Goverse (2007)

(continued)

Table 11.4. (continued)

Parameter	Values	References
Southeast coast of Florida	Mean: 67 days	Stewart and Johnson (2006)
Nest pivotal temperature: French Guiana and Suriname	Mean: 29.5 °C	Hulin et al. (2009)
Sex ratio of hatchlings from nests (proportional female)		
St. Croix, U.S. Virgin Islands	Estimated mean: 0.65	Dutton et al. (1985)
Suriname	Mean: 0.53	Godfrey et al. (1996)
Tortuguero, Costa Rica	Estimated mean: 0.67	Leslie et al. (1996)
Emergence success of hatchlings from nests		
St. Croix, U.S. Virgin Islands	Mean: 0.64	Eckert and Eckert (1990)
Awala Yalimapo, French Guiana	Mean: 0.38	Caut et al. (2006)
Southeast coast of Florida	Mean: 0.47	Stewart and Johnson (2006), Stewart (2007)
Size of hatchling		
Culebra Island, Puerto Rico	Mean: 9.07 cm SCL[a], Range: 7.91–9.90 cm SCL	Tucker (1988)
Matura and Paria Bays, Trinidad	Mean: 6.50 cm SCL	Bacon (1970)
Suriname	Mean: 5.91 cm SCL (Babunsanti)	Hilterman and Goverse (2007)
	Mean: 5.95 cm SCL (Matapica)	
Tortuguero, Costa Rica	Mean: 6.28 cm SCL	Carr and Ogren (1959)
Duration of hatchling stage	Estimated value: 1 year	Spotila et al. (1996)
Size of oceanic juveniles: Juno Beach, Florida	Range: 10–134.7 cm SCL[b]	Tucker (1988), Stewart et al. (2007)
Duration of oceanic juvenile stage: St. Croix, U.S. Virgin Islands	Estimated range: 11–13 years	Dutton et al. (2005)
Diet of oceanic juveniles: Offshore from Boynton Beach, Florida	*Aurelia* sp., *Ocryopsis* sp., warty comb jellyfish, and tunicates	Salmon et al. (2004)
Age at sexual maturity		
St. Croix, U.S. Virgin Islands	Range: 12–14 years	Dutton et al. (2005)
Western North Atlantic	Range: 24.5–29 years	Avens et al. (2009)
Size of sexually mature adult females		
U.S. Virgin Islands	Range: 127.4–172.7 cm SCL[b]	Boulon et al. (1996)
Juno Beach, Florida	Mean: 147.7 cm SCL[b]; Range: 134.7–160.7 cm SCL	Stewart et al. (2007)
Diet of adults		
Offshore from Port Aransas, Texas	Cannonball jellyfish	Leary (1957)
North Sea	Hydrozoans, Siphonophorans, Scyphozoans, *Cyanea* sp., *Aurelia* sp., *Stomolophus* sp., comb jellies, tunicates, cephalopods, and gastropods	den Hartog and van Nierop (1984)

[a]*SCL* straight carapace length, *cm* centimeters
[b]To convert from curved carapace length (*CCL*), the following equation was used: SCL = (0.9781 × CCL) − 0.7714 (Avens et al. 2012)

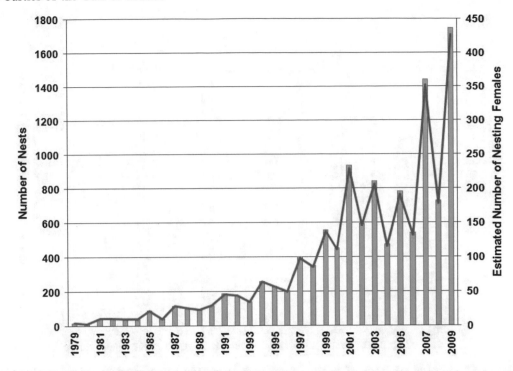

Figure 11.36. Annual number of nests (*bars*) and estimated number of nesting females (*line*), assuming 4.1 nests per female (Stewart 2007), for leatherback sea turtles for all surveyed beaches in Florida from 1979 through 2009. Note that the survey effort was not consistent among years (FFWCC FWRI 2011a).

information for leatherbacks was available specifically for the Gulf of Mexico, values are from nesting beaches in the Caribbean or northwest Atlantic Ocean or leatherback populations from other locations.

Leatherback nesting data for Florida presented below includes beaches on both the Gulf and Atlantic coasts, since separating Gulf of Mexico data was not possible. There is an increasing trend in the number of leatherback nests recorded on Florida beaches (Figure 11.36). From 1979 through 2009, the number of leatherback nests recorded each year in Florida has increased significantly, from 18 recorded nests in 1979 to 1,747 nests in 2009, representing about 5 to 420 nesting females, respectively. In 2010, 1,334 leatherback nests were recorded on Florida beaches (FFWCC FWRI 2011a).

Some of the increased leatherback nesting in Florida could be the result of increased survey and documentation efforts since the late 1970s. Leatherbacks begin their nesting season in Florida early in the year (Table 11.4), and because the leatherback nesting season starts before most nesting surveys begin, the number of nests reported in Florida is considered to be a minimum (Meylan et al. 1995).

No trend is indicated in the number of leatherback nests recorded at Tortuguero, Costa Rica from 1998 through 2009 (Figure 11.37). The lowest number of leatherback nests was recorded in 1998 (94 nests), and nest numbers have ranged from 481 to 1,107 nests, representing about 115 to 264 nesting females, from 1999 through 2009.

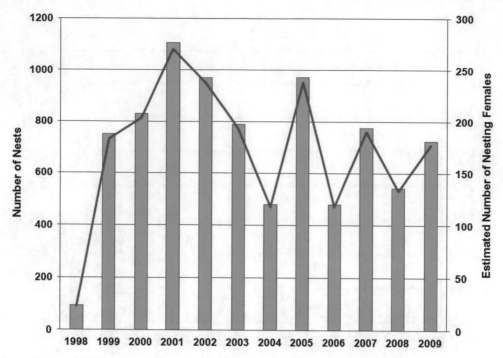

Figure 11.37. Annual number of nests (*bars*) and estimated number of nesting females (*line*), assuming 4.1 nests per female (Stewart 2007), for leatherback sea turtles in Tortuguero, Costa Rica from 1998 through 2009 (Troëng 1998, 2000; Troëng and Cook 2000; Reyes and Troëng 2001; Harrison and Troëng 2003a, b, 2004b; Haro and Troëng 2006b; Haro and Harrison 2007b; Nolasco del Aguila et al. 2008b; Debade et al. 2009; Sarmiento Devia and Harrison 2010).

11.5.1.2   Hatchling, Post-Hatchling, and Oceanic Juvenile Life History and Distribution for Gulf of Mexico Leatherbacks

In addition to being the largest species of sea turtle, leatherback hatchlings are also larger than the hatchlings of other species (Figure 11.38, Table 11.4). Immediately after emerging and crawling to the water, hatchling leatherbacks go through the swim frenzy stage, similar to hatchlings of other sea turtle species, and swim continuously for the first 24 h (Wyneken and Salmon 1992). In contrast to loggerhead and green sea turtle hatchlings, which eventually stop all swimming activities during the night, leatherback hatchlings begin a daily swimming pattern after the first 24 h and decrease swimming to 15–45 % of nighttime (Wyneken and Salmon 1992).

Hatchling leatherbacks are capable of diving soon after entering the ocean. Hatchlings between 2 and 8 weeks of age have been documented to dive deeper and longer with age (Eckert et al. 2012), while foraging exclusively on gelatinous prey throughout the water column (Table 11.4). The post-hatchling habitat remains obscure, and nothing is known about the dispersal or distribution of post-hatchling leatherbacks in the open ocean (Eckert et al. 2012). In contrast to other species of sea turtles, there is no evidence that young leatherbacks associate with *Sargassum* or epipelagic debris (Carr 1987).

Little is known about the life history or distribution of juvenile leatherbacks (Table 11.4) (Eckert et al. 2012). Leatherbacks have a unique diet that consists primarily of jellyfish, salps, and other soft-bodied coelenterates that inhabit the mid-water column in the open ocean (Table 11.4). The distribution of juvenile, as well as adult, leatherbacks is likely to be closely

Figure 11.38. Leatherback sea turtle hatchlings leaving the nesting beach (photograph by Scott R. Benson, NMFS Southwest Fisheries Science Center) (NOAA 2011).

linked to the distribution and abundance of their prey—jellyfish and other soft-bodied invertebrates—as well as their preferred temperature tolerances (Eckert et al. 2012).

A study by Eckert (2002) indicated that juvenile leatherbacks were found exclusively in waters warmer than 26 °C, but larger juveniles and subadults would venture into waters as cold as 8 °C. Leatherbacks, therefore, spend the first portion of their lives in tropical waters, venturing into cooler latitudes only after reaching a size of 97 cm (38.2 in) SCL (Eckert 2002; Avens et al. 2009). The restriction of smaller leatherbacks to warmer waters suggests that size may play a role in the ability of the species to exist in colder waters. The warm water restrictions also suggest that the onset of thermogenerating capability, which is not found in younger or smaller turtles, occurs after reaching a size of about 97 cm (38.2 in) SCL (Eckert 2002).

11.5.1.3   Adult Life History, Distribution, and Abundance for Gulf of Mexico Leatherbacks

Adult leatherbacks, which have the most extensive range of any living reptile, are primarily pelagic and are generally only seen in coastal waters when nesting (Eckert et al. 2012). While the life history information for adults is incomplete (Table 11.4), male and female leatherbacks generally return to their native nesting locales to mate and nest. They presumably mate in the waters adjacent to the nesting beaches (Reina et al. 2005; Eckert et al. 2006). Males may migrate to the area annually or may mate opportunistically at foraging grounds or nesting areas other than their own native area (Eckert and Eckert 1988).

While reproductively active females and males arrive seasonally at preferred subtropical and tropical nesting locations, nonbreeding adults range further north and south into temperate zones seeking areas containing oceanic jellyfish and other soft-bodied invertebrates (Eckert et al. 2012). Nearly 25 % of leatherback sightings away from nesting areas are associated with aggregations of jellyfish, suggesting that jellyfish distribution may drive the distribution of leatherback foraging areas (Houghton et al. 2006). Foraging occurs on both the continental shelf and in pelagic waters, and nonrandom, long-distance migrations between foraging and nesting grounds are typical (Eckert et al. 2012).

In the Gulf of Mexico, information regarding the distribution and abundance of leatherback sea turtles, which is summarized in the following paragraphs, is incomplete. However, adult leatherback distribution in the Gulf of Mexico—whether in deep-sea, pelagic waters, along the continental shelf, or in nearshore waters—has historically been associated with dense concentrations of jellyfish (Leary 1957; Fritts et al. 1983a, b; den Hartog and van Nierop 1984; Lohoefener et al. 1989; Eckert et al. 1989a, 2012; Houghton et al. 2006).

In the 1950s, Florida fishermen reported occasional sightings of leatherbacks off the coast of Sarasota and claimed they had been seen occasionally in the area since at least the 1930s (Yerger 1965). A leatherback carcass was found in Copano Bay, Texas in 1951 (Gunter 1951). In 1956, 100 leatherbacks were reported only 75 yards from the beach in Port Aransas, Texas. This group of leatherbacks appeared to be associated with a large abundance of jellyfish in the area (Leary 1957). Multiple sightings and captures of leatherbacks occurred along the Florida Gulf coast during the early 1960s (Yerger 1965). In fact, leatherbacks were once reported to be seasonally abundant off the coast of Panama City, Florida (Pritchard 1976). In 1975, leatherbacks were observed off the coast of Alabama (Mount 1975).

In 1979, aerial surveys of the Texas and Florida Gulf coasts reported more than 97 % of all turtle sightings to be in Florida, four of which were leatherbacks (Fritts and Reynolds 1981). Aerial surveys conducted in the Gulf of Mexico from May 1980 to April 1981 reported a total of 47 leatherbacks across all areas (except off the coast of south Texas), and found leatherbacks to be most common off the coast of Florida (Fritts et al. 1983a, b). During the survey, leatherbacks were more conspicuous on the continental shelf than in adjacent deeper waters (Fritts et al. 1983a, b).

Between 1988 and 1990, NMFS aerial surveys in the Gulf of Mexico reported infrequent sightings of leatherbacks, with most sightings occurring in July through November 1989 (Lohoefener et al. 1990). A survey of sea turtles in southeastern Louisiana in 1988 reported that the only evidence of leatherbacks in the area was one report from a diver in July 1988 (Fuller 1989). During the summer of 1989, six leatherback sightings were reported in south Louisiana, five of which were by divers most likely diving offshore at oil platforms (Fuller 1989). During a 1989 cruise in the Gulf of Mexico by the University of West Florida to the head of De Soto Canyon to collect neuston and *Sargassum*, eight leatherbacks were sighted near a coastal-subtropical water mass boundary region that contained high densities of jellyfish (Collard 1990).

Beginning in 1992, the U.S. Geological Survey performed a 3-year aerial and ship survey of the entire northern Gulf of Mexico. The study estimated an overall abundance of 168 leatherbacks in the continental slope area and also estimated leatherbacks to be 12 times more abundant in the winter than in the summer (Davis et al. 2000). Aerial surveys by the Southeast Fisheries Science Center during the fall of 1992, 1993, 1994, and 1996 sighted a total of three leatherbacks in the western Gulf of Mexico and eight in the eastern Gulf; only two of the turtles sighted were in water less than approximately 18 m (59 ft) deep (Epperly et al. 2002).

In CMR studies, the low recapture rate of leatherbacks, compared to other species of sea turtles, is likely due to their highly pelagic lifestyle and uncommon occurrence in coastal waters. However, captures and tag returns of leatherbacks have indicated that adult females use the Gulf of Mexico as a foraging ground. From 1970 to 1973, researchers in Suriname and French Guiana tagged more than 2,000 nesting female leatherbacks. Two of the turtles that were tagged on French Guiana beaches in 1970 and 1972, respectively, were recaptured in western Gulf of Mexico waters in 1973 (Pritchard 1976). Also, a nesting female tagged on the Costa Rican coast in 1985 was captured a year later by a shrimp fisherman off the Mississippi

Gulf coast (Hirth and Ogren 1987). In addition, tag returns from turtles tagged while nesting at index beaches in Costa Rica from 1976 to 2003 demonstrated that the northern Gulf of Mexico is a common location of dispersal for post-nesting females, as five of 21 tag returns were from this region (Troëng et al. 2004).

From 2003 through 2006, 12 adult female leatherbacks were satellite tracked from their nesting beaches at Tortuguero and Gandoca in Costa Rica and Chiriquí Beach in Panama (Evans et al. 2007). Of the four turtles that migrated to the Gulf of Mexico, three stayed within the eastern part of the Gulf off the Florida and Alabama coasts and the fourth leatherback stayed within the western Gulf. This research suggested that the Gulf of Mexico may represent a significant year-round foraging ground for leatherbacks from the Caribbean coast of Central America and not just a seasonal feeding area or pass-through region for migrating leatherbacks. For unknown reasons, jellyfish populations in the Gulf of Mexico have been increasing in recent years; it is possible that year-round foraging has increased as a response to increased jellyfish densities (Evans et al. 2007).

To assess the potential determinants of intra- and interpopulation variability in migratory patterns over the north and south Atlantic Ocean, the movements and diving behavior of 16 Atlantic leatherback turtles from different nesting sites (Chiriquí Beach in Panama, Samsambo Beach in Suriname, Awala-Yalimapo Beach in French Guiana, and Kinguere Beach in Gabon) and one foraging site (waters off Uruguay) were satellite tracked during their post-breeding migrations between 2005 and 2008 (Fossette et al. 2010). Two of the three turtles from Panama migrated to the Gulf of Mexico. After crossing the Caribbean Sea in 1 month, one turtle explored the eastern side of the Gulf, spending 2 months (September and October 2005) along the northeastern continental slope and 4 months (November 2005 through March 2006) south of the Loop Current (Fossette et al. 2010). The second turtle first moved toward the northern continental shelf of the Gulf of Mexico, and then traveled to the western and southwestern shelves from August through September 2006 toward an area between Veracruz and Yucatán, Mexico, where she remained for 6 months until March 2007. These turtles spent most of their time along the continental slope of the Gulf of Mexico, possibly foraging on gelatinous zooplankton aggregated along the front of the shelf break. By monitoring turtles from different nesting sites and one foraging area over the Atlantic Ocean, this study clearly illustrated that the general dispersal patterns and temporary residence areas used by the leatherback turtles may vary among individuals of the same nesting population and among populations (Fossette et al. 2010).

11.6 HAWKSBILL SEA TURTLE (*ERETMOCHELYS IMBRICATA*)

Linnaeus originally described the hawksbill in 1766 as *Testudo imbricata*; it was later transferred to its own genus, *Eretmochelys*, in 1843 by Fitzinger (Meylan and Redlow 2006). The specific name, *imbricata*, refers to the overlapping nature of the carapace scutes (Amorocho 2001). Unlike other sea turtles, the scutes of the hawksbill's beautiful shell or carapace are overlapping, and the rear edge of the carapace is almost always serrated (Figure 11.39) (NMFS and USFWS 1993; Meylan and Redlow 2006). The scutes are often richly patterned with irregularly radiating streaks of brown, black, orange, or red on an amber background. The small- to medium-sized hawksbill sea turtle is named for its strongly hooked beak.

Its beauty cursed the hawksbill sea turtle. As the sole source of commercial tortoiseshell, it has been exploited for centuries (Mortimer 2008). Tortoiseshell from the attractive carapace of the hawksbill can be used to make products, such as jewelry, combs, embellishments on furniture, and rims for eye glasses (Witzell 1983). Millions of hawksbills have been killed for

Figure 11.39. Hawksbill sea turtle using a coral reef (photograph by Caroline Rogers, USGS) (NOAA 2011).

the tortoiseshell markets of Asia, Europe, the Caribbean, and the USA over the past 100 years (NMFS and USFWS 2007e). Japan was historically the major importer of tortoiseshell or *bekko* from the Caribbean (NMFS and USFWS 1993). They agreed to stop importing bekko in 1993 (USFWS 2012). Although the volume of international trade has declined significantly in the past 20 years, it remains active, especially in Southeast Asia and the Americas (Mortimer 2008). In Southeast Asia, the extensive practice of selling whole, stuffed hawksbills is a relatively new threat (Mortimer 2008).

Hawksbills are the most tropical of the sea turtle species and typically nest at low densities throughout their range (NMFS and USFWS 1993, 2007e). In the past, hawksbill sea turtles were considered to be naturally rare and to have a more dispersed nesting pattern than the other sea turtle species (Groombridge and Luxmoore 1989). However, the dispersed pattern currently observed is now believed to be the result of overexploitation of previously large colonies (Meylan and Donnelly 1999). Hawksbill sea turtles are often associated with coral reefs (Meylan 1988; Meylan and Redlow 2006).

There are many gaps in the understanding of hawksbill sea turtle biology, and the oceanic phase of the post-hatchlings remains one of the most poorly understood aspects of hawksbill life history (NMFS and USFWS 2007e). Because of an almost total lack of long-term trend data at hawksbill foraging sites, nesting beach data are the primary information source used to evaluate trends in hawksbill populations. Few data are available on the at-sea mortality of hawksbills in fisheries (NMFS and USFWS 2007e).

11.6.1 Hawksbill Sea Turtle Life History, Distribution, and Abundance

Hawksbill sea turtles occur in the tropical and subtropical seas of the Atlantic, Pacific, and Indian oceans and are widely distributed in the Caribbean Sea and western Atlantic Ocean (Figure 11.40) (NMFS and USFWS 1993; USFWS 2012). They were once abundant in the tropical

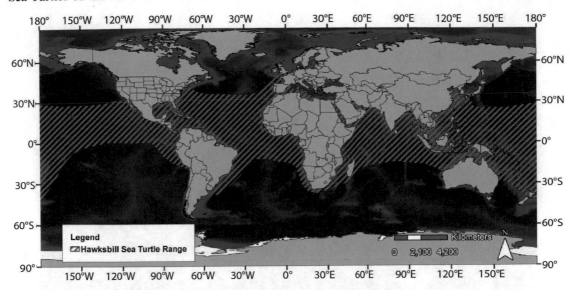

Figure 11.40. Range of the hawksbill sea turtle (from NOAA 2009d).

and subtropical regions of the world and now occur at only a fraction of their historical distribution (NMFS and USFWS 2007e). Unlike other sea turtle species, hawksbills nest in low densities on scattered, small beaches (USFWS 2012). Throughout their range, hawksbills nest in at least 70 countries on insular and mainland sandy beaches and inhabit coastal waters of more than 100 countries NMFS and USFWS (1993, 2007e). In 2007, at 83 nesting sites distributed among ten ocean regions around the world, approximately 21,212–28,138 hawksbills were estimated to nest each year (NMFS and USFWS 2007e; USFWS 2012).

Although greatly depleted compared to historical levels, nesting populations in the Atlantic Ocean are generally doing better than those in the Indo-Pacific (NMFS and USFWS 2007e). Hawksbill nesting in the Caribbean accounts for approximately 20–30 % of the world's hawksbill sea turtle population, and the number of hawksbill sea turtles living in the Caribbean was estimated at 27,000 in 2003 (Lutz et al. 2003; USFWS 2012). In the Atlantic, more population increases have been recorded in the Insular Caribbean, as compared to populations on the western Caribbean mainland or in the eastern Atlantic Ocean (NMFS and USFWS 2007e). Historically, Panama supported the single most important nesting population in the Caribbean, but hawksbill nesting at Chiriquí Beach in Panama has declined by more than 95 % during the past 50 years (Carr 1979; Meylan and Donnelly 1999). The Yucatán Peninsula in Mexico now supports the largest nesting hawksbill population in the area (Figure 11.41) (Cuevas et al. 2010).

The Insular Caribbean has eight nesting concentrations of hawksbills: Antigua/Barbuda (especially Jumby Bay), Bahamas, Barbados, Cuba (Doce Leguas Cays), Jamaica, Puerto Rico (especially Mona Island), Trinidad and Tobago, and the U.S. Virgin Islands (especially Buck Island Reef National Monument [BIRNM], an uninhabited island about 2.4 km [1.5 mi] north of the northeast coast of St. Croix) (Figure 11.41) (NMFS and USFWS 2007e). In 2002, 400–833 nesting females, or 2,000–2,500 nests, were estimated for Doce Leguas Cays, Cuba (NMFS and USFWS 2007e; Mortimer and Donnelly 2008; USFWS 2012). From 2001 through 2005, 199–332 nests were recorded each year on Mona Island, Puerto Rico, and 51–85 nests were recorded each season for Culebra Island, Caja de Muertos, and Humacao, Puerto Rico (NMFS and USFWS 2007e; USFWS 2012). From 2001 through 2006, an average of 56 nests were laid each year at BIRNM in the U.S. Virgin Islands, while 30–222 nests were laid in 2006 on U.S. Virgin Islands beaches outside BIRNM (NMFS and USFWS 2007e; USFWS 2012).

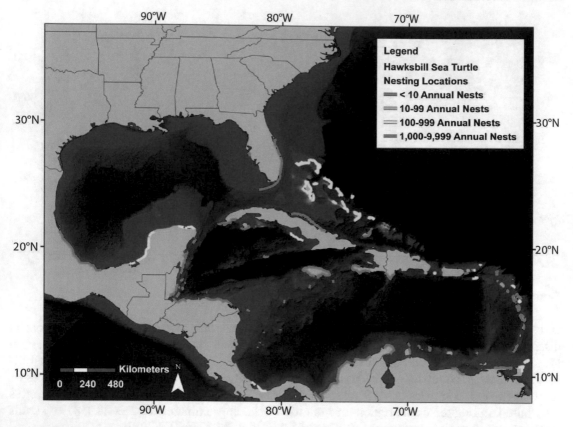

Figure 11.41. Generalized nesting beach locations of the hawksbill sea turtle in the Gulf of Mexico, Caribbean, and northwest Atlantic Ocean (interpreted from Dow et al. 2007; SWOT 2008).

11.6.1.1   Nesting Life History, Distribution, and Abundance for Gulf of Mexico Hawksbills

In the Gulf of Mexico, hawksbill nesting occurs mostly on Yucatán Peninsula beaches in the southern Gulf (Figure 11.41), and these beaches are also the location of the most important hawksbill rookery in the Atlantic (Cuevas et al. 2010). Hatchlings from this rookery are likely to be carried by the current through the Yucatán Channel north into the Gulf of Mexico (Blumenthal et al. 2009). For example, a genetic contribution of 93 % from Yucatán Peninsula nesting beaches was estimated for a group of 42 juvenile hawksbills, ranging in size from 5.2 to 36.8 cm (2 to 14.5 in) SCL, that were stranded on Texas beaches (Bowen et al. 2007). In addition, strandings of juvenile hawksbills along the Florida Gulf coast, which commonly occur, are most likely from this rookery (Meylan and Redlow 2006). Life history information for the hawksbill sea turtle is summarized in Table 11.5, including all available information for specific Gulf of Mexico beaches or regions.

The hawksbill nesting population of the Yucatán Peninsula was declining until 1978 when local and regional protection was implemented (Meylan and Donnelly 1999). As a result of an analysis of long-term trends of hawksbill nesting from 1977 through 1996 on Yucatán Peninsula beaches, the steady improvement in monitoring effort was determined to be the major explanation for the gradual increase in nesting from 1977 to 1992 (Garduño-Andrade et al. 1999). However, the average annual increase of 270 nests per year from 1992 through

Table 11.5. Summary of Life History Information for the Hawksbill Sea Turtle

Parameter	Values	References
Nesting season: Yucatán Peninsula, Mexico	April through September	Cuevas et al. (2010)
Remigration interval		
Yucatán Peninsula, Mexico	Range: 2–3 years	Garduño-Andrade (1999)
Isla Aguada, Campeche, Yucatán Peninsula, Mexico	Range: 2–4 years	Guzmán-Hernández et al. (2006)
Nesting interval		
El Cuyo, Yucatán, Yucatán Peninsula, Mexico	Mean: 18 days	Xavier et al. (2006)
Isla Aguada, Campeche, Yucatán Peninsula, Mexico	Mean: 14 days, Range: 11–15 days	Guzmán-Hernández et al. (2006)
Laguna de Términos, Campeche, Yucatán Peninsula, Mexico	Range: 14–16 days	Guzmán-Hernández and García-Alvarado (2010), Amorocho (2001)
Yucatán Peninsula, Mexico	Mean: 21 days (Isla Contoy, Quintana Roo)	Najera (1990)
	Mean: 21 days (Isla Holbox, Quintana Roo)	
	Mean: 23 days (Rio Lagartos, Yucatán)	
Number of nests/season		
Las Coloradas, Quintana Roo, Yucatán Peninsula, Mexico	Mean: 2.1 nests	Garduño (1998)
El Cuyo, Yucatán, Yucatán Peninsula, Mexico	Mean: 2.3 nests	Xavier et al. (2006)
Campeche, Yucatán Peninsula, Mexico	Mean: 3.1 nests	Guzmán et al. (1996)
Isla Aguada, Campeche, Yucatán Peninsula, Mexico	Mean: 2.8 nests, Range: 2.5–3.2 nests	Guzmán-Hernández et al. (2006)
Laguna de Términos, Campeche, Yucatán Peninsula, Mexico	Mean: 3 nests	Guzmán-Hernández and García-Alvarado (2010)
Number of eggs/nest		
El Cuyo, Yucatán, Yucatán Peninsula, Mexico	Mean: 149 eggs	Xavier et al. (2006)
Las Coloradas, Quintana Roo, Yucatán Peninsula, Mexico	Mean: 157 eggs	Garduño-Andrade (2000)
Isla Aguada, Campeche, Yucatán Peninsula, Mexico	Mean: 137 eggs	Frazier (1993)
Isla del Carmen, Chenkan, and Isla Aguada beaches, Campeche, Yucatán Peninsula, Mexico	Range: 96–183 eggs	Cuevas et al. (2008)
Yucatán Peninsula, Mexico	Mean: 148 eggs	Echeverría-García and Torres-Burgos (2007)
	Mean: 159 eggs (Telchac Puerto, Yucatán)	Echeverría-García et al. (2008)
	Mean: 161 eggs (Sisal, Yucatán)	
	Mean: 140 eggs, Range: 46–244 eggs	Frazier (1993)

(continued)

Table 11.5. (continued)

Parameter	Values	References
Yucatán Peninsula, Mexico	Mean: 149 eggs, Range: 47–194 eggs (Isla Contoy, Quintana Roo)	Najera (1990)
	Mean: 152 eggs, Range: 100–188 eggs (Isla Holbox, Quintana Roo)	
	Mean: 153 eggs, Range: 19–229 eggs (Rio Lagartos, Yucatán)	
	Mean: 140 eggs, Range: 60–247 eggs (Celestun, Yucatán)	Pérez-Castañeda et al. (2007)
	Mean: 142 eggs, Range: 60–257 eggs (Isla Holbox, Quintana Roo)	
	Mean: 145 eggs, Range: 62–241 eggs (El Cuyo, Quintana Roo)	
Egg incubation time		
Isla Aguada, Campeche, Yucatán Peninsula, Mexico	Mean: 57 days, Range: 51–64 days	Guzmán-Hernández et al. (2006)
Yucatán Peninsula, Mexico	Mean: 62 days, Range: 51–83 days (Celestun, Yucatán)	Pérez-Castañeda et al. (2007)
	Mean: 63 days, Range: 50–80 days (El Cuyo, Quintana Roo)	
	Mean: 65 days, Range: 50–80 days (Isla Holbox, Quintana Roo)	
Nest pivotal temperature		
Antigua, West Indies	29.3 °C	Hulin et al. (2009)
Bahia, Brazil	29.6 °C	Godfrey et al. (1999)
Sex ratio of hatchlings from nests (proportional female)		
Mona Island, Puerto Rico	Mean: 0.44	Diez and van Dam (2003)
Bahia, Brazil	Range: 0.91–1.0	Godfrey et al. (1999)
Emergence success of hatchlings from nests		
El Cuyo, Yucatán, Yucatán Peninsula, Mexico	Mean: 0.81	Xavier et al. (2006)
Yucatán, Yucatán Peninsula, Mexico	Mean: 0.76	Echeverría-García and Torres-Burgos (2007)
	Mean: 0.83 (Sisal, Yucatán)	Echeverría-García et al. (2008)
	Mean: 0.86 (Telchac Puerto, Yucatán)	
	Mean: 0.59 (Dzilam de Bravo, Yucatán)	Echeverría-García et al. (2009)
	Mean: 0.67 (Telchac Puerto, Yucatán)	
	Mean: 0.68 (Sisal, Yucatán)	
Yucatán Peninsula, Mexico	Mean: 0.82 (El Cuyo, Quintana Roo)	Pérez-Castañeda et al. (2007)
	Mean: 0.85 (Celestun, Yucatán)	
	Mean: 0.88 (Isla Holbox, Quintana Roo)	

(continued)

Table 11.5. (continued)

Parameter	Values	References
Size of hatchlings		
Tortuguero, Costa Rica	Mean: 4.24 cm SCL[a], Range: 3.91–4.60 cm SCL	Carr and Ogren (1966)
Wider Caribbean Region	Mean: 4.20 cm SCL, Range: 3.90–4.60 cm SCL	Amorocho (2001)
Mustang Island, Texas	Range: 5–21 cm SCL	Carr (1987)
Diet of hatchlings		
Caribbean	*Sargassum*, manatee grass, crab chela, eggs of flying fish, half-beaks, and needlefish	Meylan (1984)
Florida	Sargassum	Meylan and Redlow (2006)
Size of oceanic juveniles		
Rio Lagartos Sea Turtle Sanctuary, Yucatán Peninsula, Mexico	Range: 20–30 cm SCL	Cuevas et al. (2007)
Padre Island National Seashore, Mustang Island, and Port Aransas, Texas	Range: 20.1–29.1 cm SCL[b]	Amos (1989)
East and west coast of Florida	Mean: 20.6 cm SCL, Range: 13.4–24.8 cm SCL	Witherington et al. (2012)
Diet of oceanic juveniles		
Caribbean	*Sargassum,* manatee grass, crab chela, eggs of flying fish, half-beaks, and needlefish	Meylan (1984)
Florida	Sargassum	Meylan and Redlow (2006)
Size of oceanic juveniles at recruitment to neritic juvenile stage		
Rio Lagartos, Las Colorados, Quintana Roo, Yucatán Peninsula, Mexico	Range: 20–65 cm SCL	R. Pérez-Castañeda, Universidad Autonoma Tamaulipas, unpublished data, cited in Garduño-Andrade et al. (1999)
Size of oceanic juveniles at recruitment to neritic juvenile stage		
Florida Keys	Range: 21.4–69 cm SCL	M. Bressette, Inwater Research Group, unpublished data, cited in Witherington et al. (2012)
Broward County to St. Lucie Nuclear Plant, Florida	Range: 25.7–34 cm SCL	M. Bressette and R. Wershoven, Quantum Resources, Inc. and Broward County Audubon Society, personal communication, cited in Meylan and Redlow (2006)
Diet of neritic juveniles: Rio Lagartos Sea Turtle Sanctuary, Quintana Roo, Yucatán Peninsula, Mexico	Sponges, including *Chondrilla* sp., *Dictyopteris* sp., *Hypnea* sp., *Jania* sp., *Laurencia* sp., *Ceramium* sp., *Codium* sp., and *Gracilaria* sp.	Cuevas et al. (2007)

(continued)

Table 11.5. (continued)

Parameter	Values	References
Age at sexual maturity		
Las Coloradas, Quintana Roo, Yucatán Peninsula, Mexico	Mean: 24 years, Minimum: 14 years	Garduño (1998)
Yucatán Peninsula, Mexico	Mean: 31.2 years	IUCN (2012)
Size of sexually mature adult females		
Las Coloradas, Quintana Roo, Yucatán Peninsula, Mexico	Mean: 90 cm SCL, Minimum: 80 cm SCL	Garduño (1998)
El Cuyo, Yucatán, Yucatán Peninsula, Mexico	Mean: 94.4 cm SCL[b]	Xavier et al. (2006)
Isla Aguada, Campeche, Yucatán Peninsula, Mexico	Mean: 92.1 cm SCL[b], Range: 85.7–98.6 cm SCL	Guzmán-Hernández et al. (2006)
Isla del Carmen, Chenkan, and Isla Aguada beaches, Campeche, Yucatán Peninsula, Mexico	Range: 82.7–95.6 cm SCL[b]	Cuevas et al. (2008)
Yucatán, Yucatán Peninsula, Mexico	Mean: 93.1 cm SCL (Telchac Puerto, Yucatán), Mean: 96.5 cm SCL (Sisal, Yucatán)	Echeverría-García et al. (2008)
	Mean: 92.7 cm SCL (Sisal, Yucatán)	Echeverría-García et al. (2009)
	Mean: 99.6 cm SCL (Telchac Puerto, Yucatán)	
Yucatán Peninsula, Mexico	Mean: 94 cm SCL[b], Range of modes: 94.6–98.6 cm SCL (Celestun, Yucatán);	Pérez-Castañeda et al. (2007)
	Mean: 93.7 cm SCL[b], Range of modes: 94.6–98.6 cm SCL (El Cuyo, Quintana Roo)	
	Mean: 94.3 cm SCL[b], Range of modes: 89.6–93.6 cm SCL (Isla Holbox, Quintana Roo)	
Diet of adults		
Caribbean	Sponges, demosponges, and button polyp, *Ricordea florida*	Meylan (1984)
	Sponges, including chicken liver sponge, *Ancorina* sp., *Geodia* sp., *Placospongia* sp., *Suberites* sp., *Myriastra* sp., *Ecionemia* sp., *Chondrosia* sp., *Aaptos* sp., and *Tethya actinia*	Meylan (1988)

[a]*SCL* straight carapace length, *cm* centimeters
[b]To convert from curved carapace length (*CCL*): SCL = (0.9927 × CCL) + 0.2782 (Garduño 1998)

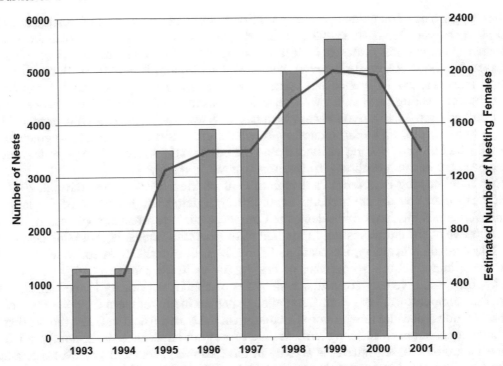

Figure 11.42. Annual number of nests (*bars*) and estimated number of nesting females (*line*), assuming 2.8 nests per female (Guzmán-Hernández et al. 2006), for the hawksbill sea turtle for selected Mexican Gulf of Mexico beaches along the coasts of Tamaulipas, Veracruz, Campeche, Yucatán, and Quintana Roo from 1993 through 2001 (Márquez-M. 2004).

1996 was considered to be indicative of real population increases, since beach coverage had peaked, with consistent monitoring efforts each year. In 1996, 4,522 nests were reported, which was equivalent of up to 2,200 nesting females; the Yucatán Peninsula hawksbill nesting population was estimated to range from 1,900 to 4,300 turtles during 1996, the largest in the western Atlantic Ocean. Most nesting in 1996 occurred on Campeche beaches; average nesting densities for the 16 monitored beaches ranged from 0.8 to 64 nests/km (0.5 to 40 nests/mi) (mean = 16.7 nests/km [10.4 nests/mi]). The increase in the nesting population after 1993 was thought to be a result of increased survival rates of juveniles and adults due to regional conservation measures and increased recruitment into the breeding stock from protected Yucatán Peninsula beaches (Garduño-Andrade et al. 1999). Similar trends in hawksbill nesting were observed along the entire Mexican Gulf coast (Figure 11.42).

In 1999, more than 6,000 nests were recorded on Yucatán Peninsula beaches (Garduño-Andrade et al. 1999; Cuevas et al. 2008). However, nesting numbers declined by 63 % from 1999 through 2004. The cause of this decline in nesting numbers was suspected to be the taking of turtles and/or impacts to the hawksbill's marine habitats (Abreu-Grobois et al. 2005). At three Yucatán Peninsula beaches—Celestun, El Cuyo, and Isla Holbox—hawksbill nesting decreased significantly from 2000 to 2001 (Pérez-Castañeda et al. 2007). Nesting numbers on Yucátan Peninsula beaches have increased since 2004 but are still below 1999 numbers (NMFS and USFWS 2007e; del Monte-Luna et al. 2012). From 2001 through 2006, 2,672 nests or 534–891 nesting females were recorded in the Yucatán Peninsula each year (NMFS and USFWS 2007e; Mortimer and Donnelly 2008; USFWS 2012).

A study was conducted on El Cuyo Beach to evaluate hawksbill nesting activity from 2002 through 2004 (Xavier et al. 2006). This beach is one of the most important hawksbill nesting beaches on the Yucatán Peninsula. During the 3-year study period, hawksbill nesting decreased by 40 %; there were 373 nests in 2002, 311 in 2003, and 217 nests in 2004. Hawksbill nesting on this beach averaged 659 nests/km (409 nests/mi) in 1996 (Garduño-Andrade et al. 1999) and decreased to about 300 nests between 1999 and 2001 (Salum-Fares 2003). Between nesting seasons, there was no difference in the size of nesting females. High site fidelity was demonstrated by the nesting hawksbills, with an average distance of 3 km (1.9 mi) between nests. Predation has risen on El Cuyo Beach, which has affected hawksbill nests (Xavier et al. 2006). El Cuyo Beach is also an important nesting beach for green turtles. Compared to green turtles, hawksbills had a wider nesting distribution on the beach and seemed to have a wider range of preferences for beach morphological features (Cuevas et al. 2010).

Hawksbill nesting data from the southern Gulf of Mexico from 1980 through 2010 were recently evaluated by del Monte-Luna et al. (2012) to determine the cause of the long-term decline. Since nesting hawksbills along the Campeche coast can reasonably be considered as representative of the entire nesting population of the Yucatán Peninsula, hawksbill nesting data from Isla Aguada, Savancuy, Chencan, and Punta Xen, Campeche, Mexico, were analyzed. A 7-year cycle in annual relative number of nesting turtles in the southern Gulf of Mexico was found, which was inversely correlated with cycles of similar periodicity in the North Atlantic sea surface temperature. Long-term population dynamics in the southern Gulf were related to a basin-wide, quasi-decadal temperature fluctuation in the North Atlantic. Other threats that also may have contributed to the long-term decline of nesting hawksbills in the southern Gulf of Mexico include increased hurricane activity in the Caribbean, regional sea-level rise, and the constant expansion of beach development (del Monte-Luna et al. 2012).

In addition to the rookery in the Yucatán Peninsula, hawksbill sea turtles lay a small number of nests each year along the Gulf coasts of Mexico and Florida (Figure 11.41). In Florida, hawksbill nesting also occurs along the east coast from Volusia through Miami-Dade counties, including Soldier Key in Biscayne Bay and the Florida Keys (Figure 11.41) (Dalrymple et al. 1985; Meylan 1992). From 1979 through 2003, 31 nests were documented in Florida and were distributed along the Atlantic coast from Volusia County to Monroe County, with a single record on the Gulf coast in Manatee County and a maximum of four nests recorded in any year (Meylan and Redlow 2006). Hawksbill nesting on both the Florida Atlantic and Gulf coasts is most likely underestimated for the following reasons: (1) beaches in areas known to be used by hawksbills are incompletely surveyed (e.g., Florida Keys), (2) beach surveys are typically not conducted during the fall months, (3) hawksbill and loggerhead sea turtle tracks are similar, (4) hawksbills nest in or under vegetation and sometimes on narrow beaches, and (5) hawksbill and loggerhead hatchlings are similar in appearance (Meylan and Redlow 2006). Low levels of hawksbill nesting are also suspected to occur in the Marquesas and Dry Tortugas (NMFS and USFWS 1993). In 1998, the first hawksbill nest was recorded on the Texas coast at PAIS (Mays and Shaver 1998).

The nesting season of hawksbills is longer than that of other sea turtle species (Table 11.5), and the small, agile females have the ability to climb over reefs and rocks to nest in beach vegetation (Figure 11.43) (NMFS and USFWS 1993; USFWS 2012). Tagging and genetic studies have demonstrated that female hawksbills have strong site fidelity and return to nest in the vicinity where they hatched (Witzell 1983; Bass 1999). Although rare daytime nesting is known, hawksbill sea turtles typically nest at night (Meylan and Redlow 2006).

Figure 11.43. Hawksbill sea turtle returning to the sea (from Scarygami 2012).

11.6.1.2   Hatchling, Post-Hatchling, and Oceanic Juvenile Life History and Distribution for Gulf of Mexico Hawksbills

Similar to other species of sea turtles, hatchling hawksbills enter an oceanic phase (Figure 11.44). This phase may involve long-distance travel carried by surface gyres, with eventual recruitment to neritic foraging habitat (NMFS and USFWS 2007e). Hatchlings entrained in the Loop Current could be expected to remain in the Gulf of Mexico for differing periods of time, depending on which branch of the Loop Current they enter (Meylan and Redlow 2006). Both newly hatched and early juvenile hawksbills have been found in association with *Sargassum* and floating weed in the Atlantic and Caribbean (Table 11.5) (Carr 1987; Mellgren and Mann 1996; Musick and Limpus 1997; Meylan and Redlow 2006). No post-hatchlings but six juvenile hawksbill sea turtles, with an average size of 20.6 cm (8.1 in) SCL, were captured from the *Sargassum*-dominated, surface-pelagic drift community in the eastern Gulf of Mexico and Atlantic Ocean from 2005 through 2011; some of these hawksbills were large enough to be on the cusp of recruitment into the neritic zone, most likely to foraging habitat in the Florida Keys (Witherington et al. 2012). Weedlines in the Gulf of Mexico likely serve as habitat for post-hatchling hawksbills from nesting beaches in Mexico and Central America (NMFS and USFWS 1993).

Between 1972 and 1984, 77 strandings of post-hatchling and juvenile hawksbills were recorded in Texas, with most occurring near Corpus Christi; these turtles most likely originated from nesting beaches on the Yucatán Peninsula in Mexico (Amos 1989; Bowen et al. 2007). Limited tagging data indicates that some post-hatchling hawksbills from the western Gulf of Mexico disperse into the Atlantic Ocean, most likely through the Florida Straits, and move northward along Florida's east coast (Meylan and Redlow 2006).

Figure 11.44. Hawksbill sea turtle hatchling moving across the beach toward the sea (from Serge_Vero 2007).

11.6.1.3 Neritic Juvenile Life History and Distribution for Gulf of Mexico Hawksbills

Oceanic juvenile hawksbills recruit to foraging habitat in the neritic zone starting at around 20 cm (7.9 in) SCL (Table 11.5). The origin of juveniles found in neritic foraging areas is related to nesting population size, geographic distance from the nesting areas, and ocean currents. Juveniles typically occupy a series of habitats as they increase in size, with larger turtles often inhabiting deeper sites (Bowen et al. 2007). Large juveniles may be associated with the same feeding location for more than 10 years (Musick and Limpus 1997). Neritic juvenile hawksbills may occupy a range of habitats, including coral reefs, rocky areas, other hard bottom habitats, seagrass and algae beds, shallow coastal areas, lagoons and oceanic islands, narrow creeks, and mangrove bays and are rarely found in water deeper than about 20 m (66 ft) (Musick and Limpus 1997; USFWS 2012). Throughout their range, neritic juvenile hawksbill sea turtles typically feed on sponges (Table 11.5). However, hawksbills are not always mainly spongivorous; for example, in a recent long-term study in the northern Great Barrier Reef, Australia, hawksbills fed mainly on algae (*Laurencia* sp. and *Gelidiella* sp.; 72 % of ingesta), along with some sponges (10 %), soft corals, and other prey (12 %) (Bell 2012).

Cuevas et al. (2007) evaluated the benthic foraging habitat of juvenile hawksbills in the Rio Lagartos Sea Turtle Sanctuary in Yucatán, Mexico, an important feeding and development area for juvenile hawksbill sea turtles. Hawksbills were found to be distributed mainly on hard bottom sites covered by octocorals, such as *Pseudopterogorgia*, and sponges of the genera *Chondrilla* and *Spheciospongia*. Based on tracking data, the average home range of the turtles was larger during the day (0.123 km^2 [0.048 mi^2]) than that used at night (0.021 km^2 [0.008 mi^2]). In addition, there were differences in habitat preferences between day and night. During the day, hawksbills mainly occupied habitats with 20–40 % octocoral cover; at night they tended to occupy bare substratum areas (Cuevas et al. 2007).

Along the U.S. Gulf Coast, juvenile hawksbills are associated with stone jetties in Texas (Amos 1989). Hawksbills are rare along the north Florida Gulf coast; juveniles are more abundant in Gulf of Mexico waters off west-central Florida than anywhere else along the Florida Gulf coast (Meylan and Redlow 2006). Hard-bottom communities on the west Florida Shelf, the southern Pulley Ridge, and the Florida Middle Ground reef complex represent potential hawksbill foraging habitat in the Gulf of Mexico off the Florida west coast; the distribution of post-pelagic hawksbills corresponds closely to the Florida Reef Tract (Meylan and Redlow 2006). During surveys conducted at KWNWR from 2002 through 2006, almost 60 hawksbills were sighted, and 19 juveniles, ranging in size from 28.2 to 69 cm (11.1 to 27.2 in) SCL and averaging 46.4 cm (18.3 in) SCL, were captured (Eaton et al. 2008).

There is little to no evidence that hawksbills use Florida's major Gulf coast estuaries, such as Tampa Bay and Charlotte Harbor (Meylan and Redlow 2006). Neritic juvenile hawksbills were captured during a study conducted from 1997 through 2004 in the nearshore waters of Gullivan Bay, in the Ten Thousand Islands off the southwest Florida Gulf coast (Witzell and Schmid 2004; Eaton et al. 2008). Three juvenile hawksbills, averaging 49.8 cm (19.6 in) SCL in size (range of 38.2–58.1 cm [15–23 in] SCL) were captured during surveys conducted in Florida Bay from 2000 through 2006 (Eaton et al. 2008).

Three large juvenile hawksbills, ranging in size from 51.9 to 69.8 cm (20.4 to 27.5 in) SCL (mean = 61.5 cm or 24.2 in SCL), were captured within DTNP in the Gulf and tracked from August 2008 through January 2011 to determine patterns of habitat use (Hart et al. 2012b). Core use areas within the park ranged from 9.2 to 21.5 km^2 (3.6 to 8.3 mi^2) and were concentrated around the flats surrounded by Garden Key, Bush Key, and Long Key. The turtles were more active during the day than at night, which could indicate active foraging during the day and resting behavior at night. After between 263 and 699 days residing within the park, two turtles migrated to Cuba, while the third hawksbill migrated toward Key West, Florida. The turtles that migrated to Cuba ceased transmitting after 320 and 687 days, while the turtle that migrated toward Key West stopped transmitting after 884 tracking days. This study highlighted unknown regional connections for hawksbills, possible turtle harvest incidents, and fine-scale habitat use of juvenile turtles (Hart et al. 2012b).

11.6.1.4   Adult Life History and Distribution for Gulf of Mexico Hawksbills

Most of the adult life history information available for hawksbills in the Gulf of Mexico is for the Yucatán Peninsula in Mexico (Table 11.5). Most hawksbills have slow growth rates, which vary within and among populations (NMFS and USFWS 2007e); growth rates of 2–4 cm (0.8–1.6 in) per year are typical for the Caribbean (Boulon 1994). Adult hawksbills may reach up to 1 m (3.3 ft) in length and weigh up to 140 kg (308.6 lb); however, they typically average about 0.75 m (2.5 ft) in length and weigh around 80 kg (176.4 lb) or less (USFWS 2012).

Recent satellite tracking studies conducted to determine the migratory patterns and feeding ground locations for hawksbills nesting on Yucatán Peninsula beaches have indicated that turtles remain in Mexican waters (Cuevas et al. 2008, 2012). In 2006 and 2007, three post-nesting females were tracked for up to 510 days from three of the major nesting beaches in Campeche; two migrated to foraging grounds off the coast of Campeche, and one migrated to the Mexican Caribbean (Cuevas et al. 2008). Ten post-nesting hawksbill sea turtles were tracked from nine different nesting beaches on the Yucatán Peninsula in 2006 and 2007; turtles that nested on the western side of the peninsula migrated to the east, while those that nested on the eastern side migrated to the west (Cuevas et al. 2012). In a second, similar study conducted in 2011, tracked hawksbills also stayed in Mexican waters; however, one female migrated north to the border of the continental shelf, while a male stayed close to the nesting beaches (Cuevas et al. 2012).

While hawksbills are not encountered in the Gulf of Mexico as frequently as some of the other species of sea turtles (Thompson et al. 1990), they regularly occur in U.S. Gulf of Mexico waters off the southern Florida coast and in the northern Gulf, especially in Texas coastal waters (NMFS and USFWS 1993). Hawksbills have been recorded in waters of all U.S. Gulf Coast states and are regularly observed in the Florida Keys (Lund 1985; NMFS and USFWS 1993; Meylan and Redlow 2006). The distribution and abundance of hawksbill sea turtles in the Florida Keys is largely unknown, and few studies have been conducted to document their distribution and abundance; however, the Florida Keys National Marine Sanctuary, KWNWR, and DTNP contain important hawksbill habitat (Meylan and Redlow 2006).

Adult hawksbills are often associated with coral reefs, where they typically forage on a limited number of sponge species (Table 11.5); however, as already mentioned, hawksbills are not always mainly spongivorous (Bell 2012). The ledges and caves of coral reefs provide important shelter for resting hawksbills both during the day and night (NMFS and USFWS 1993). Similar to other species of sea turtles, hawksbills are integral components of marine and coastal food webs (Bouchard and Bjorndal 2000). In addition, because they eat sponges, they help keep coral reefs healthy (Bjorndal and Jackson 2003).

11.7   THREATS TO GULF OF MEXICO SEA TURTLE POPULATIONS

Many anthropogenic and natural threats affect all ecosystem zones used by sea turtle populations in the Gulf of Mexico (Table 11.6). These threats, which occur either in the Gulf of Mexico or within the distribution range of the sea turtle species that occur in the Gulf, are discussed below. In addition, if available, examples of quantified impacts associated with some of these threats for each of the sea turtle species that occurs in the Gulf of Mexico are presented.

11.7.1   Incidental Capture of Sea Turtles in Commercial and Recreational Fisheries

In 1990, the incidental capture of sea turtles in shrimp trawls was identified by the U.S. National Academy of Sciences as the major cause of turtle mortality associated with human activities; in fact, this incidental capture was determined to kill more sea turtles than all other human activities combined (Magnuson et al. 1990). In addition, most of the sea turtles that are killed in shrimp trawls are neritic juveniles—the life stage most critical to the stability and recovery of sea turtle populations (Crouse et al. 1987; Crowder et al. 1994). In the first cumulative estimates of sea turtle bycatch across fisheries of the United States between 1990 and 2007, the southeast U.S./Gulf of Mexico shrimp trawl fishery was estimated to be responsible for up to 98 % of all sea turtle interactions and for more than 80 % of all sea turtle mortality. However, due to the lack of observer coverage, estimates of bycatch for this fishery are highly uncertain (Epperly et al. 2002; Finkbeiner et al. 2011). The Gulf of Mexico portion of the fishery was estimated to comprise 73 % of total interactions and 96 % of mortality; the shrimp trawl fishery was estimated to account for about 69,300 lethal takes of sea turtles before the 2003 TED enlargement requirements, and approximately 3,700 mortalities following the TED enlargement requirements and the reduction of fishing effort in the Gulf of Mexico (Finkbeiner et al. 2011).

By the late 1970s, before TEDs were developed to prevent turtles from entering the back of shrimp trawl nets and provide for escape (Figure 11.45), the only major nesting population of Kemp's ridleys was close to extinction (Henwood et al. 1992; Frazier et al. 2007). The NMFS

Table 11.6. Summary of Anthropogenic and Natural Threats Affecting the Various Ecosystem Zones used by Sea Turtle Populations in the Gulf of Mexico (NMFS and USFWS 2008; Bolten et al. 2011; NMFS et al. 2011)

Threat	Terrestrial Zone[a]	Neritic Zone[a]	Oceanic Zone[a]
Incidental capture in commercial and recreational fisheries			
Trawls		X	X
Gill nets		X	X
Dredges		X	X
Pelagic and bottom long lines		X	X
Seines		X	
Pound nets and weirs		X	
Pots and traps		X	
Hook and line		X	X
Illegal harvest			
Eggs	X		
Juveniles		X	
Adults	X	X	
Nesting beach alterations			
Cleaning	X		
Human presence	X		
Driving on beach (cars and off-road vehicles)	X		
Artificial lighting	X	X	
Construction	X		
Nourishment and restoration	X	X	
Sand mining	X	X	
Armoring and shoreline stabilization (drift fences, groins, jetties)	X		
Other anthropogenic impacts			
Channel dredging and bridge building		X	
Boat strikes		X	X
Oil and gas exploration (including seismic activity), development, and production	X	X	X
Stormwater runoff		X	X
Oil and chemical pollution and toxins	X	X	X
Algal blooms, including red tides		X	
Hypoxia		X	

(continued)

Table 11.6. (continued)

Threat	Terrestrial Zone[a]	Neritic Zone[a]	Oceanic Zone[a]
Marine debris ingestion and entanglement	X	X	X
Military activities and noise pollution	X	X	X
Industrial and power plant intake, impingement, and entrainment		X	
Dams and water diversion		X	
Sea level rise due to climate change	X		
Temperature change due to climate change	X	X	X
Trophic changes due to fishing and benthic habitat alteration		X	X
Natural impacts			
Predation	X	X	X
Beach erosion and vegetation alteration	X		
Habitat modification by invasive species	X	X	
Pathogens and disease	X	X	X
Hurricanes and severe storms	X	X	
Droughts		X	
Hypothermic stunning		X	

[a]Terrestrial zone = Nesting beach where females excavate nests and lay eggs, where embryos develop; Neritic zone = inshore marine environment from the surface to the seafloor, including bays, sounds, and estuaries, as well as the continental shelf, where water depths do not exceed 200 m (656.2 ft); and Oceanic zone = open ocean environment from the surface to the seafloor where water depths are greater than 200 m (656.2 ft)

published final regulations requiring TEDs in shrimp trawlers in June 1987; however, implementation was delayed as a result of legal and congressional action (NMFS and USFWS 2008). By the early 1990s, TEDs were required at all times of the year in all U.S. waters where the southeast U.S. shrimp fishery operated (Epperly 2003). Turtle excluder devices were also required in Mexican waters beginning in 1995 (Crowder and Heppell 2011).

An evaluation of monthly sea turtle stranding data and shrimp fishing effort from 1986 through 1989 for the northwestern Gulf of Mexico demonstrated a significant relationship: turtle strandings increased as fishing effort increased in waters landward of 15 fathoms or 9.1 m (29.9 ft) (Caillouet et al. 1991). Despite the requirement of TEDs beginning in 1990, there was no change in the relationship of monthly sea turtle stranding rates and monthly shrimp fishing intensities in the northwestern Gulf of Mexico when data collected in 1986 through 1989 and 1990 through 1993 were compared (Caillouet et al. 1996). This lack of change in sea turtle stranding rates indicated that the problem of sea turtle mortality at sea had not been solved and

Figure 11.45. Loggerhead sea turtle escaping a net equipped with a turtle excluder device (TED) (from NOAA 2011).

that further efforts were necessary. An analysis by Epperly and Teas (2002) indicated that the minimum openings of TEDs were too small to exclude large leatherback, loggerhead, and green sea turtles, but they were effective at excluding Kemp's ridleys and juvenile loggerheads. Therefore, the NMFS enacted new regulations in 2003, which required that TED openings be large enough to allow all sea turtles to escape. While TED regulations already in place were likely effective for reducing fishery-induced mortality of smaller turtles (e.g., Kemp's ridley), 2003 was considered the beginning of effective reduction of sea turtle bycatch for the shrimp trawl fishery in the Gulf of Mexico, especially for loggerheads and leatherbacks (Finkbeiner et al. 2011). Among the Gulf States, the State of Louisiana stands out because in the late 1980s, it enacted legislation prohibiting state authorities from enforcing the federal law requiring the use of TEDs by the shrimp fisheries in State waters (Louisiana Revised Statutes 1987). It is unknown how many sea turtles may have drowned in shrimp nets due to this lack of enforcement. Fortunately, the Louisiana legislature approved in 2015 a bill to repeal the old prohibition of TED enforcement law. Thus, Louisiana authorities are now able to enforce the use of TED by the Louisiana shrimp industry (Hill 2015).

While compliance and enforcement has been spotty, the correct and consistent use of TEDs in the United States and Mexican Gulf of Mexico shrimp fisheries has been effective (Lewison et al. 2003). In addition, both shrimp fisheries have declined in recent years because of many factors, including the decline in shrimp abundance, increased fuel costs, reduced shrimp prices, competition with farmed and imported shrimp, and the recent, active hurricane seasons (NMFS 2007; Caillouet et al. 2008; NMFS et al. 2011; Ponwith 2011). Therefore, the decline in shrimp fishing effort in the Gulf of Mexico since the early 1990s, as well as the spatial and temporal closures of Gulf shrimp fisheries, has also reduced bycatch mortality from shrimp trawling (Lewison et al. 2003; Shaver 2005; Caillouet et al. 2008; Crowder and Heppell 2011).

In the Gulf of Mexico, TEDs are not required for many trawl fisheries that could kill sea turtles; however, tow times are often restricted to reduce the probability of sea turtle mortality (Epperly et al. 2002). For example, skimmer trawls are allowed to use restricted tow times

(55 min from April through October and 75 min from November through March) in lieu of TED requirements as a sea turtle bycatch mitigation measure. However, recent observations have indicated that the tow times are often exceeded (Price and Gearhart 2011). Because a mass sea turtle stranding event that occurred in late spring 2010 along the Mississippi coast was attributed to skimmer trawl activity, the feasibility of using TEDs in these fisheries was recently investigated (Price and Gearhart 2011). A rule was proposed by the NMFS requiring the use of TEDs on skimmer, pusher-head, and wing-net trawls in May 2012 (NMFS 2012); however, the rule was withdrawn in late November 2012 because data gathered from a recent investigation did not support the implementation of the rule (Pulver et al. 2012; NOAA 2012).

Both pelagic and bottom longline fisheries in the Gulf of Mexico are known to incidentally take sea turtles. These longline fisheries include the U.S. Gulf of Mexico yellowfin tuna fishery, the U.S. distant water (outside the U.S. Exclusive Economic Zone) swordfish fishery, the Mexican Gulf of Mexico tuna fishery, the U.S. Gulf of Mexico shark fishery, the U.S. Gulf of Mexico grouper/snapper/reef fish/tilefish fishery, and the Mexican Gulf of Mexico shark fishery (NMFS et al. 2011). Gill net fisheries operate off the U.S. Gulf Coast as well as in nearshore state waters and incidentally capture sea turtles; however, gill nets have been banned in Florida, Louisiana, and Texas (NMFS et al. 2011). Sea turtles have also become entangled in dredges, seines, pound nets, and weirs, as well as pots and traps that are used to capture crabs, lobster, eels, and fish (NMFS et al. 2011). In addition, sea turtles are known to bite a baited hook; they have been hooked in both commercial and recreational fishing (TEWG 2000).

In addition to TED requirements, the banning of gill nets, and the permanent and temporary spatial and temporal closures of fisheries, additional measures have been implemented since the 1980s to reduce the incidental bycatch of sea turtles in fisheries. These efforts have included observer programs; developing gear solutions, such as circle hooks and bait combinations; modifying gear, such as reduced pound net mesh sizes and chain mats to prevent turtles from entering the dredge bag; and implementing careful release protocols (Conant et al. 2009; Stokes et al. 2012).

11.7.1.1 Incidental Capture of Kemp's Ridley Sea Turtles in Commercial and Recreational Fisheries

The documentation of the incidental capture of Kemp's ridleys during commercial shrimping operations, particularly in the northern Gulf of Mexico, began in 1973 when the shrimp fishery in the U.S. Gulf of Mexico was becoming highly mechanized (Frazier et al. 2007). Between 500 and 5,000 Kemp's ridleys were estimated to be killed each year prior to the requirement that the offshore shrimping fleet in the southeast United States and Gulf of Mexico use TEDs (Magnuson et al. 1990). Largely because of shrimp trawling in the southeast United States and Gulf of Mexico, 2,700 juvenile and adult Kemp's ridleys were estimated to die annually from interactions with the fisheries even after TED enlargement requirements were implemented in 2003. However, these bycatch estimates are highly uncertain due to the lack of observer coverage (Finkbeiner et al. 2011). Nevertheless, because up to 5,000 Kemp's ridleys were estimated to be killed each year prior to the requirement of TEDs in the shrimp trawl fishery (Magnuson et al. 1990), the estimate by Finkbeiner et al. (2011) may represent a significant reduction to Kemp's ridley annual mortality from shrimp trawling. The decline in shrimp fishing effort in the Gulf of Mexico since the early 1990s, as well as the spatial and temporal closures of Gulf of Mexico shrimp fisheries, has reduced bycatch mortality from shrimp trawling and has contributed to the Kemp's ridley population increase (Lewison et al. 2003; Shaver 2005; Caillouet et al. 2008; Crowder and Heppell 2011).

In the Gulf of Mexico, Kemp's ridleys rarely interact with, or are incidentally captured in low numbers by fisheries, other than shrimp trawling. From 1994 through 2006, 11 Kemp's ridleys were incidentally captured during 4,096 trips of the Mexican pelagic longline tuna fishery (Ramirez and Ania 2000; J. Molina, Instituto Nacional de Pesca, personal communication, 2007, cited in NMFS et al. 2011). No Kemp's ridleys were observed as bycatch for the U.S. shark bottom longline fishery in the Gulf from 1994 through 2002 or from July 2005 through 2010 (Hale and Carlson 2007; NMFS et al. 2011). In the Gulf of Mexico bottom longline fishery for shark, grouper, snapper, tilefish, and reef fish, no Kemp's ridleys were observed incidentally captured from 2005 through 2010 (Hale and Carlson 2007; Hale et al. 2009; NMFS et al. 2011). In the gill net fisheries operating off the U.S. Gulf Coast, no Kemp's ridleys were observed taken from 2000 through 2008, and only one Kemp's ridley, which was released alive and uninjured, was incidentally captured in 2009 (Garrison 2007; Baremore et al. 2007; Passerotti and Carlson 2009; Passerotti et al. 2010).

Kemp's ridleys are caught in both commercial and recreational hook-and-line fisheries along the U.S. Gulf Coast (Cannon et al. 1994; Seney 2008; NMFS et al. 2011). From 1980 through 1992, 112 Kemp's ridleys interacted with recreational hook-and-line gear at piers along the Texas coast, with 39 turtles documented in 1992 alone; 62 of these live captures were between the Bolivar Peninsular in Galveston County and the Texas-Louisiana border, and 63 % of the Kemp's ridleys were turtles that had been head-started (Cannon et al. 1994). These turtles, which were primarily juveniles ranging in size from 25 to 45 cm (9.8 to 17.7 in) SCL, represented a cost-effective means of gathering important Kemp's ridley data, while providing an opportunity for rehabilitation, if necessary (Seney 2008). A total of 170 Kemp's ridley hook-and-line encounters, which included 154 live captures, was recorded along the upper Texas coast from 1980 through 1995, with almost 90 % (135 turtles) occurring from 1992 through 1995 (C.W. Caillouet, NOAA NMFS, unpublished memo, 1996, cited in Seney 2008). The increased reports of hook-and-line captures during the mid-1990s may have been due to public education efforts targeting anglers, better survival of juvenile Kemp's ridleys due to shrimping regulations, and/or the initial stages of recovery exhibited by the overall population (Cannon et al. 1994; Cannon 1995; Lewison et al. 2003).

From 2003 through 2007, 42 Kemp's ridleys, with an average size of 34.6 cm (13.6 in) SCL, were captured on hook and line along the upper Texas coast in Galveston (45 turtles) and Jefferson (two turtles) counties (Seney 2008). Most of the captures (74 %) were reported from a single pier on Galveston Island. No hatchling or oceanic juveniles (less than 25 cm [9.8 in] SCL) or subadults and adults (greater than 45 cm [17.7 in] SCL) were captured on hook and line, which was similar to what occurred for Kemp's ridleys caught on hook and line along the Florida Panhandle from 1991 through 2003 (Rudloe and Rudloe 2005). However, in contrast to the ridleys caught on hook and line at piers in the Florida Panhandle, the Kemp's ridleys caught by hook and line along the upper Texas coast demonstrated low site fidelity. The hook-and-line captures were retrieved between March and October, with 81 % occurring from April through June. Forty of the hook-and-line captures were successfully rehabilitated and released (Seney 2008).

11.7.1.2 Incidental Capture of Loggerhead Sea Turtles in Commercial and Recreational Fisheries

The incidental capture of loggerheads in fisheries was concluded to be the most important threat to the northwest Atlantic Ocean loggerhead population in the most recent recovery plan (Bolten et al. 2011) and was determined to be a significant threat to loggerheads in the northwest Atlantic in the most recent status review completed by the NMFS and USFWS (Conant et al. 2009). In observed U.S. fisheries, loggerheads that are taken range from moderately sized juveniles through adults. Loggerhead hatchlings and small size classes are rarely seen as bycatch (TEWG 2009).

An estimated 63,500 loggerheads died annually from fishery interactions before bycatch mitigation strategies were mandated; an estimated 1,400 loggerheads died each year after TED opening enlargements were implemented in 2003 (Finkbeiner et al. 2011). The southeast U.S./ Gulf of Mexico shrimp trawl fishery (responsible for an estimated 23,300 annual interactions) was responsible for the most loggerhead interactions on an annual basis. In addition, due to their large nesting assemblages in the southeast United States, including the Gulf of Mexico, and their annual migrations to higher latitudes (Plotkin and Spotila 2002), loggerheads interact with more fisheries than any other sea turtle species in the United States (17 of 18 analyzed fisheries). New loggerhead bycatch estimates available for the southeast U.S./Gulf of Mexico shrimp trawl fishery, updated using 2009 data, suggest that 28,200 interactions occur each year in the Gulf, resulting in 785 loggerhead deaths (Ponwith 2011).

The estimated number of loggerhead sea turtles caught by the U.S. pelagic longline fishery in the Gulf of Mexico from 1993 and 2009 is summarized in Figure 11.46. The highest loggerhead bycatch was estimated to occur in 2002; the second highest occurred in 2003. Loggerhead bycatch estimates resulting from the Gulf of Mexico pelagic longline fishery have declined in recent years, and levels from 2005 through 2009 appear to be similar to those estimated for the 1990s (Figure 11.46).

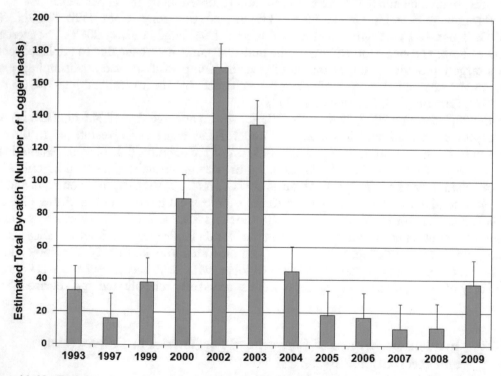

Figure 11.46. Estimated number of loggerhead sea turtle incidental captures by the U.S. pelagic longline fishery in the Gulf of Mexico from 1993 through 2009; no estimates available for 1994, 1995, 1996, 1998, or 2001; error bars = standard error (Johnson et al. 1999b; Yeung 1999, 2001; Garrison 2003, 2005; Garrison and Richards 2004; Fairfield-Walsh and Garrison 2006, 2007; Fairfield and Garrison 2008; Garrison et al. 2009; Garrison and Stokes 2010).

11.7.1.3 Incidental Capture of Green Sea Turtles in Commercial and Recreational Fisheries

Green turtles are killed as bycatch in coastal fisheries, including drift nets, longlines, set nets, pound nets, and trawl fisheries (Magnuson et al. 1990). Prior to the implementation of TED requirements, an estimated 57 % of green turtle mortalities occurred in the Gulf of Mexico as a result of shrimping activity; most of the turtles killed were juveniles (less than 60 cm [23.6 in] SCL), with the majority occurring in the central, northern Gulf (Henwood and Stuntz 1987; Thompson 1988). By the early 1980s, an estimated 229 green turtles drowned in shrimp nets annually, and most captures were in neritic Gulf of Mexico waters with a depth of less than 25 m (82 ft) (Henwood and Stuntz 1987). After TED opening enlargements were mandated in 2003, 300 green turtles were estimated to die each year (Finkbeiner et al. 2011); this study demonstrated that green turtles interacted primarily with the southeast U.S./Gulf of Mexico shrimp trawl fishery (11,300 bycatch events).

11.7.1.4 Incidental Capture of Leatherback Sea Turtles in Commercial and Recreational Fisheries

Since the mid-1900s, numerous instances of leatherbacks becoming hooked or tangled on longlines, buoy anchor lines, and other ropes and cables, leading to injury and/or death have been documented (NMFS 1992). Of 30 sea turtles caught in the Gulf of Mexico by the Japanese tuna longline fishery from 1978 to 1981, 12 (40 %) were leatherback sea turtles (Witzell 1984). Between 1992 and 1995, the U.S. pelagic longline fleet caught 73 leatherbacks in the Gulf of Mexico (Witzell 1999). Of the 621 turtles taken in the U.S. pelagic longline fishery in the Gulf of Mexico from 1992 through 2005, more than 85 % of them were leatherbacks (Kot et al. 2010).

A summary of the estimated number of leatherbacks captured by the U.S. pelagic longline fishery in the Gulf from 1993 and 2009 is presented in Figure 11.47. Although a careful statistical analysis is not available, the highest numbers of leatherback bycatch appeared to have occurred from 2002 through 2004. Leatherback bycatch resulting from the Gulf of Mexico pelagic longline fishery appears to have declined in recent years.

The first edition of the *U.S. National Bycatch Report* summarized sea turtle bycatch from 2001 through 2006 for the Atlantic and Gulf of Mexico fisheries (NMFS 2011c). The report estimated the following for bycatch of leatherback sea turtles: 63 killed annually in the Gulf due to capture in shrimp trawls; 83 caught annually by Atlantic and Gulf of Mexico shark bottom longline fisheries; and 351 caught annually by Atlantic and Gulf of Mexico pelagic longline fisheries.

While high uncertainty is associated with these estimates, approximately 2,300 leatherbacks were estimated to have died annually from fisheries interactions before bycatch mitigation strategies were mandated, and an estimated 40 leatherbacks died each year after TED opening enlargements were mandated in 2003 (Finkbeiner et al. 2011). For leatherbacks, the Atlantic/Gulf of Mexico pelagic longline fishing was estimated to be responsible for the most interactions, followed by the southeast U.S./Gulf of Mexico shrimp trawl fishery (Finkbeiner et al. 2011). For 2009, the NMFS estimated 623 interactions between leatherbacks and shrimp trawls in the Gulf, 18 of which were estimated to result in mortality (Ponwith 2011).

11.7.1.5 Incidental Capture of Hawksbill Sea Turtles in Commercial and Recreational Fisheries

Hawksbill sea turtles are caught much less frequently as bycatch than the other four species of sea turtles in the Gulf of Mexico. Nevertheless, hawksbills are susceptible, particularly in

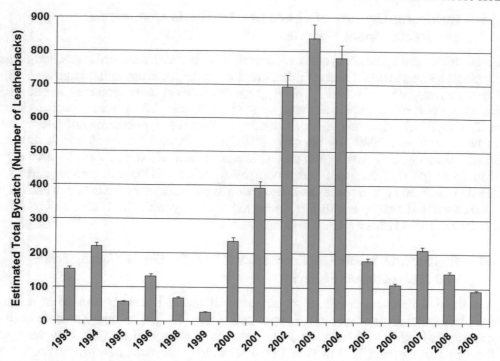

Figure 11.47. Estimated number of leatherback sea turtle incidental captures by the U.S. pelagic longline fishery in the Gulf of Mexico from 1993 through 2009; no estimates available for 1997; error bars = standard error (Johnson et al. 1999b; Yeung 1999, 2001; Garrison 2003, 2005; Garrison and Richards 2004; Fairfield-Walsh and Garrison 2006, 2007; Fairfield and Garrison 2008; Garrison et al. 2009; Garrison and Stokes 2010).

nearshore fisheries, such as drift netting, longlining, set netting, pound netting, gill netting, and trawl fisheries (Magnuson et al. 1990; NMFS and USFWS 2007e). For example, hawksbill turtle bycatch was not quantified in the 2011 national bycatch report for the southeast United States, but they may have been included in the 0.4 % of unidentified turtles (NMFS 2011c). In an effort to evaluate the effectiveness of the use of TEDs, NMFS observers documented only one hawksbill of the 13 turtles caught by U.S. shrimp fisherman in the Gulf of Mexico from March 1988 through July 1989 (Renaud et al. 1990); the same program recorded no hawksbills of two documented sea turtles caught in the Gulf of Mexico by the same fishery from September 1989 through August 1990 (Renaud et al. 1991). The sea turtle bycatch report from the 1998 pelagic longline fishery recorded no hawksbill captures in the Gulf (Yeung 1999). The limited data on hawksbill bycatch suggests that the risk of being killed or injured as bycatch in commercial and recreational fisheries in the Gulf of Mexico is low. However, given the low population abundance of hawksbills in the Gulf of Mexico, efforts must continue to ensure low bycatch impact on this species.

11.7.2   Terrestrial Zone Threats

Although uncommon, the poaching of eggs from nesting female Kemp's ridleys still occurs in Mexico and has occurred in south Texas (NMFS et al. 2011). In Florida, egg poaching does occur, and from 1980 through 2002, more than 60 arrests were made for the possession or sale of sea turtle eggs (NMFS and USFWS 2008).

Along the Gulf of Mexico coast, beaches are cleaned by mechanical raking, scraping with large machinery, hand raking, and picking up debris by hand. These activities can directly and indirectly affect sea turtles (NMFS et al. 2011). Driving is permitted on many Gulf of Mexico beaches. Nesting sea turtles have been run over and killed by vehicles, and vehicles have crushed emerging hatchlings (NMFS et al. 2011). Because the Kemp's ridley has only one primary nesting beach in Rancho Nuevo, Mexico and a secondary nesting colony at PAIS in Texas, it is particularly susceptible to beach disturbance, alteration, and destruction by natural and anthropogenic events. Beach cleaning has been documented to affect Kemp's ridleys. For example, 12 Kemp's ridley hatchlings became trapped by a sand ridge created by heavy equipment cleaning the beach on North Padre Island, Texas in 2002 and were later crushed and killed by passing vehicles (Shaver 2004).

The presence of artificial lighting on or near the beach adversely affects both nesting and hatchling sea turtles (Witherington and Martin 1996). Mortality from misdirection by artificial lighting on both the Gulf of Mexico and Atlantic coasts of Florida kills thousands of loggerhead hatchlings each year (Witherington 1997). The proportion of all emerging loggerhead hatchlings that died because of beach lighting was estimated in the early 1990s to be as high as 5–10 % (Witherington et al. 1996). Loggerheads abort nesting attempts at a greater frequency in lighted areas. Artificial lighting also deters females from emerging from the ocean to nest (Witherington 1986, 1992). Hatchling orientation of nests located at 23 representative beaches in six Florida counties was surveyed in 1993 and 1994, and approximately 10–30 % of nests showed evidence of hatchling disorientation by lighting (Witherington et al. 1996). Similar to other nocturnal nesting sea turtles, nesting leatherbacks and hatchlings can be disoriented by artificial lighting on the beach (NMFS 2011b).

A significant negative relationship was found between sea turtle nesting density and distance from the nearest of 17 ocean inlets (Witherington et al. 2005). Beach instability from both erosion and sand accretion may discourage sea turtle nesting, since the effect of inlets in lowering nesting density was found both updrift and downdrift of the inlets. When sea turtles emerged to nest in the presence of armoring structures, more returned to the water without nesting compared to turtles that emerged on nonarmored beaches (Mosier 1998; Mosier and Witherington 2002). Fewer sea turtles made nesting attempts on beaches fronted by seawalls than on adjacent beaches where armoring structures were absent (Mosier 1998). In addition, sea turtles on armored sections of beach had a tendency to wander great distances as compared to turtles that emerged to nest on adjacent natural beaches (Mosier 1998).

Since oil exploration and production occurs south of Tamaulipas and Veracruz in Mexico and at PAIS in Texas, the Kemp's ridley nesting beaches, as well as the nesting turtles, eggs, and hatchlings, could be impacted by oil spills and related activities (NMFS et al. 2011). For example, in 1979, the Ixtoc I oil well blew out and caused a fire in the Bay of Campeche in Mexico. The nesting beach at Rancho Nuevo was affected by the Ixtoc I well blowout, and large amounts of oil were released daily into the Gulf of Mexico for several months. However, the oil reached the beach after the nesting season, and nesting females were not present. Also, a loaded super-tanker, the *Mega Borg*, exploded near Galveston, Texas in 1990, causing more than 121,000 barrels of crude oil to be released into the Gulf of Mexico; sea turtles covered in oil were found after this spill (Yender and Mearns 2003).

Oil spills have affected loggerhead nesting beaches in the Gulf of Mexico. In August 1993, approximately 350,000 gal of fuel oil spilled into Tampa Bay and washed onto nesting beaches in Pinellas County, Florida (Conant et al. 2009). Impacts to loggerheads resulting from the spill included 31 dead hatchlings, 176 oil-covered nests, and 2,177 eggs and hatchlings exposed to oil or disturbed by response activities (FDEP et al. 1997).

Tropical coastlines are rapidly being developed, often leading to the destruction of hawksbill nesting habitat (Mortimer and Donnelly 2008). For example, critical hawksbill habitats are quickly being impacted by the development along the Gulf coast of the Yucatán Peninsula (Garduño-Andrade et al. 1999). Beachfront development and the clearing of dune vegetation significantly affect hawksbill sea turtles because they prefer to nest under vegetation (Mortimer and Donnelly 2008).

Global climate change may affect sea turtle nesting beaches in several ways, including sea level rise, higher ambient temperatures, and changes in hurricane/cyclone activity (Hawkes et al. 2009; Witt et al. 2010). Higher water levels associated with sea level rise will gradually and directly decrease the availability of suitable nesting sites (Witt et al. 2010). Increasing temperatures may open up areas that were previously unavailable for nesting (Witt et al. 2010), but recent studies have suggested that up to half of the currently available sea turtle nesting areas could be lost with predicted sea level rise (Fish et al. 2008). Sea turtle nesting is significantly affected by temperature, and incubation temperatures can affect incubation success, duration, and the sex of hatchlings (Mrosovsky and Yntema 1980; Witt et al. 2010; Valverde et al. 2010). Therefore, increasing temperatures have the potential to change current nest incubation regimes, as well as to skew sex ratios (Hawkes et al. 2007a). Changes in the global climate are predicted to increase the frequency and intensity of hurricanes (Webster et al. 2005), which can significantly affect the reproductive success of sea turtles.

On Gulf of Mexico beaches, predation of sea turtle eggs and hatchlings can be significant. Known predators of sea turtle eggs and hatchlings include raccoons, ghost crabs, coyotes, foxes, armadillos, domestic dogs and cats, feral pigs, skunks, bobcats, badgers, gulls, fish crows, and larval insects (Witherington et al. 2006a; NMFS et al. 2011). Invasive fire ants are also significant predators of sea turtle eggs and hatchlings on Gulf coast nesting beaches (Witherington et al. 2006a; NMFS et al. 2011). Invasive plant species are known to invade and desiccate eggs, trap hatchlings, interfere with nest construction, and lower nest incubation temperatures because of shading (Conant et al. 2009). For example, the invasive Australian pine has caused shading of beaches, lowered sea turtle nest incubation temperatures, limited accessibility to suitable nest sites, entrapped nesting turtles, interfered with nest construction, and caused sea turtle nesting activity to decline on a remote nesting beach in Everglades National Park along the Florida Gulf coast (Davis and Whiting 1977; Schmeltz and Mezich 1988; Reardon and Mansfield 1997; Hanson et al. 1998).

The hurricane season for the Gulf overlaps closely with the sea turtle nesting season (Magnuson et al. 1990). While sea turtles have evolved to deal with erosion, flooding, storm surges, and other disturbances caused by hurricanes and other severe storms by laying large numbers of eggs and distributing their nests spatially and temporally, hurricanes can affect the reproductive success of sea turtles, since they rely on specific beaches for reproduction (Carr and Carr 1972). In addition, the effects of hurricanes vary by species (Pike and Stiner 2007). For example, in the southeastern United States, leatherback turtles nest the earliest, and most hatchlings emerge before the hurricane season starts, while loggerhead turtles nest intermediately, and only nests laid late in the season would be at risk. However, green turtles nest the latest, and their entire nesting season occurs during the hurricane season; therefore, their developing eggs and nests are extremely vulnerable to hurricanes (Pike and Stiner 2007).

In 1989, the effects of Hurricane Gilbert were documented on the Kemp's ridley nesting beach in Mexico. When debris was deposited, the beach was eroded, and coral rock was exposed along the central portion of Rancho Nuevo. About 20 % of the Kemp's ridley nesting activity was displaced to the north that year (Márquez-M. 1990).

Hurricane Andrew, a Category 4 hurricane that hit south Florida on August 24, 1992, provided an opportunity to quantify the impacts of a major hurricane on six beaches where loggerheads of the Peninsular Florida subpopulation nest (Milton et al. 1994). Sea turtle nests on more than 145 km (90 mi) of beaches on the Gulf and Atlantic coasts of Florida were affected by the hurricane. The associated storm surge produced the greatest mortality through nest flooding. Loggerhead egg mortality was 100 % on beaches closest to the eye; mortality decreased with distance from the eye (Milton et al. 1994). Hurricane Andrew affected about 68 km (42 mi) of beach on the Florida Gulf coast. Within this zone, there was about a 40–50 % mortality of loggerhead eggs and hatchlings, and about 22 % of the 2,762 loggerhead nests were partially or completely destroyed (Milton et al. 1994).

Although hurricanes periodically remove the sand from the typically small hawksbill nesting beaches, the sand is usually replaced by wind and wave action. However, hurricanes may cause trees to fall and debris to be deposited on beaches; this debris hinders or prevents hawksbills from reaching their nesting habitat.

11.7.3 Neritic and Oceanic Zone Threats

The illegal poaching of juvenile and adult sea turtles in the marine environment is uncommon in the Gulf of Mexico (NMFS et al. 2011). However, sea turtles that use the Gulf could be harvested legally in nearby countries (e.g., Turks and Caicos Islands) (NMFS and USFWS 2008).

In the past, Gulf of Mexico loggerhead sea turtles could have been taken in Cuba, since an active harvest continued through the mid-1990s (Moncada Gavilan 2000). The estimated harvest of loggerheads for meat in Cuba was as follows: (1) from 1968 through 1975, at least 4,300 turtles were harvested each year; (2) from 1976 through 1987, at least 2,600 turtles were harvested each year; and (3) from 1988 through 1994, an initial level of at least 1,750 turtles were harvested each year, declining to at least 660 turtles in later years (TEWG 2009). Interestingly, there was a concurrent increase in the number of loggerhead nests laid on Florida Index Beaches, including beaches on the Florida Gulf coast, as fewer loggerheads were harvested in Cuban waters from the late 1980s through the late 1990s. In addition, the annual number of nests in Florida was still increasing when the loggerhead fishery ended in Cuba in 1996, with little sign of the decline that has characterized loggerhead nesting in Florida in recent years (TEWG 2009).

While it has declined dramatically over the last 20 years, the most significant threat to hawksbill populations is the continued illegal trade of hawksbill products. While the legal hawksbill tortoiseshell trade ended in 1993 when Japan, historically the major importer of bekko from the Caribbean, agreed to cease the imports, a significant illegal trade continues (USFWS 2012). Because of the migratory nature of hawksbill sea turtles, this trade threatens hawksbills that occur in the Gulf of Mexico.

In addition, hawksbills may still be captured illegally in the general Gulf of Mexico and Caribbean region. While hawksbills were harvested in Cuba from at least the 1500s and thousands of nesting females were captured annually through the nineteenth and twentieth centuries, a seasonal closure was introduced in 1936, and the Cuban government prohibited egg collection and the disturbance of nesting females beginning in 1961 (Carrillo et al. 1999; McClenachan et al. 2006; Moncada et al. 2012). In Cuba, the annual legal foraging ground exploitation of 5,000 hawksbills was reduced to 3,000 in 1993, 1,000 in 1994, and 500 turtles in 1995 (Carrillo et al. 1999). Cuba closed their hawksbill harvest in 2008 (Moncada et al. 2012); however, illegal subsistence fishing may still be occurring (Hart et al. 2012b).

The Gulf of Mexico is an area of high-density offshore oil and gas exploration, development, and production, and related activities affect all life stages of sea turtles. In addition to oil and natural gas being released into the Gulf from natural seeps and low-level spills, large events do occur on occasion (NMFS et al. 2011). The impacts of seismic surveys associated with oil exploration activities on sea turtles have not been fully studied (Cuevas et al. 2008). The few available studies suggest responses of sea turtles include an alarm reaction, subsequent avoidance, and sometimes temporary or permanent hearing loss (MMS 2004).

Explosives are used to remove oil platforms in the Gulf of Mexico when they are no longer in operation. These explosions have affected sea turtles in the past (Klima et al. 1988). However, an intensive observer program has been in place since 1987 to minimize sea turtle impacts from these explosions (Gitschlag 1992). The program has been successful in mitigating impacts to sea turtles associated with the explosive removal of offshore structures (Viada et al. 2008).

Leatherbacks have infrequently been observed near offshore developments associated with oil and gas operations. For example, in a survey of turtles around energy structures off the coasts of Texas and Louisiana in 1992, only two leatherbacks of 47 individual turtles were observed (Gitschlag and Herczeg 1994). During a 2-year survey, 15 leatherbacks were sighted within 8,000 m (2,625 ft) of petroleum platforms (Lohoefener et al. 1990).

Eight oiled, dead sea turtles, including one juvenile Kemp's ridley, washed up on Texas beaches after the Ixtoc I spill in the Gulf in 1979 (Rabalais and Rabalais 1980). Chronic oil exposure may have led to their poor body condition and, ultimately, their death (Hall et al. 1983). Juvenile Kemp's ridleys that were released into *Sargassum* patches offshore Padre and Mustang Islands, Texas in 1983 were found stranded on Padre and Mustang Island beaches with oily residues in their mouth, esophagus, and stomach; the residues were later determined to most likely be a result of tanker cleaning operations (Overton et al. 1983).

Van Vleet and Pauly (1987) determined that crude oil tanker discharge was significantly affecting sea turtle populations in the eastern Gulf of Mexico based on an analysis of oil residues from Kemp's ridley, loggerhead, green, and hawksbill sea turtles stranded on U.S. Gulf of Mexico beaches. There is evidence that oil pollution has a greater impact on hawksbills than on other species of sea turtles (Meylan and Redlow 2006). For example, approximately 12 % of stranded hawksbills on Florida beaches had evidence of fouling by oil, compared to 1.1 % of the strandings of other species. In addition, 22.4 % of stranded hawksbills smaller than 22 cm (8.7 in) SCL had evidence of oil (Meylan and Redlow 2006). However, from 1980 through 2002, no oil-affected turtles were found stranded on Florida Gulf coast beaches (Meylan and Redlow 2006). Ingested tar or tar on their bodies has been documented for Kemp's ridleys found along the Texas coast (Shaver 1991).

In addition to oil, other contaminants could affect sea turtles. For example, polychlorinated biphenyls (PCBs), as well as dichlorodiphenyltrichloroethane (DDT) and its breakdown products, dichlorodiphenyldichloroethylene and dichlorodiphenyldichloroethane (DDE and DDD, respectively), have been found in tissues of stranded Kemp's ridley and loggerhead sea turtles (Rybitski et al. 1995). Polycyclic aromatic hydrocarbons (PAHs), DDD, PCBs, and metals were found in loggerhead eggs that failed to hatch that were collected from 20 nests on Florida Gulf coast beaches in 1992. The failure of the eggs to hatch may have been related to the additive or synergistic toxicity of the low levels of the contaminants found in the eggs (Alam and Brim 2000). Exposure to organochlorines, such as PCBs and pesticides, has been suggested to modulate immunity in loggerhead sea turtles (Keller et al. 2006).

A large number of persistent organic pollutants (POPs) were measured in egg yolks of unhatched loggerhead eggs collected from 11 nests along the Gulf coast in Sarasota in 2002 (Alava et al. 2011). Levels of POPs were lowest in eggs collected from the Gulf coast compared to eggs collected from loggerhead nests along the Atlantic coast. Foraging ground locations used by the nesting females may have caused the differences. Data from a satellite tracking study conducted by Girard et al. (2009) indicated that females nesting on Sarasota beaches forage in the Gulf of Mexico and Caribbean Sea, where the prey is less contaminated than from sites used by females nesting along the Atlantic coast (Alava et al. 2011).

Although the effects of exposure to POPs in sea turtles are unknown, POPs have been measured in tissue, blood, and eggs from leatherbacks on the U.S. east coast (Stewart et al. 2011). In addition, trace metals have been found in blood and eggs from females in French Guiana (Guirlet et al. 2008). Since nesting females from French Guiana are known to spend time in the Gulf of Mexico (Pritchard 1976), it is possible that contaminant exposure occurs there.

Sea turtles in the Gulf are differentially affected by the ingestion of and entanglement in marine debris (Bjorndal et al. 1994; Witzell and Schmid 2005). However, all five species of sea turtles that occur in the Gulf of Mexico are significantly affected by the ingestion of and, to a lesser extent, by entanglement in marine debris, typically plastic, in the northwestern Gulf (Plotkin and Amos 1990). Compared to other species of sea turtles, the ingestion of marine debris by Kemp's ridleys is thought to be minimal because they eat more active prey, and their foraging areas are in locations where wind and currents do not concentrate marine debris (Bjorndal et al. 1994; Witzell and Schmid 2005). However, Kemp's ridleys have been documented to ingest plastic, rubber, fishing line and hooks, tar, string, Styrofoam, and aluminum (Shaver 1991; Werner 1994; Witherington et al. 2012). They also have been killed as a result of entanglement in plastic, fishing line, discarded netting, and other debris (Plotkin and Amos 1988).

Because post-hatchling and small oceanic juvenile loggerheads occupy convergences, rips, and driftlines in the open ocean, the likelihood of becoming caught in and consuming marine debris is significant; marine debris accumulates in these same areas (Carr 1986; Witherington 2002; Witherington et al. 2012). In addition to leatherbacks, loggerhead sea turtles appear to ingest more debris in all of its life stages because of habitat choice and feeding behavior (Lutcavage et al. 1997); the ingestion of debris, including plastic, Styrofoam, balloons, and tar balls occurs when debris is mistaken for or associated with prey items (Conant et al. 2009).

Since at least 1970, there have been numerous reports of leatherbacks ingesting plastic debris (Mrosovsky 1981; Mrosovsky et al. 2009). Researchers have suggested that leatherbacks are unable to distinguish between floating plastic debris and gelatinous prey; therefore, they purposely ingest plastic as though it were a prey item (Balazs 1985). In a recent analysis of autopsy records of more than 400 leatherbacks spanning 23 years, plastic was reported in 34 % of these cases, and blockage of the gastrointestinal tract by plastic was documented in multiple cases (Mrosovsky et al. 2009).

The increasing frequency of red tides and other harmful algae blooms in the Gulf of Mexico and the increasing number, geographic extent, and duration of anoxic and hypoxic dead zones caused by agricultural runoff in Mississippi River outflow to the Gulf directly and indirectly affect sea turtles (NMFS et al. 2011). Since 1995, red tide blooms, caused by the marine alga, *Karenia brevis*, have been detected every year in the Gulf of Mexico near southwest Florida. This algal species produces brevetoxin, which kills sea turtles (Pierce and Henry 2008; FFWCC unpublished data, cited in TEWG 2009). Since the early 1990s, at least 100 Kemp's ridleys found dead on Florida Gulf coast beaches have been associated with red tide events (NMFS et al. 2011).

The absence of dissolved oxygen, which is essential to most animals and plants that inhabit the Gulf of Mexico, in the dead zones kills benthic invertebrates, including crabs, the main prey item for Kemp's ridleys. Due to the reduced abundance of food, Kemp's ridley sea turtles are not likely to forage in or inhabit hypoxic areas for any length of time (McDaniel et al. 2000). Aerial surveys have indicated an absence of sea turtles in these zones (Craig et al. 2001).

The frequency of fibropapilloma tumors, which are often linked to debilitation and death, is much higher in green turtles than for any of the other sea turtle species (Witherington et al. 2006b; NMFS and USFWS 2007c). In an analysis of STSSN data for 4,328 green turtles found dead or debilitated from Massachusetts to Texas from 1980 through 1998, fibropapillomatosis was reported in green turtles only in the southern half of Florida, and 22 % of turtles in this region had tumors. The disease was more prevalent in turtles found on the Florida Gulf coast (52 %) as compared to turtles found along the Florida Atlantic coast (12 %) (Foley et al. 2005). The disease was more common in coastal waters characterized by habitat degradation and pollution, large shallow water areas, and low wave energy. In addition, 22 % of the 6,027 green turtles stranded on both the Gulf and Atlantic coasts of Florida from 1980 through 2005 had external fibropapilloma tumors (FFWCC FWRI 2011b).

In the Gulf of Mexico, sea turtles have been documented to entrain in and impinge on cooling water intake structures associated with power plant operations, and while the majority of turtles are released unharmed, some turtles are injured and killed (NMFS 2002). For example, in 1998, approximately 40 Kemp's ridleys were trapped in intake structures associated with the Crystal River Energy Complex located near the foraging grounds in the Cedar Keys, Florida area (NMFS 2002), and 92 Kemp's ridleys were entrapped from 1999 through 2004 (Eaton et al. 2008). When a significant increase in the number of Kemp's ridleys stranded on the intake bar racks of the power plant was observed in 1998, sea turtle protection measures were implemented (Eaton et al. 2008). In addition, 38 juvenile green turtles, ranging in size from 24 to 50 cm (9.5 to 19.7 in) SCL, and some loggerhead turtles were trapped in the intake structures at the energy complex during the same time period (Eaton et al. 2008).

Crabs are the main prey of Kemp's ridleys, and blue crab populations in the Gulf of Mexico have declined significantly since the late 1960s/early 1970s (TPWD 2006; Murphy et al. 2007). Crab populations off the Texas and Florida Gulf coasts reached lows in 2000 and have not rebounded. Many causes for the decline have been suggested, including drought and the subsequent reduction of freshwater flow into bays and estuaries, overharvesting, and coastal wetland loss (TPWD 2006). Recovering the blue crab stocks is challenging because of significant gaps in knowledge not only about the commercial and recreational blue crab fishery, but also about blue crab life history (Murphy et al. 2007). This decline could have significant impacts on the recovering Kemp's ridley population, especially as the ridley population continues to increase.

In the marine environment, the effects of climate change on sea turtles are difficult to study because sea turtles range across entire ocean basins, are late maturing, and are long lived (Zug et al. 2002). Temperature is known to influence the distribution and behavior of sea turtles (Hawkes et al. 2007b). Ocean currents, which are important for dispersing hatchling sea turtles, could change in magnitude or direction (Bolten 2003); these changes could influence the duration of future juvenile development (Hamann et al. 2003). Sea turtle trophic dynamics and juvenile growth and development could be altered as a result of changes to the pelagic community resulting from climate change (Bjorndal 1997; Bjorndal et al. 2000; Witt et al. 2010). Adult foraging habitat and the location and sizes of home ranges and diet could be altered as a result of changes to thermal regimes and sea surface currents resulting from climate change (Bjorndal 1997; Polovina et al. 2004; Witt et al. 2010).

Of all coastal marine habitats, seagrass beds—the primary foraging areas for green turtles—are the most susceptible and have low resiliency to disturbance. Seagrass beds occur along sheltered coasts with good water quality, are often located in areas of port development, and are downstream from point and nonpoint source discharges (Waycott et al. 2005). The hawksbill's dependence on coral reefs for shelter and food links its well-being to the conditions of coral reefs, and coral reefs are one of the most endangered ecosystems in the world (NMFS and USFWS 1993, 2007e). Because they prefer hard bottom habitats, hawksbills are thought to frequent underwater structures and hard platforms, such as those provided by offshore continental shelf development in the Gulf of Mexico. In addition, habitat loss in the Caribbean could cause more hawksbills to move into the Gulf of Mexico (Hopkins and Richardson 1984; Thompson et al. 1990).

11.7.3.1   Hypothermic (Cold) Stunning of Sea Turtles

Deep freeze events, in which water temperatures drop below 10 °C, are uncommon and brief in the northern Gulf of Mexico and typically last for only one to a few days; however, when these events occur, they can have a significant impact on coastal sea turtles, especially the juveniles (Henry et al. 1994). Most hypothermic or cold-stunning events, in which sea turtles become incapacitated and lose their ability to swim and dive as a result of rapidly dropping water temperatures, occur in shallow coastal lagoons and bays (Ogren and McVea 1982; Turnbull et al. 2000). The decreases in water temperature occur so rapidly that the turtles have little time to move to warmer waters (McMichael et al. 2008). Low water temperatures become lethal between 5 and 6.5 °C (Schwartz 1978; Moon et al. 1997). Hypothermic-stunned sea turtles may exhibit metabolic and respiratory acidosis (Innis et al. 2007). If hypothermic-stunned turtles are left untreated, they may perish in the water or when stranded on the beach, since air temperatures are even lower than water temperatures.

Adult but mainly juvenile sea turtles that overwinter in inshore waters of the Gulf of Mexico, such as along PAIS in Texas and St. Joseph Bay in Florida (Shaver 1990b; Foley et al. 2007), are most susceptible to hypothermic stunning because the temperature change is most rapid in shallow waters (Witherington and Ehrhart 1989b). A recent study suggested that it is not the rate of water temperature decrease but the duration and magnitude of the drop that drives the severity of the phenomenon (Roberts et al. 2014). Sea turtles often die from hypothermia, unless rescued, rehabilitated, and later released.

Kemp's ridleys, loggerheads, green turtles, and hawksbills have been affected by hypothermic stunning in waters along the northern Gulf of Mexico in Texas and Florida (Hildebrand 1982; Shaver 1990b; Foley et al. 2007; Roberts et al. 2014). However, historically, green turtles, especially juveniles, have been most affected by hypothermic stunning (Figure 11.48). Leatherbacks are not susceptible to hypothermic-stunning events due to their unique ability to maintain a higher body temperature than the surrounding water temperature (Frair et al. 1972) and to the fact that they spend most of their lives in deep waters. Leatherbacks have never been reported stranded due to hypothermic stunning in the Gulf of Mexico.

Historically, severe freezes in the late 1800s are thought to have contributed to the decline in the green turtle fishery in Texas (Hildebrand 1982). More recently, juvenile green turtles, and to a lesser extent Kemp's ridleys, loggerheads, and hawksbills were hypothermic stunned and stranded in Texas inshore waters during severe winters in 1971, 1979, 1983, and 1989 (Hildebrand 1982; Shaver 1990b). During a severe cold front in February 1989, 46 sea turtles were found stranded as a result of hypothermic stunning in Laguna Madre near Port Mansfield, Texas (Shaver 1990b; Figure 11.48); the hypothermic-stunned turtles included 45 green turtles (31 were found dead) and one dead loggerhead. Hypothermic-stunning events occurred in Port Isabel and South Padre Island, Texas during the winters of 2007 and 2010, in which 150 and 200 green turtles, respectively, were hypothermic stunned (Figure 11.48).

During late December 2000 and early January 2001, an unprecedented hypothermic-stunning event occurred in St. Joseph Bay along the Florida Gulf coast (Foley et al. 2007). More than 400 turtles were hypothermic stunned, which included 388 green turtles (55 found dead) (Figure 11.48), ten Kemp's ridleys (four found dead), and three loggerhead sea turtles (one found dead) (Foley et al. 2007; McMichael et al. 2008). All of the hypothermic-stunned green turtles were neritic juveniles, and most were from the Florida and Yucatán populations

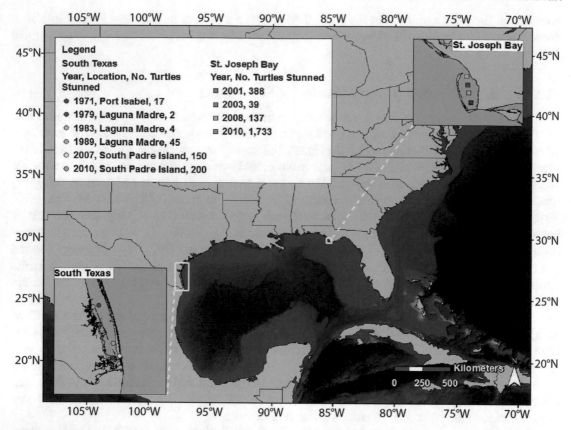

Figure 11.48. Hypothermic stunning locations of green sea turtles in the northern Gulf of Mexico from 1971 through Winter 2010 (interpreted from Shaver 1990b; H. Hildebrand, personal communication, cited in Shaver 1990b; McMichael et al. 2003, 2008; Foley et al. 2007; Texas A&M University 2011; Avens et al. 2012; Roberts et al. 2014).

(Foley et al. 2007). The 10 Kemp's ridleys ranged in size from 26.5 to 46 cm (10.4 to 18.1 in) SCL, with an average SCL of 33.4 cm (13.1 in), and one of the four Kemp's ridleys found dead was a head-started turtle from the 1998 year class. Most (337 of 401) of the sea turtles survived and were later released (Foley et al. 2007). The rehabilitated and released sea turtles included 329 green turtles, six Kemp's ridleys, and two loggerheads.

In 2003, a small hypothermic-stunning event occurred in St. Joseph Bay (McMichael et al. 2003, 2008). Forty-two turtles (39 green turtles, two Kemp's ridleys, and one loggerhead) were hypothermic stunned (Figure 11.48), and 30 of the sea turtles survived. In January 2008, another moderate hypothermic-stunning event occurred in St. Joseph Bay in which more than 100 sea turtles, mostly green turtles, were hypothermic stunned (Roberts et al. 2014).

During January 2010, Florida experienced below freezing temperatures for 12 consecutive days, resulting in a hypothermic-stunning event in St. Joseph Bay of unprecedented magnitude (Avens et al. 2012). A total of 1,733 sea turtles (mostly green sea turtles) were hypothermic stunned. While the majority of the 1,670 green turtles that were hypothermic stunned survived, 434 green turtles died from the hypothermic-stunning event (Avens et al. 2012). Air temperatures below 10 °C along with strong winds were responsible for the mass hypothermic stunning event, in which some of the turtles died when water temperatures remained between 5 and 6 °C for 3 days or so (Roberts et al. 2014; Figure 11.49).

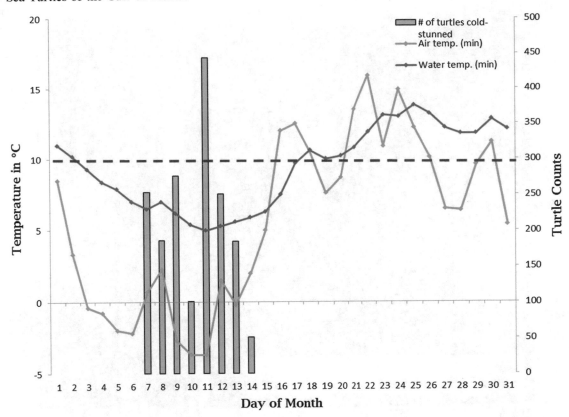

Figure 11.49. Air and water temperatures in relation to hypothermic stunned green sea turtles in Saint Joseph Bay, FL in January of 2010. The horizontal dashed red line indicates the hypothermic stunning temperature threshold for sea turtles (Roberts et al. 2014).

The frequency of occurrence, duration, and severity of hypothermic-stunning events in the Gulf of Mexico are unpredictable; however, as sea turtle populations continue to recover in the Gulf, the number of individuals impacted by hypothermic stunning may likely increase. However, because the majority of the hypothermic-stunned sea turtles are rescued, rehabilitated, and later released (as seen in hypothermic-stunning events in the northern Gulf of Mexico during the winters of 2011 and 2013/2014), the impacts of these events to the sea turtle populations are most likely minimal.

11.7.4   Sea Turtle Stranding Data

Occasionally, sea turtles wash up or strand (dead or alive) on the beaches of the Gulf of Mexico and of the U.S. Atlantic Ocean. The stranding of sea turtles is caused by many factors, including incidental capture in shrimping and fishing operations, entanglement, ingestion of marine debris, boat strikes, disease, storms, and hypothermic stunning. The STSSN was established in 1980 to collect information on and document strandings of live and dead sea turtles along the U.S. Gulf of Mexico and Atlantic coasts. The area includes the coasts from Maine through Texas, as well as portions of the U.S. Caribbean (STSSN 2012). As part of STSSN methodology, the United States coastline was divided into statistical stranding zones. Zones 1 through 21 include the Gulf coast from the Florida Keys through Texas (Figure 11.50). Important information regarding species composition, stock structure, life-history stage,

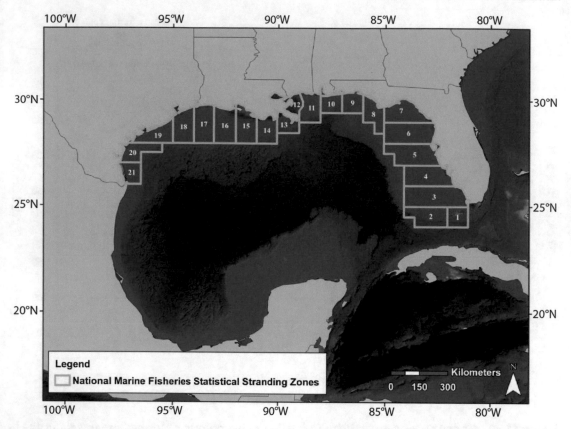

Figure 11.50. National Marine Fisheries Service statistical stranding zones for the U.S. Gulf of Mexico (from STSSN 2012).

distribution, migratory patterns, seasonality, habitat use, and causes of mortality may be inferred from the stranding data collected by the STSSN (NMFS et al. 2011). However, stranding data must be interpreted with caution, as the number of strandings recorded depends on many factors, including the surface currents, winds, time that has passed since the turtle died or was affected by the stressor that caused it to strand, and search effort (Epperly et al. 1996). In addition, while stranding data may represent a biased sample of the population, stranding distribution patterns can provide useful information on the distribution and abundance of juvenile and adult sea turtles, especially when large sample sizes and long time series are available (Meylan and Redlow 2006).

Available STSSN data prior to the Deepwater Horizon incident have been summarized for the Gulf of Mexico for each of the sea turtle species. The data available include preliminary data on the number of sea turtles stranded in each zone by species from 1986 through 2009. Strandings are defined as turtles that wash ashore, dead or alive, or are found floating dead or alive (generally in weak condition). In addition, per STSSN methodology, stranded hatchlings, as well as head-started turtles are excluded from the dataset, since their stranding may be an artifact of captive rearing and release (Teas 1993). Since the only data available from the STSSN were the number of strandings by species, an additional dataset was summarized; this dataset included sea turtle stranding data available for Florida from 1986 through 2006 and contained county, month, and size information for each stranded turtle, in addition to the number of strandings by species (FFWCC FWRI 2011b). When possible, the stranding data were summarized to demonstrate the data variability.

11.7.4.1 Kemp's Ridley Sea Turtle Strandings

From 1986 through 2009, 4,960 live and dead Kemp's ridleys were reported stranded on U.S. Gulf of Mexico beaches (STSSN 2012). More Kemp's ridley strandings (432 turtles) were reported on Gulf of Mexico beaches in 1994 than in any other year from 1986 through 2009 (Figure 11.51). The number of reported strandings along U.S. Gulf of Mexico beaches is not increasing in conjunction with the Kemp's ridley population increases based on the increased Gulf of Mexico nest counts (Figure 11.51), which illustrates the shortcomings of using strandings data to describe population abundance. Alternatively, this lack of increase in strandings, along with the increase in the nesting population in recent years, suggests that general conservation measures in the Gulf of Mexico are effective.

Of the 4,960 live and dead Kemp's ridleys reported stranded on Gulf of Mexico beaches from 1986 through 2009, approximately 45 % of them (2,242 turtles) were reported from Texas beaches (STSSN 2012). More Kemp's ridleys were typically reported stranded on Texas beaches each year from 1986 through the early 2000s than on Gulf coast beaches in Louisiana, Mississippi, Alabama, and Florida (Figure 11.52). While there is a lot of variability and a careful and detailed analysis is not available, the number of Kemp's ridley strandings reported for Texas and Louisiana beaches appears to be decreasing since a high in 1994, and Kemp's ridley strandings reported for Florida Gulf coast beaches appears to be significantly higher than in 1986 (Figure 11.52).

Along the Florida Gulf coast, few Kemp's ridleys were reported stranded along the Nature coast from Pasco through Jefferson counties. More Kemp's ridley strandings were reported on the southwest Gulf coast than on Panhandle beaches (Figure 11.53). Similar to what has occurred along the Florida Gulf coast beaches as a whole (Figure 11.52), Kemp's ridley strandings

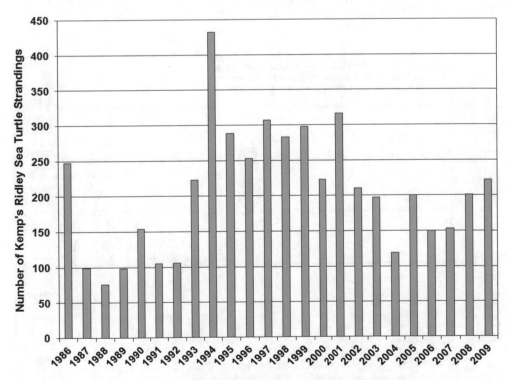

Figure 11.51. Number of reported Kemp's ridley sea turtle strandings (both live and dead turtles) on U.S. Gulf of Mexico beaches from 1986 through 2009 (from STSSN 2012).

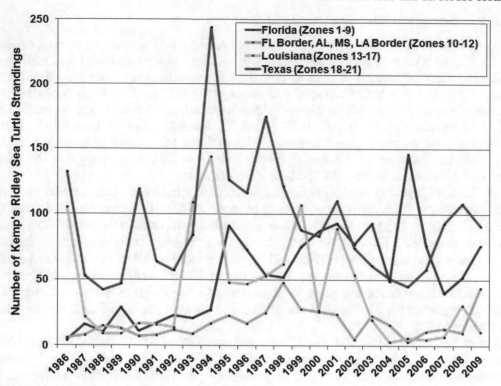

Figure 11.52. Number of reported Kemp's ridley sea turtle strandings (both live and dead turtles) on Florida, Alabama, Mississippi, Louisiana, and Texas Gulf coast beaches from 1986 through 2009 (from STSSN 2012).

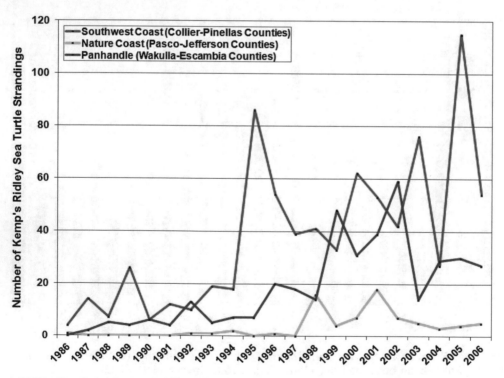

Figure 11.53. Number of reported Kemp's ridley sea turtle strandings (both live and dead turtles) along the Florida Gulf coast by region from 1986 through 2006 (from FFWCC FWRI 2011b).

reported for Florida Gulf coast regions appear to be increasing, with most of the strandings driven by southwest and panhandle regions of Florida (Figure 11.53).

Shaver (2005, 2006a) reported that more adult Kemp's ridleys were found stranded in Texas (mostly dead) than in any other state in the United States during each year from 1986 to 2003, and most of the turtles found on south Texas Gulf beaches occurred when Gulf of Mexico waters were open to shrimp trawling during the spring and summer. Despite the reported high compliance with TED regulations since they were implemented in 1990, a relationship continued on the Texas coast between shrimping and strandings on beaches through 2003 (Shaver 1997; Lewison et al. 2003). In late 2000, an annual shrimp-trawling closure of Gulf waters off south Texas beaches out to 8 km (5 mi) from shore from December 1 through mid-May was established; therefore, since 2001, South Texas nearshore waters have been closed to shrimp trawling during the entire Kemp's ridley mating and nesting seasons (Shaver 2006b). This closure, as well as the existing annual Texas closure, which extends out to 200 nautical miles from mid-May until mid-July each year, may be contributing to the lower numbers of strandings in Texas in recent years (Figure 11.52), as well as to the recent significant increase in nesting on Texas beaches (Figure 11.12).

A comparison of Kemp's ridley strandings reported for the Florida Gulf coast indicates that most turtles that stranded from 1986 through 2006 were neritic juveniles between 20 and 64 cm (7.9 and 25.2 in) SCL (Figure 11.54). The highest average numbers of strandings were reported for the 30–34 cm (11.8–13.4 in) SCL size class, followed by the 40–44 cm (15.7–17.3 in) SCL size class. From 1986 through 2006, Kemp's ridleys appeared to strand along the Florida Gulf coast most often during May and April, while the lowest average numbers of strandings

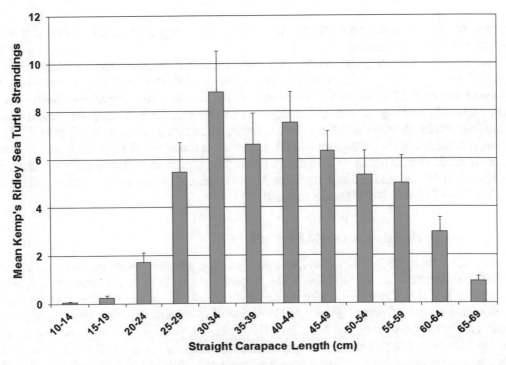

Figure 11.54. Number of reported Kemp's ridley sea turtle strandings (both live and dead turtles) along the Florida Gulf coast (Collier to Escambia counties) by size class from 1986 through 2006; error bars = standard error (from FFWCC FWRI 2011b).

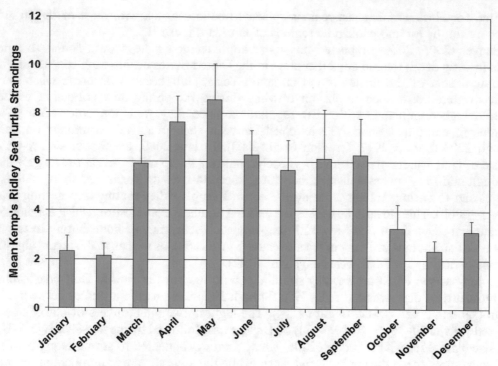

Figure 11.55. Number of reported Kemp's ridley sea turtle strandings (both live and dead turtles) along the Florida Gulf coast (Collier to Escambia counties) by month from 1986 through 2006; error bars = standard error (from FFWCC FWRI 2011b).

were reported for the winter months (Figure 11.55), which suggests that cold weather is not the main force driving strandings.

Prior to 2001, stranding data in Mexico were anecdotal or were reported only during the months of nesting activity at the main nesting beaches (NMFS et al. 2011). However, since 2001, year-round surveys of dead turtles stranded on the beaches of Tamaulipas have occurred (Figure 11.56). Strandings of Kemp's ridleys are most likely lower during the nesting season because Mexico implements a shrimp closure during the nesting season (NMFS et al. 2011). The increase in dead turtle strandings on Mexican beaches is thought to be due to an increase in the number of turtles available to be caught and killed. The data trends have been interpreted as an indication that TEDs and shrimp closures have decreased mortality of incidentally caught Kemp's ridleys (TEWG 2000; Frazier et al. 2007).

11.7.4.2 Loggerhead Sea Turtle Strandings

From 1986 through 2009, 9,289 live and dead loggerheads were reported stranded on U.S. Gulf of Mexico beaches (STSSN 2012). Approximately 15 % of all stranded loggerheads reported for the U.S. Gulf of Mexico from 1997 through 2005 were documented as having sustained some type of propeller or collision injury (NMFS and USFWS 2008). Loggerheads become caught in many materials, including fishing line, rope sacks, netting, and trap line, and from 1997 through 2005, almost 2 % of loggerheads reported stranded on Gulf of Mexico beaches were entangled in fishing gear, primarily monofilament line (NMFS and USFWS 2008). In the Gulf, high numbers of reported loggerhead strandings were associated with strong red tides in southwest Florida during some years (1995, 1996, 2000, 2001, 2003, and 2005) from

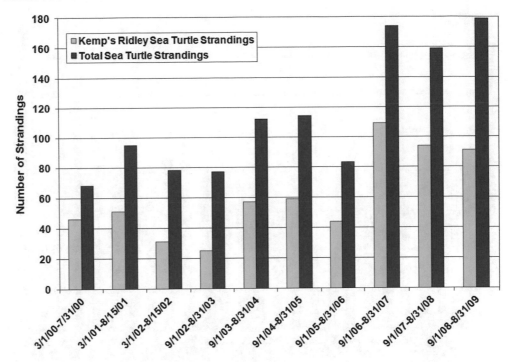

Figure 11.56. Number of dead turtles reported stranded along the coast of Tamaulipas, Mexico from March 2000 through August 2009 (from NMFS et al. 2011).

1995 through 2005 (FFWCC unpublished data, cited in TEWG 2009). From 1986 through 2009, more loggerhead sea turtles were reported stranded on Gulf of Mexico beaches in 2009 (790 turtles), than in prior years (Figure 11.57). Although no detailed and careful analysis is available, loggerhead strandings reported for the U.S. Gulf Coast appeared to be increasing, particularly since 1992 (Figure 11.57).

Of the 9,289 live and dead loggerhead strandings reported for U.S. Gulf of Mexico beaches from 1986 through 2009, very few were reported stranded on Alabama, Mississippi, or Louisiana beaches (Figure 11.58). From 1986 through 2009, loggerhead strandings reported for Texas beaches were within a similar range each year, approximately 100–200 turtles (Figure 11.58). The majority of loggerhead strandings for the U.S. Gulf of Mexico were reported for Florida beaches (Figure 11.58), with a high of 579 stranded loggerheads reported in 2009. Almost no loggerheads were reported stranded on beaches from Pasco to Jefferson counties, and fewer than 100 loggerheads were reported stranded each year on Florida Panhandle beaches from 1986 through 2006 (Figure 11.59). Fairly similar numbers of loggerheads were reported stranded along the southwest coast of Florida from 1986 through 2004; however, more than 300 loggerhead strandings were reported for the southwest coast in 2005, and almost 400 turtles were stranded on southwest coast beaches in 2006 (Figure 11.59).

From 1986 through 2006, most of the loggerheads reported stranded along the Florida Gulf coast ranged in size from 85 to 99 cm (33.5 to 38.9 in) (Figure 11.60); all of these size classes are considered adults (TEWG 2009). More loggerheads appeared to be stranded along the Florida Gulf coast during May and April, while the lowest average numbers of strandings occurred during January and December (Figure 11.61).

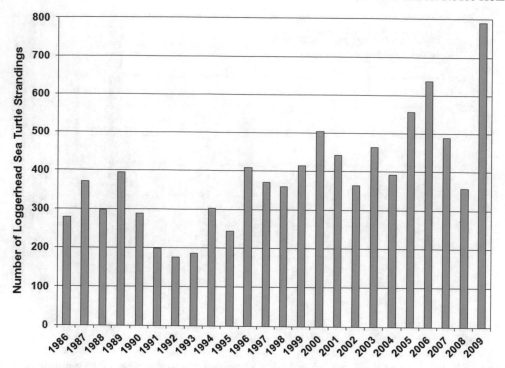

Figure 11.57. Number of reported loggerhead sea turtle strandings (both live and dead turtles) on U.S. Gulf of Mexico beaches from 1986 through 2009 (from STSSN 2012).

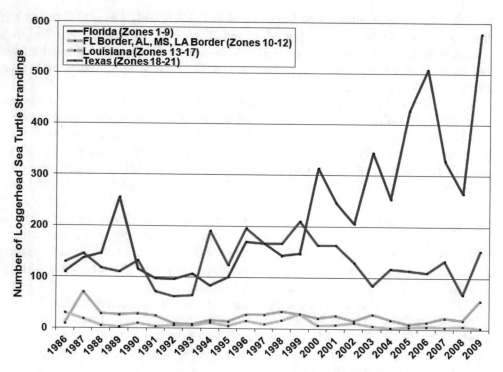

Figure 11.58. Number of reported loggerhead sea turtle strandings (both live and dead turtles) on Florida, Alabama, Mississippi, Louisiana, and Texas Gulf coast beaches from 1986 through 2009 (from STSSN 2012).

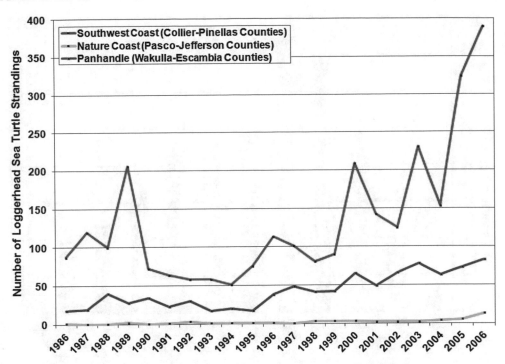

Figure 11.59. Number of reported loggerhead sea turtle strandings (both live and dead turtles) along the Floria Gulf coast by region from 1986 through 2006 (from FFWCC FWRI 2011b).

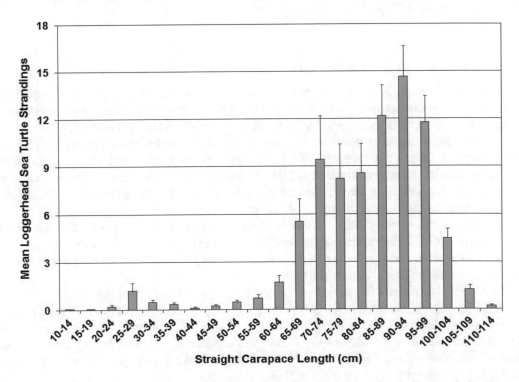

Figure 11.60. Number of reported loggerhead sea turtle strandings (both live and dead turtles) along the Florida Gulf coast (Collier to Escambia counties) by size class from 1986 through 2006; error bars = standard error (from FFWCC FWRI 2011b).

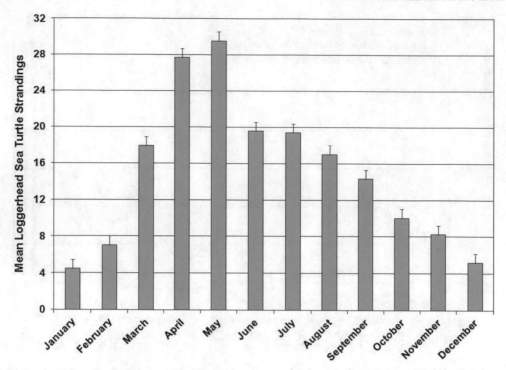

Figure 11.61. Number of reported loggerhead sea turtle strandings (both live and dead turtles) along the Florida Gulf coast (Collier to Escambia counties) by month from 1986 through 2006; error bars = standard error (from FFWCC FWRI 2011b).

11.7.4.3   Green Sea Turtle Strandings

From 1986 through 2009, 4,222 live and dead green sea turtles were reported stranded on U.S. Gulf of Mexico beaches (STSSN 2012). Green turtle strandings were most numerous in 2001 (almost 600 reports), followed by 2007 (more than 500 reports) (Figure 11.62).

Approximately equal proportions of the 4,222 green turtles stranded on U.S. Gulf of Mexico beaches from 1986 through 2009 were reported for Florida and Texas, and with few exceptions, similar numbers stranded on the beaches in both states were reported each year (Figure 11.63). Almost no green turtles were reported stranded on Alabama, Mississippi, or Louisiana beaches from 1986 through 2009 (Figure 11.63).

The majority of green turtles that were reported stranded on Florida Gulf coast beaches from 1986 through 2006 stranded on the southwest coast (Figure 11.64). Very few green turtles, typically fewer than ten turtles per year, were reported stranded on Florida Nature coast or Florida Panhandle beaches (Figure 11.64).

The highest numbers of reported green sea turtle strandings along the Florida Gulf coast from 1986 through 2006 were in the 30–44 cm (11.8–17.3 in) SCL size classes (Figure 11.65); all of these sizes classes are juvenile green turtles (Table 11.3). More green sea turtles were reported stranded along the Florida Gulf coast during January through March than any other period of the year from 1986 through 2006, while the lowest average numbers of reported strandings occurred during June (Figure 11.66).

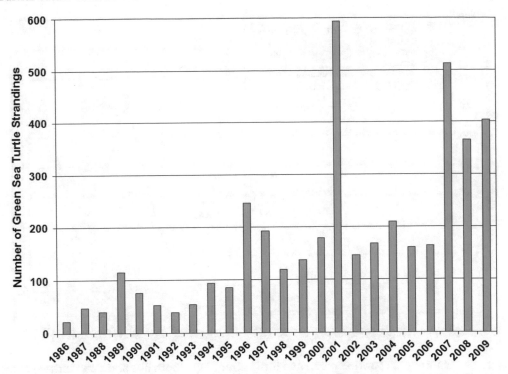

Figure 11.62. Number of reported green sea turtle strandings (both live and dead turtles) on U.S. Gulf of Mexico beaches from 1986 through 2009 (from STSSN 2012).

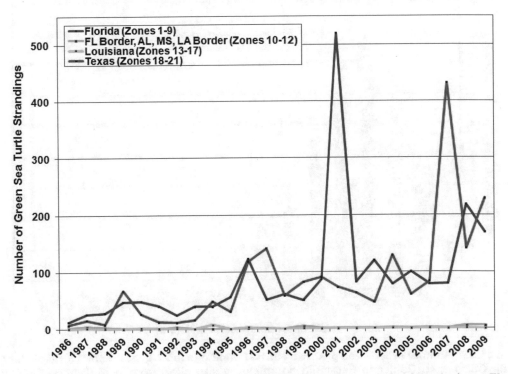

Figure 11.63. Number of green sea turtle strandings (both live and dead turtles) on Florida, Alabama, Mississippi, Louisiana, and Texas Gulf coast beaches from 1986 through 2009 (from STSSN 2012).

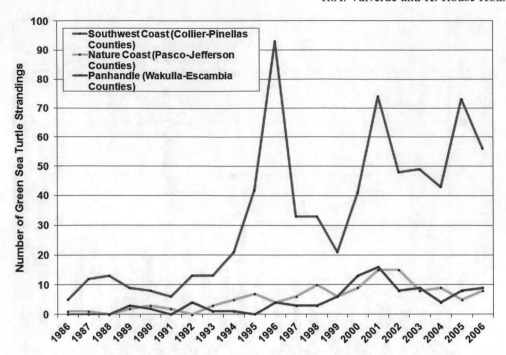

Figure 11.64. Number of reported green sea turtle strandings (both live and dead turtles) along the Florida Gulf coast by region from 1986 through 2006 (from FFWCC FWRI 2011b).

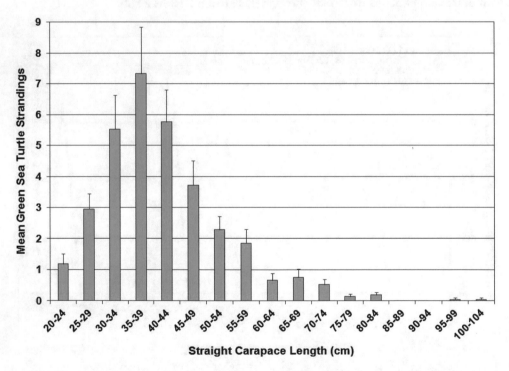

Figure 11.65. Number of reported green sea turtle strandings (both live and dead turtles) along the Florida Gulf coast (Collier to Escambia counties) by size class from 1986 through 2006; error bars = standard error (from FFWCC FWRI 2011b).

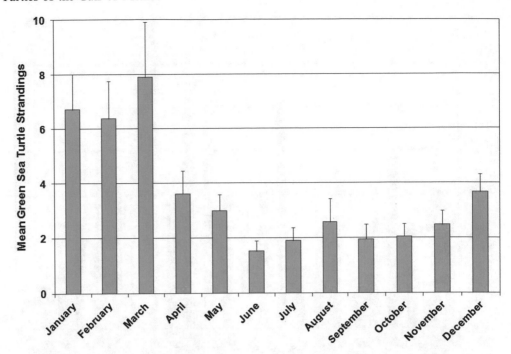

Figure 11.66. Number of reported green sea turtle strandings (both live and dead turtles) along the Florida Gulf coast (Collier to Escambia counties) by month from 1986 through 2006; error bars = standard error (from FFWCC FWRI 2011b).

11.7.4.4 Leatherback Sea Turtle Strandings

In the U.S. Gulf of Mexico, strandings of live and dead leatherbacks are far less frequent than strandings of other sea turtle species (STSSN 2012). For example, from 1986 to 2009, fewer than 30 leatherbacks typically stranded each year in the Gulf of Mexico (Figure 11.67). In addition, leatherback sea turtles represented about 2 % of the U.S. Gulf of Mexico strandings from 1986 through 2009. The yearly proportion of total strandings that consisted of leatherbacks ranged from 0.35 to 4.38 % (STSSN 2012).

The causes of leatherback strandings are most frequently attributed to boat strikes; interactions with fisheries, including entanglement in line, nets, and other gear; and ingestion of marine debris (NMFS SEFSC 2001). More frequent leatherback strandings in the Gulf tend to occur during the spring, coinciding with nearshore shrimp trawling activity (NMFS SEFSC 2001). Between 1980 and 1981, three leatherback carcasses washed ashore in Louisiana and their deaths appeared to be a result of interactions with shrimp trawls (Fritts et al. 1983a, b). In May and June 1993, 107 turtles were stranded around Grand Isle, Louisiana; the stranded turtles included two leatherbacks. These strandings were attributed to fatal interactions with the offshore longline ground fishery (Thompson 1993). In 1994, a total of 16 dead leatherback strandings in the Gulf of Mexico were attributed to interactions with shrimp trawlers (Steiner 1994).

While records of leatherback boat strikes were nonexistent prior to the late 1980s, 10 % of the 231 leatherback strandings involving boat strikes from 1980 through 1999 occurred in states on the Gulf of Mexico. Whether or not those strandings were caused directly by the boat strikes is unknown (NMFS SEFSC 2001).

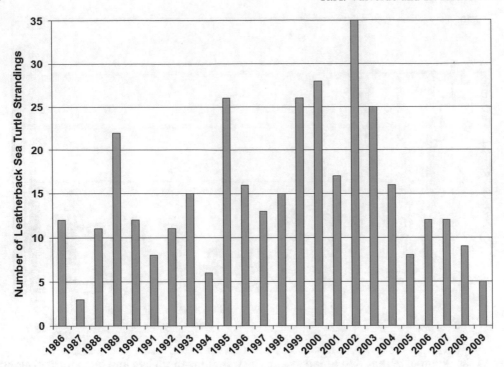

Figure 11.67. Number of reported leatherback sea turtle strandings (both live and dead turtles) on U.S. Gulf of Mexico beaches from 1986 through 2009 (from STSSN 2012).

From 1986 through 2009, 363 leatherbacks were reported to have stranded on U.S. Gulf of Mexico beaches (STSSN 2012). The highest number of leatherbacks (35) that were reported stranded on Gulf of Mexico beaches occurred in 2002 (Figure 11.67). Of the 363 leatherbacks reported stranded on U.S. Gulf of Mexico beaches, the majority was reported for Texas beaches, followed by beaches on the Florida Gulf coast (Figure 11.68). Fewer than five leatherbacks were typically reported stranded on Louisiana, Alabama, and Mississippi beaches each year (Figure 11.68).

From 1986 through 2006, more leatherbacks were reported stranded on Florida Panhandle beaches than on the remainder of the Florida Gulf coast (Figure 11.69). For each year, no more than seven leatherbacks were reported stranded in a given Florida Gulf coast region (Figure 11.69).

From 1986 through 2006, the highest average numbers of leatherback sea turtle strandings along the Florida Gulf coast were in the 140–149 cm (57.1–58.7 in) SCL size classes (Figure 11.70); this size range represents adult leatherbacks (Table 11.4). No leatherbacks were reported stranded along the Florida Gulf coast in September from 1986 through 2006, and the most reported average strandings occurred in March, followed by April (Figure 11.71).

11.7.4.5 Hawksbill Sea Turtle Strandings

Hawksbills are stranded on Gulf of Mexico beaches during all months of the year (NMFS and USFWS 1993). From 1986 through 2009, 474 live and dead hawksbill sea turtles were reported stranded on U.S. Gulf of Mexico beaches (STSSN 2012). Unlike other species of sea turtles, hawksbill strandings are often live turtles (Amos 1989). Hawksbills represented 1–5.6 % of the total turtles stranded each year on U.S. Gulf beaches. The 474 hawksbills that stranded on

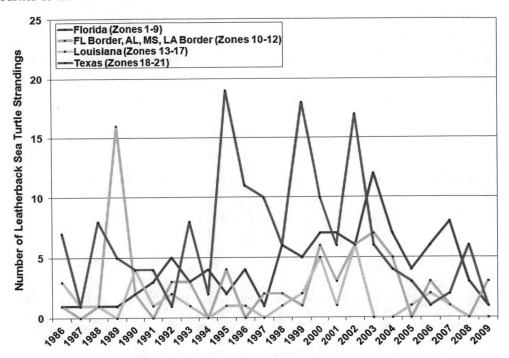

Figure 11.68. Number of reported leatherback sea turtle strandings (both live and dead turtles) on Florida, Alabama, Mississippi, Louisiana, and Texas Gulf coast beaches from 1986 through 2009 (from STSSN 2012).

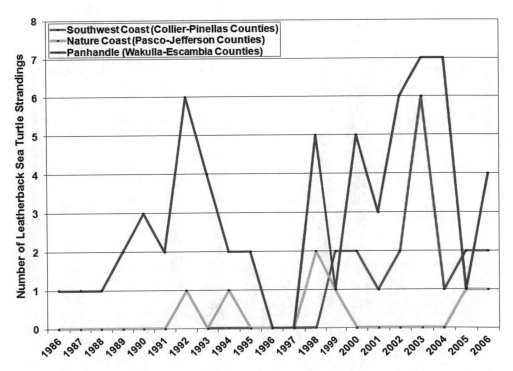

Figure 11.69. Number of reported leatherback sea turtle strandings (both live and dead turtles) along the Florida Gulf coast by region from 1986 through 2006 (from FFWCC FWRI 2011b).

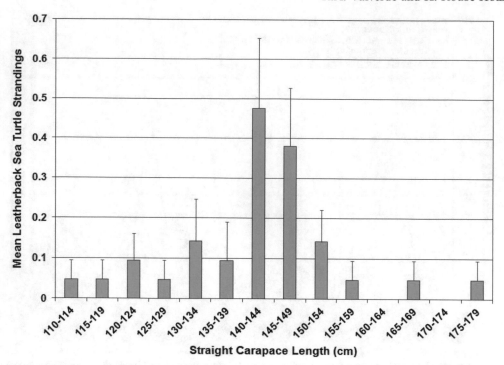

Figure 11.70. Number of reported leatherback sea turtle strandings (both live and dead turtles) along the Florida Gulf coast (Collier to Escambia counties) by size class from 1986 through 2006; error bars = standard error (from FFWCC FWRI 2011b).

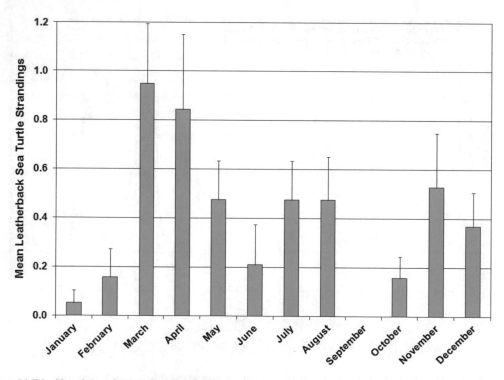

Figure 11.71. Number of reported leatherback sea turtle strandings (both live and dead turtles) along the Florida Gulf coast (Collier to Escambia counties) by month from 1986 through 2006; error bars = standard error (from FFWCC FWRI 2011b).

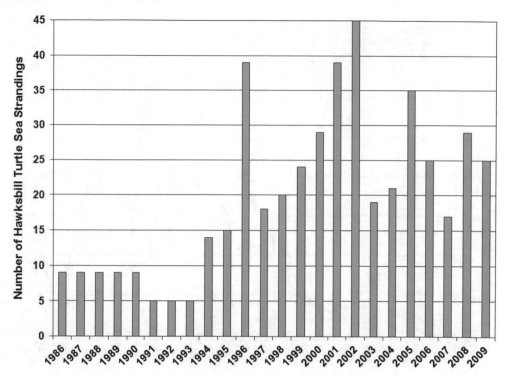

Figure 11.72. Number of reported hawksbill sea turtle strandings (both live and dead turtles) on U.S. Gulf of Mexico beaches from 1986 through 2009 (from STSSN 2012).

Gulf of Mexico beaches from 1986 through 2009 represented 2.5 % of the total strandings. Fewer than ten hawksbills were reported stranded each year on U.S. Gulf Coast beaches from 1986 through 1993. However, in 1994, the number of hawksbill strandings began to increase, with a high of 45 turtles in 2002 (Figure 11.72).

Most of the 474 hawksbill strandings reported on U.S. Gulf of Mexico beaches from 1986 through 2009 occurred in Texas (301 turtles) (Figure 11.73), with a high of 35 stranded hawksbills reported in 2002. No hawksbills were reported stranded in Alabama or Mississippi, and few to none were stranded on the Louisiana coast (Figure 11.73). On the Florida Gulf coast, 166 hawksbill strandings were reported from 1986 through 2009, and with the exception of 2001 when 27 hawksbill strandings were recorded, fewer than 15 hawksbills were reported stranded each year (Figure 11.73).

With the exception of 2001 when 23 hawksbills strandings were recorded, fewer than ten hawksbills were reported stranded each year on the southwest Florida coast from 1986 through 2006 (Figure 11.74). Few hawksbill sea turtles were reported stranded on the Florida coast from Pasco through Escambia counties from 1986 through 2006 (Figure 11.74).

The majority of hawksbill strandings reported along the Florida Gulf coast from 1986 through 2006 were juveniles (less than 60 cm [23.6 in] SCL); the highest average number of reported hawksbill strandings was in the 25–29 cm (9.8–11.4 in) SCL size class (Figure 11.75). From 1986 through 2006, more hawksbill sea turtles were reported stranded along the Florida Gulf coast during March than any other month of the year; the lowest average numbers of reported strandings occurred during November (Figure 11.76).

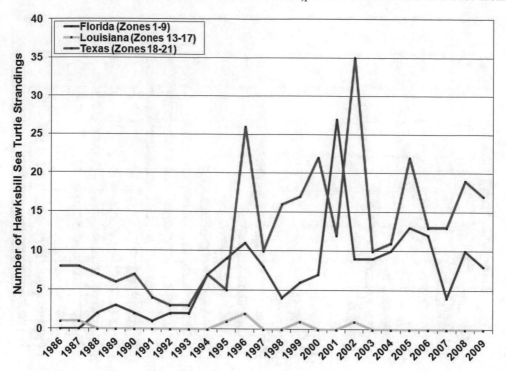

Figure 11.73. Number of reported hawksbill sea turtle strandings (both live and dead turtles) on Florida, Louisiana, and Texas Gulf coast beaches from 1986 through 2009. No hawksbill strandings were reported for Alabama or Mississippi (from STSSN 2012).

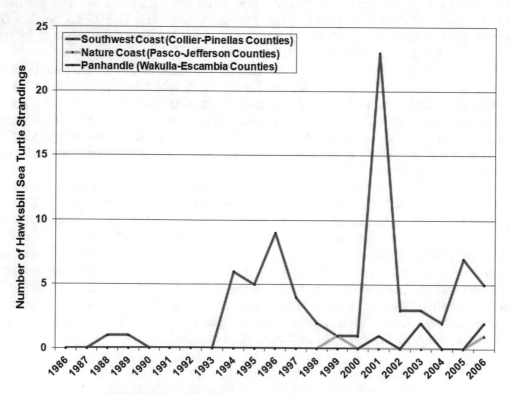

Figure 11.74. Number of reported hawksbill sea turtle strandings (both live and dead turtles) along the Florida Gulf coast by region from 1986 through 2006 (from FFWCC FWRI 2011b).

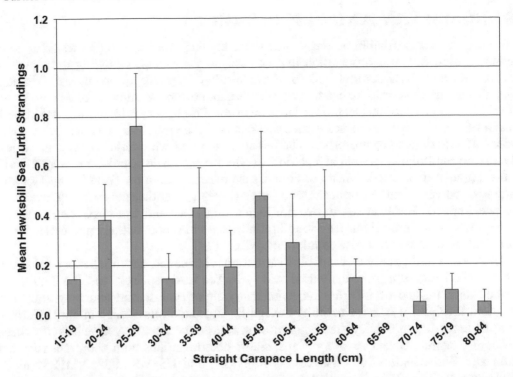

Figure 11.75. Number of reported hawksbill sea turtle strandings (both live and dead turtles) along the Florida Gulf coast (Collier to Escambia counties) by size class from 1986 through 2006; error bars = standard error (from FFWCC FWRI 2011b).

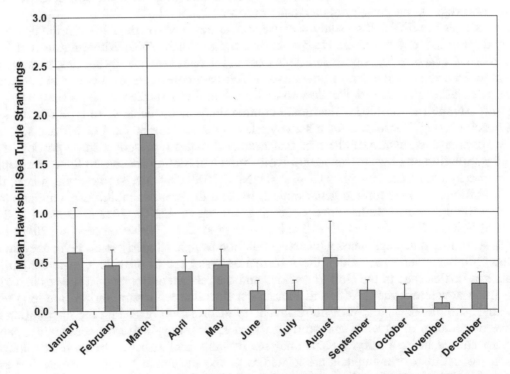

Figure 11.76. Number of reported hawksbill sea turtle strandings (both live and dead turtles) along the Florida Gulf coast (Collier to Escambia counties) by month from 1986 through 2006; error bars = standard error (from FFWCC FWRI 2011b).

11.8 SUMMARY AND DISCUSSION

In this chapter, available nesting beach data, tagging studies, satellite tracking studies, genetics studies, in-water observation and capture program data, stranding data, and other types of data have been summarized for Kemp's ridley, loggerhead, green, leatherback, and hawksbill sea turtles in order to characterize the distribution and abundance of sea turtles in the Gulf of Mexico prior to the Deepwater Horizon event. Life history information was also given for each of the five species of sea turtles that occur in the Gulf, and when available, Gulf of Mexico-specific data were presented. The threats associated with anthropogenic activities, as well as from natural events, that affect Gulf of Mexico sea turtle populations were described, and the impacts quantified, when possible, by synthesizing data on fisheries bycatch, turtle strandings, power plant entrapment and impingement, hypothermic-stunning events, and hurricane impacts. Because sea turtles are highly migratory and use terrestrial, neritic, and oceanic ecosystems throughout their long lifetimes, activities both within and outside the Gulf affect Gulf of Mexico sea turtle populations (NRC 2010).

The amount and quality of available distribution, abundance, and life history data, as well as information regarding threats, varied widely by sea turtle species, and the data are best described as a mosaic of information, with many significant spatial and longitudinal gaps throughout the Gulf of Mexico basin. For example, large amounts of high-quality information are available for Kemp's ridleys and loggerhead sea turtles because they have been the focus of significant research, recent recovery plans, status determinations, and comprehensive books (Bolten and Witherington 2003; Plotkin 2007c; NMFS and USFWS 2008; NMFS et al. 2011; USFWS and NMFS 2011). The amount of sea turtle data available for nesting beaches and nearshore waters along the U.S. Gulf Coast also varied; more information was available for Texas and Florida than for Louisiana, Mississippi, and Alabama. In addition, data available for sea turtle nesting beaches, as well as nearshore waters, along the Mexican Gulf coast varied by species and region, and long-term datasets are few and limited.

Sea turtles are difficult to study because they spend most of their lives in the ocean, are widely distributed, and have long life spans (Holder and Holder 2007; Witherington et al. 2009; NRC 2010). Consequently, significant gaps exist in the data available by species, as well as by life stage. In addition, the time frames over which data collection must occur to adequately assess population parameters for slow-to-mature, long-lived species, such as sea turtles, are daunting (Heppell et al. 2003). The most common datasets available were typically for long-term beach nesting. These data were heavily relied upon to describe Gulf of Mexico sea turtle nesting populations, aware of the fact that nesting females represent a small portion of the overall population and that extrapolating from individual nesting beaches to the entire population is inadequate at best (Heppell et al. 2007; NRC 2010). In-water studies, which have their own limitations, such as limited geographic scope and differential habitat use, continue to be lacking, although recent studies are encouraging (Cuevas et al. 2008, 2012; Fossette et al. 2010; Hart and Fujisaki 2010; Hart et al. 2012a, b; Avens et al. 2012; Witherington et al. 2012). Even though stranding data represents a biased population sample (Epperly et al. 1996; Meylan and Redlow 2006; Frazier et al. 2007), the information was used to provide a basic view of the threats sea turtles face in the Gulf of Mexico and the relative presence of a species in a given geographic area. Both the STSSN and Florida datasets that were summarized are long term, have large sample sizes, and provide rudimentary information regarding the distribution and relative abundance of juvenile and adult sea turtles, migratory patterns, seasonality, habitat use, and causes of mortality. Recent analyses of long-term fisheries bycatch data make it possible to estimate the impacts associated with the incidental bycatch of sea turtles in commercial fisheries, although the uncertainty associated with the estimates is high (Conant et al. 2009; Kot et al. 2010; Bolten et al. 2011; Finkbeiner et al. 2011; Ponwith 2011). Despite the data gaps and limitations associated with specific datasets, the current conditions of the five species of sea turtles that occur in the Gulf of Mexico have been described qualitatively, and sometimes quantitatively.

Kemp's Ridley Sea Turtles. In the Gulf of Mexico, the Kemp's ridley sea turtle has made a remarkable recovery from the brink of extinction (Heppell et al. 2007; Crowder and Heppell 2011), although the current nesting populations remain far below historical levels. The increased abundance and nesting of Kemp's ridleys in the Gulf of Mexico in recent years is most likely a result of many activities that affect all Kemp's ridley life stages—conservation and education efforts in both Mexico and the United States, the Kemp's Ridley Recovery Program, the Kemp's Ridley Head Start Experiment, the elimination of direct harvest of eggs and adult turtles, nest and hatchling protection, TED use, shrimp fishery closures, and the reduced Gulf of Mexico shrimp trawling effort in both the United States and Mexico (Shaver and Wibbels 2007; NMFS et al. 2011; Crowder and Heppell 2011). Despite the partial success story, there are threats to Kemp's ridleys in the Gulf that could cause significant impacts to the population and affect its continued recovery, particularly the incidental bycatch. The future of Kemp's ridleys is optimistically bright, and nesting numbers at Rancho Nuevo and PAIS continue to increase; however, because of their mass nesting at a single site in the western Gulf and their restricted Gulf of Mexico distribution, a significant event or the synergistic effects of multiple threats in the Gulf could have catastrophic effects on the recovering Kemp's ridley population.

Loggerhead Sea Turtles. The life history, distribution, and abundance of the loggerhead sea turtle is probably better understood than all other sea turtle species that inhabit the Gulf of Mexico (Bolten and Witherington 2003; USFWS and NMFS 2011). While the loggerhead is distributed globally, the goal of this chapter was to describe contemporary conditions of Gulf sea turtles; information for populations that occur in the Gulf of Mexico during some portion of their life cycle was the focus. These populations included the Peninsular Florida, northern Gulf of Mexico, Dry Tortugas, and Greater Caribbean subpopulations of the Northwest Atlantic Ocean DPS (USFWS and NMFS 2011). The annual nest counts from 2001 through 2009 on Florida Gulf coast and Atlantic beaches showed a decreasing trend, but in 2010 and 2011, loggerhead nest counts on surveyed Florida beaches were back to numbers similar to those recorded in 2000, indicating that the population of nesting females may have stabilized. Given the lack of a long historical record, it is not known whether these types of fluctuations are common or if the recent temporary decrease observed was caused by an acute factor. A recent analysis of annual loggerhead nest counts for the northern Gulf of Mexico subpopulation indicated a significant declining trend; however, a longer time series of nesting data is needed for an adequate evaluation (NMFS and USFWS 2008). More years of data are also needed in order to detect a trend in loggerhead nest counts for the Dry Tortugas subpopulation. Nesting for the Greater Caribbean subpopulation appears to be stable; however, a trend analysis is challenging because standardized surveys are few, survey efforts have changed, and low-level nesting occurs in many locations (NMFS and USFWS 2008).

High cumulative anthropogenic threat levels were estimated for oceanic and neritic juveniles and adults of the Northwest Atlantic Ocean DPS during the most recent loggerhead status review (Conant et al. 2009; Bolten et al. 2011). The loggerhead mortalities associated with these high threat levels resulted primarily from fisheries bycatch. Significant loggerhead mortality occurred in longline fisheries, bottom and mid-water trawl fisheries, dredge fisheries, gillnet fisheries, and pot/trap fisheries that occur not only in U.S. waters (Conant et al. 2009; Bolten et al. 2011; Ponwith 2011), but also in the Azores, the Mediterranean, on the Grand Banks, in Canadian waters, and in other locations throughout the range of the northwest Atlantic Ocean loggerhead population. Loggerhead sea turtles interacted with more U.S. fisheries (17 of 18 analyzed fisheries) than any other sea turtle species in a recent cumulative estimate of U.S. fisheries bycatch from 1990 through 2007 (Finkbeiner et al. 2011). This high level of interaction was thought to be due to their large nesting assemblages in Florida (as well as throughout the southeast and along the Gulf of Mexico) and their annual migrations to higher latitudes (Plotkin and Spotila 2002). The large range of northwest Atlantic Ocean neritic and oceanic juvenile and adult loggerheads overlaps significantly with coastal and oceanic areas where many fisheries occur, which unfortunately results in the death of thousands of logger-heads each year.

Green Sea Turtles. Despite being greatly depleted in the past, green turtle populations in the Gulf of Mexico are increasing. Green turtle nesting along the Mexican Gulf coast has increased in recent years and remains relatively stable. In addition, nesting at major rookeries in the region, such as Tortuguero, Costa Rica, and the Florida east coast, including ACNWR, has increased significantly since the 1970s. While fibropapilloma tumors have been reported for all sea turtle species, the frequency of these tumors is much higher in green turtles than in other species of sea turtles; this disease remains a threat to green sea turtles (Witherington et al. 2006b; NMFS and USFWS 2007c). Green turtles are also dependent on the continued maintenance of healthy seagrass meadows in their foraging areas. Although the impacts to green turtles resulting from the incidental bycatch in fisheries are not as significant as those for loggerheads, many green turtles die each year from fisheries interactions.

Leatherback Sea Turtles. Because leatherback sea turtles spend most of their lives in the oceanic zone, distribution and abundance data for leatherbacks in the Gulf of Mexico are incomplete. In addition, large life history data gaps still exist, especially for post-hatchlings and juveniles. However, the available data do verify that leatherbacks use the Gulf as a foraging area, and recent tracking studies have demonstrated that the Gulf of Mexico may be a significant year-round foraging ground for leatherbacks that nest along the Caribbean coast (Evans et al. 2007; Fossette et al. 2010). In the Gulf of Mexico, leatherbacks are often found in areas containing an abundance of jellyfish, their main prey; and they are less abundant than other sea turtle species, such as Kemp's ridleys, loggerheads, and hawksbills. Because of the low numbers of leatherback nesting in the Gulf, the significant gaps in available leatherback data and information, as well as their extensive migrations and large home ranges, determining the status of the Gulf of Mexico leatherback population is uncertain. However, leatherback nesting has increased significantly in Florida since the late 1970s, which may indicate that the leatherback population in the general Gulf of Mexico, Caribbean, and northwest Atlantic Ocean area is stable or increasing. A major threat to leatherbacks in the Gulf continues to be mortality as bycatch in pelagic longline fisheries (Kot et al. 2010; Finkbeiner et al. 2011).

Hawksbill Sea Turtles. Because millions of hawksbill sea turtles have been killed globally for the tortoiseshell markets of Asia, Europe, and the United States over the past 100 years, the current abundance of hawksbills in the general Gulf of Mexico, Caribbean, and northwest Atlantic Ocean area is only a fraction of what occurred historically (NMFS and USFWS 2007e). In the Gulf, the most important hawksbill rookery is located on Yucatán Peninsula beaches in Mexico (Abreu-Grobois et al. 2005; NMFS and USFWS 2007e; del Monte-Luna et al. 2012). Analyses of hawksbill nesting data from 1980 through 2010 for the Mexican Gulf of Mexico/ Caribbean coasts attributed the apparent recent declines in nesting to the low-level taking of turtles, impacts to the hawksbill's marine habitats, constant expansion of beach development, and increases in sea surface temperature, hurricane activity, and regional sea level rise (Abreu-Grobois et al. 2005; del Monte-Luna et al. 2012).

The destruction of nesting and foraging habitat is affecting hawksbills that nest along the beaches and use the nearshore areas of the Gulf of Mexico (Garduño-Andrade et al. 1999). Hawksbills are also dependent on coral reefs—one of the world's most endangered ecosystems—for food and shelter (NMFS and USFWS 1993, 2007e). In addition, although the trade in hawksbill products has declined significantly compared to historical levels, both the illegal and legal trade is still active and significant (Mortimer 2008). The low hawksbill bycatch data levels suggest that the threat of death or injury from bycatch in Gulf of Mexico fisheries is not substantial. However, given the low abundance of these populations, which renders them significantly vulnerable, it is crucial to ensure the continued protection of this species from all threats in general, and from fisheries bycatch, specifically.

Overview. This chapter has presented a significant body of data and literature on sea turtle nesting populations in the Gulf of Mexico. Most of that information focused on the northern and eastern Gulf coast, particularly on Florida beaches (e.g., Meylan et al. 1995). This information showed that north-central Gulf of Mexico beaches are essentially devoid of any significant nesting, in particular the beaches of east Texas, Louisiana, Mississippi, and Alabama. Sea turtle nesting in general increases east, west, and southwest from this north-central location and reaches its zenith around the Florida and Yucatán Peninsulas (Renaud 2001). Data available on nesting females showed that sea turtle populations in the Gulf generally exhibit a very low abundance relative to Atlantic regions outside the Gulf. This is particularly true for the loggerhead, whose nesting on Florida west coast beaches amounted to only 8.6 % of statewide nesting from 2001 through 2006 (Witherington et al. 2009). The only obvious exception to this general rule is the Kemp's ridley, whose main rookery is located in Rancho Nuevo, in northeast Mexico (NMFS et al. 2011). The very low nesting numbers indicate that all sea turtle populations in the Gulf of Mexico are particularly vulnerable to natural and anthropogenic impacts, perhaps more so than populations outside the Gulf. Moreover, the current abundance of sea turtles is so low in the Gulf mainly due to the myriad of impacts associated with the anthropogenic activities summarized in Table 11.6.

It is difficult to ascertain the abundance and trends of Gulf of Mexico sea turtle populations given the lack of long-term, in-water, systematic studies. Indeed, the Gulf may arguably be the most data-deficient basin in terms of its sea turtle populations. Efforts to determine the presence and abundance of all species in U.S. waters seem to have concentrated in the states of Florida and Texas, likely due to the presence of nesting beaches in these states, with Florida boasting by far the largest numbers (e.g., Meylan et al. 1995). Aerial surveys over Gulf of Mexico waters frequently have been used to address this deficiency of data. However, no reports exist regarding the southern and southwest Gulf of Mexico, and most of the reports available for the northern Gulf are point in time studies (e.g., over a season or a year) and lack the benefit of long-term systematic records that could be used to establish population trends.

11.9   FUTURE CONSIDERATIONS AND RESEARCH NEEDS

It has been recognized that commercial sea turtle fisheries, habitat destruction, and pollution, among other causes, have played a critical role in the decimation of sea turtles in U.S. waters (Witzell 1994). Because sea turtle nesting currently is scattered and not abundant throughout most of the rim of the Gulf of Mexico basin, future work should focus on collecting information on in-water populations in order to characterize the status of sea turtle populations in the basin. For example, large amounts of essential data on population abundance, particularly with regard to juveniles and adult turtles, can be gathered at foraging grounds, such as in the Bay of Campeche, off the Louisiana coast, and in northwest Florida waters in the case of the Kemp's ridley (Márquez-M. 1999).

A series of index sites have been identified in Florida waters that have yielded valuable information about the biology, distribution, and abundance of sea turtle populations in that state (Eaton et al. 2008). The concept of in-water index sites should be expanded to the entire Gulf of Mexico by creating a basin-wide network. Many techniques and approaches have been discussed to conduct this type of in-water work (Bjorndal and Bolten 2000). Work on in-water index sites may be complemented by aerial surveys (e.g., Witzell and Azarovitz 1996). A few major technical hurdles will need to be overcome if this gigantic task is to be pursued. Two of these hurdles are the cost of undertaking such a project in a large body of water like the Gulf of Mexico and the logistical constraints associated with flight altitude, typically around 150 m (492 ft), that complicate the detection of turtles comparable in size to adult Kemp's ridleys (Witzell and Azarovitz 1996). To address these two significant problems, a possible approach would be to use high-resolution cameras mounted on surveyor airplanes. This would keep surveying crews to a minimum, while allowing for the possibility of zooming in on the images for positive species identification and the creation of an archive of important data. Future technological and legislative advances should make it possible for high-resolution imagery for the entire Gulf of Mexico from geostationary satellites, which would provide information in real time on relative abundance on every corner of the Gulf, not only of sea turtles but also of every major vertebrate that inhabits the Gulf, including birds, cetaceans, and schools of commercial fish. Until these advances occur, a more realistic approach might be to set up observers at a fraction of the more than 3,800 oil rigs that currently exist in the Gulf of Mexico, which are mostly concentrated in Louisiana and Texas waters. Such a program may provide a reasonably good idea of the seasonal relative abundance of sea turtles, at least in the north-western Gulf. For other areas of the Gulf, index sites, such as the Chandeleur Islands in Louisiana and Waccasassa Bay in Florida, could be monitored.

Because of the many significant data gaps discussed above, the information needed for accurate abundance assessments, as well as for the calculation of important demographic parameters, for most sea turtle populations is typically not available (NRC 2010). Additional challenges to accurately estimate sea turtle populations include the limited spatial and temporal scopes of most sea turtle research projects and the lack of coordination of sea turtle data for a given species or region (e.g., the lack of comprehensive databases) (NRC 2010). Bjorndal et al. (2011) recently recommended that the following seven elements be included in strategic plans for the collection of essential sea turtle data: (1) integrate demography with abundance trends for multiple life stages and determine environmental effects on those parameters; (2) emphasize analyses of cumulative effects; (3) elucidate links among and within populations with new tools in genetics, statistical models, and tracking; (4) revise the permitting processes that now hinder peer-reviewed studies of critical processes and management alternatives for protected species; (5) encourage data sharing; (6) improve assessment tools for evaluation of anthropogenic impacts on populations by fostering interdisciplinary research among scientists, students, and managers; and (7) prioritize investments for research and monitoring.

In addition to revealing data gaps, this compilation of information for Gulf of Mexico sea turtle populations shows that sea turtle data are highly variable from year to year; an excellent example of this variability is the interannual fluctuations in green turtle nesting data. This variability highlights the importance of long-term datasets and explains why long-term trends, not year-to-year fluctuations, are critical in determining changes in sea turtle nesting popula-tions. While determining if changes to sea turtle populations have occurred is extremely difficult, it is also not possible to determine the cause of population variability. Because multiple human-made and natural threats affect all life stages of sea turtles, each threat may affect a life stage or species differently. The effects of multiple threats may be synergistic, and impacts to a specific year class may not appear on the nesting beach for many years. These issues will continue to be a challenge in the future as new threats emerge and attempts are made to quantify impacts on sea turtle health and populations associated with environmental change due to urbanization and coastal development, global warming and sea level rise, fisheries exploitation and regulation, oil and gas exploration and production, and other anthropogenic effects, such as nutrient enrichment in the Gulf of Mexico. These factors and many others interact to challenge the development of effective strategies and measures to promote sea turtle conservation.

ACKNOWLEDGMENTS

BP sponsored the preparation of this chapter. This chapter has been peer reviewed by anonymous and independent reviewers with substantial experience in the subject matter. We thank the peer reviewers, as well as others, who provided assistance with research and the compilation of information. Completing this chapter would not have been possible without the tireless work of Jonathan Ipock, ENVIRON International Corporation, in obtaining documents, compiling data and information, preparing maps and graphs, and compiling references. We also thank Kimberly Smelker for her help in compiling information for this chapter.

REFERENCES

Abreu-Grobois FA, Guzmán V, Cuevas E, Alba Gamio M (eds) (2005) Proceedings, Memorias del Taller Rumbo a la COP 3: diagnóstico del estado de la tortuga carey (*Eretmochelys imbricata*) en la Peńinsula de Yucatán y determinación de acciones estratégicas. Telchac Puerto, Yucatán, Mexico, Marzo, pp xiv–75

Ackerman RA (1997) The nest environment and the embryonic development of sea turtles. In: Lutz PL, Musick JA (eds) The biology of sea turtles, vol 1. CRC Press, Boca Raton, FL, USA, pp 83–106

Addison DS (1996) Mean annual nest frequency for renesting loggerhead turtles (*Caretta caretta*) on the southwest coast of Florida. Mar Tur Newsl (MTN) 75:13–15

Addison DS (1997) Sea turtle nesting on Cay Sal, Bahamas, recorded June 2–4, 1996. J Bahamas Sci 5:34–35

Addison DS, Morford B (1996) Sea turtle nesting activity on the Cay Sal Bank, Bahamas. J Bahamas Sci 3:31–36

Alabama Department of Conservation and Natural Resources [ADCNR] (2012) Homepage. http://www.dcnr.state.al.us/. Accessed 20 November 2013

Alam SK, Brim MS (2000) Organochlorine, PCB, PAH, and metal concentrations in eggs of loggerhead sea turtles (*Caretta caretta*) from northwest Florida, USA. J Environ Sci Health B35:705–724

Alava JJ, Keller JM, Wyneken J, Crowder L, Scott G, Kucklick JR (2011) Geographical variation of persistent organic pollutants in eggs of threatened loggerhead sea turtles (*Caretta caretta*) from southeastern United States. Environ Toxicol Chem 30:1677–1688

Allard MW, Miyamoto MM, Bjorndal KA, Bolten AB, Bowen BW (1994) Support for natal homing in green turtles from mitochondrial DNA sequences. Copeia 1994:34–41

Alonso GT (2009) Situación Actual de la Tortuga Lora (*Lepidochelys kempii*) en el Golfo de Mexico. Proceedings, Memorias de la Reunión Nacional sobre Conservación de Tortugas Marinas. Veracruz, Mexico, Noviembre, pp 33–38

Amorocho DF (2001) Status and distribution of the hawksbill turtle, *Eretmochelys imbricata*, in the wider Caribbean region. In: Eckert KL, Abreu Grobois FA (eds) Marine turtle conservation in the wider Caribbean region—a dialogue for effective regional management. Santo Domingo, Dominican Republic, 16–18 November 1999, WIDECAST, IUCN-MTSG, WWF, and UNEP-CEP, pp 41–45

Amos AF (1989) The occurrence of hawksbills *(Eretmochelys imbricata)* along the Texas Coast. Proceedings, 9th Annual Workshop on Sea Turtle Conservation and Biology, Jekyll Island, GA, USA, August, Part 1, pp 9–12

Anonymous (1992) First Kemp's ridley nesting in South Carolina. MTN 59:23

Atencio DE (1994) Marine turtle nesting activity on Eglin AFB, Florida, 1987–1992. Proceedings, 13th Annual Symposium on Sea Turtle Biology and Conservation, NOAA Technical Memorandum NMFS-SEFSC-341. National Technical Information Service, Springfield, VA, USA, pp 201–204

Atkinson C, Berrondo L, Harrison E (2010) Report on the 2009 Green turtle program at Tortuguero, Costa Rica. Caribbean Conservation Corporation, Gainesville, FL, USA and Ministry of Environment and Energy of Costa Rica, San Pedro, Costa Rica. 66 p

Avens L, Taylor JC, Goshe LR, Jones TT, Hastings M (2009) Use of skeletochronological analysis to estimate the age of leatherback sea turtles *Dermochelys coriacea* in the western North Atlantic. Endang Species Res 8:165–177

Avens L, Goshe LR, Harms CA, Anderson ET, Hall AG, Cluse WM, Godfrey MH, Braun-McNeill J, Stacy B, Bailey R, Lamont MM (2012) Population characteristics, age structure, and growth dynamics of neritic juvenile green turtles in the northeastern Gulf of Mexico. Mar Ecol Prog Ser 458:213–229

Bacon PR (1970) Studies on the leatherback turtle, *Dermochelys coriacea* (L) in Trinidad, West Indies. Biol Conserv 2:213–217

Balazs GH (1985) Impact of ocean debris on marine turtles: Entanglement and ingestion. Proceedings, Workshop on the Fate and Impact of Marine Debris, Honolulu, HI, USA, November, Session II, pp 387–429

Baremore IE, Carlson JK, Hollensead LD, Bethea DM (2007) Catch and bycatch in U.S. Southeast Gillnet Fisheries, 2007. NOAA Technical Memorandum NMFS-SEFSC-565. National Marine Fisheries Service, Panama City, FL, USA, December. 20 p

Barichivich WJ (2006) Characterization of a marine turtle aggregation in the big bend of Florida. MS Thesis, University of Florida, Gainsville, FL, USA. 46 p

Barichivich WJ, Sulak KL, Carthy RR (1998) Feeding ecology and habitat affinities of Kemp's ridley sea turtles (*Lepidochelys kempi*) in the Big Bend, Florida. Southeast Fisheries Science Center, NMFS, Panama City, FL, USA. December. 18 p

Bass AL (1999) Genetic analysis to elucidate the natural history and behavior of hawksbill turtles (*Eretmochelys imbricata*) in the Wider Caribbean: A review and re-analysis. Chel Conserv Biol 3:195–199

Bass AL, Witzell WN (2000) Demographic composition of immature green turtles (*Chelonia mydas*) from the east central Florida coast: Evidence from mtDNA markers. Herpetologica 56:357–367

Bass AL, Epperly SP, Braun-McNeill J (2006) Green turtle (*Chelonia mydas*) foraging and nesting aggregations in the Caribbean and Atlantic: Impact of currents and behavior on dispersal. J Hered 97:346–354

Bell BA, Spotila JR, Paladino FV, Reina RD (2003) Low reproductive success of leatherback turtles, *Dermochelys coriacea*, is due to high embryonic mortality. Biol Conserv 115:131–138

Bell IP (2012) The hawksbill turtle, *Eretmochelys imbricata*, (Linnaeus 1766): Ecological insights of a resident population in the northern Great Barrier Reef, Queensland, Australia. PhD Thesis, James Cook University, Queensland, Australia. 143 p

Bjorndal KA (1980) Nutrition and grazing behavior of the green turtle *Chelonia mydas*. Mar Biol 56:147–154

Bjorndal KA (1982) The consequences of herbivory for the life history pattern of the Caribbean green turtle, *Chelonia mydas*. In: Bjorndal KA (ed) Biology and conservation of sea turtles. Smithsonian Institution Press, Washington, DC, USA, pp 111–116

Bjorndal KA (1985) Nutritional ecology of sea turtles. Copeia 1985:736–751

Bjorndal KA (1997) Foraging ecology and nutrition of sea turtles. In: Lutz PL, Musick JA (eds) The biology of sea turtles, vol 1. CRC Press, Boca Raton, FL, USA, pp 199–232

Bjorndal KA, Bolten AB (1992) Spatial distribution of green turtle (*Chelonia mydas*) nests at Tortuguero, Costa Rica. Copeia 1992:45–53

Bjorndal KA, Bolten AB (eds) (2000) Proceedings of a Workshop on Assessing Abundance and Trends for In-Water Sea Turtle Populations. NOAA Technical Memorandum NMFS-SEFSC-445. National Marine Fisheries Service, Miami, FL, USA. October. 83 p

Bjorndal KA, Carr A (1989) Variation in clutch size and egg size in the green turtle nesting population at Tortuguero, Costa Rica. Herpetologica 45:181–189

Bjorndal KA, Jackson JBC (2003) Roles of sea turtles in marine ecosystems: Reconstructing the past. In: Lutz PA, Musick JA, Wyneken J (eds) The biology of sea turtles, vol 2. CRC Press, Boca Raton, FL, USA, pp 259–273

Bjorndal KA, Meylan AB, Turner BJ (1983) Sea turtle nesting at Melbourne Beach, Florida, 1. Size, growth, reproductive biology. Biol Conserv 26:65–77

Bjorndal KA, Bolten AB, Lagueux CJ (1994) Ingestion of marine debris by juvenile sea turtles in coastal Florida habitats. Mar Pollut Bull 28:154–158

Bjorndal KA, Wetherall JA, Bolten AB, Mortimer JA (1999) Twenty-six years of green turtle nesting at Tortuguero, Costa Rica: An encouraging trend. Conserv Biol 13:126–134

Bjorndal KA, Bolten AB, Martins HR (2000) Somatic growth model of juvenile loggerhead sea turtles *Caretta caretta*: duration of pelagic stage. Mar Ecol Prog Ser 202:265–272

Bjorndal KA, Bolten AB, Koike B, Schroeder BA, Shaver DJ, Teas WG, Witzell WN (2001) Somatic growth function for immature loggerhead sea turtles, *Caretta caretta*, in southeastern U.S. waters. Fish Bull 99:240–246

Bjorndal KA, Bolten AB, Martins HR (2003) Estimates of survival probabilities for oceanic-stage loggerhead sea turtles (*Caretta caretta*) in the North Atlantic. Fish Bull 101:732–736

Bjorndal KA, Bowen BW, Chaloupka M, Crowder LB, Heppell SS, Jones CM, Lutcavage ME, Policansky D, Solow AR, Witherington BE (2011) Better science needed for restoration in the Gulf of Mexico. Science 331:537–538

Blair K (2005) Determination of sex ratios and their relationship to nest temperature of loggerhead sea turtle (*Caretta caretta*, L.) hatchlings produced along the southeastern Atlantic coast of the United States. Florida Atlantic University, Boca Raton, FL, USA. 89 p

Blumenthal JM, Abreu-Grobois FA, Austin TJ, Broderick AC, Bruford MW, Coyne MS, Ebanks-Petrie G, Formia A, Meylan PA, Meylan AB, Godley BJ (2009) Turtle groups or turtle soup: Dispersal patterns of hawksbill turtles in the Caribbean. Mol Ecol 18:4841–4853

Bolten AB (2003) Active swimmers-passive drifters: The oceanic juvenile stage of loggerheads in the atlantic system. In: Bolten AB, Witherington BE (eds) Loggerhead sea turtles. Smithsonian Books, Washington, DC, USA, pp 63–78

Bolten AB, Witherington BE (eds) (2003) Loggerhead sea turtles. Smithsonian Books, Washington, DC, USA. 352 p

Bolten AB, Bjorndal KA, Martins HR, Dellinger T, Biscoito MJ, Encalada SE, Bowen BW (1998) Transatlantic developmental migrations of loggerhead sea turtles demonstrated by mtDNA sequence analysis. Ecol Appl 8:1–7

Bolten AB, Crowder LB, Dodd MG, MacPherson SL, Musick JA, Schroeder BA, Witherington BE, Long KJ, Snover ML (2011) Quantifying multiple threats to endangered species: An example from loggerhead sea turtles. Front Ecol Environ 9:295–301

Bouchard SS, Bjorndal KA (2000) Sea turtles as biological transporters of nutrients and energy from marine to terrestrial ecosystems. Ecology 81:2305–2313

Boulon RH Jr (1994) Growth rates of wild juvenile hawksbill turtles, *Eretmochelys imbricata*, in St. Thomas, United States Virgin Islands. Copeia 1994:811–814

Boulon RH Jr, Dutton PH, McDonald DL (1996) Leatherback turtles (*Dermochelys coriacea*) on St. Croix, U.S. Virgin Islands: Fifteen years of conservation. Chel Conserv Biol 2:144–147

Bowen BW, Avise JC (1995) Conservation genetics of marine turtles. In: Avise JC, Hamrick JL (eds) Conservation genetics: Case histories from nature. Chapman and Hall, New York, NY, USA, pp 190–237

Bowen BW, Meylan AB, Avise JC (1989) An odyssey of the green sea turtle: Ascension Island revisited. Proc Natl Acad Sci Biol 86:573–576

Bowen BW, Meylan AB, Ross JP, Limpus CJ, Balazs GH, Avise JC (1992) Global population structure and natural history of the green turtle (*Chelonia mydas*) in terms of matriarchal phylogeny. Evolution 46:865–881

Bowen BW, Grant WS, Hillis-Starr Z, Shaver DJ, Bjorndal KA, Bolten AB, Bass AL (2007) Mixed-stock analysis reveals the migrations of juvenile hawksbill turtles (*Eretmochelys imbricata*) in the Caribbean Sea. Mol Ecol 16:49–60

Braun-McNeill J, Epperly SP (2002) Spatial and temporal distribution of sea turtles in the western North Atlantic and the U.S. Gulf of Mexico from Marine Recreational Fishery Statistics Survey (MRFSS). Mar Fish Rev 64:50–56

Broderick AC, Godley BJ, Hays GC (2001) Trophic status drives interannual variability in nesting numbers of marine turtles. Proc Roy Soc Lond B Bio 268:1481–1487

Broderick AC, Coyne MS, Glen F, Fuller WJ, Godley BJ (2006) Foraging site fidelity of adult green and loggerhead turtles. Proceedings, 26th Annual Workshop on Sea Turtle Conservation and Biology, Island of Crete, EL, Greece. 83 p

Brongersma LD (1972) European atlantic turtles. Zoological treatises. National Museum of Natural History, Leiden, ZH, The Netherlands. 318 p

Brongersma LD, Carr AF (1983) *Lepidochelys kempi*, (Garman) from Malta. Proc Roy Neth Acad Art Sci Series C 86:445–454

Burchfield PM, Byles R, Vicente Mongrell MVZJ, Bartlett M, Rostal D (1988) Report on Republic of Mexico/United States of America: Conservation effort on behalf of Kemp's ridley sea turtle at Playa de Rancho Nuevo, Tamaulipas, Mexico, 1988. Gladys Porter Zoo, Brownsville, TX, USA. 83 p

Byles RA (1989) Satellite telemetry of Kemp's ridley sea turtle, *Lepidochelys kempi*, in the Gulf of Mexico. Proceedings, 9th Annual Workshop on Sea Turtle Conservation and Biology, Jekyll Island, GA, USA, August, Part 1, pp 25–26

Byles RA, Plotkin PT (1994) Comparison of the migratory behavior of the congeneric sea turtles *Lepidochelys olivacea* and *L. kempii*. Proceedings, 13th Annual Symposium on Sea Turtle Biology and Conservation, Jekyll Island, GA, USA, January, Part 1. 39 p

Caillouet CW Jr (1995) Egg and hatchling take for the Kemp's ridley headstart experiment. MTN 68:13–15

Caillouet CW Jr, Duronslet MJ, Landry AM Jr, Revera DB, Shaver DJ, Stanley KM, Heinly RW, Stabenau EK (1991) Sea turtle strandings and shrimp fishing effort in the northwestern Gulf of Mexico, 1986–1989. Fish Bull 89:712–718

Caillouet CW Jr, Fontaine CT, Manzella-Tirpak SA, Shaver DJ (1995) Survival of head-started Kemp's ridley sea turtles (*Lepidochelys kempii*) released into the Gulf of Mexico or adjacent bays. Chel Conserv Biol 1:285–292

Caillouet CW Jr, Shaver DJ, Teas WG, Nance JM, Revera DB, Cannon AC (1996) Relationship between sea turtle stranding rates and shrimp fishing intensities in the northwestern Gulf of Mexico: 1986–1989 versus 1990–1993. Fish Bull 94:237–249

Caillouet CW Jr, Hart RA, Nance JM (2008) Growth overfishing in the brown shrimp fishery of Texas, Louisiana, and adjoining Gulf of Mexico EEZ. Fish Res 92:289–302

Caldwell DK (1959) On the status of the Atlantic leatherback sea turtle, *Dermochelys coriacea*, as a visitant to Florida nesting beaches, with natural history notes. Q J Florida Acad Sci 21:285–291

Caldwell DK, Carr A, Hellier TR Jr (1955) A nest of the Atlantic leatherback turtle, *Dermochelys coriacea* (Linnaeus), on the Atlantic coast of Florida, with a summary of American nesting records. Q J Florida Acad Sci 18:279–284

Campbell CL (1996) Capture of Juvenile Kemp's ridleys in the nearshore waters of Apalachee Bay, Florida. Proceedings, 16th Annual Symposium on Sea Turtle Biology and Conservation, Hilton Head, SC, USA, February, pp 28–30

Cannon AC (1995) Incidental catch of Kemp's ridley sea turtles (*Lepidochelys kempii*) by Hook-and-Line along the Upper Texas Coast. Task final report for fiscal year 1994. NMFS Marine Entanglement Research Program, Galveston, TX, USA

Cannon AC, Fontaine CT, Williams TD, Revera DB, Caillouet CW Jr (1994) Incidental catch of Kemp's ridley sea turtles (*Lepidochelys kempi*), by hook and line, along the Texas Coast, 1980–1992. Proceedings, 13th Annual Symposium on Sea Turtle Biology and Conservation, Jekyll Island, GA, USA, February, Part I, pp 40–42

Carballo JL, Olabarria C, Garza OT (2002) Analysis of four macroalgal assemblages along the Pacific Mexican coast during and after the 1997–1998 El Niño. Ecosystems 5:749–760

Carr A (1952) Handbook of turtles. Comstock Publishing Associates, Cornell University, Ithaca, NY, USA. 542 p

Carr A (1957) Notes on the zoogeography of the Atlantic sea turtles of the genus *Lepidochelys*. Rev Biol Trop 5:45–61

Carr AF (1963) Panspecific reproductive convergence in *Lepidochelys kempi*. Ergeb Biol 25:297–303

Carr AF (1979) The windward road: Adventures of a naturalist on remote Caribbean shores, Revised edn. University of Florida Press, Gainsville, FL, USA. 258 p

Carr AF Jr (1984) So excellent a fishe: A natural history of sea turtles, Revised edn. Scribner's, New York, NY, USA. 280 p

Carr A (1986) Rips, FADS, and little loggerheads. BioScience 36:92–100

Carr A (1987) New perspectives on the pelagic stage of sea turtle development. Conserv Biol 1:103–121

Carr AF, Caldwell DK (1956) The ecology and migrations of sea turtles, 1. Results of field work in Florida, 1955. Am Mus Novit 1793:1–23

Carr A, Carr MH (1972) Site fixity in the Caribbean green turtle. Ecology 53:425–429

Carr A, Hirth H (1962) The ecology and migrations of sea turtles, 5. Comparative features of isolated green turtle colonies. Am Mus Novit 2091:1–42

Carr A, Ogren L (1959) The ecology and migrations of sea turtles, 3. *Dermochelys* in Costa Rica. Am Mus Novit 1958:1–29

Carr A, Ogren L (1966) The ecology and migrations of sea turtles, 6. The hawksbill turtle in the Caribbean Sea. Am Mus Novit 2248:1–29

Carr A, Carr MH, Meylan AB (1978) The ecology and migrations of sea turtles, 7. The west Caribbean green turtle colony. B Am Mus Nat Hist 162:1–46

Carreras C, Pont S, Maffucci F, Pascual M, Barceló A, Bentivegna F, Cardona L, Alegre F, San Félix M, Fernández G, Aguilar A (2006) Genetic structuring of immature loggerhead sea turtle (*Caretta caretta*) in the Mediterranean Sea reflects water circulation patterns. Mar Biol 149:1269–1279

Carrillo E, Webb GJW, Manolis SC (1999) Hawksbill turtles (*Eretmochelys imbricata*) in Cuba: An assessment of the historical harvest and its impacts. Chel Conserv Biol 3:264–280

Carriol RP, Vader W (2002) Occurrence of *Stomatolepas elegans* (Cirripedia: Balanomorpha) on a leatherback turtle from Finmark, northern Norway. J Mar Biol Assoc UK 82:1033–1034

Cato JC, Prochaska FJ, Pritchard PCH (1978) An analysis of the capture, marketing, and utilization of marine turtles. Environmental Assessment Division, National Marine Fisheries Service, St. Petersburg, FL, USA. 238 p

Caut S, Guirlet E, Jouquet P, Girondot M (2006) Influence of nest location and yolkless eggs on the hatching success of leatherback turtle clutches in French Guiana. Can J Zool 84:908–915

Chávez H (1968) Marcado y recaptura de individuos de tortuga lora, *Lepidochelys kempi* (Garman). Instituto Nacional de Investigaciones Biológico Pesqueras 19:1–28

Chávez H (1969) Tagging and recapture of the lora turtle (*Lepidochelys kempii*). Int Turtle Tortoise Soc J 2:16–19, 32–36

Chávez H, Contreras GM, Hernández DE (1967) Aspectos biologicos y proteccion de la tortuga lora, *Lepidochelys kempii* (Garman), en la costa de Tamaulipas, Mexico. Publicación Número 17. Instituto Nacional de Investigaciones Biológico Pesqueras, Ciudad de Mexico, 40 p

Collard SB (1987) Review of oceanographic features relating to neonate sea turtle distribution and dispersal in the pelagic environment: Kemp's ridley (*Lepidochelys kempi*) in the Gulf of Mexico. Final Report. Contract Number 40-GFNF-5-00193. National Marine Fisheries Service, Panama City, FL, USA, Dec. 42 p

Collard SB (1990) Leatherback turtles feeding near a water mass boundary in the eastern Gulf of Mexico. MTN 50:12–14

Collard SB, Ogren LH (1990) Dispersal scenarios for pelagic post-hatchling sea turtles. Bull Mar Sci 47:233–243

Conant TA, Dutton PH, Eguchi T, Epperly SE, Fahy CC, Godfrey MH, MacPherson SL, Possardt EE, Schroeder BA, Seminoff JA, Snover ML, Upite CM, Witherington BE (2009) Loggerhead sea turtle (*Caretta caretta*) 2009 status review under the U.S. endangered species act. A report of the loggerhead biological review team to the National Marine Fisheries Service. NMFS, Silver Spring, MD, USA. 222 p

Cornelius SE, Arauz R, Fretey J, Godfrey MH, Márquez-M R, Shanker K (2007) Effect of land-based harvest of *Lepidochelys*. In: Plotkin PT (ed) Biology and conservation of ridley sea turtles. The Johns Hopkins University Press, Baltimore, MD, USA, pp 231–252

Coyne MS (1994) Feeding ecology of subadult green sea turtles in South Texas Waters. MS Thesis, Texas A&M University, College Station, TX, USA. 76 p

Coyne MS (2000) Population sex ratio of the Kemp's ridley sea turtle (*Lepidochelys kempii*): problems in population modeling. PhD Thesis, Texas A&M University, College Station, TX, USA. 124 p

Coyne MS, Landry AM Jr (2000) Plasma testosterone dynamics in the Kemp's ridley sea turtle. Proceedings, 18th International Sea Turtle Symposium, Mazatlán, Sinaloa, Mexico, March, Part VI. 286 p

Craig JK, Crowder LB, Gray CD, McDaniel CJ, Henwood TA, Hanifen JG (2001) Ecological effects of hypoxia on fish, sea turtles, and marine mammals in the northwestern Gulf of Mexico. Coast Estuar Stud 58:269–292

Crouse DT, Crowder LB, Caswell H (1987) A stage-based population model for loggerhead sea turtles and implications for conservation. Ecology 68:1412–1423

Crowder L, Heppell S (2011) The decline and rise of a sea turtle: How Kemp's ridleys are recovering in the Gulf of Mexico. Solutions 2:67–73

Crowder LB, Crouse DT, Heppell SS, Martin TH (1994) Predicting the impact of turtle excluder devices on loggerhead sea turtle populations. Ecol Appl 4:437–445

Cuevas E, de los Ángeles Liceaga-Correa M, Garduño-Andrade M (2007) Spatial character-istics of a foraging area for immature hawksbill turtles (Eretmochelys imbricata) in Yucatán, Mexico. Amphibia-Reptilia 28:337–346

Cuevas E, Abreu-Grobois FA, Guzmán-Hernández V, Liceaga-Correa MA, van Dam RP (2008) Post-nesting migratory movements of hawksbill turtles Eretmochelys imbricata in waters adjacent to the Yucatán Peninsula, Mexico. Endang Species Res 10:123–133

Cuevas E, de los Ángeles Liceaga-Correa A, Mariño-Tapia I (2010) Influence of beach slope and width on hawksbill (Eretmochelys imbricata) and green turtle (Chelonia mydas) nesting activity in El Cuyo, Yucatán, Mexico. Chel Conserv Biol 9:262–267

Cuevas E, González-Garza BI, Guzmán-Hernández V, van Dam RP, García-Alvarado P, Abreu-Grobois FA, Huerta-Rodríguez P (2012) Tracking turtles off Mexico's Yucatán Peninsula. In SWOT (State of the Worlds Sea Turtles) Report Volume VII. Arlington, VA, USA, pp 8–9

Dahlen MK, Bell R, Richardson JI, Richardson TH (2000) Beyond D-0004: Thirty-four years of loggerhead (Caretta caretta) research on Little Cumberland Island, Georgia, 1964–1997. Proceedings, 18th International Sea Turtle Symposium, Mazatlán, Sinaloa, Mexico, March, Part VI, pp 60–62

Dalrymple GH, Hampp JC, Wellins DJ (1985) Male-biased sex ratio in a cold nest of a hawksbill sea turtle Eretmochelys imbricata. J Herpetol 19:158–159

Davis GE, Whiting MC (1977) Loggerhead sea turtle nesting in Everglades National Park, Florida, USA. Herpetologica 33:18–28

Davis RW, Evans WE, Wursig B (2000) Cetaceans, sea turtles and seabirds in the Northern Gulf of Mexico: Distribution, Abundance and Habitat Associations. Texas A&M at Galveston, National Marine Fisheries Service OCS Study MMS 2000-002. Minerals Management Service, Gulf of Mexico OCS Region, New Orleans, LA, USA. 28 p

Debade X, Nolasco del Aguila D, Harrison E (2009) Report on the 2008 leatherback program at Tortuguero, Costa Rica. Caribbean Conservation Corporation, Gainesville and Ministry of Environment and Energy of Costa Rica, San Pedro, Costa Rica. 46 p

del Monte-Luna P, Guzmán-Hernández V, Cuevas EA, Arreguín-Sánchez F, Lluch-Belda D (2012) Effect of North Atlantic climate variability on hawksbill turtles in the Southern Gulf of Mexico. J Exp Mar Biol Ecol 412:103–109

Dellinger T, Freitas C (2000) Movement and diving behaviour of pelagic stage loggerhead sea turtles in the North Atlantic: Preliminary results obtained through satellite telemetry. Proceedings, 19th Annual Symposium on Sea Turtle Conservation and Biology, South Padre Island, TX, USA, March, pp 155–157

den Hartog JC, van Nierop MM (1984) A study on the gut contents of six leathery turtles Dermochelys coriacea (Linnaeus) (Reptilia: Testudines: Dermochelyidae) from British waters and from the Netherlands. Zoologische Verhandelingen 209:1–36

Diez CE, van Dam RP (2003) Sex ratio of an immature hawksbill sea turtle aggregation at Mona Island, Puerto Rico. J Herpetol 37:533–537

Dobie JL, Ogren LH, Fitzpatrick JF Jr (1961) Food notes and records of the Atlantic ridley turtle (Lepidochelys kempi) from Louisiana. Copeia 1961:109–110

Dodd CK Jr (1988) Synopsis of the biological data on the loggerhead sea turtle *Caretta caretta* (Linnaeus 1758). USFWS Biological Report 88-14. U.S. Fish and Wildlife Service, Washington, DC, USA. 110 p

Dodd CK Jr, Byles R (1991) The status of loggerhead, *Caretta caretta*; Kemp's ridley, *Lepidochelys kempi*; and green, *Chelonia mydas*, sea turtles in U.S. waters: A reconsideration. Mar Fish Rev 53:30–33

DOF (Diario Oficial de la Federación) (1977) Acuerdo que establece como Zona de Refugio y de Veda para la proteccion de la Tortuga lora. Department de Pesca, Ciudad de Mexico, pp 1–2

DOF (1986) Decreto por el que se determinan como Zonas de Reserva y Sitios de Refugio para la proteccion, conservacion, repoblacion, desarrollo y control de las diversas species de Tortuga marina, los lugares donde anida y desovan dichas species. Department de Pesca, Ciudad de Mexico, pp 1–5

DOF (1990) Acuerdo pro el que se establece veda total para todas las species y subspecies de Tortugas marinas en aguas de jurisdicción nacional de los litorales del Océano Pacífico, Golfo de Mexico y Mar Caribe. Department de Pesca, Ciudad de Mexico, pp 21–22

DOF (2002) Acuerdo por el que se determinan como áreas naturales protegidas, con la categoría de santuarios, a las zonas de reserva y sitios de refugio para la protección, conservación, repoblación, desarrollo y control de las diversas especies de tortuga marina, ubicadas en los estados de Chiapas, Guerrero, Jalisco, Michoacán, Oaxaca, Sinaloa, Tamaulipas y Yucatán, identificadas en el decreto publicado el 29 de Octubre de 1986. Department de Pesca, Ciudad de Mexico, pp 268–269

Doughty RW (1984) Sea turtles in Texas: A forgotten commerce. Southwestern Historical Quarterly 88:43–70

Dow W, Eckert K, Palmer M, Kramer P (2007) An atlas of sea turtle nesting habitat for the wider Caribbean region. WIDECAST Technical Report No. 6. Wider Caribbean Sea Turtle Conservation Network, Beaufort, NC, USA. 267 p

Dutton PH, Whitmore CP, Mrosovsky N (1985) Masculinisation of leatherback turtle, *Dermochelys coriacea*, hatchlings from eggs incubated in Styrofoam boxes. Biol Conserv 31:249–264

Dutton DL, Dutton PH, Chaloupka M, Boulon RH (2005) Increase of a Caribbean leatherback turtle *Dermochelys coriacea* nesting population linked to long-term nest protection. Biol Conserv 126:186–194

Eaton C, McMichael E, Witherington B, Foley A, Hardy R, Meylan A (2008) In-water sea turtle monitoring and research in Florida: Review and recommendations. NOAA Technical Memorandum NMFS-OPR-38, St. Petersburg, FL, USA, June. 233 p

Echeverría-García A, Torres-Burgos E (2007) 14 Años de Monitoreo de Tortugas Marinas en Yucatán. Memorias de la Reunión Nacional sobre Conservación de Tortugas Marinas, Veracruz, VE, Mexico, Noviembre. 21 p

Echeverría-García AW, Torres-Burgos E, Quezada-Domínguez C, López-Alonzo MA (2008) Informe Técnico de la Temporada 2007 en los Centros para la Protección y Conservación de Tortugas Marinas, operados por la Secretaría de Ecología del Gobierno del Estado de Yucatán, Mexico. Centros Para la Protección y Conservación de las Tortugas Marinas, Gobierno de Estado de Yucatán, Merida, Yucatán, Mexico. 17 p

Echeverría-García AW, Torres-Burgos E, Quezada-Domínguez C, López-Alonzo MA (2009) Informe Técnico de la Temporada 2008 de los Centros para la Protección y Conservación de Tortugas Marinas (Campamentos Tortugueros), Operados por la Secretaría de Desarrollo Urbano y Medio Ambiente (Seduma) del Gobierno del Estado de Yucatán, Mexico.

Centros Para la Protección y Conservación de las Tortugas Marinas, Gobierno de Estado de Yucatán, Merida, Yucatán, Mexico

Eckert KL (1987) Environmental unpredictability and leatherback sea turtle (*Dermochelys coriacea*) nest loss. Herpetologica 43:315–323

Eckert SA (2002) Distribution of juvenile leatherback sea turtle *Dermochelys coriacea* sightings. Mar Ecol Prog Ser 230:289–293

Eckert KL, Eckert SA (1988) Pre-reproductive movements of leatherback sea turtles (*Dermochelys coriacea*) nesting in the Caribbean. Copeia 1988:400–406

Eckert KL, Eckert SA (1990) Embryo mortality and hatch success in *in situ* and translocated leatherback sea turtle *Dermochelys coriacea* eggs. Biol Conserv 53:37–46

Eckert SA, Eckert KL, Ponganis P, Kooyma GL (1989a) Diving and foraging behavior of leatherback sea turtles (*Dermochelys coriacea*). Can J Zool 67:2834–2840

Eckert KL, Eckert SA, Adams TW, Tucker AD (1989b) Inter-nesting migrations by leatherback sea turtles (*Dermochelys coriacea*) in the West Indies. Herpetologica 45:190–194

Eckert SA, Bagley D, Kubis S, Ehrhart L, Johnson C, Stewart K, DeFreese D (2006) Internesting and postnesting movements and foraging habitats of leatherback sea turtles (*Dermochelys coriacea*) nesting in Florida. Chel Conserv Biol 5:239–248

Eckert SA, Moore JE, Dunn DC, van Buiten RS, Eckert KL, Halpin PN (2008) Modeling loggerhead turtle movement in the Mediterranean: Importance of body size and oceanography. Ecol Appl 18:290–308

Eckert KL, Wallace BP, Frazier JG, Eckert SA, Pritchard PCH (2012) Synopsis of the biological data on the leatherback sea turtle (*Dermochelys coriacea*). U.S. Department of Interior, Fish and Wildlife Service, Biological Technical Publication BTP-R4015-2012. Washington, DC, USA. 160 p

Ehrhart LM (1980) A survey of marine turtles nesting at the Kennedy Space Center Cape Canaveral Air Force Station, North Brevard County, Florida. Report to the Division of Marine Resources, Florida. 216 p

Ehrhart LM, Bagley DA, Redfoot WE (2003) Loggerhead turtles in the atlantic ocean: Geographic distribution, abundance, and population status. In: Bolten AB, Witherington BE (eds) Loggerhead sea turtles. Smithsonian Books, Washington, DC, USA, pp 157–174

Epperly SP (2003) Fisheries-related mortality and turtle excluder devices. In: Lutz PA, Musick JA, Wyneken J (eds) The biology of sea turtles, vol 2. CRC Press, Boca Raton, FL, USA, pp 339–353

Epperly SP, Teas WG (2002) Turtle excluder devices—are the escape openings large enough? Fish Bull 100:466–474

Epperly SP, Braun J, Chester AJ, Cross FA, Merriner JV, Tester PA, Churchill JH (1996) Beach strandings as an indicator of at-sea mortality of sea turtles. Bull Mar Sci 59:289–297

Epperly S, Avens L, Garrison L, Henwood T, Hoggard W, Mitchell J, Nance J, Poffenberger J, Sasso C, Scott-Denton E, Yeung C (2002) Analysis of sea turtle bycatch in the commercial shrimp fisheries of the southeast U.S. Waters and the Gulf of Mexico. NOAA Technical Memorandum NMFS-SEFSC-490. National Marine Fisheries Service, Miami, FL, USA, November. 88 p

Evans D, Ordonez C, Troëng S, Dreus C (2007) Satellite tracking of leatherback turtles from Caribbean Central America Reveals Unexpected Foraging Grounds. Proceedings, 27th Annual Symposium in Sea Turtle Conservation and Biology, Myrtle Beach, SC, USA, April, pp 40–41

Fairfield CP, Garrison LP (2008) Estimated bycatch of marine mammals and sea turtles in the U.S. Atlantic Pelagic Longline Fleet During 2007. NOAA Technical Memorandum. NMFS-SEFSC-572. National Marine Fisheries Service, Miami, FL, USA. August. 62 p

Fairfield-Walsh C, Garrison LP (2006) Estimated bycatch of marine mammals and sea turtles in the U.S. Atlantic Pelagic Longline Fleet During 2005. NOAA Technical Memorandum. NMFS-SEFSC- 539. National Marine Fisheries Service, Miami, FL, USA. May. 52 p

Fairfield-Walsh C, Garrison LP (2007) Estimated bycatch of marine mammals and sea turtles in the U.S. Atlantic Pelagic Longline Fleet During 2006. NOAA Technical Memorandum. NMFS-SEFSC-560. National Marine Fisheries Service, Miami, FL, USA. June. 54 p

FDEP (Florida Department of Environmental Protection), NOAA (National Oceanic and Atmospheric Administration), USDOI (U.S. Department of the Interior) (1997) Damage assessment and restoration plan/environmental assessment for the August 10, 1993 Tampa Bay Oil Spill: Volume I—Ecological injuries—Final. Florida Department of Environmental Protection, Tallahassee, FL, USA. June. 91 p

FFWCC (Florida Fish and Wildlife Conservation Commission) (2012) Homepage. http://myfwc. com/. Accessed 20 November 2013

FFWCC FWRI (Fish and Wildlife Research Institute) (2011a) Marine turtle program: Statewide nesting totals 1979–2009. http://myfwc.com/media/1313557/Statewide-1979-2010.pdf. Accessed 20 November 2013

FFWCC FWRI (2011b) Marine turtle program: Sea turtle stranding. http://ocean.floridamarine. org/TRGIS/Description_Layers_Terrestrial.htm. Accessed 20 November 2013

FFWCC FWRI (2012) Statewide nesting beach program: Loggerhead nesting data, 2007–2012. http://myfwc.com/media/2078432/LoggerheadNestingData.pdf. Accessed 20 November 2013

Finkbeiner EM, Wallace BP, Moore JE, Lewison RL, Crowder LB, Read AJ (2011) Cumulative estimates of sea turtle bycatch and mortality in USA fisheries between 1990 and 2007. Biol Conserv 144:2719–2727

Fish MR, Cote IM, Horrocks JA, Mulligan B, Watkinson AR, Jones AP (2008) Construction setback regulations and sea-level rise: Mitigating sea turtle nesting beach loss. Ocean Coast Manage 51:330–341

Foley AM, Peck SA, Harman GR, Richardson LW (2000) Loggerhead turtle (*Caretta caretta*) nesting habitat on low-relief mangrove islands in southwest Florida and consequences to hatchling sex ratios. Herpetologica 56:436–447

Foley AM, Schroeder BA, Redlow AE, Fick-Child KJ, Teas WG (2005) Fibropapillomatosis in stranded green turtles (*Chelonia mydas*) from the eastern United States (1980–1998): Trends and associations with environmental factors. J Wildlife Dis 41:29–41

Foley AM, Singel KE, Dutton PH, Summers TM, Redlow AE, Lessman J (2007) Characteristics of a green turtle (*Chelonia mydas*) assemblage in northwestern Florida determined during a hypothermic stunning event. Gulf Mex Sci 2007:131–143

Foley AM, Schroeder BA, MacPherson SL (2008) Post-nesting migrations and resident areas of Florida Loggerhead Turtles (*Caretta caretta*). Proceedings, 25th Annual Symposium on Sea Turtle Biology and Conservation, Savannah, GA, USA, January, pp 75–76

Fontaine C, Shaver D (2005) Head-starting the Kemp's ridley sea turtle, *Lepidochelys kempii*, at the NMFS Galveston Laboratory, 1978–1992: A review. Chel Conserv Biol 4:838–845

Fontaine CT, Manzella SA, Williams TD, Harris RM, Browning WJ (1989) Distribution, growth and survival of head started, tagged and released Kemp's ridley sea turtles (*Lepidochelys kempi*) from year-classes 1978–1983. Proceedings, 1st International Symposium on Kemp's Ridley Sea Turtle Biology, Conservation and Management, Galveston, TX, USA, conservation and management, Galveston, October, Session III, pp 124–144

Fossette S, Girard C, Lopez-Mendilaharsu M, Miller P, Domingo A, Evan D, Kelle L, Plot V, Prosdocimi L, Verhage S, Gaspar P, Georges JY (2010) Atlantic leatherback migratory pathways and temporary residence areas. PLoS One 5(e13908):1–12

Fowler LE (1979) Hatching success and nest predation in the green sea turtle, *Chelonia mydas*, at Tortuguero, Costa Rica. Ecology 60:946–955

Frair W, Ackman RG, Mrosovsky N (1972) Body temperature of *Dermochelys coriacea*: Warm turtle from cold water. Science 177:791–793

Frazer NB, Ehrhart LM (1985) Preliminary growth models for green, *Chelonia mydas*, and loggerhead, *Caretta caretta*, turtles in the wild. Copeia 1985:73–79

Frazier J (1993) Una evaluación del manejo de nidos de tortugas marinas en la Península de Yucatán. Memorias del IV Taller Regional Sobre Programas de Conservación de Tortugas Marinas en la Península de Yucatán, Mérida, Yucatán, Mexico, Marzo, pp 37–76

Frazier J, Arauz R, Chevalier J, Formia A, Fretey J, Godfrey MH, Márquez-M R, Pandav B, Shanker K (2007) Human-turtle interactions at sea. In: Plotkin PT (ed) Biology and conservation of ridley sea turtles. The Johns Hopkins University Press, Baltimore, MD, USA, pp 253–296

Fretey J (2001) Biogeography and conservation of marine turtles of the Atlantic Coast of Africa. CMS Technical Series 6. United Nations Environment Programme, Convention on Migratory Species, Bonn, Germany. 429 p

Fretey J, Bour J (1980) Redécouverte du type de *Dermochelys coriacea* (Vandelli) (Testudinata, Dermochelydae). Bolletin di Zoologia 47:193–205

Fritts TH, Reynolds RP (1981) Pilot study of the marine mammals, birds and turtles in OCS Areas of the Gulf of Mexico. FWS/OBS-81/36. U.S. Fish and Wildlife Service and Bureau of Land Management Gulf of Mexico OCS Office, New Orleans, LA, USA. 139 p

Fritts TH, Hoffman W, McGehee MA (1983a) The distribution and abundance of marine turtles in the Gulf of Mexico and nearby Atlantic waters. J Herpetol 17:327–344

Fritts TH, Irvine AB, Jennings RD, Collum LA, Hoffman W, McGehee M (1983b) Turtles, birds and mammals in the Northern Gulf of Mexico and Nearby Atlantic Waters. FWS/OBS-82/65. U.S. Fish and Wildlife Service, Washington, DC, USA. 455 p

FuelFix (2011) Despite BP oil spill, Kemp's ridley turtle numbers at record levels in Texas. http://fuelfix.com/blog/2011/07/19/despite-bp-oil-spill-kemps-ridley-turtle-numbers-at-record-levels-in-texas. Accessed 20 November 2013

Fuller DA (1988) Occurrences of sea turtles in Eastern Louisiana. LSU-CFI-88-12. National Marine Fisheries Service, Panama City, FL, USA. 10 p

Fuller DA (1989) Sea turtle strandings and sightings in southeastern Louisiana. LSU-CFI-89-05. National Marine Fisheries Service, Panama City, FL, USA. 11 p

Garduño M (1998) Ecología de la tortuga carey (*Eretmochelys imbricata*) en la zona de Las Coloradas, Yucatán, Mexico. Universidad de Colima, Colima, Mexico

Garduño-Andrade M (1999) Nesting of the hawksbill turtle, *Eretmochelys imbricata*, at Río Lagartos, Yucatán, Mexico, 1990–1997. Chel Conserv Biol 3:281–285

Garduño-Andrade M (2000) Fecundidad de la tortuga de carey *Eretmochelys imbricata* en las Colorados, Yucatán, Mexico. Ciencia Pesquera 14:67–70

Garduño-Andrade M, Guzmán V, Miranda E, Briseño-Deuñas R, Abreu-Grobois FA (1999) Increases in hawksbill turtle (*Eretmochelys imbricata*) nestings in the Yucatán Peninsula, Mexico, 1977–1996: Data in support of successful conservation? Chel Conserv Biol 3:286–295

Garrison LP (2003) Estimated bycatch of marine mammals and turtles in the U.S. Atlantic Pelagic Longline Fleet during 2001–2002. NOAA Technical Memorandum NMFS-SEFSC-515. National Marine Fisheries Service, Miami, FL, USA. December. 52 p

Garrison LP (2005) Estimated bycatch of marine mammals and turtles in the U.S. Atlantic Pelagic Longline Fleet during 2004. NOAA Technical Memorandum NMFS-SEFSC-531. National Marine Fisheries Service, Miami, FL, USA. June. 57 p

Garrison LP (2007) Estimated marine mammal and turtle bycatch in shark gillnet fisheries along the Southeast U.S. Atlantic Coast: 2000–2006. NMFS Southeast Fisheries Science Center Contribution PRD-07/08-02. National Marine Fisheries Service, Silver Spring, MD, USA. November. 22 p

Garrison LP, Richards PM (2004) Estimated bycatch of marine mammals and turtles in the U.S. Atlantic Pelagic Longline Fleet during 2003. NOAA Technical Memorandum NMFS-SEFSC-527. National Marine Fisheries Service, Miami, FL, USA. September. 57 p

Garrison LP, Stokes L (2010) Estimated bycatch of marine mammals and sea turtles in the U.S. Atlantic Pelagic Longline Fleet During 2009. NOAA Technical Memorandum NMFS-SEFSC-607. National Marine Fisheries Service, Miami, FL, USA. November. 63 p

Garrison LP, Stokes L, Fairfield C (2009) Estimated bycatch of marine mammals and sea turtles in the U.S. Atlantic Pelagic Longline Fleet during 2008. NOAA Technical Memorandum NMFS-SEFSC-591. National Marine Fisheries Service, Miami, June, 63 p

Girard C, Tucker AD, Calmettes B (2009) Post-nesting migrations of loggerhead sea turtles in the Gulf of Mexico: Dispersal in highly dynamic conditions. Mar Biol 156:1827–1839

Girondot M, Godfrey MH, Ponge L, Rivalan P (2007) Modeling approaches to quantify leatherback nesting trends in French Guiana and Suriname. Chel Conserv Biol 6:37–47

Gitschlag G (1992) Offshore oil and gas structures as sea turtle habitat. Proceedings, 11th Annual Workshop on Sea Turtle Biology and Conservation, Jekyll Island, GA, USA, Jekyll Island, March, Part 1, pp 49–50

Gitschlag G, Herczeg B (1994) Sea turtle observations at explosive removals of energy structures. Mar Fish Rev 56:1–8

Godfrey MH, Mrosovsky N (1997) Estimating the time between hatching of sea turtles and their emergence from the nest. Chel Conserv Biol 2:581–585

Godfrey MH, Barreto R, Mrosovsky N (1996) Estimating past and present sex ratios of sea turtles in Suriname. Can J of Zool 74:267–277

Godfrey MH, D'Amato AF, Marcovaldi MA, Mrosovsky N (1999) Pivotal temperature and predicted sex ratios for hatchling hawksbill turtles from Brazil. Can J Zool 77:1465–1473

Godley BJ, Broderick AC, Hays GC (2001) Nesting of green turtles (*Chelonia mydas*) at Ascension Island, South Atlantic. Biol Conserv 97:151–158

González-C V, Álvarez KC, Prosdocimi L, Inchaurraga MC, Dellacasa RF, Faiella A, Echenique C, González R, Andrejuk J, Mianzin HW, Campagna C, Albareda DA (2011) Argentinean coastal waters: A temperate habitat for three species of threatened sea turtles. Mar Biol Res 7:500–508

Gorga CCT (2010) Foraging ecology of green turtles (*Chelonia mydas*) on the Texas Coast, as Determined by Stable Isotope Analysis. MS Thesis, Texas A&M University, College Station, TX, USA. 53 p

Goshe LR (2009) Age at Maturation and Growth Rates of Green Sea Turtles (*Chelonia mydas*) along the Southeastern U.S. Atlantic Coast Estimated Using Skeletochronology. University of North Carolina Wilmington, Wilmington, NC, USA. 69 p

Greer AE, Lazelljun JD, Wright RM (1973) Anatomical evidence for a counter-current heat exchange in the leatherback turtle (*Dermochelys coriacea*). Nature 244:181

Gregory LF, Schmid JR (2001) Stress responses and sexing of wild Kemp's ridley sea turtles (*Lepidochelys kempii*) in the northeastern Gulf of Mexico. Gen Comp Endocr 124:66–74

Griffin RB, Griffin NJ (2003) Distribution, habitat partitioning, and abundance of Atlantic spotted dolphins, bottlenose dolphins, and loggerhead sea turtles on the eastern Gulf of Mexico continental shelf. Gulf Mex Sci 2003:23–34

Groombridge B, Luxmoore R (1989) The green turtle and hawksbill (Reptilia: Cheloniidae): World status, exploitation, and trade. Convention on International Trade in Endangered Species of Wild Fauna and Flora, Cambridge, England. 573 p

Guirlet E, Das K, Girondot M (2008) Maternal transfer of trace elements in leatherback turtles (*Dermochelys coriacea*) of French Guiana. Aquat Toxicol 88:267–276

Gunter G (1951) Destruction of fishes and other organisms on the South Texas coast by the cold wave of January 28-February 3, 1951. Ecology 32:731–736

Guzmán V, Rejón JC, Gómez R, Silva J (1996) Informe Final del Programa de Investigación y Protección de las Tortugas Marinas de Campeche, Mexico. Temporada 1996. Situación Actual. Bol Téc 1 INP-CRIP Carmen. 40 p

Guzmán-Hernández V, García-Alvarado PA (2010) Informe Técnico 2009 del Programa de Conservación de Tortugas Marinas en Laguna de Términos, Campeche, Mexico. Contiene Informacion de: 1. CPCTM Xicalango-Victoria, 2. CPCTM Chacahito, 3. CPCTM Isla Aguada y 4. Reseña estatal regional. APFFLT/RPCyGM/CONANP. 67 p

Guzmán-Hernández V, García Alvarado PA, Rejón JC, Gómez C, Gómez J (2006) Informe Técnico Final del Programa de Conservación de Tortugas Marinas de Campeche, Mexico en 2005. Programa Nacional de Tortugas Marinas. 36 p

Guzmán-Hernández V, Cuevas-Flores EA, Márquez-Millán R (2007) Occurrence of Kemp's ridley (*Lepidochelys kempii*) along the coast of the Yucatán Peninsula, Mexico. Chel Conserv Biol 6:274–277

Hale LF, Carlson JK (2007) Characterization of the shark bottom longline fishery: 2005–2006. NOAA Technical Memorandum NMFS-SEFSC-554. National Marine Fisheries Service, Silver Spring, MD, USA. 28 p

Hale LF, Gulak SJB, Carlson JK (2009) Characterization of the shark bottom longline fishery: 2008. Panama City, Florida. NOAA Technical Memorandum NMFS-SEFSC-586. National Marine Fisheries Service, Silver Spring, MD, USA. April. 23 p

Hall RJ, Belisle AA, Sileo L (1983) Residues of petroleum hydrocarbons in tissues of sea turtles exposed to the Ixtoc I oil spill. J Wildlife Dis 19:106–109

Hamann M, Limpus CJ, Owens DW (2003) Reproductive cycles of males and females. In: Lutz PA, Musick JA, Wyneken J (eds) The biology of sea turtles, vol 2. CRC Press, Boca Raton, FL, USA, pp 135–162

Hanson J, Wibbels T, Martin RE (1998) Predicted female bias in sex rations of hatchling loggerhead sea turtles from a Florida nesting beach. Can J Zool 76:1850–1861

Haro A, Harrison E (2007a) Report on the 2006 green turtle program at Tortuguero, Costa Rica. Caribbean Conservation Corporation, Gainesville and Ministry of Environment and Energy of Costa Rica, San Pedro, Costa Rica. 48 p

Haro A, Harrison E (2007b) Report on the 2006 leatherback program at Tortuguero, Costa Rica. Caribbean Conservation Corporation, Gainesville and Ministry of Environment and Energy of Costa Rica, San Pedro, Costa Rica. 33 p

Haro A, Troëng S (2006a) Report on the 2005 green turtle program at Tortuguero, Costa Rica. Caribbean Conservation Corporation, Gainesville and Ministry of Environment and Energy of Costa Rica, San Pedro, Costa Rica. 49 p

Haro A, Troëng S (2006b) Report on the 2005 Leatherback program at Tortuguero, Costa Rica. Caribbean Conservation Corporation, Gainesville and Ministry of Environment and Energy of Costa Rica, San Pedro, Costa Rica. 25 p

Harrison AL, Bjorndal KA (2006) Connectivity and wide-ranging species in the ocean. In: Crooks KR, Sanjayan M (eds) Connectivity conservation. Cambridge University Press, Cambridge, MA, USA, pp 213–232

Harrison E, Troëng S (2003a) Report on the 2002 leatherback program at Tortuguero, Costa Rica. Caribbean Conservation Corporation, Gainesville and Ministry of Environment and Energy of Costa Rica, San Pedro, Costa Rica. 32 p

Harrison E, Troëng S (2003b) Report on the 2003 leatherback program at Tortuguero, Costa Rica. Caribbean Conservation Corporation, Gainesville and Ministry of Environment and Energy of Costa Rica, San Pedro, Costa Rica. 29 p

Harrison E, Troëng S (2004a) Report on the 2003 green turtle program at Tortuguero, Costa Rica. Caribbean Conservation Corporation, Gainesville and Ministry of Environment and Energy of Costa Rica, San Pedro, Costa Rica. 52 p

Harrison E, Troëng S (2004b) Report on the 2004 leatherback program at Tortuguero, Costa Rica. Caribbean Conservation Corporation, Gainesville and Ministry of Environment and Energy of Costa Rica, San Pedro, Costa Rica. 30 p

Harrison E, Troëng S (2005) Report on the 2004 green turtle program at Tortuguero, Costa Rica. Caribbean Conservation Corporation, Gainesville and Ministry of Environment and Energy of Costa Rica, San Pedro, Costa Rica. 51 p

Hart KM, Fujisaki I (2010) Satellite tracking reveals habitat use by juvenile green sea turtles *Chelonia midas* in the Everglades, Florida, USA. Endang Species Res 11:221–232

Hart KM, Lamont MM, Fujisaki I, Tucker AD, Carthy RR (2012a) Common coastal foraging areas for loggerheads in the Gulf of Mexico: Opportunities for marine conservation. Biol Conserv 145:185–1948

Hart KM, Sartain AR, Fujisaki I, Pratt HL Jr, Morley D, Feeley MW (2012b) Home range, habitat use, and migrations of hawksbills turtles tracked from Dry Tortugas National Park, Florida, USA. Mar Ecol Prog Ser 457:193–207

Hawkes LA, Broderick AC, Godfrey MH, Godley BJ (2007a) Investigating the potential impacts of climate change on a marine turtle population. Glob Chang Biol 13:923–932

Hawkes LA, Broderick AC, Coyne MS, Godfrey MH, Godley BJ (2007b) Only some like it hot—quantifying the environmental niche of the loggerhead sea turtle. Diversity Distrib 13:447–457

Hawkes LA, Broderick AC, Godfrey MH, Godley BJ (2009) Climate change and marine turtles. Endang Spec Res 7:137–154

Hays GC, Houghton JDR, Myers AE (2004) Endangered species: Pan-Atlantic leatherback turtle movements. Nature 429:522

Hays GC, Hobson VJ, Metcalfe JD, Righton D, Sims DW (2006) Flexible foraging movements of leatherback turtles across the North Atlantic Ocean. Ecology 87:2647–2656

Henry JA, Portier KM, Coyne J (1994) The climate and weather of Florida. Pineapple Press, Inc., Coconut Grove, FL, USA. 216 p

Henwood TA (1987) Sea turtles of the Southeastern United States, with emphasis on the life history and population dynamics of the turtle, *Caretta caretta*. PhD Thesis, Auburn University, Auburn, AL, USA. 170 p

Henwood TA, Stuntz WE (1987) Analysis of sea turtle captures and mortalities during commercial shrimp trawling. Fish Bull 85:813–817

Henwood TA, Stuntz W, Thompson N (1992) Evaluation of U.S. turtle protective measures under existing TED regulations, including estimates of shrimp trawler related turtle mortality in the wider Caribbean. NOAA Technical Memorandum NMFS-SEFSC-303. National Marine Fisheries Service, Spring, MD, USA. 21 p

Heppell SS (1997) On the importance of eggs. MTN 76:6–8

Heppell SS, Crowder LB, Crouse DT, Epperly SP, Frazer NB (2003) Population models for atlantic loggerheads: Past, present, and future. In: Bolten AB, Witherington BE (eds) Loggerhead sea turtles. Smithsonian Books, Washington, DC, USA, pp 255–276

Heppell SS, Crouse DT, Crowder LB, Epperly SP, Gabriel W, Henwood T, Márquez-M R, Thompson NB (2005) A population model to estimate recovery time, population size, and management impacts on Kemp's ridley sea turtles. Chel Conserv Biol 4:767–773

Heppell SS, Burchfield PM, Peña LJ (2007) Kemp's ridley recovery: How far have we come, and where are we headed? In: Plotkin PT (ed) Biology and conservation of ridley sea turtles. The Johns Hopkins University Press, Baltimore, MD, USA, pp 325–336

Hickerson EL (2000) Assessing and tracking resident, immature loggerheads (Caretta caretta) in and around the flower garden banks, Northwest Gulf of Mexico. MS Thesis, Texas A&M University, College Station, MD, USA. 102 p

Hickerson EL, Peccini MB (2000) Using GIS to study habitat use by subadult loggerhead sea turtles (Caretta caretta) at the Flower Garden Banks National Marine Sanctuary in the Northwest Gulf of Mexico. Proceedings, 19th Annual Symposium on Sea Turtle Conservation and Biology, South Padre Island, TX, USA, March, pp 179–181

Hildebrand HH (1963) Hallazgo del area de anidacion de la Tortuga marina "lora", Lepidochelys kempii (Garman), en la costa occidental del Golfo de Mexico (Rept., Chel.). Science 22:105–112

Hildebrand H (1982) A historical review of the status of sea turtle populations in the Western Gulf of Mexico. In: Bjorndal K (ed) Biology and conservation of sea turtles. Smithsonian Institution Press, Washington, DC, USA, pp 447–453

Hilterman ML, Goverse E (2007) Nesting and nest success of the leatherback turtle (Dermochelys coriacea) in Suriname, 1999–2005. Chel Conserv Biol 68:87–100

Hill, D. 2015. FISHING/SHRIMP: Repeals the prohibition on enforcement of the federal TEDS in shrimp nets requirement. HLS 15RS-1313. 2 pp

Hirth HF (1971) Synopsis of the biological data on the green turtle Chelonia mydas. FAO Fisheries Synopsis 85. Food and Agriculture Organization of the United Nations, Rome, Italy. 84 p

Hirth HF (1980) Some aspects of the nesting behavior and reproductive biology of sea turtles. Am Zool 20:507–523

Hirth HF (1997) Synopsis of biological data on the green turtle Chelonia mydas (Linnaeus 1758). U.S. Fish and Wildlife Biological Report 97(1). U.S. Fish and Wildlife Service, Washington, DC, USA. 120 p

Hirth HF, Ogren LH (1987) Some aspects of the ecology of the leatherback turtle Dermochelys coriacea at Laguna Jalova, Costa Rica. NOAA Technical Report NMFS 56. National Marine Fisheries Service, Spring, MD, USA. 14 p

Holder KK, Holder MT (2007) Phylogeography and population genetics. In: Plotkin PT (ed) Biology and conservation of ridley sea turtles. The Johns Hopkins University Press, Baltimore, MD, USA, pp 107–117

Hopkins SR, Richardson JI (1984) Recovery plan for marine turtles: Agency review draft. National Marine Fisheries Service, Silver Spring, MD, USA. 355 p

Hopkins-Murphy SR, Owens DW, Murphy TM (2003) Ecology of immature loggerhead on foraging grounds and adults in internesting habitat in the Eastern United States. In: Bolten AB, Witherington BE (eds) Loggerhead sea turtles. Smithsonian Books, Washington, DC, USA, pp 79–92

Houghton JD, Doyle TK, Wilson MW, Davenport J, Hays GC (2006) Jellyfish aggregations and leatherback turtle foraging patterns in a temperate coastal environment. Ecology 87:1967–1972

Hulin V, Delmas V, Girondot M, Godfrey MH, Guillon JM (2009) Temperature-dependent sex determination and global change: Are some species at greater risk? Oecologia 160:493–506

Innis CJ, Tlusty M, Merigo C, Weber ES (2007) Metabolic and respiratory status of cold-stunned Kemp's ridley sea turtles (*Lepidochelys kempii*). J Comp Phys B 177:623–630

IUCN (International Union for the Conservation of Nature) (2004) Green turtle (*Chelonia mydas*), Version 2012.1. http://www.iucnredlist.org/apps/redlist/details/4615/0. Accessed 20 November 2013

IUCN (2012) IUCN Red list of threatened species, Version 2012.2. http://www.iucnredlist.org/. Accessed 20 November 2013

Jackson JBC (1997) Reefs since Columbus. Coral Reefs 16:S23–S33

Jimenez MC, Filonov A, Tereshchenko I, Márquez-M R (2005) Time-series analyses of the relationship between nesting frequency of the Kemp's ridley sea turtle and meteorological conditions. Chel Conserv Biol 4:774–780

Joergensen K (2012) Green sea turtle hatchling. iStockphoto, Calgary, Alberta, Canada. http://www.istockphoto.com/stock-photo-22302729-green-sea-turtle-hatchling.php. Accessed 20 November 2013

Johnson SA, Ehrhart LM (1994) Nest-site fidelity of the Florida green turtle. Proceedings, 13th Annual Symposium on Sea Turtle Biology and Conservation, Jekyll Island, GA, USA, February, Part I. 83 p

Johnson SA, Bass AL, Libert B, Marshall M, Fulk D (1999a) Kemp's ridley (*Lepidochelys kempi*) nesting in Florida. Florida Scientist 62:194–204

Johnson DR, Yeung C, Brown CA (1999b) Estimates of marine mammal and marine turtle bycatch by the U.S. Atlantic Pelagic Longline Fleet in 1992–1997. NOAA Technical Memorandum NMFS-SEFSC-418. National Marine Fisheries Service, Miami, FL, USA. April. 70 p

Karnauskas M, Schirripa MJ, Kelble CR, Cook GS, Craig JK (2013) Ecosystem status report for the Gulf of Mexico. NOAA Technical Memorandum NMFS-SEFSC-653. National Marine Fisheries Service, Southeast Fisheries Science Center, Miami, FL, USA. December. 52 p

Keller JM, McClellan-Green PD, Kucklick JR, Keil DE, Peden-Adams MM (2006) Effects of organochlorine contaminants on loggerhead sea turtle immunity: Comparison of a correlative field study and *in vitro* exposure experiments. Environ Health Perspect 114:70–76

Klima EF, Gitschlag GR, Renaud ML (1988) Impacts of the explosive removal of offshore petroleum platforms on sea turtles and dolphins. Mar Fish Rev 50:33–42

Kot CY, Boustany AM, Halpin PN (2010) Temporal patters of target catch and sea turtle bycatch in the U.S. Atlantic pelagic longline fishing fleet. Can J Fish Aquat Sci 67:42–57

Lamont MA, Percival HF, Pearlstine LG, Colwell SV, Carthy RR (1998) Sea turtle nesting activity along Eglin Air Force Base on Cape San Blas and Santa Rosa, Island, Florida from 1994 to 1997. Technical Report 59. U.S. Geological Survey, Biological Resources Division, Florida Cooperative Fish and Wildlife Research Unit, Gainsville, FL, USA. 52 p

Landry AM Jr, Seney EE (2006) Nest site selection and post-nesting movements of Kemp's ridley sea turtles along the Upper Texas Coast. Final Report to the Texas General Land Office Coastal Impact Assessment Program GLO Contract 05-233N. Texas A&M University, College Station, TX, USA

Landry AM Jr, Seney EE (2008) Movement and behavior of Kemp's ridley sea turtles in the Northwestern Gulf of Mexico during 2006 and 2007. Final Report to the Schlumberger Excellence in Educational Development Program, Sugar Land, Texas. Texas A&M University, College Station

Landry A Jr, Costa D, Williams B, Coyne M (1992) Turtle capture and habitat characterization study. Final Report. U.S. Army Corps of Engineers, Galveston District, Galveston, TX, USA. 118 p

Landry A Jr, Costa D, Williams B, Coyne M (1993) Sea turtle capture and habitat characterization: South Padre Island and Sabine Pass, TX Environs. Final Report. U.S. Army Corps of Engineers, Galveston and New Orleans Districts. Galveston, TX, USA. 119 p

Landry AM Jr, Costa DT, Coyne MS, Kenyon FL, Werner SA, Fitzgerald PS, St. John KE, Williams BB (1994) Sea turtle capture/population index and habitat characterization: Bolivar Roads and Sabine Pass, TX and Calcasieu Pass, Louisiana. Final Report. U.S. Army Corps of Engineers, Galveston District. Galveston, TX, USA. 332 p

Landry AM Jr, Costa DT, Coyne MS, Kenyon FL, Werner SA, Fitzgerald PS, St. John KE, Williams BB (1995) Sea turtle capture/ population index and habitat characterization: Bolivar Roads and Sabine Pass, Texas, and Calcasieu Pass, Louisiana. Final Report. U.S. Army Corps of Engineers, Galveston District. Galveston. Texas A&M University, College Station, TX, USA. 173 p

Landry AM Jr, Costa DT, Kenyon FL, Coyne MS (2005) Population characteristics of Kemp's ridley sea turtles in nearshore waters of the upper Texas and Louisiana coasts. Chel Conserv Biol 4:801–807

LDWF (Louisiana Department of Wildlife and Fisheries) (2012) Homepage. http://www.wlf. louisiana.gov/. Accessed 20 November 2013

Leary TR (1957) A schooling of leatherback turtles, *Dermochelys coriacea*, on the Texas coast. Copeia 1957:232

LeBuff CR Jr (1974) Unusual nesting relocation in the loggerhead turtle, *Caretta caretta*. Herpetologica 30:29–31

LeBuff CR Jr (1990) The loggerhead turtle in the Eastern Gulf of Mexico. Caretta Research, Sanibel, 216 p

Leslie AJ, Penick DN, Spotila JR, Paladino FV (1996) Leatherback turtle, *Dermochelys coriacea*, nesting and nest success at Tortuguero, Costa Rica, in 1990–1991. Chel Conserv Biol 2:159–168

Lewison RL, Crowder LB, Shaver DJ (2003) The impact of turtle excluder devices and fisheries closures on loggerhead and Kemp's ridley strandings in the western Gulf of Mexico. Conserv Biol 17:1089–1097

Limpus CJ, Nicholls N (1988) The Southern Oscillation regulates the annual numbers of green turtles (*Chelonia mydas*) breeding around northern Australia. Aust Wildlife Res 15:157–161

Lohmann KJ, Lohmann CMF (2003) Orientation mechanisms of hatchlings loggerheads. In: Bolten AB, Witherington BE (eds) Loggerhead sea turtles. Smithsonian Books, Washington, DC, USA, pp 44–62

Lohmann KJ, Putman NF, Lohmann CMF (2012) The magnetic map of hatchling loggerhead sea turtles. Curr Opin Neurobiol 22:336–342

Lohoefener R, Hoggard W, Mullin K, Roden C, Rogers C (1989) Are sea turtles attracted to petroleum platforms? Proceedings, 9th Annual Workshop on Sea Turtle Conservation and Biology, Jekyll Island, GA, USA, August, Part 1, pp 103–104

Lohoefener R, Hoggard W, Mullin K, Roden C, Rogers C (1990) Association of sea turtles with petroleum platforms in the North-Central Gulf of Mexico. OCS Study/MMS 90-0025. Minerals Management Service, New Orleans, LA, USA. 90 p

Louisiana Revised Statutes (1987) RS 56:57.2. http://www.legis.state.la.us/lss/lss.asp? doc=105394. Accessed 20 November 2013

Lund PF (1985) Hawksbill turtle *Ermetmochelys imbricata* nesting on the east coast of Florida. J Herpetol 19:164–166

Luschi P, Hays GC, Papi F (2003) A review of long-distance movements by marine turtles, and the possible role of ocean currents. Oikos 103:293–302

Lutcavage ME, Plotkin P, Witherington B, Lutz PL (1997) Human impacts on sea turtle survival. In: Lutz PL, Musick JA (eds) The biology of sea turtles, vol 1. CRC Press, Boca Raton, FL, USA, pp 387–409

Lutz PL, Musick JA, Wyneken J (eds) (2003) The biology of sea turtles, vol 2. CRC Press, Boca Raton, FL, USA. 472 p

Lyn H, Coleman A, Broadway M, Klaus J, Finerty S, Shannon D, Solangi M (2012) Displacement and site fidelity of rehabilitated immature Kemp's ridley sea turtles (*Lepidochelys kempii*). MTN 135:10–13

Mager AM Jr (1985) Five-year status reviews of sea turtles listed under the endangered species act of 1973. National Marine Fisheries Service, St. Petersburg, FL, USA. 90 p

Magnuson JJ, Bjorndal KA, DuPaul WD, Graham GL, Owens DW, Pritchard PCH, Richardson JI, Saul GE, West CW (1990) Decline of the sea turtles, causes and prevention. National Academy Press, Washington, DC, USA. 260 p

Mannix P (2012) A leatherback turtle covering her eggs, Turtle Beach, Tobago stock photo. Flickr.com. http://www.flickr.com/photos/paulmannix/2633233033/. Accessed 20 November 2013

Manzella SA, Williams JA, Caillouet CW Jr (1990) Radio and sonic tracking of juvenile sea turtles in inshore waters of Louisiana and Texas. Proceedings, 10th Annual Workshop on Sea Turtle Biology and Conservation, Hilton Head Island, SC, USA, February, pp 115–120

Márquez-M R (1972) Resultados prelimiares sobre edad y crecimiento de la tortuga lora, *Lepidochelys kempii* (Garman). Memorias IV Congreso Nacional de Oceanografia, Mexico, pp 419–427

Márquez-M R (1990) FAO species catalogue: An annotated and illustrated catalogue of sea turtles known to date. FAO Fisheries Synopsis No. 125, Volume 11. Food and Agriculture Organization of the United Nations, Rome, Italy. 81 p

Márquez-M R (2004) Las tortugas marinas del Golfo de Mexico: abundancia, distribución y protección. In: Caso M, Pisanty I, Ezcurra E (eds) Diagnóstico ambiental del Golfo de Mexico. Instituto Nacional de Ecología (INE-SEMARNAT), Mexico City, DF, Mexico, pp 173–198

Márquez-M R, Villanueva OA, Sánchez PM (1982) The population of the Kemp's ridley sea turtle in the Gulf of Mexico—*Lepidochelys kempii*. In: Bjorndal KA (ed) Biology and conservation of sea turtles. Smithsonian Institution Press, Washington, DC, USA, pp 159–164

Márquez-M R, Carrasco-A M, Jiménez-O C, Byles R, Burchfield P, Sanchez-P M, Diaz-F J, Leo-P A (1996) Good news! Rising numbers of Kemp's ridley nest at Rancho Nuevo, Tamaulipas, Mexico. MTN 73:2–5

Márquez-M R, Jiménez MC, Carrasco MA, Villanueva NA (1998) Comments concerning the population tendencies of the marine turtle genus *Lepidochelys* after the year 1990. Oceánides 13:41–62

Márquez-M R, Díaz J, Sánchez M, Burchfield P, Leo A, Carrasco M, Peña J, Jiménez C, Bravo R (1999) Results of the Kemp's ridley nesting beach conservation efforts in México. MTN 85:2–4

Márquez-M R, Díaz-F J, Guzmán-H V, Bravo-G R, Jimenez-Q M (2004) Marine turtles of the Gulf of Mexico. Abundance, distribution and protection. In: Caso M, Pisanty I, Ezcurra E (eds) Environmental analysis of the Gulf of Mexico. Harte Research Institute for Gulf of Mexico Studies, Corpus Christi, TX, USA, pp 89–107

Márquez-MR (1994) Synopsis of biological data on the Kemp's ridley sea turtle, *Lepidochelys kempi* (Garman, 1880). NOAA Technical Memorandum NMFS-SEFSC-343. National Marine Fisheries Service, Southeast Fisheries Science Center, Miami, FL, USA. March. 91 p

Márquez-MR (1999) Status and distribution of the Kemp's ridley turtle, *Lepidochelys kempii*, in the wider Caribbean region. Proceedings, The Regional Meeting: Marine Turtle Conservation in the Wider Caribbean Region: A Dialogue for Effective Regional Management, Santo Domingo, Dominican Republic, pp 16–18

Márquez-MR, Ríos OD, Sánchez PJM, Díaz J (1989) Mexico's contribution to Kemp's ridley sea turtle recovery. Proceedings, 1st International Symposium on Kemp's Ridley Sea Turtle Biology, Conservation and Management, Galveston, TX, USA, October, Session 1, pp 4–6

Mays JL, Shaver DJ (1998) Nesting trends of sea turtles in national seashores along Atlantic and Gulf Coast Waters of the United States. Final report for Natural Resources Preservation Program, Project Number 95-15. National Park Service and U.S. Geological Survey, Washington, DC, USA. 61 p

McClellan CM, Read AJ (2007) Complexity and variation in loggerhead sea turtle life history. Biol Lett 3:592–594

McClenachan L, Jackson JBC, Newman MJH (2006) Conservation implications of historic sea turtle nesting beach loss. Front Ecol Environ 4:290–296

McDaniel CJ (1998) A spatial analysis of sea turtle abundance and shrimping intensity in the Gulf of Mexico: Recommendations for conservation and management. Duke University, Durham, NC, USA. 141 p

McDaniel CJ, Crowder LB, Priddy JA (2000) Spatial dynamics of sea turtle abundance and shrimping in the U.S. Gulf of Mexico. Conserv Ecol 4:15

McMichael E, Carthy RR, Seminoff JA (2003) Ecology of Juvenile Sea Turtles in the Northeastern Gulf of Mexico. Proceedings, 23rd Annual Symposium on Sea Turtle Biology and Conservation, Kuala Lumpur, Malaysia, March, pp 20–21.

McMichael E, Seminoff J, Carthy R (2008) Growth rates of wild green turtles, *Chelonia mydas*, at a temperate foraging habitat in the northern Gulf of Mexico: Assessing short-term effects of cold-stunning on growth. J Nat Hist 42:2793–2807

MDWFP (Mississippi Department of Wildlife, Fisheries, & Parks) (2012) Homepage. http://www.mdwfp.com/. Accessed 20 November 2013

Mellgren RL, Mann MA (1996) comparative behavior of hatchling sea turtles. Proceedings, 15th Annual Symposium on Sea Turtle Biology and Conservation, Hilton Head, SC, USA, February, pp 52–53

Mendonca MT (1981) Comparative growth rates of wild immature *Chelonia mydas* and *Caretta caretta* in Florida. J Herpetol 154:444–447

Mendonca MT (1983) Movements and feeding ecology of immature green turtles (*Chelonia mydas*) in a Florida lagoon. Copeia 1983:1013–1023

Mendonca MT, Pritchard PCH (1986) Offshore movements of post-nesting Kemp's ridley sea turtles (*Lepidochelys kempi*). Herpetologica 42:373–381

Merrill MW, Salmon M (2011) Magnetic orientation by hatchling loggerhead sea turtles (*Caretta caretta*) from the Gulf of Mexico. Mar Biol 158:101–112

Metz TL (2004) Factors influencing Kemp's ridley sea turtle (*Lepidochelys kempii*) distribution in nearshore waters and implications for management. Texas A&M University, College Station, TX, USA. 163 p

Meylan A (1984) Feeding ecology of the hawksbill turtle *Eretmochelys imbricata*: spongivory as a feeding niche in the coral reef community. PhD Thesis, University of Florida, Gainsville, FL, USA. 117 p

Meylan A (1988) Spongivory in hawksbill turtles: A diet of glass. Science 22:393–395

Meylan A (1992) Hawksbill turtle *Eretmochelys imbricata*. In: Moler P (ed) Rare and endangered biota of Florida. University Press of Florida, Gainesville, FL, USA, pp 95–99

Meylan AB, Donnelly M (1999) Status justification for listing the hawksbill turtle (*Eretmochelys imbricata*) as critically endangered on the 1996 IUCN Red List of Threatened Animals. Chel Conserv Biol 3:200–224

Meylan A, Redlow A (2006) *Eretmochelys imbricata*—hawksbill turtle. In: Meylan PA (ed) Biology and conservation of florida turtles. Chelonian Research Foundation, Lunenburg, MA, USA, pp 105–127

Meylan AB, Bjorndal KA, Turner BJ (1983) Sea turtles nesting at Melbourne Beach, Florida. II. Post-nesting movements of *Caretta caretta*. Biol Conserv 26:79–90

Meylan A, Castaneda P, Coogan C, Lozon T, Fletemeyer J (1991) First recorded nesting by Kemp's ridley in Florida, USA. MTN 48:8–9

Meylan A, Schroeder B, Mosier A (1995) Sea turtle nesting activity in the State of Florida 1971–1992. Florida Marine Research Publications Number 52. Florida Marine Research Institute, St. Petersburg, FL, USA. 51 p

Milton SL, Leone-Kabler S, Schulman AA, Lutz PL (1994) Effects of Hurricane Andrew on the sea turtle nesting beaches of South Florida. Bull Mar Sci 54:974–981

MMS (Minerals Management Service) (2004) Geological and geophysical exploration for mineral resources on the Gulf of Mexico Outer Continental Shelf. OCS EIA/EA MMS 2004-054. U.S. Department of the Interior, Gulf of Mexico OCS Region, Washington, DC, USA. 487 p

Moncada Gavilan F (2000) Impact of regulatory measures on cuban marine turtle fisheries. Proceedings, 18th International Sea Turtle Symposium, Mazatlán, Sinaloa, Mexico, March, Part VI, pp 108–109

Moncada FG, Hawkes LA, Fish MR, Godley BJ, Manolis SC, Medina Y, Nodarse G, Webb GJW (2012) Patterns of dispersal of hawksbill turtles from the Cuban shelf inform scale of conservation and management. Biol Conserv 148:191–199

Moon DY, MacKenzie DS, Owens DW (1997) Simulated hibernation of sea turtles in the laboratory: I. Feeding, breathing frequency, blood pH, and blood gases. J Exp Zool 278:372–380

Morgan PJ (1989) Occurrence of leatherback turtles (*Dermochelys coriacea*) in the British Isles in 1988 with Reference to a Record Specimen. Proceedings, 9th Annual Workshop on Sea Turtle Conservation and Biology, Jekyll Island, FL, USA, February, Part I, pp 119–120

Morreale SJ, Standora EA (1999) Migration patterns of Northeastern U.S. sea turtles. New York Department of Environmental Conservation NOAA Award NA36FL0430. National Marine Fisheries Service, Silver Spring, MD, USA

Morreale SJ, Standora EA (2005) Western North Atlantic waters: Crucial developmental habitat for Kemp's ridley and loggerhead sea turtles. Chel Conserv Biol 4:872–882

Morreale SJ, Plotkin PT, Shaver DJ, Kalb HJ (2007) Adult migration and habitat utilization: Ridley turtles in their element. In: Plotkin PT (ed) Biology and conservation of ridley sea turtles. The Johns Hopkins University Press, Baltimore, MD, USA, pp 213–230

Mortimer JA (1981) The feeding ecology of the west Caribbean green turtle (*Chelonia mydas*) in Nicaragua. Biotropica 13:49–58

Mortimer JA (2008) The State of the World's Hawksbills. In SWOT (State of the Worlds Sea Turtles), Report, Volume III, Arlington, VA, USA, pp 10–13

Mortimer JA, Donnelly M (2008) Hawksbill turtle (*Eretmochelys imbricata*), Marine Turtle Specialist Group 2008 IUCN Red List Status Assessment. 108 p

Mosier A (1998) The impact of coastal armoring structures on sea turtle nesting behavior at three beaches on the East Coast of Florida. MS Thesis, University of South Florida, Tampa, FL, USA. 112 p

Mosier AE, Witherington BE (2002) Documented effects of coastal armoring structures on sea turtle nesting behavior. Proceedings, 20th Annual Symposium on Sea Turtle Biology and Conservation, Orlando, FL, USA, pp 304–306

Mount RH (ed) (1975) The reptiles and amphibians of alabama. Auburn Printing Company, Auburn, AL, USA. 347 p

Mrosovsky N (1981) Plastic jellyfish. MTN 17:5–7

Mrosovsky N, Pritchard PCH (1971) Body temperatures of *Dermochelys coriacea* and other sea turtles. Copeia 1971:624–631

Mrosovsky N, Yntema CL (1980) Temperature dependence of sexual differentiation in sea turtles: Implications for conservation practices. Biol Conserv 18:271–280

Mrosovsky N, Ryan GD, James MC (2009) Leatherback turtles: The menace of plastic. Mar Pollut Bull 58:287–289

Mullin KD, Hoggard W (2000) Visual surveys of cetaceans and sea turtles from aircraft and ships. In: Davis RW, Evans WE, Würsig B (eds) Cetaceans, sea turtles and seabirds in the Northern Gulf of Mexico: Distribution, abundance and habitat associations, Volume II. Technical Report. Texas A&M at Galveston, National Marine Fisheries Service OCS Study MMS 200-2003. Minerals Management Service, Gulf of Mexico OCS Region, New Orleans, LA, USA, pp 111–172

Murphy MD, McMillen-Jackson AL, Mahmoudi B (2007) A STOCK ASSESSMENT FOR BLUE CRab, *Callinectes sapidus*, in Florida Waters. In House Report 2007-006. Florida Fish and Wildlife Conservation Commission, Division of Marine Fisheries Management, St. Petersburg, FL, USA. June. 85 p

Musick JA, Limpus CJ (1997) Habitat utilization and migration in juvenile sea turtles. In: Lutz PL, Musick JA (eds) The biology of sea turtles, vol 1. CRC Press, Boca Raton, FL, USA, pp 137–163

Najera JJD (1990) Nesting of three species of sea turtles in the northeast coast of the Yucatán Peninsula, Mexico. Proceedings, 10th Annual Workshop on Sea Turtle Biology and Conservation, Hilton Head, SC, USA, pp 29–33

Nelson DA (2000) Winter movements of sea turtles. Proceedings, 19th Annual Symposium on Sea Turtle Biology and Conservation, South Padre Island, TX, USA. March. 26 p

NMFS (National Marine Fisheries Service) (1992) Recovery plan for leatherback turtles in the U.S. Caribbean, Atlantic and Gulf of Mexico. National Marine Fisheries Service, Silver Spring, MD, USA. 65 p

NMFS (2002) Endangered species act-section 7 consultation biological opinion on the cooling water intake system at the crystal river energy complex. National Marine Fisheries Service, Silver Spring, MD, USA. 26 p

NMFS (2007) Report to Congress on the impacts of Hurricanes Katrina, Rita, and Wilma on Alabama, Louisiana, Florida, Mississippi, and Texas Fisheries. National Marine Fisheries Service, Silver Spring, MD, USA. July. 133 p

NMFS (2011a) Green turtle (*Chelonia mydas*) species account. http://www.nmfs.noaa.gov/pr/species/turtles/green.htm. Accessed 20 November 2013

NMFS (2011b) Leatherback turtle (*Dermochelys coriacea*) species account. http://www.nmfs.noaa.gov/pr/species/turtles/leatherback.htm. Accessed 20 November 2013

NMFS (2011c) U.S. National Bycatch Report. NOAA Technical Memorandum NMFS-F/SPO-117C. National Marine Fisheries Service, Silver Spring, MD, USA. September. 508 p

NMFS (2012) Sea turtle conservation; Shrimp trawling requirements. Fed Regist 77:27411–27415

NMFS SEFSC (Southeast Fisheries Science Center) (2001) Stock assessments of loggerhead and leatherback sea turtles and an assessment of the impact of the pelagic longline fishery on the loggerhead and leatherback sea turtles of the western North Atlantic. NOAA Technical Memorandum NMFS-SEFSC-455. National Marine Fisheries Service, Miami, FL, USA. March. 343 p

NMFS SEFSC (2009) An assessment of loggerhead sea turtles to estimate impacts of mortality reductions on population dynamics. NMFS Southeast Fisheries Science Center Contribution PRD-08/09-14. National Marine Fisheries Service, Miami, FL, USA. July. 45 p

NMFS, USFWS (U.S. Fish and Wildlife Service) (1991) Recovery plan for U.S. population of atlantic green turtle *Chelonia mydas*. National Marine Fisheries Service, Washington, DC, USA and Southeast Region, U.S. Fish and Wildlife Service, Atlanta, GA, USA. October. 52 p

NMFS, USFWS (1993) Recovery plan for hawksbill turtles in the U.S. Caribbean Sea, Atlantic Ocean, and Gulf of Mexico. National Marine Fisheries Service, Washington, DC and Southeast Region, U.S. Fish and Wildlife Service, Atlanta, GA, USA. December. 52 p

NMFS, USFWS (2007a) Kemp's ridley sea turtle (*Lepidochelys kempii*): 5-year review Summary and evaluation. Office of Protected Resources, National Marine Fisheries Service, Silver Spring and Southwest Region, U.S. Fish and Wildlife Service, Albuquerque, NM, USA. August. 50 p

NMFS, USFWS (2007b) Loggerhead sea turtle (*Caretta caretta*): 5-year review: Summary and evaluation. Office of Protected Resources, National Marine Fisheries Service, Silver Spring and Jacksonville Ecological Services Field Office, U.S. Fish and Wildlife Service, Jacksonville, FL, USA. August. 55 p

NMFS, USFWS (2007c) Green sea turtle (*Chelonia mydas*): 5-year review: Summary and evaluation. Office of Protected Resources, National Marine Fisheries Service, Silver Spring, MD, USA and Jacksonville Ecological Services Field Office, U.S. Fish and Wildlife Service, Jacksonville, FL, USA. August. 102 p

NMFS, USFWS (2007d) Leatherback sea turtle (*Dermochelys coriacea*): 5-year review: summary and evaluation. Office of Protected Resources, National Marine Fisheries Service, Silver Spring and Jacksonville Ecological Services Field Office, U.S. Fish and Wildlife Service, Jacksonville, FL, USA. August. 79 p

NMFS, USFWS (2007e) Hawksbill sea turtle (*Eretmochelys imbricata*): 5-year review: Summary and evaluation. Office of Protected Resources, National Marine Fisheries Service, Silver Spring and Jacksonville Ecological Services Field Office, U.S. Fish and Wildlife Service, Jacksonville, FL, USA. August. 90 p

NMFS, USFWS (2008) Recovery plan for the Northwest Atlantic population of the loggerhead sea turtle (*Caretta caretta*): Second revision. Office of Protected Resources, National Marine Fisheries Service, Silver Spring, MD, USA. January. 325 p

NMFS, USFWS, SEMARNAT (Secretary of Environment and Natural Resources, Mexico) (2011) Bi-national recovery plan for the Kemp's ridley sea turtle (*Lepidochelys kempii*), 2nd Rev. National Marine Fisheries Service, Silver Spring, MD, USA. 174 p

NOAA (National Oceanic and Atmospheric Administration) (2007) Kemp's ridley sea turtle range. http://www.nmfs.noaa.gov/pr/species/turtles/. Accessed 20 November 2013

NOAA (2009a) Loggerhead sea turtle range. http://www.nmfs.noaa.gov/pr/species/turtles/. Accessed 20 November 2013

NOAA (2009b) Green sea turtle range. http://www.nmfs.noaa.gov/pr/species/turtles/. Accessed 20 November 2013

NOAA (2009c) Leatherback sea turtle range. http://www.nmfs.noaa.gov/pr/species/turtles/. Accessed 20 November 2013

NOAA (2009d) Hawksbill sea turtle range. http://www.nmfs.noaa.gov/pr/species/turtles/. Accessed 20 November 2013

NOAA (2011) Sea turtle photos. National Oceanic and Atmospheric Administration, Silver Spring, MD, USA. http://www.nmfs.noaa.gov/pr/species/turtle/photos.htm. Accessed 20 November 2013

NOAA (2012) New data prompts NOAA fisheries to withdraw proposed rule to require turtle excluder devices in certain shrimp trawls. http://sero.nmfs.noaa.gov/news_room/press_re-leases/2012/press_release_skimmer_trawl_proposed_rule.pdf. Accessed 20 November 2013

NOAA Fisheries OPR (Office of Protected Resources) (2013) Sea turtles. http://www.nmfs.noaa.gov/pr/species/turtles/. Accessed 20 November 2013

Nolasco del Aguila D, Debade X, Harrison E (2008a) Report on the 2007 green turtle program at Tortuguero, Costa Rica. Caribbean Conservation Corporation, Gainesville, FL, USA and Ministry of Environment and Energy of Costa Rica. San Pedro, Costa Rica, 63 p

Nolasco del Aguila D, Debade X, Harrison E (2008b) Report on the 2007 leatherback program at Tortuguero, Costa Rica. Caribbean Conservation Corporation, Gainesville and Ministry of Environment and Energy of Costa Rica, San Pedro, Costa Rica. 37 p

Nolasco del Aguila D, Debade X, Harrison E (2009) Report on the 2008 green turtle program at Tortuguero, Costa Rica. Caribbean Conservation Corporation, Gainesville and Ministry of Environment and Energy of Costa Rica, San Pedro, Costa Rica. 66 p

NPS (National Park Service) (2013a) Padre Island: Current sea turtle nesting season. http://www.nps.gov/pais/naturescience/current-season.htm. Accessed 20 November 2013

NPS (2013b) Padre Island: Sea turtle recovery project. http://www.nps.gov/pais/naturescience/strp.htm. Accessed 20 November 2013

NPS (2013c) Padre Island National Seashore: Young Kemp's ridley hatchlings entering water. http://www.nps.gov/pais/naturescience/kridley.htm. Accessed 20 November 2013

NPS (2013d) The leatherback sea turtle. http://www.nps.gov/pais/naturescience/leatherback.htm. Accessed 20 November 2013

NRC (National Research Council) (2010) Sea turtle status and trends: Integrating demography and abundance. Prepared for the National Ocean and Atmospheric Administration, National Academies Press, Washington, DC, USA. 162 p

Ogren LH (1989) Distribution of Juvenile and Subadult Kemp's ridley turtles: Preliminary results from the 1980–1987 surveys. Proceedings, 1st iInternational Symposium on Kemp's Ridley Sea Turtle Biology, Conservation and Management, Galveston, TX, USA. October, Session VI, pp 116–123

Ogren L, McVea CJ (1982) Apparent hibernation by sea turtles in North American Waters. In: Bjorndal KA (ed) Biology and conservation of sea turtles. Smithsonian Institution Press, Washington, DC, USA, pp 127–132

Overton EB, Byrne CJ, McFall JA, Antoine SR, Laseter JL (1983) Results from the chemical analyses of oily residue samples taken from stranded juvenile sea turtles collected from Padre and Mustang Islands, Texas. Prepared for the Minerals Management Service, Special Report 1983-32. Center for Bio-Organic Studies, New Orleans, LA, USA. December. 25 p

Owens D, Crowell D, Dienberg G, Grassman M, McCain S, Morris Y, Schwantes N, Wibbels T (eds) (1983) Western Gulf of Mexico Sea Turtle Proceedings, January 13–14, 1983. TAMU-SG-84-105. Sea Grant College Program, Texas A&M University, College Station, TX, USA. October. 74 p

Parsons JJ (1962) The green turtle and man. Rose Printing Company, Tallahassee, FL, USA. 126 p

Passerotti MS, Carlson JK (2009) Catch and bycatch in the U.S. Southeast Gillnet Fisheries, 2008. NOAA Technical Memorandum NMFS-SEFSC-583. National Marine Fisheries Service, Panama City, FL, USA. February. 19 p

Passerotti MS, Carlson JK, Gulak SJB (2010) Catch and bycatch in the U.S. Southeast Gillnet Fisheries, 2009. NOAA Technical Memorandum NMFS-SEFSC-600. National Marine Fisheries Service, Panama City, FL, USA. February. 20 p

Pelletier D, Roos D, Ciccione S (2003) Oceanic survival and movements of wild and captive-reared immature green turtles (Chelonia mydas) in the Indian Ocean. Aquat Living Resour 16:35–41

Penick DN, Spotila JR, O'Connor MP, Steyermark AC, George RH, Salice CJ, Paladino FV (1998) Thermal independence of muscle tissue metabolism in the leatherback turtle, Dermochelys coriacea. Comp Biochem Phys A 120:399–403

Pérez-Castañeda R, Salum-Fares A, Defeo O (2007) Reproductive patterns of the hawksbill turtle Eretmochelys imbricata in sandy beaches of the Yucatán Peninsula. J Mar Biol Assoc UK 87:815–824

Pierce RH, Henry MS (2008) Harmful algal toxins of the Florida red tide (Karenia brevis): natural chemical stressors in South Florida coastal ecosystems. Ecotoxicology 17:623–631

Pike DA, Stiner JC (2007) Sea turtles vary in their susceptibility to tropical cyclones. Oecologia 153:471–478

Plotkin PT (1989) The feeding ecology of the loggerhead sea turtle in the Northwestern Gulf of Mexico. Texas A&M University, Station, TX, USA. 124 p

Plotkin PT (1995) National Marine Fisheries Service and U.S. Fish and Wildlife Service status reviews for sea turtles listed under the endangered species act of 1973. National Marine Fisheries Service, Silver Spring, MD, USA. 139 p

Plotkin PT (1996) Occurrence and diet of juvenile loggerhead sea turtles, Caretta caretta, in the northwestern Gulf of Mexico. Chel Conserv Biol 2:78–80

Plotkin PT (2007a) Introduction. In: Plotkin PT (ed) Biology and conservation of ridley sea turtles. The Johns Hopkins University Press, Baltimore, MD, USA, pp 3–5

Plotkin PT (2007b) Near extinction and recovery. In: Plotkin PT (ed) Biology and conservation of ridley sea turtles. The Johns Hopkins University Press, Baltimore, MD, USA, pp 337–340

Plotkin PT (2007c) Biology and conservation of ridley sea turtles. The Johns Hopkins University Press, Baltimore, MD, USA. 368 p

Plotkin P, Amos AF (1988) Entanglement in and ingestion of marine debris by sea turtles stranded along the south Texas coast. Proceedings, 8th Annual Workshop on Sea Turtle Conservation and Biology, Fort Fisher, NC, USA, February, pp 79–82

Plotkin P, Amos AF (1990) Effects of anthropogenic debris on sea turtles in the Northwestern Gulf of Mexico. Proceedings, 2nd International Workshop on Marine Debris, Honolulu, HI, USA, April, pp 736–743

Plotkin PT, Spotila JR (2002) Post-nesting migrations of loggerhead turtles Caretta caretta from Georgia, USA: Conservation implications for a genetically distinct subpopulation. Oryx 36:396–399

Plotkin PT, Wicksten MK, Amos AF (1993) Feeding ecology of the loggerhead sea turtle, Caretta caretta, in the northwestern Gulf of Mexico. Mar Biol 115:1–15

Polovina JJ, Balazs GH, Howell EA, Parker DM, Seki MP, Dutton PH (2004) Forage and migration habitat of loggerhead (Caretta caretta) and olive ridley (Lepidochelys olivacea) sea turtles in the central North Pacific Ocean. Fish Oceanogr 13:36–51

Ponwith B (2011) Data analysis request: Update of turtle bycatch in the Gulf of Mexico and Southeastern Atlantic Shrimp Fisheries. NMFS Memorandum for Roy E. Crabtree, January 5, 2011. 10 p. http://seaturtles.org/downloads/UPR_SEFSC_shrimp_bycatch_2011.pdf. Accessed 20 November 2013

Price AB, Gearhart JL (2011) Evaluations of Turtle Excluder Device (TED) Performance in the U.S. Southeast Atlantic and Gulf of Mexico Skimmer Trawl Fisheries. NOAA Technical Memorandum NMFS-SEFSC-615. National Marine Fisheries Service, Beaufort, NC and Pascagoula, MS, USA. May. 15 p

Pritchard PCH (1969) Studies of the systematics and reproductive cycles of the genus *Lepidochelys*. University of Florida, Gainsville, FL, USA. 226 p

Pritchard PCH (1976) Post-nesting movements of marine turtles (Cheloniidae and Dermochelyidae) tagged in the Guianas. Copeia 1976:749–754

Pritchard PCH (1982) Nesting of the leatherback turtle, *Dermochelys coriacea* in Pacific Mexico, with a new estimate of the world population status. Copeia 1982:741–747

Pritchard PCH (2007a) Evolutionary relationships, osteology, morphology, and zoogeography of ridley sea turtles. In: Plotkin PT (ed) Biology and conservation of ridley sea turtles. The Johns Hopkins University Press, Baltimore, MD, USA, pp 45–58

Pritchard PCH (2007b) Arribadas I have known. In: Plotkin PT (ed) Biology and conservation of ridley sea turtles. The Johns Hopkins University Press, Baltimore, MD, USA, pp 7–22

Pritchard PCH (2010) The most valuable reptile in the world–the green turtle. In SWOT (State of the World's Sea Turtles), Report Volume VI, Arlington, VA, USA, pp 24–29

Pritchard PCH, Márquez-M. R (1973) Kemp's ridley or Atlantic ridley, *Lepidochelys kempii*. IUCN Monograph No. 2 (Marine Turtle Series), Morges, VD, Switzerland. 30 p

Pulver JR, Scott-Denton E, Williams JA (2012) Characterization of the U.S. Gulf of Mexico Skimmer Trawl Fishery Based on Observer Coverage. NOAA Technical Memorandum NMFS-SEFSC-636. National Marine Fisheries Service, Galveston, TX, USA. October. 27 p

Putman NF, Shay TJ, Lohmann KJ (2010) Is the geographic distribution of nesting in the Kemp's ridley turtle shaped by the Migratory Needs of Offspring? Proceedings, integrative and comparative biology, Seattle, WA, USA, January, pp 305–314

Rabalais SC, Rabalais NN (1980) The occurrence of sea turtles on the South Texas coast. Contrib Mar Sci 23:123–129

Ramirez PA, Ania LV (2000) Incidence of marine turtles in the Mexican Long-line Tuna Fishery in the Gulf of Mexico. Proceedings, 18th Annual Workshop on Sea Turtle Biology and Conservation, Mazatlan, Sinaloa, Mexico, March, Part XI. 110 p

Reardon R, Mansfield K (1997) Annual report-1997 season: Dry Tortugas National Park, Sea Turtle Monitoring Program, Monroe County, Florida. National Park Service, Key West, FL, USA. 37 p

Rebel TP (1974) Sea turtles and the turtle industry of the West Indies, Florida, and the Gulf of Mexico. University of Miami Press, Coral Gables, FL, USA. 250 p

Reeves RR, Leatherwood S (1983) Autumn sightings of marine turtles (Cheloniidae) off South Texas. Southwest Nat 28:281–288

Reich KJ, Bjorndal KA, Bolten AB, Witherington BE (2007a) Do some loggerheads nesting in Florida have an oceanic foraging strategy? An assessment based on stable isotopes. Proceedings, 24th Annual Symposium on Sea Turtle Biology and Conservation, San Jose, Costa Rica. February. 32 p

Reich KJ, Bjorndal KA, Bolten AB (2007b) The 'lost years' of green turtles: Using stable isotopes to study cryptic life stages. Biol Lett 3:712–714

Reina RD, Abernathy KJ, Marshall GJ, Spotila JR (2005) Respiratory frequency, dive behaviour and social interactions of leatherback turtles, *Dermochelys coriacea* during the internesting interval. J Exp Mar Biol Ecol 316:1–16

Renaud ML (1995) Movements and submergence patterns of Kemp's ridley turtles (*Lepidochelys kempii*). J Herpetol 29:370–374

Renaud ML (2001) Sea turtles of the Gulf of Mexico. Proceedings, Gulf of Mexico Marine Protected Species Workshop, June 1999. New Orleans, LA, USA, June, pp 41–48

Renaud ML, Carpenter JA (1994) Movements and submergence patterns of loggerhead turtles (*Caretta caretta*) in the Gulf of Mexico determined through satellite telemetry. Bull Mar Sci 55:1–15

Renaud ML, Williams JA (1997) Movement of Kemp's ridley sea turtles (*Lepidochelys kempii*) and green (*Chelonia mydas*) sea turtles using Lavaca Bay and Matagorda Bay, 1996–1997. Final Report Submitted to the U.S. Environmental Protection Agency, Office of Planning and Coordination, Region 6, Dallas, TX, USA. 54 p

Renaud ML, Williams JA (2005) Kemp's ridley sea turtle movements and migrations. Chel Conserv Biol 4:808–816

Renaud M, Gitschlag G, Klima E, Shah A, Nance J, Caillouet C, Zein-Eldin Z, Koi D, Patella K (1990) Evaluation of the impacts of turtle excluder devices (TEDs) on Shrimp Catch Rates in the Gulf of Mexico and South Atlantic, March 1988 through July 1989. NOAA Technical Memorandum NMFS-SEFSC-254. National Marine Fisheries Service, Galveston, TX, USA. May. 165 p

Renaud M, Gitschlag G, Klima E, Shah A, Koi D, Nance J (1991) Evaluation of the impacts of turtle excluder devices (TEDs) on Shrimp Catch Rates in Coastal Waters of the United States along the Gulf of Mexico and Atlantic, September 1989 through August 1990. NOAA Technical Memorandum NMFS-SEFSC-288. National Marine Fisheries Service, Galveston, TX, USA. September. 80 p

Renaud M, Gitschlag G, Manzella S, Williams J (1992) Tracking of Green (*Chelonia mydas*) and Loggerhead (*Caretta caretta*) sea turtles using radio and sonic telemetry at South Padre Island, Texas, June-September 1991. Final Report. National Marine Fisheries Service, Galveston, TX, USA. 52 p

Renaud ML, Carpenter JA, Williams JA, Manzella SA (1993a) Telemetric tracking of Kemp's ridley sea turtles (*Lepidochelys kempii*) in Relation to Dredged Channels at Bolivar Pass and Sabine Pass, TX and Calcasieu Pass, LA, May 1993 through February 1994. Final Report. National Marine Fisheries Service, Galveston, TX, USA. 76 p

Renaud ML, Carpenter JA, Manzella SA, Williams JA (1993b) Telemetric tracking of green sea turtles (*Chelonia mydas*) in Relation to Dredged Channels at South Padre Island, Texas July through September 1992. Final Report. National Marine Fisheries Service, Galveston, TX, USA. 56 p

Renaud M, Carpenter J, Williams J, Carter D, Williams B (1995a) Movement of Kemp's ridley sea turtles (*Lepidochelys kempii*) near Bolivar Roads Pass and Sabine Pass, Texas and Calcasieu Pass, Louisiana. Final Report. National Marine Fisheries Service, Galveston, TX, USA. 75 p

Renaud ML, Carpenter JA, Williams JA, Manzella-Tirpak SA (1995b) Activities of juvenile green turtles, *Chelonia mydas*, at a jettied pass in south Texas. Fish Bull 93:586–593

Renaud ML, Carpenter JA, Williams JA, Landry AM Jr (1996) Kemp's ridley sea turtle (*Lepidochelys kempii*) tracked by satellite telemetry from Louisiana to nesting beach at Rancho Nuevo, Tamaulipas, Mexico. Chel Conserv Biol 2:108–109

Reyes C, Troëng S (2001) Report on the 2000 leatherback program at Tortuguero, Costa Rica. Caribbean Conservation Corporation, Gainesville, and Ministry of Environment and Energy of Costa Rica, San Pedro, Costa Rica. 31 p

Richards PM, Epperly SP, Heppell SS, King RT, Sasso CR, Moncada F, Nodarse G, Shaver DJ, Medina Y, Zurita J (2011) Sea turtle population estimates incorporating uncertainty: a new approach applied to western north Atlantic loggerheads *Caretta caretta*. Endang Species Res 15:151–158

Roberts K, Collins J, Paxton CH, Hardy R, Downs J (2014) Weather patterns associated with green turtle hypothermic stunning events in St. Joseph Bay and Mosquito Lagoon, Florida. Physical Geography 35:134–150

Rosman I, Boland GS, Martin LR, Chandler CR (1987) Underwater sightings of sea turtles in the Northern Gulf of Mexico. OCS Study MMS 87/0107. Minerals Management Service, Washington, DC, USA. 37 p

Ross JP, Beavers S, Mundell D, Airth-Kindree M (1989) The status of Kemp's ridley. Center for Marine Conservation, Washington, DC, USA. 51 p

Rostal DC (1991) The reproductive behavior and physiology of the Kemp's ridley sea turtle, *Lepidochelys kempii* (Garman, 1880). PhD Thesis, Texas A&M University, College Station, TX, USA. 138 p

Rostal DC (2005) Seasonal reproductive biology of the Kemp's ridley sea turtle (*Lepidochelys kempii*): Comparison of captive and wild populations. Chel Conserv Biol 4:788–800

Rostal DC, Grumbles JS, Byles RA, Márquez-M R, Owens DW (1997) Nesting physiology of Kemp's ridley sea turtles, *Lepidochelys kempi*, at Rancho Nuevo, Tamaulipas, Mexico, with observations on population estimates. Chel Conserv Biol 2:538–547

Rudloe A, Rudloe J (2005) Site specificity and the impact of recreational fishing activity on subadults endangered Kemp's ridley sea turtles in estuarine foraging habitats in the northeastern Gulf of Mexico. Gulf Mex Sci 2005:186–191

Rudloe A, Rudloe J, Ogren L (1991) Occurrence of immature Kemp's ridley turtles, *Lepidochelys kempi*, in coastal waters of northwest Florida. Northeast Gulf Science 12:49–53

Rybitski MJ, Hale RC, Musick JA (1995) Distribution of organochlorine pollutants in Atlantic sea turtles. Copeia 1995:379–390

Safina C, Wallace B (2010) Solving the "Ridley Riddle." In SWOT (State of the Worlds Sea Turtles). Report Volume V. Arlington, VA, USA, pp 26–30

Salmon M, Jones TT, Horch KW (2004) Ontogeny of diving and feeding behavior in juvenile sea turtles: Leatherback sea turtles (*Dermochelys coriacea* L) and green sea turtles (*Chelonia mydas* L) in the Florida Current. J Herpetol 38:36–43

Salum-Fares AD (2003) Analisis demográfico-reproductivo de la tortuga *Eretmochelys imbricata* (Linnaeus 1766) en playas arenosas de la Península de Yucatán: un estudio de largo plazo. Centro de Investigación y de Estudios Avanzados del Instituto Politécnico Nacional, Merida, Yucatán, Mexico

SEMARNAT (Secretaria de Medio Ambiente y Recursos Naturales) (2012) Homepage. http://www.semarnat.gob.mx/Pages/Inicio.aspx. Accessed 20 November 2013

Sánchez-Pérez JM, Vasconcelos-Pérez J, Díaz-Flores J (1989) Campamentos para Investigación, Proteccion y Fomento de Tortugas Marinas. In Tortugas: Golfina, Laud, Prieta, Blanca, Cahuama y Carey, Compilacion 1999. Instituto Nacional de Ecología, DF, Mexico, pp 17–28

Sarmiento Devia R, Harrison E (2010) Report on the 2009 leatherback program at Tortuguero, Costa Rica. Caribbean Conservation Corporation, Gainesville and Ministry of Environment and Energy of Costa Rica, San Pedro, Costa Rica. 39 p

Scarygami (2012) Hawksbill turtle stock photo. Flickr.com. http://www.flickr.com/photos/scarygami/5254129116/. Accessed 20 November 2013

Schmeltz GW, Mezich RR (1988) A preliminary investigation of the potential impact of Australian pines on the nesting activities of the Loggerhead Turtle. Proceedings, 8th Annual Workshop on Sea Turtle Conservation and Biology, Fort Fisher, NC, USA, December, pp 63–66

Schmid JR (1998) Marine turtle populations on the west-central coast of Florida: Results of tagging studies at the Cedar Keys, Florida, 1986–1995. Fish Bull 96:589–602

Schmid JR (2011) Assessment of marine turtle aggregations in the Coastal Waters of Lee County, Florida. Final Report to Sea Turtle Grants Program for Contract No. #09-050R,

Submitted by Conservancy of Southwest Florida to Sea Turtle Conservancy, Gainsville, FL, USA. May. 28 p

Schmid JR, Barichivich WJ (2005) Developmental biology and ecology of Kemp's ridley turtles in the eastern Gulf of Mexico. Chel Conserv Biol 4:828–834

Schmid JR, Barichivich WJ (2006) *Lepidochelys kempii*—Kemp's ridley turtle. In: Meylan PA (ed) Biology and conservation of Florida turtles. Chelonian Research Foundation, Lunenburg, MA, USA, pp 128–141

Schmid JR, Ogren LH (1990) Results of a tagging study at Cedar Key, Florida, with comments on Kemp's ridley distribution in the southeastern U.S. Proceedings, 10th Annual Workshop on Sea Turtle Biology and Conservation, Hilton Head, SC, USA, August, pp 129–130

Schmid JR, Witzell WN (1997) Age and growth of wild Kemp's ridley sea turtles, *Lepidochelys kempi*: Cumulative results of tagging studies in Florida. Chel Conserv Biol 2:532–537

Schmid JR, Witzell WN (2006) Seasonal migrations of immature Kemp's ridley turtles (*Lepidochelys kempii* Garman) along the west coast of Florida. Gulf Mex Sci 24:28–40

Schmid JR, Woodhead A (2000) Von Bertalanffy growth models for wild Kemp's ridley turtles: Analysis of the NMFS miami laboratory tagging database. In: Group TEW (ed) Assessment update for the Kemp's ridley and loggerhead sea turtle populations in the Western North Atlantic. National Marine Fisheries Service, Silver Spring, MD, USA, pp 94–102

Schmid JR, Bolten AB, Bjorndal KA, Lindberg WJ (2002) Activity patterns of Kemp's ridley turtles, *Lepidochelys kempii*, in the coastal waters of the Cedar Keys, Florida. Mar Biol 140:215–228

Schmid JR, Bolten AB, Bjorndal KA, Lindberg WJ, Percival HF, Zwick PD (2003) Home range and habitat use by Kemp's ridley turtles in west-central Florida. J Wildl Manage 67:196–206

Schroeder B, Murphy S (1999) Population surveys (Ground and Aerial) on nesting beaches. In: Eckert KL, Bjorndal KA, Abreu-Grobois FA, Donnelly M (eds) Research and Management Techniques for the Conservation of Sea Turtles. IUCN/SSC Marine Turtle Specialist Group Publication 4. Washington, DC, USA, pp 45–55

Schroeder BA, Foley AM, Witherington BE, Mosier AE (1998) Ecology of marine turtles in Florida Bay: Population structure, distribution, and occurrence of Fibropapilloma. Proceedings, 17th Annual Turtle Symposium, Orlando, FL, USA, pp 281–283

Schroeder BA, Foley AM, Bagley DA (2003) Nesting patterns, reproductive migrations, and adult foraging areas of loggerhead turtles. In: Bolten AB, Witherington BE (eds) Loggerhead sea turtles. Smithsonian Books, Washington, DC, USA, pp 114–124

Schwartz FJ (1978) Behavioral and tolerance responses to cold water temperatures by three species of sea turtles (Reptilia, Cheloniidae) in North Carolina. Fla Mar Res Publ 33:16–18

SeaTurtle.org (2011) News coverage: Scientist letter to Louisiana's Governor Makes Waves. http://seaturtles.org/article.php?id=2188. Accessed 20 November 2013

Seney EE (2008) Population dynamics and movements of the Kemp's ridley sea turtle, *Lepidochelys kempii*, in the Northwestern Gulf of Mexico. PhD Thesis, Texas A&M University, College Station, TX, USA. 168 p

Seney EE, Landry AM Jr (2008) Movements of Kemp's ridley sea turtles nesting on the upper Texas coast: Implications for management. Endang Species Res 4:73–84

Seney EE, Landry AM Jr (2011) Movement patterns of immature and adult female Kemp's ridley sea turtles in the northwestern Gulf of Mexico. Mar Ecol Prog Ser 440:241–254

Serge_Vero (2007) Sea turtle hatchling. iStockphoto, Calgary, Alberta, Canada. http://www.istockphoto.com/stock-photo-4557351-sea-turtle-hatchling.php?st=874d172.   Accessed 20 November 2013

Share the Beach (2013) Nesting season statistics. http://www.alabamaseaturtles.com/nesting-season-statistics/. Accessed 20 November 2013

Shaver DJ (1990a) Sea turtles in South Texas Inshore Waters. Proceedings, 10th Annual Workshop on Sea Turtle Biology and Conservation, Hilton Head Island, SC, USA, August, pp 131–132

Shaver DJ (1990b) Hypothermic stunning of sea turtles in Texas. MTN 48:25–27

Shaver DJ (1991) Feeding ecology of Kemp's ridley in South Texas waters. J Herpetol 25:327–334

Shaver DJ (1992) Kemp's ridley sea turtle reproduction. Herpetol Rev 23:59

Shaver DJ (1994) Relative abundance, temporal patterns, and growth of sea turtles at the Mansfield Channel, Texas. J Herpetol 28:491–497

Shaver DJ (1996) Head-started Kemp's ridley turtles nest in Texas. MTN 74:5–7

Shaver DJ (1997) Kemp's ridley turtles from international project return to Texas to Nest. Proceedings, 16th Annual Gulf of Mexico Information Transfer Meeting, New Orleans, LA, USA, December, pp 38–40

Shaver DJ (1999) Kemp's ridley sea turtle project at Padre Island National Seashore, Texas. Proceedings, 17th Annual Gulf of Mexico Information Transfer Meeting, New Orleans, LA, USA, December, Session 1G, pp 342–347

Shaver DJ (2000) Distribution, residency, and seasonal movements of the green sea turtle, *Chelonia mydas* (Linnaeus, 1758) in Texas. PhD Thesis. Texas A&M University, College Station, TX, USA. 237 p

Shaver DJ (2001a) Padre Island National Seashore Kemp's ridley sea turtle project and Texas sea turtle nesting and stranding 2000 report. U.S. Geological Survey, Reston, VA, USA. 61 p

Shaver DJ (2001b) U.S. Geological Survey/National Park Service Kemp's ridley sea turtle research and monitoring programs in Texas. Proceedings, Gulf of Mexico marine protected species workshop, New Orleans, LA, USA, June, pp 121–124

Shaver DJ (2004) Kemp's ridley project at Padre Island National Seashore and Texas Sea Turtle Nesting and Stranding 2002 Report. Proceedings, 17th Annual Gulf of Mexico Information Transfer Meeting, New Orleans, LA, USA, December, Session 1G, pp 342–347

Shaver DJ (2005) Analysis of the Kemp's ridley imprinting and headstart project at Padre Island National Seashore, Texas, 1978–1988, with subsequent nesting and stranding records on the Texas coast. Chel Conserv Biol 4:846–859

Shaver DJ (2006a) Kemp's ridley sea turtle habitat use in Mexico. Final Programmatic Report to the National Fish and Wildlife Foundation (2003-0212-009). National Park Service, Washington, DC, USA. 59 p

Shaver DJ (2006b) Kemp's ridley sea turtle nesting increasing in Texas. In: Harmon D (ed) People, places, and parks: Proceedings, 2005 George Wright Society Conference on Parks, Protected Areas, and Cultural Sites, Philadelphia, PA, USA, April, pp 391–397

Shaver DJ (2007) Texas sea turtle nesting and stranding 2006 report. National Park Service, Washington, DC, USA. 33 p

Shaver DJ (2008) Texas sea turtle nesting and stranding 2007 report. National Park Service, Washington, DC, USA. 36 p

Shaver DJ, Caillouet CW (1998) More Kemp's ridley turtles return to South Padre Island to nest. MTN 82:1–5

Shaver DJ, Rubio C (2008) Post-nesting movement of wild and headstarted Kemp's ridley sea turtles *Lepidochelys kempii* in the Gulf of Mexico. Endang Spec Res 4:43–55

Shaver DJ, Wibbels T (2007) Head-starting the Kemp's ridley sea turtle. In: Plotkin PT (ed) Biology and conservation of ridley sea turtles. The Johns Hopkins University Press, Baltimore, MD, USA, pp 297–324

Shaver DJ, Owens DW, Chaney AH, Caillouet CW Jr, Burchfield PB, Márquez-M. R (1988) Styrofoam box and beach temperatures in relation to incubation and sex ratios of Kemp's ridley sea turtles. Proceedings, 8th Annual Workshop on Sea Turtle Conservation and Biology, Fort Fisher, NC, USA, December, pp 103–108

Shaver DJ, Schroeder BA, Byles RA, Burchfield PM, Peña J, Márquez-M R, Martinez HJ (2005) Movements of home ranges of adult male Kemp's ridley sea turtles (*Lepidochelys kempii*) in the Gulf of Mexico investigated by satellite telemetry. Chel Conserv Biol 4:817–827

Shaver DJ, Hart KM, Fujisaki I, Rubio C, Sartain AR, Peña J, Burchfield PM, Gomez Gamez D, Ortiz J (2013) Foraging area fidelity for Kemp's ridleys in the Gulf of Mexcio. Ecol Evol 3:2002–2012

Spotila JR, Standora EA, Morreale SJ, Ruiz GJ (1987) Temperature dependent sex determination in the green turtle (*Chelonia mydas*) effects on the sex ratio on a natural beach. Herpetologica 43:74–81

Spotila JR, Dunham AE, Leslie AJ, Steyermark AC, Plotkin PT, Paladino FV (1996) Worldwide population decline of *Dermochelys coriacea*: are leatherback turtles going extinct? Chel Conserv Biol 2:209–222

Spotila JR, Reina RD, Steyermark AC, Plotkin PT, Paladino FV (2000) Pacific leatherback turtles face extinction. Nature 405:529–530

Standora EA, Spotila JR, Keinath JA, Shoop CR (1984) Body temperatures, diving cycles, and movement of a subadult leatherback turtle, *Dermochelys coriacea*. Herpetologica 40:169–176

Steiner T (1994) Shrimpers implicated as strandings soar in the USA. MTN 67:2–5

Stevenson CH (1893) Report on the coast fisheries of Texas. Report of the Commissioner for 1889 to 1891. U.S. Commission of Fish and Fisheries, Washington, DC, USA, pp 373–420

Stewart KR (2007) Establishment and growth of a sea turtle rookery: The population biology of the leatherback in Florida. PhD Thesis. Duke University, Durham, NC, USA. 129 p

Stewart K, Johnson C (2006) *Dermochelys coriacea*—leatherback sea turtle. In: Meylan PA (ed) Biology and conservation of Florida Turtles. Chelonian Research Monographs No. 3, Lunenburg, MA, USA, pp 144–157

Stewart K, Johnson C, Godfrey MH (2007) The minimum size of leatherbacks at reproductive maturity, with a review of sizes for nesting females from the Indian, Atlantic and Pacific ocean basins. J Herpetol 17:123–128

Stewart KR, Keller JM, Templeton R, Kucklick JR, Johnson C (2011) Monitoring persistent organic pollutants in leatherback turtles (*Dermochelys coriacea*) confirms maternal transfer. Mar Pollut Bull 62:1396–1409

Stokes LW, Epperly SP, McCarthy KJ (2012) Relationship between hook type and hooking location in sea turtles incidentally captured in the United States Atlantic pelagic longline fishery. Bull Mar Sci 88:703–718

STSSN (Sea Turtle Stranding and Salvage Network) (2012) STSSN database. Southeast Fisheries Science Center, Miami. http://www.sefsc.noaa.gov/species/turtles/strandings.htm. Accessed 20 November 2013

SWOT (State of the World's Sea Turtles) (2007a) A global glimpse of loggerhead nesting. Report Volume II. Arlington, VA, USA. 50 p

SWOT (2007b) Burning issues in conservation. Leatherback sea turtles of the world. Report Volume I. Arlington, VA, USA. 38 p

SWOT (2008) Where the hawksbill are. Report Volume III. Arlington, VA, USA. 44 p

SWOT (2010a) Kemp's and olive ridleys: Small turtles, big secrets. Report Volume V. Arlington, VA, USA. 54 p

SWOT (2010b) The green turtle: The most valuable reptile in the world. Report Volume VI. Arlington, VA, USA. 62 p

Taquet C, Taquet M, Dempster T, Soria M, Ciccione S, Roos D, Dagorn L (2006) Foraging of the green sea turtle *Chelonia mydas* on seagrass beds at Mayotte Island (Indian Ocean), determined by acoustic transmitters. Mar Ecol Prog Ser 306:295–302

Teas WG (1993) Species composition and size class distribution of marine turtle strandings on the Gulf of Mexico and Southeast United States Coasts, 1985–1991. NOAA Technical Memorandum NMFS-SEFSC-315, National Marine Fisheries Service, Miami, FL, USA. January. 43 p

TEWG (Turtle Expert Working Group) (1998) An assessment of the Kemp's ridley (*Lepidochelys kempii*) and loggerhead (*Caretta caretta*) sea turtle population in the Western North Atlantic. NOAA Technical Memorandum NMFS-SEFSC-409. Southeast Fisheries Science Center, National Marine Fisheries Service, Miami, FL, USA. March. 96 p

TEWG (2000) Assessment update for the Kemp's ridley and loggerhead sea turtle populations in the Western North Atlantic. NOAA Technical Memorandum NMFS-SEFSC-444. Southeast Fisheries Science Center, National Marine Fisheries Service, Miami, FL, USA. November. 115 p

TEWG (2007) An assessment of the leatherback turtle population in the Atlantic Ocean. NOAA Technical Memorandum NMFS-SEFSC-555. Southeast Fisheries Science Center, National Marine Fisheries Service, Miami, FL, USA. April. 116 p

TEWG (2009) An assessment of the loggerhead turtle population in the Western North Atlantic Ocean. NOAA Technical Memorandum NMFS-SEFSC-575. Southeast Fisheries Science Center, National Marine Fisheries Service, Miami, FL, USA. 131 p

Texas A&M University (2011) GEOSCIENCES NEWS: To the rescue. Texas A&M, College Station, TX, USA. http://geonews.tamu.edu/latestnews/775-to-the-rescue.html. Accessed 20 January 2014

Thompson NB (1988) The status of loggerhead, *Caretta caretta*; Kemp's ridley, *Lepidochelys kempi*; and green, *Chelonia mydas*, sea turtles in U.S. waters. Mar Fish Rev 50:16–23

Thompson N (1991) A response to Dodd and Byles. Mar Fish Rev 53:31–33

Thompson NB (1993) Preliminary analysis of sea turtle stranding event: Northern Gulf of Mexico, May/June 1993. Miami Laboratory Contribution MIA-93/94-20. National Marine Fisheries Service, Miami, FL, USA. 19 p

Thompson N, Henwood T, Epperly S, Lohoefener R, Gitschlag G, Ogren L, Mysing J, Renaud M (1990) Marine turtle habitat plan. NOAA Technical Memorandum NMFS-SEFSC-255. National Marine Fisheries Service, Miami, FL, USA. March. 20 p

Townsend CH (1899) Statistics of the fisheries of the Gulf States. In Division of Statistics and Methods of the Fisheries. U.S. Government Printing Office, Washington, DC, USA, pp 105–169

TPWD (Texas Parks and Wildlife Department) (2006) Fewer crabs—fewer fish. http://www.tpwd.state.tx.us/fishboat/fish/didyouknow/bluecrabdecline.phtml. Accessed 20 November 2013

TPWD (2012) Homepage. http://www.tpwd.state.tx.us/. Accessed 20 November 2013

Troëng S (1998) Report on the 1998 leatherback program at Tortuguero, Costa Rica. Caribbean Conservation Corporation, Gainesville and Ministry of Environment and Energy of Costa Rica, San Pedro, Costa Rica. 30 p

Troëng S (2000) Report on the 1999 leatherback program at Tortuguero, Costa Rica. Caribbean Conservation Corporation, Gainesville and Ministry of Environment and Energy of Costa Rica, San Pedro, Costa Rica. 29 p

Troëng S, Cook G (2000) Report on the 2000 leatherback program at Tortuguero, Costa Rica. Caribbean Conservation Corporation, Gainesville and Ministry of Environment and Energy of Costa Rica, San Pedro, Costa Rica. 28 p

Troëng S, Rankin E (2005) Long-term conservation efforts contribute to positive green turtle *Chelonia mydas* nesting trend at Tortuguero, Costa Rica. Biol Conserv 121:111–116

Troëng S, Chacon D, Dick B (2004) Possible decline in leatherback turtle *Dermochelys coriacea* nesting along the coast of Caribbean Central America. Oryx 38:395–403

True FW (1887) The turtle and terrapin fisheries. In: Goode GB (ed) The fisheries and fishery industries of the United States. U.S. Government Printing Office, Washington, DC, USA, pp 493–504

Tucker AD (1988) A summary of leatherback turtle *Dermochelys coriacea* nesting at Culebra, Puerto Rico, from 1984–1987 with Management Recommendations. Report. U.S. Fish and Wildlife Service, Vero Beach, FL, USA. 34 p

Tucker AD (2010) Nest site fidelity and clutch frequency of loggerhead turtles are better elucidated by satellite telemetry than by nocturnal tagging efforts: implications for stock estimation. J Exp Mar Biol Ecol 383:48–55

Tuncer B (2009) New born *Caretta* (loggerhead) sea turtle crawling on golden sands. iStockphoto. Calgary, Alberta, Canada. http://www.istockphoto.com/stock-photo-8799774-new-born-caretta-loggerhead-sea-turtle-crawling-on-golden-sands.php. Accessed 20 November 2013

Turnbull BS, Smith CR, Stamper MA (2000) Medical implications of hypothermia in threatened loggerhead (*Caretta caretta*) and endangered Kemp's ridley (*Lepidochelys kempi*) and green (*Chelonia mydas*) sea turtles. Proceedings, Annual Conference-American Association of Zoo Veterinarians, 1998, Omaha, NE, USA, October, pp 31–35

USFWS (U.S. Fish and Wildlife Service) (1991) Endangered and threatened wildlife and plants; 5-year review of listed species. Fed Regist 56:56882–56900

USFWS (2006) Preliminary report on the Mexico/United States of America Population Restoration Project for the Kemp's Ridley Sea Turtle, *Lepidochelys kempii*, on the Coasts of Tamaulipas and Veracruz, Mexico. Report Prepared by Gladys Porter Zoo and CONANP/SEMARNAT. U.S. Fish and Wildlife Service, Washington, DC, USA. 8 p

USFWS (2012) Hawksbill sea turtle (*Eretmochelys imbricata*) fact sheet. http://www.fws.gov/northflorida/SeaTurtles/Turtle%20Factsheets/hawksbill-sea-turtle.htm. Accessed 20 November 2013

USFWS, NMFS (1992) Recovery plan for the Kemp's ridley sea turtle (*Lepidochelys kempii*). U.S. Fish and Wildlife Service, Washington, DC and National Marine Fisheries Service, Silver Spring, MD, USA. 40 p

USFWS, NMFS (1996) Policy regarding the recognition of distinct vertebrate population segments under the endangered species act. U.S. Fish and Wildlife Service, Washington, DC and National Marine Fisheries Service, Silver Spring, MD, USA. Federal Register 61:4722–4725.

USFWS, NMFS (2011) Endangered and threatened species: Determination of nine distinct population segments of loggerhead sea turtles as endangered or threatened. U.S. Fish and Wildlife Service, Washington, DC and National Marine Fisheries Service, Silver Spring, MD, USA. Federal Register 76:58868–58952.

Valverde RA, Wingard S, Gómez F, Tordoir MT, Orrego CM (2010) Field lethal incubation temperature of olive ridley sea turtle *Lepidochelys olivacea* embryos at a mass nesting rookery. Endang Spec Res 12:77–86

Valverde RA, Orrego CM, Tordoir MT, Gómez FM, Solís DS, Hernández RA, Gómez GB, Brenes LS, Baltodano JP, Fonseca LG, Spotila JR (2012) Olive ridley mass nesting ecology and egg harvest at Ostional Beach, Costa Rica. Chel Conserv Biol 11:1–11

van Buskirk J, Crowder LB (1994) Life-history variation in marine turtles. Copeia 1994:66–81

van Houtan KS, Pimm SL (2007) Assessment of the Dry Tortugas National Park Sea Turtle Monitoring Program 1982–2006: Ecological Trends & Conservation Recommendations. NPS Report #H5299-05-1010. National Park Service, Washington, DC, USA. 29 p

Van Vleet ES, Pauly GG (1987) Characterization of oil residues scraped from stranded sea turtles from the Gulf of Mexico. Carib J Sci 23:77–83

Viada ST, Hammer RM, Racca R, Hannay D, Thompson MJ, Balcom BJ, Phillips NW (2008) Review of potential impacts to sea turtles from underwater explosive removal of offshore structures. Environ Impact Assess 28:267–285

Wallace BP, DiMatteo AD, Hurley BJ, Finkbeiner EM, Bolten AB, Chaloupka MY, Hutchinson BJ, Abreu-Grobois FA, Amorocho D, Bjorndal KA, Bourjea J, Bowen BW, Briseño D, Paolo C, Choudhury BC, Costa A, Dutton PH, Fallabrino A, Girard A, Girondot M, Godfrey MH, Hamann M, López-Mendilaharsu M, Marcovaldi MA, Mortimer JA, Musick JA, Nel R, Pilcher NJ, Seminoff JA, Troëng S, Witherington B, Mast RB (2010) Regional management units for marine turtles: A novel framework for prioritizing conservation and research across multiple scales. PLoS One 5, e15465

Watson JW, Foster DG, Epperly S, Shah A (2004) Experiments in the Western Atlantic Northeast distant waters to evaluate sea turtle mitigation measures in the pelagic longline fishery: Report on experiments conducted in 2001–2003. NMFS Technical Report. National Marine Fisheries Service, Silver Spring, MD, USA. 123 p

Waycott M, Longstaff BJ, Mellors J (2005) Seagrass population dynamics and water quality in the Great Barrier Reef region: A review and future research directions. Mar Pollut Bull 51:343350

Webster PJ, Holland GJ, Curry JA, Chang H-R (2005) Changes in tropical cyclone number, duration, and intensity in a warming environment. Science 309:1844–1846

Werler J (1951) Miscellaneous notes on the eggs and young of Texan and Mexican reptiles. Zoologica 36:37–48

Werner SA (1994) Feeding ecology of wild and head started Kemp's ridley sea turtles. MS Thesis, Texas A&M University, College Station, TX, USA. 65 p

Wibbels T (2007) Sex determination and sex ratio in ridley sea turtles. In: Plotkin PT (ed) Biology and conservation of ridley sea turtles. The Johns Hopkins University Press, Baltimore, MD, USA, pp 167–189

Wibbels T, Geis A (2003) Evaluation of hatchling sex ratios of Kemp's ridley in situ nests and egg corral nests in the Kemp's ridley recovery program during the 2002 nesting season. Final Report. National Marine Fisheries Service. National Marine Fisheries Service, Silver Spring, MD, USA

Wibbels TR, Owens DW, Limpus CJ, Reed PC, Amoss MS Jr (1990) Seasonal changes in serum gonadal steroids associated with migration, mating, and nesting in the loggerhead sea turtle (Caretta caretta). Gen Comp Endocr 79:154–164

Williams KL, Frick MG, Pfaller JB (2006) First report of green, Chelonia mydas, and Kemp's ridley, Lepidochelys kempii, turtle nesting on Wassaw Island, Georgia, USA. MTN 113:8

Witherington BE (1986) Human and natural causes of marine turtle clutch and hatchling mortality and their relationship to hatchling production on an important Florida nesting beach. MS Thesis, University of Central Florida, Orlando, FL, USA. 141 p

Witherington BE (1992) Behavior responses of nesting sea turtles to artificial lighting. Herpetologica 48:31–39

Witherington BE (1995) Observations of hatchling loggerhead turtles during the first few days of the lost year(s). Proceedings, 12th Annual Workshop on Sea Turtle Biology and Conservation, Jekyll Island, FL, USA, February, pp 154–157

Witherington BE (1997) The problem of photopollution for sea turtles and other nocturnal animals. In: Clemmons JR, Buchholz R (eds) Behavioral approaches to conservation in the wild. Cambridge University Press, Cambridge, UK, pp 303–328

Witherington BE (2002) Ecology of neonate loggerhead turtles inhabiting lines of downwelling near a Gulf Stream front. Mar Biol 140:843–853

Witherington BE, Ehrhart LM (1989a) Status and reproductive characteristics of green turtles (*Chelonia mydas*) nesting in Florida. Proceedings, 2nd Western Atlantic Turtle Symposium. National Marine Fisheries Service, Mayaguez, Puerto Rico, October, pp 351–352

Witherington BE, Ehrhart LM (1989b) Hypothermic stunning and mortality of marine turtles in the Indian River Lagoon System, Florida. Copeia 1989:696–703

Witherington BE, Martin RE (1996) Understanding, assessing, and resolving light-pollution problems on sea turtle nesting beaches. Technical Report TR-2. Florida Marine Research Institute, St. Petersburg, FL, USA. 73 p

Witherington BE, Crady C, Bolen L (1996) A 'Hatchling Orientation Index' for assessing orientation disruption from artificial lighting. Proceedings, 15th Annual Symposium on Sea Turtle Biology and Conservation, Hilton Head, SC, USA, February, pp 344–347

Witherington B, Lucas L, Koeppel C (2005) Nesting sea turtles respond to the effects of ocean inlets. Proceedings, 21st Annual Symposium on Sea Turtle Biology and Conservation. Philadelphia, PA, USA, February, Part II, pp 355–356

Witherington B, Herren R, Bresette M (2006a) *Caretta caretta*—loggerhead sea turtle. In: Meylan PA (ed) Biology and conservation of Florida turtles, vol 3, Chelonian research monographs. Massachusetts Chelonian Research Foundation, Lunenburg, MA, USA, pp 74–89

Witherington B, Bresette M, Herren R (2006b) *Chelonia mydas*—green turtle. In: Meylan PA (ed) Biology and conservation of Florida turtles, vol 3, Chelonian research monographs. Massachusetts Chelonian Research Foundation, Lunenburg, MA, USA, pp 90–104

Witherington B, Kubilis P, Brost B, Meylan A (2009) Decreasing annual nest counts in a globally important loggerhead sea turtle population. Ecol Appl 19:30–54

Witherington B, Hirama S, Hardy R (2012) Young sea turtles of the pelagic *Sargassum*-dominated drift community: Habitat use, population density, and threats. Mar Ecol Prog Ser 463:1–22

Witt MJ, Hawkes LA, Godfrey MH, Godley BJ, Broderick AC (2010) Predicting the impacts of climate change on a globally distributed species: The case of the loggerhead turtle. J Exp Biol 213:901–911

Witzell WN (1983) Synopsis of biological data on the hawksbill turtle, *Eretmochelys imbricata* (Linnaeus, 1766). FAO Fisheries Synopsis 137. Food and Agriculture Organization of the United Nations, Rome, Italy. 78 p

Witzell WN (1984) The incidental capture of sea turtles in the Atlantic U.S. Fishery Conservation Zone by the Japanese Tuna Longline Fleet, 1978–1981. MTN 46:56–58

Witzell WN (1994) The origin, evolution, and demise of the U.S. sea turtle fisheries. Mar Fish Rev 56:8–23

Witzell WN (1998) Long-term tag returns from juvenile Kemp's ridleys. MTN 79:20

Witzell WN (1999) Distribution and relative abundance of sea turtles caught incidentally by the U.S. pelagic longline fleet in the western North Atlantic Ocean, 1992–1995. Fish Bull 97:200–211

Witzell WN, Azarovitz T (1996) Relative abundance and thermal and geographic distribution of sea turtles off the U.S. Atlantic Coast based on aerial surveys (1963–1969). NOAA

Technical Memorandum NMFS-SEFSC-381. National Marine Fisheries Service, Miami, FL, USA. March. 10 p

Witzell WN, Schmid JR (2004) Immature sea turtles in Gullivan Bay, Ten Thousand Islands, Southwest Florida. Gulf Mex Sci 4:54–61

Witzell WN, Schmid JR (2005) Diet of immature Kemp's ridley turtles (*Lepidochelys kempi*) from Gullivan Bay, Ten Thousand Islands, Southwest Florida. Bull Mar Sci 77:191–199

Wyneken J, Salmon M (1992) Frenzy and postfrenzy swimming activity in loggerhead, green, and leatherback hatchling sea turtles. Copeia 1992:478–484

Wyneken J, Madrak SV, Salmon M, Foote J (2008) Migratory activity by hatchling loggerhead sea turtles (*Caretta caretta* L.): Evidence for divergence between nesting groups. Mar Biol 156:171–178

Xavier R, Barata A, Cortez LP, Queiroz N, Cuevas E (2006) Hawksbill turtle (*Eretmochelys imbricata* Linnaeus 1766) and green turtle (*Chelonia mydas* Linnaeus 1754) nesting activity (2002–2004) at El Cuyo beach, Mexico. Amphibia-Reptilia 27:539–547

Yender RA, Mearns AJ (2003) Case studies of spills that threaten sea turtles. Oil and sea turtles: Biology, planning, and response. Office of Response and Restoration, National Oceanic and Atmospheric Administration, Seattle, WA, USA, pp 69–86

Yerger RW (1965) The leatherback turtle on the Gulf coast of Florida. Copeia 1965:365–366

Yeung C (1999) Estimates of marine mammal and marine turtle bycatch by the U.S. Pelagic Longline Fleet in 1998. NOAA Technical Memorandum NMFS-SEFSC-430. National Marine Fisheries Service, Miami, FL, USA. November. 26 p

Yeung C (2001) Estimates of marine mammal and marine turtle bycatch by the U.S. Atlantic Pelagic Longline Fleet in 1999–2000. NOAA Technical Memorandum NMFS-SEFSC-467. National Marine Fisheries Service, Miami, FL, USA. August. 43 p

Yntema CL, Mrosovsky N (1982) Critical periods and pivotal temperatures for sexual differentiation in loggerhead sea turtles. Can J Zool 60:1012–1016

Zimmerman R (ed) (1998) Characteristics and causes of Texas marine strandings. NOAA Technical Report NMFS-143. National Marine Fisheries Service, Galveston, TX, USA. 85 p

Zug GR, Glor RE (1998) Estimates of age and grown in a population of green sea turtles (*Chelonia mydas*) from the Indian River Lagoon system, Florida: A skeletochronological analysis. Can J Zool 76:1497–1506

Zug GR, Balazs GH, Wetherall JA, Parker DM, Murakawa SKK (2002) Age and growth of Hawaiian green sea turtles (*Chelonia mydas*): An analysis based on skeletochronology. Fish Bull 100:117–127

Open Access   This chapter is licensed under the terms of the Creative Commons Attribution-NonCommercial 2.5 International License (http://creativecommons.org/licenses/by-nc/2.5/), which permits any noncommercial use, sharing, adaptation, distribution and reproduction in any medium or format, as long as you give appropriate credit to the original author(s) and the source, provide a link to the Creative Commons license and indicate if changes were made.

The images or other third party material in this chapter are included in the chapter's Creative Commons license, unless indicated otherwise in a credit line to the material. If material is not included in the chapter's Creative Commons license and your intended use is not permitted by statutory regulation or exceeds the permitted use, you will need to obtain permission directly from the copyright holder.

CHAPTER 12

AVIAN RESOURCES OF THE NORTHERN GULF OF MEXICO

Joanna Burger[1]

[1]Rutgers–The State University of New Jersey, Piscataway, NJ 08854, USA
burger@biology.rutgers.edu

12.1 INTRODUCTION

Birds are unique among vertebrates because they can fly long distances in a short period of time, and, with few exceptions, live in three-dimensional spaces. Birds that live in the water-land interface may be equally at home on land, in the air, and in the water. Most other organisms live their entire lives, or phases of their lives, in either water (fish, whales, clams, other invertebrates) or in some other medium (soil or land surface). The ability to switch from one medium to another on a daily basis requires flexibility in physiological and behavioral adaptations. A wide diversity of birds exists in the marine-terrestrial interface at the margins of continents and offshore islands. Seabirds live mainly on the oceans (pelagic), but also nest on offshore islands or along coasts (Schreiber and Burger 2001a). Herons, egrets, and some shorebirds live primarily in the marine-land interface, foraging in coastal bays and estuaries and nesting along beaches on islands, or on adjacent uplands (Burger and Olla 1984; Lantz et al. 2010, 2011; Kushlan and Hafner 2000a, b). Several shorebird species migrate or winter along coasts, but breed in the high Arctic. Many species of ducks winter along coasts but breed in inland habitats, including the prairie pothole region of North America. Other birds live mainly in coastal marshes (rails, some Passerines) and spend most of their time there.

The Gulf of Mexico has several important features for promoting high avian use and diversity: (1) a high diversity of habitats; (2) a direct pathway for Nearctic-Neotropical migrants flying to Mexico, Central America, and South America; and (3) warm coastal waters. The Gulf of Mexico is considered the most important migratory pathway in the world for waterfowl (Gallardo et al. 2004), in North America for Nearctic-Neotropical migrants, primarily songbirds (Rappole 1995; Moore 2000a), and for migrant and wintering shorebirds (Withers 2002). The four flyways of North America join in the Gulf of Mexico. Many migrants pass through central Veracruz, while others from the Mississippi and Atlantic flyways migrate directly across the open waters of the Gulf (Moore 2000a; Gauthreaux et al. 2006).

One indication of the importance of the Gulf of Mexico is the percentage of U.S. breeding populations of several species that it hosts. The U.S. Gulf Coast has a significant portion of the world population of Reddish Egret (*Egretta rufescens*) (Lowther and Paul 2002) and nearly all the Snowy Plover (*Charadrius nivosus*) that breed east of the Rockies (Elliott-Smith et al. 2004; Page et al. 2009). It also has a significant portion of the U.S. breeding populations of Sandwich Tern (*Sterna sandvicensis*), Black Skimmer (*Rynchops niger*), Forster's Tern (*Sterna forsteri*), Laughing Gull (*Larus atricilla*), and Royal Tern (*Sterna maximus*) (Figure 12.1) (Visser and Peterson 1994).

In addition, the southern Gulf of Mexico is the northern limit for many tropical species nesting in Mexico, such as boobies and Magnificent Frigatebird (*Fregata magnificens*), while the tropical Sooty Tern (*Sterna fuscata*) and Brown Noddy (*Anous stolidus*) breed as far north

© The Author(s) 2017
C.H. Ward (ed.), *Habitats and Biota of the Gulf of Mexico: Before the Deepwater Horizon Oil Spill,*
DOI 10.1007/978-1-4939-3456-0_4

Figure 12.1. A colony of Sandwich Terns, with half-grown young. A royal chick (with a *yellow bill*) is in the center. After hatching, the chicks form crèches as protection against predators. © J. Burger.

as the Dry Tortugas (Tunnell and Chapman 2000). The Laguna Madre region from southern Texas to Tamaulipas is one of the most important shorebird wintering areas (Mabee et al. 2001; Withers 2002). The region from southern Tamaulipas to Campeche contains mainly aquatic species with Nearctic-Neotropical affinities (Correa et al. 2000a, b; Gallardo et al. 2009). Many migrants, some from southern regions, winter or occur in the Yucatán peninsula (Howell 1989; Greenberg 1992; Mackinnon et al. 2011).

12.1.1   Objectives

The purpose of this chapter is to provide an overview of avian status and trends in the northern Gulf of Mexico before the Deepwater Horizon oil spill, with special emphasis on the U.S. Gulf Coast. Specific objectives include examining the avian assemblages in the Gulf generally, exploring how birds use the marine-land interface, describing the major stressors driving avian abundance and distribution, and examining spatial and temporal trends in breeding and migrant bird populations. Depending upon the authority, about 400 species of birds use the Gulf at some time of the year or at some point in their life cycle, including brief but crucial stopovers as migrants (Gallardo et al. 2009).

This chapter mainly tracks bird populations in the northern Gulf of Mexico since the 1930s or later, using indicator species and indicator groups. Prior to this time, there are no time series data on bird populations. This time period was also selected because two of the major data sets (Audubon's Christmas Bird Counts, Bird Banding Laboratory's Breeding Bird Surveys) include data for these periods. Many local and state surveys began in the 1970s. Systematic collection of local and regional data usually spans a shorter period, and often stops before the present. Changes in avifauna undoubtedly occurred with the arrival of people from Europe (clearing of forests), with market hunting (plumes for hats, eggs for food), and the massive use of pesticides such as dichlorodiphenyltrichloroethane (DDT) (King et al. 1977). For a more in depth presentation of status and trends of birds of both the northern and southern Gulf, see Burger (2017).

12.1.2   Methods

This chapter considers birds in the Gulf of Mexico ecosystem, including associated offshore islands, barrier islands, and the complex matrix of backbays, mudflats, mangroves, salt marshes, brackish marshes, and associated freshwater marshes, swamps, and uplands. Coral reefs are located mainly in Mexico, although some reefs extend to the Florida Keys (Stedman and Dahl 2008). The Gulf of Mexico itself is approximately 1,400 kilometers (km) (870 miles [mi]) in diameter and is bordered by the United States in the north, Mexico in the

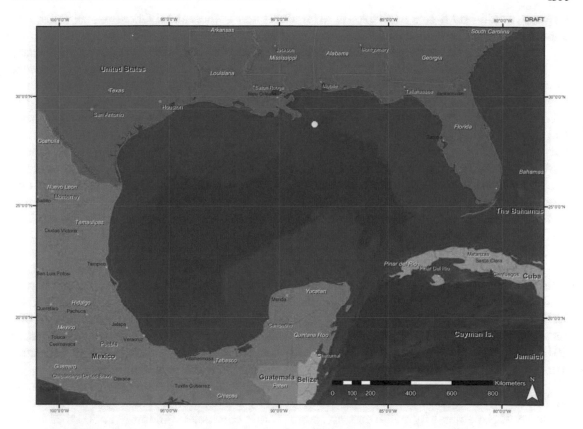

Figure 12.2. Map of Gulf of Mexico, showing the United States, Mexican, and Cuban Coasts. Photo by Wells 2013.

south, the Eastern coast of Mexico and Texas on the west, and the western coast of Florida and Cuba on the east (Figure 12.2). Three countries border the Gulf of Mexico. For many economic, ecological, ethical, and legal reasons, society should protect biodiversity in the Gulf of Mexico ecosystem (Felder et al. 2009). Understanding avian diversity in the Gulf is part of this mandate.

This chapter is derived primarily from published information in the refereed literature, in state and federal reports, and in the gray literature. All sources used are available to the public. Since it is impossible to examine the status and trends of all these species, this chapter examines selected indicators. A brief discussion of various aspects of the Gulf ecosystem and the factors that affect avian reproductive success, survival, and population dynamics are presented. This is followed by status and trends information of birds in the Gulf by individual species and species groups. Trends information is usually not available for the entire Gulf (or even for the northern coast) from the same time period. However, more complete data exist for some species, such as the Piping Plover (*Charadrius melodius*, Haig et al. 2005; Elliott-Smith et al. 2009), and comprehensive surveys of breeding and wintering Charadriiformes (gulls and terns), Anseriformes (waterfowl), and Gaviiformes through Pelecaniformes (loons through pelicans) were conducted from 1976 to 1978 by the U.S. Fish and Wildlife Service (Clapp et al. 1982a, b, 1983). These databases provide representative status and trends information for indicator species groups.

Many data gaps exist because neither the U.S. Gulf Coast nor the entire Gulf Coast has been surveyed for birds recently or completely. Different data sets are used to examine different questions. Some of these are older than others, and there may have been changes in either species composition or population levels since the data were last gathered. One of the longest-running data sets available for wintering birds is the annual Christmas Bird Counts, conducted by National Audubon Society.

Christmas Bird Counts were used to examine trends to illustrate particular points (e.g., yearly variability, differences among species, or in a given species in different Gulf States)[1] and recent trends (Niven and Butcher 2011). Niven and Butcher's (2011) analysis of the status and trends of wintering birds along the northern Gulf Coast using the Audubon Christmas Bird Counts from 1965 to 2011 is useful because it is extensive, long-term, and includes all five states. They used Christmas Counts that were centered around 7.5 miles from the Gulf coast. During this time period, the number of counts ranged from 10 to 21 (Texas), 1.7 to 6.6 (Louisiana), 2.5 to 4 (Alabama), 0 to 2 (Mississippi), and 13 to 26 (Florida). There were twice as many counts in the period from 2001 to 2010 than during 1965–1970. In general, counts were conducted by any number of people divided into parties that counted all individual birds observed during a variable period of time (limited to 24 hours (h) from mid-December to early January; Butcher 1990). The difficulty of different numbers of people, counting for different time periods, is reduced by reporting number of birds per party hour (after Link and Sauer 1999a, b).

Niven and Butcher (2011) used hierarchical log-linear models fit with Bayesian models to estimate relative abundance, relative density, and trends for the Gulf region as a whole (Sauer et al. 2009; Sauer and Link 2011). They published their findings after the Deepwater Horizon oil spill, but the trends are not reflective of this event because it occurred at the end of the time series (e.g., 2010–2011 Christmas Count); the data reflect regional trends (Niven and Butcher 2011). Christmas Bird Count data are presented, either as yearly patterns or 3-year running averages, which smooths out the temporal data, making it easier to see patterns.

Breeding Bird Surveys (BBS, Sauer et al. 2011) provide useful data for species that nest mainly along the Gulf of Mexico (e.g., Brown Pelican). Surveys conducted in June (early May in some southern states) by volunteers are point counts conducted randomly at 50 stops along preselected roadside routes. Counts start 30 minutes (min) before local sunrise, and stops are 0.8 km apart. At each stop, the observer conducts a 3-min count of all birds seen and heard within 400 meters (m). There are more than 5,000 established routes in North America, and about 2,500 are surveyed each year (Sauer and Link 2011). Data are presented as an index, which represents the mean number of birds counted per route (Sauer and Link 2011). Colonial birds present a challenge because the routes seldom pass colonies, and counts may represent birds flying around or foraging. However, since the methods are the same from year to year, they provide a useful index to assess changes in population numbers. The Bird Banding Laboratory provides information on trends by state for different species, and this information can give an overall picture of changes that can be used in conjunction with other data sets (Sauer et al. 2008).

Other methods are explained in individual sections (Green et al. 2008). The author took all photographs and all tables and figures were developed from the original data sources, unless otherwise noted. This chapter reviews current information, with three caveats: (1) Understanding population status and trends is an on-going process of new assessments, improving methods of assessment, and increasing coverage of the Gulf of Mexico, both temporally and spatially. (2) Selection of topics, indicator species and groups, and trends information was necessary. (3) The emphasis is on the northern Gulf Coast. Indicators were selected to represent avian communities and relationships, as well as different life histories and conservation status. While it is possible to write separate papers on most topics considered, the task was to provide an overview of avian communities in the Gulf of Mexico.

---

[1] http://audubon2.org/cbchist/table.html.

Finally, over the course of the last half-century, the taxonomy of North American birds has undergone several revisions (American Ornithologists' Union [AOU] Checklists), resulting in different family assignments and changes in nomenclature, particularly at the genus level. The sequence of listing families has also changed. Throughout this chapter, the nomenclature used by the authors cited was retained. The most recent AOU checklist is the 7th edition (1998), and more than 50 supplements have been published in *The Auk* since that time. Changes that are relevant to the Gulf of Mexico can be found in the individual Birds of North America Accounts (Laboratory of Ornithology, Cornell University, Ithaca, NY USA).[2]

12.2 LAWS, REGULATIONS, AND STATUS DESIGNATIONS

Laws and regulations provide the legal basis for environmental protection of birds in the Gulf of Mexico. The Migratory Bird Treaty Act (1918) and the U.S. Endangered Species Act (1973) are the main federal laws that apply to birds in the Gulf. The Migratory Bird Treaty Act protects birds that migrate between and among Canada, the United States, and Mexico. Nearly all birds that occur in the United States and Mexico are protected by this Act. The United States also signed treaties with Mexico (1936), Japan (1972), and the USSR (1976) to protect birds in those countries (Shackelford et al. 2005). The Endangered Species Act protects species listed as threatened or endangered, but the U.S. Fish & Wildlife Service also lists candidate species, those that are being considered for listing. The Convention on International Trade in Endangered Species of Wild Fauna and Florida (CITES), 1973, applies to an established list of birds that are imported, traded or sold, and where such activities threaten their populations.

In addition to international laws, and United States, Cuban, or Mexican laws, each state in the United States has laws and regulations that relate to birds. Most states have an endangered and threatened species list, and many states have a list of species of special concern. Such species are usually so designated because either their populations are in jeopardy or information is insufficient to determine status, but there is concern about their numbers or threats to their populations. Federal and state designations are given in Tables 12.1 and 12.2. Other federally listed endangered or threatened species occur along the coast, although most are

Table 12.1. Federally Listed Birds that Occur Along the Gulf Coast of the United States, Cuba, and Mexico (only non-Passerines are included)

United States
Whooping Crane (*Grus americana*)—endangered
Wood Stork (*Mycteria americana*)—threatened (Alabama, Florida, Mississippi)
Eskimo Curlew (*Numenius borealis*)—endangered
Piping Plover (*Charadrius melodus*)—threatened
Interior Least Tern (*Sterna antillarum*)—endangered
Northern Aplomado Falcon (*Falco femoralis*)—endangered
Cape Sable Seaside Sparrow (*Ammodramus maritimus mirabilis*)—endangered (Florida only)
Everglade Snail Kite (*Rostrhamus sociabilis plumbeus*)—endangered
Cuba (Earth's Endangered Species (Glenn 2006a))
Black-capped Petrel (*Pterodroma hasitata*)

(continued)

[2] Available online at http://bna.birds.cornell.edu/bna/.

Table 12.1. (continued)

Brown Pelican (*Pelecanus occidentalis*)
Cuban Black Hawk (*Buteogallus gundlachii*)
Cuban Kite (*Chondrohierax wilsonii*)
Least Tern (*Sterna antillarum*)
Sandhill Crane (*Grus canadensis*)
West Indian Whistling-duck (*Dendrocygna arborea*)
Ivory-billed Woodpecker (*Campephilus principalis*)
Mexico (Glenn 2006b)
Northern Aplomado Falcon (*Falco femoralis*)
Whooping Crane (*Grus americana*)
Elegant Tern (*Sterna elegans*)
Brown Pelican (*Pelecanus occidentalis*)
Black Rail (*Laterallus jamaicensis*)
Least Tern (*Sterna antillarum*)

Table 12.2. Endangered and Threatened Species by State for Those Breeding or Those Expected to Occur Along the Gulf of Mexico

Texas (TPWD 2004)
Brown Pelican (*Pelecanus occidentalis*)—endangered
Reddish Egret (*Egretta rufescens*)—threatened
White-faced Ibis (*Plegadis chihi*)—threatened
Wood Stork (*Mycteria americana*)—threatened
Whooping Crane (*Grus americana*)—endangered
Swallow-tailed Kite (*Elanoides forficatus*)—threatened
Bald Eagle (*Haliaeetus leucocephalus*)—threatened
Northern Aplomado Falcon (*Falco femoralis*)—endangered
Peregrine Falcon (*Falco peregrinus*)—threatened
Eskimo Curlew (*Numenius borealis*)—endangered (generally considered extinct)
Interior Least Tern (*Sterna antillarum*)—endangered
Piping Plover (*Charadrius melodus*)—threatened
Sooty Tern (*Sterna fuscatus*, now *Onychoprion fuscata*)—threatened
And a few songbirds that may be migrants (e.g., Golden-cheeked Warbler (*Dendroica chrysoparia*, endangered), Rose-throated Becard (*Pachyramphus aglaiae*, endangered), and Black-capped Vireo (*Vireo atricapillus*, threatened)). These are in coastal woodlands.
Louisiana (DWF 2012[a])
Brown Pelican (*Pelecanus occidentalis*)—endangered (also the Louisiana state bird)
Least Tern (*Sterna antillarum*)—endangered
Piping Plover (*Charadrius melodus*)—threatened/endangered
Eskimo Curlew (*Numenius borealis*)—endangered (generally considered extinct)

(continued)

Table 12.2. (continued)

Whooping Crane (*Grus americana*)—endangered
Peregrine Falcon (*Falco peregrinus*)—threatened/endangered
Bald Eagle (*Haliaeetus leucocephalus*)—endangered
Mississippi (USFWS 2012a[a])
Mississippi Sandhill Crane (*Grus canadensis*)—endangered
Least Tern (*Sterna antillarum*) (may not occur coastally)—endangered
Piping Plover (*Charadrius melodus*, except Great Lakes watershed)—threatened
Alabama (USFWS 2012b[a])
Piping Plover (*Charadrius melodus*, except Great lakes watershed)—threatened
Wood Stork (*Mycteria americana*)—endangered
Florida (FFWCC 2010)
Brown Pelican (*Pelecanus occidentalis*)—species of special concern
American Oystercatcher (*Haematopus palliatus*)—species of special concern
Marian's Marsh Wren (*Cistothorus palustris marianae*)—species of special concern
Scott's Seaside Sparrow (*Ammodramus maritimus peninsulae*)—species of special concern
Wakulla Seaside Sparrow (*Ammodramus maritimus juncicola*)—species of special concern
Wood Stork (*Mycteria Americana*)—endangered
Least Tern (*Sternula antillarum*)—threatened
Roseate Tern (*Sterna dougallii*)—threatened
Snowy Egret (*Egretta thula*)—species of special concern
Reddish Egret (*Egretta rufescens*)—species of special concern
Roseate Spoonbill (*Platalea ajaja*)—species of special concern
White Ibis (*Eudocimus albus*)—species of special concern
Tricolored Heron (*Egretta tricolor*)—species of special concern
Snowy Egret (Egretta thula)—species of special concern
Piping Plover (*Charadrius melodus*)—threatened
Snowy Plover (*Charadrius nivosus*)—threatened
Little Blue Heron (*Egretta caerulea*)—species of special concern
Osprey (*Pandion haliaetus*)—species of special concern
Black Skimmer (*Rynchops niger*)—species of special concern

Listed are all species that could get to coastal environments
[a]Earlier lists are not available
SSC species of special concern

not common in saltwater environments. The Brown Pelican (*Pelecanus occidentalis*) was listed federally until 1998 (Lindstedt 2005; USFWS 2009a). The Bald Eagle was federally delisted August 9, 2007, although they are still protected under the Eagle Act (USFWS 2010a).

Other organizations have conservation ratings or listings for many species. For example, the Audubon Society (2012) lists priority species, and the International Union for Conservation of Nature (IUCN 2011) publishes a Red List of Threatened Species. Their listings are usually similar to federal listings. The Audubon list sometimes includes species before they have been added to the federal lists (Reddish Egret, Red Knot, Marbled Godwit, and Black Skimmer) (Audubon Society 2012).

Finally, it should be mentioned that many states have designations of "species of special concern" for species with some indication that populations may have declined or lack data to indicate status. These species deserve special consideration because some may become threatened if steps are not taken to protect them.

12.3 LAND-WATER INTERFACE

Land-water interfaces usually have high species diversity and high biomass because they contain a range of different habitats. Habitats are intermixed in different patch sizes, and the interface serves as the gateway for movement into both aquatic and terrestrial environments. While it is impossible to clearly define the coastal zone, functionally it is the area on either side of the actual meeting of the land and ocean that is influenced by both marine and terrestrial inputs. The margins themselves are usually narrow, providing an opportunity for animals to move quickly from one habitat to another (Burger 1991a). Since these character-istics apply to both plant and invertebrate communities, the diversity is amplified in higher trophic levels, such as fish, birds, and mammals. The land-water interface also serves as a physical buffer for both the marine ecosystem and for the terrestrial system. Estuarine and coastal environments protect inland terrestrial habitats from excessively high tides, hurri-canes, erosion, and other severe storm events, while protecting marine environments from contamination by providing a sink for contaminants. The margin constantly changes due to the effects of wind and water.

Because it is large, the Gulf of Mexico has a long coastline with a wide range of habitats. Because of its geographical position, it has a diversity of habitats that extend from tropical to temperate and from coastal to offshore islands. The Gulf serves as a conduit or migration route to southern wintering grounds between the United States (and more northern Canada) and Mexico, Central America, and South America (Gallardo et al. 2004). The land mass to the north is larger and serves as a funnel point for birds scattered across North America that are migrating to wintering grounds along the Gulf of Mexico or farther south. Most of the birds of the Gulf of Mexico are tied to the coastal zone because of breeding constraints and foraging opportunities.

Gallardo et al. (2009) lists 395 species in 53 families as the number of bird species in the Gulf region. The main families in the Gulf are ducks (Anatidae, 46 species), gulls, terns and skimmers (Laridae, $N = 41$), herons and egrets (Ardeidae, $N = 17$), rails (Rallidae, $N = 16$), warblers (Parulidae, $N = 36$), and flycatchers (Tyrannidae, $N = 17$). The latter two groups are Passerines, but they frequently occur on coastal islands, on marshes, and in coastal forest habitats either as migrants or during the breeding season (Moore et al. 1990; Buler et al. 2007; Buler and Moore 2011). For a full list of the species, see Gallardo et al. (2009).

Coasts are impacted by weather and storm events, as well as anthropogenic factors, such as alteration of hydrological processes, introduction of toxic chemicals and nutrients, increased human population density, increased fishing and other commercial enterprises, development of wind energy, increased numbers of oil and gas platforms, and direct human disturbance. Half of the continental U.S. population resides within 50 mile of the coasts, making them the most rapidly growing areas in the United States. From the 1960s to 2015, the population density of all Gulf coastal counties is expected to increase from 187 to 327 people per square mile (NOAA 1998). Condominiums, resorts, casinos, and other commercial and industrial development already characterize large expanses of the northern Gulf Coast. Development of wind energy is ongoing, both nearshore and offshore, and has the potential to disrupt bird migration across the Gulf (Morrison 2006). Thirty-seven percent (37 %) of the population in the Gulf States lives in the Gulf Coast region (Bildstein et al. 1991; NOAA 2011). Increases in coastal and offshore development will affect birds through decreases in habitat and increased disturbance.

The potential effects of climate change are related to anthropogenic factors (Bradshaw and Holzapfel 2006), such as sea level rise and land subsidence (Daniels et al. 1993; Bayard and Elphick 2011). Increased sea level rise results in increased flooding of nests, eggs, and chicks, as well as rendering habitat on islands, beaches, or salt marshes no longer usable by nesting or foraging birds, such as Brown Pelicans, Piping Plovers, and most terns and skimmers (Daniels et al. 1993). Habitat for salt marsh species, such as Clapper Rails (*Rallus longirostris*) and Salt marsh Sparrows (Sharp-tailed Sparrow, *Ammodramus caudacutus*) (Bayard and Elphick 2011), will also be severely affected by sea level rise.

Studies suggest that habitats and species assemblages will shift considerably over the coming decades (Forbes and Dunton 2006; Greenberg et al. 2006; Day et al. 2008). Some of these changes are due to human population increases and management, and others to sea level rise or subsidence. Management of water levels in marshes can shift the salinity gradient and marsh vegetation, with consequences for marsh-nesting species. Sea level rise, storms, and hurricanes can also influence forested habitats, which in turn affects avian use by both migrants and breeding birds (Gabrey and Afton 2000; Barrow et al. 2005, 2007).

Perhaps the most important features of the Gulf of Mexico for avian populations are related to the complex interaction between natural and anthropogenic factors that result in changes in land available (losses or gains), changes in the relative amount of different habitat types (sandy beaches, marshes, mudflats), and changes in salinity. The northern Gulf coast, especially Louisiana, is losing land at a rapid rate due to complex interactions among subsidence, sea level rise, tropical and other storms, inadequate water supply, and human disturbance (Visser et al. 2005; Valiela et al. 2009). The habitats along the Gulf coast are a shifting mosaic of changing elevation and salinity gradients that result in changes in vegetation species and patterns that affect nesting. Examples of changes are given throughout this chapter, but a few examples are mentioned in Table 12.3. Some habitat shifts result in changes in populations, while others result in changes in the species of birds that are able to use that habitat.

12.3.1 Birds of the Gulf of Mexico as a Whole

There are 395 bird species that reside, migrate, or winter in the Gulf of Mexico and associated coastlines (Gallardo et al. 2009). This number may increase with time because of new information and potential range changes due to global warming. Some neotropical species may move northward into the Gulf coastal habitats (lagoons, marshes, mangroves). Semiaquatic birds (land birds feeding on aquatic species), and all land birds have been reported on islands of the Gulf or crossing its waters (Gallardo et al. 2009). Gallardo et al. (2009) drew the following conclusions: (1) approximately a third of the species occurring in the Gulf of Mexico are breeding residents with no apparent population movements; (2) about 65 % depend upon the Gulf shores for a migratory stopover, or overwintering; (3) 44 % are aquatic species and 27 % are marine; and (4) most feed on invertebrates (55 %) or vertebrates (28 %), while the others eat plants.

The recent avian update included a listing of all species by taxonomy, habitat, range, and location (Gallardo et al. 2009). These data were used to paint a picture of general avian distribution in the Gulf of Mexico, and to create a map that shows the total number of species in each of 12 sectors (Figure 12.3). The percent for each sector is the percent of the total species that is present in that sector (e.g., *N* in the sector/395 for the Gulf species list). This figure makes

Table 12.3. Examples of How Hydrological, Sea Level Changes, or Other Environmental Factors Affect Distribution and Behavior of Birds in the Gulf of Mexico

Feature	Effect on Birds
Low-lying island formations, storms, and hurricanes	Erosion of nesting islands or beach habitats in winter, or wash over of eggs and chicks of Brown Pelicans, Black Skimmers, Least Terns (*Sterna antillarum*) and other terns in colonies in Louisiana and elsewhere (Visser and Peterson 1994). Storms and hurricanes influence habitat use by migrants, as well as habitat availability for migrants and nesting birds (Barrow et al. 2005, 2007; Dobbs et al. 2009)
Changes in water flow pattern and water levels	Changes in the number and amount of shallow pools that flood periodically, and then dry down, thus concentrating prey. Reddish Egrets (*Egretta rufescens*), Roseate Spoonbills (*Platalea ajaja*), and other wading birds require a concentrated food supply of fish and invertebrates (Powell et al. 1989; Lantz et al. 2011). Low water levels limit food resources and delay breeding of Mottled Duck (*Anas fulvigula*) (Grand 1992)
Changes in salinity and influxes of freshwater	Changes in salinity result in halophytic vegetation that alters bird species composition in marshes. Clapper Rails (*Rallus longirostris*) and Seaside Sparrows (*Ammodramus maritimus*) are likely to increase, while Least Bitterns (*Ixobrychus minutus*) and Common Yellowthroats (*Geothlypis trichas*) will decrease (Rush et al. 2009a)
Sea level changes with violent storms	Changes in height of nesting beaches and islands above mean high tide result in greater washovers of beaches, with mortality of eggs and young (Visser and Peterson 1994)
Sea level changes with changes in hurricane timing, frequency, and intensity	Alteration of coastal hydrology, geomorphology, and availability of suitable nesting habitat above storm tides, causing shifts in colony locations, and declines in number of ground-nesting species (Michener et al. 1997). May also shift species composition because of habitat changes

it clear that the highest species diversity is in the southern Gulf, along the Yucatán Peninsula (although not in the sector with Cuba).

A number of non-Passerine species ($N = 93$) occurred in all 12 sectors of the Gulf of Mexico (Table 12.4). Only the non-Passerines are listed because they are more typical of the species that inhabit the coastal and marine areas. The non-Passerines that are distributed throughout the Gulf include ducks, grebes, loons, boobies, pelicans, herons, egrets, ibises, spoonbills, storks, rails, shorebirds, gulls, terns, skimmers, and a kingfisher. As might be expected, shorebirds ($N = 31$ species), ducks ($N = 10$ species), herons and egrets ($N = 10$), and gulls and terns ($N = 13$) are the most diverse groups. Scientific names in Table 12.4 are not repeated in the text that follows this section.

While the non-Passerines are normally considered the key avian component of the Gulf, Passerines are important because millions migrate around or over the Gulf each spring and fall, and others reside in the coastal environment (e.g., Seaside Sparrows, Moore 2000b). Although Gallardo et al. (2009) lists Passerine species found throughout the Gulf, their list is necessarily incomplete because the marsh, shrub, and forest habitats are continuous landward, and it is difficult to draw a suitable line for which species to include. Moreover, the distribution of Nearctic-Neotropical migrants along the southern Gulf of Mexico may be less well known than the distribution along the northern Gulf coast. Some raptors that prey on migrants may be

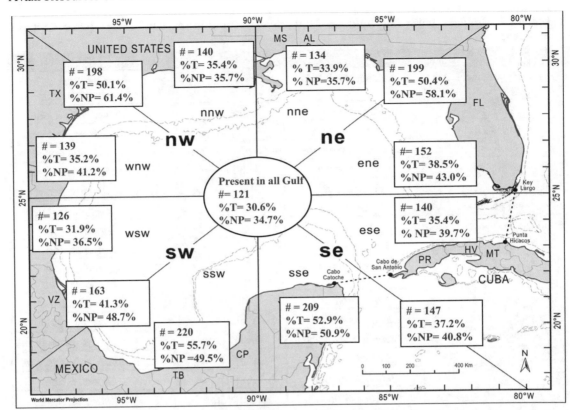

Figure 12.3. Relative avian diversity in the Gulf of Mexico. Shown are the number of species that have been recorded for that sector, the percent of total species found in the Gulf that occur in that sector (%T), and the percent of non-Passerines that are found in that sector (%NP). Data are from Gallardo et al. (2009); map made by Fabio Moretzsohn. © J. Burger.

underrepresented in species lists because they are routinely counted only at designated hawk watches (Kerlinger 1985; Woltmann 2001; Woltmann and Cimpreich 2003).

12.3.2 The Southern Gulf of Mexico Avian Community

The southern Gulf of Mexico (to the northern shores of the Yucatán) differs from the northern coast because of differences in temperature and physiognomy, which supports tropical vegetation and avifauna. From a Mexican perspective, the Gulf of Mexico is extremely important because approximately 60 % of Mexico's watersheds drain into the waters of the Gulf (Gallardo et al. 2004). Estuaries, lagoons, and other wetlands represent 30 % of the Mexican Gulf coastline; the Lagoon system at Alvarado, Veracruz has 26 % of the bird species present in all of Mexico (Gallardo et al. 2004). The extensive mangroves along the southern Gulf coast provide important habitats for foraging and nesting birds.

Lagoons and wetlands fringe the southern Gulf in Mexico, as they do in the United States, and one area, the Laguna Madre in Tamaulipas, contains 15 % of Mexico's migratory aquatic birds. About 82 % of the birds present in Laguna Madre originate in the Nearctic as it represents the southern limit of the range for several species, such as the Bald Eagle, *Haliaeetus leucocephalus*. In contrast, the region from southern Tamaulipas to Campeche

Table 12.4. Species with Distributions That Include the Entire[a] Gulf Coast (after Gallardo et al. 2009)

Common Name	Species Name	Common Name	Species Name
Fulvous Whistling-Duck	*Dendrocygna bicolor*	Semipalmated Plover	*Charadrius semipalmatus*
Wood Duck	*Aix sponsa*	Piping Plover	*Charadrius melodus*
American Wigeon	*Anas americana*	Killdeer	*Charadrius vociferus*
Mallard	*Anas platyrhynchos*	American Oystercatcher	*Haematopus palliatus*
Blue-winged Teal	*Anas discors*	Black-necked Stilt	*Himantopus mexicanus*
Pintail	*Anas acuta*	America Avocet	*Recurvirostra americana*
Green-winged Teal	*Anas crecca*	Greater Yellowlegs	*Tringa melanoleuca*
Ring-necked Duck	*Aythya collaris*	Lesser Yellowlegs	*Tringa flavipes*
Lesser Scaup	*Aythya affinis*	Solitary Sandpiper	*Tringa solitaria*
Masked Duck	[b]*Nomonyx dominicus*	Willet	*Catoptrophorus semipalmatus*
Common Loon	*Gavia immer*	Spotted Sandpiper	*Actitis macularius*
Pied-billed Grebe	*Podilymbus podiceps*	Upland Sandpiper	[b]*Bartramia longicauda*
Wilson's Petrel	[b]*Oceanites oceanicus*	Whimbrel	*Numenius phaeopus*
Masked Booby	*Sula dactylatra*	Long-billed Curlew	*Numenius americanus*
Brown Booby	*Sula leucogaster*	Marbled Godwit	*Limosa fedoa*
American White Pelican	*Pelecanus erythrorhynchos*	Ruddy turnstone	*Arenaria interpres*
Brown Pelican	*Pelecanus occidentalis*	Red Knot	*Calidris canutus*
Double-crested Cormorant	*Phalacrocorax auritus*	Sanderling	*Calidris alba*
American Anhinga	*Anhinga anhinga*	Semipalmated Sandpiper	*Calidris pusilla*
Magnificent Frigatebird	*Fregata magnificens*	Western Sandpiper	*Calidris mauri*
Great Blue Heron	*Ardea herodias*	White-rumped Sandpiper	*Calidris fuscicolis*
Great Egret	*Ardea alba*	Least Sandpiper	*Calidris minutilla*
Snowy Egret	*Egretta thula*	Pectoral Sandpiper	*Calidris melanotos*
Little Blue Heron	*Egretta caerulea*	Dunlin	*Calidris alpina*
Tricolored Heron	*Egretta tricolor*	Stilt Sandpiper	*Calidris himantopus*
Reddish Egret	*Egretta rufescens*	Buff-breasted Sandpiper	[b]*Tryngites subruficollis*
Cattle Egret	*Bubulcus ibis*	Short-billed Dowitcher	*Limnodromus griseus*
Green Heron	*Butorides virescens*	Wilson's Snipe	*Gallinago delicata*
Black-crowned Night Heron	*Nycticorax nycticorax*	Wilson's Phalarope	*Phalaropus tricolor*
Yellow-crowned Night Heron	*Nyctanassa violacea*	Pomarine Jaeger	*Stercorarius pomarinus*
White Ibis	*Eudocimus albus*	Parasitic Jaeger	*Stercorarius parasiticus*
Glossy Ibis	[b]*Plegadis falcinellus*	Laughing Gull	*Larus atricilla*
Roseate Spoonbill	*Platalea ajaja*	Franklin's Gull	[b]*Larus pipixcan*

(continued)

Table 12.4. (continued)

Common Name	Species Name	Common Name	Species Name
Wood Stork	*Mycteria americana*	Ring-billed Gull	*Larus delawarensis*
Osprey	*Pandion haliaetus*	Herring Gull	*Larus argentatus*
Black Rail	*Laterallus jamaicensis*	Lesser Black-backed Gull	*Larus fuscus*
Clapper Rail	*Rallus longirostris*	Gull-billed Tern	*Gelochelidon nilotica*
King Rail	*Rallus elegans*	Caspian Tern	*Hydroprogne caspia*
Sora	*Porzana carolina*	Royal Tern	*Thalasseus maxima*
Purple Gallinule	*Porphyrio martinica*	Sandwich Tern	*Thalasseus sandvicensis*
Common Gallinule	*Gallinula chloropus*	Common Tern	*Sterna hirundo*
American Coot	*Fulica americana*	Forster's Tern	*Sterna forsteri*
Sandhill Crane	*Grus canadensis*	Least Tern	*Sterna antillarum*
American Golden Plover	*Pluvialis squatarola*	Black Tern	*Chlidonias niger*
Black-bellied Plover	*Pluvialis dominica*	Black Skimmer	*Rynchops niger*
Snowy Plover	*Charadrius nivosus*	Belted Kingfisher	*Ceryle alcyon*
Thick-billed Plover	*Charadrius wilsonia*		

The scientific names are those used by Gallardo et al. (2009), not necessarily the most current
[a]The author does not agree with the designation of "entire" for these rare and/or local species
[b]Species may be very rare in Gulf of Mexico

contains mainly aquatic species with neotropical affinities (Correa et al. 2000a, c; Gallardo et al. 2004).

The continental platform off the coasts of Campeche and Yucatán contains reefs and keys (cays or small islands) used by nesting seabirds, including Red-footed Booby (*Sula sula*) and Least Tern, which are both on the Mexican endangered species list (Gallardo et al. 2004). While this region contains neotropical affinities, it is also influenced by the Caribbean (Gallardo et al. 2004). Thus, the Mexican coast has high species diversity because it contains both nearctic resident species (at the end of their southern range) and neotropical species (at the end of their northern range). This parallel pattern has not been given the credit it deserves (Jahn et al. 2004). Both migrants from the north (that pass through the Gulf of Mexico on their way south) and austral migrants from the south (that may migrate as far north as the Gulf in winter) share a common neotropical avifauna (Jahn et al. 2004).

Many Nearctic-Neotropical migrants pass through on their way farther south. Coastal Veracruz is a major migratory pathway for raptors (Ruelas et al. 2000), and the corridor from Texas, through Mexico to the Yucatán, is a major Nearctic-Neotropical migrant route (Rappole 1995). There is also a healthy population of breeding Mottled Ducks along the coast (Perez-Arteaga and Gaston 2004).

As is clear from Figure 12.3, there are more species on the southern Gulf of Mexico coast to Campeche Bank and the Yucatán, than on the northern U.S. Gulf coast. The Campeche Bank is an extensive, submarine continuation of the plateau that forms the Yucatán Peninsula, extending for about 650 km (404 mi) along the western and northern coasts of the Yucatán in the southeastern Gulf of Mexico. The islands used for nesting are located more than 120 km (75 mi) from the mainland and are rarely disturbed by fishermen or recreationists (Tunnell and Chapman 2000). Several species with more tropical ranges nest there, such as Masked Booby, Brown Booby, Red-footed Booby, Magnificent Frigatebird, and Brown Noddy, as well as several other species (Laughing Gull and terns, Tunnell and Chapman 2000). Tunnell and Chapman (2000) suggested that these colonies have remained fairly stable, but they require monitoring and protection. The Campeche Banks is also a stopover site for migrants, and more than a half century ago scientists were concentrating on the number of North American migrants using Veracruz (Loetscher 1955). A fuller description of the ornithology of the Yucatán can be found in Paynter (1955).

12.4 AVIAN USES OF MARINE-LAND INTERFACES

12.4.1 Functional Avian Uses

Birds use marine and coastal habitats in a variety of ways, resulting in overlapping activities, both within and among seasons. Definitions used in this chapter are shown in Table 12.5. A given species can have multiple listings. For example, Laughing Gulls breed on islands along the Gulf coast, and some may remain all year (i.e., residents). However, Laughing Gulls also breed along the Atlantic coast up to New York (Burger 1996a), and in the fall, some migrate through the Gulf of Mexico to Mexico (migrants), while others migrate to the Gulf and remain there as winter residents. They are residents, migrants, and winter visitors. In some cases, status is less clear. Red Knots breed in the Arctic and migrate through the Gulf of Mexico on their way to the Caribbean or South America (Niles et al. 2008): they were spring and fall migrants in Texas (Eubanks et al. 2006). However, recent information indicates that some knots remain the entire winter in Texas and in Florida (Burger et al. 2012a).

12.4.2 Temporal and Spatial Constraints

Birds are constrained by seasonality; most breed in the spring when food supplies are optimal (Weimerskirch 2001) and remain as residents, or migrate when conditions (food, temperature) deteriorate. Seasonal patterns have evolved over time, and there are variations even within a species. More northern members of a species that breed north of the Gulf of Mexico may be migrants that move south through the Gulf, while conspecifics that are resident in the Gulf may remain as year-round residents.

Spatial constraints often have to do with habitat suitability, whether for foraging, courting, breeding, migrating, or overwintering. With few exceptions (such as grebes and others that build floating nests), birds need dry land to breed because they lay eggs and are constrained to their nests during incubation, and often during the chick-rearing phase. Habitat suitability depends on the type and qualities required for each activity, and the stability of the habitats involved.

The most important habitat gradient in the Gulf of Mexico for birds is from open water to upland terrestrial habitats. Because birds are highly mobile, many species can be found anywhere along the gradient. "Normal" distributions change during the year, and can be altered during hurricanes or other inclement weather events. Nevertheless, species show preferences for particular habitats that meet their needs for foraging, roosting, nesting, migrating, and

Table 12.5. Definitions of Terms Used in this Chapter

	Definition of Terms
Breeding	Includes courtship, nest site selection, mate selection, egg laying, incubation, and chick rearing
Migrant	A bird that regularly moves from one region to another and back
Resident	A species that is present throughout the year and thus breeds (when it reaches adult status) and winters in the GoM
Visitor	A bird that may be present in spring, summer, fall, or winter

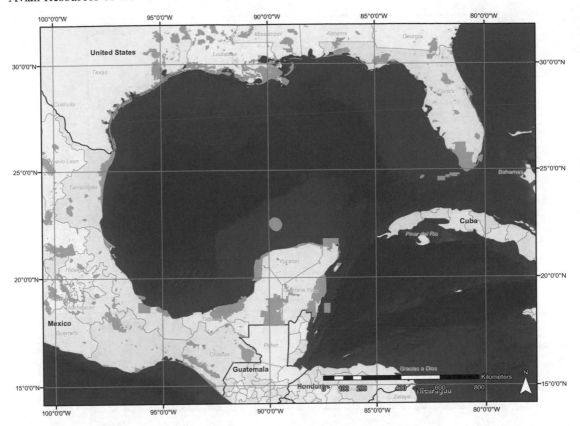

Figure 12.4. Protected coastal areas of the Gulf of Mexico, shown in green. Map courtesy of Wells (2013).

overwintering. Species composition varies along the gradient, and certain species are most likely found in specific habitats. There are also gradients in prey abundance and availability along transects from open water to shallow water, from the water surface to depths, and from the surface into the soil/sediment, depending upon moisture content and salinity. Both spatial and seasonal changes in infauna density determine prey availability for foraging birds. The available habitats, however, are also a function of how much land is protected (Figure 12.4).

12.4.2.1 Habitat Availability

The habitat types available on barrier islands and mainlands include sandy beaches, salt marshes, brackish marshes, freshwater marshes, shrub/scrub, and forests. The National Land Cover Database (2006) has several categories of interest for birds. Maps showing the habitats in each state are presented in Appendix A. In this chapter, they were combined into 11 categories. Most are self-explanatory, but barren land includes rock, sand, and clay, some of which are used by many beach-nesting birds. The three forest types (deciduous, evergreen, mixed forest) were combined (Appendix A). The relative amount of habitat available in each state is shown in Figure 12.5 (10 mile area from the coastline). Texas has a high percentage of woody wetlands, forests, and developed land. Louisiana has the greatest percentage of its coastal area as water and wetlands. Mississippi has mainly open water and wetlands, while Alabama (with the smallest coastal band) has primarily forest and woody wetlands. Florida, with the

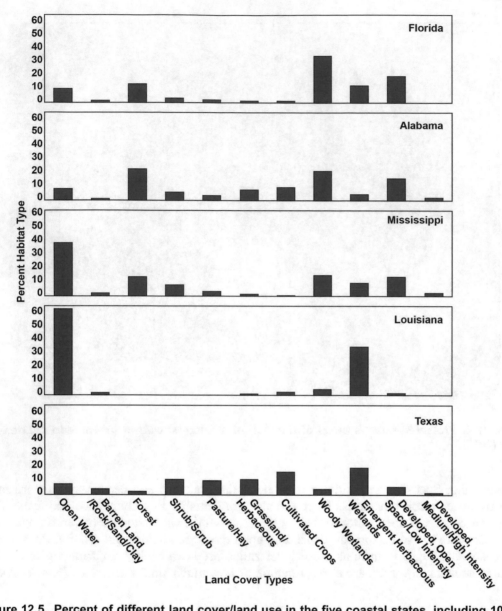

Figure 12.5. Percent of different land cover/land use in the five coastal states, including 10 mile from the Coastline (National Land Cover Database, 2006; computed from data provided by Wells 2013). © by J. Burger.

greatest coastal area, has mainly woody wetlands, developed land, and forests along its coast (Figure 12.5).

Birds have generalized niche requirements that relate to habitat availability. The open waters of the Gulf of Mexico are pelagic, and species living there are normally seabirds and some diving ducks. While winds, currents, and temperatures control the pelagic environment, the landward environments are ruled by tides. Tidal marshes are found in small, narrow pockets

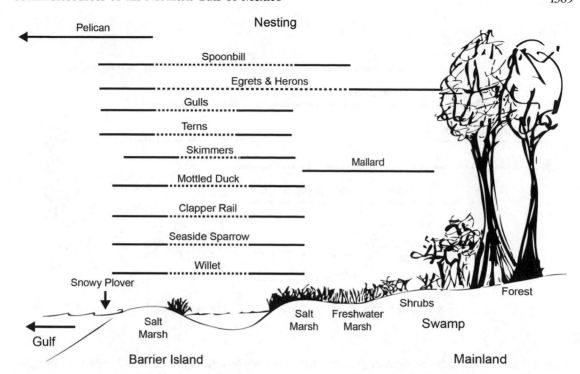

Figure 12.6. Schematic of nesting patterns of birds in the northern Gulf of Mexico. *Solid line* equals where they normally nest, and *dotted lines* connect these habitats. © J. Burger.

along coastlines, with the main vegetation being *Spartina* and *Juncus* spp. (Greenberg et al. 2006). The combination of salinity, low floristic and structural complexity, regular tidal fluctuations, catastrophic flooding, and high winds in tidal marshes creates a vulnerable, unpredictable environment, requiring flexibility and adaptability on the part of the birds living there (Greenberg et al. 2006). While tidal marshes support relatively few unique or endemic species of terrestrial vertebrates, some subspecies have differentiated (Greenberg et al. 2006), such as the Louisiana Seaside Sparrow (*Ammodramus maritimus fisheri*) (Gabrey and Afton 2000). Although birds exhibit flexibility in their choice of nesting sites, they prefer particular types of habitats (Wilson and Vermillion 2006). Gulls, terns, skimmers, and shore-birds nest on the ground, usually on bare sand or in places with sparse vegetation, or they build nests in marshes. Pelicans nest on bare ground or in vegetation that is sparse, but tall enough to allow them to maneuver their large bodies underneath it. Herons, egrets, and ibises prefer to nest on low vegetation, particularly in the Gulf, but will sometimes build nests on the ground or in shrubs and trees. Ducks, Willet, and Clapper Rail build nests low in the vegetation or on the ground, usually in marshes. Snowy Plovers and Oystercatchers build nests on open, unvege-tated sand, relying on being cryptic to camouflage their eggs. Sparrows and some other songbirds nest in marshes, scrubs, or forests (Moore et al. 1990; Buler and Moore 2011).

A schematic of nesting preferences is shown in Figure 12.6. Wintering birds also have preferred habitats. Figure 12.7 indicates the likely zonation of birds in the winter, which mainly reflects foraging and roosting sites. Habitat use is generally wider during this period as they are not restricted to nest sites.

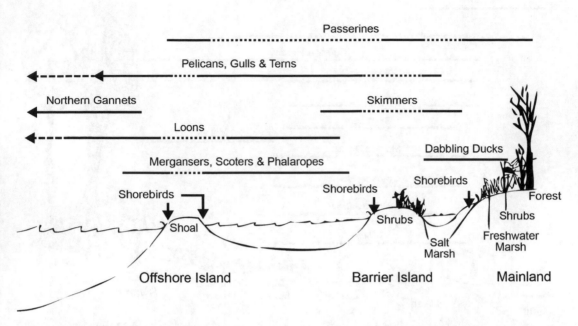

Figure 12.7. Schematic of spatial gradient for birds wintering in the Gulf of Mexico, from open water (pelagic zone) to upland habitats. *Solid line* **indicates normal habitat use,** *dotted line* **indicates area not usually used, and** *dashed line* **means frequency is less.** © **J. Burger.**

12.4.2.2 Habitat Suitability

Habitat suitability refers to whether a given habitat is usable (or suitable), considering physical, vegetative, and social features, within a context of anthropogenic factors. It is essential to distinguish both interspecific differences and those due to activities (breeding vs. migrating or overwintering; nesting vs. foraging). In the nesting season, birds are tied to their nest site during the incubation period, and non-precocial species are limited to the nest site during much of the chick-rearing phase. The chicks of precocial species (ducks and rails) are able to locomote and search for food shortly after hatching. Chicks that are not precocial (altricial) must be brooded early on because they have no feathers and cannot regulate their body temperature. They are guarded and fed until they are able to forage on their own. This imposes constraints on birds to select nest sites that are removed from the threat of tides, floods, inclement weather, and predators.

A data set for Louisiana-Alabama provides an overview of habitat use by colonial-nesting species (Portnoy 1981). Habitat preferences for common birds normally considered coastal are shown in Figure 12.8 (none with populations below 500). Most of the *Plegadis* species were White-faced Ibis (*Plegadis chihi*). This data set, because it encompassed colonies in three states, can be used to infer habitat preferences (layered upon habitat availability). The patterns reflect choices before the rapid coastal and offshore development of the last 35 years. The Brown Pelican is the only species for which the data are not typical. Because of its sharp decline in the 1950s and 1960s due to pesticides, it had not yet recovered (Wilkinson et al. 1994; Shields 2002). A similar survey in 2001 indicated that 40 % of the active pelican colonies were in saline marshes, 24 % were in freshwater marshes, 22 % were in forested wetlands, and the remainder in scrub, shrub, upland forest, or brackish marshes (Michot et al. 2003).

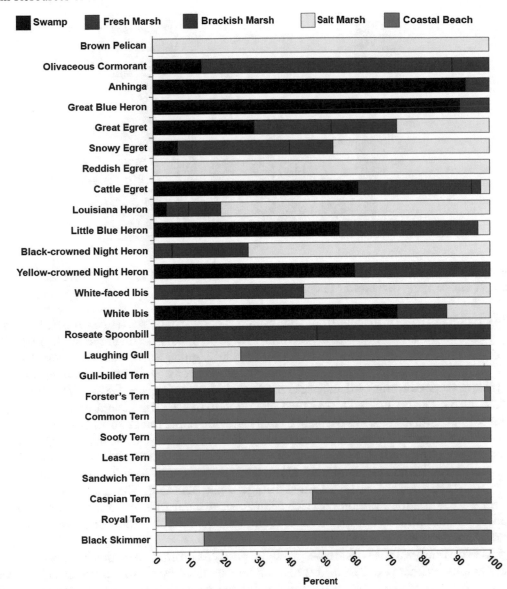

Figure 12.8. Relative habitat use by colonial nesting birds in the Louisiana, Alabama, and Mississippi Coasts of the Gulf of Mexico (after Portnoy 1981). © J. Burger.

Figure 12.8 provides a picture of horizontal nesting stratification from the Gulf landward. Most terns and Laughing Gulls nested on bare sand, and most skimmers nested on sand; although, a few nested in salt marshes. Skimmers and Laughing Gulls sometimes are forced to nest in salt marshes because of competition with other species, lack of available beaches, or human disturbance (Burger and Gochfeld 1990). Forster's Terns always nest in marshes (McNicholl et al. 2001).

Habitat use for nonbreeding birds is a function not only of habitat structure and vegetation types but also of prey types and foraging methods. Seabirds capture prey by a variety of methods, including plunge-diving for fish or invertebrates, surface-plunging, hop-plunging, hover-dipping, and picking food items off the surface of water, although gulls and some other

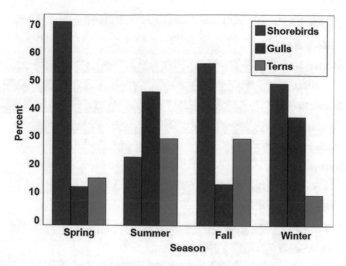

Figure 12.9. Seasonal use of beaches by shorebirds, gulls, and terns in Padre and Mustang Islands, Texas in 1979–1981 (after Chapman 1984). © J. Burger.

seabirds pick up fruit or insects from the ground, follow boats, scavenge on offal along the shore, pirate food from other seabirds, and forage at landfills (Ashmole 1971; Sealy 1973; Burger and Gochfeld 1981; Shealer 2001). In the Gulf, gulls, terns, and skimmers forage in pelagic waters, shallow tidal creeks, and behind boats or near other human activities, as well as at landfills (garbage dumps), inland lakes, and impoundments (Burger 1987a, 1988a; Burger and Gochfeld 1983a; Patton 1988). Ducks breed mainly in marshes or in distant uplands, but spend the winter in coastal areas or in nearshore environments. Some ducks form large flocks on the water and forage on the open sea (diving ducks), while others feed at the marine-land interface in bays, estuaries, marshes, fields, and other terrestrial habitats (dabbling ducks). Herons, egrets, and ibises breed on islands and along coastal areas, and feed in intercoastal habitats; they do not feed in open water as most forage while standing. Shorebirds feed along the shoreline on the mainland, along barrier islands, or around offshore islands. Their feeding method of picking up items from the sand, from shallow water, or along wrack lines, ties them to the narrow band along the shoreline.

Species diversity varies within close habitats, partly as a function of time of day, tide stage, and tide height (Withers 2002). Habitat use can be examined by season, particularly for beach habitats where birds forage and roost throughout the year, as well as during migratory stop-overs. Chapman (1984) examined seasonal use of beaches on Padre and Mustang Island barrier beaches (Figure 12.9).

This figure shows the relationship among species groups by season. Shorebirds made up the largest component in the spring, fall and winter, while gulls made up the largest component in the summer.

12.4.2.3   Mobility and Habitat Suitability

The flight abilities and inclinations to migrate or disperse are variable in birds. Seabirds are the most mobile, and are likely to fly the greatest distances from their nest sites to forage, and some circumnavigate the globe in the nonbreeding season. Many seabirds nest on offshore islands far removed from predators, such as Campeche Bank off the Yucatán (Tunnell and Chapman 2000), or on the Dry Tortugas (Dinsmore 1972), and show very high nest and colony site fidelity. Seabirds that nest on less stable coastal islands shift colony sites as conditions dictate, but have high site fidelity if colony sites remain unchanged (Buckley and Buckley 1980; Coulson 2001).

Pelicans, herons, egrets, and ibises that nest in coastal colonies use the same sites as long as they remain safe from predators and are suitable. For many species, nest site requirements drive their choice of colony site, and they will continue to nest there if the sites remain stable. In some cases, long-term stability is enhanced by habitat modification, as happened on Queen Bess Island for pelicans (Visser et al. 2005). In other cases, stability is reduced by erosion and loss of space.

For some species, choice of colony site is dependent upon foraging opportunities. Roseate Spoonbills depend upon periodic drawdown and flooding to produce pools with high prey availability (Kushlan 1979). While other herons and egrets also depend on such resources, the dependence is not as strong. White Ibis are more nomadic, both in foraging behavior and in nesting behavior (Frederick et al. 2009). They also require dry down and the concentration of suitable prey (Frederick et al. 1996). The combination of nesting and foraging habitat requirements leads to shifting colony locations for these species, and they may move hundreds of kilometers between different years. Other species are quite sedentary and are not likely to fly long distances. This has the effect of isolating populations, which can lead to subspecies. For example, Seaside Sparrows living along the Gulf are resident and do not fly long distances. Separate populations can become isolated, and if they disappear recolonization is unlikely unless there is a population nearby to provide founders (individuals to colonize).

12.5 FACTORS AFFECTING AVIAN POPULATIONS

Several factors affect populations, and provide a basis for understanding the status and trends of birds in the Gulf of Mexico. These include natural environmental factors and anthropogenic events, biological events, and interactions among them. Natural environmental events include storms, hurricanes, tidal regimes and extreme tides, extreme cold, heat or drought, and other normal or extraordinary events, such as global warming. Anthropogenic factors include contamination by oil, heavy metals, DDT, polychlorinated biphenyls (PCBs), and other pollutants (e.g., endocrine disruptors), as well as human disturbance (Coste and Skoruppa 1989). Biological stressors include social interactions (competition, cooperation, social facilitation), predation, infestations (ticks, mites), disease, and invasive species. Global change (warming, sea level rise, subsidence) is a physical change that has anthropogenic causes (Solomon et al. 2007; Edenhofer et al. 2011). Finally, intrinsic factors can affect survival and other aspects of population dynamics, including age, sex, and molt stage. For example, Common Loons are particularly vulnerable during molt while overwintering in the Gulf of Mexico (NW Florida, Alexander 1991). Coastal birds of the Gulf affected by storm events include large colonial nesting species such as Brown Pelican, beach-nesting terns and gulls (Caspian Tern, Royal Tern, Sandwich Tern, Least Tern, Laughing Gull, Black Skimmer), beach-nesting shorebirds (American Oystercatcher, Willet, Wilson's Plover, Snowy Plover), large wading birds (Reddish Egret, Roseate Spoonbill, ibises, herons, egrets), marsh birds (Mottled Duck, Clapper Rail, Black Rail, Willet, Seaside Sparrow), migratory shorebirds (Red Knot, plovers, sandpipers), and migratory songbirds on small barrier islands or coastal shrubs (warblers, orioles, buntings, flycatchers). Offshore seabirds can be affected if nesting islands are impacted (e.g., Magnificent Frigatebird) or if foraging space is reduced or rendered unusable (Northern Gannet).

The following sections are not meant to be exhaustive, but rather to illustrate the range of factors affecting birds using the Gulf of Mexico that must be considered for conservation, management, monitoring, or other purposes. More in-depth discussions can be found in chapters in Burger et al. (1980) and Schreiber and Burger (2001a) for seabirds, Kushlan and Hafner (2000a) for herons, and Moore (2000b) for Passerine migrants.

12.5.1 Habitat Loss

The availability of habitat is a prime characteristic determining nesting and foraging distribution and abundance of birds. Vegetation dispersion and land elevation determine where most birds can nest around the Gulf, while water depth and emergent vegetation influence where water birds, such as shorebirds, herons, and egrets, can forage (Lantz et al. 2010, 2011). Coastal wetlands are increasingly threatened because of development, increased use of beaches, and the continual movement of people to coasts (NOAA 2004). This has led to population declines for birds living there (Delany and Scott 2006). Many factors discussed later in this section affect habitat availability and habitat suitability. All the other threats discussed in the following sections act in concert with habitat loss, amplifying the effects of each. Overall, the U.S. coastline along the Gulf of Mexico has lost 1.2 % of intertidal wetlands (44,810 acres) in only 6 years (1998–2004, Stedman and Dahl 2008).

Louisiana provides the premier example of wetland loss. Louisiana's coasts encompass more than 9.3 million acres of barrier shorelines, swamps, and marshes (Lindstedt 2005). It contains 30 % of the remaining coastal wetlands in the continental United States, yet these wetlands are disappearing rapidly (Field et al. 1991; O'Connell and Nyman 2011). Louisiana coastal wetlands once hosted 77 % of the U.S. breeding population of Sandwich Tern, 52 % of Forster's Tern, 44 % for Black Skimmer, 16 % for Royal Tern, and 11 % for the Laughing Gull (Visser and Peterson 1994). Thus, loss of wetlands that decrease nesting habitat for species will have a significant effect on their overall populations in the United States.

The Coastal Prairie Ecosystem of east Texas and Louisiana has especially suffered losses. Many obligate grassland species breed there or stop over during migration. Losses due to degradation from fire suppression, agricultural practices, and invasive species have resulted in this habitat being globally imperiled (Barrow et al. 2005, 2007). Narrow, elongated patches embedded within these grassy marshes (oak forest patches called cheniere) provide critical stopover areas for migrant songbirds going in both directions over the Gulf of Mexico (Barrow et al. 2007). Anthropogenic and natural disturbances (hurricanes, invasive plants, industrial and residential development, and conversion to cropland) have shrunk cheniere habitat to less than 1 % of the historic presettlement area.

12.5.2 Invasive Species

Invasive species are a great concern because plant invasive species affect habitat quantity and quality, which affects avian distribution. For example, *Phragmites*, spreading into areas once dominated by salt marsh species such as *Spartina* (Greenberg et al. 2006), favors generalists over avian salt marsh specialists (Benoit and Askins 1999). In the Gulf, shifts between *Juncus* and *Spartina* stands can greatly influence the marsh-nesting birds that persist and breed successfully (Rush et al. 2009b). Increases in the nonnative Eurasian Watermilfoil (*Myriophyllum spicatus*) coincided with a 96 % decline in waterfowl populations in the Mobile-Tensaw Delta, Alabama (Goecker et al. 2006). It has largely replaced the native submerged aquatic vegetation (SAV), Wild Celery (*Vallisneria americana*), as the dominant species. Wild Celery was the preferred food of waterfowl in the region (Goecker et al. 2006). However, comparison of six surveys with historic data for waterfowl did not indicate a strong association of the invasive SAV with waterfowl declines. Another important invasive species is the Chinese Tallow tree (*Sapium sebiferum*), particularly in East Texas, Louisiana, and Mississippi (Oswalt 2010), where it forms monospecific stands (Bruce et al. 1995). Tallow seeds are spread by birds such as Red-bellied Woodpeckers (*Melanerpes carolinus*), robins (*Turdus migratorius*), and bluebirds (*Sialia sialis*) in Louisiana and elsewhere along the Gulf (Renne et al. 2002).

The Cattle Egret is one of the most invasive species in the Gulf and along the Atlantic Coast. Native to Africa, the first Cattle Egrets bred in North America in the mid-1940s. Since then, they have expanded dramatically, displacing many native egrets and herons from their traditional breeding colonies. While their spread has caused local declines in native species in traditional colony sites, it is unclear whether Cattle Egrets have generally impacted the populations of native species in the Gulf.

12.5.3 Food Resources

Food resources affect every aspect of avian life, including survival, reproduction, migration, habitat use, and even their response to inclement weather and predators. While availability of food resources is often tied to habitat availability, food will not be available if suitable habitat for the prey is not available, and food resources can be limited even when foraging habitat is not. That is, when vegetation fails to provide adequate food resources, prey can be depleted, or both vegetation types and prey types cannot be optimal or can be difficult to access or capture. For example, fish may be present for birds, but if they are unavailable because they are too deep in the water column, difficult to see or capture, or are in low densities, they may not provide an adequate food base.

Wading birds forage at different water depths, related to leg length (Powell 1987). As expected, long-legged waders forage in a greater diversity of water depths than can shorter-legged birds. The smallest species, such as the Little Blue Heron, Snowy Egret, and White Ibis, have a maximum foraging depth of 16–18 centimeters (cm), medium-sized species (Reddish Egret, Great Egret, Roseate Spoonbill) have a maximum foraging depth of 20–28 cm, and the large Great Blue Heron has a foraging depth of 39 cm (Powell 1987). Species foraging in the Gulf of Mexico exhibit both horizontal and vertical spatial patterns.

Part of foraging habitat stratification is a result of the distance birds will fly to forage away from their nest sites. Gulls and terns, for example, will fly farther than herons or egrets, and both will fly farther than Clapper Rails or Seaside Sparrows. Food resources and foraging methods differ among species as a function of species size and foraging methods, as well as age within species (Brown 1980; Burger and Gochfeld 1983b; Burger 1987a; Shealer 2001).

Songbirds depend upon microhabitats that harbor the invertebrates and fruits they consume, both during the breeding season and during migration (Barrow et al. 2007). These habitats can be destroyed not only by direct habitat destruction, but also by natural and anthropogenic forces, such as fire and hurricanes (Barrow et al. 2007).

12.5.4 Tides, Hurricanes, and Other Weather Events

Weather and unusual weather events are one of the driving forces that affect reproductive success, foraging behavior, migrating, over-wintering, and timing of life-cycle events, as well as seasonal and long-term behavior, physiology, and population trends (reviewed in Schreiber 2001). The Gulf of Mexico has relatively shallow tidal swings (generally less than 1 meter [m]; Conner et al. 1989), which makes very high tides less predictable. In most cases, birds select the highest places to nest. This is especially true for marsh nesting birds, such as solitary-nesting species (e.g., Willets; Burger and Shisler 1978; Lowther et al. 2001) and colonial species (e.g., Laughing Gulls; Burger and Shisler 1980; Burger 1996a). Very high tides, usually associated with hurricanes, other storms, or winds, reduce reproductive success by flooding out nests, eggs, and chicks in ground-nesting species. Tidal effects decrease hatching and fledging rates, and synchronize breeding behavior with lunar cycles (Shriver et al. 2007).

Hurricanes are episodic, high-energy events that accelerate routine processes (erosion, accretion) and activate others (formation of washover fans, Conner et al. 1989). Over the long term, hurricanes can create and destroy suitable habitat for nesting, foraging, and roosting. The immediate impacts of hurricanes include direct mortality from exposure to winds, rain, and storm surge (Butler 2000), as well as decreased nesting habitat for species nesting in low-lying areas, and decreased food availability for migrants, particularly songbirds in the Gulf (Dobbs et al. 2009). Some habitats are particularly vulnerable, such as low-lying barrier islands and cheniere forests. These forests suffer both short- and long-term effects, which in turn decrease foraging habitat for breeding and migrant songbirds (Barrow et al. 2007). Effects of hurricanes on habitat and substrate (leaves vs. bark) can be felt during, immediately after, and up to a year after the event (Dobbs et al. 2009).

While immediate impacts change vegetation, destroy low-lying habitats, and decrease animal populations, species can sometimes recover (Conner et al. 1989). Avian recovery from hurricanes can occur only if suitable areas are available for nesting or foraging. Immediate effects of hurricanes and other severe storms include being blown off course or forced to land (migrants; DeBenedictis 1986), and injury or death to nests, eggs, chicks, and even adults (Marsh and Wilkinson 1991).

Flying birds can flee an oncoming storm, but nests, eggs, and nonflying young are vulnerable to immediate wash-outs, cold stress, and drownings. There are often lasting effects on growing chicks that survive hurricanes. Although young Sooty Terns nesting on the Dry Tortugas (70 mile west of Key West in the Gulf) suffered abnormal growth, Brown Noddies were comparatively unaffected (White et al. 1976). Even adult Passerines can show effects following hurricanes, perhaps due to differences in prey availability (Waur and Wunderle 1992). Shorebirds can also decline following hurricanes due to habitat degradation (Marsh and Wilkinson 1991). Understanding relative vulnerability of different species to hurricanes and other severe storms may provide insights into relative population numbers, population declines, and shifts in habitat use, and can inform management and conservation.

Storms are often associated with mass mortality incidences of enroute migratory birds, including grebes (Jehl et al. 1999), eagles (Newton 2007), shorebirds (Roberts 1907), ducks (Schorger 1952), and various Passerines (Webster 1974; King 1976). One storm killed an estimated 40,000 migrant birds of 45 species on one day—the largest kill recorded for the Gulf at that time (Wiedenfeld and Wiedenfeld 1995). Weather, in conjunction with food supply, adversely affects body weight at migration time, which then affects resighting probability (indicative of survival differences), and subsequent breeding success (Newton 2006). Birds for which these effects have been found include shorebirds (Pfister et al. 1998; Baker et al. 2004), ducks (Pattenden and Boag 1989; Dufour et al. 1993), and Passerines (Smith and Moore 2003). Birds stressed by weather and a shortage of food, particularly small Nearctic-Neotropical Passerines, are often vulnerable to predators (Moore et al. 1990). Weather events, however, usually function on the large spatial scale of migration as well as affect food availability (Moore 2000b). Weather events have the potential to increase or decrease the effect of other stressors; strong winds and currents can increase the movement of pollutants and can also force oil or other contaminants further onto islands or into marshes or mangroves. Weather events, alone, however, have not caused long-term avian population declines in the Gulf because such adverse events are usually limited in space and time.

12.5.5 Climate Change, Sea Level Rise, and Land Subsidence

Climate change affects temperature, precipitation patterns, oceanic and atmospheric circulation patterns, sea level rise, and frequency, distribution, and intensity of storms, hurricanes and other weather events (Michener et al. 1997; Root et al. 2003). The Intergovernmental Panel on Climate Change (Edenhofer et al. 2011) predicts that global temperatures will rise 1.4–5.8°Celsius (°C) by 2100, an increase that is probably without precedent in the last 10,000 years. Changes can occur in the means and the extremes of temperatures and precipitation, in the length of seasons, the timing of spring, and the frequency of catastrophic events. Warmer temperatures would result in melting of glaciers and acceleration of sea level rise, which in turn would flood low-lying islands used for nesting. For example, assuming a conservative global warming scenario of only 2°C over the next century, Galbraith et al. (2005) predicted that major intertidal habitat losses for shorebirds in bays in Washington, California, Texas, and New Jersey/Delaware would range from 20 to 70 %. Such habitat losses may be large both spatially and temporally and could negatively affect avian populations in the Gulf and elsewhere if they continue. Climate change has already affected the timing of migration and breeding in some Nearctic-Neotropical migrants (Marra et al. 2005).

Changes in the timing, frequency, and intensity of storms and hurricanes can alter coastal hydrology, geomorphology, and nutrient structure, leading to changes in vegetative structure (Michener et al. 1997), which in turn will markedly affect bird use of coastal areas. Birds can adapt to slow changes more easily than to extreme events (van de Pol et al. 2010). Rush et al. (2009a) conducted censuses of birds nesting in coastal marshes of Alabama and Mississippi and found that Seaside Sparrows and Clapper Rails nested in habitats with higher salinity than did Least Bitterns (*Ixobrychus exilis*). Their models indicated that coastal alterations, sea level rise, and landward changes in habitat and salinity will lead to population increases in the former two species and declines in Least Bittern.

12.5.6 Predation, Competition, and Other Social Interactions

Social effects on survival, including competition, cooperation, and predation, are reviewed in Burger (1988b, c), Nettleship et al. (1994), and Coulson (2001). Predation pressures are often cited as the primary reason for colonial, ground-nesting species to select islands far removed from predators (Burger 1981a, 1982; Wittenberger and Hunt 1985; Coulson 2001). Predation pressures are lowest for species nesting on distant offshore islands that do not have mammalian predators, and highest for ground-nesting species on barrier islands or the mainland that are exposed to a full range of predators. Predation pressure is one of the main factors influencing colony site selection for island nesting seabirds in coastal Louisiana (Greer et al. 1988). While mammalian predators influence nesting patterns for ground- and low-nesting species, avian predators (e.g., Great Horned Owl, *Bubo virginianus*, hawks, grackles) can affect many species of birds in different habitats (Skoruppa et al. 2009).

Although birds have evolved with predators, the predator landscape has shifted with increased human occupation of the coasts. Human commensals (dogs, cats, rats) live with people in coastal communities, and people bring dogs and cats when they visit the shore: worldwide, cats are the most important predators on bird eggs and young (Nettleship et al. 1994), even on relatively remote islands such as Campeche Banks, Mexico (Howell 1989). People also inadvertently increase native predator numbers by leaving garbage out, which results in increased numbers of raccoons (*Procyon lotor*) (Burger and Gochfeld 1990), and presumably coyotes (*Canis latrans*) as well. Both are predators on some Gulf Coast barrier islands (W. Tunnell, Texas A&M University—Corpus Christi, personal communication), and if their populations increase all along the Gulf Coast, including on small, barrier islands used by nesting birds, they could seriously impact avian populations.

Competition for nest sites is often mediated by differences in arrival times, age, or size (Burger 1979a, b, 1983). Some of these factors also affect competition for foraging space or prey types (Burger 1987a; Burger and Gochfeld 1981, 1983c). Whenever prey stocks are depressed, often due to human overfishing, seabirds relying on them will also decline (Overholtz and Link 2007). Age-related differences in foraging behavior occur in many different species. For example, in the Gulf of Mexico, there were age-related differences in the success of frigatebirds pirating from Laughing Gulls in Seybaplaya, Campeche (Mexico, Gochfeld and Burger 1981), in Laughing Gulls foraging in Texas and Mexico (Burger and Gochfeld 1981, 1983c), and in Black-necked Stilts feeding in Texas (Burger 1980). Many fishery operations enable piracy because the concentrated food draws a range of species, and food items are too large to handle quickly (Furness et al. 1988).

Nesting in colonies has both negative and positive advantages (Gochfeld 1980; Burger 1981a, b; Coulson 2001). Advantages include social facilitation of breeding activities, early detection of predators, antipredator behavior, and information transfer about food sources (Ward and Zahavi 1973; Flemming and Greene 1990). Disadvantages include increased competition for food, competition for nest sites, and conspicuousness of colony members to predators (Furness and Birkhead 1984). Nesting in mixed species colonies increases the advantages (increased predator protection), while decreasing the disadvantages (competition for food resources or space; Burger 1981a, 1984a, b). Social facilitation, whereby one species derives a benefit from nesting with another, is one advantage of nesting in mixed species colonies (Gochfeld 1980; Coulson 2001). For example, Black Skimmers derive advantages from nesting with terns and gulls that mob predators to drive them from colonies, thereby protecting the nests, eggs, and chicks of skimmers from predation (Burger and Gochfeld 1990).

12.5.7 Parasites and Disease

Birds are exposed to numerous parasites and diseases, but only a few Gulf examples will be given here to illustrate possible incidences and effects. Garvin et al. (2006), examining blood parasites of Nearctic-Neotropical Passerines during spring migration in the Gulf coast, found that 21 % of 1,705 migrant Passerines were infected with one or more blood parasites. Helminth (parasitic worms) infections are quite common in Brown Pelicans along the Gulf coast, and although the effects of infections are unclear at times (Dyer et al. 2002), stressed pelicans can show the effects of parasitism (Grimes et al. 1989; Dronen et al. 2003). Similarly, 22 species of endohelminths were found in Willets collected from Texas (Dronen et al. 2002), and several platyhelminthes species (*Clinostomum* sp., *Mesotephanus* sp., *Galactosomum* sp.) were reported from shorebirds (Cormorant, Great Egret, Laughing Gull, and Pelican) in Tampa Bay and Boca Grande in Florida (Hutton and Sogandares-Bernal 1960). Nematodes (*Contracaecum* spp.) cause lesions in the proventriculus of Brown Pelicans and Double-crested Cormorants (*Phalacrocorax auritus*), and occasionally other water birds in Louisiana. The impact of harmful algal blooms (red tides) on marine bird populations has been demonstrated. Brevetoxin, a potent neurotoxin produced by the red tide dinoflagellate (*Karenia brevis*, formerly *Gymnodinium*), was found in tissues of dead Double-crested Cormorants (Kreuder et al. 2002) and in Royal Terns and Laughing Gulls (Vargo et al. 2006) in the Gulf coast region.

12.5.8 Pollutants

The land-margin interface is particularly vulnerable to pollutants, fertilizers, and wastes that flow from associated watersheds (Greenberg et al. 2006), such as from the Mississippi River (NOAA 2011). While a "dead zone" (area of hypoxia) occurs off the Louisiana and Texas Coast (NOAA 2011), its effects on overall avian populations in the Gulf have not been demonstrated.

Birds are indicators of contaminants (Sheehan et al. 1984; Fox et al. 1991; Peakall 1992; Burger 1993; Custer 2000; Burger and Gochfeld 2001, 2004a, b), because of the potential for contaminants to cause chronic effects and population declines, as well as acute mortality and other impairments (reviewed in Monteiro and Furness 1995; Rattner 2000; Burger et al. 2002). Effects have been demonstrated in both laboratory (Burger and Gochfeld 2000, 2005; Spalding et al. 2000a; Hoffman et al. 2011) and field studies (Burger and Gochfeld 1994; Frederick et al. 1999; Jackson et al. 2011). While most pollutants are anthropogenic in nature, oil and mercury also can come from natural sources. Oil seeps were known from the Gulf of Mexico long before Western colonization (Geyer 1981).

Mercury occurs naturally in seawater and also comes from anthropogenic sources (Wolfe et al. 1998; O'Driscoll et al. 2005). Comparisons of museum specimens of feathers from wading birds nesting in the Everglades from 1920 to the 1970s indicated that samples taken during the 1990s had mercury levels that were 4–5 times higher than feathers from specimens collected before 1970 (Frederick et al. 2004), indicating an anthropogenic source. Fish-eating birds are particularly vulnerable to the effects of methylmercury because it accumulates in fish. Birds that eat large fish with the highest mercury levels are most at risk (Pinho et al. 2002; Storelli et al. 2002; Burger 2009; Burger et al. 1994, 2011; Frederick et al. 1999, 2004). Common Loons (Burger et al. 1994; Burgess et al. 2005; Burgess and Meyer 2008; Evers et al. 2008), raptors (Albers et al. 2007), and songbirds (Jackson et al. 2011) are species with high mercury levels that have impaired reproduction, with possible population declines.

Ducks, such as Mallards, were once affected by seed treated with mercury (Krapu et al. 1973; Heinz 1976a, b). The toxic effects of methylmercury, particularly reproductive and neuro-behavioral deficits, have been demonstrated in the laboratory (Heinz 1979; Spalding et al. 2000b) and in the field (Frederick et al. 1999). Mercury levels in eggs from some Great Egrets in the Everglades exceeded effects levels found in the laboratory (Rumbold et al. 2001). Sensitivity to methylmercury varies greatly among species (Heinz et al. 2009). Several reviews discuss contaminants in birds in general, or of the species groups discussed in this chapter (e.g., Burger 1993; Hoffman et al. 1995; Beyer et al. 1996; Burger and Gochfeld 2001; Frederick et al. 2002; Custer 2000), but there have been no clear demonstrations that mercury levels in birds in the Gulf have affected avian population levels.

Other metals, or metalloids, including lead (Burger and Gochfeld 1994) and selenium (Ohlendorf et al. 1986, 1989) also affect bird behavior, development, and survival. Natural experimentation with Little Blue Herons in southern Louisiana wetlands (West Baton Rouge) indicated that chicks exposed to cadmium in their foods had significantly slower growth rates than nonexposed chicks, and exposure to lead was correlated with increased nestling mortality (Spahn and Sherry 1999). However, population effects from these experiments are not shown.

Brown Pelicans are the poster bird for the effects of DDT on population levels. Pelicans declined from about 5,000 individuals in Texas in the early 1960s, to fewer than 20 individuals by 1974 (King et al. 1977). Eggshell thinning, caused by the endocrine disruption effects of DDT, led to total reproductive failures (Blus et al. 1974). After DDT use was banned in the United States, pelican populations increased (King et al. 1985), and they are no longer federally listed as threatened or endangered. Similarly, high residues of organochlorine pesticides and

PCBs were found in Black Skimmers (Custer and Mitchell 1987), cormorants, and gulls (King and Krynitsky 1986), and other waterbirds from Texas (Mora 1995, 1996), and in Great Egrets from other locations (McCrimmon et al. 2011). However, population declines of gulls, skimmers, egrets, and other waterbirds from the Gulf have not been demonstrated from organochlorine pesticides. Pelican populations have increased dramatically in the Gulf since the banning of DDT (see Pelican in Indicator Species, Section 12.6.1).

Oil contributes to foraging difficulties, lowered reproductive success, and mortality, especially in seabirds (Piatt et al. 1990). The effects of oil discharges could be acute (mortality) (Dunnet 1982; Hunt 1987; Burger 1994a, 1997a, b; Lance et al. 2001; Payne et al. 2008; Wiens et al. 1996), or chronic, including the effects from operational oil discharges that affect marsh structure (McCauley and Harrel 1981; Mendelssohn et al. 1990; Fraser et al. 2006). Effects of oil include cessation of growth in chicks, osmoregulatory impairments, hypertrophy of hepatic, adrenal, and nasal gland tissue (Miller et al. 1978), reduced thermoregulation (O'Hara and Morandin 2010), reduced survival of chicks (Trivelpiece et al. 1984), and changes in hematology and blood chemistry (Newman et al. 2000). Macko and King (1980) found that oil from the Libyan crude oil spill in Redfish Bay, Texas (1976) caused significant embryo mortality in Louisiana Heron eggs, but did not affect hatchability of Laughing Gull embryos. Oil also can affect population levels of invertebrate prey, which secondarily affects birds, mammals, and even humans (Lees and Driskell 2007). However, the effects demonstrated for birds nesting along the Gulf coast are on individual birds, and not on populations or species. There is no evidence that oil in the Gulf of Mexico up to 2010 has resulted in declines in avian populations.

Because of oil development and transportation in the Gulf, birds have been exposed to both chronic and episodic spills since the 1970s. One of the first large spills was the Ixtoc I spill of June 3, 1979 in the Bay of Campeche. It released about 30,000 barrels per day, which eventually formed a thick mousse-like emulsion that floated on the surface (Energy Resources 1982). When the oil reached the southern Texas coast in August, it had broken into smaller pieces. As it reached the shore, birds moved to less suitable but unoiled places on the backshore; fewer than 20 % of shorebirds remained on the foreshore (Chapman 1981, 1984). Oiled Sanderlings and Willets spent less time foraging, and more time resting and engaged in preening than unoiled birds (Chapman 1981), which agrees with findings in shorebirds from elsewhere (Burger 1997b; Burger and Tsipoura 1998). There is no evidence, however, that such movements had long-term effects on these migrant shorebird populations in southern Texas.

Plastics and other ocean debris can cause direct mortality and injury, as well as obstruction of the gastrointestinal tract (Day et al. 1985; Azzarello and Van Vleet 1987). Vulnerability of particular birds depends upon their anatomy, methods of digestion, methods of foraging and prey identification, and their distribution geographically relative to shipping lanes, coasts, and oceanographic conditions that control the distribution of marine debris. Some birds, such as gulls, herons, and egrets, can regurgitate plastic that they ingest, although strings, plastic with jagged edges, and hooks can be caught in their esophagus or lodge in the stomach. Seabirds in the order Procellariiformes are most vulnerable to the effects of plastics because they have a small gizzard and cannot regurgitate ingested plastic (Azzarello and Van Vleet 1987). Accumulation of plastic in the stomach impedes absorption, and nonfood items may reduce food intake if the bird's stomach is full (Sturkie 1965). Plastic debris is also a problem near shore, where birds become entangled in fishing line, nets, and strings attached to kites and balloons. One bird can drag back fishing line attached to its feet, and several additional birds in the colony can then get caught in it. Although the presence of plastic debris may impact individual birds, there is no evidence that such debris has impacted avian population levels of birds nesting or migrating through the Gulf of Mexico.

Finally, birds have evolved mechanisms to deal with natural stressors (hurricanes, severe storms, native predators). These mechanisms function unless there are several years with no reproduction (e.g., Pelicans and DDT). In birds, some mortality or decreased reproduction can be compensated for by several mechanisms: (1) higher survival of remaining young or adults, (2) recruitment from elsewhere, (3) higher reproductive success of remaining birds, (4) breeding at an earlier age, and (5) breeding of birds that had not bred in previous years. For example, some young adults are unable to compete for nest sites and these do not normally breed. However, if breeding sites open (due to a mortality event), sub-adult birds, or others previously unable to breed, move in, and overall productivity remains the same.

12.5.9 Management and Physical Anthropogenic Disruptions

Many management practices are employed in coastal areas that impact birds, and many of them are designed to improve conditions for people, including dredging, shoal removal, beach nourishment, beach raking to remove debris or shells, water control, and groins or barriers (seawalls, jetties). In the nearshore and along the shore, wind energy development can impact avian use and distribution. In the Gulf itself, oil and gas development has resulted in the building of thousands of platforms in the northern Gulf of Mexico (Russell 2005). These platforms provide habitat for foraging birds that use them as roosting sites or as hunting perches (raptors). However, they also have the potential to disrupt songbird migration, especially for birds leaving the Yucatán Peninsula (Morrison 2006).

Dredging is performed to deepen channels and harbors, and the disposition of dredge spoil can have positive and negative effects on birds (Shabica et al. 1983; Guilfoyle et al. 2006). Some dredging can remove habitat, but soil deposition can create nesting habitat for Piping Plovers (Webster 2006), Least Terns (Golder et al. 2006), and Black Skimmers (Burger and Gochfeld 1990). Species of high concern with respect to dredging (both foraging and nesting) include Snowy Plover, Wilson's Plover, American Oystercatcher, Willet, Royal Tern, Least Tern, and Black Skimmer, among others (Hunter 2006).

Marshes are burned in southwestern Louisiana and Texas during the winter to favor waterfowl (Lynch 1941; Gabrey and Afton 2000). The timing of burning and the spatial extent are critical factors influencing how a given species responds to burning. For example, Louisiana Seaside Sparrows decreased in burned areas during the first breeding season, but increased during the second (Gabrey et al. 1999; Gabrey and Afton 2000).

Marsh terracing is intended to slow marsh erosion, increase marsh edge, and possibly increase bird numbers. Louisiana has 75 % more wading and dabbling birds in terraced marshes than in non-terraced marshes, but terracing did not increase bird diversity (O'Connell and Nyman 2011). Terracing slightly increased the number of herons, egrets, ibises, gulls, and terns, but it dramatically increased the number of waterfowl and Moorhens (*Gallinula chloropus*) (O'Connell and Nyman 2011).

Other managed coastal habitats in the Gulf, such as rice fields, are used by wintering waterfowl (Day and Colwell 1998; Link et al. 2011) and wading birds (Acosta et al. 1996, 2010). In Cuba, White Ibis, as well as other wading birds, concentrated in rice fields because they provided an abundance of fish, crabs, and aquatic insects (Acosta et al. 1996). Nesting on gravel rooftops, as Least Terns do in northwestern Florida and elsewhere (Gore 1991; Zambrano et al. 1997), is a prime example of using man-made habitats. Fisheries operations, such as processing, canning, and fishing itself, provide offal and other food for seabirds and coastal waterbirds (Shealer 2001; Montevecchi 2001).

12.5.10 Direct Human Activities

Habitat loss is often accompanied by increases in human activities that can affect nesting assemblages, habitat choice, foraging behavior, and reproductive success (Buckley and Buckley 1980; Erwin 1989; Burger 1994b; Carney and Sydeman 1999; Burger et al. 2004, 2007). In many cases, however, birds habituate to the presence of humans, and sometimes become more aggressive (Safina and Burger 1983; Vennesland 2010), as they do at landfills (Pons and Migot 1995). Closing landfills, however, can decrease reproductive success and survival of young birds that have difficulty foraging in other situations (Pons and Migot 1995).

The effects of increased human disturbance can be illustrated by a study of coastal birds over a three-decade period on Mustang Island, Texas (Foster et al. 2009). At the beginning of the study, an average of 19 people per day were observed on the beach, but it increased to 75 people per day by the early 1990s, and then rose to nearly 100 per day (Foster et al. 2009). Foster et al. (2009) found that some species increased significantly (Brown Pelican, Laughing Gull), but many more decreased significantly (Table 12.6). They attributed the changes to human disturbance.

Disturbance includes direct approaches, inadvertent destruction of eggs or chicks, interruption of foraging or roosting, and increased presence of dogs, as well as indirect effects, such as increased mammalian predators because of provisioning of food (Burger 1991b; Maslo and Lockwood 2009). Increased human disturbance can even delay the initiation of egg laying in Black Skimmers (Safina and Burger 1983), which has consequences if food is less available later in the season. Data on the complex interactions between species, species size, species density, and the presence of people and other disturbances bear further examination with shorebirds along the Gulf Coast. Understanding these interactions is critical for protecting the nest sites of Snowy Plover, and less so for Willet and American Oystercatcher that also nest elsewhere. Furthermore, because the Gulf is an important foraging and wintering area for more than 20 species of shorebirds, understanding how human activities affect their foraging and distribution is important for their conservation (Withers 2002). Management includes signs, fencing, wardening, and prevention of beach access by people and vehicles during the nesting season (Burger 1989; Elliott-Smith and Haig 2004), although the last method is often controversial (Mabee and Estelle 2000).

Similar data on human disturbance exist for many groups of birds, such as grebes (Keller 1989), waterfowl (Korschgen and Dahlgren 1992; Mallory and Weatherhead 1993), gulls (Hunt 1972; Burger 1981c; Burger and Gochfeld 1983b), herons (Tremblay and Ellison 1979; Parsons and Burger 1982; Fernandez-Juricic et al. 2007), pelicans (Johnson and Sloan 1975), guillemots (Cairns 1980; Ronconi and St. Clair 2002), cormorants (Kury and Gochfeld 1975; DesGranges and Reed 1981), and other colonial waterbirds (Rodgers and Smith 1995). Habitat loss amplifies the effects of human disturbance (Burger 1981d; Skagen et al. 2001). Reducing the effects of human disturbance can involve reducing the amount and types of human activities, prohibiting the presence of dogs or off-road vehicles, or habituating birds to the presence of people (Vennesland 2010).

Human disturbance, however, can also include organized human activities, such as tourist boats for diving, snorkeling, fishing, or, nature tourism. In the Yucatán, for example, two barrier peninsulas (Ria Lagartos, Celestun) are exposed to tourism boats, despite their designation as Yucatán Biosphere Reserves (Savage 1993). Disturbance comes not only from the boats and people but also from the construction of structures designed to enable tourism (Savage 1993). Presumably the effect would differ depending upon whether people are on foot, in small boats, or in large boats.

Table 12.6. Changes in Abundance of Birds on Mustang Island, Texas, from 1979 to 2007 (after Foster et al. 2009). Mean daily abundance of species ranged from 2.4 to 328.

Species	Status	Trend in Percent
Eared Grebe, Podiceps nigricollis	Winter	280.0
*Brown Pelican	Resident	586.0
Double-crested Cormorant	Winter	-82.2
*Great Blue Heron	Resident	-38.9
Cattle Egret	Resident	45.4
*Black-bellied Plover	Winter	-34.2
Piping Plover	Winter	-25.4
Snowy Plover	Winter	-3.6
*Wilson's Plover	Summer	-62.9
*American Oystercatcher	Resident	137.4
Willet	Winter	-3.4
Ruddy Turnstone	Winter	-5.5
*Red Knot	Winter	-54.0
*Sanderling	Winter	26.2
Western Sandpiper	Winter	-3.1
Least Sandpiper	Winter	-27.3
*Herring Gull, Larus argentatus	Winter	-70.3
Ring-billed Gull	Winter	10.2
*Laughing Gull	Resident	58.7
*Caspian Tern	Resident	-58.8
*Royal Tern	Resident	-68.0
Sandwich Tern	Breeding	-13.2
Common Tern	Migrant	49.0
*Forster's Tern	Resident	-87.5
Least Tern	Breeder (summer)	-35.6
Black Tern	Migrant	214.9
*Gull-billed Tern	Breeder	-53.3
*Black Skimmer	Resident	-71.3

*Before species name indicates a significant change in abundance ($p < 0.05$). Changes were attributed to human disturbance. Declines are shown in red.

All of the factors discussed in the sections above have been singly, or in combination, shown to affect bird populations in the Gulf over the short term (a storm event, a breeding season for nesting species, at migratory stopovers for Nearctic-Neotropical migrants). Long-term (decade-long) shifts in population levels of birds in the Gulf of Mexico have not been demonstrated as a result of a specific factor, except for the Brown Pelican whose population declined dramatically due to DDT. Habitat loss resulting from coastal development (and associated direct human disturbance), and sea level rise, have the potential to negatively impact avian populations along the Gulf of Mexico because they are directional and likely to continue.

12.6 STATUS OF BIRDS IN THE GULF OF MEXICO

12.6.1 Overview of Indicator Species and Groups

Because nearly 400 species reside, winter, or migrate to or over the Gulf of Mexico, it is impossible to give an account of each species. In this chapter, selected indicators are used to form a pattern to illustrate: (1) bird use in the Gulf, (2) status and trends of key species, and (3) changes of conservation concern. The Gulf of Mexico contains some of the most important habitats in North America for migrant raptors (Gallardo et al. 2009), migrant songbirds (Rappole 1995), and wintering/migrating shorebirds (Withers 2002), as well as breeding pelicans, gulls, terns, shorebirds, ibises, egrets, and herons. Indicators are used to understand the distribution and abundance of birds in the Gulf, although they are also useful as indicators of contaminants, disease, and restoration efforts (Burger 1993; Custer 2000; Erwin and Custer 2000; Frederick et al. 2009). Two kinds of indicators are considered: individual species and species groups. These indicators can serve as a baseline for future studies and for evaluating future anthropogenic effects, including restorations.

The species considered below were chosen because they were endangered or threatened (such as the Whooping Crane), species of concern, species whose major populations occur in the northern Gulf of Mexico, species that are typical of the Gulf (e.g., Reddish Egret), or were unusual in other ways (e.g., Piping Plovers winter there extensively). The rationale for the use of each species is given in Table 12.7. They were also chosen to balance migrant and resident, colonial and solitary, and different habitats. While many others could have been selected, this represents a balance for the characteristics shown in Table 12.7. Species groups were selected because the Gulf of Mexico plays an important role in their life cycle, including pelagic seabirds, waterfowl, raptors, colonial nesting birds (gulls, terns, herons, egrets, and ibises), and migrant Passerines and shorebirds, although trends data for the latter are not available. The species indicator accounts are not meant to be exhaustive or complete life history information (see *Birds of North American* [BNA], Hamer et al. 2001). Rather, the accounts give a brief description of the bird's niche and available information about their status and populations within the northern Gulf of Mexico. Information on the southern Gulf is added where available.

12.6.2 Indicator Species

12.6.2.1 Common Loon

Common Loons are large, long-lived birds with delayed maturity and low fecundity. They nest on small isolated islands in lakes in the northeastern United States and Canada. They are awkward on land, have webbed feet, are superb swimmers, and dive for fish. Their breeding range is restricted to mainland North America (Evers et al. 2010). They nest from Washington to Montana, to northwest Wyoming, north-central North Dakota, and the upper Great Lakes, and from New York to New England (Evers et al. 2010). They winter on the Pacific and Atlantic coasts, including the Gulf of Mexico and the Gulf of California. Common Loons in Mexico winter off the Texas coast (Howell and Webb 1995). They rarely winter farther south of central Mexico; some remain as far north as Newfoundland and the Aleutian Islands of Alaska (Evers et al. 2010). They also breed in Greenland, Iceland, and Northern Eurasia, and winter from the southern coast of Norway and Sweden south to the Caspian and Black Sea, China, and Formosa (Stevenson and Anderson 1994). In winter they are white below with dark gray upperparts (Figure 12.10).

Common Loons are used as indicators of environmental health in the northeast because of documented effects from acid rain and mercury (Burger et al. 1994; Nocera and Taylor 1998; Burgess et al. 2005; Burgess and Meyer 2008; Evers et al. 2008). They also are useful indicators

Table 12.7. Summary of Rationale for Selection of Indicator Species[a]

Species	Endangered & Threatened	Largely a Gulf Species	Resident	Migrant	Colonial	Solitary	Open Ocean	Mud Flat	Beach or Sand	Sand, Light Veg.	Marsh
Common Loon				X		X	X				
Brown Pelican		X	X		X		X				
Great Egret			X		X			X			
Reddish Egret		X	X		X			X			X
Roseate Spoonbill		X	X		X			X			X
Mottled Duck		X	X			X					X
Osprey		X	X			X	X				
Whooping Crane	X	X		X	X						X
Clapper Rail			X			X					X
Snowy Plover		X	X			X			X		
Piping Plover	X			X		X			X		
Laughing Gull			X		X		X	X	X	X	X
Royal Tern			X				X		X	X	
Black Skimmer			X				X		X	X	
Seaside Sparrow			X			X					X

[a]The last five columns are habitat categories

Figure 12.10. Common Loons (here in winter plumage) normally forage near the shore off the Gulf Coast, although they will forage farther out. © J. Burger.

in the Gulf of Mexico because they swim on the surface and dive for relatively large fish that are 10–15 cm long or more (Imhof 1962). Acid rain increases biomethylation of mercury in cold water, and methylmercury accumulates in fish. On the breeding grounds, mercury continues to build up in tissues as the Loons age, and increasing body burdens reduce the number of young fledged per pair (Evers et al. 2008). While they usually occur inshore, they can also range up to 100 km out into open Gulf waters (Evers et al. 2010), making them vulnerable to oceanic and Gulf coast pollutants.

Common Loons breed on small to large lakes, nesting near the edge of isolated small islets devoid of predators (Vermeer 1973a; McIntyre 1988; Barr 1996). Loons usually lay two eggs, but only fledge one chick (McIntyre 1988). Loons arrive on the northern coasts of the Gulf of Mexico by the third week of October, mainly from Minnesota and Wisconsin (Evers 2004), and numbers build up until mid- November (Alexander 1991). Mortality in Loons is due to mercury contamination, commercial fishing (Vermeer 1973b), botulism (Brand et al. 1983), and nutritional stress from high costs of plumage replacement in winter (Alexander 1991), among other factors.

Common Loon populations are probably stable to increasing in the United States (Evers et al. 2010), and the United States and Canadian population is estimated at 607,000–634,000 birds (Delany and Scott 2006). Using Christmas Bird Counts for the entire U.S. Gulf coast, Niven and Butcher (2011) computed a significant 1.6 % per year increase over the period from 1965 to 2011. Imbedded in this increase was a decrease in numbers and reproductive success in the 1980s and 1990s, partly from acid rain and mercury (Evers et al. 2010). Using the same Christmas Bird Count data, running 3-year averages were computed for Common Loon numbers from 1940 to the present (Figure 12.11). There is variation along the Gulf, with few birds recorded from Mississippi, Louisiana, and Texas, and the majority recorded off the coast of Alabama and Florida. The data show a peak in the 1980s, with a recent increase in Alabama and a decline in Florida.

Resiliency in Common Loons is low because of low clutch size (two eggs), low reproductive rate (usually raise one or fewer young per year), high mortality while at sea the first 2–3 years of life, and delayed breeding age (average age of 6 years; Evers et al. 2010). Although the loon has a long life span of around 30 years (Evers et al. 2010), it is susceptible to mercury poisoning because it eats large fish on the breeding grounds of lakes where prey fish accumulate high mercury levels (Evers et al. 2008).

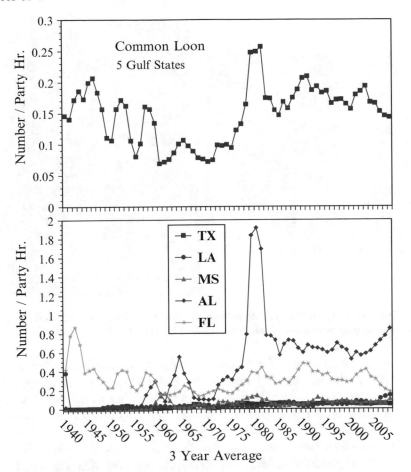

Figure 12.11. Number of Common Loons observed (number/party per hour) from 1940 to 2005, derived from Christmas Counts in the winter. Running 3-year averages were plotted to smooth out the patterns. The bottom graph shows the numbers for each state, and the top is a composite graph for all five Gulf States. © J. Burger.

12.6.2.2 Brown Pelican

Pelicans are very large, plunge-diving birds with recognizable gular pouches. They nest colonially along the Pacific, Atlantic, and the entire Gulf coasts (Figure 12.12). These iconic birds only breed along coasts, and their image is put on placemats, postcards, billboards, and signs throughout the Gulf (Eubanks et al. 2006). Their breeding range is along the Pacific coast from southern California to southern Ecuador (including the Galapagos), and along the Atlantic coast from Maryland south, around the Gulf of Mexico and Caribbean coast, to northern Venezuela and Colombia (Shields 2002).

Brown Pelicans feed on small fish (10–28 cm long), such as Menhaden (*Brevoortia patronus*) (Imhof 1962; Hingtgen et al. 1985), a major commercial fish in the Gulf. Fishermen have persecuted them because they were believed to eat commercial fish (Sprunt 1954). Pelicans dive with the bill ajar, and the force of water on impact causes the pouch to expand, trapping the fish inside. The Pelican then raises the bill above the water, pointed downward, and the water runs out, leaving prey in the pouch (Stevenson and Anderson 1994). Pelicans usually feed within 20 km of the nest site (Briggs et al. 1981), indicating the importance of having suitable nesting colonies near foraging opportunities.

Figure 12.12. Brown Pelicans nest either on the ground or in low bushes, which have to support their weight. This colony was on a small sand spit in Louisiana. © J. Burger.

Current population estimates for Brown Pelicans (*P.o carolinensis*) are 44,000–45,000 pairs; about 60 % of the 40,000 that nest in the United States do so along the Gulf Coast (Shields 2002). Pelicans are resident in most of their breeding range (Shields 2002). Pelicans breed in monospecific and mixed-species colonies, often with other ground-nesting species. They use the same colony site in successive years unless it becomes unsuitable because of habitat loss, human disturbance, or predators (Schreiber and Schreiber 1982). Colony site selection in pelicans depends upon the availability of nest sites that are free from predators and human disturbance, and are reasonably close to food. Colonies in Louisiana averaged 13 km from the mainland (Visser et al. 2005). Brown Pelicans are monogamous, mate for life, lay up to five eggs, and the young are fed predigested fish that parents deposit on the nest.

Brown Pelicans exhibited one of the most dramatic population declines ever observed in birds, which occurred between the late 1950s and the early 1970s, due to the organochlorine pesticide DDT (Shields 2002). Before the decline, populations in Louisiana and Texas were estimated at greater than 50,000 birds (Shields 2002). Lowery (1974) claimed that before the decline, most Brown Pelicans seen along the entire northern Gulf coast were produced in Louisiana. Pelicans declined from about 5,000 individuals in Texas in the early 1960s, to fewer than 20 individuals by 1974 (King et al. 1977). Populations disappeared in other places, and reintroductions were necessary. The mechanism of decline was through eggshell thinning caused by DDT; pelicans that incubated broke their eggs (Blus et al. 1974).

Brown Pelicans were reintroduced into Louisiana at Queen Bess Island in 1971 and the Chandeleur Chain in 1979 (Wilkinson et al. 1994). Before 1983, no Brown Pelicans nested in Alabama; the first ones were relocated there in 1983, and by 1990 there were 1,374 nests (Wilkinson et al. 1994). The Florida Gulf coast population of breeding Brown Pelicans declined, but remained stable in Tampa Bay after the 1990s (Hodgson and Paul 2010), while the Atlantic coast population increased (Wilkinson et al. 1994).

Trends in breeding populations have been examined in many places. Two examples are given: Queen Bess Island in Louisiana, and Galveston Bay in Texas. Breeding populations at three sites in Louisiana were followed from 1971 (when numbers had declined drastically from

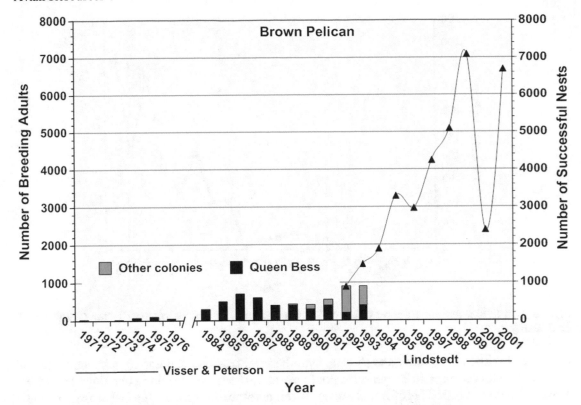

Figure 12.13. Nesting population of Brown Pelicans at Queen Bess and other colonies in Louisiana. Data from Visser and Peterson (1994) and Lindstedt (2005). © J. Burger.

DDT) to 1993 (Figure 12.13, after Visser and Peterson 1994). Pelicans were locally extirpated in Louisiana and were reintroduced at Queen Bess Island in the early 1970s (Holm et al. 2003). Subsequently, when numbers declined at Queen Bess, they increased at a nearby colony. Lindstedt (2005) reported the number of successful nests at Queen Bess and Last Islands after 1993 (Visser and Peterson 1994), and showed a small decline in the mid-1990s (Figure 12.13). Pelicans in Louisiana increased in these colonies from about 2,000 nests in 1990 to stabilize around 15,000 nests in 2003 (Holm et al. 2003; Visser et al. 2005). Pelican colonies in Louisiana are located far from the mainland and human activity, and colonies such as Queen Bess Island have required the addition of land to provide sufficient habitat (Visser et al. 2005).

Surveys of Brown Pelicans nesting in Galveston Bay, Texas, have also been made for a number of years. The number of nesting pairs has been increasing there although there were large shifts in the number of nesting pairs (Figure 12.14). The Galveston Bay Status and Trends report rated the species, used as an indicator by the program, as good—significantly increasing (GBEP 2006).

Brown Pelicans are reaching population levels on the Gulf Coast of North America that were present before the widespread use of DDT (Robinson and Dindo 2011). Pelicans are faced with severe habitat loss that might threaten their populations once again, particularly in Louisiana due to loss of available nesting sites (Visser et al. 2005). Robinson and Dindo (2011) comment that the future of Brown Pelican populations in the Gulf is unclear because of the ephemeral nature of spoil islands and natural coastal areas, as well as natural disasters, and manmade ones. Periodic reproductive failures have little effect on population levels, but recurrent breeding failures result in population declines (Schreiber 1980a). Another cause of

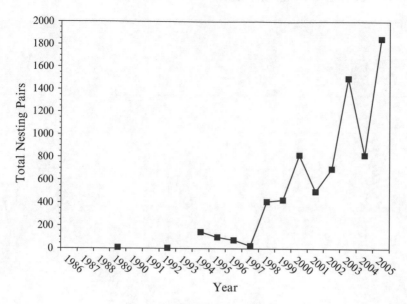

Figure 12.14. Number of nesting pairs of Brown Pelicans in Galveston Bay (after GBEP 2006). © J. Burger.

mortality is exposure to cold and storms, hypothermia, frostbite damage to gular pouches and foot webs, starvation, and longer-term cold weather effects on breeding phenology (Schreiber 1980b; Shields 2002). Therefore, changes in temperature because of global warming could increase populations of Brown Pelicans in the northern coast of the Gulf of Mexico.

Another cause of mortality in Brown Pelicans, unlike most other indicators, is from people. A study of 3,106 recoveries of Brown Pelicans banded in the Carolinas and Florida, from 1925 to 1983, indicated that more than half died from human activity, with entanglement in fishing lines as a major cause (Schreiber and Mock 1988). Pelicans are sometimes killed or maimed maliciously.

Shields (2002) plotted recovery of Brown Pelicans along the Gulf Coast as a whole, showing a steady rise in nests from the 1970s through the 1980s, with greater increases thereafter (Figure 12.15). The number of nesting Pelicans did not increase as sharply along the Atlantic coast, or along the California coast; populations in California fluctuated around 5,000 pairs since the mid-1980s (Shields 2002).

Resiliency is relatively high as evidenced by their population recovery following devastation by pesticides in the 1950s and 1960s. Pelicans reach sexual maturity at 3–5 years of age, lay up to five eggs (modal clutch is three), usually fledge one or fewer chicks, only 30 % survive the first year, and fewer than 2 % survive beyond 10 years (Schreiber and Mock 1988; Shields 2002). They probably have only an effective reproductive life span of 4–7 years although they can live for 25–30 years (Schreiber and Mock 1988). Since human disturbance and breeding habitat availability seem to be major problems, recovery from any declines will partly depend on these factors.

12.6.2.3 Great Egret

The dazzling white plumage of Great Egrets, with their long lethal yellow bill, and their motionless stance as they wait to capture prey, makes them easy to recognize (Figure 12.16). Great Egrets are cosmopolitan, inhabiting freshwater, estuarine, and marine wetlands, and are intermediate in size between the larger Great Blue Heron and the smaller egrets. Great Egrets

Figure 12.15. Populations of Brown Pelicans nesting along the northern Gulf Coast (after Shields 2002). © **J. Burger.**

breed in North and South America, in southeast Europe, northern Asia to Siberia, north China, and northern Japan, as well as in Australia (McCrimmon et al. 2011). In North America, they breed primarily along the Atlantic Coast from Maine south to all regions along the Gulf coast, to the east coast of Mexico, and down to South America, including the Caribbean Islands. On the west coast they breed in California, and on the west coast of Mexico and Central America. They also breed in scattered inland areas in the Central United States (McCrimmon et al. 2011). They winter throughout their breeding range, except for interior North America and the northeast coast (McCrimmon et al. 2011).

Egrets are useful indicators for the Gulf Coast because they are colonial, conspicuous (large and white), usually nest higher in vegetation when it will support their nests, and are key members of wading bird nesting assemblages in the coastal regions all along the Gulf of Mexico, including Mexico (Burger 1978a; Mock 1978, 1980). They feed on intermediate-size fish, as well as reptiles, amphibians (especially frogs), small mammals, birds, crustaceans, mollusks, and insects (Stevenson and Anderson 1994). They also visit inland rice fields, crawfish ponds, and wet fields to find frogs, as well as dry fields to stalk small reptiles (Eubanks et al. 2006).

Great Egrets nest in mixed-species colonies with other egrets, herons, ibises, and often Brown Pelicans. These colonies are stable as long as conditions remain viable and the habitat suitable; otherwise they switch sites (Kelly 2006a). They are monogamous, and both parents incubate and care for the young, including provisioning (McCrimmon et al. 2011). Incubation (28–29 days) begins with the first or second egg so that young hatch asynchronously; when food is in short supply, competition between siblings results in older chicks kicking eggs or younger chicks out of the nest (Mock and Lamey 1991; Stevenson and Anderson 1994).

Great Egret populations, along with other herons and egrets, declined dramatically in much of the United States during the late 1800s and early 1900s due to hunting their plumes for the millinery trade (Ogden 1978). Their plumes (called aegrettes), used in courtship displays, have a delicate, lacey appearance (Figure 12.16). The North American population of Great Egret

Figure 12.16. Great Egrets sometimes stand and wait for prey, either on logs or in shallow water.
© **J. Burger.**

declined by more than 95 % with market hunting (McCrimmon et al. 2011). Populations quickly recovered with the passage of the Migratory Bird Treaty Act in 1913, and they once again moved into breeding areas in the Northeast where they had largely disappeared (Burger 1996b; McCrimmon et al. 2011). The North American population is currently estimated at about 270,000 birds (Delany and Scott 2006). The nesting population of Great Egrets along the western Gulf coast increased from the 1930s to the 1990s. For example, numbers in Louisiana were 2,900 pairs in 1959, 11,000 pairs in 1974, and 29,000 pairs in 1990; Texas, had 5,000 pairs in 1939, 1,450 pairs in 1959, and 6,500 pairs in 1969 (McCrimmon et al. 2011).

Trends data from Shamrock Island in Texas indicate that the number of Great Egret pairs varied markedly from almost zero in 1973 to more than 160 pairs in 1999 (Gorman and Smith 2001), and thereafter numbers increased (TCWS 2012). However, there is now evidence from south Florida that numbers have declined (Figure 12.17).

Using Christmas Bird Counts (1965–2011) as a database, Niven and Butcher (2011) reported that wintering Great Egret showed a significant increase of 2.1 % per year in coastal U.S. Gulf counts. Furthermore, when Fleury and Sherry (1995) used Christmas Bird Count data to examine the effects of crayfish aquaculture on Louisiana birds, they found that Great Egrets also increased significantly from 1949 to 1989. Using Christmas Count data for all states combined also shows an increase (Figure 12.18). Using Breeding Bird Survey data, Sauer et al. (2005, 2008) shows a steady but small increase in the Great Egret population nationwide.

Great Egrets have fairly high resiliency because they were able to recover from the devastation of plume hunting. They breed when they are 1–3 years old; clutch size varies from 1 to 6; average hatching rate is about 60 %, most commonly fledge between 0.5 and 1.5 chicks per nest; and between 40 and 75 % of nests in a colony are successful (McCrimmon et al. 2011). Success can vary; Parsons and Burger (1982) reported a hatching rate of 97 %, but a fledging success of only 50 % in a Louisiana colony. These parameters do not apply if the colony is harassed, food is scarce, or they suffer hunting or other external stressors.

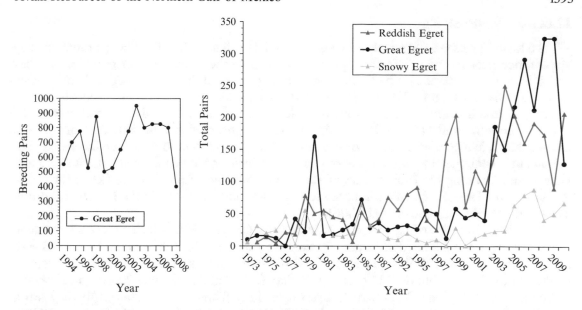

Figure 12.17. Trends in nesting pairs of Great Egret, Reddish Egret, and Snowy Egret for the Shamrock Island Colony in Texas (after Gorman and Smith 2001 and TCWS 2012, right). Left shows trends for Tampa Bay (after Hodgson and Paul 2010). © J. Burger.

Figure 12.18. Population trends in Great Egrets determined from Christmas Bird Counts from the 1950s to the present. There was a slight increase in the late 1970s. © J. Burger.

12.6.2.4 Reddish Egret

Reddish Egrets are the rarest species of heron in North America. They have a rather shaggy appearance because of the feathery plumes on both the head and back (Figure 12.19). They breed in coastal wetlands on both coasts of Florida (except in the Panhandle), Gulf of Mexico from Louisiana to south Texas and into Tamaulipas, along the Yucatán peninsula, in the Caribbean and Bahamas, and sporadically along Baja California and the Pacific coast of Mexico (Lowther and Paul 2002). The first Reddish Egret bred in Louisiana on North Island in the Chandeleur Sound in 1958 (Lowery 1974). They are resident in their breeding range, but following breeding, some birds spread out on the east coast of Mexico, down to Costa Rica and Belize, and all along the Pacific coast of Baja California and of Mexico into Central America (Lowther and Paul 2002). Significant wintering flocks can be found in the Laguna Madre in Texas and Mexico (Eubanks et al. 2006).

Reddish Egrets are of particular interest because (1) the Gulf of Mexico plays a key role in their breeding and resident distribution; (2) they are a species of special concern by the U.S. Fish and Wildlife Service (USFWS) (Bates et al. 2009); (3) they are a species of moderate concern as evaluated by the U.S. Shorebird Conservation Plan (Elliott and McKnight 2000), as well as the Southeast U.S. Regional Waterbird Conservation Plan (Hunter et al. 2006); (4) they are a priority species for habitat planning by the Gulf Coast Joint Venture (Vermillion and Wilson 2009); (5) their populations were greatly impacted by plume hunting and their populations never recovered (Paul et al. 1975; Lowther and Paul 2002; Hunter et al. 2006); and (6) they are extremely coastal. They are mainly residents, although some withdraw farther south in the Gulf of Mexico in winter (Turcotte and Watts 1999; Lowther and Paul 2002).

Reddish Egrets forage only in coastal habitats where they can appear both comical and elegant when foraging. They hunt by running, hopping, flying, and employing open-wing antics as they pursue small fish, although they sometimes stand and wait for prey. Reddish Egrets mainly forage in shallow pools where fish and invertebrates are concentrated by cyclic flooding and drying (Powell et al. 1989).

Reddish Egrets typically nest in bushes or trees in mixed species colonies along the coast and on coastal islands, and they forage in shallow, salt-water habitat (Lowther and Paul 2002), making them vulnerable to any coastal threats (Toland 1999). They also nest on dredge spoil islands (Toland 1999). They sometimes breed in small groups, and very rarely, as isolated pairs (FFWCC 2003).

Figure 12.19. Reddish Egrets often forage by waving their wings around and running about. © J. Burger.

In the 1950s, Reddish Egrets in Florida were limited mainly to the Keys, and wildlife managers experimented with transferring eggs from Texas to place them in heron nests (Sprunt 1954). Only in the last 30 years have Reddish Egret populations begun to increase in Florida Bay enough to spread up the Gulf Coast on their own (Paul et al. 1975; Powell et al. 1989). Currently, about 2,000 breeding pairs are in the United States, and 75 % of the U.S. population resides in Texas (Lowther and Paul 2002; Bates et al. 2009). The Bahamas are an important site for Reddish Egrets (Moore and Gape 2008), although surveys there indicate more than a 50 % decline in numbers since the 1980s (Green et al. 2011), which is a cause for concern. Because of their limited range, nonmigratory pattern, and colonial nesting, populations can be estimated. The breeding populations for the Gulf states are as follows: Texas 900–950 pairs, Louisiana 60–70 pairs, Alabama 5–10 pairs, and Florida 350–400 pairs, for a total of 965–1,030 pairs (>39 % of global population) (Lowther and Paul 2002; Green 2006). No Reddish Egrets breed in Mississippi. Lowther and Paul (2002) previously estimated the U.S. population to be about 2,000 pairs, but current estimates are 3,000–5,000 breeding pairs (Delany and Scott 2006). Populations are subject to considerable yearly variation. If their Gulf habitats are rendered unusable, Reddish Egrets have nowhere else to go since they are strictly a coastal species (Vermillion and Wilson 2009). Conservation concern led the Gulf Coast Joint Venture Conservation waterbird working group to designate several sites as high priority for Reddish Egret (Vermillion and Wilson 2009). These sites are centered on the south Texas coast; the waterbird working group believes they can increase breeding populations at some of these colonies by 25 %.

There are few trends data for Reddish Egrets from the Gulf States. Gorman and Smith (2001), however, tracked populations at Shamrock Island in Texas from 1973 to 1999 (Figure 12.17). While this is only one colony, it provides information on trends and variability in that colony. Reddish Egret numbers generally increased from 1973 to 1999, although the numbers were quite variable. After 1999, the numbers seemed to increase (TCWS 2012). In contrast, in Tampa Bay the numbers remained low and constant at about 100 breeding pairs (Hodgson and Paul 2010).

Fleury and Sherry (1995) used Christmas Bird Counts (1949–1988) to examine long-term population trends in Louisiana and found that populations of Reddish Egret increased 3 % per year over the 40-year period. However, from 1980 to 1988 they declined by 11.4 %. Niven and Butcher (2011) using Christmas Bird Counts for the entire U.S. Gulf coast computed a 1.6 % per year increase over the period from 1965 to 2011. A 3-year running average of Christmas Bird Count data over a longer period was computed (Figure 12.20). Variability was much greater in Texas, particularly in three time periods (early 1950s, early 1990s, and 2004–2006), which bears further examination. While Niven and Butcher (2011) show an overall increasing trend from 1965 to the present, it is not a clear consistent pattern.

Resiliency in Reddish Egrets is low as evidenced by its slow recovery from the devastation of plume hunting, particularly in relation to other egrets that recovered quickly. Most breed in the fourth year, clutch size is usually three eggs, and the maximum longevity from banded bird studies is just over 12 years.

12.6.2.5 Roseate Spoonbill

Roseate Spoonbills are stately, delicately pink birds with a greenish, flattened bill that move slowly through the water, swinging their bill from side to side (Figure 12.21). They are neotropical birds whose range extends northward to the southern United States, especially along the Gulf Coast. The main breeding area of Roseate Spoonbill is south of the United States, perhaps in Brazil (Hancock et al. 1992), and the main breeding areas in North America

Figure 12.20. Three-year running averages of Reddish Egret, computed from Christmas Counts from the late 1930s to 2008. Reddish Egrets have two color phases (*white* phase shown here). © J. Burger.

Figure 12.21. Roseate Spoonbills are the only pink species of Spoonbill (the others are all white). © J. Burger.

are along the coasts of Texas, Louisiana, and south Florida (from Tampa Bay south), with a few records from Louisiana (Dumas 2000). They rarely nest in Alabama and Mississippi. They breed sporadically along both coasts of Mexico, south to Argentina and Chile (Lewis 1983; Dumas 2000). They winter along both southern coasts of Florida, in the Gulf of Mexico, along both coasts of Mexico to Belize and Central America, and on the Pacific coast to South America (Dumas 2000). They disperse in the nonbreeding season, but mainly remain along the coasts of Louisiana and Texas, and rarely are sighted in Alabama and Mississippi (Turcotte and Watts 1999). The U.S. breeding population of this largely Gulf coast species is about 5,500 pairs, with another 3,230 pairs along the Mexican Gulf coast (Dumas 2000).

Spoonbills feed by tacto-location during day or night, at low tide (Hancock et al. 1992). While walking, they swing their slightly open bill from side to side; when it contacts prey, it snaps shut, mainly on fish, crayfish, shrimp, insects, and other aquatic invertebrates (Lewis 1983; Dumas 2000). Decline of the species in a specific area of the Gulf could be caused by loss of foraging habitat, although Spoonbills can move to other areas with suitable shallow pools for foraging.

Roseate Spoonbills nest in mixed-species colonies with other herons and egrets, although in some places they nest mainly with White Ibis (Imhof 1962). They prefer islands and keys without predator access. In a colony in Nueces Bay, Texas, they nested with Great Blue Herons, Great Egrets, Snowy Egrets, Cattle Egrets, Louisiana Herons, Black-crowned Night Herons, and Laughing Gulls (White et al. 1982), which is typical in other parts of their range. They nest low in trees or shrubs, including mangroves (Sprunt 1954; Portnoy 1977; Lewis 1983). Incubation requires 22–24 days (Lewis 1983), and the nesting season can be prolonged because nesting is not synchronous within a colony (Sprunt 1954). Postbreeding movements require more study, although some birds from Texas move a little south into Mexico (Dumas 2000), and birds from Florida move northward (Hancock et al. 1992).

Roseate Spoonbills, like many other wading birds, suffered virtual extirpation in the late 1800s to the mid-1930s because of harvesting for plumes and food. From 1865 to the late 1880s they were limited to small areas in Texas, Louisiana, and Florida (Imhof 1962). They were nearly extirpated by the late 1800s from the U.S. Gulf coast; their numbers declined to only 15 pairs during the end of the plume trade era (Rodgers et al. 1996). Spoonbill numbers gradually increased after protection from the Migratory Bird Treaty Act. They first began breeding in Texas in 1923, and Friedmann (1925) saw flocks of 75–100 feeding in shallow water and reported another flock of 1,000 feeding in southern Texas. They increased to 830 pairs by 1941 to 3,000 pairs in the 1970s, and then declined to 1,124 pairs in the 1980s (Dumas 2000). By 1996, they were up to 2,901 pairs (Dumas 2000). In Louisiana, numbers ranged up to 150 pairs in the 1940s, and then increased thereafter (while they decreased in Texas), with a 36 % increase from 1966 to 1989 (Breeding Bird Survey) (Dumas 2000). Data from Florida Bay indicated a steady increase in the number of Roseate Spoonbill colonies, and nests, but great variability among years (Figure 12.22) (Powell et al. 1989). There were fewer than 500 pairs in the 2000s (Lorenz et al. 2008). Thus their numbers appear to have declined. A recent review by the Florida Fish and Wildlife Conservation Commission (FFWCC 2011a) recommended that Roseate Spoonbills be given the status of threatened because populations are very small and restricted. Nesting colonies are affected by hydrological changes caused by management. For example, construction of canals in the Everglades reduced the flow of freshwater to the Florida Bay and decreased Roseate Spoonbills in the 1980s (Davis et al. 2005). Small fish are the primary food of Roseate Spoonbills in Florida Bay (Bjork and Powell 1994), and without a water depth threshold of 12 cm, fish are not sufficiently concentrated to provide adequate food reserves (Lorenz 2000). Thus, hydrology (salinity gradients, water depth, and dry-down) influences whether birds nest and also their reproductive success (Davis et al. 2005). Delany and Scott (2006) estimated that the number of Roseate Spoonbills in Florida and the West Indies was 3,400 individuals. Data from Shamrock Island in Texas indicated a variable, but generally stable trend from 1973 to 1999 (Gorman and Smith 2001). The number of breeding pairs varied from about 25–200 (Figure 12.23). Thereafter, numbers seemed to increase (TCWS 2012).

Using Christmas Bird Counts from 1965 to 2011, Niven and Butcher (2011) reported that wintering Roseate Spoonbills showed a significant increase of 5.9 % per year. Figure 12.24 shows populations from 1951 to 2001.

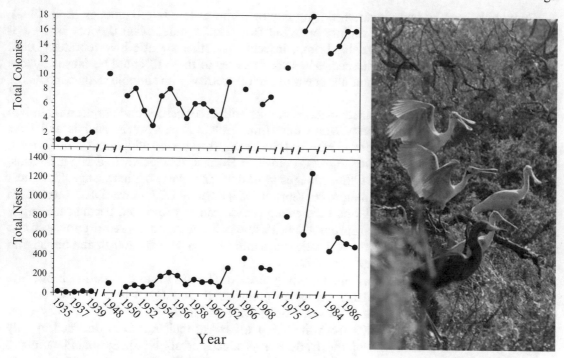

Figure 12.22. Number of colony sites and total nests of Roseate Spoonbills in Florida Bay (from Powell et al. 1989). © J. Burger.

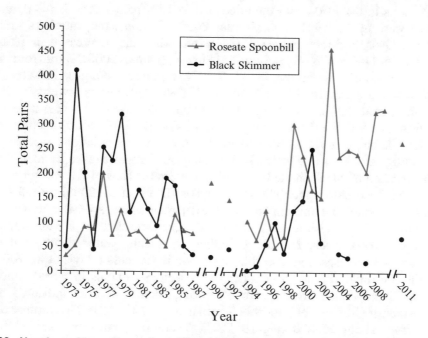

Figure 12.23. Number of breeding pairs of Roseate Spoonbills and Black Skimmers on Shamrock Island, Texas (after Gorman and Smith 2001); computed from Texas Colonial Waterbid Survey (TCWS 2012). © J. Burger.

Figure 12.24. Data from Christmas Bird Counts to illustrate changes in numbers of Roseate Spoonbills over time. © J. Burger.

Resiliency may be low in the Gulf region because historical populations were believed to be higher prior to plume hunting, and Spoonbill populations have not recovered to those levels (Dumas 2000). They usually breed at 4 years, but may breed at 3 years of age, usually lay 3–4 eggs (up to 5) and average 1–2 young fledged per nest (but this varies considerably and often colonies fail completely). Little information is available on longevity (Lewis 1983; Dumas 2000). Reproductive success partly depends upon the availability of prey that is concentrated by drying down periods (Powell et al. 1989) and varies greatly from year to year (White et al. 1982).

12.6.2.6 Mottled Duck

Mottled Ducks, a southern relative of the American Black Duck (*Anas rubripes*) and the Mallard, breed in marshes and wetlands along the Gulf of Mexico (Bielefeld et al. 2010). They resemble female Mallards, except Mallard females have a blotched bill, a whitish tail, and a white border on the front and rear of the blue speculum (Figure 12.25). They are a non-migratory resident in the Gulf Coast of the United States and into northeastern Mexico (Tamaulipas) (Howell and Webb 1995). The two disjunct populations are from western Alabama to northeast Mexico (south to Tampico), and one isolated in Florida (Johnson 2009; Bielefeld et al. 2010). The Mottled Ducks in Florida migrate north in Florida in the winter (Stevenson and Anderson 1994).

Mottled Ducks occur in near-subtropical climates of the Texas and Louisiana Gulf coast where wetlands are not subjected to near freezing temperatures (Figure 12.25) (Grand 1992). Wetland drainage, degradation of coastal marshes by saltwater intrusion and urban development pose a risk, along with hybridization with Mallards. In the nonbreeding season they concentrate in fallow-flooded fields in Florida and in harvested rice fields in the western Gulf Coast (Figure 12.26) (Bielefeld et al. 2010).

Figure 12.25. Mottled Ducks resemble female Mallards, except they lack the *white* border on the front and rear of the *blue* speculum (wing bar) © J. Burger. Shown also is the range (after Johnson 2009).

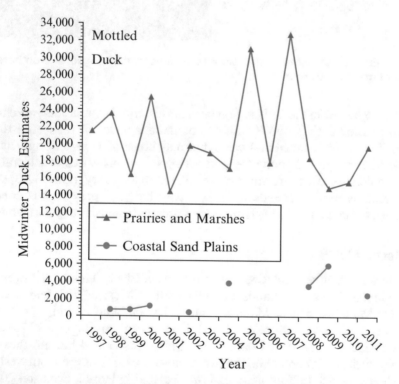

Figure 12.26. Midwinter counts of Mottled Duck from the Texas coast for two different regions (1997–2011, after Texas Parks and Wildlife Department [TPWD] 2011). © J. Burger.

The Mottled Duck prefers to feed in shallow, ephemeral wetlands (Swanson and Meyer 1977), which means they have a narrow range of habitat requirements for nesting. They feed on submerged vegetation in delta and saltwater marshes, on invasive species such as Eurasian Watermilfoil (Goecker et al. 2006), and snails and insects (Imhof 1962). They also feed on vegetation and animal matter (mollusks and crustaceans, insects) in Florida (Stevenson and

Anderson 1994). In Mississippi, more than half of their diet is animal matter (insects, snail, fish, crustaceans), as well as rice, bulrushes, pondweeds, and other aquatic vegetation (Turcotte and Watts 1999). Their dependence on estuarine habitats requires further study, as Moorman et al. (1991) found that Mottled Duck ducklings died if salinity was much greater than 12 parts per thousand (ppt), and that the tolerance may be closer to 9 ppt. This suggests that management to enhance Mottled Duck populations should take salinity into consideration when creating impoundments. Mottled Ducks breed and winter at very low densities in the fringe of the Gulf of Mexico in coastal Alabama and Mississippi, and reach their highest densities in coastal Louisiana and southeast Texas, with smaller numbers south to Veracruz (Bielefeld et al. 2010). They nest in estuarine marshes, although they have been reported nesting in farm fields (Eubanks et al. 2006). Nests and eggs are vulnerable to a wide range of predators, including raccoons, skunks, opossums, dogs, and snake; turtles and alligators (*Alligator mississippiensis*) prey on ducklings (Stevenson and Anderson 1994), and in Texas, predators also include River Otters (*Lutra canadensis*), Striped Skunks (*Mephitis mephitis*), mink (*Mustela visor*), coyotes (*Canis latrans*), Red Fox (*Vulpes vulpes*), Gray Fox (*Urocyon cinereoargenteus*), and feral dogs (*Canis familiaris*), cats (*Felis domesticus*), and Snapping Turtles (*Chelydra serpentina*) (Bielefeld et al. 2010).

Half of the Mottled Duck populations in the United States reside in Louisiana (Lindstedt 2005). Estimated Mottled Duck populations in southeastern Louisiana increased until 1994 (peak of more than 100,000), and then declined to about 18,000 by 2001. Although these populations are currently stable, estimates project further declines with loss of habitat (Lindstedt 2005). Most Gulf Coast states have designated this species of "conservation concern" (Bielefeld et al. 2010), largely because of marsh degradation and drainage. Large-scale efforts to restore hydrology of coastal marshes, as well as construction of smaller impoundments, would benefit the species (Moorman et al. 1991; Wilson 2007). Currently, the breeding population in Florida is estimated to be about 40,000 individuals, and the western Gulf Coast population is estimated at 600,000 birds (Johnson 2009), although estimates differ. For example, Delany and Scott (2006) estimated 35,000 individuals in Florida, and only 135,000 individuals in the western Gulf (Alabama to Mexico). The Florida population is stable, while the status of the western Gulf population is unclear. Breeding surveys from National Wildlife Refuges in Texas suggest a precipitous decline since 1985, when the surveys began (Johnson 2009), although breeding bird surveys show a moderate decline, and Christmas Bird counts show a decline in the same period. They increased before the 1990s (see Figure 12.27 below).

The Texas Parks and Wildlife Department (TPWD 2011) conducts annual waterfowl surveys of the central Gulf Coast Prairies and Marshes, and of the southern Coastal Sand Plains. These data indicate trends for one of the important Mottled Duck breeding areas (Figure 12.28). Mottled Ducks were much less abundant in the southern marshes of Texas, compared to the central areas.

Breeding Bird Survey data are useful for Mottled Duck because they only occur along the coasts, and although the number of routes is small, it still provides an index of numbers. Mottled Ducks showed a sharp decline in Texas, and a decline in Louisiana, although they remained stable in Florida (Figure 12.27). Trends in waterfowl populations for Mexico indicated significant long-term declines for some species, but no significant long-term trends for Mottled Duck (Figure 12.27) (Perez-Arteaga and Gaston 2004). They surveyed only two places along the Gulf Coast of Mexico, but this represented 91 % of the Mexican population. The average count last year was 49 % below the mean, and declines since the mid-1980s bear watching (Figure 12.27).

Using Christmas Bird Counts from 1965 to 2011 as a database, Niven and Butcher (2011) reported that Mottled Duck winter populations increased significantly by 1.2 % per year along

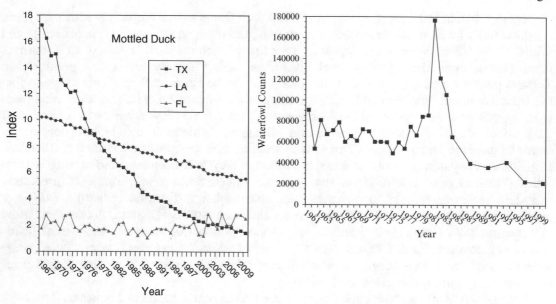

Figure 12.27. Index of Mottled Duck surveyed during the Annual Breeding Bird Surveys (shown on *left*), conducted by the U.S. Fish & Wildlife Service (Bird Banding Laboratory), with trends along the southern Gulf of Mexico coast (on *right*). Mexico counts after Perez-Arteaga and Gaston (2004). © J. Burger.

the U.S. Gulf Coast (Figure 12.28). The conflicting data (Breeding Bird Surveys vs. Christmas Counts) is troubling and bears further examination, especially given the declines at National Wildlife Refuges.

Resiliency is unclear, since the age of first breeding is unknown (perhaps at 1 year); they lay clutches of 8–12 eggs, and breeding success is unknown (Bielefeld et al. 2010). While intrinsic resiliency may be intermediate to high (based on age of breeding and clutch size), massive loss of habitat gives them few options for movement because they are an obligate estuarine species.

12.6.2.7 Osprey

These dramatic white-headed, eagle-sized hawks are familiar to coastal people throughout the United States (Figure 12.29). They are the only North American raptor to feed entirely on fish; they have strong, sharp claws and are also called "fish hawks." Ospreys are one of the most widespread species in the world; they breed or winter on all continents except Antarctica (Farmer et al. 2008). Their main breeding range in North America extends north to central Alaska, northwest Yukon, Northwest Territories, Saskatchewan, Manitoba, and Ontario, to the tree line of Newfoundland (Poole et al. 2002). They dip down into the western, central, and eastern United States, and breed along the Atlantic coast down to almost the tip of Florida, as well as sporadically along the northern Gulf Coast (Eubanks et al. 2006). A nonmigratory race breeds in Cuba (Raffaele et al. 1998). On the west coast they breed down the coast from Alaska to northern California, on Baja California coasts, and on the western Mexican mainland (Poole et al. 2002). The bulk of the North American population winters south of the United States in Central and South America (Poole et al. 2002).

They dive feet first to capture their prey from the top meter of water; this restricts them to surface-schooling fish and to shallow water. Where there is shallow water and abundant prey, they nest more densely and are considered semicolonial (Poole et al. 2002). They frequent large

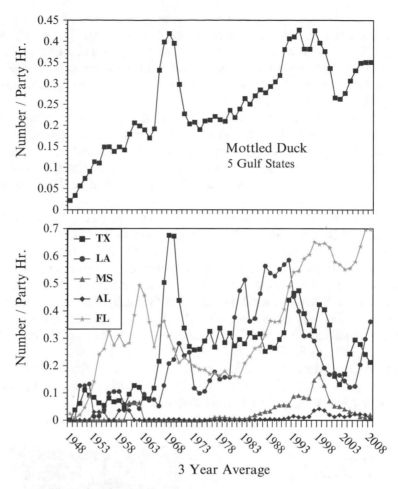

Figure 12.28. Trends data from Christmas Bird Counts to show variation in Mottled Duck counts for different states. © J. Burger.

Figure 12.29. Osprey populations declined dramatically because of exposure to DDT, and in many places, artificial nest structures and egg replacement were used to restore populations. *Right* photo shows three osprey chicks in their nest. © J. Burger.

lakes, rivers, and coastal areas where manmade structures provide perches and nesting sites. Most breed at 4 years of age. They are monogamous, lay a clutch of 1–4 eggs in April, only the female incubates (35 days), and they fledge their young in July and August. Fledging success varies greatly from region to region, perhaps as a result of prey availability. Ospreys make long migrations to Central and South America, although some remain in Florida and Mexico, and more recently, in the other northern Gulf States.

Osprey, like many other fish-eating birds, suffered population declines from the 1950s through the 1970s due to pesticides (DDT) and other contaminants (Poole et al. 2002). The chemicals resulted in eggshell thinning, depressed reproduction, and population declines. The percent of eggshell thinning was directly related to levels of dichlorodiphenyldichloroethylene (DDE), a metabolic breakdown product of DDT. Since the ban on DDT, populations have rebounded.

In 1983, there were about 8,000 breeding pairs in the United States (Henny 1983) and 16,000 to 19,000 pairs in 2001 (Poole et al. 2002). The population in the United States and Canada is about half of the world population of approximately 100,000 birds (Farmer et al. 2008). Migration counts and Breeding Bird Surveys indicate that populations have increased or remained stable; east and midwestern North America had greater increases than in the Great Lakes or western North America, based on counts in the Gulf of Mexico (Farmer et al. 2008). Surveys of Osprey at four locations around the Gulf of Mexico from 1995 to 2005 indicated that they increased significantly in the Florida Keys, in South Point and Corpus Christi, Texas, but with no significant trend in Veracruz, Mexico (Smith et al. 2008).

Christmas Bird Count data shows a clear increase in Osprey in the Gulf States, although the greatest increase was in Florida (Figure 12.30). The increase in Florida was rather steady, but increases in the other states began in the early 1980s.

Figure 12.30. Christmas bird count data for Osprey, showing steady increases. © J. Burger.

Resiliency in Osprey is low to intermediate; they breed at 4 years of age, and lay an average of three eggs, but most Osprey do not live beyond 12 years. Since they eat fairly large fish, they are vulnerable to effects from mercury, as well as to pesticides and other contaminants. They forage in shallow water and are more affected by fish abundance and behavior than other species that can fish at a greater water depth. While they adapt to the presence of people, they can suffer disturbances from approaching boats.

12.6.2.8   Whooping Crane

Whooping Cranes, the tallest North American birds (1.5 m), are snowy white with black primaries, and a brilliant carmine crown (Figure 12.31). They are a symbol of international efforts to save an endangered species, and were brought back from the brink of extinction with collaboration among Canadian and U.S. provincial and state agencies (USFWS 1986; Lewis 1995). In 1941, only 15–16 individuals wintered in Texas. They were placed on the Endangered Species List in 1967 and are one of the rarest birds in North America. Whooping Cranes are omnivorous, feeding on insects, frogs, rodents, small birds, minnows, and berries in the summer, and estuarine animal foods (blue crabs and clams) in the winter (USFWS 2010b). On their wintering grounds, in the area between the Rio Grande and Galveston Bay, Whooping Cranes feed in three habitats: (1) estuarine intertidal unconsolidated shore mud, (2) palustrine emergent persistent wetlands, and (3) estuarine intertidal emergent persistent wetlands (Anderson et al. 1996).

Whooping Cranes were once widespread, although not common in the prairie marshes of north-central United States and southern Canada, and they were a common winter resident along the coast, even in Louisiana (Lowery 1974). They are monogamous, mate for life, and nest in bulrushes (USFWS 2010b). The incubation period is 29–31 days, and chicks fledge in 80–90 days (Lewis 1995). They remain in family groups following fledging, and the young learn the

Figure 12.31. Whooping Cranes are pure *white* with a *red* crown (USFWS 2012c).

Table 12.8. World Population of Whooping Cranes in the Wild and in Captivity (after WCCA 2012)

Location	Adult	Young	Total	Adult Pairs
Aransas/Wood Buffalo	235	44	279	78
Florida non-migratory	20		20	8
Louisiana non-migratory	0	10	10	0
Wisconsin/Florida (migratory)	88	17	105	17
TOTAL BIRDS IN THE WILD	343	71	414	103
Patuxent WRC, Maryland	68	5	73	15
Crane Foundation, Wisconsin	34	1	35	11
Other Zoos	48	1	50	8
TOTAL BIRDS IN CAPTIVITY	150	7	157	34

migration route from their parents. On migration they face weather-related problems, shooting, and collisions with wires and fences, and on the wintering grounds they face shooting, disease, and predation (CWS/USFWS 2007; Gil-Weir et al. 2012). They fly some 2,600 mile (over 4,000 km) between their breeding grounds and Texas.

The Whooping Crane population from 1860 to 1870 was estimated at 1,300–1,400 (Allen 1952), although others estimated it at only 500–700 birds (Lewis 1995). In 1937, only two breeding populations remained: a sedentary population in southwestern Louisiana and the Wood Buffalo population that migrated to Aransas National Wildlife Refuge (Lewis 1995). A hurricane in 1940 reduced the Louisiana population from 12 to 6, and the last individual was taken into captivity in 1950.

There is currently only one truly wild population of Whooping Cranes; they breed in Wood Buffalo National Park in Northwest Territories and adjacent Alberta, and winter at Aransas National Wildlife Refuge (Lewis 1995; USFWS 2010b; WCCA 2012). There are, however, other Whooping Cranes at four places: (1) Grays Lake National Wildlife Refuge in Idaho, (2) Kissimmee Prairie in Florida, (3) an eastern migratory population, and (4) captive populations (USFWS 2010b, 2012c; WCCA 2012). In 2012, the total world population of Whooping Cranes was 571, of which 73 % were in the wild; the remainder were in research facilities or zoos (Table 12.8) (WCCA 2012).

The Aransas/Wood Buffalo National Park Whooping Crane population has continued to grow steadily (Figure 12.32). Resiliency in Whooping Crane is low as they start breeding when they are 4 years old, lay only two eggs, and have an estimated life span of 22–24 years (USFWS 2012c, d), although some scientists believe it is as high as 30 (Lewis 1995).

12.6.2.9   Clapper Rail

Clapper Rails are large (the size of a half-grown chicken) with gray-cinnamon buff tails and a long decurved bill. They are emblematic of salt marshes and mangrove swamps, although they are more often heard rather than seen as they skulk through the marsh (Figure 12.33). Their breeding range is from the northern United States to Brazil on the coastal fringes. They range from Massachusetts south to the Florida Keys, on the Gulf Coast from Cape Sable (Florida) west to Tamaulipas, Mexico, and from San Francisco to Baja California, as well as in some inland areas (such as the Salton Sea). Their ranges in Mexico, Central America, and South America are poorly known, but they are reported from many areas (Eddleman

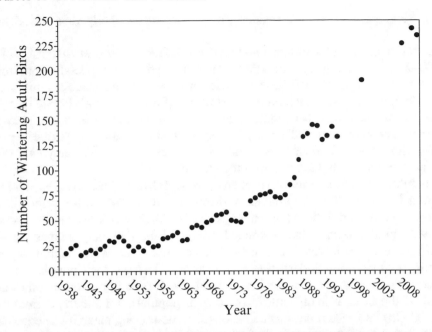

Figure 12.32. Population growth of the Arkansas National Wildlife Refuge adult population of Whooping Cranes that breed in Buffalo Wood National Park in northern Canada (after Lewis 1995; WCCA 2012). Birds are surveyed in the winter in Texas. © J. Burger.

Figure 12.33. Data on Clapper Rail from the Breeding Bird Survey, showing relative increase in Louisiana. © J. Burger. (Photo by USFWS 2012e).

et al. 1998). Most populations are resident, but the more northerly rails move south in the winter.

Clapper Rails are strictly estuarine—they very infrequently nest in freshwater marshes (Olson 1997), and are thus good indicators of marsh conditions, especially contaminants, oil, and habitat loss (Novak et al. 2011). They are a solitary species, and space out within marshes or mangroves. Clapper Rails mainly feed on crustaceans (Eddleman et al. 1998). They build nests in thick vegetation on the higher places, usually where vegetation is taller, providing some protection from aerial predators. They lay 7–9 eggs; most nest failures are due to predation on eggs, and flooding, which is likely to increase with sea level rise. Although young rails feed on their own, they remain with parents until they fledge.

Future habitat changes that result from environmental stress, sea level rise, and subsidence will result in a landward increase in salinity (McKee et al. 2004), with changes in vegetation types and nesting bird populations (Greenberg et al. 2006). Since Clapper Rails nest and forage in habitats with higher salinity than some salt marsh birds, they may increase in Gulf Coast marshes with increased salt water intrusions (Rush et al. 2009a). While habitat loss is the most critical factor, tidal flooding and hunting pressures also decrease their populations (Stevenson and Anderson 1994). Because Clapper Rails breed and forage in coastal marshes and remain hidden most of the time, it is difficult to track their population numbers, although high tides (Rush et al. 2009b) and recent advances in acoustical monitoring make it easier to count them. There are no estimates for the number of Clapper Rails in Gulf Coast marshes, although they are counted on Breeding Bird Surveys and on Christmas Bird Counts. Breeding Bird Surveys indicate that they are declining in all states, except Louisiana, with lower declines in Florida than elsewhere (Figure 12.33).

Using Christmas Bird Counts from 1965 to 2011 as a database, Niven and Butcher (2011) reported no significant trend in Clapper Rail wintering populations along the U.S. Gulf Coast. However, when these data are examined in detail, there appears to be a decline in the northern Gulf from the mid-1970s to the present (Figure 12.34).

Resiliency may be intermediate to high, depending upon the availability of a source population. If a population is extirpated from a region because of habitat loss or degradation, excess available breeders from nearby areas would be necessary to reestablish the population. They likely breed at 1 year, have average clutch sizes of more than nine eggs in the Gulf Coast, and have variable hatching success, depending upon flood losses, that can average 85 % (Eddleman et al. 1998).

12.6.2.10 Snowy Plover

Snowy Plovers are small, snowy shorebirds that blend in with their sandy habitat until they move (Figure 12.35). They were formerly considered conspecific with Kentish Plover (*Charadrius alexandrines*) in Eurasia, but are now separated as a distinct species (*C. nivosus*) (Page et al. 2009). Two subspecies of Snowy Plover breed in North America, one that nests west of the Rocky Mountains, and one that breeds east of the Rockies, primarily on the Gulf Coast (Paton 1994; Elliott-Smith et al. 2004). The west coast subspecies is already listed as threatened on the U.S. Endangered Species List, and the southeastern birds are currently being considered for listing (Brown et al. 2001). Both populations of Snowy Plover are listed as declining by the U.S. Shorebird Conservation plan (USSCP 2004). Snowy Plover has been given the highest conservation priority by the Gulf Coast Prairie Working Group (GCPWG 2000). Recently, Thomas et al. (2012) estimated the total breeding population of Snowy Plovers in North America as 23,555 (95 % confidence limits = 17,999–29,859), and noted that they may be one of the rarest of shorebirds in North America.

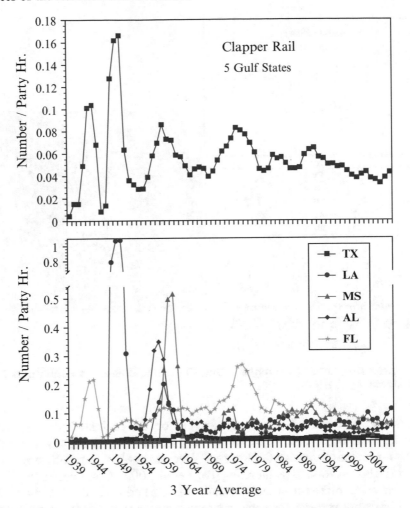

Figure 12.34. Population trends in Clapper Rail as determined by Christmas Bird Counts. Trends seem fairly constant, although the possible downturn since the 1970s bears examination. © **J. Burger.**

In the Gulf Coast, Snowy Plover are distributed sparsely along the southwest coast of Florida north to Anclote Key and along the Panhandle, in Alabama and Mississippi (mainly on offshore islands), in Louisiana—only two pair in 2001 (Zdravkovic 2005), along the lower Texas coast from Matagorda Island to the Mexican border, and south to Veracruz and the northern coast of the Yucatán Peninsula (Page et al. 2009), where they have been observed in flocks of 100 (Ornat et al. 1989). Results from a recent survey of breeding Snowy Plovers in North America, conducted in 2007 and 2008, indicated that 42 % of all breeding Plovers resided in Great Salt Lake in Utah and Salt Plains National Wildlife Refuge in Oklahoma (Thomas et al. 2012). The total population for the coast of the Gulf of Mexico was 4,515 (19 % of total). Approximately 9 % of the North American breeding population occurs in Mexico (Thomas et al. 2012).

Snowy Plover are attracted to extensive beaches with tidal pools and sand flats that provide foraging areas, although they will also feed in marshes (Withers 2002). They eat small mollusks, crustaceans, marine worms, and insects. In winter, Snowy Plovers often associate with Piping Plovers (Elliott-Smith et al. 2009) and Wilson's Plovers (*Charadrius wilsonia*) (Howell and

Figure 12.35. Three year running averages of Snowy Plovers counted on Christmas Counts for the five U.S. Gulf states. © J. Burger.

Webb 1995). Snowy Plovers nest on sandy beaches on the mainland and on barrier islands, where they make nest cups on the sand, sometimes in colonies with Least Terns (Stevenson and Anderson 1994). Incubation period is 28 days (Warriner et al. 1986); both sexes incubate, and females desert the brood soon after hatching (Page et al. 2009). Counting Snowy Plovers during the breeding season is difficult because studies of marked individuals indicate that at least twice as many birds are present for each one seen, and detection probability is 0.58 (Warriner et al. 1986; Hood and Dinsmore 2007).

Populations of Snowy Plover are likely lower in the Gulf Coast than they were in the late 1880s due to habitat loss and disturbance (Page et al. 2009). Hood and Dinsmore (2007) identified the Laguna Madre (Texas) as an important breeding area, suggesting that it be protected from development since the species is reported to be declining in the Gulf. Coordination with Tamaulipas, Mexico is critical for protection of this species; wintering Snowy Plovers use the algal mudflats there, illustrating the importance of the entire Laguna ecosystems (Mabee et al. 2001). Morrison et al. (2006) reported that the Gulf of Mexico and Caribbean race of Snowy Plover is decreasing, and likely was only 1,500 birds, although more were counted on the next census (Table 12.9) (Elliott-Smith et al. 2009).

Computing 3-year averages using Christmas Bird Count data from 1942 to 2010 indicates variability in the numbers of Snowy Plovers counted (Figure 12.35). Although there appears to be an overall increase, they have declined since 1995 (Niven and Butcher 2011). It is hard to interpret the two peaks in the early 1950s and early 1960s.

Resiliency in Snowy Plover is intermediate as they breed when 1 year old, lay a mean clutch of three eggs, and hatching success varies greatly from 12.5% to 87 % (often depending on the degree of human protection), but they can have multiple broods per season (Page et al. 2009). Paton (1994) estimated a mean adult survival of 2.7 years. Vulnerability of nests and chicks,

Table 12.9. Snowy Plover Surveyed in the Gulf of Mexico During the Winter

States	Number in 2001	Percent in 2001	Number in 2006	Percent in 2006
Texas	690	66	1,340	71
Louisiana	36	3	207	11
Mississippi	13	1	36	2
Alabama	0	0	6	<1
Florida	311	30	312	16
Total	1,050		1,895	

An additional 119 Snowy Plover were counted in 2006 in Tamaulipas, Mexico in an 89 km (55 mi) habitat survey (Elliott-Smith et al. 2009), © J. Burger

habitat losses, human disturbance, and mammalian predators all contribute to lowered success, particularly in the Gulf.

12.6.2.11 Piping Plover

Piping Plovers are a threatened and endangered shorebird that lives on open beaches, alkali flats, and sand flats. They have a distinctive dark and white pattern of bands on the head, neck, and upper breast (Figure 12.36). The black and white breaks up the outline of birds, allowing them to disappear when motionless. They were listed in 1985, and recovery plans for several regions have been developed (USFWS 1999). They are endemic to North America, with a total population of about 8,000 (Gratto-Trevor and Abbott 2011). The Piping Plover has been given the highest conservation priority by the U.S. Shorebird Conservation Plan, including for the Gulf Coast where they only winter (GCPWG 2000).

They breed on the Atlantic coast of Canada and the U.S. Great Lakes region, Great Plains, the Canadian Prairies, and St. Pierre and Miquelon (French territories off the southwest coast of Newfoundland) (USFWS 1999; Elliott-Smith et al. 2004). They do not breed in the Gulf of Mexico. They winter along the Atlantic and Gulf coasts (into Mexico), and in the Caribbean (Elliott-Smith et al. 2004; Gratto-Trevor and Abbott 2011). Piping Plovers in Texas show wintering site fidelity (Drake et al. 2001).

Piping Plovers forage on beaches, washover areas, and tidal flats on small invertebrates (Withers 2002). They frequently nest on sandy beaches with little vegetation, but with some shell or pebble cover; often near dunes (Wilcox 1959; Burger 1987b; Maslo et al. 2011). They are monogamous for one breeding season, although there are rare reports of sequential polyandry (Amirault et al. 2004). Both members of the pair incubate their four-egg clutch for 26–31 days (USFWS 1999), and both parents brood the young (Gratto-Trevor et al. 2010). The parents draw predators from their nests and chicks with very elaborate distraction displays, which involve a broken wing act (Figure 12.36).

They are dependent on management, such as restrictive access for off-road vehicles (Burger 1991b, 1994b; USFWS 1999; Maslo and Lockwood 2009). There is controversy about the effectiveness of nest protection techniques. Nest protection includes nest enclosures (predator exclosures), electrified wires on enclosures, fencing (mainly for people), wardening, predator control, and captive breeding (Burger 1987b; Murphy et al. 2003; White and McMaster 2005; Cohen et al. 2008, 2009; Maslo and Lockwood 2009). Beach nourishment increases habitat for Piping Plover (Webster 2006). Management has largely focused on reducing mortality during the breeding cycle, and it is unclear whether the sustained efforts required can be maintained (Gratto-Trevor and Abbott 2011).

Figure 12.36. Piping Plover depend upon being cryptic to avoid predators, but when faced with a predator, they begin a distraction display that ends with a full broken wing act (*right*). © J. Burger.

Table 12.10. Trends in Wintering Populations of Piping Plover Surveyed Along the Northern Gulf Coast (after Haig et al. 2005[a]; Elliott-Smith et al. 2009)

State	1991	1996	2001	2006
Florida (Gulf)	481	320	305	321
Alabama	12	31	30	29
Mississippi	59	27	18	78
Louisiana	750	420	511	226
Texas	1,904	1,333	1,042	2,090
Totals for the Gulf States	**3,206**	**2,131**	**1,906**	**2,744**

Given are number of adults recorded. In 2006, 76 Piping Plover were recorded in 89 km (55 mile) of habitat in Tamaulipas, Mexico (censuses did not cover the remainder of the southern Gulf (Elliott-Smith et al. 2009))
[a]Table 2 of Haig et al. (2005) lists 44 for the Gulf states, but this must be a typographical error, since they report only 31 from the Atlantic coast and 44 from the Gulf Coast (with a total count of 375)

As a federally endangered species, Piping Plovers are useful indicators because their populations are closely monitored, and their wintering in the Gulf Coast makes them vulnerable to coastal stressors. Populations declined during shorebird harvesting for the millinery trade in the late 1880s and early 1900s (USFWS 1999). In the latter half of the twentieth century, however, Piping Plover populations declined because of habitat loss and alterations, human disturbance, and increased nest predation (Sidle 1984). The breeding population estimate of 5,945 plovers is probably an underestimate (Elliott-Smith et al. 2009). The U.S. Fish and Wildlife Service identified human disturbance and habitat loss as the two greatest threats to Piping Plovers on the wintering ground (USFWS 2009b). Other threats during the fall and winter include hurricanes, oil spills, and red tides. Gratto-Trevor and Abbott (2011) provide the best and most comprehensive review of conservation efforts for Piping Plover.

A complete winter survey of Piping Plovers was conducted in 1991, 1996, 2001, and 2006 (Haig and Plissner 1993; Haig et al. 2005; Elliott-Smith et al. 2009). In 2001, they reported 2,389 piping plover during the winter, but 5,945 adults were counted during the breeding season (Haig et al. 2005). In 2006, the International Piping Plover Census covered more than 12,400 km of potential habitat in 2 weeks (2,470 sites, 1,300 observers) (Elliott-Smith et al. 2009; Gratto-Trevor and Abbott 2011). More Piping Plovers were recorded in 2006 in the northern Gulf than in either 1996 or 2001, but numbers were lower than in 1991. In all years, Texas had most of the Piping Plovers (Table 12.10).

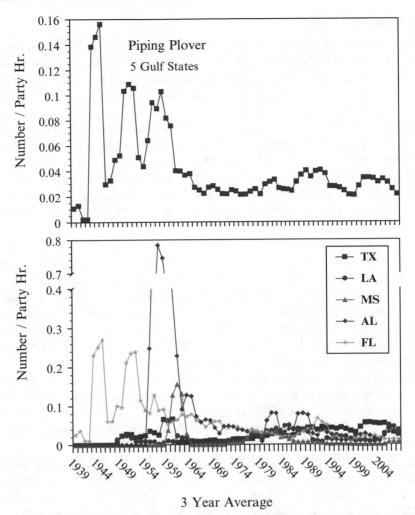

Figure 12.37. Christmas Bird Count data for Piping Plover, showing early variability, followed by much less variability in wintering plover along the Gulf Coast. © **J. Burger.**

It is also instructive to examine the Christmas Bird Count data for yearly variations (Figure 12.37). Despite recovery efforts, the number of wintering Piping Plovers counted along the Gulf Coast has remained relatively constant, and appears to have declined since the late 1990s.

Resiliency of Piping Plovers is apparently relatively low, as indicated by the great effort by the USFWS Recovery Team and state agencies to protect nesting and foraging habitats, with relatively modest success (USFWS 2003, 2009b). They breed the first spring after hatching, the number of chicks fledged per year varies from 0.3 to 2.5 per nest, and average life span is 5 years (Wilcox 1959; USFWS 1999; Elliott-Smith and Haig 2004; Gratto-Trevor et al. 2010). However, since they nest on sandy beaches exposed to human disturbance, predators, high nest failures, and habitat loss, resiliency of breeding populations is low, explaining the recent decreases in wintering birds along the Gulf of Mexico.

12.6.2.12 Laughing Gull

Laughing Gulls are small, dainty, black-hooded gulls that careen low over beaches and mudflats, or soar high in the air hawking insects (Figure 12.38). They nest in colonies of up to 25,000 pairs on sandy or rocky shores, and in salt marshes along the Atlantic and Gulf coasts of North America, as well as on some Caribbean islands, the Gulf of California, and Pacific coast of Mexico (Burger 1996a). Although they nest on the Chandeleur Islands of Louisiana, they do not nest in Mississippi (Turcotte and Watts 1999). They also nest on the islands in the Campeche Bank of the southern Gulf (Tunnell and Chapman 2000). They winter from North Carolina south through the remainder of the breeding range, along the west coast of Baja California, and along the Pacific coast from Colima, Mexico, south to Peru, the Galapagos, and Chile (Burger 1996a). They are particularly common in winter in the southern Gulf of Mexico (Howell and Webb 1995). Formerly *Larus atricilla*, Laughing Gulls are now listed as *Leucophaeus atricilla* in the latest AOU checklist and supplements (AOU 1998).

Laughing Gull populations were devastated by market hunting for plumes and eggs in the late 1880s and early 1900s (Sprunt 1954). They expanded their numbers thereafter in the northeast, but suffered competition with the larger Herring Gulls that did not breed in the region until the late 1940s (Burger 1983). Laughing Gulls did not face such competition from an expanding exotic species in the Gulf of Mexico, perhaps accounting for their large populations there. They are good indicators because they are common along the Gulf Coast, are nesting and foraging generalists, and are an integral component of nesting colonies and foraging assemblages both along the coast and in the Gulf.

Laughing Gulls are generalist foragers, eating fish, insects, other invertebrates, and garbage. They follow shrimp boats in Texas (Eubanks et al. 2006) and fishing boats off Mississippi (Turcotte and Watts 1999) in search of scraps. They dive for fish, follow boats that stir up prey, "hawk" for flying insects, and catch food thrown by beach-goers or fishermen (Stevenson and Anderson 1994; Burger and Gochfeld 1983c).

Laughing Gulls are one of the most common breeding birds along the Gulf Coast, and they are residents, migrants, and winter visitors. They do not nest in Alabama (Imhof 1962), but they do on the nearby Chandeleur Islands of Louisiana (Imhof 1962), where they are mainly coastal and rarely move inland (Lowery 1974). Along the Gulf they nest on sand and in marshes (Imhof 1962; Schreiber and Schreiber 1979; Burger 1996a, b). Most Laughing Gull colonies in coastal Louisiana are in marshes (80 %), and the remaining 20 % are on sand or in shrubs (Greer et al. 1988). They are monogamous, lay three eggs, and both sexes incubate and

Figure 12.38. Laughing Gulls often nest in marshes (*left*) and are opportunistic foragers on fish or invertebrates along the shore. © J. Burger.

Table 12.11. Changes in Abundance of Breeding Colonial Waterbirds in Galveston Bay, Texas (after Glass and Roach 1997)

Year	All Species	Laughing Gull	Black Skimmer
1973	54,645	35,860 (66 %)	1,873 (3 %)
1979	50,160	32,070 (64 %)	2,472 (5 %)
1985	39,008	22,698 (58 %)	1,101 (3 %)
1990	46,813	17,608 (38 %)	1,900 (4 %)
1996	48,126	19,052 (40 %)	1,582 (3 %)

Given are numbers of nesting pairs (with percent of total birds)

care for the young (Burger 1996a). Yearly differences in reproductive success are due to flooding, predators, and human disturbance (Burger 1996a). Nesting is a compromise between nesting on islands that are high enough to avoid flooding, while being low enough to avoid mammalian predators, and competition with other ground-nesting colonial species (Burger 1983).

A significant proportion of the North American breeding population is located in the Gulf. Many Laughing Gulls that breed along the Gulf Coast remain there, wandering along the coast in the winter, while gulls from farther north also winter along the Gulf Coast. Local surveys provide an indication of trends information. For example, periodic surveys in Galveston Bay indicate that Laughing Gulls make up a significant portion of the colonial nesting birds (Table 12.11) (Glass and Roach 1997). From 1973 to 1996 nesting pairs of Laughing Gulls declined. In 2001, there were 22,000 breeding pairs in the Galveston Bay survey (Eubanks et al. 2006). The downward trend has continued; the Galveston Bay Status and Trends Project rated Laughing Gulls as significantly decreasing (GBEP 2006).

Similarly, information from Louisiana shows declines in the number of breeding Laughing Gulls (Figure 12.39) (Visser and Peterson 1994).

There was an overall increase in Laughing Gulls on Shamrock Island in Texas from 1973 to 1999 (Figure 12.40). The trend continued until about 2007 and then declined (TCWS 2012). Information from Tampa Bay shows that the number of breeding pairs was stable until about 2008, and then they increased (Hodgson and Paul 2010). Data from Breeding Bird Surveys show a different pattern, with Laughing Gulls increasing in Texas and Alabama (Figure 12.41). This bears further examination and suggests that population trends need to be followed for a region, rather than for one colony. Furthermore, the counts are from shore and not from boats near breeding islands, which may suggest that Laughing Gulls are now foraging closer to shore.

Using Christmas Bird Counts from 1965 to 2011 as a database, Niven and Butcher (2011) reported that wintering Laughing Gulls showed a significant increase of 3.0 % per year in the Gulf region. This may reflect an increase in wintering Laughing Gulls, rather than from local residents. Using Christmas Bird Count data from 1940 to 2005, 3-year running averages were computed, showing a decline from the 1990s to the present (Figure 12.41). The 1-year increase (mainly Texas) in the last year may not reflect a real increase. These two conflicting analyses indicate the difficulty of taking different time periods to examine trends. It appears that the 40-year trend may be increasing, but the trends over the last 15 years are decreasing for both breeding populations and wintering populations along the Gulf.

Resiliency in Laughing Gulls is intermediate to high because they can breed when 3 years old, lay three eggs, have high hatching success (74–81 % in Florida) (Schreiber et al. 1979), fledge up to 1.32 chicks per nest, and live up to 19 years (Burger 1996a). Recovery potential is

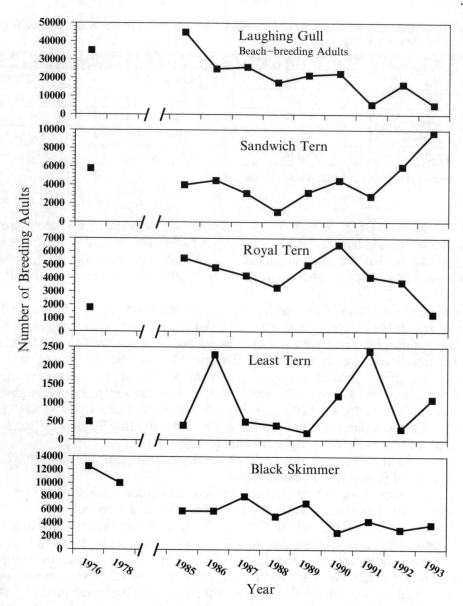

Figure 12.39. Number of breeding adults for several species nesting in Louisiana (after Visser and Peterson 1994). Photo (following page) shows Laughing Gulls at a breeding colony in Texas (photo by J. Burger). © J. Burger.

high, but in the Gulf this must be balanced against losses due to habitat loss, predators, tidal flooding, and hurricanes (Burger 1978b).

12.6.2.13 Royal Tern

Royal Terns are large, stocky terns with bright orange bills and a black crest when they start breeding. They breed primarily along the Atlantic coast from Virginia south to Florida, and along the Gulf Coast to Texas, throughout the Caribbean, along the Pacific coast of Mexico, and on the Atlantic coast of South America (Buckley and Buckley 2002). They are common

Figure 12.39. (continued)

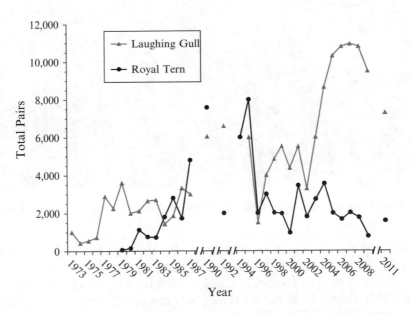

Figure 12.40. Data on pairs of Laughing Gulls and Royal Terns from 1973 to 1999 from Shamrock Island, Texas (after Gorman and Smith 2001; TCWS 2012). © J. Burger.

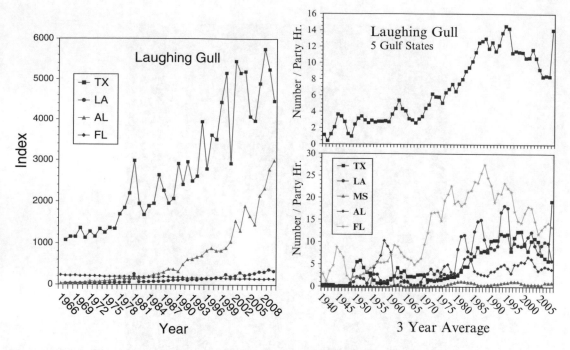

Figure 12.41. Breeding Bird Survey data for Laughing Gull showing increases in Texas and Alabama (*left*), compared to Audubon Christmas Count Data (*right*). © J. Burger.

Figure 12.42. Royal Terns have an orange bill; Caspian Terns have a red bill. © J. Burger.

breeding birds along the Gulf, nest on islands off Veracruz (Howell and Webb 1995), in the Campeche Banks (Tunnell and Chapman 2000), and in the West Indies, northern South America, and on islands in the Caribbean (Buckley and Buckley 2002). They are rarely found inland (Sprunt 1954). Formerly *Sterna maximus*, they are now listed as *Thalasseus maximus* in the latest AOU supplements (Figure 12.42).

Royal Terns are of interest because they nest on the sand in large colonies, have high colony turnover rates, and the Royal Terns along the Gulf represent about 40 % of the U.S. breeding

Figure 12.43. Royal Terns nest in dense colonies on sand. Photo by M. Gochfeld (with permission).

population (Visser and Peterson 1994; Lindstedt 2005). Abandonment of colonies is indicative of habitat quality problems that need to be addressed in the Gulf States. Royal Terns forage on small prey fish, which they capture by diving, but they also eat squid, shrimp, and crabs in Florida (Stevenson and Anderson 1994).

Royal Terns nest in dense groups, either in monospecific colonies, or with other terns (e.g., Caspian, Sandwich), Laughing Gulls, and Black Skimmers (Figure 12.43). Royal Terns nest a mere body length apart. When they nest with other species, they still nest in dense groups, which may be within or adjacent to more spaced-out gulls and terns. Royal Terns lay one egg, both incubate, and both provision the young. There is usually synchrony in egg-laying within subcolonies, which results in synchronous hatching. Parents brood very young chicks, but chicks quickly join a crèche (young birds that stay in a close-knit group), which protects them when parents are away foraging (Buckley and Buckley 2002). Buckley and Buckley (2002) estimated the number of breeding pairs for the Gulf region as follows: 1,000 in Florida (both coasts), 250 in Mississippi, 10,590 in Louisiana, and 22,463 in Texas. Breeding populations in Louisiana declined in the early 1990s (Figure 12.39) (Visser and Peterson 1994). They attributed the variability in nesting numbers to the vulnerability of their nesting habitat to storms and high tides. Delany and Scott (2006) estimated the number of Royal Terns in the United States to be 139,000. Trends data from Shamrock Island in Texas (Figure 12.40) indicates an overall increase in the number of breeding Royal Terns until the mid-1990s (Gorman and Smith 2001), and then they declined (TCWS 2012). Breeding Bird Survey data indicate a great increase in Royal Terns in Louisiana, but a slight decline in Alabama and Texas (Figure 12.44).

Analysis of Christmas Bird Count data does not indicate a significant trend (Niven and Butcher 2011), although an examination of data from Florida seems to indicate a decline, as well as a decline in the Gulf Coast overall (Figure 12.45).

Resiliency is intermediate because they do not breed until the age of 5–6 years and lay one egg, and Royal Terns can live up to 28 years, but most live fewer years (Buckley and Buckley 2002). There are few data on reproductive success because the young form crèches, making it difficult to follow families. Royal Terns have extended parental care up to the second year (Buckley and Buckley 1974), placing an additional stress on parents.

Figure 12.44. Breeding bird survey data for Royal Terns in the Gulf States. © J. Burger.

12.6.2.14 Black Skimmer

Black Skimmers, which are about the size of a Royal Tern, are familiar and striking as they fly silently just above the water with their bill dipped in—skimming (Figure 12.46). They breed from Massachusetts south along the Atlantic and Gulf coasts to southern Mexico, on islands in the Caribbean, and from southern California (and inland Salton Sea) to Nayarit, Mexico (Gochfeld and Burger 1994). A significant proportion of the world population of Black Skimmers breeds along the Gulf Coast. They winter from North Carolina south, along the Gulf Coast, south to Panama, and on the Pacific Coast to Costa Rica (Gochfeld and Burger 1994). Birds breeding in Florida are residents (Stevenson and Anderson 1994), which may also be the case for the rest of the Gulf Coast. One estimate for the number of Black Skimmers nesting along the Atlantic and Gulf coasts is 90,000 to 101,000 individuals; another 4,200 are in California and on the Pacific coast of Mexico (Delany and Scott 2006).

Black Skimmers skim across the water's surface to catch fish and invertebrates, feeding within the top 5–6 cm of water, often at dusk or at night (Erwin 1990; Burger and Gochfeld 1990; Yancey and Forys 2010). At St. Vincent National Wildlife Refuge on the Gulf Coast of Florida, 71 % of Black Skimmer foraging occurred within 2 m of the land (in water depths of 13.4 cm, Black and Harris 1983).

Black Skimmers usually nest on bare sand in colonies with gulls and terns, including Least, Sandwich, Royal and Caspian Terns, and Laughing Gulls, deriving some protection from their aggressive neighbors (Gochfeld 1978; Erwin 1979; Burger and Gochfeld 1990; Turcotte and Watts 1999). They also nest on dredge spoil or on salt marshes (Figure 12.47), and in northwestern Florida, on roofs (Gore 1991). In coastal Louisiana, Skimmers nest mainly in herbaceous vegetation (79 %), or on sand and shell beach (12 %, $N = 27$ colonies) (Greer et al. 1988). Black Skimmers are monogamous; courtship is synchronous; they lay up to six eggs, and both sexes incubate, defend the nest, and care for the chicks (Gochfeld 1980; Burger 1981e; Burger and Gochfeld 1990, 1992). The chicks are cryptically colored and blend in with sand or bleached wrack, and remain motionless until a predator or person is upon them, when they run frantically

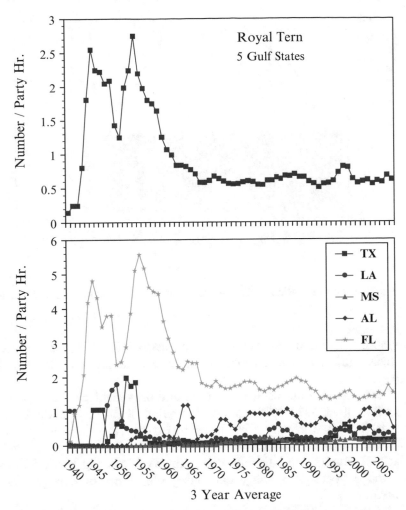

Figure 12.45. Trends in Royal Terns in the Gulf States as indicated by Christmas Bird Count data. © J. Burger.

Figure 12.46. Black Skimmers are so named because they fly just above the water surface, with the tip of their bill in the water, skimming. © J. Burger.

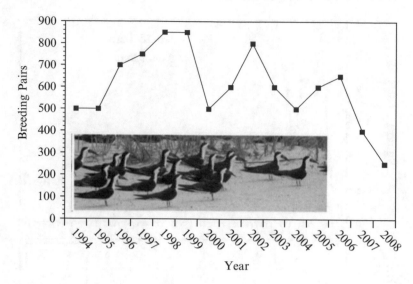

Figure 12.47. Populations of Black Skimmers nesting in Tampa Bay, showing declines. © **J. Burger.**

for cover (Burger and Gochfeld 1990; Gochfeld and Burger 1994). Skimmers often fail to brood chicks during heavy rains, and lose more chicks to cold stress than do terns (Burger and Gochfeld 1990).

Black Skimmer populations declined from oil and organochlorine chemicals in the 1960s and 1970s (Gochfeld 1973, 1974, 1979; Custer and Mitchell 1987). Contaminant levels in the Gulf, however, declined by the early 1980s to below effect levels in Galveston Bay (King and Krynitsky 1986). On the other hand, creation of dredge spoil islands has provided new nesting habitat for Black Skimmers—in Louisiana the numbers of nesting Black Skimmers steadily increased over the 5 years of a study examining the effect of dredge spoil on nesting (Leberg et al. 1995).

Waterbird surveys in Galveston Bay provide some trends information (Glass and Roach 1997). During this time Black Skimmer numbers remained relatively constant, although Laughing Gulls declined (Table 12.11). Trends information from Shamrock Island, Texas indicates an overall decline in the number of pairs on the island from 1973 to 1999, although the numbers did vary (Gorman and Smith 2001). The overall trend after 2000 for Black Skimmers, however, was downward (TCWS 2012), and there is interest in listing the species in Texas (D. Newstead, Coastal Bend Bays & Estuarine Program, Corpus Christi, TX, personal communication). Similarly, the number of breeding Black Skimmers declined in colonies in Louisiana from 1976 to the 1990s (Figure 12.39) (Visser and Peterson 1994). They attributed the decline to erosion of preferred nesting areas, human disturbance, and a reduction in the number of available sites, a recurrent theme for ground nesting colonial birds along the Gulf Coast.

Information from Florida also indicates statewide declines in the number of breeding Black Skimmers (Figure 12.47) (Hodgson and Paul 2010; FFWCC 2011b). Stevenson and Anderson (1994) refer to a single colony of 2,000 pairs in 1935, which is very large by current standards. By the late 1970s, the largest colony in the state was 1,000 pairs in Nassau County (Clapp et al. 1983). During the 2010 nesting season, the largest colony had only 450 pairs, and of the 19 colonies, 12 had fewer than 50 pairs each (FFWCC 2011b).

Breeding Bird Surveys show a steep decline in Alabama, Louisiana, and Florida, with Texas showing a slight increase (Figure 12.48). While there are few routes along the coast, the index

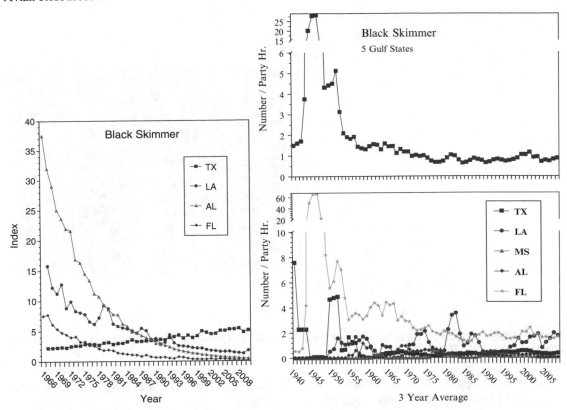

Figure 12.48. Breeding Bird Survey data for Black Skimmer for the Gulf States (*left*) and Audubon Christmas Count Data (*right*). © J. Burger.

provides information, which corroborates the breeding data reported from individual colonies. Using Christmas Bird Counts from 1965 to 2011 as a database, Niven and Butcher (2011) reported that wintering Black Skimmers showed a significant decline of 2.2 % per year along the northern Gulf. Three-year running averages show similar trends (Figure 12.48).

Resiliency in Black Skimmers is intermediate as they delay breeding until they are 3 or 4 years old, lay an average of three eggs, and probably live an average of 10–15 years (Burger and Gochfeld 1994). Since they nest on ephemeral habitats (sandy beaches) or those exposed to flood tides (marshes), reproductive success is often very low.

12.6.2.15 Seaside Sparrow

Seaside Sparrows are small, fairly nondescript brown birds that skulk in grassy vegetation, often running through the grasses (Figure 12.49). They are habitat specialists of salt and brackish marshes and occur in small, localized populations along the Atlantic and Gulf coasts (Post et al. 2009). At least five subspecies breed along the Gulf Coast: *Ammodramus maritimus mirabilis* (southern tip of Florida), *A. m. peninsulae* (northern Florida Gulf Coast), *A. m. juncicota* (Alabama), *A. m. fisheri* (Louisiana Seaside Sparrow, to Texas), and *A. m. sennetti* (southern Texas coast) (Post et al. 1983, 2009). Subspecies are separated by expanses of open water and unsuitable habitat. Seaside Sparrows breeding along the Gulf area residents, while those nesting in the northeastern United States migrate to the southern Atlantic and do not

Figure 12.49. Breeding Bird Survey data for Seaside Sparrow for the Gulf States. © J. Burger. Photo by USFWS (2012b).

migrate to the Gulf. Distribution of breeding Seaside Sparrows is not uniform, leading to the suggestion that they are semicolonial (Post et al. 2009).

Seaside Sparrows forage on grasshoppers, crickets, caterpillars, flies, moths, spiders, snails, mollusks, and small crabs, such as Fiddler Crabs, and grass seeds (Imhof 1962). They nest solitarily, occurring in relatively small, localized populations (Post et al. 2009). Although they sometimes occur with Sharp-tailed Sparrows (*Ammodramus acuticauda*), in some places, such as Alabama, they nest in wetter places than the latter species (Imhof 1962). In resident populations, such as those in the Gulf, females may remain all year on the male's territory, retaining the same mate from year to year (Post et al. 2009). Females lay 4–6 eggs, and the female incubates alone. Eggs and chicks are vulnerable to predation and tidal flooding (Post et al. 2009).

Seaside Sparrows are good indicators of the presence of healthy expanses of salt marsh; Louisiana Seaside Sparrows reside exclusively in brackish and saline marshes along the northern Gulf and are representative of the threats faced by species that breed and winter in these marshes. They are considered a species of management concern throughout their range because of habitat loss and alteration and human disturbance (Cowan et al. 1988; Greenlaw 1992). Responses to burning are unclear because the effects relate to timing (Gabrey and Afton 2000). Sparrows evolved with lightening-induced natural fires that create mosaic patterns, leaving some places unburned. In contrast, anthropogenic burning often involves large, continuous patches. Gabrey and Afton (2000) concluded that management should maintain a mosaic of burned and unburned marsh to provide adequate refuges for sparrows.

Breeding Bird Survey data for Seaside Sparrows indicate increases in Texas and Louisiana, but it is unclear how reliable these data are, given the difficulty of locating this species (Figure 12.50). Christmas Count data for the U.S. Gulf States indicates declines (Figure 12.50).

Resiliency is intermediate to high because they start breeding the spring after their hatching year, lay 2–5 eggs (modal of three in the south), and have a potential life span of 8–9 years (Post et al. 2009). Little life history information is available for *A. m. fisheri*, the species that

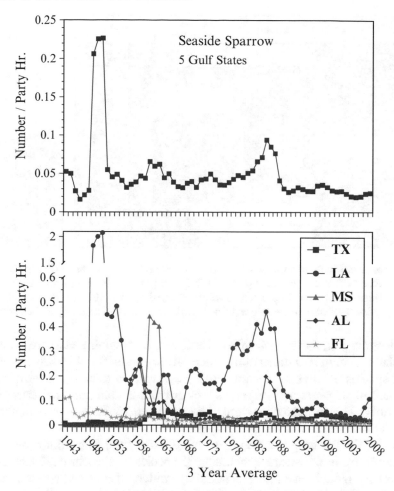

Figure 12.50. Trends in Seaside Sparrow wintering populations using 3-year running averages computed from Christmas Bird Count data. © J. Burger.

occurs from Louisiana to southern Texas. Recovery is likely low, given the loss of marshland and tidal/storm flooding while nesting.

12.6.2.16 Comparisons Among Indicator Species

It is also useful to compare indicators. Two examples illustrate variations among species: Breeding Bird Surveys for the Gulf of Mexico and habitat use by water birds at Laguna Madre in Texas (Anderson et al. 1996). These were chosen because they include many different species of birds.

The Breeding Bird Surveys, conducted by the U.S. Bird Banding Laboratory, provide useful data on trends (Figure 12.51). The Banding Laboratory produces maps that indicate whether populations are increasing or decreasing.[3] The data provide trends information that corroborates, for the most part, individual studies conducted in states and regions. A glance at Figure 12.51 shows that Osprey and Royal Tern are increasing in the U.S. Gulf, while Mottled Duck and Clapper Rail are declining in several states. There are some shifts, where populations

---

[3] Available at http://www.mbr-pwrc.usgs.gov/bbs.

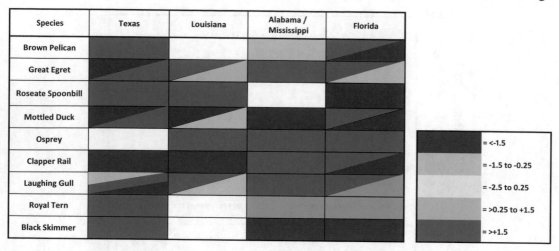

Species	Texas	Louisiana	Alabama / Mississippi	Florida
Brown Pelican				
Great Egret				
Roseate Spoonbill				
Mottled Duck				
Osprey				
Clapper Rail				
Laughing Gull				
Royal Tern				
Black Skimmer				

	= <-1.5
	= -1.5 to -0.25
	= -2.5 to 0.25
	= >0.25 to +1.5
	= >+1.5

Figure 12.51. Interspecific comparison of population trends data for the indicator species from the Breeding Bird Surveys Conducted along the Gulf Coast (Bird Banding Laboratory, from their web site). Shown are general trends, in percent change per year (from 1966 to 2010). For some regions, two trends are given because there is variation within the coastal area. © **J. Burger.**

appear to be declining in some parts of the Gulf, and increasing elsewhere, including Brown Pelican and Black Skimmer. Furthermore, there are no data for Reddish Egret and Seaside Sparrow, both species of concern for the Gulf. However, these are long-term trends, and the trend over the last 10–15 years may differ, as is clear from the temporal patterns provided in the indicator accounts. The indicator accounts show that Black Skimmer is declining and that Laughing Gull has declined over the last 20 years.

Laguna Madre (in Texas) is the second example. More than four million birds (100 species) use the Laguna Madre in mid-winter, including 35 species of shorebirds, 20 species of wading birds, and waterfowl (Muehl et al. 1994; ABC 2011). Anderson et al. (1996) examined bird use of 82 different wetland types between the Rio Grande and Galveston Bay, and found clear differences in both the diversity and types of habitats used. Some groups used many different habitat types (Gallinule and Coots), while others such as rails and grebes used few habitat types (Table 12.12). Differences among species, however, are of interest; Table 12.12 lists habitat uses of the indicator species (calculated from data in Anderson et al. 1996). There were interspecific differences in the number of habitats used, even within a group of closely related species. For example, Spoonbills used only 16 habitats, while Snowy Egret used 52. They drew the following conclusions: (1) cormorants and pelicans used wetlands with less than 30 % vegetation; (2) gulls, terns, and skimmers used estuarine and lacustrine wetlands with less than 30 % vegetation; (3) grebes and rails used palustrine aquatic-bed rooted vascular wetland types; (4) herons, egrets, and bitterns used lacustrine and estuarine wetlands; and (5) shorebirds used estuarine intertidal wetlands.

12.6.3 Indicator Species Groups

The Gulf of Mexico plays a critical role in nesting or migratory behavior of some avian groups, including pelagic seabirds, migratory hawks, wintering waterfowl, nesting colonial birds, and Nearctic-Neotropical migrants. Some of the species that make up these groups have been discussed in the previous section as indicator species, but here they are reviewed briefly because of their overall importance in the Gulf. Many of these groups are monitored separately by state agencies.

Table 12.12. Habitat Use of Waterbirds at Laguna Madre in Texas (calculated from Anderson et al. 1996)

Species	Number of Wetland Types Used	Total Birds	Total Flocks (Mean Flock Size)	Flocks/Number of Wetland Types
Brown Pelican	10	86	26 (3.3)	2.6+
Great Egret	49	1,901	631 (3.0)	12.9+
Reddish Egret	14	145	107 (1.4)	7.6
Roseate Spoonbill	16	611	81 (7.5)	5.0
Snowy Plover	8	185	15 (12.3)	1.9
Piping Plover	8	29	11 (2.6)	1.4
Laughing Gull	34	14,331	313 (45.8)	9.2+
Royal Tern	9	107	20 (11.9)	2.2
Least Tern	19	328	68 (4.8)	3.6
Black Skimmer	8	1,569	18 (87.2)	2.2−

12.6.3.1 Pelagic Seabirds

Much of the focus in this chapter and by state and federal agencies, scientists and others, deals with birds that concentrate along the coasts. It is far easier to census birds nesting there than it is to study pelagic seabirds. However, many seabirds mainly use the open, pelagic zones of the Gulf of Mexico. Seabirds do not simply migrate over or around the Gulf, but instead use the open waters of the Gulf for wintering and foraging. The Gulf waters also provide foraging habitat for more tropical-nesting species, some of which use the Campeche Banks for breeding (Tunnell and Chapman 2000), such as frigatebirds and boobies, and for North Atlantic-nesting species, such as Northern Gannet (see section below).

Of all the birds considered, the distribution and behavior of pelagic species are the most affected by prey availability and oceanic features (Hunt 1990; Schneider 1991; Ribic et al. 1997; Schreiber and Burger 2001a, b; Zuria and Mellink 2005). In the Gulf of Mexico, the Loop Current and eddy systems greatly affect distribution of seabirds in the northern region (Ribic et al. 1997). Prey is not evenly distributed over the open ocean; individual prey patches are often small and interspersed (Hunt and Schneider 1987). Foraging seabirds can search for prey by (1) looking for the presence of feeding birds (plunge-diving) as a signal of prey availability (Simmons 1972; Gotmark et al. 1986; Gochfeld and Burger 1982), (2) looking for other seabirds flying in the same direction, presumably toward a prey patch, (3) watching birds that return to a colony with food (Gaston and Nettleship 1981), (4) searching for particular oceanographic conditions, and (5) returning to known foraging areas.

Whether or not to associate with marine mammals is another decision foraging seabirds make (Burger 1988b, c). Foraging with mammals (whales, dolphins) can make prey fish more available and identify large expanses of zooplankton; it can also result in competition or interference foraging (Pierotti 1988a). Marine birds and mammals can interact in at least five ways: (1) birds can have passive associations; (2) birds can be attracted to the same resource; (3) birds can be actively drawn to marine mammals because they drive prey to the surface; (4) birds can be attracted to marine mammals to scavenge on by-products of mammal foraging; and (5) birds can actively avoid marine mammals that might prey on them; for example, Orcas eat diving birds (Pierotti 1988b).

Seabirds often forage over schools of prey fish that have been forced to the surface by predatory fish, such as bluefish (*Pomatomus saltatrix*) or tuna (Safina and Burger 1985, 1988;

Au and Pitman 1988). Such schools are usually ephemeral as prey fish soon scatter and swim away from the surface; usually there is no direct competition among the seabirds foraging above them (Burger 1988b, c).

Interactions between seabirds and fisheries include foraging near boats on fish in nets, foraging behind boats on prey churned up by fishing operations, foraging on discarded offal from factory ships, and feeding on offal near onshore facilities (Furness et al. 1992). Some seabirds panhandle food around docks. Many of the interactions among birds, marine mammals, predatory fish, and fisheries operations are described fully in chapters in Burger (1988c; 2017) and in Schreiber and Burger (2001a).

Interactions with fisheries are an ongoing concern for many diving seabirds (Forsell 1999; Gilman 2001; Gilman et al. 2005). The National Marine Fisheries Service Pelagic Observer Program observed 6,949 longline sets from 1992 to 2005 in the U.S. Atlantic longline fishery, which included the Gulf of Mexico (Hata 2006). In 52 sets, 114 seabirds were captured (69 % were dead upon retrieval). Gulls were the most common birds caught, followed by unidentified seabirds, shearwaters, and gannets (Hata 2006). Hata (2006) concluded that seabird mortality was less in Gulf waters than elsewhere.

12.6.3.1.1   Baseline Continental Shelf Surveys in 1979 and 1980–1981

A pilot study was conducted of seasonal distribution and abundance of marine mammals, sea turtles, and marine birds to make effective decisions about oil and gas development in the Outer Continental Shelf of the U.S. Gulf Coast (Fritts and Reynolds 1981). The continental shelf varies from 185 to 215 km wide off West Florida and the Yucatán coasts, to 25 km off the Rio Grande (Texas), and 13 km near Veracrúz, Mexico. They conducted aerial surveys in Florida and Texas because of the presence of major shipping lanes; surveys conducted from August to December 1979 extended 222 km perpendicular to the coast. The survey units were off Brownsville Texas; Corpus Christi, Texas; Tampa Bay, Florida; and Naples, Florida. During this time, they identified 14 bird species, and 14 categories of birds (i.e., dark terns). There were remarkably few birds on these transects. However, several conclusions were drawn: (1) terns were the most common species and were observed in all four survey areas in both August and November; (2) boobies, shearwaters, and petrels were observed mainly off Texas in August; (3) pelicans were observed off south Florida; and (4) gulls were observed mostly in November in all four survey areas.

Terns accounted for the following percentages: (1) 66 % (South Texas, August); (2) 65 % (South Texas, November); (3) 76 % (North Texas, August); (4) 35 % (North Texas, November); (5) 68 % (North Florida, August); (6) 75 % (North Florida, November); (7) 64 % (South Florida, August); and (8) 89 % (South Florida, November). Royal Terns were the most abundant of the terns (Fritts and Reynolds 1981). More birds were counted near shore than in pelagic waters.

A more extensive survey, conducted using the same methodology from May 1980 to April 1981, identified 68 bird species (Fritts et al. 1983). The four study areas extended into pelagic waters from Brownsville, Texas; Marsh Island, Louisiana; Naples, Florida; and Merritt Unit, Florida. The diversity of birds was similar in all four study sites, but was 3 times greater in the subunit off Louisiana than in Texas or South Florida (Table 12.13). These are perhaps the best data on bird distribution in the Gulf pelagic waters, and they can be used to understand species, and seasonal and geographical differences.

The most common species were Royal Tern, Laughing Gull, and Herring Gull. Numbers of Royal Terns were highest right after the breeding season, when young birds were flying. Laughing Gulls breed along the Gulf, but do not go as far offshore during the breeding season. Herring Gulls do not breed in the Gulf and they build up in February before migrating north to breeding colonies (Figure 12.52).

Table 12.13. Survey Data of Marine Birds from 1980 to 1981 in Four Study Sites in the Northern Gulf of Mexico (after Fritts et al. 1983)[a]

Species	Brownsville, Texas	Marsh Island, Louisiana	Naples, Florida	Merritt Island, S. Florida
Common Loon	0	1	130	3
Cory's Shearwater	28	7	6	149
Audubon's Shearwater	4	1	45	60
White-Tailed Tropicbird (+ unidentified)	0	1	1	3
American White Pelican	453	395	0	139
Brown Pelican	21	1	243	987
Masked Boobies	7	2	0	2
Northern Gannet	29	303	30	14
Double-crested Cormorants	4	92	219	2
Magnificent Frigatebird	1	0	96	13
Unidentified Jaegers	2	1	2	139
Herring Gull	193	1,304	193	0
Ring-billed Gull	40	133	54	29
Laughing Gull	503	2,493	221	0
Franklin's Gull (all in April)	52	0	0	225
Common (type) Terns	22	64	441	249
Sooty Tern	4	2	224	36
Bridled Tern	14	1	9	360
Royal Tern	841	1,638	2,294	22
Sandwich Tern	1	89	9	2
Black Tern	35	170	948	22
Total	**2,246**	**6,698**	**5,170**	**2,708**

Transects were out from the coast on the continental shelf. Given are the total number of each species sighted in transects from June 1980 through April 1981. © J. Burger
[a]Fritts et al. (1983) also conducted some opportunistic surveys, and recorded tropicbirds, Brown Boobies, and one unidentified Jaeger.

12.6.3.1.2 Surveys in the Northern Gulf in the Mid-1990s

Ribic et al. (1997) made four offshore cruises in the northern Gulf during four seasons ($N = 194$ transects). Data were taken between the 100 and 2,000 m isobaths in the northern Gulf of Mexico, off the coast from Texas to Florida (from about 96° west, to 88°). No species of bird was observed in all four seasons. Skuas predominated, with Pomarine Skuas being the most common. They were present in all seasons except the summer when they breed in the Arctic. During winter, the most common birds were Pomarine Skua, Herring Gull, and Laughing Gull. Overall, fewer birds were observed in the spring; Band-rumped Storm Petrel (*Oceanodroma castro*) was the most common and was observed only in the spring. Herring Gulls and Laughing Gulls were observed in winter, nearly all birds were observed in spring, and all terns were observed in summer and were more likely to be seen outside of the eddies. Pomarine Skua was more likely to be seen in the eddies (Ribic et al. 1997).

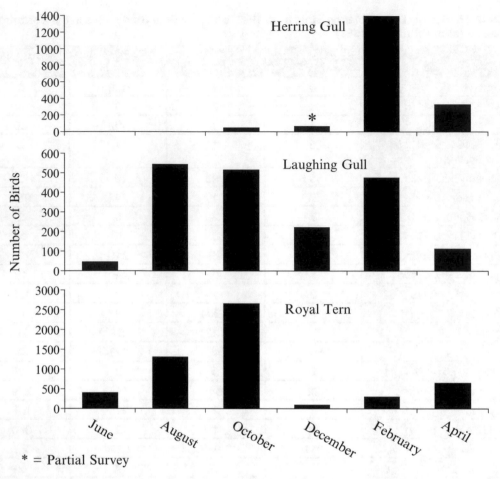

Figure 12.52. Seasonal abundance of seabirds observed from 1980 to 1981 in four transect study sites in the Northern Gulf of Mexico (after Fritts et al. 1983). © J. Burger.

12.6.3.1.3   Pelagic Surveys from 1996 to 1997 in the Northern Gulf of Mexico

As part of the GulfCed II Program, three cruises were conducted in spring, and mid- and late summer, mainly off Louisiana (Hess and Ribic 2000; Davis et al. 2000). The spring cruise spanned 44 days and covered 6,401 km of both the oceanic Gulf and continental shelf, the mid-summer cruise was 17 days and covered 2,500 km (track line in the Gulf that included eddies), and the late summer cruise was 16 days and covered 2,015 km (Hess and Ribic 2000). During the spring cruise, 5,918 seabirds were recorded during 334.8 effort hours. Terns were the most abundant group; Sooty Terns were the most abundant tern (Hess and Ribic 2000). All the other, more pelagic seabirds were much more rare in abundance. The majority of gulls observed were Laughing Gulls. About the same overall percentage of species groups were observed in the mid/late summer as in the spring. Overall the number of birds per effort hour varied by season, with the lowest level occurring in May to June (0.61 sightings per effort hour for all birds), and the highest level occurring in August to September (26.41 sightings per effort hour, mainly due to terns). Shearwater numbers were highest in August to September (1.30), Storm Petrels were highest in August (1.84), frigatebirds were highest in August (1.06), Sulids (boobies) were highest in February (0.60), tropicbirds were highest in August–September (0.03), jaegers were highest in February (1.05), gulls were highest in February (2.23), and terns were highest in August and September (24.6) (Hess and Ribic 2000).

12.6.3.1.4   Northern Gannet as an Example of a Migrant Pelagic Seabird in the Gulf

Northern Gannets, plunge-divers that breed in dense colonies on offshore rocky islands only in the North Atlantic, are a good example of a pelagic seabird that uses the Gulf of Mexico. Unlike many of the indicator species, they are only migrants to the Gulf, and adults return north to breed, although immatures may remain in the Gulf all year. And like other pelagic seabirds, it is difficult to census their numbers in the Gulf—the Gulf is vast and seabirds are spread out. Banding studies indicated that less than 15 % of gannets went to the Gulf of Mexico, but recent work with light sensitive geolocators deployed on gannets from four colonies indicated that 27 % went to the Gulf (Montevecchi et al. 2011). Thus the Gulf plays an important role in their life history and it might do so for other pelagic birds such as petrels and phalaropes.

Gannets breed in only six colonies on islands in the Gulf of St. Lawrence, and on the Atlantic coast of Newfoundland; they winter from New England south along the Atlantic into the Gulf of Mexico (Mowbray 2002). They nest in very dense, noisy colonies on offshore rocky islands, or precipitous cliffs. They are monogamous, mate for life, use the same nest site every year and lay only one egg, and both parents care for the young. Suitable nesting habitat seems to be limiting as adults remain on the colony site defending their nest sites well into October; their young depart weeks earlier. Since nest sites are scarce, there is pressure for adults to return to the colony sites (and leave the Gulf waters) as soon as possible (Figure 12.53).

They migrate south along the Atlantic coast and into the Gulf of Mexico (Mowbray 2002), south to Texas, Tamaulipas, and sometimes Veracruz (Howell and Webb 1995). Once considered a vagrant along the Texas coast, recent efforts to scour nearshore waters revealed them gliding over the Gulf in surprising numbers (Eubanks et al. 2006). Offshore from Mississippi, flocks of several hundred have been observed from boats (Turcotte and Watts 1999). Land-based sightings depend not only on observer care but also on weather and wind conditions.

Gannets are known for their high dives in which they plunge from 20 m or more straight down to the water surface. Once underwater they either catch prey directly or chase prey, using their feet and wings for propulsion (Stevenson and Anderson 1994). They are generalist and opportunistic foragers that exploit a diverse prey base along the continental-shelf waters (Mowbray 2002; Montevecchi 2008; Montevecchi et al. 2009). Foraging trips away from breeding colonies in the North Atlantic average between 196 and 452 km (122 and 280 mile)

Figure 12.53. Northern Gannets breed in dense colonies on offshore rocky islands with steep cliffs (*left*, St. Mary's Colony in Newfoundland, © J. Burger). They are strong flyers and plunge-dive for fish (flight photo by Marie C. Martin (CUNY), courtesy of NOAA 2012).

(Garthe et al. 2007), suggesting the potential for wide-ranging habitat use in the Gulf. Furthermore, changes in oceanography and the distribution and abundance of prey resources have resulted in gannets shifting diet from warm water pelagic fish to cold water fish (Montevecchi and Myers 1997). These findings imply that slight shifts in oceanographic conditions can have a massive effect on seabird distribution and foraging. While these shifts have been shown in the north Atlantic, changes in the Gulf may have similar effects on the wintering distribution of Northern Gannets.

Northern Gannets are relatively uncommon off the northeast coast of Mexico, and in Tamaulipas and Veracruz (Mowbray 2002), although global warming may result in decreasing their abundance in the southern Gulf. Most Gannets arrive in late November on the southern Atlantic Coast, and it is likely they arrive in the Gulf in December. The phenology of when breeding adults leave for colonies farther north is unclear, and some proportion of the immatures may remain year round in the Gulf, which is relevant because Northern Gannets normally breed at 5 years of age (Cramp and Simmons 1977; Mowbray 2002).

Using Christmas Bird Counts from 1965 to 2011 as a database, Niven and Butcher (2011) reported that wintering Northern Gannets showed a significant increase of 6.6 % per year in the Gulf region. Thus, the Christmas Bird Count data corresponds to the increases reported for the breeding colonies (Nettleship and Chapdelaine 1988; Chardine 2000; Mowbray 2002). Northern Gannets have a relatively low resiliency because they delay breeding until they are 3–6 years old (most are 5–6 years old); young have high mortality during the first year. Females lay only one egg, although reproductive success can be high, and they have an average life expectancy of 16 years (Mowbray 2002). Despite this, their breeding populations are increasing rapidly, and most recent estimates are that breeding Northern Gannets increased 52 % between 1984 and 1999 (Nettleship and Chapdelaine 1988; Chardine 2000; Mowbray 2002), which suggests increasing numbers in the Gulf.

12.6.3.1.5   Summary of Pelagic Seabird Use of the Northern Gulf of Mexico

Distribution of seabirds was examined in several studies involving multiple cruises to survey marine mammals, sea turtles, and seabirds. These studies form a picture of relatively low densities of seabirds out in the continental shelf of the northern Gulf of Mexico. Most seabirds foraged in flocks (except Loons), and most of these flocks contained many species. Larger flocks of seabirds fed with predatory fish and marine mammals than fed in the absence of them. The most pelagic seabirds (boobies, gannets, frigatebirds, petrels, shearwaters) were not very abundant. Royal Terns, Laughing Gulls, and Herring Gulls were the most abundant species. Seasonal use varied. Species that breed along the Gulf Coast were more common most of the year (e.g., Royal Tern, Laughing Gull), while species that breed farther north (e.g., Herring Gull, Common Loon) were not common in the spring and summer. Similar data are needed for the southern Gulf (Campeche), as well as for the pelagic waters of the Gulf.

12.6.3.2   Migratory Hawks

The U.S. Gulf Coast is not noted for migrant hawks. However, central Veracruz is the most important migratory pathway in the world for hawks (Ruelas et al. 2000; Zalles and Bildstein 2000), and some of these hawks migrate across the Gulf, while others follow the eastern Mexican coast down through Veracruz (Inzunza et al. 2010). As with many other migrants, hawk migration is partly dependent upon wind speed, wind direction, and the passage of cold fronts (Woltmann and Cimpreich 2003). Average counts (±standard deviation) from 1992 to 2004 at Veracruz were as follows (Inzunza et al. 2010): Turkey Vulture (1,897,679 ± 387,839),

Table 12.14. Counts and Significant Trends for Migrant Hawks at Four Sites Around the Gulf (after Smith et al. 2008)

Species	Florida Keys	Smith Point, Texas	Corpus Christi, Texas	Veracruz, Mexico
Turkey Vulture	–	1,529 (56)	20,996 (57)	1,988,826 (23)
		0.0	16.9	5.7
Osprey	1,154 (24)	65 (20)	167 (30)	2,969 (28)
	9.0	4.7	7.2	2.8
Mississippi Kite	19 (92)	4,320 (51)	7,020 (40)	155,651 (46)
	–	10.0	5.4	15.4
Sharp-shinned Hawk	2,971 (47)	2,913 (40)	1,076 (33)	4,542 (55)
	−12.8	−4.2	−2.6	−7.5
Cooper's Hawk, Accipter cooperi	536 (54)	1,125 (14)	663 (45)	2,529 (33)
	7.3	−1.0	3.2	1.9
Broad-winged Hawk	3,737 (28)	38,643 (73)	609,719 (45)	1,919,949 (13)
	6.1	8.2	−6.7	3.1
Swainson's Hawk	82 (60)	298 (98)	6,209 (77)	915,104 (32)
	–	10.0	18.5	13.6
American Kestrel, *Falco sparverius*	2596 (41)	1,334 (28)	506 (38)	8,252 (95)
	−8.8	−2.9	6.7	0.0
Peregrine Falcon	1,826 (28)	89 (20)	155 (37)	745 (42)
	6.9	5.8	3.2	3.2
Total raptors[a]	**13,981 (19)**	**51,275 (57)**	**639,551 (41)**	**5,260,871 (19)**

Only raptors with counts over 600 were included. Given is the mean number of hawks (coefficient of variation). The second line for each species gives percent change over the period from 1995 to 2005.
[a]Includes all raptors, even those not included in the table.

Mississippi Kite (157,199 ± 87,640), Broad-winged Hawk (1,932,255 ± 287,822), and Swainson's Hawk (819,419 ± 280,788).

Raptor populations surveyed from 1995 to 2005 at four locations (Florida Keys, Smith Point and Corpus Christi in Texas, Veracruz in Mexico) showed significant declines in some species and significant increases in others (Smith et al. 2008). Species that increased significantly at one or more sites included Turkey Vulture (*Cathartes aura*), Osprey, Swallow-tailed Kite (*Elanoides forficatus*), Mississippi Kite (*Ictinia mississippiensis*), Swainson's Hawk (*Buteo swainsoni*), Zone-tailed Hawk (*Buteo albonotatus*), and Peregrine Falcon (*Falco peregrinus*). Northern Harriers (*Circus cyaneus*) and Sharp-shinned Hawks (*Accipiter striatus*) declined at all sites (Table 12.14) (Smith et al. 2008).

It is often difficult to ascertain population trends for migrating species because shifts in migration patterns may not be readily evident without data from large geographical areas. For example, Farmer et al. (2008) reported declines of North American Kestrel, and long-term increases of Bald Eagle, Merlins, and Peregrine Falcons. However, they attributed the declines to changes in migration patterns (Farmer et al. 2008). Some species, such as Swallow-tailed Kites, are not monitored at hawk watches in North America, making counts along the Gulf Coast particularly important (Smith et al. 2008). Swallow-tailed Kites breed in scattered

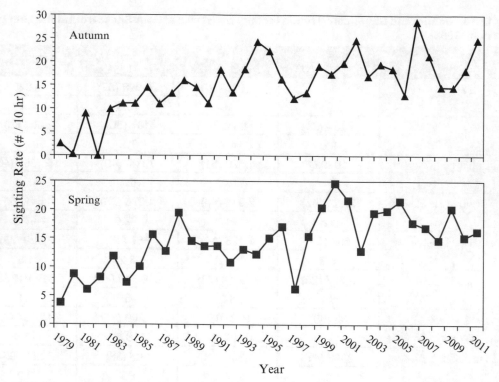

Figure 12.54. Sightings of Peregrine Falcon per 10 h of observation at the South Padre Island (Texas) hawk watch (after Seegar et al. 2011). © J. Burger.

locations in South Carolina, coastal Mississippi, Alabama and Louisiana, and southern Mexico into South America (Meyer 1995).

The south Texas and Tamaulipas coast are wintering areas for Peregrine Falcons (Enderson 1965; McGrady et al. 2002). The Padre Island Peregrine Falcon Survey, conducted since 1977, provides trends information (Figure 12.54). Because counting migrating hawks is dependent upon observation time, data are given as number of falcon sightings per 10 h time periods. Numbers have varied, but appear to have increased.

12.6.3.3   Wintering Waterfowl

One quarter of dabbling ducks once wintered in Louisiana (Palmisano 1973), and two-thirds of the Central Flyway waterfowl population also did so (Bellrose 1988). However, the Gulf Coast is no longer the chief wintering area for North American waterfowl because of coastal marsh habitat loss, sea level rise, and freshwater inputs that have reduced available habitat (Palmisano 1973; Link et al. 2011). Such steep declines require intensive study, management of hunting and habitat, and possible additional protection for nocturnal roost sites (Link et al. 2011), as well as manipulation of water levels (Bolduc and Afton 2004). Even so, 19 % of waterfowl wintering in the United States use marshes of the Louisiana Gulf Coast (Michot 1996). Migratory waterfowl also concentrate in coastal Mexico (Gallardo et al. 2004).

Texas is the top waterfowl harvest state in the Central Flyway and is in the top five hunting states in the United States Texas accounted for 42 % of the total duck harvest and 47 % of the goose harvest in the Central Flyway (TPWD 2011). Although the number of waterfowl hunters is declining nationally, Texas hunter trends have remained stable. Coastal habitat protection is a

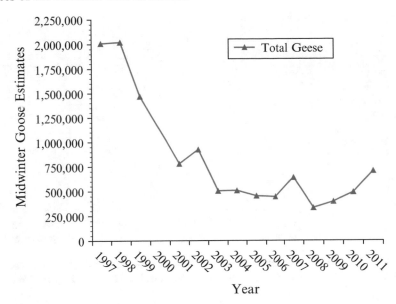

Figure 12.55. Trends in total numbers of geese counted by the Texas Parks and Wildlife Department (after TPWD 2011). Counts include Canada Geese, Snow Geese, and White-fronted Geese. © **J. Burger.**

prime concern of both the Texas Parks and Wildlife Department and the Gulf Coast Joint Venture (GCJV), a regional partnership of organizations and individuals that are concerned with waterfowl and other migratory bird populations and their habitats between Mobile Bay in Alabama and the Rio Grande in Texas (TPWD 2011). More than 1.9 million ducks winter along the Gulf Coast of Texas.

Status and trends data are available for the Texas Gulf from 1997 to 2011 (TPWD 2011). Populations of geese declined during this period (Figure 12.55). In 1997 and 1998 more than two million geese were counted, and by 2001 the numbers had declined to less than a million along the Texas coast (Figure 12.55). This area supported over 552,000 geese annually from 2001 to 2009 (TPWD 2011). Canada Geese numbers declined from over 30,000 to fewer than 7,000 in the last 3 years.

The TPWD (2011) data provide an overview of the relative numbers of ducks (Figure 12.56) as well as trends. Texas Gulf Coast prairies and marshes (mid-Texas coast) have many more ducks than the southern Coastal Sand Plains; yearly average of 1,500,000 compared to the 15-year average of 82,913 (TPWD 2011). Pintail and Gadwall were the two most common dabbling ducks in both regions, while other species, such as shovelers, were less common. Redhead and Scaup were the most common diving duck in the Prairies and Marsh region; only Scaup was most common along the southern coast. The Texas Gulf Coast also provides year-round habitat for Mottled Ducks, Black-bellied Whistling Ducks, Fulvous Whistling Ducks, and to a lesser extent, Blue-winged Teal (TPWD 2011).

The number of dabbling and diving ducks varied by year, and there appears to be a decline in the last 3 years for both groups (Figure 12.57). Although there was a decline from 1999 to 2001 for dabbling ducks, it was not as great, and no similar decline was found for diving ducks. The recent declines may be due to habitat loss caused by severe hurricanes and storms; hurricanes of 2004 and 2008 adversely affected available habitat for wintering waterfowl along the Texas and Louisiana coasts (TPWD 2011).

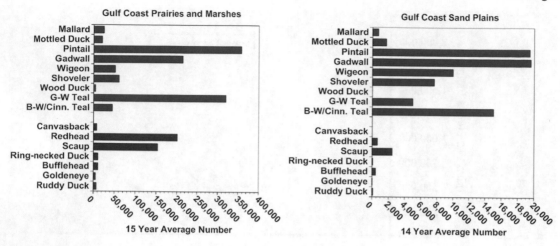

Figure 12.56. Mean number of waterfowl over a 14- or 15-year period in the Texas Gulf Coast (in winter) (after TPWD 2011). © J. Burger. G-W Teal = Green-winged Teal; Cinn. Teal = Cinnamon Teal.

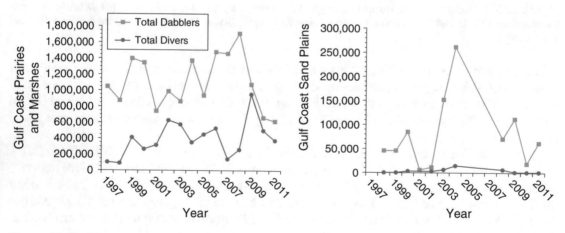

Figure 12.57. Mean number of total Dabbling Ducks and Diving Ducks surveyed along the Texas Gulf Coast (winter counts) (data from TPWD 2011). © J. Burger.

Wintering waterfowl populations have declined precipitously in the Mobile Tensaw Delta in Alabama (Figure 12.58). Surveys conducted by Lueth (1963) showed a significant decline in populations from 1939 to 1949, and data from the Alabama Wildlife and Freshwater Fisheries Department (AWFWF 2005) show these declines have continued to 2004. These data form a picture of severe waterfowl population declines in the Mobile-Tensaw Delta (Lueth 1963; Beshears 1979; Goecker et al. 2006; MBNEP 2008).

12.6.3.4 Nesting Colonial Birds

Nesting colonial waterbirds, an important component of the avifauna of the Gulf Coast, nest in abundance on the marsh islands, sandy beaches, and islands with low shrubs in coastal environments. A large number of colonial nesting birds occur throughout the Gulf, including 14 species of herons, egrets, ibises, spoonbills, and storks, as well as ten species of gulls, terns,

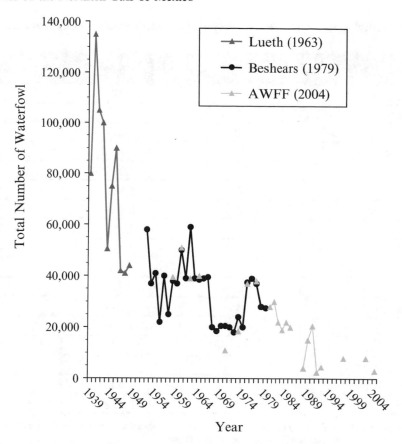

Figure 12.58. Decline in waterfowl in the Mobile-Tensaw Delta in Alabama (data from several sources, compiled in Goecker et al. 2006). © J. Burger.

and a skimmer (Table 12.4) (Gallardo et al. 2009). The northern Gulf Coast alone contains a substantial percentage of the U.S. breeding populations of a number of species, such as Reddish Egret, Sandwich Terns, and Black Skimmers, as well as significant portions of U.S. breeding gulls and other terns (Visser and Peterson 1994).

Nesting colonial birds have long been considered indicators of ecosystem health (Furness 1993; Burger 1993, 2006a, b; Erwin and Custer 2000; Burger and Gochfeld 2004b). Some colonial birds (White Ibis) are noted for their yearly shifts in colony sites and population numbers, although others use the same site for decades or longer (Frederick and Spalding 1994; Schreiber and Burger 2001a). Thus, regional estimates are most useful for species that shift sites. With marsh-nesting species, such as Forster's Tern and some Laughing Gulls, shifts due to tidal or wind flooding are usually within a local area or section of the marsh.

Two areas will serve as examples of colonial nesting birds along the northern Gulf of Mexico, one in Louisiana, and one in Texas. The Barataria-Terrebonne Estuarine system in southeastern Louisiana contains more than four million acres of wetlands and is vulnerable to wetland loss (Lindstedt 2005).

Over the last 50 years, this system has lost 47–57 km^2 (18–22 mi^2) per year (Lindstedt 2005). In 2001, 24 species of colonial birds nested in the Barataria-Terrebonne Estuaries, with as many as 37 colonies of Great Egret, 31 colonies of Snowy Egret (likely in the same colonies), and 18 of Little Blue Heron (Lindstedt 2005). Tricolored Heron (Louisiana Heron) occurred in 10 colonies,

Table 12.15. Colonial Birds That Nested in the Barataria-Terrebonne Estuary in Southeastern Louisiana

Species	Number of Colonies	Number of Pairs	Average Conspecific Pairs/Colony
Great Blue Heron	12	753	63
Little Blue Heron	18	987	55
Tricolored Heron	10	107	11
Black-crowned Night Heron	7	180	26
Great Egret	37	4,616	125
Cattle Egret	7	1,187	170
Snowy Egret	31	5,295	171
Brown Pelican	3	3,910	355
Laughing Gull	11	7,730	702
Black Skimmer	3	175	58
Roseate Spoonbill	6	352	59
Forster's Tern	6	1,230	205
Royal Tern	6	1,005	168
Gull-billed Tern	1	145	145
American Anhinga	3	25	8
White Ibis	7	7,630	1,090
Dark Ibis	12	2,922	244

Although these data are from 2001, they indicate relative abundance (after Michot et al. 2003; Lindstedt 2005). Species with fewer than 25 pairs are not included in the table. © J. Burger

Roseate Spoonbills occurred in only six colonies, and Reddish Egret was not reported (Table 12.15). Species associate with one another, so total colony size is much larger; the egrets usually nest together, along with herons and Night Herons.

Long-term trends also were examined in Galveston Bay Estuary from 1973 to 1990 (Gawlik et al. 1998). Approximately 50,000 nesting pairs of 22 species of colonial waterbirds used Galveston Bay annually since the 1970s, which represents about 30 % of the nesting colonial birds on the Texas coast. This area is also the most important wintering area for ducks and geese in the central flyway (Hobaugh et al. 1989). Laughing Gulls, a species with low annual variability, dominated the assemblage. In contrast, White Ibis and Sandwich Tern were both highly variable and abundant. Trend analysis showed that Roseate Spoonbill, Snowy Egret, and Black Skimmer declined significantly, while Neotropical Cormorant and Sandwich Tern increased significantly; the other 13 species showed no significant trends (Gawlik et al. 1998).

It seems clear from this review of the birds of the Gulf of Mexico that two types of information complement each other: local surveys and regional (up to Gulf-wide). Local surveys of one or two colonies are useful, especially when long-term data are available. Surveys from the same general area of the coast sometimes produce different results (Table 12.16), requiring careful interpretation in light of individual species characteristics. However, the two surveys were at different times, and may actually present a picture of regional shifts. Furthermore, when two different surveys from different areas produce the same result (Brown Pelican increased in both; Night Heron, Great Blue Heron decreased) it is likely to be real. In no case did a species show opposite trends. When the trends were not congruent (e.g., no trend, and increase), further investigation is necessary. For example, some species showed an increase from 1956 to 1992, but no trend thereafter. Data showing no trend should be examined further by conservationists.

Table 12.16. Trends in Colonial Waterbirds Reported for Galveston Bay (after GBST 2012) and Corpus Christi Bay (after Chaney et al. 1996) in Texas

Species	Corpus Christi Bay (1956–1992)	Galveston Bay (1990–2009)
Brown Pelican	Increase	Increase
Cattle Egret	Increase	No trend
Black-crowned Night Heron	Decrease	Decrease
Great Blue Heron	Decrease	Decrease
Snowy Egret	No trend	No trend
Tricolored Heron	No trend	Decrease
White Ibis	Increase	No trend
White-faced Ibis	No trend	Decrease
Reddish Egret	Increase	No trend
Roseate Spoonbill	Increase	No trend
Laughing Gull	Increase	No trend
Royal Tern	Increase	No trend
Forster's Tern	Increase	No trend
Least Tern	Decrease	No trend
Sandwich Tern	Increase	Decrease
Black Skimmer	Decrease	No trend
Neotropical Cormorant		Decrease

This illustrates the problem of examining colonial waterbirds in a small geographical area, and not for the same time periods. © J. Burger

Finally, understanding trends is a matter of using a "weight of evidence approach"—when all data sets indicate the same trend, it is likely real (Burger 2003). Conservation of colonial waterbirds is a matter of protection from human disturbance and predators, prevention of habitat loss, and insurance of sufficient foraging and nesting habitat, particularly for wading birds (Kushlan 2000a, b; Hafner et al. 2000; Pineau 2000; Clay et al. 2010). Active habitat management and augmentation are essential.

12.6.3.5 Neotropical Passerine Migrants

Migratory Passerines are often ignored when considering birds in the Gulf of Mexico because the focus is often on seabirds and waterbirds. Yet, the narrow band of wooded barrier islands and forested coastlines provides the departure point for Passerines crossing the Gulf of Mexico (Moore 1999). Each year billions of landbirds migrate between the northern and southern hemisphere, and many cross the Gulf of Mexico (Stevenson 1957; Moore 2000a, b). While it is an extremely important migratory pathway for Nearctic-Neotropical migrants, the Gulf of Mexico is also a formidable barrier for these migrants (Rappole 1995). Nearctic-Neotropical migrants are those that breed in the North Temperate zone and winter in the tropics (Shackelford et al. 2005), including mainly songbirds (Passeriformes), although they also include some shorebirds, terns, cuckoos, and others.

The objective of this section is to provide an overview of Nearctic-Neotropical migrants in the Gulf of Mexico and associated coastal lands, rather than in more upland habitats, such as wood plantations and bottomland hardwood forests (Wilson and Twedt 2003), and the coastal habitats of Veracruz and the Yucatán (Mackinnon et al. 2011). A number of monographs and edited books deal with neotropical migrants, including DeGraaf and Rappole (1995), Kerlinger (1995), Rappole (1995), Able (1999), Moore (2000b). Greenberg and Marra (2005), and further, Jahn et al. (2004) provide a system-wide approach to new world migration that is particularly applicable to the Gulf.

Migrants have three choices for flying between North America and Central/South America: (1) a circum-Gulf route through eastern Mexico and Texas, (2) a trans-Gulf route to the Yucatán Peninsula and the Mexican coast, and (3) a circum-Gulf route through Florida and West Indies (Stevenson 1957; Langin et al. 2009). Since migration takes place over a relatively small time scale, but a large spatial scale, different factors affect migration patterns. On a small spatial scale, habitat (amount and quality, prey availability) is critical, while on a broader scale, weather and winds become critical (Moore 2000a). Wind patterns are generally favorable for birds to migrate across the Gulf in the spring, usually determined by studies with radar (Gauthreaux 1971, 1999; Gauthreaux et al. 2006). Migration patterns are not static, but shift from year to year and season to season, depending upon prevailing winds (Barrow et al. 2005).

Analyzing data from 10 radar stations from Key West, Florida to Brownsville, Texas, Gauthreaux et al. (2006) showed that in the spring: (1) northbound migrants approached the northern Gulf Coast at between 1,000 and 2,500 m above ground; (2) the longitudes of peak arrivals were similar over the 4-year period (near longitude 75° west, northern Texas/western Louisiana); (3) wind trajectories over the Gulf of Mexico had relatively little influence on the longitude of peak arrival; (4) the longitude where the greatest density of trans-Gulf migrants arrived on the northern coast was relatively constant from year to year; and (5) on occasion, strong winds or storms displaced migrants, but migrants seemed to have a preferred route they followed. These conclusions suggest that conservation efforts should concentrate on the preferred routes and landing locations. These findings are intriguing and suggest the potential to develop conservation priorities for suitable habitat for northbound Nearctic-Neotropical migrants, especially when coupled with data from birds fitted with geolocators.

Neotropical migrants face several decisions with respect to the Gulf of Mexico, including which route to follow, when to migrate, where to make landfall before crossing the Gulf, where to make landfall after crossing the Gulf, and how long to stop at stopovers on either side of the Gulf. Passage across the Gulf is long, and birds often arrive in the Louisiana northern Gulf Coast with little fat (Yong and Moore 1997), making coastal lands critical for increasing fat stores and continued survival. Peak numbers of spring Passerine migrants occur from mid-April to early May, and radar studies indicate that nearly all the Passerine migrants arrive from directly over the Gulf of Mexico (Gauthreaux 1971). A bad storm or hurricane can kill 40,000 migrants on one day, if it occurs during a peak time when migrants are arriving from their northward flight across the Gulf (Wiedenfeld and Wiedenfeld 1995). The coastal habitats used by migratory Passerines are extremely important because estimates suggest that most Nearctic-Neotropical migrant Passerines are unable to reach northern breeding sites in a single flight without stopping (e.g., thrushes) (Yong and Moore 1997).

Passerines that are lean upon arrival often remain longer before departing for breeding grounds farther north (Moore and Kerlinger 1987). Length of stay in Louisiana after a trans-Gulf flight is related to fat-depletion upon arrival; lean birds (Parulinae warblers) remained longer than fat ones, but if weather is favorable, birds continue to migrate (Moore and Kerlinger 1987). For migrant Passerines using the northern Gulf Coast, suitable stopover habitat is a critical feature. Migrant densities were most strongly related to forest cover within a 5 km radius; this feature influenced where migrants made landfall (Buler et al. 2007). Indeed along the coast of Mississippi, northbound songbirds made landfall in resource-rich habitats within 18 km (11 mile) of the coastline (Buler and Moore 2011).

While radar is used to determine patterns of migration across the Gulf (Gauthreaux et al. 2006), data from banding stations are used to assess ecology of migrants, including timing (Moore et al. 1990; Marra et al. 2005), stopover duration (Moore and Yong 1991), and habitat use (Moore et al. 1990). Stable isotope techniques are used to connect the wintering and breeding grounds of Nearctic-Neotropical migrants (Hobson and Wassenaar 1997; Hobson et al. 2007; Hobson 2005; Kelly 2006b; Langin et al. 2009). The recent development of small geolocators makes it possible to follow migration routes of small birds, although they must be captured to remove the geolocator to obtain the data (Stutchbury et al. 2009; Burger et al. 2012b). This combination of techniques has revolutionized the understanding of migration, especially across the Gulf of Mexico and will continue to do so.

12.6.3.6 Audubon Christmas Bird Counts Along the Northern Gulf of Mexico

Niven and Butcher (2011) examined the status and trends of birds wintering along the U.S. northern Gulf of Mexico using the Audubon Christmas Counts from 1965 to 2011. Methods are described in the Methods section above. Their initial goal was to examine trends in light of the Deepwater Horizon oil spill, but there was not enough time between the spill and the counts to reflect the effects, if any, from the oil spill. To be on the conservative side, in Table 12.17 only the species with a significant decline of more than 2 % per year, and the species with a significant increase of over 2 %, are listed.

Table 12.17. Trends in Birds/Party Hours for the Northern Gulf of Mexico from 1965 to 2010 (developed from Niven and Butcher 2011)

Species	% Change	Taxonomic Group
Canada Goose	−7.0 (−2.63)	Anseriformes
Eared Grebe	−6.5 (−4.66)	Podicipediformes
Canvasback, *Aythya valisineria*	−5.6 (−3.85)	Anseriformes
American Wigeon	−5.2 (−3.62)	Anseriformes
Wilson's Plover	−4.8 (−2.93)	Charadriiformes
Northern Pintail	−4.0 (−2.67)	Anseriformes
Bonaparte's Gull, *Larus philadelphia*	−3.8 (−1.58)	Charadriiformes
King Rail	−3.2 (−1.69)	Gruiformes
Red-breasted Merganser, *Mergus serrator*	−2.8 (−1.82)	Anseriformes
Herring Gull	−2.6 (−1.85)	Charadriiformes
Red-winged Blackbird	−2.5 (−1.47)	Passeriformes
Boat-tailed Grackle	−2.3 (−0.14)	Passeriformes
Long-billed Curlew	−2.3 (−1.40)	Charadriiformes
Horned Grebe, *Podiceps auritus*	−2.3 (−0.74)	Podicipediformes
Western Sandpiper	−2.3 (−0.83)	Charadriiformes
Red Knot	−2.3 (−0.18)	Charadriiformes
Black Skimmer	−2.2 (−1.04)	Charadriiformes
American Woodcock, *Scolopax minor*	−2.1 (−0.21)	Charadriiformes
American Bittern	−2.1 (−1.12)	Pelecaniformes

(continued)

Table 12.17. (continued)

Species	% Change	Taxonomic Group
Seaside Sparrow	−2.0 (−0.71)	Passeriformes
Black-bellied Whistling Duck, *Dendrocygna autumnalis*	22.7 (18.6)	Anseriformes
Ross's Goose, *Chen rossii*	13.7 (10.5)	Anseriformes
Glossy Ibis	10.9 (7.9)	Pelecaniformes
Black-necked Stilt	10.4 (6.7)	Charadriiformes
Osprey	7.1 (6.5)	Accipitriformes
Northern Gannet	6.6 (2.2)	Suliformes
Hooded Merganser	5.9 (4.8)	Anseriformes
Roseate Spoonbill	5.9 (4.4)	Pelecaniformes
White Ibis	4.5 (3.1)	Pelecaniformes
White-faced Ibis	4.3 (1.3)	Pelecaniformes
American White Pelican	4.1 (2.4)	Pelecaniformes
Peregrine Falcon	4.0 (3.0)	Falconiformes
Bufflehead, *Bucephala albeola*	4.0 (1.9)	Anseriformes
Sandhill Crane	3.8 (2.4)	Gruiformes
Bald Eagle	3.8 (2.8)	Accipitriformes
Brown Pelican	3.7 (2.6)	Pelecaniformes
Greater White-fronted Goose, *Anser albifrons*	3.6 (1.0)	Anseriformes
Blue-winged Teal	3.5 (2.3)	Anseriformes
Anhinga	3.5 (2.9)	Charadriiformes
Marbled Godwit	3.5 (1.8)	Charadriiformes
American Oystercatcher	3.4 (2.0)	Charadriiformes
Laughing Gull	3.0 (2.1)	Charadriiformes
Palm Warbler, *Dendroica palmarum*	2.9 (1.8)	Passeriformes
Double-crested Cormorant	2.8 (2.0)	Suliformes
Wood Stork	2.8 (0.3)	Ciconiiformes
Merlin, *Falco columbarius*	2.5 (1.8)	Falconiformes
Black-crowned Night Heron, *Nycticorax nycticorax*	2.3 (1.3)	Pelecaniformes
Pied-billed Grebe	2.3 (1.6)	Gaviiformes
Common Moorhen	2.2 (0.8)	Gruiformes
Great Egret	2.1 (1.6)	Pelecaniformes
Sedge Wren	2.1 (0.9)	Passeriformes

Birds are given in decreasing order of change. For % change, the 95 % credible lower limit of decrease or increase is shown in parentheses. © J. Burger

Niven and Butcher (2011) reported that among the 20 species that declined by at least 2 % per year, 13 had the center of their ranges in the Gulf, and four declined in the Gulf, but were increasing elsewhere (Canada Geese [*Branta Canadensis*], American Wigeon, Bonaparte's Gull, Boat-tailed Grackle). These birds may be moving their wintering ranges farther north with global warming (Niven et al. 2009). Several species with winter ranges south of the northern Gulf of Mexico coast also declined (Eared Grebe [*Podiceps nigricollis*], Wilson's Plover, Long-billed Curlew, Western Sandpiper, Red Knot, Black Skimmer and American Bittern [*Botaurus lentiginosus*]). Remarkably, these were mainly shorebirds, reflecting a general decline in shorebirds worldwide (Withers 2002; Morrison et al. 2006). Surprisingly, Canada Geese showed the greatest decline.

While a 2 % per year increase or decline may not seem significant in terms of overall population dynamics, it is when an average of 2 % per year change for 40 years. Second, although there are methodological and interpretational difficulties with Christmas Bird Count data, the results are both consistent and robust over a long period of time. And finally, the changes make sense in terms of possible effects of global warming, and threats to birds that use coastal beaches (habitat losses, erosion, human disturbance, pets, and pollution).

12.7 DISCUSSION

12.7.1 Management

While this chapter does not address management specifically, management actions are discussed throughout as one of the factors affecting birds in the Gulf of Mexico. Management, however, is a complex mix of actions (dredging, hydrological control, diking, wetlands for aquaculture) aimed at improving the coastal environment for people, actions to improve ecosystem structure and function (e.g., terracing), and actions to aid particular species groups (burning for waterfowl habitat). There are management programs to restore large ecosystems in place as part of the larger Gulf of Mexico system. For example, a massive federal and state effort to restore the Everglades ecosystem features many of the prominent Gulf of Mexico species, such as White Ibis, Roseate Spoonbill, and other colonial-nesting species (Ogden et al. 2003, 2005; Frederick and Collopy 1989; Gawlik 2006). Managing for birds is difficult because it often involves trade-offs whereby a given action is positive for one species, but negative for another. Several examples have been provided in this chapter: (1) differences in salinity (affected by water control) favor some marsh species over others; (2) vegetation removal is positive for bare-sand nesting species, but not for those requiring some sparse or dense vegetation; and (3) large expanses of bare sand may encourage larger terns to nest, forcing smaller terns (e.g., Least Tern) out of otherwise optimal habitat.

Many agencies and organizations are devoted to protection and conservation of birds in general, and of coastal waterbirds in particular (e.g., SE U.S. Regional Waterbird Conservation Plan, Hunter et al. 2006; U.S. Shorebird Conservation Plan, USFWS 2004; U.S. Shorebird Conservation Plan for Lower Mississippi/Western Gulf Coast, GCPWG 2000). There are plans for particular species, such as the U.S. Fish & Wildlife Service Recovery Plans and 5-year reviews for the Piping Plover (USFWS 1999, 2003, 2009b), and the Whooping Crane Recovery Program (USFWS 1986, 2012c). Canadian provinces also produce plans for species of concern, such as the Piping Plover (White and McMaster 2005).

Status reports have been developed by the U.S. Fish & Wildlife Service for groups of birds, such as seabirds (Fritts and Reynolds 1981; Fritts et al. 1983), waterfowl (USFWS 2011), and waterbirds (Anderson et al. 1996). There are also status reports for individual species, such as Red Knot (Niles et al. 2008), American Oystercatcher (Clay et al. 2010), and Black Skimmer

(FFWCC 2011b), as well as national or international surveys (e.g., Haig et al. 2005; Elliott-Smith et al. 2009). Monitoring plans for species of concern, such as for the delisting of Brown Pelican (USFWS 2009a), have also been established. In addition, the Service develops Habitat Suitability Index Models for some species, such as Roseate Spoonbill (Lewis 1983), and evaluates the effect of offshore development on rare, threatened, and endangered species (Woolfenden 1983).

Management plans are developed for species groups such as for Colonial Waterbird Management (Coste and Skoruppa 1989), and federal agencies develop conservation strategies (MMNS 2005). Other organizations also produce assessment and trends documents. These include the Environmental Assessments by Natural Heritage Programs for the Mississippi Gulf Coast (NPS 2008), and the National Estuary Program for Texas (Tunnell and Alvarado 1996; Chaney et al. 1996) and for Alabama (MBNEP 2008), as well as for specific areas like Barataria-Terebonne (Condrey et al. 1996). Several states have breeding bird atlases (Kale et al. 1992 for Florida) and/or conduct annual surveys for waterfowl (TPWD 2011), colonial waterbirds (TCWS 2012), and shorebirds (Sprandel et al. 1997).

There are reports by management agencies whose major function is not bird protection, but have an additional mandate to protect species and the environment that relate directly to the Gulf of Mexico. The U.S. Army Corps of Engineers, tasked with beach nourishment, has incorporated creation of bird habitat in its management documents (Golder et al. 2006; Guilfoyle et al. 2006; Wilson and Vermillion 2006). The Bureau of Ocean Energy Management, Regulation, and Enforcement (formerly U.S. Minerals Management Service, U.S. Department of Interior) examined interactions between migrating birds and offshore oil and gas platforms (Russell 2005). The National Renewable Energy Laboratory considers bird movements and behavior in relation to wind energy developments (Morrison 2006).

Status evaluations are aimed at informing managers to guide policymakers and managers in making decisions. For example, the Galveston Bay Status and Trends Project (GBEP 2006) evaluates water and sediment quality, fisheries, habitat, data gaps, and indicators of bay health, using 20 years of trends data. The ratings for status of indicator species go from poor (significantly decreasing) to stable, to good (significantly increasing). Their report lists the following ratings: *poor* (Black-crowned Night Heron, Great Blue Heron, Tricolored Heron, White-faced Ibis, Laughing Gull, and Neotropical Cormorant), *stable* (Reddish Egret, Roseate Spoonbill, Snowy Egret, White Ibis, Black Skimmer, Least Tern, Royal Tern and Sandwich Tern), and *good* (only Brown Pelican). This is an excellent method of informing policymakers and the public at a glance.

All of these documents deal with status, threats, and management actions needed to restore, recover, or protect vulnerable populations. Specific methods depend upon the species, habitat, legal constraints (e.g., Endangered Species laws), geography, and species (or group) vulnerabilities.

12.7.2 Patterns of Population Changes

Several types of evidence help determine whether birds or groups of birds are increasing or decreasing, including data from Breeding Bird Surveys, Audubon Christmas Counts, Federal Species Surveys (Piping Plover, Snowy Plover), state inventories (waterfowl in Texas, colonial waterbirds in Texas), and local or refuge surveys. For some indicator species, the trends are clear and different surveys indicate the same patterns (e.g., increases in Brown Pelican, declines for Mottled Duck), but for others, the evidence is conflicting. Thus, the data in this chapter can be examined with a weight of evidence approach, whereby the different types of data are examined in total to determine population status and trends in the Gulf (Burger 2003; Krimsky 2005; Laiolo 2010). Thus, if all (or almost all) data sets suggest that a given species is increasing, it is likely that it is. Conversely, if all evidence suggests that a species is declining, it likely is. The quality of the data enters the deliberations, as does other factors, such as the temporal and spatial scale of the data, measurement error (or variability), and environmental variability.

Table 12.18 provides an overview of population trends in Breeding Bird Surveys (all North America), Breeding Bird Surveys for the Gulf States, Christmas Bird Counts for the Gulf (Niven and Butcher 2011), and from individual studies of species in the Gulf. Species in green denote an increasing population, from Gulf-based evidence (although the whole line is green), and red indicates an overall population decrease in the Gulf. Black indicates variable, and generally stable populations. A more in-depth analysis of status and trends up through 2015 can be found in Burger (2017).

Table 12.18. Comparison of Trends Data from Different Sources

Species	Breeding Bird Surveys (all na) % Change/Year	Breeding Bird Surveys from Gulf States	Audubon Christmas Bird Count % Change/Year (Gulf States)	Other Breeding Surveys or Reports from Gulf States
Common Loon	0.8	NA	1.6	Breeding populations probably stable to increasing generally, stable in the mid-west (Evers 2004; Evers et al. 2010)
Brown Pelican	6.5	Increases in Texas, and Alabama, and in part of Florida	3.7	Significant increases in Pelicans along coast 1970–2000 (Shields 2002). Increases in Galveston Bay (GBEP 2006; GBST 2012), Corpus Christi Bay (Chaney et al. 1996) and Queen Bess (Visser and Peterson 1994; Lindstedt 2005)
Whooping Crane	NG	NG	NG	Dramatic increases from 1938 to 2008 (USFWS 2012c; WCCA 2012)
Great Egret	3.1	Increases or stable in other Gulf states	2.1	Increases in Texas, but possible declines in some parts of Florida (McCrimmon et al. 2011)
Reddish Egret	NG	NG	1.6	Very variable number of breeding pairs from Shamrock Island, Texas, 1974 to 1999 (Gorman and Smith 2001), increase in Corpus Christi Bay (Chaney et al. 1996)
Roseate Spoonbill	8.8	Increases in Louisiana and Texas, with declines in Florida	5.9	Increased in Texas (Chaney et al. 1996; Dumas 2000) and in Florida Bay (Powell et al. 1989). Numbers at individual colonies very variable from 1974 to 1999 (Gorman and Smith 2001)

(continued)

Table 12.18. (continued)

Species	Breeding Bird Surveys (all na) % Change/Year	Breeding Bird Surveys from Gulf States	Audubon Christmas Bird Count % Change/Year (Gulf States)	Other Breeding Surveys or Reports from Gulf States
Mottled Duck	−3–3	Declines in all states, with small increases in parts of Texas and Florida	1.2	Midwinter counts show stable in Texas (TPWD 2011). Data from Mexico shows declines from 1983 to 2000 (Perez-Arteaga and Gaston 2004)
Osprey	2.8	Increases in gulf states with data	7.1	Only migratory counts, which indicated that they increased significantly in Florida Keys and Texas, with no increases in Veracruz (Smith et al. 2008)
Clapper Rail	−1.1	Declines in Texas, Louisiana and part of Florida (increase in part of Florida and in Alabama)	−0.2[a]	No trends data except BBS and Christmas Counts
Snowy Plover	NG	NG	−0.2[a]	Winter numbers higher in 2006 than 2001 (Elliott-Smith et al. 2009)
Piping Plover	NG	NA	−1.4[a]	Winter numbers vary, higher in 2006 than 1996 and 2001, but 2006 still lower than 1991 (Elliott-Smith et al. 2009)
Laughing Gull	4.8	Declines in Texas, both increases and declines in Louisiana, and increases in the other Gulf States	3.0	Declined in Galveston Bay from 1973 to 1996 (Glass and Roach 1997), varied from 1973 to 1999 in Shamrock Island, but highest in 1992, then declined in Texas (Gorman and Smith 2001; GBEP 2006), declined in Louisiana from 1976 to 1993 (Visser and Peterson 1994). Declines in Christmas Count data since 1990 bear examination. Recent data indicates declines in Gulf States
Royal Tern	−1.5	Increases in all Gulf states with data	0.5[a]	Declined from 1985 to 1993 in Louisiana (Visser and Peterson 1994), increased and declined on Shamrock Island (Gorman and Smith 2001), and increases at Corpus Christi Bay (Chaney et al. 1996)

(continued)

Table 12.18. (continued)

Species	Breeding Bird Surveys (all na) % Change/Year	Breeding Bird Surveys from Gulf States	Audubon Christmas Bird Count % Change/Year (Gulf States)	Other Breeding Surveys or Reports from Gulf States
Black Skimmer	−3.6	Major declines in Florida, Louisiana, and Alabama, with slight increase in Texas	−2.2	Declines in Louisiana from 1976 to 1993 (Visser and Peterson 1994). Statewide declines in Florida (FFWCC 2011b), and at Shamrock Island, Texas from 1974 to 1999 (Gorman and Smith 2001), and declines at Corpus Christi (Chaney et al. 1996)
Seaside Sparrow	3.9	NG	−2.0	No breeding trends data

NG = not given in the relevant paper(s). NA = Breeding Bird data not given for birds that do not breed in the Gulf. Green = increasing trends and Red = declines from all sources. Black = no trend or conflicting trends. © J. Burger
[a]Stable or uncertain (not significant)
Sources: U.S. Breeding Bird Survey data from Sauer and Link (2011); Gulf is from U.S. Bird Banding laboratory; Christmas Bird Count data are from Niven and Butcher (2011); other sources refers to several different papers. Sauer and Link (2011) data are given as % change/year using hierarchical models. Christmas Bird Count data are reported as % change using hierarchical models.

12.7.3 Recovery and Population Dynamics

For any complex system, it is possible to catalogue biodiversity (number of species by taxa). For some species, there are estimates of current population sizes, and perhaps trends in populations. From this review, however, it is clear that even for the key indicator species or groups, current data on population sizes for the entire Gulf are usually sparse. No Gulf-wide surveys are taken at the same time using the same methodology. However, even if sufficient surveys of populations for key species were available, this information does not necessarily provide a picture of the health of the system, predict emergent ecosystem problems, or predict future trends. This is especially true for the Gulf of Mexico because of both short-term (storms, hurricanes, tides, pollution, habitat loss, human disturbance) and long-term stressors (habitat loss, subsidence, global climate change, sea level rise). Detailed information about trophic levels, food web interactions, energy flow, and forcing functions are needed to predict emerging ecosystem change (Brown et al. 2006). This information is not available for the Gulf ecosystem, although the chapters in this series begin to bring together some of this information. However, sufficient information is available to support an integrated weight of evidence evaluation of the health of avian populations in the Gulf of Mexico.

The available database includes (1) natural history information (age of first breeding, clutch size, incubation period, parental-care period, and life span); (2) status and trends for key indicator species or species groups; (3) effects of natural and anthropogenic stressors on habitat use, and; (4) effects of social interactions (predators, competitors) on habitat use. These factors provide information on whether populations can recover quickly or not.

Information on long-term recoveries is available for the species that were devastated during the plume-hunting days of the late 1800s to the early 1900s (herons, egrets, terns), or by exposure to pesticides (Osprey, Brown Pelican), which provides insights into recovery potential. Finally, the long-term sustainability of bird populations in the Gulf is a matter of balancing the needs of people, society, economics, and the fish and wildlife ecosystems that reside there. It will ultimately depend upon the ability of governments and people to balance these different needs.

12.8 SUMMARY OF BIRDS IN THE GULF OF MEXICO

The Gulf of Mexico is a complex mosaic of habitats influenced by political, economic, sociological, and biological factors, as well as global change, sea level rise and land subsidence, tides, storms, and hurricanes. The ecosystem in the Gulf of Mexico is a matrix of tropical, subtropical and temperate habitats that include different land masses and different land margin interfaces. There are large peninsulas (Florida, Yucatán), large islands (Cuba), barrier islands, open water, and an array of offshore islands or keys, barrier beaches, sandy and gravel beaches, mangroves, salt marshes, and brackish marshes that intergrade to freshwater marshes, swamps, and more upland habitats.

The Gulf of Mexico is one of the most important places for birds in the Western Hemisphere. Birds from North America funnel over or around the Gulf of Mexico on their migratory flights; birds from both the south and the north come to winter along the Gulf or on the open water, and many species breed there. Thus, the coastal areas around the Gulf of Mexico serve as a hotspot of diversity. Several conclusions can be drawn for the Gulf as a whole:

- Most birds that use the saltwater to brackish ecosystems are seabirds, herons and egrets, shorebirds, waterfowl, gulls, terns, and specialized marsh species (Clapper Rail, Seaside Sparrow).

- About 31 % of the 395 species in the Gulf have been recorded in all areas of the Gulf.

- A higher diversity of species is found in the southern part of the Gulf compared to the north.

- A high percentage of some colonial nesting species for North America nest in Louisiana and Texas than elsewhere along the Gulf, including Reddish Egrets, Sandwich Tern, Black Skimmer, Royal Tern, Forster's Tern, and Laughing Gull, as well as Snowy Plover and Roseate Spoonbill.

- Several seabirds, such as boobies and Magnificent Frigatebirds primarily nest in the southern Gulf of Mexico, on the Campeche Banks.

- One of the greatest impacts on avian populations in the Gulf is habitat loss (either because less is available, or what is available is no longer suitable), followed by human disturbance.

- Populations of birds in the Gulf have varied in the last 50 years; some have increased and some have declined.

For the Gulf of Mexico it is necessary to distinguish between habitat availability and habitat suitability. Habitat availability is whether habitat is present and available that meets the needs of the species or species groups, such as open sandy beaches for shorebirds to feed; salt marshes for Clapper Rail and Seaside Sparrow to breed and forage; isolated islands with suitable vegetation for Brown Pelicans, terns, skimmers, herons and egrets to nest; and bare sandy beaches for Snowy Plover to breed and forage. Habitat suitability, however, refers to whether the habitat will actually meet the needs of birds with respect to providing adequate places to forage, roost, breed, and migrate, free from predators, human disturbance, high tides and storm tides, and other weather-related events. Habitat must meet the species requirements in terms of vegetation, elevation, and physiognomy, while habitat suitability relates to whether the habitat is usable in terms of predator isolation, and freedom from human disturbance. The factors that affect suitability often relate to exposure to elements (storms, tides, winds, hurricanes, floods, and over the long term, sea level rise), exposure to predators and people, degree of competition from conspecific and interspecific interactions, presence of pollutants, and physical disruptions. In short, the habitat must allow survival and reproduction. In many cases, suitable habitat is only available on islands or cays isolated from the mainland.

Habitat loss is a major factor affecting bird populations in the Gulf. Loss of habitat affects all birds, whether residents, migrants, or wintering birds. It also affects all aspects of their daily needs for breeding and nesting, foraging, and having sufficient safe places to roost. Loss of habitat is most severe at the land margin, where the land meets the sea. And it is most severe where anthropogenic activities occur, where land is modified and is no longer usable, or where land is completely developed.

Pollutants have affected behavior and populations of birds in the Gulf, although to a far lesser degree than habitat loss and modification. In the 1950s and 1960s, DDT had a great effect on fish-eating birds, such as Osprey, wading birds, and Brown Pelicans, all of which declined dramatically. Pelicans were hit especially hard, and were largely extirpated as a successful breeding bird from some regions. Mercury has affected behavior and reproduction in both resident birds (Great Egrets and other fish-eating birds), and migrants (Common Loon). Oil, while it can cause immediate mortality and chronic injury, has not been demonstrated to permanently affect any populations of birds in the Gulf. Plastics and fishing lines cause mortality, particularly in seabirds foraging in the Gulf, but the long-term effects are unclear.

Understanding avian assemblages that use the Gulf of Mexico entails examining several different factors: migrant versus resident, solitary versus colonial nester, ground versus tree nester, method of foraging, and location of foraging. The 15 indicator species examined illustrate all of these different lifestyles and behavioral patterns. Obviously nesting on the ground exposes nests, eggs, and chicks to ground predators, tidal flooding, and human disturbance, while nesting in trees exposes birds to aerial predators but usually protects them from mammalian predators. Nesting on low islands prevents mammalian predators from surviving because high tides or severe storms in the winter wash them away, but nesting there exposes birds to flooding from high tides and storms during the breeding season. Further, the indicators illustrate different life strategies; some delay breeding, have small clutch sizes, long parental care, and long life spans (Common Loon, Royal Tern). Other species breed when they are 1 year old, have large clutches, and short life spans (Mottled Duck, Clapper Rail). These factors determine how fast a species can recover from any stressor, whether natural or man-made.

The indicators illustrate a range of population trends: some are increasing; some are decreasing. In some, the variation from year to year is so great that it is difficult to ascertain trends. In others, fidelity to colony sites is so low that it is nearly impossible to census them accurately, and often their populations fluctuate wildly from year to year, depending upon water levels. Nonetheless, for the 15 indicator species, several lines of data indicate decline over the last 45 years for Mottled Duck, Clapper Rail, and Black Skimmer, and clear increases for Brown Pelican, Great Egret, Osprey, and Laughing Gulls, although data from the last 15–20 years indicate that Laughing Gull is declining.

Overall declines seem to be due to habitat loss, coupled with human disturbance and other disruptions to beach, salt marsh, and coastal environments. Dramatic increases are often due to laws and regulations (endangered species laws, cessation of the use of pesticides, e.g., Brown Pelican, Osprey), to specific management practices (Whooping Crane, Piping Plover), to habitat creation (Brown Pelican), inadvertent management (dredge spoil islands for Snowy Plover and other beach nesting species), and possibly to global warming (more northern movement of southern species, such as Roseate Spoonbill).

The avian communities of the Gulf of Mexico are varied and diverse, largely because of the diversity of habitats, the richness of the marine-land interface, the presence of a gradient from tropical to temperate, and the geography of the Gulf, which places it as the funnel point for Nearctic-Neotropical migrants. Changes in the avian community occur because of short-term and long-term stressors, which render habitat either suitable or unsuitable. Habitat loss in the Gulf, which is continuing at an alarming rate due to both natural and anthropogenic causes, will result in changes to the bird communities that can only be countered by protection and management, and that require monitoring to assess the overall health of avian communities. Finally, the needs and requirements of the avian communities must be viewed within the context of the human communities that also thrive along the Gulf Coast. And management, protection, and conservation of birds must be designed with the human dimension in mind.

ACKNOWLEDGMENTS

BP sponsored the preparation of this chapter. This chapter has been peer reviewed by anonymous and independent reviewers with substantial experience in the subject matter. I thank the peer reviewers, as well as others who provided assistance with research and the compilation of information. In particular, I want to thank my able research assistants, Taryn Pittfield, Sheila Shulka and Christian Jeitner who helped with literature reviews, data extraction and figure design, as well as in countless other ways. M. Gochfeld read drafts and provided many useful references, knowledge, and experience. I also thank Jason Wells, ENVIRON International Corporation, for developing the habitat maps of the Gulf of Mexico with emphasis on those used by birds.

REFERENCES

ABC (American Bird Conservancy) (2011) ABC protects Laguna Madre birds impacted by the Gulf oil spill. Am Bird Conservancy Newsl. p. 7

Able KP (1999) Gatherings of angels: Migrating birds and their ecology. Comstock Publishing, New York, NY, USA. 208 p

Acosta M, Mugica L, Mancina C, Ruiz X (1996) Resource partitioning between Glossy and White Ibises in a rice field system in south central Cuba. Colon Waterbird 19:65–72

Acosta M, Mugica L, Blanco D, Lopez-Lanus B, Dias RA, Doodnath LW, Hurtado J (2010) Birds of rice fields in the Americas. Waterbirds 33:105–122

Albers PH, Koterba MT, Rossmann R, Link WA, French JB, Bennett RS, Bauer WC (2007) Effects of methylmercury on reproduction in American Kestrels. Environ Toxicol Chem 26:1856–1866

Alexander LL (1991) Patterns of mortality among common loons wintering in the northeastern Gulf of Mexico. Fla Field Nat 19:73–79

Allen RP (1952) The whooping crane. Resource Report 3. National Audubon Society, New York, NY, USA. 246 p

Amirault DL, Kierstead L, MacDonald P, MacDonnell L (2004) Sequential polyandry in piping plovers, Charadrius melodus, nesting in Eastern Canada. Can Field Nat 118:444–446

Anderson JT, Tacha TC, Muehl GT, Lobpries D (1996) Wetland use by waterbirds that winter in coastal Texas. Information and Technology Report 8. U.S. Fish and Wildlife Service, Washington, DC, USA. 40 p

AOU (American Ornithologists' Union) (1998) Check-list of North American birds: The species of birds of North America from the Arctic through Panama, including the West Indies and Hawaiian Islands, 7th edn. American Ornithologist' Union, Washington, DC, USA. 769 p

Ashmole NP (1971) Seabird ecology and the marine environment. In: Farner DS, Kind JR (eds) Avian Biology, Volume 1. Academic Press, New York, NY, USA, pp 223–286

Au DW, Pitman RL (1988) Seabird relationships with tropical tunas and dolphins. In: Burger J (ed) Seabirds and other marine vertebrates: Competition, predation, and other interactions. Columbia University Press, New York, NY, USA, pp 166–204

Audubon Society (2012) Important bird areas along the Gulf Coast. New York, NY, USA. http://web4.audubon.org/bird/iba/gulfIBAs.html. Accessed 18 Feb 2013

AWFWF (Alabama Division of Wildlife and Fresh Water Fisheries) (2005) Conserving Alabama's wildlife: A comprehensive strategy. Alabama Department of Conservation and Natural Resources, Montgomery, AL, USA. 322 p. http://www.outdooralabama.com/research-mgmt/cwcs/outline.cfm. Accessed 7 Apr 2012

Azzarello MY, Van Vleet ES (1987) Marine birds and plastic pollution. Mar Ecol 37:295–303

Baker AJ, Gonzalez PM, Piersma T, Niles LJ, de Lima Serrano do Nascimento I, Atkinson PW, Clark NA, Minton CDT, Peck MK, Aarts G (2004) Rapid population decline in red knots: Fitness consequences of decreased refuelling rates and late arrival in Delaware Bay. Proc R Soc Lond B Biol Sci 271:875–882

Barr JF (1996) Aspects of common loon (*Gavia immer*) associated with mercury-contaminated waters in north-western Ontario. Can Wildl Serv Occ Pap 56:1–25

Barrow Jr WC, Randall LA, Woodrey MS, Cox J, Ruelas E, Riley CM, Hamilton RB, Eberly C (2005) Coastal forests of the Gulf of Mexico: A description and some thoughts on their conservation. USDA Forest Servc Gen Tech Rep PSW-GTR-191

Barrow WC Jr, Chadwick P, Couvillion B, Doyle T, Faulkner S, Jeske C, Michot T, Randall L, Wells C, Wilson S (2007) Cheniere forest as stopover habitat for migrant landbirds: Immediate effects of hurricane Rita. In: Farris GS, Smith GL, Crane MP, Demas CR, Robbins LL, Lavoie DL (eds) Science and the storms—the USGS response to the hurricanes of 2005. U.S. Geological Survey Circular 1306, National Wetlands Research Center Technical Report, pp 147–156

Bates EM, Deyoung RW, Ballard BM (2009) Genetic diversity and population structure of reddish Egrets along the Texas coast. Waterbirds 32:430–436

Bayard TS, Elphick CS (2011) Planning for sea-level rise: Quantifying patterns of Saltmarsh Sparrow (*Ammodramus caudacutus*) nest flooding under current sea-level rise conditions. Auk 128:393–403

Bellrose FC (1988) The adaptability of the Mallard leads to its future. In: Johnson MD (ed) Proceedings of the Mallard symposium. North Dakota Chapter, The Wildlife Society, Bismarck, ND, USA, pp 5–10

Benoit L, Askins R (1999) Impact of the spread of Phragmites on the distribution of birds in Connecticut tidal marshes. Wetlands 19:194–208

Beshears WW (1979) Waterfowl in the mobile estuary. In: Loyacano HJ, Smith J (eds) Symposium on the natural resources of mobile estuary, Alabama. U.S. Army Corps of Engineers, Mobile District, Mobile, AL, USA, pp 249–262

Beyer WN, Heinz GH, Redmon-Norwood AW (1996) Environmental contaminants in wildlife: Interpreting tissue concentrations. Society of Environmental Toxicology and Chemistry Special Publication Series. Lewis, Boca Raton, FL, USA. 494 p

Bielefeld RR, Brasher MG, Moorman TE, Gray PN (2010) Mottled duck (*Anas fulvigula*). In: Poole A (ed) The birds of North America online Cornell Lab of Ornithology, Ithaca, NY, USA. http://bna.birds.cornell.edu/bna/species/081. Accessed 13 Feb 2013

Bildstein KL, Bancroft GT, Dugan PJ, Gordan DH, Erwin MR, Nol E, Payne LX, Senner SE (1991) Approaches to the conservation of coastal wetlands in the western hemisphere. Wilson Bull 103:218–254

Bjork RD, Powell GVN (1994) Relationships between hydrologic conditions and quality and quantity of foraging habitat for Roseate Spoonbills and other wading birds in the C-111 basin. Final Report to the South Florida Research Center, Everglades National Park, Homestead, FL, USA

Black BB, Harris LD (1983) Feeding habitat of black skimmers wintering on the Florida gulf coast. Wilson Bull 95:404–45

Blus LJ, Neely BS Jr, Belisle AE, Prouty RM (1974) Organochlorine residues in brown pelicans: Relation to reproductive success. Environ Pollut 7:81–91

Bolduc F, Afton AD (2004) Relationships between wintering waterbirds and invertebrates, sediments and hydrology of coastal marsh ponds. Waterbirds 27:333–341

Bradshaw WE, Holzapfel CM (2006) Evolutionary response to rapid climate change. Science 312:1477–1478

Brand CJ, Duncan RM, Garrow SP, Olson D, Schumann LE (1983) Waterbird mortality from botulism type E in Lake Michigan: An update. Wilson Bull 95:269–275

Briggs KT, Lewis LB, William TB, Hunt GL Jr (1981) Brown pelicans in southern California: Habitat use and environmental fluctuation. Condor 83:1–15

Brown RB (1980) Seabirds as marine animals. In: Burger J, Olla B (eds) Behavior of Marine Animals, Volume 6: Shorebirds: Migration and Foraging Behavior. Plenum Press, New York, NY, USA, pp 1–31

Brown MT, Cohen MJ, Bardi E, Ingwersen WW (2006) Species diversity in the Florida Everglades, USA: A systems approach to calculating biodiversity. Aquat Sci 68:254–277

Brown S, Hickey C, Harrington B, Gill R (eds) (2001) United shorebird conservation plan, 2nd edn. Manomet Center for Conservation Sciences, Manomet, MA, USA. 64 p

Bruce KA, Cameron GN, Harcombe PA (1995) Initiation of a new woodland type on the Texas coastal prairie by the Chinese tallow tree (*Sapium sebiferum (L.) Roxb.*). Bull Torry Bot Club 122:215–225

Buckley FG, Buckley PA (1974) Comparative feeding ecology of wintering adult and juvenile Royal Terns (Aves: Laridae, Sterninae). Ecology 55:1053–1063

Buckley FG, Buckley PA (1980) Habitat selection and marine birds. In: Burger J, Olla B (eds) Behavior of Marine Animals, Volume 6: Shorebirds: Migration and Foraging Behavior. Plenum Press, New York, NY, USA, pp 69–107

Buckley PA, Buckley FG (2002) Royal Tern (*Thalasseus maximus*). In: Poole A (ed) The birds of North America online. Cornell Lab of Ornithology, Ithaca, NY, USA. http://bna.birds.cornell.edu/bna/species/700. Accessed 13 Feb 2013

Buler JJ, Moore FR (2011) Migrant-habitat relationships during stopover along an ecological barrier: Extrinsic constraints and conservation implications. J Ornithol 152:S101–S112

Buler JJ, Moore FR, Woltmann S (2007) A multi-scale examination of stopover habitat by birds. Ecology 88:1789–1802

Burger J (1978a) The pattern and mechanism of nesting in mixed species heronries. Wading birds. Research Report 7. National Audubon Society, New York, NY, USA, pp 45–58

Burger J (1978b) Nest-building behavior in laughing gulls. Anim Behav 26:856–861

Burger J (1979a) Competition and predation: Herring gulls versus laughing gulls. Condor 81:269–277

Burger J (1979b) Resource-partitioning: Nest site selection in North America herons and egrets. Am Midland Nat 101:191–210

Burger J (1980) Age differences in foraging black-necked stilts in Texas. Auk 97:634–363

Burger J (1981a) A model for the evolution of mixed species colonies of Ciconiiformes. Q Rev Biol 56:1443–167

Burger J (1981b) Aggressive behavior of black skimmers (*Rynchops niger*). Behaviour 76:207–227

Burger J (1981c) Effects of human disturbance on colonial species, particularly gulls. Colon Waterbird 4:28–36

Burger J (1981d) The effect of human activity on birds at a coastal bay. Biol Conserv 21:231–241

Burger J (1981e) Sexual difference in parental activities of breeding black skimmers. Am Nat 117:975–984

Burger J (1982) An overview of factors affecting reproductive success in colonial birds. Colon Waterbirds 5:58–123

Burger J (1983) Competition between two nesting species of gulls: On the importance of timing. Behav Neurosci 97:492–501

Burger J (1984a) Colony stability in least terns. Condor 86:61–67

Burger J (1984a) Advantages and disadvantages of mixed-species colonies of seabirds. Proceedings of 18th International Ornithological Congress, pp 905–918

Burger J (1987a) Foraging efficiency in gulls: A congeneric comparison of age differences in efficiency and age of maturity. Stud Avian Biol 10:83–89

Burger J (1987b) Physical and social determinants of nest-site selection in piping plover (*Charadrius melodus*). Condor 89:811–818

Burger J (1988a) Foraging behavior in gulls: Differences in method, prey, and habitat. Colon Waterbird 11:9–23

Burger J (1988b) Interactions of marine bird with other marine vertebrates in marine environments. In: Burger J (ed) Seabirds and other marine vertebrates: Competition, predation, and other interactions. Columbia University Press, New York, NY, USA, pp 3–31

Burger J (1988c) Seabirds and other marine vertebrates: Competition, predation, and other interactions. Columbia University Press, New York, NY, USA. 339 p

Burger J (1989) Least tern populations in coastal New Jersey: Monitoring and managing of a regionally endangered species. J Coast Res 5:801–811

Burger J (1991a) Coastal landscapes, coastal colonies and seabirds. Aquat Rev 4:23–43

Burger J (1991b) Foraging behavior and the effect of human disturbance on the piping plover (*Charadrius melodus*). J Coast Res 7:39–52

Burger J (1993) Metals in avian feathers: Bioindicators of environmental pollution. Rev Environ Toxicol 5:197–306

Burger J (1994a) Before and after an oil spill: The Arthur Kill. Rutgers University Press, New Brunswick, NJ, USA. 305 p

Burger J (1994b) The effect of human disturbance on foraging behavior and habitat use in piping plover (*Charadrius melodus*). Estuar Coast Shelf Sci 17:695–701

Burger J (1996a) A naturalist along the Jersey shore. Rutgers University Press, New Brunswick, NJ, USA. 305 p

Burger J (1996a) Laughing gull (*Leucophaeus atricilla*). In: Poole A (ed) The birds of North America online Cornell Lab of Ornithology, Ithaca, NY, USA. http://bna.birds.cornell.edu/bna/species/225. Accessed 14 Feb 2013

Burger J (1997a) Oil spills. Rutgers University Press, New Brunswick, NJ, USA. 261 p

Burger J (1997b) Effects of oiling on feeding behavior of sanderlings (*Calidris alba*) and semipalmated plovers (*Charadrius semipalmatus*) in New Jersey. Condor 99:290–298

Burger J (2003) Differing perspectives on the use of scientific evidence and the precautionary principle. Pure Appl Chem 75:2543–2545

Burger J (2006a) Bioindicators: A review of their use in the environmental literature 1970–2005. Environ Bioindic 1:136–144

Burger J (2006b) Bioindicators: Types, development, and use in ecological assessment and research. Environ Bioindic 1:22–39

Burger J (2009) Risk to consumers from mercury in bluefish (*Pomatomus saltatrix*) from New Jersey: Size, season, and geographic effects. Environ Res 109:803–811

Burger J (2017) Birds of the Gulf of Mexico. Texas A & M University Press, College Station, TX, USA (in press)

Burger J, Gochfeld M (1981) Age-related differences in piracy behavior of four species of gulls, *Larus*. Behaviour 77:242–267

Burger J, Gochfeld M (1983a) Behavior of nine avian species at a Florida garbage dump. Colon Waterbirds 6:54–63

Burger J, Gochfeld M (1983b) Behavioral responses of herring (*Larus argentatus*) and great black-backed (*Larus marinus*) gulls to variation in the amount of human disturbance. Behav Processes 8:327–344

Burger J, Gochfeld M (1983c) Feeding behavior in laughing gulls: Compensatory site selection by young. Condor 85:467–473

Burger J, Gochfeld M (1990) The black skimmer: Social dynamics of a colonial species. Columbia University Press, New York, NY, USA. 355 p

Burger J, Gochfeld M (1992) Experimental evidence for aggressive antipredator behavior in Black Skimmer (*Rynchops niger*). Aggress Behav 18:241–248

Burger J, Gochfeld M (1994) Behavioral impairments of lead-injected young herring gulls in nature. Fundam Appl Toxicol 23:553–561

Burger J, Gochfeld M (2000) Effects of lead on birds (Laridae): A review of laboratory and field studies. J Toxicol Environ Health 3:59–78

Burger J, Gochfeld M (2001) Effects of chemicals and pollution on seabirds. In: Schreiber EA, Burger J (eds) Biology of marine birds. CRC Press, Boca Raton, FL, USA, pp 485–526

Burger J, Gochfeld M (2004a) Metal levels in eggs of common terns (*Sterna hirundo*) in New Jersey: Temporal trends from 1971 to 2002. Environ Res 94:336–343

Burger J, Gochfeld M (2004b) Bioindicators for assessing human and ecological health. In: Wiersma B (ed) Environmental monitoring. CRC Press, Boca Raton, FL, USA, pp 542–566

Burger J, Gochfeld M (2005) Effects of lead on learning in herring gulls: An avian wildlife model for neurobehavioral deficits. Neurotoxicology 26:615–624

Burger J, Olla B (eds) (1984) Behavior of Marine Animals, Volume 6: Shorebirds: Migration and Foraging Behavior. Plenum Press, New York, NY, USA. 329 p

Burger J, Shisler J (1978) Nest site selection in Willets. Willson Bull 90:559–607

Burger J, Shisler J (1980) Colony site and nest selection in laughing gulls in response to tidal flooding. Condor 82:251–258

Burger J, Tsipoura N (1998) Experimental oiling of Sanderlings (*Calidris alba*): Behavior and weight changes. Environ Toxicol Chem 17:1154–1158

Burger J, Olla BL, Winn HE (1980) Behavior of Marine Animals. Volume 4: Marine Birds. Plenum Press, New York, NY, USA. 515 p

Burger J, Pokras M, Chafel R, Gochfeld M (1994) Heavy metal concentrations in feathers of Common Loons (*Gavia immer*) in the Northeastern United States and age differences in mercury levels. Environ Monit Assess 30:1–7

Burger J, Kurunthachalam K, Giesy JP, Grue C, Gochfeld M (2002) Effects of environmental pollutants on avian behavior. In: Dell'omo G (ed) Behavioural ecotoxicology. Wiley, West Sussex, UK. 433 p

Burger J, Jeitner C, Clark K, Niles L (2004) The effect of human activities on migrant shore-birds: Successful adaptive management. Environ Conserv 31(4):283–288

Burger J, Carlucci SA, Jeitner CW, Niles L (2007) Habitat choice, disturbance, and management of foraging shorebirds and gulls at a migratory stopover. J Coast Res 23:1159–1166

Burger J, Jeitner C, Gochfeld M (2011) Locational differences in mercury and selenium levels in 19 species of saltwater fish from New Jersey. J Toxicol Environ Health A 74:863–874

Burger J, Niles LJ, Porter RR, Dey AD, Koch S, Gordon C (2012a) Migration and over-wintering of red knots (*Calidris canutus rufa*) along the Atlantic coast of the United States. Condor 114:302–313

Burger J, Niles LJ, Porter RR, Dey AD (2012b) Using geolocators to reveal incubation periods and breeding biology in red knots. Wader Study Group Bulletin 119:26–36

Burgess NM, Meyer MW (2008) Methylmercury exposure associated with reduced productivity in common loons. Ecotoxicology 17:83–91

Burgess NM, Evers DC, Kaplan JD (2005) Mercury and other contaminants in common loons breeding in Atlantic Canada. Ecotoxicology 14:241–252

Butcher GS (1990) Audubon Christmas bird count. In: Sauer JA, Droege S (eds) Survey designs and statistical methods for the estimation of avian population trends. U.S. Fish Wildlife Service Biological Report 90, pp 5–13

Butler RW (2000) Stormy seas for some North American songbirds: Are declines related to severe storms during migration? Auk 117:518–522

Cairns D (1980) Nesting density, habitat structure and human disturbance as-factors-in Black Guillemot reproduction. Wilson Bull 92:352–361

Carney KM, Sydeman WJ (1999) A review of disturbance effects on nesting colonial water-birds. Waterbirds 22:68–79

Chaney AH, Blacklock GW, Bartels SG (1996) Current status and historical trends of avian resources in the Corpus Christi Bay National estuary program study area, Volume 2. CCBNEP-06B. Corpus Christi Bay National Estuary Program, Corpus Christi, TX, USA

Chapman BR (1981) Effects of the *Ixtoc* I oil spill Texas shorebird populations. Proceedings, 1981 Oil Spill Conference, American Petroleum Institute, Washington, DC, USA, pp 461–465

Chapman BR (1984) Seasonal abundance and habitat-use patterns of coastal bird populations on Padre and Mustang Island barrier beaches. Following the *Ixtoc* 1 Oil Spill. Contract 14-16-0009-80-062. U.S. Fish and Wildlife Service FWS/OBS-83/31. U.S. Fish and Wildlife Service, Washington, DC, USA

Chardine JW (2000) Census of Northern Gannet colonies in the Atlantic Region in 1999. Canadian Wildlife Service Atlantic Region Technical Report No 361

Clapp RB, Morgan-Jacobs D, Banks RC, Hoffman WA (1982a) Marine birds of the Southeast-ern United States and Gulf of Mexico. Part I. Gaviiformes through Pelecaniformes. NTIS PB82-195850. FWS Report FWS/OBS-82/01. Final Report. U.S. Fish and Wildlife Service, Minerals Management Service Gulf of Mexico OCS Office, Metairie, LA, USA. 648 p

Clapp RB, Morgan-Jacobs D, Banks RC (1982b) Marine birds of the Southeastern United States and Gulf of Mexico. Part II. Anseriformes. NTIS PB82-264995. FWS Report FWS/OBS-82/20. Minerals Management Service Gulf of Mexico OCS Office, Metairie, LA, USA. 505 p

Clapp RB, Morgan-Jacobs D, Banks RC (1983) Marine birds of the Southeastern United States and Gulf of Mexico. Part III. Charadriiformes. NTIS PB84-158773. FWS Report FWS/OBS-83/80. Minerals Management Service Gulf of Mexico OCS Office, Metairie, LA, USA. 869 p

Clay RP, Lesterhuis AJ, Schulte S, Brown S, Reynolds D, Simons TR (2010) Conservation plan for the American oystercatcher (*Haematopus palliatus*) throughout the western hemi-sphere. Version 1.1. Manomet Center for Conservation Sciences, Manomet, MA, USA. 53 p

Cohen JB, Karpanty SM, Catlin DH, Fraser JD, Fischer RA (2008) Winter ecology of piping plovers at Oregon Inlet, North Carolina. Waterbirds 31:472–479

Cohen JB, Houghton LM, Fraser JD (2009) Nesting density and reproductive success of piping plovers in response to storm and human-created habitat changes. Wildl Monogr 173:1–24

Condrey R, Kemp P, Visser J, Gosselink J, Lindstedt D, Melancon E, Peterson G, Thompson B (1996) Status and trends, and possible causes of change in living resources in the Barataria and Terrebonne estuarine systems. BTNEP Publication 21. Barataria-Terrebonne National Estuary Program, Thibodaux, LA, USA. 434 p

Conner WH, Day JW, Baumann RH, Randall JM (1989) Influence of hurricanes on coastal ecosystems along the northern Gulf of Mexico. Wetl Ecol Manag 1:45–56

Correa SJ, Berlanga MC, Wood P, Salgado JO, Figueroa EM (2000a) Río Celestún. In: Arizmendi MC, Márquez LV (eds) Áreas de Importancia para la Conservación de Aves en México. Conabio, México, pp 71–72

Correa SJ, García JB, García SP (2000b) Ría Lagartos. In: Arizmendi MC, Márquez LV (eds) Áreas de Importancia para la Conservación de Aves en México. Conabio, México, pp 73–74

Correa SJ, Salgado JO, Figueroa EM, Berlanga M (2000c) Los Petenes. In: Arizmendi MC, Márquez LV (eds) Áreas de Importancia para la Conservación de Aves en México. Conabio, México, pp 194–195

Coste RL, Skoruppa MK (1989) Colonial waterbird Rookery Island management plan for the South Texas coast. Prepared for U.S. Fish and Wildlife Report No. 14-16-0002-86-919. Corpus Christi State University, Corpus Christi, TX, USA

Coulson JC (2001) Colonial breeding in seabirds. In: Schreiber EA, Burger J (eds) Biology of marine birds. CRC Press, Boca Raton, FL, USA, pp 87–113

Cowan JH, Turner RE, Cahoon DR (1988) Marsh management plans in practice: Do they work in coastal Louisiana, USA. Environ Manage 12:37–53

Cowardin LM, Carter F, Golet C, LaRoe ET (1979) Classification of wetlands and deepwater habitat of the United States. Fish and Wildlife Service, U.S. Department of the Interior, Washington, DC, USA

Cramp S, Simmons KLE (1977) The birds of the western Paleartic. Volume 1: Ostrich to duck. Oxford University Press, Oxford, UK. 722 p

Custer TW (2000) Environmental contaminants. In: Kushlan JA, Hafner H (eds) Heron conservation. Academic Press, New York, NY, USA, pp 251–267

Custer TW, Mitchell CA (1987) Organochlorine contaminants and reproductive success of Black Skimmers in South Texas, 1984. J Field Ornithol 58:480–489

CWS/USFWS (Canadian Wildlife Service and U.S. Fish and Wildlife Service) (2007) International recovery plan for the Whooping Crane. Recovery of Nationally Endangered Wildlife, Ottawa, ON, Canada, U.S. Fish and Wildlife, Albuquerque, NM, USA

Daniels RC, White TW, Chapman KK (1993) Sea-level rise: Destruction of threatened and endangered species habitat in South Carolina. Environ Manage 17:373–385

Davis RW, Evan WE, Würsig B (2000) Cetaceans, sea turtles and seabirds in the northern Gulf of Mexico: Distribution, abundance and habitat associations. Volume 1-Executive Summary. PB2001-104839. Contracts USGS/BRD/CR-1999-0006 and OCS MMS 2000-002. MMS (Minerals Management Service), Metairie, LA, USA. 27 p

Davis SM, Childers DL, Lorenz JJ, Wanless HR, Hopkins TE (2005) A conceptual model of ecological interactions in the mangrove estuaries of the Florida Everglades. Wetlands 25:832–842

Day JH, Colwell MA (1998) Waterbird communities in rice fields subjected to different post-harvest treatments. Colon Waterbirds 21:185–197

Day JW, Christian RR, Boesch DM, Yanez-Arancibia A, Morris J, Twilley RR, Naylor L, Schaffner L, Stevenson C (2008) Consequences of climate change on the ecogeomorphology of coastal wetlands. Estuar Coast Shelf Sci 31:477–491

Day RH, Wehle DHA, Coleman FC (1985) Ingestion of plastic pollutants by marine birds. In: Shomura RS, Yoshida HO (eds) Proceedings, Workshop on the Fate and Impact of Marine Debris. U.S. Department of Commerce. Technical Memorandum NOAA-TM-NMFS_SWFC-54. NOAA (National Oceanic and Atmospheric Administration), Washington, DC, USA, pp 344–386

DeBenedictis PA (1986) The changing seasons, a hurricane fall. Am Birds 40:75–82

DeGraaf RM, Rappole JH (1995) Neotropical migratory birds: Natural history, distribution, and population change. Cornell University Press, New York, NY, USA. 676 p

Delany S, Scott D (2006) Waterbird population estimates, 4th edn. Wetlands International, Wageningen, The Netherlands, pp 1–23

DesGranges JL, Reed A (1981) Disturbances and controls of double-crested cormorants in Quebec. Colon Waterbirds 4:12–19

Dinsmore JJ (1972) Sooty Tern behavior. Bull Fla State Museum Biol Sci 16:129–179

Dobbs RC, Barrow JRWC, Jeske CW, DiMiceli J, Michot TC, Beck JW (2009) Short-term effects of hurricane disturbances on food availability for migrant songbirds during autumn stopover. Wetlands 29:123–134

Drake KR, Thompson JE, Drake K, Zonick L (2001) Movements, habitat use, and survival of nonbreeding piping plovers. Condor 103:259–267

Dronen NO, Wardle WJ, Bhuthimethee M (2002) Helminthic parasites from Willets, *Catoptrophorus semipalmatus*, (Charadriiformes: Scolopacidae), from Texas, USA, with descriptions of *Kowalewskiella catoptrophori* sp. N. *and Kowalewskiella macrospina* sp. N. (Cestoda: Dilepididae). Comp Parasitol 69:43–50

Dronen NO, Blend CK, Anderson CK (2003) Endohelminths from the Brown Pelican, *Pelecanus occidentalis*, and the American White Pelican, *Pelecanus erythrorhynchus*, from Galveston Bay, Texas, USA, and checklist of Pelican parasites. Comp Parasitol 70:140–154

Dufour KW, Ankney CD, Weatherhead PJ (1993) Condition and vulnerability to hunting among Mallards staging at Lake St-Clair. Ontario J Wildl Manage 57:209–215

Dumas JV (2000) Roseate spoonbill (*Platalea ajaja*). In: Poole A (ed) The birds of North America Online Cornell Lab of Ornithology, Ithaca, NY, USA. http://bna.birds.cornell.edu/bna/species/490. Accessed 14 Feb 2013

Dunnet GM (1982) Oil pollution and seabird populations. Philos Trans R Soc Lond A 297:413–427

DWF (Department of Wildlife and Fisheries, State of Louisiana). 2012. Species by Parish List; Rare Animal Species. [Internet]. [cited 2012 Jan 23] Available from http://www.wlf.louisiana.gov/wildlife/species-parish-list

Dyer WG, Williams EH Jr, Mignucci-Giannoni AA, Jimenez-Marrero NM, Bunkley-Williams L, Moore DP, Pence DB (2002) Helminth and arthropod parasites of the Brown Pelican, *Pelecanus occidentalis*, in Puerto Rico, with a compilation of all metazoan parasites reported from this host in the Western Hemisphere. Avian Pathol 31:441–448

Eddleman WR, William R, Conway CJ (1998) Clapper rail (*Rallus longirostris*). In: Poole A (ed) The birds of North America Online. Cornell Lab of Ornithology, Ithaca, NY, USA. http://bna.birds.cornell.edu/bna/species/340. Accessed 14 Feb 2013

Edenhofer O, Pichs-Madruga R, Sokona Y, Seyboth K, Matschoss P, Kadner S, Zwickel T, Eickemeier P, Hansen G, Schloemer S, Stechow C (eds) (2011) Intergovernmental Panel on

Climate Change (IPCC) report on renewable energy sources and climate change mitigation. Cambridge University Press, Cambridge, UK. 1075 p

Elliott L, McKnight K (eds) (2000) The U.S. shorebird conservation plan: Lower Mississippi/ Western Gulf coast regional shorebird plan. Manomet Center for Conservation Sciences, Manomet, MA, USA. 29 p

Elliott-Smith E, Haig SM (2004) Piping Plover (*Charadrius melodus*). In: Poole A (ed) The birds of North America online. Cornell Lab of Ornithology, Ithaca, NY, USA. http://bna.birds. cornell.edu/bna/species/002. Accessed 14 Feb 2013

Elliott-Smith E, Haig SM, Ferland CL, Gorman LR (2004) Winter distribution and abundance of snowy plovers in SE North America and the West Indies. Wader Study Group Bull 104:28–33

Elliott-Smith E, Haig SM, Powers BM (2009) Data from the 2006 International Piping Plover Census. USGS (U.S. Geological Survey) Data Series 426. U.S. Department of the Interior, Washington, DC, USA. 332 p

Enderson JH (1965) A breeding and migration survey of the Peregrine falcon. Wilson Bull 77:327–339

Energy Resources (1982) *Ixtoc* oil spill assessment. Final report-Executive summary. Bureau of Land Management Report AA851-CTO-71. Cambridge, MA, USA. 39 p

Erwin RM (1979) Species interactions in a mixed colony of Common Terns (*Sterna hirundo*) and Black Skimmers (*Rhynchops niger*). Anim Behav 27:1054–1062

Erwin RM (1989) Responses to human intruders by birds nesting in colonies: Experimental results and management guidelines. Colon Waterbirds 12:104–108

Erwin RM (1990) Feeding activities of black skimmers in Guyana. Colon Waterbirds 13:70–71

Erwin RM, Custer T (2000) Herons as indictors. In: Kushlan JA, Hafner H (eds) Heron conservation. Academic Press, New York, NY, USA, pp 311–330

Eubanks TL, Behrstock RA, Weeks RJ (2006) Birdlife of Houston, Galveston, and the Upper Texas coast. The Gulf Studies 10. Texas A&M University Press, Corpus Christi, TX, USA. 287 p

Evers DC (2004) Status assessment and conservation plan for the Common Loon (*Gavia immer*) in North America. U.S. Fish and Wildlife Service, Hadley, MA, USA. 88 p

Evers DC, Savoy LJ, DeSorba CR, Yates DE, Hanson W, Taylor KM, Siegel LS, Cooley JH, Bank MS, Major A, Munney K, Mower BF, Vogel HS, Schoch N, Pokras M, Goodale MW, Fair J (2008) Adverse effects from environmental mercury loads on breeding Common Loons. Ecotoxicology 17:69–81

Evers DC, Paruk JD, Mcintyre JW, Barr JF (2010) Common loon (*Gavia immer*). In: Poole A, (ed) The birds of North America online Cornell Lab of Ornithology, Ithaca, NY, USA. http://bna.birds.cornell.edu/bna/species/313. Accessed 14 Feb 2013

Farmer CJ, Goodrich LJ, Ruelas EI, Smith JP (2008) Conservation status of North Americas birds of prey. In: Bildstein BL, Smith JP, Ruelas EI, Veit RR (eds) State of North Americas birds of prey. Union Series in Ornithology 3. Nuttall Ornithological Club and American Ornithologists, Cambridge, MA, USA, pp 303–420

Felder DL, Camp DK, Tunnell JW (2009) An introduction to Gulf of Mexico biodiversity assessment. In: Felder DL, Camp DK (eds) Gulf of Mexico origin, waters, and biota: Volume 1, Biodiversity. Texas A&M University Press, College Station, TX, USA. 1393 p

Fernandez-Juricic E, Zollner PA, LeBlanc C, Westphal LM (2007) Responses of nestling Black-crowned Night Herons (*Nycticorax nyctocorax*) to aquatic and terrestrial recreational activities: A manipulative study. Waterbirds 30:554–565

FFWCC (Florida Fish and Wildlife Conservation Commission) (2003) Florida's breeding bird atlas: A collaborative study of Florida's birdlife. Reddish Egret. Tallahassee, FL, USA. http://www.myfwc.com/bba/. Accessed 14 Feb 2013

FFWCC (2010) Florida's endangered and threatened species list. Tallahassee, FL, USA. 11 p

FFWCC (2011a) Roseate spoonbill: Biological status review report. Tallahassee, FL, USA. 15 p

FFWCC (2011b) Black skimmer: Biological status review report. Tallahassee, FL, USA. 15 p

Field D, Reyer A, Genovese P, Shearer B (1991) Coastal wetlands of the United States: An accounting of a valuable national resource. National Oceanic and Atmospheric Administration, Rockville, MD, USA

Flemming SP, Greene E (1990) Making sense of information. Nature 348:291–292

Fleury BE, Sherry TW (1995) Long term population trends of colonial wading birds in the southern United States: The impact of crayfish aquaculture on Louisiana populations. Auk 112:613–632

Forbes MG, Dunton KH (2006) Response of a subtropical estuarine marsh to local climatic change in the southwestern Gulf of Mexico. Estuar Coast Shelf Sci 29:1242–1254

Forsell DJ (1999) Mortality of migratory waterbirds in Mid-Atlantic coastal anchored gillnets during March and April 1998. Administrative Report. U.S. Fish and Wildlife Service Chesapeake Bay Field Office, Annapolis, MD, USA. 34 p

Foster CR, Amos AF, Fuiman LA (2009) Trends in abundance of coastal birds and human activity on a Texas barrier island over three decades. Estuar Coast Shelf Sci 32:1079–1089

Fox GA, Gilbertson M, Gilman AP, Kubiak TJ (1991) A rationale for the use of colonial fish-eating birds to monitor the presence of development toxicants in Great Lakes fish. J Great Lakes Res 17:151–162

Fraser GS, Russell J, Von Zharen WM (2006) Produced water from offshore oil and gas installations on the Grand Banks, Newfoundland and Labrador: Are the potential effects to seabirds sufficiently known? Mar Ornithol 34:147–156

Frederick PC, Collopy MW (1989) Nesting success of five Ciconiform species in relation to water condition in the Florida Everglades. Auk 106:625–634

Frederick PC, Spalding MG (1994) Factors affecting reproductive success of wading birds (*Ciconiiformes*) in the Everglades. In: Davis S, Ogden JC (eds) Everglades: The ecosystem and its restoration. St. Lucie Press, Del Ray Beach, FL, USA, pp 659–691

Frederick PC, Bildstein KL, Fleury B, Ogden J (1996) Conservation of large, nomadic populations of White Ibises (*Eudocimus albus*) in the United States. Conserv Biol 10:203–216

Frederick PC, Spalding MG, Sepulveda MS, Williams GE, Nico L, Robins R (1999) Exposure of great Egret (*Ardea albus*) nestlings to mercury through diet in the Evergreens ecosystem. Environ Toxicol Chem 18:1940–1947

Frederick PC, Spalding MG, Dusek R (2002) Wading birds as bioindicators of mercury contamination in Florida, USA: Annual and geographic variation. Environ Toxicol Chem 21:163–167

Frederick PC, Hylton B, Heath JA, Spalding MG (2004) A historical record of mercury contamination in southern Florida (USA) as inferred from avian feather tissue. Environ Toxicol Chem 23:1474–1478

Frederick PC, Gawlik DE, Ogden JC, Cook MI, Lusk M (2009) The white ibis and wood stork as indicators for restoration of the Everglades ecosystem. Ecol Indic 9s:83–95

Friedmann H (1925) Notes on the birds observed in the lower Rio Grande valley of Texas during May, 1924. Auk 42:537–554

Fritts TH, Reynolds RP (1981) Pilot study of the marine mammals, birds, and turtles in OCS areas of the Gulf of Mexico. FWS/OBS-81/36. U.S. Department of the Interior, U.S. Fish and Wildlife Service, Belle Chase, LA, USA. 140 p

Fritts TH, Irvine AB, Jennings RD, Collum LA, Hoffman W, McGehee MA (1983) Turtles, birds, and mammals in the northern Gulf of Mexico and nearby Atlantic waters. FWS/OBS 82/65. Fish and Wildlife Service, Washington, DC, USA. 455 p

Furness RW (1993) Birds and monitors of pollutants. In: Furness RW, Greenwood JJD (eds) Birds as monitors of environmental change. Chapman & Hall, London, UK, pp 86–143

Furness RW, Birkhead TR (1984) Seabird colonies distributions suggest competition for food supplies during the breeding season. Nature 311:655–656

Furness RW, Hudson AV, Ensor K (1988) Interactions between scavenging seabirds and commercial fisheries around the British Isles. In: Burger J (ed) Seabirds and other marine vertebrates: Competition, predation, and other Interactions. Columbia University Press, New York, NY, USA, pp 232–260

Furness RW, Ensor K, Hudson AV (1992) The use of fishery waste by gull populations around the British Isles. Ardea 80:105–113

Gabrey SW, Afton AD (2000) Effects of winter marsh burning on abundance and nesting activity of Louisiana seaside sparrows in the Gulf Coast Chenier Plain. Wilson Bull 112:365–372

Gabrey SW, Afton AD, Wilson BC (1999) Effects of winter burning and structural marsh management on vegetation and winter bird abundance in the Gulf coast Chenier Plain, USA. Wetlands 19:594–606

Galbraith H, Jones R, Park R, Clough J, Herrod-Julius S, Harrington B, Page G (2005) Global climate change and sea level rise: Potential losses of intertidal habitat for shorebirds. Technical Report PSW-GTR-191. U.S. Department of Agriculture Forest Service, Washington, DC, USA

Gallardo JC, Velarde E, Arreola R (2004) Birds of the Gulf of Mexico and the priority areas for their conservation. In: Wither K, Nipper M (eds) Environmental analysis of the Gulf of Mexico, Publication Series 1. Harte Research Institute for Gulf of Mexico Studies, Texas A&M University, College Station, TX, USA, pp 18–194

Gallardo JC, Macias V, Velarde E (2009) Birds (Vertebrata: Aves) of the Gulf of Mexico. In: Felder DL, Camp DK (eds) Gulf of Mexico-origins, waters, and biota. Biodiversity. Texas A&M University Press, College Station, TX, USA, pp 1321–1342

Garthe S, Montevecchi WA, Chapdelaine G, Rail JF, Hedd A (2007) Contrasting foraging tactics by Northern Gannets (*Sula bassana*) breeding in different oceanographic domains with different prey fields. Mar Biol 151:687–694

Garvin MC, Szeil CC, Moore FR (2006) Blood parasites of nearctic-neotropical migrant passerine birds during spring trans-gulf migration: Impact on host body condition. J Parasitol 92:900–996

Gaston AJ, Nettleship DN (1981) The thick-billed Murres of Prince Leopold Island. Can Wildl Serv Monogr Ser 6:1–350

Gauthreaux SA (1971) A radar and direct visual study of passerine spring migration in southern Louisiana. Auk 88:343–365

Gauthreaux SA (1999) Neotropical migrants and the Gulf of Mexico: the view from aloft. In: Able KA (ed) Gatherings of angels: Migrating birds and their ecology. Cornell University Press, Ithaca, NY, USA, pp 27–49

Gauthreaux SA, Belser CG, Welch CM (2006) Atmospheric trajectories and spring bird migration across the Gulf of Mexico. J Ornithol 147:317–132

Gawlik DE (2006) The role of wildlife science in wetland ecosystem restoration: Lesson from the Everglades. Ecol Eng 26:70–83

Gawlik DE, Slack RD, Thomas JA, Harpole DN (1998) Long-term trends in population and community measures of colonial-nesting waterbirds in Galveston Bay Estuary. Colon Waterbird 21:143–151

GBEP (Galveston Bay Estuary Program) (2006) Galveston Bay status and trends 2004–2006: final report. Contract 582-4-650-47. Texas Commission of Environmental Quality, Houston, TX, USA

GBST (Galveston Bay Status and Trends) (2012) Galveston Bay colonial nesting waterbird indicator. 128 p, http://www.galvbaydata.org/LivingResources/ColonialWaterbirds/tabid/454/Default.aspx. Accessed 14 Feb 2013

GCPWG (Gulf Coast Prairie Working Group) (2000) U.S. shorebird conservation plan: Lower Mississippi/western Gulf coast shorebird planning region. Mississippi Alluvial Valley/West Gulf Coast Plain Working Groups, Corpus Christi, TX, USA. 64 p

Geyer RA (1981) Naturally occurring hydrocarbons in the Gulf of Mexico and the Caribbean. In: Proceedings, 1981 Oil Spill Conference, American Petroleum Institute, Washington, DC, USA, pp 445–452

Gilman E (2001) Integrated management to address the incidental morality of seabirds in longtime fisheries. Aquat Conserv Mar Freshwat Ecosyst 11:391–414

Gilman E, Brothers N, Kobayashi DR (2005) Principles and approaches to abate seabird by-catch in longline fisheries. Fish Fish 6:35–49

Gil-Weir KC, Grant WE, Slack RD, Wang HH, Fujiwara M (2012) Demography and population trend of Whooping Cranes. J Field Ornithol 83:1–10

Glass P, Roach W (1997) Status and recent trends of Galveston bays colonial waterbirds-with management implications. In: Proceedings of the State of the Bay Symposium, III, Texas Natural Resource Conservation Commission, Temple, TX, USA, pp 49–60

Glenn CR (2006a) Earth's endangered creatures-Cuba (2006–2013). Cuba. Endangered species search by area selection. http://www.earthsendangered.com/search-regions3.asp. Accessed 3 Feb 2012

Glenn CR (2006b) Earth's endangered species (2006–2013), Mexico. Endangered species search by area selection. http://www.earthsendangered.com/search-regions3.asp. Accessed 3 Feb 2012

Gochfeld M (1973) Effect of artifact pollution on the viability of seabird colonies on Long Island, New York. Environ Pollut 4:1–6

Gochfeld M (1974) Prevalence of subcutaneous emphysema in young Terns, Skimmers and Gulls. J Wildl Dis 10:115–120

Gochfeld M (1978) Colony and nest site selection by Black Skimmers. Proc Colon Waterbird Group 1:78–90

Gochfeld M (1979) Prevalence of oiled plumage of Terns and Skimmers on western Long Island, New York: Baseline data prior to petroleum exploration. Environ Pollut 20:123–130

Gochfeld M (1980) Mechanisms and adaptive value of reproductive synchrony in colonial seabirds. In: Burger J, Olla B (eds) Behavior of Marine Animals, Volume 6: Shorebirds: Migration and Foraging Behavior. Plenum Press, New York, NY, USA, pp 207–270

Gochfeld M, Burger J (1981) Age-related differences in piracy of Frigatebirds from Laughing Gulls. Condor 83:79–82

Gochfeld M, Burger J (1982) Feeding enhancement by social attraction in the Sandwich Tern. Behav Ecol Sociobiol 10:15–17

Gochfeld M, Burger J (1994) Black Skimmers. In: Poole A (ed) The birds of North America Online. Cornell Lab of Ornithology, Ithaca, NY, USA. http://bna.birds.cornell.edu/bna/species/108. Accessed 14 Feb 2013

Goecker ME, Valentine JF, Sklenar SA (2006) Effects of exotic submerged aquatic vegetation on waterfowl in the Mobile-Tensaw Delta. Gulf Mexico Sci 1:68–80

Golder W, Allen D, Cameron S (2006) Tern use of dredged materials: Designs for the creation of Tern nesting sites. In: Guilfoyle MP, Fischer RA, Pashley DN, Lott CA (eds) Summary of First Regional Workshop on Dredging, Beach Nourishment, and Birds on the South Atlantic Coast. ERDC/EL TR-06-10. U.S. Army Corps of Engineers, Washington, DC, USA, pp 37–39

Gore LA (1991) Distribution and abundance of nesting least terns and black skimmers in northwest Florida. Fla Field Nat 19:65–96

Gorman C, Smith EH (2001) Evaluation of long-term habitat and colonial waterbird dynamics of Shamrock Island, Nueces County, Texas. TAMUCC-CCS-0305. Texas A&M University-Corpus Christi, Corpus Christi, TX, USA

Gotmark F, Winkler DW, Andersson M (1986) Flockfeeding on fish schools increases individual success in gulls. Nature 319:589–591

Grand JB (1992) Breeding chronology of mottled ducks in a Texas coastal marsh. J Field Ornithol 63:195–202

Gratto-Trevor CL, Abbott S (2011) Conservation of Piping Plover (Charadrius melodus) in North America: Science, successes, and challenges. Can J Zool 89:401–418

Gratto-Trevor CL, Goossen JP, Wentworth SM (2010) Identification and breeding of yearling piping plovers. J Field Ornithol 81:383–391

Green MC (2006) Status report and survey recommendations on the Reddish Egret (Egretta rufescens). U.S. Fish and Wildlife Service, Atlanta, GA, USA. 37 p

Green MC, Luent MC, Michot TC, Jeske CW, LeBerg PL (2008) Comparison and assessment of aerial and ground estimates of waterbird colonies. J Wildl Manage 72:697–706

Green MV, Hill A, Troy JR, Holderby Z, Geary B (2011) Status of breeding Reddish Egrets on Great Inagua, Bahamas with comments on breeding territoriality and the effects of hurricanes. Waterbirds 34:213–217

Greenberg R (1992) Forest migrants in nonforest habitats on the Yucatan Peninsula. In: Hagan JH, Johnson DW (eds) Ecology and conservation of neotropical migrant landbirds. Smithsonian Institution Press, Washington, DC, USA, pp 273–286

Greenberg R, Marra PP (eds) (2005) Birds of two worlds: The ecology and evolution of migration. Johns Hopkins University Press, Baltimore, MD, USA. 488 p

Greenberg R, Maldonado J, Droege S, McDonald MV (2006) Tidal marshes: A global perspective on the evolution and conservation of their terrestrial vertebrates. Bioscience 56:675–685

Greenlaw JS (1992) Seaside Sparrow. In: Schneider KJ, Pence DM (eds) Migratory nongame birds of management concern in the northeast. U.S. Fish and Wildlife Service, Newton Corner, MA, USA, pp 211–232

Greer RD, Cordes CL, Anderson SH (1988) Habitat relationships of island nesting seabirds along coastal Louisiana. Colon Waterbird 11:181–188

Grimes J, Suto B, Greve JH, Albers HF (1989) Effect of Anthelmintics on three common helminthes in brown pelicans (Pelecanus occidentalis). J Wildl Dis 25:139–142

Guilfoyle MP, Fischer RA, Pashley DN, Lott CA (2006) Summary of first regional workshop on dredging, beach nourishment, and birds on the south Atlantic coast. ERDC/EL TR-06-10. U.S. Army Corps of Engineers, Washington, DC, USA. 68 p

Hafner H, Lansdown RV, Kushlan JA, Butler RW, Custer TW, Davidson IM, Erwin RM, Hancock HA, Lyles AM, Maddock M, Marion L, Morales G, Mundkur T, Perennous C, Pineau O, Turner D, Ulcenaers P, van Vessen J, Young L (2000) Conservation in herons. In: Kushlan JA, Hafner H (eds) Heron conservation. Academic Press, New York, NY, USA, pp 343–375

Haig SM, Plissner JH (1993) Distribution and abundance of piping plovers: Results and implications of the 1991 international census. Condor 95:145–156

Haig SM, Ferland CL, Cuthbert FJ, Dingledine J, Goossen JP, Hecht A, McPhillips N (2005) A complete species census and evidence for regional declines in Piping Plovers. J Wildl Manage 69:160–173

Hamer KC, Burger J, Schreiber EA (2001) Breeding, biology, life histories, and life history-environment interactions in seabirds. In: Burger J, Schreiber EA (eds) Biology of marine birds. CRC Press, Boca Raton, FL, USA, pp 217–262

Hancock JA, Kushlan JA, Kahl MP (1992) Storks, ibises, and spoonbills of the world. Princeton University Press, Princeton, NJ, USA. 385 p

Hata DN (2006) Incidental captures of seabirds in the U.S. Atlantic pelagic longline fishery, 1986–2005. NOAA Fisheries Service, Miami, FL, USA. 39 p

Heinz GH (1976a) Methyl mercury: Second-year feeding effects on Mallard reproduction and duckling behavior. J Wildl Manage 40:82–90

Heinz GH (1976b) Methyl mercury: Second-generation reproductive and behavioral effects on Mallard ducks. J Wildl Manag 40:710–715

Heinz GH (1979) Methylmercury: Reproductive and behavioral effects on three generations of mallard ducks. J Wildl Manage 43:394–401

Heinz GH, Hoffman DJ, Klimstra JD, Stebbins KR, Kondrad SL, Erwin CA (2009) Species differences in the sensitivity of avian embryos to methylmercury. Arch Environ Contam Toxicol 56:129–138

Henny CJ (1983) Distribution and abundance of nesting Ospreys in the United States. In: Bird DM (ed) Biology and management of bald eagles and ospreys. MacDonald Raptor Research Centre, McGill University, Montreal, Quebec, Canada. 325 p

Hess NA, Ribic CA (2000) Seabird ecology. In: Davis RW, Evans WE, Würsig B (eds) Cetaceans, sea turtles and seabirds in the Northern Gulf of Mexico: Distribution, abundance, and habitat associations: Executive summary, Volume 1. USGS/BRD/CR-1999-0005. U.S. Geological Survey, Washington, DC, USA, pp 275–315

Hingtgen TM, Mulholland R, Zale AV (1985) Habitat suitability index models: Eastern brown pelican. FWS/OBS 82/10.90. U.S. Fish and Wildlife Service, Washington, DC, USA. 20 p

Hobaugh WC, Stutzenbaker CD, Flickinger EL (1989) The rice prairies. In: Smith LM, Pederson RL, Kaminski RM (eds) Habitat management for migrating and wintering waterfowl in North America. Texas Tech University Press, Lubbock, TX, USA, pp 367–383

Hobson KA (2005) Stable isotopes and the determination of avian migratory connectivity and seasonal interactions. Auk 122:1037–1048

Hobson KA, Wassenaar LI (1997) Linking breeding and wintering grounds of neotropical migrant songbirds using hydrogen isotopic analysis of feathers. Oecologia 109:142–148

Hobson KA, Wilgenburg SV, Wassenaar LI, Moore F, Farrington J (2007) Estimating origins of three species of neotropical migrant songbirds at a Gulf coast stopover site: Combining isotopes and GIS tools. Condor 109:256–267

Hodgson AB, Paul AF (2010) 25 Years after BASIS: An update on the current status and recent trends in colonial waterbird populations of Tampa Bay. In: Cooper ST (ed) Proceedings, Tampa Bay Area Scientific Information Symposium, BASIS 5:20–23 October 2009. St. Petersburg, FL, USA, pp 233–247

Hoffman DJ, Rattner BA, Burton AG, Cairns J Jr (1995) Handbook of ecotoxicology. Lewis Publishing, Boca Raton, FL, USA. 755 p

Hoffman DJ, Eagles-Smith CA, Ackerman JT, Adelsbach TL, Stebbins KR (2011) Oxidative stress response of Forster's terns (*Sterna forsteri*) and Caspian terns (*Hydroprogne caspia*) to mercury and selenium bioaccumulation in liver, kidney, and brain. Environ Toxicol Chem 30:920–929

Holm GO, Hess TJ, Justic D, McNease L, Linscombe RG, Nesbitt SA (2003) Population recovery of the brown pelican following its extirpation in Louisiana. Wilson Bull 115:431–437

Hood SL, Dinsmore SJ (2007) Abundance of snowy and Wilson's Plovers in the lower Laguna Madre region of Texas. J Field Ornithol 78:362–368

Howell SNG (1989) Additional information on the birds of the Campeche Bank, Mexico. J Field Ornithol 60:504–509

Howell SNG, Webb S (1995) A guide to the birds of Mexico and Northern Central America. Oxford University Press, New York, NY, USA. 849 p

Hunt GL (1972) Influence of food distribution and human disturbance on the reproductive success of Herring Gulls. Ecology 53:105–106

Hunt GL (1987) Offshore oil development and seabirds: The present status of knowledge and long-term research needs. In: Boesch DF, Rabalais NN (eds) Long-term environmental effects of offshore oil and gas development. Elsevier Applied Science, London, UK, pp 539–586

Hunt GL (1990) The pelagic distribution of marine birds in a heterogeneous environment. Polar Res 8:43–54

Hunt GL, Schneider DC (1987) Scale-dependent processes in physical and biological environment of marine birds. In: Croxall JP (ed) Seabirds: Feeding biology and role in marine ecosystems. Cambridge University Press, Cambridge, UK, pp 7–41

Hunter WC (2006) Conservation priority birds species of the Atlantic coast. In: Guilfoyle MP, Fischer RA, Pashley DN, Lott CA (eds) Summary of first regional workshop on dredging, beach nourishment, and birds on the South Atlantic Coast. ERDC/EL TR-06-10. U.S. Army Corps of Engineers, Washington, DC, USA, pp 3–7

Hunter WC, Golder W, Melvin S, Wheeler J (2006) Southeast United States regional waterbird conservation plan. U.S. Fish and Wildlife Service, Washington, DC, USA. http://www.waterbirdconservation.org/southeast_us.html. Accessed 14 Feb 2013

Hutton RF, Sogandares-Bernal F (1960) Studies on helminth parasites from Coast of Florida. Trans Am Microsc Soc 79:287–292

Imhof TA (1962) Alabama birds. State of Alabama, Department of Conservation, Game and Fish Division, University of Alabama Press, Tuscaloosa, AL, USA. 530 p

Inzunza ER, Goodrich LJ, Hoffman SW (2010) North American population estimates of waterbirds, vultures and hawks from migration counts in Veracruz, Mexico. Bird Conserv Int 201:124–133

IUCN (International Union for Conservation of Nature) (2011) The IUCN red list of threatened species. Version 2011.2. Gland, Switzerland. http://www.iucnredlist.org. Accessed 14 Feb 2013

Jackson AK, Evers DC, Etterson MA, Condon Folsom SB, Detweiler J, Schmerfeld J, Cristol DA (2011) Mercury exposure affects the reproductive success of a free-living terrestrial songbird, the Carolina Wren (*Thryothorus ludovicianus*). Auk 128:759–769

Jahn AE, Levy DJ, Smith KG (2004) Reflections across hemispheres: A system-wide approach to new world bird migration. Auk 121:1005–1013

Jehl JR, Henry AE, Bond SI (1999) Flying the gauntlet: Population characteristics, sampling bias, and migrating routes of Eared Grebes downed in the Utah desert. Auk 116:178–183

Johnson FA (2009) Variation in population growth rates of mottled ducks in Texas and Louisiana. U.S. Geological Survey Administrative Report. U.S. Fish and Wildlife Service, Washington, DC, USA. 26 p

Johnson RF, Sloan NF (1975) The effects of human disturbance on the white pelican colony at Chase Lake National Wildlife Refuge, North Dakota. Inl Bird-Banding News 47:163–170

Kale HW, Pranty M, Smith BM, Biggs CW (1992) The atlas of the breeding birds of Florida. Final Report. Florida Game and Fresh Water Fish Commission, Tallahassee, FL, USA

Keller V (1989) Variations in the response of Great-crested Grebes Podiceps Cristatus to human disturbance- a sign of adaptations? Biol Conserv 49:31–45

Kelly JP (2006a) What is the lifespan of a heronry? Habitat protection and nesting colonies. The Ardeeid 6–7

Kelly JF (2006b) Stable isotope evidence links breeding geography and migration timing in wood warblers (Parulidae). Auk 123:431–437

Kerlinger P (1985) Water crossing behavior of raptors during migration. Wilson Bull 97:109–113

Kerlinger P (1995) How birds migrate. Stackpole Press, Mechanicsburg, PA, USA. 228 p

King KA (1976) Bird mortality, Galveston Island, Texas. Southwest Nat 21:414

King KA, Krynitsky AJ (1986) Population trends, reproductive success, and organochlorine chemical contaminants in waterbirds nesting in Galveston Bay, Texas. Arch Environ Contam Toxicol 15:367–376

King KA, Flickinger EL, Hildebrand HH (1977) The decline of brown pelicans on the Louisiana and Texas gulf coasts. Southwest Nat 21:417–431

King KA, Blankinship DR, Payne E, Krynitsky AJ, Hensleer GL (1985) Brown pelican populations and pollutants in Texas 1975–1981. Wilson Bull 97:201–214

Korschgen CE, Dahlgren RB (1992) Human disturbances of waterfowl: Causes, effects, and management: Waterfowl management handbook. U.S. Fish and Wildlife Service, Washington, DC, USA. 8 p. http://digitalcommons.unl.edu/icwdmwfm/12. Accessed 14 Feb 2013

Krapu GL, Swanson GA, Nelson HK (1973) Mercury residues in Pintails breeding in North Dakota. J Wildl Manage 37:395–399

Kreuder C, Mazet J, Bossart GD, Carpenter T, Holyoak M, Elie M, Wright S (2002) Clinico-pathologic features of suspected brevetoxicosis in Double-crested Cormorants along Florida Gulf coast. J Zoo Wildl Med 33:8–15

Krimsky S (2005) The weight of scientific evidence in policy and law. Am J Public Health 95: S129–136

Kury CR, Gochfeld M (1975) Human interference and gull predation in cormorant colonies. Biol Conserv 8:23–24

Kushlan JA (1979) Feeding ecology and prey selection in the White Ibis. Condor 81:376–389

Kushlan JA (2000a) Heron feeding habitat conservation. In: Kushlan JA, Hafner H (eds) Heron conservation. Academic Press, New York, NY, USA, pp 219–235

Kushlan JA (2000b) Research and information: Needs for heron conservation. In: Kushlan JA, Hafner H (eds) Heron conservation. Academic Press, New York, NY, USA, pp 331–341

Kushlan JA, Hafner H (eds) (2000a) Heron conservation. Academic Press, New York, NY, USA. 480 p

Kushlan JA, Hafner H (2000b) Reflections on heron conservation. In: Kushlan JA, Hafner H (eds) Heron conservation. Academic Press, New York, NY, USA, pp 377–379

Laiolo P (2010) The emerging significance of bioacoustics in animal species conservation. Biol Conserv 143:1635–1645

Lance BK, Irons DB, Kendall SJ, McDonald LL (2001) An evaluation of marine bird population trends following the *Exxon Valdez* oil spill, Prince William Sound, Alaska. Mar Pollut Bull 42:289–309

Langin KM, Marra PP, Nemeth Z, Moore FR, Kyser TK, Ratcliffe LM (2009) Breeding latitude and timing of spring migration in songbirds crossing the Gulf of Mexico. J Avian Biol 40:309–316

Lantz SM, Gawlik DE, Cook MI (2010) The effects of water depth and submerged aquatic vegetation on foraging habitat selection and foraging success of wading birds. Condor 112:460–469

Lantz SM, Gawlik DE, Cook MI (2011) The effects of water depth and emergent vegetation on foraging success and habitat selection of wading birds in the everglades. Waterbirds 34:439–447

Leberg PL, Deshotels P, Pius S, Carloss M (1995) Nest sites of seabirds on dredge islands in coastal Louisiana. Proc Annu Conf SEAFWA 49:356–366

Lees DC, Driskell WB (2007) Assessment of bivalve recovery on treated mixed-soft beaches in Prince William Sound. *Exxon Valdez* Oil Spill Restoration Project Final Report 040574. *Exxon Valdez* Oil Spill Trustee Council, Anchorage, AK, USA. 111 p

Lewis JC (1983) Habitat suitability index models: Roseate spoonbill. FWS/OBS-82/10.50. U.S. Fish and Wildlife Service, Washington, DC, USA. 16 p

Lewis JC (1995) Whooping crane (*Grus americana*). In: Poole A (ed) The birds of North America online Cornell Lab of Ornithology, Ithaca, NY, USA. http://bna.birds.cornell.edu/bna/species/153. Accessed 14 Feb 2013

Lindstedt DM (2005) Renewable resources at stake: Barataria-Terrebonne estuarine system in Southeast Louisiana. J Coast Res 44:162–175

Link WA, Sauer JR (1999a) On the importance of controlling for effort in analysis of count survey data: Modeling population change from Christmas bird count data. Vogelwelt 120 (Suppl):119–124

Link WA, Sauer JR (1999b) Controlling for varying effort in count surveys: An analysis of Christmas bird count data. J Agric Biol Environ Stat 4:116–125

Link PT, Afton AD, Cox RR, Davis BE (2011) Use of habitats by female Mallards wintering in southwestern Louisiana. Waterbirds 34:429–438

Loetscher FW (1955) North American migrants in the state of Veracruz, Mexico: A summary. Auk 72:14–54

Lorenz JJ (2000) Impacts of water management on roseate spoonbills and their piscine prey in coastal wetlands of Florida Bay. PhD Thesis, University of Miami, Coral Gables, FL, USA

Lorenz JJ, Langan B, Hodgson AB (2008) Roseate spoonbill nesting in Florida bay three-year report: 2004–2005, 2005–2006, and 2006–2007. Cooperative Agreement Final Report 401815G010. U.S. Fish and Wildlife Service, Vero Beach, FL, USA

Lowery GH Jr (1974) Louisiana birds. Louisiana State University Press, Baton Rouge, LA, USA. 651 p

Lowther PE, Paul RT (2002) Reddish egret (*Egretta rufescens*). In: Poole A (ed) The birds of North America Online. Cornell Lab of Ornithology, Ithaca, NY, USA. http://bna.birds.cornell.edu/bna/species/633. Accessed 14 Feb 2013

Lowther PE, Douglas HD III, Gratto-Trevor CL (2001) Willet (*Tringa semipalmata*). In: Poole A (ed) The birds of North America Online. Cornell Lab of Ornithology, Ithaca, NY, USA. http://bna.birds.cornell.edu/bna/species/579. Accessed 14 Feb 2013

Lueth FX (1963) Mobile delta waterfowl and muskrat research. Pittman-Robinson Project, 7-R, Final Report. Alabama Department Conservation, Montgomery, AL, USA

Lynch JJ (1941) The place of burning in management of Gulf coast wildlife refuges. J Wildl Manage 5:454–457

Mabee TJ, Estelle VB (2000) Assessing the effectiveness of predator exclosures for Plovers. Wilson Bull 112:14–20

Mabee TJ, Plissner JH, Haig SM, Goossen JP (2001) Winter distributions of North American Plovers in the Laguna Madre regions of Tamaulipas, Mexico and Texas, USA. Wader Stud Group Bull 94:39–43

Mackinnon B, Deppe JL, Celis-Murillo A (2011) Birds of the Yucatan Peninsula in Mexico: An update on the status and distribution of selected species. N Am Birds 65:538–552

Macko SA, King SM (1980) Weathered oil: Effect on hatchability of heron and gull eggs. Bull Environ Contam Toxicol 25:316–320

Mallory ML, Weatherhead PJ (1993) Observer effects on common goldeneye nest defense. Condor 95:467–469

Marra PP, Francis CM, Mulvihill RS, Moore FR (2005) The influence of climate on the timing and rate of spring bird migration. Oecologia 142:307–315

Marsh CP, Wilkinson PM (1991) The impact of hurricane Hugo on coastal bird populations. J Coast Res 8:327–334

Maslo B, Lockwood JL (2009) Evidence-based decisions on the use of predator exclosures in shorebird conservation. Biol Conserv 142:3213–3218

Maslo B, Handel S, Todd P (2011) Restoring beaches for Atlantic coast Piping Plovers (*Charadrius melodus*): A classification and regression tree analysis of nest-site selection. Restor Ecol 19:194–203

MBNEP (Mobile Bay National Estuary Program) (2008) State of Mobile Bay: A status report on Alabama's coastline, from the delta to our coastal waters. Mobile, AL, USA, pp 1–46

McCauley CA, Harrel RC (1981) Effects of oil spill cleanup techniques on a salt marsh. Proceedings, 1981 Oil Spill Conference, American Petroleum Institute, Washington, DC, USA, pp 401–407

McCrimmon DA, Ogden JC, Bancroft GT (2011) Great egret. In: Poole A (ed) The birds of North America online. Cornell Lab of Ornithology, Ithaca, NY, USA. http://bna.birds.cornell.edu/bna/species/570. Accessed 14 Feb 2013

McGrady MJ, Maechtle TL, Vargas JJ, Seegar WS, Pena MCP (2002) Migration and ranging of Peregrine Falcons wintering on the Gulf of Mexico coast, Tamaulipas, Mexico. Condor 104:39–48

McIntyre JW (1988) The common loon: Spirit of the northern lakes. University of Minnesota Press, Minneapolis, MN, USA. 230 p

McKee KL, Mendelssohn IA, Materne MD (2004) Acute salt marsh dieback in the Mississippi river deltaic plain: A drought-induced phenomenon? Glob Ecol Biogeogr 13:65–73

McNicholl MK, Lowther PE, Hall JA (2001) Forster's tern (*Sterna forsteri*). In: Poole A (ed) The birds of North America online Cornell Lab of Ornithology, Ithaca, NY, USA. http://bna.birds.cornell.edu/bna/species/595. Accessed 15 Feb 2013

Mendelssohn AA, Hester MW, Sasser C, Fischel M (1990) The effect of a Louisiana crude oil discharge from a pipeline break on the vegetation of a Southeast Louisiana brackish marsh. Oil Chem Pollut 7:1–15

Meyer KD (1995) Swallow-tailed kite (*Elanoides forficatus*). In: Poole A (ed) The birds of North America online Cornell Lab of Ornithology, Ithaca, NY, USA. http://bna.birds.cornell.edu/bna/species/138. Accessed 15 Feb 2013

Michener WK, Blood ER, Bildstein KL, Brinson MM, Gardner LR (1997) Climate change, hurricanes and tropical storms, and rising sea level in coastal wetlands. Ecol Appl 7:770–801

Michot TC (1996) Marsh loss in coastal Louisiana: Implications for management of North American Anatidae. Gibier Faune Sauvage Game Wildl 13:941–957

Michot TC, Jeske CW, Mazourek JC, Vermillion WG, Kemmerer RS (2003) Atlas and census of wading bird and seabird nesting colonies in south Louisiana, 2001. Barataria-Terrebonne National Estuary Program Report 32, Thibodaux, LA, USA. 93 p

Miller DS, Peakall DB, Kinter WB (1978) Ingestion of crude oil: Sublethal effects in Herring Gulls. Science 199:315–317

MMNS (Mississippi Museum of Natural Science) (2005) Mississippi's comprehensive wildlife conservation strategy. Jackson, MS, USA. 428 p

Mock DW (1978) Pair-formation displays of the great egret. Condor 80:159–172

Mock DW (1980) Communication strategies of great blue herons and great egrets. Behaviour 72:156–170

Mock DW, Lamey TC (1991) The role of brood size in regulating Egret sibling aggression. Am Nat 138:1015–1026

Monteiro LR, Furness RW (1995) Seabirds as monitors of mercury in the marine environment. Water Air Soil Pollut 80:831–870

Montevecchi WA (2001) Interaction between fisheries and seabirds. In: Schreiber EA, Burger J (eds) Biology of marine birds. CRC Press, Boca Raton, FL, USA, pp 527–558

Montevecchi WA (2008) Binary dietary responses of Northern Gannets (*Sula bassana*) indicate changing food web and oceanographic conditions. Mar Ecol Press Ser 352:213–220

Montevecchi WA, Myers RA (1997) Centurial and decadal oceanographic influences on changes in northern gannet populations and diets in the north-west Atlantic: Implications for climate change. J Mar Sci 54:608–614

Montevecchi WA, Benvenuti S, Garthe S, Davoren GK, Fifield D (2009) Flexible foraging tactics by a large opportunistic seabird preying on forage and large pelagic fishes. Mar Ecol Press Ser 385:295–306

Montevecchi WA, Fifield D, Burke C, Garthe S, Hedd A, Rail JF, Robertson G (2011) Tracking long-distance migration to assess marine pollution impact. Biol Lett 8:218–221

Moore FR (1999) Cheniers of Louisiana and the stopover ecology of migrant landbirds. In: Able KA (ed) Gatherings of angels: Migrating birds and their ecology. Cornell University Press, New York, NY, USA, pp 51–62

Moore FR (2000a) Preface. In: Moore FR (ed) Stopover ecology of Nearctic-Neotropical landbird migrants: Habitat relations and conservation implications. Studies in Avian Biology Series 20. Cooper Ornithological Society, Norman OK, USA, pp 1–3

Moore FR (ed) (2000b) Stopover ecology of Nearctic-Neotropical landbird migrants: Habitat relations and conservation implications. Studies in Avian Biology Series 20. Cooper Ornithological Society, Norman OK, USA. 133 p

Moore FR, Kerlinger P (1987) Stopover and fat deposition by North American Wood-warblers (Parulinae) following spring migration over the Gulf of Mexico. Oecologia 74:47–54

Moore FR, Yong W (1991) Evidence of food-based competition among passerine migrants during stopover. Behav Ecol Sociobiol 28:85–90

Moore FR, Kerlinger P, Simons TR (1990) Stopover on a Gulf coast barrier island by spring trans-gulf migrants. Wilson Bull 102:487–500

Moore P, Gape L (2008) Report waterbirds in the Bahamas. In: Wege DC (ed) BirdLife international important bird areas in the Caribbean: Key sites for conservation. BirdLife Conservation Series BirdLife International, Cambridge, UK

Moorman AM, Moorman TE, Baldassarre GA, Richard DR (1991) Effects of saline water on growth and survival of mottled duck ducklings in Louisiana. J Wildl Manage 55:471–476

Mora MA (1995) Residues and trends of organochlorine pesticide and polychlorinated biphenyls in birds from Texas, 1965–88. Fish Wildl Res 14:1–26

Mora MA (1996) Organochlorines and trace elements in four colonial waterbird species nesting in the lower Laguna Madre, Texas. Arch Environ Contam Toxicol 31:533–537

Morrison ML (2006) Bird movement and behaviors in the Gulf coast region: Relation to potential wind energy developments. NREL/SR-500-39572. U.S. Department of Energy, Department of Wildlife Fisheries and Sciences, Texas A&M University, College Station, TX, USA. 35 p

Morrison RIG, McCaffery BJ, Gil RE, Skagen SK, Jones SL, Page GW, Gratto-Trevor CL, Andres BA (2006) Population estimates of North American shorebirds. Wader Stud Group Bull 111:67–85

Mowbray TB (2002) Northern gannet (Morus bassanus). In: Poole A (ed) The birds of North America online Cornell Lab of Ornithology, Ithaca, NY, USA. http://bna.birds.cornell.edu/bna/species/693. Accessed 1 Feb 2012

Muehl GT, Tacha TC, Anderson JT (1994) Distribution and abundance of wetlands in coastal Texas. Tex J Agric Nat Resour 7:85–106

Murphy RK, Michaud IMG, Prescott DRC, Ivan JS, Anderson BJ, French-Pombier ML (2003) Predation on adult piping plovers at predator exclosure cages. Waterbirds 26:150–155

Nettleship DN, Chapdelaine G (1988) Population size and status of the northern gannet (Sula bassanus) in North America, 1984. J Field Ornithol 59:120–127

Nettleship DN, Burger J, Gochfeld M (eds) (1994) Threats to seabirds on islands. International Council for Bird Preservation, Cambridge, UK

Newman SH, Anderson DW, Ziccardi MH, Trupkiewicz JG, Tseng FS, Christopher MM, Zinkl JG (2000) An experimental soft-release of oil-spill rehabilitated American coots (Fulica americana): II. Effects of health and blood parameters. Environ Pollut 107:295–304

Newton I (2006) Can conditions experienced during migration limit the population levels of birds? J Ornithol 147:146–166

Newton I (2007) Weather-related mass mortality events in migrants. Ibis 149:453–467

Niles LJ, Sitters HP, Dey AD, Atkinson PW, Baker AJ, Bennett KA, Carmona R, Clark KE, Clark NA, Espoz C, González PM, Harrington BA, Hernández DE, Kalasz KS, Lathrop RG, Matus RN, Minton CDT, Morrison RIG, Peck MK, Pitts W, Robinson RA, Serrano IL (2008) Status of the red knot, Calidris canutus rufa, in the Western Hemisphere. Studies in Avian Biology Series 36. Cooper Ornithological Society, Norman, OK, USA, pp 1–185

Niven DK, Butcher GS (2011) Status and trends of wintering coastal species along the Northern Gulf of Mexico, 1965–2011. Am Birds 65:12–19

Niven DK, Butcher GS, Bancroft GT (2009) Christmas bird counts and climate change: Northward shifts in early winter abundance. Am Birds 63:10–15

NLCD (National Land Cover Database) (2006) Completion of the 2006 National Land Cover Database for the Conterminous United States, United States Geological Survey 77:858–864

NOAA (National Oceanic and Atmospheric Administration) (1998) Population: Distribution, density, and growth. By Culliton CJ. NOAA's state of the coastal report, Silver Spring, MD, USA. http://state_of_coast.noaa.gov/bulletins/html/pop_01/pop.html. Accessed 28 Sept 2012

NOAA (2004) Population trends along the Coastal United States: 1980–2008. http://oceanservice.noaa.gov/programs/mb/pdfs/coastal_pop_trends_complete.pdf. Accessed 1 Jan 2013

NOAA (2011) The Gulf of Mexico at a glance: A second glance. U.S. Department of Commerce, Washington, DC, USA. 51 p

NOAA (2012) Ecology of the Northeast U.S. continental shelf–protected species, seabirds. http://www.nefsc.noaa.gov/ecosys/ecology/ProtectedSpecies/SeaBirds/. Accessed 9 Apr 2012

Nocera JJ, Taylor PD (1998) In situ behavioral responses of common loons associated with elevated mercury (HG) exposure. Conserv Ecol 2:10–22

Novak JM, Gaines KF, Cumbee JC Jr, Mills GL, Rodriguez-Navarro A, Romanek CS (2011) Clapper rails as indicator species of estuarine marsh health. Stud Avian Biol 32:270

NPS (National Park Service) (2008) Mississippi Gulf coast national heritage area environmental assessment. Mississippi Department of Marine Resources, Biloxi, MS, USA, pp 28, 109–111

O'Connell JL, Nyman JA (2011) Effects of marsh pond terracing on coastal wintering waterbirds before and after hurricane Rita. Environ Manage 48:975–984

O'Driscoll N, Rencz A, Lean D (eds) (2005) Mercury cycling in a wetland-dominated ecosystem: A multidisciplinary study. Society of Environmental Toxicology and Chemistry, Pensacola, FL, USA. 416 p

O'Hara PD, Morandin LA (2010) Effects of sheens associated with offshore oil and gas development on the feather microstructure of pelagic shorebirds. Mar Pollut Bull 60:672–678

Ogden JC (1978) Recent population trends of colonial wading birds on the Atlantic and Gulf coastal plain. In: Sprunt A, Ogden JC, Winckler S (eds) Wading birds, Research Report 7. National Audubon Society, New York, NY, USA, pp 137–153

Ogden JC, Davis SM, Brandt LA (2003) Science strategy for a regional ecosystem monitoring and assessment program: The Florida Everglades example. In: Busch DE, Trexler JC (eds) Monitoring ecosystems. Island Press, Washington, DC, USA, pp 135–163

Ogden JC, Davis SM, Jacobs KJ, Barnes T, Fling HE (2005) The use of conceptual ecological models to guide ecosystem restoration in south Florida. Wetlands 25:795–809

Ohlendorf H, Hothem RL Bunck CM, Aldrich TW, Moore JR (1986) Relationship between selenium concentrations and avian reproduction. Transactions of the 51st North American Wildlife Research Conference 51:330–342

Ohlendorf H, Hothem RL, Walsh D (1989) Nest success, cause-specific nest failures and hatchability of aquatic birds at selenium contaminated Kesterson reservoir and a reference site. Condor 91:787–796

Olson SL (1997) Towards a less imperfect understanding of the systematics and biogeography of the Clapper and King Rail complex (*Rallus longirostris* and *R. elegans*). In: Dickerman RW (ed) The era of Allan R. Phillips: A festschrift. Horizon Communications, Albuquerque, NM, USA, pp 93–111

Ornat AL, Lynch JF, MacKinnon B (1989) New and noteworthy records of birds from the eastern Yucatan Peninsula. Wilson Bull 101:390–409

Oswalt SN (2010) Chinese tallow (Triadica sebifera (L.) Small) population expansion in Louisiana, east Texas, and Mississippi. [Internet] U.S. Dept of Agriculture, research note SRS-20. [cited 2016 Jan 4]. Available from. http://www.srs.fs.usda.gov/pubs/rn/rn_srs020.pdf

Overholtz WJ, Link JS (2007) Consumption impact by marine mammals, fish, and seabirds on the Gulf of Maine-Georges Bank Atlantic Herring (*Clupea harengus*) complex during the years 1977–2002. ICES J Mar Sci 64:83–96

Page GW, Stenzel LE, Page GW, Warriner JS, Warriner JC, Paton PW (2009) Snowy plover (*Charadrius nivosus*). In: Poole A (ed) The birds of North America online, Cornell Labs of Ornithology, Ithaca, NY, USA. http://bna.birds.cornell.edu/bna/species/154. Accessed 18 Feb 2013

Palmisano AW (1973) Habitat preference of waterfowl and fur animals in the northern Gulf Coast marshes. In: Chabreck RH (ed) Proceedings of the second coastal marsh and estuary management symposium. Louisiana State University, Baton Rouge, LA, USA, pp 163–190

Parsons KC, Burger J (1982) Human disturbance and nestling behavior in black-crowned night herons. Condor 84:184–187

Paton PWC (1994) Survival estimates for snowy plovers breeding at Great Salt Lake, Utah. Condor 96:1106–1109

Pattenden RK, Boag DA (1989) Effects of body mass on courtship, pairing and reproduction in captive Mallards. Can J Zool 67:495–501

Patton SR (1988) Abundance of gulls at Tampa Bay landfills. Wilson Bull 100:431–442

Paul RT, Meyerriecks AJ, Dunstand FM (1975) Return of reddish egrets as breeding birds in Tampa Bay, Florida. Fla Field Nat 3:9–10

Payne JR, Driskell WB, Short JW, Larsen ML (2008) Long term monitoring for oil in the *Exxon Valdez* spill region. Mar Pollut Bull 56:2067–2081

Paynter RA (1955) The ornithogeography of the Yucatan Peninsula. Peabody Museum of Natural History. Yale University, New Haven, CT, USA. 347 p

Peakall D (1992) Animal biomarkers as pollution indicators. Chapman and Hall, London, UK

Perez-Arteaga A, Gaston KJ (2004) Wildfowl population trends in Mexico, 1961–2000: A basis for conservation planning. Biol Conserv 115:343–355

Pfister C, Kasprzyk MJ, Harrington BA (1998) Body fat levels and annual return in migrating Semipalmated Sandpipers. Auk 115:904–915

Piatt JF, Carter HR, Nettleship DN (1990) Effects of oil pollution on marine bird populations: Research, rehabilitation, and general concerns. Proceedings: The oil symposium, 16–18 October 1990. Sheridan Press, Herndon, VA, USA. 210 p

Pierotti R (1988a) Associating between marine birds and mammals in the northwest Atlantic Ocean. In: Burger J (ed) Seabirds and other marine vertebrates: Competition, predation, and other interactions. Columbia University Press, New York, NY, USA, pp 31–58

Pierotti R (1988b) Interactions between Gulls and Otariid pinnipeds: Competition, commensalism, and cooperation. In: Burger J (ed) Seabirds and other marine vertebrates: Competition, predation, and other interactions. Columbia University Press, New York, NY, USA, pp 205–231

Pineau O (2000) Conservation of wintering and migrating habitats. In: Kushlan JA, Hafner H (eds) Heron conservation. Academic Press, New York, NY, USA, pp 237–249

Pinho AP, Guimaraes JRD, Marins AS, Costa PAS, Olavo G, Valentin J (2002) Environ Res 89:250–258

Pons JM, Migot P (1995) Life-history strategy of herring gull: Changes in survival and fecundity in populations subjected to various feeding conditions. J Anim Ecol 64:592–599

Poole AF, Bierregaard RO, Martell MS (2002) Osprey (*Pandion haliaetus*). In: Poole A (ed) The birds of North America online Cornell Lab of Ornithology, Ithaca, NY, USA. http://bna.birds.cornell.edu/bna/species/683/articles/introduction. Accessed 21 Feb 2013

Portnoy JW (1977) Nesting colonies of seabirds and wading birds-coastal Louisiana, Mississippi, and Alabama. FWS/OBS-77/07. U.S. Fish and Wildlife Service Biological Services, Washington, DC, USA

Portnoy JW (1981) Breeding abundance of colonial waterbirds on the Louisiana-Mississippi-Alabama coast. Am Birds 35:868–872

Post W, Greenlaw JS, Merriam TL, Wood LA (1983) Comparative ecology of northern and southern populations of the Seaside Sparrow. In: Quay TL, Funderburg JB, Lee DS, Potter EF, Robbins CS (eds) The seaside sparrow, its biology and management. Occasional Papers of North Carolina Biological Survey, Raleigh, NC, USA, pp 123–136

Post W, Post W, Greenlaw JS (2009) Seaside sparrow (*Ammodramus maritimus*). In: Poole A (ed) The birds of North America online, Cornell Labs of Ornithology, Ithaca, NY, USA. http://bna.birds.cornell.edu/bna/species/127. Accessed 18 Feb 2013

Powell GN (1987) Dynamics of habitat use by wading birds in a subtropical estuary: Implications of hydrography. Auk 104:740–749

Powell GN, Bjork RD, Ogden JC, Paul RT, Powell AH, Robertson WB (1989) Population trends in some Florida Bay wading birds. Wilson Bull 101:436–457

Raffaele H, Wiley J, Garrido O, Keith A, Raffaele J (1998) A guide to the birds of the West Indies. Princeton University Press, Princeton, NJ, USA. 511 p

Rappole JH (1995) The ecology of migrant birds: A neotropic perspective. Smithsonian Institute Press, Washington, DC, USA. 288 p

Rattner B (2000) Environmental contaminants and colonial waterbirds. USGS Patuxent Wildlife Research Center, Laurel, MD, USA. http://www.waterbirdconservation.org/plan/rpt-contaminants.pdf. Accessed 18 Feb 2013

Renne IJ, Barrow WC Jr, Randall LA, Bridges WC Jr (2002) Generalized avian dispersal syndrome contributes to Chinese tallow tree (*Sapium sebiferum*, Euphorbiaceae) invasiveness. Divers Distrib 8:285–295

Ribic CA, Davis R, Hess N, Peake D (1997) Distribution of seabirds in the northern Gulf of Mexico in relation to mesoscale features: Initial observations. J Mar Sci 54:545–551

Roberts TS (1907) A Lapland longspur tragedy: Being an account of a great destruction of these birds during a storm in southwestern Minnesota and northwestern Iowa in March, 1904. Auk 24:369–377

Robinson OJ, Dindo JJ (2011) Egg success, hatching success, and nest-site selection of Brown Pelicans, Gaillard Island, Alabama, USA. Wilson J Ornithol 123:386–390

Rodgers JA, Smith HT (1995) Set-back distances to protect nesting bird colonies from human disturbance in Florida. Conserv Biol 9:89–99

Rodgers JA, Kale HW, Smith HT (1996) Roseate spoonbill. In: Rodgers JA, Kale HW, Smith HT (eds) Rare and endangered biota of Florida, Volume 5. University of Florida Press, Gainesville, FL, USA, pp 281–294

Ronconi RA, St. Clair CC (2002) Management options to reduce boat disturbance on foraging black guillemots (*Cepphus grylle*) in the Bay of Fundy. Biol Conserv 108:265–271

Root TL, Price JT, Hall KR, Schneider SH, Rosenzweig C, Pounds JA (2003) Fingerprints of global warming on wild animals and plants. Nature 421:57–60

Ruelas I, Hoffman SW, Goodrich LJ, Tingay R (2000) Conservation strategies for the world's largest known raptor migration flyway: Veracruz the river of raptors. In: Chancellor RD, Meyburg BU (eds) Raptors at risk. World Working Group on Birds of Prey and Hancock House, Blaine, WA, USA, pp 591–596

Rumbold DG, Niemczyk SL, Fink LE, Chandrasekhar T, Harkanson B, Laine KA (2001) Mercury in eggs and feathers of great egrets (*Ardea albus*) from the Florida Everglades. Arch Environ Contam Toxicol 41:501–507

Rush SA, Soehren EC, Stodola KW, Woodrey MS, Cooper RJ (2009a) Influence of tidal height on detection of breeding marsh birds along the northern Gulf of Mexico. Wilson J Ornithol 121:399–405

Rush SA, Soehren EC, Stodola KW, Woodrey MS, Graydon CL, Cooper RJ (2009b) Occupancy of select marsh birds within northern Gulf of Mexico tidal marsh: Current estimates and projected change. Wetlands 29:798–808

Russell RW (2005) Interaction between migrating birds and offshore oil and gas platforms in the northern Gulf of Mexico. Final Report. MMS 2005-009. U.S. Department of the Interior, Minerals Manage Service, Gulf of Mexico OCS Region, New Orleans, LA, USA. 348 p

Safina C, Burger J (1983) Effects of human disturbance on reproductive success in the Black Skimmer. Condor 85:164–171

Safina C, Burger J (1985) Common tern foraging: Seasonal trends in prey fish densities, and competition with bluefish. Ecology 66:1457–1463

Safina C, Burger J (1988) Ecological dynamics among prey fish, bluefish, and foraging common tern in an Atlantic coastal system. In: Burger J (ed) Seabirds and other marine vertebrates: Competition, predation, and other interactions. Columbia University Press, New York, NY, USA, pp 93–165

Sauer JR, Link WA (2011) Analysis of the North American breeding bird survey using hierarchical models. Auk 1:87–98

Sauer JR, Hines JE, Fallon J (2005) The North American breeding bird survey, results and analysis. USGS Patuxent Wildlife Research Center, Laurel, MD, USA. http://www.mbr-pwrc.usgs.gov/bbs/bbs2007.html. Accessed 8 Feb 2012

Sauer JR, Hines JE, Fallon J (2008) The North American breeding bird survey, results and analysis. USGS Patuxent Wildlife Research Center, Laurel, MD, USA. http://www.mbr-pwrc.usgs.gov/bbs/bbs2004. http://www.mbr-pwrc.usgs.gov/bbs/bbs2010.html. Accessed 8 Feb 2012

Sauer JR, Niven DK, Link WA, Butcher GS (2009) Analysis and summary of Christmas Bird Count data. Avocetta 33:13–18

Sauer JR, Hines JE, Fallon JE, Pardieck KL, Ziolkowski DJ Jr, Link WA (2011) The North American breeding bird survey, results and analysis 1966–2010. Version 12.07.2011. USGS Patuxent Wildlife Research Center, Laurel, MD, USA

Savage M (1993) Ecological disturbance and nature tourism. Geogr Rev 83:290–300

Schneider DC (1991) The role of fluid dynamics in the ecology of marine birds. Oceanogr Mar Biol Annu Rev 29:487–521

Schorger AW (1952) Ducks killed during a storm at hot springs, South Dakota. Wilson Bull 64:113

Schreiber RW (1980a) The brown pelican: An endangered species? Bioscience 30:742–747

Schreiber RW (1980b) Nesting chronology of the eastern brown pelican. Auk 97:491–508

Schreiber EA (2001) Climate and weather effects of seabirds. In: Burger J, Schreiber EA (eds) Biology of marine birds. CRC Press, Boca Raton, FL, USA, pp 179–216

Schreiber EA, Burger J (eds) (2001a) The biology of marine birds. CRC Press, Boca Raton, FL, USA, pp 687–744

Schreiber EA, Burger J (2001b) Appendix 2: Table of seabird species and life history characteristics. In: Burger J, Schreiber EA (eds) Biology of marine birds. CRC Press, Boca Raton, FL, USA, pp 179–216

Schreiber RW, Mock PJ (1988) Eastern brown pelicans: What does 60 years of banding tell us? J Field Ornithol 59:171–182

Schreiber RW, Schreiber EA (1979) Notes on measurements, mortality, molt, and gonad condition in Florida west coast laughing gulls. Fla Field Nat 7:19–23

Schreiber RW, Schreiber EA (1982) Essential habitat of the brown pelican in Florida. Fla Field Nat 10:917

Schreiber EA, Schreiber RW, Dinsmore JJ (1979) Breeding biology of laughing gulls in Florida. Part 1: Nesting, egg, and incubation parameters. Bird-Band 50:304–321

Sealy SG (1973) Interspecific feeding assemblages of marine birds off British Columbia. Auk 90:796–802

Seegar WS, Yates MA, Doney GE (2011) Peregrine Falcon migration studies at south Padre Island, Texas. Earthspan, www.earthspan.org, pp 1–20. Accessed 18 Feb 2013

Shabica SV, Cofer NB, Cake EW (1983) Proceedings of the northern Gulf of Mexico estuaries and barrier islands research conference. U.S. Department of the Interior, National Park Service, SE Regional Office, Atlanta, GA, USA. 191 p

Shackelford CE, Rozenburg ER, Hunter WC, Lockwood MW (2005) Migration and the migratory birds of Texas: Who they are and where they are going. PWD BK W7000-511. Texas Parks and Wildlife, Austin, TX, USA. 34 p

Shealer DA (2001) Foraging behavior and food of seabirds. In: Schreiber EA, Burger J (eds) Biology of marine birds. CRC Press, Boca Raton, FL, USA, pp 137–198

Sheehan PJ, Miller DR, Butler GC, Bourdeau P (1984) Effects of pollutants at the ecosystem level. John Wiley and Sons, Chichester, UK. 429 p

Shields M (2002) Brown pelican (*Pelecanus occidentalis*). In: Poole A (ed) The birds of North America online Cornell Lab of Ornithology, Ithaca, NY, USA. http://bna.birds.cornell.edu/bna/species/609. Accessed 18 Feb 2013

Shriver WG, Vickery PD, Hodgman TP, Gibbs JP (2007) Flood tides affect breeding ecology of two sympatric Sharp-tailed Sparrows. Auk 124:552–560

Sidle JG (1984) Piping plover proposed as an endangered and threatened species. U.S. Fish Wildlife Series. Fed Regist 49:44712–44715

Simmons KEL (1972) Some adaptive features of seabird plumage types. Br Birds 65:465–521

Skagen SK, Melcher CP, Muths E (2001) The interplay of habitat change, human disturbance, and species interactions in a waterbird colony. Am Midl Nat 145:18–28

Skoruppa MK, Wood MC, Blacklock G (2009) Species richness, relative abundance, and habitat associations of nocturnal birds along the Rio Grande in southern Texas. Southwest Nat 54:317–323

Smith RJ, Moore FR (2003) Arrival fat and reproductive performance in a long-distance passerine migrant. Oecologia 134:325–331

Smith J, Farmer CJ, Hoffman SW, Lott CA, Goodrich LJ, Simon J, Riley C, Inzunza ER (2008) Trends in autumn counts of raptors around the Gulf of Mexico, 1995–2000. In: Bildstein KL, Smith JP, Inzunza ER (eds) The State of North American Birds of Prey. Series in Ornithology 3. American Ornithologists' Union and Nuttall Ornithological Club, Cambridge, MA, USA, pp 253–278

Solomon S, Qin D, Manning M, Chen Z, Marquis M, Averyt KB, Tignor M, Miller HL (eds) (2007) Intergovernmental Panel on Climate Change (IPCC) report on climate change 2007: The physical science basis. Contribution of working group 1 to the fourth assessment report of the Intergovernmental Panel on Climate Change. Cambridge University Press, Cambridge, UK. 1007 p

Spahn SA, Sherry TW (1999) Cadmium and lead exposure associated with reduced growth rates, poorer fledging success of little blue herons chicks (*Egretta caerulea*) in south Louisiana wetlands. Arch Environ Contam Toxicol 37:377–384

Spalding MG, Frederick PC, McGill HC, Bouton SN, McDowell LR (2000a) Methylmercury accumulation in tissues and its effects on growth and appetite in captive great egrets. J Wildl Dis 36:411–422

Spalding MG, Frederick PC, McGill HC, Bouton SN, Richwy LJ, Schumacher IM, Blackmore SGM, Harrison J (2000b) Histologic, neurologic, and immunologic effects of methylmercury on appetite and hunting behavior in juvenile great egrets (*Ardea albus*). Environ Toxicol Chem 18:1934–1939

Sprandel GL, Gore JA, Cobb DT (1997) Winter shorebird survey final performance report. Fla Game Fresh Water Fish Comm, Tallahassee, FL, USA. 162 p

Sprunt A Jr (1954) Florida bird life. Coward-McCann, New York, NY, USA. 527 p

Stedman S, Dahl TE (2008) Status and trends of wetlands in the coastal watersheds of the eastern United States 1998 to 2004. NOAA, U.S. Fish and Wildlife Service, Washington, DC, USA. 33 p

Stevenson HM (1957) The relative magnitude of the trans-Gulf and circum-Gulf spring migrations. Wilson Bull 69:39–77

Stevenson HM, Anderson BH (1994) The bird life of Florida. University Press of Florida, Tampa, FL, USA. 891 p

Storelli MM, Stuffler RG, Marcotrigiano GO (2002) Total and methylmercury residues in tunafish from the Mediterranean Sea. Food Addit Contam 19:715–720

Sturkie PD (1965) Avian physiology. Cornell University Press, New York, NY, USA. 766 p

Stutchbury BJM, Tarof SA, Done T, Gow E, Kramer PM, Tautin J, Fox JW, Afanasyev V (2009) Tracking long-distance songbird migration by using geolocators. Science 323:896

Swanson GA, Meyer MI (1977) Impact of fluctuating water levels on feeding ecology of breeding blue-winged teal. J Wildl Manage 41:426–433

TCWS (Texas Colonial Waterbird Survey) (2012) Colonial waterbird survey data from 1973–2011, Texas Fish and Wildlife Department, Austin, TX, USA. Unpublished data.

Thomas SM, Lyons JE, Andres BA, Elliott-Smith E, Palacios E, Cavitt JF, Royle JA, Fellows SD, Maty K, Howe WH, Mellink E, Melvin S, Zimmerman T (2012) Population size of snowy plovers breeding in North America. Waterbirds 35:1–14

Toland B (1999) Population increase, nesting phenology, nesting success, and productivity of reddish egrets in Indian River County, Florida. Fla Field Nat 27:59–61

TPWD (Texas Parks and Wildlife Department) (2004) Endangered and threatened birds in Texas and the United States. http://www.tpwd.state.tx.us/huntwild/wild/species/endang/animals/birds/. Accessed 18 Feb 2012

TPWD (2011) Waterfowl strategic plan, spring 2011. http://www.tpwd.state.tx.us/publications/pwdpubs/media/pwd_bk_w7000_1691_07_11.pdf. Accessed 18 Feb 2012

Tremblay J, Ellison LN (1979) Effects of human disturbance on breeding of Black-crowned Night Herons. Auk 96:364–369

Trivelpiece WZ, Butler RG, Miller DS, Peakall DB (1984) Reduced survival of chicks of oil-dosed adult leach's Storm Petrels. Condor 86:81–82

Tunnell JW, Alvarado SA (eds) (1996) Checklist of species within Corpus Christi Bay National Estuary Program Study area: References, habitats, distribution, and abundance. In Current status and historical trends of the estuarine living resources within the Corpus Christi Bay National Estuary Program study area. CCBNEP-06D. Corpus Christi Bay National Estuary Program, Texas Natural Resource Conservation Commission, Austin, TX, USA. 298 p

Tunnell JW, Chapman BR (2000) Seabirds of the Campeche Bank Islands, southeastern Gulf of Mexico. Center for Coastal Studies Contract 52. Atoll Research Bulletin 482. Smithsonian Institution, Washington, DC, USA. 55 p

Turcotte WH, Watts DL (1999) Birds of Mississippi. Mississippi Department of Wildlife, Fisheries and Parks, Jackson, MS, USA. 455 p

USFWS (U.S. Fish and Wildlife Service) (1986) Whooping crane recovery plan. U.S. Fish and Wildlife Service, Albuquerque, NM, USA. 283 p

USFWS (1999) South Florida multi-species recovery plan: Piping plover. South Florida Ecological Services Field Office. Accessed 28 Feb 2012

USFWS (2003) Recovery plan for the Great Lakes piping plover (Charadrius melodus). Department of Interior, Ft. Snelling, MN, USA. 141 p

USFWS (2004) United States Shorebird Conservation Plan High Priority Shorebirds – 2004. Unpub. Report. U.S. Fish and Wildlife Service, Arlington, VA. 5 p

USFWS (2009a) Draft post-delisting monitoring plan for the Brown Pelican. Ventura Fish and Wildlife Office, Ventura, CA, USA. 19 p

USFWS (2009b) Piping plover (Charadrius melodus) 5-year review: Summary and evaluation. Northeast Region, Hadley, MA, USA. 206 p

USFWS (2010a) Post-delisting monitoring plan for the bald eagle (Haliaeetus leucocephalus) in the contiguous 48 states. Divisions of Endangered Species and Migratory Birds and State Programs. Midwest Regional Office, Twin Cities, MN, USA. 75 p

USFWS (2010b) Whooping crane: Species status and fact sheet. North Florida Ecological Services Office, Southeast Region, Jacksonville, FL, USA. http://www.fws.gov/northflorida/WhoopingCrane/whoopingcrane-fact-2001.htm. Accessed 4 Mar 2012

USFWS (2011) Waterfowl population status. 2011. U.S. Department of the Interior, Washington, DC, USA. 80 p

USFWS (2012a) Species report: Listings and occurrences for Mississippi. http://www.fws.gov/ecos/ajax/tess_public/pub/stateListingAndOccurrence.jsp. Accessed 3 Jan 2012

USFWS (2012b) Species report: Listing and occurrences in Alabama. http://www.fws.gov/ecos/ajax/tess_public/pub/stateListingAndOccurrence.jsp. Accessed 18 Feb 2013

USFWS (2012c) Aransas National Wildlife Refuge: First whooping cranes at Aransas NWR. http://www.fws.gov/refuge/Aransas/multimedia/whooping_cranes.html. Accessed 9 Apr 2012

USFWS (2012d) North Florida Ecological Service Office: Species status and fact sheet: Whooping crane. http://www.fws.gov/northflorida/WhoopingCrane/whoopingcrane-fact-2001.htm. Accessed 21 Feb 2012

USFWS (2012e) National digital library. http://digitalmedia.fws.gov/cdm/search/collection/natdiglib/searchterm/clapper%20rail/order/nosort. Accessed 9 Apr 2012

USSCP (U.S. Shorebird Conservation Plan) (2004) High priority shorebirds-2004. Unpublished Report, U.S. Fish and Wildlife Service, Arlington, VA, USA. 5 p. http://www.fws.gov/shorebirdplan/USShorebird/downloads/ShorebirdPriorityPopulationsAug04.pdf. Accessed 12 Mar 2012

Valiela I, Kinney E, Culbertson J, Peacock E, Smith S (2009) Global losses of mangroves and salt marshes. In: Duarte CM (ed) Global loss of coastal habitats: Rates, causes, and consequences. Fundacion BBVA, Madrid, Spain, pp 107–133

Van de Pol M, Ens B, Heg D, Brouwer L, Krol J, Maier M, Klaus-Michael E, Kees O, Tamar L, Eising CM, Koffijberg K (2010) Do change in frequency, magnitude and timing of extreme climatic events threaten the population viability of coastal birds? J Appl Ecol 47:720–730

Vargo GA, Atwood K, Van Deventer M, Harris R (2006) Beached bird surveys on Shell Key, Pinellas County, Florida. Fla Field Nat 34:21–27

Vennesland RG (2010) Risk perception of nesting great blue herons: Experimental evidence of habituation. Can J Zool 88:81–89

Vermeer K (1973a) Some aspects of the nesting requirements of common loons in Alberta. Wilson Bull 85:429–435

Vermeer K (1973b) Some aspects of the breeding and mortality of common loons in East-Central Alberta. Can Field-Nat 87:403–408

Vermillion WG, Wilson BC (2009) Gulf Coast joint venture conservation planning for reddish egret. Gulf Coast Joint Venture, Lafayette, LA, USA. 18 p

Visser JM, Peterson GW (1994) Breeding populations and colony site dynamics of seabirds nesting in Louisiana. Colon Waterbirds 17:146–152

Visser JM, Vermillion WG, Evers DE, Linscombe RG, Sasser CE (2005) Nesting habitat requirements of the brown pelican and their management implications. J Coast Res 212:27–35

Ward P, Zahavi A (1973) The importance of certain assemblages of birds as 'information centres' for food-finding. Ibis 115:517–534

Warriner JS, Warriner JC, Page GW, Stenzel LE (1986) Mating system and reproductive success of a small population of polygamous snowy plovers. Wilson Bull 98:15–37

Waur RH, Wunderle JM (1992) The effect of hurricane Hugo on bird populations on St. Croix U.S. Virgin Islands. Wilson Bull 104:656–673

WCCA (Whooping Crane Conservation Association) (2012) Flock status: Whooping crane numbers in North America. http://whoopingcrane.com/flock-status/flock-status-2011-may/. Accessed 18 Feb 2013

Webster FS (1974) The spring migration, April 1–May 31, 1974, South Texas region. Am Birds 28:822–825

Webster WD (2006) Piping plover utilization of the Mason inlet relocation project area, North Carolina. In: Guilfoyle MP, Fischer RA, Pashley DN, Lott CA (eds) Summary of first regional workshop on dredging, beach nourishment, and birds on the south Atlantic coast. ERDC/EL TR-06-10. U.S. Army Corps of Engineers, Washington, DC, USA. 22 p

Weimerskirch H (2001) Seabird demography and its relationship with the marine environment. In: Schreiber EA, Burger J (eds) Biology of marine birds. CRC Press, New York, NY, USA, pp 115–136

Wells JB (2013) ENVIRON International Corporation, Houston, TX, USA

White CL, McMaster DG (2005) 2005 South Saskatchewan river piping plover monitoring and captive rearing project. Saskatchewan Water Authority, Regina, SK, Canada. 36 p

White SC, Robertson WB, Ricklefs RE (1976) The effect of hurricane Agnes on growth and survival of tern chicks in Florida. Bird-Band 47:54–71

White DH, Mitchell CA, Cromartie E (1982) Nesting ecology of roseate spoonbills at Nueces Bay, Texas. Auk 99:275–284

Wiedenfeld DA, Wiedenfeld MG (1995) Large kill of neotropical migrants by tornado and storm in Louisiana, April 1993. J Field Ornithol 66:70–80

Wiens JA, Crist TO, Day RH, Murphy SM, Hayward GD (1996) Effects of the *Exxon Valdez* oil spill on marine bird communities in Prince William Sound, Alaska. Ecol Appl 6:828–841

Wilcox L (1959) A twenty year banding study of the piping plover. Auk 76:129–152

Wilkinson PM, Nesbitt SA, Parnell JF (1994) Recent history and status of the Eastern Brown Pelican. Wildl Soc Bull 22:420–430

Wilson BC (2007) Northern American waterfowl management plan, Gulf Coast joint venture: Mottled duck conservation plan. North American Waterfowl Management Plan, Albuquerque, NM, USA. 33 p

Wilson RR, Twedt DL (2003) Spring bird migration in Mississippi alluvial valley forests. Am Midl Nat 149:163–175

Wilson B, Vermillion B (2006) Habitat mosaics to meet the needs of priority Gulf coast birds. U.S. Fish Wildlife Survey, Gulf Coast joint venture. http://el.erdc.usace.army.mil/workshops/06mar-dots/1Wilson.pdf. Accessed 18 Feb 2013

Withers K (2002) Shorebird use of coastal wetlands and barrier island habitat in the Gulf of Mexico. Scientific World Journal 2:514–536

Wittenberger JF, Hunt GL Jr (1985) The adaptive significance of coloniality in birds. Avian Biol 8:1–78

Wolfe MF, Schwarzbach S, Sulaiman RA (1998) Effects of mercury on wildlife: A comprehensive review. Environ Toxicol Chem 17:146–160

Woltmann S (2001) Habitat use and movements of sharp-shinned and Cooper's Hawks at Fort Morgan, Alabama. North Am Bird Bander 26:150–156

Woltmann S, Cimpreich D (2003) Effects of weather on autumn hawk movements at Fort Morgan, Louisiana. Southeast Nat 2:317–326

Woolfenden GE (1983) Rare, threatened, and endangered vertebrates of southwest Florida and potential OCS activity impacts. FWS/OBS 82/03. U.S. Fish and Wildlife Service, Washington, DC, USA. 64 p

Yancey MR, Forys EA (2010) Black Skimmers feed when light levels are low. Waterbirds 33:556–559

Yong W, Moore FR (1997) Spring stopover of intercontinental migratory thrushes along the northern coast of the Gulf of Mexico. Auk 114:263–278

Zalles JL, Bildstein KL (2000) Raptor watch: A global directory of migration sites, Bird life Conservation Series 9. Smithsonian Books, Washington, DC, USA. 438 p

Zambrano R, Robson MS, Charnetzky DY, Smith HT (1997) Distribution and status of least tern nesting colonies in southeast Florida. Fla Field Nat 25:85–116

Zdravkovic M (2005) 2004 Coastal Texas breeding snowy and Wilson's plover census and report. Coastal Bird Conservation Program, National Audubon Society, New York, NY, USA

Zuria I, Mellink E (2005) Fish abundance and the 1995 nesting season of the Least Tern at Bahia de San Jorge, Northern Gulf of Mexico. Waterbirds 28:175–180

APPENDIX A: HABITAT MAPS FOR GULF OF MEXICO, WITH EMPHASIS ON THOSE USED BY BIRDS

It is difficult to determine the amount of habitat available for birds in the Gulf of Mexico, partly because of changing habitat (quality, type, and quantity) and variable and changing requirements of birds. However, some habitat types or land cover types are unusable (such as developed lands). This appendix provides maps of habitat type (land cover) using the National Land Cover database from the National Geospatial Management Center, developed by Jason Wells, ENVIRON International Corporation, Houston, Texas in consultation with the author. Land Cover was determined for 10 mile (16.1 km) and 25 mile (40.2 km) from the coastline. Methodology is described below as well as the land cover types used (USDA, National Geospatial Management Center; accessed 28 March, 2012). The maps for each state follow, with the 10- and 25-mile area shown by dotted lines (see Figures A.1, A.2, A.3, A.4, and A.5).

Methods for GIS Maps That Follow (Wells 2013).

Data sources used:

Analytical/Operational Layers:
National Land Cover Dataset. 2006. United States Geological Survey. (raster).
    Detail County Lines. 2010. ESRI Data and Maps 2010. (dtl_cnty_ln.sdc, polyline).

Base Layers:
Detail States. 2010. ESRI Data and Maps 2010. (dtl_st.sdc, polygon). Nations data layer. 2010. ESRI Data and Maps 2010. (nation.sdc, polygon).
World Boundaries and Places. 2012. ESRI—Streaming data for ArcGIS Desktop. Copyright: © 2011, ESRI, DeLorme, NAVTEQ, TomTom.
World Transportation. 2012. ESRI—Streaming data for ArcGIS Desktop. Copyright:© 2011 ESRI, DeLorme, NAVTEQ, TomTom.
World Imagery. 2012. ESRI—Streaming data for ArcGIS Desktop. Source: Esri, i-cubed, USDA, USGS, AEX, GeoEye, Getmapping, Aerogrid, IGN, IGP, and the GIS User Community.

Software used:
ESRI ArcGIS Desktop. ArcInfo 10.0 Service Pack 4 (Build 4000) with Spatial Analyst Extension.

GIS Procedures:
For analytical purposes, the data resources were preprocessed or normalized to a consistent datum and projection. Since we were attempting to gain aerial estimates of land cover classes by state, our choice was to use the North American Datum 1983 (based on the Geodetic Reference System GRS 1980 spheroid) and NAD 1983 Albers projection using meters as the unit of measure. For mapping display purposes, the geographic coordinate system—WGS 1984 was used with projection "on-the-fly" from ArcGIS 10.

We chose operational GIS layers for this analysis as the National Land Cover Dataset (NLCD 2006) and a polyline feature class from ESRI Data and Maps 2010 called Detailed County Lines. The NLCD was used to derive comparative areal estimates of land cover classification by coastal state within 10 and 25 miles of coastal shoreline and the coastal shoreline was derived from the Detailed County Lines feature class.

The classification types comprising the NLCD:

- Developed open lands
- Developed low intensity lands
- Developed medium lands

- Developed high intensity lands
- Deciduous forest
- Evergreen forest
- Mixed forest
- Pasture/hay
- Cultivated crops
- Shrub/scrub
- Grassland/herbaceous
- Emergent herbaceous wetlands
- Woody wetlands
- Barren land
- Perennial ice/snow (not applicable)
- Open water

Datasets for NLCD 2006 were accessed by Internet download from USDA/NRCS—National Geospatial Management Center on March 28, 2012 for the states bordering the Gulf of Mexico (Texas, Louisiana, Mississippi, Alabama, and Florida). The horizontal datum referenced for these datasets was North American Datum of 1983 (NAD83) using the GRS1980 spheroid. The planar horizontal coordinate system used for the NLCD was Universal Transverse Mercator (UTM) spanning the zones 14–17 (TX-14, LA-15, MS-16, AL-16, FL-17).

We derived the coastal boundary layer for the Gulf of Mexico by extracting the relevant line type from the Detailed County Lines feature class. The attributes for this polyline feature class included line classification types of Coastline, County, International, Shoreline, and State. Using the "Select by Attributes" Tool in ArcGIS, we selected the Coastline type. The Detailed County Lines feature class included the coastlines of the Pacific and Atlantic Oceans and Gulf of Mexico. We used the "Select by Polygon" Tool to select and reduce the feature class extent representing the coastal shoreline of the Gulf of Mexico; from the Texas USA/Mexico border to the Florida Keys, along the Straits of Florida, and to the northern portion of Biscayne Bay, Miami, Florida.

The Detailed County Lines feature class has a native geographic coordinate system of WGS 1984. To reduce errors and enhance processing speed, we exported the selected and reduced shoreline feature class elements to a new feature class [dtl_cnty_ln_GOM] and converted the new feature class to NAD 1983 Albers using the transformation method NAD_1983_-To_WGS_1984_5.

This new layer formed the basis for our clipping buffer zone polygon layer of 10 and 25 miles inland from the coast, respectively. We used the ArcGIS Buffer Wizard to create new feature class polygons representing areas of 10- and 25-mile radius of the coastline.

To assist with understanding how NLCD classification types varied by state, we split the 10- and 25-mile buffer polygons at each of the state borders (e.g., Texas and Louisiana, Louisiana and Mississippi, Mississippi and Alabama) leaving new polygon features class elements of 10- and 25-mile buffer distance for Texas, Louisiana, Mississippi, Alabama, and Florida (western Florida around southern Florida to northern Miami).

The NLCD rasters were collected by state and combined into a single mosaic using the ArcGIS Raster Mosaic Tool and subsequently reprojected to NAD 1983 Albers. Once the mosaic was completed for the GoM states, the new mosaic dataset was clipped to the 25-mile buffer using the Clip Raster Tool creating a new smaller raster. This was done to reduce geoprocessing time for later operations by eliminating the majority of inland areas not relevant to this analysis.

Reclassification of the new 25-mile clipped NLCD mosaic was done to combine existing similar classifications of "Developed" areas that would be tabulated to aerial assessments by state. The classifications of Developed Open Lands (21) and Developed Low Intensity Lands (22) were combined into one Developed Open/Low Lands classification and the Developed Medium Intensity Lands (23) and Developed High Intensity Lands (24) were combined into one Developed Medium/High Intensity Lands classification. To perform the reclassification of the 25-mile clipped mosaic NLCD, the Reclassify Tool from the Spatial Analyst Extension was used. The areal extent evaluation for NLCD classification by state within 10 or 25 miles of the coastline used the following classification groups:

- Developed open/low intensity lands
- Developed medium/high intensity lands
- Deciduous forest
- Evergreen forest
- Mixed forest
- Pasture/hay
- Cultivated crops
- Shrub/scrub
- Grassland/herbaceous
- Emergent herbaceous wetlands
- Woody wetlands
- Barren land
- Open water

We assessed the areal extent of newly derived NLCD classification types by state using the Zonal Tabulate Area Tool from the Spatial Analyst Extension of ArcGIS. Inputs for geoprocessing using this tool were the 25-mile clipped/reclassified mosaic NLCD raster and the 10- or 25-mile buffer feature classes (split by state border) as the "feature zone" with attribute of STATE_NAME as the zone field. The NLCD mosaic was used as the input raster with classification "Value" as the Class Field. The result of Zonal Tabulate Area Tool is cross-tabulation containing the summation of the areas (square meters) from NLCD classification type by 10- or 25-mile buffer zone by state. This areal extent is not a true three-dimensional area since no topographic dataset was included.

For visualization purposes, the NLCD mosaic was further reclassified and reduced to the following classifications:

- Developed open/low/medium/high intensity
- Deciduous/evergreen/mixed forest
- Pasture/hay/cultivated crops
- Shrub/scrub—grassland/herbaceous
- Emergent herbaceous wetlands
- Woody wetlands
- Barren land
- Open water

Background for the National Land Cover Dataset:

National Land Cover Dataset (2006). United States Geological Survey.

"National Land Cover Database 2006 (NLCD2006) is a 16-class land cover classification scheme that has been applied consistently across the conterminous United States at a spatial resolution of 30 m. NLCD2006 is based primarily on the unsupervised classification of Landsat Enhanced Thematic Mapper + (ETM+) circa 2006 satellite data." This classification is based on the Anderson Land Cover Classification System.

Class/Value	National Land Cover Dataset—Classification Description
Water	Areas of open water or permanent ice/snow cover
11	*Open water*—areas of open water, generally with less than 25 % cover of vegetation or soil
12	*Perennial ice/snow*—areas characterized by a perennial cover of ice and/or snow, generally greater than 25 % of total cover
Developed	Areas characterized by a high percentage (30 % or greater) of constructed materials (e.g., asphalt, concrete, buildings, etc.)
21	*Developed, open space*—areas with a mixture of some constructed materials, but mostly vegetation in the form of lawn grasses. Impervious surfaces account for less than 20 % of total cover. These areas most commonly include large-lot single-family housing units, parks, golf courses, and vegetation planted in developed settings for recreation, erosion control, or aesthetic purposes
22	*Developed, low intensity*—areas with a mixture of constructed materials and vegetation. Impervious surfaces account for 20–4 % percent of total cover. These areas most commonly include single-family housing units
23	*Developed, medium intensity*—areas with a mixture of constructed materials and vegetation. Impervious surfaces account for 50–79 % of the total cover. These areas most commonly include single-family housing units
24	*Developed high intensity*—highly developed areas where people reside or work in high numbers. Examples include apartment complexes, row houses, and commercial/industrial. Impervious surfaces account for 80–100 % of the total cover
Barren	Areas characterized by bare rock, gravel, sand, silt, clay, or other earthen material, with little or no "green" vegetation present regardless of its inherent ability to support life. Vegetation, if present, is more widely spaced and scrubby than that in the green vegetated categories; lichen cover may be extensive
31	*Barren land (rock/sand/clay)*—areas of bedrock, desert pavement, scarps, talus, slides, volcanic material, glacial debris, sand dunes, strip mines, gravel pits, and other accumulations of earthen material. Generally, vegetation accounts for less than 15 % of total cover
Forest	Areas characterized by tree cover (natural or seminatural woody vegetation, generally greater than 6 m tall); tree canopy accounts for 25–100 % of the cover.
41	*Deciduous forest*—areas dominated by trees generally greater than 5 m tall, and greater than 20 % of total vegetation cover. More than 75 % of the tree species shed foliage simultaneously in response to seasonal change
42	*Evergreen forest*—areas dominated by trees generally greater than 5 m tall, and greater than 20 % of total vegetation cover. More than 75 % of the tree species maintain their leaves all year. Canopy is never without green foliage
43	*Mixed forest*—areas dominated by trees generally greater than 5 m tall, and greater than 20 % of total vegetation cover. Neither deciduous nor evergreen species are greater than 75 % of total tree cover
Shrubland	Areas characterized by natural or seminatural woody vegetation with aerial stems, generally less than 6 m tall, with individuals or clumps not touching to interlocking. Both evergreen and deciduous species of true shrubs, young trees, and trees or shrubs that are small or stunted because of environmental conditions are included
52	*Shrub/scrub*—areas dominated by shrubs less than 5 m tall with shrub canopy typically greater than 20 % of total vegetation. This class includes true shrubs, young trees in an early successional stage or trees stunted from environmental conditions

(continued)

Class/Value	National Land Cover Dataset—Classification Description
Herbaceous	Areas characterized by natural or seminatural herbaceous vegetation; herbaceous vegetation accounts for 75–100 % of the cover
71	*Grassland/herbaceous*—areas dominated by gramanoid or herbaceous vegetation, generally greater than 80 % of total vegetation. These areas are not subject to intensive management such as tilling, but can be utilized for grazing
Planted/ cultivated	Areas characterized by herbaceous vegetation that has been planted or is intensively managed for the production of food, feed, or fiber; or is maintained in developed settings for specific purposes. Herbaceous vegetation accounts for 75–100 % of the cover
81	*Pasture/hay*—areas of grasses, legumes, or grass-legume mixtures planted for livestock grazing or the production of seed or hay crops, typically on a perennial cycle. Pasture/hay vegetation accounts for greater than 20 % of total vegetation
82	*Cultivated crops*—areas used for the production of annual crops, such as corn, soybeans, vegetables, tobacco, and cotton, and also perennial woody crops such as orchards and vineyards. Crop vegetation accounts for greater than 20 % of total vegetation. This class also includes all land being actively tilled
Wetlands	Areas where the soil or substrate is periodically saturated with or covered with water as defined by Cowardin et al. (1979)
90	*Woody wetlands*—areas where forest or shrubland vegetation accounts for greater than 20 % of vegetative cover and the soil or substrate is periodically saturated with or covered with water
95	*Emergent herbaceous wetlands*—areas where perennial herbaceous vegetation accounts for greater than 80 % of vegetative cover and the soil or substrate is periodically saturated with or covered with water

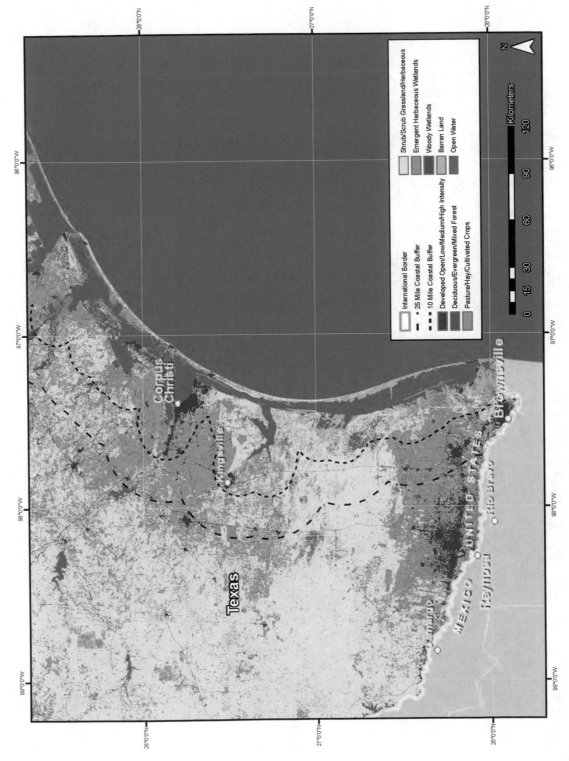

Figure A.1. Texas.

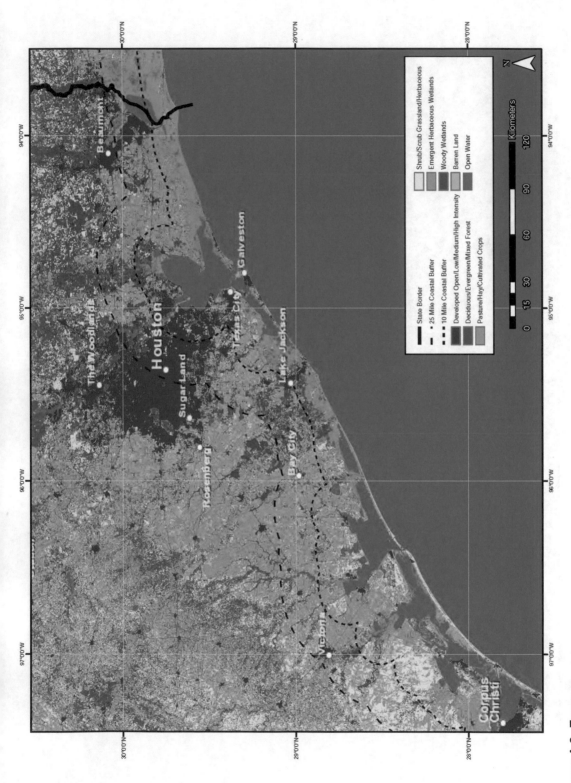

Figure A.2. Texas.

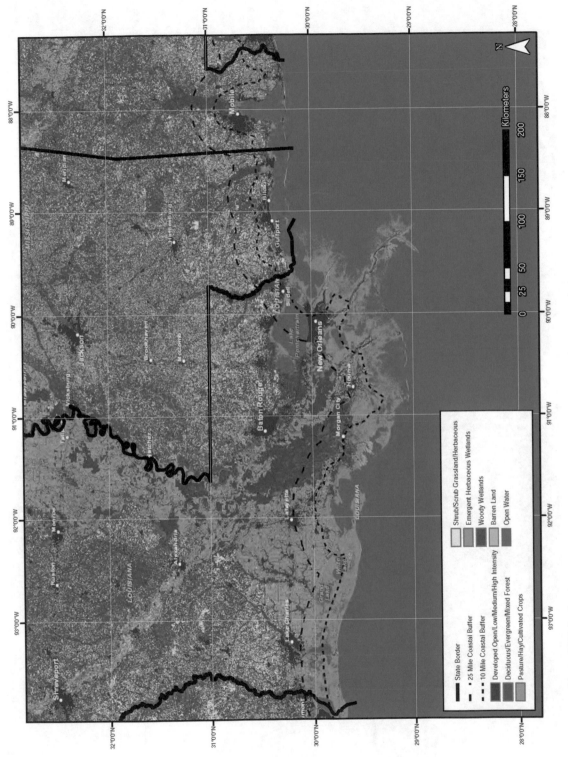

Figure A.3. Louisiana/Mississippi.

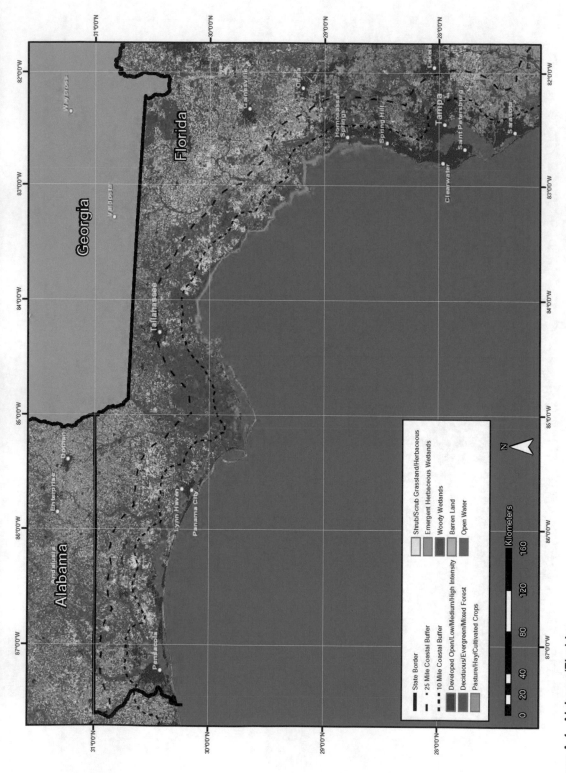

Figure A.4. Alabama/Florida.

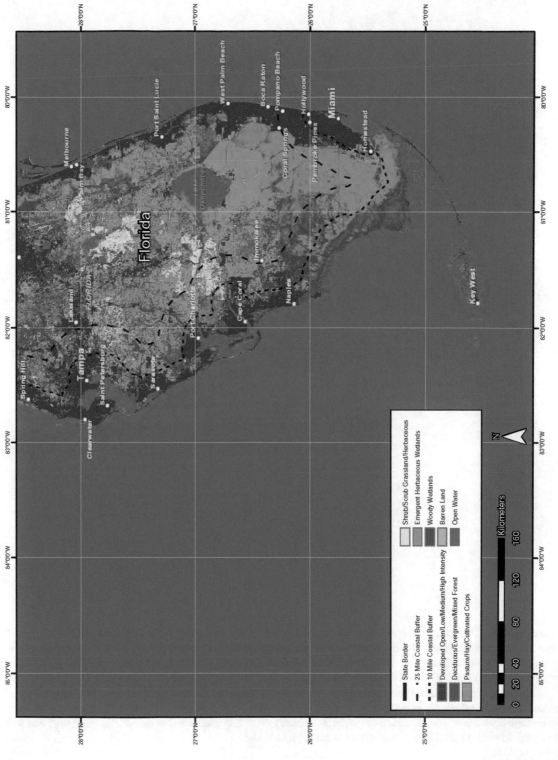

Figure A.5. Florida.

Open Access This chapter is licensed under the terms of the Creative Commons Attribution-NonCommercial 2.5 International License (http://creativecommons.org/licenses/by-nc/2.5/), which permits any noncommercial use, sharing, adaptation, distribution and reproduction in any medium or format, as long as you give appropriate credit to the original author(s) and the source, provide a link to the Creative Commons license and indicate if changes were made.

The images or other third party material in this chapter are included in the chapter's Creative Commons license, unless indicated otherwise in a credit line to the material. If material is not included in the chapter's Creative Commons license and your intended use is not permitted by statutory regulation or exceeds the permitted use, you will need to obtain permission directly from the copyright holder.

CHAPTER 13

MARINE MAMMALS OF THE GULF OF MEXICO

Bernd Würsig[1]

[1]Departments of Marine Biology and Wildlife and Fisheries Sciences, Texas A&M University, College Station, TX 77843, USA
wursigb@tamug.edu

13.1 INTRODUCTION

The marine mammals of the Gulf of Mexico consist of whales, dolphins and one species of coastal sirenian, the West Indian manatee (*Trichechus manatus*). There are no seals, sea lions, fur seals, or sea-going otters as are present in many other parts of the world. One tropical seal, the Caribbean monk seal (*Monachus tropicalis*), which was apparently never abundant in the Gulf, became extinct in the early part of the twentieth century. At about the same time, California sea lions (*Zalophus californianus*) escaped from small zoos and, for a while, appeared to be reproducing and establishing a feral population, but that population is also gone. There are no porpoises of the cetacean phocoenid family in the Gulf. Only the harbor porpoise (*Phocoena phocoena*) occurs in the North Atlantic, and waters around southern Florida are too warm for this species to have made a foray into the Gulf.

The cetaceans of the Gulf are diverse and well established, ranging from the ubiquitous, nearshore (and there is an offshore variant) common bottlenose dolphin (*Tursiops truncatus*), hereafter referred to as bottlenose dolphin unless differentiation is warranted from the Indo-Pacific bottlenose dolphin (*Tursiops aduncus*), to the sperm whale (*Physeter macrocephalus*), the largest toothed whale in the oceans and the largest toothed creature. The Gulf of Mexico is home to several species of continental shelf and deep ocean dolphins or whales of the family Delphinidae as well as deepwater beaked whales of the family Ziphiidae. There are baleen whales (infraorder Mysticeti) in the Gulf as well, members of the family Balaenopteridae from the relatively small minke (*Balaenoptera acutorostrata*) to the giant of the seas, the largest mammal on Earth, the blue whale (*Balaenoptera musculus*). Of the baleen whales, only the Bryde's whale (*Balaenoptera edeni*) is a resident of the northern Gulf.

Each major marine mammal grouping in the Gulf has evolved to make its living in quite a different way. The manatee is one of only four marine mammals (all of the order Sirenia) that feed on seagrasses and other plant material. Its common name, sea cow, is quite appropriate as manatees are indeed related to early ungulates and have a ruminant stomach somewhat similar to their terrestrial forebears. The baleen whales have a structure of keratinous material—the baleen plates—that hang from the upper jaw with finely fringed inner hairs that form a filter mat and allow for batch feeding. Although baleen whales have three different feeding methods (Heithaus and Dill 2009), the Bryde's whale, which dominates the baleen whale fauna of the Gulf, is a lunge feeder that uses a technique that has aptly been described as the greatest biomechanical action on Earth (Croll et al. 2001). The toothed whales (infraorder Odontoceti) of the Gulf, which range from the giant sperm whale to the small Clymene dolphin (*Stenella clymene*), have all evolved the ability to echolocate—use high-frequency sound—to sense their conspecifics, potential danger from sharks and killer whales (*Orcinus orca*), obstructions in the water as well as the surface and bottom, and—important at all times—potential prey. Some of

© The Author(s) 2017

C.H. Ward (ed.), *Habitats and Biota of the Gulf of Mexico: Before the Deepwater Horizon Oil Spill*,
DOI 10.1007/978-1-4939-3456-0_5

the toothed whales have also evolved extremely large, complex brains, probably due to the communication needs of sophisticated social living (Marino 2004). Large brains may also be thought of as a specialization that has allowed this group to become highly diverse and successful in a generally forbidding sea.

The present description of the cetaceans and one siren of the Gulf is an update of information in the books by Würsig et al. (2000) and Jefferson et al. (2008) as well as articles by Baumgartner et al. (2001), Davis et al. (2002), Mullin and Fulling (2004), Maze-Foley and Mullin (2006), Schmidly and Würsig (2009), Waring et al. (2010), Schick et al. (2011), and others. Few published comprehensive summaries of marine mammals of the Gulf of Mexico have been made in the past 12 years, so this update comes at a particularly opportune time. The estimated numbers of animals per species represented here for the northern (United States [U.S.]) portion of the Gulf are from the official National Oceanic and Atmospheric Administration-National Marine Fisheries Service (NOAA-NMFS) Office of Protected Resources marine mammal stock assessment reports (Waring et al. 2011),[1] unless stated otherwise. Information in the primary literature has vastly increased present knowledge, especially about bottlenose dolphins and sperm whales. However, aspects of community structure in pelagic cetaceans are only slowly coming to light. Since the publication of Würsig et al. (2000), a PhD dissertation by Ortega-Ortiz (2002) summarized information about cetaceans and cetacean habitats of the Mexican southern waters of the Gulf. This chapter does not cover much of this newer information south of U.S. waters, but a synopsis and comparison with U.S. waters is made. With the exception of several discrete areas for bottlenose dolphins in select bays and estuaries, population estimates for the Mexican and Cuban parts of the Gulf are not available.

This chapter is organized into seven major sections designed to give the reader a flavor of these charismatic megafauna, as even the smallest of dolphins in the oceans is a large—and large brained—social creature among mammals in general. Major chapter sections are as follows: (1) a general introduction; (2) history of research; (3) basic species, habitat, and number descriptions, if available; (4) discussion of anthropogenic impacts; (5) conclusions that summarize present baseline conditions; (6) references; and (7) two appendices. Throughout this chapter, cetaceans and the one sirenian of the Gulf of Mexico are described in context with their distributions and habitat preferences worldwide, as these animals are far ranging; none of the species discussed occurs only in the Gulf (i.e., none is endemic).

The Würsig et al. (2000) book describes 31 species of cetacean plus the one sirenian known or believed to be in the Gulf as of its writing in the late 1990s. It is now quite certain that three species—the more-northerly occurring long-finned pilot whale (*Globicephala melas*) and the short-beaked and long-beaked common dolphins (*Delphinus delphis* and *Delphinus capensis*, respectively)—can be taken off the list of potential Gulf inhabitants, with the recognition that quite a few species might eventually be discovered as rare vagrants. These three species are not included in this chapter. Common dolphins were quite often cited to occur in the Gulf, but apparently all such descriptions were due to confusion with other species, most commonly Clymene dolphins (Jefferson 1995; Jefferson and Schiro 1997). A further seven species have records for the Gulf but also occur so rarely that they are mentioned only in passing. Six of these are baleen whales: Atlantic right (*Eubalaena glacialis*), blue, fin (*Balaenoptera physalus*), sei (*Balaenoptera borealis*), minke (*Balaenoptera acutorostrata*), and humpback (*Megaptera novaeangliae*) whales. This leaves only the tropical Bryde's whale as a reliable baleen whale of the Gulf, which means that with modern evidence it has become apparent that the

---

[1] http://www.nmfs.noaa.gov/pr/sars/

baleen whale fauna of the Gulf is less rich than was believed even a decade ago. The seventh species is a beaked whale—Sowerby's beaked whale (*Mesoplodon bidens*). However, beaked whales are poorly described in the Gulf, and it would not be surprising to have several added to the list with further knowledge.

13.2 HISTORY OF RESEARCH IN THE GULF

13.2.1 Whaling

While whaling by itself is not research, much important detail on distribution can be gleaned from whaling records, and the Gulf of Mexico is no exception. The best known records of Gulf whaling come from the worldwide Townsend Charts (Townsend 1935), which illustrate where, when, and how many whales were taken, as recorded from logbook records of U.S. (American or Yankee) whaling ships. These are thus somewhat biased, but they provide a general review. This information has recently been updated and expanded (Reeves et al. 2011), and new insights from this re-evaluation give a modern perspective of at least some species than previously was available for the Gulf (Figure 13.1).

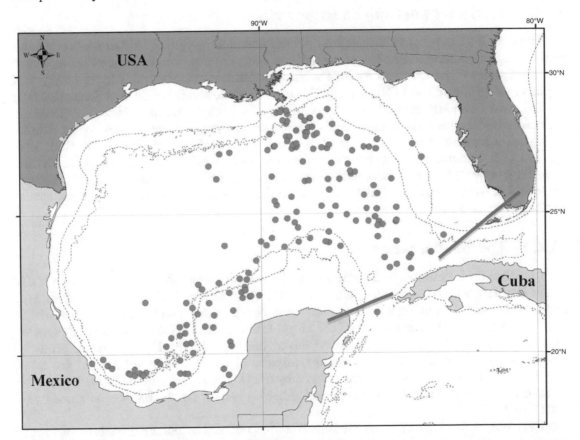

Figure 13.1. Positions of whaling ships on days when sperm whales were sighted or caught. *Dotted lines* **represent the 100 m (328 ft) and 1,000 m (3,281 ft) isobaths, and the lines between Cuba and Florida (***top right***) and the Mexican Yucatán Peninsula (***bottom left***) represent the approximate extent of the Gulf of Mexico to the southeast, as considered for marine mammals in this report. Whaling ship positions are from Reeves et al. (2011), p 44. It is instructive to compare this figure of sperm whale whaling with those of Figures 13.9 and 13.10 of modern sperm whale sightings in the northern and southern Gulf, respectively.**

Clark (1887) mentions that right whales were taken in the Gulf by eighteenth century and nineteenth century Yankee whalers, but this claim was not substantiated by Townsend (1935) or by others (Reeves et al. 2011). Instead, it is clear that the largest whaling effort—starting in 1788 and apparently ending in 1877 (with perhaps a few sporadic attempts into the twentieth century)—was on sperm whales, with occasional takes of so-called finbacks (*Balaenoptera* sp.), so-called porpoises (small delphinids), and killer whales. Besides sperm whales, another deep diving odontocete cetacean, the short-finned pilot whale (*Globicephala macrorhynchus*), was quite commonly killed. Grampus [probably Risso's dolphin (*Grampus griseus*)] was also taken. The whaling records of all three of these major whaled species overlap with what is known of present day distribution in the northern Gulf. Whaling records will be discussed in later detailed sections on the sperm whale, pilot whale, and Risso's dolphin.

The Gulf of Mexico is one of the few oceanic areas that were exploited only in early U.S. history, predominantly by takes of sperm whales. With the exception of some catches of nearshore bottlenose dolphins for the aquarium trade, there has been little exploitation for about the past 130 years. This remarkable fact allows investigation of these long-lived animals as relatively unperturbed, save for the strong buildup of shipping and industrial activities.

13.2.2 Early Opportunistic Research

Prior to 1977, reports of whales and dolphins came largely from fishermen's and other boaters' sightings and publications of strandings. These data were compiled by Gordon Gunter in a long series of papers (Gunter 1954 is but one example) and by J. C. Moore (1953), J. N. Layne (1965), Caldwell and Caldwell (1973), Lowery (1974), and Schmidly and Shane (1978). In 1980, David Schmidly started the Texas Marine Mammal Stranding Network (TMMSN) (Schmidly 1981), the first comprehensive volunteer network to obtain and report detailed biological information on strandings of live and dead marine mammals in the Gulf. Until systematic line-transect efforts from airplanes and boats began, both at about the same time, strandings provided the best information on cetaceans in the northern Gulf (with practically no information for the southern Gulf) and gave the first appreciation that there are many species of deepwater cetaceans.

13.2.3 The Modern Era

The TMMSN continues to this day and is now integrated closely with the U.S.-wide NMFS Stranding Network and specifically the Southeast Region Marine Mammal Stranding Network that includes states from Texas in the southwest to North Carolina on the north-central Atlantic coast as well as Puerto Rico and the U.S. Virgin Islands. The Office of Protected Resources, NOAA Fisheries Service Headquarters, Silver Spring, Maryland, coordinates protected resources and research programs under the auspices of the U.S. Marine Mammal Protection Act (MMPA) of 1972.[2] Manatees are covered by the U.S. Fish and Wildlife Service (USFWS), and there is some good interface in analyses of habitat use for coastal marine mammals by the NMFS and USFWS.[3]

Even before the TMMSN, there were two studies on local bottlenose dolphins—one begun by Randall Wells and colleagues in the early 1970s in the Sarasota-Bradenton area of west-central Florida (Wells 2003) and another by Susan Shane in the mid-1970s at Aransas Pass,

---

[2] http://www.nmfs.noaa.gov/pr/

[3] http://www.fws.gov/

Texas (Shane 1977). Later, Shane compared her Aransas Pass work with the results of a new study in the 1980s off Sanibel Island, Florida, almost precisely east of her former study area in Texas (Shane 2004). The Wells studies have continued and expanded and spawned several 100 publications resulting in one of the best-known cetacean populations in the world. A summary of the major findings from this set of studies is included in the section devoted to bottlenose dolphins in the Gulf.

The first systematic offshore efforts to describe marine mammals in parts of the northern Gulf of Mexico occurred during the Fritts Surveys, 1979–1981 aerial surveys led by T. H. Fritts (Fritts et al. 1983). These surveys established that several dolphins of the genus *Stenella* occur off continental shelf and slope waters, but these early surveys (originally designed for sea turtle and sea bird assessments) suffered from small cetacean sample size, precluding abundance estimates, and some unfortunate misidentifications. For example, they did not distinguish between the two species of spotted dolphins (*Stenella attenuata* and *Stenella frontalis*), now known to have quite different distribution patterns in the Gulf; at times the surveys confused spinner and Clymene dolphins (*Stenella longirostris* and *Stenella clymene*, respectively), and sometimes they misidentified either Clymene or pantropical spotted dolphins (*Stenella attenuata*) as short-beaked common dolphins. Common dolphins do not presently occur in the Gulf, nor are there verified historical records.

The NMFS conducted a series of aerial surveys in continental and coastal waters (continental shelf aerial surveys, 1983–1986; red drum aerial surveys, 1986–1987) and established some estimates of numbers of the predominantly sighted cetacean, the common bottlenose dolphin.

Since 1989, aerial and shipboard surveys have been conducted (albeit not in every year) in both coastal and deeper oceanic U.S. waters, commencing with Keith Mullin's 1989–1990 continental slope aerial surveys that for the first time properly documented distribution and abundance of oceanic cetaceans in waters off Louisiana and Mississippi (Mullin et al. 1994; Jefferson and Schiro 1997) (Figure 13.2). In the 1990s, NMFS and Texas A&M University coordinated dedicated aerial and boat surveys in concert with descriptions of physical and

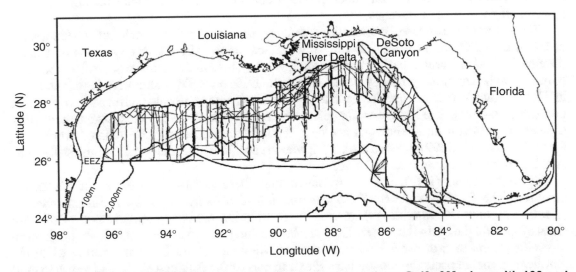

Figure 13.2. Example of the many surveys conducted in the northern Gulf of Mexico, with 100 and 2,000 m (328 and 6,562 ft) isobaths, and the boundary of the U.S. Exclusive Economic Zone (marked as EEZ). Thinner lines represent NMFS vessel surveys 1991–2001 (but not in all years). Note that there is much more survey effort between 200 and 2,000 m (656 and 6,562 ft) depths than in abyssal waters deeper than 2,000 m (6,562 ft) (from Maze-Foley and Mullin 2006).

biological oceanography—the Gulfcet surveys. These Gulfcet surveys provided a more detailed knowledge of how cetaceans are affected by seasonal and interyear climatic conditions, including the presence and positions of eddies and gyres that spin off the Loop Current in the north-central Gulf and the periodic, and at times strong, fresh water incursions from the Mississippi/Atchafalaya River system (Davis et al. 1998, 2002; Baumgartner et al. 2001; Ortega-Ortiz 2002). Such work has continued in the first 10 years of this century, although not always with similar multidisciplinary input as in the 1990s. The sperm whale seismic surveys (SWSS) of the first 6 years of the 2000s revealed much about sperm whale habitat in the north-central Gulf (Jochens et al. 2008), and Keith Mullin has continued to lead NMFS surveys (Maze-Foley and Mullin 2006).

Studies, now better integrated year to year with standardized visual (and often acoustic) data gathering and analysis techniques, are finally giving a more complete picture of the cetaceans of the northern Gulf of Mexico. Such long-term information is allowing for assessments of community structure of pelagic cetaceans at large spatial scales, as recently attempted by Schick et al. (2011). Data for the southern Gulf are much less complete, except for an analysis by Ortega-Ortiz (2002); and for western nearshore areas in the state of Veracruz, there are analyses by Galindo et al. (2009), Martínez-Serrano et al. (2011), and Valdes-Arellanes et al. (2011). The shores of Cuba are poorly represented, except for a thesis by Pérez-Cao (2004) from the Camagüey Archipelago, NE Cuba, and outside of the Gulf.

Studies of bottlenose dolphins in nearshore and offshore waters deserve special attention, since bottlenose dolphins of the Sarasota-Bradenton area of west Florida are arguably the longest continuously studied marine mammal (Wells and Scott 2009), vying for this distinction only with killer whales of several pods in the coastal Pacific near the U.S.–Canadian border (Ford 2009). Other studies also have been carried out, most notably by Shane off south Texas and southwest Florida (Shane 1977, 2004; Shane et al. 1986), Mullin (1988) in the north-central Gulf, and students of Würsig (e.g., Bräger et al. 1994; Maze 1997; Moreno 2005) off the northwest Gulf, near Galveston Texas. Most of these studies have relied heavily on photo-identification for mark-recapture numbers analyses and society descriptions, and more is now known about the generally open, fluid social systems of bottlenose dolphins than of the social systems of most other cetaceans.

The study of Caribbean (West Indian) manatee, one of only four species of the taxonomic order Sirenia (Figure 13.3) and the only vegetarians among the 122 or so species of marine mammals, had a similar early introduction by Gordon Gunter in his work with cetaceans (Gunter 1941). The monograph by Daniel Hartman (1979) on ecology and behavior of manatees in Florida (work carried out in the late 1960s) finally gave detailed information on their preferred habitats, life-history strategies, foraging techniques, and social/sexual and other aspects of behavior and behavioral ecology. While there has been much updated work (e.g., Reynolds and Odell 1991; O'Shea et al. 1995; Marsh et al. 2011), this early monograph still stands as a hallmark of marine mammal studies.

Since Hartman (1979), it is now known that there are two subspecies of Caribbean manatee—the Florida manatee (*Trichechus manatus latirostris*) and the Antillean manatee (*Trichechus manatus manatus*) (Figure 13.3). The Florida manatee regularly occurs on both sides of Florida (i.e., into the eastern Gulf of Mexico and north to about Tampa), although more rare excursions to northwest Florida, Alabama, Mississippi, and Louisiana occur, while the Antillean manatee occurs in the southern Mexican part of the Gulf and all the way south to (and even south of) the estuary of the Amazon River, where it is believed to hybridize with the much smaller-bodied Amazon manatee (*Trichechus inunguis*). Marsh et al. (2011) provide excellent up-to-date information on all Sirenia.

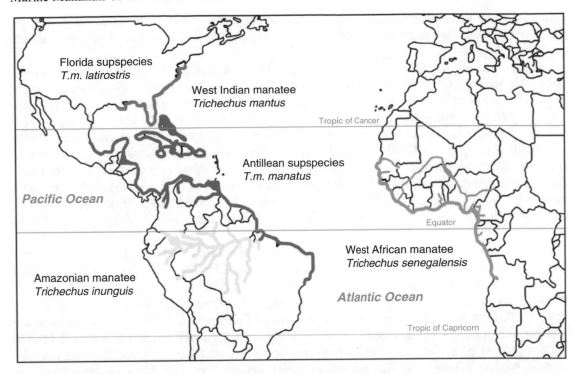

Figure 13.3. Manatees of the world. There are three species, and the Caribbean (or West Indian) manatee has the largest distribution, divided into two subspecies, shown in blue in the north and in red south of this. Both subspecies occur in the Gulf. The Florida subspecies range shown in this figure includes outlying areas, and only occasionally in summer do a few manatees occur as far north as Long Island, New York, and as far west as Louisiana, as shown here (adapted from Gonzalez-Socoloske and Olivera-Gomez 2012, after an International Union for Conservation of Nature (IUCN) species map).

In the Gulf, Florida manatees rely in large part on eight major warm-water refugia, caused by both natural warm-water springs and industrial power plants. This is a worry for wildlife managers especially in the northern part of that range such as the Tampa Bay Apollo Beach (or Big Bend) power plant (27° 48′ N). Manatees are warm-water creatures with low metabolic rates, and they quickly become cold-stressed in temperatures at or less than 17 °C; (63° Fahrenheit [F]), a common occurrence in winter off the central Florida coast and a cataclysmic event for manatees when there is a need to shut down a power plant (Laist and Reynolds 2005).

13.3   SPECIES SUMMARIES, HABITAT USE, AND NUMBERS

13.3.1   Overview of Species

Except for coastal bottlenose dolphins and Florida manatees, concepts of numbers, group sizes, seasonal and interyear distributions, and habitat use of ocean-going cetaceans in the Gulf are far from complete, and little is known about the Gulf-inhabiting Antillean manatee in Mexican waters. Nevertheless, in large part due to multidisciplinary studies carried out especially by Texas A&M University and the U.S. NMFS in the 1990s, the Gulfcet studies, and subsequent aerial and behavioral censuses by NMFS since then, some basics are known. Much of the NMFS work was conducted from research vessels and was usually in conjunction with ichthyoplankton research (e.g., Maze-Foley and Mullin 2006). However, manatee and cetacean stock assessments contain a great deal of uncertainty, and present knowledge is quite incomplete regarding what is a separate stock (i.e., population) and how many animals are contained therein.

Costly aerial and shipboard surveys in open ocean waters can rarely provide the kind of fine-scale survey lines and repetitions of surveys (in all seasons and with variabilities from year to year) needed to provide numbers estimates with low variances. Thus, surveys of the same areas in different years can provide quite different number estimates for certain species. All that can be done is present the recent ones for which the authors express greatest confidence.

While Würsig et al. (2000) described 32 potential and known species for the Gulf, several of these are likely not found at all, and several others are so rare or vagrant that they need only cursory treatment. The potential species are listed in Table 13.1 with bold highlights for those known to occur in the northern Gulf regularly enough for there to be reliable information on basic group size, sea surface temperature, and depth of occurrence (see Table 13.2 for the latter).

The North Atlantic right, blue, fin, humpback, minke, and sei whales are species that were historically present in very low numbers and are only occasionally sighted or stranded. In other words, the only baleen whale found regularly during surveys in the northern Gulf is the Bryde's whale. This is a strong departure from what was understood in the 1990s when the Würsig et al. (2000) book was being prepared, which means that the Gulf is less rich in baleen whale diversity than originally supposed. Additionally, while there are often unidentified beaked whales sighted in the Gulf, there are reliable records only for Cuvier's, Blainville's, and Gervais' beaked whales (*Ziphius cavirostris*, *Mesoplodon densirostris*, and *Mesoplodon europaeus*, respectively). Sowerby's beaked whales and several others are possibly greater in number than modern records indicate, but Sowerby's are dropped from the list during this treatment due to paucity of confirmed sightings. The colder-water long-finned pilot whale also does not seem to be present and neither do short-beaked or long-beaked common dolphins. As was mentioned before, the quite copious previous records of the genus *Delphinus* were due to misinterpretation of sightings and strandings, and they may not have occurred in the Gulf in recent (or even earlier) history.

Three species habitually occur in waters less than 200 m (656 ft) deep: the inshore/coastal manatee, inshore/coastal and shelf bottlenose dolphin, and shelf Atlantic spotted dolphin (*Stenella frontalis*); another 19 species—all but the Bryde's whale being toothed cetaceans—occur over the continental slope and into deep oceanic waters. No species in the Gulf of Mexico are endemic to the Gulf, and all but the warm-water North Atlantic endemic Caribbean manatee and Clymene and Atlantic spotted dolphins occur in other oceans. Of those that occur beyond Atlantic Ocean waters, the bottlenose dolphin and sperm and killer whales are considered cosmopolitan as they occur in tropical to quite cold waters; all others are tropical, occurring in Atlantic, Pacific, and Indian Ocean waters. While the sperm whale is considered endangered by the United States, it likely is not endangered globally. This leaves the fortunate situation that in the Gulf there are no truly threatened or endangered marine mammal species except for the manatee. Since bottlenose dolphins exist as rather discrete populations in separate nearshore systems and the sperm whale is a separate population in the Gulf, these species are of concern at the population level.

13.3.2   Species, Habitats, and Numbers

To orient the reader to how the marine mammals tend to occur worldwide and in the Gulf, a species-by-species description is in order with approximate numbers in the northern Gulf and related details. Population estimates often are approximate, even for the better-known species

Table 13.1. Potential Marine Mammal Species as Listed in Würsig et al. (2000). Those in bold are present often enough for details, as in Table 13.2.

Species	Main Reasons for Former/Present Listing
North Atlantic right whale, *Eubalaena glacialis*	One stranding, one sighting of two; reports of former hunting
Blue whale, *Balaenoptera musculus*	Two strandings
Fin whale, *Balaenoptera physalus*	Five strandings and rare sightings
Sei whale, *Balaenoptera borealis*	Five strandings
Humpback whale, *Megaptera novaeangliae*	Occasional strandings and rare sightings
Minke whale, *Balaenoptera acutorostrata*	Occasional strandings and rare sightings, Florida Keys
Bryde's whale, *Balaenoptera edeni*	**Strandings and quite common sightings**
Sperm whale, *Physeter macrocephalus*	**Common sightings**
Pygmy sperm whale, *Kogia breviceps*	**Common sightings**
Dwarf sperm whale, *Kogia sima*	**Common sightings**
Cuvier's beaked whale, *Ziphius cavirostris*	**Multiple strandings and occasional sightings**
Blainville's beaked whale, *Mesoplodon densirostris*	**Four strandings and occasional sightings**
Sowerby's beaked whale, *Mesoplodon bidens*	One stranding
Gervais' beaked whale, *Mesoplodon europaeus*	**Multiple strandings and occasional sightings**
Killer whale, *Orcinus orca*	**Common sightings**
Short-finned pilot whale, *Globicephala macrorhynchus*	**Common sightings**
Long-finned pilot whale, *Globicephala melas*	Inferred but with no confirmed records
False killer whale, *Pseudorca crassidens*	**Medium common sightings**
Pygmy killer whale, *Feresa attenuata*	**Medium common sightings**
Melon-headed whale, *Peponocephala electra*	**Common sightings**
Rough-toothed dolphin, *Steno bredanensis*	**Common sightings**
Risso's dolphin, *Grampus griseus*	**Common sightings**
Common bottlenose dolphin, *Tursiops truncatus*	**Common sightings**
Pantropical spotted dolphin, *Stenella attenuata*	**Common sightings**
Atlantic spotted dolphin, *Stenella frontalis*	**Common sightings**
Spinner dolphin, *Stenella longirostris*	**Common sightings**
Clymene dolphin, *Stenella clymene*	**Common sightings**
Striped dolphin, *Stenella coeruleoalba*	**Common sightings**
Short-beaked common dolphin, *Delphinus delphis*	Inferred due to former misidentifications
Long-beaked common dolphins, *Delphinus capensis*	Inferred but with no evidence
Fraser's dolphin, *Lagenodelphis hosei*	**Occasional sightings**
West Indian manatee, *Trichechus manatus*	**Common sightings**

Table 13.2. Group Sizes, Sea Surface Temperatures (SST), and Depths of Locations where Cetacean Species and Species Groups were Encountered During On-Effort Surveys by NMFS in the Northern Oceanic Gulf, >200 m (656 ft), 1991–2001 (adapted from Maze-Foley and Mullin 2006; all values are means, maxima, and minima; but since sample sizes and standard deviations or errors are not given here, refer to original for details).

Species	Group Size	SST (°C)	Depth (m)	General Comment
Bryde's whale	2.0, 5, 1	23.1, 25.9, 21.5	226.3, 302, 199	Upper slope
Sperm whale	2.6, 11, 1	26.02, 29.7, 21.1	1,732.4, 3,462, 198	Slope and deep ocean
Pygmy & Dwarf sperm whale	2.0, 6, 1	26.6, 29.7, 22.7	1,670.6, 3,422, 339	Slope and deep ocean
Cuvier's beaked whale	1.8, 4, 1	26.01, 28.3, 24.3	1,884.6, 3,221, 1,179	Deep ocean
Mesoplodon whale	2.3, 7, 1	26.95, 28.9, 23.1	1,291.6, 3,257, 796	Deep ocean
Ziphiid	1.7, 4, 1	26.48, 29.2, 22.5	1,876.9, 3,386, 531	Deep ocean
Killer whale	6.5, 12, 1	26.66, 28.6, 22.7	1,865.8, 2,818, 732	Deep ocean
Short-finned pilot whale	24.9, 85, 3	26.47, 28.4, 24.4	9,84.3, 2,105, 553	Slope to deep ocean
False killer whale	27.6, 70, 3	26.79, 28.7, 25.1	1,301.5, 3,294, 167	Upper slope to deep ocean
Pygmy killer whale	18.5, 84, 4	26.84, 28.2, 24.5	2,405.7, 3,422, 893	Deep ocean
Melon-headed whale	99.6, 275, 22	26.47, 28.7, 24.1	1,401.5, 3,203, 824	Deep ocean
Rough-toothed dolphin	14.1, 28, 2	25.87, 28.8, 22.3	1,572.0, 3,294, 128	Upper slope to deep ocean
Risso's dolphin	10.2, 40, 1	26.20, 29.2, 20.4	1,155.5, 3,440, 110	Upper slope to deep ocean
Common bottlenose dolphin	20.6, 220, 1	25.25, 29.5, 19.4	312.4, 2,950, 102	Upper slope and shallower waters
Pantropical spotted dolphin	71.3, 650, 3	25.94, 29.1, 21.1	1,912.2, 3,488, 280	Upper slope to deep ocean
Atlantic spotted dolphin	25.7, 68, 1	24.99, 28.3, 21.3	179.6, 362, 101	Upper slope and shallower waters
Spinner dolphin	151.5, 800, 6	25.42, 29.6, 22.2	825.7, 2,525, 275	Deep ocean to upper slope
Clymene dolphin	89.5, 325, 2	25.93, 29.2, 22.1	1,692, 3,065, 688	Deep ocean and slope
Striped dolphin	46.1, 150, 8	25.30, 28.6, 22.2	1,638.3, 3,206, 404	Deep ocean and slope
Fraser's dolphin	65.3, 117, 34	25.77, 26.5, 25.3	1,483.5, 2,141, 251	Deep ocean and slope

such as bottlenose dolphins and sperm whales. Because they exist in a huge area, there are strong sighting variations by season and sighting conditions. Line-transect sampling from airplanes and surface vessels is not an exact science, and mark-recapture population estimates can be made for only some special small and well-studied areas, such as the Sarasota-Bradenton area of west Florida for bottlenose dolphins. Most of the information for this summary comes from Würsig et al. (2000), Maze-Foley and Mullin (2006), Jefferson et al. (2008), Perrin et al. (2009), Schmidly and Würsig (2009), Marsh et al. (2011), and Waring et al. (2011), but not for data beyond March 2010; other data sources will also be cited.

It is here emphasized that population discreteness and number estimates for cetaceans and the one sirenian of the area are quite incomplete—different years yield different results, often quite widely so. While acoustic censuses along with visual ones have been touted (e.g., Davis et al. 2002), there are presently not enough data on frequency and other parameter types of sounds to say much about approximate numbers of animals beyond the possible exception of sperm whales. Recent correlations of select cetacean sounds and visual sightings are promising for improving the accuracy of future descriptions of cetacean presence (Baumann-Pickering et al. 2010).

Since about 65 % of the Gulf of Mexico is part of the nation of Mexico, it is appropriate that Spanish names are linked with English ones. The common Mexican names are in Table 13.3.

Table 13.3. English and Spanish Names of Marine Mammals in the Gulf of Mexico.

English	Spanish
Sperm whale	Cachalote
Bryde's whale	Rorcual tropical; ballena de Bryde
Pygmy sperm whale	Cachalote pigmeo
Dwarf sperm whale	Cachalote enano
Cuvier's beaked whale	Zifio de Cuvier; ballena picuda de Cuvier
Blainville's beaked whale	Zifio de Blainville; ballena picuda de Blainville
Gervais' beaked whale	Zifio de Gervais; ballena picuda de Gervais
Killer whale	Orca
Short-finned pilot whale	Calderón de aleta corta
False killer whale	Orca falsa
Pygmy killer whale	Orca pigmea
Melon-headed whale	Calderón pequeno; ballena cabeza de melón
Rough-toothed dolphin	Delfín de dientes rugosos; esteno
Risso's dolphin	Delfín de Risso
Bottlenose dolphin	Tursion o delfín naríz de botella; tonina
Pantropical spotted dolphin	Estenela moteada; delfín manchado pantropical
Atlantic spotted dolphin	Delfín manchado del Atlántico
Spinner dolphin	Delfín tornillo; estenela giradora; delfín girador
Clymene dolphin	Delfín de Clymene
Striped dolphin	Delfín listado; estenela listada
Fraser's dolphin	Delfín de Fraser
West Indian manatee	Manatí del Caribe; vaca marina del Caribe

*Spanish names reviewed by Diane Gendron, Centro Interdisciplinario de Ciencias Marinas, La Paz, Mexico, and Jaime Alvarado-Bremer, Texas A&M University at Galveston.

13.3.2.1 Bryde's Whale

The Bryde's whale (*Balaenoptera edeni*) is a small member of the Family Balaenopteridae, also called rorquals, the latter a Norwegian term referring to their throat grooves (ventral pleats) that allow them to lunge forwards into water with concentrated prey, open up the mouth/throat to prodigious volume, push the water out through a filter mat formed inside the mouth by finely fringed baleen, and swallow the euphausiid crustaceans (krill) or fish that were engulfed. Bryde's whales are approximately 13 m (43 ft) long and weigh about 12,000 kg or 12 metric tons (26,455 lbs) or 12 metric tons (13 t). As is the case for other balaenopterids and indeed baleen whales in general, females tend to be slightly larger than males. The female's larger size probably aids in gestating a 4 m (13 ft) long calf for 11 months and nursing it intensively for about 6 months; the average calf is 7 m (23 ft) long at weaning.

Bryde's whales are bluish/black above and whitish below and have a small dorsal fin that rises abruptly in front and is falcate (back-curved). The Bryde's whale's dorsal fin has a more abrupt rise and is of course taller than the dorsal fin of most delphinids. It reaches about 46 centimeters (cm) (18 in.) high in an adult. The Bryde's whale has 40–70 throat grooves that extend all the way back to the navel and three dorsal head ridges—one in the middle of the upper head and two about halfway between midline and jaw line. These ridges are a special distinguishing visual diagnostic feature, as other balaenopterids have only one. Bryde's whales are often confused with the slightly larger sei whale, and a good view of the head dorsal ridges may be needed for positive identification (Figures 13.4 and 13.5).

Bryde's whales are also termed the *tropical whale*, for they (and their recently named generic counterpart, Omura's whale (*Balaenoptera omurai*), of the tropical western Pacific and Indian Ocean) are the only baleen whale species not to be found in colder temperate waters at least part of the year. Probably as a result of their tropical nature, they tend to breed and calve year-round and do not engage in long migrations. They tend not to be highly social, and when several Bryde's whales are seen in proximity, they are likely to be in feeding, not social,

Figure 13.4. Bryde's whale (*Balaenoptera edeni*). The dorsal fin front rises abruptly from the back, allowing for quick distinction from the fin whale, whose dorsal fin rises more gently (photo by Thomas A. Jefferson, with permission).

Figure 13.5. Bryde's whale (*Balaenoptera edeni*). The top of the head has two ridges to either side of the center ridge, distinguishing it from other rorquals at close range (photo by Thomas A. Jefferson, with permission).

Table 13.4. Summary of Abundance Estimates for Northern Gulf of Mexico Bryde's Whales (*Balaenoptera Edeni*): Month, Year and Area Covered During Each Abundance Survey and Resulting Abundance estimate (N_{best}) and CV (Coefficient of Variation) (Waring et al. 2011 for these and subsequent abundance estimate tables in the Northern Gulf of Mexico).

Month/Year	Area	N_{best}	CV
Apr–Jun 1991–1994	Oceanic waters	35	1.1
Apr–Jun 1996–2001 (excluding 1998)	Oceanic waters	40	0.61
Jun–Aug 2003, Apr–Jun 2004	Oceanic waters	15	1.98

aggregations. They tend to feed not on krill but on shoals of small fishes and are capable of engulfing an entire school of fish—1 m (3 ft) or more in diameter—at or below the surface.

The worldwide population in the Atlantic, Pacific, and Indian Oceans is estimated at about 30–50 thousand whales; they are not considered endangered worldwide and are not listed as endangered under the U.S. Endangered Species Act. Until recently, best estimate for the northern Gulf was about 40, with 95 % confidence interval (CI) of 13–129 (Mullin and Fulling 2004; NMFS Stock Assessment Reports 2006–2011), but a smaller best estimate of 15 (coefficient of variation [CV] 1.98) whales (Mullin 2007) from more recent data is presently published by NMFS as the official estimate (Table 13.4). This does not necessarily mean that the population in the Gulf has declined. It simply may reflect the vagaries of sighting members of a small population depending on chance in particular line-transect surveys (Table 13.4). There is not enough information on Bryde's whales in the Gulf to determine population trend or whether there is fisheries-related or other human-caused mortality (Figure 13.6).

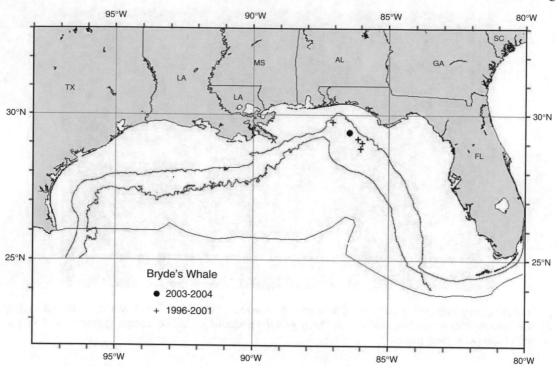

Figure 13.6. Distribution of Bryde's whale (*Balaenoptera edeni*) sightings from the Southeast Fisheries Science Center (SEFSC) spring vessel surveys during 1996 to 2001 and from summer 2003 and spring 2004 surveys. All the on-effort sightings are shown, although not all were used to estimate abundance. Solid lines indicate the 100 m (328 ft) and 1,000 m (3,281 ft) isobaths and the offshore extent of the U.S. EEZ (Waring et al. 2011 for these and subsequent figures of cetacean distribution in the northern Gulf of Mexico).

13.3.2.2   Sperm Whale

The sperm whale (*Physeter macrocephalus*), medium gray above and light gray to white below, is the largest toothed whale and, indeed, the largest toothed creature on Earth. It is highly sexually dimorphic. Males average 15 m (49 ft) in length and a prodigious 36,000 kg or 36 metric tons (79,350 lb), and females average 11 m (36 ft) and 20,000 kg or 20 metric tons (44,100 lb). The maximum size of male sperm whales is around 20 m (66 ft), although due to last mid-century's intensive worldwide whaling, there are probably few of these giants around at present. The heads of the male sperm whales grow disproportionately rapidly as they age. The male's head takes up about one-fifth to one-quarter of the body's length in young ones and up to one-third of the body's length in older males. It is obviously a secondary sexual characteristic, and males use the head for intrasex fighting and probably acoustic displays. The terrestrial analog might be deer stags with their antlers (the sperm whale male head) and roars (special male-only loud sounds that sperm whales emit). The head houses a giant structure of waxy oil—the spermaceti organ. The blowhole, placed differently from that of any other cetacean, is at the upper front of the mighty head, not along the mid-line but somewhat to the left, which results in a very distinctive, forward-tilted exhalation blow to the animal. Teeth are displayed in the lower jaw only and fit neatly into corresponding sockets in the upper jaw. The back has a dorsal ridge but no dorsal fin.

Figure 13.7. Sperm whales (*Physeter macrocephalus*) are the largest toothed whale and, indeed, the largest toothed creature on Earth. They are highly social, and all but older males are found in tight societies. There is a resident population in the northern Gulf of Mexico (photo by Thomas A. Jefferson, with permission).

Sperm whales have a matriarchal society. Females and their female young tend to stay in one or adjacent groups for many years or for life. Males leave the group as they become sexually mature, at about age 10. The matriarchy, which tends to stay in tropical and subtropical waters, allows for related animals to help each other (e.g., take turns patrolling for danger to their nondiving neonates at the surface while others dive to depth). Indeed, deep diving for food (squid and fishes) seems to have driven the evolution towards high sociality in this species. Young males tend to stay together in groups of a dozen or so animals and travel to somewhat higher latitudes than the matriarchies they have left. As they mature—males do not seem to reach social maturity for re-inserting themselves briefly into matriarchies to mate until about age 25—older males tend to be alone (probably to avoid or minimize competition for access to females), and they travel to high near-Arctic and Antarctic latitudes to feed in very deep, productive waters. The general pattern of matriarchy, maturing males, older lone males, etc., has a close analog in the matriarchal systems of the largest land mammals—African and Asian elephants (*Loxodonta* sp. and *Elephas maximus*, respectively)—in what has been termed the collosal convergence of social/sexual strategies coupled with gigantism, long lives, and extended caregiving to young (Weilgart et al. 1996; Whitehead 2003) (Figures 13.7 and 13.8).

Sperm whales occur throughout the world's oceans but generally in waters deeper than about 500 m (1,640 ft) because of their habit of seeking largely deep-diving squid and fishes. Sperm whales in the Gulf are on average 1.5–2.0 m (4.9–6.6 ft) smaller than those found elsewhere (Richter et al. 2008; Jaquet and Gendron 2009). This size difference was noted by whalers 150 years ago (Reeves et al. 2011) and strongly suggests a different population from the sperm whales of the North Atlantic, a verification of which was provided by Engelhaupt et al. (2009) from genetic analysis. Mitochondrial DNA (inherited only from the mother) shows significant differences between Gulf sperm whales and sperm whales in other parts of

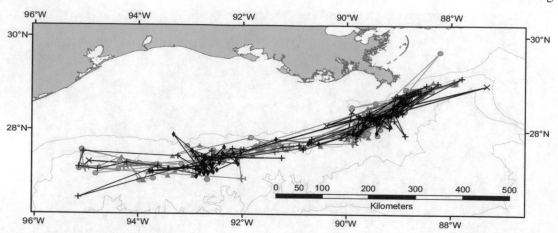

Figure 13.8. Satellite tracks of seven sperm whales (*Physeter macrocephalus*) tagged July 3, 2002, and tracked for as long as early June, 2003 (two whales) (for details see Ortega-Ortiz et al. 2012 from which this figure was taken).

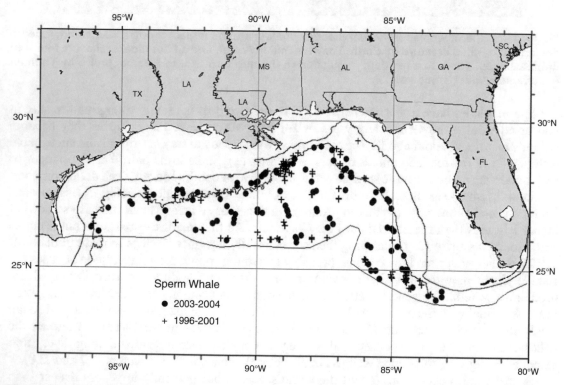

Figure 13.9. Distribution of sperm whale (*Physeter macrocephalus*) sightings from SEFSC spring vessel surveys during 1996–2001 and from summer 2003 and spring 2004 surveys. All on-effort sightings are shown, though not all were used to estimate abundance. Solid lines indicate the 100 m (328 ft) and 1,000 m (3,281 ft) isobaths, and the southern line represents the U.S. EEZ (from Waring et al. 2011).

Table 13.5. Summary of Recent Abundance Estimate for Northern Gulf of Mexico Sperm Whales (*Physeter Macrocephalus*): Month, Year and Area Covered During Each Abundance Survey and Resulting Abundance Estimate ($^N_{best}$) and CV.

Month/Year	Area	^{N}best	CV
Jun–Aug 2003, Apr–Jun 2004	Oceanic waters	16,665	0.2

the North Atlantic, while nuclear (bi-parentally inherited) DNA shows no difference. This indicates that females stay within the Gulf but that at least some males travel and breed in both the Gulf and North Atlantic. Indeed, recent satellite tracking of sperm whales showed that matriarchies stayed in waters about 200–3,499 m (656–11,480 ft) deep, generally in the area south and southwest of the Mississippi/Atchafalaya mouths, while males traveled south to Mexico's Campeche area, and one male left the Gulf but returned after about 2 months (Ortega-Ortiz et al. 2012).

Typical group size of Gulf sperm whales in the north, which is almost always of presumed matriarchies, is 8–11 animals (Richter et al. 2008), often with calves less than 3–5 years old. This is smaller than groups (24–31) in the Pacific (Coakes and Whitehead 2004), but similar to groups (about six) in the adjacent Caribbean (Gero 2005). Statistical lagged association rates (Whitehead 2009) indicate that Gulf sperm whale groups are stable for longer (about 62 days) than in the Pacific (7–19 days) (Coakes and Whitehead 2004) but similar (about 80 days) to another enclosed body of water, the Gulf of California (Jaquet and Gendron 2009). It is possible that group sizes and association rates are ecologically related and that food or other ocean-basin physical/biological variables help to define social patterns (Richter et al. 2008).

Only recently have more accurate estimates of sperm whale numbers in the northern Gulf emerged. The latest estimate is about 1,665 (CV 0.20) animals (Table 13.5) (Mullin and Fulling 2004). Sperm whales overlap strongly with shipping lanes between New Orleans and Houston, industrial seismic activities, and deepwater oil/gas rigs (Azzara 2012). They were the only large whales to be hunted in the Gulf (although apparently not into the twentieth century), and their population characteristics may still be influenced by this earlier depredation (Reeves et al. 2011). There is not enough precision to estimate population trends and current productivity rates.

Sperm whales also occur in the southern Gulf and were hunted there in the past (Reeves et al. 2011). Most sperm whales encountered during cruises in the south appear to be concentrated on the continental slope (Figure 13.10).

13.3.2.3   Pygmy and Dwarf Sperm Whales

The pygmy and dwarf sperm whales (*Kogia breviceps* and *Kogia sima*, respectively) of the family Kogiidae (which are much smaller than the sperm whale but most closely related to it) are not found together. However, they will be treated together here as they appear to have quite similar habitats and habits and are often not identified to species during surveys. Like the sperm whale, pygmy and dwarf sperm whales have a spermaceti organ of waxy oil in their heads and teeth only in the lower jaw. Again like sperm whales, the kogiids have a blunt and squarish head and an underslung lower jaw, but this is thicker than and not as long as that of the sperm whale. However, unlike the sperm whale, their blowhole is in the center top of the head, like that of dolphins, and they have a dorsal fin, also like dolphins. Both species are a steel blue/gray above and lighter below. Both species have a light colored, false gill mark just behind the eye (Figures 13.11 and 13.12).

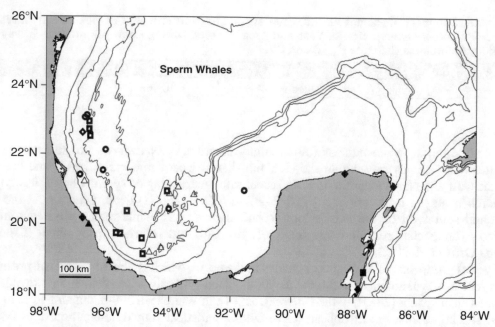

Figure 13.10. Records of sperm whales (*Physeter macrocephalus*) in the Mexican waters of the southern Gulf of Mexico and Caribbean. Display for distribution is as follows: solid symbol, strandings; hollow symbols, confirmed sightings; triangles, spring (Mar–May); squares, summer (Jun–Aug); circles, autumn (Sep–Nov); diamonds, winter (Dec–Feb); crosses, unknown dates. Thin contour lines show the 200 m, 1,000 m, 2,000 m, and 3,000 m (656 ft, 3,281 ft, 6,562 ft, and 9,843 ft) isobaths (from Ortega-Ortiz 2002).

Figure 13.11. Pygmy sperm whale (*Kogia breviceps*). They are shy and difficult to photograph, and not many good photos exist (photo by Robert L. Pitman, with permission).

The pygmy sperm whale is the larger of the two at about 2.7–3.7 m (8.8–12.1 ft) and 317–410 kg (699–904 lb). The dwarf sperm whale is the size of smaller delphinids at about 2.1–2.7 m (6.9–8.9 ft) and 136–212 kg (300–467 lb). The dorsal fin of the dwarf sperm whale is larger, relative to body size, than that of the pygmy one and is set just a bit further forward on

Figure 13.12. Dwarf sperm whales (*Kogia sima*), probably adult and young. This adult's dorsal fin is deformed. Normally the dorsal fin is not so strongly curved (photo by Robin W. Baird, with permission).

the body. Both species leap or are active at the surface only rarely, thus surfacing quite low and cryptically. It is difficult to distinguish the two species from the vantage point of a ship or an airplane except at close range and by the most expert of observers. As a result, most observations of individuals have been lumped as Kogiids. Both species dive to at least several 100 m and feed largely on squid.

Kogiids are likely much more numerous than present estimates suggest and occur in most oceans, generally in warmer waters. The slightly larger pygmy sperm whale moves to slightly higher latitudes, up to about Nova Scotia, Canada, in the western North Atlantic, as compared to about Virginia for the dwarf sperm whale. Worldwide, there are many strandings throughout their known ranges, again suggesting that the animals are more abundant than sighting records indicate, probably due to the difficulties in seeing them. Between 2003 and 2007, there were six pygmy sperm whale strandings on Florida beaches and four on Texas beaches. During the same period, seven dwarf sperm whales were stranded on Florida beaches and four on Texas beaches (with none reported for the other U.S. Gulf States). In the northern Gulf, the best estimate for both species combined is 453 (CV 0.35) (Figure 13.13 and Table 13.6) (Mullin 2007).

13.3.2.4 Beaked Whales

Beaked whales consist of 22 species worldwide. They are almost invariably in deep waters and feed on deepwater squid. Only recently has knowledge been gained about some of these species, with northern bottlenose whales (*Hyperoodon ampullatus*) studied in The Gully off Nova Scotia (e.g., Gowans et al. 2001) and tagging of Cuvier's beaked whales with short-term data tags that give details of depths, three-dimensional dive pattern, speeds (including accelerations and decelerations), and simultaneous recordings of their own click vocalizations and those of their conspecifics (Zimmer et al. 2005). This is exciting science; beaked whales are no longer ecologically unknown. The Gulf of Mexico has no known beaked whale hot spots, but detailed work has not been carried out there.

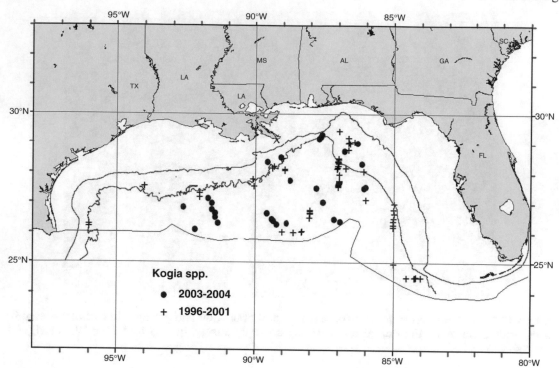

Figure 13.13. Distribution of dwarf (*Kogia sima*) and pygmy sperm whale (*Kogia breviceps*) sightings from SEFSC spring vessel surveys during 1996–2001 and from summer 2003 and spring 2004 surveys. All the on-effort sightings are shown, though not all were used to estimate abundance. Solid lines indicate the 100 m (328 ft) and 1,000 m (3,281 ft) isobaths and the offshore extent of the U.S. EEZ (from Waring et al. 2011).

Table 13.6. Summary of Combined Abundance estimates for Northern Gulf of Mexico Dwarf (*Kogia sima*) and Pygmy Sperm Whales (*Kogia Breviceps*): Month, Year and Area Covered During Each Abundance Survey, and Resulting Abundance Estimate (N_{best}) and CV.

Month/Year	Area	N_{best}	CV
Apr–Jun 1991–1994	Oceanic waters	547	0.28
Apr–Jun 1996–2001 (excluding 1998)	Oceanic waters	742	0.29
Jun–Aug 2003, Apr–Jun 2004	Oceanic waters	453	0.35

Beaked whales are often lumped during surveys in the Gulf, but there are enough data on Cuvier's, Blainville's, and Gervais' beaked whale (most beaked whales of the Ziphiidae family have traditionally been named after the person who first described them) to make some overall statements. They all occur in the open ocean, in the tropic to temperate zones, and generally in small groups, and they feed on deepwater fishes and squid. Males have only one pair of erupted lower jaw teeth, and females have none. They therefore appear to be suction feeders that can *inhale* their prey without needing teeth to bite or pierce.

Table 13.7. Summary of Abundance Estimates for Northern Gulf of Mexico Cuvier's Beaked Whales (*Ziphius Cavirostris*): Month, Year and Area covered During Each Abundance Survey, and Resulting Abundance Estimate (^{N}best) and CV.

Month/Year	Area	^{N}best	CV
Apr–Jun 1991–1994	Oceanic waters	30	0.5
Apr–Jun 1996–2001 (excluding 1998)	Oceanic waters	95	0.47
Jun–Aug 2003, Apr–Jun 2004	Oceanic waters	65	0.67

Table 13.8. Summary of Recent Abundance Estimates for Northern Gulf of Mexico *Mesoplodon* spp., a Combined Estimate for Blainville's Beaked Whale (*Mesoplodon Densirostris*) and Gervais' Beaked Whale (*Mesoplodon Europaeus*): Month, Year and Area Covered During Each Abundance Survey, and Resulting Abundance Estimate (^{N}best) and CV.

Month/Year	Area	^{N}best	CV
Apr–Jun 1996–2001 (excluding 1998)	Oceanic waters	106	0.41
Jun–Aug 2003, Apr–Jun 2004	Oceanic waters	57	1.4

Cuvier's beaked whales reach a size of 5.8 m (19 ft) for females and 5.5 m (18 ft) for males (females are larger than males in this species), and coloration can be a dark brown to slate gray. Cuvier's beaked whale heads can be quite light in color, and the erupted tooth pair of males is set far to the front of the jaw. Blainville's beaked whale males have their erupted jaw teeth on the midpoint of the jaw at a prominent upward jutting part of the lower jaw making them easy to distinguish in the field. Both males and females reach a size of about 4.7 m (15.4 ft). They occur in all tropical and temperate oceans, but apparently not in large numbers anywhere that has yet been discovered. Finally, Gervais' beaked whale, at about 4.2–5.7 m (13.8–18.7 ft) for females and about 4.2–4.6 m (13.8–15.1 ft) for the smaller males, is endemic to the tropical- and cool-temperate waters of the Atlantic (i.e., as far north as western Scotland in the East Atlantic). In the northern tropical Gulf, unidentified beaked whales (which could also be of other species not mentioned here) are estimated at 337 (CV 0.40), Cuvier's at 654 (CV 0.67), Blainville's and Gervais' beaked whale estimates are combined as *Mesoplodon* sp., and the combined best estimate is 57 (CV 1.40). Of course, it is also possible that some of these latter two species are represented in the "unidentified" category of 337 animals mentioned above. Figure 13.20 is a beaked whale in general sighting map, and Tables 13.7 and 13.8 summarize Cuvier's beaked whale and combined Blainville's/Gervais' beaked whales estimates, respectively (Figures 13.14, 13.15, 13.16, 13.17, 13.18 and 13.19).

13.3.2.5   Killer Whale

The cosmopolitan killer whales occur in all oceans from the tropics to the Arctic and Antarctic ice, although what was for many years thought to be just one species may be classified as several species in the future. They are the largest of the delphinids and have high sexual dimorphism, with adult males reaching about 9.8 m (32.2 ft) and females about 8.5 m (27.9 ft). The distinctive male dorsal fin grows throughout life, and it becomes a high erect, pointed structure in a fully mature male. Killer whales are strikingly colored, black above and white below, with a white oblong eye spot just above and behind the eye, a variably shaped saddle blaze behind and below the dorsal fin, and white undersides of the flukes, at times

Figure 13.14. Cuvier's beaked whale (*Ziphius cavirostris*), with likely conspecific tooth rake markings (photo by Thomas A. Jefferson, with permission).

Figure 13.15. Cuvier's beaked whale (*Ziphius cavirostris*), with likely conspecific tooth rake markings. The small front teeth mark this individual as a male (photo by Charlotte Dunn, with permission).

extending to a part or all of the fluke dorsum. Within a population, individuals have slightly different white marks and blaze patterns, and populations can generally be distinguished by common coloration factors within them as well. Individuals can be recognized by the distinctive natural marks, along with the pattern of scars and nicks, particularly on the dorsal fin (Figure 13.21).

Killer whale social structure is varied and complex. In the North Pacific along the shores of North America, for example, there are nearshore forms (termed *residents*) that travel little and

Figure 13.16. Blainville's beaked whale (*Mesoplodon densirostris*), adult female (photo by Robin W. Baird, with permission).

Figure 13.17. Blainville's beaked whale (*Mesoplodon densirostris*), adult male. The mid-jaw erupted teeth are crowned by barnacles (photo by Robin W. Baird, with permission).

eat salmon almost exclusively, other nearshore forms (termed *transients*) that travel over several 100 km and feed almost exclusively on marine mammals, and offshore forms that feed largely on fish but take other prey as well; the latter can move over 1,000 km (621 mi) in short time periods (days to weeks). All forms appear to be matriarchal, with youngsters staying within the mothers' group or pod for long periods or for life, but details of this vary by social grouping (Bigg et al. 1990). For example, one salmon-eating pod off Vancouver Island, British

Figure 13.18. Blainville's beaked whale (*Mesoplodon densirostris*), adult female and a large calf or subadult (photo by Robin W. Baird, with permission).

Figure 13.19. Gervais' beaked whale (*Mesoplodon europaeus*), probable mother and young (photo by Charlotte Dunn, with permission).

Columbia, Canada, is socially closed, with both female and male young staying with their mothers (but mating with others during occasional superpod congregations, which probably serve to avoid inbreeding). Such closed intergenerational social living allows for a complicated culture to be developed, due to transmission of foraging, vocal, and other information through generations (Whitehead 1998). For example, some societies of killer whales have individuals

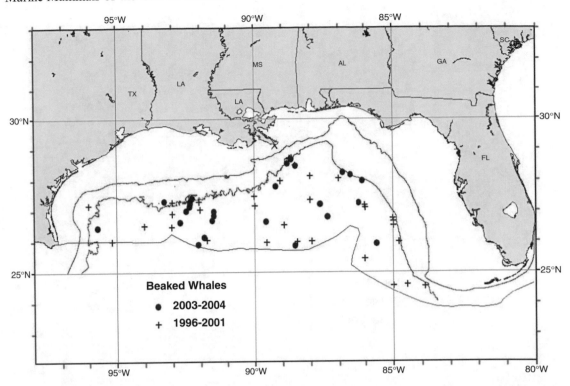

Figure 13.20. Distribution of beaked whales. Sightings from SEFSC shipboard spring vessel surveys during 1996–2001 and from summer 2003 and spring 2004 surveys. All the on-effort sightings are shown, though not all were used to estimate abundance. Solid lines indicate the 100 m (328 ft) and 1,000 m (3,281 ft) isobaths and the offshore extent of the U.S. EEZ (from Waring et al. 2011).

Figure 13.21. Killer whales (*Orcinus orca*) are the largest of the delphinids and certainly one of the most charismatic for humans (photo by Thomas A. Jefferson, used with permission).

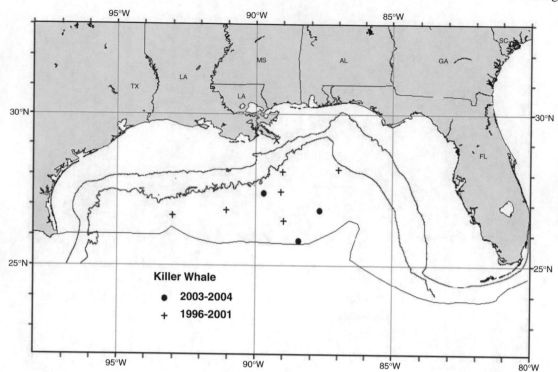

Figure 13.22. Distribution of killer whale (*Orcinus orca*) sightings from SEFSC spring vessel surveys during 1996–2001 and from summer 2003 and spring 2004. All the on-effort sightings are shown, though not all were used to estimate abundance. Solid lines indicate the 100 m (328 ft) and 1,000 m (3,281 ft) isobaths and the offshore extent of the U.S. EEZ (from Waring et al. 2011).

beaching themselves to take pinnipeds on land or in the turbulent surf zone; they cooperate in these hunts and even appear to teach youngsters the tricky business of temporarily stranding without becoming stranded forever (Guinet and Bouvier 1995).

Killer whales of the Gulf are only sporadically sighted (see Figure 13.22), and limited photo-identifications (presently 32 individuals) indicate that they travel for at least up to 1,100 km (684 mi) (O'Sullivan and Mullin 1997). It is presently unknown whether killer whales of the Gulf form a stock or population separate from those in the adjacent North Atlantic, and photo-identification comparisons and genetic data are needed. Presently, the best estimate for the northern Gulf is about 49 (CV 0.77) animals, but an earlier estimate of 133 (CV 0.49), based on data from 1996 to 2001 (Mullin and Fulling 2004), may be more appropriate, given the identification record of 32 from limited work. There are persistent reports from sport fishers that killer whales feed on tuna in the Gulf, but these have not been verified. On May 17, 2008, a killer whale became entangled in a fishing longline (Garrison et al. 2009), which suggests that killer whales take fish off longlines in the Gulf at times, as they are known to do in some other areas (Table 13.9).

13.3.2.6 Short-Finned Pilot Whales

There are two species of pilot whales. The long-finned pilot whale (*Globicephala melas*) occurs in the North Atlantic (including the western part of the Mediterranean Sea) and in the southern hemisphere but not in the North Pacific. It is a relatively cold-water species and does

Table 13.9. Summary of Recent Abundance Estimates for Northern Gulf of Mexico Killer Whales: Month, Year, and Area Covered During Each Abundance Survey and Resulting Abundance Estimate (Nbest) and CV.

Month/Year	Area	Nbest	CV
Jun–Aug 2003, Apr–Jun 2004	Oceanic waters	49	0.77

Figure 13.23. Short-finned pilot whales (*Globicephala macrorhynchus*) occur in apparent matriarchal long-term tight societal bonds (photo by Thomas A. Jefferson, with permission).

not frequent waters south of the U.S. state of Georgia, and thus does not make it (at least not regularly) into the Gulf of Mexico. The short-finned pilot whale (*Globicephala macrorhynchus*), on the other hand, occurs worldwide in the tropics and subtropics and overlaps with its congener in fringe habitats of both, including off the U.S. eastern seaboard. It can occur in groups as small as one dozen or so animals but also occurs in schools of hundreds, and before major hunting in most of its range, it even occurred in groups over a thousand. Pilot whales tend to feed on squid, but fish are also taken. Short-finned pilot whales in the Gulf at times harass groups of sperm whale matriarchies with young in them. But whether this is an attempt to feed on sperm whale newborns or perhaps to get sperm whales to regurgitate food, as is believed to have been seen, is not known (Weller et al. 1996) (Figure 13.23).

Female short-finned pilot whales become sexually mature at about age nine and are about 5.5 m (18.0 ft) in length. Male short-finned pilot whales become sexually mature at about age 15 and are about 6.0 m (19.9 ft) in length, with the male dorsal fin growing disproportionately larger and more strongly curved. The male head (or melon) becomes more bulbous and squarish than the female head as seen from the side and in older males, may even overhang the front of the jaw under the melon. Pilot whales are quite dark to black above with a light belly patch of variable shape below. They have a light chin, a grayish to white stripe or chevron dorsally behind the eye and pointing towards the dorsal fin (useful for distinguishing the species from the air), and a variably light saddle patch pattern on both sides and just behind the dorsal fin.

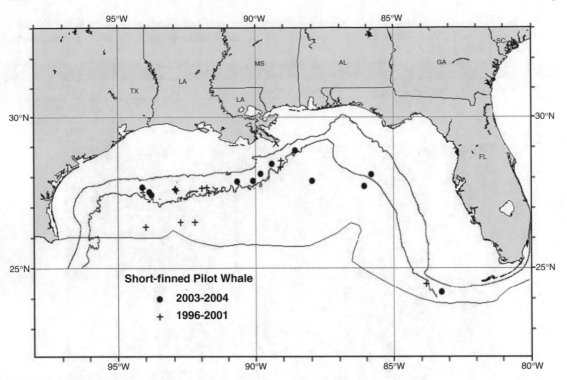

Figure 13.24. Distribution of short-finned pilot whale (*Globicephala macrorhynchus*) sightings from SEFSC spring vessel surveys during 1996–2001 and from summer 2003 and spring 2004 surveys. All the on-effort sightings are shown, though not all were used to estimate abundance. *Solid lines* indicate the 100 m (328 ft) and 1,000 m (3,281 ft) isobaths and the offshore extent of the U.S. EEZ (from Waring et al. 2011).

Data from drive fisheries off Japan indicate that social maturity of males may take considerably longer to achieve than does sexual maturity, as younger males may not be able to mate effectively with females in a generally tight matriarchal society. Generally, pilot whale females nurse their young for at least 3 years, and some evidence points to them nursing into their offsprings' teens! Females often continue to nurse their offspring even when they are no longer reproductively active (beyond about age 40). This indicates that nursing is important beyond meeting nutritional needs; it is a part of social bonding and may also extend to alo-nursing (i.e., nursing offspring that are not their own) (Kasuya and Marsh 1984). These data come from dead animals, and unfortunately long-term studies of living populations are scarce. Heimlich-Boran (1993) studied the species off Tenerife and reported that pilot whales there live in matriarchal societies that include adult males (presumably the offspring of mothers in the society) that mate outside of their immediate groupings. This system may therefore be quite similar to that of resident killer whales.

Short-finned pilot whales tend to occur in deep waters, as they feed on mesopelagic fishes and squid, but are more common over continental slopes than over the abyssal plain, and this is true for the Gulf as well (see Figure 13.24). Although there is no good, overall, worldwide population estimate, short-finned pilot whales of the eastern tropical Pacific are estimated at about 590,000 (CV 0.26) (Gerrodette and Forcada 2002) and in the northern Gulf at 716 (CV 0.34) (Table 13.10). This recent lower best estimate may be too low. Table 13.10 shows that a previous estimate was more than 2,000 animals in the same area. Short-finned pilot whales often mass strand, but only two strandings have been reported in the Gulf since the

Table 13.10. Summary of Abundance Estimates for Northern Gulf of Mexico Short-finned Pilot Whales: Month, Year and Area Covered During Each Abundance Survey, and Resulting Abundance Estimate (^{N}best) and CV.

Month/Year	Area	^{N}best	CV
Apr–Jun 1991–1994	Oceanic waters	353	0.89
Apr–Jun 1996–2001 (excluding 1998)	Oceanic waters	2,388	0.48
Jun–Aug 2003, Apr–Jun 2004	Oceanic waters	716	0.34

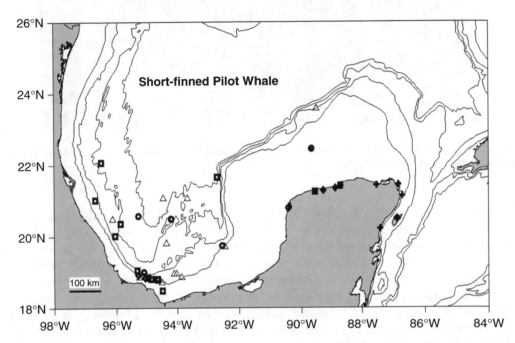

Figure 13.25. Records of short-finned pilot whales (*Globicephala macrorhynchus*) in the Mexican Waters of the Southern Gulf of Mexico and Caribbean. Display for distribution is as follows: solid symbol, strandings; hollow symbols, confirmed sightings; triangles, spring (Mar–May); squares, summer (Jun–Aug); circles, autumn (Sep–Nov); diamonds, winter (Dec–Feb); crosses, unknown dates. Thin contour lines show the 200 m, 1,000 m, 2,000 m, and 3,000 m (656 ft, 3,281 ft, 6,562 ft, and 9,843 ft) isobaths (from Ortega-Ortiz 2002).

1990s: one in 1999 (two animals) and one in 2001 (nine animals), and both strandings were off Florida; these numbers are much lower than the mass strandings that occur in many other places. Short-finned pilot whales also occur in Mexican waters and have been sighted in waters up to about 3,000 m (9,843 ft) deep (Figure 13.25).

13.3.2.7 False Killer Whale

Along with pilot, pygmy killer (*Feresa attenuata*), and melon-headed (*Peponocephala electra*) whales, the false killer whale (*Pseudorca crassidens*) species was termed a blackfish by whalers and fishermen, because of its generally very dark coloration. All of these oceanic blackfish are known for their frequent mass strandings, probably because they have tight long-term (generally matriarchal) societies, and when several animals make a navigational mistake near a shoal or headland, the integrity of the group has all others following. The false killer

Figure 13.26. False killer whales (*Pseudorca crassidens*) occur in all tropical oceans (photo by Robin W. Baird, with permission).

whale is about the same size as pilot whales, with males slightly larger than females, but its body form is considerably more slender than that of pilot whales. Group size can be just a few animals or into the hundreds. As with pilot whales, males and females travel together in apparently tight bonds. Unlike pilot whales, however, details of genetic relationships, length of maternal care, and other life history and behavioral characteristics are not yet known.

False killer whales feed on squid and fishes and also at times attack sperm whales and humpback whales. In the latter cases, it is presumed that they are attempting to isolate more vulnerable animals (i.e., old, infirm, or newborn animals) from the more robust animals, but details are unknown (Figure 13.26).

False killer whales occur in tropical and warm temperate oceans and are usually found in deep water (but not in or near the center of oceans). They may be quite close to shore where deep waters occur close to oceanic islands and atolls, such as the Hawaiian Islands. They may occur in cooler temperate waters into 50°N latitude, as well as south of the equator, probably due to their large body size (they are the third largest delphinid cetacean, after killer and long-finned pilot whales). Worldwide numbers are not available. One older estimate of about 40,000 (CV 0.64) (Wade and Gerrodette 1993) has been made for the eastern tropical Pacific. In the northern oceanic Gulf, where false killer whales occur in deep waters and not normally on the slope, the estimate is 777 (CV 0.56) (Figure 13.27 and Table 13.11).

13.3.2.8 Pygmy Killer Whale

The pygmy killer whale (*Feresa attenuata*) is slender like the false killer whale but substantially smaller, with males around 2.3 m (7.5 ft) and females 2.1 m (6.9 ft), or about dolphin size. Without adequate size reference, this species can easily be confused with false killer whales and (see below) melon-headed whales. However, Pygmy killer whales have a white patch (or goatee) at the front of their lower chin, which is more pronounced than the goatee on the melon-headed whale.

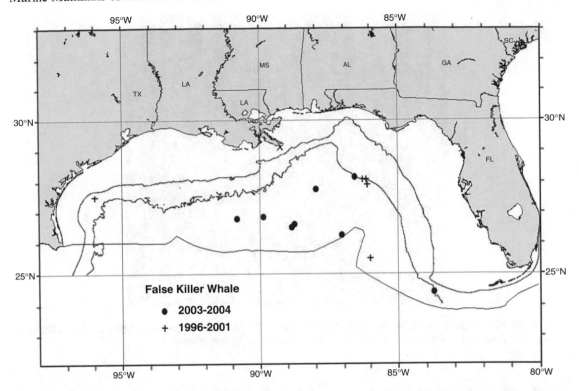

Figure 13.27. Distribution of false killer whale (*Pseudorca crassidens*) sightings from SEFSC spring vessel surveys during 1996–2001 and from summer 2003 and spring 2004 surveys. All the on-effort sightings are shown, though not all were used to estimate abundance. Solid lines indicate the 100 m (328 ft) and 1,000 m (3,281 ft) isobaths and the offshore extent of the U.S. EEZ (from Waring et al. 2011).

Table 13.11. Summary of Abundance Estimates for Northern Gulf of Mexico False Killer Whales (*Pseudorcacrassidens*): Month, Year and Area Covered During Each Abundance Survey, and Resulting Abundance Estimate (^{N}best) and CV.

Month/Year	Area	^{N}best	CV
Apr–Jun 1991–1994	Oceanic waters	381	0.62
Apr–Jun 1996–2001 (excluding1998)	Oceanic waters	1,308	0.71
Jun–Aug 2003, Apr–Jun 2004	Oceanic waters	777	0.56

Pygmy killer whales feed largely on fishes, but squid are also taken. They can be quite aggressive, and attacks on smaller as well as similar-sized delphinids, such as spotted and spinner dolphins, have been witnessed in and near tuna nets in the eastern tropical Pacific.

Pygmy killer whales occur in tropical waters worldwide in groups of about 12–50 animals, although somewhat larger groups also occur. In the eastern tropical Pacific, the population estimate is similar to that of false killer whales, slightly less than 40,000 (CV 0.64) (Wade and Gerrodette 1993). In the oceanic northern Gulf, best estimate is 323 (CV 0.60), with sightings both on the slope and in abyssal plain waters (Figures 13.28 and 13.29 and Table 13.12).

Figure 13.28. Pygmy killer whales (*Feresa attenuata*) are very small members of the blackfish group of small cetaceans (photo by Robert L. Pitman, with permission).

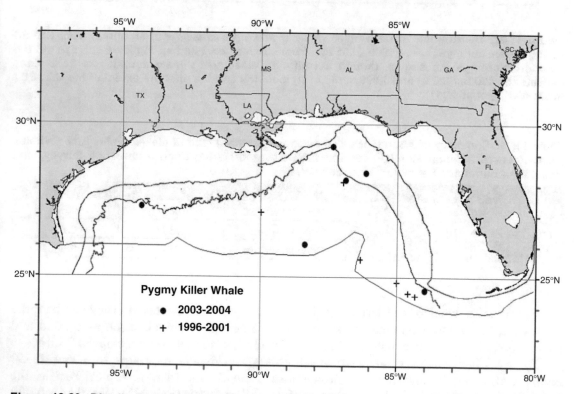

Figure 13.29. Distribution of pygmy killer whale (*Feresa attenuata*) sightings from SEFSC spring vessel surveys during 1996–2001 and from summer 2003 and spring 2004 surveys. All the on-effort sightings are shown, though not all were used to estimate abundance. Solid lines indicate the 100 m (328 ft) and 1,000 m (3,281 ft) isobaths and the offshore extent of the U.S. EEZ (from Waring et al. 2011).

Table 13.12. Summary of Abundance Estimates for Northern Gulf of Mexico Pygmy Killer Whales: Month, Year, and Area Covered During Each Abundance Survey, and Resulting Abundance Estimate (Nbest) and CV.

Month/Year	Area	N_{best}	CV
Apr–Jun 1991–1994	Oceanic waters	518	0.81
Apr–Jun 1996–2001 (excluding 1998)	Oceanic waters	408	0.60
Jun–Aug 2003, Apr–Jun 2004	Oceanic waters	323	0.60

Figure 13.30. Melon-headed whales (*Peponocephala electra*) are about half way in size between pygmy and false killer whales (photo by Robin W. Baird, with permission).

13.3.2.9 Melon-Headed Whale

The final blackfish—the melon-headed whale—is a bit larger than the diminutive pygmy killer whale, at about 2.7 m (8.9 ft) for males and 2.6 m (8.5 ft) for females. As mentioned above, chin coloration is not quite as white as that of the pygmy killer whale, but it too can have white lips. Both species (as well as the larger false killer whale) have a rounded head that is more pointed than the blunt rounded heads of pilot whales and killer whales. Melon-headed whales feed on fishes and squid (Figure 13.30).

Melon-headed whales occur throughout warm waters of the tropics and near-tropics (to about 40°N latitude and 30°S latitude), and their estimated numbers of about 45,000 (CV 0.47) are similar to those of pygmy and dwarf killer whales in the eastern tropical Pacific (Wade and Gerrodette 1993). They can occur in much larger schools (100–1,500 animals) than the false and pygmy killer whales (but not this large in the Gulf), and they are often found in multispecies aggregations with Fraser's (*Lagenodelphis hosei*) and spinner dolphins. In the Gulf, an estimated 2,283 (CV 0.76) melon-headed whales can occur in the northern oceanic area, but apparently they are more often in the western part rather than the eastern part of the Gulf (Figure 13.31 and Table 13.13).

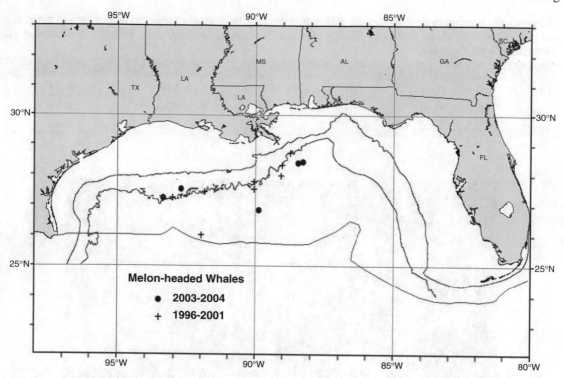

Figure 13.31. Distribution of melon-headed whale (*Peponocephala electra*) sightings from SEFSC spring vessel surveys during 1996–2001 and from summer 2003 and spring 2004 surveys. All the on-effort sightings are shown, though not all were used to estimate abundance. Solid lines indicate the 100 m (328 ft) and 1,000 m (3,281 ft) isobaths and the offshore extent of the U.S. EEZ (from Waring et al. 2011).

Table 13.13. Summary of Abundance Estimates for Northern Gulf of Mexico Melon-Headed Whales (*Peponocephala Electra*): Month, Year, and Area Covered During Each Abundance Survey, and Resulting Abundance Estimate (^{N}best) and CV.

Month/Year	Area	^{N}best	CV
Apr–Jun 1991–1994	Oceanic waters	3,965	0.39
Apr–Jun 1996–2001 (excluding 1998)	Oceanic waters	3,541	0.55
Jun–Aug 2003, Apr–Jun 2004	Oceanic waters	2,283	0.76

13.3.2.10 Rough-Toothed Dolphin

The rough-toothed dolphin (*Steno bredanensis*) is a small delphinid with a beak that tapers from the head, but is not as sharply demarked as it is in the abrupt beak of the bottlenose dolphins. This taper also gives the rough-toothed dolphin (which, indeed, has fine lateral ridges on its teeth, giving them a rough feeling) the nickname lizard dolphin. Males are about 2.7 m (8.9 ft) at maturity and females 2.3 m (7.5 ft). They are dark above and lightish below, but often with a bluish/purplish tinge of coloration, and with yellowish/white dots along the sides.

Rough-toothed dolphins do not appear to be very deep divers, preferring to feed on fishes, squid, octopuses, and often even large fishes such as mahi-mahi (*Coryphaena hippurus*) that are found in deep waters but within 100 m (328 ft) or so of the ocean's surface. Off Hawaii,

Figure 13.32. Rough-toothed dolphins (*Steno bredanensis*) generally occur in small groups. They superficially resemble bottlenose dolphins in shape and size, but their rostrum is more curved from tip to rise of head, while that of the bottlenose dolphin is very abrupt (photo by Thomas A. Jefferson, with permission).

groups of rough-toothed dolphins have gotten into the habit of taking large fishes off long-lines set by humans, and this has put them at odds with the local fishing industry.

Rough-toothed dolphins occur in tropical waters worldwide. They can be confused with bottlenose dolphins from a distance, due to similar size and general morphology, but the beak and spots should distinguish them upon closer inspection. They tend to occur in groups of ten or so animals, but larger groups of more than 100 have been seen (Figure 13.32).

We know practically nothing about the social order of rough-toothed dolphins. We know that they have a very large brain-to-body ratio, and the few that have been kept in captivity have been noted to be extremely flexible behaviorally (i.e., intelligent), with evidence for sophisticated second order learning (also called deutero-learning), which implies thought (Pryor et al. 1969). Recent work (Kuczaj and Yeater 2007) indicates that they have tight social bonds with long-term relationships.

No reliable estimates of numbers of the species worldwide are available. However, there are about 146,000 (CV 0.32) estimated for the eastern tropical Pacific (Wade and Gerrodette 1993) and 2,653 (CV 0.42) for the northern oceanic Gulf (Figure 13.33 and Table 13.14).

13.3.2.11 Risso's Dolphin

Risso's dolphin (*Grampus griseus*) is often called *grampus* by fishermen, and whaling records indicate that it was hunted for oil and meat in the Gulf in the 1700s and 1800s (Reeves et al. 2011). Males and females are about the same size, at a bit over 3 m (10 ft) in length, with no hint of sexual dimorphism. They have a prominent dorsal fin and, consequently, at a distance are sometimes confused with killer whales. However, Risso's dolphins are quite differently colored. While young, they are all gray. As they age they receive more linear scars on their bodies, until older individuals are almost entirely white. Apparently, all scrapes of their skins—presumably usually caused by intraspecific interactions of tooth rakes—disrupt dermal melanin pigments, which do not regrow or reinvade damaged skin in this species (Figure 13.34).

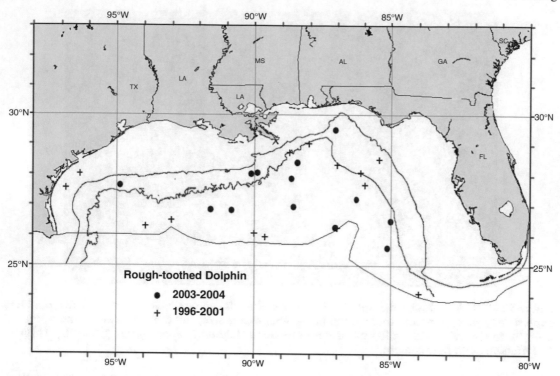

Figure 13.33. Distribution of rough-toothed dolphin (*Steno bredanensis*) sightings from SEFSC spring and fall vessel surveys during 1996–2001 and from summer 2003 and spring 2004 surveys. All on-effort sightings are shown, although not all were used to estimate abundance. Solid lines indicate the 100 m (328 ft) and 1,000 m (3,281 ft) isobaths and the offshore extent of the U.S. EEZ (from Waring et al. 2011).

Table 13.14. Most Recent Abundance Estimates (Nbest) and CV of Rough-Toothed Dolphins (*Steno Bredanensis*) in the Northern Gulf of Mexico Outer Continental Shelf (OCS) Aaters (20–200 m [66–656 ft] deep): Fall 2000–2001 and Oceanic Waters (200 m [656 ft] to the offshore extent of the EEZ) During Spring/Summer 2003–2004.

Month/Year	Area	Nbest	CV
Fall 2000–2001	Outer Continental Shelf	1,145	0.83
Spring/summer 2003–2004	Oceanic	1,508	0.39
Spring/summer and fall	OCS and oceanic	2,653	0.42

Risso's dolphins occur in tropical to cool temperate waters worldwide, and in the North Atlantic they are found as far north as Newfoundland. They are quite cold-water adapted (down to about 10 °C [50 °F]). Most occurrences are on the high seas in deep water, and numbers have generally been underestimated, it is now believed, due to the difficulty of surveying the open ocean habitat. Risso's dolphins have been described as very common cetaceans in the southern California Channel Island area, where groups can vary from 1 to approximately 100. Because they are so light colored as seen from above, individuals can be described from a circling airplane even when the animals have dived to twice their own lengths, making their social study a recently recognized plausibility. They feed on squid, but fishes and crustaceans are taken as well.

Figure 13.34. A Risso's dolphin (*Grampus griseus*) showing evidence of loss of melanin pigmentation and conspecific rake marks that stay for life (photo by Thomas A. Jefferson, with permission).

Risso's dolphins occur quite close to shore off California, Oregon, and Washington and have been designated as a subpopulation there, estimated at 16,000 (CV 0.28) (Barlow 2003). Before the intensive 1983–1984 El Niño event of the eastern Pacific, Risso's dolphins were uncommon off southern California. During the event, however, pilot whales all but disappeared (presumably because of a lack of squid), and Risso's dolphins came into the area. They are still present there in rather large numbers (Shane 1994). In the northern oceanic Gulf, they occur in some abundance both off the slope and abyssal plain, with the best estimate currently at 1,589 (CV 0.27) (Figure 13.35 and Table 13.15).

13.3.2.12 Fraser's Dolphin

Fraser's dolphin (*Lagenodelphis hosei*) is a physically robust species, with a short beak and small flippers, flukes, and dorsal fin. As adults, Fraser's dolphins have a dark tie-width stripe running along the side, demarcating the darker dorsum from the lighter ventrum. This variably appears to be much stronger in Pacific than in Atlantic animals, and it is stronger in adult males than in adult females. It is not present at all in calves and other immature animals. Males are about 2.7 m (8.9 ft) long, and females are slightly shorter, but there is no pronounced sexual dimorphism.

Fraser's dolphins feed largely on mid-water fishes and squid and may be able to dive as deep as about 600 m (1,969 ft). They are often active at the surface, frequently splashing and leaping in low arcs that create whitewater that can make their presence known from several kilometers. They occur worldwide in tropical and warm temperate waters, to about 30° north and south of the equator, but exact ranges and numbers are poorly known. There are an estimated 290,000 (CV 0.34) Fraser's dolphins in the eastern tropical Pacific (Wade and Gerrodette 1993). In the northern oceanic Gulf, no best estimate is given by NOAA because no sightings were made during the most recent surveys conducted in 2003–2004, and it is reported that sometimes none are seen for several years. Nevertheless, the most recent estimate, made from sightings in 1996–2001, is 726 dolphins (CV 0.70) (Figures 13.36 and 13.37 and Table 13.16).

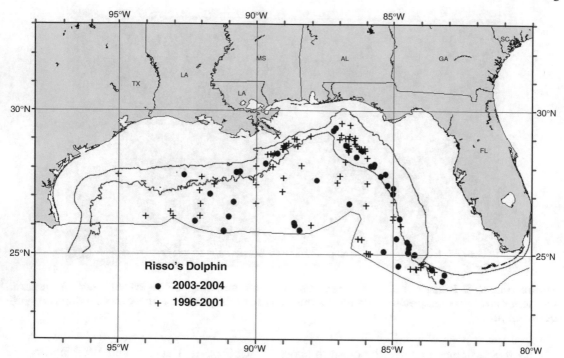

Figure 13.35. Distribution of Risso's dolphin (*Grampus griseus*) sightings from SEFSC vessel surveys during 1996–2001 and from summer 2003 and spring 2004 surveys. All the on-effort sightings are shown, though not all were used to estimate abundance. *Solid lines* indicate the 100 m (328 ft) and 1,000 m (3,281 ft) isobaths and the offshore extent of the U.S. EEZ (from Waring et al. 2011).

Table 13.15. Summary of Recent Abundance Estimate for Northern Gulf of Mexico Risso's Dolphins (*Grampus Griseus*): Month, year, and Area Covered During Each Abundance Survey, and Resulting Abundance Estimate (Nbest) and CV.

Month/Year	Area	Nbest	CV
Jun–Aug 2003, Apr–Jun 2004	Oceanic waters	1,589	0.27

13.3.2.13 Stenella Dolphins: Spinner, Clymene, Striped, Pantropical and Atlantic Spotted Dolphins

These dolphins of the *Stenella* genus are all thin-bodied and none are deep divers. They prefer to feed within the top several 100 m of the surface, although often on mesopelagic prey that comes towards the surface at night with the diurnally migrating prey of the deep scattering layer (DSL) in open oceans. While group sizes can vary greatly, stenellids are highly social animals and often occur in groups comprised of hundreds to several thousand animals. Stenellids also often occur in interspecies aggregations (e.g., spinner and pantropical spotted dolphins in the eastern tropical Pacific).

13.3.2.13.1 Spinner Dolphin

Spinner dolphins (*Stenella longirostris*) are the most numerous of the tropical oceanic cetaceans worldwide. They are thin, extremely long-beaked stenellids that occur as different morphologies (and as four different subspecies) in different parts of the tropics. Oceanic

Figure 13.36. Blunt-beaked, small bodied with very small flippers and dorsal fins, Fraser's dolphins (*Lagenodelphis hosei*) occur throughout warm waters in variable color morphs (photo by Thomas A. Jefferson, with permission).

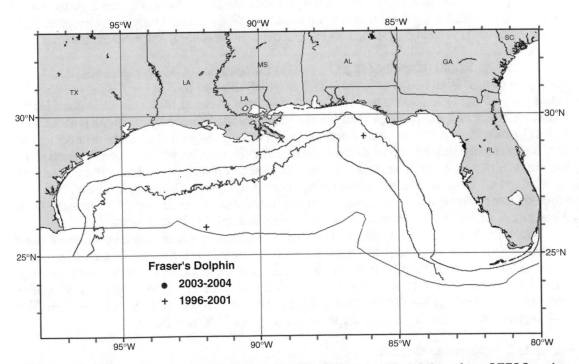

Figure 13.37. Distribution of Fraser's dolphin (*Lagenodelphis hosei*) sightings from SEFSC spring vessel surveys during 1996–2001 and from summer 2003 and spring 2004 surveys. All the on-effort sightings are shown, although not all were used to estimate abundance. Solid lines indicate the 100 m (328 ft) and 1,000 m (3,281 ft) isobaths and the offshore extent of the U.S. EEZ (from Waring et al. 2011).

Table 13.16. Summary of Abundance Estimates for Northern Gulf of Mexico Fraser's Dolphins: Month, Year, and Area Covered During Each Abundance Survey, and Resulting Abundance Estimate (Nbest) and CV.

Month/Year	Area	Nbest	CV
Apr–Jun 1991–1994	Oceanic waters	127	0.90
Apr–Jun 1996–2001 (excluding 1998)	Oceanic waters	726	0.70
Jun–Aug 2003, Apr–Jun 2004	Oceanic waters	0	-

Table 13.17. Summary of Abundance Estimates for Northern Gulf of Mexico Spinner Dolphins: Month, Year, and Area Covered During Each Abundance Survey, and Resulting Abundance Estimate (Nbest) and CV.

Month/Year	Area	Nbest	CV
Apr–Jun 1991–1994	Oceanic waters	6,316	0.43
Apr–Jun 1996–2001 (excluding 1998)	Oceanic waters	11,971	0.71
Jun–Aug 2003, Apr–Jun 2004	Oceanic waters	1,989	0.48

eastern tropical spinners show marked stripes along their flanks and reduced sexual dimorphism, and they live in huge herds of up to several thousand animals. The eastern spinner of the far eastern Pacific is almost uniformly dark gray and highly sexually dimorphic, with males having a pronounced post-anal keel and a high forward-curved dorsal fin; the eastern spinner morphology indicates a polygynous mating system, unlike the usual polygynandry (multimate or promiscuous) system of most delphinids, but detailed behavioral observations have not been carried out.

Some populations of spinner dolphins exist in the open ocean. Many others rely on daytime resting in or near island bays or in atolls and move offshore to feed at night on myctophid and squid prey of the DSL (Norris and Dohl 1980; Karczmarski et al. 2005). Despite considerable variation in size and morphology in different areas, overall spinner dolphin length is about 1.8 m (5.9 ft), making it a rather small dolphin. It is the only dolphin that spins around its axis extremely rapidly and with up to six revolutions, either in a horizontal or vertical position above the surface of the water. These spins appear to have to do with social facilitation as animals move from a resting to an alert (often highly social/sexual) state. Island spinner dolphins were studied intensively on the Kona Coast of the Big Island, Hawaii, in the late 1970s to early 1980s. Norris et al. (1994) is a detailed book of their behaviors and life-history strategies.

While there are no estimates of worldwide numbers, spinner dolphins of the eastern tropical Pacific are estimated at about 1.4 million animals for two subspecies (Gerrodette et al. 2005) and at 1,989 (CV 0.48) in the northern oceanic Gulf. However, note the large variations from other sets of surveys (Table 13.17). Almost all survey sightings of spinner dolphins in the Gulf of Mexico have been in the central and eastern Gulf, but not western Gulf, and largely, but not exclusively, in slope waters (Figures 13.38 and 13.39).

13.3.2.13.2 Clymene Dolphin

The Clymene dolphin (*Stenella clymene*) is about the same size as the spinner dolphin and has been confused with it, and also the short-beaked common dolphin. It is a bit more robust animal than the spinner dolphin, however, and has a shorter beak, black lips and a pronounced black beak tip.

Figure 13.38. Spinner dolphins (*Stenella longirostris*) occur in warm waters worldwide and in quite a few different color and body morphs (photo by Thomas A. Jefferson, with permission).

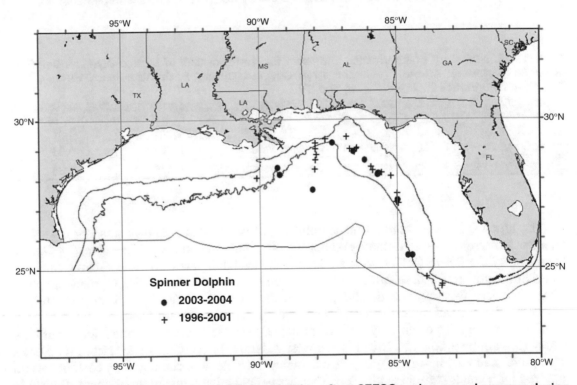

Figure 13.39. Distribution of spinner dolphin sightings from SEFSC spring vessel surveys during 1996–2001 and from summer 2003 and spring 2004 surveys. All the on-effort sightings are shown, though not all were used to estimate abundance. Solid lines indicate the 100 m (328 ft) and 1,000 m (3,281 ft) isobaths and the offshore extent of the U.S. EEZ (from Waring et al. 2011).

Figure 13.40. Clymene dolphins (*Stenella clymene*) have often been confused with spinner dolphins. They occur only in the tropical Atlantic (photo by Robert L. Pitman, with permission).

Table 13.18. Summary of Abundance Estimates for Northern Gulf of Mexico Clymene Dolphins (*Stenella Clymene*): Month, Year, and Area Covered During Each Abundance Survey, and Resulting Abundance Estimate (^{N}best) and CV.

Month/Year	Area	^{N}best	CV
Apr–Jun 1991–1994	Oceanic waters	5,571	0.37
Apr–Jun 1996–2001 (excluding 1998)	Oceanic waters	17,355	0.65
Jun–Aug 2003, Apr–Jun 2004	Oceanic waters	6,575	0.65

Not much is known about Clymene dolphins. Most of what is known was aggregated by Thomas Jefferson, a world authority on marine mammals and author of the guide *The Marine Mammals of the World* (Jefferson et al. 2008). Clymene dolphins feed largely on mesopelagic fishes and squid and take advantage of the ecological cascade of the DSL. Clymene dolphins often associate with spinner dolphins, and as the two are difficult to distinguish from a distance, this association further clouds counts of the species (Figure 13.40).

Clymene dolphins occur only in the tropical and warm temperate Atlantic from about New Jersey (in summer) down to Brazil in the west Atlantic. In the Gulf of Mexico, group sizes average 42 animals; however, group size is highly variable, and some groups contain several hundred individuals (Mullin et al. 1994). They are estimated at a minimum of about 100,000 in the Atlantic Basin, including an estimate of 6,575 (CV 0.36) in the northern oceanic Gulf; but note that previous estimates have been as high as 17,000 (Table 13.18). Unlike spinner dolphins that use mainly the eastern portion of the Gulf, Clymene dolphins largely use the abyssal part of the western section (Figure 13.41) with some overlap.

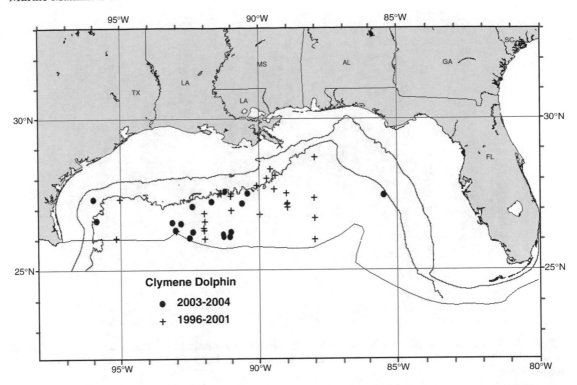

Figure 13.41. Distribution of Clymene dolphin (*Stenella clymene*) sightings from SEFSC shipboard spring surveys during 1996–2001 and from summer 2003 and spring 2004 surveys. All the on-effort sightings are shown, though not all were used to estimate abundance. Solid lines indicate the 100 m (328 ft) and 1,000 m (3,281 ft) isobaths and the offshore extent of the U.S. EEZ (from Waring et al. 2011).

13.3.2.13.3 Striped Dolphin

Striped dolphins (*Stenella coeruleoalba*) are about 2.4 m (7.9 ft) long, with little sexual dimorphism. They are a bit more robust in body form than spinner, pantropical and Atlantic spotted dolphins, but are nevertheless generally slender. They are strikingly marked with stripes along their sides, highlighted by a stripe that begins at the eye and swoops dorsally, ending just below the dorsal fin. Striped dolphins feed largely on mesopelagic fishes and squid.

This stenellid is distributed in tropical and warm temperate waters worldwide, between about 50°N latitude and 40°S latitude. In the North Atlantic, the species occurs as far north as Nova Scotia (in summer) and throughout the tropics to the southern hemisphere of Brazil and Africa. It occurs in the Mediterranean Sea and was the subject of ancient Greek frescoes. Group sizes vary from dozens of animals to hundreds and may have numbered in the thousands historically (Figure 13.42).

In the eastern tropical Pacific, the most recent population estimate of striped dolphins, which was derived from results of a 2003 line-transect survey, was about 1.5 million animals (Gerrodette et al. 2005). Abundance estimates within about 500 km (310 mi) of the U.S. West Coast have averaged about 19,000 (CV 0.28) between 1991 and 2005.

While global estimates are questionable, there are surely several million worldwide. Striped dolphins in the northern oceanic Gulf are estimated at about 3,325 animals (CV 0.48), with some sightings on the eastern Gulf slope but most in deep ocean waters (Figure 13.43 and Table 13.19).

Figure 13.42. Striped dolphins (*Stenella coeruleoalba*) are animals of the open ocean, often found in the deepest waters, including in the Gulf of Mexico (photo by Thomas A. Jefferson, with permission).

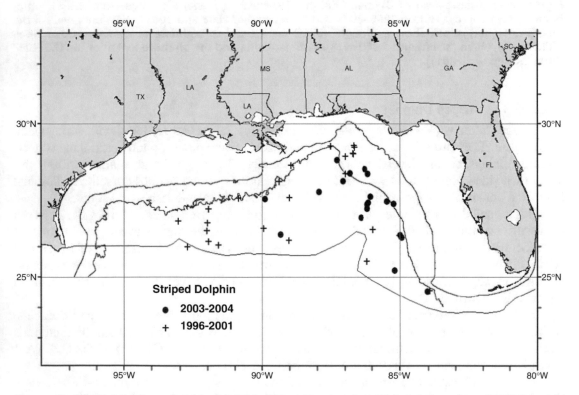

Figure 13.43. Distribution of striped dolphin (*Stenella coeruleoalba*) sightings from SEFSC spring vessel surveys during 1996–2001 and from summer 2003 and spring 2004 surveys. All the on-effort sightings are shown, although not all were used to estimate abundance. Solid lines indicate the 100 m (328 ft) and 1,000 m (3,281 ft) isobaths and the offshore extent of the U.S. EEZ (from Waring et al. 2011).

Table 13.19. Summary of Abundance Estimates for Northern Gulf of Mexico Striped Dolphins (*Stenella Coeruleoalba*): Month, Year and Area covered During Each Abundance survey, and Resulting Abundance Estimate (^{N}best) and CV.

Month/Year	Area	^{N}best	CV
Apr–Jun 1991–1994	Oceanic waters	4,858	0.44
Apr–Jun 1996–2001 (excluding 1998)	Oceanic waters	6,505	0.43
Jun–Aug 2003, Apr–Jun 2004	Oceanic waters	3,325	0.48

Figure 13.44. Pantropical spotted dolphins (*Stenella attenuata*) are likely the most numerous of the genus *Stenella* in the world's oceans and the most numerous marine mammal in the Gulf of Mexico as well (photo by Thomas A. Jefferson, with permission).

13.3.2.13.4   Pantropical Spotted Dolphin

The pantropical spotted dolphin (*Stenella attenuata*) is marked by a rather long beak, slender body, and quite strongly falcate dorsal fin. While a muted gray above and light below, it develops spots along the sides as it ages, and rough categories of age can be determined by the amount of spotting. There is much geographic variation in coloration and size by area. This dolphin occurs in all tropical and subtropical waters, worldwide, from the equator to about 40°N latitude and 30°S latitude. Sexually mature individuals are from about 1.7 m (5.6 ft) to 2.6 m (8.5 ft), with males only slightly larger than females.

Pantropical spotted dolphins occur in rather large numbers in deep waters of the world's oceans, where they feed on mesopelagic and epipelagic fishes, crustaceans, and squid often related to the DSL, but they also feed on surface-dwelling flying fishes in some areas. An estimated 640,000 still exist in the eastern tropical Pacific, but this represents probably only about 20 % of the original population(s) before intensive killing as bycatch in the tuna fishing industry during the 1950s through early 1990s (related estimates also, Gerrodette et al. 2005). In the Gulf of Mexico, pantropical spotted dolphins are the most numerous cetacean, with estimates in the northern oceanic Gulf ranging from about 34,000 (CV 0.18) to 91,000 (CV 0.16) (Figures 13.44 and 13.45 and Table 13.20). They occur on the upper slope in waters of about 100 m (328 ft), as well as their primary habitat—waters deeper than 100 m (328 ft) and into the open abyssal zone of the Gulf, including in the southern Gulf (Figure 13.46).

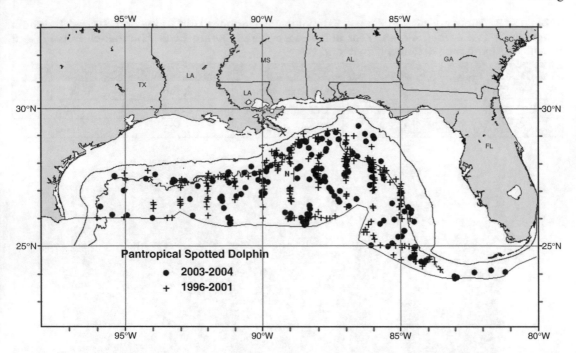

Figure 13.45. Distribution of pantropical spotted dolphin (*Stenella attenuata*) sightings from SEFSC spring vessel surveys during 1996–2001 and from summer 2003 and spring 2004 surveys. All the on-effort sightings are shown, though not all were used to estimate abundance. *Solid lines* indicate the 100 m (328 ft) and 1,000 m (3,281 ft) isobaths and the offshore extent of the U.S. EEZ (from Waring et al. 2011).

Table 13.20. Summary of Abundance Estimates for Northern Gulf of Mexico Pantropical Spotted Dolphins (*Stenella Attenuata*): Month, Year and Area Covered During Each Abundance Survey, and Resulting Abundance Estimate (^{N}best) and CV.

Month/Year	Area	^{N}best	CV
Apr–Jun 1991–1994	Oceanic waters	31,320	0.20
Apr–Jun 1996–2001 (excluding 1998)	Oceanic waters	91,321	0.16
Jun–Aug 2003, Apr–Jun 2004	Oceanic waters	34,067	0.18

13.3.2.13.5   Atlantic Spotted Dolphin

The Atlantic spotted dolphin and the Clymene dolphin are the only species of cetaceans found in the Gulf that are endemic to the Atlantic Ocean, with the Atlantic spotted dolphin occurring as far north as 50°N latitude (although more commonly only to about 40°N latitude) and about 25°S latitude. It occurs—and has been studied intensively (e.g., Herzing 1997)—on the shallows of the Bahama banks, where it socializes during daytime and (presumably) feeds on epipelagic and mesopelagic fishes and squid in the drop-off oceanic zones, feeding relative to DSL organisms in deeper waters, at night. In the Gulf, Atlantic spotted dolphins generally occur within the 200-m (656-ft) depth contour and are thus animals of the shallower waters of the oceanic and near-oceanic zones.

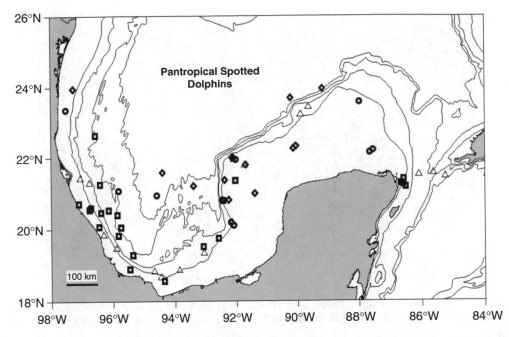

Figure 13.46. Records of pantropical spotted dolphins (*Stenella attenuata*) in the Mexican waters of the southern Gulf of Mexico and Caribbean. Display for distribution is as follows: solid symbol, strandings; hollow symbols, confirmed sightings; triangles, spring (Mar–May); squares, summer (Jun–Aug); circles, autumn (Sep–Nov); diamonds, winter (Dec–Feb); crosses, unknown dates. Thin contour lines show the 200 m, 1,000 m, 2,000 m, and 3,000 m (656 ft, 3,281 ft, 6,562 ft, and 9,843 ft) isobaths (from Ortega-Ortiz 2002).

Atlantic spotted dolphins are often confused with the pantropical species, as they also develop spots along their sides (and the rest of the body, depending on geographic area/population) as they mature. Male Atlantic spotted dolphins at full maturity are about 2.7 m (8.9 ft) in length, and females are about 2.5 m (8.2 ft). In the Atlantic Ocean, best estimate is unknown, but in the northern Gulf of Mexico, there are an estimated 37,611 (CV 0.28) animals; this estimate is not accepted as being current by the NMFS, because it is from data greater than 8 years old (Figure 13.47 and Table 13.21). Almost all of these sightings are from within the 100-m (328-ft) depth contours, especially off the Florida shelf, while a few range into the 100-m to 200-m (328-ft to 656-ft) depth area. Atlantic spotted dolphins in the Gulf do not seem to be found in deeper oceanic waters in the northern (Figure 13.48) or southern (Figure 13.49) Gulf. They co-occur in habitat with continental-slope bottlenose dolphins.

13.3.2.13.6 Common Bottlenose Dolphin

As the name implies, the common bottlenose dolphin (*Tursiops truncatus*) is a very common animal near shore in most tropical, temperate, and even cooler waters of all oceans, occurring as far north as northern Scotland in the Atlantic, as far south as mid-Patagonia in South America, and also as far south as the cold water (in winter at times slightly iced over) fjords of the South Island of New Zealand. It also occurs as separate populations in inshore bays and estuaries, alongshore barrier islands and other geographic situations, and in oceanic waters, often in quite disparate morphs of coloration and size. It is ubiquitous in nearshore areas of the Gulf of Mexico, both northern and southern, and along the continental shelf, to and beyond 200-m (656-ft) depths.

Figure 13.47. This underwater photo is of Atlantic spotted dolphins (*Stenella frontalis*) in the Bahamas, but they look very similar in the Gulf, developing spots as they age (photo by Bernd Würsig).

Table 13.21. Most Recent Abundance Estimates (N_{best}) and CV of Atlantic Spotted Dolphins (*Stenella Frontalis*) in the Northern Gulf of Mexico Outer Continental Shelf (OCS) (waters 20–200 m [66–656 ft] deep) During Fall 2000–2001 and Oceanic Waters (200 m [656 ft] to the offshore extent of the EEZ) During Spring/Summer 2003–2004.

Month/Year	Area	N_{best}	CV
Fall 2000–2001	Outer Continental Shelf	37,611	0.28
Spring/Summer 2003–2004	Oceanic	0	–
Fall and Spring/Summer	OCS and Oceanic	37,611	0.28

The bottlenose dolphin has been subdivided into over one dozen species and/or subspecies during its taxonomic history. Presently, its congener, the Indo-Pacific bottlenose dolphin (*Tursiops aduncus*), is accepted as the only other bottlenose dolphin. This designation will likely be refined with further genetic analyses. The Indo-Pacific bottlenose dolphin occurs from Cape Agulhas, South Africa, to the main island of Japan (i.e., in the Indian Ocean and the western Pacific, including most of Australia). This distribution overlaps strongly with that of the common bottlenose dolphin, and there is some confusion on species designations in certain overlap areas (Figure 13.50).

In the Atlantic, there is no argument as to species, as the Indo-Pacific bottlenose dolphin does not reach there. But there is considerable debate on population and subpopulation designations. Bottlenose dolphins typically occur in groups of about one dozen animals, as mixed age and sex groups, all-female and youngster nursery groups, and those of immature or mature males (Wells et al. 1987). However, groupings as large as 1,000 animals in the open ocean (but not in the Gulf) have been reported.

Bottlenose dolphins (*Tursiops truncatus*) occur throughout most bays, sounds, and estuaries of the Gulf of Mexico, often into quite brackish water, with salinities of less than ten parts per thousand (ppt). However, the definitions of populations or stocks is complicated by the fact

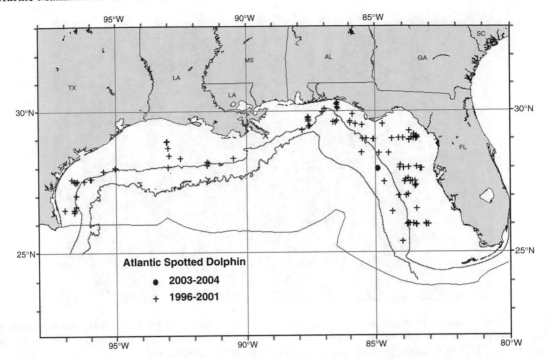

Figure 13.48. Distribution of Atlantic spotted dolphin (*Stenella frontalis*) sightings from SEFSC spring and fall vessel surveys during 1996–2001 and from summer 2003 and spring 2004 surveys. All the on-effort sightings are shown, though not all were used to estimate abundance. Solid lines indicate the 100 m (328 ft) and 1,000 m (3,281 ft) isobaths and the offshore extent of the U.S. EEZ (from Waring et al. 2011).

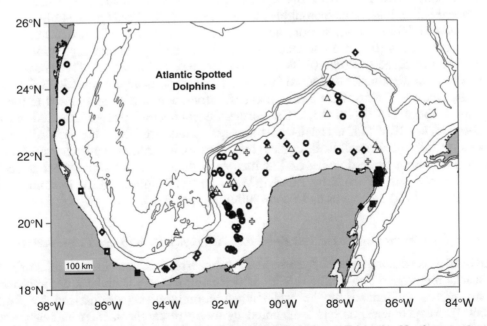

Figure 13.49. Records of Atlantic spotted dolphins (*Stenella frontalis*) in the Mexican waters of the southern Gulf of Mexico and Caribbean. Display for distribution is as follows: solid symbol, strandings; hollow symbols, confirmed sightings; triangles, spring (Mar–May); squares, summer (Jun–Aug); circles, autumn (Sep–Nov); diamonds, winter (Dec–Feb); crosses, unknown dates. Thin contour lines show the 200 m, 1,000 m, 2,000 m, and 3,000 m (656 ft, 3,281 ft, 6,561 ft, and 9,843 ft) isobaths (from Ortega-Ortiz 2002).

Figure 13.50. The ubiquitous worldwide dolphin that is most often envisioned by the nonmarine mammalogist when dolphins are mentioned. It is common in all waters except over the abyssal plain in the Gulf of Mexico (photo by Giovanni Bearzi, with permission).

that bay animals also travel outside of their major habitats and interact and mate with animals outside of these bays as well. Most populations or subpopulations have not been well studied, with the most notable exceptions being those of the Sarasota-Bradenton, Florida, area (Wells 2003) and Sanibel Island (Shane 2004) of west Florida. Similarly, dolphin populations are known in the north-central Gulf (Mullin 1988) and off Texas (Shane 1977; Moreno 2005) as well, but none has been followed for as long or as thoroughly as the subject of the Wells (2003) study.

Bottlenose dolphins of the inshore areas of the northern Gulf of Mexico number about 28–32 separate stocks (that are not necessarily distinguished as genetic populations), with an estimated total of 5,355 (no reliable CV available) animals (Table 13.22). Three coastal stocks outside of bays and estuaries and up to 20 m (66 ft) deep, total approximately 13,600 animals (see details, below). The northern continental shelf stock between 20 m and 200 m (66 ft and 656 ft) totals about 17,777 (CV 0.32). The northern oceanic stock deeper than 200 m (656 ft) totals about 3,708 (CV 0.42). Details of these groupings are given below. Anthropogenic influences on this species will be discussed, including toxins, noises, and other aspects, since more is known about these dolphins and their ecology than is known about other cetaceans in the Gulf, largely due to work by Randall Wells and colleagues, which is summarized in Reynolds et al. (2000) and Wells and Scott (2009) (Figure 13.51).

Northern Gulf of Mexico Bay, Sound, and Estuarine Stock (Often Divided into Communities)

The bottlenose dolphins seen by people inside bays, estuaries, and channels tend to be of the 28–32 stocks mentioned above. The fidelity of these animals to particular areas appears to be quite strong, as evidenced by the well-studied communities (summarized by Reynolds et al. 2000). Nevertheless, as has been noted by many researchers, there is behavioral and genetic interaction between the resident communities and members of the next category (see below) of the three NMFS-designated coastal stocks in U.S. Gulf waters to about the 20 m (66 ft) depth.

Table 13.22. Major Communities of Inshore Common Bottlenose Dolphins (Waring et al. 2010).

Blocks	Gulf of Mexico Estuary	N$_{best}$	CV	N$_{min}$	PBR	Year	Reference
B51	Laguna Madre	80	1.57	UNK	UND	1992	A
B52	Nueces Bay, Corpus Christi Bay	58	0.61	UNK	UND	1992	A
B50	Compano Bay, Aransas Bay, San Antonio Bay, Redfish Bay, Espiritu Santo Bay	55	0.82	UNK	UND	1992	A
B54	Matagorda Bay, Tres Palacios Bay, Lavaca Bay	61	0.45	UNK	UND	1992	A
B55	West Bay	32	0.15	UNK	UND	2000	E
B56	Galveston Bay, East Bay, Trinity Bay	152	0.43	UNK	UND	1992	A
B57	Sabine Lake	0[a]	–		UND	1992	A
B58	Calcasieu	0[a]	–		UND	1992	A
B59	Vermillion Bay, West Cote Blanche Bay, Atchafalaya Bay	0[a]	–		UND	1992	A
B60	Terrobonne Bay, Timbalier Bay	100	0.53	UNK	UND	1993	A
B61	Barataria Bay	138	0.08	UNK	UND	2001	D
B30	Mississippi River Delta	0[a]	–		UND	1993	A
B02–05, 29,31	Bay Boudreau, Mississippi Sound	1401	0.13	UNK	UND	1993	A
B06	Mobile Bay, Bonsecour Bay	122	0.34	UNK	UND	1993	A
B07	Perdido Bay	0[a]	–		UND	1993	A
B08	Penascola Bay, East Bay	33	0.80	UNK	UND	1993	A
B09	Choctawhatchee Bay	242	0.31	UNK	UND	1993	A
B10	St. Andrew Bay	124	0.57	UNK	UND	1993	A
B11	St. Joseph Bay	81	0.14	72	0.7	2005–2006	F
B12–13	St. Vincent Sound, Apalachicola Bay, St. George Sound	537	0.09	498	5.0	2008	G
B14–15	Apalachee Bay	491	0.39	UNK	UND	1993	A

(continued)

Table 13.22. (continued)

Blocks	Gulf of Mexico Estuary	N_{best}	CV	N_{min}	PBR	Year	Reference
B16	Waccasassa Bay, Withlacoochee Bay, Crystal Bay	100	0.85	UNK	UND	1994	A
B17	St. Joseph Sound, Clearwater Harbor	37	1.06	UNK	UND	1994	A
B32–34	Tampa Bay	559	0.24	UNK	UND	1994	A
B20, 35	Sarasota Bay, Little Sarasota Bay	160	na[c]	160	1.6	2007	B
B21	Lemon Bay	0[a]	–		UND	1994	A
B22–23	Pine Sound, Charlotte Harbor, Gasparilla Sound	209	0.38	UNK	UND	1994	A
B36	Caloosahatchee River	0[a,b]	–		UND	1985	C
B24	Estero Bay	104	0.67	UNK	UND	1994	A
B25	Chokoloskee Bay, Ten Thousand Islands, Gullivan Bay	208	0.46	UNK	UND	1994	A
B27	Whitewater Bay	242	0.37	UNK	UND	1994	A
B28	Florida Keys (Bahia Honda to Key West)	29	1.00	UNK	UND	1994	A

References: A, (Blaylock and Hoggard 1994); B, (Wells and Scott 2009); C, (Scott et al. 1989); D, (Miller 2003); E, (Irwin and Würsig 2004); F, (Balmer et al. 2008); G, (Tyson 2008)

PBR potential biological removal; refers to the maximum number of animals that may be removed from a population, not including natural mortalities, for the population to be maintained.

UNK unknown, *UND* undetermined.

[a]During earlier surveys (Scott et al. 1989), the range of seasonal abundances was as follows: B57, 0–2 (CV = 0.38); B58, 0–6 (0.34); B59, 0–182; B07 0–15 (0.43); and B36, 0–0.

[b]Blocks not surveyed during surveys reported in Blaylock and Hoggard (1994).

[c]No CV because N_{best} was a direct count of known individuals.

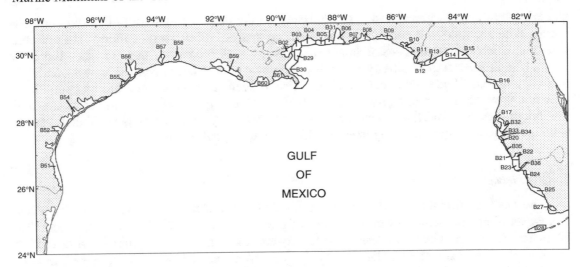

Figure 13.51. Northern Gulf of Mexico Bays and Sounds. Each of the alpha-numerically designated blocks corresponds to one of the NMFS SFSC logistical aerial survey areas listed in Table 13.1. The bottlenose dolphins (*Tursiops truncatus*) inhabiting each bay and sound are considered to comprise a unique stock for purposes of this assessment (after Waring et al. 2010). See also Table 13.22.

Analyses of mitochondrial DNA (inherited only from the mother) variations between communities and along shore indicate clinal variations among areas (Duffield and Wells 2002), and differences in the seasonality of reproduction among sites also suggest genetic differences among communities (Urian et al. 1996).

Studies by Randall S. Wells and colleagues describe the long-term structure and stability of bottlenose dolphin residents of greater Sarasota Bay, Florida, since 1970 (Irvine and Wells 1972; Scott et al. 1990; Wells 1991, 2003). By photo-identification and periodic captures for age, reproductive, and health data, presently five generations have been tracked in the area, including several first seen in the 1970s that are still living. Estimated immigration and emigration rates are about 2–3 % (Wells and Scott 1990). However, while it is rather stable and almost all individuals remain, this is not a wholly isolated, genetically closed population, and at least some calves were sired by nonresidents (Duffield and Wells 2002). While year-round residents occur in other areas as well, at least some animals can move quite long distances, as nearshore animals have been identified up to several 100 km away in Texas waters (Lynn and Würsig 2002). There is some aspect of seasonality as well. In smaller bays such as Sarasota, Florida, and San Luis Pass, Texas, some residents move into Gulf coastal waters during fall and winter and return inshore in spring and summer (Irvine et al. 1981; Maze and Würsig 1999, respectively). In larger bays, there may be even more seasonal migrations, as there is a tendency for greater numbers in northerly bays in summer and southerly bays in winter (e.g., Tampa Bay, Florida, Scott et al. 1989; and Galveston Bay, Texas, Bräger et al. 1994).

The above data must not obscure the fact that most bottlenose dolphin populations or communities of the Gulf are not thoroughly described, and much more information is needed for proper identifications and numbers. Only four populations are sufficiently well known with data from the past 8 years or less for reliable numbers estimates: Sarasota Bay (160 animals, direct count), Choctawhatchee Bay (179 animals, best estimate), Apalachicola Bay (537 animals, best estimate), and St. Joseph Bay (146 animals, best estimate). In total, an estimated 5,355 inshore/nearshore bottlenose dolphins reside in the U.S. waters of the Gulf in 28–32 bays and estuaries from the Laguna Madre, south Texas, to the Florida Keys, south Florida. Unfortunately, data are not good enough for overall trend analysis of numbers, although the one area with good, long-term data—Sarasota Bay—shows a rather constant number since the early 1970s. Maximum net productivity rate is also unknown but has been assumed to be around 0.04 (=4 %) per year (Wade 1998), based on theoretical modeling showing that cetacean populations may not grow at rates much greater than 4 % due to their typical large-mammalian pattern of low pregnancy rate and the production with each pregnancy of only a single calf that exhibits slow growth and requires a long time to achieve sexual/social maturity.

Coastal Stocks

The Gulf of Mexico coastal stocks are divided into eastern, northern, and western bottlenose dolphins that generally occur outside of bays and estuaries, but in Gulf waters less than 20 m (66 ft) deep. There is much contact between these and inshore animals, and contact between these and the greater than 20 m (66 ft) depth dolphins as well. Thus, these should be considered stocks for management purposes, not as separate or distinct behavioral or genetic entities. As mentioned above, genetic clinal-like variations exist, with animals further apart showing greater genetic dissimilarity (e.g., Duffield and Wells 2002). The three coastal stocks have approximate numbers of 7,702 (CV 0.19), 2,473 (CV 0.25), and 3,499 (CV 0.21) for eastern, northern, and western stocks, respectively. Maps of sightings per stock are provided below (Figures 13.52, 13.53, and 13.54).

Gulf of Mexico Eastern Coastal Stock

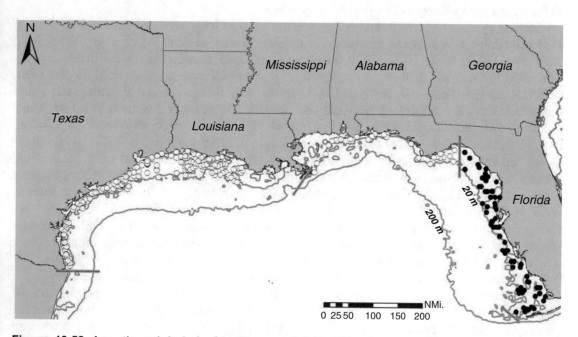

Figure 13.52. Locations (*circles*) of bottlenose dolphin (*Tursiops truncatus*) groups sighted in coastal waters during aerial surveys conducted in the western coastal stock area in 1992 and 1996 and in the northern coastal stock and eastern coastal stock areas in 2007. *Dark circles* indicate groups within the boundaries of the Eastern Coastal stock. The 20 and 200 m (66 and 656 ft) isobaths are shown (from Waring et al. 2013).

Gulf of Mexico Northern Coastal Stock

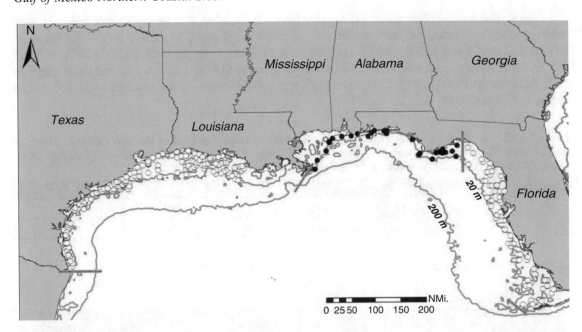

Figure 13.53. Locations (*circles*) of bottlenose dolphin (*Tursiops truncatus*) groups sighted in coastal waters during aerial surveys conducted in the western coastal stock area in 1992 and 1996, and in the northern coastal stock and eastern coastal stock areas in 2007. *Dark circles* indicate groups within the boundaries of the Northern Coastal Stock. The 20 and 200 m (66 and 656 ft) isobaths are shown (from Waring et al. 2013).

Gulf of Mexico Western Coastal Stock

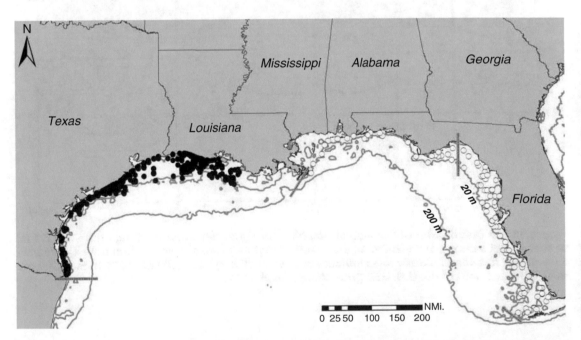

Figure 13.54. Locations (*circles*) of bottlenose dolphin (*Tursiops truncatus*) groups sighted in coastal waters during aerial surveys conducted in the western coastal stock area in 1992 and 1996, and in the northern coastal stock and eastern coastal stock areas in 2007. *Dark circles* indicate groups within the boundaries of the western coastal stock. The 20 and 200 m (66 and 656 ft) isobaths are shown. Apparent gaps between stock areas are likely due to inadequate aerial survey coverage (from Waring et al. 2013).

Continental and Oceanic Stocks

Again, there is no clear stock delineation between the shelf and more oceanic animals, with management estimates of the northern continental shelf stock at 17,777 (CV 0.32) between 20 and 200 m (66 and 656 ft); and the northern oceanic stock deeper than 200 m (656 ft) at 3,708 (CV 0.42) (Figures 13.55 and 13.56). Note that bottlenose dolphins overlap on the shelf strongly with Atlantic spotted dolphins (Figure 13.48) and neither species frequents waters deeper than about 1,000 m (Table 13.23).

Table 13.23. Summary of Abundance Estimates for the Northern Gulf of Mexico Oceanic Stock of Bottlenose Dolphins (*Tursiops Truncatus*): Month, Year, and Area Covered During Each Abundance Survey, and Resulting Abundance Estimate (^{N}best) and CV.

Month/Year	Area	^{N}best	CV
Apr–Jun 1996–2001 (excluding 1998)	Oceanic waters	2,239	0.41
Jun–Aug 2003, Apr–Jun 2004	Oceanic waters	3,708	0.42

Northern Gulf of Mexico Continental Shelf Stock

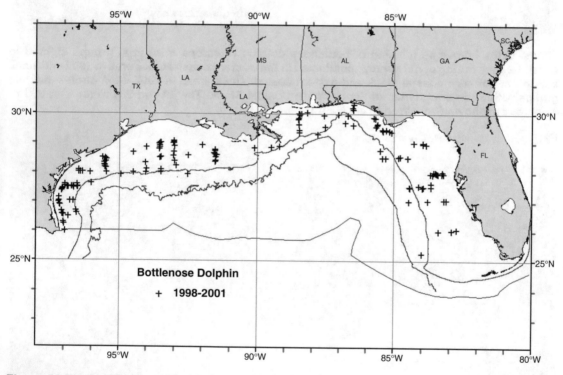

Figure 13.55. Distribution of bottlenose dolphin (*Tursiops truncatus*) sightings from SEFSC fall vessel surveys during 1998–2001. All the on-effort sightings are shown, though not all were used to estimate abundance. *Solid lines* indicate the 100 m (328 ft) and 1,000 m (3,281 ft) isobaths and the offshore extent of the U.S. EEZ (from Waring et al. 2011).

Northern Gulf of Mexico Oceanic Stock

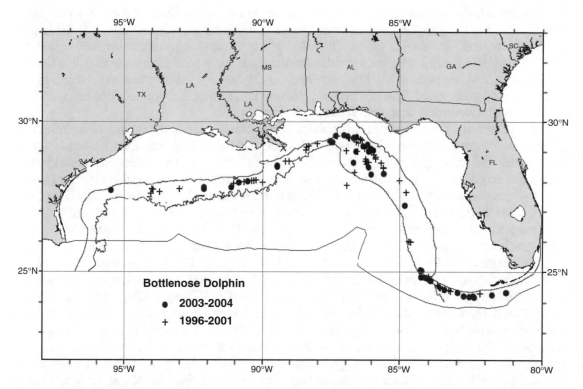

Figure 13.56. Distribution of bottlenose dolphin (*Tursiops truncatus*) sightings from SEFSC ship-board surveys during spring 1996–2001 and from summer 2003 and spring 2004 surveys. All the on-effort sightings are shown, though not all were used to estimate abundance. *Solid lines* indicate the 100 m (328 ft) and 1,000 m (3,281 ft) isobaths and the offshore extent of the U.S. EEZ (from Waring et al. 2011).

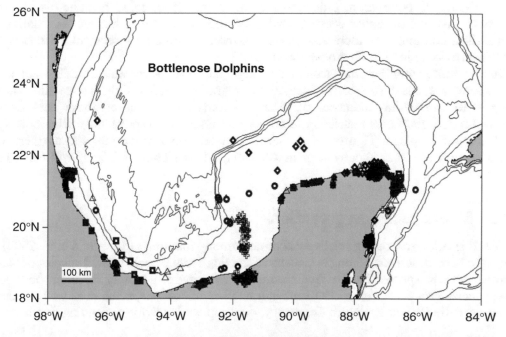

Figure 13.57. Records of bottlenose dolphins (*Tursiops truncatus*) in the Mexican waters of the southern Gulf of Mexico and Caribbean. Display for distribution is as follows: solid symbol, strandings; hollow symbols, confirmed sightings; triangles, spring (Mar–May); squares, summer (Jun–Aug); circles, autumn (Sep–Nov); diamonds, winter (Dec–Feb); crosses, unknown dates. Thin contour lines show the 200 m, 1,000 m, 2,000 m, and 3,000 m (656 ft, 3,281 ft, 6,562 ft, and 9,843 ft) isobaths (from Ortega-Ortiz 2002).

13.3.2.14 West Indian Manatee

The West Indian manatee (*Trichechus manatus*) is but one of three manatees, which along with the dugong (*Dugong dugon*) make up the small mammalian order Sirenia. It is the largest of the manatees, at about 4.6 m (15.1 ft) in length and approximately 600–1,000 kg (1,323–2,205 lb) in weight. It has a paddle-shaped flattened dorsal fin, somewhat like the tail of a beaver (*Castor canadensis*). Manatees and the dugong are the only true vegetarians of the entire grouping of more than 122 marine mammals, feeding on tropical seagrasses, water hyacinth, and even fruits that drop from vegetation above.

Manatees have been hunted for centuries for meat and hide and continue to be hunted in Central and South America. However, collisions with speeding motorboats, especially in Florida, are the most constant source of manatee fatalities in U.S. waters. In west Florida, deaths due to cold spells are also a major problem, as manatees use natural springs and (more often) warm power plant outfalls as refugia, and the latter can become death traps if a power plant is accidentally or purposefully shut down in winter (Laist and Reynolds 2005).

Two subspecies are currently recognized: the Florida manatee and the Antillean manatee. Both species occur in the Gulf of Mexico, with the Florida manatee in the northeast and the Antillean manatee in Mexican waters in the south (Figure 13.3). The Florida manatee subspecies is protected in Gulf waters by the U.S. Endangered Species Act, Convention on International Trade in Endangered Species of Wild Fauna and Flora (CITES) Appendix 1, and the International Union for Conservation of Nature (IUCN), and they all list it as endangered. Furthermore, the Florida Manatee Sanctuary Act of 1978, the Manatee Recovery Plan, and the Save the Manatee Club all help to create awareness of problems to manatees in U.S. waters and ways to mitigate these problems. While the population worldwide (that is, from northern Florida south to Brazil) is estimated at somewhat fewer than 10,000 mature individuals, about 2,800 occur off east Florida and 2,300 off west Florida in the Gulf (Marsh et al. 2011). No good estimate exists for the Antillean manatee of Mexico in the southern Gulf (Figure 13.58).

In April 2007, the USFWS announced that the West Indian manatee population of Florida was doing well and advised that the species be reclassified as threatened rather than endangered. However, computer models by a federal study showed a 50 % chance that the statewide manatee population, estimated then as about 3,300, could dwindle over the next 50 years to just 500 on either coast if further depredations such as habitat degradation and vessel strikes continued or increased. Presently, there is some disagreement as to how well manatees are doing in U.S. (and other) waters.

During winter months, manatees often congregate near warm-water outflows of power plants along the coast of Florida instead of migrating south as they once did, causing biologists to worry that manatees may have become too reliant on these human-made, warm-water refugia. Laws restricting temporary closures of power plants during cold spells have been put in place (Figure 13.59). Some manatees also move into the northern Gulf of Mexico and are sporadically seen in Alabama, Mississippi, Louisiana, and even Texas (Fertl et al. 2005).

13.3.3 Multispecies Aggregations

Mixed species groups are relatively common among mammals (Stensland et al. 2003) and often have been described among cetaceans as ranging from closely related species or species of similar size to species from different orders or having remarkably different body sizes. Mixed species groups of cetaceans occur in a number of habitats, oceanic as well as coastal, and vary greatly in their structure, frequency, duration, and activity, depending on the species

Figure 13.58. West Indian manatee. Note the rounded flippers and tail, large nostrils and small eyes. The light speckling on the back is the reflection of wavelets on the surface (photo by Christopher Marshall, with permission).

involved and the habitat. As happens more generally among mammals, mixed species groups tend to occur because of foraging advantages, predator avoidance, or both. However, there could be additional social or reproductive benefits that contribute to group formation and stability. These advantages do not need to be equal among the participating species and can vary over time (Stensland et al. 2003).

Although most cetacean groups are monospecific, several species often or regularly associate with other species for variable periods of time. For instance, bottlenose dolphins have been recorded to associate with more than 20 different cetaceans (Ballance 2009), including much larger species such as the humpback whale (Rossi-Santos et al. 2009). Spotted and spinner dolphins occur regularly in mixed schools, especially in the eastern tropical Pacific. Risso's, Pacific white-sided (*Lagenorhynchus obliquidens*), and northern right whale (*Lissodelphis borealis*) dolphins also commonly occur in association (Ballance 2009). Somewhat surprisingly, associations between dolphins and their natural cetacean predators may also take place. For instance, rough-toothed dolphins and bottlenose dolphins sometimes associated with false killer whales off the Hawaiian Islands (Baird et al. 2008).

Mixed species groups also occur in the Gulf of Mexico, but the percentage of such groups in the northern portion of the Gulf has reportedly been low (Maze-Foley and Mullin 2006). For instance, of 736 cetacean groups observed between 1992 and 1998, only 9 (1.4 %) were of mixed species (Mullin and Hoggard 2000; Mullin et al. 2004). No pantropical spotted and spinner dolphin mixed species associations were documented for the northern Gulf, despite the fact that this is a common occurrence in other ocean basins.

Association with birds and fishes is also not as common in the Gulf as it is in some other areas of the world. This may be in part because much of the information has been gathered on deep water cetacean species that tend to feed nocturnally on DSL-related organisms such as

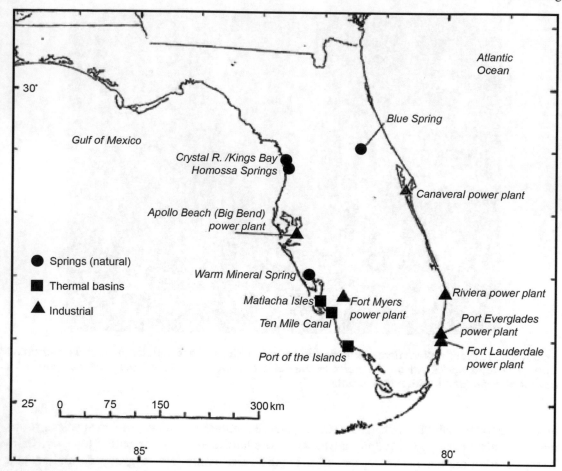

Figure 13.59. Warm-water refugia of Florida manatees (modified drawing by Adella Edwards from Marsh et al. 2011).

myctophids and squid. Nevertheless, Maze-Foley and Foley (2006) summarized the available data and found that about 2.4 % of their cetacean sightings in water deeper than 200 m (656 ft) were associated with birds and 1.1 % were with birds and fish, including surface-dwelling tunas. The most abundant of the oceanic birds in the northern Gulf are terns, the smaller shearwaters, and storm petrels (Hess and Ribic 2000), and while these make up most of the cetacean/bird associations, these associations are also not as prevalent in the Gulf as in much of the Pacific, Atlantic, and Indian oceans (Maze-Foley and Mullin 2006). While it is not certain why this is so, perhaps cetacean feeding is less often at or near the surface in the Gulf and therefore does not as often attract marine birds.

13.3.4   Cetacean Occurrences Relative to Oceanographic Features

The Gulf of Mexico consists of about 1.5 million km² (579,153 mi²) in waters of the United States, Mexico (whose borders incorporate about 65 % of the Gulf), and the western coast of Cuba. While almost totally enclosed, the Gulf is open to the Caribbean through the Yucatán Channel (about 2,000 m [6,562 ft] deep) and the shallower (about 800 m [2,625 ft] deep) Florida Straits. The Gulf has extensive continental shelf areas less than 180 m (591 ft) deep, with the shelf about 160–240 km (99–149 mi) wide off central and southern Texas, along the Florida west coast, and the Campeche Bank north of the Yucatán Peninsula, covering about 35 % of the Gulf. However, this shelf is only about 32–48 km (20–30 mi) wide south of the Mississippi River and is very narrow off Tampico, Mexico. The intermediate area between slope and abyss, at depths of 180–3,000 m (590–9,843 ft), covers about 40 % of the Gulf. The abyssal plain, at depths greater than 3,000 m (9,843 ft)—Sigsbee Plain in the west and parts of the Lower Mississippi Fan—makes up the remaining 25 %. Since cetacean habitats are in large part influenced by basic depth characteristics, it is no surprise that bottlenose dolphins are the only cetaceans normally seen off the shallow coast of Texas, out to many dozens and even more than 100 km (62 mi) from shore, while a host of more deepwater species, including sperm whales, are found within tens of kilometers of shore south of Louisiana (Davis and Fargion 1996).

Warm water from the Caribbean Sea flows into the Gulf through the Yucatán Channel, forms a Loop Current in the mid-eastern part of the Gulf, and flows out of the Gulf through the Florida Straits into the Atlantic Ocean. In the central and western Gulf, warm anticyclonic eddies that have shed off the Loop Current move slowly towards the west with adjacent cold cyclonic eddies (Sturges and Leben 2000). Upwelling cold eddies and interfaces between cold and warm eddies are areas where elevated chlorophyll levels and higher productivity occur, such as in estimated mean biomass, EMB, for example, Wormuth et al. (2000). These areas provide for a more rich, near-surface flora and fauna than outside of these zones, and therefore at least some cetaceans are attracted to them. Also, the Mississippi/Atchafalaya River complex and other rivers provide much nutrient-rich fresh water to the Gulf, draining about two-thirds of the continental U.S. watershed and one-half that of Mexico. Besides nutrients, associated land-runoff pollutants and sediments also influence the slope waters in the northern Gulf. River discharge is quite seasonal—highest flows March–May, and the lowest flows August–October—and provides rich shelf areas for spawning and juvenile fishes. For example, much Gulf menhaden (*Brevoortia patronus*) spawning occurs off the Mississippi Delta (Christmas and Waller 1975).

While a thorough analysis of physical and biological features of the northern Gulf relative to primary/upper level productivity is beyond the scope of this chapter, some interesting comparisons can be made of the dynamic nature of oceanography as related to cetacean occurrence patterns. For that discussion, Davis et al. (2002) separate (1) sperm whales as squid feeders; (2) smaller squid feeders (i.e., dwarf and pygmy sperm whales, melon-headed whales, pilot whales, pygmy killer whales, Risso's dolphins, rough-toothed dolphins, and beaked whales of the taxonomic family Ziphiidae); (3) oceanic dolphins of the genus Stenella (i.e., Clymene, pantropical spotted, spinner, and striped dolphins); and (4) upper slope bottlenose and Atlantic spotted dolphins. A final grouping is that of the nearshore and inshore living bottlenose dolphins and West Indian manatees. Killer whales were not seen during many surveys used to generate this information, but they are known to travel widely through shelf and deeper waters (Ortega-Ortiz 2002).

Sperm whale sightings in the 1990s Gulfcet studies (Davis et al. 1998, 2002) and in subsequent work (for example, Ortega-Ortiz et al. 2012) show that they are consistently present in lower slope waters south and west-south-west of the Mississippi River outfall, in a mean depth of 1,580 m (5,184 ft), and with high EMB, in colder gyre and interface cyclonic/anticyclonic waters. It is likely that the river outflows from Louisiana are especially important given that productivity is high directly south of the Delta. Although sperm whales are present in these lower slope waters year-round, they move according to productivity as measured by remote sensing, with an average time lag of about 2 weeks from primary productivity to sperm whale presence (O'Hern and Biggs 2009). Sperm whales have been in this lower slope area for a long time, as indicated by whaling records of the 1700s and 1800s (Reeves et al. 2011).

Squid feeders of the slope and deeper parts of the Gulf tend to be associated with higher salinity (less riverine influence) waters over waters of the lower, deeper slope, and in conjunction with cold-core (cyclonic) eddies or confluence zones but not in the warm anticyclonic areas.

Oceanic *Stenella* dolphins also occur most often in cold-core cyclonic rather than anticyclonic zones. They do not generally occur on the upper slope or in the abyssal, greater than 3,000 m (9,843 ft) deep zone. The pantropical spotted dolphin is the most numerous of any cetacean in the Gulf of Mexico and presents a striking feature of large group size—often more than 100 dolphins per group—and general abundance.

The more important use of cold-core rather than warm-core eddies/gyres can be found in numerous examples of data presented by Davis et al. (2002) and Ortega-Ortiz (2002) as well as subsequent studies (O'Hern and Biggs 2009). One snapshot of this general situation for the northwest central Gulf is presented in Figure 13.60.

Atlantic spotted dolphins and the offshore ecotypes of the bottlenose dolphin prefer upper-slope, continental shelf waters, and they frequent waters with a mean depth of about 200 m (656 ft). Bottlenose dolphins of nearshore and inshore areas form a separate population grouping and prefer productive, river outflow-influenced waters less than 20 m (66 ft) deep. Besides manatees (largely of west Florida), bottlenose dolphins are the only marine mammal

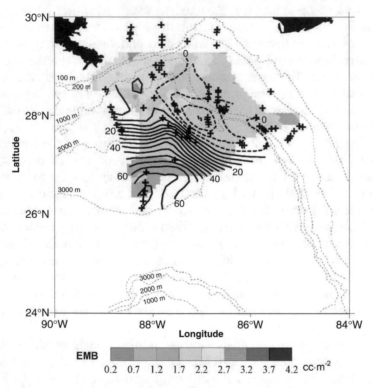

Figure 13.60. Cetacean (all identified species, summarized) sightings (+) during the mid-summer 1997 Gulfcet cruise. *Thin dashed lines* represent isobaths. *Bold solid* (positive) and *bold dashed* (negative) lines are sea surface dynamic height anomaly (DHA), relative to the 105-cm (41-in.) mean. Contour intervals are 5 cm (2 in.). The cyclone is between 0 and 13 cm (0 and 5 in.), the confluence between 0 and 24 cm (0 and 9 in.), and the anticyclone between 25 and 71 cm (9 and 28 in.). The color contours indicate nighttime EMB—estimated mean biomass (cm³m⁻²) (see scale above to the right) in the interval 10–50 m (33–164 ft) (data and figure legend from Davis et al. 2002).

species to occur in very shallow waters and the only one to do so in the western Gulf. Bottlenose dolphins outside of bays and inlets and within the about 20 m (66 ft) depth contour also engage in a partial migration, tending to move further north in summer and further south in winter. However, these movements are not likely to be much greater than about 100 km (62 mi) in most cases and are presently poorly understood (Shane et al. 1986).

Manatees prefer riverine and shallow nearshore waters where temperatures are above 17 °C (63 °F) and where seagrasses, water hyacinth, and aquatic weeds are abundant (Marsh et al. 2011). Along the western Florida coast, especially north of Warm Mineral Springs (Figure 13.59), warm-water refugia provided by waste cooling water from power plants are of special importance.

13.4 ANTHROPOGENIC IMPACTS

After a summary of the effects of physical and biological factors that influence numbers and habitat use of marine mammals in the Gulf, it is appropriate to consider human influences (i.e., anthropogenic impacts). Major anthropogenic impacts with the potential to affect cetacean and sirenian individuals and populations defined below have been modified and are based on Bearzi et al. (2011). Not all have been documented for the Gulf of Mexico. Data for each are summarized in this section:

- *Prey depletion*—Depletion of food resources caused directly or indirectly by fishing
- *Incidental mortality and injury in fisheries (bycatch)*—Mortality or injury from accidental entanglement in gear of various types including passive and active nets, longlines, traps and discarded or lost nets and lines, and illegal fishing practices (e.g., use of high explosives such as dynamite)
- *Intentional and direct takes*—Killing or capture to obtain products for animal or human consumption, live capture for public display facilities, acts of retaliation for actual or perceived damage to fish catches or gear, and shooting for sport
- *Vessel strikes*—Accidental mortality or injury from contact with a vessel, particularly the hull or propeller
- *Disturbance*—Behavioral disruption through intentional or unintentional approaches, with the potential to induce long-term effects on dolphin populations
- *Acoustic pollution (noise)*—Mortality, injury, or chronic disturbance from exposure to human-made sounds
- *Chemical contamination*—Accumulation in the body tissues, mostly through the food web, of chemicals known to adversely affect mammalian functions and health, in particular, persistent organic pollutants (POPs)
- *Ingestion of solid debris*—Mortality or injury from the ingestion of foreign objects and materials (e.g., plastic, wood, textiles) obstructing part of the digestive tract
- *Oil pollution*—Health problems or mortality deriving from contamination, contact or ingestion of hydrocarbons derived from oil spills and oil derivatives at sea
- *Ecosystem change*—Reduced habitat quality due to effects of coastal or other development (e.g., eutrophication, harmful algal blooms, alien species invasions)
- *Climate change*—Changes, potentially due to natural or human-caused climate change in, for example, prey abundance or distribution, shifts in distribution of competitors and exposure to novel diseases

13.4.1 Prey Depletion

Excessive fishing pressure and the resulting decline in fish stocks and loss of marine biodiversity is a growing concern worldwide (Pauly et al. 1998, 2002; Jackson et al. 2001; Worm et al. 2006, 2009; Swartz et al. 2010; Anticamara et al. 2011). Jackson et al. (2001) argue that "ecological extinction caused by overfishing precedes [i.e., is more important than] all other pervasive human disturbance to coastal ecosystems, including pollution, degradation of water quality, and anthropogenic climate change." Overfishing may contribute to the decline of marine mammal populations by affecting the availability of key prey (Bearzi et al. 2008). Several marine mammal populations around the world have declined rapidly, and overfishing has been suggested as one of the reasons behind their collapse (Crowder et al. 2008; Heithaus et al. 2008; Read 2008).

Pauly and Palomares (2005) analyzed landings data from fisheries in the Gulf of Mexico (as well as in the U.S. Atlantic Ocean south of Chesapeake Bay) and conclude that these regions were severely overfished and had badly degraded food webs, as evidenced by a low and declining mean trophic index, which assesses the trophic levels at which fisheries are operating. However, de Mutsert et al. (2008) subsequently point out that fisheries landings in these regions historically would be expected to have low indices because the fisheries have been dominated by menhaden and shrimp, both of which feed at low trophic levels. These authors argue that low indices derived from landings data are driven by large landings of commercial species of low trophic level, particularly Gulf menhaden and penaeid shrimp species. De Mutsert et al. (2008) also question the predictions of near-future collapses of fish populations in the Gulf of Mexico made by Worm et al. (2006), and show that—although several taxa of fish and shellfish, as well as several fisheries, have indeed collapsed in this region—the overall scenario is less dramatic than depicted in the earlier article by Worm et al. (2006). However, because the conclusions cited above are based on different datasets (i.e., the ones by the Food and Agriculture Organization of the United Nations, FAO, and NMFS), which also refer to different geographic areas, it is presently difficult to endorse either scenario.

Little information is available to permit an assessment of the effects of possible past and current impacts on marine mammals resulting from prey depletion caused by fishing in the Gulf of Mexico. A study of the potential effects of hurricane Katrina suggests that calving by bottlenose dolphins in the years following the storm may have increased for reasons including higher resource availability resulting from reduced fishing pressure, since much of the fishing fleet was destroyed by the hurricane. Calving also might have increased because many young calves were likely lost due to the hurricane, and therefore pregnancies increased afterwards. A combination of these or other factors may have been involved (Miller et al. 2010).

13.4.2 Incidental Mortality and Injury in Fisheries (Bycatch)

Fishing can deplete populations of marine mammals and other endangered megafauna, such as sharks and marine birds, in many parts of the globe through incidental bycatch in fishing gear (Lewison et al. 2004; Read 2008). Most of the following information applies primarily to bottlenose dolphins, the species that is closest to shore and has been most studied in the Gulf of Mexico. Relatively little is known even about bottlenose dolphins and the importance of impacts from fisheries; potential issues with other species are even less clear.

According to the total analyzed records from the southeastern U.S. Atlantic, including the Gulf, from 1990 to 2008, 112 (2.8 %) of the 4,029 animals stranded bore signs of fishery interaction, defined as "wounds related to fishing gear, or fishing gear attached to the animal" (Southeast U.S. Marine Mammal Stranding Network 2011). In addition to the animals reported as bearing signs of fishery interaction, a further 123 animals (3.1 %) were reported to have borne signs of human interaction, defined as "ingested plastic, debris entanglement, wounds from other weapons besides firearms (arrows, harpoons, etc.)." Side notes and specifications added to the records show that a large part of these human interactions also included signs of fishery bycatch: entanglement, amputations, cuts, and other signs that are often related to bycatch, but that at times may also be due to vandalism after stranding (Kuiken 1996; Read and Murray 2000).

Because of body decomposition at the time of inspections and other difficulties implicit in such assessments as the one described above (e.g., parts of the body not visible during the inspection, lack of necessary expertise, etc.), the numbers reported should be considered a minimum indicative estimate of the occurrence of fishery interaction among stranded animals. For example, the Southeast U.S. Marine Mammal Stranding Network (2011) reports that in 2,949 of 4,029 records, the occurrence of human interaction could not be determined due to decomposition or other problems. Many of these stranded animals are likely to have succumbed due to fisheries or other human-related interactions, but exact numbers are unknown.

Wells and Scott (1994) found that of 146 bottlenose dolphins handled during scientific catch and release live captures, about 11 % showed signs of previous gear entanglement (rope cuts, marks, etc.). However, evidence of mortality in the Sarasota-Bradenton area of west Florida was extremely low. A detailed review of their study animals yielded the conclusions that while many dolphins survive human interactions, swallowing of hooks and body constrictions by lines more often led to mortality. However, no clear numbers were available (Wells et al. 1998, 2008). Garrison (2007) describes incidents of pilot whales and Risso's dolphins becoming entangled in pelagic longlines and being released by fishermen with lines and hooks still embedded, with the supposition that many of these animals were subject to eventual mortality due to the line and hook interactions.

There is also a scarcity of information in the southern Gulf, with sporadic reports of confirmed bycatch problems. For example, of 15 records of stranded cetaceans inspected on the Veracruz coast, southwestern Gulf of Mexico, two pygmy sperm whales and one bottlenose dolphin died as a result of entanglement in gillnets (Ortega-Argueta et al. 2005). Vidal et al. (1994) surmise from limited data that especially bottlenose, Atlantic spotted, and spinner dolphins are at risk of entanglement due to gillnets in the southern Gulf. Bycatch of bottlenose dolphins is reported to be "practically zero" by the Cuban CITES administrative authority but is likely to occur in unknown numbers (Van Waerebeek et al. 2006).

Commercial fisheries that may interact with bottlenose dolphins are shrimp trawling, blue crab trap/pot fishing, stone crab trap/pot fishing, menhaden purse seining, gillnetting, and shark bottom longline fishing (Waring et al. 2010). Lack of (complete) observer program data for some of these fisheries means that the information reported below, which is limited to bottlenose dolphins, should be treated as indicative.

- *Shark Bottom Longline Fishery*—Three interactions with bottlenose dolphins have been recorded since the fishery started being observed in 1994: one mortality in 2003 and two hooked animals in 1999 and 2002 (Burgess and Morgan 2003a, b). No interactions were observed between 2004 and 2008 (Hale and Carlson 2007; Hale et al. 2007, 2009; Richards 2007). Bottlenose dolphin mortalities were estimated at 58 (CV 0.99) for 2003, but none for 2004–2008 (Richards 2007).

- *Shrimp Trawl Fishery*—Information recorded since 1992 shows that a few dozen animals have died in this fishing gear or have been caught in turtle excluder device or trawl line (Waring et al. 2010).

- *Blue and Stone Crab Trap/Pot Fisheries*—A few stranded bottlenose dolphins had polypropylene rope around their flukes (Waring et al. 2010) suggesting possible entanglement with crab pot lines.

- *Menhaden Purse Seine Fishery*—Bottlenose dolphins have died incidentally in this fishery (Reynolds 1985) with numerous self-reported kills in northern Gulf coastal and estuarine waters from the 1970s to the 1990s. The fishery was observed to take (in this sense, take means caught, including animals released) nine bottlenose dolphins, with three killed, between 1992 and 1995. Extrapolation of takes from 1992 through 1995 that considered the total number of sets indicates that up to about 172 bottlenose dolphins could have been harmed and up to 57 animals could have been killed by menhaden purse seining (Waring et al. 2010).

- *Gillnet Fishery*—Stranding data for this fishery suggest that there is probably a low frequency of takes that occur. For example, five research-related gillnet mortalities were documented between 2003 and 2008 in Texas and Louisiana (Waring et al. 2010). This is suggestive of a potential for incidental mortality in this fishery.

13.4.3 Intentional and Unintentional Direct Takes

The capture of animals from a wild population removes them from that population, and in terms of recruitment, population dynamics, and conservation value, they are effectively dead. This loss is exacerbated if certain animals are preferentially removed, such as young females, as is often the case (Van Waerebeek et al. 2006).

Since captures ceased in U.S. waters in 1989, intentional takes of bottlenose dolphins off the northeast coast of Cuba are an isolated case of removal in the Gulf of Mexico area. Nevertheless, social group effects could still be present in nature, as these dolphins can live more than 50 years (Urian et al. 2009). Bottlenose dolphins have been targets of live-capture fishery off Cuba since at least 1982. Removals occur off Sabana-Camagüey Archipelago, and 238 animals were exported from Cuba between 1986 and 2004 for the global captive dolphin industry. Twenty-eight animals were captured in 2002 alone (Van Waerebeek et al. 2006); this may be an ongoing serious problem, since global demand for aquarium dolphins is also increasing over time (Fisher and Reeves 2005). Van Waerebeek et al. (2006) recommend that the Cuban live trade in bottlenose dolphins cease until evidence of no detriment can be substantiated, but numbers within communities/populations have not been well documented, with the regional exception of Pérez-Cao's (2004) master's thesis regarding northeast Cuban waters.

Until such takes stopped in 1989, bottlenose dolphins were live-captured from several northern Gulf bays and sounds, to supply the U.S. Navy and aquarium trade (Waring et al. 2010). Between 1972 and 1989, 490 dolphins, at an average of 29 per year, were taken from several bays in the north-central Gulf as well as Tampa Bay, Charlotte Harbor to the south, and the Florida Keys. Many captures occurred in Mississippi Sound, with 202 dolphins taken. Of the dolphins captured from 1982 to 1988, 73 % were females (Waring et al. 2010), and because these animals are long-lived and slow in reproducing, population and social effects from those removals could still be present to this day.

Intentional and direct takes, such as purposeful wounding or killing of animals that are perceived to be in conflict with fisheries activities, may also take place by illegal means. Before 1988, fishermen were permitted to use almost any method, including lethal means, to protect their gear and catch. But in 1988, Congress amended the Marine Mammal Protection Act (MMPA) to forbid the lethal taking of cetaceans. Despite their protected status, cetaceans are still being shot. The Southeast U.S. Marine Mammal Stranding Network (2011) indicates that at least some strandings are of bottlenose dolphins, and in one instance, a dwarf sperm whale had been shot or otherwise wounded or killed. While this kind of interaction is presumably infrequent, actual numbers are not known. Fishermen occasionally kill or harm dolphins in retaliation for depredation of recreational and commercial fishing gear. Three cases were documented between 2006 and 2008 (Waring et al. 2010).

Two dolphin research-related mortalities occurred in western Florida in 2002 and 2006. Four others resulted from entanglements in gillnet fisheries research gear off Louisiana and Texas from 2003 through 2007. Five incidents—four of which were mortalities—involving bottlenose dolphins occurred during sea turtle relocation trawling activities by the Army Corps of Engineers. Overall, intentional or otherwise unintentional but direct takes occur now and then. However, they are not perceived to be a large problem that would endanger populations or species in the Gulf of Mexico.

13.4.4 Vessel Strikes

Vessel strikes can injure or kill a variety of marine mammals including large whales (Laist et al. 2001; Panigada et al. 2006) and dolphins (Wells and Scott 1997; Van Waerebeek et al. 2007; Wells et al. 2008). In the coastal waters of the Gulf of Mexico, marine mammals share habitat with large and increasing numbers of boats. Bottlenose dolphins suffer boat-related injuries (e.g., Wells and Scott 1997), and injuries and deaths may be high in areas of high boat traffic. Of 637 total bottlenose dolphin strandings during 2004–2008, seven showed signs of boat collision (Waring et al. 2010). Bottlenose dolphins often survive propeller strikes if these involve only soft tissue and not bone (Wells et al. 2008). Extrapolation from studies of bottlenose dolphins to other dolphin species must be done with caution, however, and Wells et al. (2008) recommend against it, especially since the very nearshore interactions tend to be with small pleasure craft, whereas offshore interactions are more likely to involve medium-size and large vessels.

A recent study by Azzara (2012) compared the known occurrence of sperm whales off New Orleans with the pattern of major shipping lanes related to the ports of New Orleans and Houston and found a high overlap between critical sperm whale habitat and vessel traffic. Because extremely high shipping traffic occurs in the areas that are most often used by sperm whales, the potential for collision exists, but these have not been reported. It is known that ship strikes increase mortalities with vessels longer than 80 m (262 ft) and traveling at speeds greater than 26 km/h or 14 knots (kts; 16 mi/h) (Laist et al. 2001).

Most living Florida manatees bear scars from vessel collisions, and it is estimated that about 30 % of all mortalities in U.S. waters are caused by such collisions (Wright et al. 1995). Manatees are generally aware of approaching vessels and attempt to evade the vessel by orienting into deeper waters of channels, and by swimming faster. However, since a rapid boat is likely to be in deepest water close to the center of channels, this behavior does not always decrease the collision potential (Nowacek et al. 2004). It is well established that slower boat speeds in manatee habitat can greatly reduce injuries and mortalities, and some progress has been made on stricter regulations, enforcement, and public awareness in recent years (Marsh et al. 2011).

13.4.5 Disturbance and Acoustic Pollution (Noise)

Sound is probably the most important sensory modality for cetaceans. All species communicate by sound. Baleen whales communicate at great distances, with fin and blue whales communicating across many tens of kilometers (Payne and Webb 1971); and toothed-whales echolocate in sophisticated fashion (Au 1993). Prolonged direct (or physical) disturbance caused by boat traffic can affect the behavior and habitat use of cetaceans. Alterations of surfacing patterns, swimming speed, directionality, group cohesion, and group fluidity have been related to boat disturbance in a number of cetacean species—several of which live in the Gulf of Mexico. Vessel traffic can also affect habitat use, which can include displacement for periods of hours to days and longer-term avoidance of areas, and reduce reproductive success (Nowacek et al. 2007).

Some of the short-term effects of boat disturbance on coastal dolphins in the Gulf of Mexico were investigated in the waters of Sarasota Bay, where dolphin whistle rates were found to increase at the onset of vessel approaches. This could be because of heightened arousal, increased motivation of animals to come close together, or to compensate for signal masking due to noise (Buckstaff 2004). Behavioral responses including changes in grouping patterns and headings as well as increased swimming speed occurred more often during experimental vessel approaches than during control (no disturbance) periods (Nowacek et al. 2001). Another study in the Mississippi Sound found short-term changes in dolphin behavior, including a decrease in feeding behavior, following the passing of speedboats (Miller et al. 2008). The long-term effects of such disturbances and the possible impacts on populations remain unknown. However, the Sarasota Bay population is known to have remained rather stable in size over the past four decades (Wells and Scott 2009).

The Gulf of Mexico is home to two of the world's ten busiest ports by cargo volume: New Orleans and Houston. In 2008, these ports hosted a combined total of 14,000 ships. An indication of behavioral disruption in sperm whale communication caused by shipping noise was found in a recent study by Azzara (2012). Azzara used recordings from one hydrophone recording buoy situated at a depth of approximately 1,000 m (3,281 ft), close by a junction of both the New Orleans and Houston waterways for major shipping. The study found significant differences in sperm whale vocalization patterns before, during, and after the passing of a ship. While this study does not show that shipping noise can negatively affect survival or reproduction, the findings are suggestive of potentially important behavioral alterations that can alter communication and foraging success and potentially cause stress. Industrial noises have been shown to change calls of several species of whales; a recent example is of blue whales (Melcón et al. 2012). Physiological stress (measured from stress hormones contained in scat samples) has been indicated relative to shipping noise and North Atlantic right whales off the U.S. Atlantic seaboard (Rolland et al. 2012).

The oceans have become much louder during the industrial (propeller-driven) age than before, especially in the lower frequencies to which baleen whales are most sensitive (McCarthy 2004). Of these noises, industrial seismic airguns for oil and gas exploration and military sonars are especially loud, and have increased ocean noise in several areas. An experimental study investigating the impact of airguns on sperm whales suggested that sperm whales in a highly exposed area of the Gulf of Mexico may not exhibit avoidance reactions to airguns, but the animals may be affected at ranges well beyond those currently regulated due to more subtle effects on their foraging behaviors (Miller et al. 2009). The high intensity of Navy-produced surveillance echolocation pings—low frequency active sonar (commonly referred to as LFA)—is known to cause death in especially deep-diving cetaceans such as some beaked whales and pilot whales and are therefore also a worry for the Gulf of Mexico (Southall et al. 2007).

A special form of disturbance is the intentional feeding of dolphins for purposes of tourism. Feeding wild dolphins is considered a form of *take* under the MMPA, since it changes natural behavior and has the potential for increasing injury or death or even creating dependency. Frequent provisioning was observed near Panama City Beach (Samuels and Bejder 2004), south of Sarasota Bay (Cunningham-Smith et al. 2006; Powell and Wells 2011), and in

Texas near Corpus Christi (Bryant 1994). Swimming with wild bottlenose dolphins—illegal under the MMPA—has been documented near Panama City Beach, and it is likely that dolphins were attracted by swimmers due to provisioning (Samuels and Bejder 2004).

13.4.6 Chemical Contamination

Toxic contaminants are a major concern in marine mammal populations because of their environmental persistence and the potential effects on reproduction and health (Gauthier et al. 1999; O'Shea et al. 1999; O'Hara and O'Shea 2001; Newman and Smith 2006). Because cetaceans are long-lived apex predators with extensive fat stores, they accumulate persistent organic pollutants (POPs)—including brominated flame retardants, dichlorodiphenyltrichloroethane (DDT) and associated compounds, polychlorinated biphenyls (PCBs), hexachlorocyclohexanes (HCHs), chlorobenzenes, and chlordanes—from lower trophic level prey (Ross 2006; Ross and Birnbaum 2003; Yordy et al. 2010). Causal links have been described between POPs exposure and immunological, endocrine, and reproductive disorders in cetaceans (Aguilar and Borrell 1994; Lahvis et al. 1995; Jepson et al. 1999; Schwacke et al. 2002; Hall et al. 2006).

Coastal stocks of bottlenose dolphins live in highly populated areas, of which some—Tampa Bay, Florida; Galveston Bay, Texas; and Mobile Bay, Alabama, for example—have much industry. Around the periphery of Galveston Bay, more than 50 % of all U.S. chemical products are manufactured and about 17 % of oil produced in the Gulf of Mexico is refined (Henningsen and Würsig 1991). Concentrations of anthropogenic chemicals and their metabolites vary from site to site and can reach levels of concern in bottlenose dolphins (Schwacke et al. 2002). Most studies conducted in the Gulf of Mexico focus on common bottlenose dolphins, particularly in western Florida (Wells et al. 2004). Levels of POPs in dolphins sampled in the northwestern Gulf of Mexico are known to be relatively high (Kucklick et al. 2011), variable by sex and age class, and have negative effects on health and reproduction. Similar levels of POPs have been found in melon-headed whales that stranded in the Gulf of Mexico (Davis 1993).

The high POPs burden carried by bottlenose dolphins in the Gulf of Mexico may increase susceptibility to parasitic microorganisms (Kuehl and Haebler 1995) and suppress immune function and increase recovery time postinfection in marine mammals (Kendall et al. 1992). There is a probable relationship between high concentrations of organochlorines (such as PCB and DDT metabolites) in the blood and male immune dysfunction (Lahvis et al. 1995). Increasing PCB concentrations may induce vitamin A deficiencies in marine mammals that can lead to reproductive disorders and susceptibility to infection (Brouwer et al. 1989).

In Sarasota Bay, Florida, Wells et al. (2001, 2003) found that transfer of organochlorines through placental and mother's milk was implicated in mortality of first-borns to (generally) young mothers (Vedder 1996; Wells et al. 2004). First-born bottlenose dolphin calves are estimated to accumulate approximately 80 % of their mother's organochlorines during the first 7 weeks of lactation (Cockcroft et al. 1989). Dead young dolphins had high PCB and chlorinated pesticide concentrations; these concentrations increased as males aged but declined to lower concentrations as females reared offspring (Küss 1998; Wells et al. 2004). Schwacke et al. (2010) and Wells et al. (2003, 2005) described a similar pattern based on samples of blubber and blood.

Organochlorines were higher in females whose calves died within the first 6 months compared to females whose calves survived (Reddy et al. 2001). POPs concentrations in adult female bottlenose dolphins in Sarasota, Florida, are generally greater in blubber than milk; however, there is congener-specific variation in mobilization of POPs from blubber to milk (Yordy et al. 2010). Deceased suckling calves collected throughout the Gulf of Mexico had nearly ten times higher concentrations of triphenylphosphate (TTP) in their blubber than adult males (Kuehl and Haebler 1995). Young suckling dolphins may be at higher risk for POPs-related health effects as their bodies undergo rapid development.

Based on probabilistic risk assessment, bottlenose dolphins sampled off Sarasota, Florida, and Matagorda Bay, Texas, indicated a high likelihood that reproductive success—primarily in primiparous females—is severely impaired by chronic exposure to PCBs. Excess risk of reproductive failure for primiparous females, measured in terms of stillbirth or neonatal mortality, was estimated as 79 % for the Sarasota sample, and 78 % for the Matagorda Bay sample (Schwacke et al. 2002). High levels of infertility were also found among common dolphins with the highest PCB burdens and most ovarian scars, which suggest that ovulation was occurring without the reproduction of a viable calf (Murphy et al. 2010). High POPs burdens also correlated with few ovarian scars in harbor porpoises, suggesting inhibition of ovulation cycles (Murphy et al. 2010).

Levels of trace elements may also be high in the Gulf of Mexico and increase with age class (e.g. Kuehl and Haebler 1995; Bryan et al. 2007). While the link between the concentration of these elements and cetacean population health and status needs to be further clarified, Rawson et al. (1993, 1995) related indicators of liver disease to high mercury concentrations in stranded dolphins from western Florida.

The findings summarized here for cetaceans highlight the importance of considering indirect anthropogenic stressors such as contaminant pollution in U.S. management schemes (Wells et al. 2005). Fine-scale spatial variation in POPs suggests that individual patterns of habitat usage can influence individual toxin burden profiles (Litz et al. 2007; Pulster and Maruya 2008). Concentrations of DDT decreased in bottlenose dolphins in the Gulf of Mexico following a ban on commercial use of DDT in the United States (Salata et al. 1995), suggesting that dolphins can recover from high contamination loads if exposure decreases.

Manatees tend to have low levels of organochlorine residues in their bodies, possibly because of their low position in the food web, the lowered recent pesticide levels in Gulf waters than during earlier studies such as by O'Shea et al. (1984), and possibly a more effective mechanism for metabolizing toxic compounds than is possessed by most terrestrial mammals (Ames and Van Fleet 1996).

13.4.7 Ingestion of Solid Debris

The world's oceans contain much plastic debris (Wolfe 1987; Laist et al. 1999; Derraik 2002) and obstruction of the alimentary canal or stomach/intestine due to ingested plastic is a known cause of cetacean mortality (e.g. Tarpley and Marwitz 1993). Much of plastic ingestion is probably due to investigating and testing items in the environment, especially for young calves recently weaned, although starvation or disease could also exacerbate such inappropriate attempts at feeding (Kastelein and Lavaleije 1992; Baird and Hooker 2000; Poncelet et al. 2000).

In 2006, three well-known dolphins of Sarasota Bay died from ingesting fishing gear, an unprecedented mortality level (Powell and Wells 2011). Additional scattered information for the Gulf of Mexico comes from the Southeast U.S. Marine Mammal Stranding Network (2011), where several animals are reported to have ingested plastic and other debris. The threat is generally considered as relatively minor, but potentially important for some teuthophagous cetacean species, such as Risso's dolphins (Bearzi et al. 2011).

13.4.8 Oil Pollution

Oil pollution can impact many parts of an ecosystem, from primary production to fishes and of course birds, sea turtles, and marine mammals (Loughlin 1994). Oil fouling of furred marine mammals has been studied most and is thought to be most dangerous, especially in sea otters that rely almost exclusively on aerated fur for thermoregulation, often in very cold environments. Williams and Davis (1995) provide a quite thorough review; however, no furred marine mammals exist in the Gulf of Mexico, so this is not a problem there.

Cetaceans feed at various trophic levels. The balaenid whales, such as North Atlantic right whales, feed on tiny calanoid copepods; blue whales feed on somewhat larger euphausiid crustaceans; other baleen whales (such as the Bryde's whales of the Gulf) feed largely on fishes; some odontocetes feed on small to medium fishes (bottlenose dolphins of nearshore waters); and other odontocetes feed on fishes and squid in the deeper ocean. A host of benthic organisms concentrate petroleum hydrocarbon residues in their tissues, but most mid and surface water crustaceans and teleost fishes metabolize and excrete them rapidly, and thus do not tend to become heavily (or for a long time) contaminated (Neff 1990). The cytochrome P450 system can be used as a biomarker to indicate exposure to polycyclic aromatic hydrocarbons (PAHs), as well as to a variety of other chemicals and stressors. However, this biomarker does not show a strong link to toxicity, lesions, or reproductive failure (Lee and Anderson 2005). A major problem for baleen whales that feed on the surface or gray whales (*Eschrichtius robustus*) that feed largely on in-benthic fauna is baleen fouling; feeding in oil slicks is therefore detrimental to these filter feeders (Geraci 1990). However, neither surface- nor bottom-foraging baleen whales regularly occur in the Gulf, and therefore, baleen fouling is not likely to be a problem there.

The Gulf of Mexico has many natural oil seeps, so it is likely that cetaceans are quite used to dealing with them. Natural seeps have been occurring not only throughout the evolutionary history of cetaceans but also before. Approximately 47 % of crude oil that is introduced into the marine environment occurs via natural seeps (Kvenvolden and Cooper 2003); however, these estimates involve broad extrapolations based on little data (NRC 2003). The Gulf of Mexico has more than 600 natural oil seeps, and one recent study identified at least 164 of these in the northern Gulf of Mexico (NASA 2000; Hu et al. 2009). Natural oil seeps have established meiofaunal communities comprised of deposit-feeding taxa capable of handling toxic environments with low oxygen levels (Steichen et al. 1996). Nevertheless, a major oil spill presents a larger and more intensive footprint than natural seeps, and potential large-scale deaths of cetacean prey may have strong, but at this time unknown, detrimental effects. Bottlenose dolphins studied behaviorally during the Mega Borg oil spill off Galveston, Texas, in 1990 showed no ability to travel out of volatile new oil. They showed some behavioral reactions to surface oil while in it, but no deaths or other signs of long-term damage were evident (Smultea and Würsig 1995). There are no data to indicate that oil spills to early 2010 in the Gulf of Mexico substantially impacted cetaceans.

13.4.9 Unusual Mortality Events

Several major unusual mortality events (UMEs) occurring in recent years have raised concern for the health of bottlenose dolphin populations along the coasts of the eastern and northern Gulf of Mexico, particularly in Florida and Texas. Since 1991 and through March 2010, 13 bottlenose dolphin and seven manatee UMEs were declared in the Gulf of Mexico. A more recent UME (January–March 2010) in the northern Gulf included a combination of bottlenose dolphins, Atlantic spotted dolphins, and pygmy and dwarf sperm whales (Table 13.24).

Table 13.24. Gulf of Mexico Unusual Mortality Events by Year (1991–2010; months not specified). See Tables B.1 and B.2 in Appendix B for Recent-Year Details.

Year	Species	Location	Cause (Category)
1991	Bottlenose dolphins	Florida (Sarasota)	Undetermined
1992	Bottlenose dolphins	Texas	Undetermined
1994	Bottlenose dolphins	Texas	Infectious disease
1996	Bottlenose dolphins	Mississippi	Undetermined
1996	Manatees	Florida (west coast)	Biotoxin
1999–2000	Bottlenose dolphins	Florida (Panhandle)	Biotoxin
2001	Bottlenose dolphins	Florida (Indian River)	Undetermined
2002	Manatees	Florida (west coast)	Biotoxin
2003	Manatees	Florida (west coast)	Biotoxin
2004	Bottlenose dolphins	Florida (Panhandle)	Biotoxin
2005–2006	Bottlenose dolphins	Florida (Panhandle)	Biotoxin
2005–2006	Multispecies (Manatees, bottlenose dolphins)	Florida (west coast)	Biotoxin
2006	Manatees	Florida (Everglads)	Biotoxin
2007	Manatees	Florida (SW)	Biotoxin
2007	Bottlenose dolphins	Texas and Louisiana	Undetermined
2008	Bottlenose dolphins	Florida (Indian River)	Undetermined
2008	Bottlenose dolphins	Texas	Undetermined
2010	Cetaceans	Northern Gulf of Mexico	Undetermined
2010	Bottlenose dolphins	Florida (St. John's River)	Undetermined
2010	Manatees	Florida	Ecological factors

Source: http://www.nmfs.noaa.gov/pr/health/mmume/

Mortality events can occur as a result of algal blooms and red tides involving the release of neurotoxins into the marine food web. For instance, a bloom of the dinoflagellate (*Karenia brevis*) occurred in western Florida in 2004, which resulted in the death of 107 bottlenose dolphins and 34 Florida manatees (Waring et al. 2010; Flewelling et al. 2005). A similar major UME occurred in 2008 in the coastal waters of Texas, resulting in over 100 bottlenose dolphin deaths. This second mortality event overlapped spatially and temporally with an algal bloom of the toxin-producing *Dinophysis* spp. and *Prorocentrum* spp. (Fire et al. 2011). Fish and seagrass can accumulate high concentrations of the brevetoxins produced by dinoflagellates and act as toxin vectors to marine mammals (Flewelling et al. 2005; Fire et al. 2007, 2008, 2011).

Red tide blooms occur in the Gulf of Mexico on a nearly annual basis. A recent study showed that bottlenose dolphins in western Florida were consistently exposed to brevetoxin and/or domoic acid over a 10-year study period (2000–2009), and 36 % of all animals tested positive for brevetoxin ($n = 118$) and 53 % tested positive for domoic acid ($n = 83$) (Twiner et al. 2011).

Harmful algal blooms are also known to alter dolphin behavior. A study in the coastal waters of western Florida found that bottlenose dolphins displayed a suite of behavioral changes associated with red tide blooms, including altered activity budgets, increased sociality, and expanded ranging behavior. These behavioral changes may result in more widespread population impacts and increased susceptibility to disease outbreaks. While the mechanisms behind red tide-associated behavioral effects are not well understood, they are most likely linked to changes in resource availability and distribution (McHugh et al. 2010).

13.4.10   Climate Change

Climate change may cause large-scale and long-lasting changes in physical and biological systems (Pollack et al. 1998; Barnett et al. 2001; Peñuelas et al. 2002; Parmesan and Yohe 2003; Díaz-Almela et al. 2007; IPCC 2007). Climate change effects on cetaceans are at present largely unknown, but it is believed that effects would be mediated mainly through alteration of prey distribution and abundance and could shift cetacean habitat use, foraging strategies and grouping patterns (Lusseau et al. 2004; Learmonth et al. 2006; MacLeod 2009; Simmonds and Eliott 2009). An increase of carbon dioxide levels that results in ocean acidification has perhaps the most important effect on marine biodiversity, but it is not understood how or in what timeframe (Orr et al. 2005; Whitehead et al. 2008).

Of the several (and for the most part poorly understood) ways in which climate change may negatively affect marine mammal populations, perhaps most relevant to the Gulf of Mexico is a potential link between climate change and hurricane occurrence. The occurrence of hurricanes in the Gulf of Mexico is known to have increased over time, and this increase has been suggested to be one of the consequences of global warming (Emanuel 2005). While the impacts of hurricanes on marine mammals are poorly understood, they may range from death, injury, short-term displacements, habitat degradation, changes in prey occurrence, and health effects (medium and long term). For instance, hurricanes Rita, Katrina and Wilma reportedly resulted in the displacement of seven bottlenose dolphins into inland areas where they do not normally occur (Rosel and Watts 2008), and Florida manatees had lower survival probabilities in years with intense coastal storms (Langtimm and Beck 2003). However, hurricanes might also benefit marine mammal populations in ways that may balance the negative effects (e.g., the dramatically reduced fishing effort following hurricanes may increase prey availability) (Miller et al. 2010). An increase of bottlenose dolphin calves was recorded approximately 1.5 years after hurricane Katrina. This increase has been suggested to have been both a consequence of the decreased fishing effort by humans and the increase in the number of females that lost calves, and hence became reproductively receptive following the storm (Miller et al. 2010).

13.4.11   Strandings

Most strandings of single cetaceans in the Gulf are by bottlenose dolphins. It is not always possible to tell whether the live or dead animals come from the bay/estuary, nearshore, or offshore stocks. Most strandings are for undetermined reasons, although emaciation, disease, possible cold exposure, and other reasons not necessarily related to human interactions are documented (Southeast U.S. Marine Mammal Stranding Network 2011). Note in Table 13.25 below that the overwhelming number of single strandings did not have a determination of cause of death, due to decomposition, vague or absent indicators, or inexperience of personnel.

The number of total stranded marine mammals (1,392) does not equal the sum of the five categories listed above (totaling 1,475), because a number of strandings were included in more than one category (e.g., Human Interaction and Fisheries).

Table 13.25. Marine Mammal Strandings in the Gulf of Mexico (2004–2008). See Tables A.1 and A.2 in Appendix A for Recent-Year Details (from Waring et al. 2010).

Strandings	Gulf of Mexico Stocks 2004–2008				
	Eastern Coastal	Northern Coastal	Western Coastal	Inshore	Total
Total stranded	86	139	526	641	1,392
Human interaction	5	3	20	55	83
Fisheries	4	1	4	31	40
Other	1	2	16	24	43
No human interaction	18	22	113	141	294
Not determined	63	114	393	445	1,015

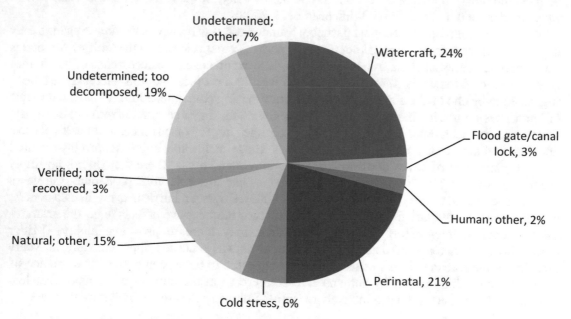

Figure 13.61. Causes of mortalities of Florida manatees (*Trichechus manatus latirostris*) (data from the Florida Fish and Wildlife Conservation Commission, 2010; drawn by Shane Blowes, and reproduced from Marsh et al. 2011).

13.4.12 Global Status and U.S. Population Trends

Manatee causes of mortality are listed as a special case in Figure 13.61. Mortalities from speeding watercraft have been the highest overall source of mortality, but there is some indication that with greater education and enforcement, these are on the wane. Cold stress due to periodic shutdown of power plants used as refuges by manatees has also decreased in recent years, due to concerted human action. Many carcasses are too decomposed for accurate assessments of cause of death, and human causes are probably larger than indicated in Figure 13.61 and Table 13.26.

Table 13.26. Marine Mammal Global Status and U.S. Population Trends for Species Found in the Gulf of Mexico.

Genus Species	Common Name	IUCN 2011* Status	IUCN 2011* POP Trend	U.S. FWS* ESA* Status	NMFS* MMPA Status	NMFS* POP Trend
Balaenoptera edeni	Bryde's whale	DD	UNK	None given	None given	Insufficient data
Physeter macrocephalus	Sperm whale	VU	UNK	Endangered	Depleted	None given
Kogia breviceps	Pygmy sperm whale	DD	UNK	None given	None given	Insufficient data
Kogia sima	Dwarf sperm whale	DD	UNK	None given	None given	Insufficient data
Ziphius cavirostris	Cuvier's beaked whale	LC	UNK	None given	None given	Gulf of Mexico (GoM) and WNA listed as "strategic"
Mesoplodon densirostris	Blainville's beaked whale	DD	UNK	None given	None given	Insufficient data
Mesoplodon europaeus	Gervais' beaked whale	DD	UNK	None given	None given	Insufficient data (NGoM listed as "strategic")
Orcinus orca	Killer whale	DD	UNK	Washington State Southern residents Endangered	Unk except: Washington State Southern residents depleted AT1 group depleted	None given
Globicephala macrorhynchus	Short-finned pilot whale	DD	UNK	None given	None given	Insufficient data
Pseudorca crassidens	False killer whale	DD	UNK	None given except: Insular Hawaii Proposed endangered	None Given	Insufficient data
Feresa attenuata	Pygmy killer whale	DD	UNK	None given	None given	Insufficient data
Peponocephala electra	Melon-headed whale	LC	UNK	None given	None given	Insufficient data
Steno bredanensis	Rough-toothed dolphin	LC	UNK	None given	None given	Insufficient data
Grampus griseus	Risso's dolphin	LC	UNK	None given	None given	Insufficient data

(continued)

Table 13.26. (continued)

Genus Species	Common Name	IUCN 2011*		U.S. FWS*	NMFS*	
		Status	POP Trend	ESA* Status	MMPA Status	POP Trend
Tursiops truncatus	Common bottlenose dolphin	LC	UNK	None given	None given except: Western North Atlantic Coastal stock Depleted	UNK
Stenella attenuata	Pantropical spotted dolphin	LC	UNK	None given	Depleted	Northeastern stock: flat Coastal stock: none given
Stenella frontalis	Atlantic spotted dolphin	DD	UNK	None given	None given	Insufficient data
Stenella longirostris	Spinner dolphin	DD	UNK	None Given	None given except: Eastern Tropical Pacific stock—Depleted	UNK
Stenella clymene	Clymene dolphin	DD	UNK	None given	None given	Insufficient data
Stenella coeruleoalba	Striped dolphin	LC	UNK	None given	None given	Insufficient data except: Western North Pacific and Mediterranean serious decline
Lagenodelphis hosei	Fraser's dolphin	LC	UNK	None given	None given	None given
Trichechus manatus	West Indian manatee	VU	Decreasing	Endangered	None given	Increasing or stable throughout much of Florida

Sources: http://www.iucnredlist.org/; http://www.nmfs.noaa.gov/pr/species/esa/mammals.htm; http://www.nmfs.noaa.gov/pr/species/mammals/cetaceans/; http://ecos.fws.gov/speciesProfile/profile/speciesProfile.

*See IUCN (International Union for Conservation of Nature, ESA (Endangered Species Act), NMFS (National Marine Fisheries Service), MMPA (Marine Mammal Protection Act) and USFWS (U.S. Fish and Wildlife Service) websites for Endangered and other definitions of status. Overall, *DD* data deficient, *LC* least concern, *VU* vulnerable, *UNK* unknown.

13.5 CONCLUSIONS: BRIEF SUMMARY OF PRESENT KNOWLEDGE

The 22 species of marine mammals of the Gulf of Mexico appear to be doing rather well, with no evidence of strong declines (or rapid increases) for any populations. Manatees and sperm whales are listed as endangered by the United States, but there is no evidence of declines in sperm whales, and manatees—while small in numbers in the Gulf—also appear now to be doing better than they were in the 1990s.

Determination of stocks (or biological populations) and numbers of marine mammals is not an exact science, except for one population of bottlenose dolphins in the Tampa Bay, Florida, area where almost each individual is known; aerial, shipboard, and acoustic censuses provide basic information on minimum numbers. However, there are broad variances (or estimates of error) depending on sighting factors, ease of acoustically identifying recorded animals, and chance of encounters with generally limited survey numbers due to costs involved. Broad variances are a common problem especially for cetacean studies in open oceans.

Much has been learned in the past 20 years about cetacean distribution in the Gulf. Bottlenose dolphins are ubiquitous in most bays that are well flushed by tides as well as along the outer shoreline. They also occur—along with Atlantic spotted dolphins—on the continental shelf in waters generally less than 200 m deep. In waters deeper than 200 m, the most commonly sighted species during multiple surveys, in descending orders of numbers, tend to be pantropical spotted, Clymene, spinner, and striped dolphins (all members of the Genus *Stenella*). Sperm whales occur generally in waters greater than 800 m deep, as do pygmy and dwarf sperm whales, pilot whales, and other members of the blackfish clade. Sperm whales occur throughout the U.S. EEZ, with a usual concentration within about 100 km south of Louisiana in an area strongly influenced by nutrient rich waters from the Mississippi River. Bryde's whales, the only common baleen whale in the Gulf, occur mainly in shallower waters (100–200 m deep) of the northeastern Gulf, largely south of Alabama and the western part of the Florida Panhandle. There is evidence that deepwater cetaceans tend to be more numerous in or near cold-core cyclones and the confluence of cyclone/anticyclone eddy pairs where primary productivity and zooplankton to higher trophic levels also tend to be more abundant than in warm-core anticyclonic eddies.

While quite a few potential anthropogenic impacts have been identified for the Florida manatee—especially cold spells and boat collisions—and cetaceans, the major impacts are probably from chemical contamination, especially POPs that are known to decrease immune responses and increase reproductive disorders in mammals in general, with building evidence for cetaceans.

REFERENCES

Aguilar A, Borrell A (1994) Abnormally high polychlorinated biphenyl levels in striped dolphins (*Stenella coeruleoalba*) affected by the 1990–1992 Mediterranean epizootic. Sci Total Environ 154:237–247

Ames AL, Van Fleet ES (1996) Organochlorine residues in the Florida manatee, *Trichechus manatus latirostris*. Mar Pollut Bull 32:374–377

Anticamara JA, Watson R, Gelchu A, Pauly D (2011) Global fishing effort (1950–2010): Trends, gaps, and implications. Fish Res 10:131–136

Au WWL (1993) The sonar of dolphins. Springer-Verlag, New York, NY, USA. 277 p

Azzara A (2012) Impacts of vessel noise perturbations on the resident sperm whale population in the Gulf of Mexico. PhD Thesis. Texas A&M University, College Station, TX, USA. 104 p

Baird RW, Hooker SK (2000) Ingestion of plastic and unusual prey by a juvenile harbour porpoise. Mar Pollut Bull 40:719–720

Baird RW, Gorgone AM, McSweeney DJ, Webster DL, Salden DR, Deakos MH, Ligon AD, Schorr GS, Barlow J, Mahaffy SD (2008) False killer whales (*Pseudorca crassidens*) around the main Hawaiian Islands: Long-term site fidelity, inter-island movements, and association patterns. Mar Mamm Sci 24:591–612

Ballance LT (2009) Cetacean ecology. In: Perrin WF, Würsig B, Thewissen JGM (eds) Encyclopedia of marine mammals, 2nd edn. Academic/Elsevier, San Diego, CA, USA, pp 196–201

Balmer BC, Wells RS, Nowacek SM, Nowacek DP, Schwacke LH, McLellan WA, Scharf FS (2008) Seasonal abundance and distribution patterns of common bottlenose dolphins (*Tursiops truncatus*) near St. Joseph Bay, Florida, USA. J Cetacean Res Manag 10:157–167

Barlow J (2003) Preliminary estimates of the abundance of Cetaceans along the U.S. West Coast: 1991–2001. Southwest Fisheries Science Center Administrative Report LJ-03-03. La Jolla, CA, USA. 31 p

Barnett TP, Pierce DW, Schnur R (2001) Detection of anthropogenic climate change in the world's oceans. Science 292:270–274

Baumann-Pickering S, Roch MA, Wiggins SM, Schnitzler H-U, Hildebrand JA (2010) Classification of echolocation signals of melon-headed whales (*Peponocephala electra*), bottlenose dolphins (*Tursiops truncatus*), and Gray's spinner dolphins (*Stenella longirostris longirostris*). J Acoust Soc Am 128:2212–2224

Baumgartner MF, Mullen KD, May LN, Leming TD (2001) Cetacean habitats in the northern Gulf of Mexico. Fish Bull 99:219–239

Bearzi G, Agazzi S, Gonzalvo J, Costa M, Bonizzoni S, Politi E, Piroddi C, Reeves RR (2008) Overfishing and the disappearance of short-beaked common dolphins from western Greece. Endang Species Res 5:1–12

Bearzi G, Reeves RR, Remonato E, Pierantonio N, Airoldi S (2011) Risso's dolphin *Grampus griseus* in the Mediterranean Sea. Mamm Biol 76:385–400

Bigg MA, Olesiuk PF, Ellis GM, Ford JKB, Balcomb KC (1990) Social organization and genealogy of resident killer whales (*Orcinus orca*) in the coastal waters of British Columbia and Washington State. In: Hammond PS, Mizroch SA, Donovan GP (eds) Individual recognition of Cetaceans: Use of photoidentification and other techniques to estimate population parameters. Report of the International Whaling Commission Cambridge Special Issue 12:383–405

Blaylock RA, Hoggard W (1994) Preliminary estimates of bottlenose dolphin abundance in southern U.S. Atlantic and Gulf of Mexico continental shelf waters. NOAA Technical Memorandum NMFS-SEFSC-356. U.S. Department of Commerce, Washington, DC, USA. 10 p

Bräger S, Würsig B, Acevedo A, Henningsen T (1994) Association patterns of bottlenose dolphins (*Tursiops truncatus*) in Galveston Bay, Texas. J Mammal 75:431–437

Brouwer A, Reijnders PJH, Koeman JH (1989) Polychlorinated biphenyl (PCB) contaminated fish induces vitamin A and thyroid hormone deficiency in the common seal *(Phoca vitulina)*. Aquat Toxicol 15:99–106

Bryan CE, Christopher SJ, Balmer BC, Wells RS (2007) Establishing baseline levels of trace elements in blood and skin of bottlenose dolphins in Sarasota Bay, Florida: Implications for noninvasive monitoring. Sci Total Environ 388:325–342

Bryant L (1994) Report to Congress on results of feeding wild dolphins: 1989–1994. NMFS (National Marine Fisheries Service), Office of Protected Resources, Washington, DC, USA. 22 p

Buckstaff KC (2004) Effects of watercraft noise on the acoustic behavior of bottlenose dolphins, *Tursiops truncatus*, in Sarasota Bay, Florida. Mar Mamm Sci 20:709–725

Burgess G, Morgan A (2003a) Commercial shark fishery observer program. Renewal of an observer program to monitor the directed commercial shark fishery in the Gulf of Mexico and South Atlantic: 1999 fishing season. Final Report. Highly Migratory Species Management Division Award NA97FF0041. U.S. National Marine Fisheries Service, Washington, DC, USA

Burgess G, Morgan A (2003b) Commercial shark fishery observer program. Renewal of an observer program to monitor the directed commercial shark fishery in the Gulf of Mexico and the South Atlantic: 2002(2) and 2003(1) fishing seasons. Final Report. Highly Migratory Species Management Division Award NA16FM0598. U.S. National Marine Fisheries Service, Washington, DC, USA

Caldwell D, Caldwell M (1973) Marine mammals of the eastern Gulf of Mexico. In: Jones JI, Ring RE, Rinkel MO, Smith RE III (eds) A summary of knowledge of the Eastern Gulf of Mexico. State University System of Florida, Gainesville, FL, USA, pp 1–3

Christmas JY, Waller RS (1975) Location and time of menhaden spawning in the Gulf of Mexico. National Marine Fisheries Report, Gulf Coast Research Laboratory, Ocean Springs, MS, USA. 20 p

Clark AH (1887) History and present condition of the fishery. In: Goode GB (ed) The fisheries and fishery industries of the United States, vol 2, Section V. History and methods of the fisheries, Part XV, The Whale fishery. U.S. Government Printing Office, Washington, DC, USA, pp 3–218

Coakes AK, Whitehead H (2004) Social structure and mating system of sperm whales off northern Chile. Can J Zool 82:1360–1369

Cockcroft VG, De Kock AC, Lord DA, Ross GJB (1989) Organochlorines in bottlenose dolphins *Tursiops truncatus* from the east coast of South Africa. S Afr J Mar Sci 8:207–217

Croll DA, Clark CW, Calambokidis J, Ellison WT, Tershy BR (2001) Effect of anthropogenic low-frequency noise on the foraging ecology of *Balaenoptera* whales. Anim Conserv 4:13–27

Crowder L, Hazen E, Avissar N, Bjorkland R, Latanich C, Ogburn M (2008) The impacts of fisheries on marine ecosystems and the transition to ecosystem-based management. Annu Rev Ecol Evol Syst 39:259–278

Cunningham-Smith P, Colbert DE, Wells RS, Speakman T (2006) Evaluation of human interactions with a provisioned wild bottlenose dolphin (*Tursiops truncatus*) near Sarasota Bay, Florida and efforts to curtail the interactions. Aquat Mamm 32:346–356

Davis JW (1993) An analysis of tissues for total PCB and planar PCB concentrations in marine mammals stranded along the Gulf of Mexico. MS Thesis. Texas A&M University, College Station, TX, USA. 139 p

Davis RW, Fargion GS (eds) (1996) Distribution and abundance of cetaceans in the north-central and western Gulf of Mexico. Final Report. MMS (Minerals Management Service), Vol 2. U.S. Department of the Interior, MMS, Gulf of Mexico OCS Region, New Orleans, LA, USA. 375 p

Davis RW, Fargion GS, May N, Leming TD, Baumgartner M, Evans WE, Hansen LJ, Mullin K (1998) Physical habitats of cetaceans along the Continental Slope in the northcentral and western Gulf of Mexico. Mar Mamm Sci 14:490–507

Davis RW, Ortega-Ortiz JG, Ribic CA, Evans WE, Biggs DC, Ressler PH, Cady RB, Leben RR, Mullin KD, Würsig B (2002) Cetacean habitat in the northern oceanic Gulf of Mexico. Deep-Sea Res Pt I 49:121–142

De Mutsert K, Cowan JH Jr, Essington TE, Hilborn R (2008) Reanalyses of Gulf of Mexico fisheries data: Landings can be misleading in assessments of fisheries and fisheries ecosystems. Proc Natl Acad Sci USA 105:2740–2744

Derraik JGB (2002) The pollution of the marine environment by plastic debris: A review. Mar Pollut Bull 44:842–852

Díaz-Almela E, Marbà N, Duarte CM (2007) Consequences of Mediterranean warming events in seagrass (Posidonia oceanica) flowering records. Global Change Biol 13:224–235

Duffield DA, Wells RS (2002) The molecular profile of a resident community of bottlenose dolphins, Tursiops truncatus. In: Pfeiffer CJ (ed) Cell and molecular biology of marine mammals. Krieger Publishing, Melbourne, FL, USA, pp 3–11

Emanuel K (2005) Increasing destructiveness of tropical cyclones over the past 30 years. Nature 436:686–688

Engelhaupt D, Hoelzel AR, Nicholson C, Frantzis A, Mesnick S, Gero S, Whitehead H, Rendell L, Miller P, De Stefanis R, Cañadas A, Airoldi S, Mignucci-Giannoni AA (2009) Female philopatry in coastal basins and male dispersion across the North Atlantic in a highly mobile marine species, the sperm whale (Physeter macrocephalus). Mol Ecol 18:4193–4205

Fertl D, Schiro AJ, Regan GT, Beck CA, Adimey N, Price-May L, Amos A, Worthy GAJ, Crossland R (2005) Manatee occurrence in the northern Gulf of Mexico west of Florida. Gulf Caribb Res 17:69–94

Fire SE, Fauquier D, Flewelling LJ, Henry M, Naar J, Pierce R, Wells RS (2007) Brevetoxin exposure in bottlenose dolphins (Tursiops truncatus) associated with Karenia brevis blooms in Sarasota Bay, Florida. Mar Biol 152:827–834

Fire SE, Flewelling LJ, Wang Z, Naar J, Henry MS, Pierce RH, Wells RS (2008) Florida red tide and brevetoxins: Association and exposure in live resident bottlenose dolphins (Tursiops truncatus) in the eastern Gulf of Mexico, USA. Mar Mamm Sci 24:831–844

Fire SE, Wang Z, Byrd M, Whitehead HR, Paternoster J, Morton SL (2011) Co-occurrence of multiple classes of harmful algal toxins in bottlenose dolphins (Tursiops truncatus) stranding during an unusual mortality event in Texas, USA. Harmful Algae 10:330–336

Fisher SJ, Reeves RR (2005) The global trade in live cetaceans: Implications for conservation. J Int Wildl Law Pol 8:315–340

Flewelling LJ, Naar JP, Abbott JP, Baden DG, Barros NB, Bossart GD, Bottein MJD, Hammond DG, Haubold EM, Heil CA, Henry MS, Jacocks HJ, Leighfield TA, Pierce RH, Pitchford TD, Rommel SA, Scott PS, Steidinger KA, Truby EW, Van Dolah FM, Landsberg JH (2005) Red tides and marine mammal mortalities. Nature 435:755–756

Ford JKB (2009) Killer whale, Orcinus orca. In: Perrin WF, Würsig B, Thewissen JGM (eds) Encyclopedia of marine mammals, 2nd edn. Academic/Elsevier Press, Amsterdam, The Netherlands, pp 650–657

Fritts TH, Irvine AB, Jennings RD, Collum LA, Hoffman W, McGehee MA (1983) Turtles, birds, and mammals in the northern Gulf of Mexico and nearby Atlantic waters. Final Report. U.S. Fish and Wildlife Service, Washington, DC, USA. 455 p

Galindo JA, Serrano A, Vásquez-Castán L, González-Gándara C, López-Ortega M (2009) Cetacean diversity, distribution, and abundance in Northern Veracruz, Mexico. Aquat Mamm 35:12–18

Garrison LP (2007) Interactions between marine mammals and pelagic longline fishing gear in the U.S. Atlantic Ocean between 1992 and 2004. Fish Bull 105:408–417

Garrison LP, Stokes L, Fairfield C (2009) Estimated bycatch of marine mammals and sea turtles in the U.S. Atlantic pelagic longline fleet during 2008. NOAA (National Oceanic and Atmospheric Administration) Technical Memorandum NMFS—SEFSC-591. Southeast Fisheries Science Center, NOAA, Washington, DC, USA. 63 p

Gauthier JM, Dubeau H, Rassart E, Jarman WM, Wells RS (1999) Biomarkers of DNA damage in marine mammals. Mutat Res 444:427–439

Geraci JR (1990) Physiologic and toxic effects on cetaceans. In: Geraci JR, St. Aubin DJ (eds) Sea mammals and oil: Confronting the risks. Academic Press, New York, NY, USA, pp 167–197

Gero S (2005) Fundamentals of sperm whale societies: Care for calves. MS Thesis, Dalhousie University, Halifax, Nova Scotia, Canada. 99 p

Gerrodette T, Forcada J (2002). Estimates of abundance of striped and common dolphins, and pilot, sperm and Bryde's whales in the eastern tropical Pacific Ocean. SWFSC (Southwest Fisheries Science Center) Administrative Report, LJ-02-20. La Jolla, CA, USA. 24 p

Gerrodette T, Watters G, Forcada J (2005) Preliminary estimates of 2003 dolphin abundance in the eastern tropical Pacific. SWFSC Administrative Report, La Jolla, CA, USA. http://swfsc.nmfs.noaa.gov. Accessed 19 Feb 2013

Gonzalez-Socoloske D, Olivera-Gomez LD (2012) Gentle giants in dark waters: Using side-scan sonar for manatee research. Open Remote Sens J 5:1–14

Gowans S, Whitehead H, Hooker SK (2001) Social organization in northern bottlenose whales (*Hyperoodon ampullatus*): Not driven by deep water foraging? Anim Behav 62:369–377

Guinet C, Bouvier J (1995) Development of intentional stranding hunting techniques in killer whale *Orcinus orca* calves at Crozet Archipelago. Can J Zool 73:27–33

Gunter G (1941) Occurrence of the manatee in the United States, with records from Texas. J Mammal 22:60–64

Gunter G (1954) Mammals of the Gulf of Mexico. Fish Bull 55:543–551

Hale LF, Carlson JK (2007) Characterization of the shark bottom longline fishery: 2005–2006. Technical Memorandum NMFS-SEFSC-554. NOAA, Washington, DC, USA. 28 p

Hale LF, Hollensead LD, Carlson JK (2007) Characterization of the shark bottom longline fishery: 2007. Technical Memorandum NMFS-SEFSC-564. NOAA, Washington, DC, USA. 25 p

Hale LF, Gulak SJB, Carlson JK (2009) Characterization of the shark bottom longline fishery, 2008. Technical Memorandum NMFS-SEFSC-586. NOAA, Washington, DC, USA. 23 p

Hall AJ, McConnell BJ, Rowles TK, Aguilar A, Borrell A, Schwacke L, Reijnders PJ, Wells RS (2006) Individual-based model framework to assess population consequences of polychlorinated biphenyl exposure in bottlenose dolphins. Environ Health Perspect 114:60–64

Hartman DS (1979) Ecology and behavior of the manatee (*Trichechus manatus*) in Florida. Special Publication 5. American Society of Mammalogists, Lawrence, KA, USA. 153 p

Heimlich-Boran JR (1993) Social organization of the short-finned Pilot Whale, *Globicephala macrorhynchus*, with special reference to the comparative social ecology of Delphinids. PhD Thesis. University of Cambridge, Cambridge, UK. 197 p

Heithaus MR, Dill LM (2009) Feeding strategies and tactics. In: Perrin WF, Würsig B, Thewissen JGM (eds) Encyclopedia of marine mammals, 2nd edn. Elsevier Press, San Diego, CA, USA, pp 414–423

Heithaus M, Frid A, Wirsing A, Worm B (2008) Predicting ecological consequences of marine top predator declines. Trends Ecol Evol 23:202–210

Henningsen T, Würsig B (1991) Bottlenose dolphins in Galveston Bay, Texas: Numbers and activities. In: Proceedings of the European Society Conference, Cambridge University Press, Cambridge, UK, pp 36–38

Herzing DL (1997) The life history of free-ranging Atlantic spotted dolphins (*Stenella frontalis*): Age classes, color phases, and female reproduction. Mar Mamm Sci 13:576–595

Hess NA, Ribic CA (2000) Seabird ecology. In: Davis RW, Evans WE, Würsig B (eds) Cetaceans, Sea Turtles and Seabirds in the Northern Gulf of Mexico: Distribution, abundance and habitat associations, Vol 2, Technical Report. OCS Study MMS 2000-003. MMS, Gulf of Mexico OCS Region, New Orleans, LA, USA, pp 275–315

Hu C, Xiaofeng L, Pichel WG, Muller-Karger FE (2009) Detection of natural oil slicks in the NW Gulf of Mexico using MODIS imagery. Geophys Res Lett 36:1–5

IPCC (Intergovernmental Panel on Climate Change) (2007) Climate change 2007: The physical science basis. Summary for policy makers. WMO/UNEP Intergovernmental Panel on Climate Change. http://www.ipcc.ch. Accessed 19 Feb 2013

Irvine AB, Wells RS (1972) Results of attempts to tag Atlantic bottlenose dolphins (*Tursiops truncatus*). Cetology 13:1–5

Irvine AB, Scott MD, Wells RS, Kaufmann JH (1981) Movements and activities of the Atlantic bottlenose dolphin, *Tursiops truncatus*, near Sarasota, Florida. Fish Bull 79:671–688

Irwin LJ, Würsig B (2004) A small resident community of bottlenose dolphins, *Tursiops truncatus*, in Texas: Monitoring recommendations. Gulf Mex Sci 22:13–21

Jackson JBC, Kirby MX, Berger WH, Bjorndal KA, Botsford LW, Bourque BJ, Bradbury RH, Cooke R, Erlandson J, Estes JA, Hughes TP, Kidwell S, Lange CB, Lenihan HS, Pandolfi JM, Peterson CH, Steneck RS, Tegner MJ, Warner RR (2001) Historical overfishing and the recent collapse of coastal ecosystems. Science 293:629–638

Jaquet N, Gendron D (2009) The social organization of sperm whales in the Gulf of California and comparisons with other populations. J Mar Biol Assoc UK 89:975–983

Jefferson TA (1995) Distribution, abundance, and some aspects of the biology of cetaceans in the offshore Gulf of Mexico. PhD Thesis. Texas A&M University, College Station, TX, USA. 232 p

Jefferson TA, Schiro AJ (1997) Distribution of cetaceans in the offshore Gulf of Mexico. Mamm Rev 27:27–50

Jefferson TA, Webber M, Pitman R (2008) Marine mammals of the world: A comprehensive guide to their identification. Academic Press, London, UK. 573 p

Jepson PD, Bennett PM, Allchin CR, Law RJ, Kuiken T, Baker JR, Rogan E, Kirkwood JK (1999) Investigating potential associations between chronic exposure to polychlorinated biphenyls and infectious disease mortality in harbour porpoises from England and Wales. Sci Total Environ 243–244:339–348

Jochens A, Biggs D, Benoit-Bird K, Engelhaupt D, Gordon J, Hu C, Jaquet N, Johnson M, Leben R, Mate B, Miller P, Ortega-Ortiz J, Thode A, Tyack P, Würsig B (2008) Sperm whale seismic study in the Gulf of Mexico: Synthesis Report. OCS Study MMS 2008-006. U.S. Department of the Interior, New Orleans, LA, USA. 323 p

Karczmarski L, Würsig B, Gailey GA, Larson KW, Vanderlip C (2005) Spinner dolphins in a remote Hawaiian atoll: Social grouping and population structure. Behav Ecol 16:675–685

Kastelein RA, Lavaleije MSS (1992) Foreign bodies in the stomach of a female harbour porpoise (*Phocoena phocoena*) from the North Sea. Aquat Mamm 18:40–46

Kasuya T, Marsh H (1984) Life history and reproductive biology of the short–finned pilot whale, *Globicephala macrorhynchus*, off the coast of Japan. Rep Int Whal Comm Sp Issue 6:259–310

Kendall MD, Safieh B, Harwood J, Pomeroy PP (1992) Plasma thymulin concentrations, the thymus and organochlorine contamination levels in seals infected with phocine distemper virus. Sci Total Environ 115:133–144

Kucklick J, Schwacke L, Wells R, Hohn A, Guichard A, Yordy J, Hansen L, Zolman E, Wilson R, Litz J, Nowacek D, Rowles T, Pugh R, Balmer B, Sinclair C, Rosel P (2011) Bottlenose dolphins as indicators of persistent organic pollutants in the western North Atlantic Ocean and Northern Gulf of Mexico. Environ Sci Technol 45:4270–4277

Kuczaj SA II, Yeater DB (2007) Observations of rough-toothed dolphins (Steno bredanensis) off the coast of Utila, Honduras. J Mar Biol Assoc UK 87:141–148

Kuehl DW, Haebler R (1995) Organochlorine, organobromine, metal, and selenium residues in bottlenose dolphins (Tursiops truncatus) collected during an unusual mortality event in the Gulf of Mexico, 1990. Arch Environ Contam Toxicol 28:494–499

Kuiken T (1996) Diagnosis of by-catch in cetaceans. In: Proceedings, 2nd European Cetacean Society Workshop on Cetacean Pathology (European Cetacean Society Special Issue 26). Montpellier, France, 2 March 1994, pp 1–43

Küss KM (1998) The occurrence of PCBs and chlorinated pesticide contaminants in bottlenose dolphins in a resident community: comparison with age, gender and birth order. MS Thesis. Nova Southeastern University, Fort Lauderdale, FL, USA

Kvenvolden KA, Cooper CK (2003) Natural seepage of crude oil into the marine environment. Geo Mar Lett 23:140–146

Lahvis GP, Wells RS, Kuehl DW, Stewart JL, Rhinehart HL, Via CS (1995) Decreased lymphocyte responses in free-ranging bottlenose dolphins (Tursiops truncatus) are associated with increased concentrations of PCBs and DDT in peripheral blood. Environ Health Perspect 103:67–72

Laist DW, Reynolds JE III (2005) Influence of power plants and other warm-water refuges on Florida manatees. Mar Mamm Sci 21:739–764

Laist DW, Coe JM, O'Hara KJ (1999) Marine debris pollution. In: Twiss JR, Reeves RR (eds) Conservation and management of marine mammals. Smithsonian Institution Press, Washington, DC, USA, pp 342–366

Laist DW, Knowlton AR, Mead JG, Collet AS, Podestà M (2001) Collisions between ships and whales. Mar Mamm Sci 17:35–75

Langtimm C, Beck C (2003) Lower survival probabilities for adult Florida manatees in years with intense coastal storms. Ecol Appl 13:257–268

Layne JN (1965) Observations on marine mammals in Florida waters. Bull Florida State Mus 9:131–181

Learmonth JA, MacLeod CD, Santos MB, Pierce GJ, Crick HQP, Robinson RA (2006) Potential effects of climate change on marine mammals. Oceanogr Mar Biol Ann Rev 44:431–464

Lee RF, Anderson JW (2005) Significance of cytochrome P450 system responses and levels of bile fluorescent aromatic compounds in marine wildlife following oil spills. Mar Pollut Bull 50:705–723

Lewison RL, Crowder LB, Read AJ, Freeman SA (2004) Understanding impacts of fisheries bycatch on marine megafauna. Trends Ecol Evol 19:598–604

Litz JA, Garrison LP, Fieber LA, Martinez A, Contillo JP, Kucklick JR (2007) Fine-scale spatial variation of persistent organic pollutants in bottlenose dolphins (Tursiops truncatus) in Biscayne Bay, FL. Environ Sci Technol 41:7222–7228

Loughlin TR (1994) Marine mammals and the Exxon Valdez. Academic Press, San Diego, CA, USA. 395 p

Lowery GH (1974) The mammals of Louisiana and its adjacent waters. Louisiana State University Press, Baton Rouge, LA, USA. 565 p

1572   B. Würsig

Lusseau D, Williams R, Wilson B, Grellier K, Barton TR, Hammond PS, Thompson PM (2004) Parallel influence of climate on the behaviour of Pacific killer whales and Atlantic bottlenose dolphins. Ecol Lett 7:1068–1076

Lynn SK, Würsig B (2002) Summer movement patterns of bottlenose dolphins in a Texas Bay. Gulf Mex Sci 20:25–37

MacLeod CD (2009) Global climate change, range changes and potential implications for the conservation of marine cetaceans: A review and synthesis. Endang Spec Res 7:125–136

Marino L (2004) Cetacean brain evolution: Multiplication generates complexity. Int J Comp Psychol 17:1–16

Marsh H, O'Shea TJ, Reynolds JE III (2011) Ecology and conservation of the Sirenia. Dugongs and Manatees. Cambridge University Press, Cambridge, UK. 536 p

Martínez-Serrano I, Serrano A, Heckel G, Schramm Y (2011) Distribution and home range of bottlenose dolphins (*Tursiops truncatus*) off Veracruz, Mexico. Cienc Mar 37(4A):379–392

Maze KS (1997) Bottlenose dolphins of San Luis Pass, Texas: Occurrence patterns, site fidelity, and habitat use. MS Thesis. Texas A&M University, College Station, TX, USA. 79 p

Maze KS, Würsig B (1999) Bottlenose dolphins of San Luis Pass, Texas: Occurrence patterns, site fidelity, and habitat use. Aquat Mamm 25:91–103

Maze-Foley K, Mullin KD (2006) Cetaceans of the oceanic northern Gulf of Mexico: Distributions, group sizes and interspecific associations. J Cetacean Res Manage 8:203–213

McCarthy E (2004) International regulation of underwater sound: Establishing rules and standards to address ocean noise pollution. Kluwer Academic Press, Dordrecht, The Netherlands. 287 p

McHugh KA, Allen JB, Barleycorn AA, Wells RS (2010) Severe *Karenia brevis* red tides influence juvenile bottlenose dolphins (*Tursiops truncatus*) behavior in Sarasota Bay, Florida. Mar Mamm Sci 27:622–643

Melcón ML, Cummins AJ, Kerosky SM, Roche LK, Wiggins SM, Hildebrand JA (2012) Blue whales respond to anthropogenic noise. PLoS One 7(2):e32681

Miller C (2003) Abundance trends and environmental habitat usage patterns of bottlenose dolphins (*Tursiops truncatus*) in lower Barataria and Caminada Bays, Louisiana. PhD Thesis. Louisiana State University, Baton Rouge, LA, USA. 125 p

Miller LJ, Solangi M, Kuczaj SA II (2008) Immediate response of Atlantic bottlenose dolphins to high speed personal watercraft in the Mississippi Sound. J Mar Biol Assoc UK 88:1139–1143

Miller PJO, Johnson MP, Madsen PT, Biassoni N, Quero M, Tyack PL (2009) Using at-sea experiments to study the effects of airguns on the foraging behavior of sperm whales in the Gulf of Mexico. Deep-Sea Res 56:1168–1181

Miller LJ, Mackey AD, Hoffland T, Solangi M, Kuczaj SA II (2010) Potential effects of a major hurricane on Atlantic bottlenose dolphins (*Tursiops truncatus*) reproduction in the Mississippi Sound. Mar Mamm Sci 26:707–715

Moore JC (1953) Distribution of marine mammals to Florida waters. Am Midl Nat 49:117–158

Moreno MPT (2005) Environmental predictors of bottlenose dolphin distribution and core feeding densities in Galveston Bay, Texas. PhD Thesis. Texas A&M University, College Station, TX, USA. 88 p

Mullin KD (1988) Comparative seasonal abundance and ecology of bottlenose dolphins (*Tursiops truncatus*) in three habitats of the north-central Gulf of Mexico. PhD Thesis. Mississippi State University, Starkville, MS, USA. 135 p

Mullin KD (2007) Abundance of cetaceans in the oceanic Gulf of Mexico based on 2003–2004 ship surveys. NMFS, SEFSC, Pascagoula, LA, USA. 26 p

Mullin KD, Fulling GL (2004) Abundance of cetaceans in the oceanic northern Gulf of Mexico. Mar Mamm Sci 20:787–807

Mullin KD, Hoggard W (2000) Visual surveys of cetaceans and sea turtles from aircraft and ships. In: Davis RW, Evans WE, Würsig B (eds) Cetaceans, Sea Turtles, and Seabirds in the Northern Gulf of Mexico: Distribution, abundance and habitat associations, vol 2. Technical Report. OCS Study MMS 96-0027. MMS, Gulf of Mexico OCS Region, New Orleans, LA, USA, pp 111–172

Mullin K, Hoggard W, Roden C, Lohoefener R, Rogers C, Taggart B (1994) Cetaceans on the upper continental slope in the north-central Gulf of Mexico. Fish Bull 92:773–786

Mullin KD, Hoggard W, Hansen LJ (2004) Abundance and seasonal occurrence of cetaceans in outer continental shelf and slope waters of the north-central and northwestern Gulf of Mexico. Gulf Mex Sci 22:62–73

Murphy S, Pierce GJ, Law RL, Bersuder P, Jepson PD, Learmonth JA, Addink M, Dabin W, Santos MB, Deaville R, Zegers BN, Mets A, Rogan E, Ridoux V, Reid RJ, Smeenk C, Jauniaux T, López A, Alonso Farré JM, González AF, Guerra A, García-Hartmann M, Lockyer C, Boon JP (2010) Assessing the effect of persistent organic pollutants on reproductive activity in common dolphins and harbour porpoises. J Northw Atl Fish Sci 42:153–173

NASA/Goddard Space Flight Center–EOS Project Science Office (2000) Scientists find that tons of oil seep into the Gulf of Mexico each year. *ScienceDaily*. http://www.sciencedaily.com/releases/2000/01/000127082228.htm. Accessed 3 Apr 2013

National Research Council (NRC) (2003) Oil in the Sea III: Inputs, fates, and effects. The National Academies Press, Washington, DC, USA. 280 p

Neff JM (1990) Composition and fate of petroleum and spill treating agents in the marine environment. In: Geraci JR, St. Aubin DJ (eds) Sea mammals and oil: confronting the risks. Academic Press, New York, NY, USA. 259 p

Newman SJ, Smith SA (2006) Marine mammal neoplasia: A review. Vet Pathol 43:865–880

NOAA (National Oceanic and Atmospheric Administration) (2004) Interim report on the bottlenose dolphin (*Tursiops truncatus*) unusual mortality event along the panhandle of Florida March–April 2004. NOAA Fisheries Office of Protected Resources, Washington, DC, USA. 36 p

Norris KS, Dohl TP (1980) Behavior of the Hawaiian spinner dolphin, *Stenella longirostris*. Fish Bull 77:821–849

Norris KS, Würsig B, Wells RS, Würsig M (1994) The Hawaiian Spinner Dolphin. University of California Press, Berkeley, CA, USA. 436 p

Nowacek SM, Wells RS, Solow AR (2001) Short-term effects of boat traffic on bottlenose dolphins, *Tursiops truncatus*, in Sarasota Bay, Florida. Mar Mamm Sci 17:673–688

Nowacek SM, Wells RS, Owen ECG, Speakman TR, Flamm RO, Nowacek DP (2004) Florida Manatees, *Trichechus manatus latirostris*, respond to approaching vessels. Biol Conserv 119:517–523

Nowacek DP, Thorne LH, Johnston DW, Tyack PL (2007) Responses of cetaceans to anthropogenic noise. Mamm Rev 37:81–111

O'Hara TM, O'Shea TJ (2001) Toxicology. In: Dierauf LS, Gulland FMD (eds) CRC handbook of marine mammal medicine, 2nd edn. CRC Press, Boca Raton, FL, USA, pp 471–520

O'Hern J, Biggs DC (2009) Sperm whale (*Physeter macrocephalus*) habitat in the northern Gulf of Mexico: Satellite observed ocean color and altimetry applied to small-scale variability in distribution. Aquat Mamm 35:358–366

O'Shea TJ, Moore JF, Kochman HI (1984) Contaminant concentrations in manatees in Florida. J Wildl Manage 48:741–748

O'Shea TJ, Ackerman BB, Percival HF (1995) Population biology of the Florida manatee: An overview. In: O'Shea TJ, Ackerman BB, Percival HF (eds) Population biology of the Florida manatee. National Biological Service Information and Technology Report I. Department of the Interior, Washington, DC, USA, pp 280–287

O'Shea TJ, Reeves RR, Long AK (1999) Marine mammals and persistent ocean contaminants. In: Proceedings of the Marine Mammal Commission Workshop, Keystone, 12–15 October 1998. Marine Mammal Commission, Bethesda, MD, USA. 150 p

O'Sullivan S, Mullin KD (1997) Killer whales (Orcinus orca) in the northern Gulf of Mexico. Mar Mamm Sci 13:141–147

Orr JC, Fabry VJ, Aumont O, Bopp L, Doney SC, Feely RA, Gnanadesikan A, Gruber N, Ishida A, Joos F, Key RM, Lindsay K, Maier-Reimer E, Matear R, Monfray P, Mouchet A, Najjar RG, Plattner G-K, Rodgers KB, Sabine CL, Sarmiento JL, Schlitzer R, Slater RD, Totterdell IJ, Weirig M-F, Yamanaka Y, Yool A (2005) Anthropogenic ocean acidification over the twenty-first century and its impact on calcifying organisms. Nature 437:681–686

Ortega-Argueta A, Perez-Sanchez CE, Gordillo-Morales G, Gordillo OG, Perez DG, Alafita H (2005) Cetacean strandings on the southwestern coast of the Gulf of Mexico. Gulf Mex Sci 23:179–185

Ortega-Ortiz JG (2002) Multiscale analysis of cetacean distribution in the Gulf of Mexico. PhD Thesis. Texas A&M University, College Station, TX, USA. 170 p

Ortega-Ortiz JG, Engelhaupt D, Winsor M, Mate BR, Hoelzel AR (2012) Kinship of long-term associates in the highly social sperm whale. Mol Ecol 21:732–744

Panigada S, Pesante G, Zanardelli M, Capoulade F, Gannier A, Weinrich M (2006) Mediterranean fin whales at risk from fatal ship strikes. Mar Pollut Bull 52:1287–1298

Parmesan C, Yohe G (2003) A globally coherent fingerprint of climate change impacts across natural systems. Nature 421:37–42

Pauly D, Palomares ML (2005) Fishing down marine food web: It is far more pervasive then we thought. Bull Mar Sci 76:197–211

Pauly D, Christensen V, Dalsgaard J, Froese R, Torres F Jr (1998) Fishing down marine food webs. Science 279:860–863

Pauly D, Christensen V, Guénette S, Pitcher TJ, Sumaila UR, Walters CJ (2002) Towards sustainability in world fisheries. Nature 418:689–695

Payne R, Webb D (1971) Orientation by means of long range acoustic signaling in baleen whales. Ann NY Acad Sci 188:110–142

Peñuelas J, Filella I, Comas PE (2002) Changed plant and animal life cycles from 1952 to 2000 in the Mediterranean region. Global Change Biol 8:531–544

Pérez-Cao H (2004) Abundance and distribution of the bottlenose dolphin Tursiops truncatus (Montagu, 1821) in two areas of the Sabana-Camagüey Archipelago, Cuba. MS Thesis. National Aquarium of Cuba, Havana, Cuba. 67 p. (English translation)

Perrin WF, Würsig B, Thewissen JGM (2009) Encyclopedia of marine mammals, 2nd edn. Academic Press, Amsterdam, The Netherlands. 1352 p

Pollack HN, Huang S, Shen PY (1998) Climate change record in subsurface temperatures: A global perspective. Science 282:279–281

Poncelet E, Van Canneyt O, Boubert JJ (2000) Considerable amount of plastic debris in the stomach of a Cuvier's beaked whale (Ziphius cavirostris) washed ashore on the French Atlantic coast. Eur Res Cetaceans 14:44–47

Powell JR, Wells RS (2011) Recreational fishing depredation and associated behaviors involving common bottlenose dolphins (Tursiops truncatus) in Sarasota Bay, Florida. Mar Mamm Sci 27:111–129

Pryor KW, Haag R, O'Reilly J (1969) The creative porpoise: Training for novel behavior. J Exp Anal Behav 12:653–661

Pulster EL, Maruya KA (2008) Geographic specificity of Aroclor 1268 in bottlenose dolphins (*Tursiops truncatus*) frequenting the Turtle/Brunswick River Estuary, Georgia (USA). Sci Total Environ 393:367–375

Rawson AJ, Patton GW, Hofmann S, Pietra GG (1993) Liver abnormalities associated with chronic mercury accumulation in stranded Atlantic bottlenose dolphins. Ecotoxicol Environ Saf 25:41–47

Rawson AJ, Bradley JP, Teetsov A, Rice S, Haller EM, Patton GW (1995) A role for airborne particulate in high mercury levels of some cetaceans. Ecotoxicol Environ Saf 30:309–314

Read A (2008) The looming crisis: Interactions between marine mammals and fisheries. J Mammal 89:541–548

Read AJ, Murray KT (2000) Gross evidence of human-induced mortality in small cetaceans. NOAA Technical Memorandum NMFS-OPR-15. NOAA, Washington, DC, USA. 20 p

Reddy ML, Reif JS, Bachand A, Ridgway SH (2001) Opportunities for using navy marine mammals to explore associations between organochlorine contaminants and unfavorable effects on reproduction. Sci Total Environ 274:171–182

Reeves RR, Lund JN, Smith TD, Josephson EA (2011) Insights from whaling logbooks on whales, dolphins, and whaling in the Gulf of Mexico. Gulf Mex Sci 29:41–67

Reynolds JE III (1985) Evaluation of the nature and magnitude of interactions between bottlenose dolphins, *Tursiops truncatus*, and fisheries and other human activities in areas of the southeastern United States. PB86-162203. National Technical Information Service, Springfield, VA, USA

Reynolds JE III, Odell DK (1991) Manatees and Dugongs. Facts on file, New York, NY, USA. 192 p

Reynolds JE III, Wells RS, Eide SD (2000) The bottlenose dolphin: Biology and conservation. University Press of Florida, Gainesville, FL, USA. 304 p

Richards PM (2007) Estimated takes of protected species in the commercial directed shark bottom longline fishery 2003, 2004, and 2005. NMFS/SEFSC Contribution PRD-06/07-08. U.S. National Marine Fisheries Service, Washington, DC, USA. 21 p

Richter C, Gordon J, Jaquet N, Würsig B (2008) Social structure of sperm whales in the northern Gulf of Mexico. Gulf Mex Sci 26:118–123

Rolland RM, Parks SE, Hunt KE, Castellote M, Corkeron PJ, Nowacek DP, Wasser SK, Kraus SD (2012) Evidence that ship noise increases stress in right whales. Proc R Soc B 279:2363–2368

Rosel PE, Watts H (2008) Hurricane impacts on bottlenose dolphins in the northern Gulf of Mexico. Gulf Mex Sci 25:88–94

Ross PS (2006) Fireproof killer whales *(Orcinus orca):* Flame-retardant chemicals and the conservation imperative in the charismatic icon of British Columbia, Canada. Can J Fish Aquat Sci 63:224–234

Ross PS, Birnbaum LS (2003) Integrated human and ecological risk assessment: A case study of persistent organic pollutants (POPs) in humans and wildlife. Hum Ecol Risk Assess 9:303–324

Rossi-Santos MR, Santos-Neto E, Baracho CG (2009) Interspecific cetacean interactions during the breeding season of humpback whale (*Megaptera novaeangliae*) on the north coast of Bahia State, Brazil. J Mar Biol Assoc UK 89:961–966

Salata G, Wade TL, Sericano JL, Davis JW, Brooks JM (1995) Analysis of Gulf of Mexico bottlenose dolphins for organochlorine pesticides and PCBs. Environ Pollut 88:167–175

Samuels A, Bejder L (2004) Chronic interaction between humans and free-ranging bottlenose dolphins near Panama City Beach, Florida, USA. J Cetacean Res Manage 6:69–77

Schick RS, Halpin PN, Read AJ, Urban DL, Best BD, Good CP, Roberts JJ, LaBrecque EA, Dunn C, Garrison LP, Hyrenbach KD, McLellan WA, Pabst DA, Palka DL, Stevick P (2011) Community structure in pelagic marine mammals at large spatial scales. Mar Ecol Prog Ser 434:165–181

Schmidly DJ (1981) Marine mammals of the southeastern United States and the Gulf of Mexico. FWS/OBS-80/41. U.S. Fish and Wildlife Service, Washington, DC, USA. 165 p

Schmidly DJ, Shane SH (1978) A biological assessment of the cetacean fauna of the Texas coast. Nat Tech Inf Ser PB-281-763. National Technical Information Service, Springfield, VA, USA. 38 p

Schmidly DJ, Würsig B (2009) Mammals (Vertebrata: Mammalia) of the Gulf of Mexico. In: Felder DL, Camp DK (eds) Gulf of Mexico—origins, waters, and biota biodiversity. Texas A&M Press, College Station, TX, USA, pp 1343–1353

Schwacke LH, Voit EO, Hansen LJ, Wells RS, Mitchum GB, Hohn AA, Fair PA (2002) Probabilistic risk assessment of reproductive effects of polychlorinated biphenyls on bottlenose dolphins (*Tursiops truncatus*) from the southeast United States coast. Environ Toxicol Chem 21:2752–2764

Schwacke LH, Twiner MJ, De Guise S, Balmer BC, Wells RS, Townsend FI, Rotstein DC, Varela RA, Hansen LJ, Zolman ES, Spradlin TR, Levin M, Leibrecht H, Wang Z, Rowles TK (2010) Eosinophilia and biotoxin exposure in bottlenose dolphins (*Tursiops truncatus*) from a coastal area impacted by repeated mortality events. Environ Res 110:548–555

Scott GP, Burn DM, Hansen LJ, Owen RE (1989) Estimates of bottlenose dolphin abundance in the Gulf of Mexico from regional aerial surveys. National Marine Fishery Service, Southeast Fisheries Science Center, Miami, FL, USA. 24 p

Scott MD, Wells RS, Irvine AB (1990) A long-term study of bottlenose dolphins on the west coast of Florida. In: Leatherwood S, Reeves RR (eds) The bottlenose dolphin. Academic Press, San Diego, CA, USA, pp 235–244

Shane SH (1977) The population biology of the Atlantic bottlenose dolphin, *Tursiops truncatus*, in the Aransas Pass area of Texas. MS Thesis. Texas A&M University, College Station, TX, USA. 238 p

Shane SH (1994) Occurrence and habitats use of marine mammals at Santa Catalina Island, California from 1983–91. Bull Calif Acad Sci 93:13–29

Shane SH (2004) Residence patterns, group characteristics, and association patterns of bottlenose dolphins near Sanibel Island, Florida. Gulf Mex Sci 2004:1–12

Shane SHR, Wells RS, Würsig B (1986) Ecology, behavior, and social organization of the bottlenose dolphin: A review. Mar Mamm Sci 2:34–63

Simmonds MP, Eliott WJ (2009) Climate change and cetaceans: Concerns and recent developments. J Mar Biol Assoc UK 89:203–210

Smultea MA, Würsig B (1995) Behavioral reactions of bottlenose dolphins to the *Mega Borg II* oil spill, Gulf of Mexico 1990. Aquat Mamm 21:171–181

Southall BL, Bowles AE, Ellison WT, Finneran JJ, Gentry RL, Greene CR Jr, Kastak D, Ketten DR, Miller JH, Nachtigall PE, Richardson WJ, Thomas JT, Tyack PL (2007) Marine mammal noise exposure criteria: Initial scientific recommendations. Aquat Mamm 33:411–521

Southeast U.S. Marine Mammal Stranding Network—NMFS Southeast Region Stranding Database (2011) Office of Protected Resources, NOAA. http://www.nmfs.noaa.gov/pr/health/networks.htm#southeast. Accessed 3 Apr 2013

Steichen DJ Jr, Holbrook SJ, Osenberg CW (1996) Distribution and abundance of benthic and demersal macrofauna within a natural hydrocarbon seep. Mar Ecol Prog Ser 138:71–82

Stensland E, Angerbjörn A, Berggren P (2003) Mixed species groups in mammals. Mamm Rev 33:205–223

Sturges W, Leben R (2000) Frequency of ring separations from the loop current in the Gulf of Mexico: A revised estimate. J Phys Oceanogr 30:1814–1819

Swartz W, Sala E, Tracey S, Watson R, Pauly D (2010) The spatial expansion and ecological footprint of fisheries (1950 to present). PLoS One 5(12):e15143

Tarpley RJ, Marwitz S (1993) Plastic debris ingestion by cetaceans along the Texas coast: Two case reports. Aquat Mamm 19:93–98

Townsend CH (1935) The distribution of certain whales as shown by logbook records of American whale ships. Zoologica 19:1–50

Twiner MJ, Fire S, Schwacke L, Davidson L, Wang Z, Morton S, Roth S, Balmer B, Rowles TK, Wells RS (2011) Concurrent exposure of bottlenose dolphins (*Tursiops truncatus*) to multiple algal toxins in Sarasota Bay, Florida, USA. PLoS One 6:1–15

Tyson RB (2008) Abundance of bottlenose dolphins (*Tursiops truncatus*) in the Big Bend of Florida, St. Vincent Sound to Alligator Harbor. MS Thesis. Florida State University, Tallahassee, FL, USA. 74 p

Urian KW, Duffield DA, Read AJ, Wells RS, Shell ED (1996) Seasonality of reproduction in bottlenose dolphins, *Tursiops truncatus*. J Mammal 77:394–403

Urian KW, Hofmann S, Wells RS, Read AJ (2009) Fine-scale population structure of bottlenose dolphins (*Tursiops truncatus*) in Tampa Bay, Florida. Mar Mamm Sci 25:619–638

Valdes-Arellanes MP, Serrano A, Heckel G, Schramm Y, Martínez-Serrano I (2011) Abundancia de dos poblaciones de toninas (*Tursiops truncatus*) en el norte de Veracruz, México. Rev Mex Biodiv 82:227–235

Van Waerebeek K, Sequeira M, Williamson C, Sanino GP, Gallego P, Carmo P (2006) Live-captures of common bottlenose dolphins *Tursiops truncatus* and unassessed bycatch in Cuban waters: evidence of sustainability found wanting. Latin Am J Aquat Mamm 5:39–48

Van Waerebeek KA, Baker N, Félix F, Gedamke J, Iñiguez M, Sanino GP, Secchi E, Sutaria D, Van Helden A, Wang Y (2007) Vessel collisions with small cetaceans worldwide and with large whales in the southern hemisphere, an initial assessment. Latin Am J Aquat Mamm 6:43–69

Vedder JM (1996) Levels of organochlorine contaminants in milk relative to health of bottlenose dolphins (*Tursiops truncatus*) from Sarasota, Florida. MS Thesis. University of California, Santa Cruz, CA, USA

Vidal O, Van Waerebeek K, Fidley LT (1994) Cetaceans and gillnet fisheries in Mexico, Central America, and the wider Caribbean: A preliminary review. In: Perrin WF, Donovan GP, Barlow J (eds) Gillnets and cetaceans, Report of the International Whaling Commission Special Issue, 15:221–233

Wade PR (1998) Calculating limits to the allowable human-coast mortality of cetaceans and pinnipeds. Mar Mamm Sci 14:1–37

Wade PR, Gerrodette T (1993) Estimates of cetacean abundance and distribution in the Eastern Tropical Pacific. Rep Int Whal Comm 43:477–493

Waring GT, Josephson E, Maze-Foley K, Rosel PE (2010) U.S. Atlantic and Gulf of Mexico Marine Mammal Stock Assessments 2010. NOAA Technical Memorandum NMFS-NE-219. http://www.nefsc.noaa.gov/nefsc/publications/ (original publication Dec. 2010). Accessed 3 Apr 2013

Waring GT, Josephson E, Maze-Foley K, Rosel PE (2011) U.S. Atlantic and Gulf of Mexico Marine Mammal Stock Assessments, 2011. NOAA Technical Memorandum NMFS-NE-

221. http://www.nefsc.noaa.gov/nefsc/publications/ (Original publication Dec. 2011). Accessed 3 Apr 2012

Waring GT, Josephson E, Maze-Foley K, Rosel PE (eds) (2013) U.S. Atlantic and Gulf of Mexico Marine Mammal Stock Assessments—2012. NOAA Tech Memo NMFS NE 223, National Marine Fisheries Service, Woods Hole, MA, USA. 419 p. http://www.nefsc.noaa.gov/nefsc/publications/. Accessed 3 Jun 2013

Weilgart L, Whitehead H, Payne K (1996) A colossal convergence. Am Sci 84:278–287

Weller D, Würsig B, Whitehead H, Norris J, Lynn S, Davis R, Clauss N, Brown P (1996) Observations of an interaction between sperm whales and short-finned pilot whales in the Gulf of Mexico. Mar Mamm Sci 12:588–593

Wells RS (1991) The role of long-term study in understanding the social structure of a bottlenose dolphin community. In: Pryor K, Norris KS (eds) Dolphin societies: Discoveries and puzzles. University of California Press, Berkeley, CA, USA, pp 199–225

Wells RS (2003) Dolphin social complexity: Lessons from long-term study and life history. In: de Waal FBM, Tyack PL (eds) Animal social complexity: Intelligence, culture, and individualized societies. Harvard University Press, Cambridge, MA, USA, pp 32–56

Wells RS, Scott MD (1990) Estimating bottlenose dolphin population parameters from individual identification and capture-release techniques. Rep Int Whal Comm Sp Issue 14:407–415

Wells RS, Scott MD (1994) Incidence of gear entanglement for resident inshore bottlenose dolphins near Sarasota, FL. In: Perrin WF, Donovan GP, Barlow J (eds) Gillnets and Cetaceans. Report of the International Whaling Commission Special Issue 15. IWC Press, Cambridge, UK. 629 p

Wells RS, Scott MD (1997) Seasonal incidence of boat strikes on bottlenose dolphins near Sarasota, Florida. Mar Mamm Sci 13:475–480

Wells RS, Scott MD (2009) Common bottlenose dolphin Tursiops truncatus. In: Perrin WF, Würsig B, Thewissen JGM (eds) Encyclopedia of marine mammals, 2nd edn. Elsevier, Amsterdam, The Netherlands, pp 249–255

Wells RS, Scott MD, Irvine AB (1987) The social structure of free-ranging bottlenose dolphins. In: Genoways HH (ed) Current mammalogy, vol 1. Plenum Press, New York, NY, USA, pp 247–305

Wells RS, Hofmann S, Moors TL (1998) Entanglement and mortality of bottlenose dolphins, Tursiops truncatus, in recreational fishing gear in Florida. Fish Bull 96:647–650

Wells RS, Duffield D, Hohn AA (2001) Reproductive success of free ranging bottlenose dolphins: Experience and size make a difference. In: 14th Biennial Conference on the Biology of Marine Mammals, Vancouver, Canada, 28 November–3 December 2001. 231 p

Wells RS, Rowles TK, Borrell A, Aguilar A, Rhinehart HL, Jarman WM, Hofmann S, Hohn AA, Duffield DA, Mitchum G, Stott J, Hall A, Sweeney JC (2003) Integrating data on life history, health, and reproductive success to examine potential effects of POPs on bottlenose dolphins (Tursiops truncatus) in Sarasota Bay, FL. Organohalogen Compd 62:208–211

Wells RS, Rhinehart HL, Hansen LJ, Sweeney JC, Townsend FI, Stone R, Casper DR, Scott MD, Hohn AA, Rowles TK (2004) Bottlenose dolphins as marine ecosystem sentinels: Developing a health monitoring system. Ecosyst Health 1:246–254

Wells RS, Tornero V, Borrell A, Aguilar A, Rowles TK, Rhinehart HL, Hofmann S, Jarman WM, Hohn AA, Sweeney JC (2005) Integrating life-history and reproductive success data to examine potential relationships with organochlorine compounds for bottlenose dolphins (Tursiops truncatus) in Sarasota Bay, Florida. Sci Total Environ 349:106–119

Wells RS, Allen JB, Hofmann S, Bassos-Hill K, Fauquier DA, Barros NB, DeLynn RE, Sutton G, Socha V, Scott MD (2008) Consequences of injuries on survival and reproduction of common bottlenose dolphins (*Tursiops truncatus*) along the west coast of Florida. Mar Mamm Sci 24:774–794

Whitehead H (1998) Cultural selection and genetic diversity in matrilineal whales. Science 282:1708–1711

Whitehead H (2003) Sperm whales: Social evolution in the ocean. University of Chicago Press, Chicago, IL, USA. 431 p

Whitehead H (2009) Estimating abundance from one-dimensional passive acoustic surveys. J Wildl Manage 73:1000–1009

Whitehead H, McGill B, Worm B (2008) Diversity of deep-water cetaceans in relation to temperature: Implications for ocean warming. Ecol Lett 11:1198–1207

Williams M, Davis RW (1995) Emergency care and rehabilitation of oiled sea otters: A guide for oil spills involving fur-bearing marine mammals. University of Alaska Press, Fairbanks, AK, USA. 300 p

Wolfe DA (1987) Persistent plastics and debris in the ocean: An international problem of ocean disposal. Mar Pollut Bull 18:303–305

Worm B, Barbier EB, Beaumont N, Duffy JE, Folke C, Halpern BS, Jackson JBC, Lotze HK, Micheli F, Palumbi SR, Sala E, Selkoe KA, Stachowicz JJ, Watson R (2006) Impacts of biodiversity loss on ocean ecosystem services. Science 314:787–790

Worm B, Hilborn R, Baum JK, Branch TA, Collie JS, Costello C, Fogarty MJ, Fulton EA, Hutchings JA, Jennings S, Jensen OP, Lotze HK, Mace PM, McClanahan TR, Minto C, Palumbi SR, Parma AM, Ricard D, Rosenberg AA, Watson R, Zeller D (2009) Rebuilding global fisheries. Science 325:578–584

Wormuth JH, Ressler PH, Cady RB, Harris EJ (2000) Zooplankton and micronekton in cyclones and anticyclones in the northeast Gulf of Mexico. Gulf Mex Sci 18:23–34

Wright SD, Ackermarn BB, Bondec K, Beckand A, Banowetz DJ (1995) Analysis of watercraft-related mortality of manatees in Florida, 1979–1991. In: O'Shea TJ, Ackermarn BB, Percival HF (eds) Population biology of the Florida manatee, National Biological Service Information and Technology Report 1. National Technical Information Service, Springfield, VA, USA, pp 259–268

Würsig B, Jefferson TA, Schmidly D (2000) Marine mammals of the Gulf of Mexico. Texas A&M University Press, College Station, TX, USA. 232 p

Yordy JE, Wells RS, Balmer BC, Schwacke LH, Rowles TK, Kucklick JR (2010) Life history as a source of variation for persistent organic pollutant (POP) patterns in a community of common bottlenose dolphins (*Tursiops truncatus*) resident to Sarasota Bay, FL. Sci Total Environ 408:2163–2172

Zimmer WMX, Johnson MP, Madsen PT, Tyack PL (2005) Echolocation clicks of free-ranging Cuvier's beaked whales (*Ziphius cavirostris*). J Acoust Soc Am 117:3920–3927

USEFUL MARINE MAMMAL WEBSITES

http://ecos.fws.gov/speciesProfile/profile/speciesProfile
http://www.fws.gov/
http://www.iucnredlist.org/
http://www.nmfs.noaa.gov/pr/
http://www.nmfs.noaa.gov/pr/species/esa/mammals.htm
http://www.nmfs.noaa.gov/pr/species/mammals/cetaceans/

APPENDIX A

Marine mammal strandings recorded with human interaction in the Gulf of Mexico, 2000–March 2010 (human interactions document the presence or absence of signs of human interaction, not necessarily the cause of death).

Table A.1. Boat Collision by Month.

Year	Month	Species	Monthly Total	Annual Total	State
2000	March	*Tursiops truncatus*	1	4	FL
	May	*Tursiops truncatus*	1		FL
	August	*Tursiops truncatus*	1		FL
	September	*Tursiops truncatus*	1		TX
2001	October	*Tursiops truncatus*	1	2	TX
	November	*Tursiops truncatus*	1		FL
2002	February	*Tursiops truncatus*	1	5	TX
	March	*Tursiops truncatus*	1		TX
	April	*Tursiops truncatus*	1		TX
	May	*Tursiops truncatus*	1		TX
	June	*Tursiops truncatus*	1		TX
2003	January	*Tursiops truncatus*	1	3	FL
	April	*Tursiops truncatus*	1		TX
	November	*Tursiops truncatus*	1		TX
2004	March	*Tursiops truncatus*	2	4	MS, TX
	April	*Tursiops truncatus*	1		TX
	August	*Tursiops truncatus*	1		TX
2005	September	*Tursiops truncatus*	1	1	FL
2006	April	*Tursiops truncatus*	1	3	TX
	June	*Stenella coeruleoalba*	1		FL
	September	*Tursiops truncatus*	1		TX
2007	March	*Tursiops truncatus*	1	2	FL
	July	*Tursiops truncatus*	1		FL
2009	February	*Tursiops truncatus*	1	7	TX
	July	*Tursiops truncatus*	2		FL
	August	*Tursiops truncatus*	1		TX
	October	*Tursiops truncatus*	1		TX
		Balaenoptera edeni	1		FL
	November	*Tursiops truncatus*	1		TX
2010	February	*Tursiops truncatus*	1	1	TX
Total				32	

Source: Southeast U.S. Marine Mammal Stranding Network 2011.

Table A.2. Bullet Wounds by Month.

Year	Month	Species	Monthly Total	Annual Total	State
2003	April	*Tursiops truncatus*	1	1	TX
2004	March	*Tursiops truncatus*	2	3	TX
	May	*Tursiops truncatus*	1		LA
2006	April	*Tursiops truncatus*	1	3	FL
	October	*Tursiops truncatus*	1		FL
	December	*Kogia sima*	1		FL
2007	June	*Tursiops truncatus*	1	1	TX
Total				8	

Source: Southeast U.S. Marine Mammal Stranding Network 2011

Table A.3. Fishery Interaction by Month.

Year	Month	Species	Monthly Total	Annual Total	State
2000	January	*Tursiops truncatus*	2	10	FL, TX
	February	*Tursiops truncatus*	2		FL, TX
	June	*Tursiops truncatus*	1		FL
	July	*Tursiops truncatus*	2		FL, MS
	August	*Tursiops truncatus*	2		TX
	October	*Tursiops truncatus*	1		FL
2001	December	*Tursiops truncatus*	1	1	FL
2002	March	*Tursiops truncatus*	1	7	TX
	May	*Tursiops truncatus*	1		FL
	July	*Tursiops truncatus*	1		FL
	September	*Tursiops truncatus*	1		FL
	October	*Tursiops truncatus*	2		TX
	December	*Tursiops truncatus*	1		FL
2003	January	*Tursiops truncatus*	1	9	TX
	March	*Stenella longirostris*	1		TX
		Kogia spp.	1		MS
	April	*Tursiops truncatus*	2		FL
	June	*Tursiops truncatus*	2		FL
	October	*Tursiops truncatus*	1		FL
	November	*Tursiops truncatus*	1		FL
2004	January	*Tursiops truncatus*	1	8	FL
	February	*Tursiops truncatus*	3		FL, TX
	March	*Tursiops truncatus*	2		FL, TX
	April	*Stenella frontalis*	2		AL
2005	January	*Tursiops truncatus*	1	4	TX
	October	*Tursiops truncatus*	1		FL
	November	*Tursiops truncatus*	1		FL
	December	*Tursiops truncatus*	1		TX

(continued)

Table A.3. (continued)

Year	Month	Species	Monthly Total	Annual Total	State
2006	January	*Tursiops truncatus*	1	15	FL
	March	*Tursiops truncatus*	1		FL, TX
	April	*Tursiops truncatus*	1		FL
	May	*Tursiops truncatus*	2		FL
	June	*Tursiops truncatus*	1		FL, TX
		Unidentified cetacean	1		
	July	*Tursiops truncatus*	6		FL, TX
	September	*Tursiops truncatus*	1		TX
	December	*Tursiops truncatus*	1		FL
2007	January	*Tursiops truncatus*	1	5	FL
	May	*Tursiops truncatus*	1		FL
	September	*Tursiops truncatus*	1		FL
	October	*Tursiops truncatus*	1		FL
	December	*Tursiops truncatus*	1		FL
2008	January	*Tursiops truncatus*	1	12	FL
	February	*Tursiops truncatus*	3		FL
	April	*Tursiops truncatus*	1		FL
	June	*Tursiops truncatus*	2		FL
	November	*Tursiops truncatus*	1		FL
	December	*Tursiops truncatus*	3		FL, TX
		Unidentified cetacean	1		
2009	February	*Tursiops truncatus*	1	6	FL
	June	*Tursiops truncatus*	1		FL
	July	*Tursiops truncatus*	1		TX
	September	*Tursiops truncatus*	2		TX
	October	*Tursiops truncatus*	1		AL
2010	January	*Tursiops truncatus*	2	3	FL
	March	*Tursiops truncatus*	1		TX
Total				80	

Source: Southeast U.S. Marine Mammal Stranding Network 2011.

Table A.4. Other Human Interactions* by Month.

Year	Month	Species	Monthly Total	Annual Total	State
2000	January	*Tursiops truncatus*	2	28	FL, TX
	February	*Tursiops truncatus*	3		FL, TX
	March	*Tursiops truncatus*	4		FL, LA, MS, TX
	April	*Steno bredanensis*	1		FL
		Tursiops truncatus	2		TX
	May	*Tursiops truncatus*	1		FL
	June	*Tursiops truncatus*	5		FL, LA
	July	*Tursiops truncatus*	2		FL, MS
	August	*Tursiops truncatus*	4		FL, TX
	September	*Tursiops truncatus*	1		TX
	October	*Tursiops truncatus*	1		FL
	November	*Tursiops truncatus*	1		TX
	December	*Tursiops truncatus*	1		TX
2001	January	*Tursiops truncatus*	1	8	AL
	March	*Tursiops truncatus*	5		AL, TX
	April	*Tursiops truncatus*	1		TX
	November	*Tursiops truncatus*	1		TX
2002	February	*Tursiops truncatus*	2	13	TX
	March	*Tursiops truncatus*	8		TX
	June	*Tursiops truncatus*	1		TX
	September	*Tursiops truncatus*	1		FL
	October	*Tursiops truncatus*	1		TX
2003	February	*Tursiops truncatus*	2	10	TX
		Unidentified cetacean	1		
	March	*Kogia* spp.	1		MS
		Tursiops truncatus	3		AL, TX
	April	*Tursiops truncatus*	1		TX
	June	*Tursiops truncatus*	1		TX
	November	*Tursiops truncatus*	1		TX
2004	January	*Tursiops truncatus*	3	9	TX
	February	*Tursiops truncatus*	2		LA, TX
	March	*Tursiops truncatus*	3		FL, TX
	October	*Tursiops truncatus*	1		TX
2005	February	*Tursiops truncatus*	1	5	TX
	August	*Tursiops truncatus*	1		LA
		Unidentified cetacean	1		
	October	*Tursiops truncatus*	1		FL
	December	Unidentified cetacean	1		FL

(continued)

Table A.4. (continued)

Year	Month	Species	Monthly Total	Annual Total	State
2006	January	*Tursiops truncatus*	1	13	FL
	March	*Tursiops truncatus*	2		FL
	April	*Stenella frontalis*	1		AL, FL
		Tursiops truncatus	1		
	July	*Tursiops truncatus*	3		FL
	August	*Tursiops truncatus*	1		TX
	September	*Tursiops truncatus*	3		FL, LA, TX
	December	*Tursiops truncatus*	1		FL
2007	February	*Tursiops truncatus*	2	10	FL, LA
	April	*Tursiops truncatus*	2		TX
	June	*Tursiops truncatus*	1		TX
	August	*Tursiops truncatus*	1		TX
	October	*Tursiops truncatus*	2		FL, TX
	November	*Tursiops truncatus*	2		TX
2008	October	*Feresa attenuata*	1	1	TX
2009	February	*Tursiops truncatus*	1	12	TX
	March	*Tursiops truncatus*	3		MS, TX
	June	*Tursiops truncatus*	6		LA
	August	*Kogia* spp.	1		FL
		Tursiops truncatus	1		
2010	–	–	–	–	–
	–	–	–		–
Total				114	

Source: Southeast U.S. Marine Mammal Stranding Network 2011.
*The Southeast Region stranding category "Other Human Interactions" encompasses a wide variety of interactions including direct physical contact and indirect impacts. Direct physical contact includes deep, clean lacerations to various body parts, removal of tail flukes and dorsal fins, gouge wounds, decapitation, broken vertebrae, ropes tied around the body, and several occurrences of incidental takes. Indirect anthropogenic impacts include ingestion of plastic debris and golf balls and entrapment due to artificial barriers such as levies. These interactions do not necessarily indicate the cause of death, but rather document the presence or absence of signs of human interaction.

There is overlap in the database under Human Interactions. Entries listed in more than one category include:

- Other human interaction + Fishery interaction = 19 entries
- Other human interaction + Shot = 1 entry
- Other human interaction + Boat collision = 8 entries
- Fishery and boat collision = 1 entry

APPENDIX B

Marine mammal unusual mortality events in the Gulf of Mexico, 2004–March 2010.

Table B.1. Marine Mammal Unusual Mortality Events in the Gulf of Mexico by Year.

Year	Species	No. of Animals	State
2004	*Tursiops truncatus*	4	FL
2005	*Stenella frontalis*	107	FL
	Tursiops truncatus		
	Unidentified cetacean		
2006	*Stenella coeruleoalba*	177	FL
	Stenella frontalis		
	Tursiops truncatus		
	Unidentified cetacean		
2007	*Tursiops truncates*	64	TX
	Unidentified cetacean		
2008	*Tursiops truncates*	99	TX
	Peponocephala electra		
2009	*Grampus griseus*	1	FL
2010	*Stenella frontalis*	73	AL, FL, LA, MI
	Tursiops truncatus		
	Unidentified cetacean		
Total		525	

Source: Southeast U.S. Marine Mammal Stranding Network 2011.

Table B.2. Marine Mammal Unusual Mortality Events in the Gulf of Mexico by Month.

Year	Month	Species	No. of Animals	State (County)
2004	March	*Tursiops truncatus*	4	Florida (Bay, Escambia, Gulf, Walton)
2005	July	*Tursiops truncatus*	7	Florida (Manatee, Pinellas, Sarasota)
	August	*Tursiops truncatus*	12	Florida (Hillsborough, Lee, Manatee, Pinellas, Sarasota)
	September	*Tursiops truncatus*	17	Florida (Charlotte, Collier, Gulf, Hillsborough, Lee, Manatee, Pinellas, Sarasota)
		Unidentified cetacean	2	
	October	*Tursiops truncatus*	25	Florida (Bay, Charlotte, Collier, Franklin, Gulf, Lee, Manatee, Pinellas, Sarasota)
		Unidentified cetacean	1	
	November	*Tursiops truncatus*	18	Florida (Bay, Franklin, Hillsborough, Lee, Manatee, Okaloosa, Pinellas, Walton)
		Unidentified cetacean	2	
	December	*Tursiops truncatus*	21	Florida (Bay, Citrus, Gulf, Lee, Manatee, Okaloosa, Pinellas, Sarasota, Walton)
		Unidentified cetacean	2	

(continued)

Table B.2. (continued)

Year	Month	Species	No. of Animals	State (County)
2006	January	*Tursiops truncatus*	22	Florida (Charlotte, Collier, Franklin, Gulf, Hillsborough, Lee, Okaloosa, Pinellas, Walton)
		Unidentified cetacean	6	
	February	*Tursiops truncatus*	28	Florida (Bay, Escambia, Franklin, Gulf, Hillsborough, Lee, Manatee, Okaloosa, Pinellas, Walton)
		Unidentified cetacean	2	
	March	*Tursiops truncatus*	36	Florida (Bay, Citrus, Collier, Escambia, Franklin, Gulf, Hillsborough, Levy, Manatee, Okaloosa, Pinellas, Sarasota, Walton)
		Unidentified cetacean	6	
	April	*Stenella frontalis*	1	Florida (Bay, Escambia, Franklin, Gulf, Lee, Manatee, Okaloosa, Sarasota, Walton)
		Tursiops truncatus	11	
		Unidentified cetacean	3	
	May	*Tursiops truncatus*	4	Florida (Lee, Manatee, Sarasota)
		Unidentified cetacean	1	
	June	*Tursiops truncatus*	5	Florida (Franklin, Manatee, Okaloosa, Pinellas, Sarasota)
		Stenella coeruleoalba	1	
	July	*Tursiops truncatus*	18	Florida (Hillsborough, Lee, Pinellas, Sarasota, Wakulla)
		Unidentified cetacean	1	
	August	*Tursiops truncatus*	10	Florida (Bay, Hillsborough, Lee, Manatee, Pinellas, Sarasota)
	September	*Tursiops truncatus*	2	Florida (Pinellas, Sarasota)
		Unidentified cetacean	1	
	October	*Tursiops truncatus*	12	Florida (Collier, Hillsborough, Lee)
		Unidentified cetacean	1	
	November	*Tursiops truncatus*	6	Florida (Hillsborough, Lee, Pinellas)
2007	February	*Tursiops truncatus*	5	Texas (Brazoria, Galveston)
	March	*Tursiops truncatus*	56	Texas (Brazoria, Chambers, Galveston, Jefferson)
		Unidentified cetacean	1	
	April	*Tursiops truncatus*	2	Texas (Brazoria, Galveston)
2008	February	*Tursiops truncatus*	24	Texas (Brazoria, Calhoun, Galveston, Jefferson, Kleberg, Nueces)
	March	*Tursiops truncatus*	67	Texas (Brazoria, Galveston, Jefferson, Matagorda)
		Peponocephala electra	1	
	April	*Tursiops truncatus*	7	Texas (Brazoria, Galveston, Jefferson)

(continued)

Table B.2. (continued)

Year	Month	Species	No. of Animals	State (County)
2009	January	*Grampus griseus*	1	Florida (Wakulla)
2010	February	*Stenella frontalis*	1	Alabama (Baldwin, Mobile); Florida (Escambia, Gulf, Santa Rosa); Louisiana (Jefferson, New Orleans, St. Martin); Mississippi (Jackson)
		Tursiops truncatus	9	
		Unidentified cetacean	1	
	March	*Stenella frontalis*	1	Alabama (Baldwin, Mobile); Florida (Escambia, Franklin, Gulf, Okaloosa, Santa Rosa, Walton); Louisiana (Cameron, Jefferson, New Orleans, St. Bernard, St. Tammany, Terrebonne); Mississippi (Hancock, Harrison, Jackson)
		Tursiops truncatus	54	
		Unidentified cetacean	7	
Total			525	

Source: Southeast U.S. Marine Mammal Stranding Network 2011.

Open Access This chapter is licensed under the terms of the Creative Commons Attribution-NonCommercial 2.5 International License (http://creativecommons.org/licenses/by-nc/2.5/), which permits any noncommercial use, sharing, adaptation, distribution and reproduction in any medium or format, as long as you give appropriate credit to the original author(s) and the source, provide a link to the Creative Commons license and indicate if changes were made.

The images or other third party material in this chapter are included in the chapter's Creative Commons license, unless indicated otherwise in a credit line to the material. If material is not included in the chapter's Creative Commons license and your intended use is not permitted by statutory regulation or exceeds the permitted use, you will need to obtain permission directly from the copyright holder.

CHAPTER 14

DISEASES AND MORTALITIES OF FISHES AND OTHER ANIMALS IN THE GULF OF MEXICO

Robin M. Overstreet[1] and William E. Hawkins[1]

[1]University of Southern Mississippi, Ocean Springs, MS 39564, USA
robin.overstreet@usm.edu

14.1 INTRODUCTION

How could the environmental health of the Gulf of Mexico (Gulf or GoM) be described in regard to fish kills, diseases, and parasitic infections before the Deepwater Horizon oil spill? This might seem like a simple question, but the answer is complex and difficult to report. Nevertheless, reporting some of that complexity is the aim of this chapter.

Many people consider the GoM to be restricted to the offshore waters, but studies on diseases in offshore water deeper than 200 meters [m] (656 feet [ft]) are few, not very detailed, incomplete, and not necessarily representative of the entire Gulf. Looking at offshore fishes in the Gulf, one should also consider the nursery grounds for those species. These grounds are usually the estuaries of the Gulf. Moreover, these estuaries are independent habitats that provide important aspects of understanding mortality and diseases of animals throughout the entire Gulf. This chapter treats the entire GoM.

Several factors have been impediments to research on this topic. An important reason for not being able to fully understand disease and its importance to the GoM involves the fact that when fishes become ill or otherwise stressed, they usually get preyed upon by predators and microbial consumers. Moreover, many diseases primarily affect the larval stages of the hosts, and individuals of these stages are so small that they are usually overlooked and seldom critically examined. When someone sees infected individuals, disease conditions may not necessarily be apparent. Infections in larval fishes by one or few individuals of a parasite species can have a role in disease epidemics. Observations of such cases can be supported by experimental laboratory studies. One critical reason that diseases in the Gulf of Mexico prior to the oil spill are not well-known is because some cutting-edge molecular diagnostic tools have been developed only recently and have not been widely applied for diagnosis of marine animal diseases. Another reason why diseases of offshore fishes as well as shellfishes, other invertebrates, marine mammals, sea turtles, and birds have not been studied relates to the high expense of collecting and studying the material. For example, charges for daily rental of large research ships can be as much as $50,000; the 29.7 m (97.4 ft) USM-GCRL R/V Tommy Munro, on which research involving the continental shelf is conducted, presently costs $7,200 plus the cost of fuel. Some of our past research has avoided high collecting costs because we took advantage of various inshore and offshore studies of fishes conducted by the National Marine Fisheries Service (NMFS) by partaking in its cruises, by accompanying other biologists on their trips, by obtaining parasites from some fishes collected by others, by conducting numerous inshore and offshore collection activities from our own small vessels, by assessing diseases, and most importantly, by utilizing the vast array of literature reflecting research conducted by others.

© The Author(s) 2017
C.H. Ward (ed.), *Habitats and Biota of the Gulf of Mexico: Before the Deepwater Horizon Oil Spill*,
DOI 10.1007/978-1-4939-3456-0_6

Information reported in this chapter covers research conducted from riverine to offshore habitats that relate to the health of marine organisms, with an emphasis on fishes.

In this chapter, parasites that can cause diseases and mortalities will be discussed, but the harmless ones also will be considered as part of the biodiversity of the GoM. Since biodiversity is widely considered to correlate with ecosystem health, the presence or abundance of parasites becomes part of that positive biodiversity. Typically, the fewer the parasites observed, the worse the environmental conditions and thus the biodiversity. Knowledge involving biodiversity of pathogens, parasites, and hosts from the GoM over the last half century has increased greatly. In fact, because about half the living fauna are symbiotic, the biodiversity of infectious agents often outweighs that of hosts and potential hosts.

Because such an enormous number of mortalities of marine life in the GoM results from eutrophication, toxins from algal blooms like the red tide, and low temperature, those conditions are treated in some detail. Moreover, stress in animals resulting from those factors plays an important role in understanding infectious diseases. Consequently, we include some of that information involving mortalities and infections in this chapter.

The purpose of this chapter involves our understanding of the environmental health of the GoM prior to April 2010. When we consider the geographic scope of the entire Gulf of Mexico, we also consider three nearby regions to be included within its borders. We do this because of the immediately adjacent identical habitat, the interchange of water across the borders recognized by Felder and Camp (2009), and the similarity of their fauna to that encountered in the Caribbean portion of the Gulf of Mexico. The fish and parasite populations in three studied locations have a strong Caribbean influence as do the birds; however, not many Caribbean-Gulf collections have actually been made. Consequently, both the checklists by Overstreet et al. (2009) and this chapter include species and diseases extending slightly outside the designated GoM borders (Felder and Camp 2009). Those include habitats north through Biscayne Bay on the Atlantic side of Florida, those located off Cancún and Cozumel (slightly south of the Gulf border of Cabo Catoche, Yucatán, Mexico), and those off Havana, Cuba. As more fish and birds from the northern Gulf of Mexico as well as elsewhere in the GoM are examined, they surely will be found to be infected with new and unreported species and diseases. Consequently, we consider it important to include all fauna indicated above to best understand the fauna of the Gulf of Mexico as reported in a compendium edited by Felder and Camp (2009).

14.2 DEFINITIONS

When reading reviews such as this, a reader must understand that different authors in the literature can either (1) use different terms for the same subject/situation or (2) use one term for different situations. The clearest way to allow a reader to understand what is being written is for the author to carefully define a term or describe the subject of investigation. A few definitions provided below will guide the reader.

- *Allelopathy* refers to a biological phenomenon by which an organism produces one or more compounds that influence the growth, reproduction, or survival of other organisms.

- *Allochthonous* in this chapter refers to an organism that obtains energy/organic matter originating from outside the system, such as point-source discharges from rivers, but also from watershed runoff and coastal tidal inlets.

- *Autochthonous* means belonging to a particular place by birth or origin. For purposes of this chapter, it refers to an organism that generates organic matter within the system produced primarily through photosynthesis, by phytoplankton productivity, or by benthic regeneration.

- *Coherence* refers to shifting baselines relative to temporal collections; one reference site is not sufficient to capture random/natural variability.

- *Disease* as a simple term refers to any alteration from the normal state of health. In medical cases, this often refers to a dose of the causative agent above a threshold value that results in harm. The term "disease" differs from the term "syndrome" in different ways by different experts; however, Dorland's Medical Dictionary (1974) defines disease as "a definite morbid process, often with a characteristic train of symptoms" and syndrome as "a combination of symptoms [signs] resulting from a single cause or so commonly occurring together as to constitute a distinct clinical entity."

- *Epizootic* as an adjective refers to a rapidly spreading disease that is temporarily prevalent and widespread in an animal population, and as a noun, refers to an outbreak of an epizootic disease. *Epidemic* has the same definition, although it is restricted by some users to cases where the animal is human; a *pandemic* is an epidemic covering a large area.

- *Infection* as defined by American parasitologists usually refers to an internal association or a combination of internal and external associations, whether that relationship results in harm or not regardless of the size of the organism. For an internal association to be called an infection by a microbiologist, the organism is usually restricted to viruses, bacteria, protists, and fungi (compare infestation). To some microbiologists, microorganisms such as bacteria that live naturally in the mouth or elsewhere in a body without causing harm are not considered infections or infectious agents by many microbiologists, but the organisms are symbionts.

- *Infestation* refers to a variety of associations, depending on the author or country of origin. These meanings are (1) an external association, the definition we prefer, (2) a metazoan symbiont (parasite), (3) a parasite's colonization, utilization, or both of the host; (4) a host being colonized, utilized, or both by parasites, (5) an environment being colonized, utilized, or both by pests, (6) a population rather than incorporated individuals, and (7) an action (as opposed to the term "infection," which would suggest a condition or a state).

- *Neoplasm/tumor* also has confusing definitions. Tumor refers to an abnormal mass of tissue. It can be benign or malignant (cancerous). Consequently, not all tumors are neoplastic; they can even be a response to inflammation or constitute a parasitic infection. However, all cancers are neoplastic. A widely used but not always accepted definition of neoplasm by the British oncologist R.A. Willis states: "A neoplasm is an abnormal mass of tissue, the growth of which exceeds and is uncoordinated with that of the normal tissues, and persists in the same excessive manner after cessation of the stimulus which evoked the change."

- *Symbiosis* defined herein refers to an association (mutualism, commensalism, or parasitism) between organisms of different species involving a unilateral or bilateral exchange of material or energy. A symbiont is any member, usually the smallest of a pair of organisms, involved in this symbiotic relationship. *Commensalism*: a symbiotic relationship in which one of two partner species benefits and the other shows no apparent beneficial or harmful effect. *Mutualism*: a symbiotic relationship in which two or more partners gain reciprocal benefits, usually mutual ones. *Parasitism*: a symbiotic relationship in which a symbiont lives all or part of its life in or on a living host, usually benefiting while harming the host in some way and usually having a higher reproductive potential than the host. For purposes of this chapter, not all parasites harm their host or not all components of a parasite's life cycle harm its hosts. Moreover, the symbiont can be facultative or obligate, and it can infect either a natural or accidental host.

14.3 MASS MORTALITIES, PRIMARILY INCLUDING FISH KILLS

Mass mortalities of fish are usually caused by very specific conditions or agents. The following general mass mortalities occur commonly or at least expectedly are treated first and typically involve a variety of fish species. Additional diseases and die-offs caused by more host-specific agents and conditions are usually restricted to one or few host species and will be reported below in Sections 14.4, 14.5, and 14.6.

14.3.1 Eutrophication

Most recognized animal mortalities in the GoM result as an undesired product from eutrophication, normally the production of organic matter that forms the basis of aquatic food webs. While this process of eutrophication and of dying animals constitutes a natural progression, its rate depends on a complex of many factors. However, some aspects of this process are influenced by human input, the natural environment, or disease agents that weaken the victim. Most animals that die during this process die from oxygen depletion, from toxins produced by specific harmful algal blooms (HABs), or from predators taking advantage of the weakened condition of the prey.

Harmful concentrations of dissolved oxygen (DO) fit different categories. For example, the following categories used by the National Oceanic and Atmospheric Administration (NOAA) and others are (1) anoxia (0 milligrams per liter [mg/L]), (2) hypoxia (>0 and ≤ 2 mg/L), and (3) biologically stressful conditions (>2 and ≤ 5 mg/L). All the categories typically occur in June–October in bottom waters of most estuaries in all Gulf States.[1] Note that corresponding stressful levels have been defined more stringently by others such as Livingston (2001) as >2 and ≤ 4 mg/L. In a recent comprehensive survey of the trophic status of estuaries in the continental United States, Bricker et al. (2008) concluded that 84 estuaries, representing 65 % of the total estuarine surface area, presently showed signs of moderate to high eutrophic conditions. The main constituent of the organic matter is carbon, with the rate of eutrophication usually expressed as grams of carbon per square meter per year (gram carbon/square meter/year [g carbon/m^2/year]). A eutrophic rate in an estuary is 300–500 g carbon/m^2/year, with its effect on the ecosystem dependent on export rates of flushing, microbial respiration, and denitrification as well as on recycling/regeneration rates. Ecologically, eutrophication involves scales of both time and space. Autochthonous organic matter loading involves that matter generated within the system and is produced primarily through photosynthesis by phytoplankton productivity or by benthic regeneration. The resulting phytoplankton blooms depend on the amount of light and nutrients consisting primarily of total dissolved nitrogen and phosphorous, including both inorganic and organic forms, but also silicates and other nutrients. In estuarine habitats, primary producers other than phytoplankton include mostly benthic microalgae, epiphytes, seagrasses, and other submerged aquatic vegetation (Figures 14.1 and 14.2).

Allochthonous organic matter, originating from outside the estuary, can usually be traced from rivers but also from watershed runoff and coastal tidal inlets. The nutrient sources

---

[1] http://www.noaa.gov/factsheets/new%20version/dead_zones.pdf.

Figure 14.1. A moderate die off of the Gulf menhaden, *Brevoortia patronus*, caused by oxygen depletion in a Mississippi bayou on June 1984.

Figure 14.2. One of thousands of floating striped mullet, *Mugil cephalus*, having died from oxygen depletion near Galveston, Texas, in July 1993, exhibiting muscular degeneration and an extensive number of dipteran maggots feeding on the decomposing flesh.

include point source discharges such as from wastewater treatment plants, industrial plants, and logging operations and nonpoint discharges from agriculture, residential lawns, and gardens. Both sources can consist of particulate matter such as plant debris, detritus, and phytoplankton and of dissolved matter including humic substances such as humic acid, mucopolysaccharides, peptides, and lipids.

Total dissolved nitrogen, phosphorous, and silicates, including both inorganic and organic forms, are basic to production of phytoplankton, which plays a central role in carbon, nutrient (primarily nitrogen and phosphorous), and oxygen cycling in estuarine and coastal waters. Phytoplankters grow rapidly, often doubling in number each day. Members of some taxonomic groups proliferate so rapidly that they form dense blooms that can affect water quality as they die, decompose, and sink, utilizing and depleting oxygen from the bottom waters. Moreover, a complex relationship exists between phytoplankton and those animals that feed on them. These animals, zooplankton and benthic filter feeders as well as some larval, juvenile, and adult fishes, graze on, depend on, and control coastal phytoplankton. When phytoplankton blooms, primarily those blooms limited by nutrient supply and light during winter and early spring when water temperatures are too low to support rapid growth of the zooplankton grazers, the excess plankton dies, resulting in oxygen depletion and fish kills. The effects of temperature on phytoplankton growth and photosynthesis are similar for most algal species, with a relatively rapid decline in production at temperatures in excess of their optimum, for example 20 (68 °F) to 25 °C (77 °F). Moreover, the photosynthesis cycle can be influenced for different variants by temperature and toxicants (Cairns et al. 1975). Also, an important aspect of photosynthesis sometimes forgotten is that while oxygen is produced during the light portion of the photosynthesis cycle, oxygen is used up during the dark or evening portion of the cycle. Consequently, a net loss of oxygen production can occur during overcast days, when the upper layer of an extensive bloom blocks the light from reaching the lower level of phytoplankton, or other conditions decreasing light to the plant community.

In contrast to the case with abundant phytoplankton in a system, zooplankton growth rates are enhanced and their biomass increases when water temperatures warm up in the spring and summer. Also, concurrent availability of nutrients for the phytoplankton can decrease because freshwater runoff, the primary source of nutrients to an estuary, often decreases, resulting in the increased grazing on and control of the phytoplankton biomass by zooplankton, benthic filter feeders, larval fishes, and some juvenile and adult fishes.

Mass mortalities of the GoM's most important natural resources in coastal and estuarine sites usually occur ephemerally, but can occur continually or seasonally. Those mortalities that occur offshore like in the "dead zone" can take place for extended periods even though boundaries of the zone can change seasonally and annually. The influence on the decline in water quality and fragile habitat health associated with nearshore events responds to factors such as rapidly growing and diversifying anthropogenic inputs associated with agriculture, aquaculture, urbanization, coastal development, and industrial expansion. Fish gills often provide an indicator of degraded water quality (Figures 14.3 and 14.4). The mass mortalities will be discussed below, but the brief background on eutrophication in general requires some attention. More detailed treatments can be found in numerous publications such as NOAA (1997), Livingston (2001), Pinckney et al. (2001), and Paerl and Justić (2011). NOAA (1997) provides specific eutrophication data for all coastal water bodies in all five Gulf of Mexico U.S. states; the book by Livingston (2001) provides continuous analysis of various rivers in the Florida coastal systems in the northeastern Gulf of Mexico from 1970 to 2000; and Pinckney et al. (2001) and Paerl and Justić (2011) describe the role of nutrient loading and eutrophication in estuarine ecology. When forecasting the future hypoxia status, the model of Justić

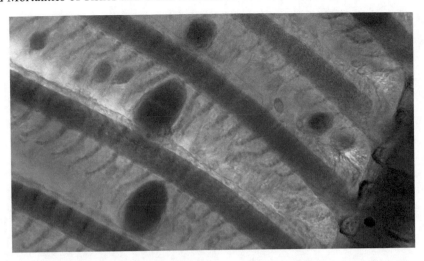

Figure 14.3. Secondary gill lamellae of the Atlantic croaker, *Micropogonias undulatus*, exhibiting progressive phases of telangiectasia, the swelling of weakened blood vessels, resulting from an early infection by the dinoflagellate *Amyloodinium ocellatum* in August 1998.

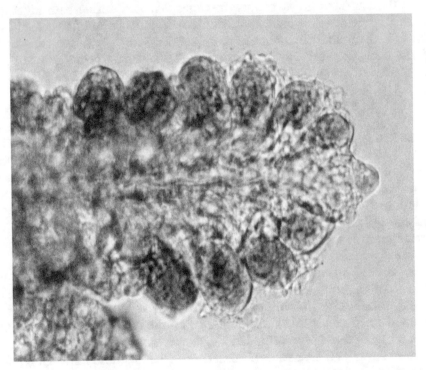

Figure 14.4. Telangiectasia, the reversible swollen blood vessels in the secondary gill lamellae of the inland silverside, *Menidia beryllina*, responding to harsh petroleum contamination.

et al. (2007) suggests that a reduction in riverine nitrogen of 40–45 % may be necessary to reach the goal of their action plan.

As a brief review, the common phenomenon of oxygen depletion usually results in mass mortalities in confined areas. For example, after a few days of overcast skies, photosynthesis with the accompanying production of oxygen during the daylight hours reduces. Without

Figure 14.5. Bluegill, *Lepomis marginatus,* with light infection of red sore disease caused by a combination of the bacterium *Aeromonas hydrophila* and the colonial peritrich ciliate *Heteropolaria colisarum* Mississippi, July 1974.

sunlight, such as during overcast or during night hours, the process of photosynthesis uses rather than produces oxygen. Consequently, if decaying matter such as dying algae, dead plant material carried from the rivers or marshes, or domestic waste from overflowing septic tanks accumulate in the confined areas, oxygen becomes depleted during both night and daylight periods. Fish will often try to avoid these conditions. Some species of fish such as menhaden, other clupeids, mullets, and catfishes are more sensitive to oxygen depletion than are other species and less likely to migrate away from areas with depleted oxygen. Some of these fish die and further reduce the amount of available oxygen, causing widespread and extensive fish kills, especially in harbors, dead-end canals, bayous, and small bays.

Fish involved in these kills are readily recognized by their pale or even whitish gills. These kills can be exacerbated by infestations of ectoparasites and bacteria. For example, centrarchid fishes such as the bluegill in Mississippi estuaries can be infested by the peritrich ciliate *Heteropolaria colisarum*. It, in turn, typically has a large concentration of the attached bacterium *Aeromonas hydrophila*. The ciliate feeds on free bacteria and organic debris, and *A. hydrophila* produces a series of proteolytic enzymes, some of which cause aesthetically displeasing lesions (Overstreet 1988) (Figure 14.5). These and other "red sore" lesions will be discussed in more detail under Section 14.4.1.1. Differentially expressed genes allow some organisms to tolerate low oxygen conditions such as in the grass shrimp *Palaemonetes pugio* exposed to cyclic hypoxia (Li and Brouwer 2013).

Mortalities caused by eutrophication occur so commonly on a seasonal and annual basis that reports seldom get published for individual cases in many areas other than in local media unless they are associated with specific bacteria, algae, toxins stresses, or other features. All harmful events do not necessarily kill the fish, and quite often the source is unknown (e.g., non-point source) (Figure 14.6). Compilations, however, are available such as for Florida (Table 14.1) and Texas and Louisiana (Zimmerman 1998; Thronson and Quigg 2008). Most mortalities seem to be caused by nuisance algae.

14.3.2 Hypoxia "The Dead Zone"

Eutrophication events discussed up to this point have dealt mostly with fish kills caused by oxygen depletion that occurred in bays and confined near-shore coastal habitats. However, other than areas in the Black Sea and Baltic Sea, a region in the northern Gulf of Mexico Continental shelf represents the largest coastal zone of hypoxia in the world. Even though this

Figure 14.6. Striped mullet, *Mugil cephalus*, caught by a commercial fisherman in December 1996 from the Pascagoula River, submitted to the National Marine Fisheries Service, and brought to us for evaluation. This case of unknown etiology affected a few of many present mullet with the pinkish-violet discoloration near a chemical plant.

Table 14.1. Dissolved Oxygen Fish Kills in Florida from 1973 to 2009 Based on Florida Fish and Wildlife Conservation Commission Fish Kill Database, http://research.myfwc.com/fishkill/[a].

Date Reported	County	City	Water Body Name	Specimen Count
11-Sep-02	Bay	Panama City	Mexico Beach, Beacon Hill area	2,000—species unknown
8-Jan-05	Bay	Panama City Beach	Crooked Creek	1,000s—mullet, shiner, bass
18-Feb-01	Charlotte	Bokeelia	Jug Creek on Pine Island	2,000—shrimp, sheepshead, species unknown
6-Nov-03	Charlotte	Placida	Placida Bay	1,000s—flounder, pinfish, pigfish, redfish
26-Oct-05	Charlotte	Port Charlotte	Vizcaya Lakes and Redwood Community	1,000s—bluegill, bass, crappie, perch, species unknown
8-Nov-05	Charlotte	Don Pedro Island	Don Pedro Island—Gulf Side	1,000s—coquinas
10-Jul-09	Charlotte	Port Charlotte	Strasburg Drive	1,000s—bass, minnow
1-Sep-09	Charlotte	Port Charlotte	Hog Island	1,000s—baitfish
23-Oct-09	Charlotte	Port Charlotte	Rossmere Road	1,000s—tilapia, bass
3-Jul-01	Citrus	Floral City	Tzalapopka Lake	1,000—bass, bluegill
16-Oct-02	Citrus	Hernando	Vanness Lake	2,000—bass, bluegill, catfish, minnows
2-Jan-04	Citrus	Crystal River	Crystal River Discharge Canal	1,000s—catfish, spotted eagle ray, sheepshead
4-Jan-04	Citrus	Citrus Springs	Barge Canal—Marker 33	1,000s—catfish, gar

(continued)

Table 14.1. (continued)

Date Reported	County	City	Water Body Name	Specimen Count
11–13-Sep-04	Citrus	Dunnellon	Withlacoohie, Withlacoochee	1,000s—bass, carp, brim, shiner, carp, catfish, shad
29-May–1-Jun-07	Citrus	Crystal River, Dunnellon	Lake Rousseau	3,000s—shad, catfish, shiner, catfish, tilapia
22-Apr-04	Collier	Naples	Lake Trafford	1,000s—catfish, bluegill, tilapia
9-Jun-04	Collier	Naples	I-75 and Pine Ridge Road	1,000s—species unknown
18-Aug-77	Escambia	Pensacola	Bayou Chico	30,000—menhaden
25-Jun-00	Escambia	Century	Unknown Water Body	2,000—shiners
27-Aug-01	Hernando	Webster	Withlacoocthee River	2,000—species unknown
26-Sep-98	Hillsborough	Tampa	Carrolwood Apts	2,000s—brim, species unknown
29-Sep-98	Hillsborough	Carrollwood	Colonial Guard Apts	2,000—brim, carp
6-May-00	Hillsborough	Riverview	Alafia—Buckhorn Springs	2,000—minnows, perch, redfish
2-Jun-00	Hillsborough	Seffner	Shangrila Subdivision	2,000—bass, crappie, tilapia
1-Aug-01	Hillsborough	Tampa	Davis Island	2,000—blue crabs, stone crabs, flounder, other species
25-Aug-01	Hillsborough	Tampa	Hillsborough Bay	2,000—All fish
17-Oct-02	Hillsborough	Lutz	Lake Hannah	2,000—bass, carp, skipjacks
10–11-Aug-03	Hillsborough	Tampa	Country Run Subdivision, Eagles Residential	1,000s—largemouth bass, shiner, bass
3-Oct-03	Hillsborough	Tampa	Bird Lake	1,000s—shiner, nile perch
2–6-May-04	Hillsborough	Apollo Beach	Masters Canal and area	1,000s—glass minnow, snook, catfish, tilapia, jack
18-Jun-04	Hillsborough	Gibsonton	Bullfrog Creek	1,000s—menhaden, sheepshead, pinfish, sand perch
13-Jul-04	Hillsborough	Tampa	Sleigh and Hillsborough River	1,000s—shad, tilapia, baitfish
1-Jun-05	Hillsborough	Apollo Beach	Masters Canal	1,000s—mullet
4–5-Sep-05	Hillsborough	Tampa	Waters Avenue, Woodlands Subdivision	1,000s—largemouth bass, bluegill, catfish, tilapia, species unknown
3-Aug-99	Lee	Fort Myers	Unknown Water Body	1,000—shiners
22-Aug-99	Lee	Fort Myers	Caloosahatchee River	1,000—baitfish
2-Apr-02	Lee	Fort Myers	Caloosahatchee River	2,000—bass, catfish, bluegill

(continued)

Table 14.1. (continued)

Date Reported	County	City	Water Body Name	Specimen Count
19–20-Apr-02	Lee	Fort Myers Beach	Estero Island	2,000s—baitfish
20-Apr-02	Lee	Fort Myers Beach	Moss Marina	2,000—baitfish
20–23-Apr-02	Lee	Sanibel	Sanibel River, Sanibel Island	2,000s—baitfish, minnows
12-Apr-04	Lee	Fort Myers	Private	1,000s—species unknown
5-May-04	Lee	Fort Myers	Lexington Country Club	1,000s—bluegill, carp
28-May-04	Lee	Sanibel	Clam Bayou	1,000s—redfish, snapper, mullet
11-Aug-05	Lee	Fort Myers	Grand Daza—Espero	1,000s—species unknown
6-Sep-05	Lee	Fort Myers	Hendry Creek	1,000s—mullet, sand perch, tilapia
27-Oct-06	Lee	Bonita Springs	Sheridan Run	1,000s—bass, carp
30-Aug-00	Manatee	Holmes Beach	Grand Canal	2,000—mullet, drum, shiners
31-May-06	Manatee	Palmetto	86 St. East	1,000s—bluegill, shiner
26-Jun-06	Manatee	Bradenton	Braden River	1,000s—bluegill, species unknown
19-Jun-09	Manatee	Bradenton	Braden River—Riverfront Drive	1,000s—snook, redfish, mullet, shiner
10-Aug-09	Manatee	Bradenton	Tidewater Preserve	1,000s—species unidentified
10-Mar-02	Miami-Dade	Miami	Venice Ave Bridge	2,000—mullet, catfish
14-Sep-02	Miami-Dade	Miami	Unknown Water Body	2,000—baitfish
22-Aug-05	Miami-Dade	Miami Springs	Miami Canal, Miami River	1,000s—species unknown
16-Nov-05	Miami-Dade	Miami	Saga Bay	1,000s—largemouth bass, peacock bass, shiner
20-Oct-09	Miami-Dade	Miami	NE 25th Court	1,000s—largemouth bass, bream, tilapia, snook
2-Oct-94	Monroe	Tavernier	Blackwater Sound	2,000—pilcards, barracuda, boxfish, puffers, spotted seatrout
20–22-Sept-03	Monroe	Key Largo	Largo Marina, Key Largo Marina	1,000s—Herring
2-Oct-06	Monroe	Lower Matecombe	Sleepy Lagoon	1,000s—glass minnow
19-Sep-09	Monroe	Marathon	Vaca Cut—109th Street, 109th St. Gulf	1,000s—sardine, pilchard, species unidentified
19-Jul-03	Okaloosa	Destin	Residential	1,000s—baitfish, softshell turtle

(continued)

Table 14.1. (continued)

Date Reported	County	City	Water Body Name	Specimen Count
11-Jul-98	Pasco	New Port Richey	Retention Pond	2,000—pinfish, sheepshead, whitebait
17-Aug-01	Pasco	Holiday	Horseshoe Lake	2,000s—species unknown
25-Aug-06	Pasco	Port Richey	Lake List to Lake Chrissy	1,000s—sunfish, brim, largemouth bass, species unknown
17-Dec-07	Pasco	Holiday	Rock Royal Dr.— Triangle Lake	1,000s—species unknown
2-Oct-09	Pasco	Odessa	Near Lake Josephine	1,000s—bream, crappie, tilapia, carp
6-Jun-94	Pinellas	St. Petersburg	Unknown Water Body	1,400—threadfin, herring
8-Nov-98	Pinellas	Clearwater	Off Exit 16	1,000—species unknown
13-Mar-00	Pinellas	St. Petersburg	Bartlett Park	2,000—snook, redfish
15-May-00	Pinellas	Clearwater	Feather Club Road	2,000—baitfish, brim
28-Jun-00	Pinellas	St. Petersburg	Lake Placid Mobile Home Park	2,000—species unknown
16-Nov-00	Pinellas	St. Petersburg	Spring Lake Apt	2,000—brim, bass
12-Jul-01	Pinellas	Oldsmar	Canal S of Lake Tarpon	1,000—shad
25-Jul-01	Pinellas	Pinellas Park	Residential Lake	1,000—bass, bluegill, shiners
19-Jun-02	Pinellas	Clearwater	Clearwater	1,000—baitfish
20-Aug-02	Pinellas	Oldsmar	Fillipi Park	2,000—catfish, trout, flounder, spade fish
8-Sep-03	Pinellas	St. Petersburg	Residential	1,000s—baitfish
14-May-04	Pinellas	St. Petersburg	County Lake	1,000s—tilapia, bluegill, crappie, golden
14-Oct-06	Pinellas	St. Petersburg	Lake Overlook	1,000s—needlefish, puffer, baitfish, greenback, pelican, blue heron
20-Aug-07	Pinellas	Seminole	Tides Golf Course	1,000s—tilapia
17–24-Aug-09	Pinellas	St. Petersburg	57th Avenue North and 112th Avenue NE	1,000s—bass, baitfish, bluegill, carp
5-Aug-78	Santa Rosa	Apalachicola	Garcon Point	10,000—alewives, croakers
2-Aug-82	Santa Rosa	Gulf Breeze	Behind Holiday Inn	15,000—alewives
5-Oct-04	Santa Rosa	Milton	Escambia—By Sandy Landing	1,000s—bass, bream, catfish
1-Nov-03	Sarasota	Venice	Residential	1,000s—species unknown
9-Jul-07	Sarasota	Sarasota	River Plantation	1,000s—bass, bluegill, species unknown
11-Jul-09	Sarasota	Osprey	Willowbend Pond	1,000—species unidentified

[a]Based on cases reported with more than 1,000 fish from a total of 1,198 cases.

zone, up to 20,700 square kilometers (km^2) (about 8,000 square miles [mi^2]) and reaching down to 30 m (about 100 ft) in depth, is called the "dead zone," it contains some life that can tolerate less than 2 mg/L oxygen. Rabalais et al. (2002) provided a good review of this seasonally and annually fluctuating zone. The zone typically occurs offshore from Louisiana between the mouth of the Mississippi River and the Texas border, but infrequently during some years it occurs off Texas, Mississippi, Alabama, and Florida. The zone receives high freshwater discharge from the nutrient-rich Mississippi and Atchafalaya rivers, and those nutrients and other organic matter help stimulate phytoplankton growth and create a stratified water column, differing in temperature, salinity, or both. The seasonally warmed surface waters establish a thermocline, with the less dense riverine fresh water further creating stratification with the saltier, cooler, denser water masses near the bottom. The phytoplankton not incorporated into the food web as well as fecal matter generated by the food web sink into bottom waters where the anaerobic bacteria decompose the matter, causing oxygen depletion. A well-defined seasonal cycle resulting from the strength and phase of river discharge, wind-mixing, regional circulation, and air–sea heat exchange processes usually generates maximum stratification during the summer and the weakest during the winter months. Because of these factors, the area comprising the zone fluctuates year to year (e.g., Fotheringham and Weissberg 1979).

In May–July 1979 after a heavy spring runoff and a diatom bloom, hypoxic bottom water developed in the upper Texas coast (Harper et al. 1981). Samples trawled from 6 m (about 20 ft) and 17 m (about 55 ft) depths consisted of only one fish species (hardhead catfish, *Ariopsis felis*), all individuals of which were dead or moribund as were many invertebrates, including the dominant polychaete population of *Paraprionospio pinnata*. Most Texas populations recovered in 1980; a few species of polychaetes that remained in low populations during the hypoxic period such as *Nereis micromma* and *Lumbrineris verrilli* increased in abundance immediately after the hypoxia abated probably because of larval recruitment; whereas others including *P. pinnata* with different life histories took much longer to reestablish.

The typical hypoxic zone, even though not as extensive as the one described above, appears from sedimentary evidence to have been present in the early 1900s and began to increase dramatically after about 1950. That is the time when the Mississippi Basin underwent a large human population increase with its increased nitrogen output through municipal wastewater systems as well as channelization and flood control of the Mississippi River along with associated deforestation, conversions of wetlands to cropland, loss of riparian zones, and expansion of agricultural discharge (e.g., Rabalais et al. 2002).

Life in the hypoxic zone differs according to the species and the oxygen concentration. Most fish are absent, some actually killed, in water with oxygen less than 2 mg/L; mantis shrimp and penaeid shrimps can tolerate 1.5 mg/L; epibenthic starfish and brittle stars die at <1.0 mg/L; and anemones, gastropods, and polychaetes die at <0.5 mg/L. At minimal levels of 0.2 mg/L, just above anoxia, sulfur-oxidation and bacteria form white mats on the sediments; at 0.0 mg/L oxygen, only black anoxic sediments exist without aerobic life. Demersal fish and invertebrates, those that live near the bottom, leave hypoxic areas and then re-occupy them by October–November. The distribution of sea turtles and cetaceans that prey upon those demersal animals seems to be somewhat dependent on the hypoxic zone (Craig et al. 2001). The oxygenated refuge habitats near the edge of the zone allow some animals to congregate (Craig 2012). The brown shrimp (*Farfantepenaeus aztecus*) and fish such as the Atlantic croaker (*Micropogonias undulatus*), spot (*Leiostomus xanthurus*), Atlantic bumper (*Chloroscombrus chrysurus*), and seatrouts collected with benthic trawls showed low DO avoidance thresholds and patterns of aggregations near these refuges. The brown shrimp, spot, and croaker showed a consistency between bottom DO avoidance thresholds and abundance in both catch per unit effort and laboratory experiments. Hazen et al. (2009) did not find that strong aggregation throughout the entire hypoxic edge of the water column, but they did find a greater biomass in the upper 7 m (23 ft) and much less biomass below 13 m (43 ft) in their hypoxic stations compared with their non-hypoxic ones.

Specific events related to oxygen depletion such as the *jubilee phenomenon* serve as a local one of those conditions that result in edible fish for those lucky enough to take advantage of the resulting "kill." Jubilees are well known in specific areas in Alabama and Mississippi and result from specific conditions. Depending on those conditions, they can be spread out over 25 km (about 15 mi) or just a few hundred meters of beach. In Alabama, most occur in the upper Eastern shore of Mobile Bay from Great Point Clear to just north of Daphne, and in Mississippi, most occur off Bellefontaine Beach and Gulfport, although they can occur elsewhere. In Alabama, where jubilees are known from as far back as the 1860s (even though documents were searched dating back to 1821), the specific set of conditions involves early morning hours before sunrise in the summer, and overcast or cloudy previous day, a gentle wind from the east, a calm or slick bay water surface, and a rising tide. These conditions produce a stratified layer of salty Gulf water accumulating in the deepest part of the northern portion of Mobile Bay overlain by lighter, fresher river water. During the calm conditions, the salty water stagnates because of decomposing plant material washed into the bay from the upstream marshes and swamps as well as supplementation by domestic wastes and becomes low in oxygen concentration. The rising tide and gentle wind-driven surface current causes an upwelling of this stagnant bottom water, forcing some species of bottom fishes and crustaceans to move ashore (Loesch 1960; May 1973; Turner et al. 1987). In Mississippi, Charles Lyles (from Overstreet 1978; Gunter and Lyles 1979), who observed them since the late 1930s, found several conditions in common. Jubilees occurred during neap tide (tides with a small difference between high and low tide occurring after the first and last quarters of the moon) at night between late June and early September, usually with rain preceding them and water with a well-defined tea color, presumably resulting from a specific phytoplankton organism. Affected animals usually include flounder, stingrays, croaker, spot, eels, blue crabs, and shrimp plus a lot of usually inedible anchovies, needlefish, and catfish. Seldom do these fish die, but they occur in extremely dense groups gulping for air; the eels usually burrowed tail first into the moist sand with their mouths wide open. Since these occur in early morning hours, neighbors often tell other neighbors about the event so they can collect large quantities of fresh seafood in wash tubs for their freezers after being caught with nets and gigs. When the sun rises, the tide changes, or the wind direction changes, the phenomenon stops, and most of the affected fish swim away. Conditions for this phenomenon, such as the role of carbon dioxide, still require scientific attention.

Phytoplankton constitute the most abundant and widespread primary producers in GoM and world waters and therefore support the bulk of marine food webs. Several of the phytoplankton species, including members of toxic algae in addition to nuisance algae, also cause animal illness and mortality of fish and other animals.

14.3.3 Nuisance Algae

Numerous species of nuisance algae commonly produce mortality events throughout the Gulf of Mexico region. Along the West Florida Coast, primary species include *Synechococcus* spp., *Anabaena* spp., *Chlorococcus minutus*, *Microcystis aeruginosa*, and other cyanobacteria (previously referred to as blue-green algae) and dinoflagellate species. These events tend to occur from April to November and last weeks to months. In the Florida Panhandle, nuisance algal events are mostly episodic, with a duration of days, and occur between July and September. Species include *Anacystis* spp., *Anabaena* spp., *Cladophora* spp., *Enteromorpha* spp., *Chlamydomonas* spp., and *Aphanocapsa* spp. In the Mississippi Delta/Louisiana Coast subregion, mortality events are mostly episodic, last from days in some estuaries to seasons in others, and generally occur between May and September in Mississippi Sound but also occur in January and February; in Barataria Bay, Louisiana, cyanobacterial blooms occur persistently throughout the year. Species in the subregion include *Exuviella* spp., *Prorocentrum minimum, Alexandrium* spp., *Anabaena circinalis, Katodinium rotundatum, Microcystis aeruginosa, Anacystis* spp., *Akashiwo sanguinea*, and others. Nuisance algal mortalities along the Texas coast occur mostly as day to month episodes between May and September except in the Upper Laguna Madre, Baffin Bay, and part of Lower Laguna Madre where *Aureoumbra lagunensis* occurs throughout the year. The latter alga produces brown tides, which occasionally block out sunlight and kill seagrasses; the blooms also occur in Florida and Mexico. During the period 1970–1995, the frequency and duration of events increased in Tampa Bay and Galveston Bay. Blooms of the dinoflagellate *Noctiluca scintillans* appear reddish orange during the day and can produce bioluminescence at night. Even though not a toxic alga, it can accumulate and emit ammonia in concentrations high enough to produce fish kills.

14.3.4 Toxic Algae: HABs, Including Red Tide

Some of the most prevalent toxic algae and associated toxins that cause animal mortalities in the Gulf of Mexico include *Alexandrium monilatum* (goniodomin A), *Karenia brevis* (brevetoxins), *Karlodinium veneficum* (karlotoxins), *Prymnesium parvum* (prymnesins), and *Akashiwo sanguinea* (surfactants). Other potential ichthyotoxic species are *Cochlodinium polykrikoides* (ichthyotoxins) and raphidophyte species such as *Chattonella marina, Heterosigma akashiwo*, and *Fibrocapsa japonica* that produce hemolysins, reactive oxygen species, polyunsaturated fatty acids, and possibly brevetoxins (Lewitus et al. 2014).

Toxic algal events in the GoM estuaries are variable in duration, lasting days to weeks in some estuaries and months to seasons in others. Impacts generally occur between June and October, except in Florida Bay and Apalachee Bay, where impacts occur between January and March. Occasionally, however, unpredictable toxic algal events may occur during any month of the year (NOAA 1997).

14.3.4.1 Red Tides, *Karenia brevis*

Most dinoflagellates are photosynthetic, possessing chlorophyll *a* and accessory pigments, and not toxic; they constitute an important and at times the dominant group of primary producers sustaining the food web. When some toxic species bloom, they cause massive fish kills. Red tide serves as the most well-known HAB in the Gulf of Mexico, with the best known species being *Karenia brevis* (previously known as *Gymnodinium breve*). A heavy bloom produces a reddish color in the water and is responsible for spectacular mass mortalities. Importantly, aerosols from a heavy bloom usually affect human respiration and occasionally cause contact dermatitis, which, in turn, provides considerable more incentive and support for research than would be received from fish kills alone. The U.S. population continued to increase between 1960 and 2010 and is projected to increase further, most significantly in coastal states, putting stress on the coasts and estuaries. Between 1965 and 1976, the number of confirmed worldwide red tide outbreaks increased sevenfold concurrent with a twofold increase in nutrient loading mainly from untreated sewage and industrial waste (Hallegreaff 1995). The threat to animals from red tide blooms is predicted by the number of dinoflagellate cells of *K. brevis*/L from a table by Lewitus et al. (2014) as (1) 1,000 cells or less (none anticipated), (2) >1,000 to 10,000 (very low, with possible human respiratory irritation, and shellfish harvesting closures when >5,000 cells/L), (3) >10,000 to 100,000 (low, human respiratory irritation, possible fish kills, and bloom chlorophyll probably detected by satellites), (4) >100,000 to 1,000,000 (medium, human respiratory irritation and probable fish kills), and (5) >1,000,000 (high, as above plus discolored water).

Blooms of the toxic alga *K. brevis* occur almost annually in the Eastern Gulf of Mexico, most frequently in Southwest Florida waters. Consequently, blooms are commonly referred to as "Florida red tides" and, as indicated above, have attracted research dollars for several decades. In fact, the University of Miami's initial Marine Laboratory, now known as the Rosenstiel School of Marine and Atmospheric Science, was established by F.G. Walton Smith to investigate red tides (e.g., Gunter et al. 1947, 1948). Also, Sammy Ray, along with Albert Collier and William Wilson, established the Galveston Laboratory of Texas A&M to investigate red tide and culture of *K. brevis* (see Zimmerman 2010). Gunter (1947) provided a short history of the Florida red tide in which he deduced that the phenomenon had been reported since 1844. He considered the death of the fish most spectacular because the dead fish floated, a diagnostic feature for fish killed by brevetoxin. He estimated the 1946–1947 red tide killed an estimated half billion fish; he said that such catastrophic kills may cover >25,000 hectare (ha) (hundreds of square miles), and the number of fish killed may even approach 1 billion. He also considered that few, if any places, on earth can produce such vast destruction of life so quickly as the dinoflagellate blooms of the shallow sea with the possible exception of fish kills along the Peruvian coast caused by El Niño. Blooms of *K. brevis* typically occur in the Gulf of Mexico; however, they can be entrained in the loop current and transported east through the Florida Straits and then north by the Gulf Stream as far as North Carolina. Quick and Henderson (1975) investigated the pathology of fish from a 1973 to 1974 Florida kill, and their evidence suggested that dehydration, hemolysis, and interference in blood-clotting mechanisms also caused fish-death in addition to neurointoxication, the previously assumed sole cause.

Brevetoxins from *K. brevis* are indeed complicated. There are several non-proteinaceous, lipid-soluble neurotoxins as well as hemolysins. For example, Baden and Mende (1982) investigated the toxicity of two of those toxins, using Swiss white mice and the western mosquitofish as assay animals. In the mice injected with one of the toxins, hypersalivation was the most pronounced sign, although copious urination and defecation commonly occurred as well as tremors, followed by marked muscular contractions. The mice exhibited compulsory chewing motions and rhinorrhea at higher doses. When given the other toxin, a distinct compound but with related chemical structure, no hypersalivation or chewing was expressed and muscular contraction was less pronounced. Mouse bioassays were used to determine the correlation between acute intraperitoneal injections and oral toxicity of shellfish extracts, and the oral assay was not recommended. The disease in humans eating brevetoxin-contaminated mollusks that goes by the name "neurotoxic shellfish poisoning" (NSP) can be debilitating but apparently non-fatal. The first toxin tested seemed to be the predominant agent responsible for the disease; the second at the dose tested produced subacute manifestations that occur in the human disease such as labored breathing, loss of appetite, and motor incoordination. Signs of the disease generally subside in 2–3 days. These signs from both toxins are typical of muscarnic stimulants, as found in *Amanita muscaria* (a poisonous mushroom), as opposed to nicotine, another stimulant acting on acetylcholine receptors and bind to voltage-sensitive sodium channels involved in the propagation of nerve impulses. Binding opens the sodium channels at a normal resting potential and consequently inhibits sodium channel inactivation, which can result in repetitive firing in nerves. Further studies described by Baden et al. (2005) characterized additional brevetoxins, each with its own specific toxicity and based on one of two different structural features (six toxins known with one and three, thought to be more potent, with the other). More importantly, these multiple brevetoxins activate brevetoxin metabolites, which can be modulated by the different, shorter, trans-fused polyether antagonist brevenals. Brevenal, obtained from either the environment or the dinoflagellate culture, binds receptors and inhibits brevetoxin binding and activity, counteracting the toxic effects on both mice and fish. The pulmonary receptor for both brevetoxins and brevenal seems to be distinct from the neuronal binding site. In other words, the multiple biotoxins and antagonists interact with at least neuronal, pulmonary, and enzymatic regulatory systems of animals, generating a complex combination of acute and chronic signs in animals, including humans, exposed to aerosolized bioactive substances produced by *K. brevis*.

Most data on Florida red tide fish kills acquired up to the last decade or so were anecdotal and qualitative but useful. Gannon et al. (2009) investigated the effects of the algal blooms on nearshore fish communities in five habitats in Sarasota Bay and adjacent areas. They looked at the cell density of *K. brevis* as well as data on fish density, fish species composition, water temperature and salinity, dissolved oxygen, and turbidity. The clupeid (herring-like fish) trophic guild (a guild [or ecological guild] consists of any group of species that exploit the same resources) was not affected by the cell density of the toxic algae as were all other eight fish trophic guilds. Fish density as measured by catch per unit effort (CPUE) and species richness of those other eight guilds all had a negative association with the algal cell density; 96 % of the local fish kills from 2003 to 2007 (ranging from 4 in 2007 to 72 in 2005, with more nearby) occurred during red tides. The guild consisting of the demersal invertebrate feeders was the most sensitive to the effects of the red tide, whereas the clupeids were the least sensitive, and, when excluding the clupeids, the difference between CPUE in red tide period versus non-red tide period ranged from 57 % in the mangrove habitat to 88 % in the GoM habitat. Fisheries-independent monitoring data (as opposed to fishery-dependent data, which are data collected directly from commercial and recreational fisheries sources) from the Tampa Bay area collected from 1996 through 2006, with an emphasis on the persistent red tide of 2005, analyzed by Flaherty and Landsberg (2011) showed that in the spring of 2006 there was a decline in the annual recruitment of juvenile spotted seatrout (*Cynoscion nebulosus*), sand seatrout (*Cynoscion arenarius*), and red drum (*Sciaenops ocellatus*). However, the subadult and adult abundance values for these fishes remained consistent with those of previous years. The respective recruitment periods of some of the other fishes did not correspond with the major red tide event. The importance of clupeid fishes such as Spanish sardines, thread herrings, and Atlantic shad in the understanding of fish kills has been recognized by Walsh et al. (2009).

The dinoflagellate *K. brevis* requires nutrients to form the catastrophic blooms. A nitrogen isotope budget of the coastal food web shows that diazotrophs (nitrogen fixers, primarily the filamentous cyanobacteria *Trichodesmium* spp.) form the initial nutrient source of red tides and clupeiformes (decomposing dead sardines, herrings, and bay anchovies) serve as the major recycled nutrient source for the maintenance of those blooms. In 2001, the dinoflagellate "harvested" >90 % of the clupeids along the West Florida Shelf rather than being harvested by fishermen. Fish kills typically originate when *K. brevis* cells lyse and release their toxins, which become absorbed directly across the gill membranes. Fish may also die after ingesting the dinoflagellate cells or toxins in the water, or after consuming contaminated biota (Landsberg et al. 2009).

The Center for Prediction of Red Tides (CPR) in Florida (Walsh et al. 2009) has developed models to assess and predict red tides based on nitrogen isotope ratios in portions of the food web that maintain *K. brevis*. The food web associated with *K. brevis* has shown to be extremely complicated and differs somewhat in different areas based on currents and winds. Some model components are based on features such as temperature. At summer temperatures, as much as 50 % of some Florida fish decay to inorganic forms of phosphorus and nitrogen within 1 day (Stevenson and Childers 2004; Walsh et al. 2009). Some clupeids can provide about 50 % of the nitrogen supply for red tides. Of equal concern is the nearly equal inclusion of the diatom-based food web, including flagellates, that also feeds the herbivores (harpacticoid and calanoid copepods and certain other members of the zooplankton), in turn feeding phytoplankton-feeding fishes (clupeiformes mentioned above including the Gulf menhaden *Brevoortia patronus*, which feeds on both phytoplankton and zooplankton, plus the mugilid [striped mullet, *Mugil cephalus*] that feeds additionally on bacterial degraded phytodetritus) and the piscivorous fish like mackerel, snappers, and groupers that feed on them. Isotope data and animal kills suggest the kills in Florida and the northern Gulf in one year, like 2006, can show how the kills decreased on the West Coast of Florida and then increased on the East Coast of Florida in 2007. The tides have "downstream" consequences up to 1,000 km (621.4 mi) from the Florida Panhandle to Cape Hatteras on the Atlantic coast.

Small fish kills can also be related to dust and associated nutrients blown into the Gulf from African and occasionally Asian deserts (Garrison et al. 2003), and those kills can include related toxic dinoflagellates in addition to *K. brevis*. Actually, in the Gulf of Mexico, there are at least nine known established species in the Kareniaceae, and most produce ichthyotoxins such as brevetoxins, karlotoxins, and gymnodimines (Steidinger et al. 2008). These include five species of *Karenia* (*K. brevis*, *K. papilionacea*, *K. mikimotoi*, *K. selliformis*, *K.* cf. *longicanalis*), three of *Takayama* (*T. pulchella*, *T. helix*, and *T. tasmanica*), and *Karlodinium veneficum*, the latter confirmed as cause of fish kills in estuarine ponds. *Karenia brevis* typically occurs in high salinity waters. In 1996, a bloom occurred in inshore waters of Alabama, Mississippi, and Louisiana, contaminating oyster beds. This bloom consisted of a complex of *Karenia* species, some of which can tolerate low salinities (5–40 parts per thousand) (ppt), but *K. brevis* was the most prominent species. Maier Brown et al. (2006) examined preserved specimens from this bloom, and they also investigated salinity tolerances of three clones of *K. brevis* and compared them with a fourth. For the three clones, the experimental minimum salinity at which growth occurred ranged between 17.5 and 20.0 ppt and optimum salinity range from 20–25 to 37.5–45 ppt, depending on the clone. In the northern Gulf of Mexico bloom, the concentration of cells/milliliter (mL) for the complex was high enough to close oyster beds in salinity as low as 14 ppt. Some agents occurred in salinities less than 10 ppt in both the northern Gulf and in Florida. Brevetoxins measured in the *K. brevis* cultures were found to be higher during the stationary phase of growth and approaching senescence, regardless of salinity, suggesting that as a natural bloom ages, it could potentially become more toxic and pose an increased threat to public health. The specific 1996 bloom seemed to originate in the Florida panhandle and move westward, rather than the typical eastern movement, into Mississippi Sound because of the unusual effects of Tropical Storm Josephine (Maier Brown et al. 2006).

Fish kills resulting from *K. brevis* also occur in Texas and Mexico. Gunter et al. (1948) reported on such massive fish kills, Zimmerman (1998) edited a report covering such mortalities of a variety of animals in Texas and Louisiana in 1994, and Magaña et al. (2003) tabularized and discussed a series of referenced reports of fish kills from various locations along the Texas coast as well as Tamaulipas-Veracruz and Yucatán, Mexico, which occurred from 1935 until 2002. Because of the severe respiratory events involving irritation, stinging eyes and nose, accompanied by a dry, choking cough, resulting from inhalation of air-borne brevetoxins, historic references provide information on Mexican events occurring from 1648 to 1875 (Magaña et al. 2003) and earlier. One case in 1792 chronicled by a government official and reported by Lerdo de Tejada (1850) indicated that sales and consumption of dead fish collected from the mass mortality of fishes on Veracruz beaches resulted in violent human mortalities. Nuñez Ortega (1878) and later others suggested that the human deaths actually resulted from bacterial contamination of or ciguatera toxins in spoiled fish. Fish kills probably resulting from *K. brevis* along the Texas shelf occurred during 1529–1534 (Adorno and Pautz 2003). Cabeza de Vaca was a survivor of the Narvaez Expedition and reported that the Capoque and Han Indians avoid fish and suspend oyster harvesting seasonally around Galveston Island; during 1534, the Avavares Indians near the Nueces River, Texas, apparently estimated seasonal changes by "the times when the fruit comes to maturity and when the fish die" (Walsh et al. 2009).

Bony fishes constitute most of the commonly killed animals, and, as indicated above, some are important sources of stored brevetoxin necessary for future blooms. They can build up to high dangerous levels in living fish tissues by being in the water with *K. brevis*, by feeding on contaminated mollusks and other invertebrates, or by feeding on contaminated fish; toxins can be abundant in the entire food web (Naar et al. 2007; Landsberg et al. 2009). Until 2000, no mass mortality of sharks or rays caused by red tide had been reported from Florida. Flewelling et al. (2010) reported the mortality of large numbers of blacktip sharks (*Carcharhinus limbatus*) and fewer Atlantic sharp nose sharks (*Rhizoprionodon terraenovae*), mostly juveniles, from the Florida Panhandle. They also examined tissues from 22 species of sharks and rays collected between 2000 and 2008 from animals both in and not associated with red tides along the West Coast of Florida and the East Coast, where some of the animals also accumulated the toxins. The amount of accumulated toxins differed among species, tissue sites, and geographical locations, and in-utero embryos also had accumulated brevetoxins. The brevetoxin concentrations in animals do not necessarily relate to being from or near blooms, and levels are not harmful for human consumption unless the liver is eaten. Large sharks seem to avoid the toxin.

Waterfowl can also be affected by red tide blooms. For example, several thousand individuals of the lesser scaup (*Aythya affinis*) and lesser numbers of other birds were found dead associated with the red tide fish kill in the Tampa Bay area. Not all birds present died. Examination for bacteria, parasites, pesticide residues, and acutely toxic material did not suggest that any was associated with the mortalities. White Peking ducklings experimentally exposed to the red tide toxins in seawater, either in addition to force-fed contaminated clams (*Mercenaria campechiensis*) or given non-contaminated clams, became lethargic, developed spastic movements of the head, and died (some individuals in the toxic seawater with non-exposed clams did not die) (Forrester et al. 1977). When Ray and Aldrich (1965) force-fed three doses of experimentally exposed oyster tissue to baby chicks, all doses produced in the chicks a loss of equilibrium, and the two higher doses produced death within 22 h. Shorebirds, including sanderlings (*Calidris alba*) and ruddy turnstones (*Arenaria interpres*), scavenged on beached individuals of the thread herring, scaled sardine, and mullets during a red tide kill. High concentrations of brevetoxin in those fish tissues corresponded with high levels in livers of shorebirds that were collected dead along the local beaches and from rehabilitation centers during the red tide event, suggesting that brevetoxin exposure serves as a risk factor for bird mortality (van Deventer et al. 2012).

Since red tide blooms have been known in the Gulf of Mexico, they have been associated with mortality of numerous animals at higher trophic levels, such as marine birds, sea turtles, and marine mammals (Gunter et al. 1948; Quick and Henderson 1974; Forrester et al. 1977; and others). Because of the ability for fishes and invertebrates (see list of maximum brevetoxin concentrations in bivalves listed by Landsberg et al. (2009)) to bioaccumulate the toxins, blooms do not necessarily have to be present to kill animals. Landsberg et al. (2009) listed hundreds of manatees (*Trichechus manatus*) and bottlenose dolphins (*Tursiops truncatus*) killed in both reported and unpublished mass mortalities and not necessarily concurrent with blooms. Even though presently impossible to determine specific lethal concentrations of the toxins and their metabolites, the presence of high levels in the animals was either solely responsible for the deaths or in combination with other harmful factors. Twiner et al. (2012) critically investigated bottlenose dolphin mortalities from the Florida Panhandle and found high levels as they also did for the clupeid *Brevoortia* sp., which was found abundant as a dietary prey in their stomach. When dead manatees from the 1996 red tide bloom were necropsied, Bossart et al. (1998)

observed severe nasopharyngeal, pulmonary, hepatic, renal, and cerebral congestion in all cases. Some exhibited pulmonary edema and hemorrhage. Immunohistochemical staining using a polyclonal primary antibody to brevetoxin exhibited intense positive staining of lymphocytes and macrophages in the lung, liver, secondary lymphoid tissues, nasal mucosa, and meninges. These data suggest that manatee mortality may occur after chronic inhalation and ingestion rather than responding in an acute event. Local rehabilitation centers have successfully recovered several species of birds, turtles, and manatees that would otherwise probably have died from the red tide. The reason humans do not die or become severely ill from inhaling aerosols or ingesting brevetoxin accumulated in fish or bivalves probably relates to their ability to avoid lethal doses. This contrasts to ciguatera toxin, which is a similar compound acting in the same manner; however, its toxin from the epibenthic dinoflagellate *Gambierdiscus toxicus* can be bioaccumulated in fishes to a much more harmful concentration without causing mortality of the fish (Naar et al. 2007).

14.3.4.2 Fish Kills From Algal Agents Other than *K. brevis*

Additional investigations on pathology of fish will show other related agents being responsible for fish mortalities. When fish kills occurred in estuarine aquaculture facilities in Maryland, they were determined to be caused by at least two isolated karlotoxins from the dinoflagellate *Karlodinium veneficum* (as *K. micrum*) by Deeds et al. (2006). *Karlodinium veneficum* has been reported from Florida in the Gulf of Mexico, has caused fish kills in Maryland and South Carolina, and is considered a cosmopolitan species. Fish from kills near Perth, Western Australia, examined by the senior author had diagnostic epithelial necrosis and shortening or loss of the secondary lamellae of the gills, the primary signs observed in the sheepshead minnow (*Cyprinodon variegatus*), a common fish in the northern Gulf of Mexico. Concentrations of toxins in filtered water from fish kills rapidly killed the experimental fish.

Also, the dinoflagellate *Pfiesteria piscicida* can produce lesions, and at one time was considered the cause of ulcerated mycosis of Atlantic menhaden, resulting in fish kills along the Atlantic coast to the GoM (Dykstra and Kane 2000) (Figure 14.7). Considerable research has gone into the cause of these lesions, and now Blazer et al. (1999) and Vandersea et al (2006) have determined that the primary agent is the pathogenic oomycete *Aphanomyces invadans*. *Pfiesteria piscicida* and later *Pseudopfiesteria shumwayae* (see Litaker et al. 2005) were originally thought to secrete potent exotoxins that caused the lesions, acute fish kills, and human disease in the mid-Atlantic estuaries. However, bioassays with *P. shumwayae* and larval fish revealed no toxin was emitted and mortality occurred only in treatments where fish and

Figure 14.7. Atlantic menhaden (*Brevoortia tyrannus*) from St. Johns River, Florida, in June 1985, exhibiting typical lesions now recognized as caused by the oomycete fungus *Aphanomyces invadans*. Fish collected and photographed by Harry Grier of the Florida Department of Natural Resources. Permission to reprint granted by H. Grier to R.M. Overstreet.

dinospores demonstrated physical contact. Dinospores swarmed toward and attached to the skin, actively feeding on and denuding fish of their epidermis and killing them by micropredation (Vogelbein et al. 2002).

Some dinoflagellates produce a toxin harmful and even deadly to humans and marine mammals, but not recognized as causing fish kills. One of these toxins is saxitoxin (STX) puffer fish poisoning, which can also on occasion include paralytic shellfish poisoning (PSP). The signs of eating toxins accumulated in puffers progress from tingling and numbness of the mouth, lips, tongue, face, and fingers; to paralysis of extremities, nausea, vomiting, and ataxia; to decreasing breathing and possibly to death by asphyxiation. The toxin occurs in the Gulf of Mexico as determined by Landsberg et al. (2006) and Deeds et al. (2008). The toxins can be produced by *Pyrodinium* by means of the shellfish, *Alexandrium cohorticula*, *A. minutum*, *A. ostenfeldii*, *Gymnodinium catenatum*, and some freshwater cyanobacteria, all of which occur in the Gulf of Mexico, but verified cases caused by the toxins in the GoM come from the bioluminescent *Pyrodinium bahamense*. Within all puffer species, they are stored in the skin, muscle, and viscera with an emphasis on ovary, making those structures a risk for human consumption. The toxin in the southern puffer (*Sphoeroides nephelus*) from the Gulf side of Florida is much less in quantity than in fish from the Atlantic side, where it can remain not depurated for over a year. However, one should realize that toxin produced by one strain of a species often does not represent that production for the species. For example, the toxin for PSP produced by 17 strains of the dinoflagellate *Alexandrium tamarense* had a wide range in the amount based on mouse bioassays. Furthermore, 15 sub-strains taken from one of those strains also had a considerable range in the amount, and that toxin from two different strains differed in the derivatives produced (see Thessen et al. 2009).

The golden alga *Prymnesium parvum* occurs worldwide, but it is best known from inland waters of Texas and estuaries of the Gulf of Mexico. Under certain environmental stresses, it produces massive fish kills, including kills of mussels and clams. Even though the alga has been identified from many locations, it often does not cause mortalities. Allelopathy has been shown to be one reason. That is a biological phenomenon by which an organism, in this case a concurrent cyanobacteria (a prokaryotic phytoplankton that has bacteria-like cellular features such as lacking a well-defined nucleus and membrane-bound organelles), produces one or more substances that influence the growth, survival, or reproduction of another organism. James et al. (2011) showed that one substance, the cyanotoxin microcystin-LR, inhibited growth of *P. parvum*, but the necessary concentration could also kill a number of other aquatic organisms.

Another non-dinoflagellate alga that attracts attention is a complex of diatoms responsible for "amnesic shellfish poisoning" (ASP). Most diatoms constitute highly proactive phytoplankton in estuaries, supporting both planktonic and benthic food webs, but the colonial *Pseudo-nitzschia* spp. produce a domoic acid toxin (DA) that causes ASP. The toxin accumulates in bivalves, but ASP is most common along the Pacific Coast in upwelling systems where seabirds and marine mammals die from it, and, consequently, marine resource management agencies both along the Pacific Coast and the GoM close shellfish beds when DA levels are high because ASP causes loss of memory in humans.

Diatoms in the genus *Pseudo-nitzschia* occur frequently in the northern Gulf in offshore and estuarine plankton, on sediments, and in both shellfish tissues and seawater in Mississippi Sound as well as in Alabama, Louisiana, and Texas (Dortch et al. 1997; Macintyre et al. 2011). Although not all species of this genus are toxic, and no case of human ASP has been reported from the GoM, when counts of the diatom and concentrations of DA in oyster tissue exceed federal guidelines, oyster reefs are temporarily closed. However, because DA occurs in GoM shellfishes, because it is produced by several species in the genus, and because it imposes a major human threat, it is presently being investigated in some detail. Even though the disease is considered a problem in high salinity waters, various species occur over a salinity range of 1 to >35 ppt in Louisiana, where oysters are typically harvested in 10–20 ppt (Thessen et al. 2005). These authors identified seven species in low salinity waters, and some are toxigenic. Much has been learned about these diatoms from laboratory work as well as from species around the world. Experimental studies have shown that the problem is extremely complex. Different strains of one species isolated from the same water sample exhibited broad differences in growth rate and toxin content when cultures contained different nitrogen sources, ammonia, nitrite, and urea (Thessen et al. 2009). Two clones of one species produce toxins; however, they preferentially utilized different nitrogen sources. Two of nine isolates of another species and two of five of still another produced DA, but the content varied by orders of magnitude. If that does not exemplify the complexity of the problem, then it should be noted that DA, in addition to being accumulated in bivalves, also occurs in tissues of zooplankton, crustaceans, echinoderms, echiurans, tunicates, and fishes; it also occurs in tissues of marine mammals, birds, and humans, all of which could be killed by it, as well as occurring in sediments, demonstrating stable transfer through the marine food web and abiotically to the benthos (Trainer et al. 2012). The latter review included considerably more information on the cosmopolitan nature and complexity in taxonomy, toxin production, toxin storage/release, bloom initiation/retention, and nutrient requirements for some of the 14 recognized species, and also mentioned that preliminary work suggested the necessity for the presence of an epibiont bacterium before sexual reproduction could occur in some clones of one species grown in axenic culture.

To reiterate the aspect of human illnesses from HABs, those that occur worldwide result from harmful algal toxins and their derivatives including saxitoxins (STX, including some paralytic shellfish poisoning [PSP]), okadaic acid (diarrheic shellfish poisoning), brevetoxins amnesic shellfish poisoning [ASP] (neurotoxic shellfish poisoning (NSP)), ciguatoxins (ciguatera fish poisoning), domoic acid (am/domoic acid poisoning), azaspiracid toxins (azaspiracid poisoning), and hepatoxins and microcystins. Dinoflagellates produce all these toxins except for domoic acid, which as discussed above is produced primarily by diatom species of the genus *Pseudo-nitzschia*, and hepatoxins and microcystins produced by cyanobacteria such as species of *Anabaena* and *Microcystis*. In addition to being produced by dinoflagellates, saxitoxin can be produced by several species of cyanobacteria, and brevetoxin can be produced by some species of raphidophytes. Deadly phycotoxins include domoic acid, saxitoxins, and ciguatoxins. Perhaps the deadliest of the phycotoxins are the STXs because of the rate of human mortality associated with exposure and the broad geographic range of distribution of STX-producing organisms. Saxitoxins produced by multiple dinoflagellate species as well as several species of cyanobacteria and can cause PSP. Moreover, the toxins can be transferred and bioaccumulate throughout aquatic food webs and therefore be vectored to terrestrial biota, including humans (Deeds et al. 2008). Ciguatera is more common in Mexican and eastern Caribbean reefs than in the northern Gulf (e.g., Okolodkov et al. 2007).

At least 15 species of *Prorocentrum*, *Dinophysis*, and *Phalacroma* are known to produce okadaic acid (OA) or its derivatives in the world's oceans, and those species occur in the GoM. However, only isolates of *Dinophysis* cf. *ovum*, *Prorocentrum texanum*, *P. hoffmannianum*, and *P. lima* have been demonstrated to produce OA in the Gulf. The toxin accumulates in bivalves, and the human disease associated with eating such bivalves is termed "diarrhetic shellfish poisoning"; conclusive evidence pointing to OA by itself causing fish disease has not been established.

14.3.5 Cold Kill

Cold kills appear conspicuous after a period of low temperature. They, however, are restricted to shallow waters and not as common as one might believe. Under normal conditions, when a cold front passes through an area, most fishes and invertebrates bury or migrate to more tolerable areas and do not die. Typically, it is the rate at which the temperature drops rather than the temperature per se that kills fish. Fish kills are more prevalent in the typically warmer southern waters of Texas and Florida than in the more temperate northern Gulf of Mexico where the rate change during freezing conditions is not as great and the fishes are more able to acclimate. A good example of this situation occurred in Mississippi in January 1973 and was studied in some detail by Overstreet (1974). During the evenings of January 13–14, 1973, a thin sheet of ice covered the surface of Paige and Cooper bayous in Jackson County, Mississippi. On the 15th, these bayous, approximately 1 to 5 m (about 3 to 16 ft) deep and completely fresh during this time in the year, became covered by a layer of the striped mullet, *Mugil cephalus*, which had surfaced and died. By January 16, the 0.6 m (2 ft) tide washed out the majority of fish, but a minimal estimation of a few hundred thousand carcasses still remained. A large number of shellcrackers, bream, bass, and catfish actively fed when local residents, who frequently fed them, placed food in the water, suggesting that no toxin occurred in the water and no low concentration of oxygen existed. In the morning of the 16th, several coastal habitats were inspected for dead and living fish, and corresponding values were obtained for salinity, chlorosity, and calcium in the water. A few other bayous also contained dead striped mullet such as the Ocean Springs Small Craft Harbor, which contained additional dead species of the striped mullet such as white mullet (*Mugil curema*), Atlantic tarpon (*Megalops atlanticus*), and fat sleeper (*Dormitator maculatus*).

Fishermen caught striped mullet during and after January 13 and 14 in nearby Graveline Bay, Bayou Porteaux, and other areas where no dead fish was observed. The unusual thing about the areas from which the mullet survived was that the water had a salinity greater than 6 ppt. At least the dying fish from Paige Bayou also exhibited starvation, had distended gallbladders with associated leaking bile (Figure 14.8), and demonstrated high levels of dichlorodiphenyltrichloroethane (DDT) metabolites and endrin pesticide residues unlike mullet samples from where no fish had died. An average-sized dead mullet was 230 millimeters

Figure 14.8. Striped mullet, *Mugil cephalus*, exhibiting viscera showing enlarged leaking gallbladder and intestine devoid of food and representative of moribund mullet during a cold kill on November 16, 1973.

(mm) (9 inches [in]) standard length, with a weight of 255 grams (g) (9 ounces [oz]), and large fish such as these are more susceptible to a variety of stresses. Foci of hepatic necrosis and an abundance of lipid material but not glycogen were demonstrated in hepatocytes of the fish livers from the mass mortalities relative to control samples. Far fewer ciliate protozoan parasites and no monogenoid or copepod infested the gills of dying mullet, and those parasites were also common in Davis Bayou where there was no mortality.

Experimental studies (e.g., Cummings 1955; McFarland 1965) have shown that when the striped mullet is gradually transferred from seawater to freshwater, it can regulate serum ion concentration, muscle ion concentration, and osmolarity and surface permeability may be reduced by prolactin in relationship with temperature. At least those dying mullet in water less than 6 ppt salinity with 4.5 g chloride/L and 94 parts per million (ppm) calcium probably had a failing ion-osmoregulatory mechanism and were unable to acclimate to the rapidly dropping temperature.

On January 25 along Cooper and Paige bayous and on January 22 in a canal off Mary Walker Bayou, each location had a few thousand bloated and decomposing floating dead fish with attached filamentous algae as long as 4 cm (1.6 in). At the same time, healthy mullet without any indication of attached algae were present (see later comment on pseudo-fish kills). In his lengthy discussion about all aspects of the mortalities, Overstreet (1974) discounted with adequate evidence several hypotheses for the mortalities, presented by interviews with longtime residents of the area.

The most severe cold fronts appear to affect the coastal biota of Florida, Texas, and occasionally in between. Severe cold fronts, presumably with air temperature less than −12 °C (10 °F), recorded for coastal Mississippi include at least January 1899, February 1914, January 1985, and December 1989 (Bergeron 2015). Cold fronts passing over the shallow waters of the Gulf in western Florida occasionally result in chilled and helpless or dead fish with massive numbers washed ashore. In waters of Cedar Keys and north, most fish in a 3-day 1917 cold wave left the coastal waters for protection from the rapid temperature drop. Dead fish were usually small, 5–8 cm (2–3 in), accompanied by crabs and small shrimp. Near Tampa, mullet, grunts, and jacks died, and further south toward Key West, "tons" of fish became numb, washed ashore, and had to be buried to avoid the stench (Finch 1917). Finch (1917) even quoted a Federal fisheries biologist as saying that a benefit of that cold spell to oysters was the near eradication of a parasite that had previously been killing the oysters near Cedar Key and Port Inglis.

Willcox (1887) reported that thousands of smelly fish killed in bays and rivers from Cedar Keys to the mouth of the Caloosahatchee River at Punta Rassa by an 1886 freeze. The numbers and species of dead fish, including oysters, differed by location, but few actually occurred along the shore of the Gulf, and those that did occurred near inlets and probably resulted from tidewater carrying them out from the bays. Nine freezing episodes at Sanibel Island, Florida, from 1886 through 1936 were reported by Storey and Gudger (1936), who listed the 1886 one as the worst for both fishes and vegetation. The air temperature in Fort Myers was −4.4 °C (24 °F) and that near the salt water was −2.2 °C (28 °F) and lasted for a day; water temperature never reached 0 °C (32 °F). About 1.3 cm (0.5 in) of ice formed in the cisterns and rainbarrels; the weather turned warm and it rained after the freeze. Generally, the local common fish species often died, but in some cases the larger fishes became lethargic and recovered before they washed ashore. Only the hardiest of fish at Sanibel Island can tolerate a water temperature rapidly dropping below 15.6 °C (60 °F). Lethargic fish have often been gathered and eaten in Florida as well as Mississippi. Those in Florida, especially those already putrefying, are often gathered and used as fertilizer. Apparently, fishing typically recovers within 2 to 3 weeks after a freeze. Another major fish kill in southern Florida occurred during January 27 through 29, 1940, when minimum air temperatures ranged from −0.6 °C (31 °F) in Miami to 10 °C (50 °F) in Key

West. Most of the killed fish included bonefish, moon fish, several different snappers, grunts, porgies, mullet, and jacks. Lesser numbers of several fishes also died or became lethargic (Miller 1940). An estimate of nearly 450,000 kilograms (kg) (1,000,000 pounds [lb]) of stunned but good edible specimens were gathered and sold by fishermen from Key Largo to Key West. Digital thermal infrared data acquired by a NOAA-5 meteorological satellite followed three consecutive cold fronts which crossed South Florida and northern Bahamas in January 1977 (Roberts et al. 1982). The third and most severe frontal system crossed the shallow, carbonate Florida Bay and depressed water temperature for 7–8 days below 16 °C (61 °F), a thermal threshold for most reef corals. Water temperature in Florida Bay decreased to at least 13 °C (55 °F). Coral and fish kills occurred along the Florida Reef Tract, with mortality at Dry Tortugas estimated at 91 %. Low water temperature was suggested as the major factor-inducing stress in this reef system. Roberts et al. (1982) discussed works by others indicating extensive drowned and killed Holocene coral reefs in the southeastern Florida shelf margin during the first stages of shallow, widespread flooding of the shelf during the sea level rise occurring approximately 7000 years Before Present. They considered the topography of the southeastern Florida shelf and other high latitude reef areas as probably being dramatically affected by the combination of reef growth and severe cold water stress.

Texas is probably the most vulnerable area on earth to cold kills. It occupies approximately 900,000 ha (3,400 mi^2) of bay waters with offshore depths being only 1.8–2.4 m (6–8 ft) deep. Polar fronts push south to the southern part of the state and occasionally are strong enough to cross the Gulf of Mexico and over the Isthmus of Tehuantepec down to Nicaragua on the Pacific coast. Fish kills extend south into Mexico (Gunter 1947). The shallow bays of Texas are connected to the GoM by typically narrow passes; consequently, the rapid drop in temperature often traps the fishes within the bays. Gunter and Hildebrand (1951) described animal kills occurring in 1951 in and around Aransas Pass. The storm with winds up to 64 km/h (40 mi/h) dropped temperatures below freezing on January 29 and remained there for 5 days, with air temperatures as low as −8 °C (18 °F). Gunter (1941) also described that the animals killed from a front passing through the same general area during January 18–22, 1940. In both cases, there were several million fish and other animals killed by the cold and numerous others numbed. Those two papers considered the freezes somewhat equivalent catastrophes to those of 1924, 1899, and 1886, but certainly not as severe as those in 1941 and 1949, although considerable mortality occurred in the 1947 freeze. Gunter (1947) considered a catastrophic cold kill to occur on the average of every 14 years from 1856 to 1940 with less damaging ones occurring at shorter intervals. Biologists of the Game, Fish, and Oyster Commission estimated that the amount of fish killed in 1951 ranged from 27 to 41 million kg (30,000–45,000 tons). The dead species differed in different areas, but most included the hardhead catfish, spotted seatrout, red drum, black drum (*Pogonias cromis*), mullets, silver perch, spot, Atlantic croaker, bay anchovy (*Anchoa mitchilli*), striped anchovy (*Anchoa hepsetus*), Atlantic cutlassfish (*Trichiurus lepturus*), toadfish, and other fishes as well as the brown shrimp, a variety of crabs, bivalves including oysters, and the occasional brown pelican, lesser scaup, white egret, and other birds plus loggerhead turtle. Based on photos of windrows roughly 0.4 km (approximately 1,500 ft) long of mass mortalities in Laguna Madre, it was concluded that southern area also incurred heavy fish kills extending for some 50 km (30 mi) along the upper Laguna shore, with lesser damage in the lower Laguna. Many of the fish as well as clams, gastropods, and starfish that died along the shore of the Gulf of Mexico became lethargic and rolled up on the beach by the heavy surf caused by the norther. In the 1980s, Texas coasts experienced three winter mass mortalities with 14 million fish killed in December 1983, 11 million in February 1989, and another 6 million in December 1989 (McEachron et al. 1994). McEachron et al. (1994) used a stepwise, standardized approach to sampling, which they admitted caused an underestimated mortality count, especially for small (<200 mm [8 in]) animals as well as the illegal activities of fishermen removing dead and dying fish prior to the census. The composition of the fish species accounting for over 50 % in each freeze were striped mullet (*Mugil cephalus*), pinfish (*Lagodon rhomboides*), Gulf menhaden (*Brevoortia patronus*), and bay anchovy. They noted that the size classes of a species affected varied between some of the freezes, but not all species, e.g., pinfish. This observation they felt led to an "instantaneous picture" of the species population structure at the time of the kill. Hence, the recommendation to fisheries managers was to respond to mass mortalities by imposing regulations to reduce fishing efforts immediately following the event to allow recruitment and compensatory mechanisms to take place.

Commercial fishing after the cold kills, at least the 1940 episode, showed a dramatic decline (Gunter 1941). While there was some difference in the decline among commercial catches from the regions of Galveston, Matagorda, Aransas, and Laguna Madre, the red drum, spotted seatrout, and black drum all declined by 78 % while that of the southern flounder declined by 95 %. However, dead flounder do not float and because of their shape they are not easily trawled or dredged, so some mortalities could have easily escaped notice. He also tabulated data for catches after the 1940 freeze from both the year of the freeze and of the prior year and noted that there was no difference in decline from catches in the Gulf of Mexico. But there was in the bays, where the water was much shallower. It took about 3 years for the commercial catch to recover. Texas fishermen seem to agree that fish will be scarce for a few months after severe cold spells; whereas those from Florida and Mississippi estimate a 2- to 3-week period. As suggested above, this difference can be explained primarily by many of the fish in Florida, Mississippi, and offshore Texas migrating to more tolerable water or recovering after a water temperature rise after being affected but not killed.

The nice thing about cold kills is that residents as well as numerous animals such as piscivorous birds and raccoons make a healthy feast of the freshly dead or numbed fish! Such is not the case for fish killed by most other causes.

14.3.6 Pseudo-Fish Kills

"Pseudo-fish kills" is our term for fish that had died a week or two earlier and submerged, only to undergo bacterial degeneration during the warm period following the cold weather. Cases of pseudo-fish kills also can occur from fish kills other than those caused by low temperatures. Metabolic byproducts or gases consisting of methane, hydrogen sulfide, and carbon dioxide are produced, becoming trapped in the body cavity. Once enough gas accumulates, the body becomes lighter and the dead fish floats to the surface. Many local citizens are sure they see fish in the state of dying and report a fish kill to state agencies, research facilities, and newspapers. Our examination of such fish, always of decayed and smelly fish usually with attached green algae, indicated that the actual mortality had taken place during the prior cold snap or other mortality event.

14.3.7 Heat Kills

Because the rapid rate change from normal to low temperatures is usually what kills fish during a freezing period, one might expect a rapid rate change from normal to high temperature to be the cause of fish kills. Theoretically that could happen, except that even when the temperature of water starts increasing at a relatively rapid rate, most fish readily escape to relatively cooler habitats. There are situations, usually those resulting from anthropogenic changes involving heated effluents, which can kill aquatic organisms. Also, there are cases where fish get inadvertently washed up into a vulnerable position and are inescapably trapped in a body of water that rapidly increases in temperature before reduction of acceptable oxygen levels, killing fish. That situation is rare. Usually when the water temperature is increased, it takes a long enough period that eutrophication takes place and fish actually die from oxygen depletion rather than from a high temperature. Probably most important, this increased temperature reduces resistance to diseases and allows the agents to become established and to become more pathogenic. In such cases, the weakened fish become readily devoured as prey items before they die from disease; they are not witnessed as dead bodies. Complicated changes in normal parasite life cycles can occur such that a parasite such as a trematode or nematode will produce an infection at a different time of year than it is typically found in the normal environment. Because of the seasonal biology of the fish, infections might take place in an unpredictable situation during an atypical season, resulting in fish death or morbidity.

We know of no marine example in the Gulf of Mexico involving parasites relative to thermal pollution; however, Khan and Hooper (2007) evaluated the effects of thermal discharge on the parasites of the winter flounder (*Pseudopleuronectes americanus*) near the coastal fossil fuel generating plant at Holyrood, Newfoundland, Canada. The water discharged into Conception Bay, but the temperature change extended up to only 1 m (3 ft) below the surface, which was 3–4 °C (37–39 °F) in May and 7–8 °C (45–46 °F) in June compared with benthic water at 0 °C (32 °F). Only summer samples were taken, although sampling occurred at a few reference sites. Biological features such as condition factor and organ indices sampled below the plume revealed no significant difference with reference samples except that the male somatic index was significantly greater than that of the other two sampled groups. The parasites, however, were another matter. Metacercariae of the heterophyid *Cryptocotyle lingua* had a greater prevalence and mean intensity of infection than reference samples, whereas the mobile peritrich ciliate *Trichodina jadranica* and the monogenoid *Gyrodactylus pleuronecti* occurred significantly less on the gills of fish samples beneath the plume when compared with those from the reference sites. Additionally, four internal parasites, one myxosporidian and three trematodes, were significantly more abundant in the reference samples, suggesting an environmental change-affected transmission of the parasites when exposed to the thermal effluent. The salmonid brown trout (*Salmo trutta*) showed an attraction to the hot water effluent from the Forsmark nuclear power plant located in the low salinity coast of the Bothnian Sea associated with the Baltic Sea, Sweden. Thulin (1987) reported a few of 401 Swedish fish with skeletal abnormalities, 22 % with the leech *Piscicola geometra*, and 9 % with the copepod *Caligus lacustris*, all apparently unexpected findings. Khan and Hooper (2007) provided several freshwater examples involving parasite indicators of thermal effluents, but we only mention one (Camp et al. 1982) in which the prevalence and abundance of metacercaria of the trematode *Ornithodiplostomum ptchocheilus* in the western mosquitofish (*Gambusia affinis*) over a 53-month study expressed a fluctuating difference in infections in thermal effluent and ambient temperatures during 31 of those months in a thermal reservoir in South Carolina. The thermal effluent initiated the trematode life cycle a few months earlier than that occurred in the ambient water, with both shedding of the cercariae and recruitment of the metacercariae being affected; but so were the nesting and foraging activities of the waterfowl definitive hosts that tend to prefer the warmer water in winter and cooler water in the summer. Of course the biology of the fish and snail were also affected.

To take complications and thermal interactions one step further, the effect of viruses on cyanobacteria should be considered. For example, *Microcystis aeruginosa* is a common species responsible for blooms and for producing HAB toxins. This situation is usually considered as part of the relatively simple eutrophication process. However, there are several undescribed viruses that infect the "blue-green algae" including *M. aeruginosa*, and which probably influence mortality. They are likely a primary factor in determining plankton crashes;

moreover, it is temperature that seems to control the crashes (Honjo et al. 2006). Paerl and Huisman (2008) stress the point that rising temperatures, often above 25 °C (77 °F), favor cyanobacteria over diatoms, green algae, and other phytoplankton species. The resulting stratification earlier in spring and destratification later in autumn increased residence times and reduced vertical mixing. More intense precipitation with associated increased nutrient discharge in conjunction with the viral infections all ultimately promoted blooms and crashes with their associated fish kills.

Hot water pouring into a cooling canal from the Florida Power & Light Company fossil fuel generating plant at Turkey Point in 1969 killed thousands of fish. The company was operating under a special permit to discharge water into Biscayne Bay at temperatures higher than allowed by pollution laws. These mortalities occurred before the company finished construction of two nuclear generating plants. In addition to the dead fish, there was a variety of dead crustaceans, mollusks, corals, and algae occurring up to 1.4 km (1 mi) from the outfall during early summer in 1969. A virtually complete kill of aquatic organisms occurred over an area of about 8.4×10^5 m^2 (200 acres) (Laws 2000). Because of those expected harmful effects, Roessler and Tabb (1974) conducted an extensive survey and determined that average temperature elevations above ambient summer water temperatures caused the depletion in the biota. The area which was elevated above 4 °C (39 °F) was approximately 30 ha (75 acres), the area between 3 (37 °F) and 4 °C (39 °F) was approximately 40 ha (100 acres), and the area between 2 (36 °F) and 3 °C (37 °F) was approximately 50 ha (125 acres). A total area of about 120 ha (300 acres) showed a decline in abundance of biota that was statistically measurable for at least part of the year. A relatively rapid recolonization took place during the winter in a portion of this area. The inner barren zone of about 20 ha (50 acres) recovered slowly because of the death of the rhizomes of the turtle grass and changes in the sediment. The optimal temperature for maximum biodiversity was between 26 (79 °F) and 28 °C (82 °F); about 50 % of the animals were excluded when the temperature was between 30 (86 °F) and 34 °C (94 °F). About 75 % were excluded when the temperature was between 35 (95 °F) and 39 °C (102 °F), over the thermal tolerance range (TTR). Most of the animals occurred where the red algae complex of species of *Laurencia* and *Digenea* was abundant; whereas few animals occurred where no algae or seagrasses occurred. These two authors predicted that an increased temperature with increased water flow in conjunction with the nuclear generators would harm an increased area without implementation of alternate methods of cooling. No indication of pollution as measured by standard chemical indicators resulted from secondary treated sewage from a local sewage treatment plant, suggesting that the kill in 1969 was caused entirely by elevated temperatures resulting from the power plant and not eutrophication. Another kill of about 2,000 fish occurred in June 1971 when the discharge water from the fossil fuel plant again reached 40 °C (104 °F) (Associated Press 1971).

The above case uses both reports of counted or estimated dead fish and reports documenting lack of or less abundant catches of fish relative to other reference collections. More often when dealing with temperature, nothing exists but circumstantial or theoretical data. This is especially true with temperature because numerous factors influence other factors, and conditions are seldom repeated so they can be adequately compared. For example, Cairns et al. (1975) provides a good article on the effects of temperature upon toxicity of chemicals to aquatic animals. The amount of literature is extensive but mostly based on laboratory studies and not adequate to make "scientifically justifiable generalizations." The number of variables is extensive, especially when including the interaction of temperature with what type or degree of toxicity; what chemical, state of compound, or mixture of compounds; what organism, in what life stage, and under what physiological condition; and environmental influences such as salinity, pH, and alkalinity. Temperature can be both a lethal factor and a controlling factor but without consistent thresholds. Temperatures outside the zone of tolerance fall in the zone of resistance, and the length of time before death is useful for trying to determine the cause of thermal death. For example, because tissue anoxia occurs at high temperatures, the toxicants act differently: copper increases metabolic demand and zinc blocks oxygen uptake by the gills, and either may be rendered more active physiologically by an increase in temperature. The body temperature of most fishes and other animals, except marine mammals and birds, corresponds almost exactly with the water temperature, taking about 3 min/1 °C for heat exchange; and the rate of metabolism undergoes an approximately twofold increase with every 10 °C (50 °F) rise in temperature, commonly referred to as Q_{10}. Cairns et al. (1975) provide and discuss the effects of temperature on a wide variety of toxicants; however, few are examples from the Gulf of Mexico and most deal with experimental studies. DeLorenzo et al. (2009) showed that temperature, salinity, and life stages of grass shrimp all affect the degree of toxicity of common pesticides to this shrimp. Lloyd (1987) provided more detail on interactions and modification of the response caused by variation of physiochemical conditions. An experimental study (Bao et al. 2008) also demonstrated that the effects of temperature above the TTR significantly increased the toxicity and hence the ecological risks of two common anti-fouling biocides (chlorothalonil and copper pyrithione) to a copepod (*Tigriopus japonicus*) and a dinoflagellate (*Pyrocystis lunula*). Male copepods were more sensitive to both compounds than the females, and the toxicity of the two biocides differed.

14.3.8 Hypersalinity (Over-Salinity)

In areas of the GoM where levels of salinity can become great enough to kill fish, the levels seldom actually reach those high concentrations. The best known is Laguna Madre of the Texas coast; this lagoon, approximately 210 km (131 mi) long by 6.5 km (4 mi) wide, is separated from the Gulf of Mexico by a narrow barrier known as Padre Island. Under normal conditions, commercial fisheries production is greater in this lagoon that in any other region of Texas (Gunter 1947). During dry years, the salinity may reach three times that of normal seawater (about 32 ppt), killing vast numbers of fish (Gunter, on our reprint, crossed out the words "specific gravity" and replaced it with "salinity"). Even though not well documented, large kills occur approximately every 10 years, but minor kills occur every year. The number of kills probably increases as the lagoon fills. We are not aware of the development of any recent artificial pass leading to the Gulf of Mexico.

Every few years in Mississippi (because of the prevailing winds), the salinity in the bays and bayous will vary between 0 and ≥40 ppt. During the periods of high salinity, we have not seen fish kills, but we note that the components of the biota differed dramatically from that expected. For example, we have occasionally seen fishes such as the Spanish mackerel (*Scomberomorus maculatus*), lookdown (*Selene vomer*), and Atlantic moonfish (*Selene setapinnis*) as well as the bottlenose dolphin abundant in the Back Bay of Biloxi.

14.3.9 Sulfate Reduction and Anaerobic Methane Oxidation

A different environment occurs in deepwater bathyal areas of the Gulf of Mexico associated with hydrate-bearing sediments where crude oil and methane advect through fault conduits to the seafloor. These areas can be considered toxic to the surrounding aerobic fauna that occupies most of the seafloor, but the occupants come in contact with different chemical compositions. The oil and gas seeps located at 590–630 m (1,900–2,100 ft) at 6–9 °C (43–48 °F) are typically overlain by chemo-synthetic communities consisting of thiotrophic bacterial (e.g., *Beggiatoa* spp.) mats, methane atrophic mussels (*Bathymodiolus* spp.), and other symbiotic associations. These well-established geologic areas are dynamic, resulting in fluctuating faunas within and surrounding them. For example, bottom waters contain about 29 mM/L of dissolved sulfate, but pore fluids from oil and gas seeps are depleted down to 0.3 mM/L sulfate. That ambient bottom water contains less than 1 μM/L sulfide, but the sediments contain 1 mM/L and all pore fluids from seeps contain up to 20 mM/L, with those concentrations in gas seeps generally higher than those in oil seeps at the same depth. This inverse relationship between sulfate and sulfide results from bacterial consumption of sulfate and concomitant production of hydrogen-sulfide during anaerobic sulfate reduction. Bacterial mats of *Beggiatoa* spp. at the seawater–sediment interface obtain their energy by oxidizing hydrogen-sulfides and producing molecular sulfur. In contrast, the *Bathymodiolus* spp. mussels from the hydrocarbon seeps, but not hydrothermal vents, contain mostly methanotropic bacterial symbionts (e.g., Aharon and Fu 2000). Gas and crude oil escape by venting and seepage is ongoing, but major expulsion events are estimated to take place at a frequency of every 300 or even less than a hundred years (Roberts 2011). Gas hydrate, an ice-like substance comprised of a gas molecule like methane surrounded by a crystalline cage of water molecules as ice, concentrates vast amounts of methane in water at depths greater than 300–500 m (980–1640 ft) and contributes to seafloor hazards. In the GoM, the gas hydrates undergo repeated near-surface formation and dissociation varying seasonally and with warm-water loop eddies. However, the largest natural geohazard associated with hydrates and methane release involves periodic landslides (Hutchinson et al. 2011). Anaerobic oxidation of methane is a microbial process taking place in anoxic marine sediments where oxidization occurs with different terminal electron acceptors such as sulfate, nitrate, nitrite, and metals.

14.3.10 Sediments and Drilling Fluids

Sedimentation resulting from storms and river flow can have a major effect on mortality of fish and covers invertebrates. Examination of gills of morbid or dead fish allows differentiation among fish killed by cold or toxins (usually reddish unless acid, nitrate in freshwater fishes), oxygen depletion (pink or white), or sediments (mud, sand, or eroded tissue).

Drilling fluids (muds) are used by offshore petroleum-drilling operations. These aqueous suspensions consist of a variety of components that are pumped down the center of the drill bit and have a composition that varies with the needs of the drilling operation. Examples are lubrication, cooling, prevention of intrusion of seawater into the borehole, antibacterial action, suspension of drill cuttings, and capture of hydrogen sulfide. The fluid may be partially or entirely discharged into the surrounding water during and after the procedures. Because of the toxic nature of some of the fluids, the U.S. Environmental Protection Agency (USEPA) conducted some laboratory tests to determine whether some substances interfered with fertilization and normal development of fish and invertebrates (Crawford 1983). Different fluids had different toxic effects. As an example, a concentration of 10 ppt of some fluids caused the diminution of heartbeat rate of the model killifish; concentrations of 1 ppt had an effect on hatching and coordination of swimming of the fry of that fish. Some fluids had no effect on the fish, and no one fluid could be considered typical. Consequently, it is hard to evaluate the effect of drilling fluids on wild populations, but animal communities surrounding drilling operations can be compared with those on nearby reference sites not undergoing such activities.

14.4 FISHES

The following sections treat mortality, health, and indicators of specific groups of animals, starting with fishes and separate from the above sections that treated general mass mortalities.

14.4.1 Infectious Parasites and Diseases

The status of fishes and the fish communities in the GoM needs to be understood relative to natural and contaminated conditions so that microbial and parasitic diseases and fish mortalities can be assessed critically. Helpful background information on many of the important model fishes such as Atlantic croaker, bay anchovy, hardhead catfish, Gulf menhaden, spot, and pinfish has been provided by Lewis et al. (2007). Comparative data are also available for the effects of hurricanes on these fish assemblages (Lewis et al. 2011).

Because the majority of the fish that die in the GoM usually are not seen or sampled since they become part of the food web, the impact of parasite-induced mortality on a population is difficult to determine (Scott and Dobson 1989; McCallum and Dobson 1995; Rousset et al. 1996). Consequently, statistical approaches have been established to estimate mortality. Lester (1984) presented six methods to estimate mortality caused by parasites in wild fish populations. Shaw et al. (1998) reviewed 49 published wildlife host-macroparasite systems and determined that in 90 % of the data sets, the negative binomial distribution provided the statistically satisfactory fit. This statistical analysis has been used by many to account for estimates of the mortality in populations due to parasite infections (May and Anderson 1978; Dietz 1982; Kennedy 1984; Scott 1987; Shaw and Dobson 1995).

Statistical evidence of parasites controlling the abundance of a population has been demonstrated by Hudson and colleagues (Hudson 1986; Hudson and Dobson 1990, 1997a, b; Hudson et al. 1985, 1992, 1998; Dobson and Hudson 1992) using the interaction of the nematode *Trichostrongylus tenuis* on the red grouse, *Lagopus lagopus scoticus*. Bruning et al. (1992) developed a population-dynamic model for phytoplankton and the impact of fungal parasites on their populations. They calculated that four parameters are needed to determine the loss-rate: prevalence of infection, developmental time for the parasite, specific growth rate of the uninfected host, and the difference between the infected and uninfected host mortality due to factors other than parasitism. Knudsen et al. (2002) used a long-term study with indirect methods to indicate that the nematode *Cystidicola farionis* increased the mortality rate in its final host (Arctic charr, *Salvelinus alpinus*). They indicated parasite-induced host mortality in hosts older than 10 years. The convex age-abundance curve indicates that heavily infected fish disappear from the population. This loss of fish has been demonstrated in other studies (Pennycuick 1971; Kennedy 1984; Esch 1994). Krkošek et al. (2007) used mathematical and empirical data to relate the influence of the copepod *Lepeophtherius salmonis* on farmed salmon, which get repeatedly infected. Not only does the parasite cause mortality of these farmed salmon, it also causes over 80 % mortality of wild juvenile salmon (*Oncorhynchus gorbuscha*) in the waters adjacent to the farms. Those authors postulated that local extinction of wild pink salmon could occur if parasite-outbreaks abate the ecosystem's ability to support the wild population.

14.4.1.1 Bacterial

During certain periods throughout most years in most areas of the GoM, lesions occur commonly on specific fishes. These sores or ulcers usually occur in fins, tail, mouth, or near

Figure 14.9. Southern flounder, *Paralichthys lethostigma*, exhibiting relatively common bacterial lesion on blind side of specimen from Pascagoula estuary, Mississippi, 1987.

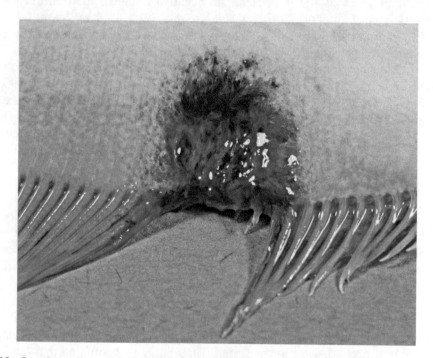

Figure 14.10. Southern flounder, *Paralichthys lethostigma*, exhibiting relatively common bacterial lesion on fin and blind side from specimen in Pascagoula estuary, Mississippi, May 1987.

anus, and regulatory agencies typically refer to those lesions or to the infected host fish as "red sore," "tail rot," "fin rot," or "mouth rot" (Figures 14.9, 14.10, and 14.11). Diagnoses usually include positive results for *Vibrio, Pseudomonas*, and other bacteria, and, if the salinity is low, counterpart infections involve *Aeromonas, Flexibacter*, and other bacteria. Causes can include

Figure 14.11. Southern flounder, _Paralichthys lethostigma_, exhibiting severe bacterial tail rot of specimen from Pascagoula estuary, Mississippi, in 1997.

injury from predator wounds, trawling or other fishing activities, harsh water quality conditions, internal or external parasites, plasmid or viral infections, co-infection with other bacteria, or even primary bacterial infections. Most often, aggravation by any one of those stressors impairs the fish skin or alimentary tract, enhancing bacterial invasion and growth. To take this process further, sometimes a wound with associated necrotic host tissue supports establishment and growth of barnacles and other fouling agents. Couch and Nimmo (1974) observed over a 10-year period in Escambia Bay, FL, a high prevalence of fin rot syndrome associated with mortalities in Atlantic croaker and spot during warm weather and oxygen depletion. They were able to demonstrate experimentally the induction of fin rot syndrome in 90 % of the exposed spot to a polychlorinated biphenyl (PCB) (3–5 µg/L of Aroclor 1254) with an associated 80 % mortality, but no attempt was made to isolate the bacteria. In most cases in the northern Gulf, the lesions undergo repair in the absence of stressors, and mortalities are not conspicuous.

The most common lesions in marine and estuarine GoM fishes involve _Vibrio_ and related bacteria and some form of stress playing a role in the disease process. _Vibrio alginolyticus_, _Vibrio anguillarum_, _Vibrio carchariae_, _Vibrio cholerae_, _Vibrio damselae_, _Vibrio ordalii_, _Vibrio parahaemolyticus,_ and _Vibrio vulnificus_ all have been reported to cause disease in marine fish and require salt to grow (Colwell and Grimes 1984) (Figures 14.9, 14.10, 14.11, 14.12, and 14.13). As methods for identification become more sophisticated, strains and new species are recognized. For an example in wild fish, a bacterial mortality event was restricted to menhaden and striped mullet in the Galveston Bay (TX) area in November 1968. The cause was attributed to _Photobacterium damselae piscicida,_ formerly known as _Pasteurella piscicida_ (Lewis et al. 1970; Panek 2005). This bacterium occurs ubiquitously in the gut of marine fishes. Thune et al. (2003) reported that in Louisiana from 1990 to 1995, heavy mortalities (32 cases) were reported in coastal hybrid striped–bass farms, with four farms closing as a direct result of _Photobacterium damselae piscicida._ This pathogen has also become a pathogen of significance in cultured cobia (_Rachycentron canadum_), with 80 % mortality at some sites

Figure 14.12. Southern flounder exhibiting ulcers and rake marks thought to be produced by an Atlantic bottlenose dolphin "playing with fish" or after escape from an unsuccessful attempt to capture the prey in Mississippi Sound in 1997.

Figure 14.13. Atlantic croaker, *Micropogonias undulatus*, exhibiting relatively common bacterial fin lesions and tail rot from shrimp trawling grounds in Mississippi Sound during October 1996.

(McLean et al. 2008). In fact, isolates from cage-cultured cobia have included *V. alginolyticus, V. parahaemolyticus, V. vulnificus,* and *V. harveyi* (McLean et al. 2008). All stages of cobia can succumb to vibriosis, and this disease can account for 45 % mortality in cage-stocked juveniles.

Vibrios not associated with overt disease in marine organisms are ubiquitous in the northern Gulf of Mexico, where *Vibrio vulnificus* has been isolated from the intestines of sheepshead (*Archosargus probatocephalus*), red snapper (*Lutjanus campechanus*), little tuna

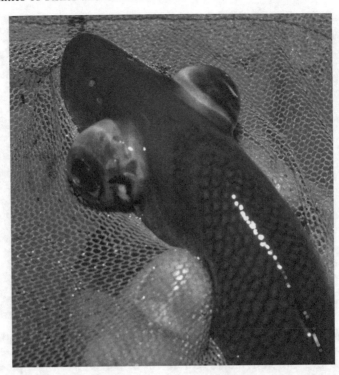

Figure 14.14. Red snapper, *Lutjanus campechanus*, exhibiting bacterial exophthalmos.

(*Euthynnus alletteratus*), Atlantic croaker, Atlantic spadefish (*Chaetodipterus faber*), Atlantic stingray (*Dasyatis sabina*), black drum, crevalle jack (*Caranx hippos*), gafftopsail catfish (*Bagre marinus*), Gulf menhaden (*Brevoortia patronus*), Gulf toadfish (*Opsanus beta*), pigfish (*Orthopristis chrysoptera*), pinfish (*Lagodon rhombodies*), scaled sardine (*Harengula jaguana*), hardhead sea catfish (*Ariopsis felis*), southern kingfish (*Menticirrhus littoralis*), Spanish mackerel (*Scomberomerus maculatus*), and white mullet (*Mugil curema*) (DePaola et al. 1994). DePaola et al. (1994) noted that the prevalence of *V. vulnificus* collected offshore in the Gulf (32–35 ppt) was 11.8 % compared with 13 % in Galveston open Gulf beaches (18.9 ppt) and 68 % in the Galveston Bay estuary (11.3 ppt). Tao et al. (2012) found that a statistically significant ($p < 0.0001$) inverse correlation between *V. vulnificus*-positive fish and salinity existed as did a positive correlation ($p < 0.03$) between water temperature and *V. vulnificus*-positive fish in Gulf locations of Ocean Springs, Mississippi, Gulf Shores and Dauphin Island, Alabama. In addition to the fish listed above by DePaola et al. (1994), Tao et al. (2012) also reported ladyfish (*Elops saurus*), striped mullet, silver perch (*Bairdiella chrysoura*), sand weakfish (*Cynoscion arenarius*), spotted weakfish (*Cynoscion nebulosus*), and red drum (*Sciaenops ocellatus*) to have *V. vulnificus* on the body surface. Buck (1990) isolated *V. alginolyticus, V. damselae*, and *V. parahaemolyticus* from fish from the GoM and adjacent Sarasota Bay, Florida. These isolates came from the gills, intestinal tract, mouth, surface skin, spines, and teeth.

Exophthalmia, a condition known as "bugeye," commonly affects fishes in the GoM. The most common cause is bacterial as shown in the red snapper maintained in culture (Figure 14.14). Figure 14.15 shows a case in the sheepshead minnow that also was probably bacterial in nature. The condition, however, can result from a variety of causes. When a trematode metacercaria infects the eye, especially in a semi-enclosed locality, a large percentage of the fish intermediate host population can be infected. Several species of diplostomoids produce this effect, and some

Figure 14.15. Sheepshead minnow, *Cyprinodon variegatus*, exhibiting exophthalmos, or bugeye, from Mississippi bayou in 1996, probably caused by bacterial infection, but similar condition occurs in eyes of several different fishes as a result of bacterial, viral, diplostomoid trematode metacercariae, or cestode metacestode as well as in cultured fish as nutritional deficiency or gas-bubble disease.

species can number 30 or more large encysted individuals in the vitreous humor or several hundred other small ones in the lens of a single eye. Presumably, heavily infected fish in the Gulf, the same as demonstrated elsewhere or in experimental studies, are vulnerable to predation by the appropriate bird or mammal final host. Metacestodes occur less commonly, but we have seen them in the eyes of Florida pompano (*Trachinotus carolinus*) and puffers. Lymphocystis, a viral disease reported later, can usually be recognized in living fish. Nutritional deficiency and gas-bubble disease usually affect fish in culture under poor husbandry conditions.

What is termed "red sore" disease in Mississippi and elsewhere in the northern Gulf is an ulcerative condition common in euryhaline fish and can occur at epizootic levels greater than 50 % prevalence, based on fishermen's comments to us, in sheepshead (*Archosargus probato-cephalus*) (Figure 14.16), black drum (*Pogonias cromis*), and centrarchids. Most cases in low salinity water are associated with *Aeromonas hydrophila*; but some lesions, especially those from fish in 15–20 ppt, had *Pseudomonas* spp. and *Vibrio* spp. Overstreet (1988) considered most of the cases he investigated as resulting from contamination, but they were associated with some secondary infections resulting from natural factors and mechanical damage from fishing or other activities. He updated prior reports from Mississippi Sound (Overstreet and Howse 1977) as approximating red sore lesions occurring in 10 % of spot and southern flounder in summer months and a lower percentage during the rest of the year. The Atlantic croaker and southern kingfish (*Menticirrhus littoralis*) also exhibit lesions, often associated with fishing activities. Overstreet also cited literature reporting 35–40 % of striped mullet in Punta Gorda, Florida, in late summer and several species in West Florida with *Vibrio damselae*. *Aeromonas hydrophila* and *Vibrio anguillarum* are also known to cause a bacterial hemorrhagic septicemia

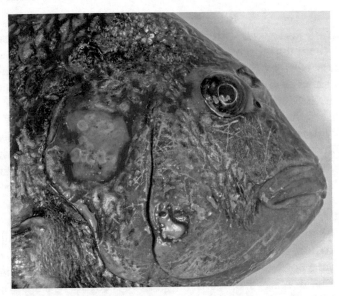

Figure 14.16. Sheepshead, *Archosargus probatocephalus*, showing ossified lesion with secondary bacterial infection; this condition can be common in low salinity conditions; Back Bay of Biloxi, Mississippi, April 1977.

and fin rot in the Florida pompano (*Trachinotus carolinus*) and striped bass (*Morone saxatilis*) in brackish water of Alabama (Hawke 1976).

The influence of environmental stressors such as low salinities or pollution with high organic content may initiate and exacerbate this disease (Overstreet 1978, 1988). *Aeromonas hydrophila* and the colonial peritrich ciliate *Heteropolaria colisarum* interact to produce ulcerating lesions in centrarchid fishes occurring in freshwater and low salinity habitats (Figure 14.5). The bacterium associates with the ciliate, and, when the organic load in the habitat increases in amount, the ciliate increases in number. Different proteolytic enzymes produced by the motile bacterium (Barrett et al. 2012) cause erosion of the epithelium, lysis of the skeletal musculature, hemolysis, and hemorrhagic septicemia (Overstreet and Howse 1977; Cipriano et al. 1984). Ultrastructural study of the ciliate by Hazen et al. (1978) assumed that since the point of the stalk attachment as shown by scanning electron microscopical images to the fish surface was not the site of pathologic changes, the relationship with the ciliate was benign. In our (Overstreet and Howse 1977) ultrastructural transmission electron microscopical images, the point of attachment was a dense granular layer overlying the spreading fibrillar attachment. Sinuous spiral fibers running longitudinally down the stalk attach directly to the collagenous lamellae of the fish scale. No cellular membrane separated host tissue from the attachments of fibrillar of granular layers. We suggested the ciliate could invade the collagenous layer. Whether this invasion could occur without the interaction of bacterial enzymes is unknown, but bacteria alone, with accompanying ciliates, could readily cause red sore lesions in Hazen's material as well as ours. In fish with severe septicemia, the liver and kidneys served as foci for toxic products produced by the bacteria. The structural integrity of both organs was destroyed, leaving minor pathologic changes in the spleen and heart (Huizinga et al. 1979).

Interesting questions concerning aspects of these red sores involve the seriousness, longevity, and cause of red sore lesions. We think the infections in sheepshead from Mississippi (Figure 14.16) provide examples of lesions associated with low salinity. During occasional periods of low salinity, sheepshead with lesions occurred and were caught by hook and line commonly from under the U.S. 90 Highway Biloxi-Ocean Springs Bridge. Because of the lesions, some calcified and some with attached barnacles, fishermen were concerned about keeping and eating them and threw them back into the water. We examined several critically, and some individuals with special marks (tattoos) were captured over and over. Most other species migrated a short distance to higher salinity waters. The lesions in some fish covered more than a third of the body surface and exhibited large areas of skeletal muscles along the flank. Because some fishermen recognized the individuals, they reported to us that the fish remained alive with lesions as long as the salinity concentration remained low, sometimes a few months. The presence of low salinity acorn barnacle supports this longevity in low salinity. Once the salinity concentration increased, fish did not exhibit lesions. We examined some and could detect scale and epithelial regeneration, suggesting recovery from the infections; regenerating tissues remained detectable for a couple of months.

We often captured, using 5-min long trawls, specimens of the Atlantic croaker that had been trawled and tossed overboard, often multiple times, by shrimp fishermen. Many of these had red sores, fin-erosion, abrasion, tail-rot, and other lesions. Presumably, some became prey of seabirds, fishes, and bottlenose dolphin, but after shrimp season we could see a time series of regenerating lesions in many specimens, suggesting recovery in a significant proportion of released fish. Few scientific studies have been conducted to understand the multiple causes of bacterial lesions.

Fin erosion can result from anthropogenic factors. These lesions can be chemical contamination, fishing techniques as indicated above, or a variety of other activities. For example, Sherwood and Mearns (1977) provided an example using strong observational and experimental evidence linking chlorinated hydrocarbon pollutants (e.g., DDT) with fin-erosion in a southern California flatfish (Dover sole, *Microstomus pacificus*) exposed to discharged municipal wastewater. In addition to lesions, the ratio of liver to body weight of laboratory-exposed fish was higher than in controls and similar to that recorded from fish from a heavily contaminated study-location.

The striped mullet, *Mugil cephalus*, in Figure 14.17 should not be confused with a fish exposed to a toxicant. It became trapped in a nearly fresh water pond at Ingalls West Bank Overpass near Pascagoula, Mississippi, and could not escape unlike the sheephead above that could have left the low salinity habitat.

Figure 14.17. Striped mullet, *Mugil cephalus*, exhibiting bacterial (*Aeromonas hydrophila*) ulcerated lesions, abraded fins, hemorrhaging, and concurrent *Saprolegnia*-like fungal infection. A series of similarly affected mullet became trapped in nearly fresh water pond at Ingalls West Bank Overpass, Pascagoula, Mississippi, March 1992.

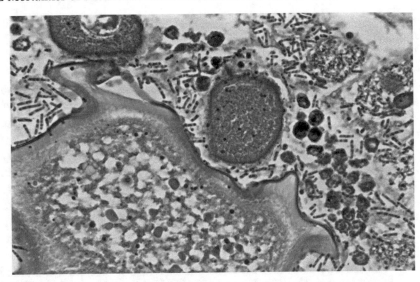

Figure 14.18. Red drum, *Sciaenops ocellatus*, with histological section of ovary showing *Streptococcus* infection in September 1991.

This was one of many such fish exhibiting a bacterial (*Aeromonas hydrophila* and *Shewanella putrefaciens* [numerous colonies on live fish, not a contaminant]) ulcerated lesion, abraded fins, hemorrhaging, and concurrent *Saprolegnia*-like fungal infection. No septicemia occurred, but the spleen exhibited a large number of macrophage aggregates, indicating the tissue damage (Overstreet 1997).

Large-scale mortality events involving bacteria as a primary causative or suspected agent have occurred, even though rarely, in the GoM. One such documented event involving *Streptococcus* sp. non-hemolytic Group B, type I_b occurred during August–September 1972 along the Alabama and northwestern Florida coastlines (Plumb et al. 1974; Wilkinson et al. 1973). Tens of thousands of fish died, mostly Gulf menhaden (*Brevoortia patronus*), but also hardhead catfish, striped mullet, Atlantic croaker, pinfish (*Lagodon rhomboides*), spot, stingray (*Dasyatis* sp.), and sand seatrout (*Cynoscion nothus*); no abnormal environmental condition was apparent during the time of the event (e.g., Figure 14.18). Experimental studies conducted with isolates obtained from two separate sites in Alabama, Soldier Creek and Bon Secour, caused 70 and 90 % mortality, respectively (Plumb et al. 1974). An isolate obtained from a "fish" from Mobile Bay, AL in September 1972 was also identified as *Streptococcus* sp. non-hemolytic Group B, type I_b (Wilkinson et al. 1973). Cook and Lofton (1975) conducted pathogenicity studies with an isolate obtained from the kidney of the menhaden from Soldier's Creek referred to as *Streptococcus* 922. Using intraperitoneal injection of the isolate, they found a mortality rate of 40–90 % in Atlantic croaker, 33–40 % in Gulf menhaden, 100 % in striped mullet, and 57–100 % in spot. Rasheed and Plumb (1984) also performed an experimental study with an isolate that was serologically identical to those above obtained from the Gulf killifish (*Fundulus grandis*) and concluded that in the killifish, disease occurred only when the portal of entry was an injured area of the body. Panek (2005) attributed a massive 1999 fish kill in the northeast Caribbean to the β-hemolytic *Streptococcus iniae*.

Fish kills due to meningitis caused by the gram-positive anaerobic bacteria *Eubacterium* sp. have also been observed in the coastal areas of the northern Gulf. This anaerobic bacterium has been cultured from striped mullet and red drum involved in an extensive mortality event near Port Aransas, Galveston, and Orange, Texas, in 1973 (Henley and Lewis 1976). Fish kills in the Biscayne Bay, Florida, region have occurred repeatedly, and Udey et al. (1976) isolated *Eubacterium* sp. from the brain tissue of all the dead and moribund fish sampled. The fish involved in the large Biscayne Bay kills included striped mullet, snook (*Centropomus undecimalis*), Gulf flounder (*Paralicthys albigutta*), and striped mojarra (*Diapterus plumieri*). Udey et al. (1976) also reported that experimental studies conducted on an isolate did not produce a toxin, nor was the agent pathogenic for mammals. It did produce mortality in channel catfish after 14 days when injected intraperitoneally. They found that this bacterium was present in every mullet tested, and they also isolated it from six other species of fishes from the Biscayne and Florida Bay areas. This bacterium can be present in a species without any display of disease. Udey et al. (1977) classified these bacteria as *Eubacterium tarantellas*, and they noted that marine fishes not entering the bays did not have *E. tarantellas*. The organism would not grow at salt concentrations above 2 %. Lewis and Udey (1978) indicated that several species of estuarine and marine fishes act as reservoir hosts. *Eubacterium tarantellas* has also been identified from the ovaries of black drum and red drum (Nieland and Wilson 1995) (also compare Figure 14.18).

The same or related bacteria from infections in fish also infect other animals in the same waters as the fishes. For examples, Cook and Lofton (1973) found *Beneckea* sp. type I to be an opportunistic infection when the crab shell has been damaged causing shell disease. The chitinoclastic bacteria in the genera *Beneckea*, *Pseudomonas*, and *Vibrio* have been isolated from the lesions of blue crab and penaeid shrimp. Shields and Overstreet (2007) listed bacteria isolated from the shell and hemolymph of the blue crab and indicated there was no relationship between black spot lesions and bacteria in the hemolymph. Lightner and Lewis (1975) observed a septicemic bacteria disease with *Vibrio alginolyticus* in wild brown, white, and pink shrimps obtained from commercial bait dealers in Galveston, Texas, in 1972 and 1973. These moralities ranged from "a few a day" to nearly 100 %. One bait camp had both *Vibrio alginolyticus* and *V. anguillarum* isolated from white shrimp that underwent 50 % mortality. Witham (1973) described a *Bacteroides* sp. infection in 140 loggerhead turtle (*Caretta caretta*) hatchlings (1–3 months old) that appeared in September of 1970 at a mariculture tank at Hutchinson Island,

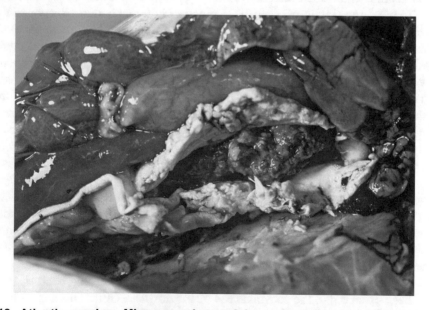

Figure 14.19. Atlantic croaker, *Micropogonias undulatus*, from Pascagoula estuary exhibiting systemic granuloma in September 1980 primarily involving swim bladder and probably caused by nocardiosis or mycobacteriosis; condition can result in fish cultured with a deficiency in Vitamin B or C.

Florida. The disease presented with necrotic, spreading skin lesions causing death within 3–7 days over a period of 3 months with a cumulative 98 % mortality. *Bacteroides* sp. was considered the primary pathogen, although *Pseudomonas aeruginosa* and *Staphylococcus epidermis* were also present.

Many other bacteria infect fishes, some hard to detect or determine without culturing blood or tissues. On the other hand, some infections produce obvious disease. Figure 14.19 exhibits systemic granuloma in a wild fish. Similar cases caused by bacteria and dietary deficiencies occur in cultured fishes.

14.4.1.2 Viral

The disease lymphocystis constitutes an interesting virus infection for a variety of reasons. It does not occur in an abundance of individuals, but various strains do infect a variety of host fishes in the GoM (Figures 14.20, 14.21, 14.22, 14.23, 14.24, and 14.25). It typically infects connective tissue cells in the skin of the body and fins, and each infected cell hypertrophies

Figure 14.20. Atlantic croaker, *Micropogonias undulatus*, with lymphocystis mass in and above eye in fish caught in June 1992 off Marsh Point, Ocean Springs, Mississippi.

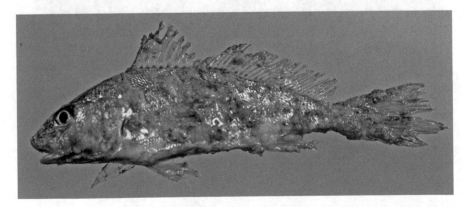

Figure 14.21. Atlantic croaker with extensive lymphocystis infection, Mississippi, 1996.

Figure 14.22. Atlantic croaker showing moderate infection of lymphocystis, Mississippi, April 1985.

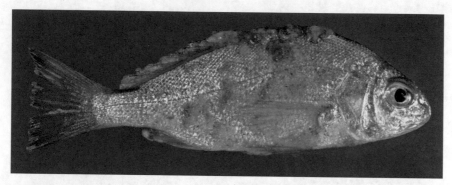

Figure 14.23. Spot, *Leiostomus xanthurus*, with heavy infection of different strain of lymphocystis, Mississippi, October 1978 (similar infection in October 1987).

Figure 14.24. Atlantic spadefish, *Chaetodipterus faber*, exhibiting still another strain of lymphocystis, Mississippi, November 1979.

Figure 14.25. Silver perch, *Bairdiella chrysoura*, **with internal infection of lymphocystis obvious in spleen; spleen located parallel to intestine, large and pinkish because of hypertrophied lymphocystis-infected cells rather than being relatively small and dark brownish.**

and can be observed as a nodule with careful examination; accumulation of these cells provides a raspberry-like tumorous growth, a cluster which is readily observable and defined as a "pseudo" tumor by Anders (1989) that is unnecessary here because we consider non-cancerous lesions as tumors. Because this viral disease does not cause mass mortalities, an epizootic can be readily followed. Lymphocystis was first reported from the GoM in a few individuals by Christmas and Howse (1970) in the Atlantic croaker and sand seatrout, *Cynoscion arenarius*, from relatively polluted locations during winter months. Individual cells can be recognized because they become greatly enlarged and possess an alcianophilic hyaline capsule that stains positive for PAS (periodic acid-Schiff). Icosahedral-shaped viral particles remain confined to the cytoplasm of the host cell, which for some strains can increase in size up to 2 mm (0.1 in) in diameter with a volume about one million-fold that of the normal host fibroblast (Anders 1989). The double-stranded DNA virions have been placed in the genus *Cystivirus* (Iridoviridae). Infections in Mississippi are most common in the Atlantic croaker and silver perch, *Bairdiella chrysoura*, and we will discuss later the unique details of this latter infection, which can occur in internal organs.

Overstreet and Howse (1977) and Overstreet (1988) reported on the history of epizootics in the Mississippi area that reached the peak of as much as 50 % of the croaker population in the mid-1970s. Between 1966 and 1969, a total of 12 of 32,688 croaker and sand seatrout exhibited infections. During the following 18 months, Overstreet and Howse (1977) observed 15 of 2,500 croaker infected. By the mid-1970s, croaker examined by them and reported by commercial shrimp fishermen had increased to hundreds of cases, with as many as half of the croaker in a trawled sample observed to be infected on several occasions. The shrimpers who had been catching croaker in Mississippi coastal waters for years recalled seeing infected individuals only during that period. Specific cases in February 1971 and summer of 1973 were mentioned (Edwards and Overstreet 1976). In June 1984 when thousands of various-sized boats trawled for shrimp, each discarded several thousand young-of-the-year croaker from their by-catch. Many declared to R. Overstreet that about 20 % of their juvenile croaker catches exhibited lymphocystis, especially in Biloxi Bay, and he verified several of those observations. Earlier on June 28, 1976, 11 out of 80 croaker in Mississippi Sound immediately north of Dauphin Island, Alabama, exhibited infections restricted to less than 2 % of their body surface. In contrast, 2 of

174 croaker samples from the higher salinity Gulf water south of the island exhibited 60 and 80 % of their surface area covered, but they could have previously acquired the infection in Mississippi Sound. What is apparently the same strain ranges from at least Texas to Sapelo Island, Georgia (Smith 1970).

We have conducted experimental studies with a few of the strains occurring in Mississippi waters. Studies by us and by Cook (1973) have shown that the enlarged cells became apparent within 5–7 days after inoculation when maintained at 25 °C, but within 9–11 days at 20 °C, with salinities at least between 10 and 35 ppt not having an effect. The disease cell clusters typically sloughed off between 20 and 30 days after inoculation. Infections could not be produced in about 2 % of fish tested from the wild. Cook (1973) reported the strain from croaker could be used to infect croaker, sand seatrout, and black drum, *Pogonias cromis*, but not spot, spotted seatrout, or bluegill. We also showed that the strain from the silver perch would not infect Atlantic croaker or spot, and the strain from spot would not infect Atlantic croaker.

The atypical strain from the silver perch produced internal infections in the heart, behind the eye, in the kidney, mesentery, spleen (Figure 14.25), liver, and ovary as well as external in the skin, fin, and gills. The presence of the cymothoid isopod *Lyroneca ovalis* causing lesions to the gill or the presence of damaged gills suggesting a prior isopod infestation indicated that such lesions might allow the virus to enter the bloodstream and infect specific cells in internal organs (Howse et al. 1977). Ultrastructural investigation of infections in the heart of the silver perch demonstrated infections similar to those in the skin in the epicardium, trabecular spaces, and sub-endocardium, but not in the adjacent myocardial cells. Wharton et al. (1977) established a fibroblast-like cell line from the swim bladder of the silver perch in which growth of the lymphocystis virus was supported but not of other viruses from fishes and mammals. They did not observe formation of the hyaline capsule in vitro, although frozen virus from their cell lines did produce tumor cells in vivo that contained this structure. They suggested that the L-15 growth medium that they used did not have sufficient muco-polysaccharide to produce the capsule. About 10,000 early stage, 200 mg (0.007 oz) juvenile red drum were imported from Texas to Israel, where they were to be reared. When the fish reached 20 g (0.7 oz), some cutaneous lymphocystis lesions appeared, and within 2 months, several hundred displayed severe infections. Some of these fish contained internal lesions, most prominently in the spleen. The origin of the virus was not indicated (Colorni and Diamant 1995).

Internal infections with the lymphocystis-virus seemed to have resulted from experimental infections until Dukes and Lawler (1975) reported on naturally occurring ocular lesions in the silver perch from Mississippi and the sand seatrout from Texas. The cells occurred in or behind the eye as well as on the cornea or adjacent skin surfaces. Figure 14.20 shows a previously unreported case of the Atlantic croaker with a lymphocystis mass occurring in and above the eye.

Most lymphocystis-infected fishes in the Gulf are sciaenids, and several strains exist, as discussed above; but members in other families have also been observed from the Gulf with lymphocystis in low prevalence. The common snook, *Centropomus undecimalis*, from Campeche, Mexico, has an infection with the hyaline capsule thicker than those reported for the species from sciaenids (Howse 1972). An infection in a 27-cm (11 in) standard length specimen of the Spanish mackerel from Venice, Florida, collected on July 11, 1986, was described by Overstreet (1988). The Atlantic spadefish, *Chaetodipterus faber*, from Mississippi in November 1979 exhibits (Figure 14.24) another new record of what is probably another strain of lymphocystis. The bluegill in freshwater habitats as well as low salinity estuaries of Mississippi occasionally exhibits what appears to be a different strain and probably what was reported by Weissenberg (1945). Our unpublished experimental infections showed that it did not become established in local sciaenid species.

Because of the spotty distribution and other factors, the expression of lymphocystis appears to have a relationship with specific toxicants or conditions. Christmas and Howse (1970) found the few affected fish occurring in industrially contaminated areas of Mississippi. Overstreet (1988) hypothesized that one or few specific toxicants, rather than general stress, may have enhanced infections by lowering host resistance.

Other circumstantial evidence occurs for induction of infections outside the Gulf of Mexico. Perkins et al. (1972) suggested that PCBs may have been responsible for lymphocystis in adults of two species of flatfishes, common hosts for lymphocystis in areas more temperate than the Gulf of Mexico, in the Irish Sea at the same time that young individuals in an uncontaminated area did not develop the disease; however, Shelton and Wilson (1973) considered hydrographic conditions such as low salinity to be a better explanation. Wolthaus (1984) reported infections in the dab, *Limanda limanda*, a flatfish in the southern North Sea, to be associated with acid iron waste from titanium dioxide production, regardless of season, but others (Möller 1985) questioned the validity of that cause. Mellergaard and Nielsen (1995) also studied the dab from 1984 to 1993. Because a severe oxygen depletion occurred in the late summer of 1986 and 1988, they were able to observe peak prevalences of lymphocystis and epidermal papilloma, another viral disease, of 14.7 and 3.3 % in 1989, respectively. They suggested that the stress caused by the oxygen depletion triggered outbreaks of both viral diseases. We know of no good experimental work relating lymphocystis infections to specific toxicants or stresses.

A massive fish kill assumed to be caused by "hardhead catfish virus" involved millions of hardhead catfish (*Ariopsis felis*), occurred in 1996, and spread from Texas to southwest Florida as followed by the Gulf of Mexico Aquatic Mortality Response Network (GMNET) sponsored by the USEPA and Gulf state agencies. Most beaches, bays, and river mouths contained thousands of dead and dying hardhead catfish and an occasional related gafftopsail catfish

Figure 14.26. Hardhead catfish, *Ariopsis felis*, from Back Bay of Biloxi, Mississippi, with hemorrhaging lesion on fin, viral die-off of May 28, 1996.

(*Bagre marinus*). For example in Mississippi, we observed or collected and examined many specimens inshore from Mississippi Sound, Back Bay of Biloxi, Davis Bayou, Biloxi Channel, and the Pascagoula River as well as specimens floating offshore from Horn and Ship islands, starting with a major kill with tons of dead and dying fish on May 28–31, 1996 and followed by collections of moribund fish from smaller kills at least on June 24, July 17, and August 6, 1996, in water with salinity ranging from 7 to 32 ppt. Initially, fish measured 25–35 cm (10–14 in) in total length, but later, some smaller fish also died. What appeared to be a recurrence of the kill occurred on November 3, 1998, with about 1,000 fingerlings dying in the mouth of the Pascagoula River among many already dead catfish. Most fish grossly exhibited hemorrhaging lesions in the gills and pectoral fins (Figure 14.26), and some had lesions of the mouth, lip, pelvic fins, and anus. Sections showed extensive hyperplasia in the gills, and the adjacent, non-hyperplastic, pale-appearing areas demonstrated an abundance of mobile peritrich tricho-dinid ciliates. Sections of visceral organs demonstrated an abundance of melanin-macrophage centers and appeared abnormal. A light red area in the liver in a few fish was shown to be infected with *Vibrio fluvialis*, an infection that did not occur in the corresponding kidneys, spleen, or systemically in the blood. This bacterium identified for us by Dawn Rebarchik at GCRL is known to be pathogenic to humans and crustaceans (Eyisi et al. 2013).

From moribund specimens from Biloxi Bay and Mississippi Sound, we (R. Overstreet and Eugene Foor) collected tissues from anterior and posterior kidney, spleen, liver, and brain and prepared them for electron microscopic observation. Intranuclear paracrystalline arrays of viral particles occurred abundantly in all preparations. The center-to-center spacing of the individual particles measured 35–50 nanometers (nm) (1.4–2×10^{-6} in). In all tissues, the particles appeared to be a DNA icosahedral virus, and it showed little selectivity in the cell type parasitized, indicating the virus had a wide host cell range. Whether or not the virus was the lethal agent, acted synergistically with other causative agents, or otherwise became expressed in dying cells only remains to be determined. However, based on the host specificity, high density of virions in dying cells, and the fact that Jan Landsberg also found the virus in moribund catfish in Florida, the virus appears to be the primary causative agent. As indicated above, an infection, not necessarily lethal, typically transforms into a disease when interacting with a stress. Since the disease is highly host-specific, the stress probably is one specifically associated with the catfish, such as one dealing with reproduction. Jan Landsberg and R. Overstreet planned to produce an extensive joint report on the mortality and the agent. We had saved considerable material at -70 °C (-94 °F) for later analyses and experiments, but all thawed during Hurricane Katrina, and we consequently had to destroy it.

We conducted a gillnet survey prior to the mortalities, and the hardhead catfish was the most abundantly captured fish. Using other methods, we would probably have found that the bay anchovy and Gulf menhaden were just as abundant, but, regardless, the catfish constituted much of the local biomass. The catfish population decreased considerably after the period of mortalities, and specimens were rare for the next few years. Additional mortalities of thousands of catfish occurred later in Mobile Bay, Alabama, and Mississippi Sound in May 2009, and these fish probably also had the viral infection. Perhaps these catfish represented specimens without an acquired immune response. Thousands of hardhead catfish also washed up dead in a lagoon in Brevard County in the East Coast of Florida in September 2005, but the cause of those mortalities is apparently unknown.

What may be the same disease as we encountered in the northern portion of the Gulf of Mexico produced 50,000 dead Mayan sea catfish, *Arius assimilus*, in Chetumal Bay in southern

Mexico from June to mid-October 1996. The catfish were also large, of 10–35 cm (4–14 in) total length (Suárez-Morales et al. 1998). The purpose of the report was to report a species of *Argulus*, but the cause of the catfish mass mortality was unknown. Additionally, about 50 tonnes (110,000 lb) of dead marine ariid catfish also related to the hardhead catfish (large individuals of *Netuma barba, Cathorops spixii, Genidens genidens*, and *Sciadeichthys luniscutis*), spreading along 1,800 km (11,184 mi) of beaches, estuaries, and lagoons of Uruguay and southern Brazil in 1994 (Costa 1994). A series of episodes occurred over 16 months during 1994 and 1995. Some of the dead fish exhibited hemorrhaging on the ventral surface and necrosis in the liver and kidney. Virus-like particles in the kidney measuring 32–42 nm in diameter were suggested to be a herpesvirus and associated with spawning stress. Other similar mortalities occurred along the coast of Sierra Leone, Western Africa, in 1980–1981 and more intensely in 1990–1993 (Ndomahina 1994). Another mass mortality of the marine catfish *Arias maculatus* (listed as the junior synonym *Tachysurus maculatus*) was reported from a 120 ha (0.5 mi^2) area near Therespuram, Tuticorin, India, in 1–2 weeks of January 1980 (Natarajan et al. 1982). Young individuals became entrapped when the water retreated after the monsoon. Whether a virus was present was not indicated, but the catfish was the only species mentioned; the dead and dying fish were stressed from a combination of high salinity, low dissolved oxygen concentration, and the presence of hydrogen sulfide.

As questioned in discussing the Gulf mortalities, whether the virus directly killed the catfish or was induced by some stress or other condition has not been established for the catfish infections from any locality, but the virus definitely caused necrosis of the infected cells. Few such massive marine fish kills have been associated with a host-specific virus, and the catfish virus and a herpesvirus infecting clupeids are great examples. Jones et al. (1997) reviewed the Australasian pilchard mortalities of 1995, which started in South Australia and spread to Geraldton, Western Australia; Noosa, Queensland; and New Zealand. A rapid spread of about 25–30 km/day (15–18 mi/day) was suggested to be caused by seabirds or other animals eating dead or dying fish and then defecating. This pandemic occurred in the Australasian pilchard, *Sardinops sagax*, and individuals died within a few minutes after clinical signs of respiratory distress occurred. Acute to subacute inflammation of the gills followed by epithelial hypertrophy and hyperplasia, and the herpesvirus in the gills was not observed in unaffected pilchards, and no correlation existed with oceanographic conditions or the presence of plankton. Another review of the same and the later 1998/1999 pilchard epidemic suggests that the origin may have resulted from importing large quantities of the pilchard into Australia to feed cage-cultured tuna (*Thunnus maccoyii*) (Gaughan 2002). Other large-scale mortalities of clupeids have also occurred such as 1,000 tonnes (2.2 million lb) of Pacific herring (*Clupea harengus pallasi*) in British Columbia in 1949; Meyers et al. (1986, 1994) described viral hemorrhagic septicemia virus associated with epizootic hemorrhages of the skin of Pacific herring in Alaska, which may have been the same virus infecting prior epidemics in British Columbia as well as in Australia.

Probably an unrelated disease in the hardhead catfish is represented by X-cell epidermal lesions not involving any visceral organs and described by Diamant et al. (1994). The description was based on a single specimen captured from Lake Pontchartrain, Louisiana. What may be the same disease was found in three of 434 sampled specimens of a related bagrid catfish (*Chrysichthys nigrodigitatus*) in the Cross River Estuary of Nigeria in 1984 and 1985 (Obiekezie et al. 1988). The cases were identified as epidermal papillomas, and transmission electron microscopical studies gave no indication of the virus or other microorganism.

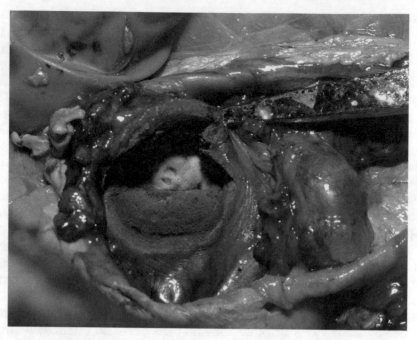

Figure 14.27. Red snapper, *Lutjanus campechanus*, **exhibiting dual deuteromycete fungal infections with** *Penicillium corylophilum* **(*white cottony* appearance of mycelium) and** *Cladosporium spaerospermum* **in the dissected swim bladder. The infection was presumably acquired when the swim bladder was deflated with hypodermic needle after the fish was rapidly raised up from relatively deep water.**

14.4.1.3 Fungal

Fungal infections can be seen more readily in freshwater fishes because the hyphae extend externally. What is probably freshwater *Saprolegnia* occurred in a marine fish trapped in freshwater (Figure 14.17). The related fungus *Aphanomyces invadans* occurs as an ulcerative skin lesion in the Atlantic menhaden (*Brevoortia tyrannus*) (Figure 14.7) along the Atlantic coast; the lesion and associated mortalities had been previously attributed to the dinoflagellate *Pfiesteria piscida* until the critical investigations by Blazer et al. (1999) and Vogelbein et al. (2001). Similar ulcerative lesions have been reported from the striped mullet, silver mullet, black drum, sheepshead, and silver perch in the Gulf from Florida, so Sosa et al. (2007) conducted an experimental study with *Aphanomyces invadans* and other oomycete fungi in the striped mullet and determined that only it caused the lesion. Other fungi such as *Lacazia loboi* causes infections (lobomycosis) in offshore bottlenose dolphin extending into the Gulf (Rotstein et al. 2009).

Little has been reported on harmful fungal infections in fish from Gulf high salinity waters. Figure 14.27 shows a dissected swim bladder of a red snapper (*Lutjanus campechanus*) that had been captured and held in a large tank for many months. It, but not 13 other cohabiting individuals, demonstrated erratic swimming behavior and was therefore necropsied. Both the swim bladder and posterior kidney of that fish exhibited dual deuteromycete fungal infections with *Penicillium corylophilum* and *Cladosporium spaerospermum*. The infection, which was not systemic, was presumably acquired when the inflated swim bladder was deflated with a hypodermic needle after the fish was rapidly raised up from relatively deep water. When cultured fungi were injected into the non-related Gulf killifish, no infection was apparent after 1 month (Blaylock et al. 2001).

14.4.1.4 Protozoan Diseases

Fish dying from a protozoan disease are difficult to obtain because (1) fish seldom die in mass mortalities but rather a few fish at a time, (2) infected fish are usually stressed and consequently become prey for predators before they can be collected, (3) unless morbid fish can be obtained, the agents, unless cyst-formers or those that produce specific gross pathologic alterations, typically deteriorate and become difficult or impossible to identify, and (4) infected or infested fish are actually responding to a toxin or other stress, which additionally allows predilection to a protozoan infection. Helpful general and specific books that treat protozoans and myxosporans of fish (Lom and Dyková 1992; Dyková and Lom 2007) provide a starting place for understanding those groups.

The holophryid *Cryptocaryon irritans* could be an important pathogenic ciliate in the Gulf, but it seldom has been known to cause mortality in wild fish. It has been reported from red drum cultured in ponds in Palacios, Texas (Overstreet 1983b), and caused problems in some marine aquaria. It is the counterpart of the well-known freshwater *Ichthyophthirius multifiliis*, and both have a similar cycle involving feeding trophonts that inhabit the basal layer of the epithelial cells on the skin and gills, a free-living tomont, and an encysted tomont that produces tomonts, which in turn develop into infective theronts that bore through the gelatinous cyst wall and infect a variety of fishes. The histophagus trophonts feed on the epidermis. Dickerson (2006) discusses both ciliates. A few different strains of *C. irritans* have been differentiated. Diggles and Adlard (1997) reported sequence differences among isolates from Moreton Bay and Heron Island, Queensland, Australia; Israel; and the United States. The strain from wild fish in Moreton Bay remains unchanged with that maintained in the laboratory for over 10 years. In Queensland, Australia, when fish are brought into the laboratory aquaria, infections often build up and fish die. The ciliate was considered rare in nature until Diggles and Lester (1996b) showed with critical, sensitive examination for encysted tomonts that 13 of 14 fish species exhibited infections with no seasonality in prevalence or intensity of infection in water temperature between 15 (59 °F) and 27 °C (80 °F). They (Diggles and Lester 1996a), however, showed in experimental temperatures of 20 (68 °F) and 25 °C (77 °F) that trophonts stayed on fish longer and tomonts took longer to excyst, producing larger theronts at 20 °C (68 °F) than at 25 °C (77 °F). The host of origin played a role in tomont incubation period and tomont size. These data suggest that *C. irritans* might play a more important role in the health of Gulf fish, if it was not for the abundance of the equally pathogenic dinoflagellate *Amyloodinium ocellatum* discussed below.

Trichodinid mobiline peritrich ciliates occur on the gills of marine and estuarine fishes as well as on the skin of freshwater fishes in the entire Gulf. When hosts are under stress, they also infect renal tubules. These are disc-shaped or hemispherical ciliates with a cytostome for feeding on bacteria and organic detritus located on the aspect facing away from the host. They usually do not occur in high enough numbers to cause mortality except in aquaculture and habitats with an excess of organic matter, or on stressed fish such as the hardhead catfish infected with the virus mentioned earlier. Infections on juvenile red drum from Mississippi are figured by Overstreet (1983b). They have a simple lifecycle and reproduce when the host is stressed or food is abundant; consequently, their presence has been and can be used as a biological indicator.

The dysteriid ciliate *Brooklynella hostilis* (Figure 14.28) infects several reef fish and the lethargic individuals, with the histophagic ciliate infecting their gills and skin, conspicuously sloughing the infected skin. Consequently, the infection described by Lom and Nigrelli (1970) from a public aquarium in Brooklyn, New York, is sometimes termed "slime-blotch disease" or "clownfish disease" because it often infects clownfishes in marine aquaria. It used to be a

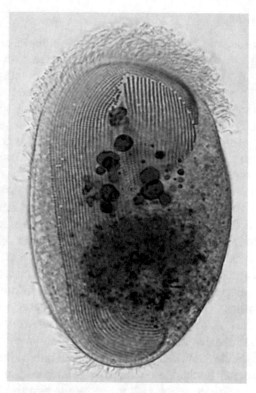

Figure 14.28. *Brooklynella hostilis*, relatively common ciliate in Mississippi waters and elsewhere in the Gulf, Atlantic coast, and Caribbean where it has been associated with fish kills. Staining and photograph by Hongwei Ma (permission to reprint granted by H. Ma to R.M. Overstreet).

problem for aquaculture or public display facilities only. In past years, R. Overstreet could seldom place a yellow-headed jawfish, *Opisthognathus aurifrons*, from the Florida Keys into an aquarium without the disease becoming conspicuous. However, within the last 30 years, infections have been detected on wild dying fish and have been associated with a Caribbean-wide mass mortality event in 1980, as well as repeated mass mortalities in South Florida and the eastern Caribbean (Williams and Bunkley-Williams 2000). But Landsberg (1995) hypothesized a more reasonable hypothesis that infections and mortalities resulted from a synergistic relationship of *B. hostilis*, *Uronema marinum*, amebae, and pathogenic bacteria with biotoxins. For example, a change in abiotic and biotic factors such as Hurricane Andrew or flooding in the Mississippi Delta may have led to changes in the successional colonization or cover of microalgae on the coral reefs. The macroalgae *Caulerpa* spp. temporarily replaced turtlegrass and other sea grasses on which many fish and invertebrates fed, and they produce the toxin caulerpenyne. That toxin as well as indirect consumption of toxic epiphytic dinoflagellates (such as *Gambierdiscus toxicus*, *Prorocentrum* spp., or *Ostreopsis*) by herbivorous fish suppresses their resistance to the protozoan diseases, causing chronic toxicity and disease as well as bioaccumulation, chronic toxicity, and mortality of those predators that fed on the herbivores. *Brooklynella hostilis*, which in many respects resembles its counterpart in fresh water *Chilodonella*, undergoes rapid multiplication by simple binary fission and then weakens and kills the host.

The scuticociliate ciliate *Uronema marinum* occurs as a free-living component of many marine systems, but it can also become a facultative parasite and infect the gills, skin, viscera, and body muscle in Pacific and Atlantic marine fishes, including those in the Gulf (Figure 14.29).

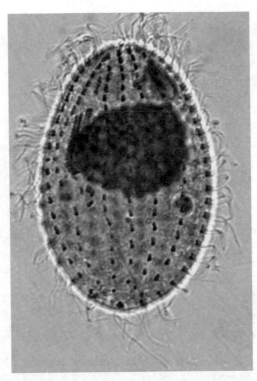

Figure 14.29. *Uronema marinum*, **a cosmopolitan ciliate found in Mississippi estuaries and in blue crab hemolymph and capable of causing disease in cultured and wild fishes and invertebrates. Staining and photograph by Hongwei Ma (permission to reprint granted by H. Ma to R.M. Overstreet).**

Occasionally, the species will enter a wild, wounded, or otherwise stressed fish and replicate rapidly, but it is more likely to cause problems in pen-reared or other cultured fish. What has been identified as *Uronema marinum* has been well studied in aquacultured olive flounder, *Paralichthys olivaceus*, in Korea, but Song et al. (2009) have shown that probably more than one species was involved in the several studies. By obtaining isolates of different scuticociliates from Korea and Japan, cloning and identifying them, and conducting experimental infections with the isolates in flounder, they determined that *Uronema marinum* did not invade the gills, skin, or brain but did produce mortality in up to 30 % of the flounder. Some strains of *Miamiensis avidus* (syn. *Philasterides dicentrarchi*), however, readily invaded the tissues and produced about 100 % mortality and others produced 70 % mortality when those fish were immersed with the ciliate. *Pseudocohnilembus persalinus* and *Pseudocohnilembus hargisi* did not produce mortalities. There are a variety of scuticociliates that can invade tissues of marine and estuarine fishes and invertebrates.

Gulf fundulids, or killifishes, primarily the Gulf killifish, *Fundulus grandis*, commonly exhibited heavy infections of the coccidian *Calyptospora funduli*. The liver (Figure 14.30) and associated pancreatic nodules serve as the primary sites, but infections also occur in the mesentery, gonads, and other tissues. The killifish acquires infections by feeding on grass shrimp in which development to the infective stage occurs (Solangi and Overstreet 1980; Hawkins et al. 1984; Fournie et al. 2000). When a fish that is not a good, susceptible host feeds on the infected grass shrimp, it may acquire the parasite but without normal development or with a strong host inflammatory response; seldom was the accidental atheriniform host killed (Fournie and Overstreet 1993). In *Fundulus grandis*, the parasite occupied up to 95 % of the

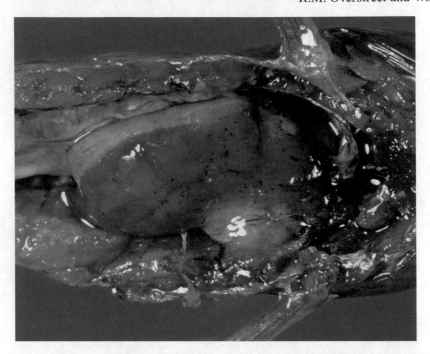

Figure 14.30. Gulf killifish, *Fundulus grandis*, with exposed liver exhibiting chalky appearing area infected by the apicomplexan *Calyptospora funduli,* typically highly prevalent, May 1980.

liver tissue without apparent harm. However, the liver stores glycogen as well as other nutrients, enzymes, and minerals for use by the killifish host when under stress. Consequently, in freezing conditions and presumably in other stressful conditions that require the stored products that, under normal conditions, do not kill uninfected killifish, heavily infected fish die readily (Solangi et al. 1982). Studies on infections in contaminated and pristine habitats in Mississippi are underway.

Most members of the genus *Eimeria*, morphologically similar to *Calyptospora*, have a direct lifecycle in which the infective stage is transmitted directly to the definitive host without developing in an intermediate host. Various species are highly pathogenic and cause mortalities in domestic and zoo animals. Whether this is true for *Eimeria southwilli* that infects the cownose ray has not been determined. However, when wild rays (*Rhinoptera bonasus*) were sampled from Pamlico Sound, North Carolina, 34/37 exhibited oocysts of this coccidian in the coelomic fluid of the seemingly healthy rays (Stamper et al. 1998). When cownose rays were placed in captivity for public display, all those except for the few specimens treated for coccidial infections died from or were associated with a highly pathogenic infection of *E. southwelli*, which was not present in other sympatric species of rays.

The parasitic dinoflagellate *Amyloodinium ocellatum* probably represents the most harmful parasite in aquaculture and display aquaria associated with the Gulf of Mexico. It has a simple lifecycle, with the feeding trophont attached to the gills (Figures 14.31 and 14.32). Once it reaches a certain size or undergoes stress, it retracts its rhizoid from the host epithelium, drops off the host, produces a thin cyst wall, and undergoes a series of synchronous divisions by binary fission. Tomites resulting from this division sporulate to form up to 256 free-living, infective, "*Gymnodinium*-like" dinospores. Under normal conditions, a host in nature would have few, if any, trophonts. However, if restricted to a confined area like an aquarium, pond, or raceway, the number of feeding trophonts on the gills increases logarithmically, and the fish

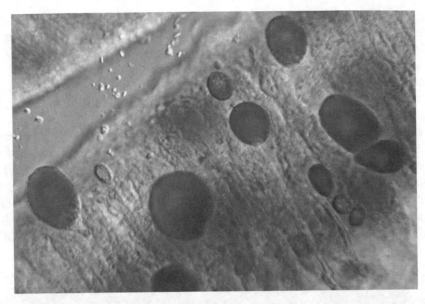

Figure 14.31. A moderate infection of the parasitic dinoflagellate *Amyloodinium ocellatum*, typically found in low mean intensity of infection in gills of the Atlantic croaker, July 1981.

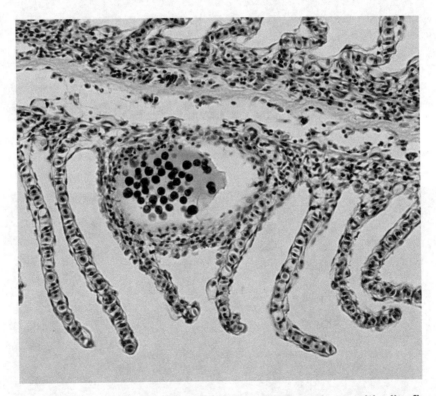

Figure 14.32. Histological section of gill of Atlantic croaker through parasitic dinoflagellate *Amyloodinium ocellatum* and associated minimal hyperplastic host response.

host becomes weakened and dies. In fact, in extremely heavy infections, trophonts may cover much of the body. Since this opportunity to encounter numerous dinospores does not usually occur in nature, mass mortalities as seen in culture are improbable. Overstreet (1993), however, reported a mass mortality on October 31 and November 1, 1984, of fish in a marina and adjoining canal in Alabama; a histological section of the spot gill illustrated the heavy infection with both large and small feeding trophonts. Of the nearly 47,000 dead fish, nearly all consisted of 15–20 cm (6–8 in) long spot.

Not all fishes are susceptible to infection with *A. ocellatum*. Lawler (1980) surveyed fish in Mississippi Sound for natural infections of parasitic dinoflagellates on fishes and found four species which R. Oversrreet confirmed. Lawler recorded 16 of 43 fish species from 28 families with *A. ocellatum*. A raceway was continually stocked with an abundant number of encysted trophonts and dying infected individuals, and specimens of 79 fishes were introduced to the heavy concentration of dinospores. Of those, 71 died and had trophonts covering the body and body-openings as well as in the intestine, allowing for protection during periods of treatment.

Figure 14.33. Sheepshead minnow, *Cyprinodon variegatus*, exhibiting tumorous-like growths caused by the invading myxosporidian *Myxobolus lintoni*, Mississippi, June 1976.

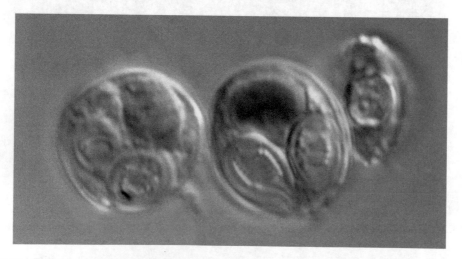

Figure 14.34. Spores of myxosporidian *Myxobolus* stained with Lugol's solution showing positive response for iodinophilous vacuole.

Resistant fishes included *Anguilla rostrata, Opsanus beta*, three species of *Fundulus, Cyprinodon variegatus, Poecilia latipinna, Menidia beryllina*, two species of *Dormitator*, and *Gobionellus hastatus*; some acquired a few trophonts if placed separately with the dinospores, but none showed erratic behavior or died and most produced an abundance of mucus.

The Myxozoa has historically been considered a protozoan taxon, but evaluation of the development of members clearly shows the group to be multicellular. Sequence data still requires additional data from select taxa as indicated by Evans et al. (2010), but the phylogenetic placement tends to relate the highly divergent group more with Cnidaria than with Bilateria. The histozoic myxosporidian *Myxobolus lintoni* invades tissues in the sheepshead minnow, *Cyprinodon variegatus*, and produces protruding neoplastic-like growths (Figures 14.33 and 14.34). The infection was described from the Gulf in Mississippi and Louisiana by Overstreet and Howse (1977), who considered infections to indicate polluted habitats. Unidentified infections of what is surely the same species had been earlier reported from polluted areas of Galveston Bay by Rigdon and Hendricks (1955). Additional cases were noted from Mississippi and Louisiana by Overstreet (1988), even though no suspect toxicant was suggested. We attempted to conduct experimental studies suggested by Overstreet (1993) to determine conditions necessary to induce invasion, but no infected fish were available when we had the presumed tubificid oligochaete hosts in culture. The infection probably can have a detrimental effect on the fish.

Many species of *Myxobolus* and *Kudoa* exist in the marine, estuarine, and riverine system associated with Mississippi Sound. For example, species of *Myxobolus* infect the bulbus arteriosus and gills of centrarchid fishes and tissues (Cone and Overstreet 1997, 1998) and those of *Kudoa* infect muscle tissue of several fishes (Dyková et al. 1994). These are all rather host-specific, but, under the proper conditions, they could cause a weakened condition in the hosts. *Kudoa hypoepicardialis* infects the space between the epicardium and compact myocardium of several marine fishes (Blaylock et al. 2004), and, unlike most species, it is associated with an inflammatory response and shows close affinity to species that cause myoliquefaction of muscle tissue.

Figure 14.35. Striped mullet, *Mugil cephalus*, infected with *Myxobolus* cf. *episquamalis* under the epithelium of scales. Fish was captured in Mississippi waters on June 3, 1996, but a species similar or identical to that infecting mullet was restricted to petroleum contaminated waters of Akko, Israel; Overstreet and Howse (1977) mentioned that such an infection at that time had yet to be seen in Mississippi.

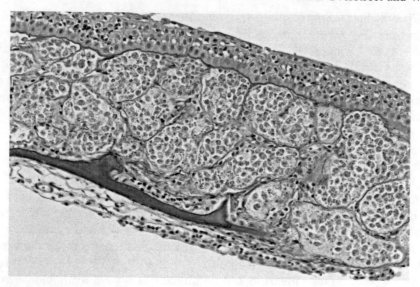

Figure 14.36. Histological section showing Myxobolus cf. episquamalis under epithelium of scales of fish shown in Figure 14.35.

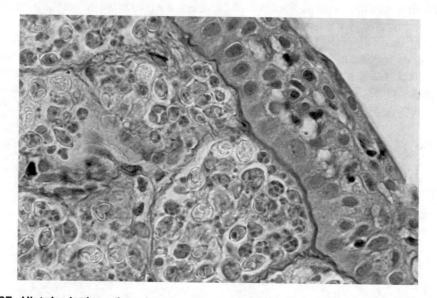

Figure 14.37. Histological section showing close up of *Myxobolus* cf. *episquamalis*.

Overstreet and Howse (1977) reported infected *Mugil cephalus* from the Mediterranean Sea with *Myxobolus* sp. on the scales. Mullet fishermen from along the Israeli coast indicated that the relatively common infection was restricted to the petroleum-contaminated waters of Akko, Israel. Consequently, they blamed infections on the oil. When we reported the infection, we indicated that we had not seen it in the Gulf, but infections would appear to be a good indicator of petroleum contamination. Overstreet (1997) reported the species in that same case as *Myxobolus* cf. *episquamalis*. In 1996, we collected what appears to be the same or similar species in a few striped mullet in Mississippi and call it *Myxobolus* cf. *episquamalis* until the myxosporidian from the Mediterranean, the Gulf, and elsewhere can be sequenced and compared (Figures 14.35, 14.36, and 14.37). Figures 14.36 and 14.37 show histological sections of the infection and the spores from the Mississippi infection.

14.4.1.5 Helminth (Worm) Parasites

In most cases, metazoan symbionts that are acquired by their fish definitive hosts serve as part of a complicated life history and do not harm their fish hosts. For example, adult trematodes, all considered as true parasites in textbooks, typically do not accumulate in large enough numbers to harm their hosts. A few trematodes such as blood flukes are more likely to be exceptions than other taxa. These typically mature in blood vessels without competition with other parasites. A few species that occur in kidneys, mesenteric vessels, and heart tissues can hypothetically accumulate in high enough numbers to harm their hosts. On the other hand, seldom is the abundance of the infected molluscan hosts present in large enough numbers in the habitat to overwhelm the definitive host. Exceptions occur when the mollusk concentrates below or near a net-pen or other aquaculture facility. The blood fluke lifecycle does not incorporate a second intermediate host; the cercaria penetrates the definitive fish host, and the metacercaria, or schistosomula, develops within the fish, usually not in the circulatory system. Bullard and Overstreet (2002) discuss details about how heavy infections of adults in the heart and in vessels associated with mesentery, kidney, and other organs could cause mortalities in cultured fish from the Gulf and are already known to cause mortalities in freshwater culture systems. In Southeast Asia, where marine aquaculture occurs more frequently, the blood-fluke *Cruoricola lates* causes mortality in the centropomid sea-bass *Lates calcarifer* (see Herbert et al. 1994). Numerous described and undescribed blood flukes occur in potential culturable fish. While most blood flukes occur in the lumen of the vessels, *Psettarium anthicum* and *Littorellicola billhawkinsi* thread themselves within the myocardial lacunae of the ventricle and atrium of the heart of cobia and Florida pompano (*Trachinotus carolinus*) and could directly harm the heart tissues if present in high intensity (Bullard and Overstreet 2006; Bullard 2010). Both hosts are popular recreational fishes, and heavy infections would influence the ability of fishermen to reel-in these fishes from relatively deep water; the cobia has unique, ubiquitous, perivenous, smooth muscle cords in viscera which may allow it to counter the effects of the trematode infection (Howse et al. 1992). The adults of other blood-fluke species could also harm their hosts, and, if in high enough intensity, the miracidia hatching from their eggs lodged in the gill filaments could destroy enough gill tissue so that the hosts would bleed to death. Some fish of importance in the sushi and sashimi market as well as cage culture such as the northern bluefin tuna, *Thunnus thynnus*, have blood flukes in the heart such as *Cardicola forsteri* that require investigation (Bullard et al. 2004).

In cases involving numerous life cycles of other non-blood fluke trematodes, the fish can serve as the intermediate host and can accumulate large numbers of or in some cases few metacercariae (larvae or juveniles). These stages develop from the infective, usually free-living cercariae, agents produced by asexual development and shed from the first intermediate molluscan host. The metacercariae either can cause physical damage or influence the behavior of the fish host so that it is more likely to be eaten by the predatory definitive host. A small number of metacercariae usually do not harm the fish, but when they do, infections can result in predation or mortality. An example involves *Bolbophorus damnificus* and its ability to cause multimillion dollar economic losses to the catfish industry. Overstreet and Curran (2004) report that seldom does a catfish fingerling harbor more than 48 encysted metacercariae, even though adults can harbor more than that (Figure 14.38), and metacercariae of other diplostomoids can occur in nervous tissue infections surpassing 2,000 metacercariae. Presumably, more than

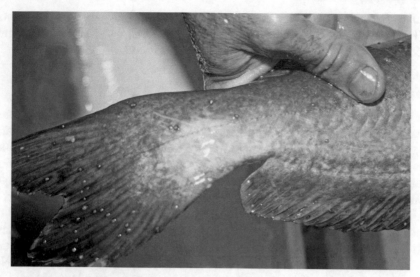

Figure 14.38. Adult channel catfish (*Ictalurus punctatus*) with metacercaria of the diplostomoid *Bolbophorus damnificus*; in the tail; infections of these in fingerling fish in aquaculture kills millions of dollars' worth of fish yearly because the American white pelican, the definitive host, feeds on the fish and transmits the infection, October 1997.

48 kills the fish because we experimentally killed fish exposed to low numbers of cercariae. We also observed kidney damage in pond-reared catfish with over 40–80 metacercariae. Furthermore, Labrie et al. (2004) determined experimentally that fish with approximately four metacercariae died when also exposed to the bacterium *Edwardsiella ictaluri* as long as the exposure was during the first 28 days of exposure, the time it takes for the protective cyst wall to form around the metacercariae. No fish died if exposed to either the trematode or bacterium when not in combination. Other trematodes maturing in either the American white pelican, *Pelecanus erythrorhynchos*, the only known final host of *B. damnificus*, or double-crested cormorant, *Phalacrocorax auritus*, can also kill the catfish in pond-culture and other fishes in the pond, if present in large enough numbers (Overstreet and Curran 2004, 2005; Overstreet et al. 2002).

As indicated elsewhere, infections by juvenile trematodes and other helminths and other parasites may debilitate or otherwise alter the behavior of a fish host so that it is more readily eaten by the definitive host. Usually, only under exceptionally stressful conditions do these parasites cause mass mortalities. An example of a trematode that both debilitates and alters host behavior is the heterophyid *Ascocotyle pachycystis* in the lumen of the bulbus arteriosus of the sheepshead minnow in the northern Gulf. The prevalence and intensity of infections of this trematode are specific to both habitat and season. In low-salinity pools, the fish is unparasitized; along the shorelines open to bays and large estuaries, the intensity may be high but in very few individuals; in isolated or semi-enclosed estuarine sloughs, the infected snail host (*Litterodinops monroensis*) and the fish remain in close proximity during the period of cercarial release, resulting in both high prevalence and intensity of infection. As many as 6,800 of these spherical, thick-cysted, approximately 250 μm (0.01 in) in diameter metacercaria can increase the size of the bulbus arteriosus by mechanical blockage to several times that of its uninfected size. Recruitment of the metacercaria is highest during the first year, with an accumulation of 100–300 parasites per month throughout the year. The rates were only 50–200 metacercaria per month during peak months and less during their third and final year. Larger fish accumulated parasites at higher rates than smaller ones and were more heavily parasitized (Coleman and Travis 1998). Those field studies plus experimental swimming performance studies (Coleman

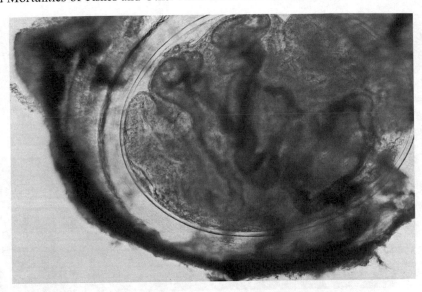

Figure 14.39. Metacercarial cyst of the heterophyid trematode *Scaphanocephalus* cf. *expansus* acquired from skin of the gray snapper, *Lutjanus griseus*, in Florida Keys (March 11, 1997); these cysts, referred to as black spots, help attract osprey or eagle definitive hosts.

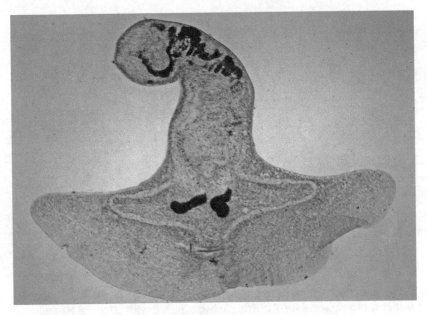

Figure 14.40. Metacercaria of *Scaphanocephalus* cf. *expansus* released from cyst figured as 14.39.

1993) showed that swimming was most detrimentally affected at low temperatures, with low dissolved oxygen levels reducing survivorship of the fish most during the winter when fed on by raccoons, other mammals, and wading birds in which the trematode matured.

The metacercaria of *Scaphanocephalus* cf. *expansus* with melanization surrounding the cyst (Figures 14.39 and 14.40) represents a mechanism allowing their osprey and eagle definitive hosts to more easily prey on infected individuals. Such responses are called "black spots" or

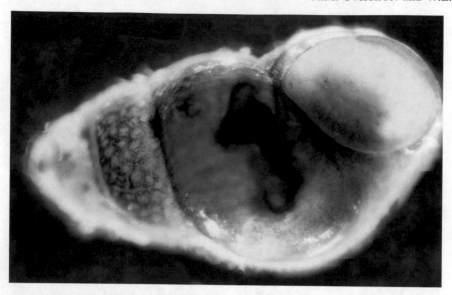

Figure 14.41. Hydrobiid snail (henscomb hydrobe, *Littoridinops tenuipes*) exhibiting microphallid trematode metacercaria of *Atriophallus minutus* in upper whorls infective to ducks (being studied with Richard Heard), Mississippi, December 1997.

Figure 14.42. The potamidid *Cerithidea scalariformis* (ladder hornsnail) showing notocotylid metacercaria attached to operculum. Florida, being studied with Richard Heard.

other names, and trematodes in several families use this strategy to be preyed on and perpetuate the species, usually in bird definitive hosts.

Whereas most metacercariae infect tissues of fishes and crustaceans, a few occur in or on snails. Various examples are there in Figures 14.41 and 14.42. The large chapter by Bullard and Overstreet (2008) treats all aspects of digeneans and even discusses aspects of Gulf species not available elsewhere.

Cestodes also can influence the behavior of intermediate hosts so that their final host more readily feeds on them. Also, they can be overburdened with infections or the cestode can injure

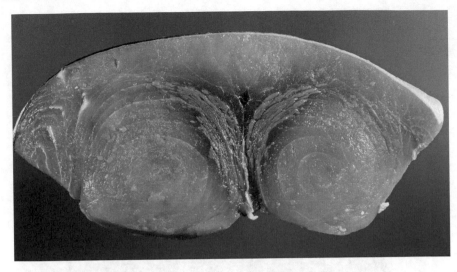

Figure 14.43. Trypanorhyncha cestode in flesh of swordfish, *Xiphias gladius;* the harm to the fish caused by such an infection is unknown.

Figure 14.44. Western mosquitofish, *Gambusia affinis*, exhibiting the invasive Asian fish tapeworm (*Bothriocephalus acheilognathi*), which has spread further across coastal Mississippi after Hurricane Katrina; specimen from Escatawpa River, Mississippi, May 1997.

visceral organs. Small cestode species are abundant and many are reported by Jensen and Bullard (2010). Relatively large species in the flesh gain the attention of fishermen, seafood consumers, and others. Several of those Gulf species have been reported (Overstreet 1977, 1978; Palm 2004; Palm and Overstreet 2000b). Even though *Otobothrium cysticum* is small, it can heavily infect butterfishes and economically affect the seafood fishery (Palm and Overstreet 2000a). Trypanorhychs in swordfish flesh (Figure 14.43) influence consumers and markets, if they realize this object is a parasite and not part of the fish. In any event, it does not harm consumers, even if eaten raw. Others such as *Poecilansystrum caryophullum* in sciaenid fishes

Figure 14.45. Red snapper, *Lutjanus campechanus*, after being fixed showing infestation of monogenoid *Neobenedenia mellini* that was causing mortalities in snapper specimens after being captured, January 2000; these worms are clear and difficult to see on live fish.

also have an unsatisfactory response by many seatrout fishermen and consumers (see questionnaire given by Overstreet 1983a). That cestode species seems to live for 2–3 years and establish an immune response by the fish against further infections. Because only about one-half of a population becomes infected, most spotted seatrout acquire a mean intensity of infection of only 1.5–4.4 worms per fish, depending on location, salinity, and presence of its shark final host (Overstreet 1977). Individual seatrout <14 cm (5.5 in) are seldom seen infected, and, if worms occur, they often associate with vulnerable organs rather than the flesh, where they occur in larger individuals. Consequently, they may kill a few hosts, but not affect the population.

When an introduced species infects a host, it can harm the host in some cases. The invasive Asian fish tapeworm (*Bothriocephalus acheilognathi*) has spread throughout much of the United States, including Gulf habitats with low-salinity or fresh water, but does not seem to have caused noticeable mortalities in the mosquitofish (Figure 14.44) or several other hosts.

Monogenoids consist of helminths that readily harm their hosts, but, as in most infections with trematodes, those usually need to occur in confined habitats like ponds or display aquaria. This group differs from the trematodes because they have a direct life cycle that does not incorporate intermediate hosts. Some species produce two types of eggs, one that will hatch soon and another that can over-winter or otherwise undergo a long period before hatching. The hatched oncomiracidium larva can infest the same individual or another of the same species or, in a few cases, other species. The capsalid *Neobenedenia mellini* provides a good example of a species that infests and can kill a wide range of non-related fishes over a wide geographic area. Figure 14.45 shows specimens of it on a red snapper after the fish was captured in the wild and transferred to an aquarium, where it killed many snapper. Bullard et al. (2000b) list many wild and captive hosts for this parasite from the Gulf and Caribbean. The species probably causes mortality of wild stressed fishes, primarily reef species. It occurs on the gills in lightly infested individuals and can cover most of the body including under the eyelids in heavy infestations. Paperna and Overstreet (1981) and Paperna et al. (1984) reported on a related species in the Gulfs of Elat and Suez, where it infested and caused lesions in a wide range of mullets. Initially when transferred into the laboratory, fish periodically died for up to 2 months, then infested individuals did not die during the next month, and those surviving finally were free of infestation after 3 months, apparently having undergone a "self-cure." On the eastern shore of the Gulf of Suez in El Bilaim Lagoon, large numbers of heavily infested, emaciated individuals of the keeled mullet, *Liza carinata*, died in April 1974 and again in February 1975. Living individuals became so lethargic that they could be captured by hand.

Lesions caused by relatively large Gulf monogenoids in sharks have been investigated in some detail. For example, the affect of the hexabothriid *Erpocotyle tiburonis* on the gills of the bonnethead shark, *Sphyrna tiburo*, produced intense hyperplastic lesions in the epithelium and resulted in the death of sharks, when reared in public aquaria. The same species of worm occurred in lower intensity and produced relatively minor lesions in wild shark individuals investigated (Bullard et al. 2001). The same authors (Bullard et al. 2000a) also studied skin lesions caused by the microbothriid *Dermophthirius penneri* on the wild blacktip shark, *Carcharhinus limbatus*. Lesions appeared as multifocal, well-demarcated, light gray patches on the skin, but they were chronic conditions not associated with secondary bacterial infections or any debilitating disease.

Three small species of *Rhabdosynochus* infesting gill lamellae of three species of snook (common snook, *Centropomus undecimalis*; swordspine snook, *Centropomus ensiferus*; and fat snook, *Centropomus parallelus*) create problems for managing aquaculture in South Florida, especially for the common snook (Kritsky et al. 2010). *Rhabdosynochus rhabdosynochus* restricts itself to hosts in fresh and low salinity brackish water, but *Rhabdosynochus hargisi* and *Rhabdosynochus hudsoni* tolerate wider salinity concentrations, and, therefore, seem to survive typical migrations of the hosts between marine and riverine systems. Consequently, some species of *Rhabdosynochus* are always present, and treatment with freshwater dips ineffectively controlled infestations in culture facilities.

Figure 14.46. Stomach of one of several water birds that exhibit ulcers containing *Contracaecum* spp., the juveniles of which can harm fish intermediate hosts, if in abundance, December 1997.

Nematodes play an important role in the parasite community. Adults have separate males and females as opposed to being hermaphroditic like the flatworms discussed above. In some cases, the adults can cause mortality, but these usually involve situations where the habitat is confined. The Gulf contains numerous species of *Contracaecum*, and the adults occur in piscivorous birds. The American white pelican hosts five species (Overstreet and Curran 2005), and in some cases, juvenile specimens embedded in the proventriculus may cause disease and even mortality. But adults, typically embedded in encapsulated ulcers, can number over 1000 per bird and rather than harming the bird, they often leave the protective ulcer, embed in the prey fish, and help break down the fish tissue. Presumably, they are helpful rather than harmful. Pelicans, herons, egrets, and many other water birds serve as definitive hosts (Figure 14.46). Fish and crustaceans are the typical second intermediate host after a copepod or other crustacean first intermediate host. When in high enough numbers, the third stage infective juvenile can weaken or cause mortality, but infections seldom cause mass mortalities. If the juvenile infects larval or postlarval fish, it can kill the fish that probably have little effect on the fish population. The effect that juveniles embedded in the liver, kidney, spleen, or other organs may have on the reproductive ability of some fish hosts should be investigated.

The nematode *Eustrongylides ignotus* obtains considerable attention because the red worm as long as a human finger can cause human peritonitis when people eat raw second intermediate fish hosts (Overstreet 2003, 2013). Small fish such as the western mosquitofish and small Gulf killifish obtain infections by feeding on the oligochaete first intermediate host. Larger fish that feed on infected mosquitofish and young killifish serve as paratenic hosts, in which the worm grows considerably but remains the same stage infective to the heron or egret bird final host. The striped bass, *Morone saxatilis*, is known to die from an association with the infection in harsh conditions (Mitchell et al. 2009).

Figure 14.47. Western mosquitofish, *Gambusia affinis*, exhibiting nymphs of the pentastomid *Sebekia mississippiensis*, in the body cavity; these nymphs mature in the lungs of the American alligator, Mississippi Bayou, May 1997.

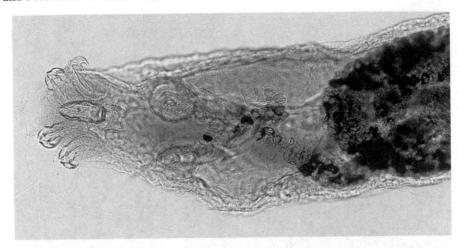

Figure 14.48. Adult female of *Sebekia mississippiensis* from the lungs of the American alligator.

Figure 14.49. Hardhead catfish, *Ariopsis felis*, with the cymothoid isopod *Nerocila acuminata*, which can occasionally be found infesting its dorsal fin in Mississippi estuaries.

14.4.1.6 Other Metazoan Parasites

Western mosquitofish that occur in bodies of water with the American alligator often contain nymphs of the pentastomid *Sebekia mississippiensis* in their body cavity (Overstreet et al. 1985) (Figure 14.47). Fish such as the Atlantic croaker, Gulf killifish, and bluegill that feed on the mosquitofish can become infected, probably debilitated, and serve as indicators of habitats containing the alligator (Figure 14.48) as well as potentially producing human infections (Overstreet 2013).

Figure 14.50. The cymothoid isopod, *Anilocra acuta,* feeding in lesion of ill-spotted gar, *Lepisosteus oculatus*, in small boat basin, Ocean Springs, Mississippi, December 1976; not sure if isopod is responsible for weakened condition of gar or if already weakened condition attracts isopods and argulids to attach and feed.

Figure 14.51. Lane snapper, *Lutjanus synagris,* from off Horn Island, Mississippi, showing exposed cymothoid isopod (in the *Livoneca redmanii-ovalis* complex), June 1997.

Isopods probably have a major influence on fish health and mortality in the Gulf, but they do not seem to cause mass fish mortalities. Members of the Cymothoidae are conspicuous because of their relatively large size. This is apparent for *Nerocila acuminata*, which occasionally infests the dorsal fin of the hardhead catfish in Mississippi estuaries and elsewhere throughout the Gulf (Figure 14.49). A species that seems more pathogenic is *Anilocra acuta* shown feeding in a lesion of an ill spotted gar (*Lepisosteus oculatus*) in the small boat basin at GCRL in Ocean Springs, Mississippi (Figure 14.50). We do not know if the isopod in several such cases caused a weakened condition in the gar or if an already weakened condition attracted isopods and argulids to attach and feed. In any event, we could bend over and cradle the gars, successfully lifting them out of the water. There are cases where crustaceans can transmit viruses and other pathogenic agents to their hosts. One such case discussed earlier (Section 14.4.1.2) suggests internal lymphocystis virus in the silver perch being transmitted from a skin infection through gill lesions produced by the cymothoid *Livoneca ovalis*. The taxonomy of the *Livoneca redmanii-ovalis* complex needs attention, but the Gulf species is common in the Gulf and infests a wide range of hosts. Figure 14.51 shows a lane snapper, *Lutjanus synagris*, from off Horn Island, Mississippi, with a typically exposed specimen. Usually a single specimen occurs on the gills, and the effect that individual has on the host or the population is difficult to establish. Blaylock and Overstreet (2002), however, provided photographs (Figures 12.38 and 12.39 from Blaylock and Overstreet (2002)) showing six individuals causing a flaring of the operculae and heavily eroded gill filaments of an anemic year-old juvenile spotted seatrout in Mississippi. *Livoneca ovalis* (most likely a separate species of the *Livoneca redmanii-ovalis* complex) was noted by both Pearson (1929) for seatrout from Texas and Overstreet (1983a) for seatrout from Mississippi to occur on runted fish during their first 2 years and was more than likely to cause a gradual mortality of those fish. The infested spotted seatrout in Mississippi ranged between 10 and 17 cm (4–7 in) in length from November through May and often hosted two specimens of the isopod. The same species also caused extensive erosion of the gills of juvenile red drum in Mississippi marshes (Overstreet 1983b). Cymothoids are also known to cause mass mortalities in cultured fish, but the majority of cases involve fishes infested by cymothoids that are not on their natural hosts (Smit et al. 2014).

Careful field and laboratory studies in the Great Barrier Reef on the cymothoid *Anilocra pomacentri* attached laterally on the pomacentrid reef fish *Chromis nitida* demonstrated an association often thought to be relatively harmless because the isopod stays attached for a long period. However, observations on a single cohort of fish showed the significantly depressed growth, reproduction, and survivorship of the infested fish. A parasitized female fish produced only 12 % of the number of eggs produced by a non-parasitized counterpart of the same size. Mortality of juvenile fish was estimated to be 88 % relative to 66 % during the first 70 days after recruitment. Within 48 h of attachment, the isopod penetrates through the skin into the muscle (Adlard and Lester 1995). In laboratory trials, fish mortality from infestation ranged from 78 % for small fish compared with 28 % for larger fish within 4 days of experimental infestation (Adlard and Lester 1994).

Most cymothoid life cycles have been based on speculation; two have been described with considerable detail: *Glossobius hemiramphi* from the ballyhoo, *Hemiramphus brasiliensis*, from South Florida (Bakenhaster et al. 2006) and *A. pomacentri* mentioned above from Australia (Adlard and Lester 1995). The final marsupial stage on the female known as a manca immediately swims to the appropriate final host and develops into a male. Then in turn, this protandric hermaphrodite develops into a female, which feeds heavily on host blood, apparently only during periods related to onset of vitellogenesis. The adult female occupies the buccal cavity of the ballyhoo or externally dorsal and posterior to one of the eyes of the chromis. Some other species probably differ because juvenile stages have been collected from hosts other than the recognized final host.

Another of the several Gulf cymothoids that attracts attention of biologists is *Olencira praegustator*, a species that infests the tongue of the Gulf menhaden. Guthrie and Kroger (1974) point out that their data, which also covers infestations of the Atlantic menhaden along the Atlantic coast, suggest that injured or infested adult menhaden school with juvenile menhaden in estuarine nursery areas rather than with their own year-class of fish in offshore waters. When enough infested individuals were present, they formed schools independent of the uninfested juveniles and remained longer in the estuary. Because male and female *Cymothoa excisa* attached to the tongue of three Caribbean snappers from seagrass beds and occupied so much space in the mouth cavity, Weinstein and Heck (1977) thought the isopod would affect the condition factor of the fish; but they found no such effect.

Other species of isopods play important roles in fish health, especially if in any way restrained. For example, most cirolanids such as *Cirolana parva* are actually micropredators that will devour restrained fishes in a short period; in fact, it and other cirolanids have been used by the shark cartilage industry for cleaning the shark carcasses of flesh prior to processing (Poore and Bruce 2012). Poore and Bruce (2012) also cited how the Florida shark industry over one summer collapsed when cirolanids (*Natalolana* spp.) swarmed and ate their way into living sharks, killing them by destroying their vital organs. Others such as *Rocinela signata* have temporary fish hosts to obtain blood meals. Excorallanids such as *Excorallana* spp. in their early stages parasitize bony fishes as well as sharks and rays and then may retire to sponges to molt and reproduce between blood meals. However, *Excorallana delaneyi* in sponges from the Gulf was not seen on fishes (Stone and Heard 1989). Many species of three genera (*Excorallana*, *Alcirona*, and *Lanocira*) in the Gulf occur temporarily on fishes and then associate in cryptic habitats with sponges, ascidians, tube-molluscs, corals, and mangroves (Delaney 1989). The tridentellid *Tridentella ornata* infests the nasal cavity of several grouper species and the red porgy, *Pagrus pagrus*, in the Gulf off Florida where they are thought to "pounce" on a fish just long enough to get a blood meal (Kensley and Heard 1997). Schotte et al. (2009) list just two named gnathid isopods from the Gulf, but we have seen adults and larvae of a few unidentified species, so more species exist. The late larvae (pranizae) of species that use elasmobranchs remain attached in the oral or buccal cavity until ready to mature, but species obtaining blood-meals from bony fish drop off and molt after each blood meal. Except for a few species that permanently attach to fish, one rarely observes larvae of the teleost feeders. Paperna and Overstreet (1981) report on one such Red Sea species that provides a good example. Mullet were placed in floating cages located in water 1–2 m (3–7 ft) and 6–8 m (20–26 ft) deep in the Gulf of Elat. Larvae from the benthos attacked fish at night and fed on their blood. Those fish in cages closest to the bottom, especially when not accustomed to their cages, became anemic from heavy infestations. The larva fed three separate times and after each, it molted and increased in size before reaching maturity. The larvae molted into male or female adults 6–8 days after their last blood meal at 24 °C (75 °F). After 22–24 additional days, the eggs developed, the young free-living larvae searched out fish for a blood meal, and the full-sized, free-living female, after producing about 90 larvae, died. Many of the mullet in the shallow cages died, but few larvae became attached and fed on mullet from the cages in deeper water.

Caligid copepods, often referred to as sea lice, commonly occur in large numbers on Gulf fishes such as flounders, hardhead catfish, drums, and seatrouts, but the effect of adult copepods on these and other hosts has not been established. Frasca et al. (2004) described in good detail the operculum lesion in wild black drum infested by *Sciaenophilus tenuis*. This and other caligids, however, would certainly have a detrimental role in cage culture, if it became more prominent in the Gulf. Ho (2000) discussed nine species reported to have caused mortality in Asian non-salmonid fishes. Salmonid aquaculture in the North and South Pacific, Scandinavian countries, and elsewhere are plagued by *Lepcophtheirus salmonis*, *Caligus elongatus*, and related species. Hewitt (1971) reported that the presumed introduced *Caligus epidemicus* killed several wild fishes in southern Australia. One reason caligids cause mortalities in aquaculture is because there is no intermediate host. Different species have different number of larval stages, but the non-feeding naupliar stages produce a copepodid that attaches to the fish by a frontal filament as chalimus stages. We have observed unidentified chalimus stages attached to fry and

Figure 14.52. Bay anchovy, *Anchoa mitchilli*, from Mississippi Sound with copepod *Lernaeenicus radiatus*, attached *within* blood vessels; infections probably weaken adult fish that more than likely will attract predators; when in postlarval fish, the host would probably be much more stressed and vulnerable, June 1981.

young juveniles from planktonic and other collections. We can only speculate that one such attached copepod can either make that fish vulnerable to predation or produce mortality.

Felley et al. (1987) examined over 27,000 fish from 33 taxa from the ichthyoplankton of Calcasieu Estuary, Louisiana, and reported caligid chalimus stage copepods attached to the dorsum of what was probably the bay anchovy (*Anchoa mitchilli*) and attached to the ventrum of the Gulf menhaden. Only 3.6 % of the postlarval anchovies and 0.2 % of the postlarval menhaden were infested. Additionally, 4.4 % of unidentified gobies had infestations of pre-adult copepods. Infestations were not observed on larval fish, and the length of the copepods on the postlarvae averaged about 15 % and reached almost 40 % of the host length. The authors concluded that the substantial hydrodynamic drag probably produced considerable stress. Overstreet (1978, 1983b) provided figures of a chalimus stage attached to the fin of the Florida pompano and postlarval red drum from off Mississippi.

The pennellid copepod *Lernaeenicus radiatus* probably has a major impact on fishes in the Gulf. Both the Gulf menhaden and bay anchovy as well as other anchovies are susceptible to heavy infestations during some years. Also the Atlantic croaker, seatrouts, Gulf killifish, and gobies often contain one or two of this embedded copepod. The lifecycle in Mississippi (Overstreet 1978) involves chalimus stages both attached to and free on the gills of the intermediate host, rock sea bass (*Centropristis philadelphica*), primarily in high salinity waters. Then the male breaks its frontal filament and transfers a spermatophore to the female, leaves the sea bass within 3–5 days, infects another fish, matures, and produces eggs within a week. Its anterior extends into the fish flesh so that its head, which forms antler-like appendages, clings around a vertebra or some other structure adjacent to a rich blood supply that serves as its food source. When vital organs of the fish are disturbed or, if too many individuals infect a host, the host can die. We used the Gulf killifish as an experimental host, and death usually occurred

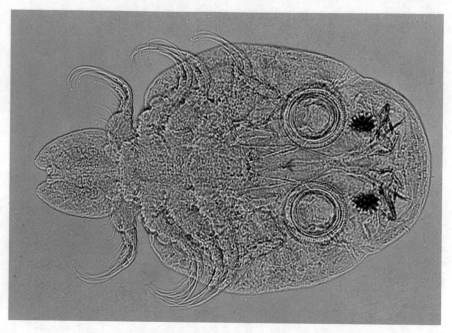

Figure 14.53. Yellow fishlouse, *Argulus flavescens*, a potentially harmful branchiuran.

when more than one or two individuals infected it. The abundance of the parasite typically depended on the abundance of the infested sea bass. Figure 14.52 shows the copepod embedded in a bay anchovy a short distance posterior to its eye.

Skin erosion and ulcerations caused by species of *Argulus* (Branchiura, a separate crustacean group superficially similar to caligid copepods) occur commonly in aquaculture and in fish on public display. Species from the Gulf (for example, Figure 14.53 showing a species from the southern flounder, but that argulid species also occurs on Atlantic croaker, *Mugil cephalus*, dasyatid rays, and other fishes, including freshwater ones) are reported by Overstreet (1978) and Overstreet et al. (1992). A species also infested the Mayan sea catfish, *Arius assimilis* in Chetumal Bay, Mexico (Suárez-Morales et al. 1998). We have seen few cases where harm has resulted in fish from the wild. Two are caused by *Argulus lepidostei* infesting ill spotted gar in conjunction with the isopod *Anilocra acuta* mentioned above and another species causing deep excavations in the skin of the gizzard shad in a Louisiana estuary. Kolipinski (1969) also reported on an infestation on the Florida spotted gar in a pond in the Everglades National Park of what was later described as *Argulus meehani* and may be the same species as occurred in Mississippi. He reported and figured the mass mortality of more than 2,000 gar. This and other aggressive argulids species such as *Argulus catostomi* can kill a fish overnight. He thought the reason that the gar died was because of a prior oxygen depletion that affected animals other than the gar, resulting in the loss of some animals that normally feed on the argulids. An experimental study showed that the flagfish, *Jordanella floridae*, and golden topminnow, *Fundulus chrysotus*, as well as a water scorpion can eat as many as 30 individuals in 28 h. Overstreet et al. (1992) also provided the microscopic anatomy of this interesting group.

Symbiotic barnacles, like those infesting the blue crab, other crustaceans, sea turtles, and marine mammals, do not infest fishes. However, some festering and marine barnacles that attach to a wide variety of substrata do embed in the skin of a few fishes, especially drums, sheepshead, and a few other species such as the hardhead catfish (Figure 14.54). Overstreet (1983b) figured a rare infestation of *Balanus improvisus* in the skin of the red drum from

Figure 14.54. Hardhead catfish, *Ariopsis felis* (SL 207 mm), with abnormal growth of acorn barnacles and secondary bacterial infection, Graveline Bayou, Mississippi, May 11, 1988.

Mississippi Sound, but he had seen it much more commonly up to 2010 in the red drum, black drum, and sheepshead.

14.4.2 Use of Parasites to Evaluate the Effects of Catastrophic Events

Many articles exist using the parasites in or on fishes as indicators to solve problems. Some parasites or group of parasites indicate general heath of the fish model, general ecosystem health, migration of the fish, stock of the fish, trophic level of the fish, feeding behavior of the fish, stress on the fish, and other features. As stressed in the article by Overstreet (1997), the ability to achieve a good indicator/answer depends on the fish species and the parasites chosen. These will differ according to the question asked, and most certainly every fish species and every parasite species will not answer a specific question. The more that is known about parasite species, the better the indications. In other words, the longevity of the larval or adult stages of the worm and the seasonality of the worm may be critical when determining when and where the infection took place. In some cases, the habitat of the first or second intermediate host will indicate in what general or specific habitat the infection was required. For example, the Atlantic croaker typically provides a good model to answer many questions. It uses the estuary to develop, and even though the juveniles move through the passes into high salinity Gulf waters sometime between late spring and early autumn depending on location, temperature changes, and other factors, some individuals remain in the estuary throughout the year. On the other hand, when the salinity of an estuary increases, mature fish from the Gulf move into the estuary and this migration can be detected by the species of parasites present. Species that occur in the croaker inshore often differ from those occurring offshore. The number of species present, or species richness, especially when the origin of the infection is known provides an indicator of the diet of the fish and general health of the environment. A healthy environment includes a large number of infected intermediate hosts, and the adult worms in the croaker indicate a healthy fish rather than one being harmed by the parasites. Moreover, the presence of long-lived species in croaker from the Gulf may indicate from which estuary it was derived. Unlike a croaker, which has a fairly large feeding habitat in the estuary, the Gulf killifish has a very restricted home range. Consequently, the species present in the killifish from a specific habitat indicate the health of that specific habitat. For example, after a catastrophic event, many of the parasites are not available to infect the killifish because their hosts no longer inhabit the area. Both the croaker and the killifish harbor each about 70 different species in the Mississippi estuary, with many in the croaker being adult stages and many in the killifish being larval stages. Many of those larval stages may occur encysted within tissues for over a year. Therefore, when monitoring juvenile killifish that were born after the catastrophic event, the absence of the specific parasites will indicate the absence of either the corresponding interme-diate or definitive hosts. When monthly or seasonally infections of the specific parasites become established in the killifish, the sampled habitat with all of its hosts is becoming healthier. The western mosquitofish, which also has a restricted home range, has many more larval stages than adult stages of parasites, and it also has 70 or so different parasites in Mississippi; it provides a good indicator of environmental health in both freshwater and low salinity habitats. Overstreet (1997) illustrates the large number of invertebrates and vertebrates such as birds, snakes, turtles, alligator, raccoon, and others that have to inhabit the location sampled at least for a period long enough to transmit the infection. Many of the occupants of a habitat, either the individuals or their offspring, become reestablished after a catastrophe. However, it takes much longer for the occupant hosts to become infected and transmit infective stages to the three example model fish. That is why the biodiversity of parasites in those models indicates the rate of restoration of a healthy environment.

Not all parasites in the model fish are harmless or absent after a catastrophic event. Many of these have a direct lifecycle in which reproduction takes place in or on the model fish rather than an indirect cycle that requires feeding on or being in the general area of infective stages. A polluted environment may be conducive to replication of these parasites with the direct cycle. Furthermore, toxic components in the environment may stress the model fish so that it is more susceptible to extensive reproduction of the parasite, and it becomes harmed. Conse-quently, examination of parasites provides a variety of indications of both host and environ-mental health.

Sometimes the use of all the parasites or helminths from all the hosts in a community or restricted habitat provides the best indication of harm or restoration. The example below on the use of parasites to assess hurricanes serves as a good example.

Hurricanes have the potential to cause considerable mass mortalities of fish and other animals, and this damage depends on the date/temperature, the amount of and direction of wind, the length of time the winds hover over the habitat, the amount of tidal amplitude, the local geography, the habitat, and other features. In many cases during and after severe storms, researchers are not available to survey the conditions because the power is out, generators and boats are destroyed, roads are blocked keeping away researchers and passage to surrounding locations, time or ability to attend to destroyed or damaged laboratory and personal facilities is curtailed, and of course the relative importance of human fatalities outweighs those of fish. In a few cases, animal mortalities are reported or described, but the most useful information regarding the effects of storms relates to indicators, which will be discussed below.

Storms are responsible for many fish kills, as well as kills of other organisms. Mortalities and strandings associated with a few storms in southern Florida have been surveyed. Several days or weeks of onshore winds in regions where such winds are not common, such as quiet waters over broad shoals and behind sandbars, can produce heavy seas with a turbulence resulting in harmful sands and other sediments that accumulate and clog gills of the fish as well as erode the gill filaments and cover otherwise healthy habitat. Robins (1957) noted that hurricanes and lesser storms produced accumulations of dead and dying fish along Marco Beach and Sanibel Island, Florida, in the Gulf of Mexico. During most storms, fishes common in the area probably escaped to safer waters; however, such storms provided ichthyologist

Robins the opportunity to collect rare or previously unknown fishes that washed up on the beaches. Under the proper conditions, massive fish kills of common species can occur. Hurricane Donna in September 1960 provided the opportunity to assess loss of biota from the Cape Sable region of northern Florida Bay including specific portions of the Everglades National Park by Tabb and Jones (1962) because they were very familiar with the biota of the area. Winds up to 241 km/h (150 mph) produced tides ranging from 0.46 m (1.5 ft) below to 3.7 m (12.1 ft) above mean water and destroyed mangrove trees, created a massive drift line of turtle grass (*Thalassia testudinum*), and churned up the calcium carbonate marl with associated hydrogen sulfide, all in different regions. Direct action of the storm with its turbulence stressed and suffocated many fish, resulting in massive fish kills in the shallow waters. The dead fish and decomposing vegetation resulted in oxygen depletion, which in turn resulted in post-hurricane fish kills and absence of several common fishes. Some fish species were scarce after the storm, but recreational fishes were abundant in deep water off Cape Sable soon after the storm; the effects in the shallow estuary were also temporary. Landings of the pink shrimp by the Tortugas fishery were about six times greater than prior landings during the same period, suggesting that the storm caused this shrimp to move from its nursery to the fishing grounds earlier than usual and at a smaller size. The healthy blades of turtle grass continually fragment, especially with extremes of temperature. The amount of turtle grass washed ashore in Biscayne Bay, just north of the area with massive fish kills, was massive; however damage to the grass beds and associated fauna due to freshwater runoff in nearshore areas could have been more severe than that caused by the physical storm damage (Thomas et al. 1961).

Hurricane Andrew (August 16–28, 1992, Category 5, with winds up to 282 km/h (175 mph)) (Tilmant et al. 1994) was a brief but extremely strong storm, with unidirectional currents and onshore tidal surges. Seagrass beds remained remarkably untouched unlike during Hurricane Donna when they were heavily destroyed. Andrew produced massive fish kills in the mangrove zone of Everglades National Park, in which the perturbed bottoms left a hydrogen sulfide smell but no sign of fish kills after 4 weeks. Also, upland forest communities were destroyed (Smith et al. 1994), but little damage occurred underwater with the exception of some submerged hardbottom communities of Biscayne Bay that encountered a loss of sponges, corals, and algae. Coral damage on upper reef surface of a few reefs (Elkhorn Reef) displaced some fish that used reefs as a protective habitat. There were few other mortalities; about half of the sea turtle eggs had already hatched, no dead manatee or crocodile occurred, and their habitats were not destroyed.

Strong storms like Hurricanes Katrina and Rita resulted in some locally restricted fish kills such as reported from the Pascagoula River floodplain lakes (Alford et al. 2008, 2009), but, for the most part, no data support the overall short-term mortality events. A few newspaper articles and photos exhibited exceptional fish kills, but most common species left the area prior to the harmful waves and returned soon after the storms. The following years typically produced good catches by recreational fishermen.

We consider a powerful method to define losses of biodiversity and faunal abundance to be through parasitological indicators. This tool is useful for damage and recovery caused by hurricanes as well as other destructive events. The perfect example is the activity caused by the Category 5 Hurricane Katrina, which occurred on August 29, 2005, with the final landfall in Mississippi near the Louisiana state line; it blew as a 280 km/h (174 mi/h) storm with gusts as high as 433 km/h (269 mi/h), with surges over 9 m (29 ft) high. The surges penetrated 10 km (6 mi) inland in Mississippi and 20 km (12 mi) along the bays and rivers. The storm devastated an area equal to the size of Great Britain. It washed away much of Southwest Pass, Louisiana, where the Mississippi River meets the Gulf of Mexico. It also washed away 25 % of the footprint of the barrier islands off Mississippi and Louisiana. It scoured 1 m (3.3 ft) of sediments in water 25 m (82 ft) to 30 m (98 ft) deep, redistributing sediments and depositing 30 cm (1 ft) of sediment at a depth of 50 m (164 ft) and killing most of the infauna but without bodies visible to the human eye. A 6-month drought following the hurricane also influenced biodiversity, mostly the terrestrial fauna and flora that was visible to the local human population. The way that parasites were used to assess the damage was reported by Overstreet (2007) and updated in this chapter.

The presence or absence of parasites in local fishes, especially resident fishes, provides an indication of the loss and recovery of the overall aquatic biota after a hurricane. Based on the species of parasite, with emphasis on its lifecycle, the length of time it takes to become reestablished, the longevity of the parasite, and the effect of the parasite on the host, one can evaluate the effects of perturbation, surge, and presence of toxicants. Most helminths have complicated lifecycles that include two to four hosts, involving a variety of invertebrates and vertebrates that may be specific, closely related, or general. Consequently, by understanding the lifecycle of a specific parasite, one can conclude that all host members of that specific cycle occur in the environment in order for the parasite to be present in the fish being investigated. It is also important that the fish being monitored fits some of the established criteria such as having a restricted home range, a relatively short lifespan, or other features necessary to indicate the specific problem being investigated (Overstreet 1997). When trying to determine how long it takes for reestablishing specific members of the infauna that were lost to perturbation, one has to examine for parasites that have cycles that require those specific hosts as members of the infauna. This is done by noting the date that individuals of the parasite show up in the specific model fish host species, preferably juvenile fish individuals born after the storm.

In the case of perturbations, some trematodes include bivalves, part of the infauna, as first or second intermediate hosts. As an example for an early reestablishment of both the parasite and the bivalve, *Diplomonorchis leiostomi* (Monorchiidae) occurred as immature specimens temporarily in spot (a sciaenid fish) from a few locations in estuaries in Mississippi in February 2006, 6 months after the storm. That trematode reoccurred in spot from a few habitats in January 2007. By March 2007, it occurred in the Atlantic croaker, the common host for that adult parasite. It occurred in low mean intensity throughout specific habitats in patchy distributions for 2.5 years post-Katrina after which time it occurred commonly in higher numbers in more individual fish in more habitats. Another trematode example from Mississippi was the monorchiid *Lasiotocus* cf. *minutus*, which matures in the Gulf killifish (*Fundulus grandis*) and is acquired from the Florida marshclam, *Cyrenoida floridana*, in marsh habitats. Even though the killifish, similar to the spot, returned to its normal habitat within days to weeks after the storm, the trematode did not show up in fish born after the storm until 19 months and then was common after 2 years. The nematode *Eustrongylides ignotus* infects the killifish as a second intermediate host after the fish feeds on the benthic oligochaete intermediate host or on a mosquitofish that had previously fed on the oligochaete, and in that case making the killifish a paratenic host. The juvenile nematode was absent from the killifish for 1.5 years and remained relatively uncommon and patchy until 2010. This conspicuous red nematode takes a long time to develop in the oligochaete and ultimately matures in the proventriculus of a few herons. In contrast with most of the parasites that occurred in infauna, some acquired from copepods were relatively common in fish hosts by 7 months. At that time, the Atlantic croaker was infected with juvenile nematodes of *Hysterothylacium reliquens*, adult nematodes of *Spirocamallanus cricotus*, and adult specimens of the hemiurid trematode *Lecithaster confusus*, which requires a snail before infecting the copepod prey.

In the freshwater area perturbed by the storm, there was a loss of the sphaeriid clam infauna, which hosted members of the gorgoderid trematodes *Phyllodistomum* spp. It took

20 months for the first species to show up in the fish, and it was very rare; by 28 months, multiple species were relatively common in the urinary bladder of catfishes, sunfishes, and fundulids.

The storm surge from Hurricane Katrina gradually encroached the rivers, bayous, marshlands, and uplands of Mississippi as well as lesser surges along Louisiana and Alabama coasts, overflowing the banks and flooding the entire area for at least 20 km (12 mi) along the bays and rivers. In an area with approximately a half meter diurnal tides, over 8 m (26 ft) of surge devastated most of this area. Much of this devastation resulted from the much greater retreat of the water relative to the ebb. From the point of view of the parasites, low salinity and freshwater was rapidly replaced with 32 ppt water, which remained in some ponds, low lands, and other areas. Consequently, snail hosts for haploporid and other trematodes had difficulty reestablishing and were sometimes outcompeted by related species. Two of these haploporid trematodes, *Culuwiya beauforti* (usually common in winter months) and *Dicrogaster fastigata* (now known as *Xiha fastigata*, usually common in summer months), are transmitted by hydrobiid snails as their first and only intermediate host, occurred but were rare 11 months later, and then disappeared from their striped mullet final host. They reappeared 8 months later but disappeared again, finally becoming common in June 2007. Their mean intensities were low in March 2008, but finally became common again in 2009. The striped mullet continually migrates from the shallower Gulf of Mexico and Mississippi Sound into the lower salinity estuarine areas, and the snails had disjunct stocks. Several species of heterophyid trematodes also utilize hydrobiid snails, but they use specific sites in their second intermediate host fishes and mature in a variety of birds and mammals. One initially showed up as a metacercaria (larval stage) in its fish host in June 2006, but others were more delayed, one metacercaria showing up in February 2008 and those of several others by 2009. In freshwater rivers in coastal Mississippi counties, numerous collections of centrarchids, catfishes, catostomids, fundulids, and other fishes yielded numerous trematode species pre-Katrina, but not until August 2006 did any fish exhibit any trematode. The first case of *Megalogonia ictaluri*, an allocreadiid presumably hosted by a sphaeriid (fingernail clam) or unionid clam to an insect or crustacean, appeared in a channel catfish in August 2006 and more were present in May 2007. *Plagiocirrus loboides,* an opecoelid hosted by a snail to a crustacean, did not show up in its fundulid final hosts in spite of numerous attempts by Steve Curran to collect it until early 2011, except for a single infection in April 2007. A few other freshwater parasites appeared by mid-2007 and many more by March 2008.

The Atlantic croaker is also one of our model fish species for parasite infections, and it usually has an abundance of parasites. One of these is the adult cryptogonimid *Metadena spectanda,* which has a typical snail-fish-fish trematode cycle to be published soon and which was first seen after the storm in July 2006 but uncommon until reestablished in 2009. Another helminth is the adult acanthocephalan *Dollfusentis chandleri*, which was not reestablished until early 2009.

In regard to parasites in the Gulf killifish and Atlantic croaker, those that do not have an indirect complicated lifecycle like monogeneans showed up shortly after the storm. The coccidian protozoan *Calyptospora funduli* in the liver and other visceral tissues of the Gulf killifish was extremely common pre-Katrina, but it did not show up until August 2006 and then occurred only patchily along the western coast of Mississippi where all properties were destroyed; it also occurred in a patchy distribution along the eastern Mississippi coastline but not until 2010. This difficult reestablishment seems strange to us because the intermediate host is the daggerblade grass shrimp (*Palaemonetes pugio*), which is extremely abundant along the entire coastline.

In contrast with the above cases, the haploporid *Intromugil mugilicolus* from the striped mullet had not been seen for decades before Hurricane Katrina occurred; it appeared commonly in March 2007 and March 2008. That worm is typically found in the mullet during winter months and absent by May or June. We do not know the snail intermediate host, but it is probably a hydrobiid that was rare in our Mississippi estuaries pre-Katrina.

Parasites in migratory fish are another matter. Adults of the bucephalid *Prosorhynchoides ovatus* in the Atlantic tripletail (*Lobotes surinamensis*) occurred abundantly inshore and offshore both before and after the storm as did other bucephalids from other offshore migratory fishes such as scombrids. The bucephalid lifecycle involves a bivalve to a fish to the final fish host.

Fauna of the sandy barrier islands off Mississippi include a variety of crustaceans, clams, snails, polychaetes, acorn worms, brittle stars, and other invertebrates. Even though those beaches were lost or reshaped, most of the nearshore beach fauna was reestablished, according to Richard Heard of USM, within 6–12 months. Some of the species on the Gulf side took longer, up to 2 years, especially those species without planktonic larvae such as pericaridean crustaceans (like amphipods and isopods). Reestablishment of these animals was easier to determine than that of those from the muddy benthos, but what is important to remember in regard to parasites is that once the invertebrate has reestablished, worms from an infected final host have to deposit eggs and the resulting larvae have to infect those invertebrate hosts. As indicated above, this infection of a parasite may take several years, and the final monitoring results provide information on the presence of all the hosts for a particular parasite in the ecosystem and the overall environmental health of that ecosystem.

Parasites are also good indicators of toxicants in the ecosystem. The myxosporidian *Henneguya gambusi* probably infects *Gambusia affinis*, the western mosquitofish, in extremely low intensity and is difficult to detect. However, when in a stream along coastal Mississippi that was contaminated with the heavy metals chromium, copper, and arsenate, the mosquitofish exhibited heavy infections that involved the intestine, gonads, kidney, and even brain tissue (Overstreet and Monson 2002). Infections presumably kill the mosquitofish in nature since when we transferred samples of fish from this location into the laboratory and maintained them in aquaria, a little more than half the fish died within 6 months when compared with none dying nor expressing an infection in samples collected from non-contaminated locations. No infected fish survived 12 months, but non-infected ones had a high survival. After the hurricane, infections were absent; however, they occurred again in August 2007. The surge apparently flushed out the contaminated water and the infected oligochaete intermediate hosts. The mosquitofish from this location before the hurricane contained no adult tapeworm infection. After the storm, the invasive Asian fish tapeworm (*Bothriocephalus acheilognathi*) showed up in the mosquitofish (Figure 14.44); this adult cestode had previously been collected from mosquitofish in locations numerous kilometers from that contaminated location.

Parasites have been shown to be good indicators of pollution, especially polycyclic aromatic hydrocarbons (PAHs) and PCBs, but usually as non-point sources. Some studies treat the Gulf (e.g., Overstreet and Howse 1977; Skinner 1982; Overstreet 1988; Landsberg et al. 1998; Vidal-Martínez et al. 2014), but most studies treat other areas (e.g., Khan and Thulin 1991; MacKenzie et al. 1995; Austin 1999; Broeg et al. 1999; Dzikowski et al. 2003). Most of the studies have shown that the parasites of fish are more sensitive biomarkers to environmental stressors than the fish by themselves. Landsberg et al. (1998) related specific natural and chemical stressors to specific parasites in the silver perch in Florida. Skinner (1982) showed that the gills of the yellow fin mojarra (*Gerres cinereus*), grey snapper, and timucu (*Strongylura timucu*) from a polluted but not a non-polluted area in Biscayne Bay, Florida, expressed excessive mucus production, epithelial hyperplasia, fused lamellae, and telangiectasia. Three species of monogenoids on the gills of fish from the polluted area occurred in significantly greater mean intensity, presumably because of gill pathology and altered host resistance to the parasites. Pech et al. (2009)

examined the effects PAHs and other chemicals on the parasites of the checkered puffer (*Spheroides testudineus*) in Yucatán lagoons. Vidal-Martínez et al. (2003) restricted one Mexican study to metazoan parasites of the Mayan catfish in Chetumal Bay; their most significant finding was that DDT concentration affected the presence of the trematode *Mesostephanus appendiculatoides* more than the PAHs. Another study showed that parasites from the pink shrimp (*Farfantepenaeus duorarum*) also responded to the chemical contamination in Campeche Sound, Mexico (Vidal-Martínez et al. 2003). Khan (1990), who studied the effects of the Exxon Valdez oil spill, determined the oil affected the presence and intensity of infections of internal parasites. More parasites occurred in the sparid *Boops boops* in Spain after the Pestige oil spill than before, indicating that different intermediate hosts became established after the spill (Pérez-del Olmo et al. 2007). Sures (2004) investigated the sensitivity of various parasites in accumulating heavy metals, and many accumulated more than the fish host. Experimental work with exposures of known PAHs to investigate the induced lesions and effects on parasites has revealed good parasite indicators. When fishes were exposed to oil for a lengthy period and then depurated in oil-free water, the fishes, with gills expressing hyperplasia, demonstrated an increase in the prevalence and intensity of both trichodinids and monogenoids (e.g., Khan and Kiceniuk 1984, 1988; Khan 1990). Water-soluble fractions of crude oil seemed to have a more toxic effect on internal helminths than oil-contaminated sediments; the prevalence and intensity of infections in both types of exposure were less than in reference controls (Khan and Kiceniuk 1983). Sediments contaminated with PAHs and PCBs were also exposed to fish and determined to have an effect on their parasites (Marcogliese et al. 1998; Moles and Wade 2001). Data collected from wild fish, especially when compared with laboratory studies, become even more significant when additional data on bioaccumulation and biomarkers such as molecular, immunological, endocrine, histological, anatomical, and others can be used in conjunction with parasite data (e.g., Van der Oost et al. 2003; Monserrat et al. 2007).

Histopathological information also provides good biomarkers for contamination because both specific and general lesions reflect specific contamination. Meyers and Hendricks (1982) summarize the literature and lesions in experimentally exposed aquatic animals caused by PAHs, PCBs, heavy metals, and numerous other compounds. Solangi and Overstreet (1982) describe lesions resulting from exposure of whole crude oil and water-soluble fractions to the tidewater silverside and hogchoker; they also showed recovery from the lesions when the oil exposure was removed. Misdiagnoses are common in the literature, and future studies can benefit from studying the review by Wolf et al. (2015).

14.4.3 Biodiversity

The role of biodiversity is not a category of disease, and biodiversity can be considered a tool to evaluate catastrophic events treated above in Section 14.4.2. We treat it separately because of its relationship with health. Since parasites comprise about half the Earth's biota, they form a critical component of biodiversity of the Gulf. An abundance of parasites, with an emphasis on helminths, indicates a healthy ecosystem or a healthy host species. This method provides an especially powerful marker because most helminths have three, plus or minus two hosts, in a specific cycle. Consequently, the presence of that specific helminth in a habitat indicates that all members of the cycle are or had recently been present in the habitat.

Because of this cycle, the absence of helminths provides indicators of disruption. The reason for the indication is that adverse impacts on the corresponding intermediate host (or final host) for the species or the population result in fewer species or smaller parasite populations. Fewer parasite species or parasite populations show there was a disruption, even if it was not otherwise apparent.

Table 14.2. Numbers of Helminth Species and Their Hosts Reported from the Gulf of Mexico Based On Felder and Camp (2009).

Vertebrate Host Helminth Group	No. Potential Host Species	No. of Reported Hosts	No. of Reported Species in 2007	No. of Reported Species in 1954	No. Listed as Endemic, 2007	No. Listed as Endemic, 1954
Chondrichthyes (Elasmobranchs)	128					
Trematodes (adult)		4	3	0	3	–
Cestodes (adult and metacestodes)		27	65	37	–	
Acanthocephalan		3	1			
Actinopterygii (bony fishes)	1,409				–	
Trematodes		347	371	198	130	164
Cestodes (adult and metacestodes)		128	41	13	–	
Acanthocephalans (adult)		67	16	7	5	
Reptiles	9					
Trematodes		8	40	15	6	7
Aves	395				–	
Trematodes		57	122	3	38	2
Acanthocephalans			16		3	
Marine mammals	31					
Trematodes		6	9	0	2	–
Cestodes (adult and metacestodes)		4	4	0	–	
Acanthocephalans		1	1		0	
Totals	1,972	652	689	273	187	173

Some protozoans (actually, Protozoa constitutes several independent phyla) have complicated life cycles with multiple hosts and can also provide indications of ecosystem or host health. Unlike the adult helminths, except for the Monogenoidea that can reproduce on their hosts, individual protozoans can produce offspring in or on the fish host. The protozoans such as coccidians, myxosporidians, microsporidians, ciliates, and others as well as monogenoids can replicate in or on the host. Consequently, a disruption in the system can stress the host or otherwise make it more susceptible to excessive replication resulting in a prolific increase in parasite numbers, a harmed host, and an indication of a disrupted and unhealthy ecosystem.

We will use biodiversity of parasites both as individuals and as communities to provide exemplary information on healthy and harmed model systems. The associated biota also goes through seasonal and long-term alterations, and these alterations can also be modified in detectable ways by both anthropogenic and natural environmental events.

Understanding biodiversity of parasites and their hosts in the Gulf of Mexico is accompanied by numerous problems. Table 14.2 helps us start understanding some of those problems regarding helminths. A large volume on Gulf biodiversity by Felder and Camp (2009) provides checklists of all named species of most animal taxa reported by the various authorities of the different taxa. The interesting aspect of the volume is that it updates the checklists occurring in an earlier bulletin listing species known at that time, also by experts in their fields (Galtsoff 1954). Not all parasite taxa were listed in either volume, but the table lists data on three helminth groups, including described adult trematodes (flukes) by Overstreet et al. (2009), adult and some undescribed metacestodes (tapeworms) by Jensen (2009), and acanthocephalans (spiny headed worms) by Salgado-Maldonado and Amin (2009). Nevertheless, the baseline of known parasitic biota is increasing, but assessments can be difficult in some cases because some major groups were not included, not all species in those three groups are known or described, and not all specific or general life cycles for listed species have been determined.

Defining the parasite fauna from the Gulf of Mexico may be difficult because some fish definitive or intermediate host species swim into and out of freshwater, and bird hosts fly to and from coastal or marine areas from adjacent freshwater or from localities other than the Gulf of Mexico. Probably, the best way to define a trematode as a Gulf species would be to determine if its molluscan host was a Gulf resident; however, even that restriction creates confusion because seasonal and yearly dynamics of the infection involve salinity and because details of the life history are usually lacking (the molluscan host(s) of many trematodes have not been discovered) (Overstreet et al. 2009). Also, some mollusks tolerate or thrive in low salinity water with as little as 1–2 ppt, yet the fish or bird host might spend most of its life in high salinity waters.

Many of the fish and parasite populations have a strong Caribbean influence as do the birds, but not many Caribbean Gulf collections have actually been made. Consequently, the checklists of Overstreet et al. (2009) include species extending slightly outside the designated Gulf (Felder and Camp 2009) up through Biscayne Bay on the Atlantic side of Florida and those located off Cancún and Cozumel (slightly south of the Gulf border of Cabo Catoche, Yucatán, Mexico, as will those off Havana, Cuba) as indicated in the introduction. As more fish and birds from the northern Gulf of Mexico as well as elsewhere in the Gulf are examined, they surely will be found to be infected with new and unreported species. Consequently, it is important to include all fauna indicated above to best understand the fauna of the Gulf of Mexico.

Assessing information from Table 14.2, we see that there were totals of 1,541 fishes and 395 birds presently described and reported from the Gulf of Mexico. Of those hosts reported to have trematodes, and not all did, there were 351 (23 %) of the fishes and 57 (14 %) of the birds. The interesting thing about this is that in 1954 there were only 198 fishes and 3 birds that were reported to be infected with adult trematodes.

If one looks at the adult trematodes listed from fishes as being endemic, there were 164 species (83 %) reported in 1954 and 133 (36 %) in 2007. The reason for this decrease reflects the increase in the number of fishes and localities examined, the improvement in identifications, and recent recognition of many species being widespread. The values for all listed adult trematodes from all definitive hosts provide similar data, 186/577 is 32 % in 2007 compared with 173/216 = 80 % as indicated by Manter (1954). Since 1954, numerous new records of fish trematode species have been reported, including those from deepwater fishes and other hosts. However, since only 23 % of possible fish hosts have been examined, and those were infected with 1.1 trematode species per fish, we predict there may be an additional thousand adult trematodes to be discovered.

Reasons for the relatively few fish species examined include the very high cost to collect fish from offshore waters, the difficulty to obtain good quality fresh specimens from the fish when aboard vessels or even on land, and the paucity of taxonomists to identify and describe the parasites. Up to now, the fish that have been examined for trematodes and other parasites have been examined from few geographical areas during a single season and in small numbers by authorities that live and work near the collection sites, primarily in Texas, Mississippi, and Florida.

In addition to trematodes from hosts not yet examined, there are many trematodes (and cestodes) that look superficially like other species, making them hard to identify. These cryptic species are now easier to identify using molecular methods. An example is a complex of species that are reported in the older literature as or close to *Homalometron pallidum* (Trematoda: Apocreadiidae) from a wide range in North America and infecting several hosts. As it turns out, numerous species exist in the complex. In the northern Gulf, the common species is *Homalometron palmeri*, which infects at least four sciaenid fishes, two fundulids, and a gerreid. It is sympatric with *Homalometron manteri* in another sciaenid and appears similar to *Homalometron pseudopallidum* in Argentina (Curran et al. 2013a). On the other hand, the actual *H. pallidum* occurs in fundulids in New England (Curran et al. 2013a). There also occur a few freshwater species from fundulids not inhabiting the southeast, but they are being described and they appear similar to *H. pallidum*. In freshwater in Mississippi and Louisiana, there are two cryptic species similar to *Homalometron armatum*, which occur northward to Lake Erie and Ontario, Canada, as well as in Mississippi, and can be separated by molecular means but with unreliable morphological differentiation (Curran et al. 2013b). A similar Gulf and Caribbean species, *Homalometron elongatum*, infecting a gerreid and appearing more elongated than *H. pallidum* and also possessing three pairs of relatively large opposing oral papillae projecting from near the mouth, makes it distinguishable from *H. pallidum* and another species by both molecular and minor morphological differences. That species, *Homalometron lesliorum*, also infects a gerreid but occurs in Costa Rica and Nicaragua on the Pacific Ocean side (Parker et al. 2010).

Similar appearing complexes also occur for nematodes (Fagerholm et al. 1996; and others) and for cestodes (Jensen 2009; Caira and Healy 2004; Caira et al. 2001; and others). Right now the published library of gene sequences involving trematodes and cestodes is relatively small, but it is gradually expanding and will be extremely helpful in identifying species in the future, thereby creating a more realistic and usable baseline. Once molecular means are used more routinely to compare similar or identical specimens from the Gulf of Mexico, southeastern coasts of the United States, Western Caribbean Sea, and the Pacific Ocean adjacent to the Panama Canal, several of the identifications will be found to be wrong, and several that are suspected to be incorrect will be found to be correct. All in all, we expect that the number of actual species will be considerably more than the presently reported number.

On the other hand, many of the trematodes and cestodes found in birds will be found to be species acquired in the northern or southern ranges of their migratory patterns and do not truly represent Gulf of Mexico species. To make the matter more confusing, the migratory pattern of many seabirds is inadequate, incorrect, or not known at all. Moreover, few cestodes have been reported from true Gulf seabirds (Hoberg 1996; Hoberg and Klassen 2002) and Jensen (2009) did not include birds in her review of Gulf cestodes. Moreover, the lack of knowledge about all seabird parasites in the Gulf makes understanding the history, ecology, and biogeography in marine systems difficult.

In the case of marine mammals and marine turtles from the central Gulf of Mexico, few parasites have been reported. Many of the animals that have been examined are those that get sick and migrate to shore or nearshore habitats to recuperate or die, and the records of their parasites actually reflect records of transient species rather than true Gulf residents with parasites originating in the Gulf. On the other hand, ill fishes and birds usually get eaten before reaching coastlines. To examine stranded or dead marine mammals and marine turtles for parasites and diseases, a researcher requires a Federal permit, which historically has been difficult to obtain. Moreover, most obscure species are unavailable for examination until several days after death of the host. Also, because of the migratory behaviors of many of the mammals and turtles, few of the parasites and diseases originate in or are endemic to the Gulf of Mexico. Also, every few years, ocean currents shift for a short period of time allowing pelagic fishes and other animals to locate near areas they seldom occupy and where few animals get periodically examined for parasites or diseases. Once again, this can provide misleading data on Gulf of Mexico parasites and diseases.

Some nematodes can survive a few days in a dead marine mammal or at least be identified, whereas trematodes and cestodes are much more difficult to find and usually rapidly degenerate after the death of their host. Consequently, recent molecular tools are more likely to detect more nematodes. Moreover, more nematodes, trematodes, cestodes, and other parasites that have complex life cycles will be listed when identified based on sequence data (when a library is or becomes available) based on larval and juvenile stages.

The Atlantic croaker (*Micropogonias undulatus*) from coastal Mississippi has been periodically examined for trophically transmitted parasites over the last 40 years. These parasites may serve as bioindicators of biodiversity, food web structure, prey utilization by hosts, and environmental health. In this chapter, we restrict data to those of the camallanid nematode

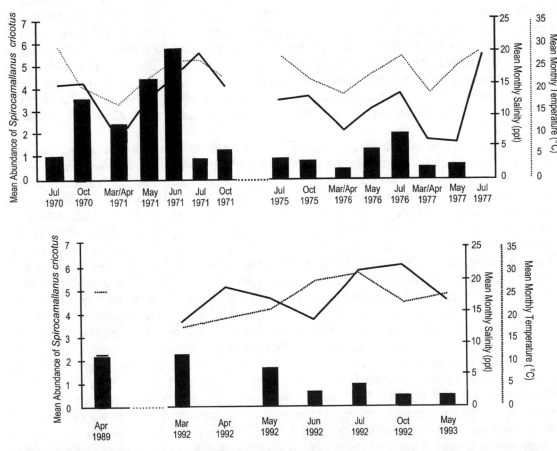

Figure 14.55. Mean abundance (number of individual nematodes recorded/number of individual fish examined) of the nematode *Spirocamallanus cricotus* from the Atlantic croaker, *Micropogonias undulatus,* from Ocean Springs, Mississippi, over several time periods of varying salinity and water temperature. The *solid line* represents salinity (ppt), and the *dotted line* represents temperature (°C). Graphs constructed by Andrew T. Claxton.

Spirocamallanus cricotus from relatively pristine Ocean Springs, Mississippi, to show historic patterns of annual and seasonal variation in abundance. This nematode is acquired by the croaker as well as numerous other fishes (Fusco and Overstreet 1978) when it feeds on copepods such as harpacticoids, the white shrimp (*Litopenaeus setiferus*) (see Fusco 1980), Atlantic brief squid (*Lolliguncula brevis*), and probably other hosts. Patterns of mean abundance, the number of the nematodes in all the croaker examined, whether infected or not, over time suggest the populations of this parasite exhibit extreme seasonal and annual variability, with some time periods exhibiting heavily infected croaker, whereas other periods exhibit few, if any, specimens of *S. cricotus* (Figure 14.55). Over the course of decades, both prevalence (% infected) and mean intensity of this parasite infecting croaker varied. In total, 1,307 croaker examined for this chapter had 2,193 individual *S. cricotus* from the 1970s and 1990s. In the early 1970s, prevalence of a sample reached as high as 74.6 %. In contrast, prevalence of *S. cricotus* in Atlantic croaker from the later 1970s was as low as 1.7 %. Low prevalence was also generally encountered during the 1990s when some collections, such as that in April of 1992, demonstrated no *S. cricotus*.

Fluctuations of *S. cricotus* over time do not appear to be strongly driven by either water temperature or salinity. The period of highest mean abundances occurred during the early 1970s (Figure 14.55). In subsequent years, this parasite was less abundant, even when environmental conditions were similar to the period when abundances were high. The lack of *S. cricotus* in later time periods suggests a decrease in either the presence or consumption of infected intermediate hosts by Atlantic croaker, different susceptibilities to infection, differences in host density, or different climatic conditions after the early 1970s. The potential mechanisms for this difference could be numerous; however, this lack of coherence, a reliable estimate of mean abundance under normal conditions, over time would hinder before-and-after comparisons of parasite population structure in relation to environmental disruptions (see review by Underwood 1994).

While the mechanism driving patterns of *S. cricotus* mean abundance in croaker remains unknown, the presence of a decline following the early 1970s was still apparent in the 1990s (Figure 14.55). In addition, abundance of *S. cricotus* often peaked within a time period during the early spring and summer months during the two periods in the 1970s and 1990s. The changes in abundance that occur on a seasonal basis would represent a further impediment to before–and–after comparisons of environmental disruption since this may necessarily involve comparing different seasons within a year. The decrease in *S. cricotus* over time is most clearly illustrated when comparing late spring and early summer months from the three time periods with one another and using that comparison to control for possible confounding seasonal effects. Even in cases where similar salinities occurred during the same months such as the spring and early summer months of the early 1970s and 1990s, mean abundance of *S. cricotus* was different, and that would suggest mechanisms besides those directly or indirectly related to either salinity or seasonality. In addition to abundance, prevalence and mean intensity also varied among time periods. Between March and June of 1971, prevalence ranged from 71 to 75 % with mean intensity varying from 3.3 to 7.2 worms/infected host. In the same months in 1976, prevalence was from 10 to 50 %, with mean intensity ranging from 2.6 to 3.9. In the 1990s, prevalence during the spring and early summer ranged between 0 and 44 %, with infected croaker harboring 3.2–3.7 worms/host. Thus, the shift in *S. cricotus* that is the result of changes in prevalence, intensity, and abundance suggests either changes in food web structure among and within time periods or physiological alterations in host immune response, but the pathway remains unresolved.

Currently, a study is in progress that will include more data that will allow for comparisons between polluted and non-polluted areas of coastal Mississippi, including the relatively

Figure 14.56. Least puffer, *Sphoeroides parvus*, exhibiting teratoma comprised of liver tissue, July 1981.

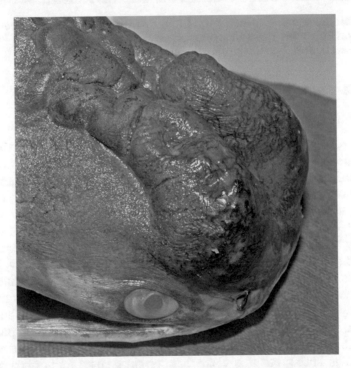

Figure 14.57. Gafftopsail catfish, *Bagre marinus,* head exhibiting disfiguring ossification, West Pascagoula River, Mississippi, June 1992.

non-polluted Ocean Springs locality. Those comparisons will examine both the effects of pollution on food web structure and whether or not the decrease in mean abundance over time is a local phenomenon.

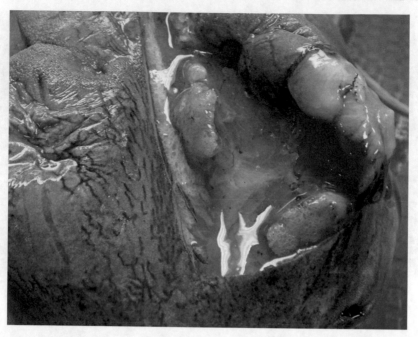

Figure 14.58. Gafftopsail catfish, *Bagre marinus,* head of different specimen exhibiting same disfiguring ossification, with head dissected to show *whitish* and *pinkish* "vacuolated" bony projections, off Horn Island, Mississippi.

14.4.4 Noninfectious Diseases and Conditions

14.4.4.1 Neoplasms

Terminology relating to cancer or cancerous conditions can be confusing because terms often overlap in meaning. For this report, we will use the term "neoplasm" which is defined as an abnormal growth of tissue that is not controlled by the surrounding tissue and continues to grow even after the stimulus that initiated it is removed (e.g., Groff 2004). A neoplasm can be a discrete structure such as a hepatocellular carcinoma or epidermal papilloma, or a neoplasm can be disseminated within tissues of the body as with a lymphoma, for example. The term "tumor" generally refers to a swelling (Figures 14.56, 14.57, and 14.58) and can be synonymous with a neoplasm but not necessarily. The term "cancer" implies malignancy, is a clinical term, and is probably best restricted to use when referring to higher animals and humans. The debate as to the role that environmental factors versus spontaneous genetic mutations or other factors such as viruses play in the initiation of neoplasms has not been fully settled, but it is well-known that environmental conditions, including life style choices such as smoking, poor diet, excessive sunlight, and certain workplace exposures might account for as many as two thirds of all cancers (Anonymous 2003); but it is clear that certain genetic traits make individuals more susceptible to exposure to environmental carcinogens (Perera 1997). These factors hold as well for fishes exposed to cancer-causing agents in the wild or the laboratory.

Perhaps no biologic condition in wildlife evokes concern as does the occurrence of clusters of neoplastic lesions particularly if the neoplasms turn out to be caused by environmental conditions or exposure to chemical carcinogens. In the last few decades, fish have been shown to be susceptible to developing neoplastic lesions from both environmental and genetic stimuli. Groff (2004) has reviewed neoplasia in fishes in general. Overviews by Harshbarger and Clark

(1990) and Harshbarger et al. (1993) concluded that epizootics (clusters) of neoplasms have occurred in fish from over 40 locations in North America. The neoplasms arose from a variety of cells and tissues including nervous, connective, reproductive, and digestive tissues as well as blood. With the exception of neoplasms involving the liver (hepatocellular and biliary adenomas and carcinomas) and skin (mainly epidermal papillomas), most of the neoplasms were unrelated to exposure to environmental conditions. Skin and liver neoplasms that occurred in 14 mostly bottom-dwelling fish species were strongly associated with exposure to PAH contaminants in sediments. Prominent among reports of contaminant-induced neoplasia in fishes from North American waters are liver neoplasms in English sole (*Parophrys vetulus*) from the Puget Sound (Myers et al. 1990), brown bullhead (*Ictalurus nebulosus*) from tributaries of Lake Erie (Baumann et al. 1990), and winter flounder (*Pseudopleuronectes americanus*) from Boston Harbor, Massachusetts (Murchelano and Wolke 1991). Both liver and epidermal neoplasms in white suckers (*Catostomus commersoni*) and brown bullhead from western waters of Lake Ontario were associated with chemical contaminants (Hayes et al. 1990). The reports above list bottom-dwelling species from cold water sites. Vogelbein et al. (1990) reported a high prevalence of liver neoplasm in mummichog (*Fundulus heteroclitus*) from a creosote-contaminated site in the Elizabeth River, Virginia, demonstrating that contaminant-induced liver neoplasia in fishes is not limited to bottom dwelling species from higher latitudes. Nevertheless, epizootics of hepatic neoplasia have not been reported from tropical or subtropical locations. Experimental studies have confirmed that environmental carcinogens such as the PAHs can cause hepatic neoplasia in laboratory-reared fish species (Hawkins et al. 1988, 1990; Fabacher et al. 1991). Furthermore, mechanisms by which chemicals cause carcinogenesis in fish relate closely, if not mirror, similar mechanisms in mammals at the organismic, tissue, cellular, and molecular levels (Ostrander and Rotchell 2005; Bailey et al. 1987; Ostrander et al. 2007).

Although clusters or epizootics of piscine neoplasms are rare in the Gulf and related systems, nerve sheath neoplasms in the bicolor damselfish (*Pomacentrus partitus*) from the Florida Keys are the best studied. They affect 5–10 % of the individuals in populations throughout the Caribbean (Schmale et al. 1983, 1986). The lesion has been considered as a

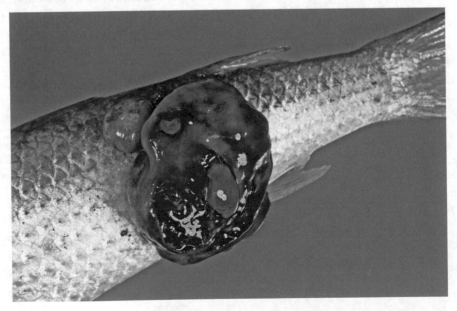

Figure 14.59. Striped mullet, *Mugil cephalus*, exhibiting a fibrosarcoma, Mississippi Sound, September 1979.

model for neurofibromatosis (von Recklinghausen's disease) in humans. Subsequently, it was found that the disease was transmissible from tumor-bearing to non-tumor-bearing specimens through tissue homogenates (Schmale and Hensley 1988) and that a virus-like agent, probably a retrovirus (Schmale et al. 1996), was responsible for the disease in damselfish (Schmale et al. 2002). Lucke (1942) reported similar nerve sheath neoplasms at a prevalence of 0.5–1.0 % in three snapper species from the Dry Tortugas near Key West. No etiologic factor for these tumors has been identified.

Otherwise, reports of neoplastic lesions in fishes from the Gulf of Mexico are of single or low numbers of cases from a variety of species and none have been clearly linked to exposure to any kind of carcinogen. Single case reports or those involving a small number of cases include squamous cell carcinoma in gulf menhaden (*Brevoortia patronus*) by Fournie et al. (1987), capillary hemangiomas in a scamp (*Mycteroperca phenax*) (Fournie et al. 1985), a hepatocellular neoplasm in a wild-caught sheepshead minnow (*Cyprinodon variegatus*) (Oliveira et al. 1994), several cases of subcutaneous fibrosarcomas (fibromatoses or fibromas) in striped mullet (*Mugil cephalus*) (Edwards and Overstreet 1976; Overstreet 1988) (note more severe neoplasms in September 1979; Figure 14.59), and subcutaneous fibromas in southern flounder (*Paralichthys lethostigma*) and the sea catfish (*Arias felis*) (Overstreet and Edwards 1976). Overstreet (1983b) illustrated an epidermal papilloma in a red drum (*Sciaenops ocellatus*). McCain et al. (1996) found hepatic neoplasms (adenomas) and preneoplastic hepatic lesions (basophilic, eosinophilic, and clear cell foci) in several specimens of hardhead catfish from chemically contaminated areas of Tampa Bay.

Few broadly based sampling programs have been conducted in the GoM using biomarkers of fish health as an indicator of the condition of the environment from which the fish were collected. The most robust and comprehensive sampling program was the USEPA's Environmental Monitoring and Assessment Program (EMAP) that examined tens of thousands of fish

Table 14.3. RTLA Tumor Specimens in Fishes Affiliated with the Gulf of Mexico.

Diagnosis	Scientific and (Common) Name	RTLA No.
Adenocarcinoma stomach	*Ocyurus chysurus* (yellowtail snapper)	7673, 7674
	Unknown serranid	7670
Adenocarcinoma stomach; Carcinoma in situ stomach	*Ocyurus chysurus* (yellowtail snapper)	7677
Angiosarcoma	**Kryptolebias marmoratus* (mangrove rivulus)	6136-7, 6139-6146, 6148-56
Carcinoma in situ stomach	*Epinephelus morio* (red grouper)	7671
	Ocyurus chysurus (yellowtail snapper)	7672, 7676
Cholangiocarcinoma	*Seriola* sp. (type of amberjack)	3820
Chondrofibroma	*Ariopsis felis* (hardhead catfish)	1113
Chondroma	*Squalus acanthias* (spiny dogfish)	3144
Dermal fibrosarcoma	*Brevoortia gunteri* (finescale menhaden)	1848
Epidermal papilloma	*Lepisosteus platostomus* (shortnose gar)	3913
	Mugil cephalus (striped mullet)	5469
	Sciaenops ocellatus (red drum)	1904
Esthesioneuroblastoma of the lateral line	*Cyprinodon variegatus* (sheepshead minnow)	3102

(continued)

Table 14.3. (continued)

Diagnosis	Scientific and (Common) Name	RTLA No.
Fibrolipoma	*Pogonias cromis* (black drum)	1662
Fibroma	*Eugerres plumieri* (striped mojarra)	678
	Lagodon rhomboides (pinfish)	3626
	Mugil cephalus (striped mullet)	807, 821
	Paralichthys lethostigma (southern flounder)	1112
	Seriola sp. (type of amberjack)	3628
Follicular cell carcinoma thyroid	*Ocyurus chysurus* (yellowtail snapper)	7664
Ganglioneuroblastoma	*Pomatomus saltatrix* (bluefish)	5241
Hemangioendothelioma	*Mycteroperca phenax* (scamp)	3179
Hemangiopericytoma	*Cyprinodon variegatus* (sheepshead minnow)	3808, 3809
Hepatocellular carcinoma	*Kryptolebias marmoratus* (mangrove rivulus)	2348, 2430-1, 2434-8, 3390, 5446-7
Hepatocytic adenoma	*Bagre marinus* (gafftopsail catfish)	5528
Iridophoroma; neurilimmoma	*Lutjanus apodus* (schoolmaster)	2289
Leiomyoma	*Mugil cephalus* (striped mullet)	5470
Lipoma	*Amia calva* (bowfin)	6397, 6399
	Brevoortia gunteri (finescale menhaden)	1664
	Eugerres (Diapterus) plumieri (striped mojarra)	596
	Paralichthys dentatus (summer flounder)	710
	Seriola sp. (type of amberjack)	4885
Mixed germ cell-sex cord stromal tumor	*Rachycentron canadum* (cobia)	7754
Myxoma	*Mugil cephalus* (striped mullet)	3131
Neurilemmal sarcoma	*Lutjanus griseus* (gray snapper)	1481
Neurilemmoma	Snapper (unidentified)	1378
Neurofibroma	*Lutjanus griseus* (gray snapper)	3892
	Stegastes partitus (bicolor damselfish)	3177
Neurofibrosarcoma	*Stegastes partitus* (bicolor damselfish)	3176
Ocular chondrosarcoma	*Kryptolebias marmoratus* (mangrove rivulus)	3973-4
Ossifying fibroma	*Caranx hippos* (crevalle jack)	1233
Rectal adenocarcinoma	*Balistes vetula* (queen triggerfish)	6435
Reticulum cell sarcoma (spleen)	*Carcharhinus plumbeus* (sandbar shark)	523
Schwannoma	*Mugil cephalus* (striped mullet)	3963
Squamous cell carcinoma	*Brevoortia patronus* (Gulf menhaden)	3618
Thyroid (?) carcinoma	*Abudefduf saxatilis* (sergeant major)	5918

* indicates that the *Kryptolebias marmoratus* (mangrove rivulus) possibly is from experimental laboratory studies. Some of these data were provided by Jeffrey C. Wolf, DVM, DACVP, Chief Scientific Officer, Manager of Virginia Pathology.

from estuarine locations along the coasts of the Gulf (see Summers 1999). The EMAP program examined over 64,000 fish specimens from the Louisianan Province from 1991 to 1994 for gross abnormalities, including tumors and lesions on the skin, malformations of the eye, gill abnormalities, and parasites. Total gross pathologies were seen in 408 specimens for an overall incidence of 0.6 %. Parasites accounted for 61 % of all gross pathologies. Nevertheless, there was a positive correlation between the occurrence of gross pathologies and sediment-contaminant concentrations (Fournie et al. 1996).

Although not a sampling program per se, the Registry for Tumors in Lower Animals (RTLA) served the environmental and comparative pathology community well for many years (Harshbarger 1977). Supported by the National Cancer Institute and housed in the Smithsonian Institution in Washington, DC, the registry for the most part depended on independent contributions from scientists and lay persons all over the world and provided diagnostic services on the accessions. Perusal of nearly 8,000 records of specimen accessions by the RTLA (Harshbarger 1965–1981 and Jeff Wolf personal communication) yielded around 75 cases of tumors in fishes from the Gulf and nearby waters or in species known to inhabit Gulf waters (Table 14.3).

Overall, neoplastic lesions have rarely been reported from fishes from the Gulf. Furthermore, to our knowledge, no epizootics of chemically induced tumors comparable to those reported above have occurred in wild fishes from the Gulf. Overstreet (1988) reviewed the occurrence of neoplasms and related histopathological conditions in fishes from the coasts of the southeastern United States, particularly the Gulf of Mexico, and found scattered examples of neoplastic lesions in individual fish species but no epizootic of neoplasia. The paucity of reports of neoplasms in Gulf fishes might be related to several factors. These include the fact that the Gulf is a rather large body of water to study, especially when compared with water bodies in other regions, and except for a few locations, it is relatively free of industrial pollution. Probably most importantly, however, the Gulf has not been studied as intensively as some other North American aquatic systems. Not all fish species are equally susceptible to chemically induced neoplasms (Hawkins et al. 1985) and susceptibility depends on habitat preferences with bottom dwelling species more susceptible than pelagic species. Most importantly, in the case of PAH exposure, the ability of the fish to convert the compounds to carcinogenic intermediates determines in large part their susceptibility to develop neoplasia (Ostrander et al. 2007). Clearly there are fish species in the Gulf that are susceptible to chemically induced carcinogenesis. For example, Atlantic croaker has been shown to be capable of metabolizing polynuclear aromatic hydrocarbons to their carcinogenic intermediates (Willett et al. 2009). Also, indigenous Gulf species including sheepshead minnow (*Cyprinodon variegatus*), Gulf killifish (*Fundulus grandis*), inland silverside (*Menidia beryllina*), and mangrove rivulus (*Kryptolebias marmoratus*) all developed liver tumors following exposure to the direct acting carcinogen methylazoxymethanol acetate (Hawkins et al. 1985). To date, the only chemically induced epizootic of neoplasia in a fish from a warm water system remains the case of mummichog (*Fundulus heteroclitus*) exposed to creosote residues in Virginia (Vogelbein et al. 1990). Nevertheless, it is likely that geographically broadly based, multi-species, sampling programs focused on examining specimens from contaminated sites will yield neoplasm prevalences in line with other well-studied systems.

14.4.4.2 Developmental Abnormalities

Abnormalities can be the result of genetics, environmental conditions, biological conditions, anthropogenic activities, and other causes. Manipulation of fish regarding reproductive activities to produce offspring utilized in aquaculture, research, and display provides good

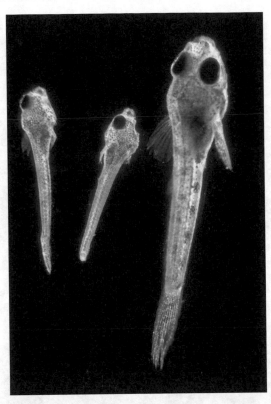

Figure 14.60. Sheepshead minnow, *Cyprinodon variegatus*, examples of abnormal 6-day-old fish resulting from experimental stripped mating; offspring from natural mating under experimental conditions seldom produce abnormal fish.

Figure 14.61. Striped mullet, *Mugil cephalus*, with atretic and fibrotic ovaries that had undergone hydration but not spawning and consequently underwent atresia after cold snap, December 1988.

Figure 14.62. Striped mullet, *Mugil cephalus*, abnormal hermaphroditic specimen from Mississippi Sound showing pinkish testes and orangish ovaries, December 1997.

Figure 14.63. Striped mullet, *Mugil cephalus*, exhibiting scoliosis and lordosis, Mississippi Sound, June 1981.

examples of abnormalities. Figure 14.60 demonstrates different-sized fish 6 days after eggs and sperm have been artificially brought together; similar abnormalities can result from suboptimal temperatures and salinities. Under normal conditions involving light-dark cycles, nutrition, and temperature, ova become hydrated, deposited, and then fertilized. Figure 14.61 shows the two ovaries containing an abundance of ova that became hydrated but not deposited, presumably because temperature or other conditions inhibited that process. After a short period, the hydrated ova hardened and started to undergo atresia, the process of their degeneration and that of the ovarian follicle. Rather than being soft and pliable, each ovary was hard and crunchy. Figure 14.62 demonstrates the gonads of a hermaphroditic striped mullet. The presence of both ovary and testes can occur naturally in some species of fish simultaneously or one at a time. This is not the case with the striped mullet, and such a condition can be a genetic abnormality or induced by a specific contaminant. When an affected individual occurs, the cause is most likely genetic. When a significant proportion of the population exhibits hermaphroditism, the cause may result from a group of chemicals in the habitat known as endocrine-disrupting compounds, such as steroids, hormones like estrogen, and some detergents and pesticides.

Figure 14.64. Spanish mackerel, *Scomberomorus maculatus*, from off barrier islands in Mississippi exhibiting a rare case of scoliosis and lordosis, November 1981.

Figure 14.65. Series of longnose killifish, *Fundulus similis*, from Mississippi Sound, showing scoliosis and lordosis, July 1981.

Three main types of spinal column abnormalities occur in fish: scoliosis (lateral deformity, zig-zag shape), lordosis (dorsal deformity, V-shape, or loss of normal curvature of the lower or posterior spine), and kyphosis (ventral deformity, inverse V-shape, front to back deformity, or hunched back); consequently, some authors prefer to join lordosis and kyphosis together as lordosis, referring to dorso-ventral deformity or as scoliosis when combining all three abnormalities.

Figures 14.63 and 14.64 show a combination of scoliosis and at least lordosis in the striped mullet and Spanish mackerel. The mackerel abnormality represents a unique case, but we have

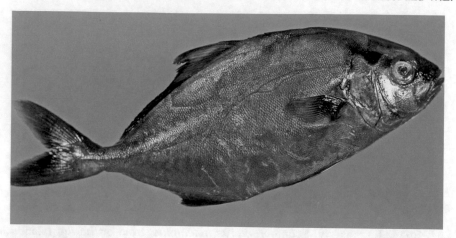

**Figure 14.66. Horse-eye jack, *Caranx latus*, with abnormal partial double lateral line, off Missis-
sippi barrier islands, November 1971.**

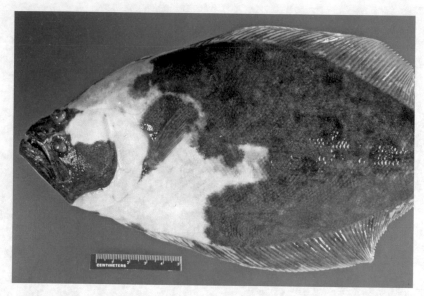

**Figure 14.67. Southern flounder, *Paralichthys lethostigma*, exhibiting partial albinism on eyed
dorsal side, Bayou Caddy, Bay St. Louis, Mississippi, August 16, 1993.**

seen several such distorted mullet over previous years (e.g., Overstreet 1978). Figure 14.65
exhibits a variety of skeletal abnormalities in the longnose killifish. Afonso et al. (2000) discuss
causes of skeletal abnormalities involving genetics, development, and environmental situations.
Such cases are relatively common in aquaculture and when rearing fish for research. As an
example, when we (Overstreet et al. 2000) reared small fish for carcinogenicity studies,
scoliosis and other abnormalities ultimately became obvious in several older specimens unless
the fry were fed an adequate diet of algae, ciliates, or immature nematodes during their initial
three or so days of culture.

Abnormalities of the lateral line system in fish seem to be rather unusual. A previously
unreported case of a partial double lateral line in the horse-eye jack was seen once only
(Figure 14.66).

Figure 14.68. Hogchoker, *Trinectes maculatus*, from Mississippi Sound showing ambicolorate pigment pattern, blind side.

Figure 14.69. Hogchoker, *Trinectes maculatus*, from Mississippi Sound, also showing abnormal pigmentation pattern on the eyed side of same fish as in Figure 14.68.

Flatfishes in coastal and estuarine waters less than 5 m (16 ft) in depth represent good models for developmental (typically reversal) and pigmentation (typically albinism and ambicoloration) abnormalities (Figure 14.67), and the foundations for discussing them were established by Norman (1934) and Gudger (1934). During development of the normal flatfish fry, one eye rotates to the opposite side so that both eyes end up on a predetermined side. Moreover, pigmentation that responds to light and background colors develops on the upper, eyed side, leaving the blindside pale. This behavior and the ability to burrow in the substrate allow flatfishes to avoid most predators from both above and below them. Abnormal cues involving light and temperature (Gartner 1986) during early development result in skeletal (primarily head and fin) and pigmentation abnormalities. Dawson (1962, 1967, 1969), Moore and Posey (1974), and Gartner (1986) described and reviewed many abnormalities in the hogchoker and other

Figure 14.70. Atlantic croaker, *Micropogonias undulatus*, exhibiting pugnose abnormality, Mississippi Sound, November 1979.

Figure 14.71. Atlantic croaker, *Micropogonias undulatus*, with top specimen exhibiting abnormal micro-eye condition, Mississippi Sound, September 1979.

Figure 14.72. Hardhead catfish, *Ariopsis felis*, conjoined twins from mouth of male brooding fish, Mississippi Sound.

flatfishes, and these hogchoker abnormalities occur more commonly (Figures 14.68 and 14.69) than similar abnormalities in most other flatfishes in the Gulf. Overstreet (1978) figured and discussed a rare case of reversal of a fringed flounder (*Etropus crossotus*). Partial albinism of the southern flounder is shown in Figure 14.67, and ambicoloration, or pigmentation on the blindside of the hogchoker shown in Figure 14.68, with abnormal pigmentation also on the eyed-side of the same individual (Figure 14.69).

Dawson (1964, 1966, 1971) and Dawson and Heal (1976) provide a series of very useful bibliographies of anomalies of fishes; each includes an index for fishes and a separate index for anomalies, allowing a reader to find nearly all abnormalities reported before 1976. Most of those from the Gulf are rare. Some examples of abnormalities include the pugnose condition in the Atlantic croaker (Figure 14.70) that also occurs in other sciaenid fishes such as the spotted seatrout in the Gulf (Overstreet 1983a). Another involving the Atlantic croaker is the lack of an eye or a micro-eye (Figure 14.71), a condition also reported for the red drum in Texas (Overstreet 1983b). The red drum from Texas is also known to exhibit scale disorientation (Gunter 1948), a common abnormality in pinfish in Biscayne Bay, Florida. Figure 14.72 shows conjoined twins still containing a yolk sac when taken from the mouth of a wild male brooding hardhead catfish. Both of these twins and those of a Japanese medaka obtained in culture (Overstreet et al. 2000) were maintained alive for a few weeks in a culture dish. More than likely, such twins would become easy prey if not carefully protected.

Rubber and plastic trash can also encircle or otherwise harm fish as well as birds, marine mammals, and seabirds. An Atlantic croaker apparently swam through a rubber band and ultimately grew around it (Overstreet 1978; Overstreet and Lyles 1974), similar to a situation where mackerel became ringed with condoms occurring near sewage effluents. Sharks also become encircled in plastic packing straps (e.g., Overstreet 1978). Because of the serious problem with trash in the past, there has been a recent attempt not to contaminate the seas.

Abnormalities also constitute good indicators of polluted environments. The best investigated area for this chapter consists of sites within Biscayne Bay, Florida. Skinner and

Kandrashoff (1988) observed over 10,000 fishes within 45 species caught by gill nets throughout the Bay over a 10-year period from 1970 to 1982. They found the most heavily-affected species were the Western Atlantic seabream (*Archosargus rhomboidalis*) with skin hemorrhaging and scale disorientation; yellow mojarra (*Gerres cinereus*) with fin erosion and eye abnormalities; Florida pompano (*Trachinotus carolinus*) with emaciation; and pinfish (*Lagodon rhomboides*) with scale disorientation. The striped mullet and Atlantic croaker were caught in significant enough numbers in 1973 and 1974 to compare prevalence of fin and skin hemorrhaging. Such bacterial infections affected all individuals in 26 of 43 collections of the mullet and more than half of the individuals in the remaining 17 collections; all individuals of the croaker in 24 of 30 collections were affected, with more than half of the fish in the remaining six collections affected. From the same general area in 1991 and 1992, Gassman et al. (1994) caught 3,650 fish of over 60 species by hook and line, but with 70 % of those belonging to one of four target species, the seabream, blue striped grunt (*Haemulon sciurus*), pinfish, and gray snapper (*Lutjanus griseus*). Missing or deformed dorsal fin rays were the most common abnormalities in the snapper (4.6 %), scale disorientation in pinfish (7.3 %), and both in the seabream (3.0 % and 3.8 %, respectively). The prevalence of these abnormal maladies was correlated with the concentration of total and aromatic hydrocarbons in sediment samples from locations within 2 km (1.2 mi) of the survey sites, but not with sediment concentrations of aliphatic hydrocarbons, polychlorinated biphenyls, or heavy metals. The grunt had a low-frequency of a variety of abnormalities, but they appear to be associated with sediment copper levels. Most of the abnormalities occurred in two locations in the more contaminated northern part of the Bay. Skin and fin hemorrhaging and eye abnormalities were seldom observed in this study as they were in that by Skinner and Kandrashoff (1988), but this difference may result from method of collection or an increase in water quality. A similar study to that by Gassman et al. (1994) conducted from November 1989 to June 1990 by Browder et al. (1993) emphasized a depression in the dorsal profile, known as "saddleback," and accounted for 76 % of all the abnormalities that they observed. A study by Corrales et al. (2000) focused on scale disorientation in pinfish from the same area. This abnormality consisting of discrete patches of scales rotated dorsally or ventrally away from the normal scale position was also reported by Overstreet (1988). Corrales et al. (2000) also found the abnormality, affecting as much as 34 % of the body surface, most prevalent in the northern part of the Bay in the pinfish, on which they also conducted experimental studies. Acute and chronic exposure to physical traumas was insufficient to induce formation of the disorientation; however, the condition could appear spontaneously in some normal juvenile and adult specimens maintained in the laboratory for 5.5 months. Their observations suggest that development occurs rapidly and is most likely the result of a sudden change in growth characteristics of cells in the affected area.

Abnormalities in fish from Biscayne Bay, Florida, even in areas of high input of sewage and urban runoff, occurred considerably less than those in fish caused by hydrocarbon contamination in the Hudson River estuary (Smith et al. 1979) and Puget Sound, Washington (Malins et al. 1984).

Clearly, recording lesions and abnormalities in fish provides good indicators of environmental health, but relating specific abnormalities to specific contaminants is usually difficult. We (Sun et al. 2009) examined hybrid tilapia from six stations from four rivers in southern Taiwan during spring and autumn from 1994 through 1996. All stations were contaminated from different non-point sources; it is important to point out that the areas were so extensively polluted that few fish other than tilapia inhabited the rivers, and consequently those locations were quite different from those with an abundance of species occurring in the less-polluted Biscayne Bay, Florida. Nevertheless, tilapia-complex provided a useful sentinel. Therefore, examples of deformities provide this dramatic difference. Contamination was derived from agriculture, industry, and domestic wastes, and specific contaminants were recorded although specific ones could not be related to specific deformities of which we noted 20 different categories. In the Kao-Ping River, scale disorientation occurred in 18 versus 2 % of the fish in the autumn of 1994 and 1995, respectively, compared with 8 and 10 % of the fish affected with disoriented scales in the spring periods. This could be compared with 33 % of the fish in the Tongkong River in the autumn of 1995 when 23 % of the fish had a bent jaw; none of the fish at that time in the Kao-Ping River had such a deformity. The percentage of fish with an opaque cornea declined from 55 to 0 % between autumn 1995 and spring 1996, and 12 % of the fish in autumn had exophthalmia. But none was blind as were a few fish in the other rivers. Never did skeletal deformities determined with radiographs occur in more than 7 % of the samples from any river during any season. The percentage of fish with frayed fins in autumn (57 %) contrasted with 27 % of those in the spring from the Kao-Ping River compared with 37 versus 1 % in the Tongkong River. Autumn was the rainy season with increased river flow and suspended sediments.

Another example from freshwater is provided because of the large number of fish examined and the abundance of abnormalities (Slooff 1982). It treats the bream (*Abramis brama*) in the Rhine River and its branches running into the North Sea. Nearly 7,000 fish were divided into males and females, examined for skeletal anomalies, and example prevalence values for specific locations were 22.7 % with deformed fins, 3.0 % with pugheadedness, 1.2 % with lack of fins or girdle, 0.7 % with spinal curvature, 0.9 % with asymmetric cranium, 1.5 % with shortened operculae, and 5.9 % with fusion of vertebrae. The prevalence of deformed fins and pugheadedness in both males and females increased in the 12 years of fish life. As with other studies, specific abnormalities could not be attributed to specific contaminants.

Figure 14.73. Green turtle (*Chelonia mydas*) from Florida Keys exhibiting fibropapillomatosis, especially in tissue around neck, near anus, and in anterior flippers; note especially on soft tissue ventral to pelvic girdle and on flippers the abundance of attached turtle barnacle, *Chelonibia testudinaria*, which has a morphological form specific to marine turtles but has recently been shown to be molecularly the same as *Chelonibia patula* on carapace of the blue crab and many other hosts (Cheang et al. 2013).

14.5 OTHER VERTEBRATE REPRESENTATIVES

14.5.1 Sea Turtles

Sea turtles have numerous parasites and diseases. Overstreet et al. (2009) reported the trematodes from the Gulf, and Herbst (1999) provided a review of all the infectious diseases. For purposes of this chapter, we will restrict ourselves to the single example of tumorous growths.

Fibropapillomatosis (FP) (Figure 14.73), tumorous growth, of marine turtles affects primarily the green turtle (*Chelonia mydas*), but has also been reported from other turtles such as loggerhead (*Caretta caretta*), olive ridley (*Lepidochelys olivacea*), hawksbill (*Eretmochelys imbricata*), leatherback (*Dermochelys coriacea*), and flatback (*Natator depressus*) (Huerta et al. 2002). Foley et al. (2005) examined data collected by the U.S. Sea Turtle Stranding and Salvage Network based on 4,328 dead or debilitated green turtles in the eastern half of the United States from Massachusetts to Texas from 1980 to 1998 and found that 22.6 % (682/3,016) of the turtles in the southern half of Florida had tumors. During that period, the percentage of turtles in southern Florida with tumors progressively increased from about 8 to over 30 %. Most of these were in the GoM, with 39 % in inshore areas and 15 % in offshore areas. Most cases were found in coastal waters characterized by habitat degradation and pollution, a large extent of shallow water area, and low wave energy during fall and winter months, and the occurrence of tumors occurred mostly in the intermediate-sized (48–70 cm [19–27 in] curved carapace length) animals. Many were emaciated or tangled in fishing line, but they showed about an equal percentage of attack by sharks as those without tumors. Historical data reported by Smith and Coates (1938) showed that in the early 1930s, less than 2 % of green turtles captured in southern Florida exhibited tumors compared with later prevalences as high as 92 % (Herbst 1994). Fibropapillomatosis spread to elsewhere in the Gulf of Mexico, Caribbean Sea, and the western Atlantic by the mid-1980s, occurring in 10 % of the stranded green turtles in the early 1980s and increasing to 30 % in the late 1990s for those found below the 29°N latitude. The presence of FP in stranded green turtles found in Florida increased at the rate of 1.2 % per year from 1980 to 1998 (Foley et al. 2005).

The size, location, and number of tumors contribute to progressive debilitation and eventual death. Tumors, ranging from 0.1 to greater than 30 cm (12 in), are typically observed externally in the inguinal and axillary regions, at the base of the tail, around the neck, in the mouth, and on the conjunctiva of the eye (Smith and Coates 1938). Gross lesions as large as 20 cm (8 in) in diameter occur internally in the lungs, kidney, heart, gastrointestinal tract, and liver (Herbst 1994).

With an increase in the number of studies on FP, a causative agent including at least the chelonid fibropapilloma-associated herpesvirus (CFPHV) (Family *Herpesviridae*, Subfamily *Alphaherpesvirinae*, proposed genus *Chelonivirus*) was determined (Stacy et al. 2008; Davidson 2010; Bicknese et al. 2010). It has also been classified as Chelonid herpesvirus 5, also restricted to marine turtles (Lu et al. 2003) in which fibropapillomas, fibromas, lung-eye-trachea disease, grey patch disease, and loggerhead genital-respiratory and orocutaneous diseases have been reported. Development of these herpesvirus infections can be acute, latent and quiescent, or appear as a disease causing highly pathogenic and life-threatening conditions to occur (Aguirre et al. 1998, 2002; Herbst 1994; Quackenbush et al. 1998, 2001; Stacy et al. 2008; Ariel 2011; Alfaro-Núñez et al. 2014). In Florida, four distinct viral variants of CFPHV have been described (A–D) (Ene et al. 2005). In the Gulf of Mexico, *C. mydas*, *C. caretta*, and *L. kempii* all have the C variant, with the A variant, the most commonly detected one, being present in both the green and loggerhead turtle populations. The D variant was found only in the loggerhead.

Green turtle (*Chelonia mydas*) fibropapillomatosis with or without mortality has been associated with herpesvirus, retrovirus, natural tumor-promoter okadaic acid, arginine, external parasites, trematode egg interaction, and environmental factors demonstrating that these combinations of conditions can be pathogenic and life-threatening (Aguirre et al. 1998; Herbst 1994; Casey et al. 1997; Landsberg et al. 1999; Dailey and Morris 1995; Alfaro-Núñez et al. 2014; Foley et al. 2005; Ene et al. 2005; Work et al. 2004; Van Houtan et al. 2010). There does not seem to be any single factor inducing infections or causing mortality. Work et al. (2001) concluded that turtles with severe FP were immunosuppressed, but immunosuppression was not a prerequisite for development of FP and neither were trace metals nor organic contaminants (Aguirre et al. 1994). Induction of FP by herpesviruses seems to be promoted by a metabolic influx of the amino acid arginine; lysine inhibits the virus and proline aids the viral infection. Moreover, eutrophication spurs nuisance algal blooms where arginine would be elevated, consequently promoting the FP tumors in *C. mydas*. Okadaic acid also promotes tumors and is produced by toxic benthic dinoflagellates in the genus *Prorocentrum*, which are fed on heavily by *C. mydas* exhibiting FP in Hawaii (Landsberg et al. 1999).

There appears to be an interesting relationship of FP with blood flukes (in the Gulf, at least two different described species of Spirorchiidae infect *C. mydas*, and three infect *C. caretta* (e.g., Overstreet et al. 2009)). Adult flukes and their eggs were initially assumed to be related to FP in relatively large specimens of *C. mydas*, but later considered not to be the immediate cause of the disease. In the mid 1960s, R. Overstreet removed tumors from many caged 20 cm (8 in) and larger specimens of *C. mydas* maintained by R.E. Schroeder in Marathon, Florida Keys, Florida, and saw some spirorchiid eggs in larger turtles, but all the tumors contained considerable cyanobacteria, diatoms, and other algae appearing to cause some cellular response. After one or more viruses had been implicated in the etiology of FP, Aguirre et al. (1998) studied the relationship between blood fluke infections and fibropapillomatosis in *C. mydas* in Hawaii. A generalized thickening and hardening of major vessels (aortic, pulmonary, mesenteric, and hepatic) and thrombosis with complete or partial occlusion occurred in turtles containing both FP and spirorchiidiasis and were considered primary causes of mortality. Similar pathogenesis has been reported from both wild and cultured sea turtles worldwide (e.g., Aguirre et al. 1998). Chen et al. (2012) provided several references on fluke infections in *C. mydas* associated with mortality, but they considered the mortality in the stranded juvenile turtles, apparently without FP, in Taiwan probably from fishery by-catch rather than the fluke-associated pathogenic alterations. When a large number of young *C. mydas* exhibiting a robust nutritional condition became moribund or died from a single hypothermic event in Florida, Stacy et al. (2010) took advantage of them to assess the pathological responses to spirorchiid infections. They determined that the responses differed relative to trematode species, species of turtle (*C. mydas* and *C. caretta*), and size of turtle host. Even though some turtles exhibited severe pathological alterations, only one specimen of *C. mydas* died from a worm infection even though infections probably contributed to poor health in others. In a report of Australian strandings in which size of turtles was not given, blood flukes caused death in 10 of 96 and contributed to death in 29 more of the 96 (Gordon et al. 1998). In those, many were chronically ill, whereas in Florida, examined turtles died from a variety of acute insults. An apparent undescribed quite pathogenic species of *Neospirorchis* sp. was most common as adult worms and associated egg masses in large *C. mydas* and *C. caretta* and infected the leptomenenges, thymic gland, and thyroid gland rather than heart and major arteries like *Neospirorchis pricei* and the other blood flukes. In general, adults of most species produce proliferative endarteritis, with parasitic granulomas and thrombosis inhibiting blood flow; eggs typically become trapped in capillaries, including those of the highly vascularized fibropapillomas, and sometimes associated with inflammation. The important point is that spirorchiids seldom cause acute infections with death, but they are often associated with mortalities when in combination with other diseases like FP. Stacy et al. (2010) also reported *C. mydas* (7 of 15) with anemia to have a severe leech infestation including egg cases involved with FP, presumably *Ozobranchus margoi*, which is discussed by Sawyer et al. (1975) as a leech occurring externally on sea turtles and harboring up to 900 individuals with lesions associated with their attachment sites. Three of those 15 turtles had the talitroidean amphipod *Hyachelia tortugae* in the skin and fibropapillomas.

The protection and population recovery of sea turtles on a global scale has had increasing attention during the past 35 years (Raustiala 1997; Wright and Mohanty 2006; Campbell 2007; Hamann et al. 2010). One of the global research priorities for marine turtles is still the etiology and epidemiology of the pandemic FP and the management of this disease (Hamann et al. 2010). Long-term studies have already shown that the disease has peaked in some regions, remained constant in others, and increased elsewhere (Van Houtan et al. 2010).

14.5.2 Birds

Birds need to adopt a strategy optimizing the use of energy for activities like reproduction and host defense. This strategy requires a "trade off" of physiological choices for both the host and pathogenic agents to maintain genetic fitness. For example, during one season or one year, a bird or group of birds may be in poor nutritional health and therefore have to direct all resources to staying alive with little or no ability to mount a defense against parasites or to grow or to reproduce. During another time of year, this same bird may have enough resources to effectively resist parasites, grow, and reproduce (Wobeser 2008). An example of a heavy infection of helminths in a group of 45 lesser scaup ducks comprised almost one million individuals, including 52 different species (Bush and Holmes 1986). Each helminth species has a life cycle with multiple hosts, making infections problematic to fully understand without knowing the complete cycle. Additionally, chicks may become infected with numerous large adult nematodes from regurgitation by parent birds. Fagerholm et al. (1996) suggest this based on observations of patent specimens of *Contracaecum magnipapillatum* in adults, chicks, and dead chicks in the breeding habitat of the black noddy (*Anous minutus*). Interactions among two or more species might be additive, synergistic, or antagonistic, resulting in host mortalities, and little is known about these or even effects of high numbers of single species in wild birds. Even though studies have focused on a few dead birds killed by parasites (see Atkinson et al. 2008), rarely do parasitic infections result in "piles of dead birds" because highly pathogenic ones tend not to impact a host population, since rapid mortalities would limit transmission to others. Populations are more detrimentally affected by the sublethal effects of chronic infections mediating reduced fecundity (Hudson and Dobson 1997b). One should also expect true seabirds, those that derive all food from the sea, defecate in the sea, and die at sea, to suffer less than coastal birds from microbial, protozoan, and probably helminth agents (Lauckner 1985).

Table 14.4 lists avian mortalities in the Gulf from 1999 to 2010, and most cases result from bacterial, fungal, and viral infections as well as toxicants. The book edited by Thomas et al. (2007) explains the agents in detail. Tropical Storm Arlene moved through the Gulf into Breton National Wildlife Refuge in June 2005 at a time when birds were vulnerable as reflected in the table. Vargo et al. (2006) summarized beached bird surveys in Pinellas County, Florida, mostly reporting mortality resulting from brevetoxin. The harmful effects of red tides on birds were discussed earlier, including an informative report by Forrester et al. (1977).

Table 14.4. Epizootics for Birds in the Northern Gulf of Mexico from 1999 to 2010 as Reported by the National Wildlife Health Center, Quarterly Mortality Reports[a].

State	Locality	Dates	Species	Number of Mortalities (e = estimate)	Cause of Death
FL	Peace River, Charlotte Harbor	01/08/99–01/14/99	Lesser scaup	50(e)	Hepatitis
TX	Colorado, Frio, Matagorda, Waller Co.	11/26/99–02/01/00	Snow goose, white-fronted goose	3,189	Avian cholera
TX	Waller Co.	01/09/00–01/31/00	Wood duck, American coot, mottled duck, green-winged teal, gadwall	291	Avian cholera
TX	Laguna Atascosa NWR	01/10/00–02/10/00	Snow goose, green-winged teal, American avocet, sandhill crane	200(e)	Open
AL	Baldwin Co., Gulf Shores	07/01/00–12/30/00	Unidentified pelican, common loon, double-crested cormorant, unidentified gull, northern gannet	100(e)	Open
TX	San Bernard NWR	11/20/00–12/01/00	Snow goose	75(e)	Open
FL	Monroe Co., Florida Keys	12/31/00–05/17/01	Brown pelican, common loon, great blue heron	250(e)	Open
TX	Nueces Co., Gulf Beach	01/28/01–03/15/01	Double-crested cormorant	100(e)	Salmonellosis
FL	Lee Co., Gasparilla Is	10/30/01–12/30/01	American white pelican	20(e)	Toxicosis: brevetoxin
LA	Offshore Louisiana	01/05/02–01/10/02	Brown pelican	50(e)	Exposure: Hypothermia
AL	Gulf Shores State Park	02/25/02–03/10/02	Unidentified loon, brown pelican, herring gull, mallard, northern gannet	20(e)	Aspergillosis
TX	Willacy Co.	07/23/02–08/01/02	Black-bellied whistling duck, eared grebe	80(e)	Toxicosis: salt
TX	Cameron Co., Harlingen	08/19/02–08/22/02	Laughing gull	32	Salmonellosis
FL	Okaloosa Co., Destin Harbor	02/01/02–07/10/02	Brown pelican, common loon, osprey, American white pelican, wood duck	60(e)	Open

(continued)

Table 14.4. (continued)

State	Locality	Dates	Species	Number of Mortalities (e = estimate)	Cause of Death
TX	Aransas NWR	12/01/03–12/05/03	Snow goose	80(e)	Open
FL	Volusia, Orange, Brevard, Martin, Palm Beach, and Broward Counties	03/08/03–04/15/03	Northern gannet, unidentified cormorant	2,500(e)	Emaciation
LA	Jefferson Parish	01/07/04–01/15/04	Eastern brown pelican	50(e)	Open
FL	Manatee Co.	07/01/04–07/31/04	Wood stork; white ibis; great blue heron; roseate spoonbill; unidentified pelican	24(e)	Open; toxicosis suspect
FL	Pinellas Co.	05/11/04–06/12/04	Mallard; muscovy; American coot; unidentified cormorant	70(e)	Botulism suspect
LA	Breton NWR	06/12/05–06/23/05	laughing gull, brown pelican, ring-billed gull, little blue heron	7,200(e)	Trauma: storm toxicosis: Oil
FL	Panama City	09/26/05–11/15/05	American coot	30	Toxicosis: suspect
FL	North Miami Beach	09/04/06–10/01/06	Muscovy, unidentified egret, NOS heron, white ibis, tricolored heron	48	Botulism suspect
FL	Key West	02/04/07–02/15/07	Unidentified seabird, brown pelican	40(e)	Toxicosis: domoic acid (red tide) suspect
TX	Galveston County Beaches, Aransas, and Nueces Counties	06/07/07–06/30/07	Northern gannet	100(e)	Emaciation: Starvation suspect
FL	St. Marks NWR	04/04/08–04/11/08	Common loon, red-breasted merganser	100(e)	Undetermined

[a]http://www.nwhc.usgs.gov/publications/quarterly_reports/index.jsp, accessed February 10, 2015.

Figure 14.74 shows a few specimens of the chewing louse *Piagetiella peralis* in the gular pouch of the American white pelican associated with a combination of louse excrement and blood located next to petechial hemorrhaging. Heavy infestations commonly cause severe ulcerating lesions covering much of the naked body of pelicans <1 week old before the lice enter the throat of older juveniles. Because of these heavy infestations and associated

Figure 14.74. A few of the many pouch lice, *Piagetiella peralis,* infesting the gular pouch of an American white pelican.

Figure 14.75. Chick of nesting least tern, *Sternula antillarum*, died of hyperthermia (heat stroke) along beach in Gulfport, Mississippi, July 1980.

Figure 14.76. The brown pelican, *Pelecanus occidentalis*, with a malformed bill; the bird had an excessive infestation of the chewing lice *Colpocephalum unciferum and Pectinopygus tordoffi* on the feathers because of its lack of ability to successfully preen, September 1993.

secondary bacterial infections, Samuel et al. (1982) suggested that the louse can have a significant effect on juvenile pelican populations.

Based on histopathological evidence in June 1980, nearly all of several hundred nesting least terns (*Sternula antillarum*) died of hyperthermia (Figure 14.75) along a narrow beach in Gulfport, Mississippi (Overstreet and Rehak 1982). The unusual heat-stroke apparently occurred because of the temperature-humidity complex arising from delayed hatching because of 39 cm (15.4 in) of rain in mid-May and because the nearby waters had low salinity, resulting in extended time for parents having to leave their nests while foraging for the bay anchovy. This fish species comprises a major dietary item of the tern and its near-nest population was small compared with most years.

Abnormalities have been seen in birds, and the best example is that of the brown pelican that had its bill so malformed that it depended on viscera from fish tossed to it by fishermen cleaning their fish in an Ocean Springs, Mississippi, harbor (Figure 14.76). Because the bird could not preen, its feathers contained an enormous number of lice. When a seagull on the end of the GCRL pier was ill, one could slowly approach it and pick it up, only to have many lice migrate up his or her arms.

14.5.3 Marine Mammals

Every few years, mortalities of marine mammals have occurred in the Gulf of Mexico. To examine dead or dying stranding animals, one required a permit or special permission and such was seldom obtained before the animal decomposed too much to be evaluated. A few cases of single stranded animals appeared to result from an ectopic parasite. Trematodes and nematodes or their eggs have been found in the brains of these animals along the Pacific coast, and, in the Gulf during the 1980s, we observed in stranded bottlenose dolphins lung infections of metastrongyle nematodes *Halocercus lagenorhynchi* and *Skrjabinalius cryptocephalus* associated with pneumonia. Those findings compared favorably with those reported by Fauquier et al. (2009). Dailey et al. (1991) and the latter article additionally found *H. lagenorhynchi* lungworms in neonates, suggesting this nematode crosses the placenta of pregnant females. Because of the lack of pinnipeds in the Gulf, we do not see fish with larval nematodes of

Figure 14.77. Flipper of stranded common bottlenose dolphin, _Tursiops truncatus_, exhibiting at least two tooth rake marks acquired during aggressive interactions between dolphins, 1997.

Figure 14.78. Stranded oceanic spotted dolphin, _Stenella frontalis_, March 15, 1993, not commonly seen stranded on Mississippi beaches; most strandings on beaches in the northern Gulf are the bottlenose dolphin.

species that mature in pinnipeds, but we do see larvae in offshore fish that mature in dolphins and other cetaceans. Because of the ability to examine the bottlenose dolphin (_Tursiops truncatus_), we use it as an example of marine mammal in the Gulf. Most dolphin exhibit external lesions. Individual bottlenose dolphin can be identified by wounds, tooth rake marks, secondary infections, and even symbiotic barnacles as figured by Overstreet (1978) and Figure 14.77. Other species of dolphin strand along Gulf beaches, but they are rare as exemplified by the stranded oceanic spotted dolphin, *Stenella frontalis,* identified from a Mississippi beach by the number of teeth (Figure 14.78).

To assess microorganisms of the bottlenose dolphin as a component of biodiversity as well as potential dolphin health and public health risks, Buck et al. (2006) reported aerobic microorganisms associated with free-ranging bottlenose dolphins in coastal waters off Florida, Texas, and North Carolina during 1990–2002. We examined blowhole and fecal samples of some of those and other dolphins for similar purposes. We also examined microorganisms from captive dolphins in Mississippi before and after Hurricane Katrina and discovered many of the same organisms reported by Buck et al. (2006) and by Williams and Barker (2001).

Evaluations of mass mortalities of bottlenose dolphin since 1990 have tentatively established that brevetoxin and morbillivirus have been at least partially responsible (Waring et al. 2007). From January through May 1990, a total of 367 bottlenose dolphin stranded in the northern Gulf of Mexico, but the cause was not established. In March and April 1992, 111 stranded in Texas, and some of these animals tested positive for previous exposure to cetacean morbillivirus. The NOAA Fisheries Working Group on Unusual Marine Mortality Events was formalized in 1992 and has evaluated several mortality events. Morbillivirus was diagnosed on the basis of histopathologic lesions, immunohistochemical chemical demonstrations of the morbilliviral antigen, and detection of morbillivirus RNA by RT-PCR in 35 of 67 stranded dolphins that occurred in the Florida Panhandle and spread west to Alabama and Mississippi, with most of the dolphins dying in Texas between 1993 and 1994 (Lipscomb et al. 1996); 29 additional dolphins exhibited advance postmortem autolysis and diagnosis was equivocal. That was a follow-up study of one (Lipscomb et al. 1994) reporting morbillivirus from a 1987–1988 epizootic from stranded Atlantic Coast bottlenose dolphins that was the first such report outside Europe. Mortalities from other types of events included dolphins in Mississippi in 1996, 120 in the Florida Panhandle between August 1999 and February 2000, and 107 from the same location in March and April 2004; all these dolphin died concurrent with red tide blooms and red tide fish kills and were assumed to be killed by brevetoxin. The West Indian manatee (*Trichechus manatus*) also succumbs to brevetoxin as indicated earlier in the section on red tides.

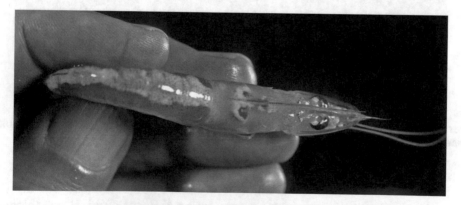

Figure 14.79. White shrimp, *Litopenaeus setiferus,* with the microsporidian *Agmasoma penaei* in the cephalothorax and along the dorsum, superficially appearing like developing gonads.

Figure 14.80. The microsporidian *Perezia nelsoni* in the skeletal musculature of the tail (abdomen) of the lower shrimp (brown shrimp, *Farfantepenaeus aztecus*) causing cotton shrimp compared with an uninfected white shrimp on top, February 1977.

Figure 14.81. A microsporidian infection (cotton shrimp) in top shrimp tail; about four species of microsporidians cause this condition in the Gulf, with *Perezia nelson* and *Tuzetia weidneri* being the most common; both tails come from the white shrimp, *Litopenaeus setiferus,* July 1977.

14.6 INVERTEBRATES

We treat invertebrates with essentially the same approach as for fishes, but we use representative-selected examples from penaeid shrimps, the blue crab, the eastern oyster, and corals. Each group is presented separately.

14.6.1 Shrimps

Like most invertebrates, shrimps, notably penaeid shrimps, succumb to a variety of diseases and other agents. Because penaeids are cultured commercially worldwide, considerable data on shrimp health have been published (e.g., Overstreet 1973, 1983c, 1987; Lightner 1996; Sindermann and Lightner 1988; Lotz and Overstreet 1990; Overstreet and Lotz 2016). For purposes of this chapter, we will discuss select ciliates, microsporidians, and viruses as examples of pathogenic organisms.

About five or six microsporidian species infect penaeids in the Gulf, and individuals with infections in the abdominal (tail) muscle are colloquially called cotton shrimp or milk shrimp (Canning et al. 2002), sometimes including *Agmasoma penaei* and sometimes not (Figures 14.79, 14.80, and 14.81). Microsporidia is a phylum of unicellular spore-forming parasites now recognized as related to fungi. *Agmasoma penaei*, primarily restricted to the white shrimp, *Litopenaeus setiferus,* does not infect striated muscles but rather the muscles lining blood vessels, foregut, hindgut, and germinal tissue of the gonads. It can occur in multiple infections in a single host (Overstreet 1973). Until recently, it has been difficult to critically separate species (Canning et al. 2002; Sokolova et al. 2015). In any event, historical data sometimes separate them and sometimes combine all species, and, in some cases, microsporidians infect a large portion of the shrimp populations. The same species occur in the South Carolina estuaries, where Miglarese and Shealy (1974) examined a total of 67,658 white shrimp on a monthly basis and found an increase in prevalence from about 15.0 % in July up to a peak of 89.5 % in November 1973. In 1919, about 90 % of the white shrimp along the Louisiana coast had their gonads destroyed by *A. penaei*; however, the largest known white shrimp crops during that general period were produced in 1920 and 1921 (Viosca 1945). Viosca (1945) stressed that evidence showed that with a prolific species like the white shrimp, the food supply and other ecological factors represented more important factors for production than the actual number of eggs laid. The most severe epizootic of this microsporidian in the Gulf of Mexico was recorded in 1929, resulting in the prevalence of 90 %, mass mortality of this shrimp, loss of 99 % of the shrimp egg production, and, in contrast with the 1919 event, an unprofitable white shrimp fishery for several following years (Gunter 1967; Muncy 1984). Lightner (1996) indicated

Figure 14.82. Penaeid shrimp showing bacterial infection in abdominal musculature.

Figure 14.83. White shrimp, *Litopenaeus setiferus,* exhibiting heavy infestation of fouling peritrich ciliate and detritus; this condition appears superficially similar to black gill disease caused by two apostomes ciliates and occasionally seen in an abundance of shrimp from Louisiana and Texas, September 1979.

the prevalence in wild populations normally did not exceed 1 %, and whereas we do not disagree, we have observed prevalence values as high as 25 % in inshore white shrimp from Mississippi and Louisiana on about three occasions during the last 45 years. Perhaps, lactic acid buildup in infected shrimp keep them from migrating offshore, similar to what was suggested for the blue crab infected with *Ameson michaelis* (see Shields and Overstreet 2007). Unlike the lack of a known lifecycle for other shrimp microsporidians from the Gulf, the cycle for *A. penaei* has been achieved experimentally. Iverson and Kelley (1976) fed pink shrimp (*Farfantepenaeus duorarum*) infected with *A. penaei* to spotted seatrout and then fed the seatrout feces to uninfected hatchery-reared postlarvae (about 1 cm [0.39 in] long) pink shrimp for 3–5 weeks. When examined grossly, no sign of infection was apparent, but histological sections demonstrated spores or other signs of infection.

Penaeid shrimps worldwide exhibit infections with more than 25 species of viruses, and most of these infections are in cultured shrimp. Lightner (2011) provided a mini-review of all the viruses in the Americas, and Lightner et al. (2012) reviewed the history of all shrimp pathogens in the Americas (Figure 14.82). The first known virus, *Baculovirus penaei*, occurs in the natural environment (Overstreet 1994; Overstreet and Lotz 2016) and provides a good contrast to the introduced viruses. Overstreet and Lotz (2016) indicated how both *B. penaei* and the viruses introduced into the Gulf can revert from being relatively harmless to the shrimp population to becoming highly pathogenic.

Black gill disease causes concern in the shrimp fishery in the Gulf and along the Atlantic coast. Actually, the disease is a syndrome because blackish or brownish gills can be caused by a variety of agents and conditions. These include several chemical irritants such as cadmium, copper, crude oil, and ammonia; microbial agents such as the virus IHHNV, bacteria *Vibrio* spp., fungi *Fusarium* sp.; ciliates such as peritrichs, apostomes, and scuticociliates; ascorbic acid deficiency; and other causes. We have seen several epidemics of apostomes, and infected shrimp with melanistic responses from Texas, Louisiana, and Mississippi were so conspicuous that the commercial product was not acceptable for the market. The conspicuous shrimp gills contained *Hyalophysa chattoni* and *Gymnodinioides inkystans*, even though other species occurred. *Hyalophysa chattoni* was the most common but least pathogenic. Moreover, it also infected grass shrimp in which no melanistic response occurred. Figure 14.83 shows a brownish version of black spot, and that shrimp was infested with attached colonial peritrichs. The effects of this disease on the population have not been established but probably depend on the stress allowing the ciliates to invade and establish on the gills. Stocks of heavily infested shrimp

Figure 14.84. "Golden shrimp" an abnormally discolored white shrimp on top with normal specimen of white shrimp underneath, Mississippi Sound, May 1985.

Figure 14.85. The palaemonid river shrimp *Macrobrachium ohione* comes down the river to encounter salt water for its larvae. Specimens can take on a variety of colors, so this "golden river shrimp" from the Pascagoula River may not be portraying a genetic phenotype as we attribute to the golden penaeid.

in aquaculture died from oxygen-depletion probably because the ciliates competed with the shrimp for oxygen (Overstreet 1973).

Other "protozoans" not normally infecting penaeids have the ability to influence the seafood market. For example, an aseptate gregarine infected *Litopenaeus vannamei* in the commercial "seed-production" facility in Texas and caused considerable economic loss until a presumed lifecycle of the coccidian could be established (Jones et al. 1994).

As with most animals, abnormalities in shrimps become apparent as more individuals are examined. Overstreet and Van Devender (1978) observed a hamartoma (non-neoplastic growth) primarily in postlarval brown and white shrimps that occurred near a harbor in Ocean Springs, Mississippi, but not elsewhere (except 2 of 33 that were near). This growth, with a 100 %

Figure 14.86. "Violet shrimp" a white shrimp on top with reference white shrimp on bottom; we saw this case once only and it involved several shrimp in the Pascagoula River, Mississippi, December 1981.

prevalence in some samples, was probably induced by a heavy metal contaminant. Also, shrimp on occasion from multiple locations have been observed with abnormal discoloration. Figure 14.84 shows a golden shrimp, perhaps a genetic anomaly. On the other hand, a palaemonid river shrimp that migrates to the estuary to spawn can acquire a variety of colors, including a golden color (Figure 14.85). A violet discoloration (Figure 14.86) more than likely is a response to a contaminant.

14.6.2 Crabs

Not all animals respond to disease agents similarly, and the susceptibility to specific agents has an influence on population structure. The blue crab provides an example model that we think responds strongly to disease as well as to predation and annual variation in salinity, temperature, and winds. The blue crab differs in its life history from many other crab species. When females mate, they seldom molt again, and they then migrate from the estuary into higher salinity Gulf water to spawn and die, while the males continue to molt, grow, and thrive in the estuary. In the warm Gulf, females can spawn multiple times, producing millions of eggs. From the disease point of view, we think that different specific disease agents can have a detrimental effect on specific Gulf crab populations, and these agents differ annually and interact strongly with predation of weakened, infected crabs. Crab stocks in the Gulf differ in recruitment from those stocks along the East U.S. coast. The megalopae settling from the plankton appeared from collections using similar sampling methods to be 10–100 times more abundant from Alabama to Texas than from Delaware to South Carolina (Heck and Coen 1995). In contrast, these authors and Perry et al. (1998) reported the abundances of juveniles as similar on both coasts and explained the difference in the extra loss of young crabs reaching carapace width of 30 mm (1.2 in) in the Gulf to be caused by predation. Such may be true, but even then, that loss could also result from disease killing a significant portion of the crabs or weakening them and allowing for additional predation.

We do not believe the disease agents necessary to control the crab population to be the same every year or years. Shields and Overstreet (2007) reviewed numerous agents that have the ability to kill or weaken the crabs. We will mention a few, and the first will exemplify one that occurs irregularly. The barnacle *Loxothylacus texanus* has been abundant for a few years during a few periods only during the last 45 years. This parasite is internal but has an externa (the female, protruding, brood chamber) located under the abdominal flap and does not look like a barnacle. Shields and Overstreet (2007) described the complicated life history, which will be briefly stated here because of its importance in understanding the prevalence of infections. There are separate male and female dispersal naupliar larvae that are attracted to light and to high salinity; they develop into relatively small female and larger male cypris larvae. The female cyprid metamorphoses into a kentrogon, which penetrates through the thin membrane between appendage joints of postmolt crabs when less than 18 mm (0.7 in) wide. A wormlike vermigon is released from the kentrogon and wraps around the crab midgut, producing the interna. The interna forms a complex of root-like branches that drain nutrition from the host, and, under appropriate environmental conditions following the host's final molt, it extrudes a virgin externa under the crab abdomen. Numerous young crabs with an equal number of males and females can be infected simultaneously under appropriate temperature and salinity conditions. Male cyprids have a weaker phototactic response, and therefore fertilize the virgin externae in crabs when in the benthos. The cyprid larvae are not viable below 12 ppt, and a mortality of 10 % still occurs at 15 ppt. Infected crabs cannot tolerate low salinity water and survive best in 25–30 ppt. Seldom does the water temperature and salinity in this estuarine environment accommodate the combination of production of barnacle larvae, fertilization, and availability of young crabs simultaneously. When these conditions are appropriate at the same time that infected crabs are present and producing barnacle naupli, over half the young crab population can become infected. Not only does the parasite kill or weaken the infected crabs when the salinity is not high enough, but infected crabs, stunted and castrated from the infection, compete with non-infected individuals for space, food, and sexual partners.

Figure 14.87. Blue Crab, *Callinectes sapidus*, covered with relatively large specimens of the Florida rocksnail, *Stramonita haemastoma floridana*; we have seen crabs commonly feeding heavily on younger specimens of the drill, October 1980.

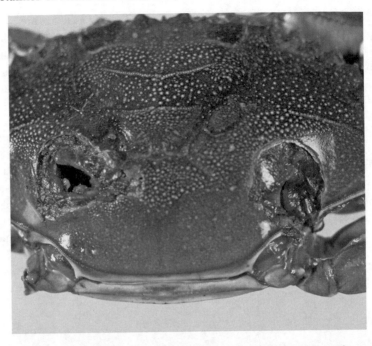

Figure 14.88. Blue Crab, *Callinectes sapidus*, showing lesions extending through the carapace; we see such cases irregularly, May 1999.

Figure 14.89. The portunid speckled swimming crab, *Arenaeus cribrarius*, exhibiting abnormal right chelipeds, July 1981; similar cases also occur with the blue crab.

Loxothylacus texanus is just one of several parasites that can control the population of juvenile blue crabs in Mississippi and elsewhere in the Gulf in addition to predation by other blue crabs and fishes. Shields and Overstreet (2007) review these in detail. Three other that can infect young crabs include the microsporidian *Ameson michaelis*, the parasitic dinoflagellate *Hematodinium perezi*, and the introduced white spot syndrome virus, or WSSV, assigned by the International Committee on Taxonomy of Viruses as the only member of the genus

Whispovirus within the family Nimaviridae. We experimentally killed crabs with the virus, but penaeid shrimps seem much more susceptible. We (Juan Carrillo, Janet Wright, and R. Overstreet) are presently investigating the effect of each of these three agents on the health and mortality of the blue crab and should be able to determine if the blue crab population is continually being controlled by at least one in a series of several disease agents.

We have seen an abundance of relatively small specimens of the Florida rocksnail, *Stramonita haemastoma floridana*, in the stomach of feeding crabs. We have seen large crabs commonly with an abundance of larger rocksnail drills on the crab carapace (Figure 14.87). This drill feeds heavily on young oysters. Blue crabs, especially those collected from contaminated locations or trapped in crab cages for several days, often exhibit shell disease. This disease results when chitonoclastic bacteria or fungi gain entrance to the shell. Such crabs appear orangish, brownish, or blackish in small to large lesions. Crabs with extensive shell disease indicate stress, keeping the individuals from molting. The lesions, caused by any of several bacteria or fungi, do not exhibit a distinct relationship with the abundance of any of the several bacteria infecting the crab hemolymph (Shields and Overstreet 2007). In some cases, lesions in the carapace will extend into the body cavity (Figure 14.88). Moreover, the blue crab and related portunid crabs occasionally exhibit abnormal chelipeds (Figure 14.89) or other structures.

14.6.3 Oyster

The Eastern oyster, *Crassostrea virginica*, exhibits high susceptibility to mortality resulting from the interaction of salinity, temperature, diseases, and predation. For purposes of this chapter, we will first provide information on *Perkinsus marinus*, previously known as *Dermocystidium marinum* or short as "dermo." This agent, initially described in 1949, probably kills more oysters in the Gulf, including Mexico and Venezuela, than any other agent; it is typically classified in the protozoan Phylum Apicomplexa, but genetic sequencing places it closer to the dinoflagellates (Ford 2011). Moreover, polymerase chain reaction (PCR) analyses can identify the species. Consequently, some infections reported from oysters in the Caribbean, Cuba, Brazil, and elsewhere using Ray's Fluid Thioglycollate Medium may involve related infectious species. Acquisition of infections usually occurs during the warm months of the year, during which the agent proliferates at water temperatures above 18 °C (64 °F) and salinities greater than 15 ppt; experimental infections have been achieved at 10 °C (50 °F) and 3 ppt and proliferation is most rapid at about 25–30 °C (77–86 °F) (Ford 2011). Mortalities between 5 and 30 % typically occur during the first year of an epizootic, reaching 60–80 % by the end of the second year, with mortalities commonly averaging 20–30 % in enzootic waters. While light infections typically influence the host little, advanced stages result in reduced feeding, growth, and reproduction, leaving oysters weak and emaciated before they die from the infection, some other infection, or predation because the infection weakened the host making it vulnerable to predators.

Wilson et al. (1990) examined oysters for prevalence and intensity of infection from 48 locations ranging from Laguna Madre in southern Texas to the Everglades in Florida as part of NOAA's Status and Trends Mussel Watch Program and found the prevalence exceeded 75 % at 25 locations. The intensity of infection did not vary with either sex or reproductive stage of the oyster; however, the distribution of infections was affected by latitude, total PAH content, and industrial and agricultural land use. PAH and pesticide concentrations were dependent on point sources, with the highest concentration values being in St. Andrews Bay, Florida; Vermillion Bay, Louisiana; and Galveston Bay, Texas. Soniat (1996) summarized data from Tabasco, Mexico, to the Everglades and found infections responded to similar factors, suggesting a combination of temperature and salinity was important but did not explain much of the variation in levels of infection. He also discussed the possibility of increased susceptibility caused by pollutants. Gold-Bouchot et al. (1995, 1997) investigated PAH fractions in oyster tissues from Tabasco and concluded the hydrocarbon concentrations were not responsible for oyster mortality, and histopathological lesions responded more to cadmium and salinity, and mortalities were confounded by the presence of *P. marinus*. A more extensive study by Noreña-Barroso et al. (1999) from Campeche, Mexico, found and was concerned by higher levels of PAHs than previously reported. MacKenzie and Wakida-Kusunoki (1997) summarized the oyster industry of eastern Mexico from Texas to Campeche.

Recent studies have shown that considerable variation exists among strains from 86 clonal cultures derived from 76 parental cultures originating from the Atlantic coast to the Gulf coast (Reece et al. 2001). They determined that 12 different composite genotypes existed, but only one was unique to Gulf Coast isolates. A single oyster can be infected with multiple strains, and virulence differs with genotypic differences. Based on earlier data from fewer isolates, Bushek and Allen (1996) found that two isolates from the mid-Atlantic region produced heavier infections in a shorter period of time than did two from the Gulf. Moreover, isolates differ in the production and activity of some extracellular proteases (La Peyre et al. 1998), and protease inhibitors differ between species of oysters (Faisal et al. 1998). These and other papers help explain why the host defense mechanisms differ in response to infections.

Two diseases in addition to *P. marinus* that cause devastating epizootics in oysters from the Northeast United States are *Haplosporidium nelsoni* (MSX) and *Haplosporidium costale* (SSO). These and other diseases and the defense mechanisms against them are detailed by Ford and Tripp (1996). Most interesting is the finding by Ulrich et al. (2007) that PCR amplification of the ribosomal rRNA detected MSX in 30 of 41 oysters sampled from Florida to the Gulf of Mexico south to Venezuela, even though an epizootic had never been reported.

A histopathological survey for infectious and noninfectious diseases in oysters as well as fishes at Pascagoula, Mississippi; Mobile, Alabama; and Pensacola, Florida; was conducted by Couch (1985); and, in most cases, the three Pascagoula locations indicated a higher prevalence of diseases. However, *P. marinus* was only apparent in 4 % from that location, even though we have seen much heavier infections in the past. He considered epithelial atrophy of the digestive gland to be the best indicator of environmental health. The normal digestive diverticula exhibited deep, thick epithelium forming triradiate and quadriradiate lumina. A total of 35.4 % of the oysters in the Pascagoula harbor demonstrated atrophy compared with 12.5 % in Mobile and 10.2 % in Pensacola. The values from the Pascagoula were heavily influenced by nearly 100 % of the oysters exhibiting failing diverticular epithelia in January and May of 1980. Twenty of the 4,496 oysters from the Pascagoula exhibited proliferative hemocyte (blood cell) disorders. These and other rare neoplastic disorders in oysters have not been evaluated using modern methods. Like Couch (1985), we have recognized thigmotrich and other ciliate protozoans as common symbionts in oysters. He mentioned they were usually nonpathogenic; however, they occasionally occluded water tubules in the gills and digestive tubules. We believe ciliates may be good indicators of water quality and environmental health. Many parasites and disease agents not involved with mass mortalities were reported by Couch (1985), Ford and Tripp (1996), and others and will not be discussed here.

Deserving of some attention are the symbionts, pests, and fouling agents that occur on or in the shell of oysters, often making them brittle and susceptible to predation. Overstreet (1978) provided illustrations and White and Wilson (1996) gathered together considerable literature on the organisms. As examples, the oyster drill (*Stramonita haemastoma*) killed more than 80 % of young oysters in 9 months. In Mobile Bay (Figure 14.87), this gastropod becomes active when the temperature reaches about 12 °C (54 °F) in the water when salinity is about 15–20 ppt (see

Garton and Stickle 1980). The blue crab feeds abundantly on young oysters, and polyclad turbellarian flatworms (*Stylochus ellipticus* and *Stylochus frontalis*) become a major threat in water with salinity above 15 ppt. Seven species of sponges are in the family Clionidae; the primary ones are *Pinone truitti*, which thrives at a salinity of 10–15 ppt and forms small holes in the shell, and *Cliona celata*, which prefers higher salinities and serves as the only species to form large coarse holes in the shell. The clam *Diplothyra curti* bores into and weakens the shell. A polychaete, *Polydora websteri*, uses a chemical agent to penetrate all layers of the shell and resides in mud-filled blisters. Another species, *Polydora cornuta*, also lives subtidally until it enters the oyster and resides in tubes consisting of mud particles held together by the oyster's mucus. In addition to weakening the shells, the symbionts produce dark lesions that can be esthetically displeasing to one eating oysters on the half shell. These and other symbionts can weaken the shell such that the black drum (*Pogonias cromis*) and cownose ray (*Rhinoptera bonasus*) can easily crush the shell of an oyster 8 cm (3.2 in) long; smaller oysters with weakened shells are consumed by a variety of other fishes.

Vibrio bacteria have commonly caused rapid epizootic mortalities of larval oysters in hatcheries, but because the density of the larvae in natural waters is much lower, mortalities are probably not common. Nevertheless, vibrio infections in adult oysters demand attention because they are known to cause human disease in those that eat raw oysters. For example, a survey of 575 laboratory-confirmed cases of vibrio gastroenteritis from Florida to Texas from 1988 to 1997 produced patients with illness that lasted a median of 7 days, produced fever in half of them, and produced bloody stools in 25 % of them. A total of 53 % of the 445 patients for whom data were available had eaten raw oysters in the week before disease-onset (Altekruse et al. 2000). A total of 31 % of the cases involved *Vibrio parahaemolyticus*, 24 % involved *Vibrio cholerae* (non-O1, non-O139), 12 % involved *Vibrio mimicus,* and the others involved six other species of Vibrio. The presence and density of *Vibrio parahaemolyticus* depends

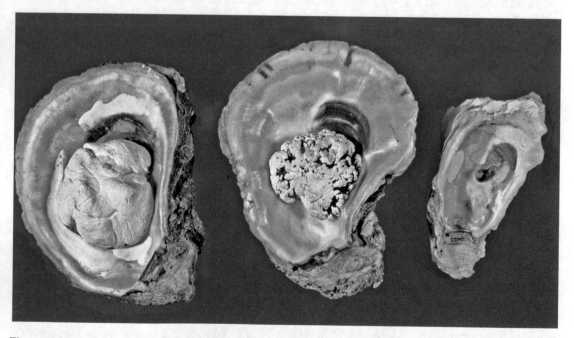

Figure 14.90. Valves of eastern oyster, *Crassostrea virginica*, with two on left exhibiting unusual abnormal nacreal bodies on internal valve surface; perhaps these resulted from repetitive covering of an abscess or foreign body. The valve on right shows myostracum similar to that historically reported in Maryland oysters as *maladie du pied* attributable to an infestation of the fungus *Ostracoblabe implexa*, October 1978.

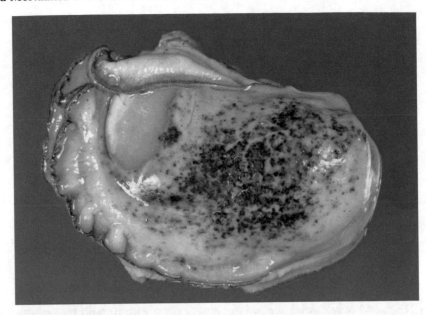

Figure 14.91. Nacreal mantle of eastern oyster, *Crassostrea virginica*, containing unusual punctate melan-like pigmentation response, January 1981.

primarily on temperature, but the bacterium requires salt to survive. It is most common during the summer, and, as an example, 44 % of the oyster samples and 30 % of the water samples in Mississippi from April to August contained the bacterium; while the total densities of the bacterium may be informative, the authors do not recommend densities as a good means to predict risk of human infection (Zimmerman et al. 2007). The presence of *Vibrio vulnificus* also requires attention because in certain high-risk individuals, those with a history of liver disease, alcoholism, and immune deficiencies, infection by means of consumption of raw shellfish or through exposed wounds may result in primary septicemia, meningitis, pneumonia, and death. The bacterium occurs regularly year-round in tropic and subtropical areas such as Charlotte Harbor, Florida, where the temperature remains moderate throughout the year, and salinity strongly controls the seasonal distribution of this bacterium between the sediment and water column. The bacterium occurred most commonly in summer months when the salinity was about 15 ppt (Lipp et al. 2001).

All aspects of the anatomy, biology, and fisheries of the eastern oyster occur in a lengthy book edited by Kennedy et al. (1996), which includes chapters on diseases (Ford and Tripp 1996) and pests (White and Wilson 1996) as indicated above. Overstreet (1978) provided a booklet including symbionts of the oyster. It figured a number of them, possibly including the valve on the right in Figure 14.90 with what appears to be a fungal infection (*Ostracoblabe implexa*) that is rare in Mississippi but more common on the Atlantic coast in cooler waters. Chris Dugan of the Maryland Department of Natural Resources (personal communication) estimated infections of perhaps 1 % in Maryland. Neither Chris or Dorothy Howard of the Oxford NOAA Laboratory (personal communication) have seen the other cases, but other melanistic conditions in the mantle of oysters along the Atlantic coast exist that have a more amorphous appearance rather than punctate (Figure 14.91). The cases figured here are rare in contrast with the more common conditions (Overstreet 1978).

Figure 14.92. Diseased maze coral, *Meandrina meandrites*. Photograph by Stephen Spotte (permission to reprint granted by S. Spotte to R.M. Overstreet).

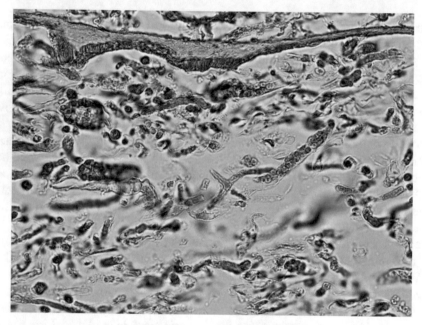

Figure 14.93. Fungal infection from diseased *Meandrina meandrites*. Photograph by Juan Carrillo (permission to reprint granted by J. Carrillo to R.M. Overstreet).

14.6.4 Corals

Corals in the tropical and subtropical areas of the Gulf of Mexico form a significant habitat for a large number of fishes. Consequently, loss of these corals by bleaching, disease, or other causes such as hurricanes can have a major impact on these fishes and other animals, and even change the habitat from a coral-dominated state to an algae-dominated state (Hughes 1994). Bleaching refers to the loss or degradation of photosynthetic symbiotic agents from the

Figure 14.94. Diseased starlet coral *Siderastrea* sp. Photograph by Stephen Spotte (permission to reprint granted by S. Spotte to R.M. Overstreet).

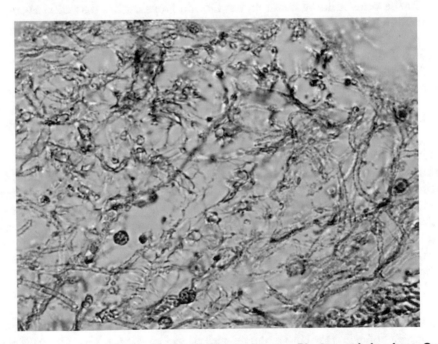

Figure 14.95. Fungal infection from diseased *Siderastrea* sp. Photograph by Juan Carrillo (permission to reprint granted by J. Carrillo to R.M. Overstreet).

endodermis of their hosts. These hosts include hard corals (stony corals), soft corals (gorgonians), and hydrozoans such as fire corals as well as other associated animals. The symbionts comprise dinoflagellates, red and green algae, and cyanobacteria or the pigments from these symbionts. Bleaching constitutes a clear sign of an unhealthy coral environment. Corals can survive being bleached for several months. In the case of the boulder star coral (*Montastraea annularis*), colonies remained bleached for 7 months after the 1987 bleaching event; they could recover and reestablish their symbionts, but those without their symbionts survived but failed to undergo gametogenesis (Szmant and Gassman 1990).

A variety of specific diseases infects corals and cause mortality in the Gulf and associated waters. The cyanobacteria *Phormidium corallyticum* causes black-band disease (BBD), expressed by a narrow dark band of its filaments encircling the coral and capable of destroying an entire colony. White band disease (WBD) has an unconfirmed etiology, but is possibly caused by adverse environmental conditions associated with a primary bacterial agent, either with or without secondary bacterial infections. Other diseases exist that can be recognized by their gross appearance. Corals are known to be infected by a variety of other internal microscopic symbionts, and we have observed a coccidium, various fungi (Figures 14.92, 14.93, 14.94, and 14.95), and bacteria in sections of both bleached and unbleached tissues. Ultrastructural investigation by Renegar et al. (2008) provided more detail of fungal infections in *Siderastrea siderea*, and it showed the affected tissue had less integrity with more degranulation and vacuolization than could be determined with regular bright-field histopathology. They also determined that identification of the fungal species was difficult. Williams and Bunkley-Williams (1990) provided a review of worldwide coral mortalities occurring from 1969 through the 1987–1988 event, including the periodic 1979–1980, 1981–1983, and 1986–1988 events. Coral disease in the Florida Keys spread rapidly from 1996 to 1998, and the different diseases exhibited different patterns of spread (Porter et al. 2001).

The event in 2005 was especially catastrophic, so we will emphasize it. Brandt and McManus (2009) provided important information on the relationship between coral bleaching and disease in the reef-building corals in the Florida Keys during that 2005 event, which also included infections in the Caribbean Sea. Both features had a positive correlation with high temperatures and with each other, but specific interactions between the two differed. White plague infections developed in the mountainous star coral (*Montastraea faveolata*), following heavy bleaching on those colonies. On the other hand, colonies of the massive starlet coral (*Siderastrea siderea*) with dark spots disease (DSD) bleached more extensively than the assumed healthy colonies. Co-occurrence of bleaching and BBD on the boulder brain coral (*Colpophyllia natans*) was apparent throughout the entire bleaching event. Bleaching, white plague (WP), and BBD each can alter structure of the coral populations by means of death of the living tissue, and DSD seems to be the most important indicator of overall reef health. Yellow band syndrome (YBS) as well as WP and BBD do not always cause mass mortalities, and environmental deterioration is considered to be responsible for the morbidity and accompanying spread of diseases (e.g., Porter et al. 2001). White pox causes great epidemic losses of the elkhorn and staghorn corals (*Acropora palmata* and *Acropora cervicornis*) in Florida and the Caribbean Sea. Lesions occurring in 1998 and spreading an average of 2.5 cm²/day (0.4 in²/day) resulted from the common human fecal enterobacterium *Serratia marcescens* (see Patterson et al. 2002).

The lack of a clear understanding of what harmful conditions actually occur on the reef seems to be partly associated with the poor characterization of diseases, syndromes, and the different stages of each as well as the fact that some reports are restricted to a single reef site and single coral species (Jordán-Dahlgren and Rodríguez-Martínez 2004). These authors studied two reefs on the western edge of the Campeche Bank in the southeastern Gulf of Mexico off the Yucatán Peninsula and not part of the primary surface circulation patterns of the Gulf. They found that of 24 coral species, only 10 included some affected colonies. Over 97 % of those affected in both reefs belong to only six coral species, *Montastraea annularis*-complex, great star coral (*Montastraea cavernosa*), knobby brain coral (*Pseudodiploria clivosa*), symmetrical brain coral (*Pseudodiploria strigosa*), mustard hill coral (*Porites*

Figure 14.96. Specimens of the leech *Calliobdella vivida,* a micropredator that feeds on blood from several fish hosts and transmits specific blood parasites (trypanosomes and hemogregarines) to the striped mullet in Mississippi.

astreoides), and *Siderastrea siderea*. Therefore, only 5.6 % of the examined colonies exhibited disease conditions, and only 3 % of those corresponded to characterized diseases and syndromes. They found only the diseases WP, WBD, and BBD and the syndromes YBS and a new one termed "thin dark-line syndrome" (TDLS). They concluded that the sensitivity of specific coral species was not the most important factor influencing disease in the shallow reef habitat nor were the type or prevalence of the conditions nor the site or density of the colonies. When these authors (Jordán-Dahlgren et al. 2005) tested the relationship of local industrial pollutants and local urban pollution with the same disease conditions in colonies of *Montastraea annularis*-complex on from one to three reefs in 1996, 1998, and 2001, they found no direct relationship. Rather than finding that the presence of disease related to environmental quality, they suggested that the reasons for disease, predominated by TDLS and YBS, resulted from their relationship with the Caribbean Sea and the warming surface water. When fragmentation of corals occurred during passage of a hurricane, WP occurred most commonly on the unattached colony fragments, especially those in contact with the sediment (Brandt et al. 2013). Coral mortalities caused by infectious disease and temperature stress both respond to cellular responses, emphasizing granular acidophilic amebocytes. Mydlarz et al. (2008) studied this cellular response in the mesoglea (connective tissue) of the common sea fan (*Gorgonia ventalina*) in the Florida Keys to the fungus *Aspergillus sydowii* and to temperature stress and concluded that this inflammatory response may allow survival of the sea fan and other corals during stressful climatic events.

Disease and corals from the East and West Flower Garden Banks in the northwestern Gulf of Mexico, an area created by uplift of underlying salt domes of Jurassic origin that rose from 100 m (330 ft) to within 17 m (55 ft) of the water's surface, were studied by Hickerson et al. (2008). Historically, the prevalence of disease in those banks was low until February 2005, when the banks experienced widespread coral disease. The plague-like disease (WP) continued to be surveyed after 2005 and was found to be most prominent during the winter months rather than in the warmer months as it occurs in the Caribbean Sea. These authors noted no WBD, which was common elsewhere in the tropical Western Atlantic.

14.6.5 Micropredators

There is only a semantic difference between a micropredator and a parasite in some cases. Some adult trematodes in the intestine of a fish may engulf host tissues without causing disease but are always considered a parasite. Leeches, isopods, argulids, and other animals obtain a blood meal from fishes and often are not considered a symbiont. Not only do they depend on the host, but they often transmit one or more blood parasites to the host. For the leech *Calliobdella vivida* on flatfishes, it transmits *Typanoplasma bullocki* to the summer flounder in the Chesapeake Bay and the hogchoker in Mississippi. When the proper alignment of low temperature, fish with a corresponding reduced immune response, and optimal salinity of 15–22 ppt for an abundance of the leech occur, the flatfishes get infected, develop splenomeg-aly, and often die (Overstreet 1982; Burreson and Zwerner 1982, 1984; Burreson and Frizzell 1986). The leech also infests the striped mullet (Figure 14.96), which also has blood parasites, but the effect on the host population has not been determined.

14.7 WHAT IS NECESSARY FOR A GOOD BASELINE FOR THE FUTURE?

Clearly, the pathobiological effects of oil spills in the Gulf of Mexico could be evaluated better if current baseline data on parasites and diseases of Gulf organisms were available. First, we lack a good and current baseline for data on parasites and diseases of Gulf organisms. This is underscored by the obvious research gaps on species and organisms as well as the large number of references from decades old studies that form the basis for much of this contribu-tion. Second, there needs to be increased acuity in recognizing and diagnosing biologically relevant pathological lesions and distinguishing them from the range of normal changes in tissue architecture. This acuity is necessary before accurately ascribing biologic effects to natural or manmade causes. Third, we need prospective knowledge of potential manmade and natural impacts to organisms in the Gulf ecosystem. Below we expand on these three points and offer some possible solutions.

In a non-intuitive way, parasites in marine organisms can reflect the health or completeness of the ecosystem in which they are found. On the surface, it might appear that parasites infect host organisms that are weakened or injured, but because many parasites depend on multiple hosts, the absence of one of those hosts can indicate a level of ecological damage. Long-term studies focused on identifying and quantifying parasite burden as well as range-extensions and identification of new parasites or new hosts. Understanding species in these host-parasite relationships at different trophic levels would be invaluable in assessing large scale impacts on the Gulf. However, two elements to achieve this are missing. First is the commitment of marine management and regulatory agencies to fund long-term, broadly-based studies to establish a robust baseline data set. Second, the scientific workforce needed to accomplish those studies is dwindling. Few traditional marine parasitology programs or programs that more broadly deal with marine pathobiology remain. The Gulf oil spill brings to the forefront the need for scientists who are competent in general marine biology and ecology, parasitology, pathology, and bacteriology as well as in the associated molecular tools that accompany those disciplines.

The public concern around the reported occurrence of pathological lesions and malforma-tions in marine organisms often outweighs the real biological significance of those findings. As with parasites, there is a thin database for lesions and malformations and a general lack of trained scientists to interpret those changes. Accurate diagnosis of lesions and malformations is key to determining their etiology. At the histologic level, a lesion represents a point in time of a dynamic process and diagnosis is often subjective. To achieve the *best* diagnoses from histological samples, the National Toxicology Program of the Department of Health and Human Services instituted the "Pathology Working Group" process wherein a panel of trained and knowledgeable pathologists evaluates contributed histopathological cases from environmental or laboratory studies and develops consensus diagnoses for those cases. A similar process needs to be applied to marine samples and conducted under the auspices of a relevant federal management or regulatory agency. This approach, along with discouraging scientists from releasing findings before they are vetted by peer review, will help maintain the integrity of the science and the confidence the public has in the scientific process.

Finally, we could better evaluate the toxicological impacts of oil spills or similar events in the Gulf if high quality baseline data were available to make before and after comparisons. This, again, is partly due to a poor baseline of data from which to make comparisons, but more broadly, it is due to relevant agencies focusing on long-term environmental events from a point-source perspective rather than concentrating on ecosystem-level effects. Yet every toxicological event begins at the lowest biochemical or molecular level of organization before it proceeds to higher level effects. A case in point is the information needed to evaluate how the oil dispersant Corexit would behave when applied in large quantities over a long period of time. All we basically knew was the acute toxicity of the compound to a small number of species. Long-term laboratory toxicological studies conducted at near "real world" toxicant concentrations are difficult and expensive to carry out but could lead to a valuable understanding of the fate and effects of potentially harmful agents in the marine environment.

14.8 CONCLUSIONS

Our knowledge of the state of health of the Gulf of Mexico fauna prior to 2010 remains based on diverse sources of academic and gray literature as well as our own unpublished investigations in which anecdotal or single-case incidents play a large role. A few long-term datasets exist, but much of those data is spotty and uneven. Long-term studies into the future will really be necessary to interpret the frequency, periodicity, intensity, and causes of disease and mortality events. We think the loss of many coral habitats from uncertain causes and the loss of estuaries because of increased populations near coastlines have a detrimental effect on the Gulf. On the other hand, an increased interest in the environment can have a positive effect on animal health.

Before 2010, episodes of fish kills, infections, and abnormalities had been documented. Acute, mass mortalities attracted attention, but when such an event occurred, attempts were made to ascribe single causes for them. Elevated mortalities are usually due to a convergence of factors, with interacting hosts, agents, and environmental conditions producing a "perfect storm." Such a balance is constantly present to some degree. At least some microbial agents, parasite infections, and environmental conditions occur in large cycles of several decades; whether this results from some underlying periodicity or from random co-occurrence of contributing factors is not certain.

Physical and chemical factors most frequently trigger large-scale mortalities. Eutrophication occurs throughout the Gulf where high nutrient input occurs. Low oxygen levels from eutrophication produce a major stress leading to fish mortality and also lead to disease and parasite-caused mortality. Red tides have a major influence on the health of fishes and other animals from the West Coast of Florida and occasionally elsewhere in the Gulf. Cold kills, which occur primarily inshore where it is hard for some animals to escape, are more disastrous in South Texas and South Florida because species there are not as well acclimated to tolerate rapid temperature changes as they are in higher latitudes of the Gulf. Heat kills, hypersalinity, sulfate reduction, sediments, and drilling fluids all have been implicated in mortality events, but they produce more localized effects. Hurricanes can occur anywhere in the Gulf, but resulting fish kills depend on the geography of the areas that the hurricanes pass through and on damage to the environment. As with most catastrophic events, the presence and absence of specific parasites can provide a good indication of environmental health and its restoration.

Few diseases cause mass mortality. When investigated, the cause usually involves one or more stresses, with an interaction between host, disease agent, and the environment. Most diseases involving infectious agents are usually shown to be highly restricted to certain geographic areas or to certain species. The most obvious infectious disease-caused mass mortality came from a catfish die-off occurring in 1996 and more cases later from Florida to Texas caused directly or indirectly by the virus. We do not know if that virus becomes intermittently introduced or if it always occurs in the habitat in low numbers until some threshold is surpassed, triggering a pandemic. Some event such as reproductive activity of the catfish may have served as the stressor, but no catastrophic event coincided with the mortality. What seems to be the same agent infects fishes in the southern Gulf of Mexico, South America, Africa, and India.

Parasites often cause disease conditions and mortalities in hosts, usually intermediate hosts, as a part of the parasitic strategy to complete the parasite life history. However, these effects tend to be ongoing at a low level without harm to the ecosystem. In cases where mass mortality occurs, changes in anthropogenic or natural environmental conditions are involved. Major stress can affect resistance of hosts to disease organisms, especially bacterial or protozoal agents. Diseases caused by a few species seem to serve as a means of host population control. Parasites, even when not harming their hosts, can be extremely useful as bioindicators in providing information about stock assessment, biological activities of hosts such as migration, feeding, and restoration of habitats as well as habitat and ecosystem health.

Neoplasms, some virally induced, have seldom been observed or reported in Gulf of Mexico fishes, although their occurrence has likely been underestimated; elsewhere, neoplasms have served as good indicators of various contaminants, particularly sediment-bound polynuclear aromatic hydrocarbons. Consequently, more attention to documenting them is warranted. Developmental abnormalities and histopathological alterations, which have been seen in many Gulf species, can indicate levels of stress from a variety of environmental factors. More quantitative data would allow researchers to tease out what factors may be involved.

Data on disease conditions in non-fish vertebrates are uneven. The best known condition in sea turtles is fibropapillomatosis, and it appears to have multiple causes. Bird mortality events are sometimes ascribed to bacterial, fungal, and viral infections, but the effects of these agents can be exacerbated by environmental conditions that sap energy and deplete needed resources. Brevetoxins and morbillivirus have been implicated in periodic marine mammal mortalities, but the cause of others is unclear, and most data are based on skewed samples from strandings. Diseases of penaeid shrimps and the blue crab have been well documented, but the effect of these diseases on host populations in the Gulf remains unclear. In the eastern oyster, the protozoan disease known as "dermo" has received a great deal of research attention. We know that its impact on oyster populations varies widely according to salinity, temperature, genotype of the infectious agent, and perhaps interaction with specific contaminants, but its variation in severity from location to location in the Gulf has not been adequately explained. Other agents and fouling agents affect oysters, but their impacts and interactions are less well studied. Loss of corals by bleaching and disease has had a major influence on tropical and subtropical Gulf communities because with their loss has come the loss of the associated fishes and invertebrates. Definitions of symbionts, parasites, and micropredators differ according to different fields and different schools of thought. Nevertheless, described associations have allowed the accumulated information to be helpful in understanding disease in the Gulf. For example, some leeches are not considered symbionts because they obtain blood meals from their hosts. However, when obtaining these meals, some species transmit protozoan parasites, which cause debilitating disease and mortality. Moreover, optimal environmental conditions promote heavy infestations of the leeches, having a significant influence on the host population.

To better understand diseases and mortalities in the Gulf, there is a need for monitoring both diseases and mortalities; for conducting more long-term, broad-scaled field work; for acquiring more expertise; and for developing more critical tools for evaluating health of the animals and health of the ecosystem.

ACKNOWLEDGMENTS

We are extremely grateful for the assistance of Jean Jovonovich, Janet Wright, Andrew Claxton, Juan Carrillo, and Kim Overstreet. Funding was acquired from numerous sources for collecting data, including U.S. Department of Agriculture; U.S. Department of the Interior; U.S. Fish and Wildlife Service and Resources and Coastal Impact Assistance Program; U.S. EPA; U.S. Food and Drug Administration; The U.S.-Israel Binational Science Foundation; Department of Health, Education, and Welfare, National Cancer Institute, and National Institutes of Health; NOAA, Oceans and Human Health Initiative and National Marine Fisheries Service; National Science Foundation; National Sea Grant Program; Mississippi-Alabama Sea Grant Program; U.S. Air Force Clinical Investigation; U.S. Army Medical Research and Development Command; The American Petroleum Institute; American Cyanamid Company; International Paper Company; State of Mississippi, Department of Marine Resources; and Mississippi Museum of Natural History. Writing the chapter was funded by BP Exploration & Production, Inc.

All images are original to the Gulf Coast Research Laboratory, The University of Southern Mississippi, unless otherwise indicated.

REFERENCES

Adlard RD, Lester RJG (1994) Dynamics of the interaction between the parasitic isopod, *Anilocra pomacentri*, and the coral reef fish, *Chromis nitida*. Parasitology 109:311–324

Adlard RD, Lester RJG (1995) The life-cycle and biology of *Anilocra pomacentri* (Isopoda, Cymothoidae), an ectoparasitic isopod of the coral reef fish, *Chromis nitida* (Perciformes, Pomacentridae). Aust J Zool 43:271–281

Adorno R, Pautz PC (eds) (2003) The Narrative of Cabeza de Vaca: Álvar Núñez-Cabeza de Vaca. University of Nebraska Press, Lincoln, NE, USA. 204 p

Afonso JM, Montero D, Robaina L, Astorga N, Izquierdo MS, Ginés R (2000) Association of a lordosis-scoliosis-kyphosis deformity in gilthead seabream (*Sparus aurata*) with family structure. Fish Physiol Biochem 22:159–163

Aguirre AA, Balazs GH, Zimmerman B, Galey FD (1994) Organic contaminants and trace metals in the tissues of green turtles (*Chelonia mydas*) afflicted with fibropapillomas in the Hawaiian Islands. Mar Pollut Bull 28:109–114

Aguirre AA, Spraker TR, Balazs GH, Zimmerman B (1998) Spirorchidiasis and fibropapillomatosis in green turtles from the Hawaiian Islands. J Wildl Dis 34:91–98

Aguirre AA, Balazs GH, Spraker TR, Murakawa SKK, Zimmerman B (2002) Pathology of oropharyngeal fibropapillomatosis in green turtles *Chelonia mydas*. J Aquat Anim Health 14:298–304

Aharon P, Fu B (2000) Microbial sulfate reduction rates and sulfur and oxygen isotope fractionations at oil and gas seeps in deepwater Gulf of Mexico. Geochim Cosmochim Acta 64:233–246

Alfaro-Núñez A, Bertelsen MF, Bojesen AM, Rasmussen I, Zepeda-Mendoza L, Olsen MT, Gilbert MTP (2014) Global distribution of chelonid fibropapilloma-associated herpesvirus among clinically healthy sea turtles. BMC Evol Biol 14:206

Alford JB, O'Keefe DM, Jackson DC (2008) Developing a framework for fisheries restoration in the Pascagoula River Basin following Hurricane Katrina. In: American Fisheries Society Symposium 64:201–218

Alford JB, O'Keefe DM, Jackson DC (2009) Effects of stocking adult largemouth bass to enhance fisheries recovery in Pascagoula River floodplain lakes impacted by Hurricane Katrina. In: Proc Annu Conf Southeast Assoc Fish and Wildl Agencies 63:104–110

Altekruse SF, Bishop RD, Baldy LM, Thompson SG, Wilson SA, Ray BJ, Griffin PM (2000) *Vibrio* gastroenteritis in the U.S. Gulf of Mexico region: The role of raw oysters. Epidemiol Infect 124:489–495

Anders K (1989) Lymphocystis disease of fishes. In: Ahne W, Kurstak E (eds) Viruses of lower vertebrates. Springer, Berlin Heidelberg, Germany, pp 141–160

Anonymous (2003) Cancer and the environment: What you need to know. National Cancer Institute, Rockville, MD, USA. 42 p

Ariel E (2011) Viruses in reptiles. Vet Res 42:100

Associated Press (1971) Hot water blamed for fish kill at Turkey Point. Daytona Beach Morning J, June 24, pp 42

Atkinson CT, Thomas NJ, Hunter DB (eds) (2008) Parasitic diseases of wild birds. Wiley-Blackwell, New York, NY, USA. 279 p

Austin B (1999) The effects of pollution on fish health. J Appl Microbiol Symp Suppl 85:234S–242S

Baden D, Mende TJ (1982) Toxicity of two toxins from the Florida red tide marine dinoflagellate, *Ptychodiscus brevis*. Toxicon 20:457–461

Baden DG, Bourdelais AJ, Jacocks H, Michelliza S, Naar J (2005) Natural and derivative brevetoxins: Historical background, multiplicity, and effects. Environ Health Perspect 113:621–625

Bailey G, Selivonchick D, Hendricks J (1987) Initiation, promotion, and inhibition of carcinogenesis in rainbow trout. Environ Health Perspect 71:147–153

Bakenhaster MD, McBride RS, Price WW (2006) Life history of *Glossobius hemiramphi* (Isopoda: Cymothoidae): Development, reproduction, and symbiosis with its host *Hemiramphus brasiliensis* (Pisces: Hemiramphidae). J Crustacean Biol 26:283–294

Bao VWW, Koutsaftis A, Leung KMY (2008) Temperature-dependent toxicities of chlorothalonil and copper pyrithione to the marine copepod *Tigriopus japonicus* and dinoflagellate *Pyrocystis lunula*. Australas J Ecotoxicol 14:45–54

Barrett AJ, Rawlings ND, Woessner JF (2012) Handbook of proteolytic enzymes, 3rd edn. Elsevier, Academic Press. 4094 p

Baumann PC, Harshbarger JC, Hartman KJ (1990) Relationship between liver tumors and age in brown bullhead populations from two Lake Erie tributaries. Sci Total Environ 94:71–87

Bergeron K (2015) Coast's 'coldest-ever' day is one for history books. Biloxi Sun Herald, January 18, pp 3F

Bicknese EJ, Childress AL, Wellehan JFX (2010) A novel herpesvirus of the proposed genus *Chelonivirus* from an asymptomatic bowsprit tortoise (*Chersina angulata*). J Zoo Wildl Med 41:353–358

Blaylock RB, Overstreet RM (2002) Diseases and parasites of the spotted seatrout. In: Bartone SA (ed) Biology of the spotted seatrout, Marine biology series. CRC Press, Boca Raton, FL, USA, pp 197–225

Blaylock RB, Overstreet RM, Klich MA (2001) Mycoses in red snapper (*Lutjanus campechanus*) caused by two deuteromycete fungi (*Penicillium corylophilum* and *Cladosporium sphaerospermum*). Hydrobiologia 460:221–228

Blaylock RB, Bullard SA, Whipps CM (2004) *Kudoa hypoepicardials* n. sp. (Mhxozoa: Kudoidae) and associated lesions from the heart of seven perciform fishes in the Northern Gulf of Mexico. J Parasitol 90:584–593

Blazer VS, Vogelbein WK, Densmore CL, May EB, Lilley JH, Zwerner DE (1999) *Aphanomyces* as a cause of ulcerative skin lesions of menhaden from Chesapeake Bay tributaries. J Aquat Anim Health 11:340–349

Bossart GD, Baden DG, Ewing RY, Roberts B, Wright SD (1998) Brevetoxicosis in manatees (*Trichechus manatus latirostris*) from the 1996 epizootic: Gross, histologic, and immunohistochemical features. Toxicol Pathol 26:276–282

Brandt ME, McManus JW (2009) Disease incidence is related to bleaching extent in reef-building corals. Ecology 90:2859–2867

Brandt ME, Smith TB, Correa AMS, Vega-Thurber R (2013) Disturbance driven colony fragmentation as a driver of a coral disease outbreak. PLoS One 8(2):e57164

Bricker SB, Longstaff B, Dennison W, Jones A, Boicourt K, Wicks C, Woerner J (2008) Effects of nutrient enrichment in the nation's estuaries: A decade of change. Harmful Algae 8:21–32

Broeg K, Zander S, Diamant A, Körting W, Krüner G, Paperna I, Westernhagen HV (1999) The use of fish metabolic, pathological and parasitological indices in pollution monitoring. Helgol Mar Res 53:171–194

Browder JA, McClellan DB, Harper DE, Kandrashoff MG, Kandrashoff W (1993) A major developmental defect observed in several Biscayne Bay, Florida, fish species. Environ Biol Fishes 37:181–188

Bruning K, Lingeman R, Ringelberg J (1992) Estimating the impact of fungal parasites on phytoplankton populations. Limnol Oceanogr 37:252–260

Buck JD (1990) Potentially pathogenic marine *Vibrio* species in seawater and marine animals in the Sarasota, Florida, area. J Coast Res 6:943–948

Buck JD, Wells RS, Rhinehart HL, Hansen LJ (2006) Aerobic microorganisms associated with free-ranging bottlenose dolphins in coastal Gulf of Mexico and Atlantic Ocean waters. J Wildl Dis 42:536–544

Bullard SA (2010) *Littorellicola billhawkinsi* n. gen., n. sp. (Digenea: Aporocotylidae) from the myocardial lacunae of Florida pompano, *Trachinotus carolinus* (Carangidae) in the Gulf of Mexico; with a comment on the interrelationships and functional morphology of inter-trabecular aporocotylids. Parasitol Int 59:587–598

Bullard SA, Overstreet RM (2002) Potential pathological effects of blood flukes (Digenea: Sanguinicolidae) on pen-reared marine fishes. Proc Gulf Caribb Fish Inst 53:10–25

Bullard SA, Overstreet RM (2006) *Psettarium anthicum* sp. n. (Digenea: Sanguinicolidae) from the heart of cobia *Rachycentron canadum* (Rachycentridae) in the northern Gulf of Mexico. Folia Parasitol 53:117–124

Bullard SA, Overstreet RM (2008) Digeneans as enemies of fishes (Chapter 14). In: Eiras JC, Segner H, Wahli T, Kapoor BG (eds) Fish diseases, Vol 2. Science Publishers, Enfield, NH, USA, pp 817–976

Bullard SA, Frasca S Jr, Benz GW (2000a) Skin lesions caused by *Dermophthirius penneri* (Monogenea: Microbothriidae) on wild-caught blacktip sharks (*Carcharhinus limbatus*). J Parasitol 86:618–622

Bullard SA, Benz GW, Overstreet RM, Williams EH Jr, Hemdal J (2000b) Six new host records and an updated list of wild hosts for *Neobenedenia melleni* (MacCallum) (Monogenea: Capsalidae). Comp Parasitol 67:190–196

Bullard SA, Frasca S Jr, Benz GW (2001) Gill lesions associated with *Erpocotyle tiburonis* (Monogenea: Hexabothriidae) on wild and aquarium-held bonnethead sharks (*Sphyrna tiburo*). J Parasitol 87:972–977

Bullard SA, Goldstein RJ, Goodwin RH III, Overstreet RM (2004) *Cardicola forsteri* (Digenea: Sanguinicolidae) from the heart of a northern bluefin tuna, *Thunnus thynnus* (Scombridae), in the northwest Atlantic Ocean. Comp Parasitol 71:245–246

Burreson EM, Frizzell LJ (1986) The seasonal antibody response in juvenile summer flounder (*Paralichthys dentatus*) to the hemoflagellate *Trypanoplasma bullocki*. Vet Immunol Immunopathol 12:395–402

Burreson EM, Zwerner DE (1982) The role of host biology, vector biology, and temperature in the distribution of *Trypanoplasma bullocki* infections in the lower Chesapeake Bay. J Parasitol 68:306–313

Burreson EM, Zwerner DE (1984) Juvenile summer flounder, *Paralichthys dentatus*, mortalities in the western Atlantic Ocean caused by the hemoflagellate *Trypanoplasma bullocki*: Evidence from field and experimental studies. Helgol Meeresunters 37:343–352

Bush AO, Holmes JC (1986) Intestinal helminths of lesser scaup ducks: An interactive community. Can J Zool 64:142–152

Bushek D, Allen SK Jr (1996) Host-parasite interactions among broadly distributed populations of the eastern oyster *Crassostrea virginica* and the protozoan *Perkinsus marinus*. Mar Ecol Prog Ser 139:127–141

Caira JN, Healy CJ (2004) Elasmobranchs as hosts of metazoan parasites. In: Carrier JC, Musick JA, Heithaus MR (eds) Biology of sharks and their relatives. CRC Press, Boca Raton, FL, USA, pp 523–551

Caira JN, Jensen K, Healy CJ (2001) Interrelationships among tetraphyllidean and lecanicephalidean cestodes. In: Littlewood DTJ, Bray RA (eds) Interrelationships of the Platyhelminthes. Taylor & Francis, London, UK, pp 135–158

Cairns J, Heath AG, Parker BC (1975) The effects of temperature upon the toxicity of chemicals to aquatic organisms. Hydrobiologia 47:135–171

Camp JW, Aho JM, Esch GW (1982) A long-term study on various aspects of the population biology of *Ornithodiplostomum ptychocheilus* in a South Carolina cooling reservoir. J Parasitol 68:709–718

Campbell LM (2007) Local conservation practice and global discourse: A political ecology of sea turtle conservation. Ann Assoc Am Geogr 97:313–334

Canning EU, Curry A, Overstreet RM (2002) Ultrastructure of *Tuzetia weidneri* sp. n. (Microsporidia: Tuzetiidae) in skeletal muscle of *Litopenaeus setiferus* and *Farfantepenaeus aztecus* (Crustacea, Decapoda) and new data on *Perezia nelsoni* (Microsporidia, Pereziidae) in *L. setiferus*. Acta Protozool 41:63–77

Casey RN, Quackenbush SL, Work TM, Balazs GH, Bowser PR, Casey JW (1997) Evidence for retrovirus infections in green turtles *Chelonia mydas* from the Hawaiian Islands. Dis Aquat Organ 31:1–7

Cheang CC, Tsang LM, Chu KH, Cheng I-J, Chan BKK (2013) Host-specific phenotypic plasticity of the turtle barnacle *Chelonibia testudinaria*: A widespread generalist rather than a specialist. PLoS One 8(3):e57592

Chen H, Kuo RJ, Chang TC, Hus CK, Bray RA, Chen IJ (2012) Fluke (Spirorchiidae) infections in sea turtles stranded on Taiwan: Prevalence and pathology. J Parasitol 98:437–439

Christmas JY, Howse HD (1970) The occurrence of lymphocystis in *Micropogon undulatus* and *Cynoscion arenarius* from Mississippi estuaries. Gulf Res Rep 3:131–154

Cipriano RC, Bullock GL, Pyle SW (1984) *Aeromonas hydrophilia* and motile aeromonad septicemias of fish. U.S. Fish & Wildlife Publications, Paper 137. http://digitalcommons.unl.edu/usfwpubs/134.

Coleman FC (1993) Morphological and physiological consequences of parasites encysted in the bulbus arteriosus of an estuarine fish, the sheepshead minnow, *Cyprinodon variegatus*. J Parasitol 79:247–254

Coleman FC, Travis J (1998) Phenology of recruitment and infection patterns of *Ascocotyle pachycystis*, a digenean parasite in the sheepshead minnow, *Cyprinodon variegatus*. Environ Biol Fish 51:87–96

Colorni A, Diamant A (1995) Splenic and cardiac lymphocystis in the red drum, *Sciaenops ocellatus* (L.). J Fish Dis 18:467–471

Colwell RR, Grimes DJ (1984) *Vibrio* diseases of marine fish populations. Helgol Meeresunters 37:265–287

Cone DK, Overstreet R (1997) *Myxobolus mississippiensis* n. sp. (Myxosporea) from gills of *Lepomis macrochirus* in Mississippi. J Parasitol 83:122–124

Cone DK, Overstreet RM (1998) Species of *Myxobolus* (Myxozoa) from the bulbus arteriosus of centrarchid fishes in North America, with a description of two new species. J Parasitol 84:371–374

Cook DW (1973) Experimental infection studies with lymphocystis virus from Atlantic croaker. In: Avault JW (ed) Proceedings of the 3rd Annual Workshop of the World Mariculture Society. Louisiana State University, Baton Rouge, LA, USA, pp 329–335

Cook DW, Lofton SR (1973) Chitinoclastic bacteria associated with shell disease in *Penaeus* shrimp and the blue crab (*Callinectes sapidus*). J Wildl Dis 9:154–159

Cook DW, Lofton SR (1975) Pathogenicity studies with a *Streptococcus* sp. isolated from fishes in an Alabama-Florida fish kill. Trans Am Fish Soc 104:286–288

Corrales J, Nye LB, Baribeau S, Gassman NJ, Schmale MC (2000) Characterization of scale abnormalities in pinfish, *Lagodon rhomboides*, from Biscayne Bay, Florida. Environ Biol Fishes 57:205–220

Costa MP (1994) Parecer tecnico n. 007/NAHP: Mortandade de bagres - Baia de Babitonga (SC) a Barra do Una (SP). CETESB, Sao Paulo, Brazil. 52 p

Couch JA (1985) Prospective study of infectious and noninfectious diseases in oysters and fishes in three Gulf of Mexico estuaries. Dis Aquat Organ 1:59–82

Couch JA, Nimmo DR (1974) Detection of interactions between natural pathogens and pollutants in aquatic animals. In: Amborski RL, Hood MA, Miller RR (eds) Proceedings of the Gulf Coast Regional Symposium on Diseases of Aquatic Animals. Wetland Res Publ No LSA-SG-74-05. Louisiana State University, Baton Rouge, LA, USA, pp 261–268

Craig JK (2012) Aggregation on the edge: Effects of hypoxia avoidance on the spatial distribution of brown shrimp and demersal fishes in the Northern Gulf of Mexico. Mar Ecol Prog Ser 445:75–95

Craig JK, Crowder LB, Gray CD, McDaniel CJ, Henwood TA, Hanifen JG (2001) Ecological effects of hypoxia on fish, sea turtles, and marine mammals in the northwestern Gulf of

Mexico. In: Robalais NN, Turner RE (eds) Coastal hypoxia: Consequences for living resources and ecosystems. Coastal and Estuarine Studies, Washington, DC, USA, pp 269–292

Crawford RB (1983) Effects of drilling fluids on embryo development. EPA Project Summary EPA-600/S3-83-021. U.S. Environmental Protection Agency, Environmental Research Laboratory, Gulf Breeze, FL, USA

Cummings EG (1955) The relation of the mullet (*Mugil cephalus*) to the water and salts of its environment: Structural and physiological aspects. PhD Dissertation, North Carolina State College, Raleigh, NC, USA. 103 p

Curran SS, Tkach VV, Overstreet RM (2013a) A new species of *Homalometron* (Digenea: Apocreadiidae) from fishes in the northern Gulf of Mexico. J Parasitol 99:93–101

Curran SS, Tkach VV, Overstreet RM (2013b) Molecular evidence for two cryptic species of *Homalometron* (Digenea: Apocreadiidae) in freshwater fishes of the southeastern United States. Comp Parasitol 80:186–195

Dailey MD, Morris R (1995) Relationship of parasites (Trematoda: Spirorchidae) and their eggs to the occurrence of fibropapillomas in the green turtle (*Chelonia mydas*). Can J Fish Aquat Sci 52(suppl 1):84–89

Dailey M, Walsh M, Odell D, Campbell T (1991) Evidence of prenatal infection in the bottlenose dolphin (*Tursiops truncatus*) with the lungworm *Halocercus lagenorhynchi* (Nematoda: Pseudaliidae). J Wildl Dis 27:164–165

Davidson AJ (2010) Herpesvirus systematics. Vet Microbiol 143:52–69

Dawson CE (1962) Notes on anomalous American Heterosomata with descriptions of five new records. Copeia 1962:138–146

Dawson CE (1964) A bibliography of anomalies of fishes. Gulf Res Rep 1:308–399

Dawson CE (1966) A bibliography of anomalies of fishes. Gulf Res Rep 2(Suppl 1):169–176

Dawson CE (1967) Three new records of partial albinism in American Heterosomata. Trans Am Fish Soc 96:400–404

Dawson CE (1969) Three unusual cases of abnormal coloration in northern Gulf of Mexico flatfishes. Trans Am Fish Soc 98:106–108

Dawson CE (1971) A bibliography of anomalies of fishes. Gulf Res Rep 3(Suppl 2):215–239

Dawson CE, Heal E (1976) A bibliography of anomalies of fishes. Gulf Res Rep 5(Suppl 3):35–41

Deeds JR, Reimschuessel R, Place AR (2006) Histopathological effects in fish exposed to the toxins from *Karlodinium micrum*. J Aquat Anim Health 18:136–148

Deeds JR, Landsberg JH, Etheridge SM, Pitcher GC, Longan SW (2008) Non-traditional vectors for paralytic shellfish poisoning. Mar Drugs 6:308–348

Delaney PM (1989) Phylogeny and biogeography of the marine isopod family Corallanidae (Crustacea, Isopoda, Flabellifera). Contributions in Science No. 409, Natural History Museum of Los Angeles County, Allen Press, Lawrence, KS, USA. 75 p

DeLorenzo ME, Wallace SC, Danese LE, Baird TD (2009) Temperature and salinity effects on the toxicity of common pesticides to the grass shrimp, *Palaemonetes pugio*. J Environ Sci Health B 44:455–460

DePaola A, Capers GM, Alexander D (1994) Densities of *Vibrio vulnificus* in the intestines of fish from the US Gulf Coast. Appl Environ Microbiol 60:984–988

Diamant A, Fournie JW, Courtney LA (1994) X-cell pseudotumors in a hardhead catfish *Arius felis* (Ariidae) from Lake Pontchartrain, Louisiana, USA. Dis Aquat Organ 18:181–185

Dickerson HW (2006) *Ichthyophthirius multifiliis* and *Cryptocaryon irritans* (Phylum Ciliophora). In: Woo PTK (ed) Fish diseases and disorders, Vol 1, CAB International, Wallingford, Oxfordshire, UK, pp 116–153

Dietz K (1982) Overall population patterns in the transmission cycle of infectious disease agents. In: Anderson RM, May RM (eds) Population biology of infectious diseases. Springer, Berlin, Germany, pp 87–102

Diggles BK, Adlard RD (1997) Intraspecific variation in *Cryptocaryon irritans*. J Eukaryot Microbiol 44:25–32

Diggles BK, Lester RJG (1996a) Influence of temperature and host species on the development of *Cryptocaryon irritans*. J Parasitol 82:45–51

Diggles BK, Lester RJG (1996b) Infections of *Cryptocaryon irritans* on wild fish from southeast Queensland, Australia. Dis Aquat Organ 25:159–167

Dobson AP, Hudson PJ (1992) Regulation and stability of a free-living host-parasite system: *Trichostrongylus tenuis* in red grouse. II. Population models. J Anim Ecol 61:487–498

Dorland (1974) Dorland's medical dictionary, 25th edn. WB Saunders, Philadelphia, PA, USA. 1748 p

Dortch Q, Robichaux R, Pool S, Milsted D, Mire G, Rabalais NN, Soniat TM, Fryxell GA, Turner RE, Parsons ML (1997) Abundance and vertical flux of *Pseudo-nitzschia* in the northern Gulf of Mexico. Mar Ecol Prog Ser 146:249–264

Dukes TW, Lawler AR (1975) The ocular lesions of naturally occurring lymphocystis in fish. Can J Comp Med 39:406–410

Dyková I, Lom J (2007) Histopathology of protistan and myxozoan infections in fishes: An atlas. Academia, Praha, Czech Republic. 219 p

Dyková I, Lom J, Overstreet RM (1994) Myxosporean parasites of the genus *Kudoa* Meglitsch, 1947 from some Gulf of Mexico fishes: Description of two new species and notes on their ultrastructure. Eur J Protistol 30:316–323

Dykstra MJ, Kane AS (2000) *Pfiesteria piscicida* and ulcerative mycosis of Atlantic menhaden—current status of understanding. J Aquat Anim Health 12:18–25

Dzikowski R, Paperna I, Diamant A (2003) Use of fish parasite species richness indices in analyzing anthropogenically impacted coastal marine ecosystems. Helgol Mar Res 57:220–227

Edwards RH, Overstreet RM (1976) Mesenchymal tumors of some estuarine fishes of the northern Gulf of Mexico. I. Subcutaneous tumors, probably fibrosarcomas, in the striped mullet, *Mugil cephalus*. Bull Mar Sci 26:33–40

Ene A, Su M, Lemaire S, Rose C, Schaff S, Moretti R, Lenz J, Herbst LH (2005) Distribution of chelonid fibropapillomatosis-associated herpesvirus variants in Florida: Molecular genetic evidence for infection of turtles following recruitment to neritic developmental habitats. J Wildl Dis 41:489–497

Esch GW (1994) Population biology of the diplostomatid trematode *Uvilifer ambloplitis*. In: Scott ME, Smith G (eds) Parasitic and infectious diseases: Epidemiology and ecology. Academic Press, San Diego, CA, USA, pp 321–336

Evans NM, Holder MT, Barbeitos MS, Okamura B, Cartwright P (2010) The phylogenetic position of Myxozoa: Exploring conflicting signals in phylogenomic and ribosomal data sets. Mol Biol Evol 27:2733–2746

Eyisi OAL, Nwodo UU, Iroegbu CU (2013) Distribution of Vibrio species in shellfish and water samples collected from the Atlantic coastline of South-East Nigeria. J Health Popul Nutr 31:314

Fabacher DL, Besser JM, Schmitt CJ, Harshbarger JC, Peterman PH, Lebo JA (1991) Contaminated sediments from tributaries of the Great Lakes: Chemical characterization and carcinogenic effects in medaka (*Oryzias latipes*). Arch Environ Contam Toxicol 21:17–34

Fagerholm HP, Overstreet RM, Humphrey-Smith I (1996) *Contracaecum magnipapillatum* (Nematoda, Ascaridoidea): Resurrection and pathogenic effect of a common parasite from the proventriculus of *Anous minutus* from the Great Barrier Reef, with a note on *C. variegatum*. Helminthologia 33:195–207

Faisal M, MacIntyre EA, Adham KG, Tall BD, Kothary MH, La Peyre JF (1998) Evidence for the presence of protease inhibitors in eastern (*Crassostrea virginica*) and Pacific (*Crassostrea gigas*) oysters. Comp Biochem Physiol B Biochem Mol Biol 121:161–168

Fauquier DA, Kinsel MJ, Dailey MD, Sutton GE, Stolen MK, Wells RS, Gulland FMD (2009) Prevalence and pathology of lungworm infection in bottlenose dolphins *Tursiops truncatus* from southwest Florida. Dis Aquat Organ 88:85–90

Felder DL, Camp DK (eds) (2009) Gulf of Mexico origin, waters, and biota: Biodiversity (Vol. 1). Texas A&M University Press, College Station, TX, USA. 1393 p

Felley SM, Vecchione M, Hare SGF (1987) Incidence of ectoparasitic copepods on ichthyoplankton. Copeia 1987:778–782

Finch RH (1917) Fish killed by the cold wave of February 2-4, 1917, in Florida. Mon Weather Rev 1917:171–172

Flaherty KE, Landsberg JH (2011) Effects of a persistent red tide (*Karenia brevis*) bloom on community structure and species-specific relative abundance of nekton in a Gulf of Mexico estuary. Estuaries and Coasts 34:417–439

Flewelling LJ, Adams DH, Naar JP, Atwood KE, Granholm AA, O'Dea SN, Landsberg JH (2010) Brevetoxins in sharks and rays (Chondrichthyes, Elasmobranchii) from Florida coastal waters. Mar Biol 157:1937–1953

Foley AM, Schroeder BA, Redlow AE, Fick-Child KJ, Teas WG (2005) Fibropapillomatosis in stranded green turtles (*Chelonia mydas*) from the eastern United States (1980-98): Trends and associations with environmental factors. J Wildl Dis 41:29–41

Ford SE (2011) Dermo disease of oysters caused by *Perkinsus marinus*. ICES Identification Leaflets for Diseases and Parasites of Fish and Shellfish, Leaflet 30. 5p

Ford SE, Tripp MR (1996) Diseases and defense mechanisms. In: Kennedy VS, Newell RIE, Eble AF (eds) The eastern oyster *Crassostrea virginica*. Maryland Sea Grant College, College Park, MD, USA, pp 581–660

Forrester DJ, Gaskin JM, White FH, Thompson NP, Quick JA, Henderson GE, Woodard JC, Robertson WD (1977) An epizootic of waterfowl associated with a red tide episode in Florida. J Wildl Dis 13:160–167

Fotheringham N, Weissberg GH (1979) Some causes, consequences and potential environmental impacts of oxygen depletion in the northern Gulf of Mexico. Offshore Tech Conf Proc OTC 3611:2205–2207

Fournie JW, Overstreet RM (1993) Host specificity of *Calyptospora funduli* (Apicomplexa: Calyptosporidae) in atheriniform fishes. J Parasitol 79:720–727

Fournie JW, Overstreet RM, Bullock LH (1985) Multiple capillary haemangiomas in the scamp, *Mycteroperca phenax* Jordan and Swain. J Fish Dis 8:551–555

Fournie JW, Vogelbein WK, Overstreet RM (1987) Squamous cell carcinoma in the Gulf menhaden, *Brevoortia patronus* Goode. J Fish Dis 10:133–136

Fournie JW, Summers JK, Weisburg SB (1996) Prevalence of gross pathological abnormalities in estuarine fishes. Trans Am Fish Soc 125:581–590

Fournie JW, Vogelbein WK, Overstreet RM, Hawkins WE (2000) Life cycle of *Calyptospora funduli* (Apicomplexa: Calyptosporidae). J Parasitol 86:501–505

Frasca S Jr, Kirsipuu VL, Russell S, Bullard SA, Benz GW (2004) Opercular lesion in wild black drum, *Pogonias cromis* (Linnaeus, 1766), associated with attachment of the sea louse *Sciaenophilus tenuis* (Copepoda: Siphonostomatoida: Caligidae). Acta Ichthyol Piscat 34:115–127

Fusco AC (1980) Larval development of *Spirocamallanus cricotus* (Nematoda: Camallanidae). Proc Helminthol Soc Wash 47:63–71

Fusco AC, Overstreet RM (1978) *Spirocamallanus cricotus* sp. n. and *S. halitrophus* sp. n. (Nematoda: Camallanidea) from fishes in the northern Gulf of Mexico. J Parasitol 64:239–244

Galtsoff PS (1954) Gulf of Mexico, its origin, and marine life. U.S. Fish and Wildlife Service. Fishery Bulletin 55(89):xiv, 604 pp

Gannon DP, McCabe EJB, Camilleri SA, Gannon JG, Brueggen MK, Barleycorn AA, Palubok VI, Kirkpatrick GJ, Wells RS (2009) Effects of *Karenia brevis* harmful algal blooms on nearshore fish communities in southwest Florida. Mar Ecol Prog Ser 378:171–186

Garrison VH, Shinn EA, Foreman WT, Griffin DW, Holmes CW, Kellogg CA, Majewski MS, Richardson LL, Ritchie KB, Smith GW (2003) African and Asian dust: From desert soils to coral reefs. BioScience 53:469–480

Gartner JV Jr (1986) Observations on anomalous conditions in some flatfishes (Pisces: Pleuronectiformes), with a new record of partial albinism. Environ Biol Fishes 17:141–152

Garton D, Stickle WB (1980) Effects of salinity and temperature on the predation rate of *Thais haemastoma* on *Crassostrea virginica* spat. Biol Bull 158:49–57

Gassman NJ, Nye LB, Schmale MC (1994) Distribution of abnormal biota and sediment contaminants in Biscayne Bay, Florida. Bull Mar Sci 54:929–943

Gaughan DJ (2002) Disease-translocation across geographic boundaries must be recognized as a risk even in the absence of disease identification: The case with Australian *Sardinops*. Rev Fish Biol Fish 11:113–123

Gold-Bouchot G, Simá-Alvarez R, Zapata-Peréz O, Güemez-Ricalde J (1995) Histopathological effects of petroleum hydrocarbons and heavy metals on the American oyster (*Crassostrea virginica*) from Tabasco, Mexico. Mar Pollut Bull 31:439–445

Gold-Bouchot G, Zavala-Coral M, Zapata-Pérez O, Ceja-Moreno V (1997) Hydrocarbon concentrations in oysters (*Crassostrea virginica*) and recent sediments from three coastal lagoons in Tabasco, Mexico. Bull Environ Contam Toxicol 59:430–437

Gordon AN, Kelly WR, Cribb TH (1998) Lesions caused by cardiovascular flukes (Digenea: Spirorchidae) in stranded green turtles (*Chelonia mydas*). Vet Pathol 35:21–30

Groff JM (2004) Neoplasia in fishes. Vet Clin Exot Anim 7:705–756

Gudger EW (1934) Ambicoloration in the winter flounder, *Pseudopleuronectes americanus*: Incomplete ambicoloration without other deformity; Complete ambicoloration with a hooked dorsal fin and with the rotating eye just over the dorsal ridge. Am Mus Novit 717:1–8

Gunter G (1941) Death of fishes due to cold on the Texas coast, January, 1940. Ecology 22:203–208

Gunter G (1947) Catastrophism in the sea and its paleontological significance, with special reference to the Gulf of Mexico. Am J Sci 245:669–676

Gunter G (1948) A discussion of abnormal scale patterns in fishes, with notice of another specimen with reversed scales. Copeia 1948:280–285

Gunter G (1967) Some relationships of estuaries to the fisheries of the Gulf of Mexico. In: Lauff GA (ed) Esturaries. Am Assoc Adv Sci, Washington, DC, USA, pp 621–638

Gunter G, Hildebrand HH (1951) Destruction of fishes and other organisms on the south Texas coast by the cold wave of January 28-February 3, 1951. Ecology 32:731–736

Gunter G, Lyles CH (1979) Localized plankton blooms and jubilees on the Gulf Coast. Gulf Res Rep 3:297–299

Gunter G, Smith FW, Williams RH (1947) Mass mortality of marine animals on the lower west coast of Florida, November 1946-January 1947. Science 105:256–257

Gunter G, Williams RH, Davis CC, Smith FW (1948) Catastrophic mass mortality of marine animals and coincident phytoplankton bloom on the west coast of Florida, November 1946 to August 1947. Ecol Monogr 18:309–324

Guthrie JF, Kroger RL (1974) Schooling habits of injured and parasitized menhaden. Ecology 55:208–210

Hallegreaff GM (1995) Harmful algal blooms: A global overview. In Hallegreaff GM, Anderson DM, Cembella AD, eds, Manual on harmful marine microalgae. IOC Manuals and Guides 33. UNESCO, Paris, France, pp 1–22

Hamann M, Godfrey MH, Seminoff JA, Arthur K, Barata PCR, Bjorndal KA, Bolten AB, Broderick AC, Campbell LM, Carreras C, Casale P, Chaloupka M, Chan SKF, Coyne MS, Crowder LB, Diez CE, Dutton PH, Epperly SP, FitzSimmons NN, Formia A, Girondot M, Hays GC, Cheng IJ, Kaska Y, Lewiston R, Mortimer JA, Nichols WJ, Reina RD, Shanker K, Spotila JR, Tomás J, Wallace BP, Work TM, Zbinden J, Godley BJ (2010) Global research priorities for sea turtles: Informing management and conservation in the 21st century. Endang Species Res 11:245–269

Harper DE Jr, McKinney LD, Salzer RR, Case RJ (1981) The occurrence of hypoxic bottom water off the upper Texas coast and its effects on the benthic biota. Contrib Mar Sci 24:53–79

Harshbarger JC (1965–1981) Activities reports. Registry of tumors in lower animals. Smithsonian Institution, Washington, DC, USA

Harshbarger JC (1977) Role of the registry of tumors in lower animals in the study of environmental carcinogenesis in aquatic animals. Ann New York Acad Sci 298:280–282

Harshbarger JC, Clark JB (1990) Epizootiology of neoplasm in bony fish of North America. Sci Total Environ 94:1–167

Harshbarger JC, Spero PM, Wolcott NM (1993) Neoplasms in wild fish from the marine ecosystem emphasizing environmental interactions. In: Couch JA, Fournie JW (eds) Pathobiology of marine and estuarine organisms. CRC Press, Boca Raton, FL, USA, pp 157–176

Hawke JP (1976) A survey of the diseases of striped bass, Morone saxatilis and pompano, Trachinotus carolinus cultured in earthen ponds. In Proceedings of the Annual Meeting-World Mariculture Society Vol 7, No. 1–4, Blackwell Publishing Ltd, Oxford, UK, pp 495–509

Hawkins WE, Fournie JW, Overstreet RM (1984) Intrahepatic stages of Eimeria funduli (Protista: Apicomplexa) in the longnose killifish, Fundulus similis. Trans Am Microsc Soc 103:185–194

Hawkins WE, Overstreet RM, Fournie JW, Walker WW (1985) Small aquarium fishes as models for environmental carcinogenesis: Tumor induction in seven fish species. J Appl Toxicol 5:261–264

Hawkins WE, Walker WW, Overstreet RM, Lytle TF, Lytle JS (1988) Dose-related carcinogenic effects of water-borne benzo[a]pyrene on livers of two small fish species. Ecotoxicol Environ Saf 16:219–231

Hawkins WE, Walker WW, Overstreet RM, Lytle JS, Lytle TF (1990) Carcinogenic effects of some polycyclic aromatic hydrocarbons on the Japanese medaka and guppy in waterborne exposures. Sci Total Environ 94:155–167

Hayes MA, Smith IR, Rushmore TH, Crane TL, Thorn C, Kocal TE, Ferguson HW (1990) Pathogenesis of skin and liver neoplasm in white suckers from industrially polluted areas in Lake Ontario. Sci Total Environ 94:105–123

Hazen EL, Craig JK, Good CP, Crowder LB (2009) Vertical distribution of fish biomass in hypoxic waters on the Gulf of Mexico shelf. Mar Ecol Prog Ser 375:195–207

Hazen TC, Raker ML, Esch GW, Fliermans CB (1978) Ultrastructure of red-sore lesions on largemouth bass (Micropterus salmoides): Association of the ciliate Epistylis sp. and the bacterium Aeromonas hydrophila. J Protozool 25:351–355

Heck KL, Coen LD (1995) Predation and the abundance of juvenile blue crabs: A comparison of selected East and Gulf Coast (USA) studies. Bull Mar Sci 57:877–883

Henley MW, Lewis DH (1976) Anaerobic bacteria associated with epizootics in grey mullet (*Mugil cephalus*) and redfish (*Sciaenops ocellata*) along the Texas Gulf Coast. J Wildl Dis 12:448–453

Herbert BW, Shaharom-Harrison FM, Overstreet RM (1994) Description of a new blood-fluke, *Cruoricola lates* n.g., n.sp. (Digenea: Sanguinicolidae), from sea-bass *Lates calcarifer* (Bloch, 1790) (Centropomidae). Syst Parasitol 29:51–60

Herbst LH (1994) Fibropapillomatosis of marine turtles. Annu Rev Fish Dis 4:389–425

Herbst LH (1999) Infectious diseases of marine turtles. In: Eckert KL, Bjorndal KA, Abreu-Grobois FA, Donnelly M (eds) Research and Management Techniques for the Conservation of Sea Turtles. IUCN/SSC Marine Turtle Specialist Group, Washington, DC, USA, pp 208–213

Hewitt GC (1971) Two species of *Caligus* (Copepoda, Caligidae) from Australian waters, with a description of some developmental stages. Pac Sci 25:145–164

Hickerson EL, Schmahl GP, Robbart M, Precht WF, Caldow C (2008) State of coral reef ecosystems of the Flower Garden Banks, Stetson Bank, and other Banks in the northwestern Gulf of Mexico. In: Waddell JE, Clarke AM (eds) The state of coral reef ecosystems of the United States and Pacific Freely Associated States: Memorandum NOS NCCOS 73. NOAA/NCCOS Center for Coastal Monitoring and Assessment's Biogeography Team. Silver Spring, MD, USA, pp 189–217

Ho J-S (2000) The major problem of cage aquaculture in Asia relating to sea lice (Southeast Asia chapter). In: Liao I, Lin C (eds) Cage aquaculture in Asia, Proceedings of the First International Symposium on Cage Aquaculture in Asia. Asian Fisheries Society/Manila and World Aquaculture Society, Bangkok, Thailand, pp 13–19

Hoberg EP (1996) Faunal diversity among avian parasite assemblages: The interaction of history, ecology, and biogeography in marine systems. Bull Scand Soc Parasitol 6:65–89

Hoberg EP, Klassen GJ (2002) Revealing the faunal tapestry: Co-evolution and historical biogeography of hosts and parasites in marine systems. Parasitology 124:3–22

Honjo M, Matsui K, Ueki M, Nakamura R, Fuhrman JA, Kawabara Z (2006) Diversity of virus-like agents killing *Microcystis aeruginosa* in a hyper-eutrophic pond. J Plankton Res 28:407–412

Howse HD (1972) Snook (Centropomus: Centropomidae): New host for lymphocystis, including observations on the ultrastructure of the virus. Am Midl Nat 88:476–479

Howse HD, Lawler AR, Hawkins WE, Foster CA (1977) Ultrastructure of lymphocystis in the heart of the silver perch, *Bairdiella chrysura* (Lacepede), including observations on normal heart structure. Gulf Caribb Res 6:39–57

Howse HD, Overstreet RM, Hawkins WE, Franks JS (1992) Ubiquitous perivenous smooth muscle cords in viscera of the teleost *Rachycentron canadum*, with special emphasis on liver. J Morphol 212:175–189

Hudson PJ (1986) The effect of a parasitic nematode on the breeding production of red grouse. J Anim Ecol 55:85–92

Hudson PJ, Dobson AP. (1990). Red grouse population cycles and the population dynamics of the caecal nematode *Trichostrongylus tenuis*. In: Lance AN, Lawton JH (eds) Red grouse population processes BES/RSPB Red Grouse Workshop. BES/RSPB Publications, pp 5-19

Hudson PJ, Dobson AP (1997a) Transmission dynamics and host-parasite interactions of *Trichostrongylus tenuis* in red grouse (*Lagopus lagopus scoticus*). J Parasitol 83:194–202

Hudson PJ, Dobson AP (1997b) Host-parasite processes and demographic consequences. In: Clayton DH, Moore J (eds) Host-parasite evolution: General principles and avian models. Oxford University Press, Oxford, UK, pp 128–154

Hudson PJ, Dobson AP, Newborn D (1985) Cyclic and non-cyclic populations of red grouse: A role for parasitism? In: Rollinson D, Anderson RM (eds) Ecology and genetics of host-parasite interactions. Academic Press, London, UK, pp 77–89

Hudson PJ, Newborn D, Dobson AP (1992) Regulation and stability of a free-living host-parasite system: *Trichostrongylus tenuis* in red grouse. I. Monitoring and parasite reduction experiments. J Anim Ecol 61:477–486

Hudson PJ, Dobson AP, Newborn D (1998) Prevention of population cycles by parasite removal. Science 282:2256–2258

Huerta P, Pineda H, Aguirre A, Spraker T, Sarti L, Barragán A (2002) First confirmed case of fibropapilloma in a leatherback turtle (*Dermochelys coriacea*). In: Mosier A, Foley A, Brost B (eds) Proceedings of the 20th Annual Symposium on Sea Turtle Biology and Conservation. U.S. Department of Commerce, Washington, DC, National Oceanic and Atmospheric Administration Technical Memorandum NMFS-SEFSC-477. 193 p

Hughes TP (1994) Catastrophes, phase shifts, and large-scale degradation of a Caribbean coral reef. Science 265:1547–1551

Huizinga HW, Esch GW, Hazen TC (1979) Histopathology of red-sore disease (*Aeromonas hydrophila*) in naturally and experimentally infected largemouth bass *Micropterus salmoides* (Lacepede). J Fish Dis 2:263–277

Hutchinson DR, Ruppel CD, Roberts HH, Carney RS, Smith MA (2011) Gas hydrates in the Gulf of Mexico. In: Buster NA, Holmes CW (eds) Gulf of Mexico origin, waters, and biota, Vol 3: Geology. Texas A&M University Press, College Station, TX, USA, pp 247–275

Iverson ES, Kelley JF (1976) Microsporidiosis successfully transmitted experimentally in pink shrimp. J Invertebr Pathol 27:407–408

James SV, Valenti TW, Roelke DL, Grover JP, Brooks BW (2011) Probabilistic ecological hazard assessment of microcystin-LR allelopathy to *Prymnesium parvum*. J Plankton Res 33:319–332

Jensen K (2009) Cestoda (Platyheminthes) of the Gulf of Mexico. In: Felder DL, Camp DK (eds) Gulf of Mexico Origin, Waters, and Biota, Vol 1: Biodiversity. Texas A&M University Press, College Station, TX, USA, pp 487–499

Jensen K, Bullard SA (2010) Characterization of a diversity of tetraphyllidean and rhinebothriidean cestode larval types, with comments on host associations and life-cycles. Int J Parasitol 40:889–910

Jones JB, Hyatt AD, Hine PM, Whittington RJ, Griffin DA, Bax NJ (1997) Australasian pilchard mortalities. World J Microbiol Biotechnol 13:383–392

Jones TC, Overstreet RM, Lotz JM, Frelier PF (1994) *Paraophioidina scolecoides* n. sp., a new aseptate gregarine from the cultured Pacific white shrimp, *Penaeus vannamei*. Dis Aquat Organ 19:67–75

Jordán-Dahlgren E, Rodríguez-Martínez RE (2004) Coral Diseases in Gulf of Mexico Reefs. In: Rosenberg E, Loya Y (eds) Coral health and disease. Springer-Verlag, Berlin Heidelberg, Germany, pp 105–118

Jordán-Dahlgren E, Maldonado MA, Rodríguez-Martínez RE (2005) Diseases and partial mortality in *Montastraea annularis* species complex in reefs with differing environmental conditions (NW Caribbean and Gulf of México). Dis Aquat Organ 63:3–12

Justić D, Bierman VJ, Scavia D, Hetland RD (2007) Forecasting Gulf's hypoxia: The next 50 years? Estuaries Coasts 30:791–801

Kennedy CR (1984) The use of frequency distributions in an attempt to detect host mortality induced by infections of diplostomatid metacercariae. Parasitology 89:209–220

Kennedy VS, Newell RIE, Eble AF (1996) The eastern oyster *Crassostrea virginica*. Maryland Sea Grant College, College Park, MD, USA. 734 p

Kensley B, Heard RW (1997) *Tridentella ornata* (Richardson 1911), new combination: Records of hosts and localities (Crustacea: Isopoda: Tridentellidae). Proc Biol Soc Wash 110:422–425

Khan RA (1990) Parasitism in marine fish after chronic exposure to petroleum hydrocarbons in the laboratory and to the Exxon Valdez oil spill. Bull Environ Contam Toxicol 44:759–763

Khan RA, Hooper RG (2007) Influence of a thermal discharge on parasites of a cold-water flatfish, *Pleuronectes americanus*, as a bioindicator of subtle environmental change. J Parasitol 93:1227–1230

Khan RA, Kiceniuk J (1983) Effects of crude oils on the gastrointestinal parasites of two species of marine fish. J Wildl Dis 19:253–258

Khan RA, Kiceniuk J (1984) Histopathological effects of crude oil on Atlantic cod following chronic exposure. Can J Zoo 62:2038–2043

Khan RA, Kiceniuk JW (1988) Effect of petroleum aromatic hydrocarbons on mongeneids parasitizing Atlantic cod, *Gadus morhua* L. Bull Environ Contam Toxicol 41:94–100

Khan RA, Thulin J (1991) Influence of pollution on parasites of aquatic animals. Adv Parasitol 30:201–238

Knudsen R, Amundsen P-A, Klemetsen A (2002) Parasite-induced host mortality: Indirect evidence from a long-term study. Environ Biol Fishes 64:257–265

Kolipinski MC (1969) Gar infested by *Argulus* in the Everglades. QJ Florida Acad Sci 32:39–49

Kritsky DC, Bakenhaster MD, Fajer-Avila EJ, Bullard SA (2010) *Rhabdosynochus* spp. (Monogenoidea: Diplectanidae) infecting the gill lamellae of snooks, *Centropomus* spp. (Perciformes: Centropomidae), in Florida, and redescription of the type species, *R. rhabdosynochus*. J Parasitol 96:879–886

Krkošek M, Ford JS, Morton A, Lele S, Myers RA, Lewis MA (2007) Declining wild salmon populations in relation to parasites from farm salmon. Science 318:1772–1775

La Peyre J, Kristensen H, McDonough K, Cooper R (1998) Effects of protease inhibitors on the oyster pathogen *Perkinsus marinus* and oyster cells in vitro. In: Kane AS, Poynton SL (eds) Proc 3rd Int Symp Aquatic Animal Health. Aquatic Pathology Center (APC) Press, Baltimore, MD, USA. 151 p

Labrie L, Komar C, Terhune J, Camus A, Wise D (2004) Effect of sublethal exposure to the trematode *Bolbophorus* spp. on the severity of enteric septicemia of catfish in channel catfish fingerlings. J Aquat Anim Health 16:231–237

Landsberg JH (1995) Tropical reef-fish disease outbreaks and mass mortalities in Florida, USA: What is the role of dietary biological toxins? Dis Aquat Organ 22:83–100

Landsberg JH, Blakesley BA, Reese RO, McRae G, Forstchen PR (1998) Parasites of fish as indicators of environmental stress. Environ Monit Assess 51:211–232

Landsberg JH, Balazs GH, Steidinger KA, Baden DG, Work TM, Russell DJ (1999) The potential role of natural tumor promoters in marine turtle fibropapillomatosis. J Aquat Anim Health 11:199–210

Landsberg JH, Hall S, Johannessen JN, White KD, Conrad SM, Abbott JP, Flewelling LJ, Richardson RW, Dickey RW, Jester ELE, Etheridge SM, Deeds JR, Van Dolah FM, Leighfield TA, Zou Y, Beaudry CG, Benner RA, Rogers PL, Scott PS, Kawabata K, Wolny JL, Steidinger KA (2006) Saxitoxin puffer fish poisoning in the United States, with the first report of *Pyrodinium bahamense* as the putative toxin source. Environ Health Perspect 144:1502–1507

Landsberg JH, Flewelling LJ, Naar J (2009) *Karenia brevis* red tides, brevetoxins in the food web, and impacts on natural resources: Decadal advancements. Harmful Algae 8:598–607

Lauckner G (1985) Diseases of Aves (marine birds). In: Kinne O (ed) Diseases of marine animals, Vol 4, Part 2, Introduction: Reptilia, Aves, Mammalia. Biologische Anstalt Helgoland, Hamburg, Germany, pp 627–643

Lawler AR (1980) Studies on *Amyloodinium ocellatum* (Dinoflagellata) in Mississippi Sound: Natural and experimental hosts. Gulf Caribb Res 6:403–413

Laws EA (2000) Aquatic pollution: An introductory text. John Wiley & Sons, New York, NY, USA. 672 p

Lerdo de Tejada MM (1850) De una noticia de los descubrimientos hechos en las islas y en el continente Americano, y de las providencias dictadas por los Reyes de España para el gobierno de sus nuevas posesiones, desde el primer viage de Don Cristobal Colon, hasta que se emprendió la conquista de Mexico. Apuntes históricos de la heróica ciudad de Vera-Cruz, Imp. de I. Cumplido, México

Lester RJG (1984) A review of methods for estimating mortality due to parasites in wild fish populations. Helgol Meeresunters 37:53–64

Lewis DH, Udey LR (1978) Meningitis in fish caused by an asporogenous anaerobic bacterium. Fish Disease Leaflet 56. U.S. Department of the Interior, Fish and Wildlife Service, Washington, DC, USA. 5 p

Lewis DH, Grumbles LC, McConnell S, Flowers AI (1970) *Pasteurella*-like bacteria from an epizootic in menhaden and mullet in Galveston Bay. J Wildl Dis 6:160–162

Lewis M, Jordan S, Chancy C, Harwell L, Goodman L, Quarles R (2007) Summer fish community of the coastal northern Gulf of Mexico: Characterization of a large-scale trawl survey. Trans Am Fish Soc 136:829–845

Lewis MA, Goodman LR, Chancy CA, Jordan SJ (2011) Fish assemblages in three Northwest Florida urbanized bayous before and after two hurricanes. J Coastal Res 27:35–45

Lewitus A, Bargu S, Byrd M, Dorsey C, Flewellin L, Flowers A, Heil C, Kovach C, Lovko V, Steidinger K (2014) Resource Guide for Harmful Algal Bloom Toxin Sampling and Analysis. Alliance, Gulf of Mexico, http://gulfofmexicoalliance.org/documents/pits/wq/goma_hab_toxin_resource_guide.pdf. Accessed on January 15, 2015

Li T, Brouwer M (2013) Gene expression profile of hepatopancreas from grass shrimp *Palaemonetes pugio* exposed to cyclic hypoxia. Comp Biochem Physiol Part D: Genomics Proteomics 8:1–10

Lightner DV (1996) A handbook of shrimp pathology and diagnostic procedures for disease of cultured penaeid shrimp. World Aquaculture Society, Baton Rouge, LA, USA

Lightner DV (2011) Virus diseases of farmed shrimp in the Western Hemisphere (the Americas): A review. J Invertebr Pathol 106:110–130

Lightner DV, Lewis DH (1975) A septicemic bacterial disease syndrome of penaeid shrimp. Mar Fish Rev 37:25–28

Lightner DV, Redman RM, Pantoja CR, Tang KFJ, Noble BL, Schofield P, Mohney LL, Navarro SA (2012) Historic emergence, impact and current status of shrimp pathogens in the Americas. J Invertebr Pathol 110:174–183

Lipp EK, Rodriguez-Palacios C, Rose JB (2001) Occurrence and distribution of the human pathogen *Vibrio vulnificus* in a subtropical Gulf of Mexico estuary. Hydrobiologia 460:165–173

Lipscomb TP, Schulman FY, Moffett D, Kennedy S (1994) Morbilliviral disease in Atlantic bottlenose dolphins (*Tursiops truncatus*) from the 1987-1988 epizootic. J Wildl Dis 30:567–571

Lipscomb TP, Kennedy S, Moffett D, Krafft A, Klaunberg BA, Lichy JH, Regan GT, Worthy GAJ, Taubenberger JK (1996) Morbilliviral epizootic in bottlenose dolphins of the Gulf of Mexico. J Vet Diagn Invest 8:283–290

Litaker W, Steidinger KA, Mason PL, Landsberg JH, Shields JD, Reece KS, Haas LW, Vogelbein WK, Vandersea MW, Kibler SR, Tester PA (2005) The reclassification of *Pfiesteria shumwayae* (Dinophyceae): *Pseudopfiesteria*, gen. nov. J Phycol 41:643–651

Livingston RJ (2001) Eutrophication processes in coastal systems: Origin and succession of plankton blooms and effects on secondary production in Gulf Coast estuaries. CRC Press, Boca Raton, FL, USA. 327 p

Lloyd R (1987) Special tests in aquatic toxicity for chemical mixtures: Interactions and modification of response by variation of physicochemical conditions. In: Vouk VB, Butler GC, Upton AC, Parke DV, Asher SC (eds) Methods for Assessing the Effects of Mixtures of Chemicals. John Wiley & Sons, New York, NY, USA, pp 491–507

Loesch H (1960) Sporadic mass shoreward migrations of demersal fish and crustaceans in Mobile Bay, Alabama. Ecology 41:292–298

Lom J, Dyková I (1992) Protozoan parasites of fishes. Elsevier Science Publishers, Amsterdam, The Netherlands. 315 p

Lom J, Nigrelli RF (1970) *Brooklynella hostilis* n.g., n.sp., a pathogenic cyrtophorine ciliate in marine fishes. J Protozool 17:224–232

Lotz JM, Overstreet RM (1990) Parasites and predators. In: Chàvez JC, Sosa NO (eds) The aquaculture of shrimp, prawn and crawfish in the world: Basics and technologies. Midori Shobo Co, Ltd, Ikebukuro, Toshima-ku Tokyo, Japan, pp 96–121 [In Japanese]

Lu YA, Wang Y, Aguirre AA, Zhao ZS, Liu CY, Nerurkar VR, Yanagihara R (2003) RT-PCR detection of the expression of the polymerase gene of a novel reptilian herpesvirus in tumor tissues of green turtles with fibropapilloma. Arch Virol 148:1155–1163

Lucke B (1942) Tumors of the nerve sheath in fish of the snapper family (Lutianidae). Arch Pathol 34:133–150

Macintyre HL, Stutes AL, Smith WL, Dorsey CP, Abraham A, Dickey RW (2011) Environmental correlates of community composition and toxicity during a bloom of *Pseudo-nitzschia* spp. in the northern Gulf of Mexico. J Plankton Res 33:273–295

MacKenzie CL Jr, Wakida-Kusunoki AI (1997) The oyster industry of eastern Mexico. Mar Fish Rev 59:1–13

MacKenzie K, Williams HH, Williams B, McVicar AH, Siddall R (1995) Parasites as indicators of water quality and the potential use of helminth transmission in marine pollution studies. Adv Parasitol 35:85–144

Magaña HA, Contreras C, Villareal TA (2003) A historical assessment of *Karenia brevis* in the Western Gulf of Mexico. Harmful Algae 2:163–171

Maier Brown AF, Dortch Q, Van Dolah FM, Leighfield TA, Morrison W, Thessen AE, Steidinger K, Richardson B, Moncreiff CA, Pennock JR (2006) Effect of salinity on the distribution, growth, and toxicity of *Karenia* spp. Harmful Algae 5:199–212

Malins DC, McCain BB, Brown DW, Chan SL, Myers MS, Landahl JT, Prohaska PG, Friedman AJ, Rhodes LD, Burrows DG, Gronlund WD, Hodgins HO (1984) Chemical pollutants in sediments and diseases of bottom-dwelling fish in Puget Sound, Washington. Environ Sci Technol 18:705–713

Manter HW (1954) Trematoda of the Gulf of Mexico. Fish Bull 55:335–350

Marcogliese DJ, Nagler JJ, Cyr DG (1998) Effects of exposure to contaminated sediments on the parasite fauna of American Plaice (*Hippoglossoides platessoides*). Bull Environ Contam Toxicol 61:88–95

May EB (1973) Extensive oxygen depletion in Mobile Bay, Alabama. Limnol Oceanogr 18:353–366

May RM, Anderson RM (1978) Regulation and stability of host-parasite population interactions: II. Destabilizing processes. J Anim Ecol 47:249–267

McCain BB, Brown DW, Hom T, Myers MS, Pierce SM, Collier TK, Stein JE, Chan S-L, Varanasi U (1996) Chemical contaminant exposure and effects in four fish species from Tampa Bay, Florida. Estuaries 19:86–104

McCallum H, Dobson A (1995) Detecting disease and parasite threats to endangered species and ecosystems. Trends Ecol Evol 10:190–194

McEachron LW, Matlock GC, Bryan CE, Unger P, Cody TJ, Martin JH (1994) Winter mass mortality of animals in Texas bays. Northeast Gulf Sci 13:121–138

McFarland WN (1965) The effect of hypersalinity on serum and muscle ion concentrations in the striped mullet, *Mugil cephalus* L. Publ Inst Mar Sci Univ Texas 10:179–186

McLean E, Salze G, Craig SR (2008) Parasites, diseases and deformities of cobia. Ribarstvo 66:1–16

Mellergaard S, Nielsen E (1995) Impact of oxygen deficiency on the disease status of common dab *Limanda limanda*. Dis Aquat Organ 22:101–114

Meyers TR, Hendricks JD (1982) A summary of tissue lesions in aquatic animals induced by controlled exposures to environmental contaminants, chemotherapeutic agents, and potential carcinogens [Toxicity to fishes]. Mar Fish Rev 44:1–17

Meyers TR, Hauck AK, Blankenbeckler WD, Minicucci T (1986) First report of viral erythrocytic necrosis in Alaska, USA, associated with epizootic mortality in Pacific herring, *Clupea harengus pallasi* (Valenciennes). J Fish Dis 9:479–491

Meyers TR, Short S, Lipson K, Batts WN, Winton JR, Wilcock J, Brown E (1994) Association of viral hemorrhagic septicemia virus with epizootic hemorrhages of the skin in Pacific herring *Clupea harengus pallasi* from Prince William Sound and Kodiak Island, Alaska, USA. Dis Aquat Organ 19:27–37

Miglarese JV, Shealy MH (1974) Incidence of microsporidean and trypanorhynch cestodes in white shrimp, *Penaeus setiferus* Linnaeus, in South Carolina estuaries. SC Acad Sci Bull 36:93

Miller EM (1940) Mortality of fishes due to cold on the southeast Florida coast, 1940. Ecology 21:420–421

Mitchell AJ, Overstreet RM, Goodwin AE (2009) *Eustrongylides ignotus* infecting commercial bass (*Morone chrysops* female × *Morone saxatilis* male) and other fishes in the southeastern USA. J Fish Dis 32:795–799

Moles A, Wade TL (2001) Parasitism and phagocytic function among sand lance *Ammodytes hexapterus* Pallas exposed to crude oil-laden sediments. Bull Environ Contam Toxicol 66:528–535

Möller H (1985) A critical review on the role of pollution as a cause of fish diseases. In: Ellis AE (ed) Fish and shellfish pathology. Academic Press, London, UK, pp 169–182

Monserrat JM, Martínez PE, Geracitano LA, Amado LL, Martins CMG, Pinho GLL, Chaves IS, Ferreira-Cravo M, Ventura-Lima J, Bianchini A (2007) Pollution biomarkers in estuarine animals: Critical review and new perspectives. Comp Biochem Physiol C Toxicol Pharmacol 146:221–234

Moore CJ, Posey CR Sr (1974) Pigmentation and morphological abnormalities in the hogchoker, *Trinectes maculatus* (Pisces, Soleidae). Copeia 1974:660–670

Muncy RJ (1984) Species profiles: Life histories and environmental requirements of coastal fishes and invertebrates (South Atlantic). White shrimp [*Penaeus setiferus*]. U.S. Fish and Wildlife Service FWS/OBS-82/11.27. U.S. Army Corps of Engineers TR EL-82-4, pp 1–19

Murchelano RA, Wolke RE (1991) Neoplasms and nonneoplastic liver lesions in winter flounder, *Pseudopleuronectes americanus*, from Boston Harbor, Massachusetts. Environ Health Perspect 90:17–26

Mydlarz LD, Holthouse SF, Peters EC, Harvell CD (2008) Cellular responses in sea fan corals: Granular amoebocytes react to pathogen and climate stressors. PLoS One 3(3):e1811

Myers MS, Landahl JT, Krahn MM, Johnson LL, McCain BB (1990) Overview of studies on liver carcinogenesis in English sole from Puget Sound: Evidence for a xenobiotic chemical etiology. 1: Pathology and epizootiology. Sci Total Environ 94:33–50

Naar JP, Flewelling LJ, Lenzi A, Abbott JP, Granholm A, Jacocks HM, Gannon D, Henry M, Pierce R, Baden DG, Wolny J, Landsberg JH (2007) Brevetoxins, like ciguatoxins, are potent ichthyotoxic neurotoxins that accumulate in fish. Toxicon 50:707–723

Natarajan P, Ramadhas V, Ramanathan N (1982) A case report of mass mortality of marine catfish. Sci Cult 48:182–183

Ndomahina ET (1994) An investigation into reported mass mortalities of catfishes of the genus *Arius* in the coastal waters of Sierra Leone. Preliminary Report, Institute of Marine Biology and Oceanography, Fourah Bay College, Freetown, Sierra Leone

Nieland DL, Wilson CA (1995) Histological abnormalities and bacterial proliferation in red drum and black drum ovaries. Trans Am Fish Soc 124:139–143

NOAA (National Oceanic and Atmospheric Administration) (1997) NOAA's estuarine eutrophication survey, Vol 4: Gulf of Mexico region. Office of Ocean Resources Conservation and Assessment, Silver Spring, MD, USA. 77 p

Noreña-Barroso E, Gold-Bouchot G, Zapata-Perez O, Sericano JL (1999) Polynuclear aromatic hydrocarbons in American oysters *Crassostrea virginica* from the Terminos Lagoon, Campeche, Mexico. Mar Pollut Bull 38:637–645

Norman JR (1934) A systematic monograph of the flatfishes (Heterosomata), Vol 1: Psettosisae, Bothidae, Pleuronectidae. British Museum (Natural History), Adlard and Son, Ltd, London, UK. 459 p

Nuñez Ortega A (1878) Ensayo de una explicación del orígen de las grandes mortandades de peces en el Golfo de México. La Naturaleza 4:188–197

Obiekezie AI, Möller H, Anders K (1988) Diseases of the African estuarine catfish *Chrysichthys nigrodigitatus* (Lacépède) from the Cross River estuary, Nigeria. J Fish Biol 32:207–221

Okolodkov YB, Campos-Bautista G, Garate-Lizarraga I, Gonzalez-Gonzalez JAG, Hoppenrath M, Arenas V (2007) Seasonal changes of benthic and epiphytic dinoflagellates in the Veracruz reef zone, Gulf of Mexico. Aquat Microb Ecol 47:223–237

Oliveira MFT, Hawkins WE, Overstreet RM, Walker WW (1994) Hepatocellular neoplasm in a wild-caught sheepshead minnow (*Cyprinodon variegatus*) from the northern Gulf of Mexico. Gulf Res Rep 9:65–67

Ostrander GK, Rotchell JM (2005) Fish models of carcinogenesis (Chapter 9). In: Mommsen TP, Moon TW (eds) Biochemistry and Molecular Biology of Fishes, Vol 6: Environmental Toxicology. Elsevier BV, Amsterdam, The Netherlands, pp 255–288

Ostrander GK, Di Giulio RT, Hinton DE, Miller MR, Rotchell JR (2007) Chemical carcinogensis. In: DiGiulio RT, Hinton DE (eds) The toxicology of fishes. CRC Press, Boca Raton, FL, USA, pp 531–597

Overstreet RM (1973) Parasites of some penaeid shrimps with emphasis on reared hosts. Aquaculture 2:105–140

Overstreet RM (1974) An estuarine low-temperature fish-kill in Mississippi, with remarks on restricted necropsies. Gulf Res Rep 4:328–350

Overstreet RM (1977) *Poecilancistrium caryophyllum* and other trypanorhynch cestode plerocercoids from the musculature of *Cynoscion nebulosus* and other sciaenid fishes in the Gulf of Mexico. J Parasitol 63:780–789

Overstreet RM (1978) Marine maladies? Worms, germs, and other symbionts from the northern Gulf of Mexico. Mississippi-Alabama Sea Grant Consortium, MASGP-78-021. 140 p

Overstreet RM (1982) Abiotic factors affecting marine parasitism. In: Mettrick DF, Desser SS (eds) Parasites—Their world and ours. Fifth International Congress of Parasitology Proceedings and Abstracts, Vol 2, August 7–14, Toronto, Canada. Elsevier Biomedical Press, New York, NY, USA, pp 36–39

Overstreet RM (1983a) Aspects of the biology of the spotted seatrout, *Cynoscion nebulosus*, in Mississippi. Gulf Res Rep Suppl 1:1–43

Overstreet RM (1983b) Aspects of the biology of the red drum, *Sciaenops ocellatus*, in Mississippi. Gulf Res Rep Suppl 1:45–68

Overstreet RM (1983c) Metazoan symbionts of crustaceans (Chapter 4). In: Provenzano AJ (ed) The biology of Crustacea: Pathobiology, Vol 6. Academic Press, New York, NY, USA, pp 155–250

Overstreet RM (1987) Solving parasite-related problems in cultured Crustacea. Int J Parasitol 17:309–318

Overstreet RM (1988) Aquatic pollution problems, southeastern U.S. coasts: Histopathological indicators. Aquat Toxicol 11:213–239

Overstreet RM (1993) Parasitic diseases of fishes and their relationship with toxicants and other environmental factors (Chapter 5). In: Couch JA, Fournie JW (eds) Pathobiology of marine and estuarine organisms. CRC Press, Boca Raton, FL, USA, pp 111–156

Overstreet RM (1994) BP (*Baculovirus penaei*) in penaeid shrimps. USMSFP 10th Anniversary Review GCRL Special Publication 1:97–106

Overstreet RM (1997) Parasitological data as monitors of environmental health. Parassitologia 39:169–175

Overstreet RM (2003) Presidential address: Flavor buds and other delights. J Parasitol 89:1093–1107

Overstreet RM (2007) Effects of a hurricane on fish parasites. Parassitologia 49:161–168

Overstreet RM (2013) Waterborne parasitic diseases in ocean. In: Kanki P, Grimes DJ (eds) Infectious diseases: Selected entries from the encyclopedia of sustainability science and technology. SpringerScience+Business Media, New York, NY, USA, pp 431–496

Overstreet RM, Curran SS (2004) Defeating diplostomoid dangers in USA catfish aquaculture. Folia Parasitol 51:153–165

Overstreet RM, Curran SS (2005) Parasites of the American white pelican. Gulf Caribb Res 17:31–48

Overstreet RM, Edwards RH (1976) Mesenchymal tumors of some estuarine fishes of the northern Gulf of Mexico. II. Subcutaneous fibromas in the southern flounder, *Paralichthys lethostigma*, and the sea catfish, *Arius felis*. Bull of Mar Sci 26:41–48

Overstreet RM, Howse HD (1977) Some parasites and diseases of estuarine fishes in polluted habitats of Mississippi. Ann New York Acad Sci 298:427–462

Overstreet RM, Lotz JM (2016) Host-symbiont relationships: Understanding the change from guest to pest (Chapter 2). In: Hurst CJ (ed) The Rasputin effect: When commensals and symbionts become parasitic, Vol 3: Advances in Environmental Microbiology. Springer International Publishing Switzerland, pp 27–64

Overstreet RM, Lyles CH (1974) A rubber band around an Atlantic croaker. Gulf Res Rep 4:476–478

Overstreet RM, Monson P (2002) Myxosporan infection expressed by exposure to heavy metals. In: Abstracts of The Tenth International Congress of Parasitology in Conjunction with the 77th Annual Meeting of the American Society of Parasitologists, Vancouver, BC, Canada, August 4–10

Overstreet RM, Rehak E (1982) Case report—heat-stroke in nesting least tern chicks from Gulfport, Mississippi, during June 1980. Avian Dis 26:918–923

Overstreet RM, Van Devender T (1978) Implication of an environmentally induced hamartoma in commercial shrimps. J Invertebr Pathol 31:234–238

Overstreet RM, Self JT, Vilet KA (1985) The pentastomid *Sebekia mississippiens* sp. n. in the American alligator and other hosts. Prac Helminthol Soc Wash 52:266–277

Overstreet RM, Dyková I, Hawkins WE (1992) Branchiura. In: Harrison FW, Humes AG (eds) Microscopic anatomy of invertebrates, Vol 9: Crustacea (Chapter 8). Wiley-Liss, New York, NY, USA, pp 385–413

Overstreet RM, Barnes SS, Manning CS, Hawkins WE (2000) Facilities and husbandry (small fish models) (Chapter 2). In: Ostrander G (ed) Handbook of experimental animals: The laboratory fish. Academic Press Limited, London, UK, pp 41–63

Overstreet RM, Curran SS, Pote LM, King DT, Blend CK, Grater WD (2002) *Bolbophorus damnificus* n. sp. (Digenea: Bolbophoridae) from the channel catfish *Ictalurus punctatus* and American white pelican *Pelecanus erythrorhynchos* in the USA based on life-cycle and molecular data. Syst Parasitol 52:81–96

Overstreet RM, Cook JO, Heard RW (2009) Trematoda (Platyhelminthes) of the Gulf of Mexico. In: Felder DL, Camp DK (eds) Gulf of Mexico—Origin, Waters and Biota, Vol 1: Biodiversity. Texas A&M University Press, College Station, TX, USA, pp 419–488

Paerl HW, Huisman J (2008) Blooms like it hot. Science 320(5872):57–58. doi:10.1126/science.1155398

Paerl HW, Justić D (2011) Primary producers: Phytoplankton ecology and trophic dynamics in coastal waters. In: Wolanski E, McLusky DS (eds) Treatise on Estuarine and Coastal Science, Vol 6, Elsevier Academic Press, Waltham, MA, USA, pp 23–42

Palm HW (2004) The Trypanorhyncha Diesing, 1863. PKSPL-IPB Press, Bogor, Indonesia. 710 p

Palm HW, Overstreet R (2000a) *Otobothrium cysticum* (Cestoda: Trypanorhyncha) from the muscle of butterfishes (Stromateidae). Parasitol Res 86:41–53

Palm HW, Overstreet R (2000b) New records of trypanorhynch cestodes from the Gulf of Mexico, including *Kotorella pronosoma* (Stossich, 1901) and *Heteronybelinia palliata* (Linton, 1924) comb. n. Folia Parasitol 47:293–302

Panek FM (2005) Epizootics and disease of coral reef fish in the Tropical Western Atlantic and Gulf of Mexico. Rev Fish Sci 13:1–21

Paperna I, Overstreet RM (1981) Parasites and diseases of mullets (Mugilidae). In: Oren OH (ed) Aquaculture of grey mullets. Cambridge University Press, Great Britain, UK pp 411–493

Paperna I, Diamant A, Overstreet RM (1984) Monogenean infestations and mortality in wild and cultured Red Sea fishes. Helgol Meeresunters 37:445–462

Parker JH, Curran SS, Overstreet RM, Tkach VV (2010) Examination of *Homalometron elongatum* Manter, 1947 and description of a new congener from *Eucinostomus currani* Zahuranec, 1980 in the Pacific Ocean off Costa Rica. Comp Parasitol 77:154–163

Patterson KL, Porter JW, Ritchie KB, Polson SW, Mueller E, Peters EC, Santavy DL, Smith GW (2002) The etiology of white pox, a lethal disease of the Caribbean elkhorn coral, *Acropora palmata*. Proc Natl Acad Sci 99:8725–8730

Pearson JC (1929) Natural history and conservation of the redfish and other commercial sciaenids on the Texas coast. Bull Bur Fish 44:129–214

Pech D, Vidal-Martínez VM, Aguirre-Macedo ML, Gold-Bouchot G, Herrera-Silveira J, Zapata-Pérez O, Marcogliese DJ (2009) The checkered puffer (*Spheroides testudineus*) and its helminths as bioindicators of chemical pollution in Yucatan coastal lagoons. Sci Total Environ 407:2315–2324

Pennycuick L (1971) Seasonal variations in the parasite infections in a population of three-spined sticklebacks, *Gasterosteus aculeatus* L. Parasitology 63:373–388

Perera FP (1997) Environment and cancer: Who are susceptible? Science 278:068–1073

Pérez-del Olmo A, Raga JA, Kostadinova A, Fernández M (2007) Parasite communities in *Boops boops* (L.) (Sparidae) after the Prestige oil-spill: Detectable alterations. Mar Pollut Bull 54:266–276

Perkins EJ, Gilchrist JRS, Abbott OJ (1972) Incidence of epidermal lesions in fish of the northeast Irish Sea area, 1971. Nature 238:101–103

Perry HM, Warren J, Trigg C, Devender TV (1998) The blue crab fishery of Mississippi. J Shellfish Res 17:425–434

Pinckney JL, Paerl HW, Tester P, Richardson TL (2001) The role of nutrient loading and eutrophication in estuarine ecology. Environ Health Perspect 109:699–706

Plumb JA, Schachte JH, Gaines JL, Peltier W, Carroll B (1974) *Streptococcus* sp. from marine fishes along the Alabama and northwest Florida coast of the Gulf of Mexico. Trans Am Fish Soc 103:358–361

Poore GC, Bruce NL (2012) Global diversity of marine isopods (except Asellota and crustacean symbionts). PLoS One 7(8):e43529

Porter JW, Dustan P, Jaap WC, Patterson KL, Kosmynin V, Meier OW, Patterson ME, Parsons M (2001) Patterns of spread of coral disease in the Florida Keys. In: Porter JW (ed) The ecology and etiology of newly emerging marine diseases. Kluwer Academic Publishers, Dordrecht, The Netherlands, pp 1–24

Quackenbush SL, Work TM, Balazs GH, Casey RN, Rovnak J, Chaves A, duToit L, Baines JD, Parrish CR, Bowser PR, Casey JW (1998) Three closely related herpesviruses are associated with fibropapillomatosis in marine turtles. Virology 246:392–399

Quackenbush SL, Casey RN, Murcek RJ, Paul TA, Work TM, Limpus CJ, Chaves A, duToit L, Perez JV, Aguirre AA, Spraker TR, Horrocks JA, Vermeer LA, Balazs GH, Casey JW (2001) Quantitative analysis of herpesvirus sequences from normal tissue and fibropapillomas of marine turtles with real-time PCR. Virology 287:105–111

Quick JA, Henderson GE (1974) Effects of *Gymnodinium breve* red tide on fishes and birds: A preliminary report on behavior, anatomy, hematology, and histopathology. In Amborski RL, Hood MA, Miller RR (eds), Proceedings of Gulf Coast Regional Symposium on Diseases of Aquatic Animals. Pub # LSU-SG-74-05. Center for Wetland Resources, Louisiana State University, Baton Rouge, LA, USA, pp 85–114

Quick JA, Henderson GE (1975) Evidences of new ichthyointoxicative phenomena in *Gymnodinium breve* red tides. In: LoCicero VR (ed) Proceedings of the First International Conference on Toxic Dinoflagellate Blooms, November 1974, Boston, MA. MA Science and Technology Foundation, Wakefield, MA, USA, pp 413–422

Rabalais NN, Turner RE, Wiseman WJ Jr (2002) Gulf of Mexico hypoxia, AKA "The dead zone". Annu Rev Ecol Syst 33:235–263

Rasheed V, Plumb JA (1984) Pathogenicity of a non-haemolytic group B *Streptococcus* sp. in Gulf killifish (*Fundulus grandis* Baird and Girard). Aquaculture 37:97–105

Raustiala K (1997) States, NGOs, and international environmental institutions. Int Studies Q 41:719–740

Ray SM, Aldrich DV (1965) *Gymnodinium breve*: Induction of shellfish poisoning in chicks. Science 148:1748–1749

Reece KS, Bushek D, Hudson KL, Graves JE (2001) Geographic distribution of *Perkinsus marinus* genetic strains along the Atlantic and Gulf coasts of the USA. Mar Biol 139:1047–1055

Renegar DA, Blackwelder PL, Miller JD, Gochfeld DJ, Moulding AL (2008) Ultrastructural and histological analysis of Dark Spot Syndrome in *Siderastrea siderea* and *Agaricia agaricites*. In: Proceedings of the 11th International Coral Reef Symposium, Ft. Lauderdale, FL, USA, July 7–11, pp 185–189

Rigdon RH, Hendricks JW (1955) Myxosporidia in fish in waters emptying into Gulf of Mexico. J Parasitol 41:511–518

Roberts HH (2011) Surficial geology of the northern Gulf of Mexico continental slope. In: Buster NA, Holmes CW (eds) Gulf of Mexico Origin, Waters, and Biota, Vol 3: Geology. Texas A&M University Press, College Station, TX, USA, pp 209–228

Roberts HH, Rouse LJ, Walker ND, Hudson JH (1982) Cold-water stress in Florida Bay and northern Bahamas: A product of winter cold-air outbreaks. J Sediment Petrol 52:145–155

Robins RC (1957) Effects of storms on the shallow-water fish fauna of southern Florida with new records of fishes from Florida. Bull Mar Sci 7:266–275

Roessler MA, Tabb DC (1974) Studies of effects of thermal pollution in Biscayne Bay, Florida. EPA-660 / 3-74-014. U.S. Environmental Protection Agency, Washington, DC, USA

Rotstein DS, Burdett LG, McLellan W, Schwacke L, Rowles T, Terio KA, Schultz S, Pabst A (2009) Lobomycosis in offshore bottlenose dolphins (*Tursiops truncatus*), North Carolina. Emerg Infect Dis 15:588–590

Rousset F, Thomas F, De Meeûs T, Renaud F (1996) Inference of parasite-induced host mortality from distributions of parasite loads. Ecology 77:2203–2211

Salgado-Maldonado G, Amin OM (2009) Acanthocephala of the Gulf of Mexico. In: Felder DL, Camp DK (eds) Gulf of Mexico Origin, Waters, and Biota, Vol 1: Biodiversity. Texas A&M University Press, College Station, TX, USA, pp 539–552

Samuel WM, Williams ES, Rippin AB (1982) Infestations of *Piagetiella peralis* (Mallophaga: Menoponidae) on juvenile white pelicans. Can J Zool 60:951–953

Sawyer RT, Lawler AR, Overstreet RM (1975) Marine leeches of the eastern United States and the Gulf of Mexico with a key to the species. J Nat Hist 9:633–667

Schmale MC, Hensley T (1988) Transmissibility of a neurofibromatosis-like disease in bicolor damselfish. Cancer Res 48:3828–3833

Schmale MC, Hensley G, Udey LR (1983) Multiple schwannomas in the bicolor damselfish, *Pomacentrus partitus* (Pisces, Pomacentridae). Am J Path 112:238–241

Schmale MC, Hensley GT, Udey LR (1986) Neurofibromatosis in the bicolor damselfish (*Pomacentrus partitus*) as a model of von Recklinghausen neurofibromatosis. Ann New York Acad Sci 486:386–402

Schmale MC, Aman MR, Gill KA (1996) A retrovirus isolated from cell lines derived from neurofibromas in bicolor damselfish (*Pomacentrus partitus*). J Gen Virol 77:1181–1187

Schmale MC, Gibbs PD, Campbell CE (2002) A virus-like agent associated with neurofibromatosis in damselfish. Dis Aquat Organ 49:107–115

Schotte M, Markham JC, Wilson GDF (2009) Isopoda (Crustacea) of the Gulf of Mexico. In: Felder DL, Camp DK (eds) Gulf of Mexico origin, waters, and biota, Vol 1: Biodiversity. Texas A&M University Press, College Station, TX, USA, pp 973–986

Scott ME (1987) Temporal changes in aggregation: A laboratory study. Parasitology 94:583–595

Scott ME, Dobson A (1989) The role of parasites in regulating host abundance. Parasitol Today 5:176–183

Shaw DJ, Dobson AP (1995) Patterns of macroparasite abundance and aggregation in wildlife populations: A quantitative review. Parasitology 111:S111–S133

Shaw DJ, Grenfell BT, Dobson AP (1998) Patterns of macroparasite aggregation in wildlife host populations. Parasitology 117:597–610

Shelton RGJ, Wilson KW (1973) On the occurrence of lymphocystis, with notes on other pathological conditions, in the flatfish stocks of the north-east Irish Sea. Aquaculture 2:395–410

Sherwood MJ, Mearns AJ (1977) Environmental significance of fin erosion in Southern California demersal fishes. Ann New York Acad Sci 298:177–189

Shields JD, Overstreet RM (2007) Diseases, parasites, and other symbionts. In: Kennedy VS, Cronin LE (eds) The blue crab, *Callinectes sapidus*. Maryland Sea Grant, College Park, MD, USA, pp 223–339

Sindermann CJ, Lightner DV (1988) Disease diagnosis and control in North American marine aquaculture. Elsevier, New York, NY, USA. 431 p

Skinner RH (1982) The interrelation of water quality, gill parasites, and gill pathology of some fishes from South Biscayne Bay, Florida. Fish Bull 80:269–280

Skinner RH, Kandrashoff W (1988) Abnormalities and diseases observed in commercial fish catches from Biscayne Bay, Florida. Water Resour Bull 24:961–966

Slooff W (1982) Skeletal anomalies in fish from polluted surface waters. Aquat Toxicol 2:157–173

Smit NJ, Bruce NL, Hadfield KA (2014) Global diversity of fish parasitic isopod crustaceans of the family Cymothoidae. Int J Parasitol Parasites Wild 3:188–197

Smith CE, Peck TH, Klauda RJ, McLaren JB (1979) Hepatomas in Atlantic tomcod *Microgadus tomcod* (Walbaum) collected in the Hudson River estuary in New York. J Fish Dis 2:313–319

Smith FG (1970) A preliminary report on the incidence of lymphocystis disease in the fish of the Sapelo Island, Georgia, area. J Wildl Dis 6:469–471

Smith GM, Coates CW (1938) Fibro-epithelial growths on the skin in large marine turtles, *Chelonia mydas* (Linnaeus). Zoologica 23:93–98

Smith TJ, Robblee MB, Wanless HR, Doyle TW (1994) Mangroves, hurricanes, and lightning strikes. BioScience 44:256–262

Sokolova Y, Pelin A, Hawke J, Corradi N (2015) Morphology and phylogeny of *Agmasoma penaei* (Microsporidia) from the type host, *Litopenaeus setiferus*, and the type locality, Louisiana, USA. Int J Parasitol 45:1–16

Solangi MA, Overstreet RM (1980) Biology and pathogenesis of the coccidium *Eimeria funduli* infecting killifishes. J Parasitol 66:513–526

Solangi MA, Overstreet RM (1982) Histopathological changes in two estuarine fishes, *Menidia beryllina* (Cope) and *Trinectes maculatus* (Bloch and Schneider), exposed to crude oil and its water-soluble fractions. J Fish Dis 5:13–35

Solangi MA, Overstreet RM, Fournie JW (1982) Effect of low temperature on development of the coccidium *Eimeria funduli* in the Gulf killifish. Parasitology 84:31–39

Song JY, Kitamura S, Oh MJ, Kang HS, Lee JH, Tanaka SJ, Jung SJ (2009) Pathogenicity of *Miamiensis avidus* (syn. *Philasterides dicentrarchi*), *Pseudocohnilembus persalinus*, *Pseudocohnilembus hargisi* and *Uronema marinum* (Ciliophora, Scuticociliatida). Dis Aquat Organ 83:133–143

Soniat TM (1996) Epizootiology of *Perkinsus marinus* disease of eastern oysters in the Gulf of Mexico. J Shellfish Res 15:35–43

Sosa ER, Landsberg JH, Kiryu Y, Stephenson CM, Cody TT, Dukeman AK, Wolfe HP, Vandersea MW, Litaker RW (2007) Pathogenicity studies with the Fungi *Aphanomyces invadans*, *Achlya bisexualis*, and *Phialemonium dimorphosporum*: Induction of skin ulcers in striped mullet. J Aquat Anim Health 19:41–48

Stacy BA, Wellehan JFX, Foley AM, Coberley SS, Herbst LH, Manire CA, Garner MM, Brookins MD, Childress AL, Jacobson ER (2008) Two herpesviruses associated with disease in wild Atlantic loggerhead sea turtles (*Caretta caretta*). Vet Microbiol 126:63–73

Stacy BA, Foley AM, Greiner E, Herbst LH, Bolten A, Klein P, Manire CA, Jacobson ER (2010) Spirorchiidiasis in stranded loggerhead *Caretta caretta* and green turtles *Chelonia mydas* in Florida (USA): Host pathology and significance. Dis Aquat Organ 89:237–259

Stamper MA, Lewbart GA, Barrington PR, Harms CA, Geoly F, Stoskopf MK (1998) *Eimeria southwelli* infection associated with high mortality of cownose rays. J Aquat Anim Health 10:264–270

Steidinger KA, Wolny JL, Haywood AJ (2008) Identification of Kareniaceae (Dinophyceae) in the Gulf of Mexico. Nova Hedwigia Beih 133:269–284

Stevenson C, Childers DL (2004) Hydroperiod and seasonal effects on fish decomposition in an oligotrophic everglades marsh. Wetlands 24:529–537

Stone I, Heard RW (1989) *Excorallana delaneyi*, n. sp. (Crustacea: Isopoda: Excorallanidae) from the northeastern Gulf of Mexico, with observations on adult characters and sexual dimorphism in related species of *Excorallana* Stebbing, 1904. Gulf Caribb Res 8:199–211

Storey M, Gudger EW (1936) Mortality of fishes due to cold at Sanibel Island, Florida, 1886-1936. Ecology 17:640–648

Suárez-Morales E, Kim IH, Castellanos I (1998) A new geographic and host record for *Argulus flavescens* Wilson, 1916 (Crustacea, Arguloida), from Southeastern Mexico. Bull Mar Sci 62:293–296

Summers JK (1999) The ecological condition of estuaries in the Gulf of Mexico. U.S. Environmental Protection Agency, EPA 620-R-98-004. Office of Research and Development, U.S. Environmental Protection Agency, Washington, DC, USA. 71 p

Sun PL, Hawkins WE, Overstreet RM, Brown-Peterson NJ (2009) Morphological deformities as biomarkers in fish from contaminated rivers in Taiwan. Int J Environ Res Public Health 6:2307–2331

Sures B (2004) Environmental parasitology: Relevancy of parasites in monitoring environmental pollution. Trends Parasitol 20:170–177

Szmant AM, Gassman NJ (1990) The effects of prolonged "bleaching" on the tissue biomass and reproduction of the reef coral *Montastrea annularis*. Coral Reefs 8:217–224

Tabb DC, Jones AC (1962) Effect of Hurricane Donna on the aquatic fauna of North Florida Bay. Trans Am Fish Soc 91:375–378

Tao Z, Larsen AM, Bullard SA, Wright AC, Arias CR (2012) Prevalence and population structure of *Vibrio vulnificus* on fishes from the northern Gulf of Mexico. Appl Environ Microbiol 78:7611–7618

Thessen AE, Dortch Q, Parsons ML, Morrison W (2005) Effect of salinity on *Pseudo-nitzschia* species (Bacillariophyceae) growth and distribution. J Phycol 41:21–29

Thessen AE, Bowers HA, Stoecker DK (2009) Intra-and interspecies differences in growth and toxicity of *Pseudo-nitzschia* while using different nitrogen sources. Harmful Algae 8:792–810

Thomas LP, Moore DR, Work RC (1961) Effects of Hurricane Donna on the turtle grass beds of Biscayne Bay, Florida. Bull Mar Sci 11:191–197

Thomas NJ, Hunter DB, Atkinson CT (eds) (2007) Infectious diseases of wild birds. Blackwell Publishing, Ames, IA, USA. 484 p

Thronson A, Quigg A (2008) Fifty-five years of fish kills in coastal Texas. Esturaries Coasts 31:802–813

Thulin J (1987) Some diseases and parasites of trout (*Salmo trutta*) attracted to a hot water effluent. In: Stenmark A, Malmberg G (eds) Parasites and diseases in natural waters and aquaculture in Nordic Countries. Zoo-Tax Naturhistoriska riksmuseet, Stockholm, Sweden. 72 p

Thune RL, Fernandez DH, Hawke JP, Miller R (2003) Construction of a safe, stable, efficacious vaccine against *Photobacterium damselae* ssp. *piscicida*. Dis Aquat Organ 57:51–58

Tilmant JT, Curry RW, Jones R, Szmant A, Zieman JC, Flora M, Robblee MB, Smith D, Snow RW, Wanless H (1994) Hurricane Andrew's effects on marine resources. BioScience 44:230–237

Trainer VL, Bates SS, Lundholm N, Thessen AE, Cochlan WP, Adams NG, Trick CG (2012) *Pseudo-nitzschia* physiological ecology, phylogeny, toxicity, monitoring and impacts on ecosystem health. Harmful Algae 14:271–300

Turner RE, Schroeder WW, Wiseman WJ (1987) The role of stratification in the deoxygenation of Mobile Bay and adjacent shelf bottom waters. Estuaries 10:13–19

Twiner MJ, Flewelling LJ, Fire SE, Bowen-Stevens SR, Gaydos JK, Johnson CK, Landsberg JH, Leighfield TA, Mase-Guthrie B, Schwacke L, Van Dolah FM, Wang Z, Rowles TK (2012) Comparative analysis of three brevetoxin-associated bottlenose dolphin (*Tursiops truncatus*) mortality events in the Florida panhandle region (USA). PLoS One 7(8):e42974

Udey LR, Young E, Sallman B (1976) *Eubacterium* sp. ATCC 29255: An anaerobic bacterial pathogen of marine fish. Fish Health News 5:3–4

Udey LR, Young E, Sallman B (1977) Isolation and characterization of an anaerobic bacterium, *Eubacterium tarantellus* sp. nov., associated with striped mullet (*Mugil cephalus*) mortality in Biscayne Bay, Florida. J Fish Board Can 34:402–409

Ulrich PN, Colton CM, Hoover CA, Gaffney PM, Marsh AG (2007) *Haplosporidium nelsoni* (MSX) rDNA detected on oysters from the Gulf of Mexico and the Caribbean Sea. J Shellfish Res 26:195–199

Underwood AJ (1994) On beyond BACI: Sampling designs that might reliably detect environmental disturbances. Ecol Appl 4:3–15

Van der Oost R, Beyer J, Vermeulen NP (2003) Fish bioaccumulation and biomarkers in environmental risk assessment: A review. Environ Toxicol Pharmacol 13:57–149

Van Deventer M, Atwood K, Vargo GA, Flewelling LJ, Landsberg JH, Naar JP, Stanek D (2012) *Karenia brevis* red tides and brevetoxin-contaminated fish: A high risk factor for Florida's scavenging shorebirds? Bot Mar 22:31–37

Van Houtan KS, Hargrove SK, Balazs GH (2010) Land use, macroalgae, and a tumor-forming disease in marine turtles. PLoS One 5(9):e12900

Vandersea MW, Litaker RW, Yonnish B, Sosa E, Landsberg JH, Pullinger C, Moon-Butzin P, Green J, Morris JA, Kator H, Noga EJ, Tester PA (2006) Molecular assays for detecting *Aphanomyces invadans* in ulcerative mycotic fish lesions. Appl Environ Microbiol 72:1551–1557

Vargo GA, Atwood K, Van Deventer M, Harris R (2006) Beached bird surveys on Shell Key, Pinellas County, Florida. Fla Field Nat 34:21–22

Vidal-Martínez VM, Aguirre-Macedo ML, Noreña-Barroso E, Gold-Bouchot G, Caballero-Pinzón PI (2003) Potential interactions between metazoan parasites of the Mayan catfish *Ariopsis assimilis* and chemical pollution in Chetumal Bay, Mexico. J Helminthol 77:173–1847

Vidal-Martínez VM, Centeno-Chalé OA, Torres-Irineo E, Sánchez-Ávila J, Gold-Bouchot G, Aguirre-Macedo ML (2014) The metazoan parasite communities of the shoal flounder (Syacium gunteri) as bioindicators of chemical contamination in the southern Gulf of Mexico. Parasites Vectors 7:541

Viosca P Jr (1945) A critical analysis of practices in the management of warm-water fish with a view to greater food production. Trans Am Fish Soc 73:274–283

Vogelbein WK, Fournie JW, Van Veld PA, Huggett RJ (1990) Hepatic neoplasms in the mummichog *Fundulus heteroclitus* from a creosote-contaminated site. Cancer Res 50:5978–5986

Vogelbein WK, Shields JD, Haas LW, Reece KS, Zwerner DE (2001) Skin ulcers in estuarine fishes: A comparative pathological evaluation of wild and laboratory-exposed fish. Environ Health Perspect 109:687–693

Vogelbein WK, Lovko VJ, Shields JD, Reece KS, Mason PL, Haas LW, Walker CC (2002) *Pfiesteria shumwayae* kills fish by micropredation not exotoxin secretion. Nature 418:967–970

Walsh JJ, Weisberg RH, Lenes JM, Chen FR, Dieterle DA, Zheng L, Carder KL, Vargo GA, Havens JA, Peebles E, Hollander DJ, He R, Heil CA, Mahmoudi B, Landsberg JH (2009) Isotopic evidence for dead fish maintenance of Florida red tides, with implications for coastal fisheries over both source regions of the West Florida shelf and within downstream waters of the South Atlantic Bight. Prog Oceanogr 80:51–73

Waring GT, Josephson E, Fairfield CP, Maze-Foley K (2007) U.S. Atlantic and Gulf of Mexico Marine Mammal Stock Assessments-2006. NOAA Technical Memorandum NMFS-NE-201, April. U.S. Department of Commerce, NMFS Southeast Fisheries Science Center, Miami, FL, USA.

Weinstein MP, Heck KL (1977) Biology and host-parasite relationships of *Cymothoa excisa* (Isopoda, Cymothoidae) with three species of snappers (Lutjanidae) on Caribbean coast of Panama. Fish Bull 75:875–877

Weissenberg R (1945) Studies on virus diseases of fish IV: Lymphocystis disease in Centrarchidae. Zoologica (NY) 30:169–184

Wharton JH, Ellender RD, Middlebrooks BL, Stocks PK, Lawler AR, Howse D (1977) Fish cell culture: Characteristics of a cell line from the silver perch, *Bairdiella chrysura*. In Vitro 13:389–397

White ME, Wilson EA (1996) Predators, pests, and competitors. In: Kennedy VS, Newell RIE, Eble AF (eds) The eastern oyster *Crassostrea virginica*. Maryland Sea Grant College, College Park, MD, USA, pp 559–579

Wilkinson HW, Thacker LG, Facklam RR (1973) Nonhemolytic group B streptococci of human, bovine, and ichthyic origin. Infect Immun 7:496–498

Willcox J (1887) Fish killed by cold along the Gulf of Mexico and coast of Florida. Bull U S Fish Commission 6:123

Willett KL, McDonald SJ, Steinberg MA, Beatty KB, Kennicutt MC, Safe SH (2009) Biomarker sensitivity for polynuclear aromatic hydrocarbon contamination in two marine fish species collected in Galveston Bay, Texas. Environ Toxicol Chem 16:1472–1479

Williams EH, Bunkley-Williams L (1990) The world-wide coral reef bleaching cycle and related sources of coral mortality. Atoll Res Bull, No. 335. NMNH Smithsonian Institution, Washington, DC, USA. 71 p

Williams EH, Bunkley-Williams L (2000) Marine major ecological disturbances of the Caribbean. Infect Dis Rev 2:110–127

Williams ES, Barker IK (eds) (2001) Infectious diseases of wild mammals, 3rd edn. Iowa State University Press, Ames, IA, USA. 558 p

Wilson EA, Powell EN, Craig MA, Wade TL, Brooks JM (1990) The distribution of *Perkinsus marinus* in Gulf Coast oysters: Its relationship with temperature, reproduction, and pollutant body burden. Int Revue Ges Hydrobiol 75:533–550

Witham R (1973) A bacterial disease of hatchling loggerhead sea turtles. Fla Sci 36:226–228

Wobeser GA (2008) Parasitism: Cost and effects. In: Atkinson CT, Thomas NJ, Hunter DB (eds) Parasitic diseases of wild birds. Wiley-Blackwell, Ames, IA, USA, pp 3–9

Wolf JC, Baumgartner WA, Blazer VS, Camus AC, Engelhardt JA, Fournie JW, Frasca S, Groman DB, Kent ML, Khoo LH, Law JM, Lombardini ED, Ruehl-Fehlert C, Segner HE,

Smith SA, Spitsbergen JM, Weber K, Wolfe MJ (2015) Nonlesions, misdiagnoses, missed diagnoses, and other interpretive challenges in fish histopathology studies a guide for investigators, authors, reviewers, and readers. Toxicol Pathol 43:297–325

Wolthaus BG (1984) Seasonal changes in frequency of diseases in dab, *Limanda limanda*, from the southern North Sea. Helgol Meeresunters 37:375–387

Work TM, Rameyer RA, Balazs GH, Cray C, Chang SP (2001) Immune status of free-ranging green turtles with fibropapillomatosis from Hawaii. J Wildl Dis 37:574–581

Work TM, Balazs GH, Rameyer RA, Morris RA (2004) Retrospective pathology survey of green turtles *Chelonia mydas* with fibropapillomatosis in the Hawaiian Islands, 1993-2003. Dis Aquat Organ 62:163–176

Wright B, Mohanty B (2006) Operation Kachhapa: An NGO initiative for sea turtle conservation in Orissa. In: Shanker K, Choudhury B (eds) Marine turtles of the Indian subcontinent. University Press, Hyderabad, India, pp 290–302

Zimmerman AM, DePaola A, Bowers JC, Krantz JA, Nordstrom JL, Johnson CN, Grimes DJ (2007) Variability of total and pathogenic *Vibrio parahaemolyticus* densities in northern Gulf of Mexico water and oysters. Appl Environ Microbiol 73:7589–7596

Zimmerman R (1998) Characteristics and causes of Texas marine strandings. NOAA Tech Rep NMFS 143. U.S. Department of Commerce, Washington, DC, USA. 85 p

Zimmerman RJ (2010) A history of the NMFS laboratory at Galveston. Gulf Mex Sci 1–2:82–96

Open Access   This chapter is licensed under the terms of the Creative Commons Attribution-NonCommercial 2.5 International License (http://creativecommons.org/licenses/by-nc/2.5/), which permits any noncommercial use, sharing, adaptation, distribution and reproduction in any medium or format, as long as you give appropriate credit to the original author(s) and the source, provide a link to the Creative Commons license and indicate if changes were made.

The images or other third party material in this chapter are included in the chapter's Creative Commons license, unless indicated otherwise in a credit line to the material. If material is not included in the chapter's Creative Commons license and your intended use is not permitted by statutory regulation or exceeds the permitted use, you will need to obtain permission directly from the copyright holder.

APPENDIX A
LIST OF ACRONYMS, ABBREVIATIONS, AND SYMBOLS

%	Percent
°C	Degree(s) Celsius
°F	Degree(s) Fahrenheit
μm	Micrometer(s)
μM	Micromole(s)
AAM	American Academy of Microbiology
ABC	American Bird Conservancy
ABC	Allowable biological catch
ac	Acre(s)
ACNWR	Archie Carr National Wildlife Refuge
ADCNR	Alabama Department of Conservation and Natural Resources
AL	Alabama
AOU	American Ornithologists' Union
ASP	Amnesic shellfish poisoning
Avg.	Average
AWFWF	Alabama Division of Wildlife and Fresh Water Fisheries
BAM	Beaufort Assessment Model
BBD	Black band disease
BBS	Breeding bird survey
BIRNM	Buck Island Reef National Monument
BNA	Birds of North America

BP	BP Exploration & Production Inc.
CCL	Curved carapace length
CFL	Curved fork length
CFPHV	Chelonid fibropapilloma-associated herpesvirus
CFR	Code of Federal Regulations
CI	Confidence interval
CITES	Convention on International Trade in Endangered Species of Wild Fauna and Flora
cm	Centimeter(s)
CMR	Capture-mark-recapture
COLREG	Collision Regulation
ComFIN	Commercial Fisheries Information Network
CPI	Consumer Price Index
CPR	Center for Prediction of Red Tides
CPUE	Catch per unit effort
CRFM	Caribbean Regional Fisheries Mechanism
CSA	Coastal Study Area
CV	Coefficient of variation
CWS	Canadian Wildlife Service
DA	Domoic acid
DDD	Dichlorodiphenyldichloro-ethane
DDE	Dichlorodiphenyldichloro-ethylene

© The Author(s) 2017
C.H. Ward (ed.), *Habitats and Biota of the Gulf of Mexico: Before the Deepwater Horizon Oil Spill*, DOI 10.1007/978-1-4939-3456-0

DDT	Dichlorodiphenyltrichloro-ethane	**g**	Gram(s)
DGoMB	Deep Gulf of Mexico Benthos	**GBEP**	Galveston Bay Estuary Program
DHA	Dynamic height anomaly	**GBST**	Galveston Bay Status and Trends
DO	Dissolved oxygen	**GCJV**	Gulf Coast Joint Venture
DoD	U.S. Department of Defense	**GCPWG**	Gulf Coast Prairie Working Group
DPS	Distinct population segments	**GCRL**	Gulf Coast Research Laboratory
DSD	Dark spots disease	**GERG**	Geochemical and Environmental Research Group
DSL	Deep scattering layer		
DTNP	Dry Tortugas National Park	**GIS**	Geographic information system
EEZ	Exclusive Economic Zone		
EMAP	Environmental Monitoring and Assessment Program	**GMFMC**	Gulf of Mexico Fishery Management Council
		GMNET	Gulf of Mexico Aquatic Mortality Response Network
EMB	Estimated mean biomass		
ESA	Ecological Society of America		
		GoM	Gulf of Mexico
ESA	U.S. Endangered Species Act	**GoMRI**	Gulf of Mexico Research Initiative
EU	European Union	**GSMFC**	Gulf States Marine Fisheries Commission
FAO	Food and Agriculture Organization (of the United Nations)		
		h/hr	Hour(s)
		ha	Hectare(s)
FCMA	Fishery Conservation and Management Act	**HAB**	Harmful algal bloom
		HCH	Hexachlorocyclohexanes
FDEP	Florida Department of Environmental Protection	**HMS**	Highly migratory species
		HMSMD	Highly Migratory Species Management Division
FFWCC	Florida Fish and Wildlife Conservation Commission		
		HRI	Harte Research Institute for Gulf of México Studies
FGBNMS	Flower Gardens Banks National Marine Sanctuary		
		ICCAT	International Commission for the Conservation of Atlantic Tunas
FL	Florida		
FL	Fork length		
FMP	Fishery/Fish Management Plan	**in**	Inch(es)
		IPCC	Intergovernmental Panel on Climate Change
FMU	Fishery management unit		
FP	Fibropapillomatosis	**IUCN**	International Union for the Conservation of Nature
ft	Foot/feet		
ft³	Cubic feet		
ftm	Fathom(s)	**kg**	Kilogram(s)
FUS	Fisheries of the United States	**km**	Kilometer(s)
		km²	Square kilometer(s)
FWRI	Fish and Wildlife Research Institute	**KWNWR**	Key West National Wildlife Refuge

L	Liter(s)	**NMFS**	National Marine Fisheries Service
LA	Louisiana		
Lb/lb	Pound(s)	**NMFS SERO**	NMFS Southeast Regional Office
LDWF	Louisiana Department of Wildlife and Fisheries		
		NMFS FSD	NMFS Fisheries Statistics Division
LJFL	Lower jaw fork length		
LSU	Louisiana State University	**NMFS SEFSC**	National Marine Fisheries Service Southeast Fisheries Science Center
m	Meter(s)		
MAFAC	Marine Fisheries Advisory Committee		
		NMS	National Marine Sanctuaries
MD	Maryland		
MDWFP	Mississippi Department of Wildlife, Fisheries, & Parks	**NOAA**	National Oceanic and Atmospheric Administration
MFMT	Maximum fishing mortality threshold	**NPS**	National Park Service
		NRC	National Research Council
mg/L	Milligram(s) per liter	**NSP**	Neurotoxic shellfish poisoning
mi	Mile(s)		
mi²	Square mile(s)	**OA**	Okadaic acid
min	Minute(s)	**OCS**	Outer continental shelf
ml; mL	milliliter(s)	**OTTF**	Oyster Technical Task Force
mm	Millimeter(s)		
mM	millimole(s)	**OY**	Optimal yield
MMNS	Mississippi Museum of Natural Science	**oz**	Ounce(s)
		PAH	Polycyclic aromatic hydrocarbon
MMPA	U.S. Marine Mammal Protection Act		
		PAIS	Padre Island National Seashore
MMS	Minerals Management Service		
		PAS	Periodic acid-Schiff
mph	mile(s) per hour	**PBR**	Potential biological removal
MRFSS	Marine Recreational Fisheries Statistics Survey		
		PCB	Polychlorinated biphenyl
MRIP	Marine Recreational Information Program	**PCR**	Polymerase chain reaction
		POLR	Private Oyster Lease Rehabilitation Program
MS	Mississippi		
MSST	Minimum spawning stock threshold	**POP**	Persistent organic pollutant
MSX	*Haplosporidium nelsoni*	**ppm**	Part(s) per million
MSY	Maximum sustainable yield	**ppt**	Part(s) per thousand
		PSP	Paralytic shellfish poisoning
NASA	National Aeronautics and Space Administration		
		RecFIN	Recreational Fisheries Information Network
NGO	Nongovernmental organization		
		RTLA	Registry for Tumors in Lower Animals
NLCD	National Land Cover Dataset		
		SAFMC	South Atlantic Fishery Management Council
nm	Nanometer(s)		

SAV	Submerged aquatic vegetation
SCAR	Scientific Committee on Antarctic Research
SCL	Straight carapace length
SCOPE	Scientific Committee on Problems of the Environment
SEAMAP	Southeast Area Monitoring and Assessment Program
SEDAR	Southeast Data, Assessment, and Review
SEFSC	Southeast Fisheries Science Center
SEMARNAT	Secretary of Environment and Natural Resources, Mexico (Ministry of the Environment and Natural Resources)
SEPM	Society for Sedimentology
SERDP	Strategic Environmental Research and Development Program
SIMB	Society of Industrial Microbiology and Biotechnology
SJRWMD	St. Johns River Water Management District
SL	Standard length
SPR	Spawning potential ratio
SSB	Spawning stock biomass
SSC	Species of special concern
SSO	*Haplosporidium costale*
SST	Sea surface temperature
STSSN	Sea Turtle Stranding and Salvage Network
STX	Saxitoxin
SWOT	State of the Worlds Sea Turtles
SWSS	Sperm whale seismic survey
TAC	Total allowable catch
TAMU	Texas A&M University
TAMU-CC	Texas A&M University at Corpus Christi

TCWS	Texas Colonial Waterbird Survey
TDLS	Thin dark-line syndrome
TED	Turtle excluder device
TEWG	Turtle Expert Working Group
TL	Total length
TMMSN	Texas Marine Mammal Stranding Network
TPWD	Texas Parks and Wildlife Department
TTP	Triphenylphosphate
TTR	Thermal tolerance range
TX	Texas
UCF	University of Central Florida
UME	Unusual mortality event
U.S.	United States
USA	United States of America
USACE	U.S. Army Corps of Engineers
USDA	United States Department of Agriculture
USEPA	U.S. Environmental Protection Agency
USF	University of South Florida
USFWS	U.S. Fish and Wildlife Service
USGS	U.S. Geological Survey
USM	The University of Southern Mississippi
USSCP	U.S. Shorebird Conservation Plan
UTM	Universal Transverse Mercator
VPA	Virtual population analysis
WAS	West Atlantic sailfish
WBD	White band disease
WCCA	Whooping Crane Conservation Association
WP	White plague
WSSV	White spot syndrome virus
YBS	Yellow band syndrome
YOY	Young-of-the-year
yr	Year

APPENDIX B
UNIT CONVERSION TABLE

Multiply	By	To obtain
Acre	0.405	Hectare
Acre	1.56 E-3	Square mile (statute)
Centimeter	0.394	Inch
Cubic feet	0.028	Cubic meter
Cubic feet	7.48	Gallon (U.S. liquid)
Cubic feet	28.3	Liter
Cubic meter	35.3	Cubic feet
Cubic yard	0.76	Cubic meter
Feet	0.305	Meter
Gallon (U.S. liquid)	3.79	Liter
Hectare	2.47	Acre
Inch	2.54	Centimeter
Kilogram	2.20	Pound (avoir)
Kilometer	0.62	Mile (statue)
Liter	0.035	Cubic feet
Liter	0.26	Gallon (U.S. liquid)
Meter	3.28	Feet
Metric ton(ne)	1.102	U.S. short ton
Mile (statue)	1.61	Kilometer
Pound (avoir)	0.45	Kilogram
Square feet	0.093	Square meter
Square kilometer	0.386	Square mile
Square mile	640	Acre
Square mile	2.59	Square kilometer

© The Author(s) 2017
C.H. Ward (ed.), *Habitats and Biota of the Gulf of Mexico: Before the Deepwater Horizon Oil Spill*,
DOI 10.1007/978-1-4939-3456-0

INDEX

A

Acoustic pollution, 1551, 1555–1557
Actitis macularius, 1364
Agricultural runoff, 1289
Aix sponsa, 1364
Albacore (*Thunnus alalunga*), 877
Algae, 884, 979, 1207, 1243, 1244, 1249, 1250,
     1253, 1274, 1289, 1596, 1602, 1612,
     1614–1616, 1661, 1680, 1687, 1707
Algal bloom, 1277, 1560, 1590, 1605, 1687
Allelopathy, 1590, 1609
Allochthonous, 1590, 1592
Allowable biological catch, 911
American Anhinga, 1364, 1438
American Avocet, 1689
American Bird Conservancy (ABC), 1426
American coot, 1365, 1689, 1690
American Golden Plover, 1365
American Ornithologists' Union (AOU), 1357,
     1414, 1419
American Oystercatcher (*Haematopus
     palliatus*), 1359, 1364
American white pelican, 1364, 1428, 1441,
     1646, 1652, 1689–1691
American Wigeon, 1364, 1441, 1442
Amnesic shellfish poisoning (ASP),
     1609, 1610
Anaerobic methane oxidation, 1617–1618
Anas acuta, 1364
Anas americana, 1364
Anas crecca, 1364
Anas discors, 1364
Anas platyrhynchos, 1364
Angelfishes, 872
Anhinga anhinga, 1364
Anthropogenic impacts, 1196, 1277, 1315, 1316,
     1490, 1551–1565, 1585
AOU. *See* American Ornithologists' Union
     (AOU)
Apalachicola Bay, 977, 1214, 1539, 1541
Aransas Bay, 1250, 1539
Archie Carr National Wildlife Refuge
     (ACNWR), 1223, 1224, 1239–1241
Ardea alba, 1364
Ardea herodias, 1364

Arenaria interpres, 1364, 1607
Armored searobins, 871
Arsenic (As), 1664
Atchafalaya Bay, 1091, 1539
Atlantic angel shark (*Squatina dumerili*), 987
Atlantic bluefin tuna (*Thunnus thynnus*), 877,
     913–923, 1005
Atlantic blue marlin (*Makaira nigricans*), 876,
     877, 923–931, 1005
Atlantic goliath grouper (*Epinephelus
     itajara*), 877
Atlantic sailfish (*Istiophorus albicans*), 876,
     877, 939–946, 1005
Atlantic sharpnose shark (*Rhizoprionodon
     terraenovae*), 986
Atlantic spotted dolphin (*Stenella frontalis*),
     1496, 1497, 1536, 1537, 1564
Atlantic sturgeon (*Acipenser
     oxyrhynchus*), 881
Atlantic swordfish (*Xiphias gladius*), 876, 877,
     931–939, 1005
Atlantic thread herring (*Opisthonema
     oglinum*), 877, 1118
Autochthonous, 1590, 1592
Aythya affinis, 1364, 1607
Aythya collaris, 1364

B

Bacterial contamination/contaminants, 1606
Bacterial disease, 1619–1629
Baffin Bay, 1603
Bald Eagle (*Haliaeetus leucocephalus*), 1358,
     1359, 1363
Banded drum (*Larimus fasciatus*), 881
Barataria Bay, 1091, 1539, 1603
Barataria-Terrebonne Estuary, 1437, 1438
Bartramia longicauda, 1364
Basking shark (*Cetorhinus maximus*), 987
Basslets, 871
Batfish, 875
Bay of Fundy, 990
BBS. *See* Breeding Bird Survey (BBS)
Beach advisory(ies), 870
Beach closing(s), 1221, 1297
Belted kingfisher, 1365

© The Author(s) 2017
C.H. Ward (ed.), *Habitats and Biota of the Gulf of Mexico: Before the Deepwater Horizon Oil Spill*,
DOI 10.1007/978-1-4939-3456-0

he United States
asters